JOHN M^C GUIRE
(ANALYZERS)
791-5425

INSTRUMENT ENGINEERS'
Handbook

INSTRUMENT ENGINEERS'
Handbook

Volume II
Process Control

BÉLA G. LIPTÁK, Editor

CHILTON BOOK COMPANY

PHILADELPHIA · NEW YORK · LONDON

Copyright © 1970 by Béla G. Lipták
First Edition   *All rights reserved*
Published in Philadelphia by Chilton Book Company
and simultaneously in Ontario, Canada,
by Thomas Nelson & Sons, Ltd.
ISBN 0-8019-5519-X
Library of Congress Catalog Card Number 73-80445
Designed by William E. Lickfield
Manufactured in the United States of America
by Graphic Services, Inc. and Vail-Ballou Press, Inc.

To the Young People of
Hungary and Czechoslovakia

CONTRIBUTORS

ROBERT J. BAKER
BS
Technical Director
  Wallace & Tiernan Division, Pennwalt Corp.
    (*Section 10.14*)

ROGER M. BAKKE
BS, MS
Senior Engineer
  IBM Corp.
    (*Sections 7.13, 8.10 and 8.11*)

HANS D. BAUMANN
IE, PE
Vice-President Engineering
  Masoneilan International, Inc.
    (*Sections 1.3 through 1.7 and A.3*)

CHESTER S. BEARD
BSEE
Engineer Specialist
  Bechtel Corp.
    (*Sections 1.13 through 1.20*)

JOHN W. BERNARD
BSChE, MSIE, PE
Manager, Systems Technology
  The Foxboro Co.
    (*Sections 8.14 and 8.15*)

BENJAMIN BLOCK
BS, MSChE, PE
Project Engineer
  Abbott Laboratories
    (*Section 10.8*)

RICHARD W. BORUT
Section Engineer, Instrument Design
  M. W. Kellogg Co.
    (*Section 5.1*)

AUGUST BRODGESELL

BSEE
Assistant Chief Instrument Engineer
Crawford & Russell, Inc.
(*Chapter II*)

ANTHONY M. CALABRESE

BSChE, MSChE, BSEE, PE
Section Engineer, Instrument Analytical
M. W. Kellogg Co.
(*Section 10.7*)

ARMANDO B. CORRIPIO

BSChE, MSChE, PhDChE, PE
Assistant Professor of Chemical Engineering
Louisiana State University
(*Section 9.5*)

SAMUEL G. DUKELOW

BSME
National Sales Manager
Bailey Meter Co.
(*Section 10.6*)

LAWRENCE S. DYSART

BSEE
Sales Manager for Control Equipment
Robertshaw Controls Co.
(*Section 3.3*)

HOWARD L. FELDMAN

BSME, PE
Product and Market Planner
The Foxboro Co.
(*Section 4.1*)

CHARLES E. GAYLER

BSChE, PE
Manager of Engineering Services
Hooker Chemical Corp.
(*Sections 1.9, 1.10 and 1.12*)

CONRAD H. HOEPPNER

BSEE, MSEE, PE, PhD
President
Industrial Electronics Corp.
(*Section 6.4*)

HAROLD L. HOFFMAN

BSChE, PE
Refining Editor
Hydrocarbon Processing
(*Section 10.3*)

PER A. HOLST

MSEE
Technical Supervisor
The Foxboro Co.
(*Sections 9.2 and 9.3*)

FRANKLIN B. HOROWITZ

BSChE, MSChE, PE
Process Engineer
Crawford & Russell, Inc.
(*Sections 10.12 and 10.13*)

STUART P. JACKSON

BSE, MSc, PhD, PE
Partner
Jackson Associates
(*Section 5.4*)

ROBERT F. JAKUBIK

BSChE
Manager, Process Control Applications
Digital Applications, Inc.
(*Sections 8.1, 8.2 and 8.3*)

VAUGHN A. KAISER

BSME, MSE, PE
Member of Technical Staff
Profimatics, Inc.
(*Sections 8.7, 8.8 and 8.9*)

DONALD C. KENDALL

BSChE, MSChE
Senior Systems Analyst
The Foxboro Co.
(*Section 10.2*)

CHANG H. KIM

BSChE
Senior Group Leader
Control Engineering Development
UniRoyal Chemical Div. of Uniroyal, Inc.
(*Sections 8.6 and 10.15*)

PAUL M. KINTNER

BSEE, MSEE, PhDEE
Manager, Digital Products and Systems Development
Cutler-Hammer, Inc.
(*Sections 4.6 and 4.7*)

BÉLA G. LIPTÁK

ME, MME, PE
Consultant and Chief Instrument Engineer
Crawford & Russell, Inc.
(Introduction, Sections 1.2, 1.21, 10.5, 10.11,
and Miscellaneous)

ORVAL P. LOVETT JR.

BSCE
Supervising Instruments Engineer
E. I. duPont de Nemours & Co.
(Sections 1.1, 1.8 and 3.1)

DALE E. LUPFER

BSME, PE
Process Automation Consultant
Engel Engineering Co., Inc.
(Section 10.4)

CHARLES L. MAMZIC

BSME
Manager, Systems and Application Engineering
Department
Moore Products Co.
(Sections 4.3, 4.4 and 6.1)

ALLAN F. MARKS

BSChE, PE
Engineering Specialist—Instrumentation
Bechtel Corp.
(Sections 6.2 and 6.3)

FRED D. MARTON

Dipl. Ing.
Former Managing Editor
Instruments and Control Systems
(Sections 1.11, 5.2 and 5.3)

CHARLES F. MOORE

BSChE, MSChE, PhDChE
Assistant Professor of Chemical Engineering
University of Tennessee
(Sections 7.1 through 7.5)

RALPH L. MOORE

BSME, MSME, MSInst. E, PE
Instrument Engineer
E. I. duPont de Nemours & Co.
(Section 3.2)

PAUL W. MURRILL

BSChE, MSChE, PhDChE, PE
Vice-Chancellor
Louisiana State University
(Section 7.12)

A. ELI NISENFELD

BA, BSChE
Senior Instrument Engineer
Bechtel Corp.
(Section 9.4)

GEORGE PLATT

BSChE
Staff Engineer—Instrumentation
Bechtel Corp.
(Sections A.1 and A.2)

HOWARD C. ROBERTS

ABEE, PE
Professor Emeritus
University of Illinois
(Section 4.5)

FRANK M. RYAN

BSChE
Editor & Associate Publisher
Instrumentation Technology
(Preface)

CHAKRA J. SANTHANAM

BSCh, BSChE, MSChE, ChE, PE
Senior Process Engineer
Crawford & Russell, Inc.
(Sections 10.9 and 10.10)

WALTER F. SCHLEGEL

BE
Assistant Chief Process Engineer
Crawford & Russell, Inc.
(Section 10.1)

FRANCIS G. SHINSKEY

BSChE
Senior Systems Design Engineer
The Foxboro Co.
(Sections 7.6 through 7.11 and 9.1)

WILLIAM L. SKAGGS

BSChE, PE
Bonner & Moore Associates, Inc.
(Sections 8.12 and 8.13)

Cecil L. Smith   BSChE, MSChE, PhDChE, PE
                                Associate Professor of Chemical Engineering
                                     Louisiana State University
                                        (*Section 7.12*)

James E. Talbot   BSEP
                                Associate Editor
                                Instrumentation Technology
                                     (*Section 4.2*)

Joseph Valentich   BSE
                                Mechanical Engineer
                                     Westinghouse Electric Corp.
                                        (*Section 6.4*)

Marvin D. Weiss   BSE, MSChE
                                Assistant Professor of Mechanical Engineering
                                     Valparaiso University
                                        (*Sections 8.4 and 8.5*)

PREFACE

F. M. Ryan

Automation, like mechanization before it, has had profound social effects, though not the ones that were anticipated. Both of these technological trends have fundamentally changed the role of the production worker, but they have not displaced him.

When assembly lines and other forms of mechanization began to appear in what has been called the second industrial revolution, there was widespread concern over mass labor displacement and unemployment. Some people were mechanized out of jobs, but from the perspective of the overall economy, this was a short-term and small-scale effect. Population expansion, a rising standard of living, the growth of new consumer demands for goods and services, and other economic factors combined to override the unemployment problem in the long run.

The significant long-range effect of mechanization was to mechanize the production worker himself to some extent. He became almost an insignificant cog in the total production machine, like Charlie Chaplin in "Modern Times." *Isolated from his end product, he lost a sense of pride and meaning in his work. He was no longer a craftsman.*

As automation spread, a similar pattern appeared. The same fears of displacement and unemployment were expressed. Again, although this effect has certainly occurred, it is not a large-scale or long-range problem, nor is it the most significant social effect of automation. Automatic control has other more important justifications than reduction of manpower. In many industries, manpower allocations are already minimal, and are based on the number of people required to deal with emergencies that may occur. They cannot be further reduced, in spite of automation.

The really problematical effect of automation has been ·a further insidious alienation of the machine or process operator from his craft. Behind a barrier of automatic control equipment, he can lose his sense of the process and his feeling of control over it. Instead of being an active manipulator, he becomes a passive monitor. Control rooms—in the absence of emergencies—can be very boring environments. And the attention and alertness of operators suffers accordingly.

The human being is particularly suited to certain types of control functions—for example, to highly adaptive reactions; and particularly unsuited to certain others—for example, observing slowly accumulating trends or waiting for alarms. When emergencies do occur and his intervention is required, the effectiveness of the operator is limited by his isolation from the process and his lack of feel for what has happened.

A good man-machine interface that takes such problems into account is one of the recurring challenges in modern control system design. Communications between the operator and the system should be designed for the convenience and efficiency of the man, not the machine. The interface should be compatible with human sensing mechanisms and the laws of perception. Perhaps we need to design special servo-mechanized intermediaries between the operator and the process to give him a better feel for process conditions, as we do in many aircraft control systems. Maybe we need fairly elaborate simulations of the process and its controls to develop the operator's proficiency at interacting with the system. In control system design, we have far to go before we meet Weiner's criterion of "the human use of human beings."

Automatic control has great significance in another area of social concern—by implication rather than by effect so far. That is its pertinence to some of the social problems brought on by technology in general: air and water pollution, mass transit, traffic control. Many of these ecological and logistical crises are amenable to solutions based on automation theory and techniques. Instrumentation can give us the means to measure adverse effects, and control theory can give us the science to minimize them. We have hardly begun to apply either to these urgent questions.

Both of these challenges, the "human engineering" of automation systems and their application to some social problems, will demand the effective utilization of all our instrumentation resources, from sensors to computers. They will probably also require the development of new hardware and the elaboration of new theory. Most of all, they will require an information exchange among specialists in many fields. A handbook such as this one, by providing a catalog of resources and an inventory of information, serves as a valuable instrument in this process.

CONTENTS

CONTRIBUTORS   vii

PREFACE (F. M. Ryan)   xiii

CONTENTS   xv

LIST OF IMPORTANT TABLES   xx

INTRODUCTION (B. G. Lipták)   1

     Measurement Accuracy   1
     Definitions, Symbols and Abbreviations   13

CHAPTER I

CONTROL VALVES (H. D. Baumann, C. S. Beard,
     C. E. Gayler, B. G. Lipták, O. P. Lovett and
     F. D. Marton)   19

     Contents   20

     Introduction   30

     1.1   Terminology and Nomenclature   31
     1.2   Control Valve Sizing   40
     1.3   Globe Body Forms   81
     1.4   Bonnet Types, Packings and Lubricators   98
     1.5   Trim Designs   104
     1.6   Control Valve Accessories   118
     1.7   Positioners, Transducers and Boosters   123
     1.8   Diaphragm and Piston Actuators   132
     1.9   Butterfly Valves   145
     1.10   Saunders Diaphragm Valves   154
     1.11   Conventional Ball Valves   161
     1.12   Characterized Ball Valves   167
     1.13   Pinch Valves   171
     1.14   Sliding Gate Valves   179

1.15   Plug Valves   185
1.16   Ball and Cage Valves   197
1.17   Expansible Tube or Diaphragm Valves   202
1.18   Fluid Interaction Valve   208
1.19   Miscellaneous Valves and Actuators   211
1.20   Valve Actuators   221
1.21   Control Valve Selection and Application   256

CHAPTER II

OTHER TYPES OF FINAL CONTROL ELEMENTS   291
    (A. BRODGESELL)

Contents   292

Introduction   294

2.1   Throttling Electrical Energy   299
2.2   Variable-Speed Drives   309
2.3   Pumps and Feeders as Final Control Elements   324
2.4   Dampers   341

CHAPTER III

REGULATORS (L. S. DYSART, B. G. LIPTÁK, O. P. LOVETT and
    R. L. MOORE)   345

Contents   346

Introduction   347

3.1   Regulators vs. Control Valves   348
3.2   Pressure Regulators   351
3.3   Temperature Regulators   364

CHAPTER IV

CONTROLLERS AND LOGIC COMPONENTS
    (H. L. FELDMAN, P. M. KINTNER, B. G. LIPTÁK,
    C. L. MAMZIC, H. C. ROBERTS and J. E. TALBOT)   383

Contents   384

Introduction   387

4.1   Pneumatic vs Electronic Instrumentation   388
4.2   Electronic Controllers   398
4.3   Pneumatic Controllers   420

4.4	Function Generators and Computing Relays	457
4.5	Relays and Timers	481
4.6	Static Logic Switching Elements	498
4.7	Synthesis and Optimization of Logic Circuits	516

CHAPTER V

CONTROL BOARDS AND RECEIVER DISPLAYS
(R. W. Borut, S. P. Jackson, B. G. Lipták and F. D. Marton)

531

Contents 532

Introduction 534

5.1	Control Rooms and Panel Boards	535
5.2	Indicators	566
5.3	Recorders	575
5.4	Stand-by Power Supply Systems	587

CHAPTER VI

TRANSMITTERS AND TELEMETERING
(C. H. Hoeppner, B. G. Lipták, C. L. Mamzic, A. F. Marks and J. Valentich)

615

Contents 616

Introduction 618

6.1	Pneumatic Transmitters and Converters	619
6.2	Electronic Transmitters	640
6.3	Signal Converters	655
6.4	Telemetering Systems	662

CHAPTER VII

CONTROL THEORY (R. M. Bakke, B. G. Lipták, C. F. Moore, P. W. Murrill, F. G. Shinskey and C. L. Smith)

709

Contents 710

Introduction 714

| 7.1 | Automatic Control Terminology and Basic Concepts | 715 |
| 7.2 | Process Variables, Dynamics, Characteristics and Degrees of Freedom | 736 |

7.3   Controllers and Control Modes   754
7.4   Transfer Functions, Linearization and Stability
        Analysis   769
7.5   Closed-Loop Response With Various Control Modes   789
7.6   Feedback and Feedforward Control   802
7.7   Adaptive Control   813
7.8   Cascade Control   819
7.9   Ratio Control   826
7.10   Selective Control   832
7.11   Optimizing Control   837
7.12   Tuning of Controllers   842
7.13   Controller Tuning by Computer   859

CHAPTER VIII

PROCESS COMPUTERS (R. M. Bakke, J. W. Bernard,
      R. F. Jakubik, V. A. Kaiser, C. H. Kim, B. G. Lipták,
      W. L. Skaggs and M. D. Weiss)   873

Contents   874

Introduction   879

8.1   Computer Terminology   882
8.2   Process Computer Installations   900
8.3   Planning of Computer Projects   907
8.4   Computer Languages   915
8.5   Control Algorithms   931
8.6   Signal Conditioning and Desirable Wiring Practices   937
8.7   Computer Interface Hardware   961
8.8   Set-Point Stations   978
8.9   Memory Devices   985
8.10   Data Logging and Supervisory Program   1006
8.11   Optimizing Program   1041
8.12   Direct Digital Control   1069
8.13   Operator Communication   1086
8.14   Hierarchical Computers   1097
8.15   Application of Process Computers   1109

CHAPTER IX

ANALOG AND HYBRID COMPUTERS (A. B. Corripio,
      P. A. Holst, A. E. Nisenfeld and F. G. Shinskey)   1137

Contents   1138

9.1   Pneumatic Analog Computers   1140

9.2   Electronic Analog and Hybrid Computers   1151
9.3   Analog and Hybrid Computer Applications   1179
9.4   Analog Computers in Distillation Column Controls   1232
9.5   Process Reactor Models and Simulation   1259

                          CHAPTER X

PROCESS CONTROL SYSTEMS (R. J. BAKER, B. BLOCK,
                A. M. CALABRESE, S. G. DUKELOW, F. B. HOROWITZ,
                H. L. HOFFMAN, D. C. KENDALL, C. H. KIM,
                B. G. LIPTÁK, D. E. LUPFER, C. J. SANTHANAM and
                W. F. SCHLEGEL)   1269

Contents   1270

Introduction   1276

10.1   Control of Chemical Reactors   1277
10.2   Computer Control of Batch Reactors   1287
10.3   Control of Distillation Towers   1319
10.4   Optimizing Control of Distillation Columns   1343
10.5   Control of Refrigeration Units   1369
10.6   Control of Steam Boilers   1385
10.7   Control of Furnaces   1420
10.8   Control of Dryers   1449
10.9   Control of Crystallizers   1466
10.10   Control of Centrifuges   1477
10.11   Control of Heat Exchangers   1486
10.12   Control of Pumps   1513
10.13   Control of Compressors   1525
10.14   Effluent and Water Treatment Controls   1540
10.15   Analog and Digital Blending Systems   1559

                          APPENDIX

            (H. D. BAUMANN, B. G. LIPTÁK and G. PLATT)   1579

Contents   1580

A.1   Instrumentation Flow Sheet Symbols   1581
A.2   Interlock Logic Symbols   1592
A.3   Estimating Valve Noise   1600

SUBJECT INDEX   1607

List of Important Tables

Alphanumeric Computer Codes 920
A/D Converter Characteristics 970
Analog Computer, Integrator Modes 1159
Analog Computer Circuits 1153
Analog Computer Symbols 1181
Analog and Digital Techniques, Advantages 1121
Assembly Program to Calculate Output Signal From Temperature Deviation 924

Binary and Other Number Systems 918
Block Diagrams, Manipulations 779
Block Diagram Symbols 778
Bonnet Characteristics 280
Boolean Algebra Equivalents 528

Cavitation Resistance of Materials 264
Central Processor Characteristics 1106
Centrifuge Characteristics, Filtering 1483
Centrifuge Characteristics, Sedimentation 1479
Color Code for Control Panel Wiring 560
Compressor Capacity Control Methods 1525
Compressor Parameters, Throttling Method 1529
Computer, Ferrite Core Memory 989
Computer Application Factors 1135
Computer Codes, Alphanumeric 920
Computer Control Program, Idealized 930
Computer Form, Idealized Fill-in-the-Blanks 927
Computer Input Form Table 927
Computer Memory Characteristics 987
Computer Optimizing, Data for Sample Problem 1061
Computer Optimizing, Sample Problem 1059
Computer Optimizing, Sample Problem Output 1062
Computerized Process Control System Design 1111
Console Functions for Computerized Process Control Systems 1126
Control Devices, Service Applicability 299
Control Mode, Conventional 755
Control Mode Response to Upsets 756
Control Valve Discharge and Recovery Coefficients 44
Control Valve Material Selection 96
Control Valve Noise Sources 267
Control Valve Orientation Table 26
Control Valve Sizing Equations 42

Controller Tuning Comparison of 3c Equations   851
Controller Tuning Criteria   843
Controller Tuning Criteria Based on Load Disturbance   853
Controller Tuning Criteria Based on Set-Point Disturbance   853
Controller Tuning Criteria for Proportional Control Systems   867
Controller Tuning Equations   850
Critical Pressure of Various Fluids   51
Cryogenic Fluids, Properties   282

ddc, Reliability   1082
ddc Systems, Hierarchical   1108
Degrees of Freedom Analysis, Direct Contact Water Heater   741
Degrees of Freedom Analysis, Simple Water Boiler   743
Diode Characteristics   501
Disc Drives, Removable   1000
Disc Packs   1001
Distillation Column Optimization Equations   1351, 1358, 1364
Dryer Characteristics, Batch   1451
Dryer Characteristics, Continuous   1452

Electropneumatic Positioner Signals   128

Ferrite Core Memory Features   989
Fixed-Head Drum Memories   999
Fortran Program for PID Algorithm   926

Gauge Factor Variation With Temperature   673

Head Flowmeters, Ratio Settings   829
Head-per-Track Disc Memories   1000
Heat Exchanger Control, Two-Way vs Three-Way Valves   1495
Heater Control System Features   1501

Identification Letters, Their Meanings   1582
Information Required in a Computer-Controlled System   1084
Integrated Circuit Characteristics   513

K-Values for Various Vapors   65

Laplace Transform Pairs   771
Logic Design Operation Symbols   518
Logic Design Operation Symbols for Binary Functions   1593–97

Magnetic Tape Storage Systems, Reel-to-Reel   1001
Memory Device Characteristics   987
Moving-Head Disc Memories   1000
Multiplexer Feature Summary   957

Natural Frequency of Control Valves 140
Noise Conditions, Average, in Industries 941

Operation Code, Assembly Language Mnemonic 923
Operation Codes, Machine Language 922
Organizing a Multi-Level Computerized Process System 1113
Orientation Tables
    Control Valves 26
    Non-Valve Final Control Elements 296

PID Control Program, Overall, for Entire Plant 924
Packing and Bonnet Selection Table 280
Pipe Noise Attenuation Factor for Valves 1603
Power Factor Constant for Saturable Core Reactors 306
Pump Control Methods 1513

Refrigerant Characteristics 1372
Regulator Selection for Various Service Conditions 358
Regulators vs Control Valves 349
Regulators Used in the Gas Industry 358
Relay Characteristics, Range 484
Relay Function Symbols 1585
Relays, Hermetically Sealed 491
Relays, Open-Contact 490
Reliability of Plant Functions, Minimum Acceptable 1123
Rotary Drives, Features and Service Applicability 310

Set-Point Station Features 980
Sizing, Control Valve 42
Software Capabilities 1107
Stand-by Power Supply System Classifications 590
Stand-by System Component Redundancy 591
Strain Gauge Resistance vs Temperature 678
Stroking Time for Various Actuator Accessory Combinations 128
Symbols, Miscellaneous 1586
Symbols for Logic Design Operations 518
System Inaccuracy Summary 11

Thermal Bulbs, Construction Materials and Features 380
Transistor Characteristics 504
Transmission Distance vs Response Time 639
Trim Material Selection 111
Tuning Equations for PI Controllers Based on Load Disturbance 868
Tuning Equations for PID Controllers Based on Load Disturbance 869
Tuning Equations for Proportional Control Based on Load Disturbance 868

Valve Actuators, Non-Pneumatic   222
Valve Coefficients   44
Valve Sizing Formulas, Summary   42
Valve Stroking Times   128

Water Treatment Control Instrumentation   1557

Ziegler-Nichols and Cohen-Coon, Comparison of 3C Equations   851
Ziegler-Nichols and Cohen-Coon Equations   850

INTRODUCTION

B. G. Lipták

The scope of Volume I of Instrument Engineers' Handbook was limited to the measurement and safety aspects of process instrumentation. Volume II treats such major subject matters as process control systems, panels and panel instruments, control theory, control valves, and computers, as well as many lesser subject areas.

As noted by Edward Teller in his preface to the first volume, it is the general nature of handbooks and reference books to require periodic revised editions. This handbook is no exception. The contributors who have prepared the material for the first edition are constantly updating their files to make sure that no new development is unnoticed. All suggestions and comments from readers concerning the content, format and organization of the material will be welcomed and whenever possible will be acted upon and incorporated in editions to come.

Accuracy

There are a large number of technical subjects which would deserve separate discussion and elaboration as part of the introduction to this handbook. No topic, however, is more deserving than the subject of accuracy. The reasons for this are multiple:

> Accurate measurement of process variables is an essential prerequisite to good control.
> The term itself is poorly defined and is widely misunderstood.
> The interrelationships between accuracy, rangeability, calibration and maintenance are not always recognized.
> There is a tendency on the part of some manufacturers to use misstatements of accuracy as a sales gimmick whereby their products appear in a more favorable light than those of more responsible suppliers.

Terminology

"Accuracy" is defined by Webster as "freedom from error or the absence of error." This, to start with, is contrary to the widespread use of this term.

1

When an accuracy statement is given as "±1 percent accuracy," in almost all cases the intended meaning is "±1 percent inaccuracy." This mistake illustrates the carelessness (lack of accuracy) which prevails in dealing with this subject.

The purpose of all measurement is to obtain the true value of the quantity being measured, and error is thought of as the difference between the measured and the true quantity. Because it is impossible to measure a value without some uncertainty, it is equally impossible to know the exact error. All that can be stated in connection with the accuracy of a measurement, therefore, is the limits within which the true value will fall.

The accuracy related terminology can be illustrated by an example of target shooting, as shown in Figure 1. The spread in the nine shots fired in a tight pattern into the upper right hand corner of the target represents the *random error* of the shooter. His shooting is repeatable and precise, but

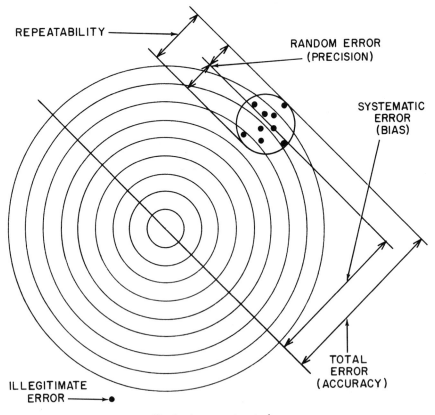

Fig. 1 Accuracy terminology

precision alone does not represent accuracy, it is only the measure of random error.

The deviation between the mean impact and the bullseye represents the *systematic error*. This error (caused by the wind or faulty adjustments of the sights) is repeatable and can be eliminated, because it is not related to the shooter's inability to duplicate his shots. Systematic error is also referred to as *bias*, which is the displacement of the measured or observed value from the true one.

The shot in the lower left corner of the target represents the *illegitimate error* which is caused by blunders, and such should be totally eliminated.

Accuracy of measurement can thus be defined as the *sum of random and systematic error*. If the purpose of an installation is to maintain process conditions at previously experienced levels, without having an interest in the true values of these conditions, then it is desirable to reduce the random error, without paying much attention to the remaining bias. In such an installation, in other words, a precise, repeatable, but inaccurate measurement is sufficient.

Inversely, if the interest is in approaching the true value of the measurement because the installation serves such absolute purposes as accounting or quality control, then a repeatable detection is insufficient, and attention must be concentrated on accuracy, which will be achieved only through the reduction of both random *and systematic* errors.

If it is impossible to determine the systematic error, it can still be corrected by calibration against a fixed standard, such as standard thermal elements, analytical samples or weights.

It is often suggested[1] to elaborate on accuracy statements, beyond the point of stating some percentage values. A good example of this is the following quotation taken from a National Bureau of Standards Calibration Certificate for a turbine flowmeter:

> The results given are the arithmetic mean of ten separate observations, taken in groups of five successive runs on two different days. The reported values have an estimated *overall uncertainty* of ±0.13 percent, based on a standard error of ±0.01 percent and an allowance of ±0.1 percent for possible systematic error.

Flow Measurement Example

In order to bring this discussion into an area of specific relevance to the process industry, an example of flow measurement will be used. Figure 2 shows three flow detection installations all serving the purpose of totalizing a process stream. All flow sensors are for a full range of 100 GPM instantaneous flow rate, and it will be our purpose to evaluate their accuracy at the flows of 20 GPM and 80 GPM. The evaluation will consider two different basic

[1]R. E. Kemp, "Accuracy For Engineers," *Instrumentation Technology.*

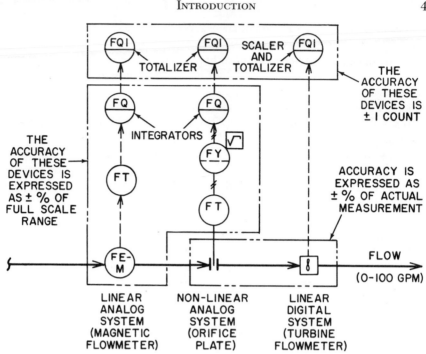

Fig. 2 Variations on possible flow totalization loops

assumptions—one, that component errors are *additive* and two, that system accuracy is likely to approach the accuracy of the least precise component in the system. Errors introduced by counter-totalizers will be neglected; devices having their accuracy based on the full scale range will be assumed to be inaccurate to ±0.5 percent FS; orifice plates will be treated as having a ±0.5 percent *of rate* inaccuracy; and a value of ±0.25 percent of rate will be used for turbine flowmeters. (Refer to Table V in Volume I for details on flow sensor characteristics.)

Analog, Linear

The ideal behavior for a linear flow sensor, such as a magnetic flowmeter, is shown in Figure 3. The line marked "actual" represents the relationship between the true flow and the output signal generated by a specific flow sensor. This deviation from the ideal is plotted in Figure 4 in terms of error, as a percentage of *full scale*, with the error limits being ±0.5 percent FS.

The same performance, if plotted as a function of the actual flow rate reading (instead of full scale), results in the relationship shown in Figure 5. The accuracy limits shown are conservative, in the sense that the specific detector performance is better, at most points of its range, than what the error limits would imply. It is also true that the meter inaccuracy increases with

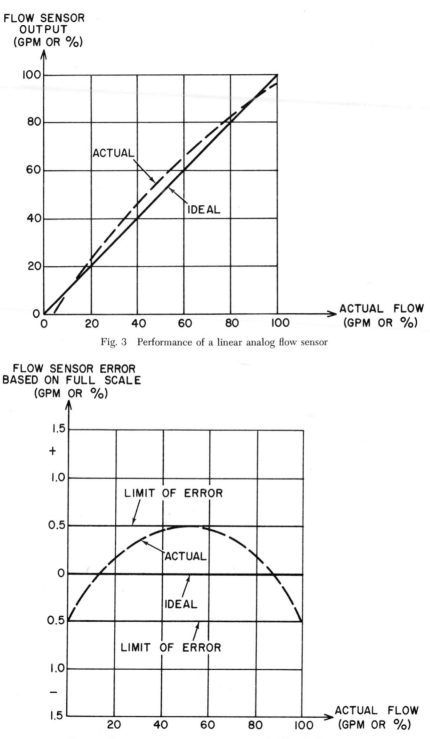

Fig. 3   Performance of a linear analog flow sensor

Fig. 4   Flow sensor error as a percentage of full-scale flow rate

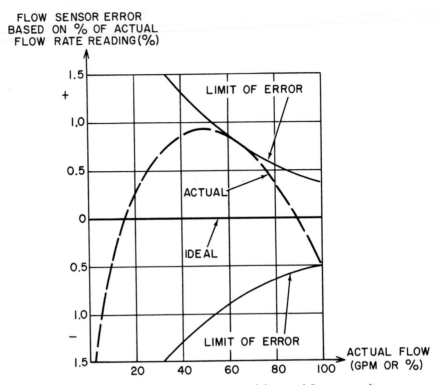

FLOW SENSOR ERROR
BASED ON % OF ACTUAL
FLOW RATE READING (%)

Fig. 5   Flow sensor error as a percentage of the actual flow rate reading

dropping flow rates if the inaccuracy is expressed as a percentage of actual reading, as shown in Figure 6.

Analog, Non-Linear

In case of an orifice type flow sensor, the actual measurement (pressure drop across the orifice plate) has a square relationship to the desired measurement (flow). Figure 7 illustrates both this ideal non-linear relationship and the actual performance of a specific instrument. Because in most cases the square root must be extracted before the signal becomes useable (in our example of Figure 2, a square-root extractor is installed before the integrator), the "gain effect" of this extraction must be recognized. As shown by Figure 8, the extraction of the square root improves the accuracy at the higher flow rates, but degrades it as the flow rate is reduced.

Figure 9 depicts the relationship between the pipeline Reynolds number and the discharge coefficient for various head type flow elements. The Reynolds number is determined by the use of equation 1.2(33):

$$\text{Re} = \frac{3.160 G_f Q_f}{D\mu} \qquad 1.2(33)$$

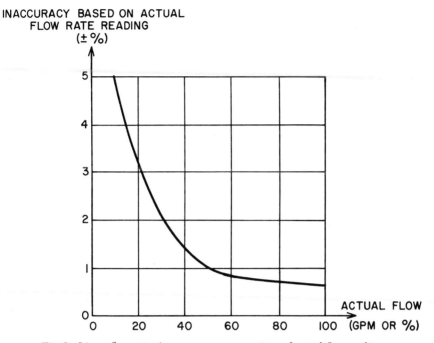

INACCURACY BASED ON ACTUAL
     FLOW RATE READING
           (± %)

Fig. 6   Linear flowmeter inaccuracy as a percentage of actual flow reading

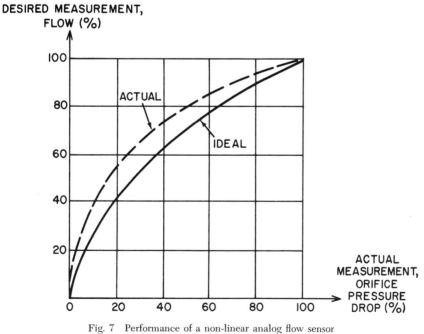

DESIRED MEASUREMENT,
      FLOW (%)

Fig. 7   Performance of a non-linear analog flow sensor

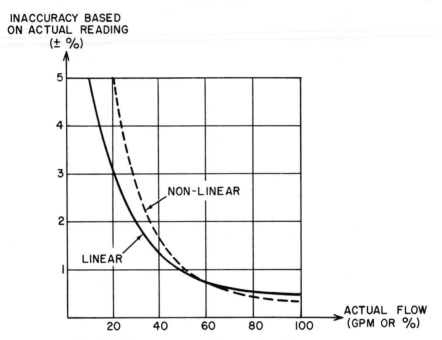

Fig. 8   Relative inaccuracies of linear and non-linear flowmeters

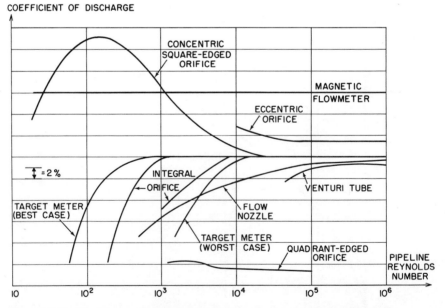

Fig. 9   Discharge coefficients as a function of sensor type and Reynolds number.
(Courtesy of The Foxboro Company.)

where G_f = process fluid specific gravity (at 60°F),
Q_f = liquid flow in GPM (at 60°F),
D = pipe inside diameter (in inches), and
μ = process fluid viscosity (in centipoises).

As shown by Figure 9 the orifice plate discharge coefficient is constant within ±0.5 percent over a Reynolds number range of 2×10^4 to 10^6. This consistency of the discharge coefficient guarantees a corresponding limitation of inaccuracies based on *actual* flow, over a range that is wider than what the d/p flow transmitter can handle. This capability of the orifice plate element is illustrated in Figure 10.

Digital, Linear

Figure 11 illustrates the calibration of a turbine meter in terms of the K factor (pulses per gallon), which is rather similar to the calibration curve of an orifice plate (Figure 9). The inaccuracy of a turbine meter is also stated as a percentage of the *actual* flow reading and not of full scale. As shown in Figure 12, turbine meter accuracy can be improved by reducing the rangeability requirement of the unit.

System Accuracy

Having reviewed the inaccuracies of the various components in the loops shown in Figure 2, the next step is to evaluate the resulting overall system accuracies. It is important to emphasize here that there is no proven basis for evaluating the accumulative effect of component inaccuracies, and only an actual system calibration can reliably establish the total inaccuracy. It should also be emphasized that the probability favors the minimum number of components in a measurement loop to result in the best accuracy system. The only exception to this is the fact that, in case of digital systems, no additional error is introduced by the addition of functional modules.

Without actual system calibration the evaluation of loop accuracy must be based on assumptions. Table 13 summarizes system inaccuracies that can be expected under various conditions. The accumulated effect of component inaccuracies has been based on one of two assumptions:

Basis 1—Here it is assumed that the inaccuracy of each component is *additive,* and therefore system inaccuracy is the sum of component inaccuracies (very conservative basis).

Basis 2—Here the assumption is that all component inaccuracies can be neglected except that of the least accurate component, and therefore system inaccuracy is the same as the inaccuracy in this one component (very optimistic assumption).

From Table 13, it is seen that if Basis 1 is accepted for system accuracy evaluation, an orifice type installation operating at 20 percent of full scale

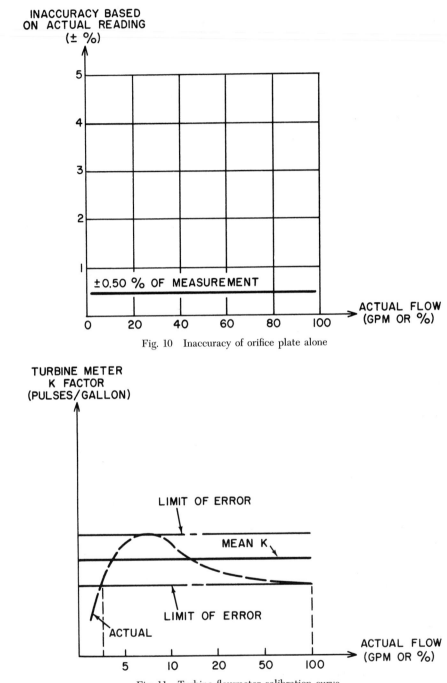

Fig. 10   Inaccuracy of orifice plate alone

Fig. 11   Turbine flowmeter calibration curve

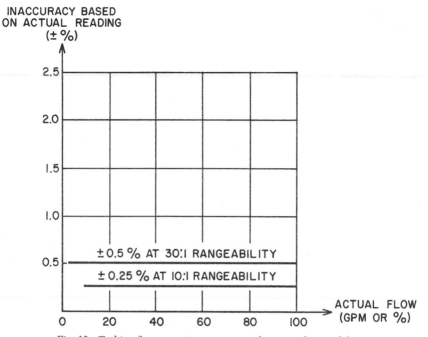

Fig. 12   Turbine flowmeter inaccuracy as a function of rangeability

Table 13
SYSTEM INACCURACY SUMMARY BASED ON
ACTUAL READINGS

Assumption Used to Estimate Accumulated System Inaccuracy	Basis 1		Basis 2	
Type of Flow Detection Loop — Operating Flow Rate (GPM)	20	80	20	80
Analog, Linear (Magnetic Flowmeter)	±9.0%	±1.5%	±3.0%	±0.5%
Analog, Non-linear (Orifice Flowmeter)	±12.0%	±2.0%	±5.0%	±0.5%
Digital, Linear (Turbine Flowmeter)	±0.25%	±0.25%	±0.25%	±0.25%

will be inaccurate to ±12 percent of the reading, although no component inaccuracy in the loop exceeds ±0.5 percent FS.

Figure 14 illustrates the system inaccuracies if they were evaluated on a basis slightly more conservative than Basis 2 and much less conservative than Basis 1.

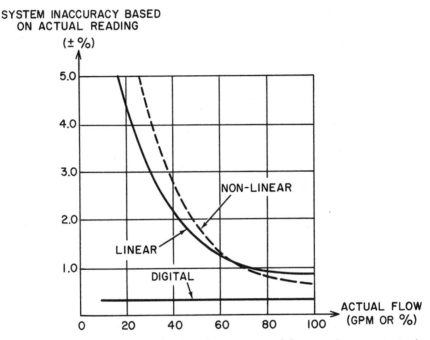

Fig. 14   System inaccuracies as a function of sensor type and flow rate. (Accuracy in simple flow measurement, TI-1-30a, The Foxboro Company.)

From the data in Table 13 and Figure 14, it can be concluded that accuracy is by no means a clearly defined single number when one wishes to evaluate the performance of a multi-component system under varying load conditions.

From the foregoing discussion, therefore, only qualitative conclusions can be drawn. These include—

Accuracy is likely to be improved by reducing the number of components in a measurement loop.

Accuracy is meaningful only in combination with rangeability. The wider the rangeability required (expected load variations), the more inaccurate the measurement is likely to be. Furthermore, the linear analog system accuracies are less affected by rangeability requirements than are non-linear analog systems, and the rangeability effect on digital systems is less than on any other type loop.

On non-accounting systems, the interest is focused on repeatability (random error) and not on accuracy. The repeatability of most measurement loops is several-fold better than their accuracy.

Instrumentation worth installing is worth calibrating. In this regard, several points should be made; namely, that the accuracy of a multi-component system is unknown unless calibrated as a system, that the calibration equipment must be *at least three times* more accurate than the system being calibrated and that periodic recalibration is an essential prerequisite to good control. (For an automatic calibration system, refer to Figure 10.121.)

Instrumentation worth installing is worth keeping in good condition. All measurement devices are process limited in the sense that their performance can be affected by corrosion, plugging, coating or process property variations, and therefore scheduled maintenance is required to guarantee reliable operation.

In summing up this discussion of accuracy, it is important that the terms used be clearly understood by those using them, that accuracy be stated as a function of rangeability, that multi-component systems be distinguished from single-component ones, that the prerequisites of calibration and scheduled maintenance be emphasized.

Definitions, Symbols and Abbreviations

The reader is referred to the following locations for definitions relating to instrumentation:

Volume I—
>Pages 6 to 17 for general definitions.
>Pages 395 and 396, relating to pyrometry.
>Pages 574 to 576, relating to viscometry.
>Pages 966 to 969, relating to relief valves.
>Pages 1060 to 1068, relating to electrical safety definitions.

Volume II—
>Section 1.1 for control valve terminology.
>Section 7.1, relating to control theory.
>Section 8.1 for computer terminology.

Listed below are some abbreviations and symbols which an Instrument Engineer is likely to use:

A   Angstrom
AC   Alternating current

ACFM	Volumetric flow at actual conditions in units of cubic feet per minute.
ACS	American Chemical Society
A/D	Analog-to-digital
AEC	Atomic Energy Commission
AGA	American Gas Association
AM	Amplitude modulation
A/M	Automatic-to-manual
ANS	American National Standard
ANSI	American National Standards Institute
APHA	American Public Health Association
API	American Petroleum Institute
°API	API degrees of liquid density
ARI	Air Conditioning and Refrigeration Institute
ASA	American Standards Association
ASCE	American Society of Civil Engineers
ASME	The American Society of Mechanical Engineers
ASRE	American Society of Refrigerating Engineers
ASTM	American Society for Testing and Materials
atm	Atmosphere
AWG	American wire gauge
bar	Barometer
°Ba	Balling degrees of liquid density
BCO	Bridge-controlled oscillator
°Be	Baumé degrees of liquid density
°Bk	Barkometer degrees of liquid density
°Br	Brix degrees of liquid density
bbl	Barrels
BSI	British Standards Institution
BTU	British thermal units
BWG	Birmingham wire gauge
°C	Degrees centigrade
cal	Calories
cc	Cubic centimeter
cgs	Centimeter-gram-second
CI	Cast iron
CIL	Canadian Industries Limited flow index
cm	Centimeter
cpm	Cycles per minute
cps	Cycles per second
cps	Centipoises
CPU	Central processor unit
CRT	Cathode ray tube

CSA	Canadian Standards Association
ctks	Centistokes
cu	Cubic
D	Derivative
D/A	Digital-to-analog
db	Decibel
DC	Direct current
ddc	Direct digital control
deg	Degrees
diam	Diameter
DO	Dissolved oxygen
d/p cell	Differential pressure transmitter (A Foxboro trademark)
DPDT	Double pole, double throw
emf	Electromotive force
Eq	Equation
°F	Degrees Fahrenheit
FC	Fails closed
FCI	Fluid Controls Institute
FIA	Factory Insurance Associations
FM	Factory Mutual
FM	Frequency modulation
FO	Fails open
FPA	Fire Protection Association
fpm	Foot per minute
fps	Foot per second
FS	Full scale
ft	Foot
ft-lb	Foot pounds
g	Acceleration due to gravity
gal	Gallons
g-cal	Gram-calories
G-M	Geiger-Mueller
GPH	Gallons per hour
GPM	Gallons per minute
hhv	Higher heating value
hr	Hours
Hz	Hertz, frequency per second (kHz = 10^3 Hz, MHz = 10^6 Hz)
I	Integral
ICE	Institute of Civil Engineers
ID	Inside diameter
IEC	International Electrotechnical Commission
IEEE	Institute of Electrical and Electronics Engineers
I/O	Input and output

in.	Inches
IPS	Iron pipe size
IPTS	International practical temperature scale
IR	Infra-red
ISA	Instrument Society of America
ISO	International Organization for Standardization
ISTM	International Society for Testing Materials
JCU	Jackson candle units
°K	Degrees Kelvin (centigrade absolute)
Kg	Kilograms
Kips	Thousands of pounds
Km	Kilometers
Kw	Kilowatts
l	Liters
lb	Pounds
LEL	Lower explosive limit
LP	Liquid petroleum
LPGA	National LP-Gas Association
LVDT	Linear variable differential transformer
m	Meters
M	Thousand; Mach number
M	Mega (10^6)
ma	Milliamperes
madc	Milliamperes, direct current
mc	Millicurie
MCA	Manufacturing Chemists' Association
MI	Melt index
ml	Milliliter
mm	Millimeters
mmf	Magnetomotive force
mph	Miles per hour
mR	Milliroentgen
ms	Milliseconds
mv	Millivolts
mvdc	Millivolts, direct current
Mw	Megawatt
Mw	Molecular weight
NBFU	National Board of Fire Underwriters
NBS	National Bureau of Standards
NDIR	Non-dispersive infrared
NEMA	National Electrical Manufacturers Association
NFPA	National Fire Protection Association
NSC	National Safety Council

OD	Outside diameter
ORP	Oxidation-reduction potential
OTS	Office of Technical Services
oz	Ounces
PB	Proportional band
PD	Proportional and derivative
pf	Pico-farad (10^{-9} farad)
pH	Acidity index (negative logarithm of hydrogen ion concentration)
PI	Proportional and integral
PID	Proportional, integral and derivative
PM	Phase modulation
ppb	Parts per billion
ppm	Parts per million
PSI	Pounds per square inch
PSIA	Pressure in pounds per square inch absolute
PSID	Differential pressure in pounds per square inch
PSIG	Above atmospheric (gauge) pressure in pounds per square inch
PVC	Polyvinyl chloride
°Q	Quevenne degrees of liquid density
rad	Radius
RC	Resistance-capacitance
rev	Revolutions
Re	Reynolds number
RF	Radio frequency
RH	Relative humidity
RI	Refractive index
rms	Root mean square
rpm	Revolutions per minute
rps	Revolutions per second
RTD	Resistance temperature detector
SAMA	Scientific Apparatus Makers Association
SCFH	Standard cubic feet per hour
SCFM	Standard cubic feet per minute
SCR	Silicone-controlled rectifier
sec	Seconds
SG	Specific gravity
SPDT	Single-pole double-throw
SPST	Single-pole single-throw
sq	Square
SSU	Saybolt univeral seconds
SWG	Standard (British) wire gauge
T_d	Rate (or derivative) time
T_i	Reset (or integral) time

°Tw	Twaddell degrees of liquid density
UEL	Upper explosive limit
UHSDS	Ultra-high-speed deluge systems
UL	Underwriters' Laboratories
UPV	Unfired pressure vessel
USAS	USA Standards
USASI	USA Standards Institute
UV	Ultraviolet
VA	Volt-ampere
VCO	Voltage-controlled oscillator
vs	Versus
wt	Weight
yd	Yards
yr	Years

GREEK ALPHABET

Alpha	A	α	Eta	H	η	Nu	N	ν	Tau	T	τ
Beta	B	β	Theta	Θ	Θ ∂	Xi	Ξ	ξ	Upsilon	Υ	υ
Gamma	Γ	γ	Iota	I	ι	Omicron	O	o	Phi	Φ	φ ϕ
Delta	Δ	δ	Kappa	K	κ	Pi	Π	π	Chi	X	χ
Epsilon	E	ε	Lambda	Λ	λ	Rho	P	ρ	Psi	Ψ	ψ
Zeta	Z	ζ	Mu	M	μ	Sigma	Σ	σ s	Omega	Ω	ω

Acknowledgments

I wish to thank my wife, Mártha, and our children for tolerating the consequences of this extra-curricular activity during the last few years; Pamela Smith not only for typing all the correspondence, but also for correcting my spelling errors; Samuel Russell of Crawford & Russell for providing me with secretarial and other help; William Carlson for reproducing manuscripts at a speed "beyond the call of duty"; and all the contributors for their spirit of cooperation and for their patience in submitting to modifications of their sections.

Finally, I wish to thank my parents for having taught me that with faith in God and with determined hard work, nothing is impossible.

BÉLA G. LIPTÁK

Stamford, Connecticut
January 13, 1970

Chapter I

CONTROL VALVES

H. D. Baumann, C. S. Beard,
C. E. Gayler, B. G. Lipták,
O. P. Lovett and F. D. Marton

CONTENTS OF CHAPTER I

	INTRODUCTION	30
1.1	TERMINOLOGY AND NOMENCLATURE	31
	General Terms	31
	Terms Relating to the Valve Body	31
	Terms Relating to the Valve Actuator	37
1.2	CONTROL VALVE SIZING	40
	Introduction	40
	Valve Coefficients	41
	Sizing for Liquid Service	45
	The Basic Equation	45
	Critical Flow and Cavitation	46
	The Pressure Recovery Factor	49
	Example 1 for Cavitating Flow	51
	Pipe Reducer Effects	52
	Example 2 for Sizing With Reducers	53
	Reducer Effects on Critical Flow	57
	Example 3 for Reducers and Critical Flow	59
	Sizing for High Viscosity	59
	Alternative Viscous Sizing Technique	62
	Example 4 for High-Viscosity Sizing	63
	Non-Newtonian Fluids	64
	Gas and Vapor Sizing	64
	Non-Critical Gas Service	64
	Non-Critical Vapor Service	65
	Critical Gas or Vapor Service	65
	Generalized Sizing Formula for Compressible Flow	67
	Example 5 for Compressible Fluid Sizing	68
	Flashing Liquids	70
	Sum of Cv's Technique	70
	Critical Pressure Method	71
	Mixture Density Technique	71
	Outlet Density Method	73
	Example 6 for Flashing Sizing	73
	Gas-Liquid Mixtures	74

Assigning Control Valve ΔP 74
Safety Factors in Valve Sizing 77
Nomenclature 78
References 79
1.3 GLOBE BODY FORMS 81
Body Forms 81
Single-Seated Valves 82
Double-Seated Valves 83
Cage Valves 84
Split-Body Valves 86
Angle Valves 87
Y-Style Valves 88
Jacketed Valves 89
Three-Way Valves 89
Small-Flow Valves 90
Valve Connections 92
Materials of Construction 96
References 97
1.4 BONNET TYPES, PACKINGS AND LUBRICATORS 98
Bolted Bonnets 98
Packings 99
Bellows Seals 100
Double Packings 101
Radiation Bonnets 102
Extension Bonnets 103
1.5 TRIM DESIGNS 104
Flow Characteristics 104
Rangeability 105
Trim Configurations 108
Trim Materials 110
Plug Stems 112
Noise and Wear Reduction 113
Multiple Orifices in Parallel 113
Multiple Orifices in Series 115
Multiple Orifices in Series and Parallel 116
References 117
1.6 CONTROL VALVE ACCESSORIES 118
Handwheels 118
Limit Switches 119
Solenoid Valves 119
Limit Stops 120
Stem-Position Indicators 121
Airsets 121

1.7   Positioners, Transducers and Boosters   123
     Pneumatic Valve Positioners   124
     Electropneumatic Positioners   127
     Electropneumatic Transducers   129
     Boosters   130
1.8   Diaphragm and Piston Actuators   132
     Definitions   133
     Actuator Performance   133
     The Steady-State Equation   134
     Actuator Sizing Example   136
     Actuator Non-Linearities   137
     Dynamic Performance of Actuators   137
     Spring-Mass System Dynamics   138
     Pneumatic System Dynamics   140
     Piston Actuators   141
     Safe Failure   142
     Relative Merits of Diaphragm and Piston Actuators   144
     Nomenclature   144
1.9   Butterfly Valves   145
     History   145
     Operation   146
     Construction   147
     T-Ring Seal   149
     Piston Ring Seal   149
     Linings   150
     Dual-Range, Multi-Disc Designs   151
     Fish-tail Discs   151
     Torque Characteristics   151
1.10   Saunders (Diaphragm) Valves   154
     Operation   154
     Construction   156
     Straight-Through Design   156
     Full-Bore Valve   157
     Dual-Range Design   157
1.11   Conventional Ball Valves   161
     Features   162
     Body   163
     Trim   164
     Connections   165
     Flow Characteristics   165
     Sizing   166
     Conclusions   166

1.12 CHARACTERIZED BALL VALVES   167
      Operation   167
      Construction   168
      Characteristics   170
1.13 PINCH VALVES   171
      Mechanical Pinch Design   173
      Pneumatic Pinch Design   175
      Shutter Closure   176
      Accurate Closure   177
1.14 SLIDING GATE VALVES   179
      Knife Gate Valves   179
      Positioned-Disc Valves   182
      Plate-and-Disc Valves   182
1.15 PLUG VALVES   185
      V-Ported Design   186
      Adjustable Cylinder Type   188
      Semispherical Plugs for Tight Closure   191
      Expanding Seat Plate Design   191
      Eccentric Shaft Design   191
      Retractable Seat Type   193
      Overtravel Seating Design   193
      Eccentric Rotating Spherical Segment Type   195
1.16 BALL AND CAGE VALVES   197
      Cage Positioned Ball Design   197
      Ball Unseated by Stem   199
      Ball Gripped by Cage   201
1.17 EXPANSIBLE TUBE OR DIAPHRAGM VALVES   202
      Expansible Tube Type   202
      Expansible Diaphragm Design   206
1.18 FLUID INTERACTION VALVES   208
1.19 MISCELLANEOUS VALVES AND ACTUATORS   211
      Expansible Element In-Line Valves   211
      Positioned Plug In-Line Valves   212
      Diaphragm-Operated Cylinder In-Line Valves   215
      Digital Control Valves   216
      Digital-to-Pneumatic Positioners   217
      Digital-to-Pneumatic Transducers   218
      Digital Valve Actuators   218
1.20 VALVE ACTUATORS   221
      Pneumatic Actuators   223
          Cylinder Type   223
          Rotation by Spline or Helix   225

Vane Type 228
Pneumo-Hydraulic Type 229
Rotary Pneumatic Type 231
Electro-Pneumatic Actuators 232
Electric Actuators 233
Linear Actuators (Solenoids) 233
Electro-Hydraulic Actuators 236
External Hydraulic Source 236
Hermetically Sealed Power Pack 237
Reversible Motor and Pump 240
Jet Pipe Systems 240
Hydraulic Control of Jacketed Pinch Valves 240
Multiple Pump Systems 241
Vibratory Pumps 241
Two-Cylinder Pumps 243
Electric Rotary Actuators for Linearized Stem Motion 243
Reciprocating Motion Type 245
Linearized Stem Motion Through Cams and
  Eccentrics 245
Drive Sleeve Type 247
Electric Quarter-Turn Actuators 252
1.21 CONTROL VALVE SELECTION AND APPLICATION 256
Leakage 256
Factors Affecting Leakage 257
Soft Seats 257
Single vs Double Ports 257
Temperature Effects 257
Pipeline Forces 258
Seating Forces and Materials 258
Rangeability 258
Leakage and Clearance Flows 259
Valve Characteristics 259
Installed Valve Behavior 260
Cavitation 262
Methods to Eliminate Cavitation 262
Revised Process Conditions 263
Revised Valve 263
Revised Installation 263
Material Selection for Cavitation 263
Noise, Vibration and Flowing Velocity 265
The Phenomena of Sound and Noise 265
Controlling Noise 265

Sources of Control Valve Noise	269
Limiting High-Velocity Gas Flow	269
Reducing Gas Regulator Noise	270
Reducing Cavitation Noise	271
Natural Frequency Vibration	271
Horizontal Plug Vibration	272
Vertical Plug Oscillation	272
High-Pressure Service	274
High-Temperature Service	276
Limitations of Metallic Parts	276
Packing Limitations	277
Jacketed Valves	280
Low-Temperature Service	281
Vacuum Service	283
Small-Flow Valves	285
Viscous and Slurry Service	288

Table I

ORIENTATION TABLE FOR CONTROL VALVES

Features	Split Body, Single Seated	Angle Valve, Single Seated	Globe Valve, Top-Bottom Guided, Double Seated, V-port	Cage Valve, Top Entry, Single Seated	Butterfly Valve 60° Open	Butterfly Valve 90° Open	Saunders Diaphragm Valve
Applicable Section of Chapter I	1.3	1.3	1.3	1.3	1.9	1.9	1.10
Capacity [$C_v = (C_d)(d^2)$] Critical	$10d^2$	$14d^2$	$10d^2$	$12d^2$	$12d^2$	$20d^2$	$14d^2$
Non-critical	$12d^2$	$28d^2$	$12d^2$	$14d^2$	$17d^2$	$35d^2$	$22d^2$
Flow Characteristics	Good	Good	Good	Good	Fair–Good	Fair	Poor[4]
Cost	Medium	Medium–High	Medium	Medium	Low	Low	Low–Medium
Available Size Range (in.)	¼–12	1–6	1–24	1–4	1–150	1–150	¼–20
Max./Min. Operating Pressure (PSIG)	6,000/vacuum	6,000/vacuum	6,000/vacuum	6,000/vacuum	2,000/vacuum	2,000/vacuum	200/<1 mm Hg[9]
Max. Pressure Drop (PSID)	3,000	3,000	2,000	1,000	1,000	1,000	100[5,9]
Max./Min. Operating Temp. (°F)	1,200/−40	1,200/−40	1,200/−40	1,200/−40	2,000[2]	2,000[2]	350/−80[7,9]
Available in Jacketed Construction	No	Yes	Yes	Yes	No	No	Yes
Applicable to Flashing Service	Yes	Yes	Limited	Yes	No	No	No
Applicable to Cavitating Service	Limited	Limited	Yes	Yes	No	No	No
Applicable to High Viscosity ($\mu > 10^4$ cps) Service	Limited	Limited	No	No	Limited	Limited	Yes–Limited
Applicable to Slurry Service	Yes	Yes–Limited	No	Limited	Limited	Limited	Yes
Tight Shut-off Available[1]	Yes[3]	Yes[3]	No (¼–2% leakage)	Yes[3]	Yes[3]	Yes[3]	Yes
Available Rangeability[10]	Up to 50:1	Up to 50:1	30:1 to 50:1	30:1	20:1	30:1	15:1[6]
Has "Fail-safe" (Open or Closed) Action	Yes	Yes	Yes	Yes	Yes	Yes	Yes
Available in Variety of Materials	Yes	Yes	Yes	Yes	Yes	Yes	Limited[7]
Flow in Either Direction	Yes	Yes	Possible	No	Some designs	Some designs	Yes
Unobstructed Streamline Flow	Not fully	Some designs	No	No	Not fully	Not fully	Almost[8]
Has Tendency to Cavitate Due to High Recovery	Some		No	No	Yes	Yes	Yes
Stem or Shaft Sealed from Process by	Packing	Packing	Packing	Packing	Packing and O-ring	Packing and O-ring	Diaphragm
Valve Trim Easily Removed	Yes	No	Yes	Yes	No	No	No
Low Maintenance	Yes	No	No	Yes	Yes	Yes	No

[1] See section 1.21 for a discussion of leakage.
[2] Metal-to-metal seated designs only.
[3] Soft-seated designs only.
[4] Fair with dual range design.
[5] Diaphragm life short at high ΔP.
[6] Better with dual range design.
[7] Limited by diaphragm material.
[8] Yes, with straight-through design.
[9] See Figure 1.10e.
[10] Refer to section 1.21 for the definition of rangeability.

26

Table I (Continued)

Features	Conventional Ball Valve	Characterized Ball Valve	Pinch Valve	Sliding Gate Valves Plate and Disk Type	Sliding Gate Valves V-insert Type	Sliding Gate Valve Positioned Disk Type
Applicable Section of Chapter I	1.11	1.12	1.13	1.14	1.14	1.14
Capacity $[C_v = (C_d)(d^2)]$ Critical	$14d^2$	$14d^2$	$25d^2$	$5d^2$–$10d^2$	$18d^2$	$8d^2$
Non-critical	$24d^2$	$22d^2$	$60d^2$	$6d^2$–$12d^2$	$22d^2$	$9d^2$
Flow Characteristics	Good	Good	Poor[12]	Fair	Fair	Fair
Cost	Medium	Medium	Low	Medium	Low	Low
Available Size Range (in.)	⅛–42	2–16	⅛–24	½–6	2–36	1–3
Max./Min. Operating Pressure (PSIG)	10,000/<1 mm Hg	350/<1 mm Hg	150/vacuum	300/vacuum	10^{14}	10,000/vacuum
Max. Pressure Drop (PSID)	1,000	200	60	300	10^{14}	5,000
Max./Min. Operating Temperature (°F)	1,000/−300	600	350	450	1,000	750
Available in Jacketed Construction	Yes	No	Yes	No	No	No
Applicable to Flashing Service	No	No	No	Yes	No	Yes
Applicable to Cavitating Service	No	No	No	Yes	No	Yes
Applicable to High Viscosity ($\mu > 10^4$ cps) Service	Yes—Limited	Limited	Yes	No	Limited	No
Applicable to Slurry Service	Yes	Yes	Yes	No	Limited	No
Tight Shut-off Available[1]	Yes	Yes	Yes	Yes[3]	Yes[3]	Yes[3]
Available Rangeability[10]	Over 50:1	Over 50:1	15:1	40:1	15:1	40:1
Has "Fail-safe" (Open or Closed) Action	Yes	Yes	No	Yes	No	Yes
Available in Variety of Materials	Yes	Limited	Limited[13]	No	No	No
Flow in Either Direction	Yes	No	Yes	No	Yes	Yes
Unobstructed Streamline Flow	Not fully	No	Yes	No	No	No
Has Tendency to Cavitate Due to High Recovery	Yes	Yes	Yes	No	No	No
Stem or Shaft Sealed from Process by	O-ring	O-ring/packing	Sleeve	Packing	Packing	Delrin seals
Valve Trim Easily Removed	Yes[11]	No	No	No	Yes	No
Low Maintenance	Yes	No	No	No	Yes	No

11 Only in top-entry design.
12 Better in second half of stroke and with the "shutter closure" design.
13 Limited by sleeve material.
14 More in sizes under 10 in.

27

Table I (Continued)

Features / Type of Control Valve Body	Plug Valve V-Ported Type	Plug Valve Adjustable Cylinder Type	Plug Valve Expanding Seat Plate Type	Plug Valve Eccentric Shaft Type	Plug Valve Retractable Seat Type	Plug Valve Eccentric Rotating Type (Camflex)	Ball and Cage Type Valves
Applicable Section of Chapter I	1.15	1.15	1.15	1.15	1.15	1.15	1.16
Capacity [$C_v = (Cd)(d^2)$] Critical	$12d^2$	$12d^2$	$10d^2$	$14d^2$	$14d^2$	$12d^2$	$14d^2$
Non-critical	$17d^2$	$14d^2$	$16d^2$	$24d^2$	$24d^2$	$14d^2$	$20d^2$
Flow Characteristics	Good	Fair	Fair	Fair	Fair	Good	Good
Cost	Low	Medium	Low	Medium	Low	Medium	Medium–High
Available Size Range (in.)	½–24	⅜–6	2–16	4–60	4–12	1–12	¼–14
Max./Min. Operating Pressure (PSIG)	700/vacuum	125	700	600/vacuum	10,000	1,200/vacuum	2,500/vacuum
Max. Pressure Drop (PSID)	400	40[15]	500	200	5,000	300[16]	2,500
Max./Min. Operating Temperature (°F)	450	400	400	450	450/−100	750/−320	1,800/−425
Available in Jacketed Construction	Yes	No	No	No	No	No	No
Applicable to Flashing Service	No	No	No	No	No	Yes	Yes
Applicable to Cavitating Service	Limited	No	Limited	No	No	Yes	Yes
Applicable to High Viscosity ($\mu > 10^4$ cps) Service	Limited	Limited	Limited	No	No	Yes—Limited	Yes—Limited
Applicable to Slurry Service	Yes	No	No	No	No	Yes	Yes
Tight Shut-off Available[1]	Yes	No	Yes	No	Yes	Yes	Yes
Available Rangeability[10]	20:1	10:1	20:1	Over 50:1	Over 50:1	Over 50:1	Over 50:1
Has "Fail-safe" (Open or Closed) Action	Yes	Yes	Yes	No	No	Yes	Yes
Available in a Great Variety of Materials	Yes	No	No	Yes	Yes	No	Average
Flow in Either Direction	No	No	Yes	No	No	Yes	No
Unobstructed Streamline Flow	No	Not Fully	No	Not fully	Not fully	Not fully	Not fully
Has Tendency to Cavitate Due to High Recovery	Some	No	Some	Yes	Yes	Some	No
Stem or Shaft Sealed from Process by	Packing	Precision fit	Packing	Packing	Packing	Packing	Packing
Valve Trim Easily Removed	Yes	No	Yes	No	No	No	Yes
Low Maintenance	Yes	Average	Yes	Average	Average	Average	Average

[15] ½ PSID on gas service.   [16] More with positioner.

28

Table I (Continued)

Features	Type of Control Valve Body	
	Expansible Tube or Diaphragm Type Valve	Fluid Interaction Valve
Applicable Section of Chapter I	1.17	1.18
Capacity [$C_v = (C_d)(d^2)$] Critical	$10d^2$	$10d^2$
Non-critical	$12d^2$	$14d^2$
Flow Characteristics	Function of pilot	Diverting
Cost	Low–Medium	Low–Medium
Available Size Range (in.)	1–12	½–4
Max./Min. Operating Pressure (PSIG)	1,500	100[17]
Max. Pressure Drop (PSID)	1,200	50[17]
Max./Min. Operating Temperature (°F)	150/0	Above 1,000[18]
Available in Jacketed Construction	No	No
Applicable to Flashing Service	Limited	—
Applicable to Cavitating Service	Limited	—
Applicable to High Viscosity ($\mu > 10^4$ cps) Service	No	No[19]
Applicable to Slurry Service	No	Limited[20]
Tight Shut-off Available[1]	Yes	No
Available Rangeability[10]	20:1	10:1
Has "Fail-safe" (Open or Closed) Action	No	No
Available in a Great Variety of Materials	No	Yes
Flow in Either Direction	No	No
Unobstructed Streamline Flow	No	Yes
Has Tendency to Cavitate Due to High Recovery	No	Yes
Stem or Shaft Sealed from Process by	Sleeve	No moving parts
Valve Trim Easily Removed	No	Has no trim
Low Maintenance	Average	Yes

[17] Operating principle places no limit; values given are those of existing installations.
[18] Can handle molten metals.
[19] Operation requires a minimum Reynolds number value.
[20] Pilot line needs to be kept from plugging.

29

This chapter covers all aspects of control valves, their features, relative merits and accessories.

Orientation Table I provides a summary of the more important characteristics of the various valve designs. The reader will find this table valuable in selecting the proper valve for a particular application.

The first section in this chapter deals with the standard terminology applicable to control valves. The second section covers the subject of control valve sizing. Sections 1.3 through 1.8 deal with the hardware aspects of the globe type control valves, separately discussing the valve bodies, bonnets, trims, accessories, positioners and diaphragm actuators. Between Sections 1.9 and 1.19 each section is assigned to a particular type of control valve. At the beginning of each of these sections a feature summary is presented, providing cost, flow characteristics and other data for the type of valve involved. Section 1.20 discusses the various non-diaphragm type valve actuators.

The last section in this chapter is devoted to selection and application. As such, it covers the subjects of leakage, rangeability, cavitation, noise and vibration together with the various special services, including high pressures and temperatures, low flows, temperatures, pressures and the slurry or high-viscosity applications.

Other types of final control elements are covered elsewhere in this volume. Self-contained regulators, where the control valve and the controller are combined into a single unit, are discussed in Chapter III of this volume in regard to *pressure* and *temperature*. Volume I covers self-contained *level* regulators in Section 1.19 and *flow* regulators in Section 5.19. Other types of final control elements, such as dampers, feeders, pumps and electric heaters are all discussed in Chapter II of Volume II.

1.1 TERMINOLOGY AND NOMENCLATURE *

General Terms

Valve   A pressure-dissipating device designed to modify flow of fluids in pipes.

Control Valve   A valve designed to modify flow of fluids in pipes and used for control purposes via an actuator responding to an external signal.

Regulator   A valve with an actuator responding to the condition of the fluids in the body.

Hand Valve   A valve with a manual actuator.

Actuator   The portion of a valve which responds to the applied signal and causes the motion resulting in modification of fluid flow.

Valve Body   The portion of the valve containing the flowing fluid and the device which modifies the flow of fluids through it.

Terms Relating to the Valve Body

Valve Body Assembly   An assembly of a body, bonnet assembly, bottom flange and trim elements. The trim includes a valve plug which opens, shuts or partially obstructs one or more ports.

Valve Body   A housing for internal valve parts having inlet and outlet flow connections (Figures 1.1a, b, c and d).

> *Several common body arrangements are employed, as follows:*
> *Single-ported means one port and one valve plug* (Figure 1.1b).
> *Double-ported means two ports and one valve plug* (Figure 1.1a).
> *Two-way means two flow connections: one inlet and one outlet.*
> *Three-way means three flow connections, two of which may be inlets with one outlet (for converging or mixing flows)* (Figure 1.1d), *or one inlet and two outlets (for diverging or diverting flows)* (Figure 1.1c).

Bonnet Assembly   An assembly including the part through which a valve plug stem moves and a means for sealing against leakage along the stem. It usually provides a means for mounting the actuator (Figures 1.1f, g, h and i).

° Some of the following terminology was used verbatim from Diaphragm Actuated Control Valve Terminology (ASME Standard 112) with the permission of the publisher, The American Society of Mechanical Engineers, United Engineering Center, New York, New York.

31

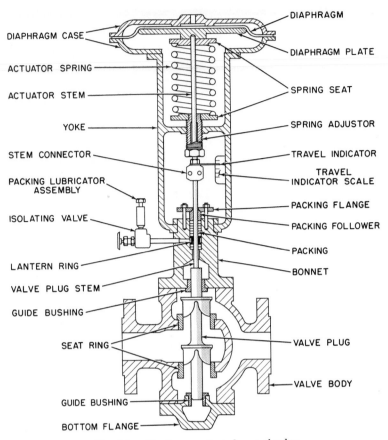

Fig. 1.1a   Diaphragm-actuated control valve

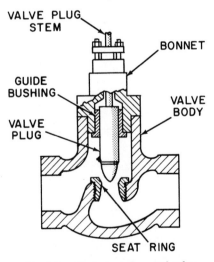

Fig. 1.1b   Single-ported control valve

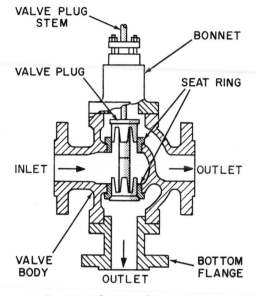

Fig. 1.1c   Three-way diverting valve

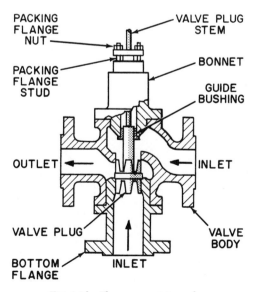

Fig. 1.1d   Three-way mixing valve

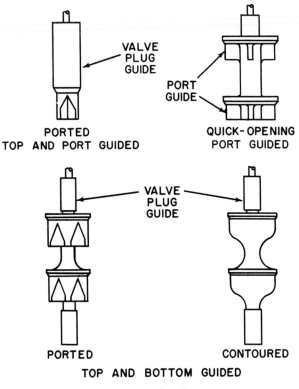

Fig. 1.1e   Valve plugs

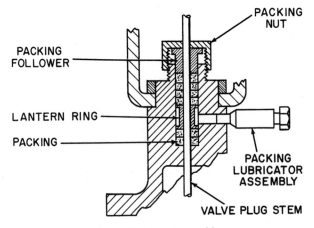

Fig. 1.1f   Bonnet assembly

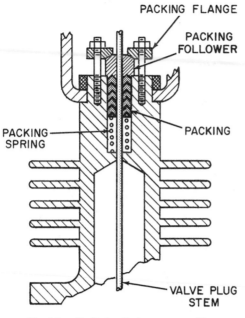

Fig. 1.1g Radiation fin bonnet assembly

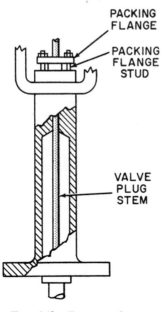

Fig. 1.1h Extension bonnet
assembly

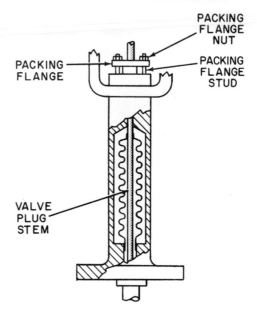

Fig. 1.1i   Bellows seal assembly

Sealing against leakage may be accomplished by means of packing or a bellows. A bonnet assembly may include a packing lubricator assembly with or without isolating valve. Radiation fins or an extension bonnet may be used to maintain a temperature differential between the valve body and sealing means.

Bonnet   The major part of the bonnet assembly, excluding the sealing means.

Radiation Fin Bonnet   A bonnet with fins to reduce heat transfer between the valve body and packing box assembly (Figure 1.1g).

Extention Bonnet   A bonnet with an extension between the packing box assembly and bonnet flange (Figure 1.1h).

Bellows Seal   A seal which uses a bellows for sealing against leakage around the valve plug stem (Figure 1.1i).

Packing Box Assembly   The part of the bonnet assembly used to seal against leakage around the valve plug stem, including various combinations of all or part of the parts shown in Figures 1.1a, f and g.

Isolating Valve   A hand-operated valve between the packing lubricator assembly and the packing box assembly to shut off the fluid pressure from the lubricator assembly (Figure 1.1a).

Bottom Flange   A part which closes a valve body opening opposite the bonnet assembly or in a three-way valve may provide an additional flow connection.

It may include a guide bushing and, in a three-way valve, may also include a seat (Figures 1.1a, c and d).

Seat Ring   A separate piece inserted in a valve body to form a valve body port (Figures 1.1a, b and c).

Seat   That portion of a seat ring or valve body which a valve plug contacts for closure (Figures 1.1a, b, c and d).

Valve Plug   A moveable part which provides a variable restriction in a port (Figures 1.1a, b, c and d).

Because of desired characteristics and for functional reasons, there are many forms of valve plugs, ported and contoured, a few of which are illustrated in Figure 1.1e.

Valve Plug Guide   That portion of a valve plug which aligns its movement in either a seat ring, bonnet, bottom flange or any two of these.

Typical examples are shown in Figure 1.1e.

Valve Plug Stem   A rod extending through the bonnet assembly to permit positioning the valve plug (Figures 1.1a, b, c and d).

Guide Bushing   A bushing in a bonnet, bottom flange, or body to align the movement of a valve plug with a seat ring (Figures 1.1a, b and d).

Guiding of a valve plug may be accomplished by an integral part of a bonnet or bottom flange or by a seat ring or seat ring extension.

Top-and Port-Guided Design   A design in which the valve plug is aligned by a guide in (a) the bonnet or body and (b) in the body port (Figure 1.1e).

Port Guided   A design in which the valve plug is aligned by the body port or ports only (Figure 1.1e).

Top and Bottom Guided   A design in which the valve plug is aligned by guides (a) in the body or (b) in the bonnet and bottom flange (Figure 1.1e).

Top Guided   A design in which the valve plug is aligned by a single guide (a) in the body adjacent to the bonnet or (b) in the bonnet.

Stem Guided   A special case of top guided in which the valve plug is aligned by a guide acting on the valve plug stem.

Terms Relating to the Valve Actuator

Diaphragm Actuator   A fluid pressure operated spring or fluid pressure opposed diaphragm assembly for positioning the actuator stem in relation to the operating fluid pressure or pressures (Figures 1.1a, j and k).

Diaphragm   A flexible pressure-responsive element which transmits force to the diaphragm plate and actuator stem (Figures 1.1a, j, and k).

Diaphragm Plate   A plate concentric with the diaphragm for transmitting force to the actuator stem (Figures 1.1a, j and k).

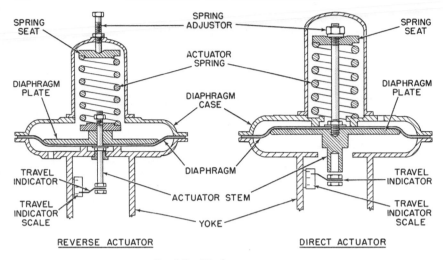

REVERSE ACTUATOR   DIRECT ACTUATOR

Fig. 1.1j   Diaphragm actuators

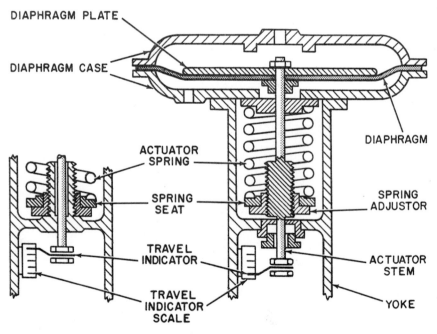

Fig. 1.1k   Reverse actuator

Diaphragm Case   A housing, consisting of top and bottom sections, used for supporting a diaphragm and establishing one or two pressure chambers (Figures 1.1a, j and k).

Actuator Stem   A rod-like extension of the diaphragm plate to permit convenient external connection (Figures 1.1a, j and k).

Yoke   A structure which supports the diaphragm case assembly rigidly on the bonnet assembly (Figures 1.1a, j and k).

Direct Actuator   A diaphragm actuator in which the actuator stem extends with increasing diaphragm pressure (Figures 1.1a and j).

Reverse Actuator   A diaphragm actuator in which the actuator stem retracts with increasing diaphragm pressure (Figures 1.1j and k).

1.2 CONTROL VALVE SIZING

Introduction

One might be justified in asking how it was possible that until recent years no reliable valve sizing formulas were developed and whether one is justified in keeping up with the modern trend in valve sizing if the old rules of thumb were good enough for decades. The answer to this very basic question has several aspects.

Part of the answer lies in the remarkable rangeability (ratio between maximum and minimum controllable flow) capability of most valves. This feature made it possible to provide for acceptable valve performance even when the sizing error reached not 10 percent, not 100 percent, but 1,000 percent.

Another major consideration in the slow evolution of sizing formulas was psychological. The valve manufacturers recognized that it was a poor policy to furnish the users' application engineers with complex equations, curves or slide rules, because many of them in response switched to a competing brand that promised valve sizing through simple, undemanding techniques.

The next logical question is, how has the situation changed in the last few years?

The changes too have been several, and they include aspects of both hardware and the human element. Probably the most outstanding factor in changing the situation is the appearance of the computer in the chemical industry. This has done two things.

First, use of the computer pointed out the optimization potentials in the various processes, which earlier were not noticed, and at the same time it indicated the necessity for a better match between process and control valve in order to improve overall control. And second, the computer did not stop at showing that using undemanding techniques was inefficient; it went further by offering its services to assist in the sizing, thereby removing the major obstacle to progress, the human element.

This really changed the situation, because it became a matter of prestige to have a computer in the design office and to have a valve sizing program in that computer. Having succeeded in channeling the work to the machine, the application engineer's attitude changed completely, and he became the one who demanded complex equations from the supplier of the competing brand so that he could feed the hungry machine.

This description is somewhat overdrawn, but the fact is that nowadays precise and close control is demanded of the control hardware, and this can be provided only if the final control element (the valve) is properly selected and sized.

The purpose of this section is to provide the reader with a "state of the art" summary of the present thinking on valve sizing while pointing out that this field is still in evolution and that the material presented should not be considered as firm and final. Table 1.2 provides a summary of the more important equations discussed.

Valve Coefficients

The first recommendations concerning the desirability of a capacity index for control valves were advanced during the second world war. Today, practically all valves manufactured are provided with capacity data in C_v units. By definition *the valve coefficient C_v is the number of GPM of $60°F$ water that will pass through the valve with a pressure drop of one PSI.*

The valve coefficients for the majority of valve designs were arrived at by actual tests. In these tests the flow and pressure drop across the valve are detected while it is installed in a *straight, valve size* pipe section. The valve coefficient is then arrived at by

$$C_v = \frac{Q_f}{\sqrt{\Delta P}} \text{(for water)} \qquad 1.2(1)$$

Depending on the particular valve design involved, the same pressure differential and valve size can result in substantially differing water flow rates, because of the internal flow paths involved. This is just one way of saying that different valve designs have different capacities.

The valve discharge coefficient, defined as

$$C_d = \frac{C_v}{d^2} \qquad 1.2(2)$$

where d = valve size in inches

is a useful indicator of relative capacity between the various designs. By the use of this coefficient a general equation is developed giving close *approximation* of control valve C_v's.

$$C_v = C_d d^2 \qquad 1.2(3)$$

Table 1.2a lists some approximate C_d values for a number of valve designs. It is apparent from the tabulation that the same size valves, exposed to the same pressure drop, will pass substantially different quantities of water as a function of their construction, and as expected, none of them will approach the capacity of an open pipe section.

When control valve capacities were tested on gas service, it was noted that different control valve designs, having the same C_v rating, can have substantially different gas capacities under the same pressure conditions. This is due to the variations in pressure recovery of the various designs. (The

Table 1.2

SUMMARY OF VALVE SIZING FORMULAS

Basis for Selecting the Sizing Formula	State of the Fluid	Liquid	Gas	Vapor (for steam, $K = 2.1$)
Basic Equations For Cold, Non-viscous, Non-cavitating or Non-flashing Liquids or for Vapors and Gases Where the Expansion at the Valve Seat Can Be Neglected. $(\Delta P \ll P_1/5)$	Volumetric Flow (GPM or SCFH)	$C_v = \dfrac{Q\sqrt{G_f}}{\sqrt{\Delta P}}$ 1.2(12)	$C_v = \dfrac{Q}{963}\sqrt{\dfrac{(G_f)(T_1)}{\Delta P(P_1+P_2)}}$ 1.2(36)	
	Mass Flow (lbm/hr)	$C_v = \dfrac{W}{500\sqrt{(\Delta P)(G_f)}}$	$C_v = \dfrac{W}{3.22\sqrt{(\Delta P)(G_f)(P_1+P_2)}}$	$C_v = \dfrac{W}{K}\sqrt{\dfrac{1}{\Delta P(P_1+P_2)}}$ 1.2(37)
Critical Flow Due to Cavitation $\Delta P > \Delta P_{allowed}$ $P_v < 0.7\,P_1$	$P_v < \dfrac{P_c}{10}$	$C_v = \dfrac{Q_f\sqrt{G_f}}{C_f\sqrt{P_1-P_v}}$ 1.2(15)		
	$P_v > \dfrac{P_c}{10}$	$C_v = \dfrac{Q_f\sqrt{G_f}}{C_f\sqrt{P_1-R_cP_v}}$ 1.2(16)		
Pipe Reducer Effect $D/d > 1.0$	Non-critical $P_v \simeq 0$	$C_v = \dfrac{F_1 Q_f\sqrt{G_f}}{\sqrt{\Delta P_t}}$ 1.2(21)	$C_v = \dfrac{F_1 Q\sqrt{(G_f)(T_1)}}{825\,C_f\,P_1\,X}$	$C_v = \dfrac{1.16 F_1 W}{(K)(C_f)(P_1)(X)}$
	Critical $\Delta P_t > \Delta P_{t\,allowed}$ $P_v < 0.7\,P_1$ $< 0.1\,P_c$	$C_v = \dfrac{Q_f\sqrt{G_f}}{C_{fr}\sqrt{P_1-P_v}}$ 1.2(24)	$C_v = \dfrac{Q\sqrt{(G_f)(T_1)}}{825(C_{fr})(P_1)}$	$C_v = \dfrac{1.16W}{(K)(C_{fr})(P_1)}$

High Viscosity Resulting in $Re_{valve} \leq 500$	$$C_v = \frac{F_2 Q_f \sqrt{G_f}}{\sqrt{\Delta P}}$$ <div align="right">1.2(31)</div>		
Superheated Non-critical Vapor Flow			$$C_v = \frac{W\left(1 + \dfrac{1,400}{T_s}\right)}{K}\sqrt{\frac{1}{\Delta P(P_1 + P_2)}}$$ <div align="right">1.2(38)</div>
Critical Vapor or Gas Flow $\Delta P > \Delta P_{allowed}$		$$C_v = \frac{Q\sqrt{(G_f)(T_1)}}{825(C_f)(P_1)}$$ <div align="right">1.2(40)</div>	$$C_v = \frac{1.16W}{(K)(C_f)(P_1)}$$ <div align="right">1.2(41)</div>
Generalized Sizing Formula for Compressible Flow		$$C_v = \frac{Q\sqrt{(G_f)(T_1)}}{825(C_f)(P_1)(X)}$$ <div align="right">1.2(45)</div>	$$C_v = \frac{1.16W}{(K)(C_f)(P_1)(X)}$$ <div align="right">1.2(48)</div>
Flashing Liquids	If percent flash is known, use the sum of equations 1.2(16) and 1.2(45) or 1.2(48). If $(G_m)(P_v - P_2)$ product curves are available, use equation 1.2(50). If neither of above is known, use 1.2(16) as qualified in text.		
Gas-liquid Mixtures	Calculate gas and liquid C_v's separately and base valve selection on their sum.		

Table 1.2a

CONTROL VALVE DISCHARGE AND RECOVERY COEFFICIENTS

Valve Type	Discharge Coefficient C_d	Pressure Recovery Coefficients C_1	C_f
Globe, Single Seated, Flow to Open, 50% Reduced Port	5	32	0.90
Globe, Double Seated, % V-Port Plug, 50% Reduced Port	6	35	0.95
Split Body, Flow to Open, 50% Reduced Port	6	33	0.90
Globe, Double Seated, Contoured Plug, 50% Reduced Port	7	31	0.80
Globe, Single Seated, Flow to Open, 50% Reduced Cage Trim	7	30	0.80
Globe, Angle, Flow to Open, 50% Reduced Port	8	36	0.95
Globe, Angle, Flow to Close, 50% Reduced Port	10	30	0.80
Globe, Single Seated, Flow to Open, Full Port	11	35	0.90
Split Body, Flow to Open, Full Ported	12	30	0.80
Globe, Double Seated, % V-Port Plug, Full Port	13	35	1.00
Globe, Single Seated, Flow to Open, Full Ported, Cage Trim	14	33	0.85
Globe, Double Seated, Contoured Plug, Full Port	14	35	0.90
Eccentric Rotating Plug Valve	14	31	0.85
Butterfly, 60° Open with $D/d = 2$	15	25	0.62
Globe, Angle, Flow to Open, Full Port	16	35	0.90
Butterfly, Line Size, 60° Open	17	23	0.70
Globe, Angle, Flow to Close, Full Port	20	31	0.80
Characterized Ball Valve, with $D/d = 1.5$	22	22	0.63
Globe, Streamlined Angle, Flow to Close, Reduced Port	23	18	0.50
Butterfly, 90° Open, with $D/d = 2$	23	21	0.50
Non-Characterized Throttling Ball, with $D/d = 1.5$	24	20	0.55
Butterfly, Line Size, 90° Open	35	17	0.56
Open Pipe Section with an L/D Ratio of 4	110	°	°

° Not applicable.

phenomenon of pressure recovery is discussed later.) If C_g is taken as a gas flow coefficient determined by air tests at the critical pressure drop, then the above phenomenon can be expressed numerically as

$$C_1 = \frac{C_g}{C_v} \qquad\qquad 1.2(4)$$

Table 1.2a lists some approximate C_1 values for a variety of valve designs. Still another indicator: C_f is listed in this same table, which allows one to convert C_v based on apparent ΔP to one based on critical liquid flow conditions.[1]

The critical flow factor C_f is also defined as the ratio between critical and non-critical C_v coefficients of the same valve.

$$C_f = \frac{C_v \ critical}{C_v} \qquad\qquad 1.2(5)$$

In other words C_f allows one to determine the extent by which the valve will actually appear to be smaller than its published C_v rating, if exposed to critical (cavitating or sonic) conditions.

$$C_v \ critical = C_f C_v \qquad\qquad 1.2(6)$$

It seems that the two factors C_1 and C_f are indicators of the same phenomenon (capability to recover pressure drop in the valve vena contracta) and differ only because the first is based on critical velocity air tests while the second is measured in cavitating (critical) liquid flow tests. An approximation of

$$C_1 = 36C_f \qquad\qquad 1.2(7)$$

describes their relationship for most cases. From the Table 1.2a one should note that the high-capacity valves ($C_v \geq 14d^2$) are also the high-recovery ones, and when they are used, the factors C_1 or C_f must be taken into consideration. As to how they are used in actual sizing, refer to the next paragraphs.

Sizing for Liquid Service

The Basic Equation
When a given amount of liquid is passing through a restriction such as a valve port, its flowing velocity must increase at the orifice. The energy for the increase in velocity is obtained from the static head, resulting in a localized pressure decrease. If the potential energy term is neglected due to assumed constant elevation of the flow path through the valve, then Bernoulli's theorem describing the conservation of energy will contain only static, velocity and friction head terms.

$$P_1 - P_2 = \frac{\rho}{2g}(\overline{V}_2^2 - \overline{V}_1^2) + \Delta P \qquad\qquad 1.2(8)$$

If the fluid is incompressible and the valve outlet has the same diameter as the inlet, then the velocity head term disappears and Bernoulli's equation reduces to a statement of static head conversion into friction head loss.

$$P_1 - P_2 = \Delta P \qquad\qquad 1.2(9)$$

Let us now investigate the relationship between flow rate and friction drop. For this purpose a brief discussion of the Reynolds number is essential. By definition, Re expresses the ratio of inertial forces to viscous forces.

$$Re = \frac{\rho \overline{V} d}{\mu} \qquad 1.2(10)$$

At low values of Re, viscous forces predominate and therefore pressure difference approaches direct proportionality to flow velocity or flow rate. At high Reynolds numbers, inertial forces predominate and the pressure drop to flow relationship is a square root one.

$$Q_f = KA \sqrt{\frac{\Delta P}{G_f}} \qquad 1.2(11)$$

If the KA product representing the area ratio, measurement units, correction factors and cross-sectional areas is replaced by the term C_v, the well-known liquid sizing formula results:

$$C_v = \frac{Q_f \sqrt{G_f}}{\sqrt{\Delta P}} \qquad 1.2(12)$$

This equation has been recommended by the Fluid Controls Institute[2] and is widely used, unfortunately with little understanding of its limitations.

The following paragraphs will attempt to outline the conditions under which correction factors should be applied to this equation.

Critical Flow and Cavitation

In the last decade, people involved in the design and use of control valves came to the realization that the measurement of apparent pressure differential across a control valve does not necessarily relate to the flow capacity. In other words, equation 1.2(12) does not necessarily apply in all cases.

It has been observed that, in accordance with Bernoulli's theorem, when a fluid stream is accelerated in passing through a control valve, the energy for this is provided by the conversion of static head into velocity head, resulting in a localized pressure decrease. The velocity reaches its maximum in the vena contracta, and therefore the static pressure is the minimum at this point.

The pressure gradients in a control valve are illustrated in Figure 1.2b when the vena contracta pressure is above the vapor pressure of the fluid, and therefore equation 1.2(12) does apply.

If the valve outlet pressure (P_2) is further lowered, this will cause an increase of velocity at the vena contracta with even more of the static pressure head being converted to velocity head, further lowering this minimum pressure in the valve. At some outlet pressure value, the pressure at the vena contracta

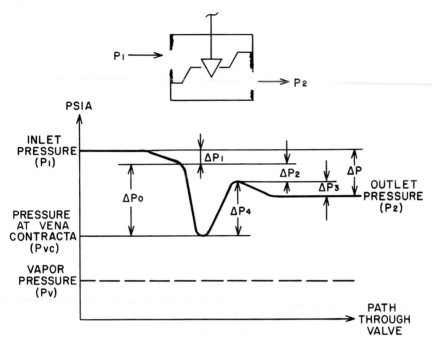

ΔP   – OVERALL PRESSURE DROP ACCROSS THE VALVE
ΔP₀   – DIFFERENTIAL OF VENA CONTRACTA AND PORT INLET PRESSURES
ΔP₁   – VALVE INLET PRESSURE LOSS
ΔP₂   – PERMANENT PRESSURE LOSS IN VALVE PORT
ΔP₃   – VALVE OUTLET PRESSURE LOSS
ΔP₄   – PRESSURE RECOVERY

Fig. 1.2b   Pressure gradients in a control valve under non-critical conditions

reaches the vapor pressure of the fluid (Figure 1.2c). At this point localized vaporization occurs, and voids, or *cavities,* appear in the fluid stream. The pressure drop across the valve corresponding to this condition is shown as point #1 in Figure 1.2d and is referred to as the *point of incipient cavitation.* The solid line in this illustration shows the actual flow curve through a valve as it differs from the relationship suggested by equation 1.2(12).

As shown by this curve after the formation of the first vapor cavities, the increase in flow rate will not be proportional to the increase in pressure drop through the valve. In other words (referring to Figure 1.2c), a further decrease in P_2 will not result in the expected increase of flow, but will have a lesser effect because of the vapor formation at the vena contracta. When the apparent pressure drop across the valve corresponds to point #2 in Figure 1.2d, a further lowering of the outlet pressure will not increase the flow at

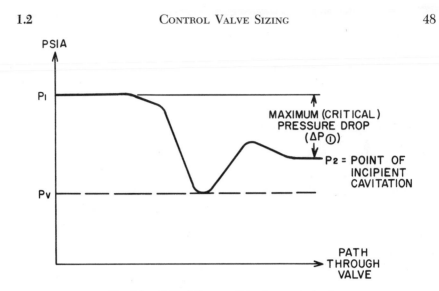

Fig. 1.2c Critical flow condition in a control valve

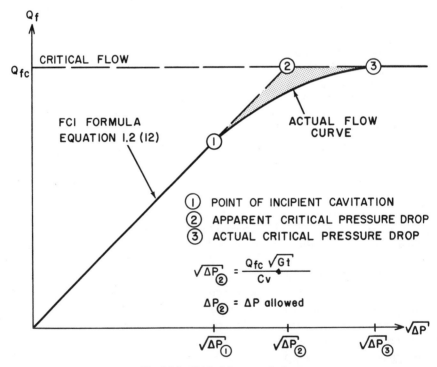

Fig. 1.2d Critical flow terminology

all. This condition is sometimes referred to as *choked flow condition*, implying that the localized vaporization at the vena contracta has caused critical flow, and a further increase in flowing velocity at that point is not possible.

As discussed in more detail in Section 1.21, the localized vapor formation is reversed after the vena contracta, and as the velocity head converts back to static pressure head, the pressure is raised. This pressure recovery is noted as ΔP_4 in Figure 1.2b and is responsible for the collapse or implosion of the previously formed cavities. For the purposes of this section, the above phenomenon need not be discussed in detail because it has no major influence on the sizing limitations.

The Pressure Recovery Factor

As was discussed earlier, the factor C_f allows one to determine the extent by which the valve will appear to be smaller than its published C_v rating if exposed to critical flow conditions. This means that the C_f factor is an indicator of the recovery capability of the particular valve, and the higher the pressure recovery the smaller the valve will *appear* under cavitating conditions relative to its standard C_v rating. Using the terminology of Figure 1.2b;

$$C_f = \sqrt{\frac{\Delta P}{(\Delta P_0 + \Delta P_1)}} \qquad 1.2(13)$$

when ΔP has reached ΔP_2 or $\Delta P_{allowed}$, as defined in Figure 1.2d. This relationship then allows for the use of the factor C_f in defining the maximum ΔP at which choked flow occurs:

$$\Delta P_{allowed} = C_f^2(P_1 - P_v) \qquad 1.2(14)$$

As shown in Table 1.2a, the more streamlined the valve design, the lower the value of C_f will be and the lower the flow capacity will be under critical conditions, relative to the standard C_v rating.

By replacing ΔP in equation 1.2(12) with $\Delta P_{allowed}$, a new sizing equation is arrived at for cavitating conditions at low vapor pressures:

$$C_v = \frac{Q_f \sqrt{G_f}}{C_f \sqrt{P_1 - P_v}} \qquad 1.2(15)$$

This equation assumes that the flow through the valve will follow equation 1.2(12) until $\Delta P_{allowed}$ is reached and that higher pressure drops will not increase the corresponding flow. With reference to Figure 1.2d this means that the equation assumes points 1 and 3 to be identical with point 2, and therefore the shaded area is neglected. In case of globe valves with reduced trims, this assumption is quite accurate, but in other valve designs it is not. (A mathematical relationship to describe the actual flow curve between points 1 and 3 is given under the discussion of gas sizing.)

If the vapor pressure of the flowing liquid at the valve inlet temperature is greater than 10% of its thermodynamic critical pressure, then it is advisable to correct the above equation by the critical pressure ratio, R_c. Values of R_c can be obtained from Figures 1.2e and f, and an incomplete tabulation of

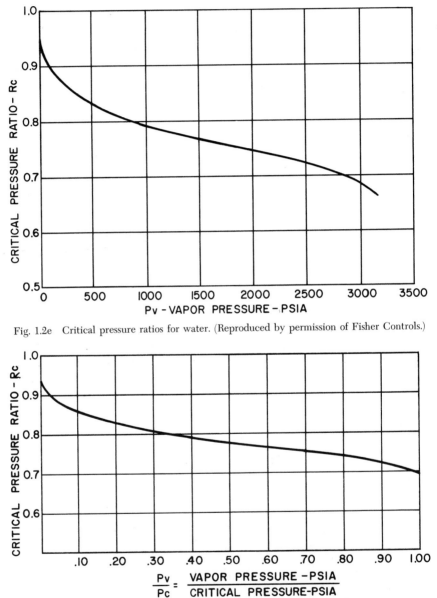

Fig. 1.2e   Critical pressure ratios for water. (Reproduced by permission of Fisher Controls.)

Fig. 1.2f   Critical pressure ratios for liquids other than water. (Reproduced by permission of Fisher Controls.)

Table 1.2g
CRITICAL PRESSURE OF
VARIOUS FLUIDS, PSIA°

Ammonia	1,636	Isobutane	529.2
Argon	705.6	Isobutylene	580
Butane	550.4	Methane	673.3
Carbon Dioxide	1,071.6	Nitrogen	492.4
Carbon Monoxide	507.5	Nitrous Oxide	1,047.6
Chlorine	1,118.7	Oxygen	736.5
Dowtherm A	465	Phosgene	823.2
Ethane	708	Propane	617.4
Ethylene	735	Propylene	670.3
Fluorine	808.5	Refrigerant 11	635
Helium	33.2	Refrigerant 12	596.9
Hydrogen	188.2	Refrigerant 22	716
Hydrogen Chloride	1,198	Water	3,206.2

° Reproduced by permission of Fisher Governor Company.

thermodynamic critical pressures is given in Table 1.2g. The effect on valve size of the R_c factor which accounts for the influence of fluid properties on cavitation *can* be neglected if $P_v < 0.1\ P_c$ without drastically influencing the precision of sizing. If the vapor pressure is more than 10% of P_c, then the following equation is recommended:

$$C_v = \frac{Q_f \sqrt{G_f}}{C_f \sqrt{(P_1 - R_c P_v)}} \qquad 1.2(16)$$

Note that both equations 1.2(15) and 1.2(16) are based on the assumption that the vapor pressure of the flowing fluid does not exceed 70 percent of the valve inlet pressure. This limitation is necessary because otherwise the error represented by the shaded area on Figure 1.2d would become too great. (Flashing and two-phase flow are covered in a later paragraph.)

Example 1 for Cavitating Flow

Process Data:	Fluid	Water
	T_1	200°F
	P_1	100 PSIA
	P_2	20 PSIA
	Q_f	1,000 GPM
	P_v	11.5 PSIA
	G_f	0.96
	C_f for a characterized ball valve from Table 1.2a	0.63

Check ΔP against

$$\Delta P_{allowed} = C_f^2\,(P_1 - P_v) = (0.63)^2\,(100 - 11.5) = 35.1\ \text{PSID}$$

and note that the available total 80 PSID should not be used as the basis for sizing and that equation 1.2(12) is not applicable.

Because of the low vapor pressure, the critical pressure ratio R_c can be neglected, and therefore equation 1.2(15) can be used.

$$C_v = \frac{Q_f \sqrt{G_f}}{C_f \sqrt{P_1 - P_v}} = \frac{1{,}000\,\sqrt{0.96}}{0.63\,\sqrt{100 - 11.5}} = 165$$

If equation 1.2(12) had been used, the resulting

$$C_v = \frac{1{,}000\,\sqrt{0.96}}{\sqrt{80}} = 109$$

would have caused serious undersizing.

Pipe Reducer Effects

Usually the sizing data are provided for control valve stations with little thought, if any, given to the fact that, if the valve is smaller than line size, then some of the available pressure drop will be consumed by the reducers and will not be available at the valve. Neglecting this phenomena in valve sizing did not cause serious errors until the various high-capacity valves had begun to gain acceptance. With these valves, it is not uncommon to see reducers with 2:1 or greater diameter ratios. For terminology used in connection with reducers, see Figure 1.2h.

One might visualize this as two resistances in series, sharing in the total pressure drop available.

$$\Delta P_t = \Delta P + \Delta P_r \qquad\qquad 1.2(17)$$

If an equivalent valve coefficient (C_{vr}) is assigned to the reducers, it is possible to relate the required control valve coefficient to the apparent coefficient of the overall valve station (C_{vt}).

$$\frac{1}{C_v{}^2} = \frac{1}{C_{vt}{}^2} - \frac{1}{C_{vr}{}^2} \qquad\qquad 1.2(18)$$

If a discharge coefficient is used to determine the equivalent valve coefficient for the reducer, then

$$C_{vr} = C_{dr}d^2 \qquad\qquad 1.2(19)$$

In accordance with Reference (3), Figure 1.2i relates C_{dr} to the reducer diameter ratio and to its swage angle. As shown in Table 1.2k under alternatives 3 and 4, it is possible to size valves using equation 1.2(18) above and the curves in Figure 1.2i.

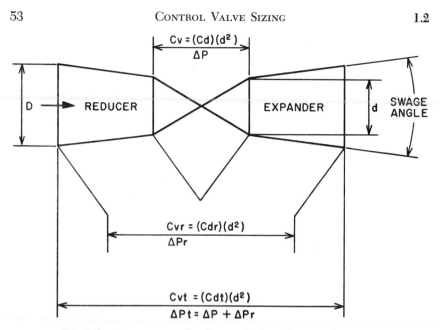

Fig. 1.2h   Terms associated with control valve-reducer combinations

Tests have shown[3] that more reliable results are obtained if the apparent discharge coefficient for the overall station (C_{dt}) is also considered. This factor is defined as

$$C_{dt} = C_{vt}/d^2 \qquad\qquad 1.2(20)$$

If the pipe reducer correction factor (F_1) is defined as the ratio of valve station capacity with reducers, to the capacity when the valve and pipe diameters are the same, then

$$C_v \text{ required} = F_1 C_{vt} = \frac{F_1 Q_f \sqrt{G_f}}{\sqrt{\Delta P_t}} \qquad\qquad 1.2(21)$$

The relationship between F_1, the diameter ratio, C_{dt} and the swage angle is given in Figure 1.2j[3]. By the use of this method (alternatives 1 and 2 in Table 1.2k), slightly different results are reached than if C_{dt} is disregarded (equation 1.2(18)).

Example 2 for Sizing With Reducers

The process data provided in Table 1.2k requires an overall valve station coefficient $C_{vt} = 190$. If the application engineer were not aware of the need for giving consideration to the reducers used, he would probably conclude that a high-capacity 3-in. valve, being less expensive than a 4-in. globe with the same C_v rating, is the logical choice.

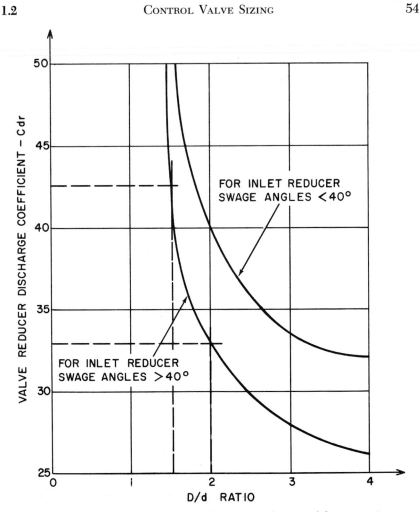

Fig. 1.2i   Valve reducer discharge coefficients as a function of diameter ratio

The results in Table 1.2k clearly show that the large-capacity 3-in. valve could not pass the required flow and that only the 4-in. globe is acceptable. This tabulation also compares the results in terms of the F_1 factor for the sizing techniques disregarding and considering the apparent overall discharge coefficient, C_{dt}. The results indicate that for standard globe valves there is no appreciable difference between the two methods, but for high-capacity valves there is and the disregarding of the C_{dt} coefficient can result in undersizing.

It is therefore recommended that the F_1 factor be obtained from data which take C_{dt} into consideration.

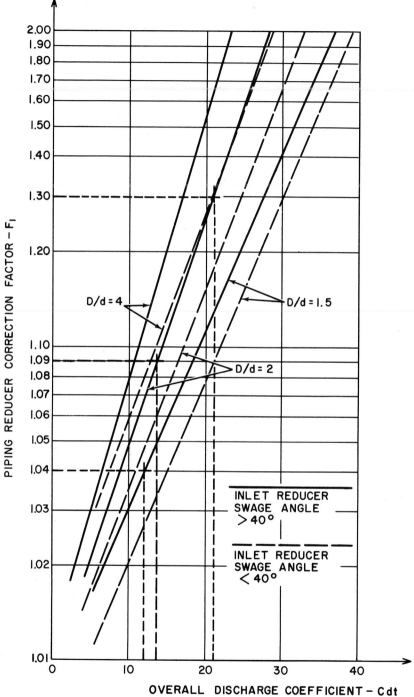

Fig. 1.2j   Pipe reducer correction factor as a function of discharge coefficient and D/d

Table 1.2k
TABULATION OF CALCULATIONS FOR EXAMPLE 2

Process Data for Control Valve and Reducer Combination:

$Q_f = 1{,}900$ GPM
$\Delta P_t = 100$ PSID
$G_f = 1.0$
$P_v = 0$ PSIA
$C_{vt} = Q_f \sqrt{G_f}/\sqrt{\Delta P_t} = 190$

Note: The selected valve will be installed in a 6″ diameter pipe, using reducers with *greater* than 40° swage angles.

Steps in Calculation	Considering the C_{dt} Factor		Disregarding the C_{dt} Factor	
	Alternative 1 4″ Globe (2-port)	*Alternative 2* 3″, 90° Butterfly	*Alternative 3* 4″ Globe (2-port)	*Alternative 4* 3″, 90° Butterfly
D/d Ratio	1.5	2	1.5	2
C_d for Valve Type Considered (from Table 1.2a)	13	23	13	23
C_{dr} from Figure 1.2i			43	33
$C_{vr} = C_{dr}d^2$			690	296
$C_{dt} = C_{vt}/d^2$	12	21	$1/C_v^{\,2} = 1/C_{vt}^{\,2} - 1/C_{vr}^{\,2}$ $1/C_v^{\,2} = = \dfrac{1}{190^2} - \dfrac{1}{690^2}$	$1/C_{vr}^{\,2}$ $1/C_v^{\,2} =$ $= \dfrac{1}{190^2} - \dfrac{1}{296^2}$
The Pipe Reducer Correction Factor F_1 Is Determined from Figure 1.2j	Yes	Yes		
C_v Available in Proposed Valve	208	207	208	207
C_v Required	198	247	198	231
C_v Req'd/$C_{vt} = F_1$	1.04	1.30	1.04	1.22

Reducer Effects on Critical Flow

The pressure recovery coefficient C_f allows one to determine the extent by which the valve appears to be smaller than its standard C_v rating when exposed to critical conditions. If in addition the valve is installed between reducers, it will appear to be even smaller.

Figure 1.21 shows some of the pressure gradients in a valve-reducer combination. When ΔP reaches $\Delta P_{allowed}$, the flow is choked. If there were no reducers, C_f would be defined as

$$C_f = \sqrt{\frac{\Delta P}{(\Delta P_0 + \Delta P_1)}} = \sqrt{\Delta P_{allowed}/(P_1 - P_v)} \qquad 1.2(13)$$

But because an additional ΔP_5 pressure drop occurs in the pipe reducer, this lowers the critical pressure recovery factor correspondingly:

$$C_{fr} = \sqrt{\Delta P/(\Delta P_0 + \Delta P_1 + \Delta P_5)} \qquad 1.2(22)$$

According to reference #3, C_{fr} can be defined as a function of the valve C_d, C_f and the reducer d/D by the following equation:

$$C_{fr} = \frac{1}{\sqrt{\dfrac{1}{C_f^2} + \dfrac{C_d^2}{900}\left(1 - \dfrac{d^4}{D^4}\right)}} \qquad 1.2(23)$$

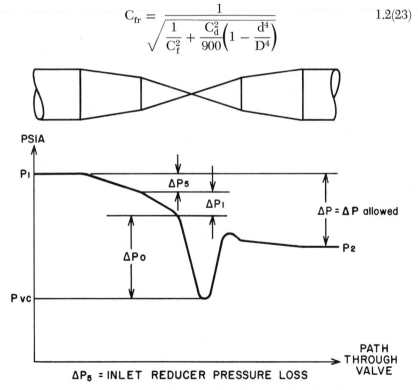

Fig. 1.21 Pressure gradients in control valve with reducers—under choked flow conditions

This equation for easier use is plotted[3] on Figure 1.2m, graphically relating C_{fr} to the D/d ratio and to the valve type involved. Once C_{fr} for the valve-reducer combination is defined, the valve can be sized by replacing C_f with C_{fr} in equation 1.2(15).

$$C_v = \frac{Q_f \sqrt{G_f}}{C_{fr} \sqrt{P_1 - P_v}} \qquad 1.2(24)$$

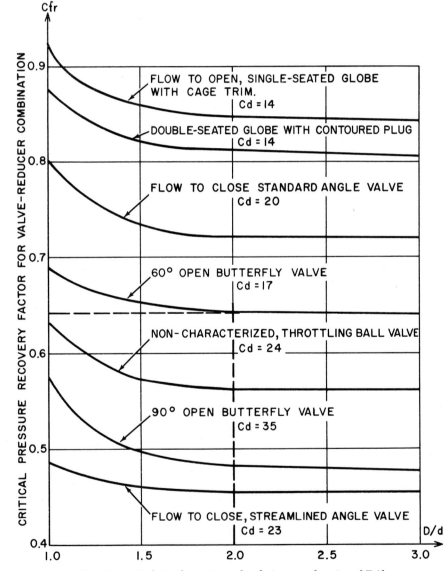

Fig. 1.2m C_{fr} factor for various valve designs as a function of D/d

Example 3 for Reducers and Critical Flow

If the possibility of critical flow conditions is overlooked, the process data presented in Table 1.2n would indicate an overall valve station coefficient $C_{vt} = 218$ and a C_v requirement of 238, which is less than the available coefficient of 272.

If the application engineer realizes that at $P_v = 11.5$ PSIA critical flow is a possibility, he will proceed to determine the critical pressure recovery coefficient for the valve-reducer combination, using Figure 1.2m. Having thus obtained $C_{fr} = 0.645$, he will then apply equation 1.2(24), which results in a C_v requirement of 352, well exceeding the available C_v of 272. Therefore, by properly considering the critical flow conditions in this case, the application engineer determined that the proposed valve is not sufficient for the application.

It might also be noted that if the reducer effects were overlooked and the sizing were performed for critical flow only, using equation 1.2(15), the result of $C_v = 326$ would still have indicated that the selection of a 4-in. butterfly is not sufficient.

Sizing for High Viscosity

As was noted in equation 1.2(10), the Reynolds number expresses the ratio between inertial and viscous forces acting in a flowing stream. In the sizing

Table 1.2n
TABULATION OF CALCULATIONS FOR EXAMPLE 3

Process Data for Control Valve and Reducer Combination			*Note:* The selected valve will be installed in an 8″ diameter pipe, using reducers with *greater* than 40-degree swage angles.
$Q_f =$	1,000	GPM	
$\Delta P_t =$	20	PSID	
$G_f =$	0.96		
$P_v =$	11.5	PSIA	
$P_1 =$	30	PSIA	

Steps in Calculation	Basis for Calculation	Consider 4″ Butterfly Having $D/d = 2$, $C_v = (C_d)(d^2) = 17 \times 16 = 272$, $C_f = 0.7$	
		Assuming Conditions to be Non-critical	Critical Conditions Assumed
$C_{vt} =$		$1,000 \dfrac{\sqrt{0.96}}{\sqrt{20}} = 218$	$1,000 \dfrac{\sqrt{0.96}}{0.7\sqrt{18.5}} = 326$
$C_{dt} = C_{vt}/d^2$		13.6	20.4
Pipe Reducer Correction Factor(F_1) from Figure 1.2j		1.09	
C_{fr} from Figure 1.2m			0.645
$C_v = Q_f \sqrt{G_f}/(C_{fr} \sqrt{P_1 - P_v})$			352
C_v Available in Proposed Valve		272	272
C_v Required		238	352

techniques presented to this point, it was assumed that the inertial forces dominate, and therefore the pressure-drop-to-flow relationship is square root. It must be realized that this is true only for turbulent flows. At low Reynolds numbers the viscous forces predominate and pressure difference approaches direct proportionality to flow velocity and to viscosity as expressed by the Poiseuille-Hagen relationship for laminar flow:

$$\Delta P = \frac{32 Q_f L \mu}{D^2 A} = (\text{constant})(Q_f)(\mu) \qquad 1.2(25)$$

The Reynolds number at the seat of conventional control valves is calculated as follows:[5]

$$\text{Single seated:} \quad \text{Re} = \frac{15{,}500 Q_f \rho}{\sqrt{C_v}\ \mu} = \frac{15{,}500 Q_f}{\sqrt{C_v}\ \nu} \qquad 1.2(26)$$

$$\text{Double seated:} \quad \text{Re} = \frac{10{,}200 Q_f \rho}{\sqrt{C_v}\ \mu} = \frac{10{,}200 Q_f}{\sqrt{C_v}\ \nu} \qquad 1.2(27)$$

where Q_f = sizing liquid flow rate (GPM),
ρ = density (slugs/ft^3),
C_v = the valve coefficient required for turbulent conditions,
μ = absolute viscosity (centipoises),
ν = kinematic viscosity (centistokes).

The lower the Reynolds number, the less flow will be passed by the same valve under the same conditions. In recent years some tests have been performed[4][5] to develop information on laminar flow sizing and to collect data for viscosity correction factors. Figure 1.2o shows the relationship[4] between the Reynolds number at the valve seat and correction factor F_2.

To size a valve using this technique involves the determination of the valve coefficient required, assuming turbulent flow (equation 1.2(12)) and then using this C_v to determine the Reynolds number at the valve seat (equations 1.2(26) and 1.2(27)). Having obtained Re, F_2 is found from Figure 1.2o and used as a multiplier to the turbulent C_v.

$$C_v(\text{laminar}) = F_2 C_v(\text{turbulent}) \qquad 1.2(28)$$

This method of viscosity correction is very appealing in its simplicity, but it also has a number of limitations, which the application engineer should be fully aware of:

1. Probably the most serious limitation of this method is the way the Reynolds number is proposed to be determined. One should realize that, by using the "turbulent" C_v in equations 1.2(26) and 1.2(27), the resulting Re has little to do with the actual Reynolds number in the "viscous" valve. In other words, the Re value entered in Figure 1.2o is *not* the Reynolds number

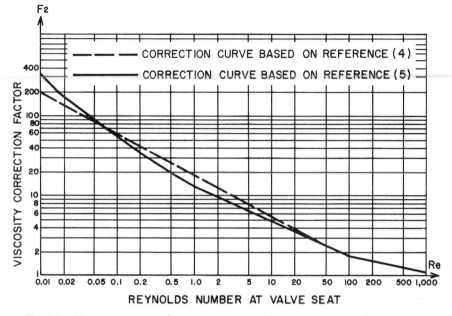

Fig. 1.2o   Viscosity correction factor as a function of Reynolds number at the valve seat

expected at the seat of the valve being sized, but is the Re in a *much smaller* valve, one which would be of sufficient size only *if* the flow were turbulent.

2. The other major limitation of this approach has to do with the testing program, which was "Reynolds number oriented" instead of being oriented in regard to viscosity. This means that during the tests only moderately viscous oils have been used, and therefore low Re values could be achieved only by reducing flow velocity. One might say that the flow pattern should not be different for a stream that has an Re = 0.1 because of high viscosity or due to low velocity, but there is a substantial difference in the *measurements* involved. In other words, to reproduce the Reynolds number corresponding to 10^5 or 10^6 cps viscosity, using water or light oils, the flow stream has to be slowed down to the extent that its rate is expressed in cubic centimeters and the drop through the valve in units of 10^{-2} PSID. Detection of such quantities is difficult to perform accurately and is partially responsible—one would assume—for the difference between the correction factor curves noted in References 4 and 5.

3. Other limitations could also be listed, including a better relationship between the particular valve design and the corresponding Re than those given in equations 1.2(26) and 1.2(27), but the main problems are those noted above.

Alternative Viscous Sizing Technique

The limitations discussed earlier caused the writer to develop an alternative valve sizing technique which attempts to give a theoretically more reliable basis for viscous sizing.[11] This method relates the Reynolds number at the valve seat to that in the pipeline, takes the valve discharge coefficient (C_d) into consideration and compares fairly accurately with the results of actual installations. The philosophy behind this technique is as follows:

Because viscosity inversely affects the Reynolds number, most high-viscosity streams follow the laminar flow pattern. Under laminar flow conditions, viscosity and pressure drop are proportional. (Doubling viscosity will double pressure drop.) If the viscosity *corresponding to the beginning of the laminar range* (μ_l) *is determined*, the ratio between that and the actual viscosity will equal the ratio between available and sizing pressure drops.

$$\mu/\mu_l = \Delta P \text{ available}/\Delta P \text{ sizing} \qquad 1.2(29)$$

If the viscosity correction factor F_2 is defined as

$$F_2 = \sqrt{\mu/\mu_l} \qquad 1.2(30)$$

then

$$(C_v) \text{ viscous} = F_2(C_v) \text{ turbulent} \qquad 1.2(31)$$

The use of this sizing method involves the following steps:

1. *Determine the* Re *value corresponding to the beginning of the laminar range.* Experience has shown that truly laminar conditions will *not* yet exist at the valve seat when Re = 2,000, and it has been found that for sizing purposes the assumption of Re = 500 for the beginning of the laminar range gives good results.

2. *Find* Re_{pipe} *corresponding to* $Re_{valve} = 500$. Due to the smaller flow area through the valve relative to the pipe and to the flow pattern across the valve, there is no question that the Reynolds number under identical flow conditions will be substantially higher in the valve than in the pipe. How substantial this difference is can be a function of pipe surface quality, of the number of upstream pipe elbows or of the general upstream history of the fluid, but a conservative relationship has been found to be

$$Re_{pipe} = \frac{d}{D}\sqrt{\frac{C_d}{110}}Re_{valve} \qquad 1.2(32)$$

This equation takes into consideration the valve size (d) relative to the pipe and also the particular valve design involved C_d.

3. *Determine the viscosity* μ_1 *that will result in* Re_{pipe}. Having determined in step 2 the Reynolds number which corresponds to the beginning of the laminar range, it is now necessary to calculate the viscosity that will result in it. The Re for a pipeline can be calculated as

$$Re_{pipe} = \frac{3{,}160 Q_f G_f}{D\mu} \qquad\qquad 1.2(33)$$

From the above equation

$$\mu_1 = \frac{3{,}160 Q_f G_f}{D Re_{pipe}} \qquad\qquad 1.2(34)$$

If equations 1.2(32) and 1.2(34) are combined and $Re_{valve} = 500$ is substituted, then

$$\mu_1 = \frac{6.31 Q_f G_f}{d\sqrt{C_d/110}} \qquad\qquad 1.2(35)$$

4. *Find viscosity correction factor F_2.* Having calculated μ_1, we can now, using equations 1.2(30) and 1.2(31), calculate both the correction factor and the required valve coefficient. If this required C_v calculates to be much less than the coefficient of the selected valve, then steps 1 through 4 are to be repeated using a valve one size smaller.

Example 4 for High-Viscosity Sizing

In Table 1.2p a set of process data is given and the sizing results are listed for both sizing techniques. The method based on References 4 and 5

Table 1.2p
TABULATION OF CALCULATIONS FOR EXAMPLE 4

$Q_f = 100$ GPM $G_f = 1.0$ $P_1 = 100$ PSIA	$\Delta P = 25$ PSID° $P_v = 0$ PSIA $\mu = 1 \times 10^5$ cps	Line size: 10″ diameter Valve type: angle globe C_d: 16	
Steps in Calculation *Basis for Calculation*	*Ref. 4 as Basis for Sizing*	*Proposed Alternative Method*	
		Assuming 8″ Valve	*Assuming 6″ Valve*
$C_{v\,turbulent} = Q_f \sqrt{G_f}/\sqrt{\Delta P}$	20	20	20
$RE_{turbulent\ valve}$ from eq. 1.2(26)	3.47		
F_2 from Figure 1.2o	9.3		
$Re_{Laminar\ Valve}$ Assumed		500	500
Find Re_{pipe} from Eq. 1.2(32)		152	114
Find μ_1 from Eq. 1.2(35)		208	278
$F_2 = \sqrt{\mu/\mu_1}$		22	19
$C_{v\,required} = F_2 C_{v\,turbulent}$	186	440	380
		$C_{v\,reqd.} \ll$ $C_{v\,available} = 1{,}024$	$C_{v\,reqd.} \simeq$ $C_{v\,available} = 576$

° ΔP is given for valve only. Drop in reducers is separately considered.

would suggest that a 4-in. valve is sufficient, while the technique recommended by the author indicates that a 6-in. valve is required. The difference between the two techniques is less pronounced at low viscosity levels and is more substantial at high viscosities.

The difference between the two techniques can be illustrated by comparison of Reynolds numbers. The actual Re at the seat of a 6-in. valve will be approximately

$$Re = \frac{15,500 Q_f}{\nu \sqrt{C_v}} = 0.65$$

whereas the correction factor of 9.3 was arrived at based on a Reynolds number of 3.47, corresponding to a valve $C_v = 20$, which bears no relation to the problem at hand. Further, if the correct Reynolds number were entered into Figure 1.2o, the difference between the two methods would diminish.

Therefore, a *third* (trial and error) sizing technique can also be considered, whereby a valve near line size is assumed, the corresponding Re is calculated using equation 1.2(26), and then F_2 is obtained from Figure 1.2o.

Non-Newtonian Fluids

The preceding discussion on viscous sizing assumes Newtonian fluid behavior. If the process fluid involved is non-Newtonian and its apparent viscosity is a function of shear rate, then the best approach is to test the valve under normal process conditions. If reliable data is available on shear rate vs. apparent viscosity, then the previously discussed sizing methods can be applied. It is usual practice to work with the apparent pipeline viscosity for pseudoplastic, thixotropic and plastic solid materials, because these fluids will exhibit a lower apparent viscosity at the higher shear rates inside the valve, and therefore the use of the higher pipeline viscosity introduces a safety factor. For dilatant (shear thickening) fluids the opposite is true, and for that reason a valve smaller than line size would seldom be considered.

Gas and Vapor Sizing

Non-Critical Gas Service

Until the early 1960s several sizing equations were in use which differed from each other in the gas density assumed. Some had their sizing based on the gas density corresponding to the upstream pressure, others (the conservative ones) had it based on the downstream pressure, and the majority would use something in between. In 1962 the Fluid Controls Institute[2] suggested that the gas density be based on the average of the upstream and downstream pressures, resulting in this well-known and widely used equation:

$$C_v = \frac{Q}{963} \sqrt{\frac{G_f T_1}{\Delta P (P_1 + P_2)}} \qquad 1.2(36)$$

The limitations of this equation will be discussed later, but it is important to note that the equation is unreliable for high-recovery valves with substantial pressure drops. It is therefore suggested that the use of equation 1.2(36) be limited to instances in which $\Delta P \ll P_1/5$.

Non-Critical Vapor Service

For saturated vapors the FCI sizing formula[2] is

$$C_v = \frac{W}{K} \sqrt{\frac{1}{\Delta P(P_1 + P_2)}} \qquad 1.2(37)$$

This equation is also limited to cases in which $\Delta P \ll P_1/5$ holds true. For values of K for a variety of saturated vapors, refer to Table 1.2q.

Table 1.2q
VALUES OF K FOR VARIOUS VAPORS

Vapor	K	Vapor	K
Steam	2.1	Freon 14	8.4
Chlorine	5.4	Freon 22	6.2
Dowtherm A	5.6	Freon 114	8.3
Ammonia	2.7	Methane	3.9
Carbon Dioxide	5.0	Propane	4.5
Methyl Chloride	4.5	Butane	4.7
Methylene Chloride	5.5	Isobutane	4.7
Sulfur Dioxide	5.1	Ethylene	4.1
Freon 11	7.4	Propylene	4.5
Freon 12	7.1		

When the vapors are not saturated, but superheated, the following equation should be used, subject to the same limitations as the preceding ones:

$$C_v = \frac{W(1 + 1,400/Ts)}{K} \sqrt{\frac{1}{\Delta P(P_1 + P_2)}} \qquad 1.2(38)$$

Critical Gas or Vapor Service

It was briefly noted earlier (Figure 1.2b) that the further P_2 is reduced the more of the static head in the valve will be converted into velocity head—in accordance with Bernoulli's theorem, equation 1.2(8)—at the vena contracta. At the same time we know that once sonic velocity is reached at the valve throat the flow is *choked*, in the sense that further lowering of the downstream pressure will not increase the flow rate. This phenomenon is very similar to that discussed in connection with cavitation (Figure 1.2d), and the pressure recovery factors listed in Table 1.2a also apply here to gas sizing. These factors indicate the extent to which the valve will appear to be smaller than its published C_v rating when exposed to critical flow conditions.

The Fluid Controls Institute, in presenting its sizing formulas, [2] assumed that critical flow occurs when $\Delta P > P_1/2$, and that if $\Delta P < P_1/2$, then equation 1.2(36) applies. Tests have proved that the pressure recovery factor of the particular valve design limits the maximum obtainable flow under critical conditions to less than that predicted by the FCI. Figure 1.2r also shows that

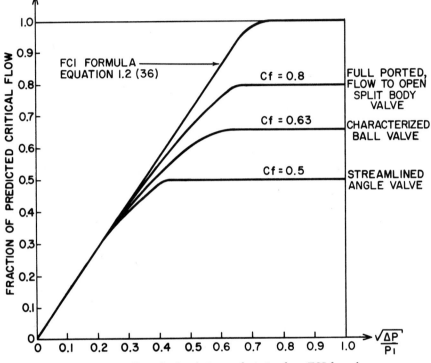

Fig. 1.2r   Effect of valve design on deviation from FCI formula

deviation from the predictions of equation 1.2(36) start at pressure drops substantially below $\Delta P_{allowed}$. The pressure drop corresponding to critical flow—further increase in ΔP will not increase flow—is defined as

$$\Delta P_{allowed} = \frac{C_f^2 P_1}{2} \qquad 1.2(39)$$

If the process conditions are such that $\Delta P > \Delta P_{allowed}$, then critical flow conditions will exist and the following equations can be used for sizing.

For gases: $C_v = \dfrac{Q\sqrt{G_f T_1}}{825 C_f P_1}$ 1.2(40)

For saturated vapors: $C_v = \dfrac{1.16W}{KC_f P_1}$ 1.2(41)

Generalized Sizing Formula for Compressible Flow

Thus far equations have been presented for compressible flow which are applicable only within limitations:

1. Equations 1.2(36), 1.2(37) and 1.2(38) are applicable only at those low pressure drops where the deviation between actual flow conditions and the FCI formula is small (Figure 1.2r). This limitation has been approximated by $\Delta P < P_1/5$, but even this can result in undersizing with "high recovery" valves, and therefore the use of these equations is generally discouraged.

2. Equations 1.2(40) and 1.2(41) are applicable only where $\Delta P > \Delta P_{allowed}$ (Figure 1.2d).

Therefore the equations discussed to this point are applicable only to the extreme (high or low) pressure drop conditions and not to the more usual cases, which are likely to fall in the area where neither FCI nor critical flow sizing is applicable. This is why there has been a need for the development of a generalized sizing formula for compressible flow.

The first such generalized empirical formula was introduced by Buresh and Schuder.[6] This equation is given below with the single modification of leaving out the correction for specific heat ratio, the effect of which is negligible considering the overall precision involved.

$$C_v = \frac{Q\sqrt{G_f T_1/520}}{C_1 P_1 X} \qquad\qquad 1.2(42)$$

where

$$X = \sin Y_2 \qquad\qquad 1.2(43)$$

and

$$Y_2 = \frac{3,420}{C_1}\sqrt{\frac{\Delta P}{P_1}} \qquad\qquad 1.2(44)$$

Generalized equation 1.2(42) involves a sine function and relates to the pressure recovery factor C_1. Boger[7] has proposed to eliminate the sine function and relate the correction to the recovery factor C_f. His basic equation is the same as 1.2(42), but he defines X and Y as noted and replaces the factor C_1 with C_f.

$$C_v = \frac{Q\sqrt{(G_f)(T_1)}}{825\ C_f P_1 X} \qquad\qquad 1.2(45)$$

$$X = Y_1 - 0.148 Y_1^3 \qquad\qquad 1.2(46)$$

and

$$Y_1 = \frac{1.62}{C_f}\sqrt{\frac{\Delta P}{P_1}} \qquad\qquad 1.2(47)$$

For the reader's convenience in obtaining values of X without time-consuming calculations, Figure 1.2s has been prepared, which relates both Y_1 and Y_2 to X in graphic form. The introduction of factor X makes equations

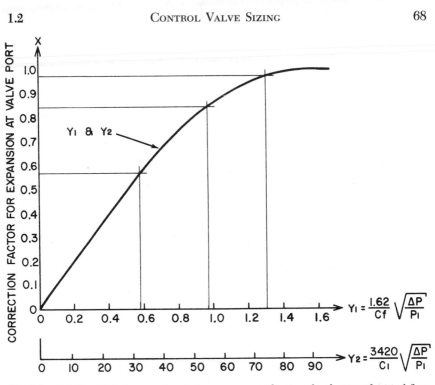

Fig. 1.2s   Relationship between valve design, pressure gradients and reduction of critical flow rate

1.2(42) and 1.2(45) applicable to all valve designs and to all values of pressure drops.

The equivalent of equation 1.2(45) for saturated vapors is

$$C_v = \frac{1.16W}{KC_fP_1X} \qquad 1.2(48)$$

Example 5 for Compressible Fluid Sizing

Table 1.2t was prepared to illustrate the following:

  a. The results obtained using the three sizing concepts:
    1. The FCI formulas.
    2. The generalized equations.
    3. The critical flow formulas.
  b. The results obtained as actual ΔP increases:
    1. $\Delta P < P_1/5$
    2. $\Delta P > P_1/5$
    3. $\Delta P > \Delta P_{allowed}$
  c. The results obtained as the pressure recovery coefficient changes.

Table 1.2t
TABULATION OF CALCULATIONS FOR EXAMPLE 5

Process Data for Split-body Valve with $C_d = 12$; $C_f = 0.8$ $(C_f = 0.65)^{\circ\circ}$

Process Fluid = Saturated Steam $\quad W = 100,000$ #/hr $\quad P_1 = \quad 1,200$ PSIA $\quad \Delta P^\circ = \quad 100$ PSID $\quad \Delta P < P_1/5 < \Delta P_{allowed}$	Same as Case 1 But $\Delta P = 280$ PSID $P_1/5 < \Delta P$	Same as Case 1 But $\Delta P = 500$ PSID $\Delta P > \Delta P_{allowed}$	
Sizing Basis	*Case 1*	*Case 2*	*Case 3*
Basis 1, Using eq. 1.2(37) for $\Delta P < P_1/5$ Cases	$C_v = 99(99)$	$C_v = 62(62)$	$C_v = 49(49)$
Basis 2, Using eq. 1.2(48), the Generalized Gas Sizing Equation	$Y_1 = 0.58(0.72)$ $X = 0.575(0.68)$ $C_v = 99(103)$	$Y_1 = 0.98(1.2)$ $X = 0.845(0.93)$ $C_v = 68(75)$	$Y_1 = 1.31(1.6)$ $X = 0.975(1.0)$ $C_v = 59(70)$
Basis 3, Using eq. 1.2(41) for Critical Flow Which Applies Only When $\Delta P > \Delta P_{allowed}$	$C_v = 57(70)$	$C_v = 57(70)$	$C_v = 57(70)$

$^\circ$ ΔP is given for valve only. Drop in reducers is separately considered.
$^{\circ\circ}$ Numbers in ( ) are based on valve design with $C_f = 0.65$.

Since sizing basis 2 is empirical, it is assumed to be correct, and the results of other sizing techniques are therefore compared with this reference.

As expected, the more pressure drop available, the smaller will be the required C_v, while the smaller the recovery coefficient (C_f), the greater the C_v necessary for the higher recovery valve involved.

In reviewing the results of the FCI formula (basis 1), it is apparent that at low pressure drops and high recovery coefficients the sizing results are accurate, but with increasing values of ΔP and decreasing values of C_f the undersizing due to the use of the FCI equations becomes pronounced.

An evaluation of the results of basis 3 indicates that the closer ΔP is to $\Delta P_{allowed}$, the less undersizing will result from using the critical flow equation. At values of $Y_1 \geq 1.5$ or $Y_2 \geq 90$, the results of basis 3 and those of the generalized equation in basis 2 are the same.

It is the authors' recommendation to use the generalized equation (basis

2) for all compressible sizing applications and thereby eliminate the necessity for judging if the other equations are applicable or not.

Flashing Liquids

When a saturated or nearly saturated liquid enters a control valve and some of its static head is converted into velocity head at the valve seat (Figure 1.2b), the resulting pressure reduction causes vaporization. After the vena contracta, some of this pressure is likely to be recovered, but if the valve outlet pressure is below the vapor pressure corresponding to the inlet temperature, then some of the vaporization will be permanent. This phenomenon is called *flashing,* and sizing control valves under flashing conditions is one of the less developed areas of valve sizing.

Over the years the following sizing techniques have been advanced for flashing:

1. Use one size larger valve than for non-flashing liquid.
2. Use line size valve.
3. Size valve for liquid, but provide oversized outlet.
4. Size valve for liquid, but use streamlined design.
5. Sum of C_v's technique.
6. Critical pressure method.
7. Mixture density technique.
8. Outlet density method.

The first four of these techniques are not really sizing methods but only "rules of thumb," and as such they will be disregarded in the following discussion.

Sum of C_v's Technique

This approach involves the determination of the stream composition at the valve outlet and then separately the calculation of the valve coefficient required for the liquid and vapor portions. The valve selection is then made, based on the sum of these separately calculated coefficients.

The results and the reliability of this technique are greatly influenced by the basis used for calculating the individual coefficients.[12] The amount of flashed vapors at the valve outlet is much less than at the vena contracta, before the pressure recovery section. Therefore, if the basic equations 1.2(12) and 1.2(37) were applied to the liquid and vapor fractions, serious undersizing could result, due to the disregarding of the critical flow condition at the vena contracta.

It is recommended that when the sum of C_v's technique is used, equations 1.2(16) and 1.2(45) or 1.2(48) be applied, so that the temporary flashing at the vena contracta is taken into consideration together with the particular valve design involved. The results of this technique tend to be more conservative as the amount of flashing increases, but as expected, the error in sizing is in the safe direction.

Critical Pressure Method

This technique is based on the assumption that valves for flashing can be sized by the same method as applied for cavitation, and that the limitation of $P_v < 0.7P_1$ noted for equation 1.2(16) can be disregarded. The advocates of this sizing approach define the critical or allowable pressure drop as

$$\Delta P_{allowed} = C_f^2(P_1 - R_c P_v) \qquad 1.2(49)$$

This technique does not consider the fact that some of the fluids flashed at the vena contracta will stay permanently in their vapor form. Basically, this method is very similar to the sum of C_v's technique, *if* the vapor C_v was disregarded. Consequently, while the sum of C_v's approach is erroneous at high quantities of flash, due to its double consideration of critical flow effects and the resulting oversizing, the critical pressure method is erroneous at low amounts of flash, because the permanent vapor portion is disregarded, thereby resulting in undersizing.

Mixture Density Technique

The mixture density technique[8] considers the gravity of the flashed mixture and reportedly finds close correlation with test results. Figure 1.2u illustrates the relationship between pressure and the gravity of the flashed mixture, and the area under this curve is termed by the author[8] the usable

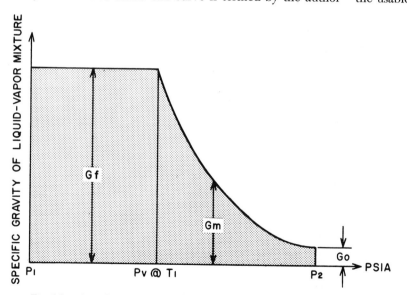

Fig. 1.2u Specific gravity of liquid-vapor mixture as a function of pressure

density-pressure drop product. The sizing equation relates to this area and because the tests substantiating this technique were performed with low recovery valves ($C_f \simeq 0.85$), the equation has been multiplied by $(0.85)/(C_f)$ to expand its coverage to valves with other recovery characteristics;

$$C_v = \frac{0.85Q_fG_f}{C_f(\sqrt{G_m(P_v - P_2)} + \sqrt{G_f(P_1 - P_v)})} \qquad 1.2(50)$$

If the fluid entering the valve is saturated, $P_v = P_1$, then the equation becomes

$$C_v = \frac{0.85Q_fG_f}{C_f\sqrt{G_m(P_v - P_2)}} \qquad 1.2(51)$$

Figure 1.2v relates the $G_m(P_v - P_2)$ product to ΔP for saturated water at various inlet pressures. Similar families of curves have been prepared for a fairly long list of industrial fluids.[8]

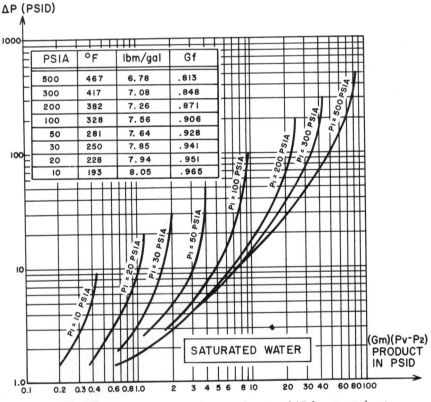

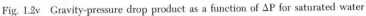

Fig. 1.2v Gravity-pressure drop product as a function of ΔP for saturated water

Outlet Density Method

This technique is included only for the sake of completeness, because there is little theoretical justification for considering this method, which is based on the following assumption:

If the gravity of the liquid-vapor mixture at the valve outlet is determined and is used as the density in a non-compressible liquid sizing formula, the result will approximate the flashing C_v requirements. The assumption that a non-compressible equation can be used when vapors are present is obviously false and will result in greater oversizing errors as the amount of flash increases.

Example 6 for Flashing Sizing

In Table 1.2w the results of all four sizing techniques have been summarized, but basis 4 will not be discussed further, and the reader is advised

Table 1.2w

TABULATION OF THE CALCULATIONS FOR EXAMPLE 6

$Q_f = 100$ GPM $G_f = \quad 0.906$	$P_1 = 100$ PSIA $P_v = 100$ PSIA	$T_1 = 328°F$ $C_f = \quad 0.85\ (0.65)°$	
Pressure Drop Through Valve; ΔP (PSID)°°		10	50
Amount Flashed at Valve Outlet (lbm/hr)		395	2,360
Basis 1 Sum of C_v's Technique	C_v required for *liquid* portion only, based on Eq. 1.2(16)	36.7(48.4)	35.2(46.3)
	C_v required for *vapor* portion only, based on Eq. 1.2(48)	4.2(4.6)	15.6(19.9)
	Total C_v required	40.9(53.0)	50.8(66.2)
Basis 2 Critical Pressure Method	$\Delta P_{allowed}$, based on Eq. 1.2(49)	6.5(3.8)	6.5(3.8)
	C_v required, calculated on the basis of Eq. 1.2(16)	37.0(49.0)	37.0(49.0)
Basis 3 Mixture Density Technique	$(G_m)(P_v-P_2)$ product, obtained from Figure 1.2v	4.8(4.8)	8.6(8.6)
	C_v required, calculated on the basis of Eq. 1.2(51)	41.2(54.0)	31.0(40.0)
Basis 4 Outlet Density Method	Mixture gravity at outlet, G_0	0.27	0.037
	$C_v = W/(500\ \sqrt{\Delta PG_f})$	56.0	67.0

° Numbers in parentheses are based on valve design with $C_f = 0.65$.
°° ΔP is given for valve only. Drop in reducers is separately considered.

not to consider it when sizing for flashing conditions. Saturated water is used with two values of pressure drops flowing through two valve designs, which differ in their pressure recovery coefficients.

Assuming that the mixture density technique is truly substantiated by test results, it can be concluded that at low amounts of flash the critical pressure technique seems to result in some undersizing, while the sum of C_v's method is quite accurate. At high amounts of flash, the mixture density technique gives the lowest results out of the three methods considered, and therefore the other two approaches can be considered safe insofar as they will not cause undersizing.

Much work remains to be done before one can say that a proven and accurate sizing technique exists for flashing conditions, and for the time being, the reader has to be satisfied with the discussion above.

Gas-Liquid Mixtures

When a liquid, well mixed with a non-condensable gas, flows through a valve, the sum of C_v's technique seems to give acceptable accuracy results. It is suggested that for the gas portion equation 1.2(45) be used, while for the liquid portion either 1.2(12) or 1.2(16), depending on the liquid vapor pressure at the inlet temperature.

Another method has also been reported[9] based on test data gathered on air and water mixtures.

$$C_v = \frac{W}{360\sqrt{\Delta P(G_{m1} + G_{mc})}} \qquad 1.2(52)$$

Where G_{m1} is the upstream mixture gravity and G_{mc} is the mixture gravity at the vena contracta. G_{mc} is found by

$$G_{mc} = \frac{G_{m1}(1 - \Delta P)}{P_1 C_f^2} \qquad 1.2(53)$$

The results using this method come very close to those obtained by the sum of C_v's technique.

Assigning Control Valve ΔP

A generalized statement would say that the more pressure drop is assigned to a valve the better it will control, because the more its influence will be relative to the overall system. This is like saying that the stronger the lead sled dog, the better control it has on the direction of travel; in other words, this is a statement appreciated by all but useful to none because of its non-quantitative nature.

Another widely accepted statement is that the control valve characteristics should match those of the process. This can mean that if the purpose of the loop is temperature control, then the valve characteristics should be so selected that a say 10 percent load change will be matched by a corresponding 10 percent change in the flow of the heating or cooling media, regardless of whether the valve is just opening or is almost fully open. For such a loop, therefore, the equal percentage valve of the increasing sensitivity type is to be used, which gives increasing flow increments for fixed increments of lift. Similarly for fast pressure loops or for level loops where the ΔP is likely to vary over a wide range, the equal percentage valve characteristic would be selected, while for the majority of standard level control applications and for flow control loops with wide ranges, the linear port is the proper selection.

Having established that the proper selection of valve characteristics is important in achieving good control, and knowing that the published valve characteristics are true *only if ΔP is constant* (Figure 1.2x), one should evaluate

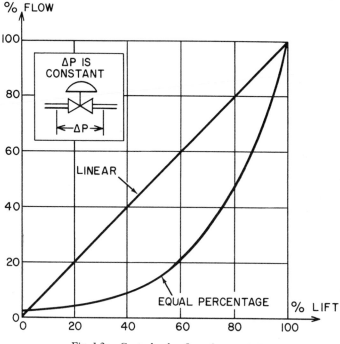

Fig. 1.2x   Control valve flow characteristics

those factors, which can cause changes in ΔP and thereby affect the valve characteristics.

One of those factors is the ratio between the pressure drop assigned to the valve and the friction drop in the rest of the system ($\Delta P/\Delta P_s$). The other factor is the head characteristics of the typical centrifugal pump or compressor, which has the same effect. This factor can be evaluated as the ratio of total pressure differential at minimum and maximum flow $[(\Delta P_t)min/(\Delta P_t)max]$. The effects of both of these factors have been graphically represented[10] as they distort the inherent flow characteristics of the valve. The value of this distortion coefficient is calculated by

$$D_c = \frac{(\Delta P_t)min \, (\Delta P)}{(\Delta P_t)max \, (\Delta P_s)} \qquad 1.2(54)$$

Figure 1.2y illustrates the pressure drop terms used in this formula.

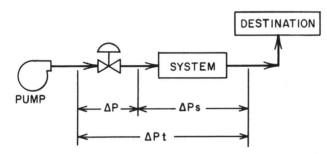

Fig. 1.2y   Terminology for system pressure drops

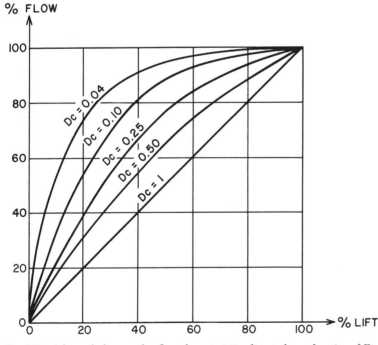

Fig. 1.2z   Inherently linear valve flow characteristics distorted as a function of D_c

From Figure 1.2z it can be seen that an inherently linear control valve will approach on-off characteristics, and Figure 1.2aa shows that an inherently equal percentage valve will be distorted toward linear characteristics if the value of D_c drops below 0.25. To prevent this from occurring on critical control loops, it is suggested that the same amount of pressure drop be assigned to the control valve as is required by friction drop in the rest of the system $(\Delta P = \Delta P_s)$.

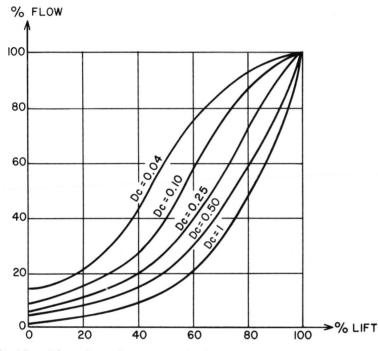

Fig. 1.2aa   Inherently equal percentage valve flow characteristics distorted as a function of D_c

Safety Factors in Valve Sizing

The purpose of this section is to allow the application engineer to calculate the size of control valves instead of having to guess it. This approach will be successful only if the process data for sizing is also accurately calculated. One common problem in this connection is the accumulation of safety factors.

It is important to realize that valve sizing should consider the minimum, normal and maximum operating loads, not only the maximum. (Section 1.21 provides a detailed discussion on rangeability.) If the valve sizing is based on maximum load only and the instrument engineer applies a safety factor to the data concerning flow, pressure drop, viscosity, flashing etc. before using it, the accumulated effect will be an inflated valve coefficient requirement which is always exceeded by the C_v of the actual valve selected. Such over-sizing is not only expensive but can result in controllability and rangeability problems.

The recommended procedure is first to calculate the minimum, normal and maximum C_v's required, without the use of safety factors. The ratio between maximum and minimum C_v's will then indicate the rangeability requirement of the particular installation, and the instrument engineer can judge if a single valve can handle this range. The valve is then selected, with the realization that its characteristics usually are best in the 20 to 90 percent lift range (Figure 1.2x). This, in case of an equal percentage characteristics trim, corresponds to a 7 to 70 percent operating C_v range.

If the rangeability requirement is *less* than 10:1, then C_v max calculated should fall at this 70 percent point of the selected valve. If the rangeability is *more* than 10:1 but less than what the selected valve can handle, then it is recommended to *locate the* C_v *min–*C_v *max range symmetrically* over the 7 to 70 percent operating range, such that *the two ranges are centered on each other*, but without allowing C_v max to exceed 80 percent of the actual valve C_v available.

This recommendation assumes that the normal load will fall more or less between C_v max and C_v min. If the normal load expected is much closer to the maximum than the minimum, then one might select a less than 70 percent point to correspond to C_v max in order to provide additional future capacity, and inversely, if the minimum and normal loads are close together, it is logical to build in more flexibility at the low end by shifting the entire operating range upward.

Nomenclature

C_d	Valve discharge coefficient (C_v/d^2)
C_{dr}	Valve reducer discharge coefficient (C_{vr}/d^2)
C_{dt}	Discharge coefficient for valve-reducer combination
C_f	Pressure recovery factor $(C_1/36)$
C_{fr}	Pressure recovery factor for valve-reducer combination
C_g	Supercritical gas flow coefficient $(C_1 C_v)$
C_v	Valve capacity coefficient
C_{vr}	Equivalent valve coefficient of the valve reducers
C_1	Pressure recovery coefficient $(36\,C_f$ or $C_g/C_v)$
d	Nominal valve size (in.)
D	Nominal pipe diameter (in.)
D_c	Flow characteristics distortion coefficient
F_1	Pipe reducer correction factor
F_2	Viscosity correction factor
g	Gravitational acceleration $(32.17\ \text{lbm-ft/lbf-sec}^2)$

G_f	Specific gravity at inlet temperature, relative to air or water
G_m	Specific gravity of liquid-gas or of liquid-vapor mixture, relative to water
G_0	Specific gravity of flowing stream at valve outlet, relative to water
K	Vapor constant (Table 1.2q)
L	Equivalent or actual pipe length
$P_{allowed}$	Valve outlet pressure that produces maximum flow at given inlet conditions (PSIA)
P_c	Pressure corresponding to thermodynamic critical point, where fluid state is undefined (PSIA)
P_v	Liquid vapor pressure at inlet temperature (PSIA)
P_{vc}	Pressure at vena contracta (PSIA)
P_1	Valve inlet pressure (PSIA)
P_2	Valve outlet pressure (PSIA)
ΔP	Valve pressure drop $(P_1 - P_2)$
$\Delta P_{allowed}$	Apparent pressure drop at which choked flow first occurs (PSID)
Q	Volumetric flow (SCFH)
Q_f	Liquid flow (GPM)
R_c	Critical pressure ratio $[f(P_v/P_c)]°$
T_s	Degrees superheat (°F)
T_1	Valve inlet temperature [°R = °F + 460]
$\overline{V}$	Velocity (ft/sec)
W	Mass flow (lbm/hr)
X	Correction factor for expansion at valve port
ρ	Density (lbm/ft^3)
μ	Viscosity (centipoises or lbf − sec/ft^2)

REFERENCES

1. H. D. Baumann, "The Introduction of a Critical Flow Factor for Valve Sizing," ISA Transactions, Vol. 2, No. 2, April 1963.
2. "Recommended Voluntary Standard Formulas for Sizing Control Valves," Fluid Controls Institute, Inc., FCI 62-1, May 1962.
3. H. D. Baumann, "Effect of Pipe Reducers on Valve Capacity," Instruments and Control Systems, December 1968.
4. G. F. Stiles, "Liquid Viscosity Effects on Control Valve Sizing," paper presented at Texas A & M 19th Annual Symposium on Instrumentation for the Process Industries, January 1964. Lab Report 3, Problem 1263, Fisher-Governor Co., January 19, 1964.
5. E. W. Singleton, "Control Valve Sizing for Liquid Viscous Flow," Engineering Report No. 7, Introl Limited.
6. J. F. Buresh and C. B. Schuder, "The Development of a Universal Gas Sizing Equation for Control Valves," ISA Transactions, Vol. 3, No. 4, October 1964.
7. H. W. Boger, "Recent Trends in Sizing Control Valves," 23rd Annual Symposium on Instrumentation at Texas A & M, January 1968.

° Boger defines the term R_c as $R_c = 0.96 - 0.28\sqrt{P_1/P_c}$

8. A. J. Hanssen, "Accurate Valve Sizing for Flashing Liquids," Control Engineering, February 1961.
9. L. R. Driskell, "Practical Guide to Control Valve Sizing," Instrumentation Technology, June 1967.
10. H. W. Boger, "Flow Characteristics for Control Valve Installations," ISA Journal, November 1966.
11. B. G. Lipták, "How to Size Control Valves for High Viscosities," Chemical Engineering, December 24, 1962.
12. B. G. Lipták, "Valve Sizing for Flashing Liquids," ISA Journal, January 1963.

1.3 GLOBE BODY FORMS

FC

THREE-WAY

FC
FAIL CLOSED
GLOBE VALVE

NOTE: THE LETTER
"S" IF MARKED INSIDE
THE VALVE SYMBOL,
REFERS TO "SPLIT
BODY" AND THE LETTER
"C" TO CAGE DESIGN.

FO
FAIL OPEN
ANGLE VALVE

Sizes:	1″ to 24″.
Design Pressure:	Up to 2,500 PSIG.
Design Temperature:	−40 to 1,200°F.
Rangeability:	Up to 50:1.
Capacity:	$C_v = 12d^2$ (non-critical flow).
Materials of Construction:	Great variety of materials.
Partial List of Suppliers:	Black, Sivalls & Bryson, Inc.; A. W. Cash Co.; Conoflow Corp.; Fisher Controls Co.; Foxboro Co.; General Controls—ITT; Honeywell, Inc., Industrial Div.; Kieley & Mueller, Inc.; Leslie Co.; Masoneilan International, Inc.; Research Controls Co.; Valtek Co.

Body Forms

The actual fluid handling portion of a control valve is called the *valve body assembly*. When properly operated by an actuator, it will modulate the flow of process fluid to help regulate the pressure, flow rate, temperature, liquid level or some other variable in a particular control system. This control function can only be performed through pressure reduction. A control valve is always a pressure-reducing device (Section 1.21). The valve body assembly consists of a pressure-tight body, a bonnet or top closure assembly, trim elements and sometimes a bottom flange. The shape and style of the valve body assembly is usually determined by the type of trim elements it contains with the exception of valves that have to fulfill particular functions or meet piping requirements such as three-way valves or angle valves. The body can have screwed or flanged ends or be welded into the piping system.

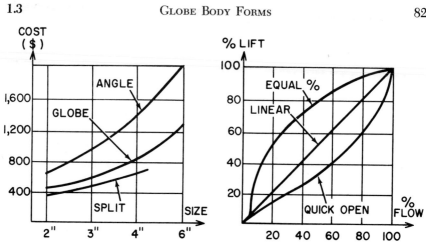

Fig. 1.3a   Cost and inherent characteristics of carbon steel globe valves

The trim always consists of a plug connected by a reciprocating stem to the actuator, one or two seat rings and guide bushings.

Most users of control valves prefer valve trims that will provide tight shutoff. Unfortunately, under the present state of the art, single-seated non-balanced valve trims must be used to provide tight shutoff under all possible service conditions. With lapped-in metal-to-metal seats, leakage figures below 0.01 percent of maximum capacities are possible (Section 1.21).

Single-Seated Valves

Single-seated valves can be provided with top-guided or top- and bottom-guided valve plugs. As shown in Figure 1.3b, the top-guided valve requires

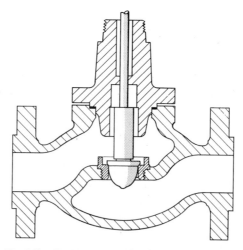

Fig. 1.3b   Top-entry, top-guided single-seated globe valve

only one body opening. In addition, the mass of the valve plug is reduced, which increases the natural undamped frequency of the trim and makes this valve less susceptible to vibration than the top- and bottom-guided valves in Figure 1.3c. Here a bottom flange has to be added to provide the second guide

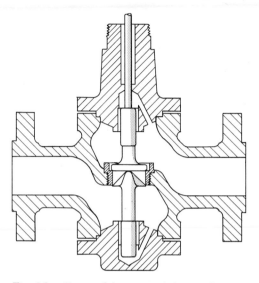

Fig. 1.3c   Top- and bottom-guided invertible single-seated globe valve

bushing. This design permits reversal of the fail-safe mode. That is, the valve body can be inverted and the bottom flange replaced by the bonnet. In this case, the valve plug will open with increase in diaphragm stroke. Thus the valve will "fail-closed" with the direct actuator instead of "fail-open" in the original body position shown in Figure 1.3c. The bottom flange can also be used to clean the valve body from sediments, and this can be an important factor, particularly after startup of a new system. A single-seated valve attached to standard diaphragm actuator should always be installed "flow tending to open" to insure dynamic stability of the trim.

Double-Seated Valves

The reversibility of the valve plug in Figure 1.3c applies equally well to a double-seated valve (Figure 1.3d), which until recently was the most commonly used control valve style. This type of valve is semi-balanced, i.e., the hydrostatic forces acting on the upper plug tend to cancel out the forces acting on the lower plug. The result is less actuator force requirement. However, there is always an unbalanced force due to a slight area difference between the upper plug and the lower plug (the lower plug is $\frac{1}{16}''$ to $\frac{1}{8}''$ smaller in diameter to permit assembly) and due also to the effect of dynamic hydrostatic forces acting respectively on the throttling area of each plug. Such

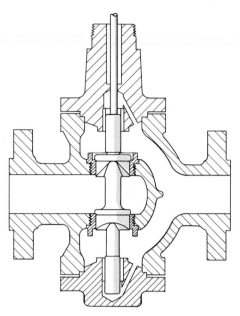

Fig. 1.3d   Top- and bottom-guided invertible double-
seated globe valve

dynamic off-balance forces can be quite high, particularly with the contoured
(lathe-turned) plugs, and can reach as much as 30 percent of such forces
of the equivalent single-seated valve plugs. Characterized V-port plugs (Section 1.5) are preferred for higher pressure drops, especially on larger valve
sizes.

Cage Valves

Another form of semi-balanced trim is provided in the cage valve (Figure
1.3e), which uses a piston with piston ring seal attached to the single-seated
valve plug. Here the hydrostatic forces acting on top of the piston or below
the valve plug tend to cancel out. The main disadvantage of these semi-
balanced trim configurations is the increase in seat leakage unless soft-seat
inserts are employed. Leakage figures of around 0.5 percent of maximum flow
capacity have to be anticipated. Of particular interest in the cage design is
the fact that the seat ring is clamped in by a cage. This is in marked contrast
to the designs shown in Figures 1.3b and c, where a metal shoulder of the
screwed in seat ring seals directly against the body wall. The threaded retention of the seat ring, while quite rugged, does not lend itself very well to
maintenance, particularly after extended service. The ease of maintenance
has made the cage-supported or cage-guided valve designs more popular in
recent years. Figure 1.3f shows such a typical design, this time with unbalanced
trim for tight shutoff requirements. In this case the flow characteristic is

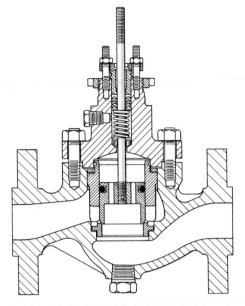

Fig. 1.3e   Piston-balanced single-seated globe valve
with cage type seat ring retainer. (Courtesy of the
Fisher Controls.)

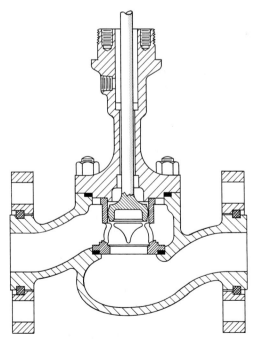

Fig. 1.3f   Single-seated ported cage valve with piston
plug

obtained by specially shaped ports placed directly in the valve cage. The valve plug has a form of a disc and serves only to provide shutoff and the function of a "curtain" to generate the desired effective flow area. Besides ease of maintenance, this type of trim provides a more turbulent flow pattern, making this particular valve less susceptible to cavitation.

Frequently built also, in contrast to the top entry cage trim configuration, is a "quick change" trim type supported by a bottom flange (Figure 1.3g). This permits drainage of the body and quick inspection of the valve trim without the need to remove the actuator portion.

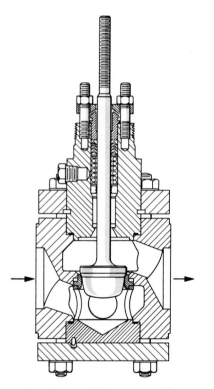

Fig. 1.3g   Quick-change trim valve designed for trim removal through bottom flange

Split-Body Valves

Another way to provide a means to replace the valve trim quickly is provided in the split-body valve (Figure 1.3h). Here the seat ring and valve plug can be removed quite rapidly by separating the two body halves. Additional advantages in this particular design come from its adaptability to slip-on

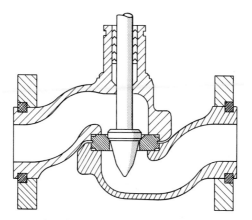

Fig. 1.3h   Single-seated split-body valve with removable flanges

flanges. These make it possible to reduce the weight of the actual valve casting, which results in a cost saving, particularly in corrosion-resistant alloys. They also bring about a reduced inventory, since only one body is required for several pressure ratings. The body shape is relatively streamlined, reducing the tendency to collect sediments. An added benefit of split-body valves is the possibility of mounting the lower body half at 90 degrees to the line axis to facilitate installation. Welded connections are impractical due to the inability to separate the two housing parts for maintenance.

Most control valve body assemblies of the globe form are interchangeable for a given size and pressure rating because their face-to-face dimensions are standardized.[1,2]

Angle Valves

Angle valves are one of the many special configurations of body assemblies. It has been recommended that they be installed in a flow-to-close position for high-pressure-drop service. This application, however, while favorable to the valve body and trim, places the major burden of pressure reduction on the downstream piping. It is for this reason that this particular trend has recently been reversed. The proper use of angle valves, other than accommodating a special piping requirement or to aid drainage, is for erosive service where impingement of solid particles is to be avoided, and for special applications such as coking hydrocarbons. Figure 1.3i shows a typical streamlined angle valve with venturi outlet designed especially for coking. The streamlining of the valve plug is done especially to avoid settling of deposits of fluid particles on the body wall. The use of this type of valve for general high-pressure drop applications is not recommended, because the streamlined flow path results in exceptionally high-pressure recovery, which in turn reduces

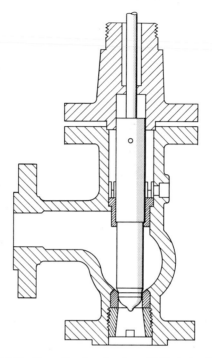

Fig. 1.3i   Streamlined angle valve with Venturi outlet

the installed flow capacity under choked flow conditions and makes the valve highly susceptible to cavitation on liquid service.

Y-Style Valves

For applications in which drainage of the body passages and a high flow capacity is required, as in certain cryogenic or molten metal applications, the use of a Y-valve is recommended (Figure 1.3j). This valve in the figure is fitted with a vacuum jacket, which provides maximum thermal insulation as would be required in cryogenic applications, handling liquid hydrogen, for example. Note the compactness of this design which permits minimum and uniform wall thicknesses, which in turn enables rapid cool-down rates. The Y-valve can be installed with the actuator on a vertical axis, thereby permitting self-draining of body cavities.

The vacuum jacketing shown in Figure 1.3j consists of a stainless steel sheet metal enclosure with attached metal bellows at the inlet and outlet ports of the valve. These bellows take up mechanical tolerances and thermal expansion and can be welded in turn to a similar jacket around the actual pipeline.

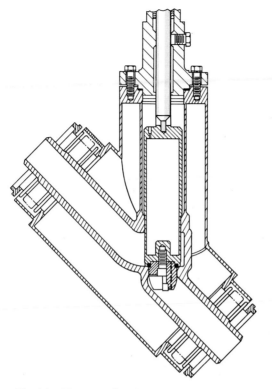

Fig. 1.3j   Vacuum-jacketed Y-valve for high flow and
self-draining

Jacketed Valves

Another type of body enclosure is shown in Figure 1.3k. This is a so-called "steam jacket" welded around the body casting. When filled with steam or any other heating medium such as Dowtherm, it keeps the body sub-assembly sufficiently hot to prevent solidification or crystallization of certain fluids.

Three-Way Valves

Three-way valves are another form of specialized valve body configurations. There are types for two basic services: (1) for mixing service, that is, the combination of two fluid streams passing to a common outlet port (Figure 1.3l), and (2) for diverting service, that is, separating a common inlet port into two outlet ports (Figure 1.3m).

A typical application for a mixing valve would be the blending of two different fluids to provide an end product with a desired property and consistency. A diverting valve might be used in temperature control. One portion

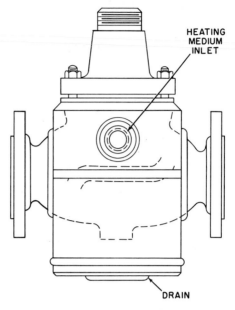

Fig. 1.3k   Steam-jacketed valve

of the passing hot fluid could pass through the heat exchanger to permit temperature control of an independent fluid stream and then regain the portion by-passed through the second outlet port.

The forces acting on the double-seated three-way plug do not balance, because in each of the three flow channels different pressure levels exist. The valve plugs in these particular illustrations are port- and top-guided. This eliminates the need for the lower stem guide and permits unrestricted flow capacity through the lower outlet flange connection.

Small-Flow Valves

Small-flow valves are used primarily in laboratories and pilot plants which essentially duplicate complete chemical processes on a reduced scale or for specialized applications such as pH control. The most common valve is the small single-seated globe configuration with screwed end connections, as shown in Figure 1.3n. The trim consists of a precision honed plug fitted into an orifice made of a hard alloy. The control area consists of a fine slot milled or scratched into the outer surface of the piston-shaped plug. The rangeability, i.e., the ratio between maximum and minimum controllable flow, is quite *limited* for this type of valve plug due to the inherent leakage flow between piston and orifice. The C_v ratings of small-flow valves should be treated with extreme caution and should be used for reference purposes only (below values of 0.01). It is quite common for a laminar flow pattern to predominate, particularly with viscous fluids or in low-pressure applications of these small-flow valves. Laminar flow means that the flowing quantity increases linearly with pressure drop (Section 1.2).

In contrast to the relatively long-stroke configurations discussed earlier there are certain short-stroke or variable-stroke devices such as the flow control valve shown in Figure 1.3o. Here a synthetic sapphire ball is allowed to lift off a metal orifice, thereby modulating minute fluid flows. The particular advantage of this valve is that the maximum diaphragm stroke can be adjusted to produce various actuator strokes with a 3–15 PSIG signal to cover C_v ranges from 0.07 to 0.00007.

A similar valve especially suitable for high-pressure service (up to 50,000 PSIG) is illustrated in Figure 1.3p. Variation in C_v rating of identical plugs and seats is achieved by mechanical adjustment of a toggle arrangement. This changes the valve stroke between 0.010 and 0.150 in. A total C_v range from 1 to 1×10^{-6} can be covered with different trim inserts.

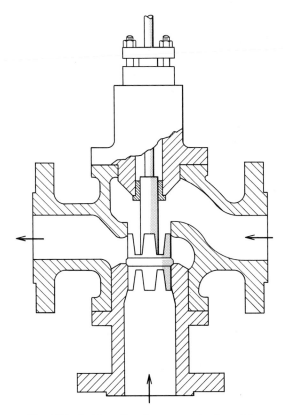

Fig. 1.31   Inside seating three-way valve for mixing
service

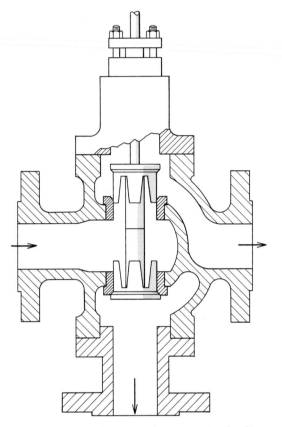

Fig. 1.3m   Outside seating three-way valve for diverting service

Valve Connections

The most popular body end connection is the machined flange. In the
United States, these flanges are standardized under a USAS Standard B16.5.[3]
The rating of the valve flange is a function of process fluid pressure and
temperature. Other countries have similar standards, such as DIN in Germany.
Their rating system however is quite different. Whereas USAS B16.5 sets the
rated pressure level at an elevated temperature, DIN defines the rated pressure
level at room temperature. Therefore, a USAS 150 PSIG flange can be used
for higher pressures than a DIN 10 flange.

For ease of identification, USAS has provided that all cast iron flanges
be machined without a raised face. Steel and alloy flanges in ratings of 150
and 300 PSIG have a $\frac{1}{16}$-in.-high raised face, and flanges including and above
400 PSIG rating have a $\frac{1}{4}$-in.-high raised face. For most high-pressure flange
connections (above 600 PSIG), USAS employs ring joint gaskets, which provide

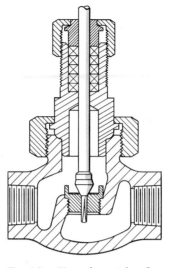

Fig. 1.3n   Union bonnet low-flow
valve with close clearance grooved
piston plug

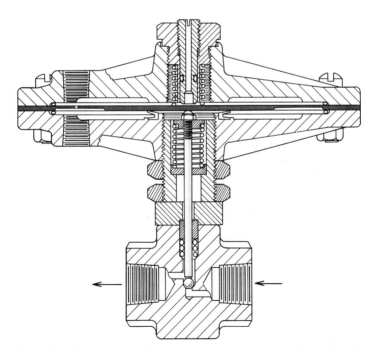

Fig. 1.3o   Short-stroke low-flow design using synthetic sapphire ball and
adjustable stroke actuator. (Courtesy of the A. W. Cash Co.)

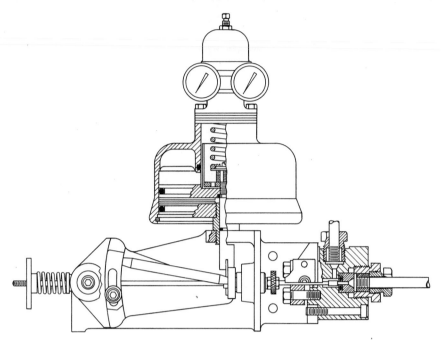

Fig. 1.3p   Special ratio linkage actuated low-flow valve for pressures to 50,000 PSIG

a more reliable gasketing between the body and the mating line flange, particularly at elevated temperatures.

Slip-on flanges, as discussed under Split-Body Valves, make it possible to use one standard body for different pressure ratings or they can reduce costs (when carbon steel slip-on flanges are used on stainless steel body for example). Two halves of a tubular ring are placed into a groove cut into the outside of each body end connection, which then key and retain the flange after bolting against the line flange (Figure 1.3h).

Flangeless bodies (clamped-in) have become more common in recent years, having originally been applied to butterfly valves. The inherent advantage of such a body style is universal applicability. A valve body with a 600 PSIG rated wall thickness, for example, can be installed between 150, 300, 400, or 600 PSIG flanges. It also permits simplified machining of the valve body out of bar stock (Figure 1.3g). Metals with high tensile strength must be selected for the tie rod materials to absorb the longitudinal expansion of the clamped-in valve body due to temperature changes. With proper care, flangeless bodies offer some advantages including the relative ease of installation, particularly if companion flanges are welded to the pipe in the field. Lining up the mating bolt holes here is not as critical as it is between line and body flanges.

Welding ends are recommended where unusually high line stresses or thermal shock conditions exist, as they do in many power plant applications. Socket-welded ends are easy to align and are more common with smaller valves (up to 2-in. size). Butt weld ends must be used where full penetration of the weld and a subsequent X-ray check is required (Figure 1.3q). A valve

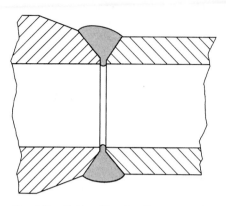

Fig. 1.3q   Butt-weld valve line connection showing full penetration weld

body material should be selected that is suitable for welding to the adjoining line flange. Body materials with high carbon or high chrome content should be avoided.

Certain chemical processes requiring pressure ratings above 5,000 PSIG utilize *lens type high-pressure fittings,* as shown in Figure 1.3r. This is a

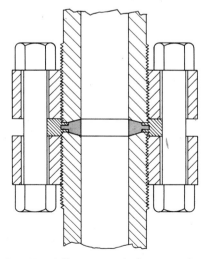

Fig. 1.3r   Self-energizing high-pressure lens type joint

self-energizing type of seal, i.e., the lens ring expands with increase in line pressure. (Further details are given in Section 1.21.)

Materials of Construction

A valve body is a pressure-carrying vessel. Hence, material selection must follow guidelines set forth in the Unfired Pressure Vessel Code published by ASME. For example, a valve body material should not be used for low-temperature service when the Charpy impact value is less than 15 PSIG.

Particular attention should be paid to the weldability of body materials. For example, to avoid intergranular corrosion due to carbon precipitation specify ASTM A351-52T GR.CF8C if a stainless steel valve body is to be welded in the line. Table 1.3s is a simplified guide for the selection of suitable body materials.

Other special trade-name alloys are available for special corrosive applications. Valve bodies of such alloys as Monel metal, Hastelloy, nickel, Inconel, Durimet 20 and others are available from most manufacturers.

Table 1.3s
BODY MATERIAL SELECTION CHART

Material	Maximum Pressure Rating (PSIG)	Rating Per USA Standard	ASTM Specification	Low-Temperature Limit (°F)	High-Temperature Limit (°F)	Service
Cast Iron	800°	B16.1 B16.4	A126 B	−20	400	Non-corrosive, non-fire-hazard
Cast Carbon Steel	2,500	B16.5	A216-WCB	−20	775°°/1,000	Non-corrosive
Forged Steel	2,500	B16.5	A105	−20	775°°/1,000	High pressure
Cast Chrome-moly Steel	2,500	B16.5	A217-WC6	−20	1,200	Elevated temperature, flashing fluids
Cast Stainless Steel (AISI 304)	2,500	B16.5	A351-CF8	−450	1,500	Corrosive, low temperature
Cast Stainless Steel (AISI 316)	2,500	B16.5	A351-CF8M	−350	1,500	Corrosive, high temperature
Cast Nickel Steel	2,500	B16.5	A352-LC3	−150	650	Non-corrosive, low temperature
Cast Bronze	300	B16.24	B62	−350	450	Low temperature, marine, oxygen
Cast Monel	2,500	B16.5	A296-M35	−350	900	Corrosive, salt water
Cast Hastelloy C	2,500	B16.5	A494		1,000	Corrosive

° 250 PSIG for steam.　　°° If welded in line.

The *bolting material* is usually carbon steel for cast iron bodies. However, higher strength studs such as specified under ASTM A193-52T GR.B7 are generally employed for carbon steel and alloy steel bodies. If the service temperature exceeds 850°F, specify the stud alloy as GR.B14. The respective material code for the nuts are ASTM A194-51 GR.2H or GR.4 above 850°F. Steel alloy studs are usually sufficient for stainless steel body sub-assemblies unless there is a danger of corrosive fluids contacting the studs or where the atmosphere is corrosive.

Bonnet and blind flange gaskets are usually metal-clad asbestos, although it is possible to use flat gaskets made of asbestos fiber sheets in about $\frac{1}{16}$-in. thickness, particularly on cast iron valve body sub-assemblies. High-pressure valves or exotic service conditions may require solid metal gaskets made from soft iron or stainless steel. However, solid metal rings or metal O-rings do require extremely fine surface finishes. Elastomer O-rings are sometimes used with threaded bonnets for temperatures to about 180°F.

REFERENCES

1. "Uniform Face to Face Dimensions for Flanged Control Valve Bodies," Instrument Society of America, RP4.1, Pittsburgh, Pennsylvania.
2. "Recommended Voluntary Standards for Face to Face Dimensions of Control Valves," Fluid Controls Institute, FCI 65-2, Pompano Beach, Florida.
3. "Steel Pipe Flanges and Flanged Fittings," The American Society of Mechanical Engineers, USAS B16.5, New York, New York.

1.4 BONNET TYPES, PACKINGS AND LUBRICATORS

The valve bonnet mounts the actuator to the valve body and seals against process fluid leakage along the stem. This section discusses the bonnet design variations that can be considered and the special-purpose bonnets, such as the radiation and extension types. On the subject of sealing the stem, the choices of packing materials and lubricators is presented together with some discussion of the more special seals, such as the dual packings and the bellows seals.

Bolted Bonnets

The bonnet assembly, according to ASME Standard 112, is "an assembly including the part through which a valve plug stem moves and a means for sealing against leakage along the stem. It usually provides a means for mounting the actuator." The bonnet is generally a separate valve part that can be removed to give access to the valve trim. This then makes the bonnet a pressure-carrying part sealing the contained fluid against the outside. The usual bonnet joint is flanged with a machined recess to provide guiding and containment of a gasket (Figure 1.4a). Bonnet and bottom flange are usually stud bolted. Through bolting is more common on high-pressure valves. The bolt

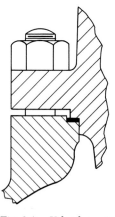

Fig. 1.4a Valve bonnet joint showing retained gasket design

98

size and flange thicknesses are calculated in accordance with the applicable ASME Code requirements.

Certain low-pressure valves, particularly in sizes below 2 in., employ a threaded bonnet, which reduces the total weight and is more economical to produce than a flanged joint. One disadvantage of this particular connection is the difficulty in removing the bonnet after extended service, particularly with corrosive fluids. The lower portion of the bonnet usually contains a guide bushing made from a hard material such as heat-treated AISI type 440-C stainless steel or 17-4PH stainless. Special guide materials such as glass-filled Teflon are available for corrosive or cryogenic applications.

Packings

The upper bonnet portion has to seal the sliding stem in a stuffing box assembly (Figure 1.4b). This assembly consists of a packing flange, packing

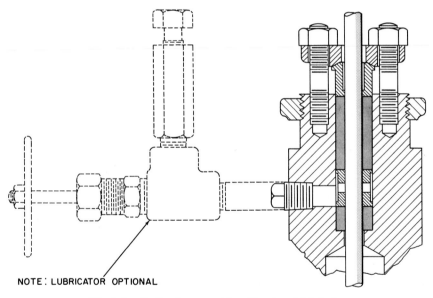

NOTE: LUBRICATOR OPTIONAL

Fig. 1.4b   Stuffing box assembly with external lubricator

follower, a lantern ring and a number of equally spaced packing rings. There are a number of packing materials on the market, each having particular advantages and disadvantages to suit different service conditions. One requirement for a control valve packing is to seal the sliding stem and yet produce minimum friction. External lubrication by means of a packing lubricator usually combined with an isolating valve for safety is frequently used. The grease is inserted between the individual packing rings in the space provided by the lantern ring. The viscosity of the particular grease is selected to provide maximum lubricity at a given service temperature. Other choices are dictated by corrosion requirements.

External lubrication works reasonably well, although it is somewhat cumbersome and requires frequent reloading of the lubricator.

To overcome this disadvantage, valve manufacturers have been utilizing the low-friction properties of Teflon, either in the form of solid Teflon chevron rings (Figure 1.4c) or as a coating on asbestos yarn. The solid Teflon packing

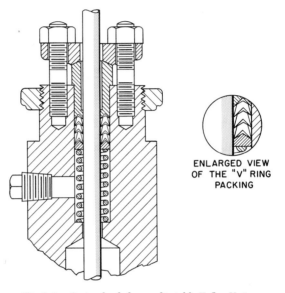

ENLARGED VIEW
OF THE "V" RING
PACKING

Fig. 1.4c   Spring-loaded non-adjustable Teflon V-ring
stuffing box assembly

is somewhat handicapped by the low "memory" of Teflon and therefore requires a constant loading pressure such as provided by the internal spring arrangement shown in Figure 1.4c. Such a packing cannot be retightened once installed.

Of particular importance in selecting this type of packing is the surface finish of the stem, which must be in the order of 5 to 8 RMS. An accurate diameter and a finish of about 16 RMS is required on the inside of the stuffing box.

Braided asbestos yarn is still the basic packing material, lubricated either with Teflon additives, external coatings, or in the case of high-temperature service, by graphite.

Bellows Seals

The packings described so far do not provide absolutely tight seals, particularly after extended service. There are instances that require a 100 percent seal, e.g., if the valve is handling toxic or radioactive fluids, or when the fluid to be handled is extremely valuable. A typical bellows seal construction is shown in Figure 1.4d. The bellows is usually made of stainless steel

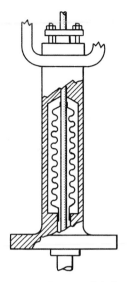

Fig. 1.4d Typical bellows seal for reciprocating valve stem

and can be hydraulically formed (as shown) or welded in individual segments (the nested type). The bellows itself is welded on top to the valve stem and on the bottom against a clamped-in fitting containing the guide bushing and an anti-rotation device. The latter prevents the bellows from being twisted during assembly, thereby disassembling the valve. To insure the tightness, a mass spectrometer test is usually specified, measuring helium leakage rates below 1×10^{-6} cc/sec, from atmospheric pressure to vacuum.

One of the basic limitations of a bellows seal is its inability to withstand high pressures. The pressure rating decreases with increasing valve size. Typical pressure ratings are 150 PSIG at 600°F. Higher pressure ratings to about 500 PSIG can be obtained by the use of special bellows having an increased wall thickness. This is usually done only at the expense of the number of life cycles obtainable. Another way of increasing the pressure rating is to load the exterior side of the bellows with pressurized inert gas. This method is usually costly and cumbersome.

Double Packings

An economical substitute for bellows seal may be a double packing consisting essentially of two packings in series with a leak-off connection in between. This type of stem seal is used quite frequently in atomic energy applications involving radioactive fluids. The leakage coming through the first set of packing rings is piped to a non-pressurized waste container. See Figure 1.4e and also refer to Section 1.21 for additional information.

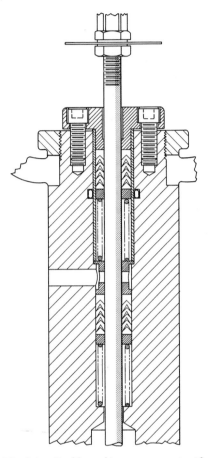

Fig. 1.4e   Double packing arrangement with leak-off connection and independent packing adjustment

Radiation Bonnets

This may be used when the fluid temperature in a valve exceeds 400° F. The two functions are, first to extend the leakage path of the heat coming from the body by conductance, and second to provide sufficient radiation surface by welded or cast-on ribs to absorb the heat leakage through radiation and convection. The fins may be omitted if the difference between fluid temperature and maximum packing temperature is less than 350°F. See the discussion, Section 1.21, on the subject of high-temperature service applications.

Extension Bonnets

Plain extension bonnets (Figure 1.4f) are required for low-temperature service, i.e., for fluid temperatures below −20°F. Its purpose is to keep the packing box from freezing. To help achieve this purpose, heating coils are sometimes wrapped around the bonnet. In any case, the extension bonnet should *not* be thermally insulated.

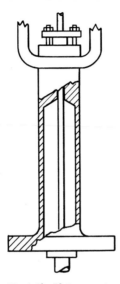

Fig. 1.4f   Plain extension bonnet design for fluid temperatures   below −20°F

In valves specially built for cryogenic service, the extension bonnet often consists of a thin-walled stainless steel tube (to reduce cooldown weight) and it is welded directly to the body casting. The bonnet flange is then attached to the top of the stainless steel tube and connects to the wall of an insulating vessel, a so-called "cold box." For further details on cryogenic service control valves refer to Low-Temperature Service in Section 1.21.

1.5 TRIM DESIGNS

The main components of the trim in a globe valve are the valve plug and the seat. A large portion of the pressure dissipated in the valve is absorbed in the trim. In addition to its pressure-dissipating function, the trim also serves to determine the inherent flow characteristics of the valve. This section describes the design variations of valve trims, including a discussion on their forms, rangeability and materials of construction. Particular attention is given to the wear and noise aspects of valve trims, which subject is further amplified in Section 1.21.

Partial List of Suppliers of Low Noise and Wear Valves: Control Components, Inc. ("Self Drag" Trim); Fisher Controls ("Whisper Trim"); Hammel Dahl, Div. of ITT ("Flash Flo" Trim); Masoneilan International, Inc. ("Lo db" Valve); Yarway Corp. ("Turbo-Cascade")

Flow Characteristics

All control valves are essentially pressure-reducing devices. In other words, they have to throttle the flowing fluid in order to achieve control. The most widely used form of throttling is with a single-stage orifice, although multiple-stage orifice devices are recently being employed at an increasing rate to combat noise and erosion problems. The most common single-step orifice consists of a seat ring and a plug member (Figure 1.5a). A typical valve plug has a cylindrical guide, a seating portion that provides tight shut-off, and a characterized portion which—in conjunction with the seat ring—provides a variable orifice area as a function of valve stroke. The relation between this area change and stroke is important because it defines the flow characteristics of the valve.

The term "flow characteristics" usually refers to the inherent flow characteristics, which means that the variation in flow as a function of valve stroke is measured with a *constant* differential pressure across the valve. In contrast, the installed flow characteristic is experienced when the valve is installed in a complete control loop (Section 1.2). In this case, the differential pressure across the valve is not constant, but varies as function of the pump characteristic, the hydrostatic resistance of elbows, reducers, and block valves, and the friction loss in the piping. It is often desired that the installed flow characteristic be as linear as possible to enable a constant proportional band

104

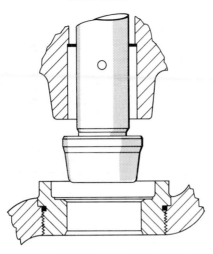

Fig. 1.5a   Top-guided plug and seat ring

setting of the controller over a wide load range, i.e., the increase or decrease in actual flow through the valve should be linearly proportional to the variation in valve stroke or instrument output signal. If the pressure drop across the control valve is nearly constant under all flow conditions (such as in pressure-reducing installations), then the inherent flow characteristics and the installed characteristics are the same and one should specify a *linear trim* (Figure 1.5b).

If, on the other hand, the pressure drop across the valve varies greatly between the closed and wide open positions, then one should specify the *equal percentage characteristic*.

The mathematical relationship between flow and valve lift of an equal percentage trim can be defined as follows:

$$m/M = R^a \qquad\qquad 1.5(1)$$

where a = x/X − 1,

    x = any valve lift,

    X = maximum lift,

    m = flow at lift x,

    M = maximum flow at lift X and

    R = rangeability.

Rangeability

Rangeability R has to be defined for this type of characteristic. For example, R could be assumed to be infinite if the flow were to be shut off

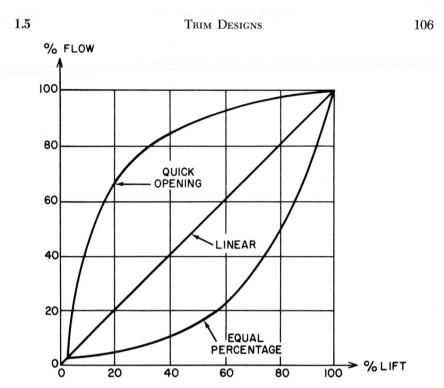

Fig. 1.5b   Inherent flow characteristics—quick-opening, linear, and equal percentage

completely at zero stroke. (Rangeability is discussed further in Section 1.21.) In practice, however, this is not the case because there is a minimum clearance between the cylindrical plug and the seat ring (Figure 1.5c). As soon as the valve plug lifts off the seat, there is a sudden increase in flow, corresponding to the annular clearance. Rangeability must therefore be defined as the ratio between maximum and minimum controllable flow, where the rate of flow change follows the desired flow characteristic. The maximum rangeability for single-seated valves is usually 50 to 1, while large double-seated valves have a rangeability of no more than 25 to 1. However, there are now certain valve types such as ball valves on the market that do exhibit rangeabilities as high as 100 to 1, since they are not limited by mechanical considerations such as the minimum clearance between valve plug and orifice. Theoretically, the higher the rangeability, the larger the permissible ratio between minimum pressure drop (100 percent lift) and maximum pressure drop (zero lift) across the valve. As a rule of thumb, one might select the equal percentage charac-teristics whenever the ratio between maximum and minimum pressure drop across the valve, over the full flow range, is more than 3 to 1.

     A quick-opening characteristic is obtained with a beveled seated disc or a plain flat disc type of valve plug as shown in Figure 1.5d. Here the inherent

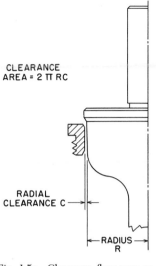

Fig. 1.5c   Clearance flow area as
determined by plug and seat ring
design

flow-to-stroke characteristic is approximately linear up to a valve lift equivalent to about 25 percent of orifice diameter. Plugs of this kind are used on
self-contained pressure regulators having a short actuator stroke or for on-off
service. Figure 1.5b shows the inherent flow characteristic of a typical quick-
opening single-seated valve. This characteristic might be used in relief applications where the valve should open as fast as possible.

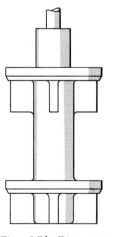

Fig.   1.5d   Disc   type
(quick-opening) plug

Trim Configurations

The desired flow characteristic is obtained by turning the outside contour of the valve plug (Figure 1.5e) or by porting an outside skirt of the plug (Figure 1.5f). The contoured plug is generally used on single-seated valves particularly with small trim sizes, while the ported plug, having a more constant off-balance area, is more common in double-seated control valves. There are inherent manufacturing advantages in a contoured plug. It can be made of bar stock and can be hard-faced more easily for erosive services. The ported plug must be cast or forged and is difficult to hard-face. On the other hand, the more abrupt change in flow pattern with a ported plug produces less pressure recovery and makes this plug dynamically more stable in the axial direction. However, a tendency of this plug in certain cases to "spin" around its axis at high-pressure drops can be troublesome. Low-pressure recovery allows an increase in differential pressure before cavitation occurs.

The actual flow area at maximum stroke is usually larger than the ported area, i.e., the plug pulls out of the seat to provide a larger opening at the end of the lift to compensate for the pressure loss in the body, which is assumed to be acting in series with the orifice.

There has been a recent trend to place the contoured portion in the seat ring instead of on the plug. This reduces the mass of the plug, which in turn increases the natural undamped frequency of the trim and makes the plug less sensitive to vibration induced by fluid turbulence. In this design too one

Fig. 1.5e   Lathe-con-
toured single-seated
valve plug

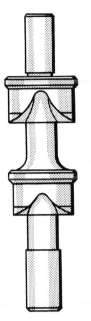

Fig. 1.5f   Cast double-
seated ported plug

can distinguish between the contoured seat ring and the ported seat ring.

Figure 1.5f illustrates a cage trim having characterized ports placed in the outer guide ring. Figure 1.5g in contrast shows a contoured seat ring that is clamped between the valve body and a cage type retainer. The additional

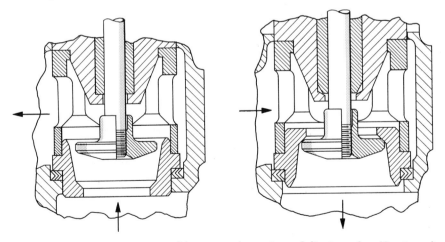

Fig. 1.5g   Valve design using invertible contoured seat ring and disc type plug. (Courtesy of
A. W. Cash Co.)

advantage of this design is that the seat ring can be inverted to provide the identical characteristic for both "fail-open" or "fail-closed" plug positions using only one direct acting actuator.

Seat rings can be screwed into the body, as is the case in most double-seated valves, or they can be retained against the body bridge by a cage type retainer. When utilizing a retainer a flexible metal gasket is placed between the seat ring shoulder and the body. Although screwed-in seat rings have been quite standard, they do have a tendency to loosen up due to temperature cycling, allowing leakage between the seat ring and body thread. Circumferential welding is used at times to overcome this problem.

Integral seats machined directly into the body bridge have been used on occasion. Hard-facing is quite difficult, however, and so is maintenance.

Trim Materials

The basic trim material is austenitic chrome nickel stainless steel, such as AISI 316 stainless steel, which covers the majority of service conditions. But if the pressure drop is high or if the fluid contains erosive particles, then a hard coating with cobalt-chrome alloys (such as some of the stellites) is recommended. For even more severe service, valve plugs made of solid tungsten carbide or ceramic are available. There are conflicting views concerning when hard coatings should be used in relation to pressure drop. Table 1.5h provides some generalized guidelines for the selection of trim materials and Section 1.21 contains additional information.

When specifying hard-faced plugs and seats, one should consider the plug and seat rings separately, using the following logic. The plug is often supplied with a layer of hard-facing alloy on the seating surface only, since this is sufficient if high-pressure drops occur primarily when the valve is closed. For continuous high-pressure-drop service, the complete plug surface, including the throttling portion and the guides, may be coated. A completely coated plug can often be used in conjunction with a seat ring which has the hard facing on the seating surface only, because the point of maximum velocity on the seat ring is thereby still protected. As the severity of the service increases, hard facing of the complete inner surface of the seat ring may be recommended. A number of cobalt and boron base hard-facing alloys are now available along with improved coating techniques such as flame spraying. This substantially increases the usage of these excellent valve trim materials.

Solid hardenable alloys for trim material include 440-C stainless steel (17 percent chrome, 1.0 percent carbon) ranging up to 60 Rockwell C in hardness and type 17-4PH stainless steel, a precipitation hardening material combining corrosion resistance and a hardness of up to about 40 Rockwell C. Type 440-C is generally used for simple shapes that can be made from bar stock or forgings and is limited in corrosion resistance. Type 17-4PH is more readily available in cast form and is a good general purpose alloy for service up to 750°F.

Table 1.5h
SELECTION OF TRIM MATERIAL

Service	Operating Temperature (°F)	Operating Pressure Drop (PSID)	Plug and Seat Material	Plug Guide Material	Guide Bushing Material
Liquid (Non-corrosive, Non-cavitating)	<750	<200	316 SS	316 SS	440-C SS
		>200	Cobalt-based alloy or 17-4PH SS	316 SS	
Flashing Liquid (Non-corrosive)	<750	<300	Hard-faced 17-4PH SS	316 SS, 17-4PH SS	440-C SS
Steam	<750	>150	440-C SS, 416 SS, 17-4PH SS	416 SS, 17-4PH SS	440-C SS
Steam	<750	<150	316 SS, 17-4PH SS	316 SS, 17-4PH SS	440-C SS
Oxygen	<−20	<150	316 SS	316 SS	316 SS nitrided
Abrasive Particles			Tungsten carbide, ceramic	Tungsten carbide, ceramic	Cobalt-based alloy
General	750–1200		Hard-faced	Hard-faced	Cobalt-based alloy
General	−20 to −320	<150	316 SS	316 SS	316 SS, nitrided

When tight shutoff is required, *soft seat inserts* can be provided for standard valve plugs in single- or double-seated configurations (Figure 1.5i). Teflon is commonly used for corrosive applications and for fluid temperatures ranging up to 450°F. Other elastomers such as Buna-N have a service limit at about 180°F. The soft inserts are usually replaceable and should be put in controlled compression, i.e., the maximum actuator thrust should be taken up by a surrounding metal lip pressing against the metal seat ring to prevent crushing of the soft insert. About 50 lb for Teflon and 25 lb for rubber seal inserts per linear inch of plug circumference should be allowed as additional seating force, when selecting the valve actuator.

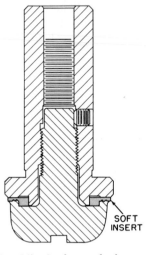

Fig. 1.5i   Single-seated plug and
seat ring with soft insert

Plug Stems

The usual form of connecting the valve stem to the valve plug is by means
of a fine thread, as shown in Figure 1.5j. Part of the stem fits tightly into
a recess on top of the plug guide to provide alignment in the axial direction.
The stem is brought up to an interference fit on the unfinished parts of the
female thread to provide a solid and vibration proof connection. To keep the

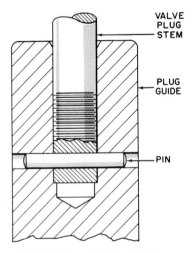

Fig. 1.5j   Specially threaded stem-
to-plug connection used for control
valves

stem from unscrewing, a pin is inserted across the bore. The pin is normally located in the plug guide section so that it cannot back out even if it becomes loosened.

Plug materials such as the cobalt base alloys, cemented carbides, or ceramics require special consideration such as welding, the use of softer alloy inserts as in the case of ceramics, shrink fits or clamping devices. In other valve types such as split-body valves the stem is an integral part of the plug sub-assembly. In this case, the plug is guided at the lower portion of the stem. This is called stem guiding and is useful in certain *slurry applications*, since it prevents build-up of deposits on top of the plug guide.

Noise and Wear Reduction

Noise limitations set by law and a continuously increasing number of high-pressure drop requirements with their attendant wear problems require special trim designs. A discussion of this subject in Section 1.21 is complimentary to its treatment below. Interestingly enough, most design considerations that result in noise reduction also diminish wear. Present designs can be classified in three groups.

> 1. Multiple orifice arrangements in parallel.
> 2. Multiple orifice or velocity head-loss elements in series.
> 3. A combination of both.

Common to all these designs is the aim of creating high hydraulic resistance (pressure loss) to the fluid stream, which in turn reduces the flow capacity (unfortunately resulting in a relatively low C_v value).

Multiple Orifices in Parallel

Figure 1.5k shows a typical valve with multiple orifices in parallel. Here a multiplicity of relatively small orifices, progressively uncovered by a moving cylindrical plug, create a turbulent flow pattern primarily due to the rapid expansion in flow area between each small orifice and the plug seat area. This results in a high-velocity head-loss coefficient, i.e., low-pressure recovery. Little pressure recovery in turn allows high pressure drops before cavitation occurs (see Section 1.2). Cavitation, the most destructive phenomena associated with flowing liquids, must be avoided to insure reasonable valve trim life no matter how hard or wear resistant the material is.

Even when service with a compressible fluid (steam and gases) requires velocities above sonic at critical pressure-drop ratios ($P_1/P_2 \geq 2$), some reduction in noise power is achieved by the decrease in the orifice diameter. Lighthill's[1] equation illustrates the effect of certain design parameters:

$$W = E \frac{dD^2V^8}{d_0C^5} \qquad\qquad 1.5(2)$$

where W = accoustical power,
         E = efficiency factor,
         d = density of flowing gas,
         D = orifice jet diameter,
         V = jet exit velocity,
        d_0 = density of atmosphere, and
        C = sound velocity in atmosphere.

For example, a 50 percent decrease in orifice diameter brings a noise reduction by a factor of 4 (6 db sound-pressure level). However, reducing the jet velocity by a factor of 2 brings a 256-fold decrease in sound power (or noise intensity). Such energy reduction results in a theoretical lowering of sound pressure level by 24 db. Lowering jet velocities is therefore a much more efficient way to reduce noise. Some of the sound energy is added by the increase in effective

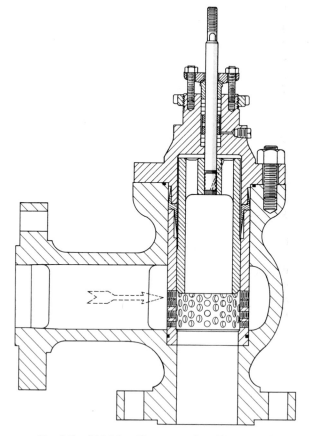

Fig. 1.5k   Multiple-orifice cage valve. (Courtesy of
Hammel Dahl, Div. of ITT.)

flow area that is necessary to compensate for the decrease in velocity in order to maintain the same flow rate, but substantial improvements are possible. For a valve noise estimating technique, refer to Section A.3.

Multiple Orifices in Series

As illustrated in Figure 1.5l, multiple orifices in series provide for a decrease in jet velocity due to employment of "multiple velocity head-loss elements" in series.

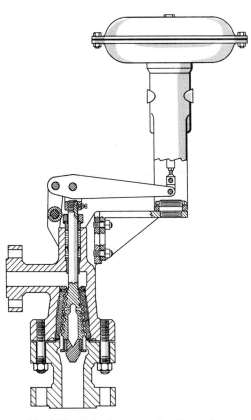

Fig. 1.5 l   Multiple-velocity head-loss valve

The friction loss in the fluid passing the multiple throttling teeth is similar to that found in a labyrinth seal. Because velocity according to Bernoulli's theorem is

$$V = \sqrt{(2g\ \Delta h)/K} \qquad\qquad 1.5(3)$$

velocity V can be significantly reduced for a given pressure drop (Δh) by the employment of a sufficiently high velocity head-loss coefficient K. This is possible only by providing fluid resistance in series. In contrast, K is always less than 1 in a single-orifice valve.

A reduction in fluid velocity leads not only to a significant reduction in acoustical energy, but also results in less wear. A K factor above 1.0 means no pressure recovery and therefore no cavitation. Little research has been done to determine the relationship between wear and throttling velocities of various gases. It is reasonable to assume however that the erosive effects of entrained particles in a gas stream are diminished as a function of the reduction in impact energy, i.e., as a function of velocity squared, V^2.

The effective flow area has to increase between the entrance and the outlet to compensate for the change in gas density. This requires an increase in the outlet flange size, for little would be accomplished by a significant reduction in noise level at the trim if the noise were then generated at the outlet of the valve due to high exit velocities.[2]

Multiple Orifices in Series and Parallel

A design with multiple orifices in series and parallel utilizes the pressure-loss-producing effects of a fluid passing through a series of sharp turns (Figure 1.5m). However, individual turns here are not formed by intermeshing teeth of plug and seat, but by a series of channels cut or etched into thin plates, which in turn are stacked to produce a great number of parallel layers of such multiple channel arrangements. A piston type valve plug passing through the center section opens up successive layers to the inner orifice and

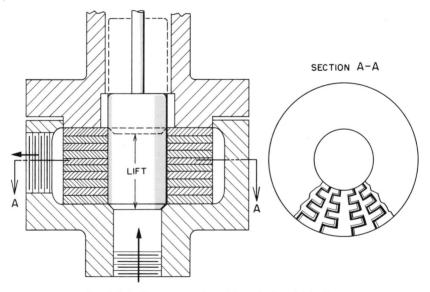

Fig. 1.5m   Piston type valve with stacked multiple discs

thereby controls the flow with an inherently linear characteristic. The total velocity head loss produced by such a trim is a function of the number of turns introduced in the fluid stream, i.e., the width of each throttling plate. It is customary to select a K factor that will bring the throttling velocity close to the pipeline velocity. The resultant valve size is of course more bulky than a conventional single-seated valve. However, cost and size of such special valves are offset by savings in trim replacement and by the omission of silencers and special pipe configurations.

REFERENCES

1. M. J. Lighthill, "On Sound Generated Aerodynamically," I General Theory, Proc. of The Royal Society of London, Vol. 211, Ser. A. (1952).
2. H. D. Baumann, "How to Reduce Control Valve Throttling Noise," Instrument Technology, September 1969.

1.6 CONTROL VALVE ACCESSORIES

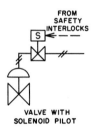

SIDE-MOUNTED
HANDWHEEL

TOP-MOUNTED
HANDWHEEL
OR LIMIT STOP

Cost: For side-mounted handwheels, $130 up to 3″ valve size, $210 up to 6″ valve size, $400 for larger sizes.

Price for top-mounted limit stop handwheels is around $40.

Explosion proof, SPDT limit switches can be obtained for $40.

Explosion proof solenoid pilots cost about $30 each.

For positioners, transducers, etc., refer to Section 1.7.

VALVE WITH
LIMIT SWITCH

FROM
SAFETY
INTERLOCKS

VALVE WITH
SOLENOID PILOT

Handwheels

One of the principal accessories on automatic control valves is an arrangement to override the pneumatic actuator manually in case of air failure or during certain maintenance operations. There are two types available:

a. The top-mounted handwheel operator.
b. The side-mounted handwheel operator.

A top mounted handwheel version is shown in Figure 1.6a. It consists primarily of a threaded yoke fastened to the top diaphragm case, a spindle with handwheel in a suitable sealing arrangement to prevent the operating air from leaking past the spindle. When not in use, it is retracted all the way so that valve operation is not hindered. When needed, the spindle is screwed down on top of the diaphragm, which in turn will push down the actuator stem and overcome the opposing force of the actuator spring. This design is relatively inexpensive, but additional headroom is required. This type of handwheel can be used as a limit stop in the upper direction.

The side-mounted handwheel, in contrast, moves the actuator stem

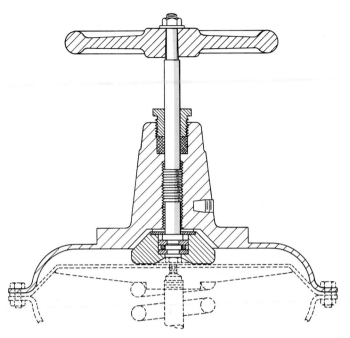

Fig. 1.6a Top-mounted handwheel for spring-opposed pneumatic actuator

directly (Figure 1.6b). Here a bell crank transmits the force from the rotating handwheel spindle to the valve stem. Lost motion, equivalent to the normal stroke of the valve, is provided in the assembly so that there is no hindrance to the automatic operation of the valve. Such side-mounted handwheels are usually found on large control valves where the required height would make it inconvenient for the operator to actuate the valve manually. This design is more expensive, but one of its advantages is the ease of maintenance of the actuator itself. It is usually possible to service and replace the diaphragm while the valve itself is held in position by the handwheel. The handwheel can also provide the function of a limit stop in either direction of travel.

Limit Switches

The limit switch is attached to the actuator yoke by a suitable bracket, which in turn senses the motion of the valve stem through a takeoff arm. Limit switches can be either single- or multi-throw, and they are used to signal when the valve stem reached a predetermined position. Such information on valve opening can be used to actuate safety and other interlocks.

Solenoid Valves

A surprising number of valves are used for safety functions in a loop, i.e., they are opened or closed in response to an electric signal from a remote

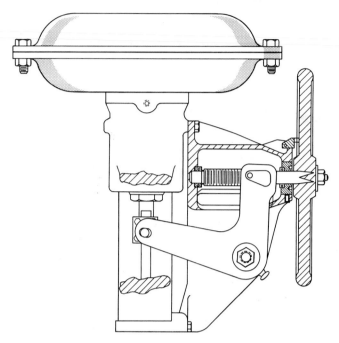

Fig. 1.6b   Side-mounted handwheel for spring-opposed pneumatic actuator

location. Solenoid valves are used to supply the signal air either directly to the actuator or indirectly by blocking the air supply to the positioner. Solenoid valves are the least expensive form of electrical actuation, although an air supply is needed. These valves come in a variety of styles and flow capacities and with explosion proof housings. It is usually specified that the valve supplier mount them directly to the diaphragm case by means of a suitable nipple.

    Certain piston actuators that do not use a spring to provide fail-safe functions utilize stored air capacity to meet the valve fail-safe requirements. A lock-up valve and check valve are placed between the air supply line, a storage bottle and the actuator or positioner. In case of failure of the air supply, the check valve will trap the air in the storage bottle, which then can be used to close or open the valve as desired by selectively loading the actuator piston. This approach to fail-safe design is *not* as positive as a spring, but is sufficient in less critical installations.

Limit Stops

    When use of the handwheel as a limit stop is not practical, other limit stops can be provided as part of the body sub-assembly. Figure 1.6c shows a typical limit stop mounted on the bottom of a globe valve which is guided

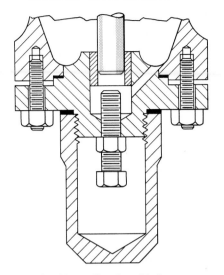

Fig. 1.6c   Externally adjustable limit stop in
valve body sub-assembly

top and bottom. This stop consists of an adjustable spindle sealed by a cap. This stop can be adjusted while the valve is in operation to meet the exact required minimum or maximum settings.

Stem-Position Indicators

Stem-position indicators show the exact valve position to operating personnel at a remote location. Such valves may be located in unmanned pumping stations or in a hazardous area closed off to operating personnel, such as near an atomic reactor. Remote position indicators can be electrical, with a linear variable resistor suitably connected to the valve stem. The electrical signal is then shown on a calibrated panel meter.

Another way to indicate stem position remotely is with a pneumatic signal. This is more desirable where fire hazard exists. Most pneumatic positioners can be inverted to work as motion transmitters. When properly modified, they will transmit a pneumatic signal as a function of valve stroke, which is indicated on a calibrated receiver gauge.

Airsets

Most common of all accessories is the *air filter regulator*. (More details on regulators are provided in Chapter III.) This is a compact self-contained regulator with a maximum flow capacity around 20 SCFM of air with an integral filter and drip valve. This air filter regulator is used to supply pressurized air to either the positioner or a yoke-mounted controller. Their main advantage is that they provide a way to set the individual pressure level of a positioner, i.e., to meet the exact air pressure requirement for a particular valve which may range from 20 to 80 PSIG. Piping the usual plant air supply of perhaps 80 PSIG directly to each valve could overstress smaller valve stems or damage receiver bellows in positioners or controllers.

1.7 POSITIONERS, TRANSDUCERS AND BOOSTERS

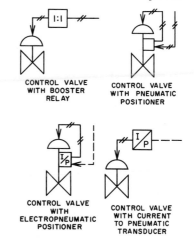

CONTROL VALVE
WITH BOOSTER
RELAY

CONTROL VALVE
WITH PNEUMATIC
POSITIONER

CONTROL VALVE
WITH
ELECTROPNEUMATIC
POSITIONER

CONTROL VALVE
WITH CURRENT
TO PNEUMATIC
TRANSDUCER

Materials of Construction:	Die-cast zinc or aluminum.
Supply Pressures:	Up to 100 PSIG; higher for boosters.
Signal Ranges:	3–9, 3–15, 6–30, 9–15 PSIG for pneumatic and 1–5, 4–20, 10–50 madc for electronic devices.
Accuracy:	Positioners are repeatable to ±0.2% and accurate to ±0.5% of span, transducers are accurate to ±1% of span and boosters are accurate to ±0.1% of span.
Cost:	Pneumatic positioner costs vary from $75 to $100, the price of explosion proof current-to-air transducers is $135 and volume boosters can be obtained for $60. Standard electropneumatic positioner can be obtained for $150.
Partial List of Suppliers:	Fairchild Hiller Corp. (boosters only); Fisher Controls Co. (all but boosters or electropneumatic positioners); Foxboro Co. (all but boosters); Honeywell, Inc. (all but boosters); Masoneilan International, Inc. (all but boosters); Moore Products (boosters and pneumatic positioners); Motorola, Inc. (transducers only); Taylor Instrument Cos. (positioners).

Pneumatic Valve Positioners

Pneumatic valve positioners are the most important accessories associated with control valves, but the reasons for using a positioner are frequently misunderstood. Positioners are sometimes over-applied, i.e., the control valve would be more stable and perform a better control function without the positioner. A positioner contributes an additional time constant to the overall control loop, and this makes it more difficult to achieve dynamic stability.

A positioner works best when its response time together with the valve is much faster than that of the process. In other words, the oscillatory frequency of the controlled loop when responding to an upset in the process should be at least 5 to 10 times lower than the bandwidth of the controller. A positioner with a single valve member (with only one additional time constant) is more stable than one with two valve members in series (a flapper nozzle—volume relay combination, for example).

Certain slow systems that fall into this category definitely call for the use of positioners. Applications include temperature control, liquid level control, gas flow control, and mixing and blending. In certain fast systems, such as for liquid pressure control or liquid flow control, a volume or ratio booster is usually more advantageous.

Assuming that the dynamics of the process allows the use of a positioner, then it should be employed when the actuator pressure (required to stroke the valve under actual service conditions) differs from the controller signal. This is usually the case in applications of high pressure drop, when the so-called "bench range" (the actual spring range of the diaphragm actuator) deviates from the standard 3–15 PSIG signal range. (Refer to Section 1.8 for a more detailed discussion.)

The second positioner requirement is the use of split range signals, i.e., when one common controller signal commands two or more control valves. A typical example is found in certain temperature control applications where the 3–9 PSIG signal portion operates a liquid coolant valve while the 9–15 PSIG signal span commands a steam valve to another heat exchanger. The liquid valve should operate in reverse, closing when signal level increases. In this case, the liquid valve will be wide open and the steam valve closed at the 3 PSIG signal level. Upon decrease in temperature, the controller signal will increase to 9 PSIG, at which point the coolant valve is completely closed and the steam valve is starting to open. (In some cases the two spans might overlap slightly for increased stability.) Such reversal between direction of valve stroke and signal can easily be obtained with most valve positioners.

Other reasons for using the positioner should be evaluated carefully. These include high stem friction due to tight stuffing boxes and solidifying liquids. It is sometimes better to allow minor dead-bands imposed by friction than to jeopardize loop stability by adding the positioner time constants.

All pneumatic positioners can be characterized as either force balance or motion balance types. *Force balance* means that the feedback derived from the valve position provides a force to balance the controller signal acting on an input diaphragm or bellows. In the *motion balance* type the stem motion is compared directly with a similar motion produced by a bellows expanded by the air signal.

Figure 1.7a shows a cross-sectional view of a force balance positioner.

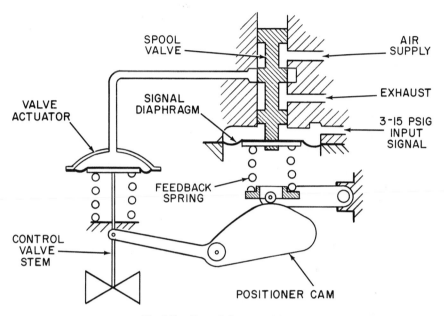

Fig. 1.7a   Force balance positioner

The controller signal acts on a diaphragm, creating a signal force which is opposed by a feedback spring. A temporary offset in the diaphragm position motivates a spool valve, which in turn allows supply air to flow to the diaphragm of the valve actuator. The resultant stem motion is sensed by a lever which rotates a cam. This cam displacement is then converted by a suitable lever arrangement into compression of the feedback spring, which in turn produces an equivalent force to match the signal level. Hence the term "force balance." The use of a cam to characterize the feedback motion has gained increasing importance. One advantage is that certain rotary valves that have a built-in characteristic (butterfly valves, ball valves, etc.) can now be modified so that their flow characteristic can match the requirements of the system. However, a cam feedback is truly effective only when the system (process loop) is *slower* than the positioner-valve combination.

A typical motion balance positioner is shown in Figure 1.7b. Here the feedback motion, sensing the movement of the actuator stem, is reduced by

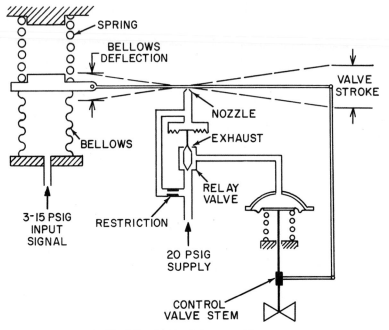

Fig. 1.7b   Motion balance positioner

an adjustable lever arrangement (which allows stroke adjustment) to produce a deflection that is equivalent to that produced by the air signal, forcing a bellows to expand against a spring.

If signal air increases, then the bellows will deflect the beam on the bellows side to move up, which uncovers a nozzle, allowing air to escape from a relay diaphragm. The motion of the relay diaphragm actuates a valve, which allows supply pressure to go to the actuator diaphragm case. The resultant actuator stem motion returns the lever to its original position, i.e., the nozzle and flapper are again brought into tangency.

A good pneumatic positioner should have a dead-band of less than 0.2 percent and an open loop gain of at least 40 to 1. It will position a valve within 0.5 percent of the lift span. The steady state air consumption is determined by the type of pilot arrangement used, and it usually ranges from 10 to 30 SCFH. However, the instrument air system *should be capable of providing* a continuous supply of at least 250 SCFH to assure good response under adverse dynamic conditions.

Most positioners have an integral by-pass valve which blocks the positioner output pressure and allows the instrument air to flow directly to the diaphragm actuator during maintenance or check-out. These by-pass valves are frequently misapplied, because a high percentage of valves require positioner output pressures in excess of 20 PSIG and therefore could not operate at all if the controller signals were directly applied to the actuator.

The dynamic response of a valve positioner with an assumed single time constant can be analyzed by considering the pneumatic actuator and positioner together in a typical block diagram,

$$I \longrightarrow \boxed{\frac{KG}{1 + KG}} \xrightarrow{O} \qquad\qquad 1.7(1)$$

where I = input (controller signal) in percent of span,
   O = stem position in percent of stroke,
   K = static gain (open loop) of positioner, and
   G = transfer function of positioner and actuator.

$$G = \frac{1}{1 + Ts} \qquad\qquad 1.7(2)$$

where T = time constant (open loop), in seconds and
   s = Laplace operator

Thus,

$$\frac{O}{I}(s) = \frac{K}{K + 1 + Ts} \qquad\qquad 1.7(3)$$

The break frequency (in radians per second) is a useful tool in evaluating the dynamic response of certain instruments. It can be defined as follows:

$$\omega = \frac{K + 1}{T} \qquad\qquad 1.7(4)$$

Open loop gains (K) range from 40 to 200, depending on the type and make of the positioner.

Table 1.7c shows stroking times (seconds) and time constants (T) for a variety of diaphragm actuators with typical side-mounted positioners in comparison with stroking times obtained when the controller output is fed directly to the actuator.

Electropneumatic Positioners

Similar in function to pneumatic positioners, these devices accept an electrical input signal, usually in analog form. Typical input signals are given in Table 1.7d.

An electromotive force is produced when an electric signal is fed through the windings of a torque motor or a voice coil. This force has to be balanced by a spring being deflected by the motion of the valve stem, as shown in Figure

Table 1.7c
STROKING TIMES IN SECONDS
FOR VARIOUS ACTUATOR-ACCESSORY COMBINATIONS

Valve Actuator Diaphragm Area (in.²)	Valve Stroke (in.)	Positioner Time Constant T (sec)			Stroking Times (sec)		
						Controller Signal to Actuator (No Positioner)	
			Valve Positioner	¼ in. Size Solenoid	With Volume Booster	With Relay in Controller	No Accessories
50	¾	4	3	2	1	4	15
75	1	8	6	2.5	2	7	32
100	1½	12	10	4	3	16	70
150	2	20	20	9	6	28	140
200	3	36	40	17	10	65	260

1.7e. An increase in signal temporarily tilts a beam covering an air nozzle, resulting in an increase in air signal, which in turn, after being amplified in a relay, drives the actuator stem to its desired position. The change in stem position increases tension in the feedback spring until the electromotive force of the voice coil is balanced.

Electropneumatic positioners as a rule are explosion proof and should be suitable to stand shock and vibration levels of up to 2 g's at frequencies below 60 Hz. If vibration at the valve location is severe, locate the electropneumatic transducer remotely from the valve.

When output signals from computers are used in place of conventional controllers, they are digital in nature and have to be transformed into an analog signal by a suitable converter. (See Sections 6.3 and 9.7.) This conversion can be avoided by the use of a *digital electropneumatic positioner* which accepts a pulsed digital input signal that drives a stepping motor. The stepping motor in turn drives a lead screw, compressing a spring whose force is balanced by the feedback spring when in equilibrium.

The resolution obtained with such a device approaches 0.1 percent as a limit, since it takes roughly 1,000 pulses to fully stroke the valve. The rate

Table 1.7d
ELECTROPNEUMATIC POSITIONER SIGNALS

Input Signals (DC)	Load Requirement of Controller	Actual Internal Resistance of Positioner
1–5 ma	0–3,000 ohms	2,300 ohms
4–20 ma	0–800 ohms	175 ohms
10–50 ma	100–600 ohms	100 ohms

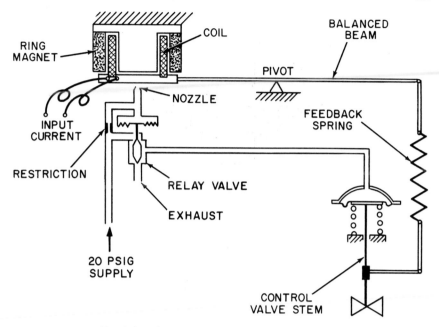

Fig. 1.7e   Electropneumatic force balance positioner

of the steps coming from the computer cannot exceed 400/sec, and power requirement is about 5.5 watts.

Electropneumatic Transducers

Electropneumatic transducers operate similarly to electropneumatic positioners, because they convert an analog electronic controller signal into a proportional air signal, usually 3–15 PSIG (Sections 6.1 and 6.3). Unlike positioners, however, they do not sense the final stem position. One of their main advantages is that they can be mounted remotely from the control valve when the vibration level at the valve is excessive.

Figure 1.7f shows a typical transducer design. The incoming electronic signal produces an electromotive force that rotates a bar magnet armature in a torque motor. The rotating armature temporarily covers a nozzle. Changes in nozzle back-pressure are then amplified through a relay whose output is piped into a feedback bellows, thus restoring the force equilibrium of the armature in the conventional manner. Span adjustment is usually obtained by varying the magnetic shunt.

The high-capacity pneumatic relay can be omitted if the transducer feeds directly into the signal bellows of a pneumatic positioner or some other low-volume system. The omission of the relay in this case *improves the stability* of the control valve. Typical accuracy obtained with an electropneumatic

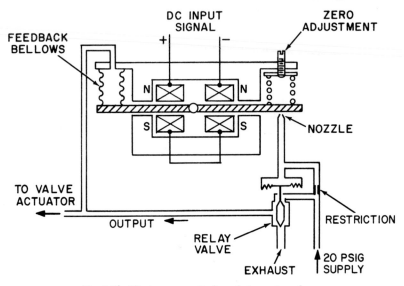

Fig. 1.7f   Electropneumatic force balance transducer

transducer is ± 1 percent of output span. This figure includes combined linearity and hysteresis. Input signals and internal resistance values are identical to the ones listed for electropneumatic positioners in Table 1.7d. The transducer case is commonly designed for Class I, Group D, Division 1 hazardous locations (explosion proof).

Boosters

Booster relays are essentially air loaded, self-contained pressure regulators (Section 4.4). They can be classified in three broad groups:

1. Volume boosters used to multiply the available volume of the air signal.
2. Ratio relays used to multiply or divide the pressure of an input signal.
3. Reversing relays which produce a decreasing output signal for an increasing input signal.

One of the main characteristics of boosters is the high exhaust capacity, i.e., the bleed rate matches closely the maximum output capacity. Figure 1.7g shows a cross section of a typical volume booster relay. Rolling diaphragms of the static balancing type are used to eliminate the effects of off-balance forces acting on the relatively large inlet and exhaust valve seats. A sensing tube connecting the outlet to the inner diaphragm chamber provides an aspirating effect under high flow conditions to provide additional valve lift. Under equilibrium conditions the signal air acting on the top diaphragm is

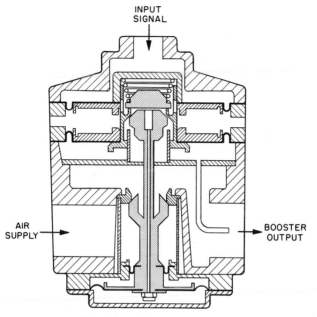

Fig. 1.7g   Volume booster relay

balanced by the outlet pressure pushing the lower diaphragm. Any imbalance of these forces causes a change in position of the valve plug.

In the case of ratio relays, there is a difference between the effective area of the signal and feedback diaphragm. Typical ratios are: 1:2, 1:3, 2:1, 3:1 and 5:1, where the first number refers to the signal and the second to the output pressure. These ratios can be maintained within approximately 1 percent. Booster relays are usually made of diecast aluminum to withstand up to 250 PSIG supply pressures. They are usually provided with $\frac{1}{4}$-in.-pipe tap connections except on high-capacity models, which may be $\frac{1}{2}$ in. or $\frac{3}{4}$ in. in size. They are highly accurate devices and respond to signal variations as low as 0.01 PSI. Typical flow capacities range from 10 to 150 SCFM. Applications include the use of volume relays to increase the frequency response of a control valve. This is sometimes preferable to the use of positioners on fast control loops.

Reversing relays are employed when two control valves, one air-to-open and the other air-to-close, are operated from the same controller. They might also be used to reverse part of the output pressure from a single-acting positioner to a double-acting piston actuator.

Ratio relays can provide for split-ranging, i.e., a 1:2 ratio relay could change a 3 to 9 controller signal to a 3 to 15 PSIG output signal.

1.8 DIAPHRAGM AND PISTON ACTUATORS

**DIAPHRAGM
ACTUATOR**

**PRESSURE
BALANCED
DIAPHRAGM
ACTUATOR**

**SINGLE-ACTING,
CUSHION LOADED
PISTON ACTUATOR**

**DOUBLE-ACTING
PISTON ACTUATOR**

**HAND
ACTUATOR**

Pneumatic valve actuators respond to an air signal by moving the valve trim into a corresponding throttling position. This section covers the two basic designs most frequently applied to globe valves: the diaphragm and the piston actuators. In connection with the performance of these actuators an analysis is presented of the various forces positioning the plug, including diaphragm, spring and dynamic forces generated by the process fluid. An understanding of the interrelationships between these forces will allow the reader to properly size these actuators and make the correct spring selection.

In addition, there are discussions on the failure safety of valve actuators and on the relative merits of diaphragm and piston actuators.

Definitions

An actuator (as defined in Section 1.1) is that portion of a valve which responds to the applied signal and causes the motion resulting in modification of fluid flow.

Thus, an actuator is any device which causes the valve stem to move. It may be a manually positioned device, such as a handwheel or lever. The manual actuator may be open–closed, or it may be manually positioned at any position between fully open and fully closed. Other actuators are operated by compressed air, hydraulics, and electricity.

The actuators to be discussed here will be those capable of moving the valve to any position from fully closed to fully open and those using compressed air for power. Of such there are two general types, the spring and diaphragm actuator and the piston actuator.

> *Spring and Diaphragm Actuator*   An actuator using a flexible diaphragm to which a variable air pressure is applied to oppose a spring. The combination of diaphragm and spring forces act to balance the fluid forces on the valve.
>
> *Piston Actuator*   An actuator using a piston within a cylinder on which a combination of fixed and variable air pressures are used to balance the fluid forces on the valve. Sometimes springs are used, usually to assist valve closure. Excluding springs, there are two variations of piston actuators: cushion loaded and double acting.
>
> *Cushion-Loaded Piston Actuator*   A piston actuator on which a fixed air pressure, known as the cushion pressure, is opposed by a variable air pressure and is used to balance the fluid forces on the valve.
>
> *Double-Acting Piston Actuator*   A piston actuator on which two opposing variable air pressures are used to balance the fluid forces on the valve.
>
> *Function of an Actuator*   1. To respond to an external signal directed to it and cause an inner valve to move accordingly. With the proper selection and assembly of components, other functions can be obtained, such as a desired fail-safe action. 2. To provide a convenient support for certain valve accessory items, including positioners, limit switches, solenoid valves and local controllers.

Actuator Performance

The actuators under discussion in this section are restricted to pneumatic types. The external signal, therefore, is an air signal of varying pressure. The air signal range from a pneumatic controller is commonly 0–18 PSIG. Signal or actuator input pressure starts at 0 PSIG, not 3 PSIG. A common mistake is to confuse the 3–15 PSIG range of transmitter output pressure with the signal to a valve. The higher value of 18 PSIG is fixed only by the air supply to the controller (or positioner), and it can easily be set to 20 PSIG or higher.

A variety of other input pressures are sometimes used, such as 0–30 or 0–60 PSIG.

Both the spring–diaphragm and the piston actuators produce linear motion to move the valve. These actuators are ideal for use on valves requiring linear travel, such as globe valves. A linkage or other form of linear-to-rotary motion conversion is required to adapt these actuators to rotary valves, such as the butterfly type.

The Steady-State Equation

In spring-and-diaphragm actuators stem positioning is achieved by a balance of forces acting on the stem. These forces are due to pressure on the diaphragm, spring travel, and fluid forces on the valve plug (Figure 1.8a).

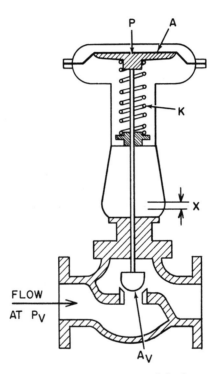

Fig. 1.8a   Forces on a spring-and-diaphragm valve

Equation 1.8(1) can be derived from a summation of forces on the valve plug adopting the positive direction downward.

$$PA - KX - P_vA_v = 0 \qquad 1.8(1)$$

Equation 1.8(1) applies to a push-down-to-close actuator and valve combination with flow under the plug. This type of actuator is commonly referred to as direct acting. Another popular actuator configuration is one causing the stem to rise on an increase of air pressure. It is commonly called a reverse actuator (Figure 1.8b).

By using the same sign convention, the equation for this valve configuration is seen to be

$$-PA + KX - P_vA_v = 0 \qquad\qquad 1.8(2)$$

If the flow direction is reversed in Figure 1.8a, the equation is

$$PA - KX + P_vA_v = 0 \qquad\qquad 1.8(3)$$

Likewise, reversing flow direction in Figure 1.8b results in equation 1.8(4):

$$-PA + KX + P_vA_v = 0 \qquad\qquad 1.8(4)$$

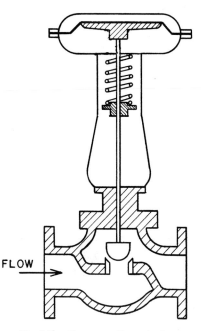

FLOW

Fig. 1.8b   Reverse acting actuator

These equations are simplified because they do not consider friction and inertia. Friction occurs in the valve stem packing, in the actuator stem guide and in the valve plug guide or guides (Figure 1.1a). Usually, for static valve actuator sizing problems, negligible error is introduced by ignoring the friction terms.

If equation 1.8(1) is plotted as signal pressure vs stem travel and if the case of no fluid forces on the plug (bench test) is assumed, then the curve shown in Figure 1.8c is obtained.

Next, consider the case of plug forces due to fluid flow, assuming that the term P_v is constant for all travel positions. This has the effect of shifting the straight line to the right to some position depending on the magnitude of P_v. Curves similar to those in Figure 1.8c can readily be drawn for the other valve configurations represented by equations 1.8(2), 1.8(3), and 1.8(4).

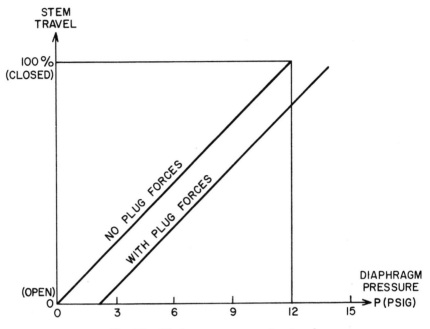

Fig. 1.8c   Diaphragm pressure vs stem travel

Actuator Sizing Example

A typical problem uses values applicable to a 1-in. single-ported valve:

$$A = 46 \text{ in.}^2,$$

$$X = \tfrac{5}{8} \text{ in. full travel, and}$$

$$K = 885 \text{ lb-force/in.}$$

If no plug forces exist, equation 1.8(1) reduces to $PA = KX$. We solve for the pressure change required for full travel as follows:

$$P = \frac{KX}{A}$$

$$= \frac{(885)(\tfrac{5}{8})}{46}$$

$$= 12.03 \text{ PSI} \tag{1.8(5)}$$

which is reasonably close to the 12 PSI desired operating span. Practical considerations of variations in spring constants and in actuator-effective areas usually prevent such a close approach to the desired span, and frequently a ±10 percent leeway is permitted.

When there are plug forces, it is seen from Equation 1.8(1) that an additional actuator force is required to maintain balance. The actuator pressure required to begin stem motion can be calculated for the case of a 1-in.-diam. plug and 100 PSID pressure drop. Equation 1.8(1) can be used to solve for P as follows (stem travel is zero, thus there are no spring forces):

$$P(46) = (885)(0) + (100)(\pi/4)$$

$$P = 1.7 \text{ PSIG}$$

This means that the diaphragm pressure must increase to 1.7 PSIG before stem travel begins. Figure 1.8c shows this with the line labeled "With plug forces."

Actuator Non-Linearities

In practice we encounter many non-linearities, and the ideal curves in Figure 1.8c are not obtained. These non-linearities occur due to several sources, such as the variable effective diaphragm areas. The effective diaphragm area varies with travel and with the pressure level on the diaphragm. Figure 1.8d illustrates this for three different sizes of diaphragms.

Another non-linearity is in valve plug forces P_vA_v. Figure 1.8e illustrates the variations in plug forces for two 4-in. valves, single ported and double ported. It also shows the effects of flow over and under the plugs of a single-ported valve.

Springs are also non-linear in that spring rates vary with travel. By judicious selection of springs, considering their spring rate and travel, the effects of non-linearity on the valve assembly can be minimized.

When all of these non-linearities are considered, a typical plot of actuator travel vs diaphragm pressure would not be a straight line as shown in Figure 1.8c but *might* be a curve such as in Figure 1.8f.

A non-linear curve, such as the one labeled "actual" in Figure 1.8f, is not necessarily objectionable. When used in an automatic control loop, the static non-linearities are compensated for by the controller. This curve is actually a part of the gain term in the valve's transfer function, and the other part is the flow characteristic. When a valve positioner is used, the positioner overcomes these non-linearities, and the result is the ideal curve.

Dynamic Performance of Actuators

Several control valve sub-systems must be analyzed in order to thoroughly evaluate their dynamic performance.

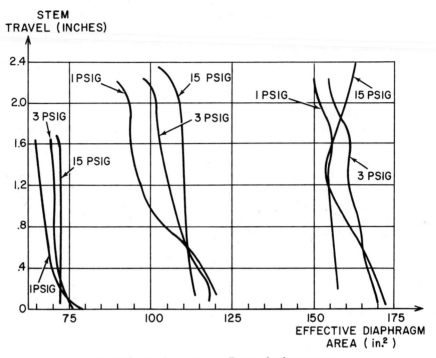

Fig. 1.8d   Non-linearities in effective diaphragm areas

The separate systems include:

1. The spring-mass system of the valve's moving parts.
2. The pneumatic system from controller output to valve diaphragm chamber.
3. If a valve positioner is used, there are two separate pneumatic systems. One from the controller output to the positioner and another from the positioner output to the diaphragm chamber.
4. The interconnecting tubing is a consideration in all of the pneumatic systems.

Spring-Mass System Dynamics

Analysis of the spring and mass system is only valid for linear systems. It is necessary either to neglect consideration of the non-linear elements or have a system wherein the non-linear effects are minor. In the case of control valves *with sufficient power* in the actuator, the latter case is approached. With such an understanding of the non-linear effects, we proceed as though valve actuators were linear devices.

The spring-mass system is represented by the following differential equation.

$$M\frac{d^2x}{dt^2} + b\frac{dx}{dt} + Kx = PA - P_vA_v \qquad 1.8(6)$$

The static, non-time-dependent terms of equation 1.8(6) are identical with equation 1.8(1). The transfer function of the valve actuator is the LaPlace transform of differential equation 1.8(6).

$$\frac{x(s)}{P(s)} = \frac{A/K}{(w/gK)s^2 + (b/k)s + 1} \qquad 1.8(7)$$

This can be written in terminology more useful to instrument engineers using the time constant τ (tau) and damping factor ζ (zeta).

$$\frac{x(s)}{P(s)} = \frac{1}{\tau^2s^2 + 2\tau\zeta s + 1} \qquad 1.8(8)$$

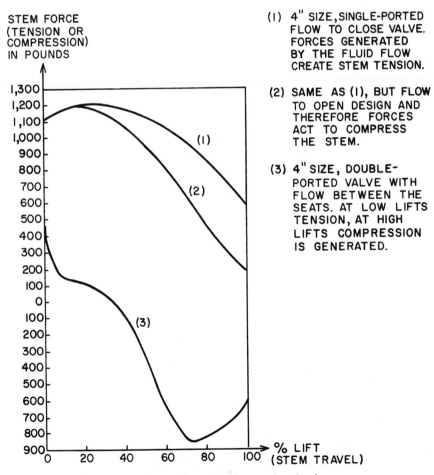

STEM FORCE (TENSION OR COMPRESSION) IN POUNDS

(1) 4" SIZE, SINGLE-PORTED FLOW TO CLOSE VALVE. FORCES GENERATED BY THE FLUID FLOW CREATE STEM TENSION.

(2) SAME AS (1), BUT FLOW TO OPEN DESIGN AND THEREFORE FORCES ACT TO COMPRESS THE STEM.

(3) 4" SIZE, DOUBLE-PORTED VALVE WITH FLOW BETWEEN THE SEATS. AT LOW LIFTS TENSION, AT HIGH LIFTS COMPRESSION IS GENERATED.

% LIFT (STEM TRAVEL)

Fig. 1.8e   Non-linearities due to valve plug forces

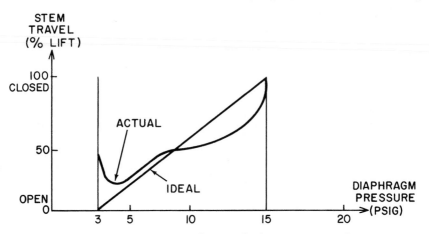

Fig. 1.8f   Ideal and actual relationship between diaphragm pressure and stem travel

The coefficient of the s^2 term in equation 1.8(7) is the square of the reciprocal of the undamped natural frequency of the spring-mass system. It is a useful number in understanding the relative importance of a control valve's dynamic components.

Table 1.8g is a list of the natural frequencies for a variety of valve sizes, considering the "average" design.

Table 1.8g
NATURAL FREQUENCY OF
CONTROL VALVES

Valve Size, (in.)	Undamped Natural Frequency (Hz)
1	32
1½	27
2	22
3	16
4	14
6	10
8	9

The real significance of Table 1.8g is in the values listed. The largest of the sizes listed has a natural frequency of nearly 10 Hz, which is ten times faster than the typical pneumatic performance of a control valve.

Pneumatic System Dynamics

The preceding discussion considered the transfer functions of the spring and diaphragm actuator from the actuator pressure to the resulting stem travel.

Next we will consider the pneumatic transfer function of the pressure from the controller to the diaphragm pressure (Figure 1.8h).

The analysis of transfer tubing dynamics involves distributed parameter systems and linear control theory analysis. Some systems, especially with short

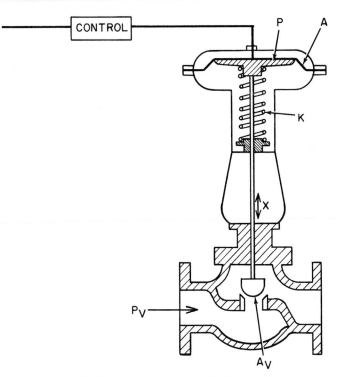

Fig. 1.8h   Forces on a spring-and-diaphragm valve

transmission lines, behave nearly like linear systems. A short tube behaves like pure resistance, and the air volume above the diaphragm is a capacitance resulting in a resistance-capacitance time constant. Some values obtained from tests with very short tubing are given in Table 1.8i. The values given are from actual tests in which the air supply was not limiting. These figures show that the valves are capable of fast response. Performance is usually limited by the controller's or positioner's ability to supply the required air fast enough. Section 4.3 contains a more detailed discussion of transmission lags and methods of boosting.

Piston Actuators

Piston actuators are either single or double acting. The single-acting actuator, Figure 1.8j, utilizes a fixed air pressure, known as the *cushion*, to oppose the controller signal. This valve does not have spring or diaphragm

Table 1.8i

TIME CONSTANTS FOR
SHORT TUBE SECTIONS

Valve Size (in.)	Time Constant (sec)
1	0.03
2	0.05
4	0.8

area non-linearities, but it is of course subject to the same plug force non-linearities (Figure 1.8e) as the spring and diaphragm actuator.

In order to use such an actuator for throttling purposes, it is *necessary* to have a positioner. The positioner senses the actuator motion and causes the valve to move accordingly. It cannot be used as a proportioning travel device without the positioner; consequently, its performance is that of the "ideal" curve in Figure 1.8f.

A *double-acting piston actuator* is one which eliminates the cushion regulator and uses a positioner with a built-in reversing relay. Thus the positioner has two air pressure outputs, one connected above the piston and the other below. The positioner receives its signal and senses travel in the same manner as a single-acting positioner. The difference is in the outputs; one pressure increases and the other decreases to cause piston travel.

Safe Failure

The valve application engineer must choose between the two readily available fail-safe schemes for control valves, either fail-open or closed. His choice will be based upon process safety considerations in the event of control

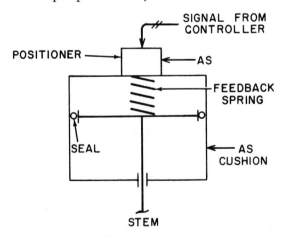

Fig. 1.8j   Single-acting piston actuator

valve air failure. He should consider complete plant air failure, controller signal failure and local air supply failure. Local failure is significant when a valve positioner is being used and when piston actuators with cushion loading are used.

The choice must be based on detailed knowledge of the valve application in the overall process or system. Two generalizations are that in a heating application, the valve should fail closed, and on a cooling application it should fail open. There are certainly applications where either failure mode is equally safe; then considerations of standardization may be used.

Fail-safe involves the selection of actions of actuator and inner valve. Both actuator and inner valve usually offer a choice of increasing air to push the stem down or up, and pushing the stem down may open or close the inner valve. The proper choice of combinations may be made by fail-safe considerations. The process application of the valve must be investigated to determine whether, on instrument air failure, it would be better to have the valve go fully open or fully closed.

There may not be much flexibility in the inner valve action. For example, a single-seated top-guided valve *must* have a push-down-to-close plug. There is freedom of choice, however, in either single- or double-seated top- and bottom-guided valves. Other valve bodies, such as the Saunders and pinch valve styles, *must* be of the push-down-to-close type. Rotary types, such as butterfly and ball valves, may be arranged either way.

The inner-valve flexibility leads to two cases: one in which either inner-valve action is permissible and one in which the inner-valve must be push-down-to-close.

When there is freedom of choice of inner-valve action, overall valve action may be obtained by selecting the suitable inner-valve action and always using increasing air to push down the actuator. This is known as a *direct actuator*. A direct actuator is preferred because of economy reasons in spring and diaphragm actuators. The savings may be in purchase cost. It is also realized in maintenance costs, because there is no stem seal to cause possible leakage and maintenance costs. The piston type actuator is equally suitable for either direct or reverse action; if it is the actuator to be used, the application engineer has complete freedom of choice of actions.

When the inner valve must be push-down-to-close, it is necessary to use both direct and reverse actuators to accomplish the desired fail-safe actions. Figure 1.8k summarizes the possibilities available.

Relative Merits of Diaphragm and Piston Actuators

When choosing between piston actuators and the spring–diaphragm type, the fail-safe consideration may be the reason for the final selection. If properly designed, the spring is the best way of achieving fail-closed action. Fail-open action is less critical.

Valve Failure (Overall)	Fail Open		Fail Closed	
Actuator	Direct	Reverse	Reverse	Direct
Inner Valve	Direct	Reverse	Direct	Reverse

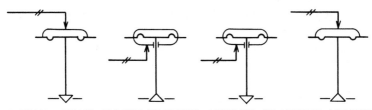

OVERALL VALVE FAILURE POSITIONS ACHIEVED BY VARIOUS ACTUATOR AND INNERVALVE COMBINATIONS

Fig. 1.8k   Relative merits of diaphragm and piston actuators

Piston actuators depend upon air lock systems to force the valve closed on air failure. Such systems may work well initially, but there are many possibilities for leaks to develop in the interconnecting tubes, fittings, and check valves, and such piston actuator systems are not considered reliable. Air lock systems also add to the actuator's cost. Piston actuators may also be specified with closure springs to provide positive failure positions.

Valve installation in the line is also a factor to consider. Flow over the plug assists in maintaining valve closure after air failure, but as explained in Section 1.3, the considerations involving dynamic stability are more important, and therefore the use of "flow-to-open" valves is recommended.

Nomenclature

    A Effective diaphragm area (in.²)

    A_v Effective inner-valve area (in.²)

    K Spring rate (lb-force/in.²)

    M Mass (lb-mass)

    P Diaphragm pressure (PSIG)

    P_v Valve pressure drop (PSID)

    b Viscous friction force (lb-force)

    g Gravitation constant $\left(\dfrac{\text{(lb-mass)in.}}{\text{(lb-force)sec}^2}\right)$

    s Laplace operator, d/dt

    w Weight of moving parts (lb-mass)

    x Stem travel (in.)

    τ Time constant (sec)

    ζ Damping factor

1.9 BUTTERFLY VALVES

Sizes:	1″ to 150″
Design Pressure:	Up to 2,000 PSIG.
Design Temperature:	From cryogenic to 2,000°F.
Rangeability:	20:1.
Capacity:	$C_v = 17d^2$ (non-critical flow in 60-degree open valve).
Materials of Construction:	Any castable or machinable material. Linings can be rubber, plastic or ceramic.
Special Features:	T-ring and piston ring seals, fishtail, U-port and other special designs.
Partial List of Suppliers:	Allis-Chalmers; Associated Control Equipment, Inc.; Conoflow Corp.; Continental Div., Fisher Controls; DeZurick Corp.; Jamesbury Corp.; Masoneilan International, Inc.; Posi-Seal International, Inc.; Rockwell Manufacturing Co.

History

Webster defines the butterfly valve as "a damper or throttle valve in a pipe consisting of a disk turning on a diametral axis" (Figure 1.9b). The simple butterfly valve is one of the oldest types of valves still in use. It was probably used for many of the earliest natural draft furnaces. The damper used in the heating and cooking stoves of yesteryear is a variety of the butterfly valve.

The butterfly valve has been widely adopted only since about 1920. This acceptance was the result of improved design and recognition by engineers of public waterworks. The butterfly valve is particularly applicable to low-pressure on-off service, such as is usually encountered in waterworks applications.

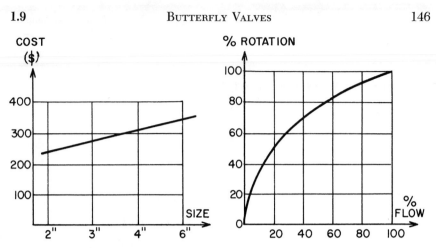

Fig. 1.9a   Cost and inherent characteristics of 60° opening standard duty, carbon steel butterfly control valves

The modern butterfly valve is suitable for high-pressure-drop and tight shutoff applications. Its self-cleaning and straight-through flow pattern holds advantages for some liquid-solid applications. This straight-through design is also advantageous due to its high capacity and also when erosion is considered.

Operation

Of major importance is the design of the disc, also called the vane, flapper, louver, and blade. One type of butterfly disc is circular and seats (closes) vertically. The other type of disc is elliptical and closes 10–15 degrees off the vertical (Figure 1.9c). This type is more expensive but gives a tighter seal.

Circular discs are usually designed to rotate 360 degrees, and a certain clearance is required between disc and body. The following discussion is in reference to this most common *swing-through* design, and exceptions are carefully noted.

As the disc moves through its 90-degree rotation the valve moves from fully closed to fully open. It is fully open when the disc is parallel to the

Fig. 1.9b   Vane positions of butterfly valve

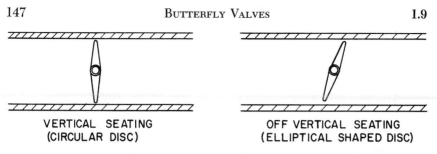

VERTICAL SEATING
(CIRCULAR DISC)

OFF VERTICAL SEATING
(ELLIPTICAL SHAPED DISC)

Fig. 1.9c   Variations in butterfly vanes

flow pattern. The free area between the disc and the wall, which is open to the process fluid is calculated based upon the supposition that the projected area of the disc is elliptical. A plot of open flow area vs percent vane rotation is very similar to the inherent characteristics curve of Figure 1.9a.

Construction

Mechanically, butterfly valves consist of the valve body, the disc and the shaft, with the necessary packing and grease fittings needed for sealing, lubrication and support (Figure 1.9d).

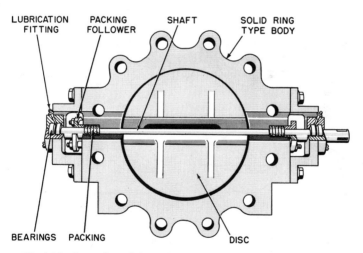

Fig. 1.9d   Swing-through butterfly valve, metal seated, with solid ring

The body is usually the solid ring type, which is mounted between pipe flanges. The disc, in the swing-through design, is cast in one piece. Disc travel distance can be limited with external stops. The thickness of the disc is a function of the maximum pressure drop required. The shaft is a one-piece rod supporting the disc. The thickness of the shaft is a function of torque required, which, in turn, is a function of pressure drop. Correct alignment eliminates binding with the swing-through disc.

Variations or special features available in butterfly valves are the design variations of the disc or the body. These adaptations include the fishtail design, the piston ring and T-ring seals. Each of these are shown and discussed in the following paragraphs.

The thickness of the disc is a function of the flowing pressure differential. Light-duty butterfly valves are made for low-pressure-drop services, and therefore the discs are thinner. The disc used in the higher-pressure-drop valve is reinforced to withstand the higher torque. When the pressure drop is appreciable, then the torque required to open the valve necessitates a heavy shaft. This also requires a heavy-duty disc much thicker at the center of the diametral axis than a light-duty one. These designs are referred to respectively as *heavy pattern* and *light pattern*.

The heavy-duty disc reduces the available flow area appreciably. Even when the valve is wide open, the flow area is reduced by the shaft and support area of the disc.

The dark areas in Figure 1.9e represent the minimum restriction, when

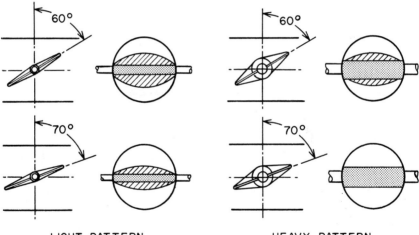

LIGHT PATTERN   HEAVY PATTERN

Fig. 1.9e   Effect of design pattern on flow area in butterfly valves

the valve is wide open. These areas are larger than the areas taken-up by the shaft alone. This difference is the area needed to attach the disc to the shaft. The shaded areas are those taken up by the wings of the disc at various opening angles.

There are only two body styles readily available. These are the solid body valve for insertion between flanges and the spool type body (Figure 1.9f). The spool valve requires more space than does the solid ring type.

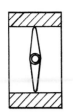 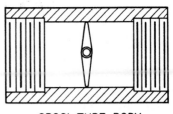

SOLID RING BODY   SPOOL TYPE BODY

Fig. 1.9f   Available butterfly valve body styles

T-Ring Seal

The purpose of the T-ring is to obtain "bubble tight" shut-off. The T-ring system operates by air pressure loading an elastic ring in the body of the valve so that the ring is expanded until it contacts the disc, thus providing bubble-tight shut-off (Figures 1.9g and h). The T-ring itself never comes into contact with the disc until the disc is in the closed position, at which point the seal is pressurized and forced into contact with the disc. The T-ring seal is available in most types of commercial elastomer selected to suit the process fluid.

Piston Ring Seal

The piston ring design is similar in operation to that of an automotive piston ring, and its purpose is to eliminate leakage around the circumference of the ring. Piston ring valves are used for high-temperature applications where other sealing methods are unsuitable—hot gas exhausts, refinery vent lines, coking ovens and testing laboratories (Figure 1.9i). Piston ring seals are suitable for temperatures from $-300°F$ to $1,500°F$.

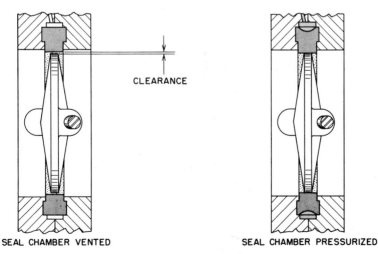

CLEARANCE

SEAL CHAMBER VENTED   SEAL CHAMBER PRESSURIZED

Fig. 1.9g   T-ring seal in butterfly valves

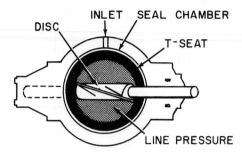

INLET   SEAL CHAMBER
DISC
T-SEAT
LINE PRESSURE

DISC OPEN, 90° POSITION. NO
PRESSURE IN CHAMBER.

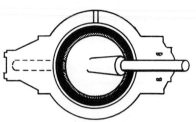

DISC ROTATES TO CLOSED, 0°
POSITION. NO PRESSURE IN
CHAMBER. NOTE CLEARANCE
BETWEEN DISC AND T-SEAT.

SEAL PRESSURE

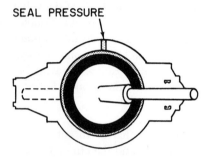

DISC CLOSED. SEAL CHAMBER
PRESSURIZED. T-SEAT IS
"PRESSURE SEALED" AGAINST DISC.
RESULT: BUBBLE TIGHT SHUT-OFF.

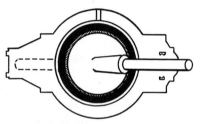

DISC CLOSED. SEAL CHAMBER
EXHAUSTED. NOTE CLEARANCE
BETWEEN DISC AND T-SEAT.

LINE PRESSURE

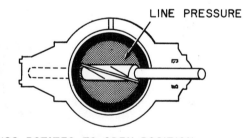

DISC ROTATES TO OPEN POSITION.
NO PRESSURE IN CHAMBER.

Fig. 1.9h   T-ring seal positions during butterfly valve operation

Linings

Both the disc and the valve are lined to suit the application. Rubber and
plastics, including Teflon, are the lining materials mostly used. Refractory type
linings are also available for the body.

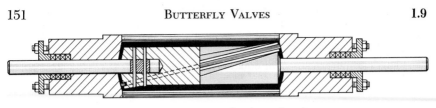

Fig. 1.9i   Piston ring seal in butterfly valve

Dual-Range Multi-Disc Designs

One method of increasing valve rangeability (turndown) is by the use of the multiple discs. At low loads with the dual-range design, a small inner disc does the throttling, and the large disc starts to open only when the capacity of the inner one is exhausted. Another design is the three-way butterfly, which consists of a single actuator and two valves mounted onto a pipe tee.

Fishtail Discs

Much research during the last few years on the design of butterfly valves has resulted in the fishtail design, which is available in both heavy and light patterns (Figure 1.9j). The torque characteristics of the fishtail are better than those of the regular butterfly valve, and it has a relatively low operating noise level.

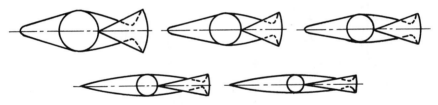

Fig. 1.9j   Fishtail butterfly discs

Torque Characteristics

Torque characteristics of butterfly valves require more careful consideration than any other valve type. The disc acts much like an air-foil or the wing of an aircraft. The aircraft wing is shaped so that a lower pressure is exerted on its upper surface than on the lower; therefore the wing is pushed up. With the standard butterfly valve the foil, or disc, is symmetrical. The pressure distribution resulting from the different flow velocities at the leading and trailing ends of the disc is unequal. This results in different pressures around the face of the disc, producing a torque which tends to close the valve.

Only at 0° and at 90° are the pressures equal on both sides of the disc. Between 0° and 90° the thrust load of the wing of the disc turned toward the upstream side is larger than that on the downstream side. This is referred to as the unbalanced or hydraulic torque. Its size is a function of the pressure drop and disc diameter.

This can be expressed in equation form as follows:

$$T_u = K(\Delta P)(\text{disc diameter})^3 \qquad 1.9(1)$$

where K is the unbalanced torque coefficient. In addition, there is a static torque, which is the torque required to overcome friction of the bearings and shaft. It resists movement of the disc in either direction. The sum of the unbalanced and the static torques is the actual torque required to change the valve's position. These torques are shown in Figure 1.9k as three lines, one for opening, one for closing and one for the unbalanced torque by itself. The torque on this 10-in.-diameter disc becomes negative on closure at approximately 35 degrees open. Thus from 35 degrees to the closed position the valve tends to close itself. The torque characteristics also indicate that at about a 75–80 degree opening the torques for both opening and closing maximize,

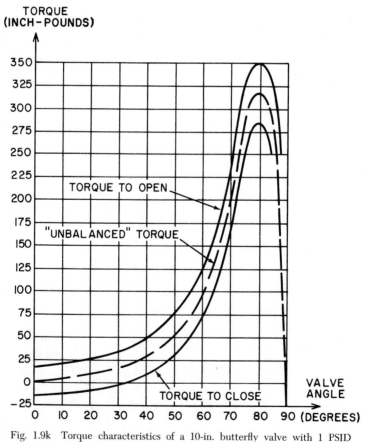

Fig. 1.9k   Torque characteristics of a 10-in. butterfly valve with 1 PSID pressure drop

and above this rotation angle the torque returns to zero. Therefore the torque requirements are non-linear and because of the difference between the opening and closing torques the butterfly valve can be said to display a torque hysteresis.

The angle at which the maximum torque requirement occurs is a function of the ratio between the pressure differential across the valve when it is closed to the pressure differential at maximum flow. As this ratio increases, the maximum torque angle decreases. Each installed butterfly valve will have a particular opening angle corresponding to its maximum torque condition, and if it is a control valve it should not be opened further than this point. It is safe practice to throttle the butterfly valves only up to their 60 degree opening.

The fishtail disc valve requires less torque than the standard disc valve (Figure 1.91).

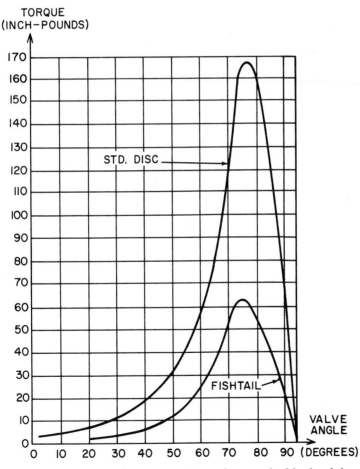

Fig. 1.91 Comparison of torque requirements between the fishtail and the standard butterfly valve

1.10 SAUNDERS (DIAPHRAGM) VALVES

Sizes:	¼″ to 20″.
Design Pressure:	Up to 200 PSIG (Figure 1.10e).
Design Temperature:	From −80 to 350°F if operating pressure is below 50 PSIG (Figure 1.10e).
Rangeability:	15:1.
Capacity:	$C_v = 22d^2$ (non-critical flow).
Materials of Construction:	Bodies available in most castable or machinable materials with a variety of linings. Diaphragm available in several rubber and plastic materials. Lining materials might include titanium; the body can be of graphite.
Special Features:	Dual-range, full-bore, and straight-through design.
Partial List of Suppliers:	Conoflow Corp.; Fisher Controls; Foxboro Co.; Grinnell Co., Inc.; Hills-McCanna Co.; Masoneilan International, Inc.; Taylor Instrument Cos.

Operation

The Saunders valve is also referred to as a diaphragm valve and occasionally as a weir valve. Conventional Saunders valves utilize both the diaphragm and the weir for control of the flow (Figure 1.10b).

The Saunders valve is opened and closed by moving a flexible or elastic diaphragm toward or away from a weir. The elastic diaphragm is moved toward the weir by the pressure of a compressor on the diaphragm. The compressor is attached to the valve stem for this purpose. The diaphragm, which is attached to the compressor at the center, is pulled away from the weir when the compressor is withdrawn.

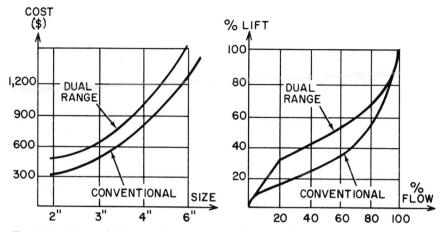

Fig. 1.10a   Cost and inherent characteristics of Saunders (diaphragm) control valves with glass-lined iron body

For high-vacuum service it is often desirable to evacuate the bonnet in order to reduce the force pulling the diaphragm away from the compressor. This is especially desirable for large valves, where otherwise the flow vacuum might be sufficient to tear the diaphragm from the compressor.

A Saunders valve can be considered as a half pinch valve. The pinch valve contains two diaphragms that move toward or away from each other, whereas the Saunders valve has only one diaphragm and a fixed weir.

Because of their design similarity, their flow characteristics are also similar, as illustrated by Figures 1.10a and 1.13a. Figure 1.10c shows the three basic positions of a Saunders valve.

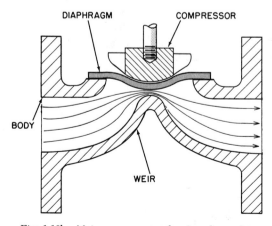

Fig. 1.10b   Main components of a Saunders valve

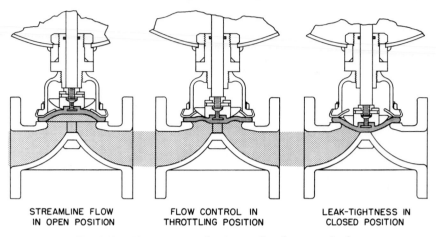

| STREAMLINE FLOW | FLOW CONTROL IN | LEAK-TIGHTNESS IN |
| IN OPEN POSITION | THROTTLING POSITION | CLOSED POSITION |

Fig. 1.10c   Main positions of weir type Saunders control valve

Construction

The body of a conventional Saunders valve (Figure 1.10d), because of its simple and smooth interior, lends itself well to lining with plastics, glass, titanium, zirconium, tantalum and other corrosion-resistant materials. Valve bodies are available in iron, stainless and cast steels, alloys and plastics. Iron bodies are lined with plastic, glass, special metals and ceramics. The flow capacity of lined Saunders valves in the smaller sizes (under 2 in.) is about 25 percent less than that of unlined ones.

The diaphragm for the conventional Saunders valve is available in a wide choice of materials. These include polyethylene, Tygon, white nail rubber, gum rubber, Hycar, natural rubber, Neoprene, Hypalon, black butyl, KEL-F, Teflon, etc., with various backings including silicone. Some contain reinforcement fibers.

Maintenance requirements of the Saunders valve depend on diaphragm life, which is determined by the diaphragm's resistance to the flow material (which may be corrosive or erosive), pressure and temperature (Figure 1.10e).

The compressor is designed to clear the finger plate, or diaphragm support plate, and to contour the diaphragm so that it matches the weir (Figure 1.10d).

The purpose of the finger plate is to support the diaphragm when the compressor has been withdrawn. The finger plate is utilized for valve sizes 1 in. and larger. For valves larger than 2 in., the finger plate is built as part of the bonnet.

Straight-Through Design

The valve seat of the straight-through diaphragm valve is not the conventional weir but is contoured into the walls of the body itself (Figure 1.10f). The longer stem stroke of the straight-through valve necessitates a very flex-

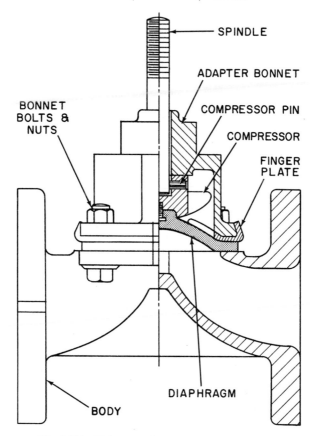

Fig. 1.10d   Main parts of a weir type Saunders valve

ible diaphragm. The increased flexure requirement tends to shorten the life of the diaphragm, but the valve's smooth, self-draining, straight-through flow pattern makes it applicable to hard-to-handle materials, such as to slurry.

The flow characteristics of the straight-through design are more nearly linear than those of the conventional Saunders valve.

Full-Bore Valve

The full-bore Saunders valve has a body design modified by special forming of the weir. As a result the internal flow path is fully rounded at all points, permitting ball brush cleaning (Figure 1.10g). This is an important feature in the food industry, which requires valve interiors with a smooth, easy-to-clean surface.

Dual-Range Design

The rangeability and flow characteristics of a conventional Saunders valve are rather poor, and as such it is not suitable for critical control applications.

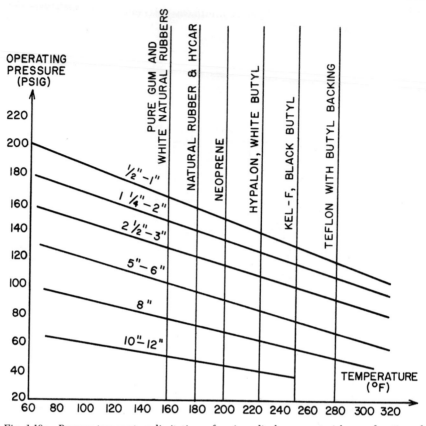

Fig. 1.10e   Pressure-temperature limitations of various diaphragm materials as a function of valve size

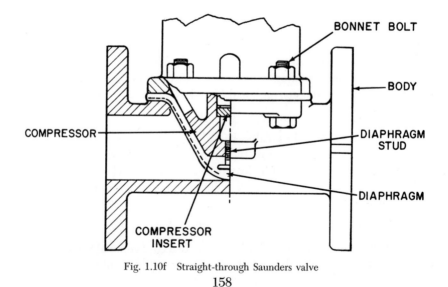

Fig. 1.10f   Straight-through Saunders valve

158

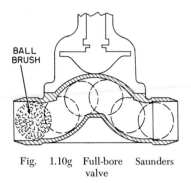

Fig. 1.10g   Full-bore   Saunders
valve

The dual-range design, with its improved flow characteristics, represents an improvement in this regard (Figure 1.10a).

The dual-range valve contains two compressors, which provide independent control over two areas of the diaphragm (Figure 1.10h).

The first increments of stem travel raise only the inner compressor from

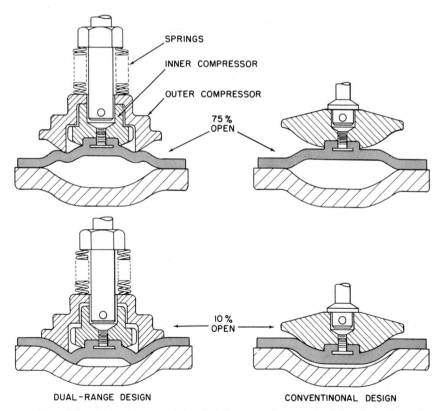

Fig. 1.10h   Stroking characteristics of dual-range and conventional Saunders valves

the weir. This allows flow through a contoured opening in the center of the valve, rather than through a slit across the entire weir.

This improvement in the shape of the valve opening helps prevent clogging and the dewatering of stock and keeps abrasion at a minimum. While springs hold the outer compressor firmly seated, the inner compressor may be positioned independently to provide accurate control over small amounts of flow.

When the inner compressor is opened to its limit, the outer compressor begins to open. From this point on, both compressors move as a unit. When wide open, this valve provides the same flow capacity as its conventional counterpart.

1.11 CONVENTIONAL BALL VALVES

THREE-WAY
BALL CONTROL
VALVE

FULL-PORTED
BALL CONTROL

Sizes:	$\frac{1}{8}''$ to 42''.
Design Pressure:	Up to 10,000 PSIG.
Design Temperature:	From $-300°$F to 1,000°F.
Rangeability:	Over 50:1.
Capacity:	$C_v = 24d^2$ (non-critical flow).
Materials of Construction:	*Body:* Cast or bar stock brass or bronze, carbon steel, stainless steel, ductile iron, aluminum, Monel, plastics, glass; also hafnium-free zirconium (for nuclear applications). *Ball:* Forged naval bronze, carbon steel (also plated), stainless steel, plastics, glass. *Seats:* Teflon, Kel-F (both tetrafluoro-ethylene), Delrin, buna-N, Neoprene, Perbunan, Hypalon, natural rubber, graphite.
Special Features:	Full-ported, three-way, split body, two-directional.
Partial List of Suppliers:	American Chain & Cable Co.; Ball Valves, Inc.; Besler Co.; Cameron Iron Works; Cooper Alloy Corp.; Crane Co.; Fisher Controls; Flodyne Controls, Inc.; General Kinetics Corp.; Hills McCanna Co.; Hoke, Inc.; Hydromatics, Inc.; Jamesbury Corp.; Lunkenheimer Co.; Clayton Mark & Co.; Pacific Valves, Inc.; Rockwell Mfg. Co.; Rockwood

Sprinkler Div., Gamewell Co.;
Tube Turn Plastics, Inc.; Walworth
Co.; Worcester Valve Co.

The ball valve contains a spherical plug that controls the flow of fluid through the valve body.

Two basic types of ball valves are manufactured at present: (1) the quarter-turn type and (2) the lift-off type. (The V-notched and other ball types are variants of the quarter-turn type.)

This section is concerned with the quarter-turn pierced ball type; the lift-off type, or ball and cage valve, is discussed in Section 1.16.

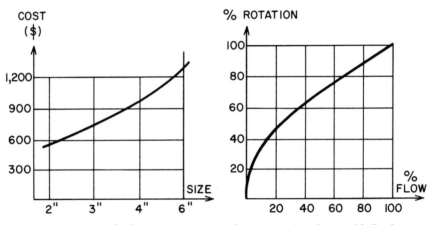

Fig. 1.11a   Cost and inherent characteristics of conventional, carbon steel ball valves

The quarter turn required (90 degrees) to fully uncover or cover an opening in the valve body can be imparted to the ball either manually, by turning a handle, or mechanically by the use of an automatic valve actuator. Actuators used for ball valves may be the same as those used to control other valve types—pneumatic, electric (or electronic), hydraulic, or combination. The latter types include electro-pneumatic, electro-hydraulic, electro-thermal, or pneumo-hydraulic actuations, as treated in Sections 1.8 and 1.20, but of special design for rotating motion. Most of the ball valves on the market are available with built-in, or integral, valve actuators. They are designed so that they can be applied with or without an actuator or so that they can be fitted with other manufacturers' actuators. This section describes ball valves as special control valves.

Features

The spherical plug lends itself not only to precise control of the flow through the valve body but also to tight shutoff. Thus the ball valve may

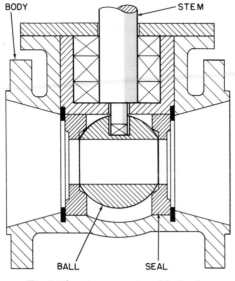

Fig. 1.11b   Top-entry pierced ball valve

assume the double role of control and block valve. (Figure 1.11b). Special materials used for valve seats help achieve these functions.

Body

    The body of a ball valve is predominantly made as a two-way globe valve type, but three-way and split-body valves are also manufactured. Figure 1.11c illustrates some porting arrangements of multi-port valves. Body materials are listed in the feature summary at the beginning of this section.

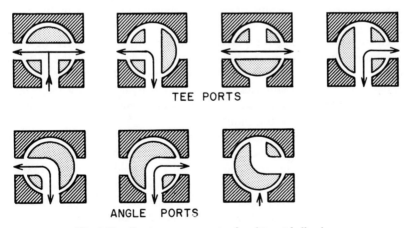

Fig. 1.11c   Porting arrangements of multi-port ball valves

Trim

The ball is cradled by seats on the inlet and the outlet side. The seats are usually made of plastic and are identical on both sides, especially in double-acting valves. Tetrafluoroethylene materials are preferred for their good resilience and low-friction properties (Figure 1.11d). Some valves have

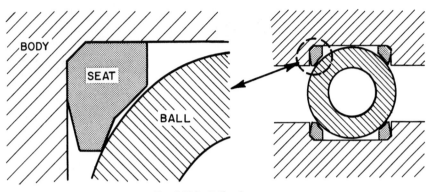

Fig. 1.11d   Ball valve seats

their plastic seats backed up by metallic seats to insure tightness in the event that the soft seat gets damaged by high temperature such as in a fire (Figure 1.11e). Such precautions are imperative on shipboard, for nuclear installations, and in cryogenic applications.

The valves are designed so that lubrication is unnecessary, and the torque required to turn the ball is negligible. Both upstream and downstream seats of the pierced ball are sometimes made freely rotatable in order to reduce wear. In some designs the seats are forcibly rotated a fraction of a turn with each quarter turn of the ball. Thus seat wear, which is concentrated at the points where the flow begins or ends on opening or closing of the valve, respectively, is distributed over the periphery of the seat.

To facilitate cleaning or replacing worn seats, in some designs, the whole seating assembly is made in the form of a tapered cartridge. If the valve has top-entry design, the cartridge can be removed without disturbing the valve arrangement. O-rings usually close off stem and seats, and thrust washers made of tetrafluoroethylene compensate for axial stem thrust due to line pressure and reduce stem friction to a minimum.

Some seats are preloaded by springs or are made tapered for wear compensation and leak-tight closure. Where fluids of high temperature are handled, seats of graphite are recommended. They hold tight up to 1,000°F.

Ball and stem are often machined from one piece. Other designs use square ends on the stem to engage in square recesses of the ball. In this case, the ball is made floating in fixed seats, while other designs provide a fixed location of ball and stem through the application of top and bottom guiding and ball bearings.

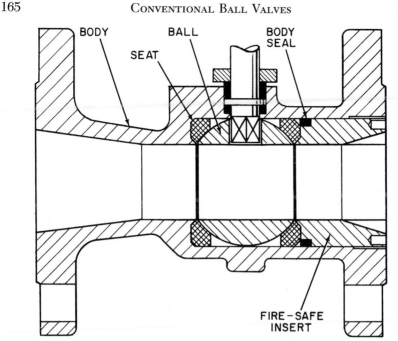

Fig. 1.11e   Fire-safe ball valve

Balls are subject to wear by friction. Where long life and dead-tight closure are of paramount importance, a design is recommended which provides for lifting the ball off its seat before it is turned. This measure also prevents freezing or galling. Lift-off is achieved by mechanical means such as an eccentric cam. This design also facilitates the handling of slurries and abrasive fluids, and it can be used for high pressures.

The proper materials for body and trim depend on the application. For handling chemicals or corrosive fluids, all wetted parts will possibly require stainless steel, plastics or glass (borosilicate glass is preferred for impact strength).

Connections

Ball valves are made with the same connections as used in all other valve types. Where screwed connections must be used, valves manufactured with ends that take the place of unions should be preferred. Threads are either NPT or AND standard.

Flow Characteristics

The flow characteristics of a ball valve approximate those of an equal-percentage plug (Figure 1.11a). Both curves compare favorably with those of other rotary-stem valves. The flow path through a ball valve includes two orifice restriction locations (Figure 1.11f). Balls characterized either by a

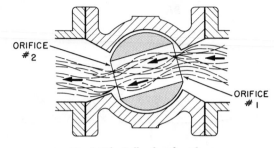

Fig. 1.11f   Ball valve throttling

notch or by a non-circular bore give somewhat better characteristics (Section 1.12). Critical flow in ball valves is encountered at $\Delta P = 0.15P_1$, far below the usual figure of 50% of absolute inlet pressure.

Sizing

Sizing of ball valves proceeds along the lines described in Section 1.2, with the possible exception that due to the essentially straight-through flow feature a ball valve can be chosen whose size is equal to the nominal pipe size, which might be an advantage on slurry service. The low-pressure loss and high recovery of a ball valve must be considered in the calculations. For discharge and recovery coefficients (C_d, C_1 and C_f, respectively) refer to Table 1.2a, and for critical pressure recovery factor C_{fr} in valve-reducer combinations, see Figure 1.2m.

Conclusions

The application of ball valves for control is a comparatively new field. Not many useful hints are published in manufacturers' literature, and it is recommended to proceed with caution, relying partly on one's experimentation. Where much noise is encountered, such as with natural gas flowing at high speed, it may be necessary to bury the valve.

The use of ball valves to control liquid oxygen in the experimental X-15 air-space vehicle as early as 1961, and later in the Atlas rocket, attests to the precise controllability of the ball valve. A special valve with dual-ball design for mixing liquid oxygen with liquid ammonia in precise proportions has also been used in the Atlas.

1.12 CHARACTERIZED BALL VALVES

Size:	2″ to 16″.
Design Pressure:	Up to 300 PSIG USAS rating.
Design Temperature:	Up to 600°F with metallic packing.
Rangeability:	Over 50:1.
Capacity:	$C_v = 22d^2$ (non-critical flow).
Materials of Construction:	Body, ball, seal ring and shaft are available in 316 stainless steel. Chrome plating available for ball and carbon steel for valve bodies.
Special Features:	Depending on "contour edge" of ball, the flow characteristics vary slightly between suppliers. Slurry design provides for continuous purging of low-activity zone of valve to prevent build-up of solids, dewatering or entrapment.
Partial List of Suppliers:	Fisher Controls; General Controls, ITT; Masoneilan International, Inc.

Operation

During the past several years there has been a considerable amount of development work on new valve types. The characterized ball valve category includes the V-notched ball valve and the parabolic ball valve. These valves were introduced in 1962, partially in an effort to solve the problem of valve clogging and dewatering in paper stock applications. Since then these valves have come into more widespread use which is the result of increased valve rangeability as well as of the shearing action at the sharp edges of the valve as it closes.

In essentially all of these characterized ball valves the "ball" has been modified so that only a portion of it is used (Figure 1.12b). The edge of the partial ball can be contoured or shaped to obtain the desired valve character-

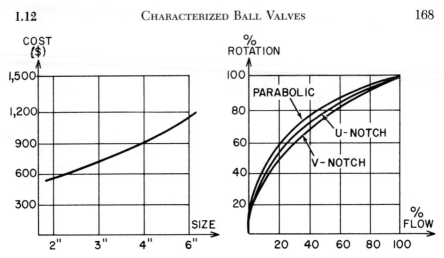

Fig. 1.12a   Cost and inherent flow characteristics of steel body, characterized ball valves

istics. The V-notching as used by one manufacturer serves this purpose as well as its shearing purpose.

This shape or contour of the valve's leading edge is the main difference between the various manufacturers' products. The ball is usually closed as it is rotated from top to bottom, although this action can be reversed.

Construction

Mechanically the characterized ball valves are very similar to their ancestor, the ball valve. However, because of the assymmetrical design, the characterized ball valve has some design problems that are not significant with the conventional ball valves. A typical characterized ball valve is shown in Figure 1.12c, in its end and side views. The main parts of a characterized ball valve are as follows:

The ball. The controlling edge of the ball can be notched or contoured to produce the desired flow characteristics. They are presently available as

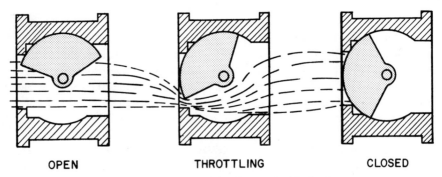

Fig. 1.12b   Positions of the characterized ball valve

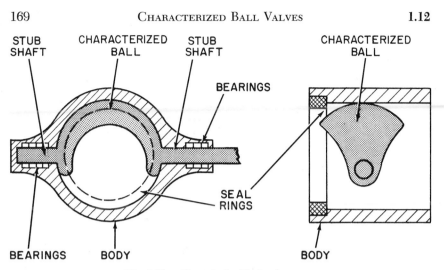

Fig. 1.12c   Characterized ball valve parts

a V-notched and as a parabolic curve. Mechanically this part can create problems by bending under pressure and thus introducing movement into the shaft seals.

The stub shafts can be distorted by the bending of the partial ball under operating loads.

Valve bodies are still not designed for high pressure or for installations other than insertion between flanges.

The seal ring and seal-retaining ring are usually held in place by companion flanges. Damage due to overtightening of flange bolts sometimes occurs. Figure 1.12d illustrates a special sealing arrangement useful in slurry applica-

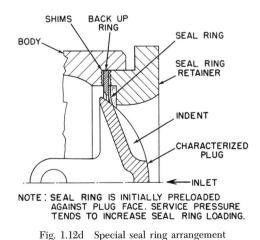

NOTE: SEAL RING IS INITIALLY PRELOADED
AGAINST PLUG FACE. SERVICE PRESSURE
TENDS TO INCREASE SEAL RING LOADING.

Fig. 1.12d   Special seal ring arrangement

tions, due to the purging effect created by the flow into the otherwise low-activity zone, through the indent in the ball plug.

Characteristics

The flow characteristics are dependent upon the shape of the edge of the partial ball and on the method of installation. The shape of the V notch at the edge of the valve varies from concave for small openings to convex for large openings. Figure 1.12e shows this together with the corresponding shapes for the parabolic ball valves.

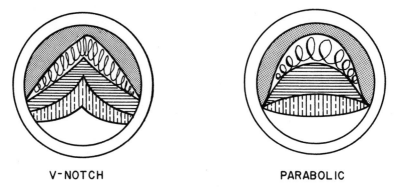

V‑NOTCH   **PARABOLIC**

Fig. 1.12e   Shapes of throttling areas of some characterized ball valves

The flow characteristics for parabolic, U- and V-notched valves are given in Figure 1.12a. These curves are based on water flow. If the characteristics were evaluated using compressible fluids at critical velocities, these curves would be flatter, closer to linear.

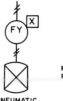

1.13 PINCH VALVES

MECHANICAL
PINCH VALVE

PNEUMATIC
PINCH VALVE

Sizes:	$\frac{1}{8}''$ to 24''.
Design Pressure:	Up to 150 PSIG.
Design Temperature:	350°F.
Rangeability:	15:1.
Capacity:	Up to $C_v = 60d^2$ (noncritical flow).
Materials of Construction:	Sleeves available in several rubber and plastic materials. Bodies are usually iron or aluminum.
Special Features:	Pneumatic or hydraulic jacketed, shutter and accurate closure designs.
Partial List of Suppliers:	Clarkson Co.; Flexible Valve Corp.; Mine & Smelter Supply Co.; RKL Controls, Inc.; Red Valve Co., Inc.

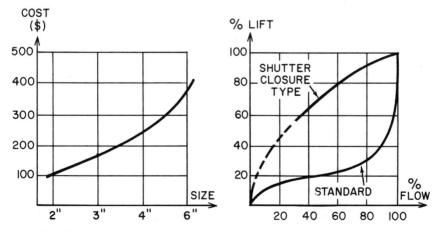

Fig. 1.13a Cost and inherent flow characteristics of pinch valve with iron body and Neoprene sleeve and without accessories

171

Pinch valves are flexible tubes that are compressed mechanically or pneumatically. They are like the rubber tube and pinchcock found in a chemical laboratory. Pinch valves characteristically have smooth walls when fully open and when partially pinched, and the elastomeric tube can close over and break up lumps of material. These qualities make this valve particularly suitable for handling slurries, minerals or gravel in suspension and *viscous* materials. Chemicals that do not attack the sleeve can be handled in a completely enclosed manner.

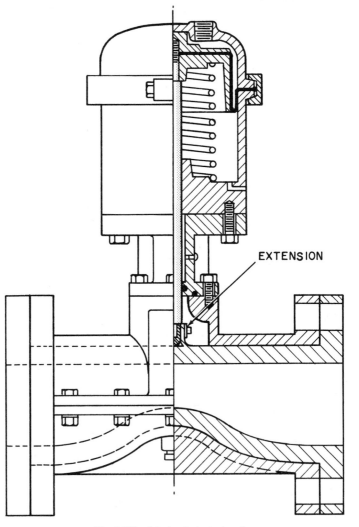

Fig. 1.13b   Mechanical pinch valve

Mechanical Pinch Design

Mechanical pinch design eliminates contact of flowing fluid with metal parts. The body may be molded, (Figure 1.13b), with the ends in the form of a complete flange which is held between the valve-body flange and a line flange. The body may also be molded with belled ends (Figure 1.13c), which

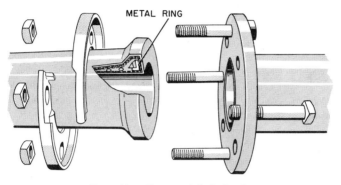

Fig. 1.13c   Sleeve with belled ends

enclose a metal ring for strength and rigidity. A split-body flange is used in this design. A wide variety of elastomeric bodies is available. Representative of these are buna-N for hydrocarbons, Hypalon for chemical and temperature resistance, PGR (conductive) for abrasion and high velocities, silicone for high and low temperatures and many others.

The sleeve may take many forms, depending upon application and means of actuation. The usual sleeve for on-off service and maximum flow capacity is a full bore that has the same size as the connected piping. Capacity in this case is the same as that of an equivalent pipe length of the same diameter using the friction factor of the material in the sleeve. The body may have a reduced port (Figure 1.13d) to increase pressure drop at the valve, and the walls of the body may be thickened for high-pressure service. The inner cir-

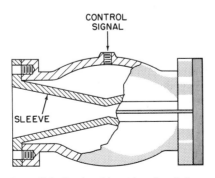

Fig. 1.13d   Pinch valve with reduced sleeve

cumference may be shaped (Figure 1.13e) to obtain tight closure without excessive stresses in the body wall or actuator force requirement.

The body may be prepinched for proportional control. The flow characteristics curve of the pinch valve (Figure 1.13a) shows practically no increase

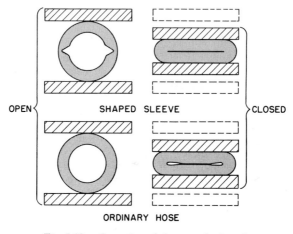

OPEN   SHAPED SLEEVE   CLOSED

ORDINARY HOSE

Fig. 1.13e   Operation of sleeve with shaped
inner circumference

in flow from 50 percent to 100 percent, thus restricting the usable opening lift for proportional control to 50 percent. The prepinched design (Figure 1.13b) starts its closure from this 50 percent point to obtain a usable flow characteristic *immediately* upon movement of the actuator closing fixture. The prepinched tube has been designed with an integrally molded, fabric-reinforced extension that connects to the upper pinch bar, relating the opening to pinch bar position without depending upon line pressure to expand the tube.

A pinch valve may consist of the sleeve alone or may be enclosed in a metal sheath. The tube may be designed for slip-on installation (Figure 1.13f) in sizes up to $2\frac{1}{2}$ in.

The actuator usually consists of a stationary bar and a movable bar positioned by a pneumatic or electric power unit. Both top and bottom bars may compress the body (Figure 1.13g) upon movement of the power unit. Units are furnished with a sanitary end-connection, which satisfies all standard specifications for use in food and drug processing.

By its nature, the pinch valve is a fail-open device. Selection of the actuator may make it either fail-open, or, by sufficient spring or auxiliary pneumatic pressure on the actuator, it may become fail-closed. Best positioning for proportional control is obtained by using a double-acting pneumatic actuator with a valve positioner, but with this design a positive (spring) failure position is not available.

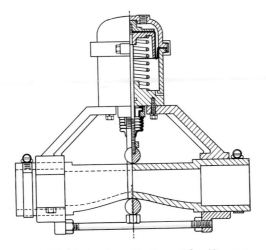

Fig. 1.13f   Pinch valve with slip-on tube. (Courtesy RKL Controls Inc.)

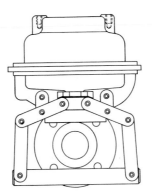

Fig. 1.13g   Pinch valve with compressor bars. (Courtesy Flexible Valve Corp.)

Pneumatic Pinch Design

By enclosing the tube in a metal spool, the unit may be operated by pneumatic or hydraulic pressure in the annular space. The form of the closure (Figure 1.13h) is similar to the mechanical closure, but is even more effective in surrounding lumps or stones and preserving leak-tight operation. Pneumatic operation has been very effective in breaking up chunks of lime or clay without detrimental abrasive or corrosive action on the gum-rubber sleeve used for this service.

Screwed bodies are available through $2\frac{1}{2}$-in. sizes and flanged bodies are being offered through 14 in. One-inch bodies can be used to 275 PSIG, and

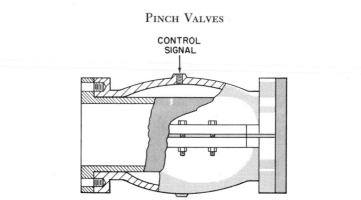

Fig. 1.13h   Pneumatic pinch valve. (Courtesy
Red Valve Co., Inc.)

the 14-in. body is rated at 100 PSIG. Metal bodies are cast iron or steel with
150 or 300 PSIG USAS flanges.

Shutter Closure

Shutter closure has been obtained by use of an elastomer sleeve in a metal
housing for pneumatic or hydraulic actuation (Figure 1.13i). The shutter is
obtained by use of a *muscle* which exerts pressure around the periphery of
the sleeve. Round-hole configuration is maintained through about 50 percent
reduction in diameter, under which condition the area has been reduced to
25 percent of the open condition. This round hole will pass larger particles
than the slotted configuration obtained with most pinch valves. The metal
housing is drilled for bolting between 150 PSIG USAS flanges. Sizes range
from 1 to 6 in. For added flexibility required for expansion, smaller sleeve
areas are also available.

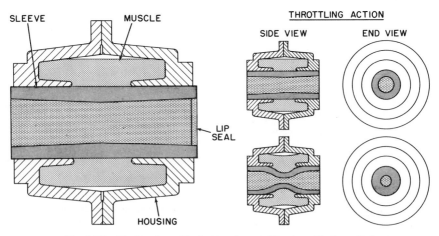

Fig. 1.13i   Pinch valve with shutter closure. (Courtesy Clarkson Co.)

A package has been developed to utilize this valve for automatic control with a number of primary elements, such as magnetic meters for flow measurement, gamma ray instruments for density measurement or diaphragm pressure transmitters for level measurement. The package includes a pneumatic-hydraulic pump unit furnishing hydraulic oil up to 270 PSIG to a non-bleed pneumatic ratio relay. The pilot accepts a 3–15 PSIG air signal and, acting as an open-loop system, supplies hydraulic pressure to position the valve sleeve. Bleed, which occurs when a reduction of pressure is required to open the valve, is returned to the pump reservoir. The valve should be sized for use between fully open and 50 percent closed, with consideration of required rangeability of the process. Its flow characteristics (Figure 1.13a) are superior to those of conventional pinch valves.

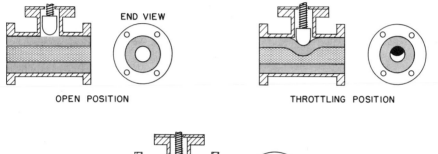

OPEN POSITION   THROTTLING POSITION

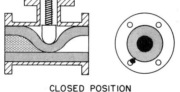

CLOSED POSITION

Fig. 1.13j   Pinch valve with "accurate closure"

Accurate Closure

A form of elastomeric sleeve valve (Figure 1.13j) employs what is technically known as "accurate closure." This may be compared to a powerful thumb that compresses the sleeve until the top wall meets the bottom wall. This closure forms a crescent. The sleeve is a thick, molded elastomer of natural rubber or plastic with an embedded metal ring in each end to maintain the circular configuration at inlet and discharge. The sleeve is inserted into a flanged metal housing. Sizes are from 1 to 10 in. with 150 PSIG USAS flanges. Shutoff pressure is limited to 50 PSIG.

The housing is designed with a mounting flange upon which a manual or a wide variety of automatic actuators can be adapted. The sleeve can be replaced quickly. Capacities are high, being dependent only on the friction factor of the sleeve and the face-to-face dimension. The C_d of the $C_v = C_d d^2$ relationship approximates 90. Required operating forces are relatively high due to the thick wall of the sleeve. About 9,000 lb is required to close a 10-in. valve against a 50-PSID differential. This force is readily available with a 12-in. cylinder or electric actuator. Smaller valves and differentials require relatively smaller actuators.

1.14 SLIDING GATE VALVES

Sizes:	½″ to 6″ with disc and plate type.
	2″ to 36″ with V-insert type.
Design Pressure:	As low as 10 PSIG with large, V-insert valves and as high as 10,000 PSIG with small, positioned-disc valves.
Design Temperature:	Up to 1,000°F with asbestos packing.
Rangeability:	50:1, lower with V-insert type.
Capacity:	$C_v = 6d^2$ to $C_v = 12d^2$ for most, but $C_v = 22d^2$ with V-insert type and non-critical flow.
Materials of Construction:	Monel, Hastelloy and alloy-20 in addition to the more standard body materials.
Partial List of Suppliers:	DeZurik Corp.; Grove Valve & Regulator Co.; Jordan Valve Div., Richard Industries Inc.; McCartney Mfg. Co.; Willis Oil Tool Co.

Knife Gate Valves

Changing the flow rate of a valve by sliding a hole or plate past a stationary hole is a very basic approach to throttling flows. The most common valve, the gate valve, is just this. Although occasionally used for automatic control, the design is not considered to be a control valve. A form of "guillotine" gate valve (Figure 1.14b) is much used in the pulp and paper industry due to its shearing ability and non-plugging body design.

Use of a "slab-type" valve with a round hole (Figure 1.14c) gives flow characteristics created by converging circles, which is similar to equal percentage for 70 percent of its flow, and then it becomes linear. Seventy percent of the flow occurs during the last 30 percent of opening.

A V-shaped insert (Figure 1.14d) in the valve opening creates a parabolic

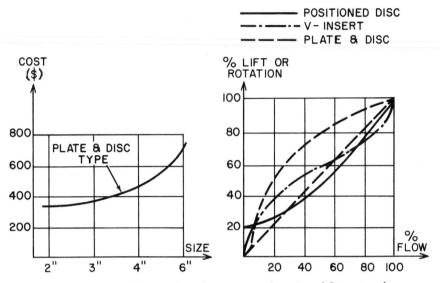

Fig. 1.14a   Cost and inherent flow characteristics of cast iron sliding gate valves

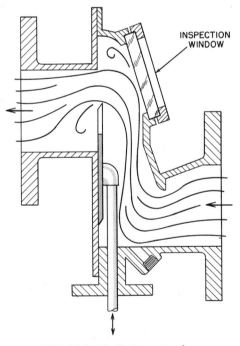

Fig. 1.14b   Guillotine gate valve

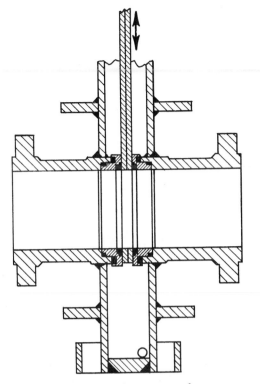

Fig. 1.14c   Slab type gate valve

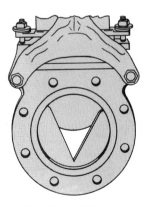

Fig. 1.14d   Sliding gate
valve with V-insert

flow characteristic somewhat similar to the V-ported globe valve. The success of these valves in proportional control is entirely dependent upon the ability of the actuator to achieve accurate positioning.

Positioned-Disc Valves

Rotation of a movable disc with two holes to progressively cover two holes in the stationary disc successfully throttles flow (Figure 1.14e). This

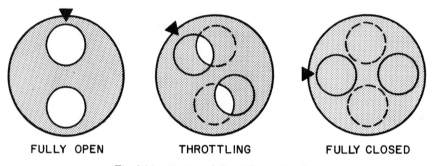

FULLY OPEN   THROTTLING   FULLY CLOSED

Fig. 1.14e   Positioned-disc sliding gate valve

variable choke was designed to control flow of high-pressure oil wells. Use of ceramic or tungsten carbide discs allows handling of pressures up to 10,000 PSIG. Such valves are presently furnished in 1- and 2-in. sizes with areas from 0.05 in.^{2} ($\frac{1}{4}$-in. hole) to 1.56 in.^{2} (two 1-in. holes).

Angle design is used for proportioning control, with an actuator capable of controlling the discharge flow at quarter-turn movement. Both linear and rotary type output actuators can be used. The relation between the discs remains in the status-quo mode without power being required from the positioner. A stepping actuator (Figure 1.14f), positions in 1-degree increments with a pneumatic input to a double-acting, spring-centered piston. Rotation occurs through a rack and pinion. Limit switches are provided, and a stepping switch can be used for position transmission and, conceivably, for feedback in control systems.

Plate-and-Disc Valves

A wide variety of flow characteristics is available using a stationary plate in the valve body and a disc made movable by the valve stem. The plate (Figure 1.14g) is readily replaceable by removing a flanged portion of the body which retains the plate with a pressure ring. Areas of the plate are undercut to reduce friction. A circumferential groove provides flexibility and allows the plate to remain flat in spite of differential pressures, expansion or contraction of the body. The stem contacts the disc by a pin through a slot in the plate.

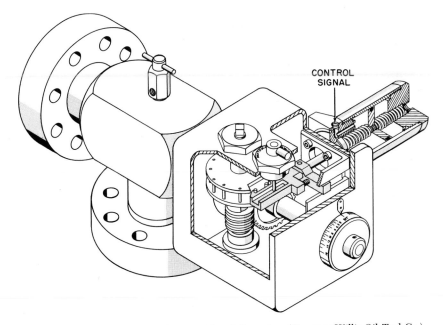

Fig. 1.14f   Stepping actuator for a positioned-disc valve. (Courtesy Willis Oil Tool Co.)

The disc is held in contact with the plate by upstream pressure and by retaining guides. The contacting surfaces of the disc and plate are lapped to light band flatness. The chrome-plated surface of the stainless steel plate has a hardness comparable to 740 Brinell to resist galling and corrosion and obtain smooth movement of the disc. The material, with the registered name of Jordanite, is reported to have an extremely low coefficient of friction, is applicable to high pressure drops and has great resistance to heat and corrosion.

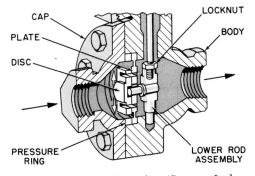

Fig. 1.14g   Plate-and-disc valve. (Courtesy Jordan-Div. Richards Industries.)

Flow occurs through mating slots in the disc and the plate. Positive shutoff occurs (Figure 1.14h), when the slots are separated. Flow increases on an approximately linear relationship until the slots are lined up for maximum flow. Capacities are about $C_v = 6.5d^2$ through the 2-in. size and about $C_v = 12d^2$ through the 6-in. size.

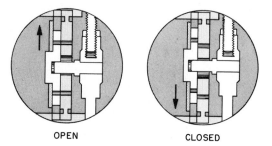

OPEN   CLOSED

Fig. 1.14h   Throttling with a plate-and-disc valve.
(Courtesy Jordan-Div. Richards Industries.)

Stem travels to obtain full flow are very short due to the slot relationship, and low-lift-diaphragm actuators can be used for positioning. Forces needed for positioning are low, requiring only sufficient power to overcome friction between the plate and disc, which is right-angle motion and not opposed to the direction of flow. Valve bodies are offered in sizes between $\frac{1}{4}$ in. and 6 in. and with ratings through 300 PSIG USAS, depending on the material, with a selection of trims and packings. Many styles of actuators are used, including one with a thermal unit and cam actuation. This body design has been adapted for extensive use as a self-contained pressure or temperature regulator.

1.15 PLUG VALVES

Sizes:	$\frac{1}{2}''$ to 60″, depending on design.
Design Pressure:	Up to 10,000 PSIG with the retractable seat type.
Design Temperature:	From −320 to 750°F with the eccentric rotating plug type.
Rangeability:	Over 50:1 with some designs.
Capacity:	Up to $C_v = 24d^2$ with some designs under non-critical flow conditions.
Materials of Construction:	Available construction materials are a function of the design involved. The V-ported plug valve, for example, can be obtained in Ni-resist, alloy-20, Monel, nickel, Hastelloy, rubber and plastic lining, in addition to the standard materials.
Partial List of Suppliers:	Allis-Chalmers; Continental Hydraulics Div., Continental Machines; Darling Valve Co.; DeZurik Corp.; Foxboro Co.; General Valve Co.; Hydril Co.; Masoneilan International, Inc.; North American Manufacturing Co.; Orbit Valve Co.

Many quarter-turn valves are not of the ball or butterfly types. The plug cock, for instance, consists of a tapered or straight cylinder containing a hole, and this is inserted into the cavity of a valve body. This design, in wooden form, was used in the water distribution system in ancient Rome and probably predates the butterfly valve. The plug cock is not commonly used for proportional control, but variations of it are, nevertheless, used both for control and for special applications.

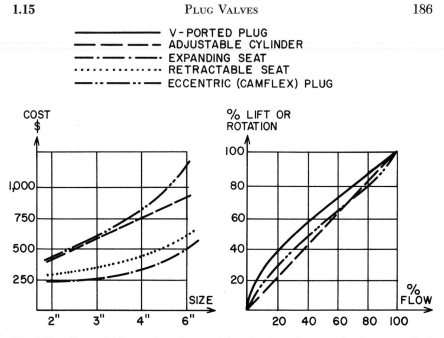

——————— V-PORTED PLUG
— — — — ADJUSTABLE CYLINDER
—·——·—· EXPANDING SEAT
·················· RETRACTABLE SEAT
—··——··— ECCENTRIC (CAMFLEX) PLUG

Fig. 1.15a   Cost and inherent flow characteristics of various plug control valves in standard materials

V-Ported Design

The V-ported plug valve (Figure 1.15b) is used for both on-off and throttling control of slurries and fluids containing solids in suspensions greater than 2 percent. These applications occur principally in the chemical or pulp and paper industries. A diamond-shaped opening is created by matching a V-shaped plug with a V-notched body. Straight-through flow occurs on 90-

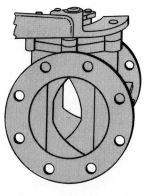

Fig.   1.15b   V-ported plug valve

degree rotation, when the plug is swung out of the flow stream. Shearing action and a pocketless body make the valve applicable to fibrous or viscous materials. The orifice configuration develops a modified linear flow characteristic with C_v capacities approximating $17d^2$. Valves are flanged from 3 to 16 in. in bronze, corrosion-resistant bronze or stainless steel. The body may be rubber-lined with a rubber-coated plug. A cylinder actuator and valve positioner are used for proportional control.

A true V-port opening is obtained (Figure 1.15c) by a rotating segment

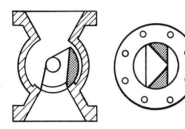
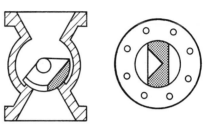

OPEN

THERE ARE NO SHARP CORNERS OR NARROW OPENINGS TO PACK WITH STOCK. LARGE PORT AREA AND CLEAN INTERIOR DESIGN ASSURE HIGH FLOW CAPACITY.

THROTTLING

CLOSE PLUG-TO-SEAT CLEARANCE REMAINS CONSTANT. THE V-ORIFICE RETAINS ITS SHAPE THROUGHOUT THE CYCLE.

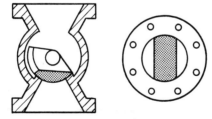

CLOSED

FLOW CONTINUES UNTIL THE V-ORIFICE IN THE LEADING EDGE OF THE PLUG ROTATES PAST THE SEAT. THROTTLING IS SMOOTH DOWN TO THE SHUTOFF POSITION.

Fig. 1.15c   Throttling with a V-ported plug valve

closing against a straight edge. The valve can be smoothly throttled on thick stock flows without the stock packing or interfering. The valve is available from 4 through 20 in. with C_v stated as more than $20d^2$. The valve is available in body materials and trims for use in the pulp and paper or in the chemical industries. A cylinder-operated rack and pinion is used for on-off service with the addition of a valve positioner for throttling services.

A variety of flow characteristics (Figures 1.15d, e, f and g) can be obtained with shaped throttling plates. Rotation of the plug is within a TFE sleeve locked into the body by metal lips. Recessed areas in the body minimize the sleeve's sealing effect. Although rangeability is stated as 20:1, this is made usable for the capacity of the valve by designing the port to handle full flow at open position. The valve is available in $\frac{1}{2}$- to 12-in. sizes and to 600 PSIG USAS rating for use to 400°F.

Adjustable Cylinder Type

In another form of quarter-turn valve flow is varied by rotating the core and by raising or lowering a *curtain* with an adjusting knob (Figure 1.15h). Proportional opening at any curtain position is made with the control handle, which may be attached to an actuator for automatic control. Various openings are obtained by these manipulations. The valve is widely used for combustion control and for mixing, in which case valves are "stacked" on a common shaft or operated by linkages from the same actuator. A port adjustment technique is based upon the use of the linear characteristic with constant pressure drop as related to the effect of downstream pressure drops through piping, orifices and burners. After installation, the curtain is closed until the pressure drop

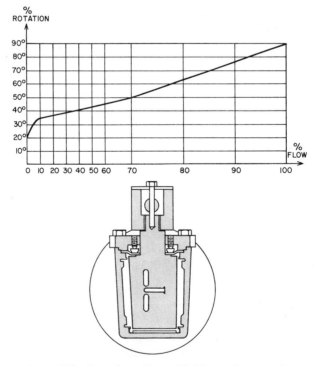

Fig. 1.15d   Plug valve with modified linear characteristics

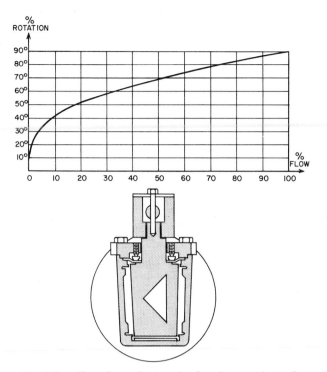

Fig. 1.15e Plug valve with triangular throttling port for modified parabolic characteristics at 25% flow

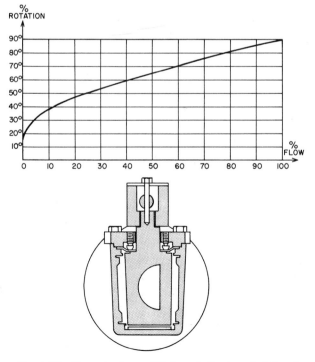

Fig. 1.15f Plug valve for modified parabolic characteristics at 50% flow

across the valve is one-sixth of the total pressure drop of the system using the control handle in wide open position. This provides a flow characteristic approximating linear without decreasing sensitivity by limiting valve stroke. Equal percentage characteristic is obtained by manipulation of the linkage to the actuator.

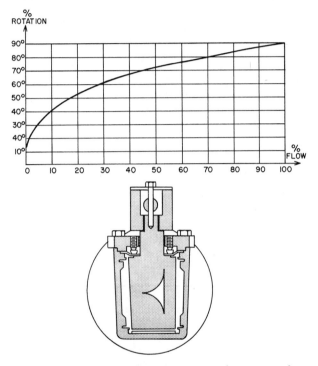

Fig. 1.15g   Plug valve with equal percentage characteristics for reduced flow

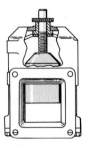

CURTAIN ½ OPEN,
CORE FULL OPEN

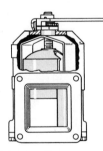

CURTAIN ¾ OPEN,
CORE ROTATED ¼

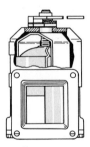

CURTAIN ¾ OPEN,
CORE ROTATED ½

Fig. 1.15h   Adjustable cylinder plug valve

Semispherical Plugs for Tight Closure

Various designs have been developed to obtain tight closure without the continuous friction of seals during rotation, as with most ball valves. The valve design illustrated in Figure 1.15i uses an eccentric ball. In the closed position,

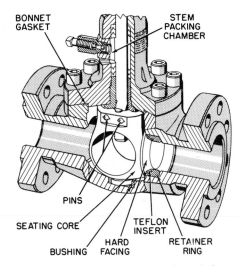

BONNET
GASKET

STEM
PACKING
CHAMBER

PINS

SEATING CORE

TEFLON
INSERT

HARD
FACING

RETAINER
RING

BUSHING

Fig. 1.15i   Plug valve with semispherical plug
for tight closure. (Courtesy Orbit Valve Co.)

the rectangular end of the stem protrudes into the ball and the closure face is wedged toward the seating surface. As the stem is rotated for opening, the closure surfaces separate and then the pin moves into a vertical slot so that rotation occurs. A non-lubricated seal is possible with a primary Teflon seal enclosed in a body seat retainer ring. The valve is adapted for automatic operation by connecting the stem to a diaphragm actuator.

Expanding Seat Plate Design

Another means of seating uses metal-to-metal or resilient seats carried in two seating segments (Figure 1.15j). These segments are carried on a rail which is tapered so that downward stem movement forces the plates against the inlet and discharge ports. To obtain opening, the first few turns of an actuator cause retraction of the seal plates and then plug rotation procedes. These plates can be removed by merely removing the bottom plate of the valve V-body. Actuators have been adapted for operation of this valve.

Eccentric Shaft Design

Double seating has been accomplished in a spherical valve using a geometric design employing an eccentric shaft (Figure 1.15k). On initial

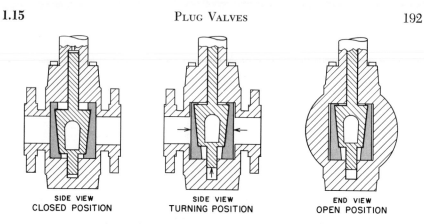

SIDE VIEW
CLOSED POSITION

SIDE VIEW
TURNING POSITION

END VIEW
OPEN POSITION

Fig. 1.15j   Plug valve with expanding seat plate

rotation of the valve, the relative offset of the seats with respect to the shaft center provides pullout action to separate the plug seat from the stationary body. This design permits building very large valves without excess weight in the spherical plug member. Standard sizes range from 4 to 48 in. for pressures up to 300 PSIG and temperatures to 450°F. The eccentric center design allows use of trunions to carry the plug.

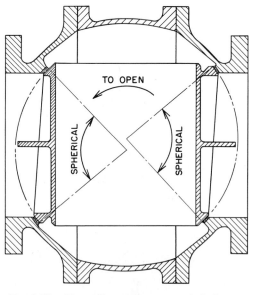

Fig. 1.15k   Plug valve with eccentric shaft design.
(Courtesy Darling Valve Co.)

Retractable Seat Type

Positioning of a movable seal after a spherical plug is in the closed position creates tight closure with sliding friction. A trunion-mounted partial sphere is operated by spur or worm gears (Figure 1.15l). The gear system

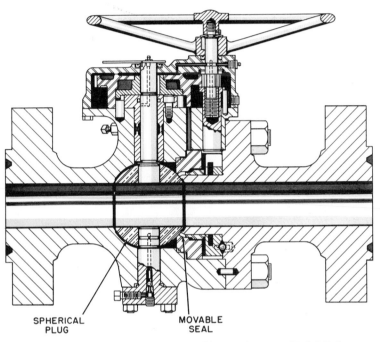

SPHERICAL   MOVABLE
PLUG   SEAL

Fig. 1.15 l   Plug valve with retractable seat. (Courtesy Hydril Co.)

rotates the plug until it is in a closed position, at which point additional rotation of the drive creates a camming operation to compress the packing ring. In the opening operation, the packing ring is released before rotation of the plug occurs. Although applicable to low pressure ranges, this valve is particularly useful up to API 5,000 PSIG rating for 10,000 PSIG service. It is used mostly in the oil fields.

Overtravel Seating Design

Use of a cylinder for a flow passage fabricated within a tapered cylindrical plug allows building a plug valve at least through 60-in. without undue weight and attendant inertia to rotation. Rotation is caused by pushing down on a rotator with the stem operated by a piston (Figure 1.15m). Continuous movement of the stem completely closes the plug, leaving a small crescent (Figure 1.15n), due to the plug being slightly raised from seating. In Figure 1.15o,

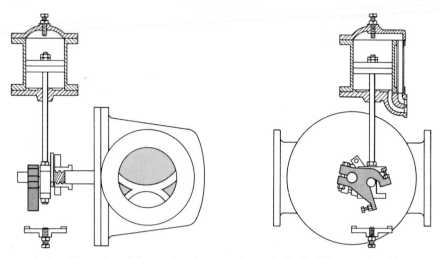

Fig. 1.15m   Plug valve of the overtravel seating design in its throttling position. (Courtesy Allis-Chalmers.)

the stem has contacted a seating adjustment to force the plug into the tapered seating surface for tight closure. The design makes possible a rapid restriction of flow area, causing about 65 percent closure with 30 percent stem travel. Complete rotation of the plug occurs at about 65 percent stroke, with the additional stroke utilized for seating. Inasmuch as rapid closure to about 20 percent area does not create serious surge pressures, this valve can be used for emergency closure without undue consideration of the piston speed. Rangeability well exceeding 50 to 1 is claimed for proportional control serv-

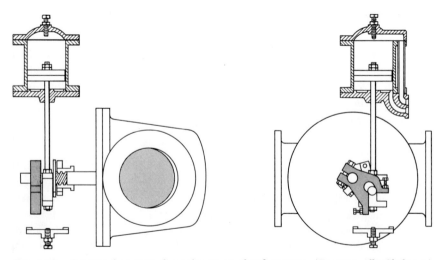

Fig. 1.15n   Overtravel seating plug valve in its closed position. (Courtesy Allis-Chalmers.)

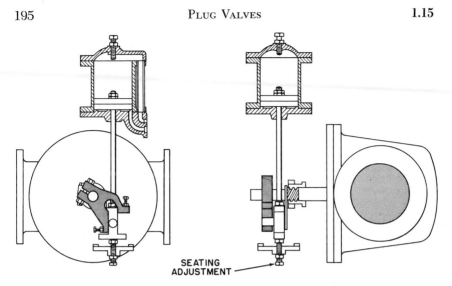

Fig. 1.15o   Overtravel seating plug tightly closed

ice. Free rotation and relatively low weight of plug per valve size contribute to low power requirements. Materials are selected by the manufacturer to suit many services.

Eccentric Rotating Spherical Segment Type

Although designed for uses comparable to those of globe or butterfly valves, the eccentric rotating spherical segment valve can be considered as semibalanced due to its low torque requirements (Figure 1.15p).

This control valve makes exaggerated use of the offset center, as used

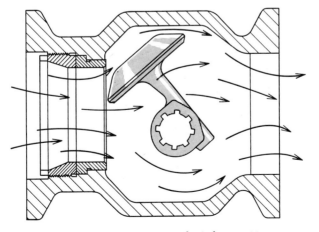

Fig. 1.15p   Eccentric rotating spherical segment type plug valve. (Courtesy Masoneilan International Inc.)

in butterfly valve designs, to obtain contact at closure without rubbing. The seat portion of the plug has the form of a spherical segment which is rotated 50 degrees for maximum opening. The support arms for the plug flex upon closure to cause tighter contact with the seat upon increase in actuator force.

Flow characteristic approaches linear (Figure 1.15a). Change in characteristic is accomplished with a cam on the positioner. Capacity is between a double-seated valve and a butterfly valve. Reduced trim is obtained with the same plug, using a reduced seat ring. High flow capacity is achieved with only moderate pressure recovery in the body, so the critical flow factor is much higher than that of a butterfly valve throughout its throttling range.

Torque requirements follow a uniform decrease from closed to open without an objectional peak. The valve is furnished with a low-convolution cylinder actuator, with or without positioner, designed to fit the valve. Since the drive shaft is normally horizontal, an extended bonnet may be furnished for use at elevated temperatures.

The valve is furnished in sizes from 1 through 12 in. with a 600 PSIG body for clamping between flanges rating from 150 PSIG to 600 PSIG. Use of a positioner allows maximum pressure drops of 600 PSID in up to 3-in. sizes or 400 PSID in up to 6-in. sizes with a high-pressure positioner. A rangeability of over 50 to 1 is claimed. The entire valve positioner unit is much lighter than the normal globe valve actuator and positioner.

1.16 BALL AND CAGE VALVES

Sizes:	$\frac{1}{4}''$ to $14''$.
Design Pressure:	Up to 2,500 PSIG.
Design Temperature:	From -425 to $1,800°F$.
Rangeability:	Over $50:1$.
Capacity:	$C_v = 20d^2$ (non-critical flow).
Materials of Construction:	Stainless steel is standard.
Special Features:	Good resistance to cavitation and vibration.
Partial List of Suppliers:	DeVar-Kinetics Div., Consolidated Electrodynamics; Otis Engineering Corp.; Powers Regulator Co.

Cage Positioned Ball Design

Positioning of a ball by a cage, in relation to a seat ring and discharge port, is used for control (Figure 1.16b). The valve consists of a venturi-ported body, two seat rings, a ball that causes closure, a cage that positions the ball, and a stem that positions the cage. Seat rings are installed in both inlet and discharge but only the discharge ring is active. The body can be reversed for utilization of the spare ring.

The cage rolls the ball out of the seat as it is lifted by the stem, positions it firmly during throttling and lifts it out of the flow stream for full opening (Figure 1.16c). The cage is contoured for unobstructed flow in the open position. Cage design includes four inclined control surfaces. The two surfaces next to the downstream seat lift the ball out of the seat and roll it over the top edge of the seat ring as the valve is opened. As the valve opens further, the ball rolls down the first two inclined surfaces to the center of the cage to rest on all four inclined surfaces. The Bernoulli effect of the flowing stream holds it cradled in this position throughout the rest of the stroke. A non-rotating slip stem is guided by a bushing at the bottom and a gland at the

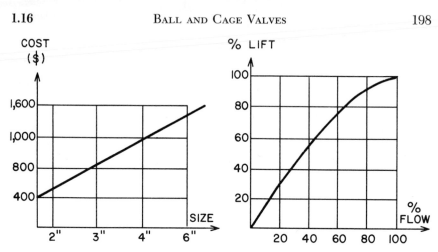

Fig. 1.16a   Cost and inherent characteristics of ball and cage valves with 1,500 PSIG rating and in stainless steel materials

top of the bonnet. A machined bevel near the base of the stem acts as a travel limit and allows for back-seating.

Ball and cage valves are furnished in sizes from $\frac{1}{2}$ in. to 14-in., with USAS ratings from 150 PSIG to 2,500 PSIG. Reported flow coefficients (C_v) are consistently high. The flow characteristic reflects the increasing enlargement of the crescent between the surface of the ball and the discharge port (Figure 1.16a). With a flow characteristic starting at zero flow, the rangeability is very high, over 50 to 1, depending only upon the ability of the actuator to position the cage.

Tight shutoff occurs over a long operating life due to the continual rotation of the ball at each operation, which offers a new seating surface each time it is closed. Closure is positive due to the wedging of the cage in addition to line pressure. Although tightly closed, the stem force for opening is approxi-

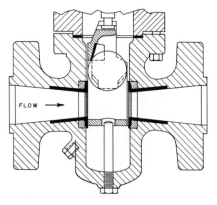

Fig. 1.16b   Cage positioned ball valve

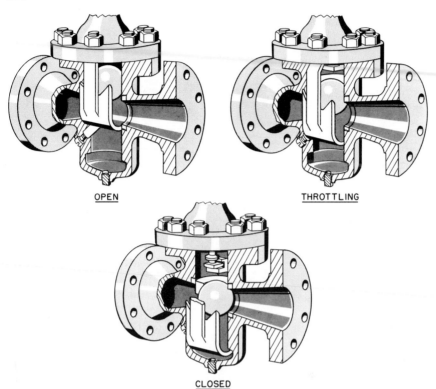

Fig. 1.16c   Throttling with cage positioned ball valve. (Courtesy DeVar-Kinetics.)

mately 25 percent of a single-seated valve due to the manner in which the inclined surfaces of the cage roll the ball away from the seat. Opening and closing force factors have been determined for all sizes of the valve. The low opening force requirement ($4.76 \times 2,000$, or 9,520 lb for a 4-in. valve at 2,000 PSIG) is beneficial in selecting the actuator. The design is conducive to minimizing cavitation effects because flow tends to follow the curve of the ball, thus reducing turbulence. Cavitation tends to occur in the venturi passage, not at the seat, allowing use of hardened or replaceable throats. The expanding venturi discharge assists in handling flashing liquids.

The bonnet design readily lends itself to the adaptation of a variety of linear or rotary actuators. Because the ball must be moved completely out of the flow stream, stem travels are at least as much as the diameter of the valve throat.

Ball Unseated by Stem

The ball cage has also been used for regulators. The ball is cradled in the cage with the valve installed in the vertical position (Figure 1.16d). The

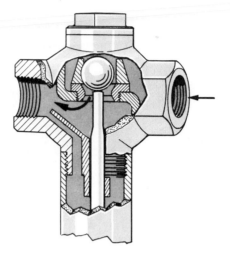

Fig. 1.16d   Ball unseated by stem. (Courtesy
Powers Regulator Co.)

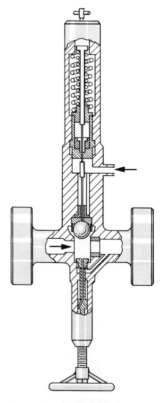

Fig. 1.16e   Ball and cage valve
for emergency closure

stem of the regulator, coming from below the ball, forces the ball away from the seat. Flow is around the ball through the annular space, similar to the flow in a single-seated valve.

Ball Gripped by Cage

A variation of the ball and cage design is used for emergency closure (Figure 1.16e). Separate springs and ejection pistons allow high and low limit settings. Pressure above the high setting pushes the piston down to eject the ball from the holder into the seat. Low pressure allows the low-pressure spring to push the piston down. The ball is held firmly on the seat by the differential pressure. An internal bypass is opened to equalize the system pressures. Rotation of the bypass handwheel moves the ball back into the holder, the reset rod is retracted, and the by-pass valve closed for normal operation.

1.17 EXPANSIBLE TUBE OR DIAPHRAGM VALVES

Sizes:	1″ to 12″.
Design Pressure:	Up to 1,500 PSIG.
Design Temperature:	150°F.
Rangeability:	20:1.
Capacity:	$C_v = 12d^2$ (non-critical flow).
Materials of Construction:	Standard metals as body materials. Rubber sleeve and plastic body coatings are available.
Special Features:	Large variety of pilots available, making it suitable to a broad range of special applications.
Partial List of Suppliers:	Fisher Controls; Grove Regulator Co.

Expansible Tube Type

Control of flow is obtained by use of an expansible tube which is slipped over a cylindrical metal core containing a series of longitudinal slots at each end and a separating barrier in between. A cylindrical, in-line jacket surrounds the tube so that the fluid line pressure can be introduced between the jacket and the sleeve to cause the sleeve to envelop the slots. With pressure shutoff to the annular space and bled to the downstream line, the line pressure in the valve body will cause the valve to open fully (Figure 1.17b). Control of the pressure on the sleeve creates a throttled flow condition by first uncovering the inlet slots and then progressively opening the outlet slots. A continuous dynamic balance between fluid pressure on each side of the sleeve makes it possible to obtain wide rangeability from a no-flow to a fully open condition.

The basic operation of the valve can be accomplished with a manually positioned, three-way valve, or by a three-way pilot valve positioned from a remote location. A variety of automatic pilots give versatility to the basic valve. For reduced pressure control, a pilot is used to modulate the jacket

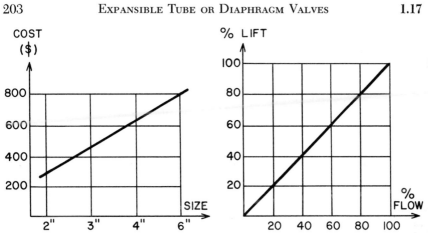

Fig. 1.17a   Cost and inherent characteristics of expansible tube valves

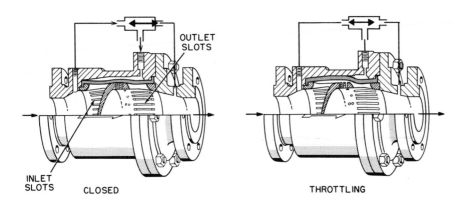

CLOSED   THROTTLING

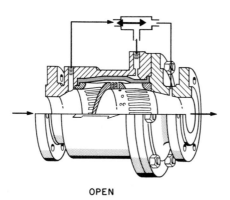

OPEN

Fig. 1.17b   Expansible tube valve

pressure in response to the sensed pressure in the downstream pipeline. As downstream pressure tends to fall below the set point, the double-acting pilot positions itself to reduce the jacket pressure. This allows the valve to open to a throttling position. Therefore downstream pressure increases to the set pressure, with attendant change in flow rate which holds the set pressure. The static sensing line is separate from the pilot discharge line, precluding a pressure drop effect in the sensing line.

Another form simulates a conventional regulator, in that system gas is bled into the jacket annular space through a fixed orifice and bled off through the pilot regulator. In this form, the static sensing line and pilot output are common. Double-acting pilot systems use seven control ranges from 2 to 1,200 PSIG with corresponding inlet pressures up to 1,500 PSIG. The fixed orifice design is available for control from 2 oz to 600 PSIG.

Back-pressure control and pressure relief are obtained in the same manner, with pilot regulators that sense upstream pressure. By using a separate sensing and bleed port, a buildup from cracking to fully open can be varied from 3 percent to 14 percent of the set pressure. Return to normal operation causes the valve to create absolute shutoff. Emergency shutoff service may use an external pressure source piloted to obtain immediate shutoff upon abnormal conditions.

A diaphragm-operated, three-way slide valve may also control jacket pressure by proportioning the inlet and outlet pressures. Application of the controller output pressure to the three-way slide valve diaphragm actuator causes the sleeve valve to open proportionally, in a manner somewhat similar to that of a conventional diaphragm control valve. In this manner, the valve becomes a control valve (Figure 1.17c).

Differential, or flow control, is accomplished by using a pilot valve in which the diaphragm is acted upon by both upstream and downstream pressures. The differential static lines may be taken at the inlet and outlet of the line valve, or, for better precision, across an orifice plate in the line.

An unlimited variety of control systems is possible with the pilots available. One of its important applications in gas distribution is pressure boosting. The line pressure losses due to increased consumption can be counteracted by automatically increasing the set pressure of the distribution control valve. Normal pressure is controlled by pilot #1 in Figure 1.17d. An increased flow is sensed at a differential pressure-producing device, orifice, or flow tube, which opens a high-differential pilot to cut in a boost-pressure-control regulator (pilot #2), which is set at a higher control pressure. Return to normal flow cuts out the boost pressure control and reinstates the normal control pilot.

The expansible tube valve is made in sizes from 1 through 12 in. with pressure ratings from 200 PSIG in iron to 1,500 PSIG in steel construction. It is made flanged or flangeless for insertion between line flanges. The flangeless body is cradled in the studs between the line flanges. Removal of the body

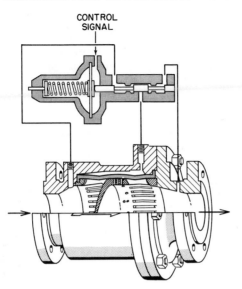

Fig. 1.17c   Expansible tube valve with throttling control pilot

is made easier by expanding the flanges about $\frac{1}{8}$ in. using nuts on the studs
inside the flanges. Tight shutoff or throttling requires a differential between
the line pressure or external source used for closure and the downstream
pressure sensed on the inside surface of the sleeve that is exposed to this
pressure. This differential requirement for a special low-pressure 2-in. valve
is 3.6 PSID, and 1.6 PSID for a 4-in. valve. The low-pressure series requires
from 21 PSID for the 1-in. size to 4.6 PSID for the 10- and 12-in. sizes.

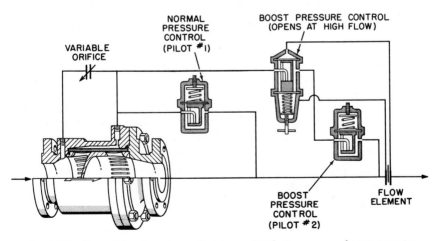

Fig. 1.17d   Expansible tube valve applied to gas distribution pressure boosting service

High-pressure models require from 58 PSID for the 1-in. size to 11 PSID for the 10- and 12-in. sizes.

The body design allows tight shutoff, even with comparatively large particles in the flow stream. Freezing of the pilot by hydrates is not common because the intermittent and small bleed occurs only to open the valve and as such is not conducive to freezing. The pilot may be heated or housed, or even located in a protected area close to the warm line. With its only moving part a flexible sleeve, this type of valve has no vibration to contribute to noise. The flow pattern also helps make this valve from 5 to 30 db more silent than most regulators. Flow capacities are comparable to those of single-seated as well as many double-seated regulators.

Expansible Diaphragm Design

An expansible element (Figure 1.17e) is stretched down over a dome-shaped grid causing shutoff of the valve when pressure above this resilient member overcomes the line pressure under the element. Line pressure is evenly directed over the expansible area by a series of pressure channels. Selection of the correct action on a pilot that supplies line or external pressure

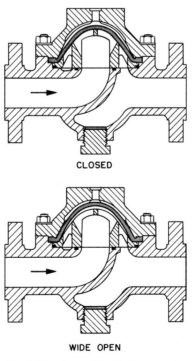

CLOSED

WIDE OPEN

Fig. 1.17e   Expansible diaphragm valve

to the exterior of the expansible element causes the valve to control as a back-pressure or reducing control valve. For back-pressure control, or pressure relief, the static line is taken upstream of the valve. Increase in line pressure will increase the bleed from the annular space between the expansible element and the metal housing. Reducing regulation is accomplished by restricting the bleed upon increase in downstream pressure and increasing it upon decrease in downstream pressure.

The valve is available in iron or steel with ratings to 600 PSIG. Models are available from $-10°F$ to $150°F$, using a molded, buna-N diaphragm. Relief valve pressures are from 30 to 300 PSIG, while reducing service varies from 5 to 150 PSIG. Capacity factors vary from $C_v = 11d^2$ in smaller sizes to $C_v = 14d^2$ in larger valves.

CONTROL
PILOT LINE

1.18 FLUID INTERACTION VALVES

Sizes:	½″ to 4″.
Design Pressure:	100 PSIG or greater. No theoretical limit.
Design Temperature:	Can handle molten metals.
Rangeability:	10:1.
Capacity:	$C_v = 14d^2$ (non-critical flow).
Materials of Construction:	Theoretically unlimited.
Special Features:	No moving parts; memory capability.
Partial List of Suppliers:	Moore Products Co.

The Coanda effect, the basis of fluidics, is used in diverting valves from ½-in. through 4-in. sizes. The Coanda effect means the attachment of a fluid stream to a nearby sidewall of a flow passage. This effect can be used in a so-called flip-flop valve for diverting a stream from one discharge port to

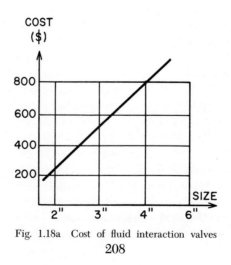

Fig. 1.18a   Cost of fluid interaction valves

208

another. Figure 1.18b shows the flow through the right-hand port due to both control ports being closed. Opening the right-hand port, to allow air or liquid to enter, will shift the flow. The industrial valve has rectangular diverting tubes, but the end connections may be circular. Control is maintained by opening or closing the control port or by injecting low-pressure air or liquid through a solenoid or other pilot valve.

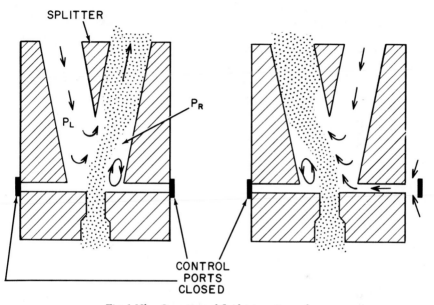

Fig. 1.18b Operation of fluid interaction valves

In this valve, with the stream flowing in one diversion tube, the flow at the inlet contains some potential (pressure) energy and some kinetic (flowing) energy. Much of the potential energy is converted to kinetic energy at the nozzle. Up to 70 percent recovery of potential energy occurs in a diffuser section. *Fifty percent recovery* is guaranteed for commercial valves, and somewhat less for gases above critical flow and for viscous fluids ($\Delta P \approx P_1/2$).

Installation must allow the recovered pressure to create the desired flow against friction effects of piping or fittings. An uninhibited flow will create some aspirating effects in the open outlet; restriction causes blocking, while *excessive restriction will cause a leak* or diversion to the open port.

Industrial valves, with their ability to divert in less than 100 msec, fill a wide variety of uses. The primary one is for level control in which the effluent not required for filling may be returned to storage. It is necessary only to use a dip tube set at the control point, as shown in Figure 1.18c. Lack of moving parts or of detrimental effects due to fast diversion action allows the system to provide very close control.

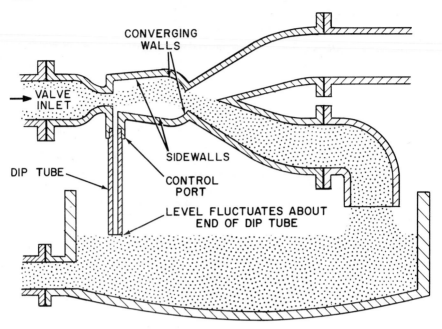

Fig. 1.18c   Level control with fluid interaction diverting valve

Numerous uses of diversion valves exist, such as tank filling, which is accomplished by using an external signal. Diversion of a process stream upon contamination, sensed by a pH or other analyzer, is important in paper mills and chemical plants. The ability to divert rapidly makes this valve applicable to oscillating flows. The valve may be used if total by-passing of a heating or cooling medium is adequate for temperature control.

This design is proposed as the basis for a 100-in.^{2} valve to control the exhaust of a high-performance turbojet engine. The fluidic valves can be used for diversion of engine exhaust gases from tailpipe propulsion nozzles to wing-mounted lift fans with ambient control flow. A four-ported valve has been used for direction control of a missile. By manipulation, a flow stream emits from each port. Upon proper injection of control streams the flow is directed to the ports required to create change in direction. A sophisticated digital control system maintains flow volume control in a given direction by controlling the accumulated time that the flow is directed to the desired port. It is evident that the ultimate use of this valve design has not been reached.

1.19 MISCELLANEOUS VALVES AND ACTUATORS

The previous sections cover a broad range of control valve designs. The purpose of this section is to complete this coverage with the discussion of some more special designs. The following types of control valves are discussed in this section: The expandable element in-line valve, which is used in the gas pipeline field, the positioned plug in-line valve, which is most adaptable to toxic fluid services and is used in the aerospace industry, and the diaphragm operated cylinder in-line valve, which is most suitable to high-pressure gas service, where a reduced noise level is desired.

Also discussed are digital control valves that can directly receive the digital signals of a process control computer and special valve accessories that are compatible with such a computer—positioners, transducers and actuators.

<table>
<tr><td>Partial List of Suppliers:</td><td>VALVES: American Meter Controls, Inc.; Belfab Co.; G. W. Dahl Co.; Eisenwerk Heinrich Schilling; Fisher Controls; Process Systems Instrumentation Inc. ACCESSORIES: Conoflow Corp.; Honeywell, Inc.; Weston Hydraulics Div., Borg-Warner Corp.</td></tr>
</table>

Expansible Element In-Line Valves

Streamlined flow of gas occurs in a valve in which a solid rubber cylinder is expanded or contracted to change the area of an annular space (Figure 1.19a). A stationary inlet nose and discharge bullet allow hydraulic pressure to force a slave cylinder against the rubber cylinder to vary its expansion. Control is from a diaphragm actuator, with the diaphragm plate carrying a piston. The piston acts as a pump to supply hydraulic pressure to the slave cylinder.

The rubber cylinder offers the seating ability of a soft seat valve. It has the capability of closing over foreign matter, and the design allows for the use of a restricted throat for reduced capacity. With this design pressure-drops as high as 1,200 PSID have been handled with a low noise level. The valve may be utilized as a pressure reducer or for back-pressure control, depending upon the system requirements.

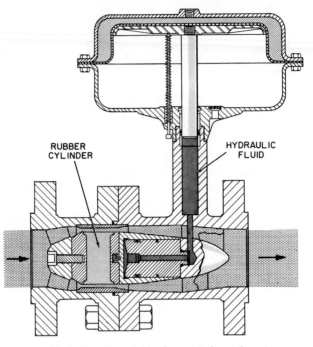

RUBBER
CYLINDER

HYDRAULIC
FLUID

Fig. 1.19a Expandable element in-line valve

Available sizes are from 1 through 6-in. The 1-in. valve can have screwed connections, while all sizes can be flanged. The body is steel with flange ratings to 600 PSIG USAS. A valve positioner can be used by calibrating stem position to annular space reduction and thereby obtaining accurate flow control relative to controller output.

Positioned Plug In-Line Valves

The positioned plug in-line valve, excluding control units, resembles a pipe spool. It is only necessary to inject pressure into its ports for positioning the valve plug. This simple design (Figure 1.19b) requires only three pressure seals. The plug is carried on a cylinder which also includes the piston. Pressure in one port causes closing, while the opposite port is used for opening. The valve has only one moving part. The valve is available in sizes from 2 in. to 8 in. for use to 350 PSIG at 400°F. Control quality is dependent upon the pilot valves and auxiliary units employed.

Another in-line valve available in small sizes (Figure 1.19c) carries the valve plug on a bridge in the operating cylinder, with the seat as part of a split body.

A spring-loaded version of this design uses the beveled end of the moving cylinder to seat on a replaceable soft seat retained in a dam held in position

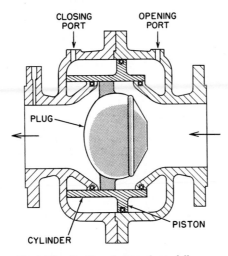

Fig. 1.19b   Positioned plug valve in fully open position. (Courtesy Eisenwerk Heinrich Schilling.)

by struts from the inside wall of the valve body. The spring loading may cause fail-close or fail-open actions, as illustrated on Figure 1.19d. The unit can be powered with line fluid or by an external pressure source. A double bleed feature can be incorporated to eliminate the possibility of actuation and line fluids combining if one of the dynamic seals should fail. All seats and seals are replaceable by separation at the body flange.

The unit is particularly adaptable to fluids that are toxic or difficult to contain, such as nitrogen tetroxide, hydrogen and others used in the aerospace industry. The unit is furnished in sizes from $1\frac{1}{2}$ in. to 18 in., with ratings

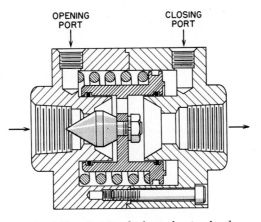

Fig. 1.19c   Positioned plug valve in closed position. (Courtesy G. W. Dahl Co.)

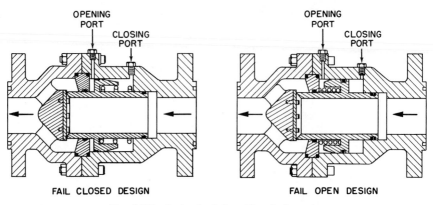

Fig. 1.19d   Spring-loaded positioned plug valve

to 2,500 PSIG USAS. All types of end connections are available, and control is dependent upon the auxiliary control components selected for the application. An explosion-proof limit switch can be furnished for position indication.

A similar valve (Figure 1.19e) has been adapted to manual operation. Rotation of the external handwheel carrying internal threads moves a cylinder with external threads. The cylinder's leading edge closes on the soft seat.

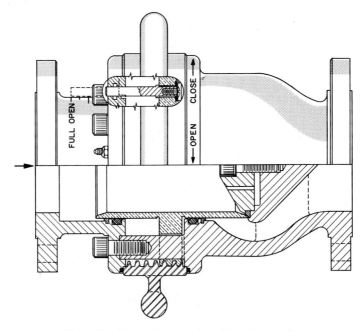

Fig. 1.19e   Manually operated positioned plug valve

Diaphragm-Operated Cylinder In-Line Valves

An in-line valve using a low convolution diaphragm for positive sealing and long travel (Figure 1.19f) is designed particularly for gas regulation. The low level of vibration, turbulence and noise of this in-line design makes it

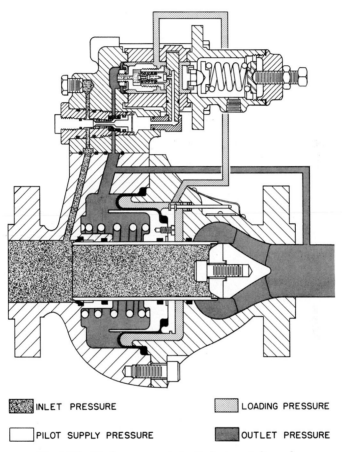

INLET PRESSURE   LOADING PRESSURE

PILOT SUPPLY PRESSURE   OUTLET PRESSURE

Fig. 1.19f   Diaphragm operated cylinder type in-line valve

suitable for high-pressure gas service. Inlet pressures to 1,400 PSIG and outlet pressures to 600 PSIG are possible in the 2-in. size. It is a high-capacity valve, as expressed by $C_v = 23d^2$. As a gas regulator the unit is supplied with a two-stage pilot to accept full line pressure. This pilot resists freeze-up and serves as a differential limiting valve. All portions of the pilot and line valve will withstand a full body rating of 600 PSIG USAS.

Digital Control Valves

An 8-bit digital system controlling a corresponding number of flow elements in a valve body (Figure 1.19g) constitutes a digital control valve with a rangeability of 255 to 1. The body is of in-line design, with an inlet

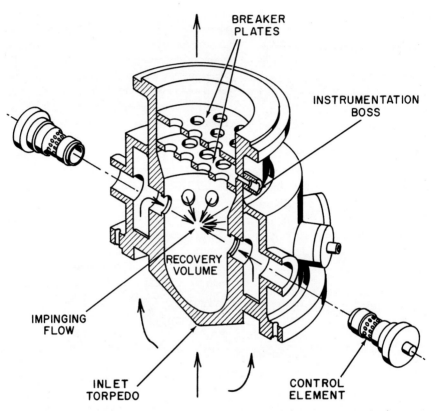

Fig. 1.19g   Digital control valve. (Courtesy Process Systems Instrumentation.)

torpedo to distribute flow to the passages in which the on-off control elements are placed. The flow through each element is directed to the center of the body. This design results in maximum pressure recovery within the body, with minimum recovery in the downstream piping. Further, the impinging-flow concept minimizes the possibility of cavitation and resulting erosion.

Breaker plates are installed within the body to damp high-frequency pressure waves present in the fluid system.

In another design, 12 control elements, Figure 1.19h, are used instead of the normal eight. The binary series of control element flow capacities is 1, 2, 4, 8, 16, 32, 32-32, 32-32, 32-32. By this method no bit in the series handles more than $12\frac{1}{2}$ percent of the total flow. The largest volume (64)

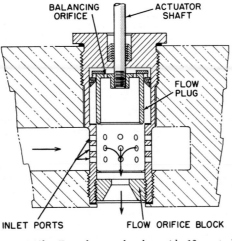

Fig. 1.19h Digital control valve with 12 control
elements

consists of two elements (32–32) located opposite each other and operating
simultaneously. Thereby the 8-bit computer word remains the same but
operates 12 control elements.

Each control element is identical except for the replaceable orifice block,
which is sized for the required flow. The element consists of a plunger working
in a multi-holed cage. The unit is designed for balancing of forces and to obtain
tight shutoff. The actuator shaft adapts itself to movement by solenoids or
pneumatic actuators.

Digital-to-Pneumatic Positioners

With the entry of digital computers into the field of process control, the
need becomes pronounced for devices that will operate on digital input signals.
Some of these components are discussed in Section 8.7 under "I/O interface
hardware," others in Section 6.3 dealing with converters and still others are
covered in these following paragraphs.

Use of direct digital control (ddc) in its fullest sense requires a digital
valve or actuator, one that can take a position dictated by the computer. This
means that an input of 120 bits to an actuator requiring 600 bits for full stroke
would result in a 20 percent opening of the corresponding valve. Many devices
can be jogged rapidly and would approximate the required opening but have
no position feedback to verify the position.

Most applications of ddc outputs use a digital to pneumatic transducer
or digital-to-pneumatic valve positioner. The positioner (Figure 1.19i) utilizes
a DC stepping motor that rotates in discrete steps (200 steps per revolution,
5 revolutions for full range). A traverse nut travels on the threaded motor
shaft to vary the tension of the range spring. Spring tension varies the position

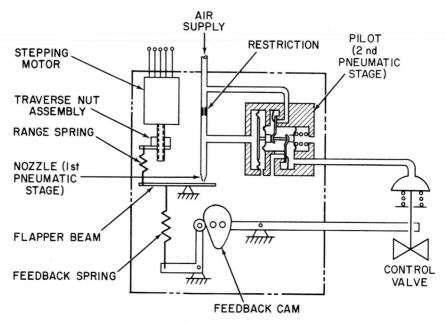

Fig. 1.19i   Digital-to-pneumatic valve positioner

of the flapper beam to increase or decrease pressure on the diaphragm of the second stage pilot. Valve stem position rotates a cam to vary the feedback spring tension to create a force balance with the range spring. The cam can be contoured to vary the valve flow characteristic.

Stepping motors can be DC bifilar (5 wire), DC synchronous (3 wire) or AC synchronous (3 wire). Operation is by single-pole, double-throw switches or transistors operated by the computer. Motor speed is 240 steps per second as a DC stepping motor, or 72 rpm as an AC synchronous motor. The 5-wire DC motor runs at a rate of 400 steps per second.

Digital-to-Pneumatic Transducers

A digital-to-pneumatic transducer (Figure 1.19j) has a non-rotating nut on the motor shaft which positions a flexible shaft to set the range spring, varying a beam-nozzle relationship to obtain a 3–15 PSIG output to the valve diaphragm or amplifier. The motor is capable of creating 1,000 increments in five revolutions at a speed of 400 steps per second, which allows full stroke change in $2\frac{1}{2}$ sec.

Digital Valve Actuators

Digital actuators can accept the output of digital computers directly without digital-to-analog converters. For their operation only simple on-off elements are needed. The number of output positions which can be achieved is equal to 2^n, where n equals the number of inputs. Accuracy of any position is a function of the manufacturing tolerances. Resolution is established by the number of inputs and by the operating code selected for a given requirement. The smallest move achievable is called a 1-bit move. The code may be binary, complementary binary, pulse or special purpose. A 3-input piston adder assembly produces 8 discrete bit positions. The adders in Figure 1.19k are shown in the 6-bit extended position. The interlocking pistons and sleeves will move when vented or filled through their selector valves. This same adder can be used to position a four-way spool valve, with a mechanical bias to sense position. The spool valve controls the position of a large-diameter piston actuator or force amplifier.

Use of a DC motor featuring a disc-armature with low moment of inertia has created another valve actuator particularly adaptable to a digital input. Brushes contacting the flat armature conduct current to the armature segments. Incremental movement is caused by half-waves at line frequency for rotation in either direction. The rotation of the armature is converted to linear stem motion by use of a hollow shaft internally threaded to match the valve stem. Actuator output is 5,000 lbf maximum for non-continuous service at a rate of 0.4 in./sec through a valve stroke of 3 in. The actuator is de-energized at stroke limits or at power overloads by thermal overload relays.

Operation of the actuator requires application of a thyristor (SCR) unit designed for this purpose. This unit accepts pulses from a computer or pulse generator. The thyristor unit consists of two SCRs with transformer, triggers

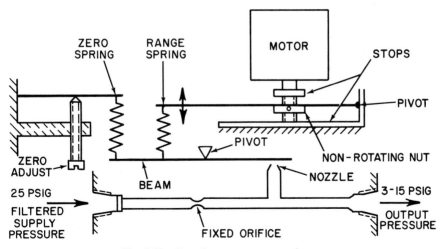

Fig. 1.19j   Digital-to-pneumatic transducer

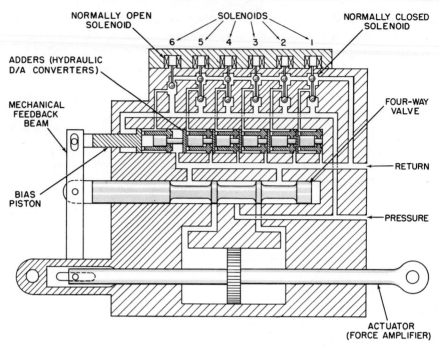

Fig. 1.19k   Digital valve actuator. (Courtesy Weston Div. Borg Warner.)

with pulse shift circuit, facility for manual actuator operation, and the previously mentioned thermal overload relays. The output consists of half-waves to pulse the armature of the actuator.

Modules are also available for process control with the necessary stem position feedback, slow pulsing for accurate manual positioning, and full-speed emergency operation.

1.20 VALVE ACTUATORS

**SPRING-LOADED
PNEUMATIC CYLINDER
ACTUATOR**

**DUAL ACTING
PNEUMATIC CYLINDER
ACTUATOR**

Valve actuators can be classified by their energy source (pneumatic, electric, hydraulic, etc.) or by the motion they generate (linear or rotary). The pneumatic-linear valve actuators, including both the cylinder and the diaphragm types, are covered in Section 1.8 in connection with the discussion on globe valves and their accessories. In this section the following additional basic actuator categories will be treated:

Pneumatic-Rotary
Electric-Linear
Electric-Rotary
Electric-Hydraulic
Electric-Pneumatic
Pneumatic-Hydraulic

Table 1.20a lists some of the main features of the presently available linear electric actuators.

**ELECTRIC SOLENOID
ACTUATOR SHOWN
WITH MANUAL RESET**

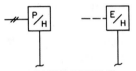

**PNEUMATIC–HYDRAULIC
AND ELECTRO–HYDRAULIC
ACTUATORS**

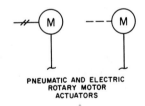

**PNEUMATIC AND ELECTRIC
ROTARY MOTOR
ACTUATORS**

Partial List of Suppliers:

PNEUMATIC: The Andale Co.; Bettis Corp.; Ledeen Div., Textron Co.; OIC Corp.; Philadelphia Gear Corp. (Limitorque); Shafer Valve Co. ELECTRIC-LINEAR, INCLUDING SOLENOIDS: Ametek Inc.; Automatic Switch Co.; H. Beck & Sons Inc.; Conoflow Corp.; G. W. Dahl Co.;

Table 1.20a
NON-PNEUMATIC VALVE ACTUATORS FOR PROPORTIONAL
AND ON-OFF APPLICATIONS

Actuator Type	Maximum Stroke (in.)	Maximum Stem Force (lbf)	Stem Motion Rate (sec/in.)	Approximative Cost ($)		Partial List of Suppliers
				Base	Cost Per 1,000 in.-lbf	
Electric	(12)°	(1,200)	(2.5)	(340)	(24)	General Controls ITT
Electric	(10)	(800)	(4)	(235)	(29)	Limitorque-Div. Philadelphia Gear
Electric	$9\frac{3}{4}$ ($9\frac{3}{4}$)	6,600 (6,600)	0.2 (0.2)	5,000 (4,500)	78 (70)	Raco Machine Co.
Electro-Hydraulic	(6)	(6,000)	(12)	(430)	(12)	Oil Dyne-Div. Racine
Electric	($4\frac{1}{2}$)	(2,000)	(6.5)	(270)	(27)	Limitorque-Div. Philadelphia Gear
Electric	($4\frac{1}{2}$)	(10,000)	(7.5)	(390)	(9)	Limitorque-Div. Philadelphia Gear
Electric	4 (4)	700 (1,200)	24 (55)	695 (385)	245 (80)	Conoflow Corp.
Electro-Hydraulic	3	5,000	2.7	1,300	84	Fisher Controls
Electro-Hydraulic	($2\frac{5}{8}$)	(3,000)	(18)	(430)	(55)	General Controls ITT
Electro-Hydraulic	$2\frac{1}{2}$	1,500	7	840	224	General Precision Systems
Electro-Hydraulic	(2)	(5,000)	(220)	(485)	(48)	Fisher Controls
Electric	2 (2)	500 (500)	7 (15)	600 (350)	600 (350)	H. Beck & Sons
Electric	2	1,400	8	665	238	G. W. Dahl Co.
Cylinder & Servo	2	5,300	0.25	1,800	170	Bocnshaff & Fuchs

° Data given in parentheses is for on-off actuators.

General Controls ITT; J. D. Gould Co.; Lawrence Co.; Marotta Valve Corp.; Philadelphia Gear Corp. (Limitorque); Ramcon Corp.; Ruggles-Klingemann; Skinner Electric Valve Div., Skinner Precision Industries Inc. E<small>LECTRIC</small>-R<small>OTARY AND</small> E<small>LECTRIC</small> H<small>YDRAULIC</small>: Clarkson Valve Co.; Crane Co.; G. W. Dahl Co.; Fisher Controls; General Controls ITT; General Precision Systems Inc.; Honeywell, Inc.; Ledeen Div., Textron; Moog Inc.; Oil Dyne Inc.; Ramcon Corp.

Pneumatic Actuators

Cylinder Type

Increased use of ball and butterfly valves or plug cocks for control has bred a variety of actuators and applications of existing actuators for powering these designs.

Positioning a quarter-turn valve with a linear output actuator using a lever arm on the valve resolves itself into a problem of mounting and linkages. The actuator can be stationary, with a bushing to restrain lateral movement of the stem. This requires a joint between the stem and a link pinned to the lever arm. The actuator can be mounted on a gimbal mechanism to allow required movement. The actuator can be hinged to allow free rotation to allow for the arc of the lever arm.

Various *Scotch yoke* designs, such as the one shown in Figure 1.20b, can be used with one, two or four cylinders. Use of rollers in the slot of the lever arm utilizes the length of the lever arm of the valve opening or closing points.

A rack and pinion can be housed with the pinion on the valve shaft and

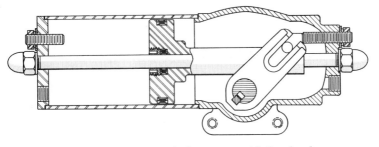

Fig. 1.20b   Pneumatic cylinder actuator with Scotch yoke

the rack positioned by almost any linear valve actuator. The rack (Figure 1.20c) can be carried by a double-ended piston or by two separate pistons (Figure 1.20d) in the same cylinder, where they move toward each other for counterclockwise rotation and away from each other for clockwise rotation. Similar action is obtained (Figure 1.20e) by two parallel pistons in separate cylinder bores. Dual cylinders are used in high-pressure actuators used to rotate ball valves as large as 16 in. in less than 0.5 sec. An actuator similar to the one shown in Figure 1.20f can be spring-loaded for emergency or positioning operation.

On-off operation of cylinders for quarter-turn valves requires solenoid or pneumatic pilots to inject pneumatic or hydraulic pressure into the cylinders. Open or closed position must be set by stops that limit shaft rotation or piston travel. Thereby the valve rotation is stopped and held in position until reverse action is initiated. Positioning action requires a calibrated spring in the piston or diaphragm actuator, a valve positioner, or a positioning valve system that loads and unloads each end of the cylinder. Rotation of the valve must be translated to the positioner by gears, direct connection, cam or by linkage. The valve positioner must be the type that includes the four-way valve. The positioning valve system can be a four-way valve with a positioner for use with a pneumatic controller. The piston is sometimes positioned by

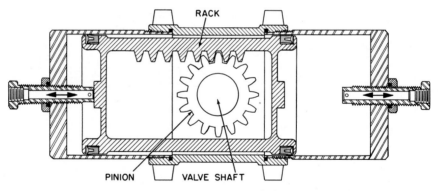

Fig. 1.20c   Rack and pinion actuator with dual-acting cylinder

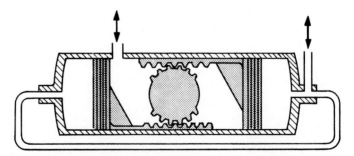

Fig. 1.20d   Rack and pinion actuator operated by two separate pistons

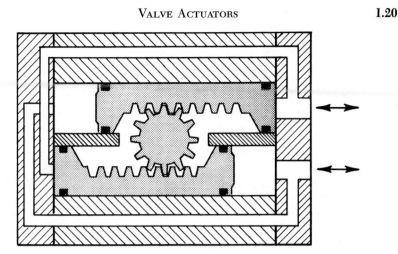

Fig. 1.20e   Rack and pinion actuator operated by two parallel pistons

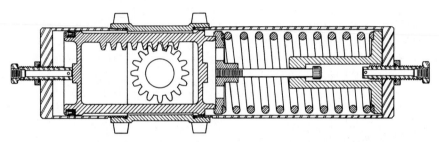

Fig. 1.20f   Rack and pinion actuator with spring-loaded (fail-safe) cylinder

a servo-system consisting of a servo-valve that accepts an electronic signal, a four-way valve to amplify and control the pressure to the cylinder and a feedback signal from a potentiometer or LVDT.

An electro-hydraulic power pack or pneumatic pressure source may be used to furnish pressure to a pair of cylinders, one for each direction of rotation, as shown in Figure 1.20g.

Rotation by Spline or Helix

A multiple helical spline rotates through 90 degrees as pressure below the piston (Figure 1.20h) moves the assembly upward. A straight spline on the inside of this piston extension sleeve rotates the valve closure member through a mating spline. The cylinder is rated at 1,500 PSIG. The actuator will rotate 1-in. and 1½-in. valve stems. The mounting configuration is designed to adapt to many quarter-turn valves.

A non-rotating cylinder (Figure 1.20i), with an internal helix to mate with a helix on a rotatable shaft, creates a form of rotating actuator. Hydraulic or pneumatic pressure in the drive end port (left) causes counterclockwise

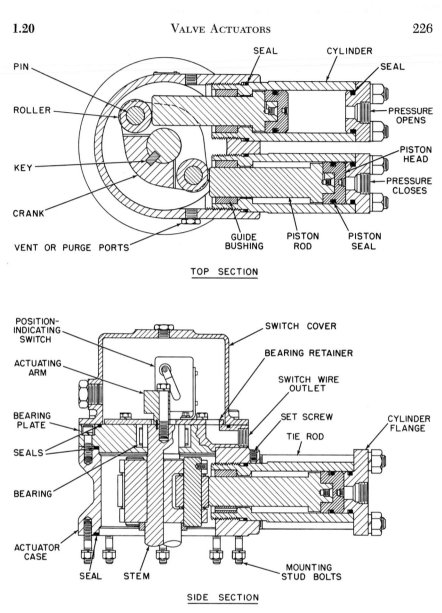

Fig. 1.20g   Crank and roller actuator operated by a pair of cylinders

rotation; clockwise rotation is caused by pressure on the opposite side of the piston. There is a patented seal between the internal and the external bores of the cylinder and the external surface of the shaft. The unit is totally enclosed by seals to protect it from contaminated atmospheres. A hydraulic pump, reservoir and necessary controls can be mounted integrally.

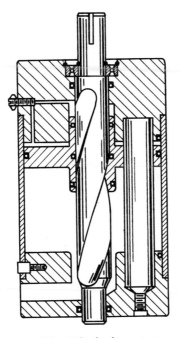

Fig. 1.20h   Helical spline actuator

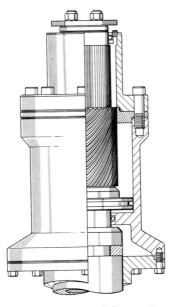

Fig. 1.20i   Rotating helix actuator

Vane Type

Injection of pressure on one side of a vane to obtain quarter-turn actuation is straightforward and obtainable with a minimum number of parts. A single vane (Figure 1.20j) can be used for 400 to 20,000 in.-lb. Units can be mounted together for double output, or a double vane design (Figure 1.20k) can also

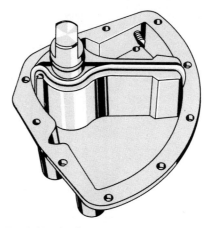

Fig. 1.20j   Single-vane quarter-turn actuator

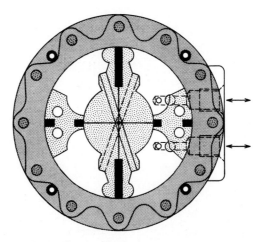

Fig. 1.20k   Double-vane quarter-turn actuator

be used. The success of the vane actuator as a control device is dependent upon the control systems. By use of an auxiliary pneumatic pressure source, all types of fail-safe actions are possible although *not* as positively as with spring loading. Use of line pressure to create hydraulic pressure on the vane, is piloted by both manual and automatic methods. Use of a rotary potentiometer to sense position and complete a bridge circuit is necessary for proportional control.

Pneumo-Hydraulic Type

An actuator with two double-acting cylinders uses an integrally designed pneumatic or electric powerpack (Figure 1.20l). This has the advantage of

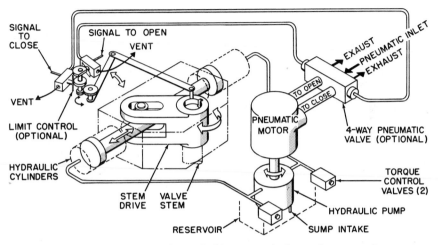

Fig. 1.20 l   Actuator with two double-acting cylinders and power pack

furnishing a constant hydraulic pressure to the cylinders regardless of the power source to the prime mover. Few cylinder sizes are needed to cover a wide range of torque outputs. The prime movers are sized and selected to obtain the actuator speeds desired with the pneumatic pressures or electric voltages available. Multiple auxiliary switches, position transmitter, and positioning devices are adapted to the unit.

Gas pressure is used to create hydraulic pressure using two bottles (Figure 1.20m). The stability of a hydraulically operated cylinder is utilized in this manner, using line gas pressure and the bottle size for amplification. The manual control valve can be replaced by a variety of electric or pneumatic pilot valves for automatic control. A hand pump is furnished which can take over the hydraulic operation in the absence of gas pressure or malfunction of the pilot controls. This self-sustaining approach to cylinder operation finds wide application for line break shutoff and for the various diverting and by-pass operations of a compressor station.

A hydrostatic system consisting of a pneumatic prime mover (Figure 1.20n) on the shaft of a hydraulic pump to run a hydraulic motor has interesting features. This actuator incorporates many of the features of other high-force geared actuators that rotate a drive sleeve. Torque control consists of

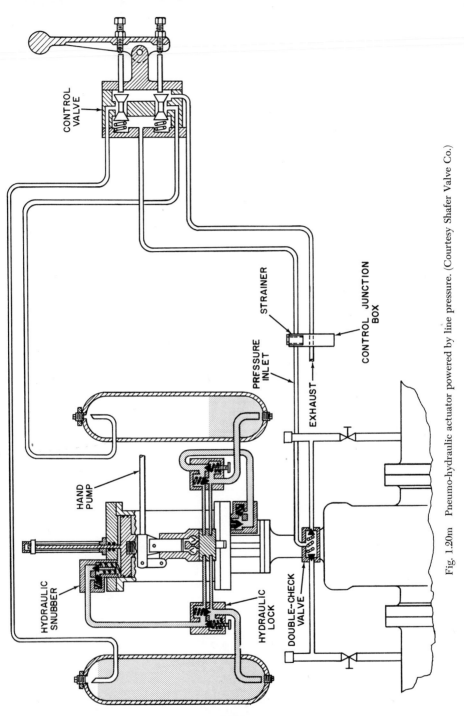

CONTROL VALVE

HAND PUMP

HYDRAULIC SNUBBER

HYDRAULIC LOCK

DOUBLE-CHECK VALVE

PRESSURE INLET

STRAINER

EXHAUST →

CONTROL JUNCTION BOX

Fig. 1.20m Pneumo-hydraulic actuator powered by line pressure. (Courtesy Shafer Valve Co.)

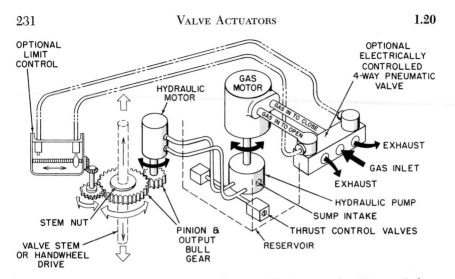

Fig. 1.20n   Actuator system with pneumatically powered hydrostatic valve. (Courtesy Ledeen Div. Textron.)

a relief valve in the hydraulic line to the motor. This eliminates the reactive force of spring-loaded torque controls. Starting torque occurs because hydraulic slippage of the pump allows the motor to reach maximum speed. Direction and deactivating control is attained with a four-way valve. As many as 16 auxiliary switches, settable at any position, are housed in the unit. Limit switches can be pneumatic or electric. A wide range of torque outputs and speeds is obtained by selection of prime mover, pump and motor combinations. Initial success of the unit was partially due to its adaptability for retrofit to existing valves. The unit can be manually operated if required.

Rotary Pneumatic Type

Pneumatic pressure is used to power a rotary motor to drive any of the large gear actuators. Control is by a four-way valve. The motor shown in Figure 1.20o is running in one position. This will continue until the valve

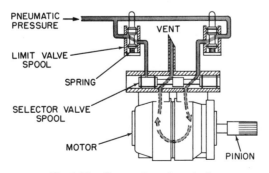

Fig. 1.20o   Rotary air motor actuator

is repositioned or until a cam operates a shutoff valve at one end of the stroke. Reversal of the four-way valve causes reverse operation. An intermediate position causes the motor to stop. The four-way valve can be operated by pneumatic or electric actuators for remote automatic control. A position transmitter will allow adaptation to closed-loop, proportional control.

Electro-Pneumatic Actuators

An actuator that defies classification, except that it is pneumatically powered and electrically controlled for proportional application, is described at this point. Operation of a threaded drive sleeve occurs when spring-loaded pawls create a jogging action on a drive gear. Pressure introduced through one of the external lines selects the pawl to become active when the rocker arm is repetitively rocked by the pneumatic motor. A lead screw positions

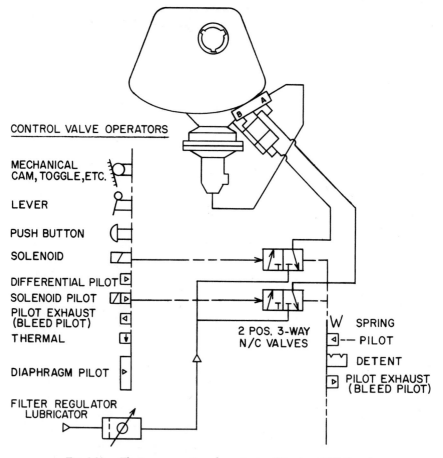

Fig. 1.20p   Electro-pneumatic valve actuator. (Courtesy OIC Corp.)

a sliding block to operate control switches, potentiometer and position indicators. The lead screw is driven from a small spur gear and bevel gears.

Air is supplied from 60 to 140 PSIG to give torque outputs up to 360 or 720 ft-lbf using two actuators. Air consumption is from 0.75 SCF to 1.70 SCF per revolution at 140 PSIG. Maximum valve stem diameter that can be rotated is 2 in. Numerous motivating combinations are used for control, including electro-pneumatic, electric, and fully pneumatic. A wide variety of components can be used to build up these systems, as shown in Figure 1.20p.

Electric Actuators

Linear Actuators (Solenoids)

Use of a solenoid (a soft iron core that can move within the field set up by a surrounding coil) is made extensively for moving valve stems. Although the force output of a solenoid may not have many electrical or mechanical limitations, its use as a valve actuator has economic and core (or stem) travel limitations, and it is expensive.

A solenoid valve consists of the valve body, a magnetic core attached to the stem and disc, and a solenoid coil (Figure 1.20q). The magnetic core

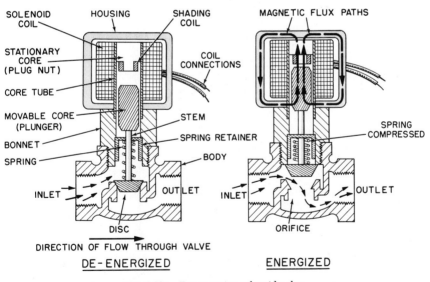

Fig. 1.20q   Direct-acting solenoid valve

of this valve moves in a tube closed at the top and sealed at the bottom, allowing the valve to be packless. A small spring assists the release and initial closing of the valve. This valve is electrically energized to open. Stronger

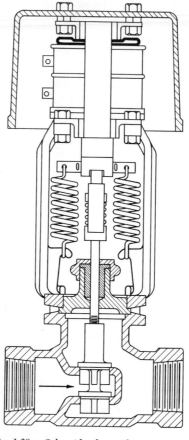

Fig. 1.20r   Solenoid valve with strong return
springs

springs are used to overcome the friction of packing when it is required (Figure 1.20r). Reversing the valve plug causes reverse action (open when de-energized). Increased stroking force is obtained by using the mechanical advantage of a lever with a strong solenoid (Figure 1.20s).

Use of the solenoid to open a small pilot valve (Figure 1.20t) increases the port size and allowable pressure drop of solenoid operated valves. A widely employed device is a small solenoid pilot valve to supply pressure to a diaphragm or piston for a wide range of output forces. Pilot operation applies pressure to a diaphragm or piston or may release pressure, allowing the higher upstream pressure to open the valve. A good example is the in-line valve (Figure 1.20u). Most solenoid valves are designed to be continually energized, particularly for emergency shut-down service. Thus the power output is limited to the current whose I^2R developed heat can be readily dissipated.

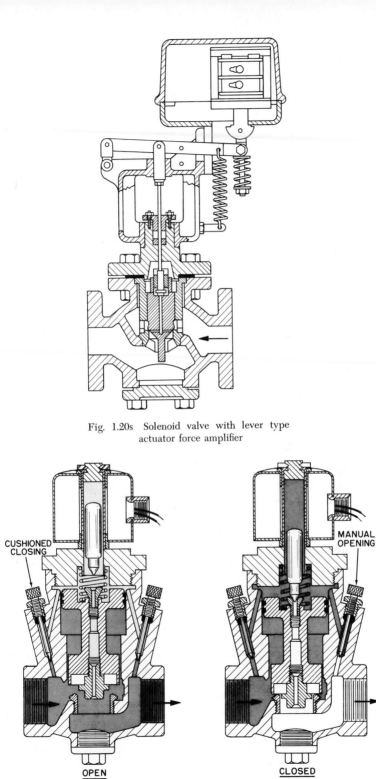

Fig. 1.20s   Solenoid valve with lever type
actuator force amplifier

CUSHIONED
CLOSING

MANUAL
OPENING

OPEN

CLOSED

Fig. 1.20t   Pilot-operated solenoid valve. (Courtesy J. D. Gould Co.)

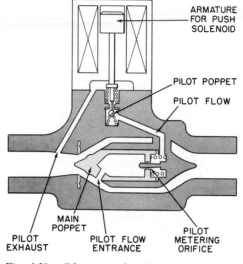

ARMATURE
FOR PUSH
SOLENOID

PILOT POPPET

PILOT FLOW

MAIN
POPPET

PILOT
EXHAUST

PILOT FLOW
ENTRANCE

PILOT
METERING
ORIFICE

Fig. 1.20u   Pilot-operated in-line solenoid valve.
(Courtesy AMETEK-APCO.)

Use of a high source voltage and a latch-in plunger overcomes the necessity for continuous current. The single-pulse valve-closing solenoid is disconnected from the voltage source by a single-pulse, delatch solenoid and hence does not heat up after it is closed. A pulse to the delatch solenoid permits the valve to be opened by a spring.

Three-way solenoid valves with three pipe connections and two ports are used to load or unload cylinders or diaphragm actuators (Figure 1.20v). Four-way solenoid pilot valves are principally used for controlling double-acting cylinders.

Electro-Hydraulic Actuators

External Hydraulic Source   Use of the term electro-hydraulic has been applied to actuator systems in which the hydraulic pressure is supplied by a hydraulic mule for one actuator or for a number of actuators close enough to each other to utilize the one source. In these cases the electro-terms refer to the application of this hydraulic power to the actuator by electrical control means. In the broad sense, the use of two three-way solenoids or one four-way solenoid externally mounted to the actuator constitutes an electro-hydraulic system.

More extensively, electro-hydraulic applies to a proportionally positioned cylinder actuator. This requires a servo-system which is a closed loop within itself. A servo-system requires one of the standard command signals, which is usually electrical but can be pneumatic. This small signal, which often requires amplification, controls a torque motor or voice coil to position a flapper or other form of variable nozzle. This positions a spool valve or comparable device to control the hydraulic positioning of a high-pressure second-stage valve. The second stage valve directs operating pressure to the cylinder for very accurate positioning. Closing the loop requires mechanical (Figure 1.20w) or electrical feedback to compare the piston position with the controller output signal.

Hermetically Sealed Power Pack   A much more useful and compact type of electro-hydraulic actuator combines the electro-hydraulic power pack with the cylinder in one package. Many of these actuators are designed as a truly integral unit. An electric motor pump supplies high-pressure oil through internal ports to move the piston connected to the stem (Figure 1.20x). The small magnetic relief valve is held closed during the power stroke until de-energized by an external control or emergency circuit to allow the spring to cause a "down" stroke. The same unit can be used to cause spring return to up position using a Bourdon switch to cause force limit.

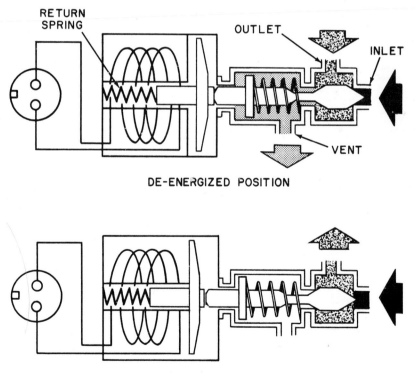

Fig. 1.20v   Normally closed three-way solenoid valve

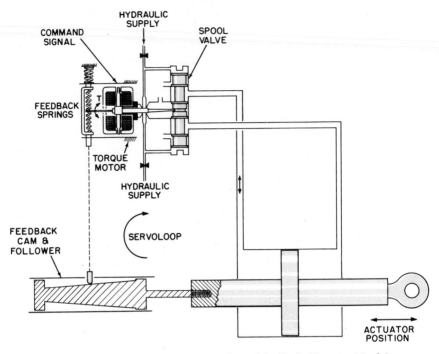

Fig. 1.20w   Two-stage servo valve with mechanical feedback. (Courtesy Mood Servo-
controls Co.)

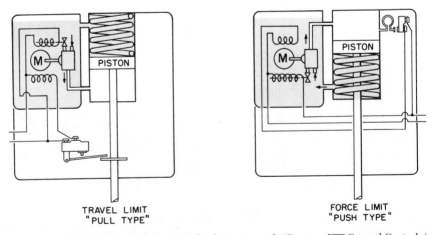

Fig. 1.20x   Hermetically sealed electro-hydraulic power pack. (Courtesy ITT General Controls.)

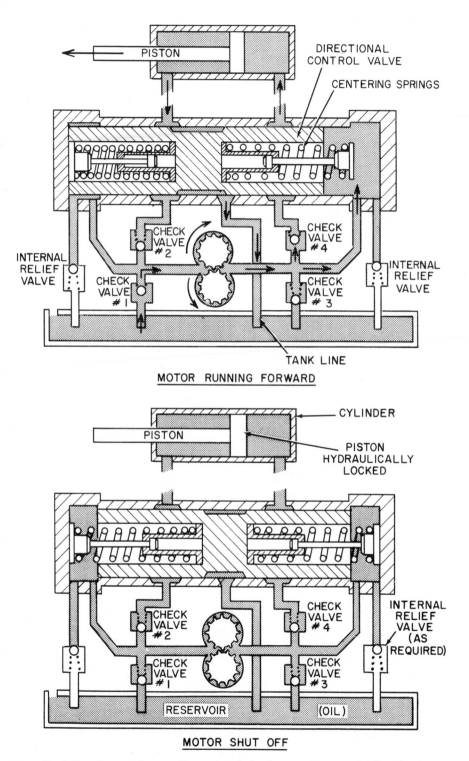

Fig. 1.20y   Pump with reversible motor and closed center. (Courtesy Oil Dyne Inc.)

Reversible Motor and Pump   A reversible motor can be used to drive a gear pump in a system to remove oil from one side of the piston and deliver it to the other side (Figure 1.20y). The check valves allow the pump to withdraw oil from the reservoir and position the directional control valve in order to pressurize the cylinder. Reversing the motor (and pump) reverses the direction. When the motor is de-energized, the system is "locked up." For proportional control, a feedback is necessary from stem position to obtain a balance with the control signal.

Jet Pipe Systems   A very old control system for a cylinder, the jet pipe, is employed in an electro-hydraulic actuator. An electro-mechanical moving coil in the field of a permanent magnet is used to position a jet that can direct oil to one end or the other of the cylinder actuator (Figure 1.20z). A force

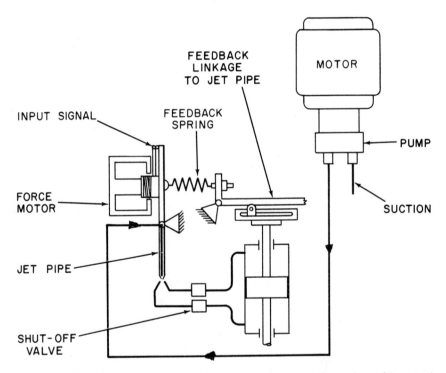

Fig. 1.20z   Electro-hydraulic actuator with jet pipe control. (Courtesy Singer-General Precision.)

balance feedback from stem position creates the balance with the controller signal.

Hydraulic Control of Jacketed Pinch Valves   Controlled hydraulic positioning of a sleeve valve is obtained with a moving coil and magnet to position the pilot (Figure 1.20aa) which controls pressure to the annular space of the

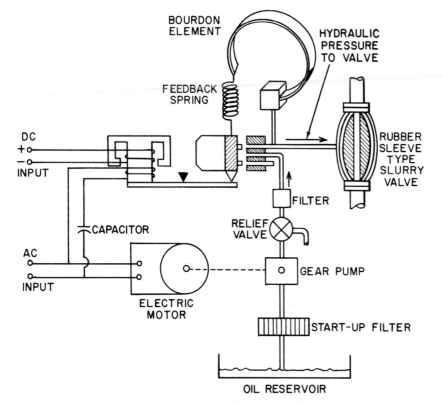

Fig. 1.20aa   Electro-hydraulic control of jacketed pinch valve. (Courtesy The Clarkson Company.)

valve. Feedback is in the form of a Bourdon tube, which senses the pressure supplied to the valve and moves the pilot valve to lock in that pressure.

Multiple Pump Systems   The multiple pump system consists of three pumps running on the shaft of one prime mover (Figure 1.20bb). There is one pump for each side of the piston and one for the control circuit. The force motor tilts a flapper to expose or cover one of two control nozzles. The flow through a restricting nozzle allows pressure to be transmitted to one side of the piston or the other. Force balance feedback is created by the effect of a ramp attached to the piston shaft, which positions a cam attached to the feedback spring. Upon loss of electric power the cylinder shutoff valves close to lock up the pressure in the cylinder and assume a status quo. A by-pass valve between the cylinder chambers allows pressure equalization to make use of the manual handwheel.

Vibratory Pumps   A vibratory pump is used in a power pack mounted on a cylinder actuator (Figure 1.20cc). A 60-Hz alternating source causes a

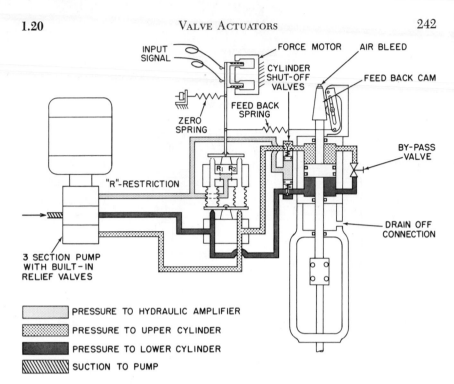

INPUT SIGNAL

FORCE MOTOR

AIR BLEED

CYLINDER SHUT-OFF VALVES

FEED BACK CAM

ZERO SPRING

FEED BACK SPRING

BY-PASS VALVE

"R"-RESTRICTION

R₁ R₂

DRAIN OFF CONNECTION

3 SECTION PUMP WITH BUILT-IN RELIEF VALVES

PRESSURE TO HYDRAULIC AMPLIFIER

PRESSURE TO UPPER CYLINDER

PRESSURE TO LOWER CYLINDER

SUCTION TO PUMP

Fig. 1.20bb   Multiple pump hydro-electric valve actuator

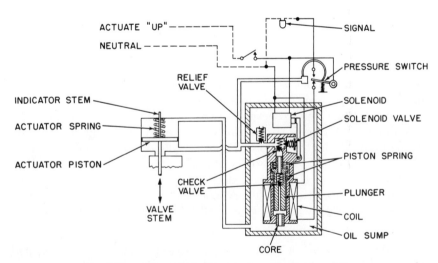

ACTUATE "UP"

NEUTRAL

SIGNAL

PRESSURE SWITCH

RELIEF VALVE

INDICATOR STEM

SOLENOID

ACTUATOR SPRING

SOLENOID VALVE

ACTUATOR PISTON

PISTON SPRING

CHECK VALVE

PLUNGER

VALVE STEM

COIL

OIL SUMP

CORE

Fig. 1.20cc   Electro-hydraulic actuator with vibratory pump

plunger to move toward the core and, upon de-energization, a spring returns it. The pump operates on this cycle and continues until the piston reaches the end of its stroke, when a pressure switch shuts it off. The solenoid that retains the pressure is de-energized by an external control circuit. Maximum stem force is 1,500 lbf at a rate of about 1-in./min, or 5,000 lbf at 0.3-in./min.

Two-Cylinder Pumps   A two-cylinder pump, driven by a unidirectional motor, injects pressure into one end of a cylinder or the other, depending upon the positions of two solenoid relief valves (Figure 1.20dd). The solenoid on the left is closed to move the piston to the right, with hydraulic pressure relieved through the other relief valve. Motion continues until the valve it is operating is seated. The buildup of cylinder pressure operates the pressure switch at a predetermined setting to de-energize the motor and both solenoid relief valves. This locks the hydraulic pressure in the cylinder. Switching of the three-way switch will start the motor and reverse the sequence of the relief valves to move the piston to the left. At full travel (which is the up position of a valve stem), a limit switch shuts down the unit. Open-centering the three-way switch at any piston position will lock the piston (and valve stem) at that point.

Remote control is accomplished by manipulation of the open-center switch. Automatic control is acknowledged by including this open-center function, which may be solid state, in the control circuit. A potentiometer or LVDT which senses the stem position is required for feedback.

Stem output is 6,000 lbf at a rate of 3 sec/in. Stem travels up to 7-in. are available, although longer travels are feasible. The entire system is designed in a very compact explosion-proof package which may be mounted on a variety of valve bonnets. The speed of response to energizing and de-energizing the control circuit makes it feasible to adapt the unit to digital impulses.

Electric Rotary Actuators for Linearized Stem Motion

Use of the rotary output of an electric motor for valve actuation requires a set of gears to reduce normal motor speeds to usable valve operating parameters, at the same time increasing the power output to fit valve operating requirements. Electric actuators vary in output from a few inch-ounces using 20 volt-amp motor to a 50-horsepower motor on an actuator developing torque in the foot-pounds range.

Use of this power requires adaptation to the operation of a valve. Actuator outputs take the following forms:

    a. *Linear*   Reciprocating motion as used for globe valves and other types.

    b. *Rotary*   Used to obtain short linear stem travels through a cam, eccentric or system of levers.

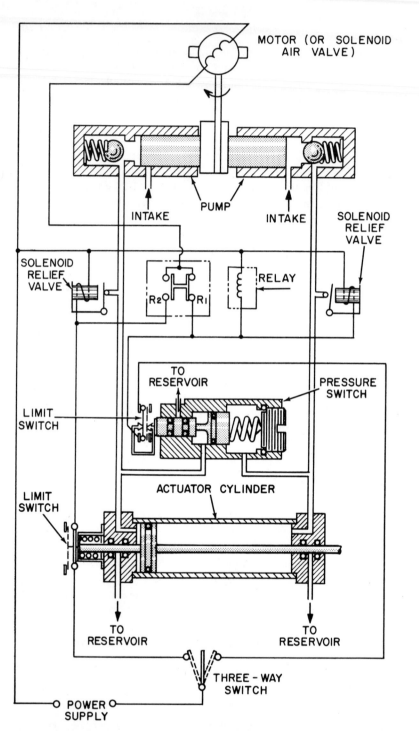

Fig. 1.20dd Two-cylinder pump type electro-hydraulic actuator. (Courtesy Ledeen Div. Textron.)

c. *Rotary drive sleeve* Used to rotate a drive sleeve, creating reciprocating motion of a stem which is not allowed to rotate.

Reciprocating Motion Type A representative of classification "a" is an actuator (Figure 1.20ee), which uses a worm and a rack and pinion to translate

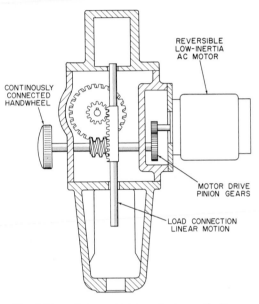

Fig. 1.20ee Rotary electric actuator with linear output

horizontal shaft motor output to vertical linear motion. Maximum force output is approximately 1,500 lbf at about 0.1-in./min. A continuously connected handwheel which must rotate the rotor of the motor, can be used when there is short stem travel and relatively low force output.

The actuator is designed with a conventional globe valve bonnet for ease of mounting. Units operate on 110 volts and have been adapted to proportional use with a 135-ohm Wheatstone bridge circuit or any of the standard electronic controller outputs.

Linearized Stem Motion Through Cams and Eccentrics. The actuator portion of classification "b" designs is commonly called a *damper motor* because its normal use is to position dampers and louvers by 90-degree or 180-degree rotation using various types of level arm combinations. For this application, a shaded-pole motor is used for non-reversing service and a capacitor-type motor is used for reversing service. The motor drives a geartrain to rotate the bull gear on the output shaft of the unit.

The drive shaft may carry a number of auxiliary switches. Two of these may be used for limit control unless separate limit switches are located at another point. The remaining auxiliary switches, which must be adjustable, are used to sequence other equipment such as valves, pumps or blowers. When used for proportional control in a Wheatstone bridge circuit, the unit has an open potentiometer with the wiper positioned by the drive shaft. The unit must contain an open-center (mousetrap) relay. Some units have speed control by use of an adjustable hydraulic reaction to motor shaft rotation.

Application of the linear requirement of a valve necessitates a cam linkage for translation from rotary to linear motion. Use of a linkage (Figure 1.20ff)

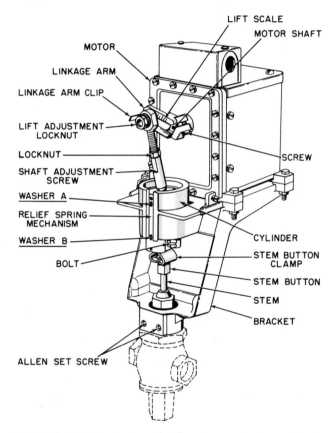

Fig. 1.20ff   Electric actuator with linkage to convert rotary motion to linear. (Courtesy Honeywell, Inc.)

allows continuous unidirectional motor rotation. The stem moves down through 180-degree rotation of the lever arm and returns during the next 180 degrees. The unit is sometimes provided with a brake to maintain position with power off.

Some units use a heavy, external coil spring for return of the shaft to its original position when de-energized. Although most models are limited to valve sizes not larger than 3 in. and to low-pressure-drop services, they are in demand for heating and air conditioning applications.

Drive Sleeve Type Classification "c" actuators rotate a drive sleeve within the actuator which is threaded to mate with the stem thread. They include many designs, features and force outputs. All basic units, such as the one shown in Figure 1.20gg, will include a prime mover, gear train,

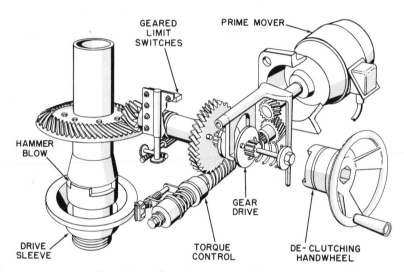

Fig. 1.20gg   Electric actuator with drive sleeve

drive sleeve, limit switches and a facility for mounting on a valve. All additional features may or may not be available on all units. In fact, many applications do not require certain of the features. Use of electric, gear-operated actuators has emphasized the necessity for refinement of this design. The following features must be considered in the selection of this type of actuator:

1. *Wide range of speeds and torque outputs.* Speed requirements vary from 10 rpm to about 160 rpm. The upper limit of available torque seems to be about 60,000 ft-lbf. Stem thrusts and rotational drive torques are limited only by the size of the motor used and the ability of the gear, bearings, shafts, etc., to carry the load. Speed of operation depends on the gear ratios, adequate prime mover power and means of overcoming the inertia of the moving system for rapid stopping. This is most important for proportional control uses. Some actuators have a limited selection of drive speeds while others are furnished in gear ratios in discrete steps of 8–20 percent between speeds.

2. *Adaptability to a wide variety of rising stem valves.* Two features are required: a drive sleeve as part of the actuator, and a drive nut to drive a sleeve which is part of the valve yoke. The first requires bearings and castings, as part of the actuator design to carry the entire force load of the valve. The second requires an elongated drive nut to mate with the configuration on the drive sleeve of the valve. A cooperative effort is required between valve and actuator manufacturers to insure identical threads in the drive sleeve and valve and to insure sufficient stem length to reach the drive sleeve.

3. *Means of manual operation.* Manual operation is sometimes necessary for normal operational procedures, such as startup, or under emergency conditions. Only units that are rotating very slowly or those with low output should have continuously connected handwheels. Most units have the handwheel on the actuator but have a clutch for demobilizing the handwheel during powered operation. Clutches are manual engage and manual disengage, or manual engage but disengage upon release of the handwheel. Others are manual engage when the handwheel is rotated, with the motor reengaging when the handwheel is not being rotated; or power reengage, which takes the drive away from the operator upon energization, leaving the handwheel freewheeling.

4. *Electrical equipment for use in any normal area or atmosphere.* Most of these actuators include much of the electric gear within the housings of the unit. Components such as limit, auxiliary and torque switches and position or feedback potentiometers are run by gearing to stem rotation, so they must be housed on the unit. Installation is optional concerning push-buttons, reversing starters, lights, control circuit transformers or line-disconnect devices in an integral housing on the unit. Any or all of these components may be located externally, such as in a transformer, switch or control house. The enclosures must be designed to satisfy NEMA requirements for the area. (See Section 10.10 of Volume I.)

5. *High breakaway force.* Resistance to opening requires a method of allowing the motor and gear system to develop speed to impart a "hammer blow" which starts motion of the valve gate or plug. Selection of motors with high starting torque is not always sufficient. The dogs of a dog-clutch rotate before picking up the load, or a pin on the drive may move within a slot before picking up the load at the end of the slot. Systems are used which delay contact for a preselected time or until the tachometer indicates the desired speed of rotation.

6. *Torque control for shutdown at closure or due to an obstruction in the valve body.* Although accomplished in numerous ways, each one uses a reaction spring to set the torque. When rotation of the drive sleeve is impeded, the spring will collapse, moving sufficiently to operate a shut-off switch.

7. *Position indication, local or transmitted.* An indicator can be geared to the stem rotating gear, but it becomes a problem when actuator rotation

varies from 90 to a possible 240 revolutions. Gearing of a cam shaft operating the position indicator or auxiliary switches must be calculated to obtain a fairly uniform angle. Upon correct gearing, an indicating arrow or transmitting potentiometer can be rotated.

8. *Maintenance of status quo position.* This is no problem when the actuator includes a worm gear or an Acme stem thread. Use of spur gears can cause instability when positioning a butterfly or ball valve. Status quo is obtained by use of a motor brake or insertion of a worm gear into the system.

9. *Protection against stem expansion.* The status quo ability of an Acme thread or worm gear is detrimental when the valve itself is subjected to temperatures high enough to expand the stem. This expansion, when restrained, can damage the seat or plug, bend the stem, or damage the actuator thrust bearings. One of the original patents for this type of actuator included Belleville springs to allow the drive sleeve to move with the thermal expansion and relieve the linear force.

10. *Versatility in mounting methods.* Industry has dictated a set of dimensions for the mounting flanges and bolt holes for newly manufactured valves. Retrofit mounting requires adaption to existing valves. Mounting requires a plate to match the existing valve, which is either screwed into the yoke upon removal of the manual drive sleeve, welded or brazed to the yoke, or, for a split yoke, bolted to the yoke.

11. *Adaptability to a wide variety of control circuits.* This feature includes adaptability to many voltages and to single or polyphase supplies. Polyphase motors of 240–480 volts and 60 Hz predominate. Single-phase motors up to about 2 horsepower are used. Reversing starters with mechanical interlocks are used for both proportional and on-off service. The coils that open and close the contacts are energized by an open-center double-pole switch, which can be incorporated in the automatic control circuit. For manual control, the starter may be the type that maintains contact, requiring the open or close button to be held in position. Some units are also wired for momentary depression so that the actuator runs until it reaches the limit switch, or until a stop button is depressed.

12. *Proportional control.* Proportional control of these large units can be accomplished by including the coils of the reversing starter in a proportional control circuit. This requires a position feedback which may be a potentiometer. For this type of control, a Wheatstone bridge circuit would be used.

The reversing starter controls any voltage or phase required by the motor. A transducer to transform any of the accepted electronic controller outputs (1–5 ma, etc.) to a resistance relative to controller output, permits use of the actuator in these systems.

A smaller unit, with a force output of 1,460 lbf at stall and 500 lbf at a rate of 0.12-in./sec, uses solid state control of the motor. At the same time it eliminates stem position feedback into the controller. A DC signal

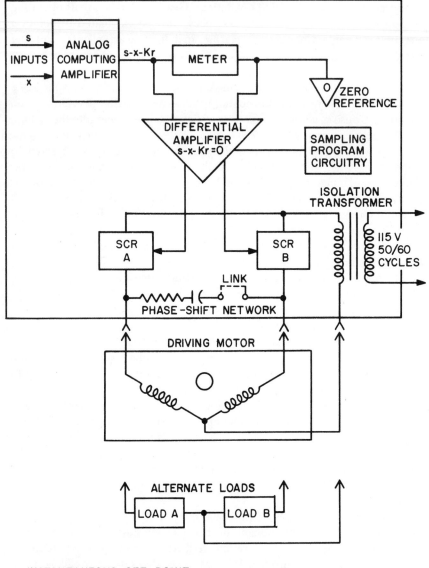

s = INSTANTANEOUS SET POINT

x = INSTANTANEOUS TRANSMITTER SIGNAL

K = ADJUSTABLE TRANSMITTER COMPENSATION

r = INSTANTANEOUS RATE OF CHANGE OF TRANSMITTER SIGNAL

Fig. 1.20hh   Proportional motor control circuit with position feedback. (Courtesy G. W. Dahl
Co., Inc.)

"x" (Figure 1.20hh) from a process transmitter provides the loop feedback. Transmitter and other lags are compensated for by the controller. Systems for ratio, cascade and feedforward control can also be incorporated.

The system is particularly adaptable to slow loop responses, such as temperature, as well as high-response systems, such as flow. A differential amplifier responds to the magnitude and the polarity of an internally modified error signal which triggers SCR's to obtain bidirectional drive. The synchronous motor can be driven in either direction, depending upon the relation to the setpoint.

An electric proportional actuator (Figure 1.20ii), is designed for con-

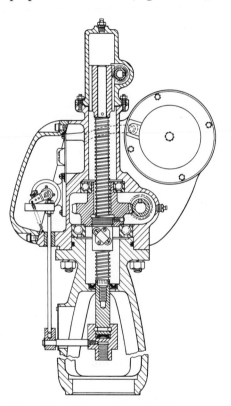

Fig. 1.20ii Electric proportional actuator
with continuously rotating drive sleeve

tinuous rotation of a drive sleeve on a ball-screw stem thread. Three thousand pounds of thrust is obtained at a stem speed of 1 in./min.

One or two DC signals are used separately or numerically added or subtracted. Triacs operate on position error to control a DC permanent magnet motor which positions a stem within an adjustable dead band. Degree of error and rate of return are sensed by a lead network to determine the direction and time that the motor must run.

Stem reversals are almost instantaneous. The back emf of the motor is used as a velocity sensor, and is fed into a circuit which allows adjustment of the speed of drive sleeve rotation. Gain can be adjusted to control oscillation of the stem. Stem position feedback is by an LVDT. Use of the DC motor allows for torque control through sensing of motor current. A manual handwheel is furnished which can only be used when the unit is de-energized by a manual/automatic switch.

13. *Rotating Armature.* An internally threaded drive sleeve in the armature of the motor is used to obtain a linear thrust up to 6,600 lbf at a rate of 10 in./min. Bearings in the end cap support the drive assembly (Figure 1.20jj).

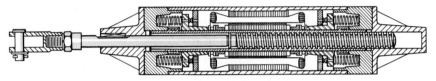

Fig. 1.20jj   Electric actuator with rotating armature

The drive stem is threaded to match the drive sleeve and is kept from rotating by a guide key. Thrust-limit switch assemblies are mounted in each end of the housing to locate the hollow shaft in mid-position. When the linear movement of the drive stem is restricted in either direction, the limit switch involved will operate to shut down the unit. Thermal cut-outs in the motor windings offer additional overload protection. Strokes are available from 2 to 48 in. The unit has been adapted for proportional control by use of an external sensing position for feedback. For use as a valve actuator, it must be mounted so that the drive stem can be attached to the valve stem, or a suitably threaded valve stem must be supplied.

Electric Quarter-Turn Actuators

Electric quarter-turn actuators are either linear or rotary, although rotary types predominate. A linear unit consists of the motor, gears, and a lead screw that moves the drive shaft (Figure 1.20 kk). A secondary gear system rotates cams to operate limit and auxiliary switches. The unit may have a brake motor for accurate positioning and a manual handwheel. The bracket on the rear end allows for the actuator to rotate on the pin of a saddle mount, so that the drive shaft can be pinned directly to the lever arm of a valve or a lever arm fixture that fits the shaft configuration of a plug cock or ball valve. Maximum output is 1,600 lbf at 5 in./min.

Rotary types are motor-gear boxes in various forms. Electrical gear must include adjustable cams to operate limit and auxiliary switches. A potentio-

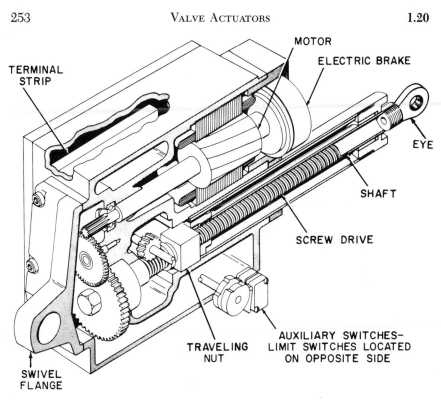

Fig. 1.20kk   Electric quarter-turn actuator with linear output

metric feedback calibrated to the rotation of the actuator is required for use with a control circuit. The unit shown, in Figure 1.20ll was developed to slide over and be keyed to the shaft of a boiler damper. Actuators of this type must be adaptable to mounting on and operating a variety of quarter-turn valves. The difficulty in setting limit switches for accurate stopping of the valve in the shut position emphasizes the advisability of incorporating a torque-limiting device to sense closure against a stop. Opening can be controlled by a limit switch. One compact unit contains the features noted with an output of 9,000 in.-lbf of torque. The unit is powered by a motor with a high-torque capacitor and includes a mechanical brake, feedback potentiometer, limit switches and a declutchable handwheel.

Actuators for larger outputs use externally mounted motors. Two motors operating a single gear train have been used to obtain 45,000 in.-lbf of torque through 90-degree rotation in 75 sec. An actuator has been designed for accurate rotary positioning that develops 5,000 ft-lbf at stall and will rotate 90 degrees in 10 sec. SCR's energize the motor as commanded by a servo-trigger assembly housed separately. Stem position is transmitted from a differential transformer for comparison with the demand input for error

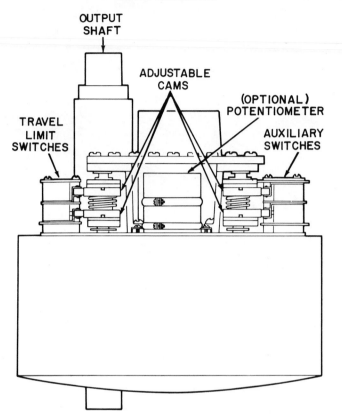

Fig. 1.20 ll   Electric quarter-turn actuator of the rotary type

determination. Large gear actuators, of the type used to drive gate valves through a threaded drive sleeve, are sometimes used for quarter-turn operation. The unit must be geared down externally, for a usable operating rate, or by use of geared or slow-speed motors.

An actuator is designed particularly for quarter-turn service (Figure 1.20mm). A permanent-magnet motor is used which is a direct-reversing, high-starting-torque type for use with AC or DC power. Use of a permanent magnet instead of wound field coils permits long life under on-off service. Instant break-away and efficient transfer of prime mover power to the valve stem are obtained with a modified, concentric planetary system consisting of a semiflexible gear within a rigid gear. A three-lobed cam transmits power to the gears by creation of a deflection wave transmission to cause a three-point mesh of the gearing teeth on 30 percent or more of the external gearing surface. A walking motion is created to amplify torque because of the engagement of a great area of the gear surface during transmission. Lubrication has been eliminated in this design.

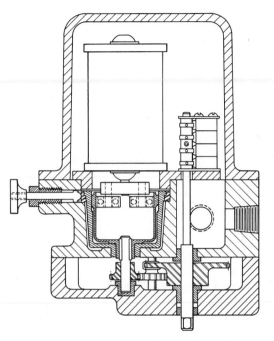

Fig. 1.20mm   Electric quarter-turn actuator with a
permanent-magnet motor

Although designed for 90-degree operation and up to 350 ft-lbf at 3-sec stroking time, the unit has been adapted to continuous operation with an output of up to 150 ft-lbf. The unit is available for use with various DC voltages and with 115 volts AC. They are equipped with positive declutching, instant manual override. Also offered are solid state variable-speed, constant torque control, battery pack, fail-safe controls, slide-wire position indication, and limit switches.

1.21 CONTROL VALVE SELECTION AND APPLICATION

The previous twenty sections dealt with the design features and operating characteristics of the various types of control valves. This section concerns the application considerations encountered by the instrument engineer in selecting and applying these devices. The basic areas discussed in this section include—

> Leakage
> Rangeability
> Cavitation
> Noise, Vibration and Flow Velocity
> High Pressure
> High Temperature
> Low Temperature
> Vacuum
> Low Flow
> Viscous and Slurry Streams

If one or more of these conditions apply to the contemplated installation, it is suggested that the corresponding paragraphs in this section be read before making the valve selection. Choosing the valve based only on Orientation Table I at the beginning of this chapter is not recommended in these cases because the table does not and cannot reflect all the intricate factors contributing to the proper selection.

Leakage

Any flow through a fully closed control valve when exposed to the operating pressure differentials and temperatures is referred to as leakage. It is expressed as a cumulative quantity over a specified time period for tight shutoff designs and as a percentage of full capacity for conventional control valves. Orientation Table I contains some quantitative data on shutoff and leakage characteristics for the various control valve designs. These will not be repeated here, but they should provide a reference to the reader on this matter.

One basic point is that control valves are usually designed considering their flow characteristics and other features related to control quality, and leakage is *not* one of these considerations. The functions of control and tight shutoff might therefore be separated and assigned to different valves rather than being combined into a single one. Emphasis on tight shutoff in a control valve would probably direct the attention of the application engineer toward the 90-degree-turn ball or plug type designs, which might not be the best selection from a control quality point of view, and control quality should not be compromised for leakage considerations. This rule too is dependent on what one means by tight shut-off. In most cases, even the average globe control valve will minimize leakage if the process fluid is viscous and has high surface tension. There would be no compromise involved in combining control and shut-off functions on such a service. At the same time, if bubble-tight shutoff is required for a hot, dry, low-molecular-weight gas with a possible vacuum on the downstream side of the valve, this would definitely call for putting control and shutoff functions into separate valves.

Factors Affecting Leakage

Some valve manufacturers list in their catalogs the valve coefficients applicable to the fully closed valve. For example, a butterfly valve supplier might list a C_v of 13.2 for a fully closed, metal-to-metal seated 24-in. valve. It should be realized that such figures apply only to new, clean valves operating at ambient conditions. After a few years of service valve leakage can vary drastically from installation to installation as affected by some of the factors to be discussed.

Soft Seats

One of the most widely applied techniques for providing tight shutoff over reasonable periods of time is by the use of soft seats. Standard materials used for such services include Teflon and Buna-N. Teflon is superior in its corrosion resistance and in its compatibility to high-temperature services up to 450°F. Buna-N is softer than Teflon but is limited to services at 200°F or below. Neither should be considered for static pressures of 500 PSIG or greater, for use with fluid containing abrasive particles, or if critical flow is expected at the valve seat.

Single vs Double Ports

New, clean, metal-to-metal seated double-ported globe valves under ambient conditions will usually leak more than their single seated counterparts. How much this leakage will be is affected by many factors, and therefore the only rule of thumb that can be offered is that if 2–3 percent leakage cannot be tolerated, then the double-ported globe should not be considered.

Leakage characteristics of other single-ported valve designs are given in orientation Table I.

Temperature Effects

It is frequently the case that either the valve body is at a different temperature than the trim or that the thermal expansion factor for the valve plug is different from the coefficient for the body material. It is usual practice in some valve designs (such as the butterfly) to provide additional clearance to accommodate the expansion of the trim, when designing for hot fluid service. The leakage will therefore be substantially greater if such a valve is used at temperatures below those for which it was designed.

Temperature gradients across the valve can also generate strains that promote leakage. Such gradients are particularly likely to exist in combining service three-way valves, when the two fluids involved are at different temperatures. This is not to imply that three-way valves are inferior from a leakage point of view. Actually their shutoff tightness is comparable to that of single-seated globe valves.

Pipeline Forces

Pipe strains on a control valve will promote leakage as a temperature gradient does. For this reason, it is important not to expose the valve to excessive bolting strains when placing it in the pipeline and to isolate it from external pipe forces by providing sufficient supports for the piping.

Seating Forces and Materials

The higher the seating force in a globe valve the less leakage is likely to occur. An average value is 50 lbf per linear inch of seat circumference. Where necessary, a much increased seating force will create better surface contact by actually yielding the seat material. Seating forces of this magnitude (about ten times the normal) are practical only when the port is small.

Seating materials are selected for compatibility with service conditions, and Stellite or hardened stainless steel is an appropriate choice for non-lubricating, abrasive, high-temperature and high-pressure-drop services. These hard surface materials also reduce the probability of nicks or cuts occurring in the seating surface, which might necessitate maintenance or replacement.

Rangeability

The terms rangeability and repeatability are probably the two most important and least understood expressions in the instrumentation profession. Rangeability as applied to control valves is usually defined in very vague terms, such as being the ratio of *maximum to minimum controllable* flows, where the word controllable implies that within this range the deviation from the specified *inherent flow characteristic* will not exceed some stated limits. The value of such a definition would be rather limited even if the limits of this deviation were internationally agreed upon. Therefore, the best thing to do is to review the subject from a common sense point of view. Some quantitative information on this subject concerning the various valve designs is given in Orientation Table I.

Rangeability is of interest for basically two reasons:

    a. It tells the point at which the valve is expected to act on-off or lose control completely due to leakage.

    b. It establishes the point at which the flow-lift characteristic starts to deviate from the expected.

Leakage and Clearance Flows

Leakage flow is the amount of flow through a fully closed valve. It is true that rangeability cannot exceed the ratio between maximum and leakage flow, but it can be and usually is *substantially less*. (Otherwise the rangeability of a soft-seated valve could be stated as infinity, which is obviously false.) Published values of leakage and clearance flows are based on tests performed on new valves at low pressures and ambient temperatures. These values can increase substantially due to seat damage or just due to the operating pressure and temperature conditions.

Clearance flow is that finite quantity of flow that occurs as the valve just barely starts to open. A soft-seated valve might not have leakage, but it too has some minimum clearance flow. Figure 1.21a illustrates the behavior of a typical linear valve with the clearance flow noted. This valve cannot maintain a flow rate *under* its clearance flow level because its behavior in this range is *inherently on-off*. It is also true, therefore, that rangeability cannot exceed the ratio between maximum and clearance flow either, but it can be and usually is markedly less. This is true because being above the clearance flow level only means that the valve will not act on-off, but it *does not yet* mean that it has the expected flow characteristics.

Valve Characteristics

As it was discussed in other parts of this chapter, it is very important in quality control that the control valve characteristics approximately match the controlled process. Therefore, if a linear valve is selected for a particular installation, the minimum controllable flow through that valve is defined as the flow which linearly relates to the corresponding lift. Figure 1.21a shows that a deviation from the design flow characteristics occurs at a much higher flow rate than clearance-flow.

Figure 1.21b schematically illustrates the flow characteristics of various other plug types. This plot on semilog coordinates allows one better to define the characteristics of the so-called *equal percentage valves*, which by definition will produce a change in flow for a unit change in lift, which is a fixed percentage of the flow rate at that point.

If unit sensitivity is defined as a 1 percent increase in flow, resulting from a 1 percent increase in lift, then in Figure 1.21b the percent ported valve is shown with a unit sensitivity of 4 percent flow change per percentage lift

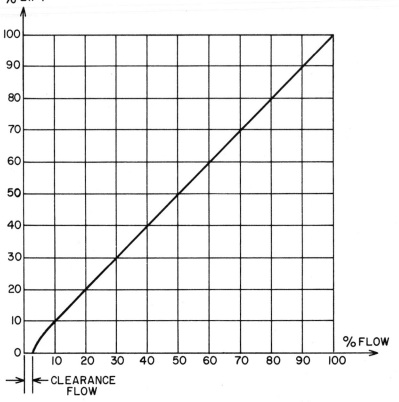

Fig. 1.21a   Linear flow characteristics

and the percent contoured with a 3 percent per percentage unit sensitivity. Figure 1.21c gives the idealized characteristics of various equal percentage valves.

By the use of unit sensitivity, it is possible to give quantitative limits of acceptable deviation from flow characteristics as a basis of defining valve rangeability. For example, if a particular process installation requires a valve with a 3 percent per percentage unit sensitivity, then the rangeability of this valve could be defined as *the ratio between those maximum and minimum flows which will stay within say 0.2% of the required unit sensitivity.* Once more, it is true that valve rangeability cannot exceed this ratio, but it can be less, because until now, this discussion has been limited to uninstalled control valves.

Installed Valve Behavior

When the control valve is installed as part of a process plant, its flow characteristics are no longer independent of the rest of the system. The fluid

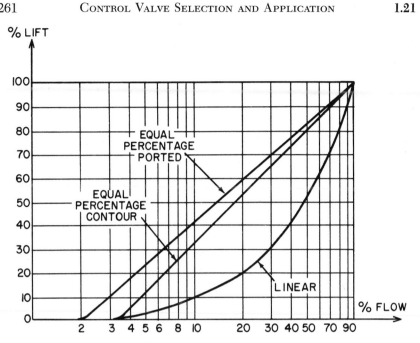

Fig. 1.21b   Various flow-lift characteristics

flowing through the valve is exposed to frictional resistances in series with that of the valve, and depending on the type of pump involved, there is a definite relationship between flow rate and pressure head generated. Equation 1.2(54) defined the resulting distortion coefficient, and Figures 1.2z and 1.2aa showed the resulting actual installed (distorted) flow characteristics.

From these curves, one can conclude that the particular installation

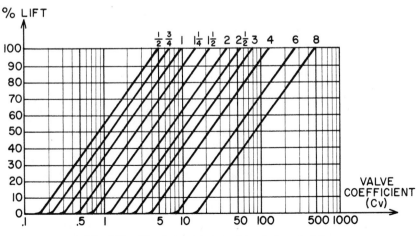

Fig. 1.21c   Equal percentage contoured characteristics

involved can have a very substantial effect on both flow characteristics and rangeability. *Clearance flow alone can increase as much as ten fold, and equal percentage characteristics can be distorted toward linear or even quick opening under adverse conditions.*

In summary, the reader is cautioned to use the type of rangeability information given in Orientation Table I with extreme care, fully analyzing the factors discussed in these paragraphs, and above all being very conservative with the *installed* rangeability expectations from those valves, which are depended on to maintain their flow characteristics for quality control. It is *not unusual* to install two valves in parallel when rangeability requirements exceed a 10 to 1 ratio and if control is of extreme importance.

Cavitation

The phenomenon of cavitation and its effect on valve capacity is discussed in Section 1.2 in connection with valve sizing. The purpose of this paragraph is to discuss means of eliminating cavitation and increasing the useful life of the valve under cavitating conditions.

There are several reasons why cavitation was given increasing attention during the 1960s. One of these is that with the development of new materials, chemical processes in general shifted toward operations using higher pressures and temperatures. Another reason is the development of high recovery valves, which tends to create problems of cavitation even under those conditions, where previously the standard globe valve did not experience such. Much research on cavitation has resulted in a better understanding of its causes and cures.

The flow velocity at the vena contracta of the valve tends to increase by obtaining its additional energy from the potential energy of the stream, causing a *localized* pressure reduction, which if it drops below the fluid's vapor pressure, results in temporary vaporization. (Fluids form cavities when exposed to tensions equal to their vapor pressure.) Cavitation damage always occurs downstream of the vena contracta when pressure recovery in the valve causes the temporary voids to collapse. Destruction is due to the *implosions*, which generate the extremely high-pressure shock waves in the substantially non-compressible stream. When these waves strike the solid metal surface of the valve or downstream piping, the damage gives a cinder-like appearance. Cavitation is usually coupled with vibration and a sound like rock fragments or gravel flowing through the valve.

Cavitation damage always occurs downstream of the vena contracta at the point where the temporarily formed voids implode. In case of flow to open valves, the destruction is almost always to the plug and seldom to the seat.

Methods to Eliminate Cavitation

Because no known material can remain indefinitely undamaged by severe cavitation, the best and the only sure solution is to eliminate it completely.

Cavitation can be reduced or eliminated by several methods, listed in the following paragraphs.

Revised Process Conditions

A slight reduction of operating temperature can usually be tolerated from a process point of view, and this might be sufficient to lower the vapor pressure sufficiently to eliminate cavitation. Similarly, increased upstream and downstream pressures, with ΔP unaffected, can also relieve cavitation. For this reason, it is essential that such control valves be installed at the lowest possible elevation in the piping system.

If cavitating conditions are unavoidable from the process conditions point of view, then it is actually preferred to have not only cavitation, but also some permanent vaporization—flashing—through the valve. This can usually be accomplished by a slight increase in operating temperature or by decreasing the outlet pressure. Flashing is preferred, because the pressure shock waves due to the cavity implosions during pressure recovery are not as damaging in a compressible medium as they are in a pure liquid.

Revised Valve

Where the operating conditions cannot be changed, it is logical to review the type of valve selected in terms of its pressure recovery characteristics. (Table 1.2a gives valve recovery coefficients.) The more treacherous the flow path through a particular valve design, the closer is its recovery coefficient (C_f) to unity and the less likelihood exists for cavitation. Equation 1.2(49) approximately defines the point of incipient cavitation. The "self-drag trim" described in Figure 1.5m is particularly suited to eliminate cavitation problems, since it takes most of the pressure drop in the etched disc passages and not in the throttling trim.

Revised Installation

In order to eliminate cavitation, it is possible to install two or more control valves in series. Cavitation problems can also be alleviated by absorbing some of the pressure drop in breakdown orifices or in partially open block valves upstream to the valve. The amount of cavitation damage is related to the 6th power of flow velocity or to the 3rd power of pressure drop. This is the reason why reducing ΔP by a factor of 2, for example, will result in an 8-fold reduction in cavitation destruction.

Material Selection for Cavitation

If it is impossible to eliminate cavitation completely, then it is important to select construction materials that will provide reasonably long life. Special consideration is usually given to the trim materials when cavitation is expected, whereas valve body material is seldom selected on this basis.

Table 1.21d compares the cavitation resistance of a number of trim and

Table 1.21d
RELATIVE RESISTANCE OF VARIOUS MATERIALS TO CAVITATION

Possible Trim and/or Valve Body Materials	Relative Cavitation Resistance Index	Approximative Rockwell C Hardness Values	Corrosion Resistance	Cost
Aluminum	1	0	Fair	Low
Synthetic Sapphire	5	Very high	Excellent	High
Brass	12	2	Poor	Low
Carbon Steel, AISI C1213	28	30	Fair	Low
Carbon Steel, WCB	60	40	Fair	Low
Nodular Iron	70	3	Fair	Low
Cast Iron	120	25	Poor	Low
Tungsten Carbide	140	72	Good	High
Stellite #1	150	54	Good	Medium
Stainless Steel, Type 316	160	35	Excellent	Medium
Stainless Steel, Type 410	200	40	Good	Medium
Aluminum Oxide	200	72	Fair	High
K-Monel	300	32	Excellent	High
Stainless Steel, Type 17-4 PH	340	44	Excellent	Medium
Stellite #12	350	47	Excellent	Medium
Stainless Steel, Type 440C	400	55	Fair	High
Stainless Steel, Type 329, Annealed	1,000	45	Excellent	Medium
Stellite #6	3,500	44	Excellent	Medium
Stellite #6B	3,500	44	Excellent	High

valve body materials. This table shows that hardness alone is no indication of cavitation resistance. While it is important to consider cavitation in selecting the materials of construction for a valve, other factors, such as resistance to corrosion, erosion and abrasion and temperature limitations, should not be overlooked.

The best overall selection for cavitation resistance is Stellite 6B (28 percent chromium, 4 percent tungsten, 1 percent carbon, 67 percent cobalt). This is a wrought material and can be welded to form valve trims in sizes up to 3 in. Stellite 6 is used for hard-facing of trims and has the same chemical composition but less impact resistance. Correspondingly, its cost is lower.

In summary, the applications engineer should first review the potential methods of eliminating cavitation. These would include adjustment of process conditions, revision of valve type or change of installation layout. Only if none of these techniques guarantee the complete elimination of cavitating conditions should the design engineer consider accepting the presence of this phenomenon and select the trim and body materials for compatibility with reasonably long life. These materials are selected only on a relative basis, because, in an absolute sense, no known material remains indefinitely undamaged by cavitation.

Noise, Vibration and Flowing Velocity

The subject matters of flowing velocity, vibration and control valve noise are all interrelated. When the application engineer is interested in limiting flowing velocity or vibration in the valve, he is concerned not only with the resulting noise, but also with other by-products, such as mechanical failure and shock wave formation. At the same time, reducing the operating noise level of a control valve is likely to increase its useful life.

The Phenomena of Sound and Noise

Sound is the sensation produced when the human ear is stimulated by a series of pressure fluctuations transmitted through the air or other medium. Sound is described by specifying the magnitude and frequency of these fluctuations. The magnitude is expressed as the level of sound pressure having the units of decibels (db). This is a logarithmic function related to the ratio between an existing sound pressure and a reference sound pressure. The reference sound pressure has been selected as 0.0002 microbars, which in more frequently used engineering units is equivalent to 0.29×10^{-8} PSI.

$$\text{decibels} = 20 \log \frac{\text{existing sound pressure}}{0.0002 \text{ microbars}} \qquad 1.21(1)$$

Figure 1.21e illustrates the relationship between loudness and the decibel readings in a certain area. This illustration can be slightly misleading, because apparent loudness to the human ear is dependent on *both* the sound pressure level and its frequency. Sounds of equal sound pressure levels will appear to be louder as the frequency approaches 2,000 Hz. Apparent loudness in phons is defined as the sound pressure in decibels of a pure tone having a frequency of 1,000 Hz. For pure tones, the equal loudness contour in Figure 1.21f shows the pressure level required for the tone to sound as loud as the corresponding reference tone of 1,000 Hz.

Table 1.21g relates the various ways of describing the frequency of a sound and also notes the approximate frequency levels usually generated by the various noise sources in a control valve. Figure 1.21h gives some approximate indication of the noise damage risk as a function of sound pressure and frequency.

Controlling Noise

The transmission of a noise requires a source of sound, a medium through which this sound is transmitted and a receiver. Each of these three factors can be changed to reduce the noise level. In some cases, such as when the

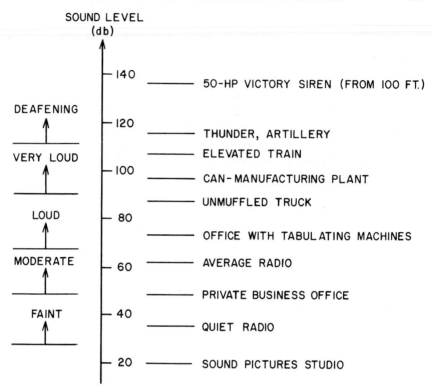

Fig. 1.21e   Relationship between loudness and decibel reading

noise is caused by vibrating control valve components, the vibrations must either be eliminated or result in valve failure. In other cases, such as when the source of noise is the hiss of a gas-reducing station, the accoustical treatment of the noise medium is sufficient. Figure 1.21i illustrates a silencer design that can be installed downstream from a gas-regulating valve; due to the resulting accoustical attenuation, it reduces the sound pressure by a factor of five (from 96 to 82 db).

The higher the frequency of vibration, the more effective are the commercially available sound absorption materials. Figure 1.21j is an example of accoustical treatment for the outside of a pipe. It is also possible to cover the inside walls of the building with sound-absorbing materials to prevent the reflection and radiation of the sound waves. It is important to seal all openings: a 12-in.-thick concrete wall will lose 95 percent of its effectiveness as a sound block if a 1-in.-diameter hole is bored through it.

Personnel can be restricted from noisy areas or required to wear ear-protective devices when in them (controlling the receiver), but this approach is unlikely to be applied in the process industry. The most effective and

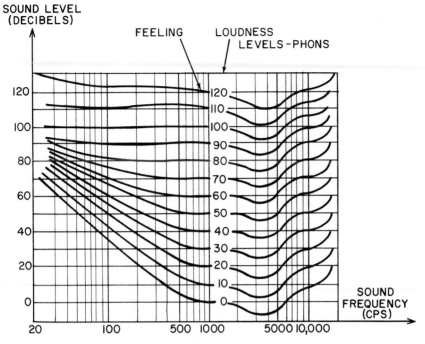

Fig. 1.21f Apparent loudness contours

Table 1.21g
SOUND FREQUENCIES AND SOURCES IN VALVES

Frequency (Hz)	Octave Band Number	Sound Description	Typical Noise Source in Valves
20–75	1	Rumble	Vertical plug oscillation
75–150	2		Cavitation°
150–300	3	Rattle	
300–600	4	Howl	Horizontal plug vibration
600–1,200	5		
1,200–2,400	6	Hiss	Flowing gas
2,400–4,800	7	Whistle	
4,800–7,000	8	Squeal	Natural frequency vibration
20,000 and up		Ultrasonic	

° The noise accompanying cavitation varies and can be much higher in frequency than noted here.

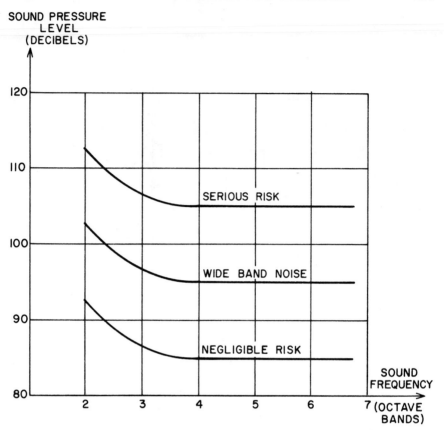

Fig. 1.21h   Hearing damage risk chart. (From "Control Valve Noise," C. F. King, Fisher Governor Co., Marshalltown, Iowa.)

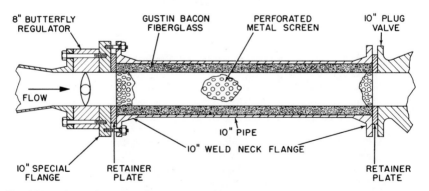

Fig. 1.21i   Silencer for gas-regulating stations. (From "Control Valve Noise," C. F. King, Fisher Governor Co., Marshalltown, Iowa.)

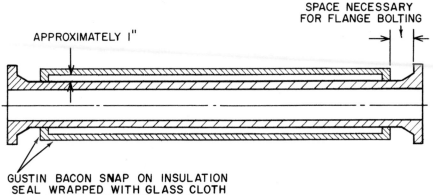

GUSTIN BACON SNAP ON INSULATION
SEAL WRAPPED WITH GLASS CLOTH
IMPREGNATED WITH RESIN

Fig. 1.21j   Accoustical treatment of pipe walls

desirable technique is the treatment of the noise source, and this subject will be covered in the following paragraphs.

Sources of Control Valve Noise

Noise can be generated in a control valve as a consequence of several conditions, including

1. High-velocity gas flow
2. Liquid cavitation
3. Natural frequency vibration
4. Horizontal plug vibration
5. Vertical plug oscillation

Limiting High-Velocity Gas Flow

Very high gas or vapor velocities not only generate noise in the control valve but also cause erosion, reduce stability, shorten life, impair general performance and frequently cause vibration and damage to the valve trim. Recognition of this led to velocity limit recommendations which were rather arbitrary and involved the setting of maximums for inlet and/or outlet velocities in regulator stations.

Later investigations suggested that it was the outlet velocity which should be kept below sonic, so that supersonic velocities would be avoided farther downstream. The argument behind this recommendation[1] was the desirability of retaining the throttling (pressure-reducing) process inside the reducing valve instead of allowing it to take place partially in the downstream reducer, i.e., the pressure reduction should not be completed in the pipe reducer.

[1] "Why Limit Outlet Velocities in Reducing Valves," Hans D. Baumann, Instruments & Control Systems, Sept. 1965.

Throttling in the reducer is undesirable for the following reasons:

a. If sonic velocity is exceeded at the valve outlet, then the static pressure at that point is going to be higher than the regulator setting, which therefore can expose the downstream piping to pressures higher than those for which it was designed. Such condition is unsafe and should be avoided.

b. When sonic velocity exists at the valve outlet, the pipe expander acts as a supersonic diffuser expanding the gas and increasing its velocity to supersonic levels. The downstream piping is usually designed for a velocity of around 6,000 fpm, and therefore near the end of the expander the supersonic gas stream must be decelerated. (Sonic velocity in air corresponds to 67,560 fpm.) This sudden conversion of velocity head into static head is accomplished by a process of turbulent friction, called *shock waves*, which besides being noisy, also can contribute to vibration damage.

Based on these considerations, it is suggested that *the regulator outlet be sized as a secondary orifice* by selecting its diameter to correspond to a subsonic valve outlet velocity. Under subsonic conditions, the downstream expander causes a decrease in velocity as flow area increases, which is just the opposite of the phenomenon described for supersonic conditions. To determine the regulator outlet diameter required for subsonic outlet velocities, the following equations can be used:

$$\text{for steam:} \quad D = \sqrt{W/(36P_2)} \qquad\qquad 1.21(2)$$

$$\text{for gas:} \quad D = \sqrt{W/(90P_2 \sqrt{G_f})} \qquad\qquad 1.21(3)$$

$$D = \sqrt{(Q\sqrt{G})/(1,100P_2)} \qquad\qquad 1.21(4)$$

where D = outlet port diameter (in.),
G = specific gravity relative to air,
G_f = specific gravity at flowing temperature,
P_2 = downstream pressure (PSIA),
Q = volumetric flow (SCFH) and
W = mass flow (1bm/hr).

Reducing Gas Regulator Noise

At subsonic velocities, the sound power of a gas jet increases with the eighth power of the velocity. At supersonic velocities the aerodynamic sound pressure level increases with the sixth power of gas velocity. The noise consideration can be another reason for limiting the valve outlet velocity as discussed in the preceding paragraph. Considering the subsonic relationship between noise and speed, if the outlet gas velocity is reduced from say 60,000 fpm to 15,000 fpm, this should reduce the corresponding sound pressure level of noise by a factor of $4^8 = 65,536$.

Noise reduction by decreasing outlet velocity is a sound approach, but the cost and control quality implications of the use of oversized regulators should not be overlooked either. The availability of oversized outlet connections on conventional regulators places some additional restrictions on this approach.

Table 1.21g shows that the frequency of this aerodynamic noise usually falls between 1,200 and 4,800 Hz, which is a hissing sound.

From the above, it is evident that one of the most effective methods of reducing regulator noise is by decreasing the gas velocity. Another potential method of "source treatment" is to reduce the number of valve ports, since single-seated valves tend to be quieter than double-ported ones.

Aerodynamic noise is usually reduced by treatment of the sound path. This might involve the use of heavy wall piping, eliminating the sudden expansions or contractions in the conduit, burying the piping underground, acoustical treatment of building and pipe walls (Figure 1.21j) and the installation of silencers (Figure 1.21i) or multiple orifice restrictors (Figure 1.21k), which are sized to keep the average velocity low.

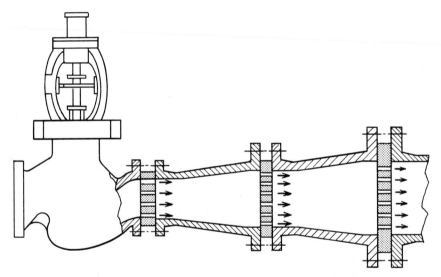

Fig. 1.21k Use of multiple-orifice restrictors for noise reduction

Reducing Cavitation Noise

The noise caused by the imploding bubbles in a cavitating liquid stream can vary from a low-frequency rumble to a high-frequency squeal. The elimination of this noise is achieved by eliminating the cavitation itself. The methods for accomplishing this were discussed earlier in this section.

Natural Frequency Vibration

The control valve plug and stem can be compared to a mass and spring assembly, where the plug represents the mass and the stem acts as the spring.

The natural frequency of such system is a function of the mass size and of the spring constant. In other words, the natural frequency of a plug and stem assembly can be changed by changing either the mass of the plug (size or type change) or the diameter of the stem.

Natural frequency vibration in a control valve usually occurs at low-pressure-drop conditions when the amount of energy dissipated by the valve is small. Its cause might be the intricate upstream piping configuration, cavitation inside the valve or other reasons. The accompanying noise is usually in the 3,000 to 7,000 Hz range.

Natural frequency vibration should be eliminated because it is likely to cause either mechanical failure or failure from overheating. If the valve plug is a solid casting or forging, the probability of natural frequency vibration is substantially reduced. Changing the plug style or, if necessary, its size usually corrects the condition. The natural frequency of the plug-stem assembly can also be affected by the type of guiding involved, such as top guiding vs top-bottom guiding, and by the number of ports in the valve.

The causes of the natural frequency vibration of the plug should be determined, because they are usually symptoms of inferior system design.

Horizontal Plug Vibration

The horizontal impingement of the flowing stream on the valve plug can cause vibrations at frequencies other than the natural frequency of the plug-stem assembly. This usually occurs below 1,500 Hz as the plug vibrates horizontally between the guide bushings. In addition to noise, this can cause mechanical damage and eventual failure.

Minimizing the clearance between guide posts and guide bushings and applying hard surfacing (such as Stellite) is practically all that the valve manufacturer can do to avoid this vibration. If that is not sufficient to eliminate it, then the upstream piping configuration, the flow direction through the valve and other factors affecting the fluid's contact with the valve plug should be investigated.

Vertical Plug Oscillation

The vertical impingement of the flow on the valve plug can also generate vibration. Its frequency is a function of the mass-spring rate relationship between the valve and its actuator and is usually less than 100 Hz.

In case of single-seated valves with flow-to-close operation, severe *hydraulic hammer* can occur on liquid service when the plug is positioned close to the seat. Flow-to-close flow patterns should be avoided, because even if they do not cause vertical vibration, they are likely to contribute to instability when the valve is in near-closed position, due to the "bathtub effect." In double-seated valves, the problem of vertical oscillation is usually less pronounced, but an axial force can still be produced by the unequal pressure drops across the two seats. In some cases the damping characteristics of the actuator prevent this type of vibration.

In installations where the damping effect of the actuator alone is not sufficient, the recommended method to suppress vertical oscillation is to install a hydraulic snubber between the yoke and the diaphragm casing. One such snubber is shown on Figure 1.21l. The damping action is produced by the

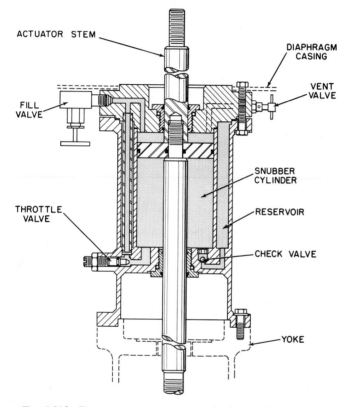

ACTUATOR STEM

DIAPHRAGM
CASING

VENT
VALVE

FILL
VALVE

SNUBBER
CYLINDER

THROTTLE
VALVE

RESERVOIR

CHECK VALVE

YOKE

Fig. 1.21l Dampener to prevent vertical plug oscillation. (From "Inner-Valve Instability," Kieley & Mueller, Inc., Middletown, N.Y.)

restricted oil flow from one side of the piston to the other when the stem speed is high. At normal stem velocity, it has practically no effect.

The following categories of valves usually require a hydraulic snubber:

    a. All flow-to-close, single-seated valves except angle valves or the ones with sizes under 1 in.

    b. Double-seated valves on gas or vapor service, if the product of inlet pressure and valve size is 1,800 or greater. (6″ valve with 300 PSIG inlet pressure, for example.)

    c. Double-seated valves on liquid service if the inlet velocity exceeds
       15 fps.

High-Pressure Service

The Orientation Table I at the beginning of this chapter lists the pressure ranges for which the various control valves are designed, and Section 1.3 provides further information on the pressure ratings of globe valves. The discussion here will complement this material in elaborating further on some aspects of control valves for high-pressure service.

High pressure has different meanings according to the industry using it. In the food or fertilizer industry 1,000 PSIG would probably be considered high pressure. In the gas or utility industry, pressures up to 10,000 PSIG are not uncommon. In the production of high-density polyethylene, pressures in the range of 50,000 PSIG are usually encountered. The demand for control valves at even higher pressure levels is rather limited, because no industrial process except diamond synthesis is known to operate at pressures exceeding 200,000 PSIG.

Design of high-pressure valves usually necessitates at least three considerations:

    a. Increased physical strength
    b. Selection of erosion-resistant material
    c. Use of special seals

Both the body and the moving parts should be strong. Valve bodies for high-pressure service are usually forged to provide homogeneous materials free of voids and with good mechanical properties. The loads and stresses on the valve stem are also high. For this reason, higher strength materials are used with increased diameter. As shown in Figure 1.21m, they are usually kept short and are well guided to prevent column action.

High operating pressure frequently involves high pressure drops. This usually means erosion, abrasion or cavitation at the trim. For the selection of the proper plug and seat materials on cavitating service, Table 1.21d can be consulted, but cavitation and erosion resistance are usually *not* properties of the same metal. Materials resistant to erosion and abrasion include 440C stainless steel, flame-sprayed aluminum oxide coatings (Al_2O_3) and tungsten carbide. On the stem, where the unit pressure between it and the packing is high, it is usually sufficient to chrome plate the stem surface to prevent galling. Special "self-energizing" seals are used with higher pressure valves (above 10,000 PSIG service) so that the seal becomes tighter as pressure rises. Popular body seal designs for such service include the delta ring closure and the Bingham closure (Figure 1.21n).

The angle valve designs illustrated in Figures 1.21m and n are not restricted to high-pressure service, and straight flow-through patterns are

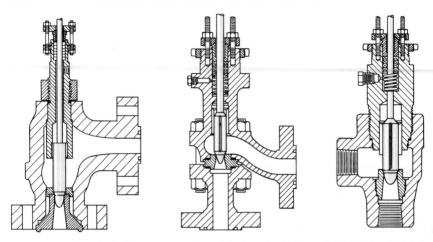

Fig. 1.21m   Valves for high-pressure service with body ratings between 2,500 PSIG and 10,000
PSIG

equally available. The design features of the self-drag trim valve are discussed
in Section 1.5.

The methods by which the self-energizing seals are used in connecting
the high-pressure valves into the pipeline are shown in Figure 1.21o. These
designs depend on the elastic or plastic deformation of the seal ring at high
pressures for self energization.

Special packing design and material are also required in high-pressure
service, because conventional packings would be extruded through the clearances. To prevent this, the clearance between stem and packing box bore is
minimized, and extrusion-resistant material, such as Teflon-impregnated glass,
is used for packing.

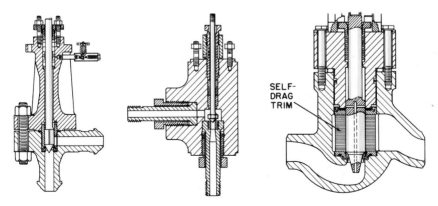

Fig. 1.21n   Valves for high-pressure service with body ratings between 10,000 PSIG and 60,000
PSIG

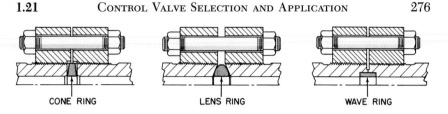

CONE RING   LENS RING   WAVE RING

Fig. 1.21o   Line connections for the pipeline installation of high-pressure valves

High-pressure services increase the probability of noise, vibration and cavitation, and the corresponding paragraphs in this section should be reviewed.

High-Temperature Service

The Orientation Table I lists the temperature limitations of various valve designs. Section 1.4 provides further information on high-temperature bonnets and packings for globe valves. The discussion below will complement this information in elaborating on the requirements of valves in high-temperature service.

All process conditions involving operating temperatures *in excess of 450°F* are considered high temperature. There is a limited number of reports concerning maximum temperatures at which control valves have been successfully installed, but they do indicate satisfactory operation at temperatures up to 2,500°F.

High operating temperatures necessitate the review of at least three aspects of valve design:

    a. Temperature limitations of metallic parts
    b. Packing temperature limitations
    c. Use of jacketed valves

Limitations of Metallic Parts

The high operating temperatures are considered in selecting materials for both the valve body and trim. For the body, it is suggested that bronze and iron be limited to services under 400°F, steel to operation below 850°F, and the various grades of stainless steel, Monel, nickel or Hastelloy alloys to temperatures up to 1,200°F.

For the valve trim, 316 stainless steel is the most popular material, and it can be used up to 750°F. For higher temperatures, the following trim materials can be considered:

    17-4 PH stainless steel, up to 900°F, tungsten carbide, up to 1,200°F, Stellite or aluminum oxide, up to 1,800°F.

At high temperatures, the guide bushings and guide posts tend to wear excessively, and this can be offset by the selection of proper materials. Up to 600°F, 316 stainless steel guide posts in combination with 17-4 PH stainless steel guide bushings give acceptable performance. If the guide posts are surfaced with Stellite, the above combination can be extended up to 750°F service. At operation over 750°F, both the posts and the bushings require Stellite.

Packing Limitations

The various types of standard packings and their temperature limitations are covered in Section 1.4. In general, the packing temperature limitation for most non-metallic materials is in the range of 400°F to 550°F, the maximum temperature for metallic packings is around 900°F, and Teflon should not be exposed to temperatures above 450°F.

The relationship between the process and the packing temperature is a function not only of the type of bonnet used, but also of the physical relations between valve and bonnet. (Bonnet designs are described in Section 1.4.) The heat transmitted by the process to the packing travels by conduction (through the metal), convection (through the process fluid) and sometimes also by radiation. Some application engineers prefer the use of bellows seal bonnets with an auxiliary stuffing box and vent connection to minimize the heat transmitted by convection through the process fluid.

Recent investigations seem to indicate that the value of using radiation fins instead of the extended bonnets is much less than the effectiveness of *mounting the bonnet below the valve.* Fig. 1.21p shows that with liquid service, with the bonnet above the valve, the packing will be exposed to the full process temperature due to the natural convection of heat in the bonnet cavity. If the bonnet is mounted below the valve, no convection occurs and the heat from the process fluid is transferred by conduction in the bonnet wall only. Therefore, by this method of mounting, the allowable process temperature is substantially increased.

Figure 1.21q provides a method for determining packing temperature, with gas or vapor service, with the bonnet above the valve. With vapor service, it is likely that vapors will initially condense on the wall of the bonnet, lowering the temperature to the saturation temperature of the process fluid, but in some cases the heat conducted by the metallic bonnet wall will be sufficient to prevent this condensation from occurring. In short, the packing temperature will be at *or above* saturation temperature (T_s) with vapor service, but if the bonnet is mounted below the valve, the packing temperature is substantially reduced, due to the accumulated condensate. Table 1.21p shows that, if the bonnet is below the valve, the relationship between process and packing temperature is not affected by the phase of the process fluid.

Table 1.21r lists some factors which relate effectiveness of the various bonnet designs in reducing packing temperature. The smaller the value of F, the more effective the bonnet will be.

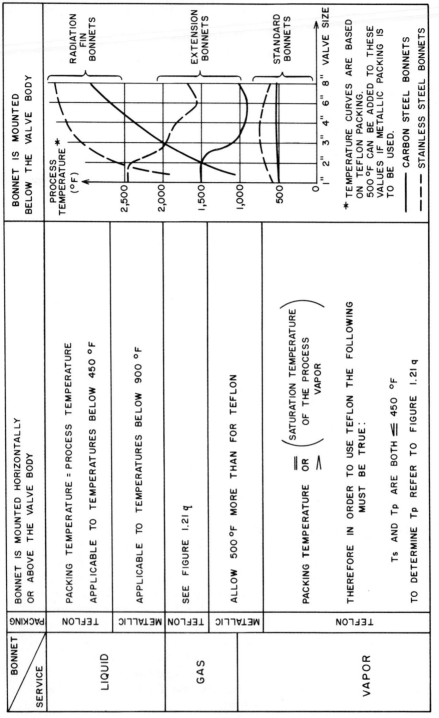

BONNET SERVICE	PACKING		
LIQUID	TEFLON	BONNET IS MOUNTED HORIZONTALLY OR ABOVE THE VALVE BODY PACKING TEMPERATURE = PROCESS TEMPERATURE APPLICABLE TO TEMPERATURES BELOW 450 °F	
	METALLIC	APPLICABLE TO TEMPERATURES BELOW 900 °F	
GAS	TEFLON	SEE FIGURE 1.21 q	
	METALLIC	ALLOW 500 °F MORE THAN FOR TEFLON	
VAPOR	TEFLON	PACKING TEMPERATURE $\;\substack{= \\ >}\;\left(\substack{\text{SATURATION TEMPERATURE} \\ \text{OR} \\ \text{OF THE PROCESS} \\ \text{VAPOR}}\right)$ THEREFORE IN ORDER TO USE TEFLON THE FOLLOWING MUST BE TRUE: T_s AND T_p ARE BOTH $\leqq$ 450 °F TO DETERMINE T_p REFER TO FIGURE 1.21 q	

Fig. 1.21p Maximum allowable process temperatures for various combinations of service, packing and bonnet types

278

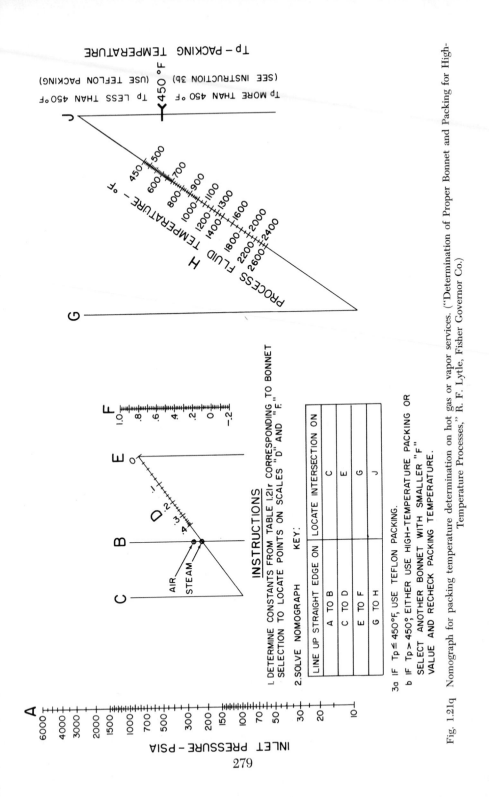

Fig. 1.21q Nomograph for packing temperature determination on hot gas or vapor services. ("Determination of Proper Bonnet and Packing for High-Temperature Processes," R. F. Lytle, Fisher Governor Co.)

INSTRUCTIONS

1. DETERMINE CONSTANTS FROM TABLE 1.21r CORRESPONDING TO BONNET SELECTION TO LOCATE POINTS ON SCALES "D" AND "F".

2. SOLVE NOMOGRAPH KEY:

LINE UP STRAIGHT EDGE ON	LOCATE INTERSECTION ON
A TO B	C
C TO D	E
E TO F	G
G TO H	J

3a IF Tp ≤ 450°F, USE TEFLON PACKING.

b IF Tp > 450°, EITHER USE HIGH-TEMPERATURE PACKING OR SELECT ANOTHER BONNET WITH SMALLER "F" VALUE AND RECHECK PACKING TEMPERATURE.

Table 1.21r
BONNET CHARACTERISTICS°

Valve Size	Bonnet Material°°	Bonnet Factors	Standard Bonnet	Extension Bonnet	Radiation Fin Bonnet
1″	CS	D	0.07	0.29	
		F	.83	.25	
	SS	D	.13	.33	
		F	.65	.15	
1½″	CS	D	.07	.24	0.25
		F	.83	.39	.35
	SS	D	.12	.30	.31
		F	.69	.22	.21
2″	CS	D	.11	.29	
		F	.72	.25	
	SS	D	.17	.33	
		F	.55	.15	
3″	CS	D	.10	.24	.31
		F	.74	.39	.20
	SS	D	.17	.31	.34
		F	.57	.21	.11
4″	CS	D	.07	.25	
		F	.83	.36	
	SS	D	.14	.31	
		F	.65	.20	
6″	CS	D	.04	.23	
		F	.90	.40	
	SS	D	.08	.30	
		F	.80	.22	
8″	CS	D	.06	.23	.33
		F	.86	.40	.14
	SS	D	.11	.30	.35
		F	.72	.22	.10

° "Determination of proper bonnet and packing for high-temperature processes." R. F. Lytle, Fisher Governor Company.
°° CS: carbon steel, SS: stainless steel.

Jacketed Valves

Orientation Table I at the beginning of this chapter shows that a number of control valve designs are available with heat transfer jackets. Those which are not standard with this feature can be traced or jacketed by the user. The purpose in installing a jacketed valve can be for either cooling or heating.

When a cooling medium is circulated in the jackets, this is usually done to lower the operating temperature of the heat-sensitive working parts. Such jacketing is particularly concentrated onto the bonnet, so that the packing temperature is reduced relative to the process. For certain operations at very high temperatures, intermittent valve operation is recommended, such that when the valve is closed it is cooled by the jacket, and when it is opened, it is kept open only long enough to prevent temperature equalization between the valve and the process.

Heating jackets with steam or hot oil circulation are usually used to prevent the formation of cold spots in the more stagnant areas of the valve or where the process fluid otherwise would be exposed to relatively large masses of cold metal. Figure 1.21s shows one of these valves, designed to prevent localized freezing or decomposition of the process fluid due to cold spots.

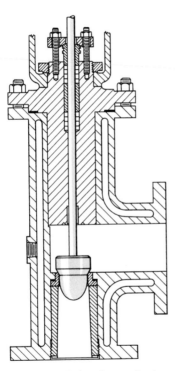

Fig. 1.21s   Jacketed control valve
for high-temperature service

Low-Temperature Service

Orientation Table I at the beginning of this chapter lists the temperature limitations of the various valve designs. The discussion below will complement this information by elaborating on the requirements of valves in cryogenic service. Cryogenic service is usually defined as temperatures below −150°F. Properties of some cryogenic fluids are listed in Table 1.21t.

Valve materials for operation at temperatures down to −450°F include copper, brass, bronze, aluminum, 300 series stainless steel alloys, nickel, Monel,

Table 1.21t
PROPERTIES OF CRYOGENIC FLUIDS

Properties Gas	Boiling Point (°F)	Critical Temperature (°F)	Critical Pressure (PSIA)	Heat of Vaporization at Boiling Point (BTU/lbm)	Density (lbm/ft³)		
					Gas at Ambient Conditions	Vapor at Boiling Point	Liquid at Boiling Point
Methane	−259	−117	673	219	0.042	0.111	26.5
Oxygen	−297	−181	737	92	0.083	0.296	71.3
Fluorine	−307	−200	808	74	0.098	—	94.2
Nitrogen	−320	−233	492	85	0.072	0.288	50.4
Hydrogen	−423	−400	188	193	0.005	0.084	4.4
Helium	−452	−450	33	9	0.010	1.06	7.8

Durimet and Hastelloy. The limitation on the various steels falls between 0°F and −150°F, with cast carbon steel representing 0°F and 3½ percent nickel steel being applicable to −150°F. Iron should not be used below 0°F.

Conventional valve designs can be used for cryogenic service with the proper selection of construction materials and with an extension bonnet to protect the packing from becoming too cold. The extension bonnet is usually installed vertically so that the boiled-off vapors are trapped in the upper part of the extension, which provides additional heat insulation between the process and the packing. If the valve is installed in a horizontal plane, a seal must be provided to prevent the cryogenic liquid from entering the extension cavity. When the valve and associated piping are installed in a large box filled with insulation ("cold box"), this requires an unusually long extension in order to keep the packing box in a warm area.

Although conventional valves with proper materials can be used on cryogenic service, there are valves designed specifically for this application. The unit shown on Figure 1.21u has some characteristics which make it superior to conventional valves on cryogenic service. Most important is its small body mass, which assures a small heat capacity and therefore a short "cool-down period." In addition, the inner parts of the valve can be removed

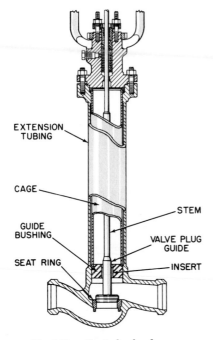

EXTENSION
TUBING

CAGE

STEM

GUIDE
BUSHING

VALVE PLUG
GUIDE

SEAT RING

INSERT

Fig. 1.21u   Control valve for cryogenic service

without removing the body from the pipeline, and if the valve is installed in a cold box, no leakage can occur inside this box because there are no gasketed parts.

The most effective method of preventing heat transfer from the environment into the process is by vacuum jacketing the valve and piping (Figure 1.21v). The potential leakage problems are eliminated by the fact that there are no gasketed areas inside the jacket.

For cryogenic services where tight shutoff is required, Kel-F has been found satisfactory as a soft seat material.

Vacuum Service

Orientation Table I lists some of the minimum pressures that the various valve designs can handle. The term "can handle" is relative and is usually qualified in terms of leakage limitations. In some processes, the in-leakage from the atmosphere results in overloading the vacuum source; in others it represents a contamination that cannot be tolerated. Potential leakage sources include all *gasketed areas* and, to an even greater extent, the locations where *packing boxes* are used to isolate the process from the surroundings.

For vacuum service, valves that do not depend on stuffing boxes to seal the valve stem generally give superior performance. Such designs include the

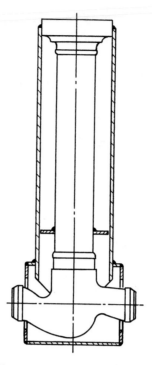

Fig. 1.21v   Vacuum jacketing of
cryogenic valve

various Saunders valves (Section 1.10) and pinch valves (Section 1.13). These designs unfortunately are limited in their application by their susceptibility to corrosion and their temperature and control characteristics. Their applicability to vacuum service is further limited by their design. The jacketed pinch valve versions, for example, require a vacuum source on the jacket side for proper operation, and the mechanically operated pinch and Saunders designs are limited in their capability to open the larger size units against high vacuum on the process side. The vacuum process tends to keep the valve closed, and this can result in the diaphragm breaking off the stem and rendering the valve inoperative.

For services requiring high temperatures and corrosion-resistant materials, in addition to good flow characteristics and vacuum compatibility, conventional globe valves can be considered, with special attention given to the type of seal used.

One approach to consider is the use of double packing, as shown in Figure 1.21w. The space between the two sets of packings is evacuated so that air leakage across the upper packing is eliminated. The vacuum pressures on the two sides of the lower packing are approximately equal, and therefore there

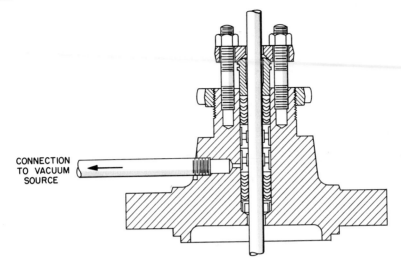

CONNECTION
TO VACUUM
SOURCE

Fig. 1.21w   Double packing to seal process under vacuum

is no pressure differential to cause leakage across it. Usually the space between
the two packings is exposed to a slightly higher vacuum than the process,
so that no *in-leakage* is possible.

Double packing provides reasonable protection against in-leakage under
vacuum, but it does not relieve the problems associated with corrosion and
high temperatures. When all three conditions exist (vacuum, corrosive flow
and high temperature), the use of bellows seals is considered (Section 1.4).
The bellows are usually made of 316 stainless steel and are mass spectrometer
tested for leakage. They not only prevent air infiltration but also can protect
some parts of the bonnet and topworks from high temperature and corrosion.
Like all metallic bellows, these too have a finite life, and therefore it is
recommended that a secondary stuffing box and a safety chamber be added
after the bellows seal. A pressure gauge or switch can be connected to this
chamber between the bellows and the packing to indicate or warn when the
bellows seal begins to leak and replacement is necessary.

Considerations similar to those noted for high-vacuum service would also
apply when the process fluid is toxic, explosive or flammable.

Small-Flow Valves

Valves with small flow rates are found in laboratory and pilot plant
applications. Even in industrial installations the injection of small quantities
of neutralizers, catalysts, inhibitors or coloring agents can involve flows in
the cubic centimeter per minute range.

Valves are usually considered miniature if their C_v is less than one. This
generally means a $\frac{1}{4}$-in. body connection and a $\frac{1}{4}$-in. or smaller trim. The

topworks are selected to protect against oversizing, which could damage the precise plug. There are at least three approaches to the design of miniature control valves:

1. The use of smooth-surfaced needle plugs.
2. The use of cylindrical plugs with a flute or flutes milled on it.
3. Positioning the plug by rotating the stem.

Needle plugs give more dependable results than the ones with grooves, scratches or notches because the flow is distributed around the entire periphery of the profile. This results in even wear of the seating surfaces and eliminates side thrusts against the seat. The trim is machined for very small clearances, and hard materials or facings are recommended to minimize wear and erosion. Needle plugs are available with equal percentage (down to $C_v = 0.05$), linear and quick opening characteristics (Figure 1.21x). Some manufacturers claim

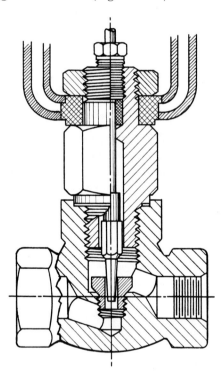

Fig. 1.21x   Needle type small-flow valve

the availability of valves with coefficients of $C_v = 0.0001$ or less. At these extremely small sizes it is very difficult to characterize the plugs (equal percentage is not available), and rangeability also suffers.

It is easier to manufacture the smaller cylindrical plugs with one or more

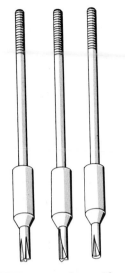

Fig. 1.21y Plugs with flute(s) milled on cylindrical surface

grooves (Figure 1.21y) and obtain the desired flow characteristic by varying the milling depth. Both the needle and the flute plugs are economical, but it is difficult to reproduce their characteristics and capacity accurately.

More reproducible control is provided by the valve illustrated in Figure 1.21z. Here the lateral motion of the plug is achieved by rotating the stem through a lead screw. The linear diaphragm motion is transferred into rotation by the use of a slip ball joint. Valve capacity is a function of orifice diameter (down to 0.02 in.), number of threads per inch in the lead screw (from 11 to 32), amount of stem rotation (from 15 to 60 degrees), and the resulting total lift, which generally varies from 0.005 to 0.02 in.

The extremely short distance of valve travel makes accurate positioning

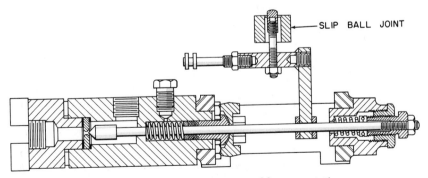

Fig. 1.21z Small-flow plug positioned by stem rotation

of the plug essential, and this necessitates a positioner. The combination of a long stem and short plug travel makes this valve sensitive to stem load and temperature effects. Because the differential thermal expansion can cause substantial errors in plug position, this valve is limited to operating temperatures below 300°F.

Viscous and Slurry Service

When the process stream is highly viscous or when it contains solids in suspension, the control valve is selected to provide an unobstructed, streamline flow path. Orientation Table I contains some recommendations concerning the applicability of the various valve designs for slurry and viscous services.

The chief difficulty encountered with heavy slurry streams is plugging, and the conditions that might contribute to this include; a difficult flow path through the valve, shoulders, pockets or dead-ended cavities in contact with the process stream. Valves with these characteristics must be avoided because they represent potential areas in which the slurry can accumulate, settle out and as a result gel, freeze, solidify, decompose or, as most frequently occurs, plug the valve completely.

The ideal slurry valve is one with the following features:

1. Provides full pipeline opening in its open position.
2. Flow is unobstructed and streamlined in its throttling position.
3. Its flow characteristic and rangeability are good.
4. Its pressure and temperature rating is high.
5. It is available in corrosion-resistant materials.
6. It is self-draining.
7. There is a "fail-safe" action.
8. There is a positive seal between process and topworks.

Unfortunately, no one valve meets all of these requirements, and the instrument engineer has to judge which features are essential and which can be compromised.

If, for example, it is essential to provide a full pipe opening when the valve is open, there are several valves that can satisfy this requirement, including the various pinch valves (Figure 1.21aa), the full-opening angle valves (Figure 1.21bb), some of the Saunders valve designs (Figure 1.21cc) and the full-ported ball valves. While these units all satisfy the requirement of fully open pipeline when open, they differ in their areas of limitations.

The pinch valves (Section 1.13), for example, are limited in their materials of construction, pressure, temperature ratings, flow characteristic, speed of response, and rangeability, but they do provide self-cleaning streamlined flow, which in some designs resembles the characteristics of a variable venturi.

The Saunders valves (Section 1.10) and pinch valves have similar features, including the important consideration that the sealing of the process fluid does

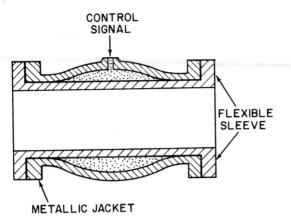

Fig. 1.21aa   Jacketed pinch valve

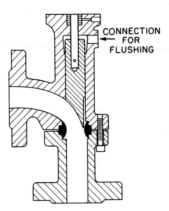

Fig.   1.21bb   Scooped-out   plug
angle valve

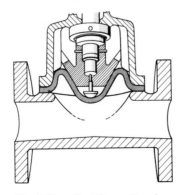

Fig. 1.21cc   Straightway Saunders
valve

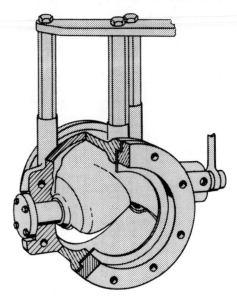

Fig. 1.21dd   Characterized ball valve

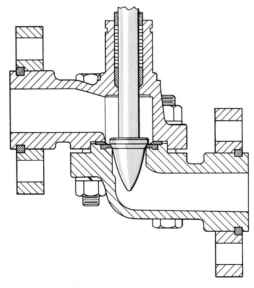

Fig. 1.21ee   Self-draining valve

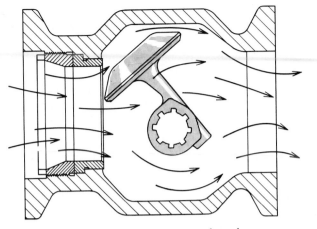

Fig. 1.21ff   Eccentric rotating plug valve

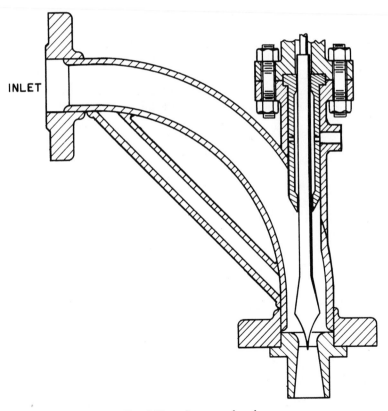

INLET

Fig. 1.21gg   Sweep angle valve

not depend on stuffing boxes. They are superior in the availability of corrosion-resistant materials, but they are inferior if a completely unobstructed streamline flow is desired.

The angle valve with a scooped-out plug satisfies most requirements except that its flow characteristics are not the best, and it is necessary to purge above the plug in order to prevent solids from migrating into that area.

Full-ported ball valves in their open position are as easy for the stream to pass as any open pipe section, but in its throttling position, both its flow path and its pressure recovery characteristics are less desirable.

Valves which do not open to the full pipe diameter but still merit consideration on slurry service include the characterized ball valves (Figure 1.21dd), various self-draining valve types (Figure 1.21ee), the eccentric rotating plug designs (Figure 1.21ff), and the sweep angle valves (Figure 1.21gg).

Each of these have some features that represent an improvement over some other design. The characterized ball valve, for example, exhibits an improved and more flexible flow characteristic in comparison with the full-ported ball type. The self-draining valve allows slurries to be flushed out of the system periodically. Complete drainage is guaranteed by the fact that all surfaces are sloping downstream.

The sweep angle valve, with its wide-radius inlet bend and its venturi outlet, is in many ways like the angle slurry valve. Its streamlined non-clogging inner contour minimizes erosion and reduces turbulence. In order to prevent the process fluid from entering the stuffing box, a scraper can be furnished, which if necessary can also be flushed with some purge fluid.

Chapter II

OTHER TYPES OF FINAL CONTROL ELEMENTS

A. Brodgesell

CONTENTS OF CHAPTER II

	Introduction	298
2.1	Throttling Electrical Energy	299
	Silicon-Controlled Rectifiers	300
	Maximum Rate of Current Rise	301
	Maximum Rate of Voltage Change	303
	Cooling	303
	Special Considerations	304
	Ignitron Tube	304
	Saturable Core Reactors	305
	Magnetic Amplifiers	306
	Power Amplifiers	307
	Relays And Contactors	308
2.2	Variable-Speed Drives	309
	Variable-Speed DC Motors	310
	Starting Circuits	313
	Dynamic Braking	314
	Reverse Rotation	315
	Control Devices	315
	Speed Changes on AC Motors	315
	Magnetic Coupling	315
	Eddy Current Coupling	315
	Magnetic Particle Coupling	317
	Magnetic Fluid Clutches	318
	Mechanical Variable-Speed Drives	318
	Stepped Speed Control	318
	Continuously Variable-Speed Drives	319
	Variable-Pitch Pulley Systems	320
	Hydraulic Variable-Speed Drives	321
2.3	Pumps and Feeders as Final Control Elements	324
	Plunger Pumps	325
	Diaphragm Pumps	326
	General Considerations for Displacement Pumps	328
	Opposed Centrifugal Pumps	332

	Ball Check Feeders	333
	Belt Type Gravimetric Feeders	335
	Self-Powered Gravimetric Feeders	337
	Roll Type Volumetric Feeders	337
	Screw Type Volumetric Feeders	338
	Belt Type Volumetric Feeders	339
	General Considerations for Solids Feeders	339
	Conclusions	340
2.4	Dampers	341
	Parallel-Blade Dampers	341
	Fan Suction Dampers	343
	Variable-Orifice Damper Valves	343
	Conclusions	344

Table II
ORIENTATION TABLE FOR NON-VALVE TYPE FINAL CONTROL ELEMENTS

Service \ Hardware	Throttling Electrical Energy					Adjustment of Drive Speeds				Pumps and Feeders as Final Control Elements								Dampers	
	SCR	Ignitron	Saturable Core-Reactor	Amplifier	Contactor	Electric Motor	Magnetic Coupling	Mechanical Variable-Speed Drive	Hydraulic Drive	Plunger Pump	Diaphragm Pump	Opposed Centrifugal Pump	Ball Check Feeder	Belt Feeder	Self-Powered Feeder	Roll Feeder	Screw Feeder	Vane Dampers	Variable-Orifice Valve
DC Output Required	E	E	G	P	P														
High Operating Voltage	P	E	G	P	P														
Reliability at Frequent Overloads and Transients	G	E	E	P	G	G	E	G	E										
Zero Output Required with Zero Input Signal	E	E	P	P	E														
Continuous Speed Adjustment Not Required						E	G	E	G										
Very Wide Speed Regulation Range						E	G	P	G										
High Output Power						G	E	P	G										
Vibration at Driver or Load						P	E	P	G										
Shock Loads						P	G	P	E										
High Output Pressure										E	G	G	G						
Low Output Pressure, Low Flow										E	E	G	E						
Coarse Slurry Service										P	P	E	P						

Table II (Continued)
ORIENTATION TABLE FOR NON-VALVE TYPE FINAL CONTROL ELEMENTS

Service \ Hardware	Throttling Electrical Energy					Adjustment of Drive Speeds				Pumps and Feeders as Final Control Elements								Dampers	
	SCR	Ignitron	Saturable Core-Reactor	Amplifier	Contactor	Electric Motor	Magnetic Coupling	Mechanical Variable-Speed Drive	Hydraulic Drive	Plunger Pump	Diaphragm Pump	Opposed Centrifugal Pump	Ball Check Feeder	Belt Feeder	Self-Powered Feeder	Roll Feeder	Screw Feeder	Vane Dampers	Variable-Orifice Valve
Very Low Slurry Flows										P	G	E	G						
Very Thick Slurry with Fine Solids										P	P	P	E						
Viscous Fluids										P	G	P	P	E	G	G	G		
Power Source Not Available														P	E	P	P		
Low Flow of Fine Powders														E	G	E	G		
Solids Blending														E	P	G	G		
Low-Pressure Gas																		E	E
Throttling Solids Flow																		P	G
Large Flow Rates																		E	E
Low Pressure Loss																		E	E
Accuracy	E	E	E	E	P	E	E	G-P	E	G	G	G	P-G	E	P	G	G	P	E

E—Excellent, G—Good, P—Poor

297

Final control elements considered in this chapter include those associated with throttling of fluid flow, electrical energy and speed. Methods for throttling electrical energy can be successfully applied to speed control by selection of appropriate hardware. The same is true of fluid flow control where speed can be used as the manipulating variable. Generally, a number of speed control devices will depend on throttling electrical energy to achieve their purpose, and the same is true for fluid flow control devices. In all such instances, the text is cross referenced to the appropriate section for a discussion of the various hardware available. The orientation table for this chapter will help in narrowing the choice of hardware for the application at hand.

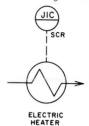

ELECTRIC
HEATER

2.1 THROTTLING ELECTRICAL ENERGY

Range: Up to 200,000 volt-amperes with standard units.

Accuracy: Better than ±1% full scale.

Cost: $2,000 for proportional control of 50-Kw load.

Partial List of Suppliers: Electronic Control Systems Inc.; General Electric Co.; R-I Controls; Westinghouse Electric Co.

Table 2.1a
SERVICE APPLICABILITY OF THE VARIOUS CONTROL DEVICES

Service	Control Element	SCR	Ignitron	Saturable Core Reactor	Amplifier	Contractor
DC Output Required		√	√		√	
Operating Voltage Above 600 Volts			√	√	√	√
Limited Mounting Space		√				
High Reliability Under Severe Overloads or Transients			√	√		

Devices for throttling electrical energy can be grouped into two classifications—on-off devices and proportional control devices. Examples of the former are power relays and contactors; examples of the latter are the ignitron tube, the silicon-controlled rectifier, the saturable core reactor and the power amplifier. On-off devices are suitable for applications involving slow process and where some cycling can be tolerated. For detailed coverage of these devices, refer to Section 4.5. With the advent of solid state devices, silicon-controlled rectifiers have gained wide acceptance for applications requiring close control, due to their inherent advantages.

Silicon-Controlled Rectifiers

The silicon-controlled rectifier (SCR) is a four-layer, solid-state silicon device consisting of alternating layers of negatively and positively doped wafers. For more details on doping and on diodes refer to Section 4.6. It is a three-element device consisting of an anode, a cathode and a gate, encapsulated and bonded to a thermally conductive base. In order for the SCR to conduct, the anode must be positive with respect to the cathode and a trigger signal at the gate must initiate conduction. (Figure 2.1b). Once the gate has

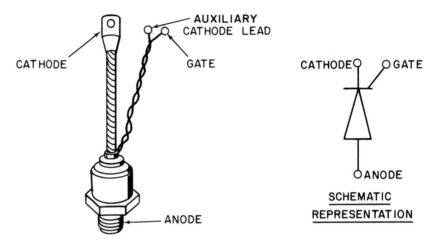

Fig. 2.1b   Silicone-controlled rectifier

been pulsed by a DC voltage, conduction is self-sustained until the anode-to-cathode voltage polarity is reversed. Modulation of power to the load is achieved by varying the firing angle of the SCR. The firing angle is defined as the point of turn-on with respect to the half cycle of line voltage (Figure 2.1c). By moving the firing angle over 0 to 180 degrees, power to the load can be varied from 100 percent to 0 percent.

For DC applications, two SCRs are combined with two diodes to provide a full-wave rectified, modulated output. For AC loads, two SCR's are connected in reverse parallel. The AC and DC configurations are shown in Figure 2.1d. During the positive half cycle of line voltage, SCR-1 is pulsed to deliver power to the load, and SCR-2 is pulsed during the negative half cycle. By timing the pulse signals with respect to the appropriate half-cycle line frequency, the power delivered to the load is modulated.

The function of the gate circuit is to provide an appropriately timed pulse signal to fire the SCR. However, the method of switching the SCR into conduction is quite critical. During turn-on, there is a relatively high forward voltage drop, resulting in localized power dissipation. The resultant temperature rise may be sufficient to destroy the device over a period of time. Most

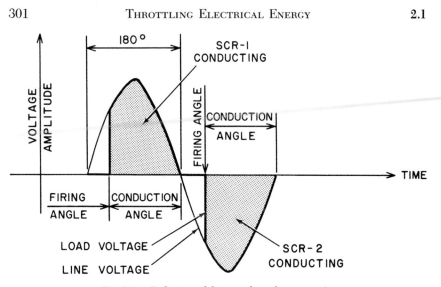

Fig. 2.1c   Definition of firing and conduction angles

gate circuits are therefore designed to provide a fast rising pulse whose magnitude is much greater than that required for firing. This insures a fast turn-on and minimum power dissipation in the SCR.

It is important that the gate circuit produce a symmetrical output for a given control signal input when the load is transformer coupled. A symmetrical output means that the firing angle for both SCRs is the same, resulting in zero DC level. An assymmetrical firing, on the other hand, will result in a net DC component in the output voltage which may saturate the transformer. Transformer saturation will cause excessive current through the SCR which may blow fuses or destroy the SCR.

The SCR has an extremely low thermal capacity and for this reason must be protected from overloads. Standard fuses or circuit breakers are not sufficiently fast to protect the SCR against transients. *High-speed fuses* developed for use with SCR are normally provided.

Maximum Rate of Current Rise

Each SCR has a maximum rate of current rise (di/dt), specified by the manufacturer, which must not be exceeded. In cases of purely resistive loads and a firing angle of 90 degrees, the current would immediately surge to a maximum before the total surface of the SCR reached conduction. A high current on a small conducting surface can produce a sufficient temperature rise to cause failure after a period of time. One method of dealing with this problem is to insert a small inductance into purely resistive load circuits. There is usually sufficient leakage reactance of the transformer in transformer-

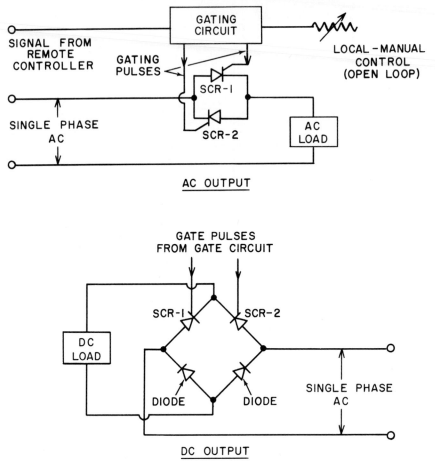

Fig. 2.1d   Control of AC and DC loads by single-phase SCR units

coupled loads to limit the maximum rate of current rise to below destructive values.

Some load circuits are designed to minimize inductance, and for these applications a high di/dt rating of the SCR is required. The di/dt rating of the SCR can be increased by using the "field-initiated" principle. SCR turn-on is achieved in two steps. The gate initiates turn-on at a small area, and the resultant load current turns on the SCR completely. The method by which this is achieved is beyond the scope of this book. However, increased di/dt rating does mean a trade-off of other device parameters. The higher di/dt rating results in a lower current rating, lower surge rating, and higher thermal resistance. Therefore the di/dt rating of the SCR should be matched to the application.

Maximum Rate of Voltage Change

An SCR *will become conductive* in the absence of a gate signal if the forward voltage drop increases very rapidly. This rapid voltage change, or dv/dt, can be detrimental in transformer coupled applications where destructively high line currents may result. Although high dv/dt may be a problem only with some applications, it is generally advantageous to prevent false firing of the SCR by using an energy-absorbing network (Figure 2.1e). Slow voltage

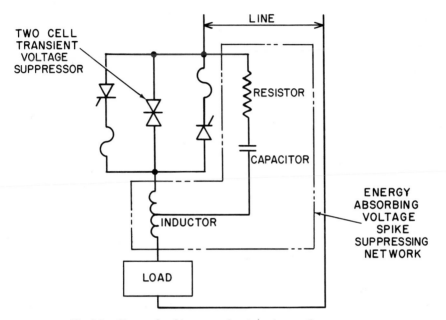

Fig. 2.1e   Energy-absorbing network using voltage spike suppressors

transients will not be absorbed by this network, and the SCR may be destroyed if its blocking voltage rating is exceeded. One method to solve this problem is to connect selenium zener type cells in parallel with the SCR.

Two cells are used back to back to by-pass transients of either polarity. The cells represent an open circuit up to the voltage rating of the SCR. At higher voltages, the cells become a closed circuit, conducting the transient *around* the SCR.

Cooling

The temperature of the SCR will increase during normal operation due to the power dissipated in it. This power loss stems from the fact that there is a voltage drop of approximately one volt during the conducting state. To prevent dangerously high temperatures, the SCR is mounted on a metallic heat sink, which is cooled by air or water. Finned heat sinks can generally be used in conjunction with natural air convection, or with forced air convection at higher power levels. Water cooling can be used effectively where the cooling by air convection is not possible.

Special Considerations

Problems can be caused by some load configurations if they are not recognized and taken into consideration at the design stage of the SCR power unit. Some transformer designs exhibit very high in-rush currents, high enough to clear the protective fuses. One way to overcome the problem is to provide signal conditioning in the gate circuit to turn the SCR on and off gradually over several cycles of line current. On 3-phase transformer-coupled loads, the load on each phase must be balanced to prevent transformer saturation. Since transformer saturation will result in high currents, three single-phase SCR's and transformers, connected to the 3-phase source, should be considered in such instances.

Purely resistive loads, such as heaters, pose the problem that their resistance when cold is very low in comparison with their resistance when hot. If full line voltage is applied to startup, the current drawn would be sufficient to clear the SCR fuses. An adjustable load current limiter can be provided to hold the load current within safe limits. Generally the limiter consists of a current transformer to sense load current and of electronic circuitry to modify the SCR command signal in such a way that the load current remains below a preset, adjustable limit. Capacitive loads may also draw excessive charging currents, and an inductor or current-limiting resistor may be necessary.

Ignitron Tube

Ignitron tubes are similar to the SCR in function, but quite different in construction. The ignitron is a *mercury arc rectifier* consisting of a graphite anode, a cathode containing a pool of mercury, and an electrode called the ignitor (Figure 2.1f). Ignitron operation is achieved by applying a DC voltage pulse to the ignitor. The arc produces a large number of mercury ions which permit current to flow from the anode to the cathode when the proper potential exists.

In general, the ignitron is inferior to the SCR in terms of size, efficiency, life expectancy, and resistance to shock and vibration. The ignitron, however, can be subjected to more than 200 percent overloads for periods in the order of one minute without damage. This factor alone does not outweigh the advantages of the SCR. However, at very high power or voltage levels the ignitron can be used to better advantage since the maximum operating voltage of the ignitron is 2,400 volts and that of the SCR is only about 600 volts. Due to the lower efficiency of the ignitron, the amount of cooling required is much greater than that of the SCR. Normally cooling water is used, or forced convection at low power ratings.

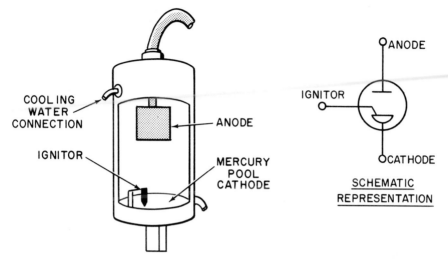

Fig. 2.1f   Ignition tube

From an operational standpoint, the ignitron can perform the same functions as the SCR without the necessity for protective circuitry. However, at low or moderate power levels the advantages of the SCR far outweigh those of the ignitron. Only at power and voltage levels above those for which SCRs are presently available does the ignitron offer a clear advantage.

Saturable Core Reactors

A saturable core reactor is a device consisting of an iron core, an AC or gate winding, and a control winding. The impedance of the gate winding can be changed by means of the magnetic flux produced by the control winding. The relationship between control voltage and output voltage is linear until magnetic saturation of the core is approached. Further increases in control voltage will no longer increase the output voltage. To permit throttling the AC power on both half cycles, a gate winding for each half cycle of voltage is used.

Saturable core reactors are relatively inefficient, large in size and require either water or air cooling, depending on installation and power rating. Their main advantage lies in their immunity to transients and overloads. SCRs with properly designed gating and compensating circuits, however, can offer similar reliability at greater efficiency and reduced size.

Saturable core reactors used with purely resistive loads should be rated for the load. If the load contains a reactive component, the reactor must be sized as follows:

$$R_{kva} = L_p \times P_{fc} \qquad\qquad 2.1(1)$$

where R_{kva} = reactor power (KVA),
     L_p = load power (Kw), and
     P_{fc} = power factor constant.

The power factor constant can be determined from Table 2.1g.

Table 2.1g
POWER FACTOR CONSTANT
FOR
SATURABLE CORE REACTORS

Power Factor of Load	Power Factor Constant (Pfc)
1.00	1.00
0.98	1.09
0.95	1.17
0.90	1.29
0.85	1.42
0.80	1.55
0.75	1.70
0.70	1.84
0.65	1.98

Saturable core reactors *cannot* be used when the load voltage must be reduced to zero or when full line voltage must be applied to the load. Typically, the load voltage is 5 percent of line voltage at zero input to the control winding, and it is 90 percent of line voltage at full control winding excitation.

The range of control voltage for a saturable core reactor is in the order of 0 to 100 volts. Since the standard electronic controller does not have the required signal range, a magnetic amplifier is commonly used in conjunction with the reactor to boost the controller output signal.

Magnetic Amplifiers

Magnetic amplifiers are essentially saturable core reactors with additional circuitry to produce the desired relationship between input and output signals. Since the output signal required is usually DC, a full-wave rectifier is included in the output. In addition to the control windings powered by an external control instrument, the magnetic amplifier may include bias and auxiliary windings. A typical magnetic amplifier is shown in Figure 2.1h.

The bias winding supply voltage is adjustable and effectively shifts the minimum output voltage over the input signal range. Thus, for a 1–5 milliampere input, the output may be at a minimum from 1 to 3.5 milliamperes and start to increase only above 3.5 milliamperes input. The auxiliary winding can be used for positive or negative feedback or to vary the output range

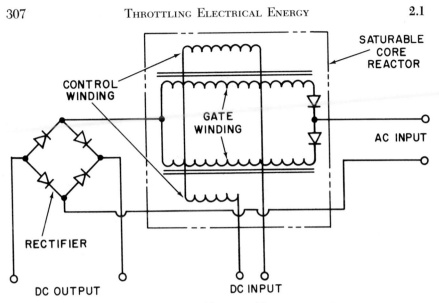

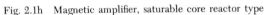

Fig. 2.1h   Magnetic amplifier, saturable core reactor type

for a given input signal span. For example, the amplifier output can be driven full scale with only half of the input signal span.

The output of the magnetic amplifier, similar to that of the saturable core reactor, cannot be driven to zero. Positive features of the magnetic amplifier are its input-to-output isolation, resulting in the fact that high transient voltages on the input do not appear in the output. Magnetic amplifiers with unity gain are therefore ideally suited for applications where the receiving device is sensitive to transient overvoltages.

Power Amplifiers

The basic element of an amplifier is the vacuum tube or transistor. The simplest tube used in amplifiers is the triode (Figure 2.1i). The triode is a three-element device consisting of anode, cathode, and grid. A heater element is also provided at the cathode to serve as a source of electrons.

The heater element emits the electrons and these occupy the space around the element in the absence of a voltage. However, the plate or anode is made positive with respect to the cathode by an external power supply. The electrons are accelerated to the anode by its positive potential and current flows. The grid, interposed between cathode and anode, consists of a fine, metallic lattice through which electrons can pass freely. The control signal is applied to the grid so that the potential of the grid is negative with respect to the cathode. This negative voltage tends to repel electrons from the plate and can therefore modulate the plate current. Variations in signal voltage at the grid are reflected as variations of plate current at increased voltage.

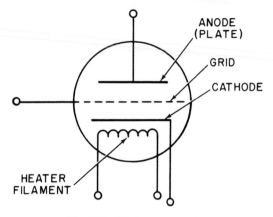

Fig. 2.1i   Triode vacuum tube

Transistors function in a different manner but the net effect is identical. The transistor is also a three-element device consisting of collector, emitter, and base. Small changes in base current will produce large changes in collector current.

The active elements of tubes or transistors can be combined with passive elements of resistors, inductors, and capacitors in various configurations to yield amplifiers with desired features of linearity, efficiency, amplification, and frequency response.

Amplifiers are *not* normally used as final control elements. Reproduction of input waveform, while important in communications, is redundant in process control, where only modulation of power is of interest. For this reason, the previously discussed devices offer more economical and efficient solutions.

Relays and Contactors

These devices offer another method of throttling electrical energy; however, only on-off control is possible. While this type of control may be satisfactory for secondary service, applications requiring close control without overshoot must be avoided. Since their operating speed is low, these devices can be used only with relatively slow processes. Additional coverage is given in Section 4.5.

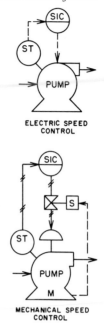

ELECTRIC SPEED CONTROL

MECHANICAL SPEED CONTROL

2.2 VARIABLE-SPEED DRIVES

Rangeability:	Between 4:1 and 40:1 with electric types, from 4:1 to 10:1 with mechanical designs and up to 40:1 with hydraulic ones.
Accuracy:	±1% full scale. Better with narrow spans.
Costs:	$1,000 for 5-hp mechanical drive, $1,200 for 5-hp motor and SCR controller, and $20,000 for 600-hp magnetic coupling drive.
Partial List of Suppliers:	Dynamatic Div., Eaton, Yale & Towne Inc.; General Electric Co.; Gerbing Manufacturing Co.; Louis Allis Co.; Philadelphia Gear Corp.; U.S. Motors Corp.; Vickers Division of Sperry Rand Corp.

The number of methods for speed adjustment of rotating machines is great. These methods fall into two categories—speed adjustment of the prime mover and speed adjustment through a transmission connecting the driver to the driven machine. Within each of these two groups several degrees of

Table 2.2a

FEATURE SUMMARY AND SERVICE APPLICABILITY
OF VARIOUS ROTARY DRIVES

Service and Feature \\ Type of Drive	Electric Motor	Eddy Current or Magnetic Couplings	Mechanical Stepped-Speed Transmissions	Continuously Variable Mechanical Drives	Hydraulic Drives
Very Wide Range of Speeds	√				
Few Speed Steps With Remote Control	√				
Few Speed Steps With Local and Manual Control	√		√		
Very High Output Power at Variable Speed		√			
Vibration at Load or Driver		√			
Small or Moderate Load with Narrow Speed Range				√	
Accurate Speed Control	√	√			√
Shock Loads, Frequent Overloads					√
Speed Reversal Required	√		√		√

sophistication are possible, ranging from manually actuated step-wise speed changes to continuously variable automatic speed changers. Each method of speed control offers certain advantages and disadvantages which must be weighed against the design criterion to permit a proper selection.

Variable-Speed DC Motors

Direct current motors lend themselves extremely well to speed control, as demonstrated by the variety of speed-load and load-voltage characteristics obtainable by means of parallel, series, or separate excitation. A typical set of speed-load characteristics for shunt, series and compound motors is shown in Figure 2.2b. The inherent speed regulation, or constancy of speed under

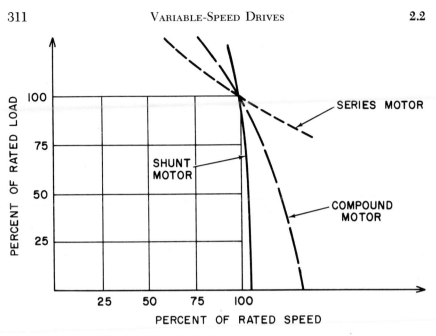

Fig. 2.2b   Speed-load characteristics of DC motors

varying load conditions, of the shunt motor is shown graphically in the diagram. The speed change of the motor is only 5 percent for a load change from zero to full load. Although the methods of speed control are applicable to all three types of DC motors, the superior speed regulation of the shunt motor accounts for its wider use on control applications.

The speed of a DC motor is a function of armature voltage, current, and resistance, the physical construction of the motor, and the magnetic flux produced by the field winding. The equation relating these variables to speed is

$$S = \frac{V - IR}{K\Phi}$$
2.2(1)

where S = motor speed,

      V = armature terminal voltage,

      I = armature current,

      R = armature resistance,

      Φ = magnetic field flux, and

      K = a constant for each motor depending on physical design.

Three methods of speed control are suggested by equation 2.2(1): adjustment of field flux Φ, adjustment of armature voltage V, and adjustment of armature resistance R.

The first of these, adjustment of field flux, involves varying the field current. This can be accomplished by means of a field rheostat, or the current can be controlled electronically in response to a set point signal through an energy-throttling device such as an SCR. Open-loop control should be utilized only when the load is constant and precise speed control is not essential. Closed-loop control is generally used in conjunction with electronic devices and offers better control than open loop, but at higher cost (Figure 2.2c). Either

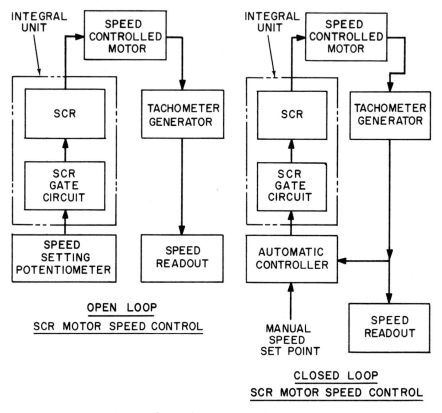

Fig. 2.2c   Speed control systems, open loop and closed loop

type of control requires a measurement of the controlled speed; in closed-loop control the measured speed is compared with the set point and the field current is adjusted automatically to bring the difference to zero. For a discussion of speed sensors refer to Section 9.3 of Volume I. In open-loop control a speed readout is required to permit set adjustments by the operator for the desired speed.

Adjustment of field flux yields a motor with constant horsepower. The allowable armature current is approximately limited to the motor rating in order to prevent overheating. The effects of changing flux and changing speed effectively cancel each other so that the allowable output horsepower—the product of armature current and induced armature voltage—is constant.

Torque, however, varies directly with field flux, and therefore this type of speed control is suited for applications involving increased torque at reduced speeds. The speed range possible with field flux adjustment is approximately 4 to 1.

Armature voltage control can also be used to control motor speed. With armature voltage control, the change in speed from zero load to full load is almost entirely due to the full-load armature resistance drop, and this speed change is independent of the no-load speed. Consider two motors with identical speed changes from zero to fully loaded, but one operating at 100 rpm and the other at 1,000 rpm at zero load. In terms of percent speed change, a 10 rpm variation may be unacceptable for the lower-speed motor, but may be insignificant for the higher-speed motor.

One method used for armature voltage control is the use of a motor-generator set to supply a controlled voltage to the motor whose speed is to be regulated. An AC motor drives a DC generator whose output voltage is variable and supplies the variable-speed motor armature. An obvious disadvantage of this method is the initial investment in three full-size machines; this method is so versatile, however, that it is often used. Armature voltage, of course, can also be controlled electronically by one of the devices discussed in Section 2.1. These are more commonly used for speed control due to the lower investment and greater efficiency than the motor-generator set. In the motor with controlled armature voltage, both the allowable armature current and field flux remain constant. The driver therefore has a *constant torque output,* as opposed to the constant horsepower output of the field-controlled motor.

The armature voltage method of speed control yields speed ranges in the order of 10 to 1. By combining field flux control and armature voltage control, speed ranges of 40 to 1 are obtainable. The base speed of the motor is set at full armature voltage and full field flux; speeds above base are obtained by field flux control, speeds below base by armature voltage control. Speed ranges greater than 40 to 1 are obtainable through the addition of special motor windings and SCR controls.

Adjustment of armature resistance is another method of speed control suggested by equation 2.2(1), and it can be used to obtain reduced speeds. An external, variable resistance is inserted into the armature circuit (Figure 2.2d). Speed regulation with this method and its variants is very poor, however, and this type of speed control is not commonly used. An added disadvantage of the armature resistance method is the decrease in efficiency due to the power consumption in the resistor.

Starting Circuits

DC motors must be protected from high in-rush currents during starting. The starting current through the armature is limited by resistors R1, R2 and R3 in Figure 2.2e. When the start button is depressed momentarily, relay M

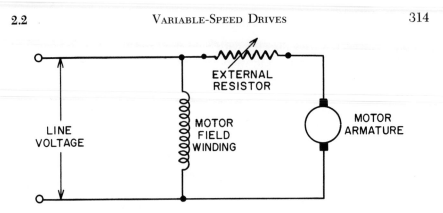

Fig. 2.2d   Variable armature resistance to control shunt motor

and time delay relays TD1, TD2 and TD3 are energized. The two M contacts close instantaneously providing power to the motor. After a time delay, contact TD1-1 closes, shunting resistor R1. Contacts TD2-1 and TD3-1 close successively until the armature is directly on the line. The number of resistors is set by the torque and current limitations of the motor and the desired smoothness of startup.

Dynamic Braking

The motor will also act as a generator, converting mechanical to electrical energy. This feature can be utilized to advantage where rapid stopping of

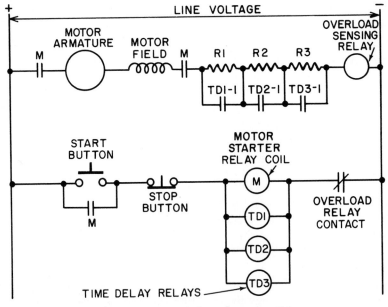

Fig. 2.2e   Starting circuit for non-reversing DC motor

the motor is required. When power is disconnected from the motor, a resistor is automatically connected across the armature and the mechanical energy of the rotating member is dissipated as heat in the resistor.

Reverse Rotation

DC motors can be run in reverse rotation by changing the polarity of the armature voltage. This can be accomplished by means of two contactors—one forward, one reverse. These must be either mechanically or electrically interlocked to inhibit closing both contactors simultaneously, and to prevent applying reverse line voltage to the motor prematurely.

Control Devices

Normally, DC power for operation of the motors is not readily available. Rectifier tubes and solid state devices such as SCRs are used extensively to convert the incoming power to DC and simultaneously to throttle the current or voltage delivered to the motor in response to an external control signal. A detailed discussion of these devices is given in Section 2.1.

Speed Changes on AC Motors

AC motors, while not readily usable for continuous speed control, can be made reversing or furnished with several fixed speeds. The squirrel cage induction motor is best suited from the standpoint of reliability, and it can be furnished with up to four fixed motor speeds. Speed changes are accomplished through motor starter contacts which reconnect the windings to yield a different number of poles. In the resulting pole motor, the sections of the stator winding are interconnected through the motor starter to provide different motor speeds. On separate winding motors, a different winding with the required number of poles is energized for each motor speed. The starter contactors, however, must be interlocked to prevent two or more contactors from closing simultaneously.

Magnetic Coupling

Eddy Current Coupling

In most electrical machinery, eddy currents are detrimental to operating efficiency, and great pains are taken to eliminate them. In the eddy current coupling, however, these currents are harnessed and are the basis for the operation of the coupling. The eddy current coupling is a non-frictional device where input energy is transferred to the output through a magnetic field. The eddy current coupling consists of a rotating magnet assembly separated from a rotating ring or drum by an air gap. In addition, a coil is wound onto the magnet assembly (Figure 2.2f), or on larger units, the coil is stationary on the coupling frame. There is no mechanical contact between the magnets and drum. When the magnet assembly is rotated, the drum remains stationary until

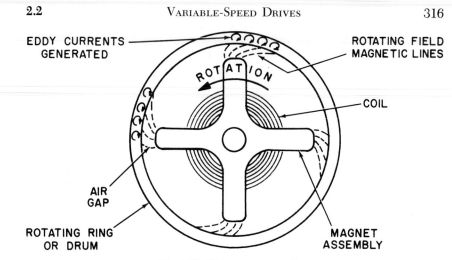

Fig. 2.2f   Eddy current coupling

a DC current is applied to the coil. Relative motion between magnet assembly and drum produces eddy currents in the drum, whose magnetic field attracts the magnet assembly. Attraction between the two magnetic fields causes the drum to follow the rotation of the magnet assembly. The attraction between the two rotating members is determined by the strength of the coil's magnetic field and by the difference in speed between the two members. Thus by controlling the coil excitation, the amount of slip and, hence, the output speed can be controlled.

Being a slip device, the eddy current coupling must of necessity develop slip and reject the slip power in the form of heat. The amount of slip loss can be determined from equation 2.2(2)

$$P_s = P_L \frac{S_s}{S_0} \qquad\qquad 2.2(2)$$

where P_s = slip loss power,
       P_L = load power,
       S_s = slip speed (rpm) and
       S_0 = output speed (rpm).

Slip devices always generate heat and eddy current couplings are either air or water cooled. Small air-cooled units below 5 horsepower can be designed to dissipate all of the rated power. Air-cooled units above 300 horsepower in size can dissipate only about 25 percent of the rated power capacity. For this reason, air-cooled units are not recommended where cooling capacity would be greater than about 20 percent of rated power. Water-cooled units are designed to dissipate the rated horsepower continuously. In addition, water-cooled units show a small efficiency advantage over air-cooled ones, particularly on "water-in-the-gap" types, where the water contributes slightly to the torque capability. Generally, water-cooled units are preferred to air-cooled types except when lack of coolant precludes their use or where very low slip losses are encountered.

Eddy current couplings require only a low percentage of transmitted power for excitation. Typically a 3-horsepower unit will require 50 watts of excitation, while a 12,000-horsepower unit will require 20 kilowatts. Eddy current couplings are readily adaptable to SCR or magnetic amplifier control, providing speed control within 1 percent accuracy, over 10 to 100 percent load change. The speed-control devices mentioned above are discussed in Section 2.1.

Efficiency of the eddy current coupling is very good, particularly in the large sizes and at full speed and torque. At lower speeds, efficiency drops considerably, but efficiencies as high as 95 percent are possible at full excitation and torque. Since there is no contact between the input and output shafts of the coupling, the unit will not transmit vibrations. On prime movers that exhibit some torsional vibration, the use of an eddy current coupling can be of advantage because it will suppress or at least attenuate these vibrations.

Integral combinations of motor, coupling, and excitation are available in small sizes of one horsepower or less. Air-cooled couplings range in size up to 900 horsepower, but the larger sizes are practical only where a relatively low cooling capacity is required. Liquid-cooled units can be as large as 18,000 horsepower capacity. Speed-control units range from simple open-loop control to precise, automatic closed-loop control with tachometer-generator speed feedback.

Magnetic Particle Coupling

The magnetic particle coupling offers another solution to adjustable speed drives. Basically the coupling consists of two concentric cylinders separated by an air gap and a stationary excitation coil surrounding the cylinders. Ferromagnetic particles fill the gap between the concentric cylinders. When a controlled amount of current is used to energize the coil, the particles form chains along the magnetic lines of flux connecting the cylinder surfaces. The shear resistance of the magnetic particles is proportional to the coil excitation and provides the basis for power transmission from input to output.

The output torque of this unit is always equal to input torque regardless of speed. This fact allows the output torque to be set at standstill by controlling the coil excitation. Whenever the torque capacity of the coupling is exceeded, slip will result with accompanying heat liberation. Whenever the coupling is used for speed control, selection of a coupling with adequate cooling capacity is of great importance. Air-cooled units are available with large heat dissipation capacity; however, water cooling is more effective and therefore preferred.

Magnetic particle couplings can be used effectively for speed control, since they can provide a constant torque output independent of speed. However, a closed-loop control system is required for effective speed control in order to cancel the effect of changing torque on slip speed.

Magnetic Fluid Clutches

The magnetic fluid clutch is similar in operating principle to the magnetic particle coupling, but either a magnetic dry powder or magnetic powder suspended in a lubricant is utilized. A typical disc type magnetic fluid clutch is shown in Figure 2.2g.

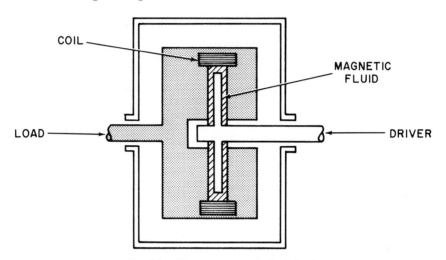

Fig. 2.2g   Disc type magnetic fluid clutch

While the operating principle is similar to that of magnetic particle coupling, the clutch can be used only for low-power applications in the order of a few horsepower. Recharging of the clutch due to fluid deterioration is a definite disadvantage; however, recharging can be fairly easily accomplished during routine maintenance.

Mechanical Variable-Speed Drives

Stepped Speed Control

Mechanical methods of speed adjustment offer a number of gear and pulley devices for both stepped and continuously variable-speed control. Stepped speed control methods provide setting a number of speeds very accurately. However, they are not readily adaptable to automatic process control. The stepped pulley system shown in Figure 2.2h is one of the earliest methods of speed adjustment. Its advantages are low cost and simplicity, but belt slippage contributes to inefficiency, high maintenance and reduced speed control. Two factors to be considered in the design of a stepped pulley system are the proper ratio of pulley diameters to obtain the desired speeds, and

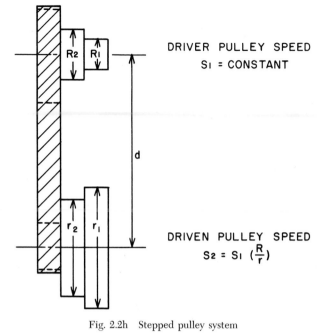

Fig. 2.2h   Stepped pulley system

pulley dimensions such as to maintain belt tension for all positions. Pulley dimensions for a system such as shown in Figure 2.2h can be determined from the relationships in equations 2.2(3) and 2.2(4).

$$\frac{\pi}{2}(R_1 + r_1) + \frac{(R_1 - r_1)^2}{4d} = \frac{\pi}{2}(R_2 + r_2) + \frac{(R_2 - r_2)^2}{4d} \qquad 2.2(3)$$

$$\frac{S_2}{S_1} = \frac{R_2}{r_2} \qquad 2.2(4)$$

The variables are defined in Figure 2.2h.

Gear transmissions offer high efficiency of power transmission at precise stepped speed control. For speed changes of gear train drives, either clutches or brakes are required or the change must be made at rest. Epicyclic gears, such as planetary gears, offer the most compact unit, operating quality, and high efficiency. However, auxiliary clutches and brakes are required and the cost is therefore higher than that of other gear drives. The complexity of control arrangements increases rapidly with the number of speeds required, making the planetary gear drive impractical above four or five speeds.

Continuously Variable-Speed Drives

Cone pulley systems of the type shown in Figure 2.2i are a natural evolution of the stepped pulley system shown in Figure 2.2h. Cone pulleys

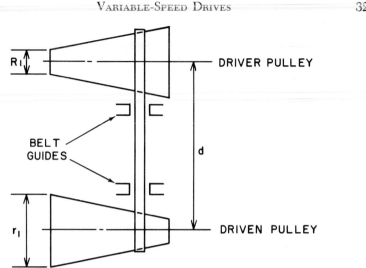

Fig. 2.2i   Cone pulley system

are designed similarly to stepped pulleys. A series of diameters is calculated equidistant on the pulley axis. The diameter end points are joined to form the pulley contour. The cone pulley is inherently inefficient, since contact surface speed varies across the belt causing slippage. Belts must be kept narrow to reduce slippage and wear, and thus the capacity for power transmission is reduced. Belt guides at the pulleys are required to hold the belt in position. These guides must move simultaneously for speed changes.

Variable-Pitch Pulley Systems

Cone pulleys were the forerunners of the more sophisticated variable pitch pulley systems, which permit continuous, automatic speed adjustments over wide ranges. Speed adjustment in all of these systems is obtained by means of sliding cone face pulleys or sheaves, whose effective diameter can be changed.

A simple variable-pitch sheave is shown in Figure 2.2j. Two flanges are mounted on a threaded hub. The flanges are set to the desired spacing and locked in place with a setscrew. Speed adjustments are accomplished by changing the flange spacing, producing in effect a pulley of different diameter. This method is very economical, but it requires stationary speed adjustment and special adjustable motor bases to maintain belt tension.

In place of the adjustable motor base, a spring-loaded flat-face idler pulley can be used to maintain belt tension when center-to-center distance must be held constant. Alignment of driving and driven sheaves is also critical, because belt wear will be severe with poor alignment. Also, speed adjustment with this drive is limited by the fact that the setscrew must engage a flat surface.

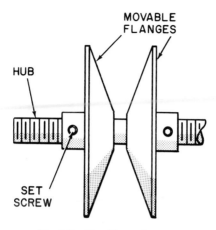

Fig. 2.2j   Variable-pitch sheave

A number of designs are available with in-motion speed adjustment. The designs vary from manual speed adjustment, through a crank or handwheel, to automatic actuators. Generally these drives use one sheave with adjustable pitch and one spring-loaded sheave which automatically adjusts itself to maintain belt tension (Figure 2.2k). Speed adjustment is obtained by means of a mechanical linkage which moves one of the flanges of the driving sheave. The opposite flange on the driven sheave is spring loaded and moves to maintain belt tension and alignment. Radial motion of the belt along the flange faces in response to speed changes is facilitated by the rotary motion of the sheaves. When speed adjustments are made at rest, the belt cannot move radially and the adjusting linkage may be damaged. On automatic actuators, an interlock should be provided which vents air off the diaphragm whenever the drive is stopped to inhibit speed changes at rest.

Variable-pitch sheave drives are available as package units including motor. Speed adjustment ranges of 4 to 1 are common but higher ranges to 10 to 1 are possible. Horsepower ranges to 100 horsepower are also possible. For very narrow speed ranges, drives to 300 horsepower are available.

Hydraulic Variable-Speed Drives

Hydraulic variable-speed drives utilize a pump with fixed or variable displacement driving a hydraulic motor with fixed or variable displacement. The pump is driven by an electric motor at fixed speed. Output speed is controlled by changing pump or motor displacement.

Pumps and motors are virtually identical, differing only in the location of the power input and output. Several designs of variable displacement units are possible, including axial piston, radial piston and vane types. An axial piston unit is shown in Figure 2.2l.

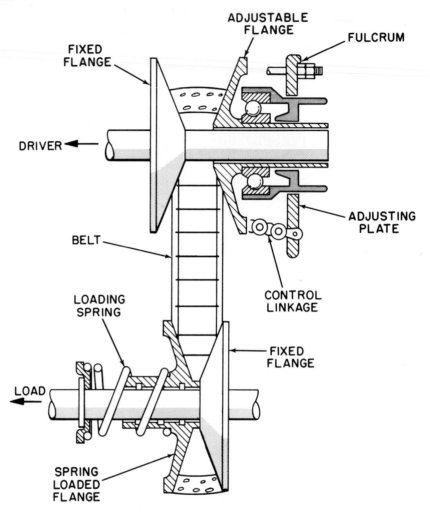

Fig. 2.2k   Continuously variable mechanical speed transmission

Of course, combinations of pumps and motors with variable or fixed displacement are possible for speed control. Pumps with variable displacement combined with motors with fixed displacement yield a variable-horsepower, *fixed-torque* drive. Motor speed reversal is possible by reversing the pump stroke.

Combinations of fixed-displacement pump with variable-displacement motors produce drives with *fixed horsepower* and variable torque. Speed reversal in this combination, however, requires the use of a valve to reverse the supply and return connections at the motor. A variable-displacement pump

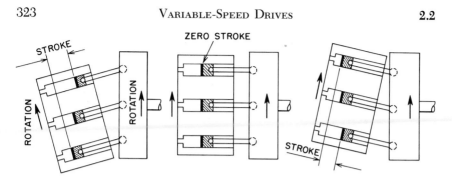

Fig. 2.2 1   Axial piston pump with variable displacement

in conjunction with a variable-displacement motor has characteristics intermediate between the two previously discussed combinations. Range of speed control, however, is widest for this combination.

Pumps and motors with fixed displacement can be used for speed control by controlling the amount of fluid delivered to the motor. This can be accomplished by means of a pressure relief valve on the pump discharge and throttling valve in the oil line to the motor. Motor reversal can also be accomplished by means of a four-way valve, reversing the oil supply and return lines at the motor. These methods, however, are comparatively inefficient and are not used with automatic control.

Hydraulic drives are available as pump-motor packages or the motor can be mounted remotely and connected to the pump by hydraulic tubing. Displacement is adjustable through push-pull rods or handwheels, and both methods are adaptable for operation by hydraulic and pneumatic actuators or by electric motors for automatic speed control. Speed ranges are adjustable up to about 40 to 1 at horsepower ratings up to 4,000.

Several attributes of hydraulic variable-speed drives are clear advantages over the other types of drives discussed. Hydraulic drives offer the fastest response in acceleration, deceleration and speed reversal. They are generally better suited than other types of drives for high shock loads, frequent speed-step changes, and reversals. A disadvantage associated with hydraulic drives is frequent fluid leaks. Besides requiring maintenance, leakage of hydraulic fluid creates safety hazards which can preclude their use.

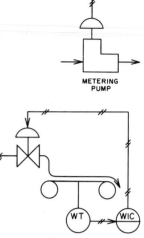

2.3 PUMPS AND FEEDERS AS FINAL CONTROL ELEMENTS

METERING
PUMP

GRAVIMETRIC FEEDER

Design Pressures:	Up to 4,000 PSIG for liquids, and atmospheric for solids.
Materials of Construction:	Cast iron, steel, stainless steel, carpenter 20, Hastelloy, plastics and glass.
Range:	Up to 40 GPM for liquids, and 3,000 lbm/hr for solids.
Rangeability:	Can exceed 100:1 with both pumps and feeders.
Accuracy:	$\pm\frac{1}{4}\%$ to $\pm2\%$ full scale.
Cost:	$1,000 for pump with positioner and $6,000 for an automated gravimetric solids feeder.
Partial List of Suppliers:	Beckman Instruments, Inc.; Fischer & Porter Co.; Milton Roy Co.; Mec-O-Matic; Seiscor Inc.; Wallace & Tiernan Div., Pennwalt Corp.

Flow control of liquids and solids can be accomplished by means of pumps and feeders which incorporate the measurement and control element in a single unit. Metering pumps and feeders are designed to provide measurement and control of the process. For a measurement-oriented discussion of these pumps, refer to Section 5.17 of Volume I. However, it is also possible to utilize standard centrifugal pumps with variable-speed control to give the desired control action.

Plunger Pumps

These pumps are suitable for use on clean liquids at high pressures and low flow rates. A typical plunger pump is shown in Figure 2.3a. The pump

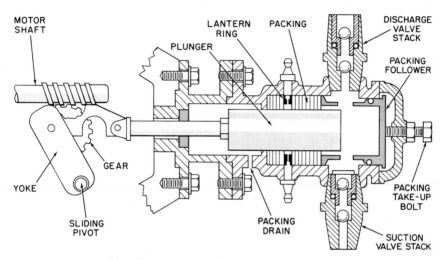

Fig. 2.3a   Plunger or piston type metering pump

consists of a plunger, cylinder, stuffing box, packing, and suction and discharge valves. Rotary motion of the driver is converted to linear motion by an eccentric. The plunger moves inside the cylinder with reciprocating motion, displacing a volume of fluid on each stroke.

Stroke length, and thus the volume delivered per stroke, is adjustable. The adjustment can be a manual indicator and dial, or for automatic control applications, a pneumatic actuator with positioner can be provided. Stroke adjustment alone offers operating flow ranges of 10 to 1 from maximum to minimum. Additional rangeability can be obtained by means of a variable-speed drive. A pneumatic stroke positioner used in conjunction with a variable-speed drive provides rangeability of at least 100 to 1. In the case of automatic stroke adjustment and variable speed, the pumping rate can be controlled by two independent variables, or the controller output can be "split-ranged" between stroke and speed adjustment.

The reciprocating action of the plunger results in a pulsating discharge flow as represented in Figure 2.3b by the dotted simplex curve. For applications where these flow pulsations cannot be tolerated, particularly if a flow

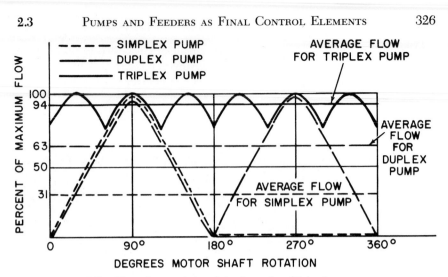

Fig. 2.3b   Flow characteristics of simplex and multiple plunger pumps

measurement is required, pumps can be run in duplex or triplex arrangements. With the duplex pump, two pumps are driven off the same motor, and the discharge strokes are phased 180 degrees apart. With a triplex arrangement, three pumps are driven by one motor and the discharge strokes are 120 degrees apart. Both the duplex and triplex pumps provide a smoother flow than the single pump, as shown by the dashed and solid curves of Figure 2.3b.

For blending two or more streams, several pumps can be ganged to one motor. Stroke length adjustment can be used to control the blend ratio, and drive speed can control total flow. However, in this case rangeability is sacrificed for ration control. Pumping efficiency is affected by leakage at the suction and discharge valves. These pumps are therefore not recommended for fluids such as slurries, which will interfere with proper valve seating or settle out in pump cavities.

Diaphragm Pumps

Diaphragm pumps, as their name implies, use a flexible diaphragm to achieve pumping action. The input shaft drives an eccentric through a worm and gear. Rotation of the eccentric moves the diaphragm on the discharge stroke by means of a push rod. A spring returns the push rod and diaphragm during the suction stroke. A typical pump is shown in Figure 2.3c.

Operation of the diaphragm pump is similar to that of the plunger pump; however, discharge pressures are much lower due to the strength limitation of the diaphragm. Their principle advantage over the plunger pumps is lower cost. Designs with two pumps driven by one motor can be used to advantage for increased capacity or to smooth out flow pulsations. By combining automatic stroke length adjustment with a variable speed drive, operating ranges can be as wide as 20 to 1. These pumps can be used only on relatively clean fluids, because solids will interfere with proper suction and discharge valve seating or may settle out in the pump cavities.

The weakness of the diaphragm pump design is in the diaphragm, which is operated directly by the push rod. The diaphragm has to be flexible for pumping and yet strong enough to deliver the pressure. The strength requirement can be reduced by using a hydraulic fluid to move the diaphragm, thereby eliminating the high differential pressures across it. This design consists basically of a plunger pump to provide hydraulic fluid pressure for diaphragm operation and the diaphragm pumping head (Figure 2.3d). The forces on the diaphragm are balanced, and discharge pressures comparable to plunger pumps are possible. The volume pumped per stroke is equal to the hydraulic fluid displaced by the plunger, and this volume is controlled by the stroke length adjustment as in the plunger pump.

A pump design using a flexible tube to achieve pumping action is shown in Figure 2.3e. Motion of the plunger displaces the diaphragm, which in turn causes the flexible tube to constrict, forcing fluid in the tube to discharge (similar to the operation of a peristaltic pump). This design is better suited for use on viscous and slurry liquids than the previously discussed types because the flow path is straight with few obstructions and no cavities, but seating of the valves can still be a problem.

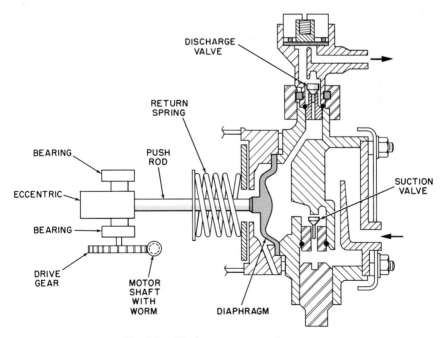

Fig. 2.3c   Diaphragm type metering pump

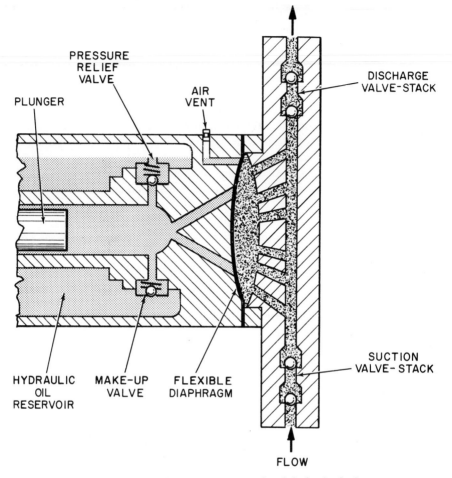

Fig. 2.3d   Diaphragm pump operated with hydraulic fluid

General Considerations for Displacement Pumps

In order to insure a properly working installation, a number of factors associated with the physical installation and with the properties of the fluid must be considered. Some factors which can contribute to a poor installation include:

    a. Long inlet and outlet piping with many fittings and valves.
    b. Inlet pressure higher than outlet pressure.
    c. Pocketing of suction or discharge lines.
    d. Low suction head or suction lift.

A tortuous flow path in the pump suction or discharge can be troublesome when the fluid handled contains solids, is of high viscosity, or has a high vapor

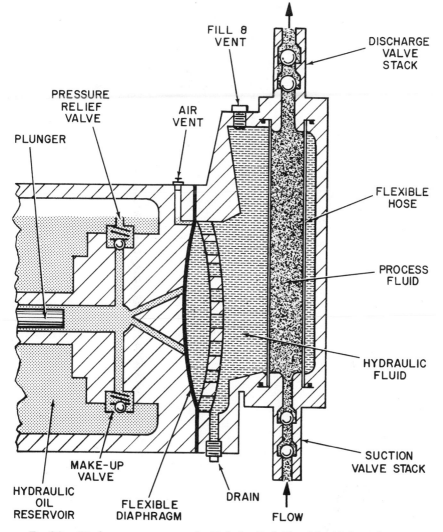

Fig. 2.3e Diaphragm pump operated with hydraulic fluid and flexible hose element

pressure, or if the suction head available is low. Generally, valves that offer a full flow path (such as ball valves) are preferred. Needle valves should be avoided. If the inlet pressure is higher than discharge, the fluid may flow unrestricted through the pump. Spring-loaded check valves at the pump are undesirable because the ball check should be free to rotate and find a new seating surface for increased valve life. For such applications the installations shown in Figures 2.3f and g offer solutions. In Figure 2.3f the piping arrangement will supply the head to prevent through flow and syphoning. Dimension "A" is a variable depending on pump capacity and fluid velocity. This dimen-

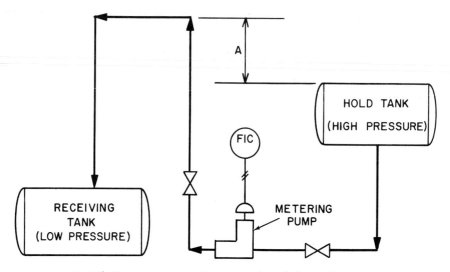

Fig. 2.3f   Piping arrangement to prevent through flow and syphoning

sion varies between 2 and 10 ft and increases with capacity and velocity. Figure 2.3g illustrates the use of a spring-loaded back-pressure valve, to overcome the suction pressure. For this installation a volume chamber to dampen pulsations should be placed between the pump discharge and the valve.

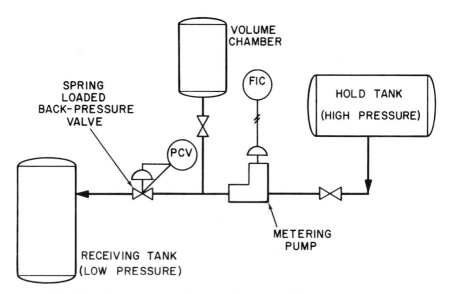

Fig. 2.3g   Metering pump with artificial head created by back-pressure valve

Dissolved or entrained gases in the fluid can destroy metering accuracy and, if they are of sufficient volume, they can stop all pumping action. Figure 2.3h illustrates an installation designed to vent entrained gases back to the fluid hold tank.

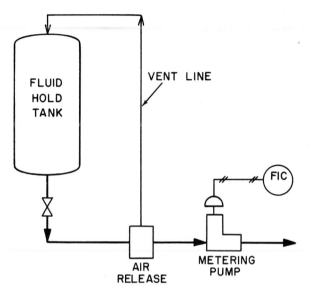

Fig. 2.3h   Elimination of entrained gases in metering pump installations

It is always desirable to locate the pump below and near the fluid hold tank. Under these conditions the fluid will flow by gravity into the pump suction and loss of prime is unlikely. If the pump cannot be located below the hold tank, other measures must be taken to prevent loss of prime.

In order for the pump to operate properly, the net positive suction head must be above the minimum practical suction pressure of approximately 10 PSIA. Net positive suction head (NPSH) is given by equation 2.3(1):

$$\text{NPSH} = P - P_v \pm P_h - \sqrt{\left(\frac{\text{lvGN}}{525}\right)^2 + \left(\frac{\text{lvC}}{980\text{Gd}^2}\right)^2} \qquad 2.3(1)$$

where P = feed tank pressure (PSIA),

P_v = liquid vapor pressure at pump inlet temperature (PSIA),
P_h = head of liquid above or below the pump centerline (PSID),
l = actual length of suction pipe (ft)
v = liquid velocity (ft/sec),
G = liquid specific gravity,
N = number of pump strokes per minute,
C = viscosity (centipoise) and
d = inside diameter of pipe (in.).

For liquids below approximately 50 centipoise, viscosity effects can be neglected, and equation 2.3(1) reduces to

$$\text{NPSH} = P - P_v \pm P_h - \frac{(\text{lvGN})}{(525)} \qquad 2.3(2)$$

The calculated value of NPSH must be above the minimum suction pressure required by the pump design.

In addition to multiple pumping heads, a pulsation dampener can be used on the pump discharge to smooth the flow pulsations. The pulsation dampener is a pneumatically charged diaphragm chamber that stores energy on the pump discharge stroke and delivers energy on the suction stroke, thus helping to smooth the flow pulses. In order to be effective, however, the dampener volume must be equal to at least five times the volume displaced per stroke.

Opposed Centrifugal Pumps

The opposed centrifugal pump is not a device specifically designed as a control element but represents an adaptation of a centrifugal pump to flow control. This method of control is particularly suitable for coarse, rapidly settling slurries at low flow rates. In such instances the conflicting requirements of control at low flow, and large free area to pass the solids, may make it impossible to find a suitable control valve. A system that requires a small quantity of slurry to be fed to a receiving vessel under controlled conditions is depicted in Figure 2.3i. Pump P_1 continuously circulates the slurry from the feed tank at high velocity. A branch line from the discharge of P_1 is run

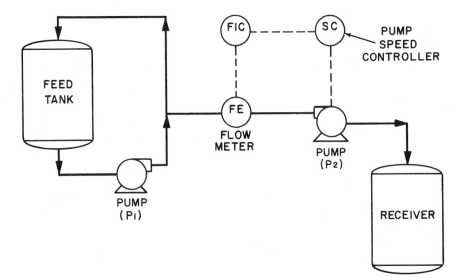

Fig. 2.3i   Opposed centrifugal pump as final control element

to the opposed centrifugal pump P_2. Pump P_2 is connected in opposition to the direction of slurry flow, and pressure drop to throttle flow is obtained by means of the mechanical energy supplied to the pump. A variable-speed driver on pump P_2 permits changing the pump pressure drop. At full speed the pressure difference across P_2 is sufficient to stop the branch line slurry flow completely. A magnetic flowmeter or some other suitable device can be used to measure the slurry flow. An SCR speed control can be used to vary pump speed in response to the flow controller output signal. A detailed discussion of variable-speed drives is given in Section 2.2.

Ball Check Feeders

Ball check feeders find their application in the control of small slurry flows. The heart of the feeder is a plug valve with a passage drilled through the plug. A ball check is held captive in the passage but is otherwise free to move. The plug can rotate through 360 degrees and can be driven by either a pneumatic or electric motor. A separate control unit contains the required timers, feed rate adjustment, feed rate recorder and counter devices.

Figure 2.3j shows the feeder in its reset position. The flow passage through

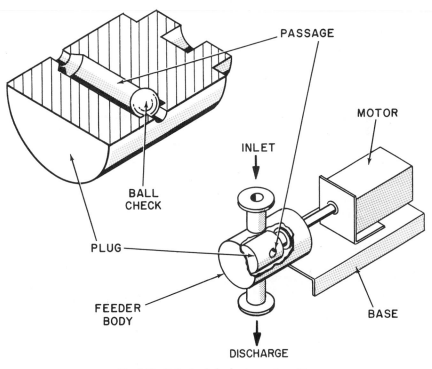

Fig. 2.3j   Ball check feeder in reset position

the plug is blocked and no flow exists. An electrical signal from the control unit starts the drive motor, and the plug rotates to the position shown in position "A" of Figure 2.3k. At this point a cam-operated limit switch cuts power to the motor to permit discharging of the feeder. The process pressure

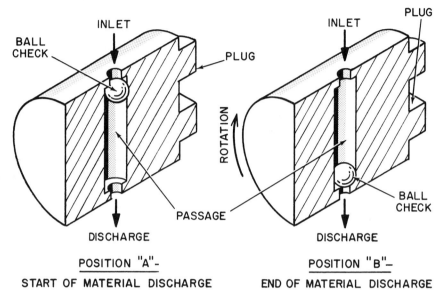

Fig. 2.3k   Position of ball check at start and completion of material discharge

on the upstream side of the feeder is greater than that on the downstream side, and the ball moves down in the passage, forcing out the material ahead of it. At the same time material enters the passage above the ball check (position "B"). After a short delay, the motor is again energized by the control unit and rotates the feeder to the position shown in Figure 2.3j. At this point the limit switch again cuts power to the motor, and the feeder remains in the reset position until an electrical signal from the control unit initiates the next cycle. The feed rate can be controlled by changing the reset time at the control unit.

Standard feeder capacities are between 2 and 8 cc/dump, and maximum feed rate is between 30 and 40 dumps/min. As mentioned above, the feeder requires a differential pressure for operation. This differential, approximately 50 PSID, serves to force material out of the feeder passage and seat the ball properly at the end of its travel. Since feed rate is inferred from dump rate, feeder leakage will affect accuracy of metering.

The feeder is well suited for fine slurries at high concentration. Due to the size of the feed chamber, coarse slurries cannot be handled.

Drivers for the ball check feeders can be electric or pneumatic motors.

Air motors offer slightly better feed rates; however, the electric motor can be operated directly from the control unit without the need for another energy source. The control unit is electric and operates a solenoid valve in the air supply line when a pneumatic drive is used.

Belt Type Gravimetric Feeders

Belt type gravimetric feeders incorporate both the feed rate sensing and control portion of the control loop. For a more measurement oriented discussion, refer to Section 7.6 of Volume I. The designs available vary considerably in terms of sophistication, and the choice of the particular design will be dictated by operating flow range, accuracy required, and flowing properties of the solid handled. Gravimetric feeders consist of a feed section, a weigh belt, a means of sensing the belt loading, and a driver.

A typical constant speed feeder is shown in Figure 2.31. The weight of

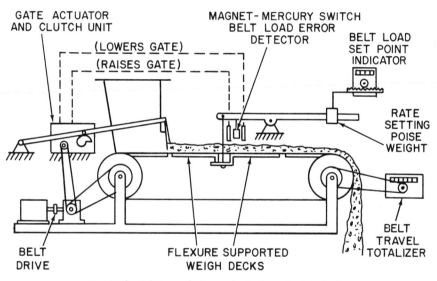

Fig. 2.31   Belt type electromechanical gravimetric feeder

material above the weigh deck is sensed by a two-piece platform. This weight represents the belt loading and is transmitted to the scale beam by a yoke. The beam is balanced by a poise weight. As long as the beam is balanced, the feed rate is at set point. When belt loading changes, the scale beam becomes unbalanced, and the magnet closes one of the two mercury switches. This signal actuates the gate positioner, which in turn operates the feed gate. The feed gate moves to change belt loading in the direction that will restore the scale beam to balance. When balance is reached, the mercury switch opens and the gate remains in its last position.

In this typical design, gate position is used to vary material depth on the belt (belt loading), thus maintaining the desired weight flow. Feed rate, however, is the product of belt loading and belt speed. By adding a variable-speed drive to this feeder, belt loading is maintained by the gate, and feed rate set-point changes are accomplished by speed changes. Belt speed can be varied through a gear box, variable-speed transmission, or variable-speed SCR-controlled motor. Either variable belt speed or variable belt loading provide operating feed rate ranges of 10 to 1.

Wider operating ranges can be accommodated by using variable belt speed and variable belt loading. Variable belt loading provides a feed range of 10 to 1, and the electric SCR drive has a feed range of 20 to 1, for a total rangeability of 200 to 1. Additional rangeability can be achieved with a two-speed gear box.

For accurate feeding of the material, particles must be fairly uniform in size and free flowing. Flow ranges obtainable with variable belt loading are less than 10 to 1 if particle size is greater than approximately ⅛ in. Materials that tend to flood the feeder cannot be handled with the feed gate. For such materials, a rotary volumetric feeder of the type shown in Figure 2.3m must be used. Each chamber of the rotary feeder deposits a fixed quantity

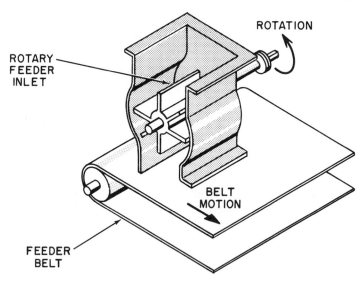

Fig. 2.3m   Volumetric feeder of the rotary type

of material on the belt, and its blades distribute the solids uniformly. Belt loading here is controlled by the speed of rotation.

It is possible to operate several feeders in parallel or in cascade from the same set point, and this mode of operation can be used effectively for blending solids. Necessary instrumentation consists of transmitters for belt loading and speed and a multiplying relay whose output is the product of belt loading and belt speed. Also available are readout instruments to measure and set feed rates, a set-point regulator, and a controller to maintain the desired feed rate by adjusting belt speed or gate opening. This instrumentation and the feeder are available as a package. For additional information see Section 7.6 of Volume I.

Self-Powered Gravimetric Feeders

Self-powered feeders can be used to advantage where reduced accuracy can be tolerated or where a power source is not readily available. The feeder operates on the force balance principle, where the weight of material on the impact pan is balanced by weights. The essential parts of the feeder are shown in Figure 2.3n. In operation, material flows through the inlet spout and strikes

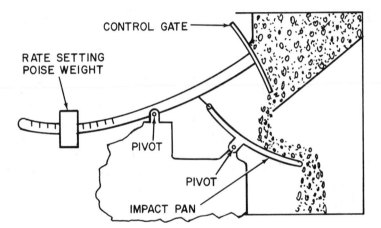

Fig. 2.3n   Self-powered gravimetric feeder

the impact pan. If its force is greater than the preset weight, then the impact pan, gate, and weight beam will rotate clockwise about the fulcrum. The gate moves to throttle material flow until the force of the weights and the force on the impact pan are in balance. The gate will open to admit more material if the force on the impact beam should decrease. Feed rate can be changed by adjusting the position of the counterweights.

Roll Type Volumetric Feeders

Roll feeders are low-capacity devices for handling dry granules and powders (Figure 2.3o). The feeder consists of a feed hopper, two feed rolls and a drive unit. Guide vanes in the hopper distribute the material and provide agitation by oscillating. The feed rolls form the material into a uniform ribbon,

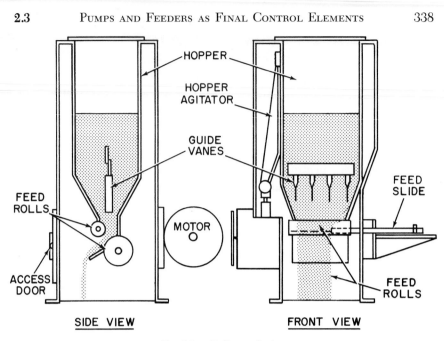

Fig. 2.3o   Roll type feeder

and feed rate is controlled either by means of a slide that varies the width of the ribbon or by means of a variable-speed drive. Operating ranges are typically 6 to 1 with the feed slide and 10 to 1 with mechanical or electrical variable-speed drive. A 3–15 PSIG air or a milliampere input can be used as the control signal for the electrical variable-speed drive. For materials that tend to cake or bridge in the hopper, agitators can be provided to maintain the material in a free-flowing state.

Screw Type Volumetric Feeders

Screw type feeders can handle higher material flow rates than roll feeders. The feeder element in this device is a screw whose rotary motion delivers a fixed volume of material per revolution. The screw is located at the bottom of a hopper so that the feed element is always flooded with material. Rotation of the screw discharges material, at one or both ends of the screw, into the receiving vessel. Screws grooved in one direction discharge material at one end only. Screws grooved in opposite directions from the middle deliver material at both ends.

For materials with a tendency to cake or clog the feed screw, the double-ended screw can be provided with a lateral oscillating motion which imparts a cleaning action. In this case material is alternately fed from one end or the other, depending on the direction of lateral motion.

In order to assure an accurate feed, the hopper on the inlet side of the feeder must be so designed as to assure a uniform supply of material at the feed screw. Vibrators can be added to the hopper to keep the solids agitated and to prevent caking or bridging.

Feeder drives are usually electric motors. With a constant-speed drive, operating feed is adjustable over a 20-to-1 range by means of a mechanical clutch that varies the operating time per cycle. At 75 percent feed rate setting, the screw will be operating over 75 percent of a clutch revolution.

Addition of a variable-speed drive can extend the operating range to 200 to 1. The variable-speed drive can be electric or mechanical. The electric type will accept any standard milliampere signal; the mechanical type will operate on a 3–15 PSIG signal.

Belt Type Volumetric Feeders

The belt type volumetric feeder is a high-capacity feeder. The design and operation of this unit are essentially the same as that of the belt gravimetric feeder, but in the belt volumetric feeder the weighing portion is omitted. Feed rate is controlled by positioning the gate to vary material depth on the feed belt. Mechanical or electrical variable-speed drives can be used to extend the operating range available with gate positioning. The operating range by gate positioning is 10 to 1 for particle sizes under $\frac{1}{8}$ in. in diameter. For larger particle sizes the operating range decreases. The largest particle size that can be handled by this feeder is approximately 1.5 in. in diameter.

A mechanical variable-speed drive using a 3–15 PSIG pneumatic control signal, or an electrical variable-speed drive using a milliampere control signal, will provide additional rangeability of up to 20 to 1. A two-speed gear box can also be used, and this provides a 10-to-1 step change in belt speed.

General Considerations for Solids Feeders

The physical properties of the material to be handled, such as angle of repose, tendency to cake, density, particle size, and flow characteristics, will affect the design of the feeder and associated equipment. Since material flow upstream of the feeder is uncontrolled, surge capacity must be provided. A feed hopper with a sloping bottom section can be used to store a supply of material for the feeder. The bottom portion of the hopper, however, must have a slope great enough to insure material flow over the entire cross section of the vessel. The minimum slope will be dictated by the angle of repose, which is determined experimentally for the material. For particulate solids with a tendency to form into a solid mass under quiescent storage condition, flexible, vibrating hopper walls are required to agitate the material and to keep it in a flowing state. Materials with poor flow characteristics may require the use of a vibrating outlet on the vessel to insure flow. Very free-flowing solids, on the other hand, will flood a feeder gate, and the use of a rotary valve in place of a gate is recommended.

Particle size will affect the operating range of the feeder. Maximum ranges are possible with powders and small granules, and operating range will decrease with increasing particle size. Flow characteristics and uniformity of particle size will affect accuracy. Generally, feeder accuracy is in the order of ± 1 percent, but higher accuracies are possible with uniform, smooth-flowing materials.

Conclusions

Pumps and feeders are available with constant-speed or variable-speed drives. The latter vary from incremental speed changers, such as gears and pulleys, to continuously variable drives, such as variable-speed motors and transmissions. Variable-speed drives are discussed in Section 2.2.

A recapitulation of the performance features and a technical evaluation of the devices discussed is provided here. Plunger pumps are characterized by fairly high capacities and high output pressures. Corrosive liquids do not present a problem, since the pump is available in a large variety of corrosion resistant materials. Their use on slurry service, however, is not recommended.

Diaphragm pumps offer lower capacities and lower output pressures than plunger pumps. The diaphragm does provide a positive seal that prevents fluid leakage from the head into the drive unit. The diaphragm, however, must be conservatively rated in relation to operating conditions for good service life. The hydraulically operated diaphragm offers an advantage in this respect since the diaphragm is pressure balanced and therefore less likely to fail. The diaphragm pumps with a flexible hose have generally the same characteristics, except that the straight-through pumping chamber makes this pump more suitable for slurry service.

Opposed centrifugal pumps and ball check feeders are specific for severe slurry service. Due to their high cost, these devices should be considered only when all other methods of control fail to meet operating requirements.

Belt type gravimetric feeders offer high capacity, wide rangeability, and high accuracy for control of solids mass flow. However, the physical properties of the solid will affect the feeder design parameters. Accuracy and rangeability decrease with increasing particle size, although rangeability can be widened by using both variable belt speed and variable belt loading.

The properties of the flow particle determine whether gate or rotary valve is used at the feeder inlet and how the surge hopper should be designed. Self-powered feeders are not as accurate as belt feeders, but they require no external power and very little space.

Volumetric feeders vary in design according to capacity. Roll type feeders with their lower capacity are suitable for powders and very fine granules. Screw type and belt type feeders offer higher capacities and ability to handle larger particle sizes.

2.4 DAMPERS

Design Pressure:	To 15 PSI differential pressure.
Materials of Construction:	Steel, galvanized steel, stainless steel with aluminum.
Range:	To 66″ × 120″.
Accuracy:	±5% to ±10% of full scale.
Cost:	$300.
Partial List of Suppliers:	Honeywell, Inc.; Lundy Electronics & Systems Inc.; Penn Vent Co.; Ruskin Manufacturing Co.; Syntron, Inc.

Dampers are used extensively for throttling fluid flows where control quality is not very critical. Typical applications are air conditioning systems, flue gas ducts, and fans. Due to the large surface areas involved, pressure differential is a limiting factor. Dampers for solids service are normally exposed to very low gas pressures, and forces due to weight or motion of the solids impose the limitations on the vanes.

Parallel-Blade Dampers

Parallel-blade dampers are a modification of the butterfly valve discussed in Section 1.9. Structurally they consist of a rotating vane or blade within the damper frame. In larger sizes multiple blades are used to reduce the torque requirement of the actuator. Flow characteristics are similar to those of the butterfly valve (Figure 2.4a). Leakage rates are comparable to those of the butterfly valve, and soft blade edging to reduce leakage at low operating temperatures is available. The blades of multiple units are linked mechanically as illustrated in Figure 2.4b. The position of the driving blade for these units must be specified to fit the location of the actuator.

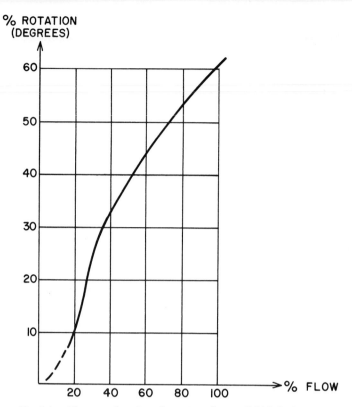

Fig. 2.4a Flow as a function of rotation of a parallel-blade damper

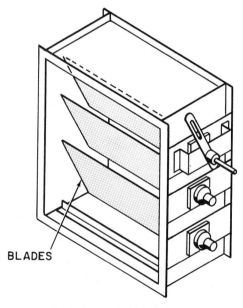

Fig. 2.4b Parallel-blade damper

Fan Suction Dampers

On blowers and fans, where throughputs must be controlled, radial vane dampers can be utilized. The damper consists of a number of radial vanes arranged to rotate about their radial axis (Figure 2.4c).

Control quality is not very good, and leakage rates are fairly high in the closed position. Control applications typically involve such secondary services as furnace draft control.

Variable-Orifice Damper Valves

Variable-orifice valves incorporate the principle of the iris diaphragm of the camera to achieve control action. The closure element moves within an annular ring in the valve body and produces a circular flow orifice of variable diameter (Figure 2.4d). The flow characteristics are similar to those of a linear valve. Tight shutoff, however, is not possible, and leakage rates are in the

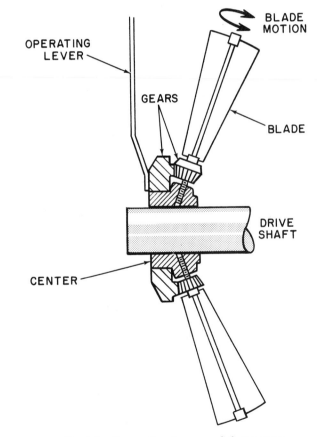

Fig. 2.4c   Fan suction damper, radial vane type

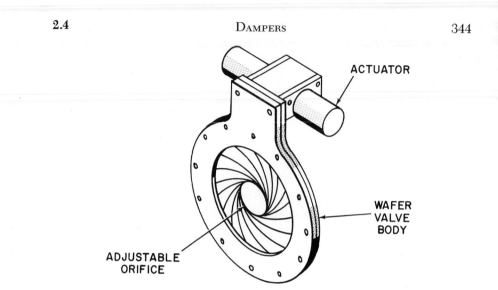

ACTUATOR

WAFER
VALVE
BODY

ADJUSTABLE
ORIFICE

Fig. 2.4d   Variable-orifice damper valve

order of or greater than those of a butterfly valve of equal size. Maximum pressure differential is limited to approximately 15 PSID. Dual valve units are available with a common discharge port for blending two streams.

On solids service, the variable orifice valve can be used for throttling. The valve, however, must be installed in a vertical line. The shutter mechanism of the valve forms a dam, making the valve unsuitable for solids service in horizontal lines.

Conclusions

Dampers are suitable for control of large flows at low pressures where high control accuracy is not a requirement. Typical applications of these units include air conditioning systems and furnace draft control. Variable-orifice valves, although smaller than dampers, offer better control quality and can be used to control vertical solids flow.

Chapter III

REGULATORS

L. S. Dysart, B. G. Lipták,
O. P. Lovett and R. L. Moore

CONTENTS OF CHAPTER III

	INTRODUCTION	347
3.1	REGULATORS vs CONTROL VALVES	348
	Advantages of Regulators	348
	Advantages of Control Valves	348
	Disadvantages of Regulators	348
	Disadvantages of Control Valves	350
	Conclusions	350
3.2	PRESSURE REGULATORS	351
	Introduction and Definitions	352
	Weight-Loaded Regulators	352
	Spring-Loaded Regulators	353
	Pilot-Operated Regulators	354
	Regulators for the Gas Industry	356
	Regulator Features	356
	Seating and Sensitivity	357
	Droop or Offset	359
	Regulator Noise	360
	Regulator Sizing and Rangeability	361
	Regulator Stability	361
	Regulator Safety	362
	Regulator Installation	362
	References	363
3.3	TEMPERATURE REGULATORS	364
	Types and Styles of Temperature Regulators	365
	The Regulator Valve Body	368
	Thermal Systems	371
	Vapor-Filled System	371
	Liquid-Filled System	375
	Hot Chamber System	376
	Fusion Type System (Wax Filled)	377
	Thermal Bulbs and Fittings	379
	Transmission Tubing	380
	Control Characteristics	381
	Special Features and Designs	381
	Conclusions	382

A regulator is a device that measures, controls and throttles a control loop. In other words, a regulator incorporates a sensor, a controller and a control valve. Theoretically, there can be as many types of regulators as there are process properties, but the following regulator types are most readily available:

Level regulators	(Volume I, Section 1.19)
Vacuum regulators	(Volume I, Section 2.7)
Flow regulators	(Volume I, Section 5.19)
Mass flow regulators	(Volume I, Section 7.6)
Pressure regulators	(Volume II, Section 3.2)
Temperature regulators	(Volume II, Section 3.3)

Although the majority of regulator types are discussed in Volume I as part of the treatment of detection techniques, the two most important types, pressure and temperature regulators, are separately covered in this chapter. This chapter is application oriented, but it also provides suggestions and recommendations as to when complete control loops should be used and when regulators will suffice.

3.1   REGULATORS vs CONTROL VALVES

Generally, a regulator is a control valve with a built-in controller. Many control applications in pressure, temperature and flow can be handled with either a regulator or control valve. What are the considerations for choosing between them and which type of valve should be selected? The advantages and disadvantages will be presented here, but the user will have to decide what is best for his application.

Advantages of Regulators

Regulators usually cost less than the combination of control valves, transmitters, and controllers. They are lower to buy, to install and to maintain. When the applications require larger valves, the economics begin to change in favor of control valves.

Regulators have a built-in controller and do not require an air supply. There are resulting savings in purchase and installation costs.

Regulators are not subject to air supply failure. The power to operate them is contained in the fluid being controlled. In critical fail-safe applications or at locations remote from a source of compressed air, this characteristic can be very important. However, diaphragm failure in a regulator usually causes valve opening, which *can* be unsafe.

Advantages of Control Valves

Control valves are used with an external controller having the flexibility of one, two, or three mode control and with remote-manual operation, and they are compatible with any measuring system. Materials of construction of the valve include any castable metal, and the valve may be lined. Fail-safe may be in any desired form. The many valve accessories include positioners, limit switches, manual handwheels, solenoids, and local controllers. They may be specified for maximum interchangeability with valves in other services, thus accomplishing maintenance savings by reducing spare parts requirements.

Disadvantages of Regulators

Set point of a regulator is provided integrally, and *remote control is not possible.* The controller is single-mode or proportional only, and therefore it suffers from having a set-point droop curve varying with throughput. It is limited in materials of construction and interchangeability and accessories cannot be applied to regulators. Regulators are limited to pressure, temperature, flow, and level type applications and cannot be used with some of the more complicated measurements, such as analysis, or with electrical signals, such as thermocouples. Even level regulators will not suffice if a large level range is required.

Table 3.1a
REGULATORS VS CONTROL VALVES

REGULATORS	CONTROL VALVES
Advantages	
Cost is lower in small sizes, and if ordinary construction materials are used.	It is used with external controller, which may be remote or local and may contain any control mode.
Size is smaller.	There is practically no limitation in material of construction.
Cost to install is less.	
The controller is built in.	When used with a remote controller, remote-manual control is possible.
	There is no limitation in fail-safe.
	A wide variety of accessories can be attached.
	Interchangeability between many service conditions is possible.
Disadvantages	
Remote control is impossible.	Cost is higher.
Set point must be set at the regulator location.	Handling is more difficult because of size and weight.
Controller is proportional only, thus exhibits a typical droop curve.	Air connections are required, increasing installation cost and making air failure a possibility.
Materials of construction are somewhat limited.	
Interchangeability in different services is limited.	
A very limited number of accessories can be used.	
Applications are fewer.	

Disadvantages of Control Valves

Control valves including the corresponding controllers and transmitters, are more expensive to purchase, install and maintain. They are generally larger, thus requiring more space for installation and making handling more difficult. They require air supplies with the resulting increased cost and increased maintenance due to the possibility of leakage.

Conclusions

There are definite applications for both regulators and control valves. Each has some advantages and disadvantages. Generally, if either is suitable, the lower-cost regulator is preferred. There undoubtedly are many potential applications where regulators could be successfully used in place of the more expensive control valve. Table 3.1a is a summary of the merits and drawbacks associated with the two basic designs.

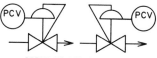

PRESSURE REDUCING AND
BACK PRESSURE
REGULATORS

3.2 PRESSURE REGULATORS

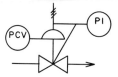

AIRSET, REGULATOR WITH
OUTLET PRESSURE RELIEF
AND INDICATOR

Regulator Types:
    a. Weight-loaded,
    b. Spring-loaded,
    c. Piloted $\frac{1}{4}''$ air regulator,
    d. Internally piloted,
    e. Externally piloted.

    Note: In the following summaries, the letters "a" to "e" refer to the designs listed above.

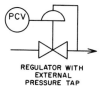

REGULATOR WITH
EXTERNAL
PRESSURE TAP

Sizes:
    $\frac{1}{2}''$ to 6" for "a," $\frac{1}{4}''$ to 4" for "b," $\frac{1}{4}''$ for "c," $\frac{1}{2}''$ to 6" for "d" and $\frac{3}{8}''$ to 12" for "e."

Design Inlet Pressure:
    Up to 6,000 PSIGs with "b," to 1,500 PSIG with "d" and "e," and to 500 PSIG with "a" and "c."

PRESSURE DIFFERENTIAL
REGULATOR WITH
INTERNAL AND EXTERNAL
PRESSURE TAPS

Minimum Regulated
Outlet Pressure:
    Down to 2 PSIG with "b" and "e," to 0.5 PSIG with "a" and "d" and to 0.1 PSIG with "c."

Droop or Offset:
    5% to 80% for "b," 2% to 10% for "e," 1% to 2% for "a" and "d" and $\frac{1}{2}$% for "c."

Cost for Size Ranges
Noted Above:
    $30 for "c," $10 to $400 for "b," $100 to $1,200 for "a" and "d" and $75 to $2,500 for "e."

351

Partial List of Suppliers:
Alcon Products Corp. (b); A. W.
Cash Valve Mfg. Co. (b); Cono-
flow Corp. (b); Fisher Controls
(a, b, e); General Controls,
ITT (b); Jordan Valve Div.,
Richards Inc. (b, e); Kieley &
Mueller Inc. (a, b); Kimray Inc.
(d); Leslie Co. (d); Masoneilan
International, Inc. (b, d); Moore
Products Co. (c); C. A. Norgren
Co. (b); Spence Engineering Co.,
Inc. (a, b, e).

Introduction and Definitions

The self-contained pressure regulator, developed late in the 19th century, has the advantages of simplicity, dependability, ruggedness, and low cost.[1] It is operated by energy from the flowing fluid itself.

Definitions of some terms commonly used in connection with regulators are listed below:

Droop   The amount by which the controlled variable; pressure, temperature, or liquid level deviates from the set value at minimum controllable flow when the flow through the regulator is gradually increased from the minimum controllable flow to the rated capacity.

Disturbance Variable   An undesired variable applied to a system which tends to affect adversely the value of a controlled variable.

Open-Loop Gain   The ratio of the change in feedback variable to the change in actuating signal.

As shown in Figure 3.2a, the diaphragm is the "brain" of the regulator. It compares the set point, which is converted into a spring force, to the regulated pressure, which is converted into a force by the diaphragm itself, and it adjusts the valve opening to reduce the error between the two. Thus the diaphragm is a feedback device, an error detecting mechanism, and an actuator. Figure 3.2a shows a spring-loaded pressure-reducing valve. Regulators are also available to control back pressure, differential pressure, and vacuum. They can be applied to gases, vapors and liquids, and they are actuated by springs, weights and gas pressures. Thousands of different regulator types and designs are available.[2] Descriptions of the typical regulator designs are given here.

Weight-Loaded Regulators

The weight represents a constant actuating force on the diaphragm, whereas a spring changes its force as it is compressed. Thus droop is minimized.

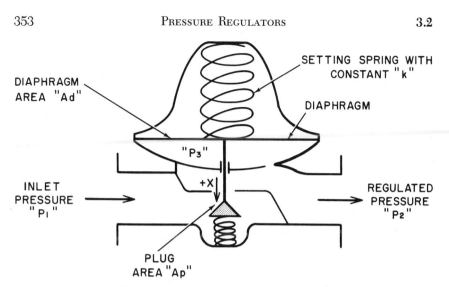

Fig. 3.2a   Spring-loaded pressure-reducing valve

Set-point changes are accomplished by adding to the weight or by changing its position (Figure 3.2b). The weight-actuated regulator is widely used to regulate gas pressures at 1 PSIG or below and where load changes are slow. It is *not* recommended for service where mechanical shock or vibration are present or for regulation of incompressible fluids.

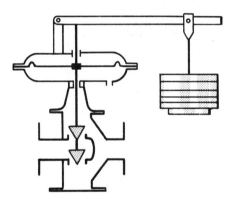

Fig. 3.2b   Weight-loaded pressure-regulating
valve with external pressure tap

Spring-Loaded Regulators

The "air regulator" is a $\frac{1}{4}$-in. air pressure regulator used to reduce the instrument air pressure to a level compatible with pneumatic instruments. The term "airset" is also used. Regulators with the ability to relieve excess regulated pressure are designated as "bleed type." This design is recommended for dead-end (no flow) service. Air regulators from various manufacturers have capacities of from 10 to 60 SCFM and are usually provided with an integral filter, hence the name "filter regulator."

One design of the spring-actuated regulator is shown in Figure 3.2c. The valve is normally closed. Compression of the spring by turning the handwheel

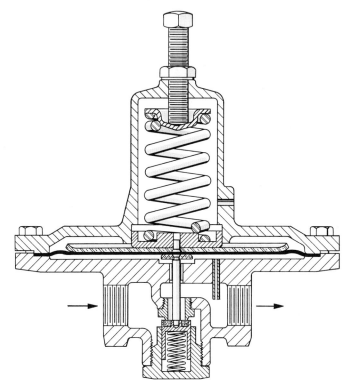

Fig. 3.2c   Spring-loaded pressure regulator

or setscrew opens the valve. Increasing downstream pressure acts beneath the diaphragm, raising it, and closing the valve. Thus spring compression can be adjusted to provide the desired downstream pressure at the flow throughput demand.

A spring-actuated vacuum pressure regulator is shown in Figure 3.2d.

The spring-actuated regulator is used more than any other type because of its economy and simplicity.[2, 3] It also has an extremely fast dynamic response. Its disadvantages include high droop, awkward set-point adjustment, and sensitivity to shock and vibration.[3]

Pilot-Operated Regulators

The pilot-loaded regulator is a two-stage device. The first stage is a spring-actuated regulator that controls pressure on the diaphragm of the main

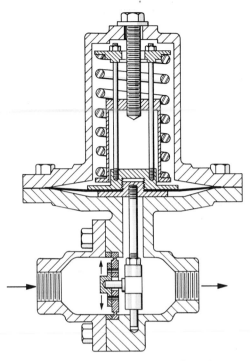

Fig. 3.2d Spring-loaded vacuum regulator. (Courtesy of Jordan Valve Div. Richards Industries.)

regulating valve. Its advantages are that the actuating fluid operating the first-stage regulator is at the upstream pressure, providing a higher force level to the actuating mechanism, and that its travel is very short, which reduces its droop. Thus more accurate regulation is possible with this design. The pilot (first-stage) regulator can be external or internal, as shown in Figure 3.2e. Here it controls pressure to a diaphragm operating the main valve. In some designs this diaphragm is replaced by a piston, which serves the same function. Since piloted regulators use the difference between upstream and downstream pressures for actuation, a minimum differential is required for proper operation.

The piloted regulator provides accurate regulation at a wide range of pressures and capacities. It is more expensive to purchase and install on the basis of valve size, but not necessarily on the basis of capacity.[1] It is more complex and requires clean flowing fluids to avoid plugging the small passages and ports. Its response is slower than that of the spring-loaded design.

The externally piloted regulator is available in the widest range of sizes and has the advantage of permitting the pilot regulator to be mounted at a distance from the main valve. Thus it can be mounted in an accessible location and away from possible pipeline shock and vibration. The external pilot also simplifies maintenance.

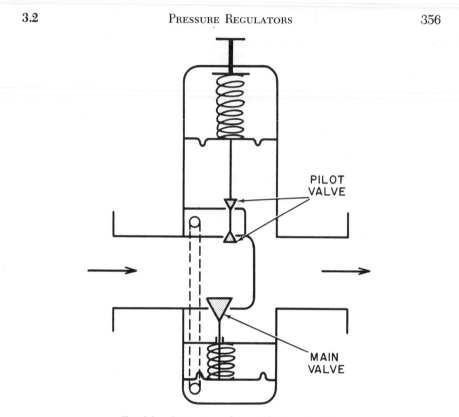

Fig. 3.2e   Pressure regulator with integral pilot

Figure 3.2f illustrates an internally piloted regulator with pneumatic set-point loading. This design features the convenience of remote set-point adjustment at distances up to 500 ft, but it is more expensive to operate because it requires compressed air. Table 3.2g provides data to help in deciding between spring-loaded and pilot-loaded regulators.

Regulators for the Gas Industry

In addition to the regulators available for general service, the gas utility industry utilizes an extensive family of regulators designed specifically for gas service.[4, 5] The types of regulators for these services are tabulated in Table 3.2h.

Regulator Features

The self-contained regulator has no adjustments, either in terms of capacity (reduced valve trim) or stability (proportional band and reset adjustments). Thus service conditions must be well defined in order to specify a regulator that will result in a satisfactory installation. Consideration must be given to the following features.

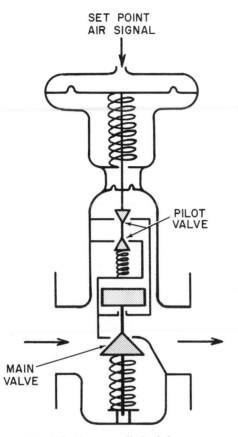

Fig. 3.2f   Pneumatically loaded remote
set-point pilot regulator

Seating and Sensitivity

A balance of forces on the valve stem (Figure 3.2a) are as follows:

$$kx = A_p(p_1 - p_2) - A_dp_3 \qquad\qquad 3.2(1)$$

where k = spring compression rate (lbf/in.),
 $\quad x$ = valve stem movement (in.),
 $\quad A_p$ = area of port (in.^{2}),
 $\quad p_1$ = inlet pressure (PSIG) and
 $\quad A_d$ = area of diaphragm (in.^{2}).

Insertion of typical values in equation 3.2(1) can result in the product $A_p(p_1 - p_2)$ being relatively large compared with A_dp_3. Thus the actuating force level is comparatively low, and a balanced plug is therefore desirable because it eliminates the term $A_p(p_1 - p_2)$. The double-seated plug shown in Figure 3.2b is essentially balanced, but it is limited to continuous service,

Table 3.2g
REGULATOR SELECTION FOR VARIOUS SERVICE CONDITIONS[3]

Recommended Regulator Type	Magnitude of Pressure Reduction	Supply Pressure Variations	Load Variations
Self-actuated	Moderate	Small	Moderate
Self-actuated	Moderate	Large	Moderate
Pilot-actuated	Moderate	Large	Large
Pilot-actuated	Large in one stage	Moderate	Large
Pilot-actuated	Large in two stages	Large	Large in first stage
Self- or Pilot-actuated			Moderate in second stage

since it *cannot* be made to shut off tightly. The sliding gate mechanism shown in Figure 3.2d is also balanced, but it cannot be used with high pressure drops. Analysis of the regulator as a feedback mechanism[6] shows that the balanced plug reduces sensitivity to variations in supply pressure.

Intermittent service requiring tight shutoff necessitates the use of a single-seated valve. Various designs utilizing balancing pistons, diaphragms and ingenious seating configurations can provide a balanced plug in single-seated designs at the expense of greater complexity and cost. In valves with

Table 3.2h
REGULATORS USED IN THE GAS INDUSTRY

Application	Inlet Pressures	Regulated Outlet Pressures
Appliance	8" H_2O	3–4" H_2O
Household	2–100 PSIG	5–8" H_2O
Community of Users	Up to 200 PSIG	3" H_2O to 60 PSIG
Transmission Line Take-off in Two Stages	Up to 1,200 PSIG	100–1,000 PSIG
Transmission Line Take-off to Individual User	Up to 1,500 PSIG	3–600 PSIG

port diameters of 1 in. or less, the ratio of diaphragm area to port area can be made large enough that balancing is not required for moderate pressure drops.

Droop or Offset

The regulator is a complete, self-contained feedback control loop with only proportional control action.[6] Thus the regulated pressure will be offset by changes in the disturbance variables (upstream pressure and flow demand in case of a pressure-reducing valve). The offset in regulated pressure with changing flow is called droop. It can be expressed[6] as follows:

$$\Delta p = \frac{(kx/q)(p_1^2 - p_2^2)}{kxp_2 + A_d(p_1^2 - p_2^2)} \Delta q \qquad 3.2(2)$$

where q = throughput and other factors are from equation 3.2(1) and Figure 3.2a.

Equation 3.2(2) shows droop to be a function of the pressures and of the regulator design parameters such as spring rate k, valve lift x, and diaphragm area A_d. Size and economic considerations limit the length of the spring (for low spring rate) and also limit the diaphragm area. Low lift (x) is necessary to reduce diaphragm fatigue as well as to minimize droop. Figure 3.2i illustrates the phenomenon of droop.[7]

Equation 3.2(2) applies to the flow range of q_1 to q_2 in this figure, which is the operating span of the regulator. While ideally linear, it takes on many shapes in practice because of valve plug flow characteristics and varying effective diaphragm areas. The valve is fully open at q_2 and acts as a fixed orifice from q_2 to q_3. The span from 0 to q_1 is dominated by flow-generated pressure forces on the plug,[8] and q_1 is considered to be the minimum controllable flow rate. The minimum flow rate is a function of plug design and is typically *5 to 10 percent of maximum capacity* (q_2).

Figure 3.2i emphasizes that maximum regulator capacity is not at full

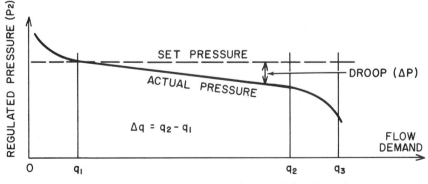

Fig. 3.2i   Regulated pressure as a function of flow demand

valve opening, but at maximum acceptable droop. Information on droop versus flow is therefore essential for satisfactory regulator performance.

Regulator designs to minimize droop are available. They include placing the feedback sensing line at a point of high velocity, either by placing it in the throat of a slight restriction or by extending it into the flowing fluid (Figure 3.2c). The latter design makes use of the aspirating effect. Droop compensation is also provided by a *moving valve seat*, called a pressure-compensating orifice, which moves with upstream pressure. The *roll-out diaphragm*, which reduces its effective area with spring compression, also acts to reduce droop. These devices make possible the specification of the droop curve over limited flow spans.

Regulator Noise

High velocities of compressible fluids are primarily responsible for noisy regulator installations, a continuing source of complaint. Aerodynamic noise caused by high-velocity gas flow in the valve body has been identified as the principal source of the irritating "screaming." The volume of a gas increases in proportion to the ratio of inlet and outlet pressures. Gas velocity increases in proportion to the volume increase if the cross-sectional area of the regulator outlet port does not also increase. Increased velocity raises pressure drop until sonic velocity is reached. A tapered pipe expander downstream from the valve has been shown to increase the sonic velocity (mach 1) to supersonic velocities of mach 2 to mach 3. Velocities above sonic result in (a) noise, (b) a major portion of the energy conversion being downstream from the valve plug, (c) a static pressure build-up in the downstream piping that can far exceed its pressure rating, and (d) a "choking" (reduced capacity) of the regulator due to the downstream static pressure.[9] A more detailed discussion of this subject is given in Section 1.21.

Since sonic velocity is known to occur when downstream pressure is less than 50 percent of upstream pressure, it does not occur at pressure reductions of less than 2 to 1. Because changes in flow path direction also contribute to noise, a maximum gas velocity of 200 fps (500 fps below ground) has been recommended. Gas velocity in regulators is difficult to calculate because of interacting velocity, pressure drop, and cross-sectional area. Many manufacturers provide tables of maximum capacities for quiet operation. (Section A.3.)

Maintaining gas flows to less than sonic velocities at the regulator outlet is recommended to avoid noisy installations. Since velocity is difficult to calculate, sonic velocity is avoided by limiting pressure reduction to less than the critical ratio of 2 to 1. High-to-low reduction should be made with regulators in series, also called stages. Two to three stages are common to reduce noise and improve regulation. Slightly less than critical reduction should be made in the second of two stages, or in the second and third of three stages, with the remainder of the reduction across the first stage.

Noise is further reduced by eliminating changes in flow direction, and thus as much straight pipe as practical on both sides of the regulator is recommended. Mufflers are also available for further noise reduction.

Noise in regulators on liquid service that is caused by the valve being opened or closed too quickly is called water hammer. This is a vibration due to the "bath tub stopper" effect of the plug operating too close to the seat because of the valve being oversized. Another noise source is cavitation due to localized vena contracta pressure drop from which vibration damage and valve metal erosion can result. Maximum velocities of 15 fps are recommended to avoid these difficulties.

Regulator Sizing and Rangeability

Oversizing is the most common error in regulator selection. The droop characteristic makes a larger valve attractive, a greater capacity being obtainable for the same droop. The larger valve also reduces noise because of its larger passages. These apparent advantages are offset by higher cost, severe seat wear, and poor regulation.

The limitation on sizing is rangeability. Rangeability varies from 4 to 1 for a steam regulator, which cannot be operated close to its seat because of wire drawing, to over 50 to 1 for an air regulator. Rangeability is illustrated by Figure 3.2i. Minimum flow q_1 is 5 to 10 percent of q_2, depending on the seat configuration. Maximum flow is not necessarily q_2 but is determined by maximum acceptable droop. Thus a typical rangeability is 10 to 1, and therefore oversizing may not permit accommodation of the minimum flow requirement of certain installations.

Regulators are sized on the basis of tabulated data or valve coefficients (C_v) provided by the manufacturer. Size must be chosen to accommodate maximum flow at minimum pressure drop. The valve should ideally operate at some 50 to 60 percent open under normal conditions. Catalog information must be used judiciously, however, since capacity and droop might be specified at different points. Rated capacity might result in too high a velocity, or the (external) pressure feedback tap might have been located at a different point during the testing than in the application.

Rangeability can be increased by two regulators in parallel. The pressure set point of one regulator is set 10 percent higher than the other. Thus one regulator will be wide open while the other modulates under normal conditions. As demand is reduced, the second regulator will close and the first will modulate. Leakage in the second regulator must be small enough to be a minor portion of the capacity of the first regulator.

Regulator Stability

A regulator chosen in accordance with the preceding information may perform very poorly because the regulated pressure does not stabilize. Stability of the regulator installation depends on the open-loop gain,[6] which for the regulator shown in Figure 3.2a is defined as

$$K_o = \frac{A_d(p_1^2 - p_2^2)}{kxp_2}$$

3.2(3)

where K_o is the open-loop gain.

The factors here are the same as those upon which the droop depends, as shown in equation 3.2(2). Comparison of the two equations indicates that parameter changes which decrease droop would increase the open-loop gain and decrease stability, as expected in a feedback mechanism. Since the regulator has no controller mode adjustments, the manufacturer must choose design parameters before fabrication, which will provide an adequate compromise between droop and stability. It can thus be expected that the open-loop gain of some applications will be too high, resulting in a noisy and cycling pressure regulator.

Guidelines for a stable installation are few. Difficulties are generally found after installation, and because the regulator has no adjustments, it is costly to alleviate the conditions. Generally, field revisions to stabilize an installation include (a) relocation of the pressure-sensing tap, (b) redesign of the downstream piping to provide more volume, and (c) restricting the pressure feedback line, either by a needle valve in an external line or by filling an internal line and redrilling a smaller hole.

Regulator Safety

Diaphragm rupture is the most common failure source in a regulator. Most regulators fail full open upon diaphragm failure, an unsafe condition. Thus, a relief valve on the regulated pressure is recommended if an increase in regulated pressure to the level of the supply pressure is considered unsafe. In installations where it is imperative that the user continue to be supplied even on regulator failure, two regulators can be placed in series. The second regulator will have its setpoint adjusted some 10 percent higher than the first, and it will thus remain wide open under normal circumstances. Upon failure of the first regulator, the second will take over the pressure regulation.

Regulator Installation

Regulator installation is generally considered to be easier than regulator selection. The following installation suggestions will help ensure satisfactory regulator performance:

1. Steam regulators should be preceded by a separator and a trap.
2. All regulators should be preceded by a filter or strainer.
3. A valve by-pass is recommended where it is necessary to service the regulator while continuing to supply the users.

4. Straight lengths of pipe upstream and downstream will reduce noise.
5. External feedback lines should be $1/4$-in. pipe or tubing.
6. Locate the pressure feedback tap at a point where it will not be affected by turbulence, line losses or sudden changes in velocity. A distance of 10 ft from the regulator has been recommended.

REFERENCES

1. Baumann, H. D., "Regulator or Control Valve?" Isa Journal, (December 1965), p. 51.
2. Kubitz, R., "Valves in Gas Regulators," Instruments and Control Systems (April 1966), p. 139.
3. O'Connor, J. P., "12 Points to Consider When Selecting Pressure Regulators," Plant Engineering (March 1958), p. 109.
4. Beard, C. S., and F. D. Marton, "Regulators and Relief Valves," Chilton Co., Philadelphia, Pennsylvania, p. 79.
5. Perrine, E. B., "Pressure Regulators and Their Application," Instruments and Control Systems (September 1965), p. 167.
6. Moore, R. L., "The Use and Misuse of Pressure Regulators," Instrumentation Technology (March 1969), p. 52.
7. Yedidiah, Sh., "Effect of Seat Configuration on Pressure Operated Control Valves," Design News (October 11, 1967), p. 114.
8. Quentzel, D., "A Balance Factor for Pressure Reducing Valves," Control Engineering (May 1965), p. 97.
9. Baumann, H. D., "Why Limit Velocities in Reducing Valves?" Instruments and Control Systems (September 1965), p. 135.

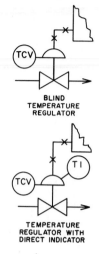

**BLIND
TEMPERATURE
REGULATOR**

**TEMPERATURE
REGULATOR WITH
DIRECT INDICATOR**

3.3 TEMPERATURE REGULATORS

Sizes:	From ¼″ to 6″.
Features:	Two- or three-way designs with direct or reverse action, only proportional mode of control and the only adjustment capability being the changing of set point. Bulb pressure may exceed 1,000 PSIG.
Ranges:	−25 to 480°F.
Spans:	Standard from 30 to 60°F, special up to 100°F.
Materials of Construction:	BODY in iron, bronze, steel or stainless steel, BULB in copper, steel, stainless steel or coated with PVC, Teflon etc., TUBE in either copper or stainless steel.
Tubing Length:	Usually 10 ft with 50 ft being the maximum.
Base Cost:	$125 for standard unit in ½″ size. $250 for same in 2″ size.
Partial List of Suppliers:	Dresser Industries, Inc. Valve Division; Lawler Automatic Controls; Leslie Co.; Powers Regulator Co.; Robertshaw Controls Co., Fulton Sylphon Division; Sarco Co.; Spence Engineering Co.; Trerice Co.

The advantages inherent in temperature regulators and the relatively wide range of choices available today will recommend them for many temperature control jobs which do not require the high performance level or sophisticated features of temperature controllers and systems of the pneumatic or electric/electronic types.

A temperature regulator is a temperature controlling device (Figure 3.3a)

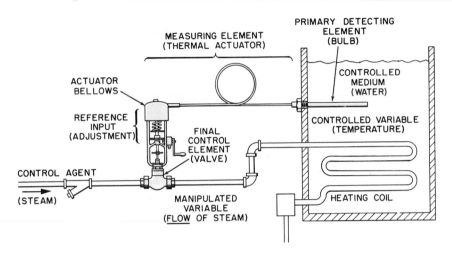

Fig. 3.3a   Self-actuated temperature regulator

including within itself a primary detecting element (bulb), a measuring element (thermal actuator), a reference input (adjustment), and a final control element (valve). It might properly be referred to as an "automatic temperature control system," but the term regulator is preferred.

Requiring no external power (electricity, air, etc.), regulators are described as "self-actuated." They actually "borrow" energy from the controlled medium for the required forces.

Types and Styles of Temperature Regulators

The two types of regulators available, direct-actuated and pilot-actuated, are distinguished by the way in which the valve (the final control element) is actuated.

In the direct-actuated type, the power unit (bellows, diaphragm etc.) of the thermal actuator is directly connected to the valve plug and develops the force and travel necessary to fully open and close the valve (Figure 3.3b). In the pilot-actuated type, the thermal actuator moves a pilot valve, internal or external (Figure 3.3c). This pilot controls the amount of pressure energy from the control agent (fluid through valve) to a piston or diaphragm, which in turn develops power and thrust to position the main valve plug. The pilot

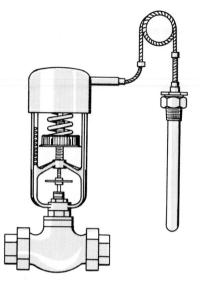

Fig. 3.3b   Direct actuated temperature regulator

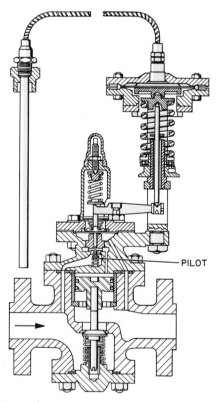

PILOT

Fig. 3.3c   Pilot-actuated temperature regulators

may be internal or external. When external, independently acting multiple pilots are also available.

Direct-actuated regulators are generally simpler, lower in cost, and more truly proportional in action (with somewhat better stability), whereas pilot-actuated regulators have smaller bulbs, faster response, shorter control (proportional) band, can handle higher pressures through the valve, and can handle possible interrelated functions through use of multiple pilots, such as temperature plus pressure plus electric interlocks, etc.

There are two styles of temperature regulators, self-contained and remote-sensing, and which one is appropriate depends on the location and structure of the measuring element (thermal actuator).

Self-contained regulators have the entire thermal actuator contained within the valve body, and the actuator serves as the primary detecting element (Figure 3.3d). Thus the self-contained style can sense only the tem-

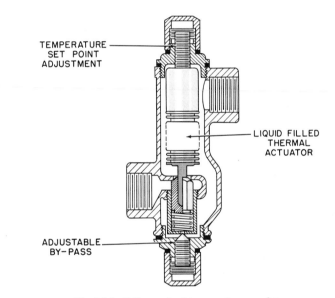

TEMPERATURE SET POINT ADJUSTMENT

LIQUID FILLED THERMAL ACTUATOR

ADJUSTABLE BY-PASS

Fig. 3.3d   Self-contained temperature regulator

perature of the fluid flowing through the valve, and that fluid might be said to be both the controlling agent and the controlled medium. The device regulates the temperature of the fluid by regulating its own flow. Self-contained regulators are generally provided with liquid expansion or fusion type thermal elements.

Remote-sensing regulators have the bulb (the primary detecting element) separate from the power element (bellows, etc.) of the thermal actuator, and generally connected to it by flexible capillary tubing (Figures 3.3a and b).

This style is able to sense and regulate the temperature of a fluid (the controlled medium) as distinct from that flowing through the valve (the control agent).

The self-contained style is simpler, frequently lower in cost, and "packless" (has no stem sliding through the valve body "envelope"), but it is limited in application to such uses as regulating the temperature of coolant (water, etc.) that is leaving the engine, compressor, exothermic process, etc.

The Regulator Valve Body

Further discussion here is limited to some special forms available and to factors relating to selection and use due to force limitation of the thermal actuator. See also the discussion on globe valves in Section 1.3, which is also applicable here.

Action refers to the relationship of stem motion to plug and seat position. A direct acting valve closes as the stem moves down into the valve. A reverse acting valve opens as the stem moves down. A three-way valve combines these, one port (or set of ports) opening and the other port(s) closing on downward movement.

Relating valve action to motion of the actuator on a temperature change gives the following combinations:

Heating control is accomplished by a direct acting regulator, using a direct acting valve to reduce the flow of heating medium on temperature rise.

Cooling control is most frequently achieved by a reverse acting regulator with a reverse acting valve to increase the flow of coolant on rising temperature. However, on systems using a secondary coolant with steady flow through the process, a direct acting regulator with its valve in a by-pass line is occasionally used to vary the flow through the heat exchanger. A three-way valve provides positive control of such a cooling system by giving a full range of flows through the two legs of the cooling circuit (exchanger and by-pass).

Mixing of two media at different supply temperatures to control the mixed temperature is also accomplished with three-way valves.

Single-seated valves are desirable for minimum seat leakage in the closed position. The pilot-actuated regulators *always* use single-seated valves (Figure 3.3e). On direct-actuated types, the use of single-seated valves is limited to a 2-in. maximum by two factors: the required full-open lift (approximately $\frac{1}{4}$ of the valve size) and the closing force required (full port area times maximum pressure drop across the valve). A single-seated valve with piston balanced plug (Figure 3.3e) eliminates the closing force problem but requires slightly higher full-open lift.

Double-seated valves are most common on direct-actuated regulators in sizes up to 4 in. The reasons for this include the feature of low lift to full open (approximately $\frac{1}{8}$ of valve size) and the minimum closing force requirement. The unbalanced area of double-seated valves is the difference between

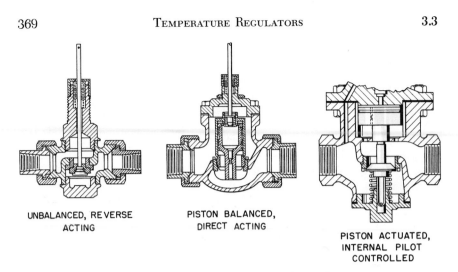

UNBALANCED, REVERSE
ACTING

PISTON BALANCED,
DIRECT ACTING

PISTON ACTUATED,
INTERNAL PILOT
CONTROLLED

Fig. 3.3e   Single-seated, two-way regulator valves

the two port areas, which is approximately 6 percent of the larger of the two—hence, the term "semi-balanced" for double-seated valves (Figure 3.3f).

Three-way valves as used with temperature regulators are of three basic designs (Figure 3.3g). Smaller valves (to 2 in. maximum) are of the "unbalanced" single-seated construction with a common plug moving between two seats. In sizes from 2 to 6 in. there are two designs. A sleeve type plug (requiring a body seal) moves between two seats and provides near balance of closing forces, but it requires typical single-seated lift. A semi-balanced design, essentially two double-seated valves for each flow path with connected plugs, gives both low lift and low seating force. This construction is somewhat more expensive, however.

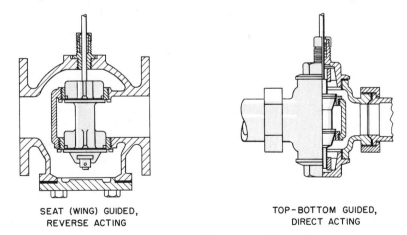

SEAT (WING) GUIDED,
REVERSE ACTING

TOP-BOTTOM GUIDED,
DIRECT ACTING

Fig. 3.3f   Double-seated, two-way regulator valves

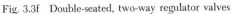

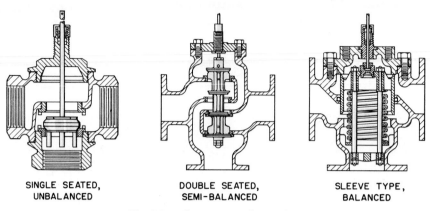

SINGLE SEATED,	DOUBLE SEATED,	SLEEVE TYPE,
UNBALANCED	SEMI-BALANCED	BALANCED

Fig. 3.3g   Three-way regulator valves

Plug guiding is considered to be a less important factor with regulators than with control valves because the applications involve relatively low pressures and velocities. Therefore, many direct-actuated single-seated valves are only stem guided. Most double-seated valves are seat ("wing") guided, but some top- and bottom-guided designs are also in use. Pilot-actuated valves are generally well guided, as are the piston balanced valves.

Flow characteristics   Most regulator valves are of the quick opening type in order to use the shortest practical lifts. On special basis, low-lift characterized plugs are available in some small single-seated valves and also in a few double-seated valves.

Materials of Construction   Standard materials used in body construction are bronze in sizes below 2 in. and cast iron in larger sizes. Steel and stainless steel are considered to be special body materials. The standard bronze valve trim is used with pressure drops below 50 PSID, while stainless steel is recommended for greater drops. On low pressure and temperature services, composition discs can be used for tight shutoff. Stellite or hardened alloy trim for extreme services is special, though some pilot-actuated designs offer it as standard. Monel and other special alloys are available for specialized services, such as seawater.

Pressure rating of a temperature regulator is frequently determined more by limitations in the thermal actuator power (seating force) than by the design and materials of the valve body and trim. Therefore, while catalogs will usually state the maximum temperature and pressure rating for the valve, they should also give a lower recommended maximum pressure limit for a direct-actuated regulator. This will usually decrease as valve size increases, and it should be given full cognizance. With pilot-actuated valves and piston-balanced valves (direct actuated) the actuator limitation is *not* a factor. However, regulators are seldom available or recommended for pressures above 250 PSIG.

End Connections   To lessen the chance of installation damage, union ends are usual for $\frac{1}{2}$ to $1\frac{1}{2}$ in. valves (in some cases 2 in.). Flanged ends are standard in sizes 2 in. and larger.

Valve Capacity   The valve capacity data published by manufacturers is reasonably accurate, having been determined by recognized test procedures. Most manufacturers publish capacity tables for common fluids, from which the full-open capacity figures can be read directly for various supply pressures and pressure drops. In a few cases, the information takes nomographic form. Some list the valve capacity factor C_v for each valve size and type.

Seat Leakage   Regulators should not be considered as positive shutoff devices. However, in "as new" condition, leakage across seats in the closed position for metal seated valves may be generally expected to be at the level of 0.02 percent of full capacity for single-seated and 0.5 percent for double-seated regulators. Single-seated valves with composition disc are almost "dead tight," or "bubble tight."

Thermal Systems

There are four classes of "filled system" thermal actuators used in temperature regulators. These are distinguished by the principle of power development. All develop power and movement proportional to the measured temperature, and hence temperature regulators are basically proportional in action. For a detailed discussion of filled thermal systems, refer to Section 4.3 of Volume I. Gas-filled systems (SAMA Class III) and bi-metal or "differential expansion" thermal elements are not presently used in the family of devices covered here.

Vapor-Filled System

The operation of the vapor pressure class (SAMA Class II) is illustrated in Figure 3.3h. The thermal actuator, having been evacuated to eliminate the "contaminating" effect of air and other gases and vapors, is partially filled

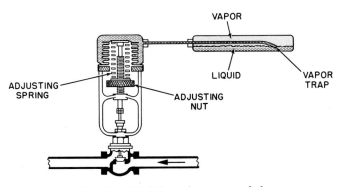

Fig. 3.3h   Vapor-filled thermal actuator and element

with a volatile liquid. Liquids used are chemically stable at temperatures well above the range used.

The bellows and tubing usually contain liquid only, although with ambient conditions *above* the set point of the regulator they may contain vapor only—*never both* under conditions of normal operation. The bulb contains the liquid-vapor interface, and so the pressure within the actuator is the vapor pressure of the charge or fill liquid at the temperature sensed by the bulb. This vapor pressure, increasing with rising temperature, acts on the bellows against the force of the adjustment to produce the power required to reposition the valve plug.

Figure 3.3i shows various charging liquids used, each having a different vapor pressure curve and atmospheric boiling point. Different ranges are achieved by using the different liquids whose curves are shown, not by changing springs.

Ranges of adjustment from −25 to +480°F are commonly available, and common spans are in the order of 40 to 60°F. In some special cases, involving small short-stroke valves, longer spans up to 100°F are furnished. Other special designs have shorter ranges of 25 to 30°F.

Adjustment of vapor pressure regulators is accomplished by changing the initial thrust of the adjusting spring (Figure 3.3h). Turning the adjusting nut toward the bellows to compress the spring increases the spring thrust. This requires an increased pressure to start valve movement and thus effects a higher temperature setting. Lowering the adjustment to decompress the spring permits motion of the stem with lower actuator pressure, thereby lowering the set temperature.

The rate of the adjusting spring (force required per inch of spring compression) is an important factor in determining the length of adjustment span as well as the proportional band of a regulator. The stiffer the spring, the longer the span or range, but also the longer the proportional band.

Weight-and-lever adjustment (Figure 3.3j) once very common, is seldom used today because of problems of position, location and general vulnerability of the moving lever. Here the force produced by a weight acting on an adjustable length lever or "movement" arm opposes the thrust produced by the thermal actuator. Increasing the length of the lever raises the setting. Since the force of the weight-and-lever system has no "rate" (being almost constant for a given setting), the proportional band of a weight-adjusted regulator is narrower than for a spring-adjusted design.

Over-temperature protection is standard in most regulators of the vapor pressure class. It consists of a special spring-loaded and compressible "over-run" section of the actuator stem. When increasing temperature produces a force beyond that for normal control action, this over-run compresses. The added movement of the actuating bellows drains the remaining liquid from the bulb, and pressure then follows the "gas law" instead of the vapor pressure curve. This provides protection up to 100°F above the top end of the range.

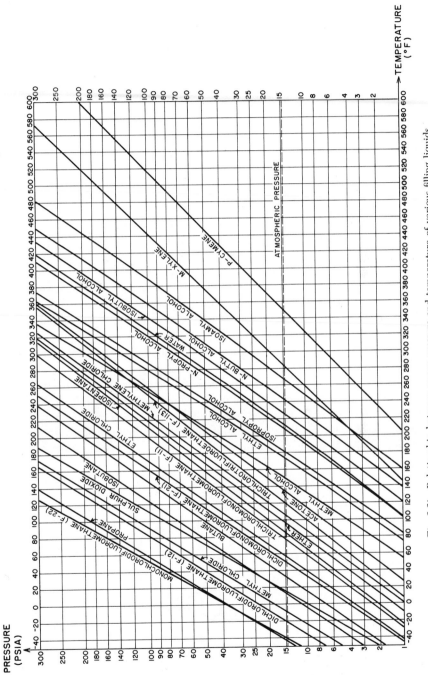

Fig. 3.3i Relationship between vapor pressure and temperature of various filling liquids

373

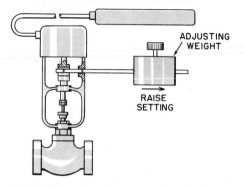

Fig. 3.3j   Weight-adjusted vapor pressure-filled regulator

Temperature indication is available as a special feature (Figure 3.3k). Generally, this is simply a compound pressure gauge, sealed in the thermal system and responding to the changing pressure of the fill. The dial is calibrated to correspond to the temperature-pressure curve of the particular fill (charge) and therefore "reads" the bulb temperature. In some designs the thermometer has its own sensing element, with a separate bulb contained in or attached to the regulator bulb. For this design improved characteristics are claimed. In either case, the "thermometer" is in the order of approximately ±2 percent accuracy. Thermometers are furnished with both Fahrenheit and Celsius calibration.

Actuator bellows design (diameter, area, length, etc.) is the major factor in determining the characteristics of the vapor pressure regulator. A small bellows, having limited power and stroke, results in a compact actuator, successful only for small valves and on moderate to low pressures. Increasing bellows size makes direct actuation of larger valves possible at higher pressure

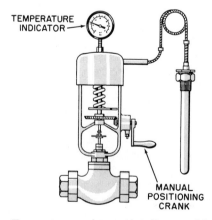

Fig. 3.3k   Temperature regulator with indicator and handwheel

levels. However, since the bellows movement requires the transfer of some liquid from the bulb to the bellows (equal to the volumetric displacement), larger bellows require larger bulbs. In general, larger bellows result in shorter adjustable ranges but also narrower proportional bands. Some manufacturers offer a choice of designs relating to bellows size in order to offer different characteristics.

Bulb size of the vapor pressure class regulators, governed by the volumetric displacement of the bellows (area × valve travel), varies with actuator design and with valve design and size. It is also affected by the possible need to operate under "cross-ambient" conditions; i.e., with ambient temperature either above or below the regulator's set point. This requirement produces a bulb substantially larger than that used where the ambient temperature will always be either above or below the range of the regulator. The use of vapor-filled thermal elements on cross-ambient applications is generally *not* recommended. Bulb sizes of this class range from ½ in. OD by 6 in. long in some very small units to 1½ in. OD by 36 in. long in large valves with cross-ambient fill.

Bulb construction for vapor pressure regulators includes a "vapor trap" (Figure 3.3h). This serves to give the best possible response by insuring liquid transfer of pressure changes from bulb to bellows. Since the end of the vapor trap tube must always be in liquid, it requires that the installation of the bulb be made in accordance with markings on the bulb. For installations where the tube connection end of the bulb is below the horizontal, the extended vapor trap must be omitted.

Advantages of the vapor pressure class regulators—
No ambient temperature effect on calibration.
Wide choices of designs, characteristics and ranges available. Many special features available.

Liquid-Filled System

The operation of the liquid expansion class (SAMA IB) is illustrated in Figure 3.3l. The thermal actuator is completely filled with a chemically stable liquid. Changing temperature at the bulb produces a volumetric change, which causes the small area bellows to move in a corresponding direction and thereby move the valve. The force and action are positive and linear, limited only by the components of the element, chiefly the actuator and adjustment bellows.

Ranges are from 0 to 300°F, with spans generally being from 30 to 50°F. The special 100°F spans are sometimes furnished with very small, short-stroke valves. The different ranges (not field changeable) are achieved by filling procedure and sealing temperature. Hydrocarbon fills are generally used. Mercury is not used as a filling fluid for regulators because of cost and hazards.

Adjustment is achieved by moving the actuator bellows relative to the valve plug (Figure 3.3a) or by use of a separate adjusting bellows (Figure 3.3l).

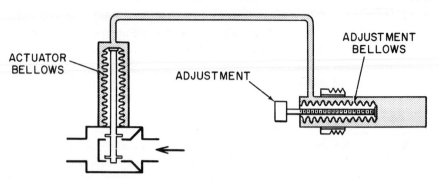

UNIFORM LIQUID EXPANSION PER DEGREE TEMPERATURE RISE
AT SENSING BULB COMPRESSES VALVE ACTUATING BELLOWS
TO CONTROL FLOW OF HEATING OR COOLING MEDIUM

Fig. 3.3 l   Liquid-filled thermal actuator and element

This adjustment bellows transfers liquid into the actuator bellows to lower
the set point, or vice versa. The adjustment may be located at the bulb or
at the valve, or it may be separate from both.

The proportional band is generally wider than with other classes of
actuators, and it is a direct function of bulb volume to actuator bellows area.

Over-temperature protection, which permits over-travel of the bellows,
allows an uncontrolled temperature to overrun the set point by up to 50°F.

Ambient temperature effects on the liquid in the tubing, bellows and
adjustment are minimized by design, but wide ambient swings do introduce
errors.

Packless construction (no moving stem through the body envelope) is a
feature of this class regulator. However, use on corrosive fluids must be
avoided.

Temperature indication is not generally available.

Advantages of the liquid expansion class regulators include a positive
actuating force, linear movement and packless design.

Hot Chamber System

The hot chamber class has no corresponding SAMA designation. This
thermal actuator is partially filled with a volatile fluid (Figure 3.3m) and has
some characteristics common to the liquid expansion class. Rising temperature
at the bulb forces liquid to the actuator bellows or "hot chamber." There
the heat of the control agent, always at a temperature substantially above
the regulator range, flashes the liquid into a superheated vapor. The pressure
increase in the bellows causes the valve to move against a return spring.

Hot chamber regulators are furnished in sizes from $\frac{1}{2}$ to $1\frac{1}{2}$ in. Because
of practical considerations, only direct acting designs are available (close on

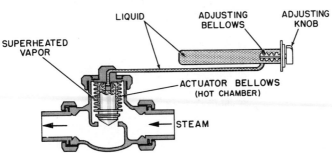

Fig. 3.3m　Hot chamber type thermal actuator and element

rising temperature) and are limited to use on steam at pressures up to 75 PSIG. Theoretically, other designs are also possible.

Ranges are from 30 to 170°F (to 200°F in special cases) with spans from 30 to 60°F.

Adjustment is achieved by the use of adjusting bellows which are usually located at the bulb but may be separate and remote.

Ambient temperature effects are minimized by design, but wide ambient swings are nevertheless detrimental. Effects of steam supply pressure variations are partially self-compensating.

Bulbs (thermostats) take several forms; for steam space heating systems the room (wall) mounting types or duct mounting types are common, while for liquid heating the cylinder types with pressure-tight bushings are utilized.

Advantages of the hot chamber class include—

Calibrated knob and dial adjustment available.

Good response, proportional characteristics.

Packless design.

Compact construction.

Fusion Type System (Wax Filled)

The fusion class does not have a SAMA equivalent. In this least common class, the compact element is filled with a special wax containing a substantial amount of copper powder to improve its rather poor heat transfer characteristics (Figure 3.3n). A considerable and positive volumetric increase occurs in the wax as temperature rises through its fusion or melting range. This volumetric change produces a force used for valve actuation through the ingenious arrangement of a sealing diaphragm acting to compress a rubber plug which "squeezes" out, the highly polished stem. This element has more hysteresis and dead band than the other classes.

Different waxes, such as natural waxes, hydrocarbons, silicones, and mixes of these are available which provide a variety of operating (control) spans between 100 and 230°F.

No adjustment of set point is provided, since the fusion (melting) span

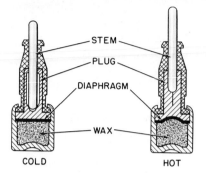

Fig. 3.3n   Wax-filled (fusion type) thermal element

of the wax is narrow, from 6 to 15°F. A heavy return spring, which is always present in the regulator, forces the rubber components of the element to follow the wax fill on falling temperature.

Remote bulb designs have not been offered since they would provide no advantages over the other three classes. Thus fusion class regulators are available only in "self-contained" styles (Figure 3.3o).

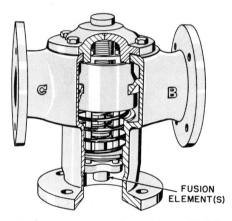

Fig. 3.3o   Self-contained temperature regulator with wax-filled (fusion type) element

Over-temperature protection up to 50°F above the operating span is provided by the flat temperature expansion curve of the wax above its melting range.

Valve sizes are from 1 to 6 in., with the majority being of the three-way valve design in sizes from 2 to 6 in.

Advantages of the fusion class regulator lie in their compact size and low cost of the thermal element, even where multiple elements are required to stroke larger valves.

Thermal Bulbs and Fittings

The "primary sensing element" of temperature regulators is usually called the bulb in remote sensing style designs. The following forms, sizes, fittings and materials are available.

Forms   The cylindrical bulb form is usually standard and is suitable for most liquid control requirements (Figure 3.3p). For faster response in air or

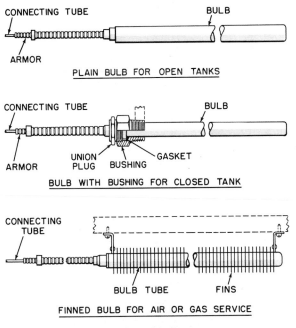

Fig. 3.3p   Thermal bulb variations

gas, a cylindrical form with metal fins is utilized. Other increased surface forms are the precoiled bulb (for "liquid expansion" and "hot chamber" classes) and the tubular bulb. The tubular bulb is made from a substantial length of flexible tubing, to be formed and supported in the field, as, for example, in and across a duct or wrapped around piping. Other forms include those with dead (inactive) extensions, flexible extensions, etc.

Size   Because of widely varying regulator designs and bulb volume requirements, no bulb dimension standards have been developed. Bulb sizes vary between the extremes of $\frac{1}{2}$ in. diameter by 9 in. long and $1\frac{1}{2}$ in. diameter by 36 in. long. Most bulbs range from $\frac{3}{4} \times 9$ in. to $1\frac{1}{4} \times 24$ in. long. Some manufacturers will vary the diameter and length combination to fit special needs.

Fittings   For open tank use, no fittings are needed. The union type screwed bushing for pressure vessel or pipeline installation is most common

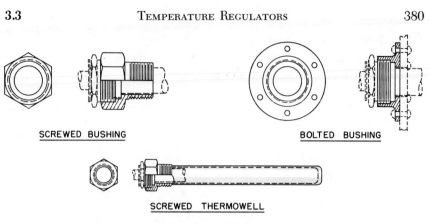

SCREWED BUSHING   BOLTED BUSHING

SCREWED THERMOWELL

Fig. 3.3q   Thermal bulb fittings

(Figure 3.3q). These are generally "ground joint" unions but are sometimes gasketed. Wells, or "sockets," are also available for higher pressure ratings, corrosion protection and to maintain the integrity of the pressure vessel or system during service. Sanitary fittings are also available.

Materials and Pressure Ratings   Bulbs are usually built-up assemblies and are available in copper or in other materials at extra cost. Table 3.3r gives some of their features.

Transmission Tubing

The tubing which transmits the pressure from the bulb to the actuator bellows is usually covered with armor (spiral or braided) for mechanical

Table 3.3r
CONSTRUCTION MATERIALS AND FEATURES OF THERMAL BULBS

			Rating	
Bulb Material	*Assembly*	*Fittings*[5]	*Pressure*[1] *(PSIG)*	*Temperature* *(°F)*
Copper, Brass	Brazed	Brass	700	300
Stainless Steel[2]	Welded	Stainless Steel	1,100	600
Steel	Welded	Steel, Stainless Steel	1,000	600
Lead[3]	Fusion	Lead	50	250
Monel	Welded	Monel	1,000	600
PVC Coated[4]	Fusion	None	—	160
Teflon Coated	Shrinking	None	—	300

[1] Average, based on 1-in.-diameter bulb. Larger diameters have lower ratings.
[2] Type #316 is available.
[3] Some lead bulbs are sheathed over copper.
[4] Coatings are usually over copper, but may be over other materials.
[5] Wells or sockets are available in all metals. The use of copper bulbs with other well materials is usual.

strength. A tube with $^3/_{16}$ in. OD and approximately 0.050 in. wall thickness is common.

The standard tubing material is copper, with brass armor, but stainless steel (tube and armor), lead-coated copper, and PVC- or Teflon-coated copper or stainless steel are also available.

Standard tubing length is 10 ft. Longer or shorter lengths are also provided, but lengths above 50 ft are not recommended because of increased mechanical hazards and slow response.

Control Characteristics

Proportional mode is the basic control mode of all temperature regulators. The proportional band is not adjustable in the field but is determined by various design factors, and it generally varies in direct proportion to the valve size and lift. For example, a typical 1-in. vapor pressure regulator, using a double-seated valve, has a mid-range proportional band (full-open to full-closed) of 7°F, and a similar 2-in. regulator has a proportional band of 11°F.

Response time of temperature regulators is much longer than that of corresponding pneumatic or electronic control systems. Time constants vary widely with design factors such as actuator bellows size (area), valve size (lift), transmission tube length, bulb size and material (heat conductivity). Typical regulators will have time constants in the range of 30 to 90 sec, although it might be reduced to under 10 sec under favorable conditions.

Special Features and Designs

Temperature indication is a feature available with most vapor pressure class regulators (Figure 3.3k). It usually consists of a pressure gauge sealed in the thermal system and calibrated to correspond to the particular range (charging liquid) used. Accuracy is about ±2 percent of indicated span.

Fail-safe design is available as a special version of the vapor pressure class regulator. Here failure (loss of fill) of the thermal actuator produces downward movement of the valve stem, i.e., the valve goes to the "safe" position, producing shutoff of the heating medium or full flow of the cooling medium, depending on valve action. Adjustable ranges are short (approximately 30°F), and valve actuating force is rather low, since the actuator works on the sub-atmospheric portion of the non-linear vapor pressure curve of the filling medium.

Manual handwheels are available in a few regulator versions (Figure 3.3k). These are useful to manually override the element, to provide a minimum flow valve position, or for emergency operation.

Combination temperature/pressure control is available in the pilot-actuated regulator. This is accomplished in two ways. In the external pilot version (the first way), two separately adjusted pilot units are piped in series in the pilot pressure line. The temperature pilot modulates the flow of heating medium (control agent) to regulate the process temperature, while the pressure pilot limits the discharge pressure from the main valve to its setting under maximum load conditions. In the internal pilot version the two pilot elements are "cascaded;" i.e., the temperature element mechanically changes the setting of the pressure element from zero at no load to the maximum pressure for which the pressure pilot is set at full load.

Conclusions

Since the characteristics of a regulator cannot be altered in the field, greater care must be taken in proper process evaluation and regulator selection than would be the case with a pneumatic or electronic controller having an adjustable control mode or modes.

In general, temperature regulators are most successful on control applications having high "capacitance" (and, therefore, slow response requirements), such as heated or cooled storage tanks, process ovens and dryers, and/or on applications with minimum load changes, such as metal treating (cleaning and plating) tanks.

Proper installation includes good bulb (primary element) location where process temperature changes can be sensed quickly, but away from potential damage by moving equipment. Regulators in sizes 2 in. and greater are installed upright in horizontal lines, and regulators of all sizes should be protected by strainers.

Regulator maintenance is minimal. Direct-actuated regulators with packed valve stems may require occasional lubrication and infrequent replacement of packing and cleaning or polishing of stem. Pilot-actuated regulators with piston-actuated valves (Figure 3.3c) may require programmed cleaning to remove "glaze" and deposits which would cause the piston to stick. Depending on the service, occasional examination and servicing of valve seats may be required to maintain needed tightness.

Relative to pneumatic control loops, the advantages and limitations of temperature regulators can be summarized as follows:

The advantages of regulators include their low installed cost, ruggedness, simplicity together with low maintenance requirements and lack of need for an external power source.

Its limitations include the fixed proportional control capability, the local set point, the limited and narrow ranges, the slow response, the size and operating pressure limitations and the large bulb size requirements.

Chapter IV

CONTROLLERS AND LOGIC COMPONENTS

H. L. Feldman, P. M. Kintner,
B. G. Lipták, C. L. Mamzic,
H. C. Roberts and J. E. Talbot

CONTENTS OF CHAPTER IV

	Introduction	387
4.1	Pneumatic vs Electronic Instrumentation	388
	General Considerations	388
	Transmission Signals and Lags	390
	Signal Converters and Computer Interfacing	392
	Hazardous Atmospheres and Intrinsic Safety	393
	Maintenance	395
	Costs	396
4.2	Electronic Controllers	398
	The Controller's Function	398
	Basic Parts of the Controller	400
	Input Variations	401
	Control Modes	402
	Special Features	406
	Displays	408
	Balancing Methods	411
	Mounting	412
	Servicing	414
	Other Electronic Controllers	415
	Feature Check List	419
4.3	Pneumatic Controllers	420
	History and Development	422
	Pneumatic Controller Principles	423
	Miniature Receiver Controllers	427
	Derivative Relay	430
	Miniature Control Stations	432
	Four-Pipe System	432
	Remote-Set Station	435
	Computer-Set Station	435
	Single-Station Cascade	435
	"No Seal" Station	437
	Procedureless Switching Station	438

Miniature High Density Stations   441
     Mid-Scale Scanning Station   441
     Procedureless Switching Station   444
Large-Case Receiver Controllers   447
Direct-Connected Large-Case Controllers   449
     Direct-Connected, Field Mounted Controllers   449
     Field-Mounted Receiver Type Controllers   451
Special Control Circuits   452
Effect of Transmission Distance on Control   454

4.4   FUNCTION GENERATORS AND COMPUTING RELAYS   457
Time Function Generators (Programmers)   459
Adjustable Ramp and Hold Programmers   460
Profile Tracer Programmers   460
Electric Line and Edge Follower Programmers   462
Step Programmers   463
Computing Relays   464
     Pneumatic Multiplying and Dividing   464
     Electronic Multiplying and Dividing   465
     Pneumatic Adding, Subtracting and Inverting   467
     Electronic Adding, Subtracting and Inverting   470
     Pneumatic Scaling and Proportioning   470
     Electronic Scaling and Proportioning   471
     Pneumatic Differentiating   472
     Electronic Differentiating   474
     Pneumatic Integrating   475
     Electronic Integrating   476
     Pneumatic Square Root Extracting   476
     Electronic Square Root Extracting   477
     High- and Low-Pressure Selector and Limiter   478
     Electronic High- and Low-Voltage Selector
     and Limiter   479

4.5   RELAYS AND TIMERS   481
Definitions and Functions of Relays   482
Characteristics   482
Relay Types   483
     Relay Contact Configurations   484
     Relay Structures   485
     Specialized Relay Structures   487
     Contact Materials and Shapes   488
     Selection of Relays   489
     Application and Circuitry   489
     Relative Costs of Relays   493

Reliability   493
Relays vs Other Logic Elements   493
Special Relays   494
Definitions and Functions of Timers   494
Interval Timers   494
Programming Timers   495
4.6   Static Logic Switching Elements   498
Semiconductors   499
Diodes   499
Transistors   501
Logic Operations   505
Diode Switching Circuits   505
Transistor Switching Circuits   507
Integrated Circuits   510
Integrated Switching Circuits   512
Medium- and Large-Scale Integration   515
4.7   Synthesis and Optimization of Logic Circuits   516
Logic Symbols and Tables   516
The Basic Logic Functions   517
Synthesis With AND/OR Logic Operations   518
Negation and DeMorgan's Theorem   519
Synthesis With NAND/NOR Operations   520
Graphic Logic Symbols   520
Graphic Translation From AND/OR to
    NAND/NOR Logic   521
Logic Redundancy   522
Logic Simplification Through Logic Maps   524
Logic Simplification With Boolean Algebra   527

This chapter covers three basic areas of process instrumentation:

1. Controllers
2. Function generators
3. Logic components and systems

The first and last sections are application-design oriented, while the others describe the available hardware.

In connection with process controllers, a discussion on the relative merits of pneumatic and electronic systems is followed by the description of the features and limitations for pneumatic and electronic controllers.

The section on computing relays and function generators is complimentary to the coverage of analog computers in Chapter IX.

The last three sections in this chapter are assigned to sequencing and logic components and systems.

The last section is devoted to the subject of optimizing these systems, in order to minimize the hardware requirements for furnishing the necessary interlock logic.

4.1 PNEUMATIC vs ELECTRONIC INSTRUMENTATION

It is not possible to make a firm recommendation in favor of either pneumatics or electronics that would be universally applicable, but it is feasible to list the considerations which tend to favor one or the other.

Pneumatic systems are at an advantage due to their standardized signals, lower installed cost, safety in hazardous locations and because of the ready availability of qualified maintenance personnel.

Electronic systems are usually favored when computer compatibility, long transmission distances, or very cold ambient conditions are considered or when long-term maintenance is to be minimized.

In the following discussion, some of these considerations are examined in detail.

Flow Sheet Symbols

PNEUMATIC

ELECTRONIC

General Considerations

Pneumatics or electronics is not a choice between controllers using different types of supply power but a choice between different concepts of the application of automatic control systems. Generally, when pneumatic controllers are used, pneumatic transmitters and pneumatic transmission of the measurement and control signals are utilized. Similarly, electronic controllers involve electronic transmitters and electric transmission. The consideration is one of control loops rather than just controllers, and the entire loop becomes of interest, including the process which closes the loop.

Since either system can be made to work properly, and since each system

388

has inherent difficulties peculiar to it, the choice between electronics and pneumatics becomes an evaluation of (1) which problems exist, (2) which problems are more easily resolved and (3) which inconveniences one is willing to accept. The major considerations are graphically illustrated in Figure 4.1a.

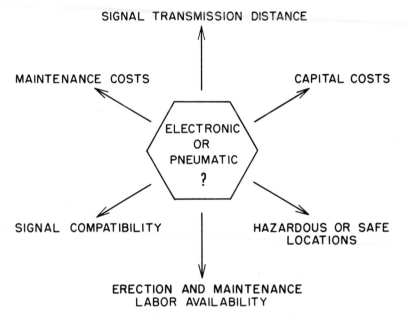

Fig. 4.1a   Electronic vs pneumatic controllers

Are the signal transmission distances too long for good control, taking into account the dead time and time constant of pneumatic transmission lag? Are initial capital costs more or less important than continuing operating costs? Are there some or many areas of the plant in which hazardous atmospheres require a very critical evaluation of electronic instruments? Does the particular area in which the plant is located have a ready supply of people capable of installing and maintaining pneumatic or electronic systems? Is it more economical to have the transmission signals compatible with data loggers and computers or with the final control element which is generally an air-operated valve?

In spite of all predictions to the contrary, pneumatic systems have continued to outsell electronic systems by a substantial margin, although electronic controllers are capturing a greater share of the market each year. The hybrid system, in which some loops are electronic and some pneumatic, is gaining popularity and represents a compromise solution to take advantage of the best qualities of both systems.

There is no definite answer to the pneumatics vs electronics question.

Pneumatic systems are not always less expensive, and electronic systems do not always produce the fastest loop response when air-operated valves are used. The important thing is to understand the nature of the problem and then to solve it for the particular case.

Most plants require some instruments to be provided with a back-up power supply to allow for the safe and orderly shutdown in the event of a general power supply failure. For the pneumatic system an air reservoir and for the electronic system a string of batteries (with an inverter if AC is required) will usually suffice. (Section 5.4 has more details.) The number of instruments which must be provided with a safe supply, the length of time this supply must be maintained and the nature of the supply should be considered along with the primary selection of instrumentation.

There are several topics that are not discussed in this section, such as ambient condition limitations, accuracy and range of controller adjustments. These subjects are not being ignored; they are simply not pertinent to an evaluation of pneumatics vs electronics because there is no significant difference between the two, although one might note some advantages associated with the use of electronic systems under very cold ambient conditions.

Transmission Signals and Lags

Electronic instruments suffer from a lack of transmission signal standardization but have the advantage over pneumatics in that electrical transmission lags are negligible. There are many signal ranges used with electronic instruments, including various milliampere DC levels, volts DC and millivolts AC, some with live zeros and some without live zeros. The most popular are the 1–5, 4–20 and 10–50 madc ranges, but even so, when manufacturers use the same signal level the power supply voltage is apt to be different. The electronic controller has input impedances (into which the transmitter must drive), its output must go into the final control element load, and all controllers have output load restrictions. What this incompatibility appears to indicate is that with electronic instruments one cannot "mix and match" between manufacturers as is commonly and easily done with pneumatic transmitters, controllers, recorders and valve positioners.

The situation is somewhat improved because some instrument manufacturers offer their electronic controllers with several choices of input and output signals. Also, current-to-current converters are available for either current conversion, such as 10–50 madc in and 4–20 madc out, or power amplification.

One great advantage of electronic control is the virtual disappearance of the transmission lag. Pneumatic transmission of the measurement signal from the transmitter to the controller, and of the control signal from the controller to the valve, contributes additional dead times and time constants to the control loop. The calculation of lag of pneumatic transmission is extremely difficult and tedious because the transmission line has resistance, capacitance and inductance, all of which are distributed along the length of the transmission distance. The capacitance is dependent on the volume of the tubing (derived from length and inside diameter) and the capacitive load at its end, commonly referred to as the terminal volume. The resistance is a function of the fluid friction involving many parameters, not the least of which are tubing length, bends, material, size and the fittings employed. Most quantitative data on pneumatic transmission lags comes from experimental results. Unfortunately the experimental results tend to be inconsistent, since the capacity of the pneumatic relay, the terminal volume and the level of the signal, which are of extreme importance, not only vary from test to test but are many times ill defined.

Electronic instruments entirely eliminate the consideration of transmission lags, but if pneumatic instruments are preferred or required, several loop modifications can be made to reduce the lag or to eliminate its importance. The severity of the lag and its effect on control is best approached from a qualitative standpoint considering both the hardware and the process. With fast-responding processes such as flow the lag in signal transmission is of greater importance than in the control of temperature, because in most systems both the process response and the primary element response are relatively slow. Pressure control systems tend to be fast. Liquid level systems can be fast if the process is characterized by high throughput and a small vessel, and therefore transmission lags should be minimized. The effect of long transmission lags in fast processes is that the controller is receiving a measurement signal that does not represent the current condition of the process because the dead time, time constant and phase lag contribution of the transmission system have resulted in poor or unstable control.

The most commonly used pneumatic tubing is $\frac{1}{4}$ in. OD. Since the resistance of the tubing varies as the fourth power of the diameter, and the capacitance of the tubing varies as the second power of the diameter, the transmission lag may be reduced by the use of larger tubing. The time constant of $\frac{3}{8}$-in.-OD tubing is roughly half that of $\frac{1}{4}$-in.-OD tubing, but the degree of improvement is also influenced by the ratio of tubing volume to terminal volume. Increasing the size of the tubing is helpful when transmitting to a bellows element terminal volume but not necessarily so when transmitting into a 50-cu.-in. valve motor. The time constant of 200 ft of $\frac{1}{4}$-in.-OD tubing into a receiving bellows is approximately 1.4 sec, and the time constant increases about 2 sec for each 100 ft up to 1,000 ft transmission distance.

Pneumatic transmission lags can be reduced by using pneumatic boosters at the field transmitting end to increase the air-delivering capacity and by using valve positioners at the field receiving end to reduce the terminal volume into which the controller is feeding. However, pneumatic boosters can introduce additional dead time, worsen frequency response of the loop or create phase shifts, all of which degrade the performance of control. For recording purposes only, transmission lags introduce no penalty unless the frequency response of the tubing attenuates the measurement signal.

Another method frequently employed to eliminate the effects of transmission lag is to locate the controller near the transmitter and valve and to provide in the central control room the recorder with set-point adjustment, auto-manual switch, manual adjustment and controller output indicator. Although this method allows satisfactory control, it has two serious disadvantages. First, *four* pneumatic lines are generally needed between the field and the control room instead of two. The extra two are needed to transmit the set-point signal to the field-mounted controller and a seal to allow manual operation from the control room. Secondly, with this scheme, all control mode adjustments must be made at the controller in the field, while the record of the process is available only on the control panel.

The American Petroleum Institute Recommended Practice 550 suggests that for general service applications for which transmission lag is not ordinarily critical, the total length of tubing from the transmitter to the controller plus that from the controller to the control valve should not exceed 400 ft; neither run should exceed 250 ft. In practice, longer runs are often used without serious effects on the process. With transmission distances less than 400–500 ft (total for both directions), practically no improvement is possible in the process recovery time by eliminating the transmission lag, because the air-operated valve response time is the limiting factor.

Signal Converters and Computer Interfacing

There are many transmitters whose normal output is an electric signal, such as turbine flowmeters, chromatographs, viscometers, etc. On the other hand, most final control elements continue to be pneumatically operated valves or power cylinders. Very often data logging or some form of computer control is required. Pneumatic systems are compatible with final control elements and electronic systems are compatible with the electric signal transmitters and computers. The ordinary pressure, level, flow and temperature transmitters do not enter into the consideration, since they are readily available with either pneumatic or electric output.

If a system is made all electric or all pneumatic, signals will often have to be converted from one form to the other in the typical multiloop plant. There are available pneumatic-to-electric, electric-to-pneumatic, voltage, and resistance-to-pneumatic converters (Section 6.3) and electro-pneumatic valve positioners to complete the loop, but these extra items increase cost and decrease reliability and loop accuracy.

The availability of special electric output transmitters should not affect the signal compatibility decision, for they are generally a very small percentage of the controller inputs. The problem is to keep controller outputs compatible with air-operated valves or with logger and computer input requirements when the latter exist or are included in future plans.

Input signals for data loggers or process control computers can be generated in an electronic transmission system by taking the voltage across a series dropping resistor. Hence the computer compatibility of electronic systems is inherently easy and inexpensive. The computer output can be fed back to the analog process controllers to modify set points or perform other functions. When pneumatic transmission is used, pneumatic-to-current converters or servo driven potentiometers must be used to produce the electrical input signals. In this case, there is a greater sacrifice in accuracy, and considerably more hardware is required. Computer compatibility (including data logging) is a key function in many large processing plants, and electronic control systems lend themselves nicely to this tie-in.

Hazardous Atmospheres and Intrinsic Safety

Very often transmitters and final control elements must be located in hazardous areas. Sometimes it is necessary to locate the controllers and associated devices such as ratio units and square root extractors in hazardous areas. With pneumatic instruments hazardous areas present no difficulty, but with electronic instruments serious problems arise which require consideration. These problems can be resolved, but generally only by extra expense and/or inconvenience to the maintenance staff.

The hazardous atmospheres with which we are concerned occur in those areas in which flammable gases or vapors, combustible dust, ignitable fibers or ignitable flyings may be present under normal or abnormal operating conditions of the process. The National Electric Code is the generally accepted authority for the definition of hazardous locations, the classification of various areas (Classes, Groups and Divisions), and the hardware requirements for satisfactory installation. Hazardous locations are discussed in Section 10.11 of Volume I of this Handbook.

Since air containing combustible material can be ignited by the release of electrical energy or by the presence of a high surface temperature, the environment in which the instruments are operated can play a major role in choosing between pneumatics or electronics. Provisions must always be made to render electronic instruments safe for use in a hazardous location. These provisions also apply to the electrical connections, wiring, junction boxes and so on.

Safety codes vary from country to country and, within a country, from one government level to another. Compliance requires familiarity with the governing codes and familiarity with the nature of the electronic instruments under consideration.

Class I locations are the most often encountered hazardous atmospheres. These are areas in which flammable gases or vapors are or may be present in the air in quantities sufficient to produce explosive or ignitable mixtures. In Division 1 locations these flammable gases or vapors are present under normal operating conditions, during maintenance, and during breakdown or faulty operation. In Division 2 locations the flammable gases or vapors are present only under abnormal conditions of the process. Generally, for Division 1 locations all electrical instruments must be enclosed in explosion-proof or flameproof (UL) housings unless they are intrinsically safe. For Division 2 locations the electrical instruments may be in a general purpose enclosure if their contacts (if any) are hermetically sealed or under normal conditions do not release sufficient energy to ignite a specific hazardous atmospheric mixture and if the maximum operating temperature of any exposed surface does not exceed 80 percent of the ignition temperature in degrees Celsius of the gas or vapor involved. If these stipulations are not satisfied, an explosion-proof housing must be used in a Division 2 location.

It is seen, then, that there are two methods available for rendering electrical instruments satisfactory for installation in Division 1 locations: explosion-proof housing and intrinsically safe apparatus. Both methods are being employed, although intrinsic safety is greatly preferred because it allows the instrument to be serviced while it is energized and does not require hazardous area wiring methods. Since the purpose of an explosion-proof housing is to prevent flame and hot gases from igniting the surrounding atmosphere should there be vapor ignition within the housing, it stands to reason that the instrument may not be operated without the housing intact if the plant area in which it is located remains hazardous. Case purging is also acceptable under the codes, but this is impractical except in special, isolated situations. Some transmitters are fitted with zero adjustments and some with both zero and span adjustments external to the explosion-proof housing so that calibration can be done with the housing intact. This facility overcomes part of the inconvenience of explosion-proof housings, but still does not allow trouble-shooting and repair in situ with the instrument energized.

Most electronic transmitters are not suitable for Division 2 locations without some adaption (generally optionally available) unless they are intrinsically safe.

Instrument manufacturers are increasingly providing intrinsic safety in their electronic models. Intrinsically safe instruments and wiring are incapable of releasing sufficient electrical energy under normal or abnormal conditions of the instrument loop to cause ignition of a specific hazardous atmospheric mixture, and, in the case of intrinsically safe loops, the provisions of Articles 500 to 517 of the National Electric Code do not apply. Intrinsically safe transmitters located in a hazardous area may be connected to non-intrinsically safe controllers located in a safe area. In the event that the controller must be located in a hazardous area, some are available that are suitable (optionally) for Division 2 locations.

Some instruments and instrument loops are designed to be inherently intrinsically safe, while others use separate barrier devices just prior to the entry of the cabling into the hazardous area. The barrier is a passive device that limits the available energy in the hazardous area. While barriers serve the purpose, their adverse effect on cost, reliability and accuracy must be considered. The advantage to be gained from the use of barriers is that *only* the instruments in the hazardous area need be considered for testing and certification rather than the entire loop. This is especially important, for the number of different loops that can be created is without limit.

British Standard CP 1003, Electrical Apparatus and Associated Equipment for Use in Explosive Atmospheres of Gas or Vapor, and the British Institute of Petroleum Electrical Safety Code have a Division 0 in which hazardous gases are always present and where the conditions normally require the *total exclusion* of any electrical equipment (except in special circumstances where this is impracticable, in which cases intrinsically safe equipment or special measures such as pressurization may be used). However, except when the only source or storage of energy is an inherently safe battery, intrinsically safe circuits invariably contain means to restrict potentially dangerous currents to within safe levels. Thus very few intrinsically safe circuits would be suitable for continuous use in Division 0 atmospheres, since the possibility of ultimate failure of the means of protection cannot be ignored.

Extreme care must be taken when evaluating the hazardous conditions, the pertinent regulations and the instrument design. Even when instruments are designated and approved as intrinsically safe, there remain, in some cases, subtle limitations such as load resistance restrictions and maximum acceptable value of capacitance and inductance of the connecting cables. In some cases only loops tested by a certifying agency are acceptable, while in other cases intrinsic safe design is sufficient. When options are available to allow the installation of an instrument in a more hazardous atmosphere than that for which the standard model is rated, they are invariably available at extra cost. Pneumatic instrumentation eliminates all of these potential problems and additional costs, and is generally applied when hazardous conditions are present unless process dynamics or other ulterior considerations require the use of electronics.

Maintenance

Costs and problems associated with maintaining the instrumentation are rarely taken into serious consideration when the initial control system decisions are being made, because operating costs, of which instrument maintenance is only one, are allocated from different budgets and are subject to different tax considerations than capital costs. However, because maintenance costs continue year after year, they deserve more thought than is generally given them. Although difficult to prove statistically, it is generally felt that electronic instruments require less maintenance after the initial aging period of a few months has transpired.

Maintenance departments depend on the local labor supply, made up of both skilled and trainable workers, whose availability and the hourly rates vary from area to area. Some plants use contract maintenance and pass this problem off to another party. While this method relieves the plant owner of the labor problem should one exist, it in no way helps solve the problem.

The cost of outfitting either a pneumatics or electronics maintenance shop is not a large budget item, but electronic test equipment is considerably more expensive and more delicate than pneumatic test equipment. If the maintenance department wishes to repair the electronic modules or cards rather than discard them or return them to the manufacturer for repair, a higher order of electronic test equipment will be required, such as a signal generator and an oscilloscope.

An adequate supply of spare parts must be stocked to allow the maintenance shop to repair and replace worn and damaged components. Spare parts generally comprise 5 to 10 percent of the total instrumentation cost. Because electronic instruments are more expensive initially, their spare parts also are more expensive.

Costs

Many determinations between pneumatics and electronics are made on the basis of installed costs. The significant cost areas of installed instrumentation are (1) instruments, (2) panels and (3) erection, and the instrument category is the most significant. Although all other areas of cost favor the electronic system, the large difference in instrument cost is never completely overcome in the uncomplicated situation.

Instrument for instrument, the electronic counterpart of the pneumatic version can cost anywhere from 25 to 100 percent more. Detailed cost comparisons for 200 loop plants show that the electronic instruments cost generally 50 to 70 percent more than the pneumatic instruments. This comparison does not include those items common to both systems such as control valves and local loops which tend to be pneumatic even when electronic systems are selected for plant-wide control. When the common items are included, the extra cost of electronics is naturally less because there is the same absolute difference between two larger financial packages.

The cost comparison of central control room panels is a function of the pneumatic tubing and electric wiring methods. A plastic tube pneumatic panel costs essentially the same as electric wiring in a plastic trunking electronic panel. Copper tubing increases the pneumatic panel costs, and rigid conduit or screening increases the electronic panel costs, but the variation either way is approximately 10 percent of the total panel fabrication costs (less instruments but including accessories).

The erection costs generally break down to 50 percent for transmitter installation and process side hookups, 35 percent for transmission side hookups and 15 percent for the air supply installation. In the largest cost category there is no difference between pneumatics and electronics. Since the cable trays account for almost half of the transmission side erection costs, there is virtually no difference here. Only in the air supply installation can substantial savings be made by using electronic instruments. While air supplies are still required for air operated final control elements and local pneumatic loops, 5 to 10 percent of the total erection material and labor costs can be saved with electronics.

Where other factors such as computer or data logging inputs and long transmission distances affecting control warrant consideration, electronics are superior to pneumatics from the cost viewpoint, but when hazardous locations must be dealt with, electronics cost more than pneumatics.

Continuing costs should not be overlooked when a new installation is being considered. With the advent of solid state circuitry and highly reliable semi-conductors coupled with good design, electronic instruments require less maintenance than pneumatic instruments and, hence, maintenance costs are less with electronic systems. Over a ten-year period this cost can be very significant in a large plant.

4.2 ELECTRONIC CONTROLLERS

Input Ranges:	1–5, 4–20, 10–50 madc; 1–5, 0–10 vdc; ±10 vdc
Output Ranges:	1–5, 4–20, 10–50 madc; 1–5, ±10 vdc; ±2 madc
Repeatability:	0.5%
Accuracy:	±1%
Control Modes:	Manual, proportional, proportional plus integral, proportional plus derivative, proportional plus integral plus derivative and integral.
Displays:	Set point, process variable, output, deviation and balance.
Cost:	$400 to $700.
Partial List of Suppliers:	Bailey Meter Co.; Beckman Instruments, Inc.; Bell & Howell; Bristol Co.; Fischer & Porter Co.; Foxboro Co.; General Electric Co.; Hagan/Computer Div. of Westinghouse; Honeywell, Inc.; Leeds & Northrup Co.; Motorola; Robertshaw Controls Co.; Sybron Corp.; Taylor Instrument Cos.

The Controller's Function

Most industrial processes require that certain variables, such as flow, temperature, level, and pressure, remain at or near some reference value, called a set point. The device that serves to maintain a process variable value at the set point is called a controller. The controller looks at a signal that represents the actual value of the process variable, compares this signal to the set point and acts on the process to minimize any difference between these two signals.

Any simple process control loop (Figure 4.2a) contains the equivalent of a sensor, transmitter, control element (usually a valve), and a controller. The sensor measures the actual value of the process variable. The transmitter amplifies this sensed signal and transforms it into a form suitable for sending

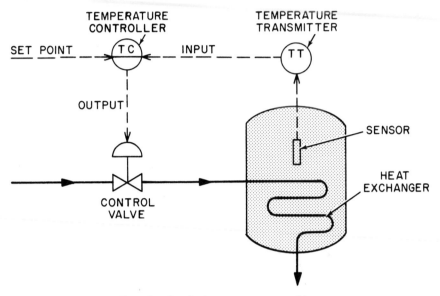

Fig. 4.2a   Standard temperature control loop

to the controller. The controller really has two inputs: this measured signal and a set-point signal. The set point, however, may be internally generated. The controller subtracts the two input signals, producing a deviation or error signal. Additional hardware in the controller shapes the deviation signal into an output according to certain control actions or modes (Section 7.3). The controller output then typically sets the position of a pneumatic control valve or other final element in a direction to decrease the error.

Although the function of the controller is easy to define, it may be implemented in many different ways. For example, the controller may work on pneumatic (Section 4.3), fluidic, electric, magnetic, mechanical, or electronic principles—or on combinations of these. This section, however, will confine itself primarily to a certain class of miniature electronic controllers that accept and generate standard process input and output current signals. Other kinds of electric and electronic controllers, having various forms of inputs and outputs, will be described briefly at the end of the section.

The general nature of their inputs, outputs, and control modes makes the electronic and pneumatic controllers applicable to a wide variety of process control situations. The same controller may control flow in a pipe, which responds nearly instantly to valve movements, or a temperature in a distillation column, which often takes hours to respond to disturbances. Any process variable is a candidate for control as long as it can be transduced to a standard current signal. The instrument engineer must choose and tune the control modes to fit the dynamics of the variable to be controlled (Sections 7.3 and 7.12).

Basic Parts of the Controller

The construction of electronic process controllers breaks down into six basic parts or sections: input, control, output, display, switching, and power supply (Figure 4.2b). The input section comprises the hardware for generating

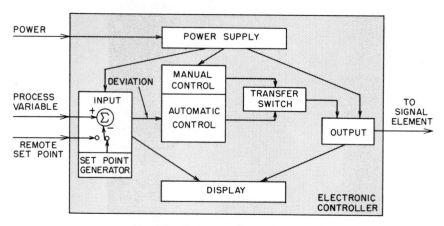

Fig. 4.2b   Basic controller components

the set-point signal, for accepting and conditioning the process signal, and for comparing the two signals to produce a deviation signal. The heart of the control section is generally a chopper-stabilized AC amplifier that acts on the deviation signal. Associated circuitry amplifies, integrates, and differentiates the deviation signal to produce an output with the necessary proportional gain, integral (reset), or derivative (rate) control actions. (In some controllers the derivative circuit acts on the process signal rather than the deviation signal.)

The controller's output has two forms: automatic and manual. When in the manual mode, the controller's output remains at a steady level, often dependent on the shaft position of a potentiometer. Switching and balancing hardware on the front panel selects the automatic or manual output. In some cases the source of the manual output is the control amplifier (modified by switching), while in other cases the source is entirely separate, requiring only the operation of a power supply.

The display section on the front panel carries information about the set point, process variable value, deviation, and controller output. The front panel also contains mechanisms for switching between manual and automatic modes and for adjusting the value of the set-point input and the manual output. A secondary indication, balance, allows the operator to equalize the manual and automatic outputs before transferring from the automatic to the manual mode.

The power supply section serves to transform the incoming AC line voltage to the proper DC levels to operate the other controller sections. Some controller power supplies also serve to energize instrumentation external to the controller. For example, the controller may power the transmitter in the same loop. In other cases the power supply itself may be external to the controller, energizing many controllers and other instruments.

In addition to these basic parts, the controller may have various special optional or standard features. Some examples include alarms, feedforward inputs, output limits, and batching aids (described later).

Input Variations

Input and output signals for electronic controllers have more or less been standardized, through general usage, to three ranges: 1–5, 4–20, and 10–50 madc. Nearly all the makes on the market will accommodate at least one of these as a standard input signal. Many offer all three as either standard or optional features. On the output side, the manufacturers generally offer only one of the signals, the most popular being 4–20 madc.

The use of a DC current amplifier for transmitting signals over great distances has one primary advantage over a voltage signal: small resistances that develop in the line (from switches and terminal connections and from the line itself) do not alter the signal value. Once the input signal reaches the controller, it is generally dropped across a suitable resistor to produce a voltage that matches the working voltage range of the controller's amplifier. The signal's live zero aids in troubleshooting since it differentiates between the real signal zero and a shorted or grounded conductor. Besides these three ranges, some manufacturers offer uncommon inputs such as 1–5 and 0–10 V DC, and outputs such as 1–5 V DC; centered-zero signals ($\pm$10 V DC and $\pm$2madc) are also available.

The controller's set-point input is either internally (local) or externally (remote) generated. Manufacturers generally provide a switch to let the user choose between the two, but some require him to choose between two different models. In some electronic controllers the two signals can be displayed and nulled to equalize them before transferring from one to another. It is also possible, but more costly, to have the unused signal *automatically* track the other, so that the two signals are always equal. If the remote and local set-point signals are not equal before transfer, the controller will see a sudden set-point change. The output will therefore also change abruptly, unnecessarily disturbing or "bumping" the process.

The local set point is often merely the output of a voltage divider that is connected to the set-point adjustment and display mechanism. The divider consists of either a multi-turn variable resistor (potentiometer) or a large fixed circular slidewire with wiper. The divider's output is compared to the voltage drop produced by the process input signal, yielding a deviation signal that corresponds to the difference between the two.

The remote set-point signal must generally be the same range as the process input signal. Manufacturers provide terminals at the back of the controller to accept this signal. Common examples in which a remote set-point signal comes into play are the cascade and ratio control systems (Sections 7.8 and 7.9, and Figure 4.2c). In the cascade system the source of the set-point signal is the output of another controller, called the primary, or master, controller. The controller that accepts this remote set point is the secondary, or slave, controller. Another common source of remote set point is a function generator or a programmer, set up to make the controlled variable follow a prescribed pattern (Section 4.4).

Blending systems are a form of ratio control. Here the set-point signal is really a second process variable from a flow transmitter. The controlled stream is forced to follow the uncontrolled (or wild) stream in some adjustable proportion (ratio). So the signal from the flow transmitter on the wild stream serves as a remote set point. The adjusting device to fix the desired ratio between the two streams is usually incorporated on the front panel of the controller, making it a special model. Calibrated ratios generally run from 0.3 to 3.0.

An optional feature pertaining to a remote set point is a servo system (motorized set point) that makes the set-point display mechanism track the remote set-point value. Otherwise the display should be blanked or marked in some way because it corresponds to the local rather than remote set point. In some controllers the set point is displayed on a galvanometer, and then the above comments do not apply. Motorized set points often find application in supervisory control systems where the remote set-point source is a computer.

Control Modes

The controller must deal with processes having wide ranges of dynamic characteristics. Adjustable control modes give it the needed flexibility. These modes act on the deviation signal, producing changes in controller output to compensate for process disturbances that have affected the controlled variable. Electronic hardware modules provide one or more of the following control actions (Sections 7.1, 7.3):

1. proportional gain
2. proportional gain + integral (reset)
3. proportional gain + derivative (rate)

4. proportional gain + integral + derivative
5. integral (proportional speed floating)

The control action with the most widespread use is proportional gain plus integral. A few manufacturers offer proportional gain alone, and fewer yet offer pure integral.

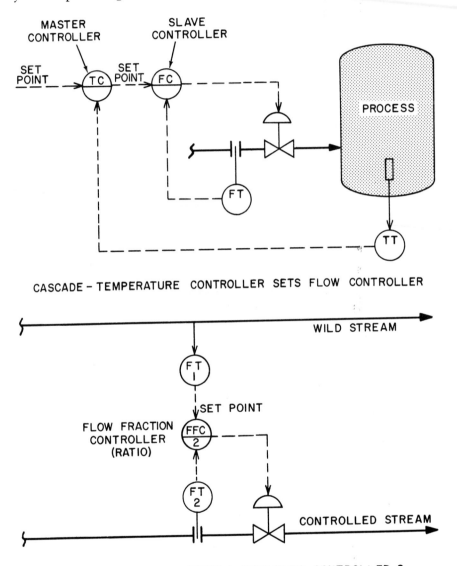

CASCADE – TEMPERATURE CONTROLLER SETS FLOW CONTROLLER

RATIO – FLOW TRANSMITTER I SETS FLOW CONTROLLER 2

Fig. 4.2c   Cascade and ratio control systems

In some cases the manufacturer provides one basic controller that has all three modes. The user then switches out derivative or integral if it is not needed. Otherwise he picks from a variety of models, depending on desired modes, fast or slow integral rates, fast or slow derivative rates, and other kinds of options. The choice depends on the dynamics of the process to be controlled.

The proportional mode varies the controller output in proportion to changes in the deviation. High gains mean that the controller output varies greatly for small changes in the deviation. Extremely high gains amount to on-off control. In processes requiring relatively low gains, the importance of the integral mode increases, because at low proportional gains the controller will be satisfied even though the controlled variable and the set point are relatively far apart (offset). The integral mode continues to drive the output until the deviation goes to zero. The derivative mode is a refinement that produces an output component proportional to the rate of change of either the deviation or the process input signal. Its effect is to anticipate changes in the process variable under control. Figure 4.2d shows a simplified circuit for implementing three-mode control.

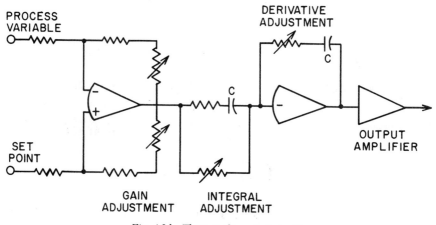

Fig. 4.2d   Three-mode control circuit

The advantage of having the derivative mode act on the process input rather than on the deviation is that the controller does not respond directly to a set-point change. If it acts on the deviation, the derivative component of the output depends, for example, on how fast the operator turns the set-point knob. Of course he could always switch to manual before making a set-point change and then switch back to automatic. A few manufacturers can provide both kinds of derivatives as optional connections, but most offer only one or the other.

Adjustments on each control-mode module let the user fit (tune) the controller to the process dynamics (Section 7.12). Continuous or multi-position knobs are located inside the case and are calibrated in terms of the degree of control-mode action. Typical units and values are:

1. proportional gain 0.2 to 50 (proportional band 500 to 2%)
2. integral, 0.04 to 100 repeats per minute
3. derivative 0.01 to 20 minutes

Higher and lower values for each mode are available, depending on the manufacturer. The term proportional band (PB) is also used; it is related inversely to proportional gain:

$$PB = \frac{100}{\text{proportional gain}} \%$$

One repeat per minute means that the integral mode will ramp the output up or down, repeating the proportional action that corresponds to a particular deviation every minute. Ramping continues until the deviation is zero. A derivative setting of one minute means that the controller output should satisfy a new set of operating conditions one minute sooner than it would without the derivative mode. Many manufacturers also have scaling switches or jumpers that multiply the effects of the integral and derivative modes by factors of 2 or 10. Also, the ability to switch the derivative and integral modes completely out of the control circuit is an aid for troubleshooting and calibrating.

The direction on control action depends on set point and process variable input connections. Direct control means that as the process variable increases, the controller output also increases. Reverse control means that the controller output responds in a direction opposite to that of the changing process variable. A direct-reverse switch in the controller reverses the input connections to bring about the desired controller action. If the derivative action is on the process measurement rather than the deviation, this connection must be reversed also.

Some circuit designs permit changes in the control settings while the controller is operating in automatic, without disturbing the output. Sometimes only the proportional gain changes have no immediate effect on output. Otherwise, the controller must be switched to manual before changing the settings. Usually the settings are somewhat interacting, meaning that a change in one affects the value of another. However, certain controllers incorporate features such as separate amplifiers for each mode to eliminate or minimize this interaction. These controllers have an advantage when tuning is critical.

Although the process controllers have standard output signal values, different makes have different limitations on the maximum load resistance in the output circuit. The load resistance is the sum of the series resistances of all the devices driven by the controller's output current. Typically the output current drives an electro-pneumatic converter that connects to a diaphragm-actuated control valve. These varying load requirements between manufacturers complicate matters somewhat if the user wants to buy parts of the control loop from different manufacturers.

A sampling of actual specifications for maximum load resistance includes: 75, 500, 600, 800, 1,500, and 3,000 ohms. Higher maximums, of course, give the user greater flexibility. A few manufacturers require a minimum load resistance as well. At least one vendor asks the user to adjust the load resistance in the output circuit to an exact value, which is a severe limitation.

Special Features

Manufacturers usually offer special features, either optional or standard, to enhance the convenience and usefulness of the controllers. The most common features involve alarm modules and output limits in some form. Electronic alarms may be set up on either the process measurement, the deviation, or both, depending on the supplier. Also, they may be specified as actuating on either the high or low side of the alarm point. Some controllers have panel lights for connecting to the alarm contacts as well as terminal connections for external alarms such as lights, bells, horns, or buzzers.

Nearly all manufactures of electronic controllers limit the high side of the output current range to the maximum standard value. Often a zener-diode arrangement serves to clamp the current at this value. Some manufacturers go a step further and place a fixed limit on the low end of the range. Still others offer adjustable output limits on both ends. Output limits serve only to constrain the range of the manipulated variable (valve stroke).

Certain limits can prevent a phenomenon called reset (integral) windup. Windup occurs if the deviation persists longer than it takes for the integral mode to drive the control amplifier to saturation. This might occur, for example, at the beginning of a batch process, under temporary loss of either controller input signal, or for a large disturbance or set-point change. Only when the deviation changes sign (process variable value crosses set point) does the integral action reverse direction. So the process variable will most likely overshoot the set point by a large margin.

If the limit acts in the feedback section of the control amplifier's integral circuit, the controller output will immediately begin to drive in the opposite direction as soon as the process signal crosses the set point. This approach is commonly referred to as *anti-reset windup*. On the other hand, if the limit acts directly on the output as discussed earlier, it essentially diverts excess current coming from the control amplifier, and therefore the output will remain at a high value for some time after the process crosses the set point. The output limit will simply divert less and less excess current. This extends the time that the output remains at a saturated value, aggravating the overshoot due to windup. Anti-reset windup circuits are usually offered only to eliminate saturated output currents at the high end of the signal range.

A few manufacturers go so far as to provide special batch controller models to minimize reset windup. Here, special circuit modifications actually begin to drive the output out of saturation *before* the process variable value crosses the set point. Reset windup can be avoided in batch applications if the operator equalizes the process variable value and setpoint in the manual mode before switching from manual to automatic operation.

Another controller feature gaining acceptance is a feedforward input connection. In these controllers a feedforward input signal adds to the normal controller output signal if desired. Feedforward control (Section 7.6) assumes that the user knows exactly how much the control valve (or other final element) should change to compensate for known changes in a certain process input variable (Figure 4.2e). This relieves the controller of compensating for this

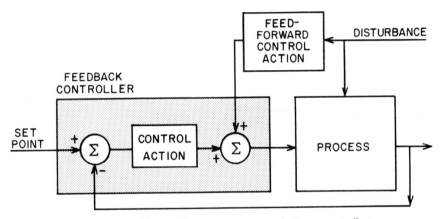

Fig. 4.2e   Feedforward input connection on electronic controllers

particular process disturbance by normal feedback control. A few manufacturers offer gain and bias modules for conditioning the feedforward signal before it sums with the controller output signal. A direct and reverse switch is necessary to match the feedforward signal to the controller action.

Other special features offered by various manufacturers include:

Communication jack: lets signal wires between the controller and transmitter serve also as an audio communication link; an AC carrier system does not disrupt the DC process variable signal. This eases calibration and troubleshooting of the control loop.

Trend record: provides jacks or pins for patching a recorder to the process variable input signals available in voltage form in the controller.

Output tracking: locks the controller output to a remote signal. This feature is useful, for example, when an electronic controller backs up a computer control installation.

Process retransmission: converts process current signal to a low impedance voltage signal for distribution to other parts of the control panel. Other devices may then be connected in parallel to this signal without loading down the controller input signal.

Special input modules: extract square root, select inputs, accept resistance inputs, excite strain gages, convert pulse inputs, integrate, isolate, condition, and limit inputs. Some manufacturers offer much wider selections than others.

Battery backup: automatically takes over when regular power mains fail.

Displays

As mentioned earlier, the controller's front panel indicates four basic signals: set point, process variable, error deviation and controller output. The deviation meter sometimes doubles as a balance meter for equalizing the manual and automatic signals or local and remote set-point signals when the operator wants to switch from automatic to manual output. The front panel has one small meter to show the controller output. Displays for the other three basic variables take on various forms differing among manufacturers and sometimes differing among models of the same manufacturer. Controllers without any panel displays are called blind controllers.

The majority of process controllers merge the set-point, deviation, and process variable indications into one display (Figure 4.2f). This display consists essentially of a deviation meter movement combined with a long steel tape or drum scale behind the meter's pointer. The scale length ranges from 6 to 10 in., depending on the manufacturer, but only a portion of this scale shows; the tape or drum may be moved so that the scale value corresponding to the desired set point rests exactly at the center of the display, marked by a hairline or a transparent green band. If used, the green band masks the red pointer for small deviations. For larger deviations the red pointer peeks glaringly from behind the band to alert the operator.

The deviation meter movement corresponds to the visible part of the scale. For example, if 50 percent of the scale shows, the deviation movement is ±25 percent of full scale. So the pointer indicates the process variable value as long as it is on scale, the hairline at the center gives the set-point value, and the distance between the hairline and pointer indicates the deviation.

The mechanical set-point scale movement also drives components for generating the corresponding electrical set point signal. Commonly, the scale movement drives a multi-turn potentiometer. In some controllers, a wiper attached to a movable scale drum contacts a large fixed circular slidewire connected as a voltage divider. Either design should avoid gearing between the set-point drive and electrical components, otherwise backlash would destroy their one-to-one correspondence.

The controllers having a large deviation display with a long, expanded scale offer high resolution and readability of each signal value. Of course the deviation must be on scale to be able to read the process value at all. Also the electrical and mechanical set-point values must be calibrated carefully or the process reading will be inaccurate.

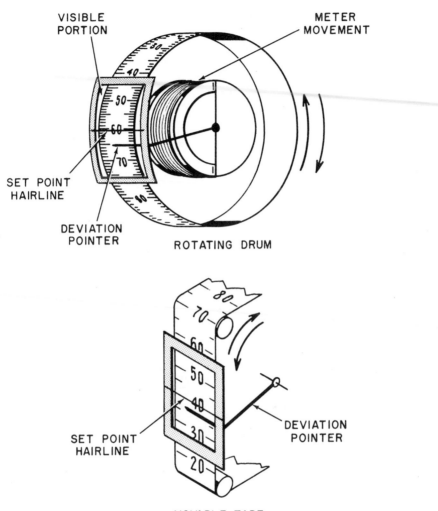

Fig. 4.2f   Expanded scale displays for showing set point, deviation, and process variable

In controllers without this kind of display, some compromise is usually made in the indication of set point, deviation, and process variable value. It is impractical and confusing to put a display for each of these signals on the controller's narrow face. The set point must be displayed, but either the process variable or deviation display can be sacrificed without great loss, since one is implied by the other once the set point is known.

The set point is either read out directly on a meter or mechanically displayed with a calibrated dial and index arrangement (Figure 4.2g). Sometimes the process variable and set-point value are shown with two pointers

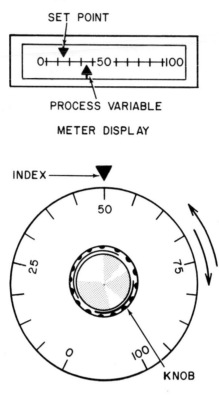

Fig. 4.2g   Standard set-point displays

on a single scale. This lets the operator compare easily the two for estimating the deviation. In another case it is the process value that must be estimated from markings that relate the deviation meter readings to a calibrated set-point dial.

One way to circumvent the limited space on the controller's front panel is to have one meter serve for indicating more than one variable. The operator then must position a switch to choose the variable he wants to read. One manufacturer indicates five basic controller signals (including a balance indication) with one meter. To avoid confusion, a set of switches inside the case determines the direction of meter movement for different switch positions so that the meter movement always drives upscale.

The controller's output meter shows either the manual or the automatic output signal, depending on the transfer switch position. Interchangeable tags at the side of the meter remind the operator of the valve positions (open or closed) for the extreme controller outputs. In certain controllers the output meter can be physically inverted so that the pointer always drives upscale as a valve opens. Occasionally the output meter may act as a deviation meter to null the automatic and manual outputs before transfer.

The front panel may also contain *alarm light displays*. These are often connected to alarm modules behind the panel that monitor deviation or process variable input. Some lights may merely run to an external connection at the rear of the controller. The position and form of the lights vary greatly. One manufacturer backlights legend plates. Another backlights the top and bottom of the deviation meter to show high and low deviation. In one case, deviation alarm lights replace the deviation meter itself. A few manufacturers provide a small standard panel light to indicate whether the local or remote set point is implemented.

Balancing Methods

All these controllers have a switch on the front panel for transferring between automatic and manual outputs. The switch takes many forms, including lever, toggle, rotary, and pushbutton. It is often important to balance the two output signals before transferring from automatic to manual to avoid a sudden change in controller output that might disturb or "bump" the process. Controllers that feature "bumpless transfer" have some provision for balancing the two output signals just before switching.

No balancing is necessary when switching from manual to automatic output because the unused automatic controller signal takes the same value of (or tracks) the output when the controller is in the manual mode. So the operator merely switches from manual to automatic and need not concern himself with balancing procedures. However, if a deviation exists between process and set-point values at the time of transfer, the integral action in the controller (if present) will immediately begin to drive the controller in a direction to eliminate this deviation. For large deviations the integral action will often produce a controller output change fast enough to bump the process.

This balanceless, bumpless transfer from manual to automatic output depends on a modified integral circuit to make the automatic signal track the manual output. So the integral function must be present if balanceless transfer is desired.

Nearly all manufacturers require a balancing procedure for bumpless transfer when the controller is switched from automatic to manual output. The transfer switch has a balance position to accommodate this procedure. In the balance position, the controller retains its automatic output, but the functions of the display meters change. The automatic and manual output signals usually feed to either side of a deviation meter. This allows the operator to equalize the two signals by adjusting the manual output until the meter reads zero at midscale. This nulling method has a high sensitivity for accurate balancing. Usually the large deviation meter, if present, serves to compare the output signals, but some manufacturers connect the output meter as a deviation meter for the controller's balance state. At least one supplier places a small separate deviation meter on the panel that always shows the condition of balance.

Still other manufacturers provide a momentary pushbutton on the front panel that, when depressed, connects the manual signal to the output meter in place of the automatic signal. To balance the two signals the operator repeatedly presses the button while adjusting the manual output. When the pointer on the output meter remains motionless, the two signals should be about equal.

A unique design features balanceless, bumpless transfer in both directions. This means that the unused output signal tracks the other signal at all times. In this case the control amplifier is switched into a simple integrator configuration when the operator transfers from automatic to manual mode. This circuit holds the latest value of the automatic output as a manual output. To change the manual value, the operator throws a momentary switch that introduces a DC input signal to the integrator circuit's input. This causes the output to ramp up or down, depending on whether the input signal is positive or negative. So the operator jogs the manual output to the desired value. The control amplifier also holds the latest manual output signal at the instant the operator switches back to automatic operation, but integral action begins immediately to eliminate any existing deviation between the set point and process variable values.

Mounting

Miniature electronic controllers are designed for panel mounting in a relatively clean and safe environment such as a control room. The room should be free from corrosive vapors and excessive vibration. A few controllers may have ambient temperature limitations ranging from 0 to 130°F, but others are more restricted. Some models may be mounted in areas designated by the National Electrical Code as Class I, Group C, Division 2. These areas require that equipment have no open sparking contacts, but do not require explosion proof cases. (See Sections 10.10 and 10.11 in Volume I.) Here the manufacturers provide hermetically sealed on-off power switches to avoid open sparking.

The controller's small frontal size allows many controllers to be installed in a small panel space, called a high-density configuration. The controller's depth (over 20 in.) and weight (often over 15 lb) usually requires additional support at the rear of the panel for multiple mountings. In general, the controllers can be mounted as multiple arrays in four ways (Figure 4.2h), in individual adjacent panel cutouts, side by side in a single large panel cutout, in large cases or packs that accept some multiple of controller chassis, or on shelves that accept some number of individual controller cases.

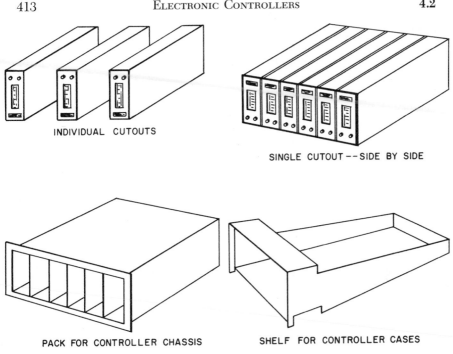

Fig. 4.2h   Panel mounting arrangements

The mounting approach depends on the particular manufacturer, but some companies offer more flexibility than others. A few require individual panel cutouts for each controller. This increases panel fabrication costs. When the controllers are placed side by side in a single cutout, the manufacturer may offer outside trims to frame the installation. The company will also suggest ways to support the back ends. Packs sometimes have both vertical and horizontal capacity, while shelves are limited to horizontal mountings. Some horizontal packs may be specified for any number of controllers, but others are limited either by a maximum number or by certain multiples of controllers. Mounting in packs or shelves generally simplifies power and signal connections.

If the panel tilts too far from the vertical, the controller may not operate properly. For example, one manufacturer specifies a maximum of 60-degree panel tilt, and another stops at 75-degree tilt from the vertical. Still others say that the controller works in any position. But some companion products, like recorders, have limitations on their mounting position.

In all these electronic controllers the chassis slides part way out of the case without interrupting the control signal. A retractable plug connects the chassis with all the necessary power, signal input and signal output leads. This permits tuning, calibration, and realignment while the controller operates in place. In a few controllers the panel display section can be mounted in a location different from that containing the control and output circuitry.

Although the controller is mounted in a relatively safe area (the control room), its signal must eventually go into the field to operate some final element, such as a control valve. Some control systems have been approved by either Factory Mutual or Underwriters Laboratories as intrinsically safe for hazardous field areas. (See Section 10.11 in Volume I.) This means that the output signal in the field cannot release sufficient electrical or thermal energy, under defined normal or abnormal conditions, to ignite a specific hazardous atmospheric mixture. Special barrier circuits and energy limiters, sometimes with zener diodes (Figure 4.2i), limit the output signal energy, even if the full voltage

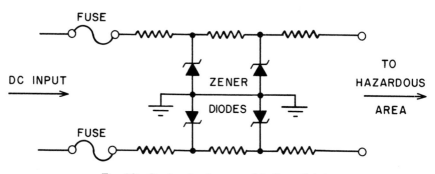

Fig. 4.2i   Barrier circuit approved in Great Britain

of the power main is somehow connected to the signal lines. Some suppliers put the barrier in the controller case; others prefer to place a barrier at the location where the signal lines leave the control room. Also, distinct mechanical separation of energy levels in the controller will add to its intrinsically safe properties.

Servicing

The electronic controller frequently is the first to receive the blame for loop malfunctions because its panel tends to give the first indication that something is wrong. Before removing the controller, however, the user should check other loop components and the operation of main plant equipment. He should also make sure that the controller output is not shorted to ground in the field.

If the controller is at fault, the course of action depends on the manufacturer's provisions for emergency servicing. In many cases the manual section of the controller will continue to work, so the operator can set the controller output at a desired DC current value. But the instrument maintenance technician must somehow remove the automatic section so he can troubleshoot it. The following items highlight the basic approaches taken by manufacturers:

1. Manual and automatic sections are completely independent, plug-in modules. One may be removed without affecting the other. They may even have separate power supplies to ease testing of the faulty unit. Sometimes only the automatic section is removable for emergency servicing.
2. Adjustable plug-in power module takes the place of manual output when the controller is removed from its case.
3. Battery-operated or auxiliary manual station replaces controller to hold output current. In some cases the substitution can be made without interrupting the signal to the process and (by a balancing procedure) without abruptly changing (bumping) the value of the controller output signal.

A few manufacturers make no provisions for emergency servicing. The faulty unit must simply be replaced regardless of signal interruption and output bumping.

Once the controller or its automatic section is in the shop, the technician can try to isolate the trouble by hooking up suitable input signals and a dummy load to the controller, closing a mock control loop. Some suppliers offer a jig that plugs into the controller to simulate a closed loop. Terminals are provided for input signals and for monitoring the deviation and output signals. Most companies at least provide easily accessible test pins or jacks for checking and calibrating key controller signals. Instruction manuals contain guides and schematics for diagnosing the fault from the symptoms. Multimeters used for voltage signal measurements should have at least 20,000 ohms per volt sensitivity.

The best buy is the controller that requires little maintenance and can be quickly replaced and repaired when it does need servicing. All the electronic controllers will perform satisfactorily in a control loop, at least in the short run. But those that frequently break down may result in costly process downtime, a factor that tends to override any differences in initial controller purchase and installation costs. Highly modular controller designs and circuit accessibility simplifies maintenance. Rugged, high-quality, solid-state circuit designs along with dust-tight encapsulation of key parts, including meter displays, help to prevent controller breakdowns. Silicon transistors are more rugged and reliable than those made of germanium. Much depends on the quality control practices implemented by the particular supplier.

Other Electronic Controllers

The electronic controllers discussed so far were all general-purpose devices with standard inputs and outputs, analogous to the 3–15 PSIG standard signals in pneumatic control systems (Section 4.3). A wide variety of other controllers on the market are tailored to specific kinds of control by specialized input or output schemes, or both.

The most common examples by far are electronic temperature controllers for driving electric heaters (Section 2.1) and motor speed controllers (Section 2.2). The temperature controllers are designed to accept directly inputs from sensors like thermocouples, resistance bulbs, or thermistors. Their outputs may drive relays or contactors, silicon-controlled rectifiers (SCRs), or saturable reactors.

Since the sensor connects directly to the controller, the input signal lines handle very low power levels; the lines are kept short to avoid interference from other sources. The input circuit will be for a particular kind of sensor and certain range of temperature. Common thermocouple inputs include iron-constantan, copper constantan, Chromel-Alumel, and platinum-rhodium platinum pairs. The controller's input circuit will compensate for variations in the thermocouple's cold junction temperature. Resistance bulb and thermistor inputs generally feed into resistance bridge circuits.

As in all feedback controllers, the input circuit will compare the incoming signal with a reference or set-point signal. Control circuitry will act on the resulting deviation. The form of control ranges from simple two- (on-off) or three-position (high-low-off) control to sophisticated three-mode control. Some manufacturers offer three-mode control on one set-point position and on-off control on another. Proportioning, however, is generally on a time basis, called time-proportional control. Here the output is driven full on or off, but the control signal varies the percentage of on-to-off time in some duty cycle, such as 10 seconds. The *average* power level determines the load temperature. This kind of control applies to both relay and SCR outputs. However, the SCR is capable of faster switching than the relay, permitting proportioning periods of a fraction of a second and giving smoother control. Some controllers trigger the SCR to chop up each AC cycle; in others the SCR proportions the number of AC cycles that are on and off in a certain period (Figure 4.2j). Available SCRs can handle loads up to 300 Kw in three line phases.

A common example of an electronic controller with relay outputs is the positioner controller (Figure 4.2k). Here the controller energizes one of two relays to operate an electric motor in either direction. The motor drives some final element to a desired position. A feedback signal from the final element tells the controller when the desired position has been reached. Essentially the controller acts as a master controller in a cascade system. The relays, motor, final element, and feedback signal constitute the slave loop, and the master controller generates the slave loop's set point by conventional control action.

A meter-relay combination provides an extremely simple way of implementing two- and three-position, high-gain control. Basically, the input signal, such as a thermocouple, drives a galvanometer's pointer to indicate temperature. Another pointer (or index) is manually set to the desired temperature on the same indicating scale. The idea is to operate a relay or an SCR when the two pointers are in the same position.

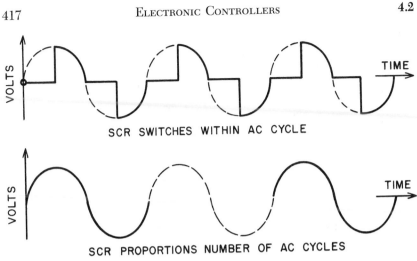

SCR SWITCHES WITHIN AC CYCLE

SCR PROPORTIONS NUMBER OF AC CYCLES

Fig. 4.2j   SCR switching techniques for driving electrical heaters

The system shown in Figure 4.2l optically senses the correspondence of the two pointers. The set-point arm carries a small photocell and light source. As long as light strikes the photocell, amplified current energizes the relay to supply power to a load. When the indicated temperature reaches the set-point temperature, a vane on the indicating pointer breaks the light beam, and the relay drops out. Some meter relays are equipped with two set-point arms and relays to give three-position control. In a meter with an SCR output the vane can be shaped to provide some degree of proportional control over a narrow temperature band. The optical system avoids any interaction between the process and set-point pointers, as would be present in mechanical or magnetic systems. Obviously the idea can be extended to other displays where the variable is indicated by a pointer's position. For example, it has been applied to electronic circular and strip-chart recorders.

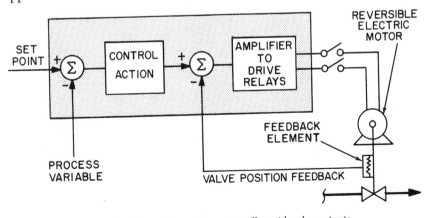

Fig. 4.2k   Electronic positioner controller with relay outputs

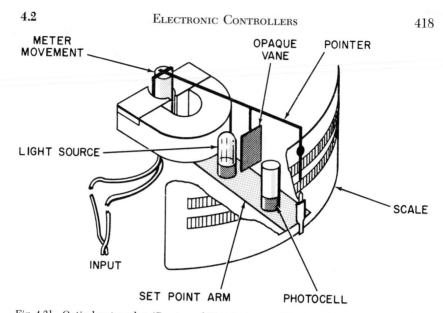

Fig. 4.21   Optical meter relay. (Courtesy of West Instrument Division of Gulton Industries Inc.)

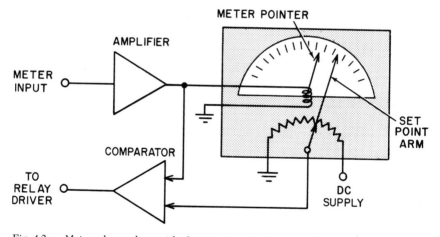

Fig. 4.2m   Meter-relay package with electronic input and output circuits (usually integrated circuits)

Meters can also be purchased with built-in electronic amplifiers to boost the input signal as in Figure 4.2m. In one meter-relay package of this type, the meter's set-point arm wipes a resistance element. This resistance element, connected as a voltage divider, provides a voltage reference signal, which is compared to the amplified input. The deviation, after further amplification, drives a transistor that operates a control relay. The meter and resistance element can be encapsulated in an inert atmosphere if the device must be installed in a relatively hostile environment.

Feature Check List

The following is a summary of the relevant features that an instrument engineer must consider when picking a controller:

Control mode selection
Input and output ranges (including set point)
Output load resistance range
Emergency service provisions
Maintenance accessibility, convenience
Panel readability and accuracy
Control repeatability and accuracy
Tuning ranges and resolution
Electrical classification
Power requirements; need for regulation
Mounting flexibility, density
Switches for local/remote set point, direct/reverse action, and manual/automatic operation
Balancing procedures and accuracy
Alarm modules and lights
Output limits
Anti-windup
Computer compatibility

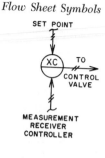

4.3 PNEUMATIC CONTROLLERS

Type:	Receiver controllers: Indicating, recording, miniature, high-density miniature and large case.
	Direct-connected controllers: Blind, indicating or recording in large or medium case for field or panel mounting.
Application:	Receiver type used to control any variable that can be measured and translated into an air pressure by a pneumatic transmitter. Includes automatic and manual control and set-point adjustment.
	Direct-connected types contain their own measuring element in contact with the process. The following are some of the measuring elements available: Pressure, differential pressure, temperature, level, pH, thermocouple, radiation pyrometer, humidity.
Typical Front Panel Size:	Miniature: 6″ × 6″
	Miniature, high-density: 3″ × 6″
	Large case: 15″ × 20″
Minimum Response Level:	Less than 0.01% of full scale
Input Ranges:	3–15 PSIG or direct connected.

Output Ranges:	3–15 PSIG.
Repeatability:	±0.5%.
Accuracy:	±1%.
Displays:	Set point, process variable, output, deviation and balance.
Maximum Frequency Response:	Flat to 30 Hz.
Maximum Zero Frequency Gain:	750.
Control Modes:	Manual, proportional, integral (reset), derivative, floating, differential gap, two-position
Cost:	Miniature indicating controller—$380. Miniature recording controller—$640. Miniature high density indicating controller—$420. Large case recording controller—$550. Large case indicating controller—$500. Direct-connected, field-mounted, pressure pilot type indicating controller—$320.
Partial List of Suppliers:	MINIATURE: Bailey Meter Co.; Beckman Instruments, Inc.; Bristol Co.; Fischer & Porter; Foxboro Co.; Hagan Controls Corp.; Honeywell, Inc.; Moore Products Co.; Taylor Instrument Cos.
	HIGH-DENSITY MINIATURE: Fischer & Porter; Foxboro Co.; Honeywell, Inc.; Moore Products Co.; Taylor Instrument Cos.
	LARGE CASE, RECEIVER AND DIRECT-CONNECTED: Bailey Meter Co.; Bristol Co.; Fischer & Porter; Foxboro Co.; Hagan Controls Corp.; Honeywell, Inc.; Masoneilan International, Inc.; Taylor Instrument Cos.
	LOCALLY MOUNTED RECEIVER OR DIRECT-CONNECTED: Bailey Meter Co.; Black-Sivalls and Bryson, Inc.; Bristol Co.; Fisher Controls; Foxboro Co.; Hagan Controls Corp.; Kieley and Mueller, Inc.; Masoneilan International, Inc.; Moore Products Co.; U.S. Gauge Div. of Ametek.

History and Development

Pneumatic controllers were first introduced at the turn of the century. They logically followed the development of diaphragm-actuated valves in the 1890s. Early types were all direct-connected, local-mounting, indicating or blind types. Large case indicating and circular chart recording controllers appeared around 1915. All early models incorporated two-position, on-off action or proportional action. It was not until 1929 that reset action was introduced. Rate action followed around 1935.

Up until the late 1930s, all controllers were direct-connected and therefore had to be located close to the process. Pneumatic transmitters were not introduced until the late 1930s. To make them compatible, the large case pressure recording and indicating controllers were easily converted into receiver controllers. This made remote mounting practicable and centralized control rooms became a reality. Because of the inherent advantages, the combination of pneumatic transmitters and receiver controllers quickly became popular. Since the recording and indicating receiver controllers were quite large, control rooms and panel boards were likewise spacious. Additionally, all control boards had a similar, monotonous look and usually came in one color—black.

A revolution in design occurred in 1948 with the introduction of miniature instruments. Here the concept of the controller evolved into a combination of a small, approximately 6×6-in. panel front, indicating and recording control station and a blind receiver controller. The station permitted the operator to monitor the measured variable, set point, and valve output; it allowed him to switch between, and operate in, either the automatic or manual control modes. Instantly, miniature controllers ushered in the era of the graphic panel in which the instruments are inserted into graphic symbols representing the attendant process apparatus. Control rooms became more compact, control boards more meaningful and colorful, and because operators quickly developed a "feel" for the process, training time was considerably reduced.

Nevertheless, graphic panels too were wasteful of space and presented major modification problems each time the process was changed. This led to the evolution of the semi-graphic panel, in which a graphic symbol diagram of the process appeared above the miniature instruments mounted in neatly spaced rows and columns.

In 1965, miniature, high-density mounting style stations appeared. The new lines brought with them the most advanced ideas in displays, operating safety and simplicity, packaging, installation simplicity and servicing facility. Along with some of the standard miniature controllers, they offer computer compatability along with some unique control capabilities which had previously been impractical.

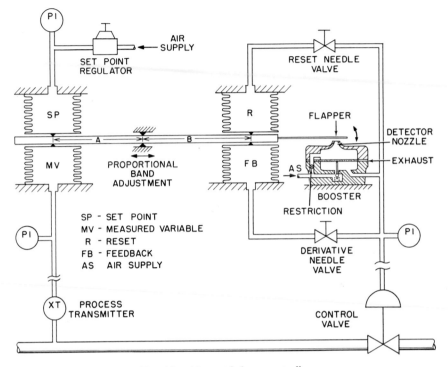

Fig. 4.3a   Moment-balance controller

Pneumatic Controller Principles

A receiver type pneumatic controller is shown schematically in Figure 4.3a. A process transmitter, lower left, senses the measured variable, which may be pressure, temperature, flow, etc., and transmits a proportional air pressure to the MV bellows of the controller. The controller compares the measured variable against the set point (SP) and sends a corrective air signal to manipulate the control valve, thereby completing the feedback control loop.

The controller consists of two sets of opposed bellows, of equal area, acting at opposite ends of a force beam which rotates about a movable pivot. Extending from the right end of the beam is a flapper which baffles the detector nozzle of the booster relay.

Booster Circuit   Supply air connects to the pilot valve of the booster and flows through a fixed restriction into the top housing and out the detector nozzle. The flapper is effective in changing the back pressure on the nozzle as long as the clearance is within one-fourth of the nozzle diameter. The restriction size is selected on the basis that the continuous air consumption will be reasonable and that it will be large enough not to clog with typical instrument air. The nozzle, on the other hand, must be large enough that when the flapper has a clearance of one-fourth nozzle diameter, nozzle back-pressure drops practically to atmospheric. It must not be so large, however, that the seating of the flapper becomes too critical. A typical size of restriction is 0.012 in. ID while the nozzle would be 0.050 in. ID. The nozzle back-pressure is a function of flapper position. (For a more detailed discussion of flapper-nozzle detector circuits, refer to Section 6.1 on Pneumatic Transmitters.)

The exhaust diaphragm senses nozzle back-pressure and acts on the pilot valve. If the back-pressure increases, it pushes down on the valve, opening the supply port to build up the underside pressure on the diaphragm until it balances the nozzle back-pressure. If the back-pressure decreases, the diaphragm assembly moves upward, allowing the valve to close off the supply seat while opening an exhaust seat in the center of the diaphragm. This allows the underside pressure to exhaust through the center mesh material of the diaphragm assembly until the pressures balance.

Proportional Response   Since a pressure range of 3–15 PSIG is an almost universal standard for representing 0 to 100 percent of the range of measured variable, set point, and output on receiver controllers, this description will assume the operation to be in this range. In explaining the proportional response, first assume that the derivative needle valve is wide open and that the reset needle valve is closed with 9 PSIG mid-scale pressure, trapped in the reset bellows R (Figure 4.3a). If the set point is adjusted to 9 PSIG, then when the measured variable equals set point, the flapper will automatically be positioned so that the booster output, acting on the feedback bellows FB, will be equal to the reset pressure, namely, 9 PSIG. The reason for this is that the force beam will only come to equilibrium when all of the moments about the pivot come to balance. Any unbalance in moments causes a rotation of the beam with attendant repositioning of the flapper and change in feedback pressure until moment balance is restored.

If the pivot is positioned centrally, where moment arm A equals B, then for every 1 PSI difference between set point and measured variable there will be a 1 PSI difference between reset and controller output, or feedback. This represents a 100 percent proportional band setting, or a gain of 1. Proportional band is defined as the input change divided by the output change times 100. Gain equals the ratio of output change over input change. See proportional band chart, Figure 4.3b.

$$PB\% = \frac{\text{change in input}}{\text{change in output}} \times 100$$

$$\text{Gain} = \frac{\text{output change}}{\text{input change}}$$

If the pivot in Figure 4.3a is shifted to the right, to the point where moment arm A is four times greater than moment arm B, then every 1 PSI

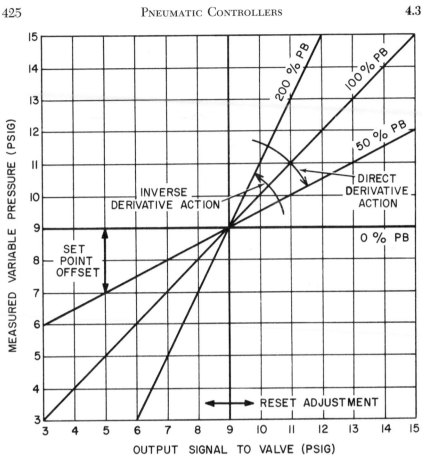

Fig. 4.3b   Graphic representation of control functions

change in measured variable results in a 4 PSI change in output. This gives a proportional band of 25 percent, or a gain of 4. If the pivot is moved to the left reversing the ratio, the proportional band would be 400 percent and the gain 0.25. If the pivot could be moved to the right to coincide with the center of the R and FB bellows, the most sensitive setting would be achieved, i.e., approaching 0 percent band or infinite gain, as the slightest difference between MV and SP would rotate the flapper to change the output to 0 PSI or full supply pressure, depending upon the direction of the error (on-off control).

Reset Response   If it were practical to use 1 or 2 percent proportional band on all processes, proportional action alone would be sufficient for most processes. However, most process loops become unstable at much wider bands than this. Noisy flow control loops, for example, may require more than 200 percent band for stability. Suppose the process in this case could tolerate a band no narrower than 50 percent. Then, according to the proportional band diagram, the controller could only control exactly at set point when valve pressure had to be 9 PSIG. If the valve pressure, because of conditions, had to be 5 PSIG, for example, it could only be so when the measured variable deviates from set point by 2 PSI (16.7 percent of scale error). In most cases this amount of error, more correctly termed offset, would be intolerable. Nevertheless, in practice, the valve pressure must change as the load changes. As an example, to hold level in a tank, the control valve on the inlet would obviously have to change opening as rate of effluent changes.

If the load were such as to require a 5 PSIG valve pressure, one way of eliminating the offset would be to *manually* change the reset pressure R to 5 PSIG. The output or valve pressure would then be 5 PSIG when MV would equal SP. This action amounts to applying manual reset to the controller and is used in some occasions. Actually, it was more popular years ago, but currently rather uncommon.

It is a simple matter to make this reset action automatic. It only requires that the reset bellows be able to communicate with the controller output pressure through some adjustable restriction such as a needle valve. The reset action must be tuned to the process in such a way as to allow the process sufficient time to respond. Too fast a reset speed, in effect, makes the controller "impatient" and results in instability. Too slow a speed results in stable operation, but the offset is permitted to persist for a longer period than necessary.

To describe the automatic reset action, assume that SP is at 9 PSIG and R is trapped at 9 PSIG, while valve pressure FB, because of load, must be at 5 PSIG. According to Figure 4.3b, with the proportional band at 50 percent, MV will be controlled at 7 PSIG. If the reset needle valve is then opened slightly, the pressure in the reset bellows will gradually decrease. As it drops from 9 to 7 PSIG, the measured variable pressure will rise from 7 to 8 PSIG. The reset pressure will continue to drop until it exactly equals output, i.e., 5 PSIG. At this point, the measured variable will exactly equal set point, 9 PSIG. With this circuit, regardless of where the valve pressure must be, the controller will ultimately provide the correct valve pressure with no offset between set point and measured variable.

Reset action can also be understood by looking at it from the opposite direction, i.e., controller configuration. The controller cannot come to equilibrium as long as there is any difference in pressure between R and FB, since the open communication through the reset needle valve will cause the reset to continue to change which in turn directly reinforces the feedback pressure. R and FB, in turn, cannot be equal unless SP and MV are equal to each other. This fact alone assures that the controller will maintain corrective action until it makes MV exactly equal to SP regardless of where FB must be. Referring to Figure 4.3b, reset in effect shifts the proportional band lines along a horizontal axis at set-point level. It makes the center of the band coincide with the required valve pressure.

Reset time is the time required for the reset action to produce the same change in output as that resulting from proportional action as the error remains constant. For example, if an error resulting in 1 PSI output change due to proportional action is applied to a controller and the error is sustained, a 1 minute reset setting would cause the output to continue to change at the rate of an additional 1 PSI per minute in the corrective direction. The term repeats per minute is also used to characterize reset. This term is the reciprocal of reset time.

Derivative Response   If a needle valve is inserted between the booster output and the feedback bellows as in Figure 4.3a, it delays the rebalancing action of the feedback bellows and causes the controller to give an exaggerated response for changes in the measured variable. The degree of exaggeration is in proportion to the speed or rate at which the measured variable changes. (The term derivative action refers to the mathematical description of rate.) Derivative action is particularly effective in the slow processes, as with most temperature control loops. It compensates for lag or inertia. For a sudden change of even small magnitude, it provides an extra "kick" to the control valve because it recognizes that, with the lag which exists, even a small sudden change in measured variable signifies that a considerable exchange of energy has taken place and that the situation is likely to get worse before getting better. Conversely, as it drives the process toward set point, it begins anticipating the inertial effect and begins cutting back the valve response accordingly.

The simple method of achieving derivative action in Figure 4.3a closely resembles some of the approaches used in practice, but it has some serious limitations. In this method, the derivative action interacts with the proportional and reset responses and, in fact, follows the proportional response. It is therefore useless in preventing overshoot on start-up and on large upsets, as will be described later under special batch controllers. The derivative also interacts with set-point changes. An independent derivative unit is shown in Figure 4.3e and is described later.

Derivative action can be considered to temporarily rotate the proportional band lines in Figure 4.3b clockwise in response to rate of change, i.e., derivative temporarily narrows the band. Derivative time is the time in minutes by which the output would lead the feedback pressure during a steady ramp input change. This is described in greater detail under Derivative Relay.

Miniature Receiver Controllers

Two actual designs of force-balance receiver controllers are shown in Figures 4.3c and d. The controller in Figure 4.3c closely resembles that of Figure 4.3a, except that the bellows all act from one side against a pivoted

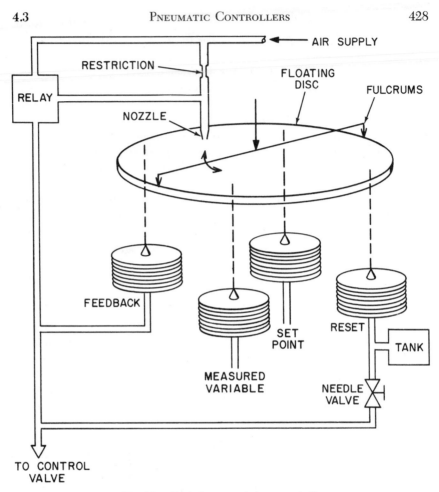

Fig. 4.3c   Typical moment-balance controller

"wobble plate." The wobble plate acts as the nozzle baffle. Rotating the pivot axis changes the proportional band. When the pivot axis coincides with the reset and feedback bellows, 0 percent proportional band results; when it coincides with the measured variable and set-point bellows, infinite band results. When the axis bisects the two bellows axis, 100 percent proportional band results. Otherwise, the operation is the same as described for Figure 4.3a.

The controller in Figure 4.3d is constructed of machined aluminum rings, separated by rubber diaphragms with bolts holding the assembly together. The lower portion of the controller forms the booster section. This is quite similar to the booster in Figure 4.3a. The detector section consists of three diaphragms. The upper and lower diaphragms have equal areas while the center diaphragm has half the effective area of the other two. Reset pressure acts in the top chamber. Assume this pressure is at midscale and that the reset needle valve is closed. The reset pressure acts on the top diaphragm, which is part of a 1:1 reproducing relay. Supply air passes through a restriction and out the exhaust nozzle. The diaphragm baffles the nozzle to make the back pressure equal to the reset pressure. Assume further that the proportional band needle valve is closed. The pressure then acting on top of the detector section will be the reset pressure as reproduced by the 1:1 relay, since the two chambers are connected via a restriction. If the measured variable signal equals the set point, the detector diaphragm assembly, with its integral nozzle seat, will baffle off the nozzle so as to make the controller output, or valve pressure, equal to the reset pressure, thus bringing all of the forces acting on the detector to balance. If the measured variable then increases by 1 PSI, the increase in pressure acts downward on the lower diaphragm as well as upward on the center diaphragm. The net effect is the same as having the

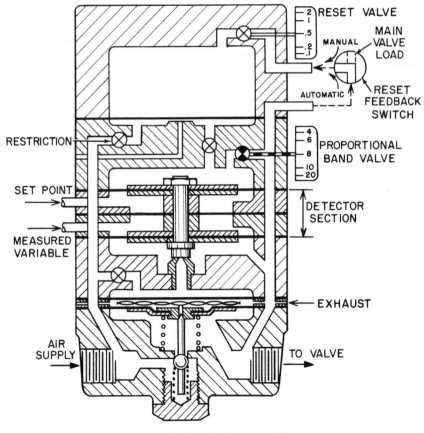

Fig. 4.3d Force balance controller

pressure act downward on a diaphragm of half the area of the lower diaphragm. Since the output pressure acts upward on the full area of the lower diaphragm, it need increase only $\frac{1}{2}$ PSI to bring the forces to balance. Since a 1 PSI change in input resulted in a $\frac{1}{2}$ PSI change in output, the proportional band is said to be at 200 percent.

If the proportional band needle valve were wide open, so that it would provide negligable resistance, then if the measured variable pressure increased ever so slightly above set point, the full effect of the resultant change in output would be felt on top of the detector stack. This would cause the output to increase further, which in turn would feed upon itself and the action would continue to regenerate until the output reached its maximum limit. Therefore, with the proportional band needle valve wide open, the narrowest proportional band is obtained. If the proportional band needle valve is set to where its resistance equals that of the restriction which separates the reproducing relay from the top of the detector section, then the following action results. If the measured variable deviates from set point by 1 PSI in an increasing direction, instantly the output will rise $\frac{1}{2}$ PSI because of the construction of the detector section. The difference in pressure between the controller output and the reproducing relay will cause a flow through the reset needle valve and the intermediate restriction. Since the resistance of the two is equal, the pressure drop will divide equally, causing a $\frac{1}{4}$ PSI increase on the top of the detector section. This $\frac{1}{4}$ PSI increase directly causes the output to increase $\frac{1}{4}$ PSI, which further causes the pressure on top of the detector to increase by $\frac{1}{8}$ PSI. The action continues until equilibrium is obtained, with the output having changed a total of 1 PSI and with the pressure on top of the detector section having increased $\frac{1}{2}$ PSI. Since a 1 PSI change in variable resulted in a 1 PSI change in output, this needle valve opening provides a 100 percent proportional band.

The reset action in this controller is similar to that of Figure 4.3a in that any change in reset pressure propogates down through the unit to directly affect the output. The controller will not come to equilibrium until all forces are balanced, i.e., the measured variable will have to equal set point and the reset pressure will have to equal controller output.

Derivative Relay

It was noted earlier that the type of derivative circuit used in Figure 4.3a had definite limitations. Derivative action is more effective if it is non-interacting and if it can be applied *ahead* of the proportional and reset action of the controller. Such a derivative unit can be built into the controller, or it can take the form of a separate relay as shown in Figure 4.3e. This relay employs diaphragms, but the design can be executed with bellows just as well. The signal from some process transmitter, typically a temperature transmitter, would be connected to the input of the relay.

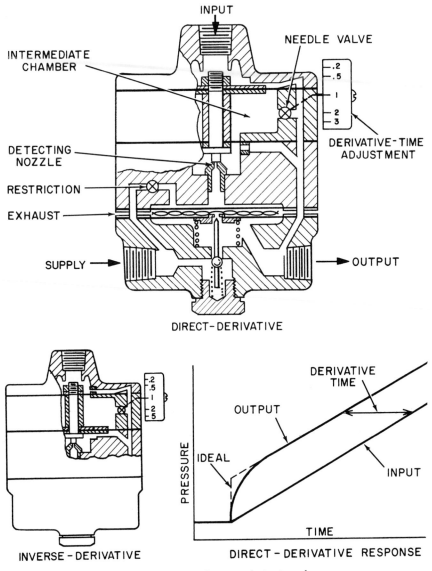

Fig. 4.3e   Direct and inverse derivative relays

Due to the difference between input- and output-diaphragm areas, a step change in input produces a largely amplified step change in output to maintain force balance.

In a steady-state condition (with no change in input), the output pressure acts on both sides of the output diaphragm—and output pressure rebalances input pressure directly.

The output pressure is connected to the intermediate chamber through the needle valve. There is therefore a lag between a change in output and a change in intermediate pressure.

With a continuous ramp change in input, the intermediate pressure will lag the output by a constant amount, proportional to the rate of change in output. Thus, the intermediate pressure will partially rebalance the input, reducing the effective gain. The result is that the output will continuously lead the input by a definite amount, proportional to the rate of change in input.

The time by which the output leads the input is the "derivative time" as set on the graduated needle valve.

An inverse derivative relay, as shown in Figure 4.3e, works in the opposite manner and attenuates high-frequency signals. It can therefore serve as a stabilizing relay in "noisy" processes.

Miniature Control Stations

Miniature control stations having a panel face of nominally 6 × 6 in. and inserted into individual cutouts having approximately 10 in. center-to-center distances are one of the common types found on central control room panels in the various process industries today. Figure 4.3f shows a typical cross section of some of the types of units available.

A typical miniature indicating control station is shown in Figure 4.3g. The measured variable is indicated on the center pointer, and the set point is indicated on the peripheral pointer of a duplex gauge. On automatic, the operator changes the set point by adjusting the set-point regulator and noting the set point on the gauge. The controller is connected to the control valve via the manual-automatic switch. The controller compares the measured variable with the set point and manipulates the control valve to bring the variable on set point. If the operator wishes to switch to manual, he notes the valve pressure by operating the upper left-hand switch on the station. He next turns the right-hand switch to "seal," which isolates the controller from the control valve. He then adjusts the regulator to match the noted valve pressure and then turns the right-hand switch to the manual position, which connects the regulator directly to the valve. In manual, the operator directly adjusts the valve while the controller reset follows whatever changes are made to the valve. In switching back to automatic, the operator goes to "seal" position and adjusts the set-point regulator to match the measured variable. If the set point and measured variable are equal at switchover, and if the reset is equal to the valve pressure, the controller output should then similarly equal the valve pressure, and the switchover is effected without a "bump."

Four-Pipe System

Since there is a lag in the transmission of pneumatic signals, the dynamic capability of a control loop can be affected with increasing distance between the controller and the process. (The subject of the effect of transmission distance on control is covered in more detail at the end of this section.) However, the transmission lag will only be significant on the fast processes such as liquid flow control and then only as the distance exceeds 300 ft. For

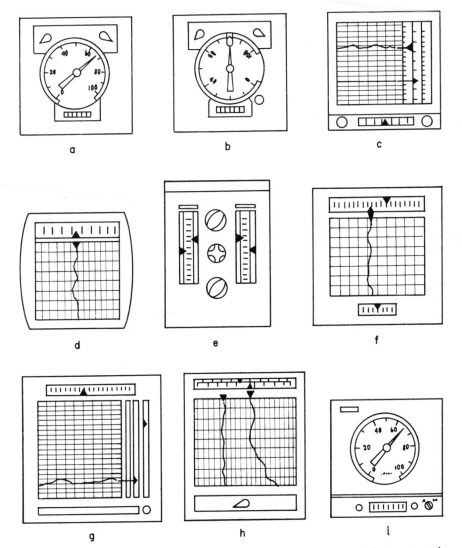

Fig. 4.3f Available types of miniature pneumatic controllers. (a) Typical indicating control station. (b) Indicating control station with 12 o'clock scanning feature. (c) Recording control station with 30-day strip-chart and vertical moving pen. (d) Recording control station with horizontal moving pen and daily chart tear-off feature. (e) Indicating control station with two duplex vertical scale indicators. (f) Recording control station with no "seal" position. (g) Recording control station with servo-operated pen. (h) Recording control station with procedureless switching. (i) Indicating control station with instant procedureless switching.

S - AIR SUPPLY
SP - SET POINT
MV - MEASURED VARIABLE
CO - CONTROLLER OUTPUT
R - RESET FEEDBACK
C - CONTROLLER
————MECHANICAL CONNECTION
—#——#—PNEUMATIC CONNECTION
M/A - MANUAL – AUTOMATIC

AUTO   SEAL   MANUAL

M/A SWITCH POSITIONS

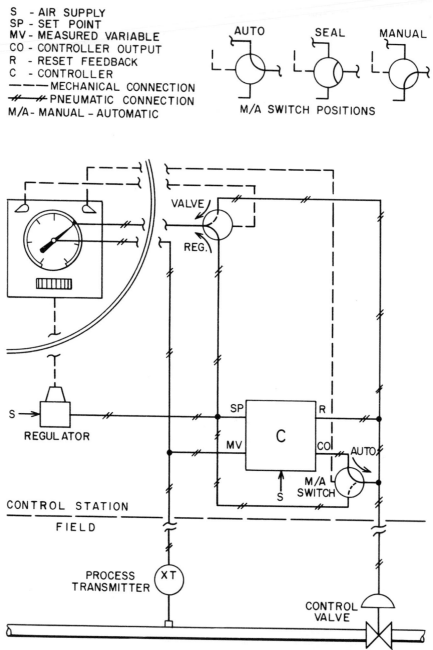

Fig. 4.3g   Miniature control station with integral-mounted controller

circuits where this lag would prove objectionable, the type of control station shown in Figure 4.3h is employed. Here the controller is mounted in the field near the measuring transmitter and control valve. Four connections are run between the control station and the field-mounted equipment. These involve the measured variable, set point, valve pressure and relay operating pressure lines. Hence, the name *four-pipe system* is most often applied to characterize this circuit, whereas the circuit in Figure 4.3g is often referred to as a *two-pipe system*. Since the lines going back to the station amount to dead-ended parallel connections, the dynamics of the control loop are the same as they would be with any closed-loop system. In switching from automatic to manual, the operator goes to the "seal" position, which actuates the cutoff relay to isolate the controller from the valve and permits the operator to change the set-point regulator to match the noted valve pressure. This pressure is then connected to the valve when the operator turns the switch to manual. Returning to automatic involves the same procedure as described for Figure 4.3g.

The disadvantages of this circuit are that (1) it is more costly to run four transmission tubes between the control station and the field-mounted equipment and (2) the controller settings can not be adjusted from the control panel.

Remote-Set Station

If a modification as shown in Figure 4.3i is added to the basic station in Figure 4.3g, the control station can accommodate a remote set-point signal from sources such as a ratio or proportioning relay, primary cascade controller, or analog computer.

Computer-Set Station

The addition of a stepping motor to the set-point regulator or to the set-point motion transmitter as in Figure 4.3j allows the control station to be set from a digital computer. The station still allows the computer to be disconnected and permits the set point to be adjusted locally and provides direct manual control.

Single-Station Cascade

Cascade control can be implemented either with two stations, such as Figure 4.3g on the primary and Figure 4.3i on the secondary, or with a single station, such as shown in Figure 4.3k. The latter scheme not only eliminates one station, but offers operating safety and convenience as well. A common problem when using two stations is that when the operator switches to manual, which he does on the secondary station, he often forgets to switch the primary station to manual as well. The primary controller then wanders around aimlessly and is not balanced for switchover back to cascade. Compounding this is the fact that cascade circuits are often employed on the most critical loops.

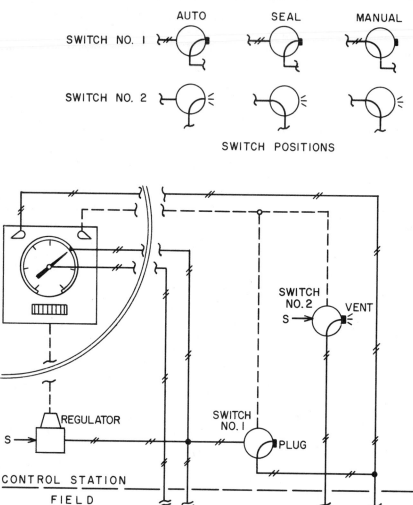

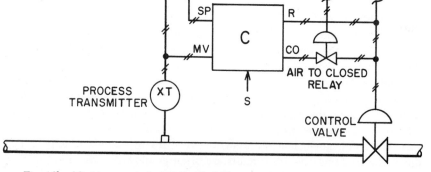

Fig. 4.3h   Miniature control station with field-mounted controller (four-pipe system)

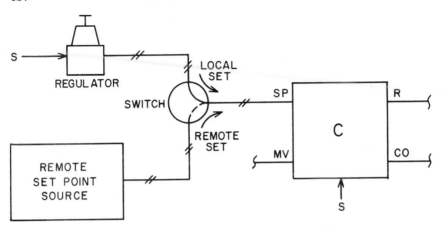

Fig. 4.3i   Control station circuit for remote set-point adjustment

In Figure 4.3k, the regulator has three functions: (1) Set point to primary controller in cascade control, (2) set point to secondary controller for independent secondary control, and (3) manual valve setting. There is a seal position between each step while the regulator is set for its upcoming function. The key to this station is the concept used in making the secondary measured variable MV2, the reset feedback of the primary controller while on manual or secondary control. Versions are also available which allow cascade, independent automatic control on the primary and manual.

"No Seal" Station

The use of two regulators or motion transmitters in a station, Figure 4.3l, eliminates the need for a "seal" position. When the operator wishes to switch

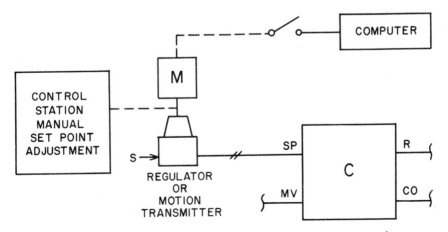

Fig. 4.3j   Control station circuit for computer adjusted set-point control

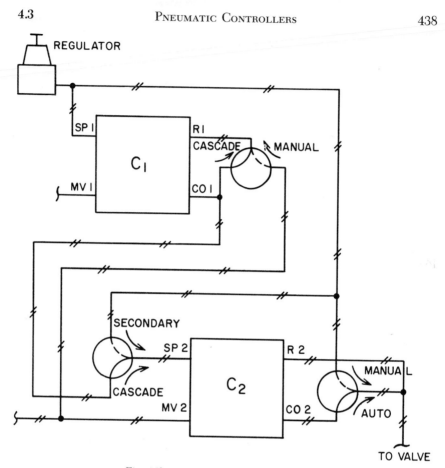

Fig. 4.3k   Single-station cascade controller

to manual, he adjusts the manual regulator to match the controller output while viewing a deviation indicator. When they are aligned, he transfers control.

Procedureless Switching Station

Two methods of procedureless switching have been introduced in conventional miniature stations. With these, the operator simply turns a switch and the station automatically takes care of pressure-balancing problems. The two approaches are shown in Figure 4.3m. The mechanism on the left is a combination motion transmitter/receiver.

The motion transmitter provides set-point pressure in automatic and valve pressure in manual—just as the regulator does in Figure 4.3g. As a motion transmitter, a friction clutch holds the index lever at whatever position the operator sets it. A restriction-nozzle circuit senses the position and converts it to a proportional pneumatic output which is fed back to the rebalancing bellows. A 0 to 100 percent movement of the index gives a 3–15 PSIG output pressure. When acting as a receiver, supply pressure is cut off the restriction nozzle circuit and the pressure to be sensed is admitted to the rebalancing bellows. The friction clutch is disconnected so that the index lever can be moved by the rebalancing bellows. The unit is so designed and calibrated that a 3–15 PSIG sensed pressure produces a 0 to 100 percent index movement. Therefore, as the operator moves the switch from automatic to manual, the following actions take place in sequence and automatically: The index is de-clutched; supply is cut off the restriction nozzle circuit; controller output (valve) pressure is admitted to the bellows causing the index lever to take a position proportional to the pressure; controller output is disconnected from the valve line; the clutch is engaged; supply pressure is readmitted to the restriction nozzle circuit and the unit again acts as a motion transmitter, now providing valve pressure. Switching back to automatic involves the same sequence, in reverse. This is the same sequence as carried out in Figure 4.3g, except that here it is automatic.

A second system for procedureless switching involves the use of *self-synchronizing regulators*, termed *syncros*, by which one regulator provides set point while the other is used for manual valve loading (Figure 4.3l). As the name implies, this regulator can synchronize itself to some varying pneumatic pressure and thereby provide automatic balancing. The regulator employs a reaction nozzle circuit which results in very low spring force (approximately 1 oz) to develop a 3–15 PSIG output. The setting spring is adjusted by rotation

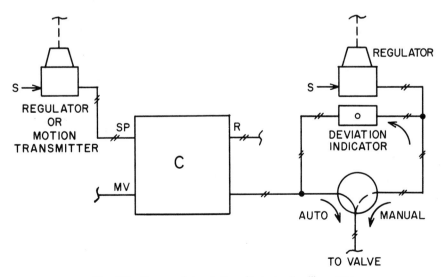

Fig. 4.31   Two-regulator station eliminates "seal" position

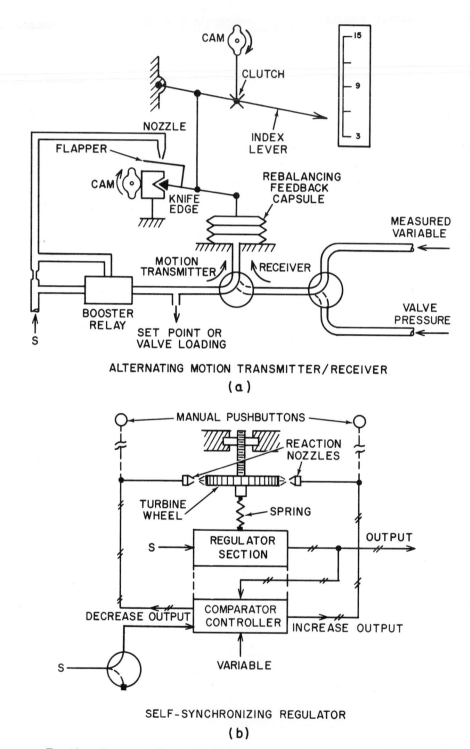

ALTERNATING MOTION TRANSMITTER/RECEIVER

(a)

SELF-SYNCHRONIZING REGULATOR

(b)

Fig. 4.3m   Two approaches to self-balancing, procedureless switching control stations

of a turbine wheel with an integral lead screw (Figure 4.3m). If supply is connected to the comparator controller section, air is transmitted to the increase-decrease nozzles to make the regulator section output match the variable input pressure. If supply is cut off from the comparator controller, the regulator section output remains locked in, with the memory being a function of lead screw position. The unit can then be driven manually by the operator.

On the station, when in automatic, the set-point syncro is manually adjusted by the operator, while the valve-operating syncro keeps itself matched to the controller output to allow instant transfer to manual. In manual, the operator adjusts the valve-loading syncro while the set-point syncro tracks the measured variable.

In addition to procedureless switching, these stations can be switched from a remote location or source, manually or automatically; they can be gang-switched and they make possible the operation of stations connected in parallel on one loop as, for example, having one station in the central control room and the other in the field local to the process. Whichever station is not in active service keeps itself fully synchronized and ready to be made active at any instant.

Miniature High-Density Stations

These represent the latest developments in pneumatics. The stations have a typical panel size of 3 × 6 in., mount adjacent to each other and allow very compact and efficient panel arrangements (Figure 4.3n). The units incorporate novel packaging features which simplify panel construction and design and facilitate servicing. Much of the design is aimed at making the job of the operator simpler, faster and safer, in line with the present trend to consolidate control rooms, to increasingly shift the plant to one operator, and to handle processes that keep getting faster, more critical and more complex.

Mid-Scale Scanning Station

Figures 4.3o, p and q show three types of high-density control stations which feature a mid-scale deviation scanning pointer. The pointer is driven either by a differential detector or differential servo which compares the measured variable against set point. If the two are equal, the red deviation pointer is positioned at mid-scale, where it is screened off by a green scan band. If there is a deviation, the red pointer stands out prominently.

In Figure 4.3o, a fixed, nominal 4-in. vertical scale is employed, and there are separate pointers to indicate set point and measured variable. The station uses the two-regulator approach to achieve "no seal" switching as in Figure 4.3l. The operator does have to balance pressures before switching, however.

In Figure 4.3p, the station employs an expanded scale, which provides

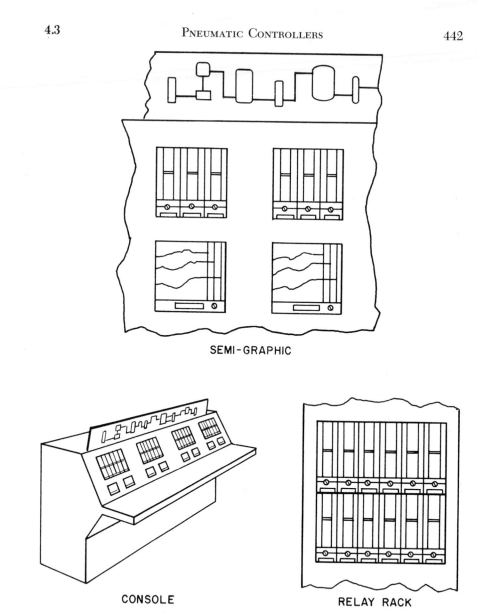

SEMI-GRAPHIC

CONSOLE   RELAY RACK

Fig. 4.3n   Typical mounting arrangements of high-density control stations

greater readability. The only indication on the scale is deviation, however, and this requires that the set-point transmitter scale and deviation servo stay in calibration relative to each other in order to provide an accurate reading of the variable. While the expanded scale gives greater readability, it does have to be moved to bring the reading on scale when the variable makes any excursions. This station and the station in Figure 4.3q, likewise, use the two-regulator approach to eliminate the need for a seal position. In both cases the operator balances pressures prior to switchover. In Figure 4.3p, the operator notes deviation on a ball-in-tube indicator. In Figure 4.3q, the valve switch has a detent action while the indicator switch operates at the mid-throw position, so that the operator moves the integral switch lever back and forth across center while manually matching pressures before switching. The controller in Figure 4.3q is a deviation type actuated by displacement of a deviation link in the indicator circuit. The controller acts to hold the link at its "zero" position.

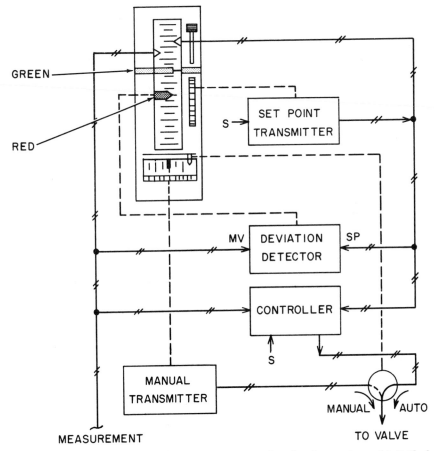

Fig. 4.3o   Functional diagram of high-density station with mid-scale scanning and individual indication of set point and measured variable

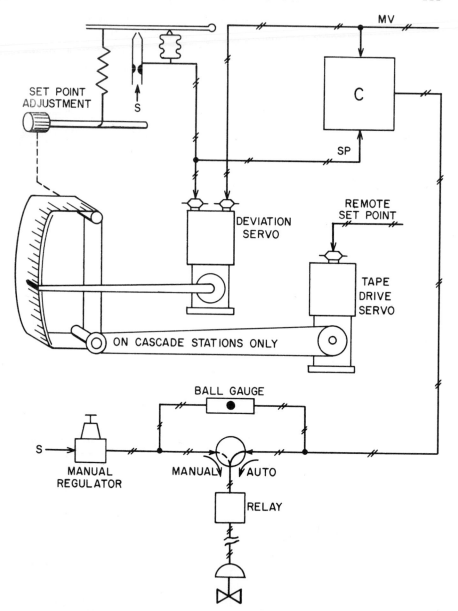

Fig. 4.3p   Mid-scale scanning station with expanded scale

Procedureless Switching Station

Stations in Figures 4.3r and 4.3s offer procedureless switching for the operator. Both have a fixed 4-in. scale, separate indication of set point and measured variable and a scanning concept which involves having the set-point indicator overlap the measured variable indicator when control is normal.

The station in Figure 4.3r employs two self-synchronizing regulators, termed syncros (Figure 4.3m), one for set point and the other for manual valve loading. On automatic, the operator manually adjusts the set-point syncro, while the valve-loading syncro automatically tracks controller output. On manual, the operator adjusts the valve-loading syncro while the set-point syncro tracks the measured variable. Thus the station is always balanced for instant transfer of control mode.

Like its 6 × 6-in. counterpart, these stations can be gang-switched, switched remotely or automatically, and operated in parallel from different

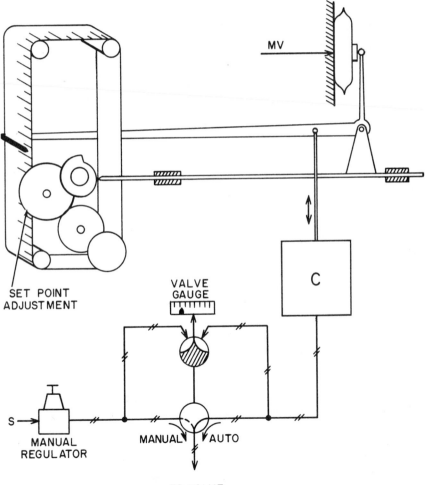

Fig. 4.3q   Mid-scale scanning station, expanded scale and deviation controller

SR- SELF-SYNCHRONIZING REGULATOR
S/S - SELF- SYNCHRONIZING ACTIVATION SIGNAL

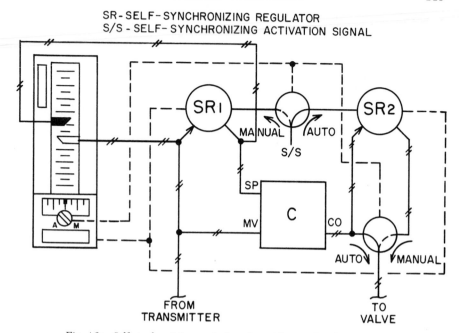

Fig. 4.3r   Self-synchronizing control station with procedureless switching

locations while maintaining themselves in syncronism, and they are available in single-station cascade arrangements.

The station in Figure 4.3s employs a dual function motion transmitter/ receiver for manual valve loading and valve pressure indication. This unit is similar to that described in Figure 4.3m. On automatic, the index lever is declutched and the feedback capsule, connected to the valve pressure line, moves the index accordingly. On manual, the clutch engages, and the mechanism reverts to a motion transmitter which provides manual valve loading. This allows procedureless switching to manual. Switching to automatic is also procedureless, assuming that the set point of the process does not change. If the operator wishes the controller to operate at some new value, compared to where he had the process on manual, he must obviously change the set point to that value prior to switching to automatic. However, to facilitate switching to automatic on loops where the set point remains fixed, the station incorporates a separate balancing controller which operates while the station is in manual. The balancing controller manipulates controller reset pressure to keep the controller output equal to the manual valve loading even though the measured variable may be off set point. This allows the operator to switch to automatic while off the intended set point and yet have the pressures balanced at switchover and the system return to set point without overshoot. This feature of the circuit is somewhat limited on narrow proportional band applications, which is usually the case with slow processes, as it takes little deviation from set point before the reset would have to be at either a vacuum or considerably above supply to obtain the balance between valve loading and controller output.

This control station also makes available as an optional extra a small integral circular chart recorder with two-speed switch for trend recording and for continuous recording of variables not requiring extreme precision or readability. Special stations are also available for batch control and selector control (Figures 4.3w and 4.3x).

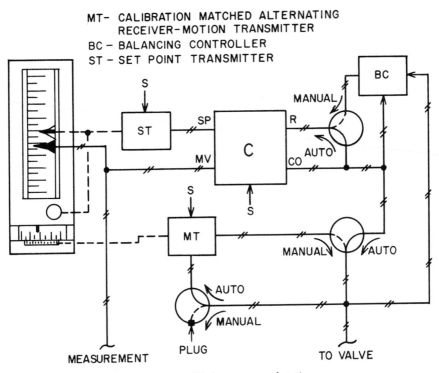

MT- CALIBRATION MATCHED ALTERNATING
RECEIVER-MOTION TRANSMITTER
BC – BALANCING CONTROLLER
ST – SET POINT TRANSMITTER

Fig. 4.3s   Self-balancing control station

Large-Case Receiver Controllers

Because of their size, large-case receiver controllers are not as popular as the miniature types. They nevertheless still find considerable use in plants and industries where the 24-hour circular chart is traditional and desirable.

Most large-case controllers operate on a displacement balance principle. The set point is a mechanical index setting. The measured variable acts on a pressure spring such as a bellows, spiral or helix which moves the recorder pen, or indicator pointer. A differential linkage detects any deviation between the index and pen position and actuates the flapper-nozzle system in an effort to bring the deviation to zero.

One example of large-case controllers is shown in Figure 4.3t. If the pen moves clockwise, the differential link moves upward and the bell-crank moves

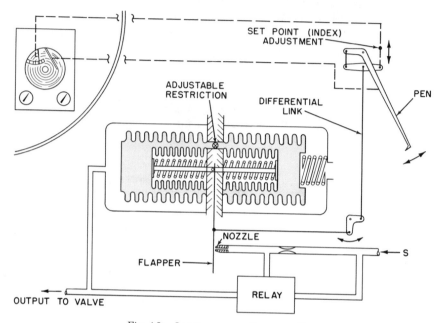

Fig. 4.3t   Large-case recording controller

the flapper toward the nozzle. The resultant increase in nozzle back-pressure is reproduced by the relay, whose output is connected to the control valve and the proportioning bellows housing. The pressure increase is transmitted through the oil to the small inner bellows, causing the two connected inner bellows to move to the right. The spring in the left inner bellows compresses while the spring in the right distends. The motion also causes the large righthand bellows to move to the right against the housed spring. As the center rod joining the inner bellows moves to the right, the flapper is moved back away from the nozzle. This negative feedback results in proportioning action. The greater this negative feedback, the greater change will be required in measured variable to obtain a given change to the valve. Adjusting the linkage to change the amount of this negative feedback changes the proportional band.

Opening the adjustable restriction between the two large bellows allows oil to flow from the bellows at higher pressure to the one at lower pressure. In this example, it would flow from the left to the right bellows, causing the inner bellows to move left, moving the flapper toward the nozzle and increasing output further. This action would continue to regenerate until the pen finally returned to the index where full balance would be achieved with the oil pressure being equal in the two large bellows and with the two inner bellows centered.

Large-case controllers are also available with remote air-operated set-point adjustment as required in cascade and ratio control.

Direct-Connected Large-Case Controllers

Direct-connected controllers have their own measuring elements which, as the term implies, are directly connected to the process. They therefore eliminate the need for a transmitter. However, because of this fact, direct-connected controllers must be located in the vicinity of the process which limits their use to local rather than remote central control rooms. Having to run process connections to the control room would be necessary with these controllers, but it is costly, troublesome, and hazardous. These units are still used, however, usually on smaller installations and on local panels.

Direct-connected large-case controllers of the indicating and recording type predate the receiver type by 20 years. In fact, receiver controllers were simply the pressure controller version of the direct-connected controller type.

The operation of direct-connected large-case controllers is the same as that of large-case receiver controllers. The pen or pointer arm, instead of being actuated by pressure from a process transmitter, is actuated by its own built-in sensing system. A cross section of some of the measuring elements available with large-case controllers is shown in Figure 4.3u. These include elements for pressure, absolute pressure, draft, vacuum, differential pressure, liquid level and filled systems for temperature measurement. Basic electrical measurements involved in thermocouples, resistance bulbs, radiation pyrometers and pH probes are accommodated in the potentiometer versions of large-case pneumatic controllers.

As with the large-case receiver controller, the direct-connected type is also available with remote set-point adjustment for cascade and ratio control.

Direct-Connected, Field-Mounted Controllers

Direct-connected controllers mounted in the field, often termed pressure and temperature pilots, are smaller than the large-case instruments, come in a weatherproof case, and are available as indicating or blind types. They can be pipe-mounted, mounted directly on a valve or surface, or flush-mounted on a local panel. They include their own measuring element. Figure 4.3v shows some typical types and mounting arrangements.

These units are considered the least expensive pneumatic controllers. Their performance, likewise, is considered less precise than those previously discussed. They find considerable use on small local installations and on the

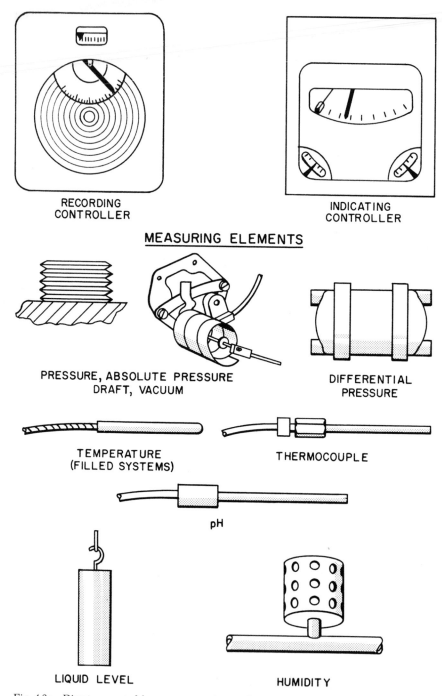

RECORDING
CONTROLLER

INDICATING
CONTROLLER

MEASURING ELEMENTS

PRESSURE, ABSOLUTE PRESSURE
DRAFT, VACUUM

DIFFERENTIAL
PRESSURE

TEMPERATURE
(FILLED SYSTEMS)

THERMOCOUPLE

pH

LIQUID LEVEL

HUMIDITY

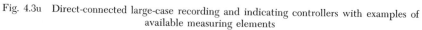

Fig. 4.3u   Direct-connected large-case recording and indicating controllers with examples of
available measuring elements

450

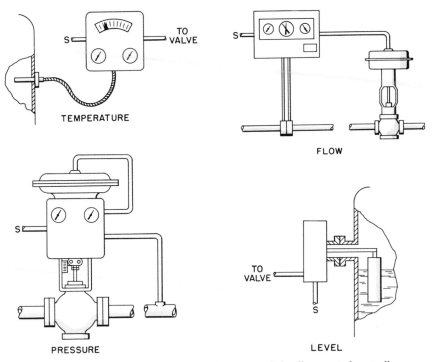

Fig. 4.3v Typical arrangements of directly connected, locally mounted controllers

many local field loops in larger plants. Every plant has such loops which are not critical, do not require that the measured variable appear on the central control board, that the set point be adjustable from the board and that it be possible for the operator to switch to manual. Combining a pressure pilot with a control valve makes a pressure regulator. Local level regulating loops are also quite common. Temperature pilots can be used to regulate temperature in preheaters, for example, where high precision is not essential.

The pilots can be had with on-off, differential gap, proportional, reset and derivative modes of control. The principle of operation is similar to that discussed in connection with Figures 4.3a and t.

Field-Mounted Receiver Type Controllers

In some applications, a field-mounted local controller is used, but it receives a signal from a transmitter rather than having its own measuring element. This would be in cases where the signal must be transmitted to the control board for recording, alarm, or indication or where a measurement is made that is more readily handled by one of the great selection of transmitters available. For these applications, there are field mounted versions of the receiver controllers shown in Figures 4.3c and d as well as receiver versions of the normally direct-connected field-mounted controllers shown in Figure 4.3v. Remote adjustment of set point is also an optional extra with these controllers.

Special Control Circuits

Feedforward, ratio, cascade and other multiple loop systems are quite straightforward to implement with pneumatic hardware. Two circuits, one involving automatic selector control and the other involving batch control, receive enough attention that controllers are sometimes packaged specially for these applications.

With automatic selector control, also called override, or limit control, two or more control loops are connected to a common valve in such a way that, under normal conditions, the normal control loop has command of the valve; however, if some abnormal condition arises, one of the other loops automatically moves in and takes over control to keep operation within safe limits. Unlike safety shutdown systems, normal control is only cut back as much as necessary to stay within safe limits. When the abnormal condition improves, the normal loop resumes control. Figure 4.3w shows a system such as would be used on a compressor station on one of the transcontinental

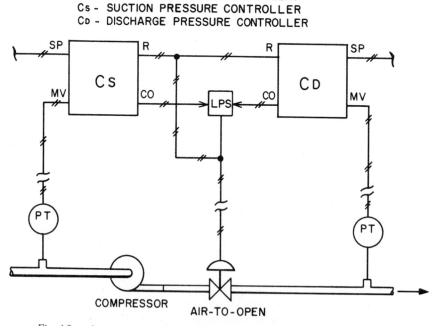

LPS- LOW PRESSURE SELECTOR RELAY
Cs - SUCTION PRESSURE CONTROLLER
Cᴅ - DISCHARGE PRESSURE CONTROLLER

Fig. 4.3w   Automatic override or selector control circuit on pipeline compressor pipelines. Normal control is on discharge pressure. If the suction pressure gets too low, however, as would be the case if the compressor upstream failed or if a line rupture occurred, the discharge controller would open the valve wide, which would lower suction beyond the safe limit, causing vapor lock and compressor burnout. To avoid this, a second loop—which senses suction pressure and whose controller set point is equivalent to the low safe limit—is coupled with the discharge pressure control loop by means of a low-pressure selector relay. Since the control valve operates air-to-open, the selector relay chooses and transmits to the valve the output of the controller which wants the valve more nearly closed. Under normal conditions, with adequate suction pressure, the discharge pressure controller will have the lower output and hence command the valve. If suction pressure drops to the set point, the suction pressure controller moves into control. The key to correct implementation of this circuit is in making the reset feedback of both controllers common with the valve pressure. In this way the controlling unit has its reset acting normally while the stand-by controller is prevented from having its reset saturate or wind-up. Its reset should exactly match valve pressure at the instant it is to take over.

Figure 4.3x shows a system for *batch control*. The problem is that when a batch is completed and the controller is left on automatic with the manual valve closed, the reset keeps acting until it saturates, or winds up. When the system starts up again, if the controller has proportional and reset action only, or if the derivative unit is the interacting type which follows proportional and reset action, as in Figures 4.3a and c, the controller makes no effort whatever to move down from full scale until the error changes sign, i.e., the process crosses over set point. A considerable overshoot will obviously result.

One solution is an anti-reset wind-up relay. This is simply a throttling relay set to operate at 15 PSIG, the wide-open position of the control valve. As long as the controller output is below 15 PSIG, the relay transmits the output to the reset feedback connection and reset acts normally. If the output goes above 15 PSIG, the relay begins exhausting the reset feedback line until it maintains the output at 15 PSIG. Thus it does not affect control, but when the system is shut down, it brings the reset down to whatever value it takes to limit output at 15 PSIG. This allows the proportional action to get into the act at start-up so that it can prevent overshoot. How effective it is depends upon proper tuning of the controller.

An alternative solution is to use either a separate derivative unit or a controller with a built-in derivative unit ahead of the proportional plus reset sections. These then act on the derivative modified signal. The derivative unit's output crosses over set point well ahead of the variable itself, and this starts the reset unwinding in time to prevent overshoot.

The effectiveness of this circuit also depends upon proper tuning. Too little derivative allows some overshoot; too much causes initial undershoot.

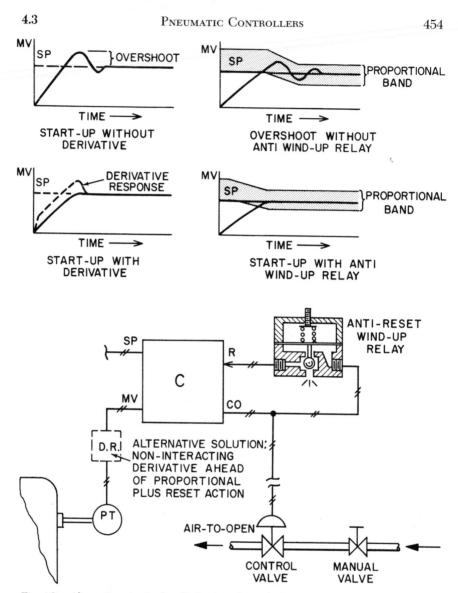

Fig. 4.3x   Alternative circuits for elimination of overpeaking in start-up of batch processes

Effect of Transmission Distance on Control

Since there is a lag with pneumatic transmission, control is affected as the distance between the process and controller increases. Figure 4.3y shows the result on recovery time as step load upsets of 10 percent are imposed on a liquid flow control system as the distance between controller and process increases. Since liquid flow control is one of the fastest processes, this amounts

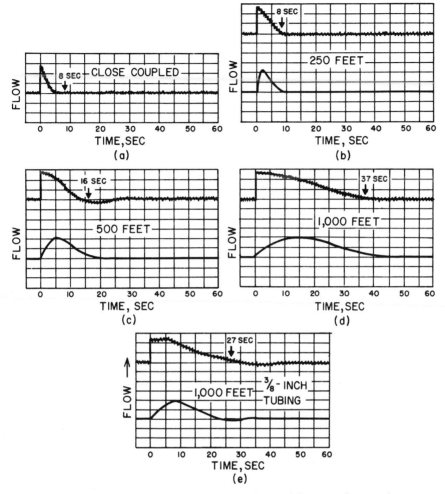

Fig. 4.3y   Effect of transmission distance on control of a liquid flow control process (worst case example) with 10 percent step upset. Upper noisy curve shows flow recorded locally, and lower smooth curve shows flow recorded remotely at controller. (J. D. Warnock, How Pneumatic Tubing Size Influences Controllability, *Instrumentation Technology*, February 1967.)

to a worst case example and the effect on slower processes is proportionately less.

From the charts it can be seen that with all instruments close-coupled—it took 8 seconds for the system to recover. At 250 ft distance, the recovery time was still approximately 8 seconds. At 500 ft, it was 16 seconds and at 1,000 feet, 37 seconds. These results were obtained using ¼-in.-OD tubing, which is conventional. When ⅜-in. tubing was used, there was a significant improvement. At 1,000 ft, the recovery time was reduced from 37 to 27 seconds. An equivalent electronic control loop had an 8-second recovery time. The upper noisy record on charts (b), (c), (d) and (e) was of the flow as recorded at the transmitter. The lower smooth record was of the flow as it appeared on the recorder located remotely with the controller.

If this lag was objectionable, the four-pipe system shown in Figure 4.3h could be used, and the dynamic performance would be equivalent to that of the closed-loop system, or the installation of booster relays in the transmission lines, could also be considered.

PROGRAMMER

4.4 FUNCTION GENERATORS AND COMPUTING RELAYS

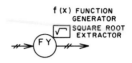

f (X) FUNCTION
GENERATOR

√‾ SQUARE ROOT
EXTRACTOR

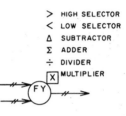

> HIGH SELECTOR
< LOW SELECTOR
Δ SUBTRACTOR
Σ ADDER
÷ DIVIDER
X MULTIPLIER

LARGE CASE, CAM-ACTUATED,
TIME FUNCTION GENERATORS
(PROGRAMMERS):

Cost: $300 to $1,000.

Partial List of Suppliers: Bailey Meter Co.; Barber Colman
                                 Co.; Bristol Co.; Foxboro Co.;
                                 Honeywell, Inc.; Leeds & Northrup
                                 Co.; Taylor Instrument Cos.

LARGE CASE, ADJUSTABLE
RANGE AND HOLD PROGRAMMERS:

Cost: $650 to $1,100.

Partial List of Suppliers: Foxboro Co.; Taylor Instrument Cos.

MINIATURE AND LARGE-CASE
PNEUMATIC PROFILE TRACERS
(PROGRAMMERS):

Accuracy: ±0.25% FS.

Cost: $800 to $1,400.

Partial List of Suppliers: Gaston County Dyeing Machine Co.;
                                 Moore Products Co.

ELECTRIC LINE AND EDGE
FOLLOWER PROGRAMMERS:

Accuracy: ±0.25% FS.

Cost:	$750 to $1,500.
Partial List of Suppliers:	Beckman Instruments, Inc.; Leeds & Northrup Co.; Research, Inc.; U.S. Gauge, Div. of Ametek.

STEP PROGRAMMERS:

Cost:	$100 to $2,000.
Partial List of Suppliers:	Agastat Div., Elastic Stop Nut Corp.; Bristol Co.; Industrial Timer Co.; Precision Products and Controls, Inc.; Sealectro Corp.; Taylor Instrument Cos.; Tenor Co.

COMPUTING RELAYS:

Accuracy:	$\pm\frac{1}{2}\%$ for all types except the differentiating and integrating relays, which are $\pm 15\%$ if uncompensated and $\pm 1\%$ if compensated in specially built units.
Cost:	Prices shown are for pneumatic; electronic costs are distinguished by (.....).
	High and low selectors—$20 to $55 ($165 to $235).
	Adding and subtracting relays—$75 to $250 ($440 to $600).
	Square-root extractors and function generators—$150 to $250 ($290 to $350).
	Scaling and proportioning relays— $150 to $250 ($430 to $600).
	Multiplying and dividing relays— $240 ($275 to $1,000).
Partial List of Suppliers:	After the manufacturer's name the letters "e" and "p" note if electronic and/or pneumatic units are offered. Bailey Meter Co. (e, p); Beckman Instruments, Inc. (e, p); Bell & Howell Control Products Div. (e); Bristol Co. (e, p); G. W. Dahl Co. (p); Fischer & Porter Co. (e, p); Foxboro Co. (e, p); General Electric Co. (e);

Honeywell, Inc. (e, p); Leeds & Northrup Co. (e); Moore Products Co. (p); Motorola Instrumentation & Control, Inc. (e); Robertshaw Controls Co. (e, p); Sorteberg Controls Co. (p); Taylor Instrument Cos. (e, p); Westinghouse Electric Corp., Hagan Div. (e).

Time Function Generators (Programmers)

The simplest, and the least expensive, analog time function generator is the camtype programmer. These are assembled in large-case circular chart recorder housings (Figure 4.4a) and consist of a motor-driven cam which moves

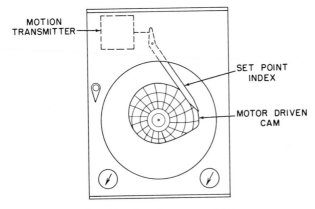

Fig. 4.4a   Cam type programmer

the set-point index to which a motion transmitter is connected. The output is usually a 3 to 15 PSIG pneumatic set-point signal. Electric outputs are also available. The time-base is a function of motor speed, and a wide selection of speeds is available. The cams can be made of plastic or metal. It is also common to incorporate an integral controller, direct-sensing element and circular chart recorder in the same housing.

Cam programmers are usually applied to batch processes, which are repeated time after time. Making up new cams and changing them is *not* a simple matter. These units are not as accurate as the profile tracer and line follower types of more recent manufacture. The cam rise is also limited for mechanical reasons to about 50 degree cam rotation for full-scale movement of the index. Curvilinear coordinates make the cams more difficult to lay out as compared with programming a device with rectilinear coordinates.

Adjustable Ramp and Hold Programmers

For batch processes in which the controlled variable must be made to rise at a controlled rate, then hold at some preset value and, possibly, fall at a controlled rate, programmers such as that in Figure 4.4b are often

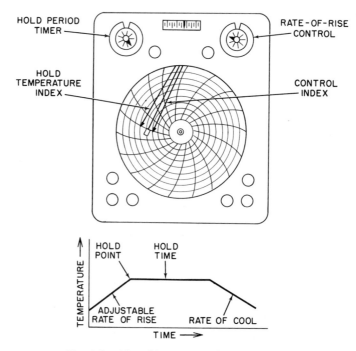

Fig. 4.4b   Adjustable ramp-and-hold programmer

preferable to cam types, particularly if the program must be changed periodically. These, too, are usually packaged as large-case circular chart recorders. In this type programmer, the set-point index is driven by a constant-speed motor. The rate of rise is set by adjustment of an interrupter timer, which makes contact for a set percentage of the basic timer cycle time. The movement of the index is, therefore, actually in steps, but the steps are so small that the operation is, for all practical purposes, continuous. The set point rises until it coincides with the hold point index, at which point the hold timer is energized while the interrupter timer is de-energized. Controlled cooling rate requires driving the set-point index in reverse.

Usually this type of programmer comes complete with a controller element and direct sensing element.

Profile Tracer Programmers

This type of programmer comes in a miniature pneumatic recorder type case, 6 × 6 in. (Figure 4.4c), or in a large case. The program is stored on

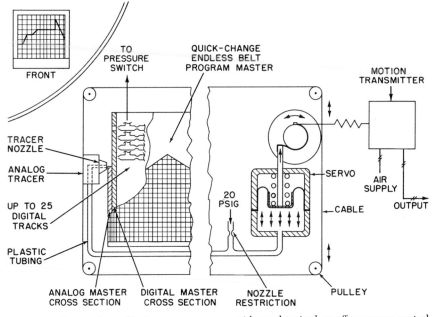

Fig. 4.4c   Pneumatic profile tracer programmer with synchronized on-off sequence control switches

a laminated, endless belt, plastic master. It combines an analog set-point program with up to 25 synchronized digital tracks for operation of logic circuits, auxiliary equipment, solenoid valves, lights, etc. There is no limit to the slope which the programmer can follow—even slopes of 90 degrees are accommodated.

Since the master program can be quickly changed, these programmers are often used where the program does require periodic change and where accurate reproduction of the program is essential, as in textile dyeing processes. The complete program is stored on the master, thus eliminating the need for having an operator make various program settings for each change and, therefore, also eliminating the chance for human error in setting the program.

These programmers are accurate to within $\frac{1}{4}$ percent of full scale, which makes them applicable when accuracy alone is the critical requirement of the operation.

The endless belt master is made up by plotting the desired analog program on the rectilinear chart and cutting the top portion away with scissors. The second layer serves as a backing and also is used to program the synchronized digital tracks. At any point of the program where a switch action is desired, a hole is punched in with a conductor's punch. The back of the analog program has a pressure sensitive adhesive which joins the two sections. A splice finishes the make-up of the master.

In operation, a motor drives the master program. A cable-mounted tracer nozzle senses the step which occurs on the analog program profile. The back-pressure of the nozzle actuates a servo, which, through the cable drive, keeps the tracer following the profile. Operating from the same servo drive is an accurate force balance type motion detector. Sensing the back side of the digital master are a series of vertically aligned nozzles. Normally, their back-pressure is high since the master baffles off the nozzles. However, if a punched hole presents itself, the back-pressure of that particular nozzle drops to zero, actuating the connected pressure switch.

Electric Line and Edge Follower Programmers

Electric line and edge follower programmers are also good for better than $\frac{1}{4}$ percent accuracy. In the electrostatic line follower type, Figure 4.4d,

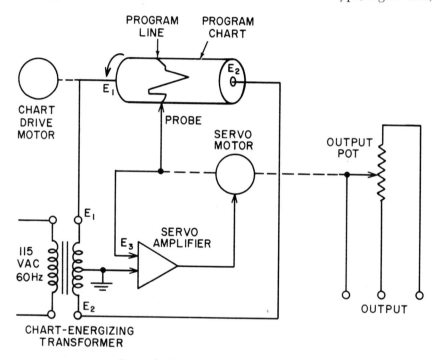

Fig. 4.4d   Electric line follower programmer

the desired program curve is etched into a conductive surface chart, dividing it into two electrically isolated surfaces. The surfaces are energized by oppositely phased AC voltages establishing a gradient across the gap. A non-contacting probe senses the electrostatic field developed by the surfaces and energizes a servo amplifier to keep the probe tracking the line, which is at zero potential. Attached to the servo drive is the wiper of a potentiometer whose output is proportional to line position. The photoelectric line follower type functions to keep the line centered between two slightly overlapping pickup heads. The detector must be manually set over the line at start-up, and slope rate is limited by the speed of the follower mechanism.

The photoelectric edge follower consists of a chart which is divided into a transparent and an opaque section at the program line. A photocell detector senses the edge and a servo system tracks it.

Up to eight digital tracks are available with the electric programmers.

Step Programmers

Step programmers are used for on-off event-sequencing. (See also Sections 4.5 and 4.6.) They do not provide an analog output. A typical type consists of a perforated drum (Figure 4.4e). Each perforation represents a step in one

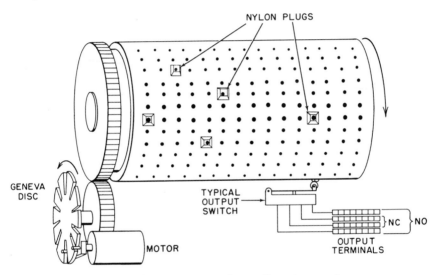

Fig. 4.4e   Drum programmer for on-off event sequencing

channel. Drums are available with from 30 to 100 steps and from 16 to 93 channels. Inserting a nylon plug into a hole results in a switch actuation on the corresponding step and channel. The stepping can be initiated by a remote sensor switch, counter, timer or pushbutton.

These units are easily programmed and can replace complex logic and interlock circuits which are commonly implemented with electromechanical relays. They not only replace such systems, but eliminate the need for their custom design and construction, as well.

Related to the step programmer is the continuous, multi-channel, cam timer in which a number of individually adjustable cams, mounted on a single drive shaft, provide event sequencing control.

Computing Relays

Pneumatic Multiplying and Dividing

In the force bridge multiplier-divider shown in Figure 4.4f, input pressures act on bellows in Chambers A, B and D. The output is a feedback

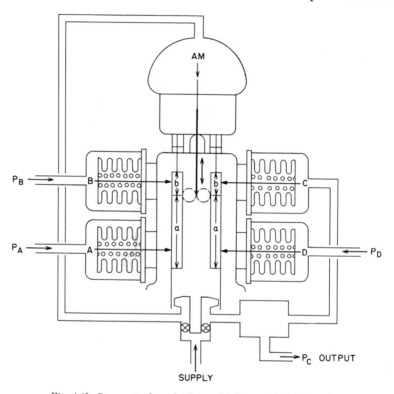

Fig. 4.4f   Pneumatic force bridge multiplying and dividing relay

pressure in Chamber C. The bridge consists of two weigh-beams which pivot on a common movable fulcrum, with each beam operating a separate feedback loop. Any unbalance in moments on the left-hand beam causes a movement of the fulcrum position until a moment-balance is restored. An unbalance in moments on the right-hand beam results in a change in output pressure until balance is restored. Equations which characterize the operation of the force-bridge are

$$A \times a = B \times b \quad \text{and} \quad D \times a = C \times b \qquad 4.4(1)$$

The equation reduces to

$$A \times C = B \times D \qquad 4.4(2)$$

or

$$C = \frac{B \times D}{A} \qquad\qquad 4.4(3)$$

Multiplication results when the two input variables are connected to Chambers B and D. Division results when the dividend is connected to either Chambers B or D, with the divisor connected to A. Simultaneous multiplication and division results when B, D and A chambers are used.

Cam-Actuated Pneumatic, Multiplying and Dividing Relays The significant advantage of a cam-actuated multiplying and dividing relay is that it can operate with practically any type of non-linear function which can be cut on a cam. This can mean operation with logarithmic functions, as in pH measurement, and computation in narrow, suppressed ranges of measurement which result in good resolution. The pure multiplier-divider as in Figure 4.4f, when used for temperature and pressure compensation for example, uses input signals proportional to the total absolute temperature and pressure range, starting with zero. Since the usable temperature and pressure range might be a small percentage of the total measurement range, the results might lack precision.

In Figure 4.4g, input pressure P_1 and output P_0 act on double diaphragm capsules, and the net resultant force in each is in the direction of the larger area diaphragm. Input P_1 creates force Y, which pulls the baffle, pivoted at A, away from the nozzle. Output pressure P_0 creates force X which moves the baffle closer to the nozzle. The θ input-output relationship is a function of the angle of the nozzle beam. When angle θ is 45 degrees, the relationship is 1:1. This can also be considered the multiplication factor or gain, K. At larger angles, K is greater than 1, and at smaller angles, smaller than 1. The multiplicand P_2 acts on the cam-positioning cylinder and thereby changes the nozzle beam angle in accordance with the cam characteristic. The zero adjusting springs subtract the 3 PSI zero from P_1 and set a 3 PSI zero on the output, respectively.

The characteristic equations are

$$(P_0 - 3) = (P_1 - 3)K \qquad\qquad 4.4(4)$$
$$\text{Cotangent } \theta = K \qquad\qquad 4.4(5)$$
$$K = f(P_2 - 3) \qquad\qquad 4.4(6)$$

Combining 4.4(4) and 4.4(6),

$$(P_0 - 3) = (P_1 - 3)f(P_2 - 3) \qquad\qquad 4.4(7)$$
$$P_0 = (P_1 - 3)f(P_2 - 3) + 3 \qquad\qquad 4.4(8)$$

Electronic Multiplying and Dividing

In Figure 4.4h, inputs e_1 and e_2 are multiplied in the diode bridge. Conduction of the diodes in the bridge is dependent upon the relative magni-

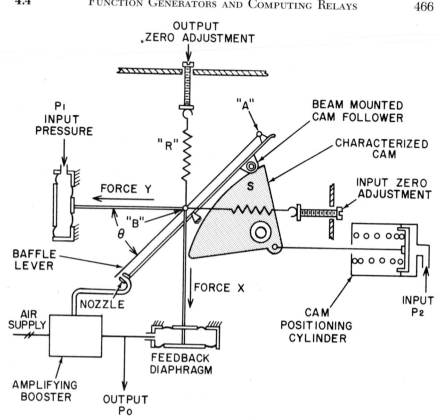

Fig. 4.4g   Pneumatic cam-characterized multiplying and dividing relay

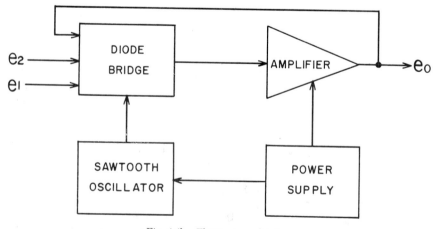

Fig. 4.4h   Electronic multiplier

tude of the inputs, with respect to the constant slope of the sawtooth input. The output of the diode bridge is a trapezoid, which has an area equivalent to

$$\text{Area} = e_1 e_2 \tan \theta \qquad\qquad 4.4(9)$$

The angle θ is established by the constant slope of the sawtooth, and thus,

$$\text{Area} = K e_1 e_2 \qquad\qquad 4.4(10)$$

The output voltage, e_o, is amplified and filtered to a DC signal, and its voltage level will, therefore, be proportional to the area, and hence, to the product of e_1 and e_2.

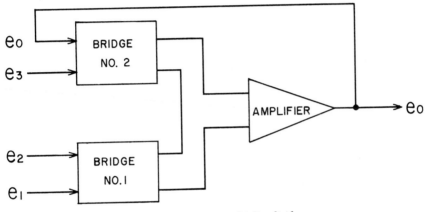

Fig. 4.4i   Electronic multiplier-divider

Adding another diode bridge to the multiplier circuit produces a multiplier/divider (Figure 4.4i). The input to the amplifier is the output difference from the two bridge networks.

$$e_o = A(K e_1 e_2 - K e_3 e_o) \qquad\qquad 4.4(11)$$

where A = gain of the amplifier.

$$e_1 e_2 = \frac{e_o}{AK} + e_3 e_o \qquad\qquad 4.4(12)$$

The term e_o/AK is very small if the amplifier gain is high, and thus,

$$e_o = \frac{e_1 e_2}{e_3} \qquad\qquad 4.4(13)$$

Pneumatic Adding, Subtracting and Inverting

In the force balance, arithmetic computing relay, Figure 4.4j, a signal pressure in chamber A acts downward on a diaphragm with unit effective

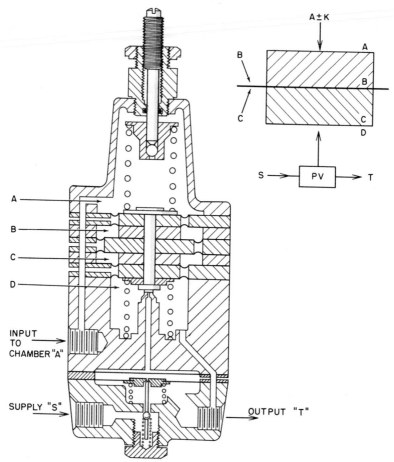

Fig. 4.4j   Pneumatic adding, subtracting, inverting and biasing relay

area. A signal in chamber B also acts downward on an annular diaphragm configuration, likewise having an effective area of unity. Signal pressures in chambers C and D similarly act upward on unit effective diaphragm areas. Any unbalance in forces moves the diaphragm assembly with its integral nozzle seat. The change in nozzle seat clearance changes the nozzle back-pressure and, hence, changes the output pressure, which is fed back into chamber D until force balance is restored. The basic equation which describes the operation of the relay is

$$T = A + B - C \pm K \qquad\qquad 4.4(14)$$

K is the spring constant. It is adjustable to give an equivalent bias of ± 18 PSI.

The relay in Figure 4.4k is a modification of Figure 4.4j in that it incorporates additional input chambers and output feedback chambers. It can be used to add and/or average up to nine inputs. Figure 4.4k is an averaging relay for five inputs. The averaging feature keeps all signals in the same standard 3–15 PSIG range.

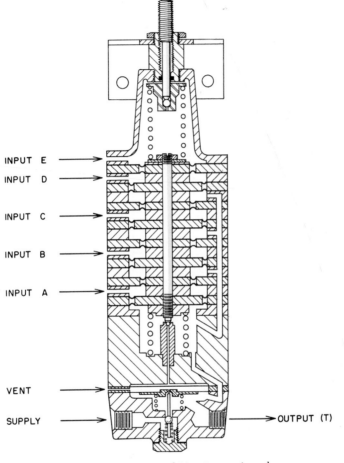

INPUT E

INPUT D

INPUT C

INPUT B

INPUT A

VENT

SUPPLY

OUTPUT (T)

Fig. 4.4k   Pneumatic multi-input averaging relay

The characteristic equation is

$$T = \frac{A + B + C + D + E}{5} \pm K \qquad\qquad 4.4(15)$$

Figure 4.4j relay provides inverting, or reversing action, by setting the bias spring loading to a maximum and connecting the input to subtracting chamber

C. If the bias is set at $+18$ PSIG, then a 3–15 PSIG signal in chamber C results in a 15–3 PSIG output.

The equation describing the operation is

$$T = K - C \qquad\qquad 4.4(16)$$

Electronic Adding, Subtracting and Inverting

In Figure 4.4l, the two input potentials e_1 and e_2 are compared in the multiple comparator, which produces a proportional output to the amplifier.

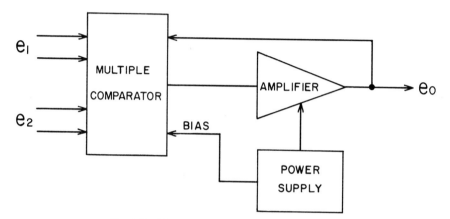

Fig. 4.41   Electronic adder, subtracter and inverter

The current paths of the two inputs may be the same or opposite, resulting in either an adding or subtracting circuit, respectively.

Inverting is accomplished by biasing the comparator to produce maximum output with no input. Applying a reverse input (i.e., a reverse current input with respect to bias current) causes the output to decrease with increasing input. The feedback signal is such that the amplifier acts as a unity gain network.

Pneumatic Scaling and Proportioning

Scaling, or proportioning, involves multiplication by a constant. Several approaches are available, namely, (1) special fixed-ratio relays, (2) pressure transmitters, (3) proportional controllers, and (4) adjustable ratio relays.

The fixed-ratio scaler is the simplest if the correct ratio is available and if adjustability and exact ratio is unnecessary. Figure 4.4m shows such a relay. The input pressure is connected to the top chamber and acts on the upper diaphragm. Output acts upward on the small bottom diaphragm. The gain is a function of the relative effective areas of the large and small diaphragm as determined by the dimensions of the diaphragm ring. The bottom spring

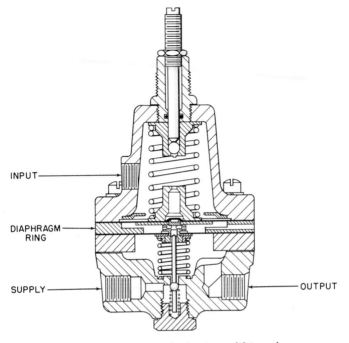

Fig. 4.4m   Pneumatic, fixed-ratio amplifying relay

applies a negative bias to the input and the adjustable top spring allows exact zero setting. The operating equation is

$$T = AP_1 + K \qquad\qquad 4.4(17)$$

where A is the gain constant and K is the spring setting.

Where the scaling must be exact and does not have to be adjusted periodically, pressure transmitters are an economical, reliable and accurate choice. Where the scaling factor must be modified occasionally, conventional ratio relays, which often consist of the proportioning section of a controller, are commonly used.

Electronic Scaling and Proportioning

Simple electronic scaling or proportioning involves combining a voltage divider circuit with an amplifier. The voltage divider circuit is connected to either the input or output side, depending upon whether the gain is to be greater or less than unity.

In Figure 4.4n, the amplifier comes to balance when Δe_i equals zero. Since the voltage divider is on the output, only a portion of the amplifier output is fed back to counterbalance the input voltage. Therefore, the output

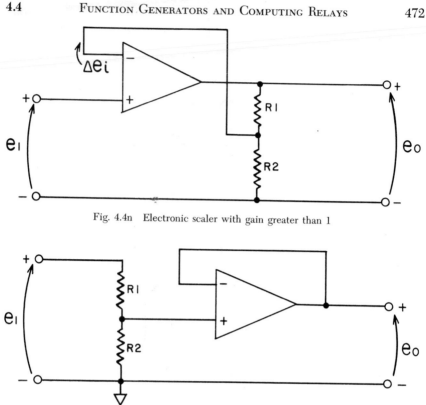

Fig. 4.4n   Electronic scaler with gain greater than 1

Fig. 4.4o   Electronic scaler with gain less than 1

will rise above e_1, resulting in gains greater than one. The operation can be expressed as

$$e_o = e_1 \frac{(R_1 + R_2)}{R_1} \qquad 4.4(18)$$

In Figure 4.4o, the voltage divider is on the input side, so that only a voltage equal to or less than e_1 is impressed across the amplifier input, resulting in gains less than one. This can be expressed as

$$e_o = e_1 \frac{R_2}{R_1 + R_2} \qquad 4.4(19)$$

Pneumatic Differentiating

A differentiating relay produces an output proportional to rate of change of input. Figure 4.4p shows an ideal pneumatic differentiating relay. The relay is basically similar in construction to relay (Figure 4.4j), except that the annular

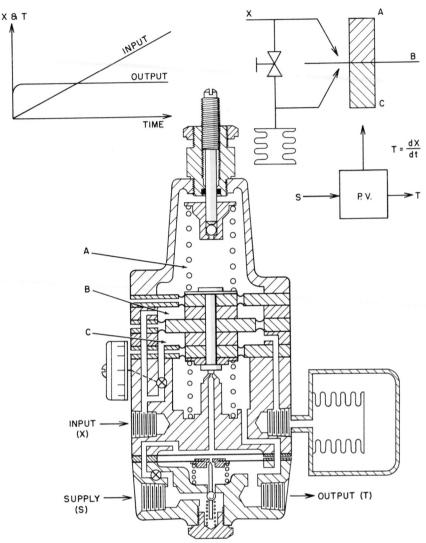

Fig. 4.4p   Pneumatic differentiating relay

effective diaphragm area between chambers B and C is more than ten times the effective area of the small diaphragms between chambers A and B—giving a gain of greater than ten. The input signal is transmitted unrestricted to chamber B and passes to chamber C through an adjustable restriction. When the input is steady, the forces resulting from pressures in B and C chambers cancel each other, so that the output equals the zero-spring setting (usually mid-scale, 9 PSIG, if both positive and negative rates are to be measured).

If the input pressure changes, a differential develops across the restriction. The relay transmits an output proportional to this differential. For accurate results, this differential must be directly proportional to the rate-of-change of input. Using a needle valve which produces laminar flow provides a linearly proportional volumetric flow, but the differential developed across the needle valve is a function of the mass flow which varies with static pressure, because of compressibility. This compressibility error is approximately ±15 percent. The effect can be fully compensated, however, by the addition of a variable volume to chamber C, the restricted chamber. As the static pressure increases, tending to make the differential smaller because of higher mass flow-rate, the volume increases proportionately to maintain a constant differential. The needle valve setting determines the rate time constant.

The compensated relay is not a standard piece of hardware. In most cases, a non-compensated differentiating relay is satisfactory.

Electronic Differentiating

The input amplifier in Figure 4.4q is capacitor coupled so that only the rate of change of the input signal is seen by the amplifier. Two diodes in the

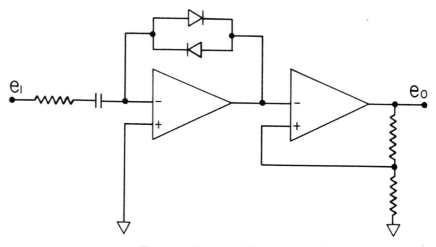

Fig. 4.4q   Electronic differentiator

feedback of the amplifier allow its output to go positive or negative (depending on the direction of the rate of change) by an amount equal to the forward drop across the diodes (only a few tenths of a volt). The output amplifier inverts and amplifies this signal by its open loop gain.

A small positive feedback is applied to the last amplifier to prevent output from "chattering" at the diodes' switching point.

Pneumatic Integrating

Integration, the reverse of differentiation, essentially involves measurement of accumulated pressure resulting from a flow which is proportional to the offset (from some chosen reference) of the input variable. Figure 4.4r shows an ideal integration relay. The input signal loads chamber B. The output, it should be noted, is the accumulated pressure in chamber A, *not* the booster pilot output. The input signal determines the pressure differential across the

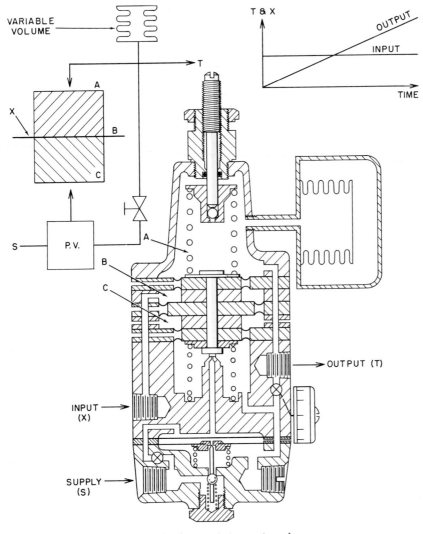

Fig. 4.4r   Pneumatic integrating relay

needle valve. As in the case of the differentiation relay, with laminar flow across the needle valve, the volumetric flow is directly related to the differential. The mass flow, however, which determines the accumulated pressure, still varies with the static pressure, because of compressibility. This effect is also compensated by connecting a variable volume to chamber A. The needle valve sets the proportionality constant of the integrator.

Neither the compensated differentiating relay, nor the compensated integrator is available as standard hardware. Usually the non-compensated relay (actually a proportional-speed floating controller) is satisfactory.

Electronic Integrating

The first amplifier in Figure 4.4s, a simple inverting type, performs the integration function as the charge accumulates across the capacitor of the

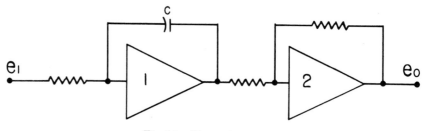

Fig. 4.4s   Electronic integrator

RC network. The second amplifier is an inverting, general purpose type, which relates the output directly to the input.

Pneumatic Square Root Extracting

This function is commonly required to linearize signals from differential-type flow transmitters. The force bridge, Figure 4.4f, provides square root extraction when the output is connected in common to the A and C chambers, giving the equation

$$C^2 = B \times D \qquad\qquad 4.4(20)$$

Other solutions are based on (1) use of a cam characterized function generator and (2) a geometric relationship, namely, change in cosine compared with the change in included angle, for small angular displacements (Figure 4.4t). Starting with the input and output at 3 PSIG, an increase in input causes the floating pilot link to restrict the pilot nozzle. This increases the output pressure and moves the output feedback bellows upward, until balance is restored. Since the length of the floating link is fixed, the angular displacement produced by movement of the output bellows follows the relationship

$$\cos \theta = 1 - \frac{X}{L} \qquad\qquad 4.4(21)$$

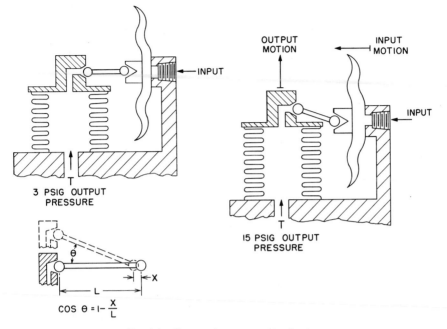

Fig. 4.4t   Pneumatic square root extractor

A plot of the angle θ (output displacement) vs X (input displacement) in this equation shows the relationship to be virtually an exact square root for small angular motion.

Electronic Square Root Extracting

The square root converter, Figure 4.4u, combines a DC amplifier with a negative feedback diode network. As current into the amplifier increases, the amplifier gain decreases with decreased feedback resistance in the diode

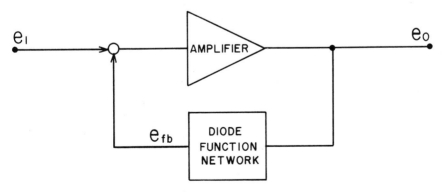

Fig. 4.4u   Electronic square root extractor

network. The gain varies according to, typically, seven straight line segments which approximate a square root function. This is accomplished by having seven diode-resistance paths, in the feedback network, automatically parallel each other with increasing input. The output stabilizes when the diode network modified feedback counterbalances the input.

High- and Low-Pressure Selector and Limiter

Selector relays are used in override systems.

The high-pressure selector relay compares two pressures and transmits the higher of the two in its full value. In Figure 4.4v, the two input pressures

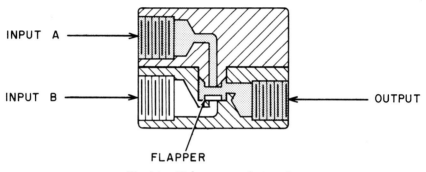

Fig. 4.4v   High-pressure selector relay

act against a free-floating flapper disc. The differential pressure across the flapper always results in closure of the low-pressure port.

In the low-pressure selector, Figure 4.4w, if input A is less than input B, the diaphragm assembly throttles the pilot plunger to make the output

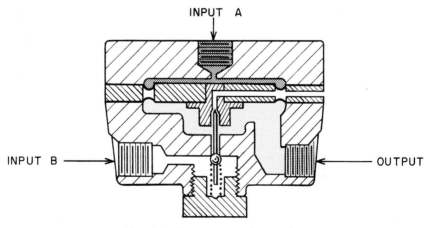

Fig. 4.4w   Low-pressure selector relay

equal to input A (the conventional action of a 1:1 booster relay). If input B is less than A, the supply seat of the pilot plunger is wide open so that pressure B is transmitted in its full value.

Limiting Function   By connecting a set reference pressure into one of the ports of the high-pressure selector, a low-limit relay results. Conversely, by connecting a set reference pressure into the low-pressure selector, a high-limit relay results. Limit relays are available with the reference-setting regulator built into the relay.

Electronic High- and Low-Voltage Selector and Limiter

The higher of the two positive inputs in Figure 4.4x causes a higher negative potential at the cathode of one of the diodes (CR1 or CR2). The

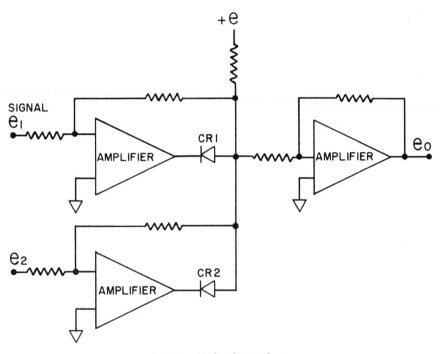

Fig. 4.4x   High-voltage selector

forward bias of this diode passes the higher input, and reverse biases the other diode to isolate the lower input. Thus if signal e_1 drops below signal e_2, CR2 is forward-biased to pass signal e_2, and CR1 is reverse-biased to isolate signal e_1. All the amplifiers are unity-gain inverter types.

Substituting a fixed input for one of the variables produces a low-limit relay.

To obtain a low-voltage selector (Figure 4.4y), the diodes are inverted

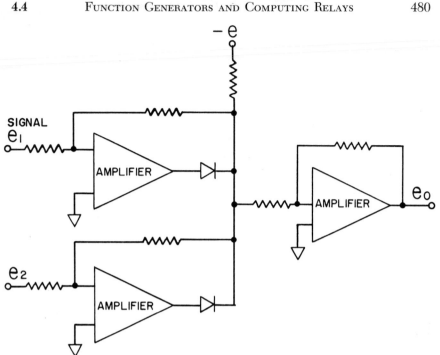

Fig. 4.4y   Low-voltage selector

and a negative supply (e) is used. Thus, the least positive input forward-biases one of the diodes (by the least negative potential applied to the anodes of the diodes). This automatically reverse biases the other diode and isolates the higher input from the output.

Substituting a fixed input for one of the variables produces a high-limit delay.

Flow Sheet Symbols

ON-OFF

$\longrightarrow$ (FY) $\dashrightarrow$

RELAY

(KJC) $\overset{\text{TO}}{\underset{}{\text{RECEIVER}}}$ $\dashrightarrow$

TIME SEQUENCE
CONTROLLER
(TIMER)

4.5 RELAYS AND TIMERS

Available Capacities:

RELAYS can handle microvolts to kilovolts, microamperes to kiloamperes, with response speeds from milliseconds to any longer period.

TIMERS can handle 1 to 100 circuits, up to 15 amperes per circuit, and time ranges from milliseconds to weeks.

Power and Frequency:

RELAYS will be actuated by the energy of milliwatts to a few voltamperes and will operate on frequencies from DC to RF.

TIMERS will usually be powered off 120-volt, 60-Hz supplies.

Environmental Limits:

Some relays can be exposed to temperatures from −50 to 400°F, vacuum to 400 PSIG and to above 100 G shock or vibration. Timers are usually more restricted.

Sizes:

Above 0.02 in.^{3} for relays and above 3 in.^{3} for timers.

Cost:

RELAYS from $3. See Tables 4.5e and f.

TIMERS from $50 for industrial units.

Partial List of Suppliers:

Relay suppliers (r) and timer suppliers (t) are thus distinguished. Adams and Westlake Co. (r); Agastat Div. of A. W. Haydon Co. (t); Allied Control Co. (r); AMF Alexandria Div. (r); Babcock Electronics (r); Barber-Colman Co. (r); E. W. Bliss Co. (t); C. P. Clare and Co. (r); Cook Electric

Co. (r); Durakool, Inc. (r); Eagle Signal Div. of Cutler-Hammer, Inc. (t); Elastic Stop Nut Corp. (t); General Electric Co. (r); Hickok Electrical Instrument Co. (t); Hi-G, Inc. (r); Honeywell, Inc. (r); Industrial Timer Corp. (t); IBM Industrial Products Marketing (r); Leach Corp. (r); Ledex, Inc. (r); Oak Electro/Netics Corp. (r); Sealectro Corp. (t); Struthers-Dunn, Inc. (r); Teledyne Co. (r); Tork Time Controls (t); Western Reserve Electronics, Inc. (t); Westinghouse Air Brake (r); Weston Instruments, Inc. (r); Zenith Controls, Inc. (t).

Definitions and Functions of Relays

An electrical relay is a device which initiates action in a circuit in response to some change in conditions in that circuit or in some other circuit. Pneumatic, fluidic, and other relays exist, as well as electrical. Most of the relays discussed here are electromechanical; solid-state and vacuum-tube devices are not discussed.

The function of a relay is simply to open or close an electrical contact, or a group of contacts, in consequence of a change in some electrical condition—such a change often being called a "signal." These contact closures are employed in the associated circuitry to select other circuits or functions, to turn on or off various operations, etc.

More generally, a relay may be considered as an amplifier and controller; it has a power gain factor which is defined as the ratio of output power to input power. Thus a relay may require a coil current of 0.005 amperes at 50 volts but can control 2,500 watts of power—a gain of 10,000. Among the many special forms of relays many other specific functions can be identified.

Characteristics

The relays described here are all either electromechanical or electrothermal. They appear in a number of forms, but the following characteristics are common to all of them.

They control the flow of electric current by means of contacts. These contacts show extremely high resistance when open (megohms) and low resistances when closed (milliohms). They may have multiple contacts (as many as eight DPDT contact assemblies are readily obtainable on stock relays) with each contact assembly electrically isolated from all others. The contacts are actuated in some definite and positive sequence.

The actuating coil and circuit can be (and usually is) completely isolated from the controlled circuit. It may be actuated by electric energy of entirely different character from the controlled one.

Each of the various mechanical structures has certain advantages and certain limitations. Some respond rapidly—in less than one millisecond—but do not handle safely great amounts of power; some handle large amounts of power but at somewhat lower speed and perhaps with some danger of contact bounce, etc. Nearly all forms can be obtained open, enclosed in dust shield, or hermetically sealed. Some are vacuum-type for handling extremely high voltages.

More specific information on the characteristics of the different relay forms will be given under the discussion of relay structures and contact materials.

Relay Types

The entire spectrum of relays extends from DC relays with contact capacities from microamperes to kiloamperes and coil energies from a few milliwatts to several watts. With meter relays and amplifier relays the actuating energy may be a small fraction of a microwatt. The spectrum also includes AC relays handling from a few watts of power to many kilowatts. Relays can control both AC and DC potentials in the thousands of volts and frequencies from DC to RF, and they can respond to specific frequencies only or indiscriminately to all frequencies alike.

Many of them are special forms, for special applications. This discussion, however, will be limited to the types and forms most used in typical instrumentation work—principally small DC and AC relays, sensitive relays, miniature and sealed relay types, and some small power relays.

For many applications, it is convenient to consider first the actuating energy needed for a relay. A rough classification, associating coil requirement with power-handling ability, is given in Table 4.5a. The values are only approximate.

Another classification is by operating function. Available energy often dictates relay choice, examples of which are given below.

Meter relays and ultra-sensitive relays, actuated by very small energies (a few milliwatts or even microwatts), are used where only a minimum signal is available, as at the output of a transducer or bridge.

General purpose or small control relays are used where no great amount of power need be handled but flexibility in application and reliability in operation are essential. Depending on the number of contacts, they usually require from 200 to 800 milliwatts, which may be either AC or DC and control perhaps 5 amperes at 120 volts per contact.

Small power relays are used in larger sizes. These usually have AC coils, which in turn may be controlled by sensitive relays. Small power relays in

Table 4.5a
THE RANGE OF RELAY CHARACTERISTICS

Type of Relay	Coil Power Required (approximate)	Contact Capability (typical)
Meter Relay	As low as 1 microwatt	Low energy only
Ultra-Sensitive Relay	Less than 10 milliwatts	0.5 to 1.0 amperes, non-inductive
Sensitive Relay	From 10 to 60 milliwatts	1.0 ampere non-inductive typical
Typical Crystal-Can Relay	100 milliwatts per form C contact	0.5 to 1.0 amperes, non-inductive
Transistor-Can (TO-5) or Miniature Relay	150 milliwatts per form C contact	0.5 ampere, non-inductive
Reed Relay	200 milliwatts per form C contact	0.5 to 1.0 amperes, non-inductive
General Purpose Relay	200 milliwatts per form C contact, or as much as 3 volt-amperes AC	10 amperes, 120 volts AC
Small Power Relay	Usually AC coils, from 1 to 10 volt-amperes.	30 amperes, 240 volts AC

this discussion have contact capabilities of 30 amperes or less, at 600 volts or lower. Much larger relays, of course, are also available.

Choices may be made on other bases, for instance, mechanical size. *Miniature relays* may be TO-5 size (the size of a TO-5 transistor case), with a volume of about 0.02 cubic inch, crystal-can size (about ⅛ cubic inch), miniature plug-in (about 1 cu in.) or general purpose plug-in (about 2 cu in.), all with about the same contact capacity. Miniature relays are used with printed-circuit construction or other compact assemblies.

In terms of their use, relays may be appliance grade, general-purpose, aerospace, military, or other; there may be little difference in their reliability when used for their prescribed purposes, but the more expensive types will operate more satisfactorily under adverse conditions.

In terms of mode of construction, there are clapper type relays, telephone types, solenoid-actuated types, reed relays, and many other forms.

Relay Contact Configurations

Electromechanical relays are produced with a wide range of contact arrangements, in various combinations. A number of these have been established as standard forms; those most used are shown in Figure 4.5b with their

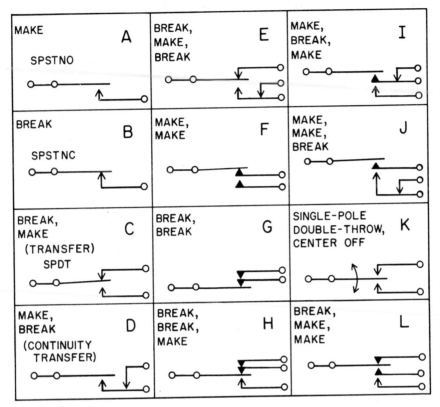

Fig. 4.5b   Standard relay contact configurations

identifying code letters. The same relay contact assembly can be described as single-pole, double-throw (SPDT), as make, break (continuity transfer), or as having form C contacts—all mean the same thing. Four form C contacts means 4PDT, and so on.

Certain mechanical structures better tolerate some contact pile-ups (the term for an assembly of leaf contacts) than others. Small clapper type relays seldom carry more than three form C contacts; telephone type relays may handle as many as eight form Cs or perhaps four form Cs and one or two form Es or form Fs. These contact assemblies permit the actuation of some circuits only after some others have been actuated, which means that the action is mechanically positive.

Electrical insulation between contacts can be very high, permitting high-voltage use with low leakage. Capacity coupling can also be kept low.

Relay Structures

Electromagnetic relays are actuated by magnetic forces produced by electric current flowing through coils of wire. In most such relays, the magnetic force moves an iron armature; in a few (in particular, meter relays) the coil itself is moved in a magnetic field.

The widely used mechanical structures are sketched in Figure 4.5c. At A in the sketch the elements of a clapper type relay are shown, and at B

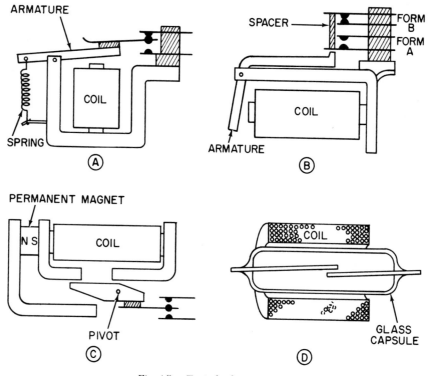

Fig. 4.5c   Typical relay structures

the elements of a telephone type relay. When no current flows in the coil, the relay armature is held away from the core by a spring. When current flows through the coil, the magnetic field produced pulls the armature toward the core, decreasing the air gap. As the air-gap decreases, the reluctance of the magnetic circuit also increases, and the pull increases. Thus, there is usually appreciably greater contact force when the relay is energized than when the contact force is only that of the return spring.

Both types—at A and at B—can be used with either direct or alternating coil current (a shading coil is normally added when AC is used). When AC is used, the added impedance due to the smaller air gap often helps to reduce coil heating. The telephone type relay has the advantage that, when used with DC, it can easily be given small time delays either on opening, or on closing, or on both, simply by adding a ring slug to the core.

Two other types now in wide use are shown in Figure 4.5c, at C and D. At C the balanced-force mechanism is illustrated. This structure has two magnetically stable positions. One of them is controlled by the permanent magnet, and the other is controlled by the magnetic force from the coil, which must be strong enough to overcome the force of the permanent magnet. These relays can be made in relatively small bulk, yet be positive in action and little affected by vibration.

The reed relay is shown in the partial sectional drawing of Figure 4.5c, at D. Two reeds, of magnetic metal, are mounted in a glass capsule, which is itself installed within a coil. Current flowing through the coil produces a magnetic field, magnetizing the reeds and causing them to attract each other, bend and make contact. The contacting surfaces are usually plated with precious metal contact alloy. The spring action required is provided by the reeds themselves. Reed relays are among the fastest available, some operating in less than 500 microseconds. They are available in several contact configurations, can be polarized, or can be made into latching relays by adding small permanent magnetic elements. They are available with dry or with mercury-wetted contacts. In addition to permitting several reeds in a single capsule, as many as a dozen capsules are sometimes operated by a single coil assembly. Reed relays are very widely used in transistorized driver systems, because of their small size, high reliability, and long life—100,000,000 operations are not unusual.

Specialized Relay Structures

The elementary magnetic structures described earlier are combined and elaborated to produce special-purpose relays.

By adding small permanent magnets to the relay, a relay armature may be forced in one direction for a signal of one polarity, and in the opposite direction if the signal has the opposite polarity; the elements of such a polarized relay are shown at A in Figure 4.5d.

Mechanical or magnetic latching devices can be applied to produce a relay which can permit either one of two circuits to be actuated, but not both at the same time. A simple form of this is shown at B in Figure 4.5d.

For certain applications, it is useful to be able to actuate one of two separate circuits depending on which of two signals is the larger—with little regard to the absolute magnitude of either signal. Two coils, with a single tilting armature, afford this ability (Figure 4.5d at C). Usually a weak spring is added to insure that when no signal at all is present the relay will go to its neutral position.

Multi-position rotary switches can be driven by magnetic ratcheting devices, permitting the selection of any desired contact position. Such stepping switches are available in a wide range of positions and number of circuits; they can also be secured with electrical reset or for continuous rotary opera-

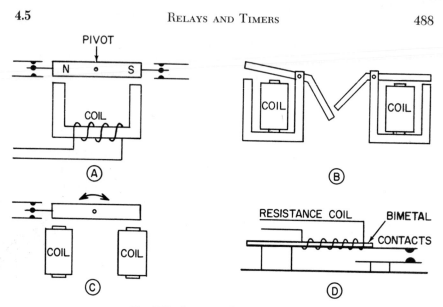

Fig. 4.5d   Some specific relay structures

tion. Stepping switches are extensively used in some communication systems and in computing devices, and they have a large number of miscellaneous applications as well.

Simple thermal time-delay relays are popular for causing one action to be delayed for one brief period after another, where accuracy of timing is not critical. One form consists of a thermal bi-metal strip, wound with a resistance coil which heats on the application of current, resulting in contact closure as soon as the bi-metal becomes warm enough to bend and establish the contact (Figure 4.5d at D). In another form, the electric current flows through a resistance wire, which expands and causes movement. These relays are somewhat affected by ambient temperature and cannot be recycled instantly. Another popular low-cost time-delay relay employs a small dashpot to delay the armature movement.

The lowest-current relay structures are the moving-coil meter type relays. These employ a d'Arsonval meter movement carrying delicate contacts. Sometimes the meter pointer is retained and sometimes contact force is supplemented by magnetic contacts or by auxiliary coils. Such relays are susceptible to vibration, shock and overloading. They are, however, capable of rather close adjustment for over-voltage or under-voltage use.

Contact Materials and Shapes

A variety of contact materials is available, with characteristics suiting them to various applications.

For very low-current, low-voltage applications (dry circuit) it is essential that materials be selected which do not oxidize, develop insulating coatings, or erode mechanically. Some combinations of precious metals (as gold and palladium) and some proprietary alloys satisfy these requirements. Such contacts are used in choppers and in meter relays where sticking can be a serious problem.

Silver and silver-cadmium contacts withstand fairly high currents without overheating, but they tend to form coatings (oxide and sulphide) which, while conductive, do have appreciable resistance.

Mercury-wetted contacts can be expected to have higher current ratings and lower contact resistances than dry contacts of the same size, and like mercury-pool contacts they are usually less noisy and display far less bounce. Dry contacts may fail by welding together; mercury contacts seldom do this. Relays with mercury contacts must usually be mounted in a nearly vertical position and are vibration sensitive.

Shapes of contacts depend upon their use. Heavy-current contacts are usually dome-shaped. Low-resistance, small-current contacts are often crossed cylinders, so placed as to wipe against each other. In wire-spring relays the round wires themselves, plated locally with contact material, also form the contacts. They thus are long-lived and inexpensive. In reed relays, the flat strips which are the reeds also form the contacts. They may be shaped at the ends and usually are plated for good contact.

Contact mounting is an important part of relay design. In multi-contact relays it is essential that all contacts can bear properly without interfering with each other. In low-voltage applications it is usually desirable that a wiping contact be provided (higher voltages can break through surface films). In nearly all relays it is desirable that contact bounce and chatter be minimized. Some forms of reed relays, in particular, are very fast, yet do not bounce.

Selection of Relays

Among the many factors affecting the selection of relays are cost, size, speed, capacity, energy required, etc., and also more restrictive items such as mounting limitations, open or sealed contacts (sometimes required for safety, sometimes for protection against unfavorable ambient conditions). The catalogs of the many relay manufacturers list dozens of types, forms, sensitivities, and though relays have been in use for more than a century, new forms are still appearing. Tables 4.5e and 4.5f list a few representative types of open- and sealed-contact relays respectively.

Application and Circuitry

Satisfactory application of relays requires that the relay functions be clearly understood, that relay characteristics be established, that the relay be selected to fit the need, and that the circuitry be designed to couple properly

Table 4.5e
TYPICAL OPEN-CONTACT RELAYS

Relay Type	Coil Description	Contact Description	Mounting	Installed Cost
Sensitive DC Relay	1,000–10,000 ohms, 20 milliwatts per form C	Up to 4 form C, $\frac{1}{2}$ amperes	Screw	$ 6
Plate-Circuit Relay	2,000–20,000 ohms, 50 milliwatts per form C	Up to 2 form C, 2 amperes	Screw	$ 4
Small Utility-Relay	AC or DC, 6–120 volts, about 250 milliwatts	Up to 2 form C, 2 amperes, 120 volts	Screw	$ 3
General Purpose Relay	AC or DC, about 200 milliwatts per form C	Up to 3 form C, 10 amperes, 120 volts	Screw or plug-in	$ 6
General Purpose Relay	AC or DC, about 200 milliwatts per form C	Up to 3 form C, 10 amperes, 120 volts	Plug-in, with dust cover	$ 8
Small Power Relay	Usually 120 volt AC, 2–10 volt-amperes	Up to 3 form C, 30 amperes, 600 volts	Screw	$10

the relay to the rest of the system. Thus, it is usual to begin by determining what amount of energy must be controlled and how much energy is available as signal. In doing this it is necessary to consider the number of contacts needed. It may be necessary to use two relays, cascaded, if signal energy is too small, or two relays paralleled to provide enough contacts.

Ambient conditions must be considered. Are *sealed relays* needed? Is there a *space* problem? Is there a *vibration*, shock or temperature problem?

If a suitable relay has been selected and is available for the projected use, the circuit problem must be considered.

In general the same design criteria may be used as in designing other circuits, yet relay circuitry has some problems of its own, as a result of some basic relay characteristics. Among the most important of these are the problems of transients across relay coils, and the problem of protecting relay contacts from sparking, arcing, and welding.

Low-current, low-voltage applications, as in most communication or logic circuits, do not usually experience serious problems with contacts, and often the most difficult problem is the maintenance of clean contact surfaces under low-current (dry-circuit) conditions. But when larger currents are handled, especially if the load is inductive or experiences an appreciable inrush for any reason, steps should be taken to protect the contacts from the effects of arcs, sparking, or welding. Under certain conditions arcs tend to develop between contact and case, or contact and mounting. This can be prevented by proper circuit design.

Welding of contacts from high-inrush currents can be minimized by using contacts large enough and of suitable materials. Occasionally more drastic means—such as parallel contacts—must be considered. Lamp loads and starting of single-phase motors are especially troublesome.

Sparking at contacts, or arcing, resulting from interrupting an inductive load, is minimized by installing spark suppressors. These may be RC circuits or special devices—diodes or other solid-state devices. Mercury-pool relay contacts are often used for heavy currents because the circulation of mercury provides a clean contact surface for each operation. Large power relays may use double-break contacts, blowout devices, or other means which are not within the scope of this section.

Most relay coils possess enough inductance to produce large transients when their currents are interrupted. It is not uncommon for a 28-volt relay coil to produce a transient as high as 1,000 volts. These transients can be

Table 4.5f
TYPICAL HERMETICALLY SEALED RELAYS

Relay Type	Contact Description	Mechanical Size (in.)	Speed of Action (milliseconds)	Installed Cost
Midget Plug-In	1 form C, 0.5 to 1.0 amperes	$\frac{3}{4} \times 2$	2	$ 8
Mercury-Wetted Plug-In	1 form C, 2 amperes	$1\frac{1}{2} \times 3\frac{1}{2}$	5	$15
Mercury-Wetted Plug-In	4 form C, 2 amperes	$1\frac{3}{4} \times 3\frac{1}{2}$	5	$40
Balanced-Armature DC, Sensitive	2 form C, 2 amperes	$1 \times 1 \times 2$	3 (shock resistant)	$15
Crystal-Can Relay	2 form C, 2 amperes non-inductive	$0.8 \times 0.8 \times 0.4$	1.5 (shock resistant)	$20
Reed Relay, Dry Contacts	1 form C, 12 volt-amperes.	$0.5 \times 0.5 \times 3$	1	$ 4
Transistor-Can Relay.	2 form C, 1 ampere non-inductive	$0.3 \times 0.3 \times 0.3$ (0.1 oz weight)	1.5	$40
Mercury-Pool Power Relay	1 form A or form B	$1.5 \times 1.5 \times 6$	Slow, mounts only in vertical, and is affected by vibration.	$15

reduced by adding a short-circuited winding on the coil as an absorption device, which is often done on small sealed relays. This method can only be used on DC relay coils. Neon lamps, semi-conductor devices and the like are also often used. The problem is a serious one and should not be ignored.

With many relay applications, it is advantageous to *apply overvoltage* momentarily to insure fast and positive action, then to reduce the coil current to a lower value to avoid overheating. Of various methods used for this, two are shown in Figure 4.5g. At A is a circuit useful only for DC relays. A 24-volt

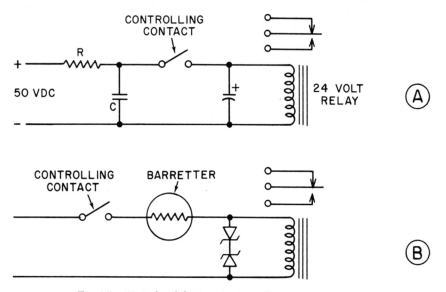

Fig. 4.5g   Typical coil-driving circuits with transient suppressors

relay is fed from a 50-volt power source, and the capacitor C is initially charged to 50 volts, so that when the controlling contacts are closed, the initial current is double the normal coil current. After the capacitor has discharged, the resistor R limits the coil current, and if R is equal to the relay-coil resistance, only normal coil current flows. In this diagram, a diode transient suppressor is shown connected across the coil.

Circuit B in Figure 4.5g can be used with either AC or DC coil excitation. It employs a positive-coefficient resistor—often a barreter or a tungsten-filament bulb—in series with the relay coil. When it is cold its resistance is quite low, but after current has passed through it for a few seconds the resistor becomes hot, its resistance greatly increases and the coil current reduces to the normal value. This diagram shows the use of two Zener diodes as transient suppressors across the coil.

Sometimes an extra relay contact is used to insert more resistance after a relay has been energized. It should be remembered that the coil current for an AC relay normally becomes smaller when the air gap is reduced by closing it.

The relay user should not hesitate to consult the relay manufacturer for assistance in selecting and applying these devices. No catalog can possibly contain all the available data, and most manufacturers are happy to supply details on relay characteristics and application.

Relative Costs of Relays

Small general purpose relays can be installed for less than any competing device—$2 to $4 each. Relays carrying more contacts, hermetically sealed, with transient protection and the like, will cost $10 and more installed. High-sensitivity relays are somewhat costly, making the use of amplified relays desirable. For some typical pricing information refer to Tables 4.5e and 4.5f.

Reliability

Relay reliability depends upon quality of relays, but even more on correct application. Relays carrying signal currents may be expected to operate 100,000,000 times or more before failure, and relays with currents in the order of amperes necessarily wear faster, but with proper contact protection lives of 100,000 to 1,000,000 operations are common.

Reliability does depend upon selecting the proper relay, contact protection, use of dust covers or hermetic seals when needed, proper circuit design, and observance of the usual rules of good practice as regards to voltages, ambient conditions, etc.

Relays vs Other Logic Elements

In the field of instrumentation, relays compete primarily with various kinds of solid-state devices, in particular silicon-controlled rectifiers, silicon switches, and transistors (Sections 4.6 and 2.1).

Electromechanical relays out-perform solid-state devices in some respects. They offer extremely low resistances when closed and extremely high resistances when open; they provide essentially complete isolation of the controlled circuit from the controlling circuit, and they can provide essentially simultaneous actuation of several circuits, or actuation at small intervals of time in succession, at no added cost. Some relays can handle higher voltages, at higher frequencies, than solid-state devices, or in other applications, higher currents and higher powers. They provide positive circuit closures and can tolerate much abuse with only a shortening of their useful life.

Some relays are as small or smaller than the competing solid-state device when installed. Still others can be furnished at a lower cost installed than competing devices, and some—such as latching relays or steppers—can offer other advantages, such as a functional memory which is not destroyed by a power interruption.

For extremely high-speed operation, solid-state devices excel, because only a few relay forms will operate in less than a millisecond. For the ultimate in compactness, disregarding cost, integrated solid-state devices are best. Many relays operate on ordinary line power, in contrast to solid-state devices, which often require some other power source. Thus if only a few relays are needed, they may be selected for this reason, while if a large number of devices are involved the per unit power supply cost becomes negligible.

Modern relays withstand extremes in ambient conditions well.

Special Relays

A multitude of special relays is available, many of them little known outside of their special field of application. Some which might be of interest to the instrumentation engineer include: sequence counters with transmitting contacts, opto-electronic relays (in which coupling between actuator and circuit closure is a light beam, permitting elaborate interlocking of functions), meter relays with 1 microwatt sensitivity, ultra-sensitive polarized DC relays (40 microwatt sensitivity), resonant-reed relays for remote-control switching, coaxial RF relays, vacuum relays for several thousand volts, multipole relays which can actuate 60 or more circuits simultaneously, polarized telegraphic and pulse-forming relays, voltage-sensing relays for over-voltage and under-voltage protection, phase-monitoring relays for protection in polyphase power circuits, impulse-actuated relays, instrumentation relays with negligible capacitance between circuits, and cross-bar switches, which can select any one of 100 or more circuits.

Definitions and Functions of Timers

In this context, timers are defined as devices which initiate action in an electrical circuit at some predetermined time or on a time schedule. Time-determining devices may be mechanical (including pneumatic, hydraulic, clock motors, or electric motors), or electronic (electronic time-measuring circuits of all kinds), or thermal (thermal time-delay devices). Primary emphasis will be placed on mechanical types.

The classification should not include simple time-delay relays, such as the thermal time-delay relays and the lagged telephone relays mentioned earlier (Figure 4.5d). They are ordinarily used only to prevent interference between operating functions.

The functions of timers are as follows: to initiate action after a known delay, to actuate a number of circuits in accurately timed succession, to time reactions or processes, to monitor a number of circuits at definite time intervals, and the like.

Interval Timers

When a single time interval (fixed or adjustable) must be established, interval timers may be used. Such timers employ a clock mechanism, a synchronous motor, a calibrated dashpot or any other timing device to produce some action at a specific time interval after the signal is applied to the timer. The range of time intervals possible varies from a few milliseconds to as much as an hour for pneumatic devices and it may be extended to as long as desired by using electric clock motors. The accuracy of time measurement is usually quite high, even in factory conditions. Most such interval timers are available with reset devices so they can be operated remotely or on a repetitive basis. Simple interval timers are often used in conjunction with more sophisticated programmers. In such arrangement they can handle specific single timing tasks under the control of the programming timer.

Programming Timers

Programming timers are capable of controlling a number of circuits on a continuing basis where many different actions are programmed at many different times during the entire operating cycle of the programmer.

Most widely used are the cam-operated programming multiple-circuit time switches, drum programmers, and band-programming timers or punched-card programmers—all quite alike in principle but varying in construction.

Cam-operated timers, or timeswitches, employ an array of mechanical switches arranged with an equal number of adjustable cams on a shaft (Figure 4.5h). A clock or a synchronous motor turns the shaft at the desired speed—one

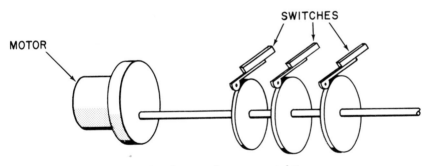

Fig. 4.5h   Elements of a cam-operated timer

revolution for the desired cycle duration. During the cycle, each of the switches is opened and closed once (occasionally two or three times is permissible), for durations established by the cam settings. Such a timer can control as many as 60 circuits, with a cycle as short as a few seconds or as long as a few weeks. Usually about 95 percent of the cycle can be programmed either "on" or "off" for each circuit; usually the contacts can handle currents as great as 10 amperes at 120 volts. These timers are available in a variety of sizes and speeds.

Cam-operated timers are often used in conjunction with (either in cascade or in parallel) interval timers, which permits much greater flexibility of action. They are reasonable in cost, reliable, and their timing accuracy is high— limited by facility in adjustment.

A limitation of cam-operated timers is that each switch can be actuated only a few times during each timing cycle. This limitation is entirely overcome in drum programmers, band-programming timers and punched-card pro- grammers.

Band-programming timers employ a belt or band of flexible plastic mate- rial, several inches wide and as long as needed to provide the required cycle length (Figure 4.5i). An array of switches is provided, each with an actuating

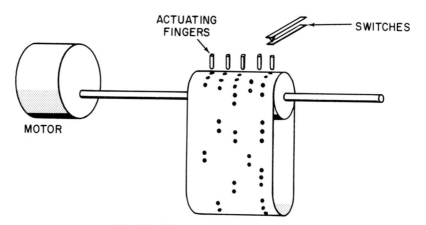

Fig. 4.5i   Elements of a band programmer

finger resting on the band. Where the band is intact the switch is held open, but where it is perforated the switch can close. As many as 80 such switches can be used and the band can be of any reasonable length. Many hours of operation can be controlled with many distinct on-off cycles for each switch during the total cycle. Quite complicated switching patterns can be provided.

Punched-card programmers operate in a similar manner. The cards are usually handled singly instead of in a continuous band. Such programmers are in fact quite similar to the Jacquard loom principle. Drum programmers are somewhat similar in that a drum in which many movable studs are placed rotates in a time cycle under an array of switches (Figure 4.5j).

In a band or drum programmer, the moving patterned band is often made to move against the switch array by a pulsed stepping motor rather than a simple synchronous motor. The stepping motor is driven by timed pulses from a separate source, perhaps varying in rate according to a subsidiary pattern. Added flexibility is thus afforded.

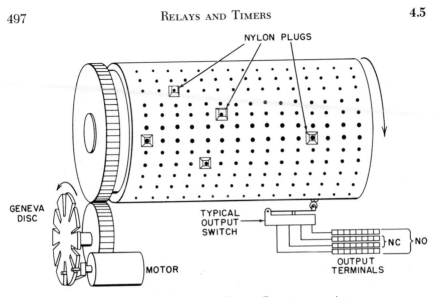

Fig. 4.5j   Drum programmer for on-off event sequencing

Other types of timers include various electronic time generators, either single-shot or producing complex time patterns, actuating relays which provide the final contact closures. These are capable of higher operating speed—shorter intervals—than most electromechanical timing devices.

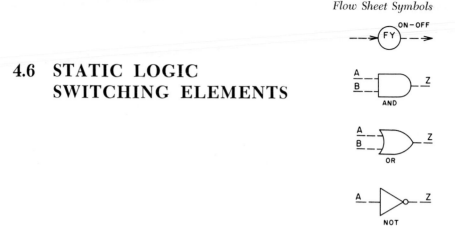

4.6 STATIC LOGIC SWITCHING ELEMENTS

Cost:	$4 to $10 for AND, OR and NOT elements. $10 to $20 for flip-flops. $10 to $30 for timers.
Partial List of Suppliers:	Cutler-Hammer, Inc., Digital Equipment Corp., General Electric Co., Square-D Co., Westinghouse Electric Corp.

Static logic switching elements are electronic devices that perform control switching operations without moving parts.

Their advantages are (1) high switching speeds, (2) life independent of number of operations, (3) operation unaffected by dirt and corrosive atmospheres, (4) small size, (5) low power consumption.

Their disadvantages are (1) sensitivity to electrical interference, (2) lack of input-output isolation, (3) low power output driving capabilities.

The choice of static logic elements over mechanical relay elements is a matter of economics and requirements. In general, solid-state elements are indicated when the following factors are important:

1. Switching speed requirements are high (above, say, 100 per second). At such speeds the life of an ordinary relay is limited to the order of a few hundred hours.
2. The system requirements are complex, involving a large number of elements for implementation.
3. The system receives signals from a digital computer which are at static switching levels.
4. Space is limited such as in portable equipment.
5. Process down-time is costly and must be minimized.

498

The choice of relays is indicated when the following factors predominate (see also Section 4.5):

1. Switching speed requirements are low.
2. The system is simple requiring only a few relays for implementation.
3. The signals received and generated by the system are at relatively high voltages and current levels.
4. The input signals must be isolated from the output signals.
5. Maintenance personnel are unsophisticated, but already familiar with relays, and retraining would be costly.

This section will begin with a brief discussion of semiconductor materials, describe the common switching devices based on material, define the basic switching actions, and describe circuits for carrying out these switching actions.

Semiconductors

Switching—the changing of a circuit from zero-conductance to full-conductance and vice versa—is accomplished with relays by the opening and closing of contacts. With solid-state switching devices, this is accomplished by utilizing the properties of a class of solids known as semiconductors. The two most significant semiconductor materials for solid-state switching devices are germanium and silicon, and of these, silicon has become the most utilized, because devices constructed of silicon can operate at higher temperatures than those of germanium.

The semiconductor material is modified to give switching properties by introducing some other material through a process called doping. Although the concentration of this material is very low (typically one part in ten million), the electrical properties of the semiconductor are significantly changed.

There are two kinds of doping. The first is called N-type and results in a N-type semiconductor. It is also called donor doping, because the effect of the doping is to make available conductance electrons in the material. Materials used for donor doping are phosphorous, arsenic and antimony. The second type of doping is termed P-type or acceptor doping, because the added material "accepts" electrons from the silicon creating conductance *holes*. (Holes are really the absence of electrons and are analogous to bubbles which are the absence of water.) Commonly used acceptor doping materials are boron, aluminum and gallium.

Diodes

A diode is formed by joining together a small amount of P-material to a similarly small amount of N-material to form what is termed a P-N junction (Figure 4.6a). Such a junction has the property of offering a comparatively

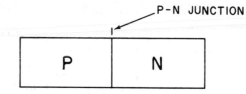

Fig. 4.6a Diagrammatic representation of P-N junction

low resistance for one direction of current flow and very high resistance for the flow of current in the reverse direction. Figure 4.6b gives a plot of the voltage-current characteristics for a typical diode. As seen in the conducting region the current increases rapidly after a small threshold voltage is exceeded. In the opposite direction, the current is small and increases slowly with voltage. The diode is said to be reverse biased with a reverse voltage. Table

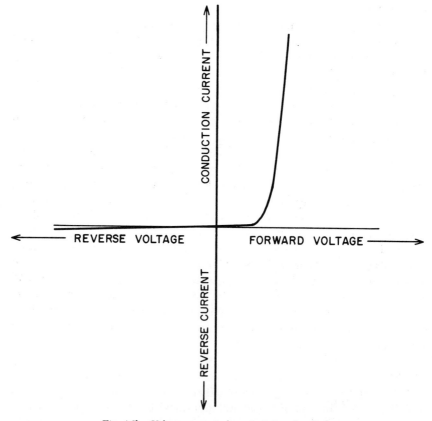

Fig. 4.6b Voltage-current characteristics of a diode

4.6c shows the typical characteristics of diodes for the two types of semi-conductor materials.

As seen, the diode formed from germanium semiconductor material has a lower forward voltage drop (lower conducting resistance), but silicon has

Table 4.6c
DIODE CHARACTERISTICS

Material	Forward Voltage Drop (Volts)	Reverse Current (10^{-6} amperes)		
		77°F	185°F	302°F
Germanium	0.25	5	300	
Silicon	0.8	0.025		5

a much lower reverse current (higher non-conducting resistance). The difference becomes quite marked at elevated temperatures.

The symbol used for the diode in circuit representations is shown in Figure 4.6d. Shown are the voltage polarities across the diode which give conduction and non-conduction and are respectively called forward voltage and reverse voltage.

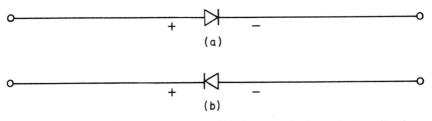

(a)

(b)

Fig. 4.6d   Symbols used for diodes in circuits. (a) Voltage polarity for conduction; (b) voltage polarity for non-conduction

Transistors

The transistor is formed by combining two P-N junctions as shown in Figure 4.6e, where position (a) shows a PNP transistor and (b) shows a NPN transistor. The voltages applied and the resultant current flow are also shown for the common-emitter connection (the emitter is common to the input and output circuits) which is the commonly used arrangement for switching circuits.

The internal action of the transistor is quite complicated, but two general characteristics (which can be observed by external measurements) represent the significant characteristics for switching action:

1. The circuit from base to emitter controlling the current I_b in Figure 4.6e closely resembles a diode.

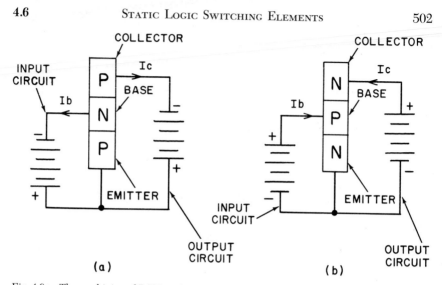

Fig. 4.6e　The combining of P-N junctions into transistors. (a) Arrangement for the PNP-transistor; (b) arrangement for the NPN-transistor. Also shown are the external voltages applied and the resultant current flows.

2. The circuit from collector to emitter (through which I_c flows) also resembles a diode, but if *reverse* voltage is applied, a significant action can be observed: If $I_b = 0$, I_c will be very small as for a normal diode with reverse voltage, but when I_b begins to flow, I_c will also begin to flow proportionally to I_b but *increased by an amplification factor*.

The relationship is expressed by the equation

$$I_c = \beta I_b \qquad\qquad 4.6(1)$$

where β is the current amplification factor of the transistor.

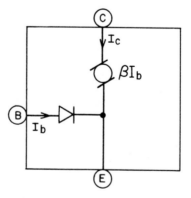

Fig. 4.6f　Equivalent circuit for NPN-transistor

Based on these two properties an equivalent circuit (somewhat simplified) can be drawn as shown in Figure 4.6f where the collector I_c is considered to be controlled by the current generator, βI_b.

Now consider the circuit in Figure 4.6g. The output voltage of the circuit is of particular interest. This can be expressed as

$$V_o = V - I_c R_1 \qquad\qquad 4.6(2)$$

$$V_o = V - \beta I_b R_1 = V - \beta V_i (R_1/R_i) \qquad\qquad 4.6(3)$$

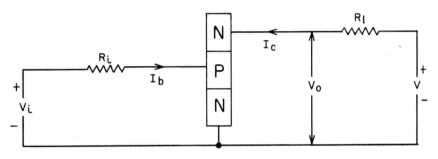

Fig. 4.6g   Basic circuit for transistor switching action

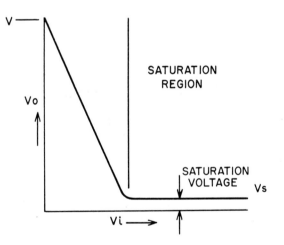

Fig. 4.6h   Plot showing variation of output voltage V_o as a function of input voltage V_i for the circuit of Figure 4.6g

A plot of 4.6(3) is shown in Figure 4.6h. The region where the collector voltage V_o becomes small is termed the saturation region.

The characteristic which makes the transistor effective as a switching device is that the current amplification is maintained even in the saturation region. The voltage in that region, called the saturation voltage, is less than

0.2 volts for typical switching transistors, and the effective resistance between collector and emitter at saturation is accordingly small (in the order of a few ohms). On the other hand when $I_b = 0$, the collector-to-emitter resistance is that of a reverse biased or non-conducting diode and is in the region of millions of ohms. The collector-to-emitter circuit thus appears to open and short under the control of I_b and, in fact, an analogy can be made to a relay as shown in Figure 4.6i, where the coil of the relay is analogous to the base-emitter

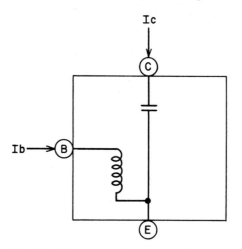

Fig. 4.6i   Diagram showing analogy of the transistor switching circuit in Figure 4.6g to a relay coil and contact

input circuit and the contact of the relay is analogous to the collector-emitter circuit. However, there is one inherent difference between the relay and transistor as switching devices, a difference which leads to different design methods. The emitter terminal is common to the input and output circuits of the transistor and they *cannot be isolated* as the coil and contact of the relay can. Table 4.6j gives the characteristics of some commonly used switching transistors.

Table 4.6j
SWITCHING TRANSISTOR CHARACTERISTICS

"JEDEC" Transistor Type Numbers	Material	Beta	Saturation Voltage, Vs	Reverse Current, I_b (10^{-6} amperes)		Switching Times (10^{-6} seconds)	
				77°F	185°F	on	off
2N404	Germanium PNP	30	0.15	0.05	100	0.34	0.57
2N2894	Silicon PNP	30	0.15	0.08	5	0.06	0.09
2N2369A	Silicon NPN	40	0.2	0.01	0.1	0.01	0.03

Table 4.6j shows the superiority of silicon for higher temperature operation. The lower switching times (higher switch speeds) of the 2N2894 and 2N2369A compared with the 2N404 are due to their method of construction and reflect an improvement in fabrication techniques in the last few years. Figure 4.6k shows the commonly used symbols for transistors which will be adopted for the discussion that follows.

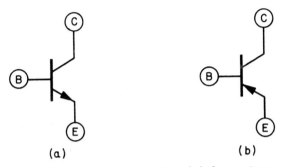

Fig. 4.6k   Commonly used circuit symbols for transistors.
(a) NPN transistor symbol; (b) PNP transistor symbol

Logic Operations

Before considering switching circuits formed from diodes and transistors, the nature of the switching operations themselves will be defined. The general requirement for switching devices is that they generate on-off output control signals from on-off input signals, in a way described as generating a *logic function*. It will be seen in the next section that logic functions are obtained by interconnecting switching circuits, where each circuit generates an elementary logic operation. The five elementary and basic logic operations generated by static switching circuits are:

1. NOT   The output is *on* if the input is *off* and vice versa.
2. AND   The output is *on* if *all* inputs are *on*.
3. OR   The output is *on* if *one or more* inputs are *on*.
4. NOR   The output is *off* if *one or more* inputs are *on*.
5. NAND The output is *off* if *all* inputs are *on*.

Diode Switching Circuits

Figure 4.6l shows how two diodes (and resistors) can be connected to give the AND-logic operation. The input signals range between two values 0 and $+V_i$ (V_a or V_b), and there is a supply voltage $+V$ which is *greater* than $+V_i$. Let us now consider how V_o will change as the input signals assume their possible values? There are four possible combinations of input signal values, and they can be listed as follows:

$$V_a = 0 \qquad +V_i \qquad 0 \qquad +V_i$$
$$V_b = 0 \qquad 0 \qquad +V_i \qquad +V_i \qquad\qquad 4.6(4)$$

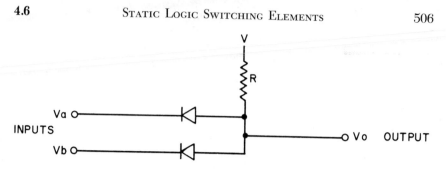

Fig. 4.61   Circuit diagram for diode type, "AND" logic generator

where V_a and V_b are the two inputs and 0 and $+V_i$ are the two signal values for each input.

For the first case where V_a and V_b both equal zero, both diodes are conducting (the voltage across the diodes is in a direction to give conduction). The output voltage V_o differs from the input only by the small voltage across the diodes required for conduction. This will also be true for the last case where V_a and V_b both equal $+V_i$, because V_i was assumed to be negative in respect to $+V$. Therefore, for the first and last columns of this tabulation, V_o is equal to 0 and $+V_i$ respectively, if the diode-conducting voltage is neglected.

For the second and third columns of the 4.6(4) one of the inputs has the value 0 and the other the value $+V_i$. The significant action here is that the diode with the input of 0 will be conducting but the diode with the value of $+V_i$ will be *non-conducting* and have, in effect, a reverse voltage applied. This is seen by inspection of Figure 4.6m. The conducting diode will maintain

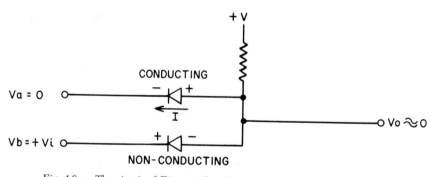

Fig. 4.6m   The circuit of Figure 4.6l with one input at 0 and the other at V_i

V_o as near zero, and the application of $+V_i$ on the other diode is the equivalent of applying a reverse voltage. The result is that V_o will be equal to 0 (neglecting the diode-conducting voltage) for both cases where one input is equal to 0 and the other equal to $+V_i$.

The action for the four cases can be summarized by adding an "output" line to the 4.6(4) tabulation

$$
\begin{array}{cccc}
V_a = 0 & +V_i & 0 & +V_i \\
V_b = 0 & 0 & +V_i & +V_i \\
\hline
V_o = 0 & 0 & 0 & +V_i
\end{array}
\qquad 4.6(5)
$$

where V_o represents the output. The AND operation is thus generated if $+V_i$ is associated with the idea of being "on." The output has the value of $+V_i$ only if both inputs are at $+V_i$.

The *OR-logic* action is obtained by reversing the diodes and having a *negative* supply voltage as shown in Figure 4.6n. The action can now be summarized as follows:

$$
\begin{array}{cccc}
V_a = 0 & +V_i & 0 & +V_i \\
V_b = 0 & 0 & +V_i & +V_i \\
\hline
V_o = 0 & +V_i & +V_i & +V_i
\end{array}
\qquad 4.6(6)
$$

or the output is $+V_i$ if either or both of the inputs have the value V_i.

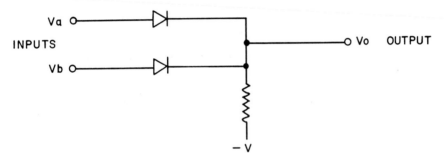

Fig. 4.6n   Diode circuit to generate OR logic

The two circuits can be cascaded to give more complex logic operations as illustrated in Figure 4.6o, where two AND operations involving V_a, V_b and V_c, V_d are the inputs to an OR operation.

However, there are two limitations to diodes as switching elements. The "inverting" logic operations, such as the NOT function, cannot be generated, and there is inherent signal deterioration due to the forward voltages necessary for diode conduction, which limits the amount of cascading and the complexity of the logic function which can be formed.

Transistor Switching Circuits

The basic transistor switching circuit for a NPN-transistor is shown in Figure 4.6p. Referring to the previously given discussion on transistor action, the output voltage V_o will be equal to $+V$ when $V_a = 0$, since both I_b and

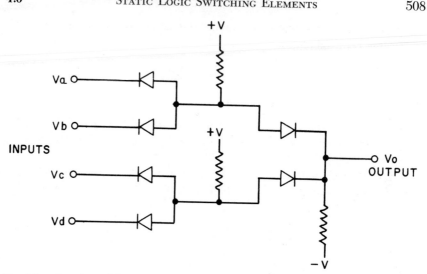

Fig. 4.6o   Cascading of diode logic circuits. Shown are two AND circuits whose outputs are to an OR circuit

$I_c = 0$. When the input voltage is equal to $+V_i$ (assuming that this causes sufficient I_b to flow to give saturation), V_o will be close to 0, differing from it only by the small value of saturation voltage. If this is neglected, a table can be prepared to describe this action:

$$
\begin{array}{ccc}
V_a = & 0 & +V_i \\
\hline
V_o = & +V & 0
\end{array}
\qquad 4.6(7)
$$

If both $+V$ and $+V_i$ are taken as "on" values, it is seen that the NOT logic operation is generated.

If a second resistor is added to the circuit of Figure 4.6q, the resultant

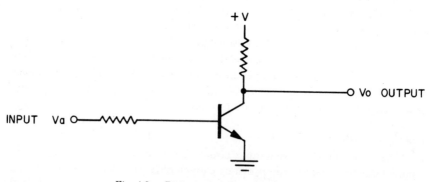

Fig. 4.6p   Basic transistor switching circuit

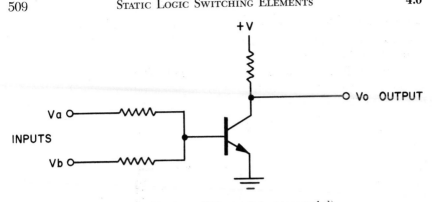

Fig. 4.6q   Two-input NOR circuit (resistor-coupled)

action is easily determined: $+V_i$ applied to *either* of the inputs will cause transistor saturation. The switching table then is

$$
\begin{array}{lcccc}
V_a = & 0 & +V_i & 0 & +V_i \\
V_b = & 0 & 0 & +V_i & +V_i \\
\hline
V_o = & +V & 0 & 0 & 0
\end{array}
\qquad 4.6(8)
$$

This is recognized as the NOR-logic operation as previously defined. The output is off (output at 0) if any of the inputs are on (input at $+V_i$).

NOR circuits are readily cascaded, as shown in Figure 4.6r, with the

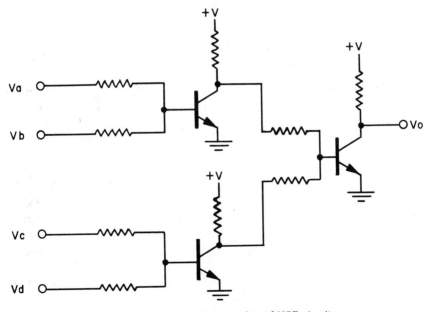

Fig. 4.6r   Example of the cascading of NOR circuits

significant advantage compared with the circuit of Figure 4.6o that there is no signal deterioration, since each transistor switching operation completely regenerates the signal.

The diode circuit of Figure 4.6l can be combined with the transistor circuit of Figure 4.6p to give the NAND-logic operation as shown in Figure 4.6s. The action is that of the AND operation generated by the diode circuits

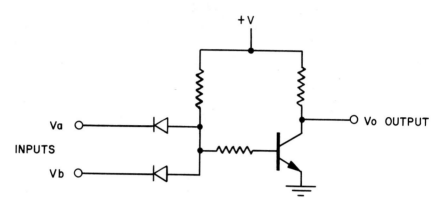

Fig. 4.6s   Circuit for obtaining NAND logic based on combining diode and transistor circuits (diode-coupled)

followed by a NOT operation generated by the transistor. The action can be tabulated as follows:

$$
\begin{array}{lcccc}
V_a = & 0 & +V_i & 0 & +V_i \\
V_b = & 0 & 0 & +V_i & +V_i \\
\hline
V_o = & +V & +V & +V & 0
\end{array}
\qquad 4.6(9)
$$

The circuits of Figures 4.6q and 4.6s are respectively called resistance-coupled and diode-coupled transistor switching circuits. Figure 4.6t shows two circuits in which a transistor is incorporated for each input. The first circuit generates the NOR operation and the second the NAND operation.

Integrated Circuits

The term "integrated circuits" refers to a way of fabricating the previously discussed circuits in such a way that the technique for fabricating a single transistor is applied to the entire circuit. Transistors, diodes and resistors are formed and interconnected with a process that is analogous to casting a metal part as against assembling the part from sub-components by welding or other techniques.

The integrated circuit process begins with a silicon wafer about 1 in. in diameter and 6 to 8 mils in thickness.

The basis of the integrated circuit technology is the application of a set

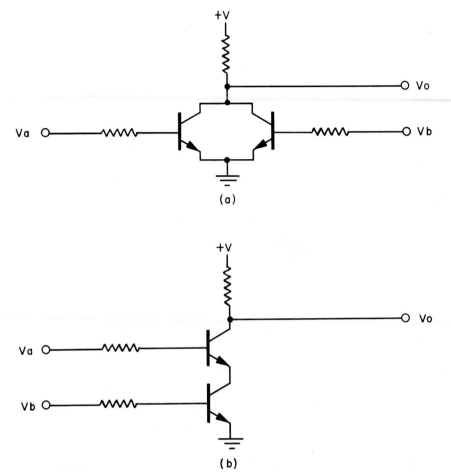

Fig. 4.6t   Transistor logic circuits based on incorporating a transistor for each input. (a) NOR circuit; (b) NAND circuit

of "masked diffusions," where a succession of photographic masks are applied to the wafer and selected areas of the wafer are doped (see discussion on semiconductors) to give a pattern of P-material and N-material joined in a way to form transistors and diodes.

The components thus formed are then interconnected by depositing (again controlled by a photographic mask) a metallic pattern.

The significant nature of the process is that this is carried out for up to several hundred circuits simultaneously, for this is the number of circuits (called dice) which a wafer can hold. The result is a very small circuit (typically 0.050 in. square). After the completion of wafer processing, the dice are separated by scribing and assembled into packages.

Integrated Switching Circuits

There are four main families of integrated circuit switching in common use for control:

1. RTL (resistance-transistor logic)
2. DTL (diode-transistor logic)
3. TTL (transistor-transistor logic)
4. HLL (high-level logic)

RTL was the first type of switching logic produced in integrated form. A typical circuit is shown in Figure 4.6u. The circuit is similar to that of Figure 4.6t, where the logic generated is the NOR operation ("high" voltage represents "on").

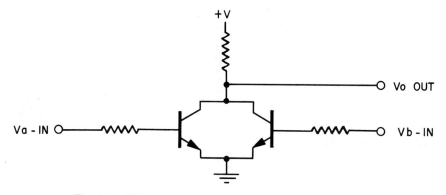

Fig. 4.6u   RTL (resistance-transistor logic) integrated logic circuit

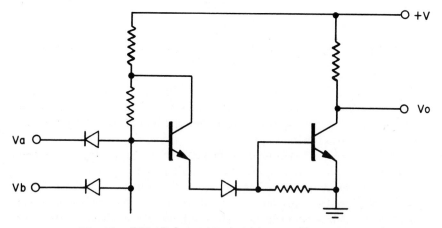

Fig. 4.6v   DTL (diode-transistor logic) integrated logic circuit

A modified form of the circuit of Figure 4.6s is the basis of the DTL circuit, shown in Figure 4.6v. The circuit, it will be recalled, generates the NAND operation.

A more complicated circuit is the TTL circuit shown in Figure 4.6w.

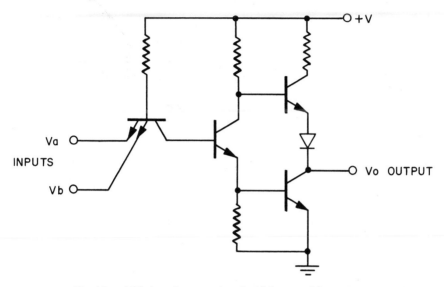

Fig. 4.6w   TTL (transistor-transistor logic) integrated logic circuit

This circuit is characterized by a high speed of switching and it generates the NAND operation.

A special kind of circuit (HLL) is one designed to discriminate against electrical interference of the type encountered in many industrial situations. The circuit is a modification of DTL where elements have been introduced to give large voltage swings and, in effect, to make the switching signals large compared with the interfering signals. A summary of integrated circuit characteristics is given in Table 4.6x.

Table 4.6x
INTEGRATED CIRCUIT CHARACTERISTICS

Type	Resistance to Noise	Switching Speed	Cost
RTL	Poor	Fair	Low
DTL	Fair	Good	Medium
TTL	Good	High	Medium
HLL	Excellent	Low	High

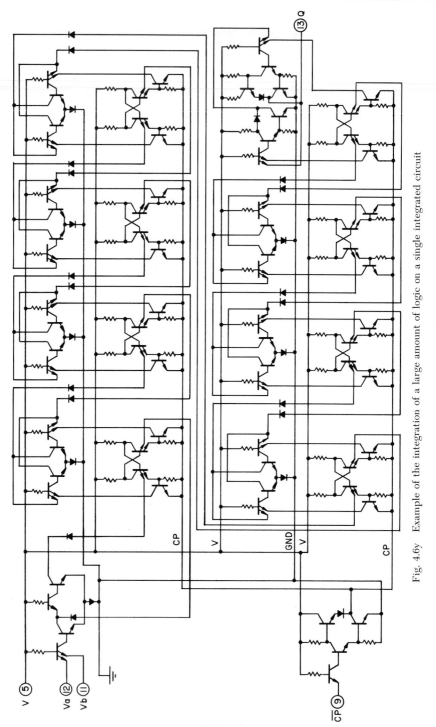

Fig. 4.6y   Example of the integration of a large amount of logic on a single integrated circuit

514

Medium- and Large-Scale Integration

Not only can a large number of circuits be produced simultaneously by the methods described, but by increasing the size of the dice on the wafer, very large and complicated logic circuits can be made. Circuits are now available where several hundred transistors have been incorporated on a single die. Figure 4.6y gives an example of a circuit for an eight-bit shift register.

4.7 SYNTHESIS AND OPTIMIZATION OF LOGIC CIRCUITS

The main purpose of logic design is to synthesize an interconnection of switching elements (static or relay type) which will satisfy the switching functions required by the process. The development of such a design involves several main steps.

Usually the first step is to functionally define the logic interlock requirements by statements of the following nature: "Agitator to start automatically if level is above 50 percent, temperature is between 170°F and 180°F and if the feed valve is closed."

The next step is to convert these statements into logic symbols, where the three conditions mentioned can become inputs A, B, C and the agitator might be represented by output Z. The following text describes the various alphabetic and graphic logic symbols, together with the truth tables and other tools, which help the instrument engineer in the synthesis of logic circuits.

Once a concise expression of the required logic operations has been arrived at using any one of the possible symbols, it is desirable to remove logic redundancies. The simplification of logic is accomplished by either logic maps or through the use of Boolean algebra. All that remains to be done after this step is to select the hardware (the type of switching elements) which will be used to fulfill the simplified logic requirements. It should be understood that logic simplification always means less hardware and therefore less cost, to satisfy the same interlock requirements.

Logic Symbols and Tables

Basic logic components and operations were defined and discussed in the preceding sections. The terms used included words such as on-off, high, low, AND, OR, NOT, etc., but logic design is more commonly based on the use of symbols.

A logic signal (or logic variable), such as one into or out of a logic switching element, can be represented by the symbols A, B, C . . . X, Y, Z. The two values of the signals (for engineering design) are represented by the symbols "1" and "0." Thus a statement such as, "The motor shall operate only if both switches are closed," can be stated symbolically as

$$Z = 1 \quad \text{only if} \quad A = 1 \quad \text{and} \quad B = 1 \qquad 4.7(1)$$

where Z is the motor, A and B are switches, and "1" means motor operating and switch closed.

Tables of switching operations in Section 4.6 show how a switching device carried out its operations. Similar tables, in logic design called truth tables, can be based on the above symbols.

The procedure is to list in a systematic way all of the possible combinations of inputs (independent variables) to a logic operation using the symbols "1" and "0." The following is a horizontal form of the table shown for inputs A, B and C.

A	0101	0101	4.7(2)
B	0011	0011	
C	0000	1111	

The manner in which the patterns are established: 01010101 for A, the first input, 00110011 for B, the second, etc., serves the purpose of representing *all* possible input signal combinations.

The Basic Logic Functions

The five logic functions previously given in terms of physical values can be defined through tabulation 4.7(2).

A	0101	0101	
B	0011	0011	
C	0000	1111	
AND	0000	0001	4.7(3)
OR	0111	1111	4.7(4)
NOR	1000	0000	4.7(5)
NAND	1111	1110	4.7(6)

As seen, the definition of the logic operation is made by listing the value ("1" or "0") generated for each input combination (column).

Such a table can also be used to define general logic functions which are to be synthesized from more basic functions. For example, tabulation 4.7(7) defines the logic operation of parity generation:

A	0101	0101	
B	0011	0011	
C	0000	1111	
Z	1001	0110	4.7(7)

which is to generate a 1 if the number of 1s in the inputs is even; otherwise 0 is generated.

Not only are logic quantities represented by symbols but also operations upon those quantities can also be represented by symbols. Table 4.7a shows the common symbols used in engineering design:

Table 4.7a

OPERATION SYMBOLS IN LOGIC DESIGN

Word Designation	Symbolic Representation
NOT A	$\overline{A}$
A AND B AND C	$A \cdot B \cdot C$
A OR B OR C	$A + B + C$
A NOR B NOR C	$\overline{A + B + C}$
A NAND B NAND C	$\overline{A \cdot B \cdot C}$

This symbology leads to concise expressions of logic operations. For example:

$$Z = A \cdot B + C \cdot D \qquad 4.7(8)$$

states that two outputs of two AND operations based on A, B and C, D respectively are the inputs to an OR operation.

Synthesis With AND/OR Logic Operations

Synthesis using AND/OR logic operations, sometimes called English or non-inverting logic, is carried out through two special functions, the *minterm* and the *maxterm*.

The minterm is a logic function where there is a single "1" in the output. For example:

$$
\begin{array}{ll}
A & 0101 \\
B & \underline{0011} \\
Z & 0100 \qquad 4.7(9)
\end{array}
$$

A minterm is implemented through an AND operation based on the column where the single "1" resides. The rule is that all inputs with "0" in that column have a NOT operation applied. For the example, then, $Z = A \cdot \overline{B}$ defines the minterm.

The maxterm is a logic function where there is a single "0" in the output. For example:

$$
\begin{array}{ll}
A & 0101 \\
B & \underline{0011} \\
Z & 1011 \qquad 4.7(10)
\end{array}
$$

A maxterm is implemented by a process which is the inverse of that for the minterm. An OR operation is used, and those inputs with a "1" for the column where the single "0" resides have the NOT operation applied. For the example, $Z = \overline{A} + B$ defines the maxterm.

A logic function can be synthesized through the OR-ing together of minterms. For example, the function:

$$
\begin{array}{ll}
\text{A} & 0101 \\
\text{B} & 0011 \\
\hline
\text{Z} & 0110
\end{array}
$$

4.7(11)

can be synthesized as $A \cdot \overline{B} + \overline{A} \cdot B$, where the two minterms 0100 and 0010 have been joined together with an OR operation. Alternatively, the function can be synthesized by the AND-ing together of maxterms. For the above example, this is $(A + B) \cdot (\overline{A} + \overline{B})$, where the two maxterms 0111 and 1110 have been joined by an AND function.

Example: The logic function for parity generation previously given in 4.7(7), was as follows:

$$
\begin{array}{lll}
\text{A} & 0101 & 0101 \\
\text{B} & 0011 & 0011 \\
\text{C} & 0000 & 1111 \\
\hline
\text{Z} & 1001 & 0110
\end{array}
$$

4.7(7)

This can be synthesized by either

$$Z = \overline{A} \cdot \overline{B} \cdot \overline{C} + A \cdot B \cdot \overline{C} + A \cdot \overline{B} \cdot C + \overline{A} \cdot B \cdot C \qquad 4.7(12)$$

or

$$Z = (\overline{A} + B + C) \cdot (A + \overline{B} + C) \cdot (A + B + \overline{C}) \cdot (\overline{A} + \overline{B} + \overline{C})$$

4.7(13)

Negation and DeMorgan's Theorem

Before considering synthesis through NOR and NAND functions, some basic principles for inverting logic functions will be established.

The first principle is that two cascaded inversions cancel, since this amounts to a double negation. Stated in symbolic form:

$$\overline{\overline{A}} = A \qquad 4.7(14)$$

where it is indicated that the logic signal A has been inverted twice, returning it to its original value.

The second is that the inversion of logic based on AND/OR operations can be obtained by inverting every input signal and at the same time interchanging AND and OR operations. (DeMorgan's theorem). For example, the inversion of

$$Z = A \cdot B + C \cdot D \qquad 4.7(15)$$

is obtained as:

$$\overline{Z} = (\overline{A} + \overline{B}) \cdot (\overline{C} + \overline{D}) \qquad 4.7(16)$$

These two principles can be used to obtain equivalent forms for both the NOR and NAND operations. For the NOR operation,

$$\overline{A + B} = \overline{\overline{\overline{A} \cdot \overline{B}}} = \overline{A} \cdot \overline{B} \qquad 4.7(17)$$

For the NAND operation:

$$\overline{A \cdot B} = \overline{\overline{\overline{A} + \overline{B}}} = \overline{A} + \overline{B} \qquad 4.7(18)$$

These two expressions are called the *DeMorgan equivalents*.

Synthesis With NAND/NOR Operations

The synthesis of a logic function by NAND-operations can be based on the minterm synthesis previously given which, it will be remembered, resulted in a set of AND operations followed by OR operations. The procedure is to replace *both* AND and OR operations by NAND operations.

It can also be readily shown that a maxterm synthesis consisting of OR operations followed by AND operations can be accomplished by NOR operations, where the NOR operation replaces both the OR and AND operations. For example:

$$Z = A \cdot B + C \cdot D \qquad 4.7(19)$$

is an example of a minterm synthesis. Following the procedure stated above, the AND operations A $\cdot$ B, C $\cdot$ D are replaced by NAND operations to give $\overline{A \cdot B}, \overline{C \cdot D}$. Next, the OR operation is also replaced by a NAND operation to give

$$Z = (\overline{\overline{A \cdot B}) \cdot (\overline{C \cdot D})} \qquad 4.7(20)$$

The fact that this is equivalent to the original minterm expression is readily shown by using the DeMorgan equivalent given in the preceding paragraph: $\overline{X \cdot Y} = \overline{X} + \overline{Y}$ for the last NAND operation to give

$$Z = (\overline{\overline{A \cdot B}}) + (\overline{\overline{C \cdot D}}) \qquad Z = A \cdot B + C \cdot D \qquad 4.7(21)$$

Graphic Logic Symbols

Logic can be represented graphically through *logic diagrams*. First, graphic symbols are adopted to show the basic logic operations. Figure 4.7b lists those which are now in prevalent use, where a symbol shape indicates the basic operations of NOT, AND and OR. The inverting operations of NAND, NOR are represented by the same shapes but with small circles which indicate inversion. Also shown are symbols for the DeMorgan equivalent forms for the NAND/NOR operations discussed earlier.

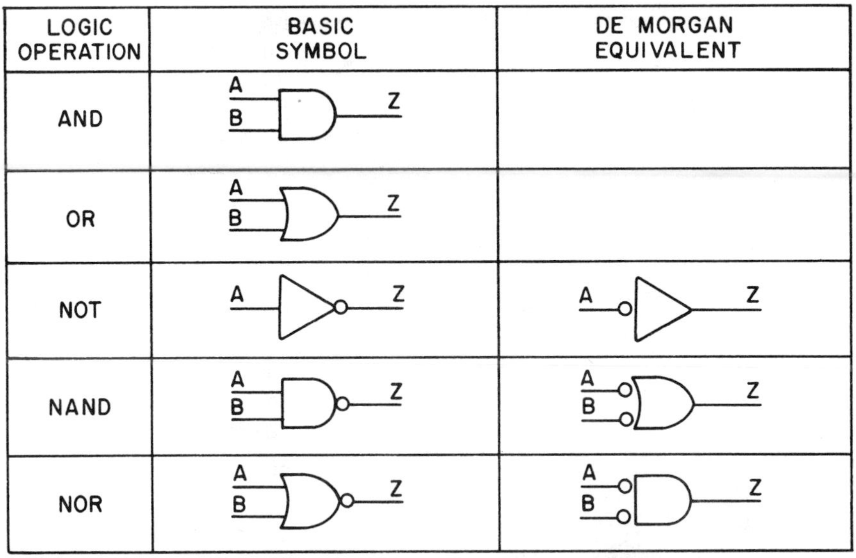

LOGIC OPERATION	BASIC SYMBOL	DE MORGAN EQUIVALENT
AND	(AND gate symbol, inputs A, B; output Z)	
OR	(OR gate symbol, inputs A, B; output Z)	
NOT	(inverter symbol with output bubble, input A; output Z)	(inverter symbol with input bubble, input A; output Z)
NAND	(NAND gate symbol, inputs A, B; output Z)	(OR gate with input bubbles, inputs A, B; output Z)
NOR	(NOR gate symbol, inputs A, B; output Z)	(AND gate with input bubbles, inputs A, B; output Z)

Fig. 4.7b   Summary of symbols used to represent logic operations on logic diagrams

Graphic Translation From AND/OR to NAND/NOR Logic

It is usually easy for the designer to express a switching problem in terms of AND/OR operations. In fact, common sense will often suffice for a solution without resorting to the formal design methods described. However, implementing that logic with NAND/NOR circuits can be difficult and confusing. The following describes a graphic technique for making a conversion which requires no particular skill or knowledge of logic design principles.

The method utilizes the symbols of Figure 4.7b and is based on the two DeMorgan equivalent symbols shown for the NAND and NOR operations. The process begins with expressing the logic with AND/OR symbols. Next the inverting circles are added from the general NAND/NOR symbols, depending upon which circuit (NAND or NOR) is being employed. The diagram now shows the inversions introduced by converting to NAND or NOR logic.

The inversions introduced by the NAND/NOR logic conversion are of three types: (1) cancelling, (2) wanted, and (3) unwanted.

Cancelling inversions appear as circles at *both* ends of a connecting line; they may be accepted as not influencing the original AND/OR logic. Wanted inversions are those which coincide with the NOT operations of the original logic and they may of course be accepted. Unwanted inversions, on the other hand, must be "cleaned out" by introducing an inverter which, in effect, cancels the unwanted inversion. The DeMorgan equivalent symbols may now be replaced.

Example: The expression

$$Z = \bar{A} \cdot (B + \bar{C})$$ 4.7(22)

will be implemented with both NAND and NOR circuits. Figure 4.7c shows successively the diagram based on AND/OR symbols, the translation to NAND

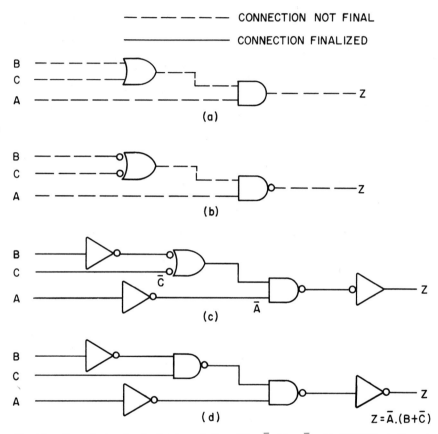

Fig. 4.7c   Example showing the implementation of $Z = \bar{A} \cdot (B + \bar{C})$ with NAND logic through a graphical transformation. (a) Beginning with AND/OR symbols; (b) adding of inverting circles for NAND operations; (c) adding of inverters; (d) final diagram using basic symbols

symbols, the addition of inverter to cancel unwanted inversion and the final diagram. Figure 4.7d shows the same steps using NOR symbols.

Logic Redundancy

Logic expressions based on minterms or maxterms previously described often are necessarily complex. For example, the following,

$$
\begin{array}{ll}
A & 0101 \\
B & \underline{0011} \\
Z & 1111
\end{array}
\qquad 4.7(23)
$$

would be defined by the minterm expression

$$
Z = \overline{A} \cdot \overline{B} + A \cdot \overline{B} + \overline{A} \cdot B + A \cdot B \qquad 4.7(24)
$$

whereas it is obvious that $Z = 1$ suffices to define the logic. Another example:

$$
\begin{array}{lll}
A & 0101 & 0101 \\
B & 0011 & 0011 \\
C & \underline{0000} & \underline{1111} \\
Z & 0001 & 0001
\end{array}
\qquad 4.7(25)
$$

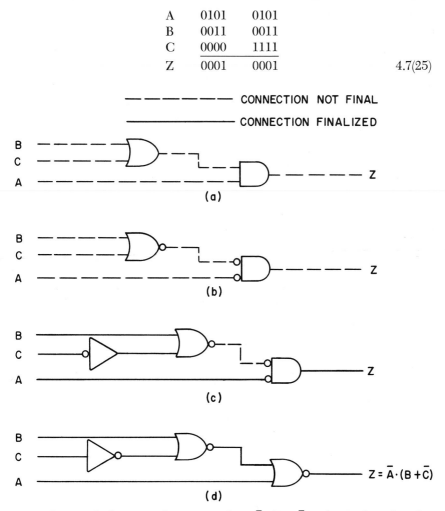

Fig. 4.7d Example showing implementation of $Z = \overline{A} \cdot (B + \overline{C})$ with NOR logic through a graphical transformation. (a) Beginning with AND/OR symbols; (b) adding of inverting circles for NOR operations; (c) adding of an inverter; (d) final diagram using basic symbols

could be expressed as

$$Z = A \cdot B \cdot \bar{C} + A \cdot B \cdot C \qquad\qquad 4.7(26)$$

but

$$Z = A \cdot B \qquad\qquad 4.7(27)$$

suffices to express the logic as can be verified by inspection of the tabulation 4.7(25), because the output is "1" whenever both A and B = "1" *regardless* of the value of C.

Unnecessary inputs in a logic expression is called logic redundancy. The rule for eliminating redundancy is that an input can be taken out from a set of minterms if it appears in those minterms with all possible combinations and if the remaining inputs in the minterms are constant. In the 4.7(24) example A, B could be eliminated because they appeared with their four possible values in four minterms. In 4.7(26), C can be eliminated because it appeared in two minterms with the two possible values of C = 0 and C = 1.

Logic Simplification Through Logic Maps

Some cases of logic redundancy are easily seen by inspection of the logic table as in the examples given previously. However, many cases of redundancy are difficult to recognize. The most frequently used tool for aiding in finding redundant terms is a graphical method known as the logic map, or Karnough map.

The logic map consists of an array of squares with each square corresponding to a column of the logic table. The best way to show how the map is constructed is to number the columns of the logic table starting from left to right. This is illustrated for three-inputs in Figure 4.7e, which shows the

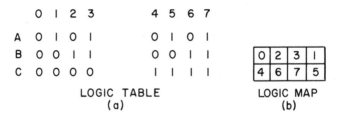

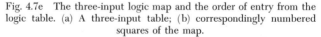

Fig. 4.7e   The three-input logic map and the order of entry from the logic table. (a) A three-input table; (b) correspondingly numbered squares of the map.

logic map array for three inputs with numbers to indicate the correspondence to the logic table. (This is only one of several forms of constructing a map.) Figure 4.7f then shows how the values for A, B, C would be entered onto the map.

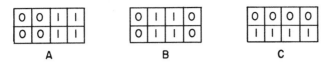

Fig. 4.7f   The three input maps resulting when each input row (A, B, C) of the table is entered onto the map

The maps of Figure 4.7f are fixed, since they correspond to the input portion of the logic table. The true implementation of the map occurs when the *output* portion of the table (the "Z" row) is mapped. For example, the previously given example of logic redundancy,

$$
\begin{array}{lll}
A & 0101 & 0101 \\
B & 0011 & 0011 \\
C & 0000 & 1111 \\
\hline
Z & 0001 & 0001
\end{array}
$$

4.7(25)

results in the map of Figure 4.7g.

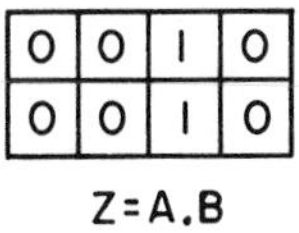

0	0	I	0
0	0	I	0

Z = A.B

Fig. 4.7g   Map resulting from the entry of Z = 0001 0001

The rule for the use of the map of Figure 4.7g is simply this: Any two *adjacent* squares with "1s" represent *an input* which can be eliminated. This is recognized by inspecting the corresponding area in the input maps on Figure 4.7f. It is seen that A, B have the constant value of 1 in this area, but C has its two possible values of 0 and 1. Note that squares on the sides of the map are considered contiguous as shown in Figure 4.7h.

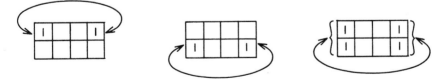

Fig. 4.7h   Illustration of contiguous squares based on "wraparound"

Further, any *four* squares with "1s" and which are symmetrical, i.e., forming either a square or rectangle, eliminate *two inputs*. Figure 4.7i gives examples together with the value of the simplified logic expressions.

A type of simplification for which the map is especially useful is shown in Figure 4.7j. Here, a "1" is shared between two different pairings and

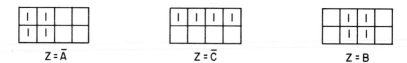

Fig. 4.7i   Examples of the elimination of two inputs by the grouping of four squares. Resultant expression is obtained by inspection of Figure 4.7e.

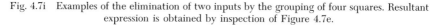

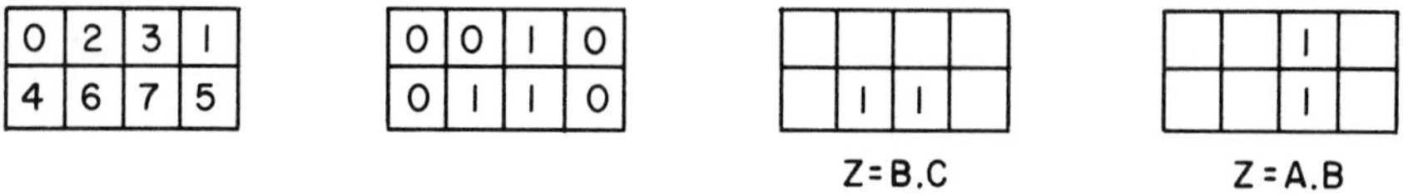

Fig. 4.7j   Example showing the sharing of a square between two groups. Square-7 is shared between B · C and A · B

expressions. This is an example of the rule that areas defining logic simplification can overlap.

Figure 4.7k shows how a four-input map can be constructed. Figure 4.7l gives the four input maps, and Figure 4.7m gives examples of the elimination of one input through the grouping of two squares.

Figure 4.7n gives examples of the elimination of two inputs by the grouping of four squares, and Figure 4.7o shows the elimination of three inputs through eight squares.

	0	1	2	3	4	5	6	7	8	9	10	11	12	13	14	15
A	0	1	0	1	0	1	0	1	0	1	0	1	0	1	0	1
B	0	0	1	1	0	0	1	1	0	0	1	1	0	0	1	1
C	0	0	0	0	1	1	1	1	0	0	0	0	1	1	1	1
D	0	0	0	0	0	0	0	0	1	1	1	1	1	1	1	1

LOGIC TABLE
(a)

0	2	3	1
8	10	11	9
12	14	15	13
4	6	7	5

LOGIC MAP
(b)

Fig. 4.7k   Four-input map and order of entry. (a) Four-input logic table with numbered columns; (b) corresponding squares on map

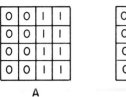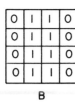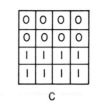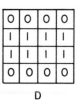

Fig. 4.7l   The input maps for the four-input (A, B, C, D) logic table

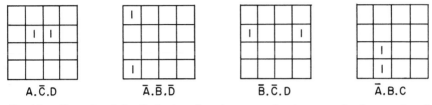

Fig. 4.7m   Examples of the elimination of one input on a four-input map by the grouping of two squares. Simplified expression is obtained by inspection of Figure 4.7l.

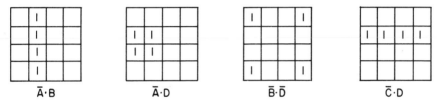

Fig. 4.7n   Example of the elimination of two inputs from a four-input map by the grouping of four squares. Resulting expression is obtained by inspection of input maps in Figure 4.7l.

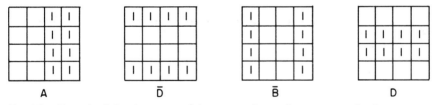

Fig. 4.7o   Example of the elimination of three inputs from a four-input map by the grouping of eight squares. Resulting expression is obtained by inspection of input maps in Figure 4.7l.

Logic Simplification With Boolean Algebra

Boolean algebra is the class of mathematics used to manipulate logic expressions. This section will give a few of the theorems of this algebra and show how it can be used to simplify logic and cast it into different forms. The particular theorems are easy to remember, since they for the most part parallel familiar arithmetic and conventional algebraic operations.

The following Boolean algebra equivalents are stated:

$$1 \cdot A = A \qquad\qquad 4.7(28)$$
$$0 \cdot A = 0 \qquad\qquad 4.7(29)$$
$$1 + A = 1 \qquad\qquad 4.7(30)$$
$$0 + A = A \qquad\qquad 4.7(31)$$
$$A + \overline{A} = 1 \qquad\qquad 4.7(32)$$
$$A + A = A \qquad\qquad 4.7(33)$$
$$A \cdot B + A \cdot C = A \cdot (B + C) \qquad\qquad 4.7(34)$$

To show the use of these relationships for logic simplification, consider the expression defined by the tabulation 4.7(25):

$$
\begin{array}{lll}
A & 0101 & 0101 \\
B & 0011 & 0011 \\
C & \underline{0000} & \underline{1111} \\
Z & 0001 & 0001
\end{array}
\qquad 4.7(25)
$$

The minterm expression is

$$Z = A \cdot B \cdot \overline{C} + A \cdot B \cdot C \qquad\qquad 4.7(26)$$

By 4.7(34) above, this can be "factored":

$$Z = A \cdot B \, (C + \overline{C}) \qquad\qquad 4.7(35)$$

By 4.7(32) above, this becomes

$$Z = A \cdot B \cdot 1 \qquad\qquad 4.7(36)$$

And by 4.7(28), it becomes

$$Z = A \cdot B \qquad\qquad 4.7(27)$$

Consider another expression:

$$Z = A \cdot B \cdot C + A \cdot \overline{B} \cdot C + A \cdot \overline{B} \cdot \overline{C} \qquad\qquad 4.7(37)$$

This is simplified by repeating the middle term, which can be done without changing the value of the logic, based on 4.7(33):

$$Z = A \cdot B \cdot C + A \cdot \overline{B} \cdot C + A \cdot \overline{B} \cdot \overline{C} + A \cdot \overline{B} \cdot C \qquad 4.7(38)$$

Then, using 4.7(34),

$$Z = A \cdot C \, (B + \overline{B}) + A \cdot \overline{B} \, (\overline{C} + C) \qquad\qquad 4.7(39)$$

which becomes, by 4.7(28) and 4.7(32),

$$Z = A \cdot C + A \cdot \overline{B} \qquad\qquad 4.7(40)$$

This can be placed in still another form by again using 4.7(34):

$$Z = A \cdot (C + \overline{B}) \hspace{3cm} 4.7(41)$$

It might be said that the repeating of the term $A \cdot \overline{B} \cdot C$ in the original expression is the equivalent of the overlapping of areas on the logic map.

Chapter V

CONTROL BOARDS AND RECEIVER DISPLAYS

R. W. Borut, S. P. Jackson,
B. G. Lipták and F. D. Marton

CONTENTS OF CHAPTER V

	INTRODUCTION	534
5.1	CONTROL ROOMS AND PANEL BOARDS	535
	Control Rooms	536
	Panel Types	537
	Flat Panels	537
	Breakfront Panels	538
	Consoles	538
	Panel Layout	538
	Large-Case Instruments	539
	Miniature Instruments	540
	High-Density Instruments	542
	Graphic Panels	543
	Semi-Graphic Panels	546
	Full Graphic Panels	546
	Variations of Graphic Panel Constructions	546
	Back of Panel Layout	550
	Panel Piping and Tubing	550
	Panel Wiring	553
	Power Distribution	555
	Battery Back-Up	556
	Wiring and Terminal Identifications	558
	Panel Materials of Construction	560
	Panel Inspection	561
	Panel Shipment	564
	Panel Specifications	564
	Conclusions	565
5.2	INDICATORS	566
	Terminology of Analog Indication	566
	Indication of Measurements	567
	Fixed-Scale Indicators	568
	Movable-Scale Indicators	571
	Parametric Indication	572
	Digital Indicators	573

5.3 RECORDERS   575
        Charts and Coordinates   576
        Circular Chart Recorders   576
        Strip Chart Recorders   578
        X-Y Recorders   579
        Multiple Recorders   579
        Recording Means   580
        Event Recorders   581
        Miniature Recorders   581
        Service Recorders (Portable)   582
        Photographic and Electrostatic Recording   582
        Magnetic Recording   584
        Digital Recorders (Printers)   584
        Facsimile Recording   584
        Holograms   584
        Video Tape Recording   584
5.4 STAND-BY POWER SUPPLY SYSTEMS   587
        Summary of Features   589
        Power Failure Classifications   591
            Source Failure   592
            Equipment Failure (Inverter)   595
            Common Bus Branch (Load) Failure   598
        System Components   600
            Rotating Equipment   601
            Batteries   603
            Battery Chargers   603
            Static Inverters   603
            Bus Transfer Switches   608
            Protective Components   609
        Stand-by Power Supply Systems   610
            Multicycle Transfer System   610
            Subcycle Transfer System   610
            "No-Break" Transfer System   611
        System Redundancy   612
        Specifications   613

This chapter deals with instrumentation systems, their organization and features as they appear in the centralized control rooms of modern chemical plants. The emphasis here is not on the features or on the operating principles of the various components, but on the overall design of control rooms, control panels, and on the display capabilities of the various receiving instruments, particularly of indicators and recorders.

The last section of this chapter is devoted to the subject of stand-by power supply systems as they apply to electronic or computerized plants, where the user could not afford to allow the temporary failure of electric power to shut down the automated controls of the plant.

Chapter IV discussed the features and operation of controllers, function generators and logic devices, and Chapter VI will concern itself with the subject of signal transmission techniques. Between these three chapters the reader should find most of the design information that he needs for the control loop components after the measuring element and prior to the final control element.

The method of displaying the various measurements and the design of operators' panels are areas where not scientific rules, but individual preferences play an important part. In this area of human engineering, some trends occur due to similar motivations, as in the field of fashion, while others have sound reasons in the new developments of our modern technology. As to what the apparent trends are in the early seventies, one might mention the continuing tendency toward high-density displays, with the operators' attention being attracted by special deviation indicators or the gradual increase of digital readouts, while the analog displays still used seem to become increasingly vertical and, when also recorded, this is frequently done on a trend basis instead of on a continuous basis.

The packaging in which a civilization likes to receive its information in many ways is indicative of society's outlook and, as such, is a fascinating area to observe. The terms "diversity" and "high density" are applicable not only to the manner in which instrumentation is packaged, but also to other aspects of our life. In the design of control rooms, more and more attention is given to aesthetic considerations and operators' comfort.

MOUNTED ON MAIN
CONTROL PANEL

MOUNTED ON LOCAL
CONTROL BOARDS

MOUNTED ON REAR
OF MAIN PANEL

MOUNTED ON REAR
OF LOCAL BOARDS

5.1 CONTROL ROOMS AND PANEL BOARDS

Cost:　　　　　　　　　　$500 to $1,000 per linear foot.

Partial List of Suppliers:°　AETRON, Div. of Aerojet-General Corp.; Customline Control Products Inc.; Custom Engineering Co.; Honeywell Inc., Special Systems Div.; Lake Erie Electric Mfg. Co.; Mercury Company Inc.; Monitor Panel Co.; Scam Instr. Corp., Panellit Div.; Swanson Engineering Co.; Sycamore Engineering & Mfg. Co.

Early control panels, from which simple or batch type processes were operated, required very little engineering and design for field fabrication. Most control boards contained several panels of large-case recorders or controllers, a few alarm units, and a potentiometric temperature indicator. The instrument lead lengths were short, and the control philosophy uncomplicated.

Today's integrated units, with their huge through-puts and minimum holding times, require centralized control panels to keep them operating at optimum level. The development of sophisticated pneumatic and electronic instrumentation has kept pace with the growth of the Gargantuas they control. This sophistication has engendered the growth of a number of control panel manufacturers specializing in fabricating, piping and wiring complete control

° As most process units are constructed by members of building trades unions, prefabricated control boards will not be permitted on the site unless they have labels certifying that they were constructed by members of the required unions. The required labels are: United Association of Journeymen and Apprentices of the Plumbing and Pipe Fitting Industry of the United States and Canada, (UA); and the International Brotherhood of Electrical Workers (IBEW).

centers. The complexity of these control centers has made field fabrication unpractical.

The design of a control panel is dependent upon many considerations, such as control room layout, panel profile, instrument type, complexity of process, sophistication of control philosophy, and aesthetics.

Control Rooms

The control room must be designed so that only those operations necessary for the direct control of the plant are performed there. The operators must not be distracted by unassociated functions. The room should have limited access and *not* act as a passageway. The panels must either be arranged so that unauthorized personnel cannot tamper with the panel instruments or with the auxiliaries mounted on the rear. Figure 5.1a shows a typical control room layout.

Air conditioning and room pressurization is now the rule. Aside from operator comfort, a constant ambient temperature at the instruments will minimize the possibilities of drift. Room pressurization is normally used where

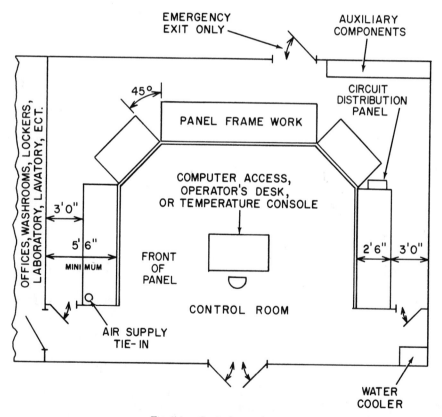

Fig. 5.1a   Control room layout

the plant atmosphere is explosive or flammable, and it is achieved by forcing fresh air through ducts, from a safe area. This permits the reduction of area classification from either "hazardous" or "semi-hazardous" to "non-hazardous," with commensurate savings in instrument and installation costs. (Section 10.11 of Volume I gives details on area classification.)

The illumination in the control room must be of a level consistent with close work. The panel should average 50 foot-candles across its face. The back of the panel area should be lighted to 30 foot-candles.

The most advantageous ratio of panel length to control room area is obtained by bending the panel to a ∪ shape. Right angle bends of the panel, as opposed to 45-degree bends, should be avoided. The slightly increased panel length that could be gained in this manner, is negated by the interferences to opening instrument doors and/or withdrawing the chassis. An operator can monitor a greater length of panel if it bends around him.

Panel Types

There are three basic types of control panel shapes, straight, breakfront and console. Each type has its family of variations. The dimensions shown in the various illustrations are only typical, and instrument heights and spacing must be adjusted to suit the particular manufacturer and application.

Flat Panels

The type of panel shown in Figure 5.1b is least expensive, easiest to construct, and the simplest to design of all other types. The straight vertical

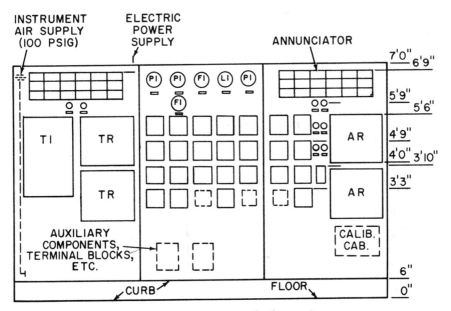

Fig. 5.1b Typical layout for flat panel

plane of the panel allows an orderly layout of tubing, electrical duct work and miscellaneous equipment. Instruments and auxiliary components can be arranged so that all are accessible for maintenance and calibration. The lower row of instruments, approximately 3 ft 3 in. from the floor, should be utilized for recording or indicating instruments, as this elevation is rather inconvenient for operation. As the maximum of four horizontal rows of miniature recorders and controllers can be used, this type of panel will require more control room space than some other, even higher density layouts.

Breakfront Panels

Breakfront panels allow greater utilization of the board's front plane, because the instruments located in the lower rows are swung upward to a convenient height (Figure 5.1c). The top portion of the panel is swung downward to an angle normal to the line of sight, allowing better visibility. The additional rows of instrumentation obtained with this layout cuts the overall panel length requirement. This higher instrument density, however, significantly reduces the space for maintenance, and for mounting of auxiliary components in the back of the panel.

Consoles

Consoles appear to be aesthetically very pleasing to most people. They are often used with high-density instrumentation in control rooms that are limited in size. Normally, their length is set by the operator unit responsibility limits, i.e., one operator per console and one console per operator. The lengths vary between 4 and 12 ft.

Auxiliary equipment such as transducers and pressure switches can be installed inside the console cabinet, but the arrangement of the flush-mounted instruments as shown in Figure 5.1d severely limits the available free space. Maintenance of instruments within the console can therefore become a real problem.

Consoles are often coupled with a conventional flat "back-up" panel, which contains the larger sized instruments and auxiliary components. As the average operator is about 5 ft 9 in. tall, the "see over" console should not rise above 5 ft 0 in. This will allow the operator to see the back-up panel over the top of the console.

Panel Layout

Currently available are three basic instrument sizes, which may be fitted to almost any panel configuration and type. From the large-cased conventional type, nominally 18 in. wide by 24 in. high, down through the miniature type, nominally 6 in. square, to the high-density type, 2 in. wide by 6 in. high, spans the range of available sizes and shapes.

The same factors which determined the panel type, very often also determine the instrument type.

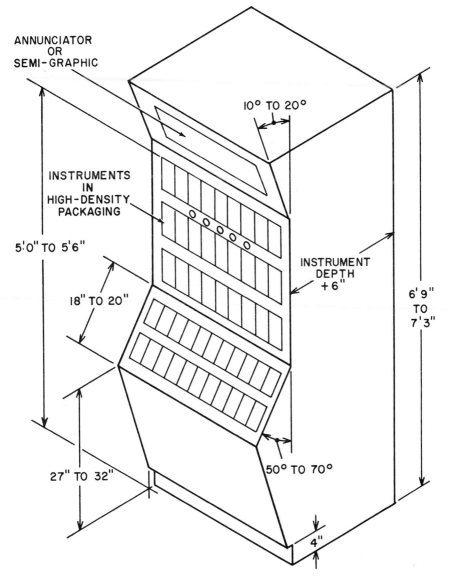

ANNUNCIATOR
OR
SEMI-GRAPHIC

10° TO 20°

INSTRUMENTS
IN
HIGH-DENSITY
PACKAGING

5'0" TO 5'6"

INSTRUMENT
DEPTH
+6"

6'9"
TO
7'3"

18" TO 20"

27" TO 32"

50° TO 70°

4"

Fig. 5.1c   Breakfront panel structure

Large-Case Instruments

Large conventional instruments can be conveniently operated when mounted two rows high (Figure 5.1e). They do not permit a great deal of flexibility in layout, but they are rugged and are suitable for almost any outdoor location as local control stations. The control systems for most of these applications are not apt to be complex or to have numerous instrument components.

Miniature Instruments

The "miniature" instruments are the most widely used process control instruments. They allow a moderate reduction in panel length (Figure 5.1f).

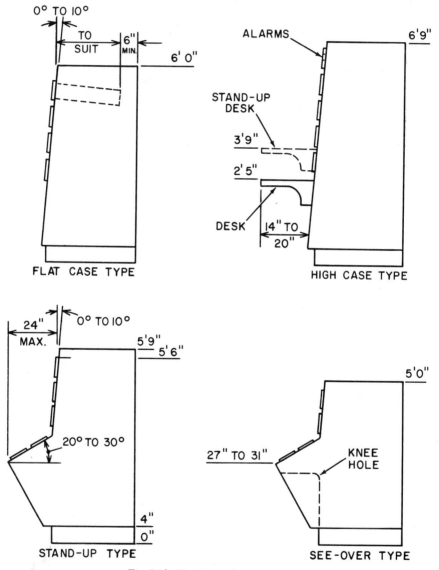

Fig. 5.1d Variations of console shapes

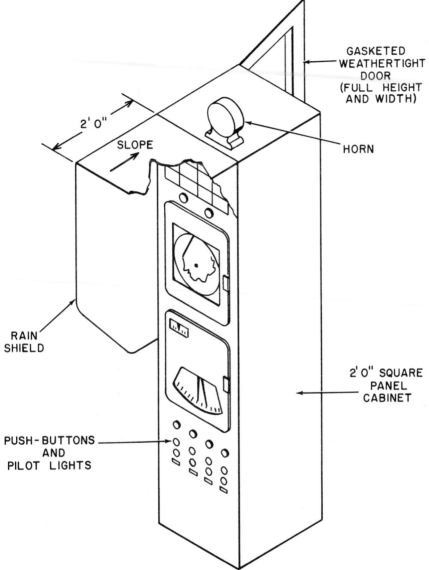

Fig. 5.1e   Local panel cabinet with large case instruments

This is sufficient for most indoor installations. They can be utilized in almost all configurations of breakfront or console designs, but there are some mounting angle limitations which should be kept in mind in designing the panel.

For estimating the length of control panels, a horizontal spacing of 9

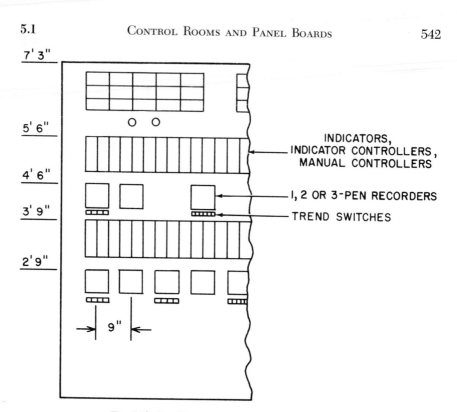

Fig. 5.1f Panel layout using miniature instruments

in. between vertical centerlines of miniature instruments may be used. Ten to twelve inches may be used between horizontal centerlines. Instrument manufacturers have not fully standardized overall dimensions, cutouts, connection locations, etc. Each instrument must therefore be checked for the manufacturer's recommended spacing and installation recommendations.

High-Density Instruments

There are several high-density type miniature instrument designs available on the market. They may be mounted in groupings to suit a particular processing unit, as shown in Figure 5.1g, or in long rows to condense panel length, as illustrated in Figure 5.1h.

This type of instrument layout requires additional space on the rear of the panel for auxiliary equipment. Additional equipment is sometimes mounted on the wall behind the panel or in a peripheral equipment room remote from the panel. High-density instruments tend to have longer chassis, requiring about six additional inches in panel depth. Some of these instruments require amplifiers or converters, which must be mounted in close proximity to the primary instrument.

Some high-density instrument designs do not include a similar size recorder. In this instance a standard two- or three-pen 4-in. strip chart recorder from the miniature (6 by 6-in.) line may be utilized (Figure 5.1f). The use of trend recorders rather than permanently wired ones will considerably reduce the overall recorder requirements.

Graphic Panels

A graphic control panel pictorially depicts a simplified flow diagram of the processing unit and of its control philosophy. The most common materials used for this depiction are of colored plastic or melamine. Back-painted translucent vinyl or other plastics are also used to some extent. The lines and symbols may be affixed to a removable steel, aluminum, or plastic plate.

The extra expense of a graphic panel may be justified for some of the following reasons:

1. To enable the panel operator to visualize a complex process flow pattern.
2. To make understandable a sophisticated control philosophy with complex interrelationships between variables.

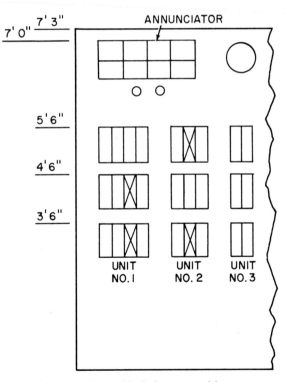

Fig. 5.1g   Grouped high-density panel layout

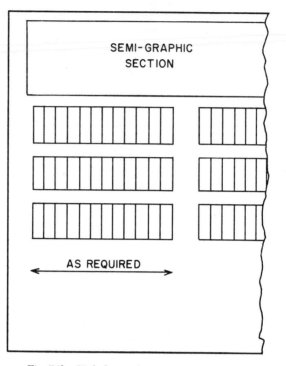

Fig. 5.1h   High-density layout without grouping

3. For the training of new operators.
4. For aesthetic enhancement of the control room (a motivation seldom admitted but frequently present).

Although the industry has not yet standardized shapes or dimensions for equipment symbols, there is some agreement in utilizing and adopting some common symbols for graphic layout purposes, such as the ones shown in Figure 5.1i. The dimensions shown here are typical, and can be adjusted to suit, as long as they are consistent throughout the panel. Symbols representing equipment, such as furnaces, vessels, drums, pumps and compressors, are usually shown in silhouette. Internals are shown only when necessary to insure complete understanding of how the equipment functions.

In general, the graphic display shows the process flow from left to right in a step-by-step sequence, where fresh feed enters from the left and the product stream exits to the right. Only the process streams and those utilities which are required for clarifying unit operations should be shown. Similarly, only important local instrumentation need be shown.

The flow diagrams should be laid out with horizontal and vertical lines. The pattern and its density should be as consistent as possible. Vessels should

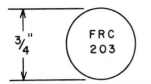

THE SYMBOL IS WHITE IF THE CONTROLLER
IS PANEL MOUNTED.
FOR LOCAL CONTROLLERS THE SYMBOL IS
THE SAME COLOR AS THAT OF THE
PROCESS LINE.

INSTRUMENT SYMBOL

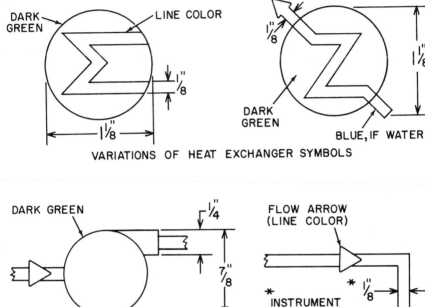

VARIATIONS OF HEAT EXCHANGER SYMBOLS

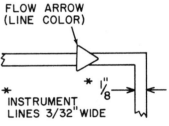

PUMP SYMBOL

PROCESS FLOW
LINE SYMBOL

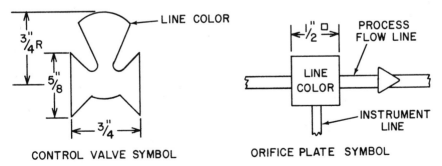

CONTROL VALVE SYMBOL   ORIFICE PLATE SYMBOL

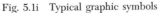

Fig. 5.1i Typical graphic symbols

utilize the middle three-fifths of the graphic section, with a common bottom line where practical. Equipment in similar service should utilize similar positions. For example, reflux drums would be aligned along the upper three-fifths division line; reboilers would rest atop the lower two-fifths line, etc. This leaves the top and bottom fifth of the panel for long horizontal lines. The diagram should be designed so that as few lines as possible cross each other. Where a crossing is necessary the horizontal line should break.

Semi-Graphic Panels

Semi-graphic panels are either flat or of the breakfront type. The top portion of the panel is utilized by a flow diagram of the process (Figures 5.1h and j). The location of each instrument should be so selected that it is installed as directly underneath its corresponding symbol in the graphic diagram as is feasible. The instrument item number is clearly noted in both locations. In some instances, alarm or running lights are also installed within the graphic section in the appropriate location. The instrument density is reduced with this type of panel, because of the desirability of locating the instruments in relative proximity to their graphic symbols.

Full Graphic Panels

Full graphics cover the full front face of the control panel with the process diagram as shown in Figure 5.1k. Instruments are positioned in the panel at the point which corresponds to their measurement or control location. As in all other panels, the instruments may be aligned in horizontal and vertical rows for ease in conduit, duct and tube layout. The instrument density for this type of panel is extremely low and can vary from one to three instruments per linear foot of board length. This significantly extends the overall length of panel.

Variations of Graphic Panel Constructions

In addition to the glued-on plastic type graphic panels, there are other designs available. The degree of difficulty in making changes to the graphic diagram should be considered if there is a likelihood of process revision in the future. The aforementioned plastic strips are relatively easy to change.

If the lines and symbols are painted on the panel, changes can be made easily when the panel is new, but as the paint ages and fades, colors become difficult to match.

A flow diagram can also be drawn full scale on reproducible paper and placed in a semi-graphic frame. It may be held in place and protected with a glass or clear rigid plastic cover. This method of graphic presentation is simple to change by revising the drawing master and acquiring a new print. This type of panel is not as aesthetically pleasing as other types because it is usually black and white rather than color-coded.

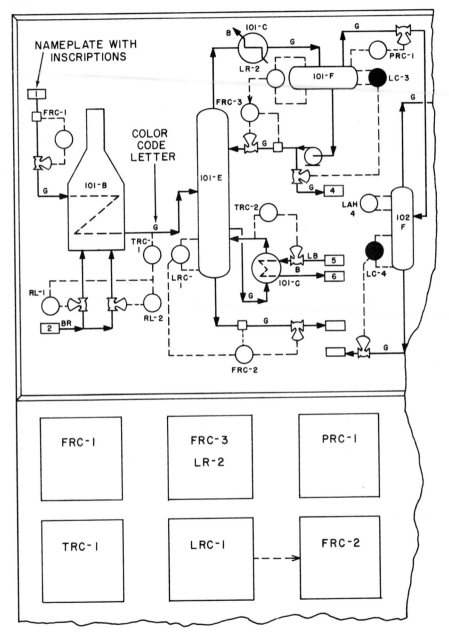

Fig. 5.1j Semi-graphic panel with miniature instruments

Another method of graphic process display is by the use of automatic slide projectors. This system is used with a rear-projection type, translucent screen. In order to gain the necessary focal length distance, a set of mirrors is strategically placed on the rear of the panel. This approach to graphics

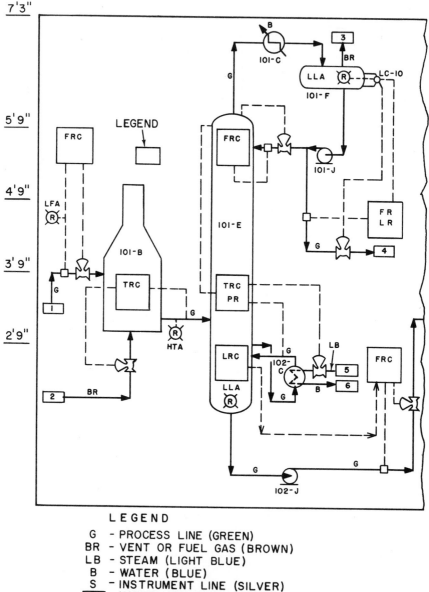

LEGEND

G – PROCESS LINE (GREEN)
BR – VENT OR FUEL GAS (BROWN)
LB – STEAM (LIGHT BLUE)
B – WATER (BLUE)
S – INSTRUMENT LINE (SILVER)
NO. – NAMEPLATE WITH INSCRIPTION

Fig. 5.1k   Full graphic panel layout with miniature instruments

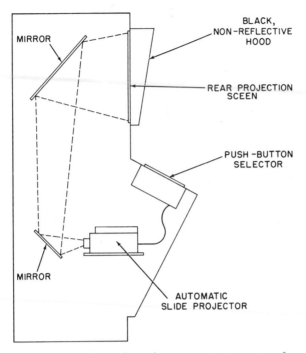

Fig. 5.11 Panel console with rear projection system for graphic presentation

allows a great deal of flexibility. As many slides can be prepared as desired, showing the process or instrument systems or sub-systems in color and in as much detail as is deemed necessary. Slides can be prepared which give instructions for emergency procedures, unit operating parameters, and optimum set points, for various product choices. One serious drawback to the system is that the bright light in the control room tends to wash out the image and reduce legibility. Therefore the lighting level must be reduced, and the fixtures must be so located as to minimize reflections on the screen. In addition, the screen should be furnished with a black shield to reduce side lighting (Figure 5.11).

Yet another type of graphic panel is fabricated by back-engraving a sheet of clear plastic. The plastic lines so engraved, are then filled with the selected colors. This type of panel is easy to maintain, as only the smooth surface is exposed.

Changes can be made by re-engraving and by refilling. This is extremely difficult, and even when expertly done, tell-tale signs of the change remain.

The front-engraved and enamel-filled graphic line work allows vivid coloring and sharp line work, making this panel one of the most impressive and pleasing to the eye. It is not feasible to make changes on this panel, and maintenance is somewhat more difficult, as the engraved lines tend to fill with dust.

Back of Panel Layout

In order for the overall panel design to be well done, sufficient attention must be given to the back of panel arrangement. It must be verified that there is sufficient room to run the conduits, ductworks, air headers, tube leads, etc. There must also be enough room to mount the various switches, relays, converters, amplifiers and other auxiliary components. A back of panel profile will sometimes assist in this (Figure 5.1m). Auxiliary equipment must be so located, that its connecting wiring or tubing does not obstruct the maintenance or calibration of, and accessibility to, the other instruments.

Panel Piping and Tubing

Most panel manufacturers stock an acceptable line of industrial grade tubing, pipe and fittings. One of the most commonly used tubing materials (and sizes) behind the panel is copper ($\frac{1}{4}$ in. OD, 0.030 in. wall thickness, ASTM B68). It is relatively resistant to corrosion, readily available, and rigid enough to require only a minimum of support, yet it is sufficiently ductile to bend to precise measurements. It is available with a tightly extruded PVC sheath for corrosive atmospheres, and with cadmium or tin plating for damp locations.

Another commonly used tubing material is aluminum ($\frac{1}{4}$ in. OD, 0.032 in. wall thickness, federal specification WWT-700/4c). Aluminum tubing offers good resistance to the attack of many types of chloride or ammonia atmospheres. It is somewhat softer to work with than copper and requires significantly more support. It has a tendency to work-harden, which makes it particularly susceptible to vibration failures. This tubing is also available with a PVC sheath.

The most frequently used plastic tubing material (and size) is polyethylene ($\frac{1}{4}$ in. OD, 0.040 in. wall thickness). This tubing should be of a material which has been environmental stress tested in accordance with ASTM D1693. Although polyethylene is markedly less expensive than the metallic tubes, additional costs are incurred in the use of an extensive support network which consists of either plastic, slotted or sheet metal duct. The ducting should extend to within one or two inches of the connected instrument. Unsupported lengths are to be avoided because the plastic tubing has a tendency to kink and also because it is almost impossible to run this soft tubing neatly. Tubing is available and may be *color coded* in colors conforming to the ISA Recommended Practice ISA RP7.2.

Other tubing materials are available, such as stainless steel, nylon, polyvinyl, rubber, and glass, but they are not as commonly used.

Fittings for all the tubing materials mentioned are available. The industry

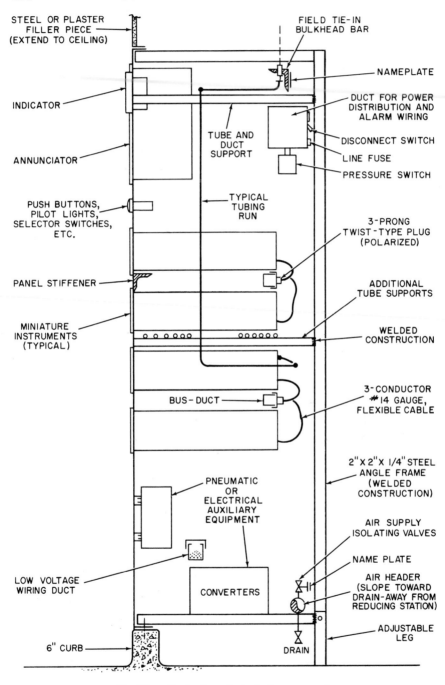

Fig. 5.1m   Back of panel—typical arrangement

has generally standardized on compression type tube fittings for panel work rather than flared, soldered, or other specialty types.

Pipe material should be seamless red brass, threaded in accordance with USAS B2.1, and soldered or brazed. The fittings should be rated to at least 125 PSIG.

Air supply isolating valves should be either straight-line needle type or packless diaphragm two-way type.

The panel air supply should consist of two sets of parallel-piped reducing valves and filters. Each set should be sized and manifolded so that either set can supply the panel requirements. In general, 0.5 SCFM of air flow per air user is sufficient. The main air header must be sized so that the pressure drop to the furthest instrument is under 1 PSID.

The use of a plugged tee and isolating valve, or a three-way air switch allows for calibration of instruments or components without disconnecting them or losing the measurement signal to other loop components (Figure 5.1n).

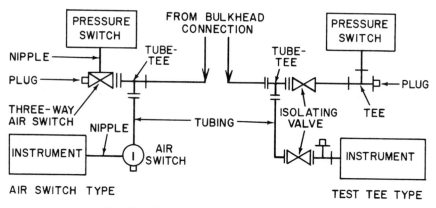

Fig. 5.1n   Typical test or calibration connections

The isolating valve need only be used when the measurement signal is branching to another instrument, or, when the instrument is not furnished with an integral spring loaded air cut-off valve, as in plug-in type devices.

The location of the instrument air supply tie-in to the board is usually noted on the panel drawings, as in Figure 5.1b.

The pneumatic signal lead tie-in location usually requires only a general description, as there is enough flexibility in the tubing installation for adjustment. Notes such as "top of panel" or "bottom of panel" are usually sufficient.

Each tubing termination for field leads should be identified with the item number of the instrument it serves and its function. For example FRC-10-V might indicate that the field lead terminates at FRC-10 control valve, while FRC-10-T would mean that the air signal tube is received from the so-identified flow transmitter.

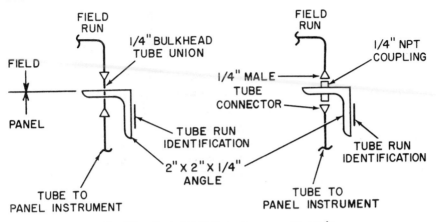

Fig. 5.1o Typical field tie-ins to pneumatic panels

There are several different types of panel-field, interface connections, some of which are shown in Figure 5.1o. To allow for future changes or additions at least 10 percent additional tie-in points should be provided.

Panel Wiring

The first step in deciding what equipment to use is to determine the nature and the degree of hazard. The National Electrical Code[1] describes hazardous locations by Class, Group and Division. The Class defines the physical form of the combustible material mixed with air:

> Class I—Combustible material in the form of a gas or vapor.
> Class II—Combustible material in the form of a dust.
> Class III—Combustible material in the form of a fiber, such as textile flyings.

The Group subdivides the Class:

> Group A—Atmospheres containing acetylene.
> Group B—Atmospheres containing hydrogen, gases or vapors of equivalent hazard, such as manufactured gas.
> Group C—Atmospheres containing ethyl ether vapors, ethylene, or cyclopropane.
> Group D—Atmospheres containing gasoline, hexane, naphtha, benzine, butane, propane, alcohol, acetone, benzol, lacquer, solvent vapors, or natural gas.
> Group E—Atmospheres containing metal dust, including aluminum,

---

[1]NFPA 70-1968 USAS C1-1968 "National Electrical Code 1968" Canadian Equivalent, CSA Standard C22.1—1966 "Canadian Electrical Code Part I"

magnesium and their commercial alloys, and other metals of similarly hazardous characteristics.

Group F—Atmospheres containing carbon black, coal or coke dust.

Group G—Atmospheres containing flour starch or grain dusts.

The Division defines the probability of an explosive mixture being present. For instance, a hazardous mixture is *normally* present in a Division 1 area, but will only be *accidentally* present in a Division 2 area.

In addition to knowing the area classifications, one should also be aware of the National Electrical Manufacturers Association (NEMA) terminology for classifying equipment enclosures.

NEMA 1	General purpose
NEMA 2	Drip tight
NEMA 3	Weatherproof
NEMA 4	Watertight
NEMA 5	Dust tight
NEMA 6	Submersible
NEMA 7	Hazardous (Class I, Groups A, B, C or D)
NEMA 8	Hazardous (Class I, Groups A, B, C or D)—oil-immersed
NEMA 9	Hazardous (Class II, Groups E, F or G)
NEMA 10	Explosionproof—Bureau of Mines
NEMA 11	Acid- and fume-resistant, oil-immersed
NEMA 12	Industrial

For additional details on this subject, refer to Sections 10.10 and 10.11 in Volume I.

Most panels are enclosed by or parallel to a wall, with a door on either or both ends to limit unauthorized access. Under such conditions, and when the area is general purpose and non-hazardous (Figure 5.1a), it is permitted to reduce the mechanical protection requirements for the wiring. Electronic transmission, power and signal wiring need not be enclosed in conduit or in thin-walled metallic tubing.

All of the wiring may be run in sheet metal or slotted plastic duct. The insulated wire may be run exposed, from the duct to the instrument, an inch or two without the necessity for a conduit nipple. Bare or exposed terminals, however, are *not* permitted.

Panels installed in hazardous or semi-hazardous areas must be installed in strict adherence to the National Electrical Code requirements. As the code does not allow much flexibility in these cases, this discussion will be limited to the non-hazardous applications, where the designer has some flexibility. For requirements in hazardous locations refer to Volume I, Sections 10.10 and 10.11.

Power Distribution

Instrument power supplies should be taken from a reliable source, with automatic switchover capability to an alternate power supply to be used upon failure of the main. The two typical stand-by power sources are a separate supply bus, fed from batteries or from a different source, and a steam- or gasoline-powered generator. A detailed coverage of back-up power supply systems is given in Section 5.4.

To reduce the cross-sectional area of the power feeders, it is often expedient to have the panel manufacturer furnish and mount a three-phase 440/208/120-volt transformer directly on the panel. Then only a 440-volt power supply is provided from the switchgear to the transformer (Figure 5.1p).

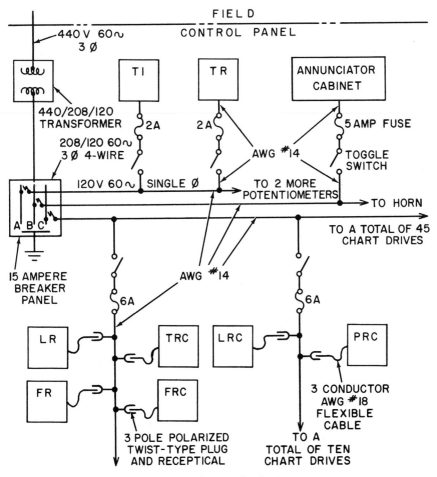

Fig. 5.1p   Typical power distribution system

A conventional lighting type circuit breaker panel may be installed on the back of the panel to provide the necessary circuit distribution. This permits the panel manufacturer to install the complete system and significantly reduces field tie-in time. Breakers, sized to trip above 15 amperes, and AWG #14 wire gauge are generally used. Three-wire power circuits having hot, grounded neutral, and ground leads are frequently used.

To avoid the possibility of overloading, circuits should be lightly loaded to approximately half their rated capacity, or, to a maximum of 850 volt-amperes.

The following groups of instruments will keep the load on 15-ampere circuits within acceptable limits.

- 4 potentiometer type temperature instruments
- 1 annunciator cabinet with horn
- 2 analyzer circuits (900 volt-amperes maximum)
- 45 miniature pneumatic recorder chart drives
- 5 miscellaneous auxiliary components, at 800 volt-amperes maximum
- 10 electronic instrument loops (500 volt-amperes maximum)
- 2 emergency trip circuits (500 volt-amperes maximum)

Secondary sub-circuits must be utilized so that a short or ground at one instrument does not trip out the 15-ampere circuit breaker but only the associated fuse, and so that each instrument (or group) can be isolated for maintenance or replacement. This isolation is accomplished with a fused disconnect device. The fuses must be coordinated so that it is not possible for the 15-ampere breaker to trip before a fuse blows. The fuse must also go before a significant voltage dip occurs. An exception to the individual fusing isolation rule is chart drive power to pneumatic instrumentation. Here, ten, or some similar number of drives, may be commoned to one fuse. A three-pronged, polarized, twist type plug can serve as a disconnect.

When instruments are internally fused, that fuse may be utilized, but the instrument circuits must be checked to verify that the fuse protects the complete chassis and not just a single critical component.

Battery Back-Up

There are several factors which must be considered in deciding whether a stand-by power supply is required. These are as follows:

1. The power source used has a history of failures with up to half an hour duration.
2. A unit which utilizes normally energized solenoids and/or relays which must be manually reset.
3. Flame safety system.
4. Extremely fast-acting process with electronic controllers in criti-

cal services, where a dip in power supply could send the unit into uncontrollable cycling.

5. Computerized process control system.

A typical uninterruptable power supply is the battery back-up system shown in Figure 5.1q. This system consists of a battery charger, a bank of

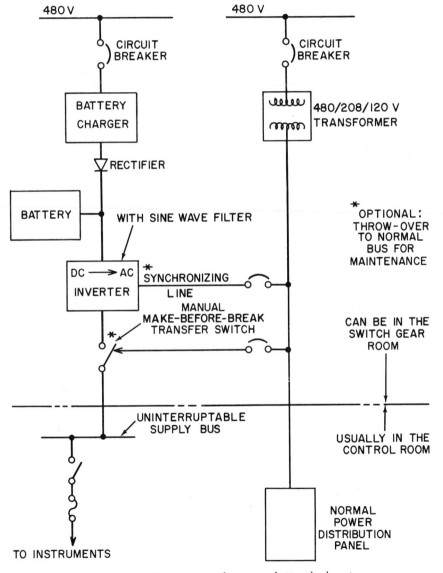

Fig. 5.1q   Stand-by power supply system—battery back-up type

batteries and an inverter. One of the main advantages to this system is that the AC power input phase does not have to be syncronized with the output phase. The instruments are normally powered directly through this system. Upon failure of the mains, the batteries, which have been floating on the charging current, start feeding the inverter.

The battery ampere-hour capacity should be sized using one or more of the following considerations.

1. Length of time of average power outage $\times$ $1\frac{1}{2}$.
2. Length of time plant will remain operable after power mains fail $\times$ $1\frac{1}{2}$.
3. Length of time to switch to alternative power supply $\times$ 2.
4. Length of time instruments will be required to bring about an orderly shutdown.

The charger must be sized so that it can simultaneously operate the unit and recharge the battery system after a discharge. A recharge time of eight hours is reasonable.

Circuit breakers and fuses downstream of the inverter must be sized and coordinated so that the available current will trip them out before a significant voltage disturbance could occur.

Wiring and Terminal Identifications

Most electronic control loops are of the "two-wire" type. This means that the locally mounted transmitter or control element does not require a separate power supply but takes its actuating force from the signal wires. Therefore the installation requires only a two-conductor cable for each transmitter or final control element.

To simplify field wiring and minimize overall field installation costs, auxiliary components are usually mounted on the panel. In this way the complex interconnecting wiring is installed by the panel manufacturer.

For flexibility in making loop changes and additions, one method of installation is the use of a centralized terminal block for each complete instrument loop on the back of the panel. In this system, each transmission and control loop is assigned a set number of terminals in the field tie-in junction box. Each group of terminals is identically marked, and the terminal marking strip carries the instrument loop number. Each component of the loop is then wired to this terminal as shown in Figure 5.1r. Spare terminals should also be included for future instrument components.

Terminal strips are available with white plastic or painted marking strips, suitable for pencil identification. At all electronic instrument field tie-ins, the junction blocks should be identified with the instrument loop number, its function, and its polarity in the following manner: FRC-10-T-(+) FRC-10-T-(−).

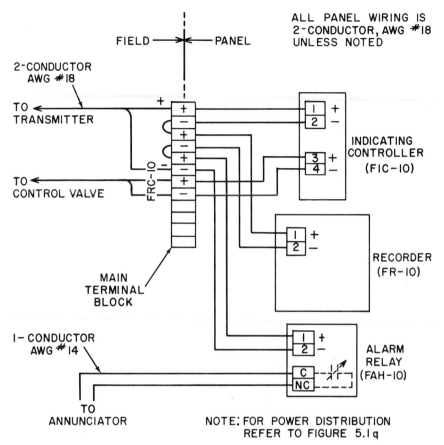

Fig. 5.1r   Typical wiring diagram for an electronic instrument loop

All other terminals are identified similarly, except that the terminal is marked with the instrument terminal designation instead of its polarity.

For equipment such as relays, switches, and other components without terminal designations, the terminal should duplicate the identification shown in the wiring diagrams, such as SV-10-VS (SV-1), or as annunciator 'A'-(S1-1).

Wire identification data should duplicate that shown on the terminal marking strips: instrument item number, function, and polarity. Function and polarity identification may be replaced with color-coded wire insulation. The *color code* to be effective must be simple, and it must be consistent. A typical color code is given in Table 5.1s.

In addition to the color coding the wire may be identified with the instrument item number by means of a preprinted marker tape or by an identified plastic sleeve. The sleeves are either sized to a snug fit over the wire insulation or are of the shrink type.

Table 5.1s
CONTROL PANEL WIRING COLOR CODE

Color of Wire Insulation	AC Service (120-volt, 60-cycle supply using AWG #14 wires)	DC Service (low voltage using 2-conductor AWG #18 wires)
Black	"A" phase—hot	Positive power, transmission or control signal
Red	"B" phase—hot	Negative power or control signal
Blue	"C" phase—hot	—
White	Neutral	Negative transmission signal
Brown	Annunciator common (H)	Annunciator common
Orange	Annunciator signal	Annunciator signal
Green	Ground	Ground
Gray	Miscellaneous interconnections and jumpers	

Wire terminations should be made with crush-type wire lugs. Special lugs are available for solid wire. A good lug to use is a flanged spade lug. This type combines the ease of installation of a spade lug, with the flanges holding the lug in place if the terminal screw becomes loose.

Where wire is run within a duct, it need not be laced. Although it gives a neater appearance lacing is time consuming and is not required where wires must be frequently added or rerouted.

Panel Materials of Construction

The materials used most frequently for the construction of panelboards are steel and various plastics. The advantages of steel are as follows:

Strength   At instrument cut-outs, where the panel is weakened, it requires less stiffening than other materials. Steel stands up better to the cantilever effect of the bezel supported, flush-mounted instruments.

Ease of Construction   Holes can be drilled or flame cut. Auxiliary equipment supports can be welded at any point. Bend-backs on straight panels and the breakfront shapes add significantly to panel stiffness.

Safety   When grounded, the panels offer an excellent path to ground if an instrument chassis or case becomes energized due to a short circuit or mistermination.

Attractive Finish   Long lengths can be finished in any color or hue desired, without visible seams.

Ready Availability   Sheet steel of good quality and with a minimum of surface pitting is readily available.

Disadvantages of steel panels include the following:

High Susceptibility to Corrosion   In corrosive atmospheres, the finish is only as durable as the paint.

Difficulty in Adding Cut-outs   Once the panel has been constructed and instruments are installed, cutting the steel for further instruments scatters steel particles that may interfere with the mechanical or electrical operation of the surrounding instruments.

The plastics available for panelboards are all melamines—Formica, Peonite, Micarta, Textolite, etc. They are used either as $\frac{1}{2}$-in. thick sheets or as laminates. The cores of the laminates are such materials as aluminum, flakeboard, steel, and plywood.

Plastic panels, like those of steel, have both good and unfavorable aspects:

Great Durability   The finish is very resistant to scratching and heat.

Availability in Colors   Sheets come in many colors, but if the desired color is not in stock, there may be a long wait for delivery. A color may vary slightly from one run at the factory to the next.

Seam Required Every Four Feet   This is true regardless of panel length, because sheets are usually 4 by 8 ft.

Steel Frame Required   The plastic panel must be bolted to a skeletal steel frame for support.

Routing May Be Necessary   Many switches, pilot lights and other flush-mounted instruments are designed for a $\frac{3}{8}$-in. maximum panel thickness. Solid Formica and some laminates will require routing in the back of the panel, around the cut-out for bezel locking rings or locknuts, because of their thickness.

Less frequently used materials used for panel construction are stainless steel, aluminum and fiberglass.

Panel Inspection

A complete panel inspection at the manufacturer's plant will usually pay dividends in ease of installation, field tie-ins, loop checkouts, and in a smoother plant start-up with fewer field manhours expended.

A panel designer, familiar with the overall instrumentation and operating philosophy of the unit, can visually inspect and functionally check a control panel most expeditiously. Together with a pipe fitter and an electrician, he can locate piping errors or wiring misterminations in a fraction of the time that it would take in the field.

Each panel has its own peculiarities. The following listing may be used as a guide in formulating a check-list prior to inspection.

1. Panel construction dimensions:
   a. Overall dimensions.
   b. Thickness of panel.
   c. Size of framing.

2. Materials of construction, including:
   a. Panel and framing material.
   b. Panel finish; smooth, unblemished, correct color.
   c. Piping materials; copper, brass, PVC etc., sizes, correct valves, fittings etc.
   d. Wiring; proper wire gauge, type and insulation.
   e. Hardware; acceptable industrial grade, rated equal to or better than service requirements.

3. Construction features:
   a. Workman-like finish and appearance of overall panel.
   b. All instruments and equipment properly aligned.
   c. Tubing, piping, and wiring neatly layed out, adequately supported and without interference to instrument maintenance.
   d. All equipment rigidly mounted.
   e. Back of panel auxiliaries and miscellaneous hardware properly identified by item numbers.
   f. Verify that all field tie-ins are identified.

4. Instrumentation:
   a. All instruments installed in their proper location on panel.
   b. Correct instruments are furnished and installed, including the charts, scales, model or type numbers, instrument nameplate inscriptions, etc.

5. Preliminary checks are performed as follows:
   a. Power distribution is checked in these steps:
      (1) Verify that no one is working on panel.
      (2) Check that panel is securely grounded.
      (3) Put all disconnect switches and circuit breakers in off position.
      (4) Pull all polarized plugs.
      (5) With a high resistance (light or other) across the input terminals, energize the panel.
      (6) If light dims, find and remove the ground or short circuit.
      (7) Energize each circuit, and check each power supply sub-circuit sequentially. De-energize the circuit after checking and prior to energizing the next one.
      (8) Check for proper voltage.
   b. Air supply is checked in the following steps:
      (1) Close all instrument air-supply isolating valves.
      (2) Close reducing station gate valves.
      (3) Connect clean, dry air supply to panel at specified pressure.
      (4) Open reducing station blocks and check downstream air pressure of each reducing station (set at 20 PSIG).
      (5) Blow-down filters and header drain valves.

(6) Increase header pressure until relief valve pops.

(7) Individual air supplies may be checked as each loop is operated.

(8) Bubble-test main air header connections for leakage.

6. Functional tests may include the following:

    a. Pneumatic instruments are tested as follows:

       (1) Simulate input signal at bulkhead fitting with a 3–15 PSIG regulator

       (2) Attach 0–30 PSIG gauge to controlled output at the bulkhead.

       (3) Turn on air supply.

       (4) Verify that bulkhead and air supply nameplates are correctly inscribed.

       (5) Vary the input signal and watch the output gauge for proper response.

    b. Alarm and 120-volt control circuits are tested as follows:

       (1) Jumper input terminals one at a time to verify that the correct annunciator light flashes. The horn may be disconnected after the first alarm checkout.

       (2) Energize relay circuits on panel by simulating input signals.

       (3) Across the terminals actuating remote solenoid valves or trips, use pilot lights.

       (4) As each item is checked, verify the tagging of equipment and of field tie-ins.

       (5) Energize all chart drives, place mark on roller and check after one hour for movement.

    c. Electronic instruments are tested in the following manner:

       (1) Energize loop for checking and de-energize when checked.

       (2) Simulate input signals at field tie-in points. Check input signal type, level and voltage. This is particularly important for special instruments.

       (3) Put proper resistance across output terminals.

       (4) As each instrument is checked, verify identification of equipment and of field tie-in terminals.

If the inspector is unable to stay at the shop and verify that all corrections, misterminations, errors, etc. have been corrected prior to panel shipment, then a "punch list" is prepared. One copy is left with panel manufacturer, the other is kept as a record, and the third copy is sent to the field, so that the panel can be checked upon arrival there.

After a proper panel inspection there should be no difficulty in hooking up the field-tie-ins. Any problems in the loop checkout and calibration will most likely be external to the panel. This will significantly ease the trouble shooting in the field.

Panel Shipment

A panel should be handled as little as possible, because the chance of damage is much higher during loading and unloading than when in motion on the carrier.

When shipping by truck and if a panel is to be installed immediately upon arrival at the job site, only skids with a light framework holding a tarpaulin are necessary. To save time and handling, the panel should go via a softly sprung, exclusive, covered van.

If it cannot be installed immediately and must be stored at the plant site, heavier crating is required and a thicker plastic sheeting should be utilized. As time is not critical, and the panel is better protected, an exclusive van need not be utilized in this case.

Shipment by train, although less expensive for large distances, requires additional handling and moving. Some trains are also severely jostled during make-up and routing.

When shipping by boat, the panel should be sent as below decks cargo. The panel crating must be especially heavy and must be cushioned within the case. The wrapping should effectively seal out the salt air. Prior to sealing, the voids in the crating should be liberally loaded with dessicants such as silica gel. Heavy, impregnated, water-resistant paper or 5-mil-thick polyethylene can be used for wrapping. All seams should be covered by waterproof tape.

When possible smaller panels should be utilized to ease handling and all panel equipment must be securely braced.

Air freight does not require any particular crating or wrapping other than those required for a non-exclusive van. The particular air line must be contacted and questioned as to weight and overall size limitations on each panel and crate. The plant site airport may also be checked to verify that it is capable of receiving the type of plane furnished or required for the shipment.

Panel Specifications

In addition to the drawings and diagrams described earlier, a written specification covering other important aspects of the panel's manufacture must be developed. This specification is the vehicle which precisely instructs the panel manufacturer as to design options and material to be used for the particular panel.

The length of the specification is unimportant as long as it includes every important design feature. Some options and material choices may be left up to the panel manufacturer to decide.

A panel specification should include at least the following subjects:

Extent   Defines the design drawing specifications and codes furnished by the purchaser which the panel manufacturer is to follow.

Engineering   Describe the extent and type of engineering drawings to be developed by the panel manufacturer and whether "as built" drawings are required. It also includes the number of prints and reproducibles required and the approval or review requirements of preliminary designs.

Construction Describe the type of panels and their fabrication. This includes NEC area classification, ambient conditions and similar requirements.

Design Specify methods of installing wiring and piping systems. Materials of construction for wire, pipe, tubing, ducts, nameplate inscriptions, etc.

Materials Describe completely all materials to be used. A generic description is usually sufficient.

Inspection Give number and types of inspections planned, which may include preliminary inspections during specific stages of construction. This section of the specification should also describe the extent of inspection required such as visual, point-to-point checks, or functional testing.

Shipping Specify type of conveyance used to ship panel to plant site, type of crating, and protection requirements.

Guarantees Understand conditions under which a panel or equipment may be rejected, and the length of time during which the panel is covered by the manufacturer's warranty.

Conclusions

Panel designs need not be limited to the basic examples discussed in this section. There is no limit to the number of design variations. Each panel may be formulated of new and different shapes specifically conceived and adapted to its own unique application.

The multiplicity of design parameters is such that drawings and specifications can not cover every particular feature. The only reasonable way to insure the development of the exact control panel desired is by close coordination between the panel manufacturer and the panel user.

INDICATOR

5.2 INDICATORS

MULTIPOINT PANEL
INDICATOR

DIGITAL
INDICATOR

MULTIPOINT SCANNING
INDICATOR

Special Features:	Analog, digital, movable pointer, movable scale, parametric, scanning, multipoint, etc.
Partial List of Suppliers:	Dresser Industries, Inc., Dresser Industrial Valve and Instrument Div.; General Electric Co.; Heise Bourdon Tube Co., Inc.; Helicoid Gage Div. of American Chain and Cable Co., Inc.; Jacoby-Tarbox Corp.; Marsh Instrument Co., Div. of Colorado Manufacturing Corp.; Wallace & Tiernan, Inc.; U.S. Gauge Div. of Ametek, Inc.; Weston Instruments, Inc., A. Schlumberger Co.; Westinghouse Electric Corp.

Terminology of Analog Indication

Indication implies a representation to the eye from which the mind infers either an individually distinct state or, in most instances, a quantity in which it is interested.

Indications can be conveyed to human perception in a variety of ways. In instrumentation, we are primarily concerned with the measurement of a quantity to which we attach a numerical value. This magnitude of measurement can be conveyed to the eye either individually by a digit, by a combination of digits, or on a graduated scale on which digits are shown in a logical sequence. In the latter case, a movable reference such as a pointer is required to indicate the digit of interest on the scale.

Visual observation also includes an indication of the existence of a variable without necessarily attaching a quantitative value to it.

The indicating apparatus most frequently used for showing the variable is some form of a meter. When a meter also traces a record, this fact is brought out by the adjective "recording," such as "a recording voltmeter."

The term gauge (or gage) is always used for an instrument that only indicates.

The suffix -scope refers to an instrument used for viewing only. An oscilloscope is an indicator; an oscillograph is a recorder. The oscillogram is the recording itself.

All of the aforementioned indicators are analog instruments. They use one physical phenomenon to indicate another one by analogy. One of the most widely used analog indicators is the liquid-in-glass thermometer, which utilizes the property of metallic media such as mercury, mercury-thallium, gallium or of various organic liquids to expand under the influence of heat to indicate the temperature of another medium.

The term "analog" has become popular with the advent of the analog computer. Here an "analog" electrical quantity is used to represent any variable, since the electrical quantity lends itself to easy computation.

Other analog measuring methods include the following examples:

One type of viscosimeter uses the time a ball requires to sink from the surface of a liquid medium in a cylinder to the bottom as an analog measure of the liquid's viscosity.

The vapor-pressure thermometer is actuated by pressure to indicate temperature.

The frequency of an alternating current is measured by the vibration of a reed in an instrument containing several reeds each tuned to a different frequency. The particular reed whose resonant frequency is closest to the current frequency swings with the greatest amplitude and indicates the frequency on an adjacent scale.

Indication of Measurements

Indications on a graduated scale necessarily require the relative motion of two elements. One of the two must be a fixed reference. In most indicators the scale is stationary and the indicating reference moves. The indicating element may be a liquid column, a float, a pointer, or a beam of light. The scale may be laid out on a straight line or on a circular arc. Figures 5.2a, b and c illustrate some of the available design variations. The required simplicity of the movements in indicating instruments usually renders any other possibility for a different shape prohibitive. The larger the radius is, the better the readability. A scale of more than 180 degrees laid out on a circular disk is usually referred to as a dial. A scale is called uniform if the graduations on it are equidistant. The increments may also be different, reflecting some

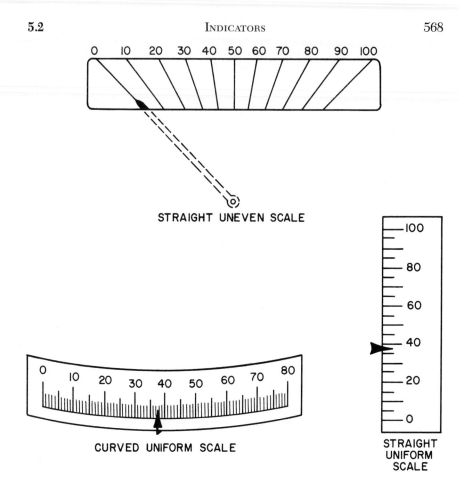

Fig. 5.2a   Straight and curved scales

other than a linear relationship. Flowmeters usually have a square root scale, vapor-pressure thermometers have an uneven scale which becomes progressively wider, calibrated according to the individual filling medium.

Fixed-Scale Indicators

A liquid column moving in a transparent, usually cylindrical tube with a stationary scale on the tube or placed inside forms the simplest fixed-scale indicator. Liquid-in-glass thermometers and manometers are examples of this. Manometer and thermometer scales are often adjustable in height for recalibration, and the tubes are usually vertical. One manometer type uses two tubes connected in the shape of a ∪. However, one leg of the ∪ may be bent on an angle which is then called an inclined-tube manometer. For very low pressures it offers the advantage of a longer, more accurately readable scale,

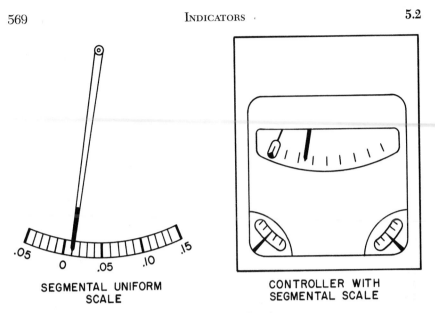

SEGMENTAL UNIFORM
SCALE

CONTROLLER WITH
SEGMENTAL SCALE

Fig. 5.2b   Segmental scale indicators

stretched out by the reciprocal of the sine of the angle which the inclined
leg forms with a horizontal line (Figure 5.2d). In some designs the inclined
leg is provided with a logarithmic scale to stretch readings in the low range
comparatively further out.

Readings in liquid columns are taken at the crest of the meniscus formed
due to capillary attraction. Mercury forms a convex meniscus, most other
liquids a concave meniscus.

A plummet (or plumb bob) floating in a calibrated tapered tube is used
in the variable-area type flowmeter (of which the rotameter is one version)

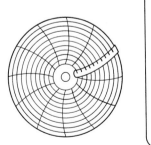

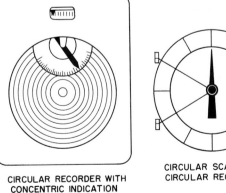

SEGMENTAL SCALE
ON CIRCULAR RECORDER

CIRCULAR RECORDER WITH
CONCENTRIC INDICATION

CIRCULAR SCALE ON
CIRCULAR RECORDER

Fig. 5.2c   Indication methods used on recorders

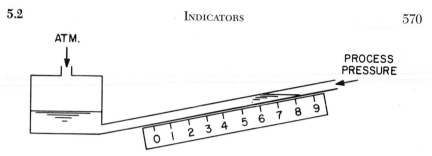

Fig. 5.2d   Inclined manometer type indicator

to indicate by its position the rate of flow through the tube. The position of a plummet in a liquid inside a graduated transparent tube is an indication of fluid density.

The most commonly used moving reference for indication is a pointer (in a timepiece called a "hand"). The pointer's color usually contrasts with the background for better readability. Pointers are frequently covered with luminescent materials to make readings in darkness possible. Transparent plastics are used where the pointer would obscure other indicators under it, e.g., when a totalizing counter is on the same dial. Incandescent, fluorescent, or neon edge-lighting are used to facilitate observation. To minimize the parallax or apparent displacement effects caused by the lens, knife-edge pointers are used. Also, mirrors are used in the same plane as the dial, and the readings are taken at an angle of observation where pointer and mirror image coincide. Sometimes the dial is raised to pointer level. In another meter, either a part of the rotatable scale or the pointer are optically projected onto a coated window, whereby parallax is completely eliminated.

Indicating instruments usually read from left to right, or clockwise. Where

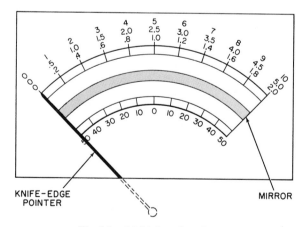

Fig. 5.2e   Multiple scale indicator

negative values have to be read, as in vacuum indications, graduations progress from right to left. Compound gauges indicating pressure and vacuum have the zero point near the middle of the scale. When readings near zero are of lesser importance, suppressed-zero ranges are used with narrower graduations at the beginning of the scale followed by wider increments. There are scales with extended sections and scales with condensed sections.

A multiple scale is illustrated in Figure 5.2e. Others are used for thermometers which indicate temperature in two units, such as Farenheit and Celsius degrees. Pressure gauges usually have a circular scale of 270 angular degrees (Figure 5.2f). Precision gauges are made with 350-degree scales and

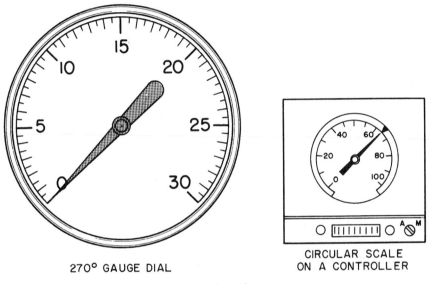

270° GAUGE DIAL   CIRCULAR SCALE
ON A CONTROLLER

Fig. 5.2f   Dials

the possibility of extending the pointer rotation to two turns which, covering a total of 660 degrees, results in a scale of 80 in. in a 16-in.-diameter gauge.

A light beam can be used as the movable reference, throwing a light spot on a scale. Readings are taken at the center of the light spot.

Multiple-scale meters show different ranges of the same variable on concentric scales. A selector switch changes the range in such electrical indicators and shows the range of the scale on which the reading is to be taken.

Movable-Scale Indicators

The alternative to the moving pointer is the moving scale and fixed reference or index (Figure 5.2g). A dip stick with a scale engraved on it used for immersion in a vessel containing a liquid is the simplest indicator in this

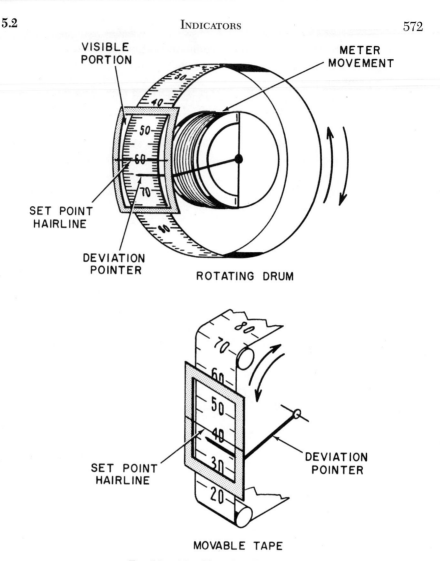

Fig. 5.2g   Movable-scale indicators

category. A large circular disk with graduations on its entire periphery, read at an index in 12 o'clock position, affords great precision in reading fine subdivisions (Figure 5.2h). The hydrometer is also in this category. The floating vertical scale projecting through the surface of a liquid shows the specific gravity or density of the liquid as indicated by the surface of the liquid on the scale.

Parametric Indication

A parametric indication tells only that a state exists, e.g., whether a fluid is flowing or static. The precise level of water in a boiler may be less important

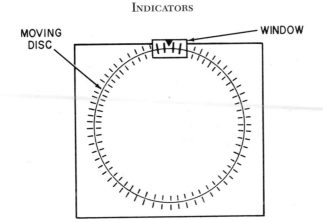

Fig. 5.2h   Rotating disc type movable-scale indicator

than the fact that it contains an adequate quantity. The sight flowmeter equipped with a vane or a paddlewheel moving in a glass enclosure, and the bubbler in which an airstream passing through a narrow glass tube monitors air flow, do not measure but do indicate, just as the full water column is a non-quantitative indicator of liquid level. In an electrical system, an indicator light is often used to show that an electric potential exists at a certain point. A Geiger counter indicates radiation by clicking or blinking.

Various other non-metering analog indicators are in use in industry. Color being a phenomenon easily distinguished by the human eye, color changes are also used in instrumentation to indicate changes of state. Chemical conditions and reactions are often indicated by color. This is exemplified by the pH meter and by litmus paper.

The change that takes place in the physical state of a compound by application of heat is used to monitor approximate temperature. Solid waxes with various admixtures show by drooping in an oven that the desired amount of heat has been absorbed by the object to be heated. Color changes brought about by heat in certain materials are likewise utilized for rough temperature indications.

Digital Indicators

Digital indicators present the readout in numerical form. Fewer mistakes are made in reading a number than in deciphering an indication on a scale. Numerical readouts are called for when a quantity is to be counted. Thus counters usually present the result in figures, even though an analog indication of a variable measured at an instant offers other advantages.

A speedometer is an analog indicator for an instantaneous rate of speed, whereas mileage traveled is digitally shown on a counter.

A flowmeter customarily uses analog indication for the rate of flow (i.e., the quantity flowing at a given instant per unit time). The total quantity having passed a given point is usually indicated numerically on a separate counter, even though a somewhat complicated conversion by an integrator is required. A positive displacement meter indicates total flow by adding liquid quantities flowing through a pipe. It is generally provided with a digital readout.

Digital clocks and thermometers are easier to read than dial clocks and scale thermometers. The necessary analog-to-digital conversion is even simpler in electrical measurements.

A digital computer requires that digital data be fed to it. Many quantities easily obtained in analog form thus require conversion. Digital meters are now made for the readout of almost any variable. Voltmeters (including panel meters with gaseous numeric display devices), thermocouple thermometers, indicators of pressure, load, strain or torque, pH meters, oscilloscopes, and stroboscopes—all have become available with digital displays.

RECORDER

5.3 RECORDERS

TWO PEN RECORDER
WITH ONE OF THE
PENS ACTUATING A
"HIGH" SWITCH

MULTIPOINT,
MULTIVARIABLE
DATA LOGGER

MULTIPOINT,
SCANNING RECORDER

Special Features:

Circular, strip, x-y, multiple, operations/events, drum, etc., chart recorders, oscillographs, photographic, electrostatic, digital, facsimile, magnetic, printing, hologram and video tape types.

Partial List of Suppliers:

Alden Electronic and Impulse Recording Equipment Co., Inc.; American Meter Co.; Bacharach Industrial Instrument Co., Div. of American Bosch Arma Corp.; Ampex Corp.; Bailey Meter Co.; Barber-Colman Co.; Beckman Instruments, Inc.; Bell & Howell, Electronic Instrumentation Group; Benson-Lehner Corp.; Bristol Div., American Chain & Cable Co.; Brush Instruments Div., Clevite Corp.; Thomas A. Edison Industries, Instrument Div.; Esterline-Angus Co.; Fairchild Instrumentation Div., Fairchild Camera & Instruments Corp.; Fischer & Porter Co.; Franklin Electronics, Inc.; Gulton Industries, Rustrak Instrument Div.; The Foxboro Co.; General Electric Co.; Genisco Technology Corp.; Hagan

Controls Corp., Div. of Westinghouse Electric Corp.; The Hays Corp.; Hewlett-Packard Co., Moseley Div.; Hogan Faximile Corp., Subsidiary Teleautograph Corp.; Honeywell, Inc.; Houston Instrument Co.; Keinath Instrument Co.; Kinelogic Corp.; Leeds & Northrup Co.; Perkin-Elmer Corp.; Precision Instrument Co.; Radio Corporation of America; Texas Instruments Inc.; Tally Corp.; Teletype Corp.; Taylor Instrument Cos.

Charts and Coordinates

The movement of the recording arm holding the scribe, usually a pen, is actuated by and shows the amplitude of the process variable. The motion of the scribe is either linear or in an arc. The chart is moved regularly and mechanically. These two motions thus produce a record of variable vs time. Any point on the continuous plot obtained in this manner can be identified by two values, called coordinates, and many coordinate systems are in use. Most industrially used charts have Cartesian coordinates. If the reference lines are straight and cross each other at right angles, they are called rectilinear coordinates. If at least one of the reference lines is an arc of a circle, the coordinates are curvilinear.

Another system, rarely used, is polar coordinates, which determine the location of a point by its vector distance from a pole and its vector angle relative to a base line.

The chart's shape gives the primary classification into (1) circular charts and (2) rectangular charts, in sheet or strip form. The strip chart, more used than sheets, can be torn off, or stored rolled up or in Z folds. Standard strip lengths are from 100 to 250 ft.

Circular Chart Recorders

The circular chart recorder is the earliest design, when recording instruments were provided with round cases. Today cases are generally rectangular even if the charts used are round. The circular chart recorder has the merit of simplicity and low price and is still used in industry despite its disadvantage of lesser readability compared with the strip chart recorder. Chart speeds are a uniform number of revolutions per unit time. Chart drives making one revolution in 24 hours are standard, but instruments are also equipped with drives of 15 minutes per revolution, 1, 4, 8, 12, or 48 hours per revolution and 7, 8 or 28 days per revolution.

The circular chart shows concentric circles crossed by circular arcs whose center, when the pen crosses the arc, is the same as that of the recording arm (Figure 5.3a). The tangents at the intersection points of the two curves should be as near to a right angle as possible for best readability. The concen-

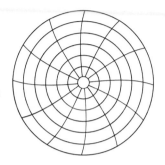

Fig. 5.3a   Circular chart

tric circles form the scale on which the variable is read. The "time arcs" laid out at convenient uniform distances divide the full circles into appropriate intervals of the total period. When the pen is moved in a straight line, as in a level meter with float and cable arrangement, the time lines are straight.

The full circles are equidistant if the variable is a linear function of time. Orifice flowmeter charts are usually square root, and as such are difficult to read in the lower 25 percent of the chart range, as the graduations become rather condensed.

Standard chart diameters are 6, 8, 10 and 12 in., although there are also 3 and 9-in. charts. The calibrated scale length is less than half of the nominal chart size, with a maximum scale of 5 in. for the 12-in. recorder. Despite the fact that there are only a few basic sizes used, chart manufacturers report stocking more than 35,000 different charts, since there are many recorder manufacturers, and every one has his own combinations of size, range, number of revolutions, etc.

Circular charts offer the advantage of a flat surface. Use of up to four pens for multiple-point recording is possible, utilizing the entire calibrated width. Since most recorders are also equipped with an indicating scale, the scale length is confined to the calibrated length of the chart, which for strip chart recorders is a maximum of approximately 11 in., while for the circular chart recorder which makes the use of concentric indicating scales possible, a maximum linear length of 34″ can be obtained. Other designs have a segmental or arc-shaped indicating scale which is necessarily shorter.

Among the special features of circular chart recorders are automatic chart changers which are intended to be better than strip chart recorders by supplying a collection of daily records in a continuous manner. Charts are usually designed with increasing values from the center out, but some recorders register on reversed charts with "upscale" toward the center.

Strip Chart Recorders

The standard strip chart recorder is characterized by the uniform linear motion of the paper, usually vertical (Figure 5.3b). The time lines, always

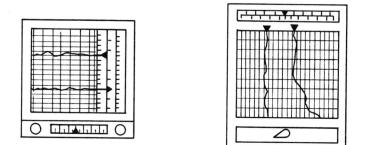

Fig. 5.3b   Miniature multi-point strip chart recorders

running perpendicular to the direction of motion, are straight, while the measurement lines can be straight or curved, furnishing either rectilinear or curvilinear recordings (Figures 5.3c and 5.3d). The curvilinear recordings requiring less complicated linkage geometry than rectilinear-recording movements offer even better readability than recordings on round charts. Charts calibrated to read directly in process variable units are most commonly used, although some read in percent of full scale. The pen position is plotted as a function of time. Scales are uniform with the exception of incremental recorders where the chart is moved only with the occurrence of an *event*. Strip charts with square-root scales are also available for orifice type flow measurements.

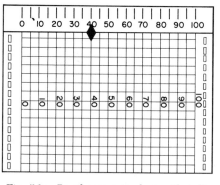

Fig. 5.3c   Rectilinear strip chart with indicator scale

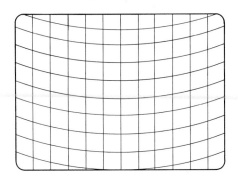

Fig. 5.3d   Curvilinear strip chart

X-Y Recorders

X-Y recorders are rectangular chart recorders which plot a function of two variables such as stress vs strain or temperature vs pressure. In one type the chart is stationary and the scriber is moved along both abscissa and ordinate by two analog signals via electromechanical servomechanisms operating independently. In another version of this type the chart is moved in the Y direction, while the stylus slides on an arm in the X direction. Plot lines may be continuous or a succession of points.

There are also combination function plotters, called X-Y-Y_1 recorders. Some allow the pen to be driven along either axis at a constant speed, thus making recordings of X vs Y, Y vs T, and X vs T possible. Recorders with three independent servo systems allow the recording of two variables against a third one.

Function plotters may also accept digital signals directly from magnetic tape, punched card, paper tape, or keyboard, yet supply an analog plot by a simple conversion of the digital signals. Scales may be linear or logarithmic along both axes.

X-Y recorders usually have flat beds in sizes from $8\frac{1}{2}$ by 11 in. to 45 by 60 in., but some record on a drum type platen.

Besides obtaining an analog plot, digital records can also be provided by X-Y recorders with analog-to-digital conversion and conventional digital printout methods.

Some X-Y recorders print from the back so as not to obstruct vision; others plot on a glass screen.

Oscillographic methods are also used for X-Y recording by a light beam and ultraviolet-sensitive paper which is immediately developed by exposure to daylight. Writing speeds of 200 fps can be obtained by this method.

Multiple Recorders

A recorder handling one variable only is a single-point recorder. A recorder automatically handling several points on the same chart, such as several temperatures measured by thermocouples at various locations, is a multiple-point recorder. A round chart recorder usually can handle up to four points due to practical design limitations. To distinguish between the various points, color or numerical coding is used. In the latter method numbers may be printed in almost uninterrupted sequence quasi forming a line. A special type recorder with one pen having six tips in six different colors, usually can plot up to 12 measurements. A full-size strip chart recorder can handle up to 24 measurements. Multi-bank recorders can record up to 200 separate inputs on a 12-in. strip.

A recorder can be equipped with more than one actuating mechanism responding to different sensors, such as pressure and associated temperature. Depending on the size of the recorder, several actuators can be placed in one case. Such a recorder is called a multi-variable recorder. Since most recorders also have indicating scales, design considerations usually limit the number of pointers to two or three since they have to cross each other. A multi-channel recorder is illustrated on Figure 5.3e.

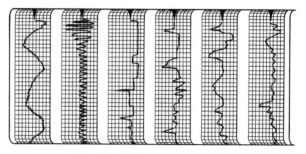

Fig. 5.3e   Multi-channel recorder

Another way to make multiple recordings in one instrument is to record several ranges of the same variable. An instrument designed for this purpose is called a multi-range recorder. It may have one measuring element and the total range may be switched to a narrower span for closer observation, or it can be actuated by two or more sensors, each with a different range.

One special design, using the sweep-balance technique, can plot up to 400 points, presenting them on charts in 4 by 6-in. frames printed in columns and rows on one panel with from one to four records per frame. The charts are held against translucent platens with transparent process symbols showing through the chart. A graphic panel can thus be presented behind the chart.

Chart papers are usually made up of various layers, with the grain running crosswise for stability against humidity.

Recording Means

Recorders use many means of scribing but pen and ink is the oldest and most frequently used method. Numerous improvements have minimized smearing, spilling of ink, heavy lines, skipping, and other drawbacks of ink writing. Some filling systems are pressurized, made splatter-proof, or completely eliminated by cartridges or closed systems which have a long useful life and are eventually thrown away without ever getting a chance of putting a blot on a chart or soiling a finger. The ink may have different chemical and physical properties than usual. Some inks are viscous, with about 50 percent dye content against about 2 percent in conventional ink. Others are heated, made to dry upon contact with the chart. "Solid-state" wax has been used as ink.

Ball points and fiber tips are in use. The width of the trace has been reduced in some systems to 0.01 in., which is no wider than a hairline.

Other means of scribing have been devised. Charts are coated with waxes, carbon, metal, oxides, or electrolytes. They have been made to react to heat, pressure, chemicals, electricity, or light intensity. A stylus is used in such systems which leaves the recording mark by such as burning away a coating by electrolytic action, arcing, or using mechanical force to remove a layer from the base by abrasion or knife scraping. No stylus is instantly ready for scribing or is insensitive to vibration.

Event Recorders

Operations or event recorders mark the occurrence, duration, and/or type of an event. Flat strip charts or drum charts are used. These recorders serve to supervise a large number of operations indicating on the chart information such as on-time, down-time, speed, load, overload, etc. The markings are usually made in the form of a bar, with interruptions in a continuous line indicating a change in the process. Other systems use a code in the line from which information can be taken as to the time of an event, its duration, the frequency of occurrence, etc. One such recorder type uses a drum rotating at a constant rate while the pen scribes short horizontal lines in a number of columns, forming a bar when the pen moves up with each rotation. Event recorders may scan a great number of points in succession, putting a record down for each point. High-speed techniques by which up to 100 points can be scanned every millisecond, for high-speed printout on teletypewriters show the trend in this field.

There are event recorders which plot combinations of analog and event records. One unit is available with eight analog channels and 64 event channels.

Miniature Recorders

Strip chart recorders originally were designed 12 in. wide. These are now designated as "full width" strip chart recorders. Their calibrated width is about 11 in., since space is used up for the perforations to transport the chart. The advent of the graphic panel stimulated the development of the miniature recorder. A miniature recorder can be placed on the graphic panel at the location where the measurement is taken. By utilizing depth instead of width for the recorder, most elements have maintained their original size, and accuracy has *not* been sacrificed. Miniature recorders occupy about one-fifth of the front panel space used up by a full size strip chart recorder. A calibrated chart width of 4 in. has become standard, and readings are as accurate as those taken on a circular chart, because the charts are mostly rectilinear or only slightly curved, and the penholding arms are long. The miniature recorder is made both as an electric servo type and as a pneumatically actuated direct-writing type.

The medium size recorder was developed as a compromise between the miniature and its full size counterpart to increase the calibrated length of the circular chart scale, yet stay below the panel space required by the full-size recorder. Recorders in this category are made in nominal chart widths of 6, 7 and 11 in. and are called "compact" by their manufacturers. Companion models are made with circular charts.

Besides being classified by the technique applied to obtain a permanent record of a variable and by the shape of the chart, recorders can also be grouped by their means of printing, such as photographic or electrostatic, and by their intended use, such as stationary and portable (service).

Service Recorders (Portable)

Service recorders are usually small portable instruments used for check-out work. They are made in both circular and strip chart types, as single- and multiple-point units (up to 12 points on a 5-in. chart), and also as multi-range and multi-variable recorders. They record practically all variables and include oscillographs. Nominal sizes are 4 in. and 5 in. for round chart or strip chart instruments.

Photographic and Electrostatic Recording

Photographic and electrostatic methods are primarily used to make a permanent record of an oscillograph trace. Records are obtained either by direct methods or by the use of separate processing units after exposure.

Galvanometer oscilloscopes make a transient display of amplitude vs time on a fluorescent screen by using a mirror attached to the galvanometer to project a beam of light onto the screen whereby light is emitted at the spot where the beam impinges. This method is confined to electric phenomena with a maximum frequency of about 6,000 Hz. The persistence of the human eye traces a curve on the screen as the dot moves along. The screen is usually equipped with scaling and indexing dots to coordinate the curve. For high-frequency response (to 100,000 Hz) a cathode-ray oscilloscope is used. Figure 5.3f illustrates a typical oscillogram.

A permanent record can be obtained with special cameras capable of recording high-speed phenomena in single frame exposure (with exposure

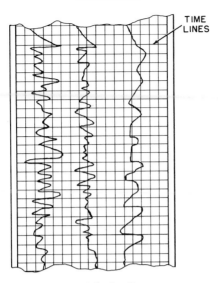

Fig. 5.3f   Oscillogram

times as short as 10 nanoseconds), on 35-mm film, or on cine film with standard rates of 8, 16, or 24 frames per second and up to 18,000 pictures per second. The pictures can be developed within minutes.

An oscillograph records directly on a chart driven at constant speed by a timing motor. Either photo-sensitive paper (or film) or modern electrostatic methods are applied. Light-beam oscillographs can record 52 or more channels with the aid of separate galvanometers.

The photographic methods applied are either conventional wet processing, using chemicals, or print-out. The wet method involves a time lag of up to 30 minutes except when the drying is accelerated by heating the platen. This method is called "develop out." The print-out method has a typical time lag of only a few seconds. Both methods require special papers, and the print-out materials give an instantaneous image only when the recording beam moves comparatively slowly. When it moves fast, the beam yields only a latent picture, which has to be made visible by subsequent exposure to room light for about 60 seconds. By using high-intensity lighting and heating the platen, the latent picture can be brought out by ultraviolet or fluorescent light in seconds.

Some photosensitive paper can be stored rolled up in cartridges after exposure, without the need for processing. There are integral processors and attachments to automatically develop and dry the record while the recorder is in operation. With direct-writing photorecording papers the writing rates of 50,000 in. per second are possible.

Electrostatic methods use fixed conducting styli which make contact with a thin plastic surface coated with a dielectric. The traces are made visible either by dry powders or liquid toners. No chart paper is used in order to avoid shifting. The coordinates are simultaneously printed on the paper.

Magnetic Recording

Magnetic recording on special tape is a highly developed and most practical method for storing information on waveforms for analysis. This can be done either in the analog or digital mode. Analog tape recording uses either the biased AC method or frequency modulation.

Digital recording on tapes can either be continuous on several tracks handled by separate recording heads, or incremental, storing many hundreds of binary digits per second. The recording thus makes direct reproduction by digital computers possible.

Digital Recorders (Printers)

Digital recording is performed by printing the output of electronic equipment on paper in digital form.

Originally this was done by attaching solenoids to the types of an electrical typewriter and printing the data on an $8\frac{1}{2}$ by 11-in. sheet. This method is still the basis for digitally recording data from on-off elements.

Printing data on tape in numerical, alphanumerical, or symbolic form is favored, because it is logged in tabular format. This is done on 3-in.-wide tape, which has become the standard, but presentation on a preprinted form facilitates evaluation of the data.

Other electromechanical means for printing, besides the typewriter, are represented by the line printer. While the automatic typewriter prints about 100 words per minute, modern high-speed line printers can handle up to 40 lines of numerical or alphanumerical data per second, recording in up to 32 columns. The adding machine type tape printer can add and subtract readings and print out the total.

Printouts can be keyed to time and show quantities handled in a certain time interval, such as gallons of gasoline pumped per minute.

Ultra-high-speed recording in digital form can be achieved by combining electronic readouts with electrostatic printing. One such instrument can present 900,000 characters per minute.

When data are received in analog form, digitizing equipment such as a digital voltmeter may have to be inserted in the circuit to provide digital input into the printer.

Where data are represented by the output of logical levels, power amplification is required for the operation of the solenoids.

Most data input is in decimal form. If coded signals are received, a decoder may be required, unless decoding equipment is built into the printer and the data are simultaneously scanned for straight decimal and/or coded input.

Tape printers receive parallel input. Serial input is required to operate remote printing equipment used in electric communication systems such as teletype equipment. Teletypewriters are, in principle, electric typewriters that are actuated by magnets on the keys when receiving a signal from the remote transmitting station. Input at the transmitting station may be either manual (by keyboard) or automatic by prepunched tape. A code, usually the Baudot code, is used having five pulses of equal length which permit 32 combinations. Decoding is required for input into the teleprinter.

Facsimile Recording

Facsimile recorders reproduce written matter and illustrations transmitted by telephone or telegraph wire or by microwaves.

The material to be transmitted is subdivided into elemental parts and is scanned (usually by a phototube). The parts are transmitted in the form of electric pulses and are reassembled by photographic, electrolytic, and heated or pressure-stylus methods.

Line drawings can be reproduced by a direct-recording process; halftone pictures require an amplitude modulated transmission system in which the pulses represent the blackness of the elementary area.

Facsimile recording finds primary use in transmitting weather maps, since a number of reproductions on dry carbon paper can be simultaneously made, but it is also used in radar sampling and oceanography.

Some plotters accept binary data and print out contour maps with symbols. There are also graphic recorders with facsimile converters.

The sweep-balance technique utilizes an electrolytic helix on a drum, rotating at high speed, and an endless electrode for recording illustrations on electrosensitive paper.

Holograms

Holograms are negatives obtained without a camera by exposure of a high-resolution photographic plate. An interference pattern is produced by highly coherent light, such as a laser beam, reflected from a subject and a reference source. Holograms record all information contained in light waves and reproduce a truly three-dimensional picture, even in color, when properly illuminated and viewed through a high-power microscope. Photographic pictures, in contrast, record only the intensity of light in the scenery.

Recent advances have made it possible to take holograms by illumination with white light through a fly's eye lens consisting of a great many small sensors which see the scene from various angles. Previously, only monochromatic light afforded the possibility of producing holograms. This recent achievement should have many applications.

Video Tape Recording

Video tape recordings (VTRs) are made similarly to magnetic tape recordings, except that electric signals record both picture and sound. The frequency-modulated signals magnetize the finely distributed iron powder particles in the tape emulsion via recording heads. The signals are recovered by reproducing heads, demodulated, and need no further processing to activate both the screen of the cathode ray tube and the sound system in a conventional manner. The method finds industrial applications in closed-circuit television.

The kinescope recording method photographs the picture on film, thus requiring further processing with attending time delay.

5.4 STAND-BY POWER SUPPLY SYSTEMS

Main Features:	For classification, refer to Tables 5.4e and 5.4f.
Cost:	For the cost of batteries, alternators, chargers and inverters, refer to Figures 5.4a, b, c and d.
Partial List of Suppliers:	Cyberex Co.; ESB Inc.; General Electric Co.; Holt Broders Co.; King-Knight Co.; Onan Div., Studebaker Corp.; Power Sources Div., Technical Operations Inc.; Rochester Instrument Systems Inc.; Solid State Controls Inc.; Westinghouse Electric Corp.

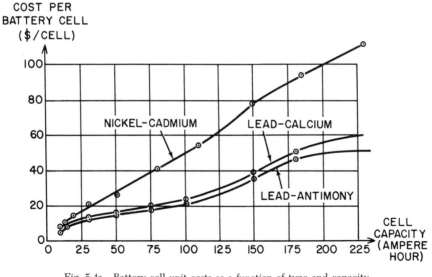

Fig. 5.4a   Battery cell unit costs as a function of type and capacity

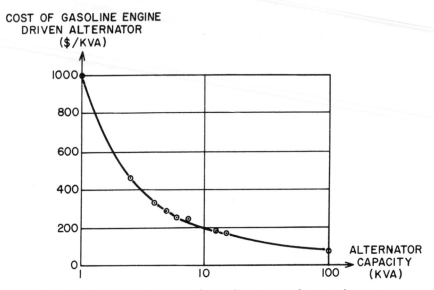

Fig. 5.4b   Cost of gasoline engine driven alternator as a function of capacity

Electric power lines are commonly assumed to be a perfectly reliable source of constant voltage. This assumption is valid when complete source reliability, particularly on a short-term basis, is not important. Control and instrument engineers frequently make this assumption for systems which do

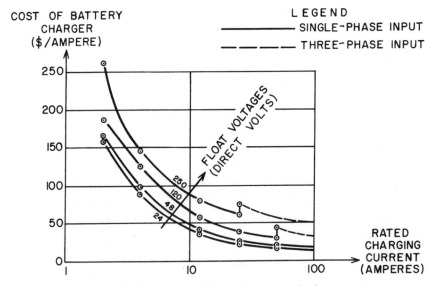

Fig. 5.4c   Battery charger cost as a function of current and voltage requirements

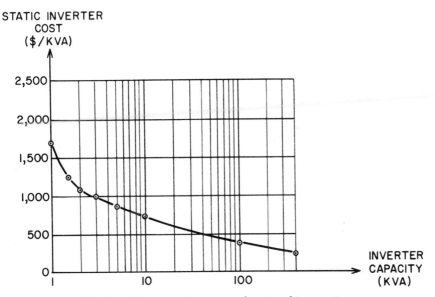

Fig. 5.4d   Static inverter cost as a function of its capacity

not satisfy this criterion. Obvious long-time power outages are fairly rare in modern power systems. Notable exceptions, however, have called attention to the complete spectrum of possibilities of power failure. Voltage dips too short to be noticed by human senses can occur frequently during the lightning season on exposed suburban and rural power systems. These systems serve many industrial plants. In addition to the obvious power failures due to weather, there are many transient variences due to electrical, mechanical and human occurrences. A system designer who assumes a perfectly reliable power source is responsible for any loss of production and damage to machines and plant facilities resulting from power irregularities.

The trend in control system design is to use faster components operating on smaller signals. This results in increasing system sensitivity to line voltage variation and in particular to short transients. As control instrumentation technology advances, the need for stand-by power sources increases.

Frequently the cause of the transients which result in control or instrument circuit complications is not the lack of voltage alone. Phase shift, a change in frequency, inadequate transient response and noise can be equally damaging. Therefore an adequate stand-by power supply system must consider all types of power failure.

Summary of Features

Stand-by systems can be quite complex. They involve the use of a number of components which may be of the static types, electromechanical types, or some combination of both. The selection of the stand-by system and the components for it should be based upon the degree of integrity required of the application.

It is frequently possible to improve stand-by system reliability *and* decrease cost. An example is that of the redundant input circuit shown in Figure 5.4h, under Source Failure. By separating system functions, it is possible to purchase the component that does the precise job required. Over-specification of components frequently results in more expensive equipment having less reliability.

Much attention has been focused on the problem of failure of the incoming power line. While this is an important consideration, it is not the only consideration. Attention should also be given to the equipment comprising the stand-by power supply system. Should critical pieces of the stand-by equipment be less reliable than the incoming power line, failure will be more frequent and little improvement will have been accomplished. Also of importance is the load circuit. If a number of load branches are connected to the output of a stand-by power supply system, the failure of any one load may result in the failure of the remaining loads. Proper design of the load system will minimize or eliminate this possibility.

The importance of system redundancy cannot be overemphasized. Once designed, the system must be evaluated to determine its weakest link. The cost of redundant equipment may then be evaluated in light of the importance of maintaining the load.

Table 5.4e summarizes the standby system classifications and lists the most commonly used components. Obviously, special arrangements may be found

Table 5.4e
SUMMARY OF STAND-BY SYSTEM CLASSIFICATIONS
AND "MOST USED" COMPONENT PARTS

System Component	Stand-by System Classification		
	Multi-Cycle	*Sub-Cycle*	*"No-Break" (Minimum Transient)*
Secondary Power Source	Engine-driven alternator or generator *starting* on primary source failure	Engine or motor driven alternator or generator *running* with flywheel	Battery
Inverter	Rotating or static	Rotating or static	Static
Bus Transfer Switches	Electromechanical	Static	Static

which extend some of the components into an additional classification. This table is intended as a means of outlining the discussion which follows.

Various degrees of *redundancy* may be designed into the standby system. This redundancy may occur on the input side of the inverter, on the output side or on both sides. A tabulation of components in a single and a double source redundant system is provided in Table 5.4f. Both input and output redundancy is included.

Table 5.4f
STAND-BY SYSTEM COMPONENT REDUNDANCY

Levels of Redundancy	Input Side	Output Side
One Source	Battery and battery charger	Inverter
One Source With Some Equipment Redundancy	Battery, battery charger and rectifier	Inverter with transfer to line
Two Sources	Battery, battery charger and engine-driven generator	Inverter with transfer to alternate inverter
Two Sources With Some Equipment Redundancy	Battery, battery charger, rectifier and engine-driven generator	Inverter with transfer to alternate inverter and then to line

Power Failure Classifications

Stand-by power systems can be characterized by the time it takes to achieve full output from the stand-by power source after failure of the primary source. This characterization by time implies that various stand-by systems can be discussed in terms of transfer time. This transfer time might be as long as many cycles—for large electromechanical switching devices or for engine-driven alternators or generators which start up on failure of the primary source—to fractions of a cycle—for some of the solid-state switching devices or for motor-alternator or generator sets with flywheels and to no-break systems. Most early stand-by systems provided a time interval during which there was no voltage to the load on transfer from the primary source to the stand-by source or on retransfer from the stand-by source to the primary source. As control and instrumentation circuits have become more critical, stand-by systems have been developed which include new techniques for transferring power sources such that essentially no transfer time occurs. For the most part, these no-break systems cost little more than those requiring a significant amount of time to transfer. In general, this characterization of the stand-by power system by transfer time is disappearing, since a large percentage of present stand-by systems are of the no-break variety.

Another means of characterizing stand-by systems is by type of failure. One's first thought is to protect the critical loads from failure of the commercial power line. A careful scrutiny into the system suggests that there are other points worthy of consideration. Among these is the failure of stand-by power supply components and the failure of the load.

Source Failure

A very simple stand-by power system is shown in Figure 5.4g. It consists of an AC power line feeding a battery charger. The battery charger, in turn,

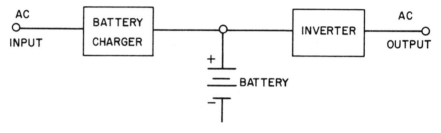

Fig. 5.4g   Basic AC stand-by system

floats a battery which provides power to the inverter. The inverter provides an AC output through a distribution panel to a number of loads. Should the AC line fail, the battery charger will cease to provide the current to the inverter. The current will then be provided by the battery which is floating on the system. In this fashion, the inverter supplies the loads until such time as the AC line is re-energized, at which time the battery charger again provides the power for the inverter, for the loads and at the same time provides recharge current to the battery. Thus, the simple stand-by system of Figure 5.4g protects against a line failure, since there is no cessation of power to the loads when the AC line fails.

A suggested improvement in this basic system is shown in Figure 5.4h. By separating the functions of the battery charger into (1) supplying steady-state running current to the inverter and (2) supplying recharge current to the battery, two rectifiers can be used. An unregulated rectifier is adequate to supply load current by means of the inverter. The battery charger rectifier must be regulated to insure long battery life. Since the unregulated supply is less likely to fail, some additional reliability is gained. This can be seen by considering the results of failure of the battery charger. In the circuit of Figure 5.4g, should the battery charger fail, the system is no longer operable after the energy stored in the battery is consumed by the load. In the case

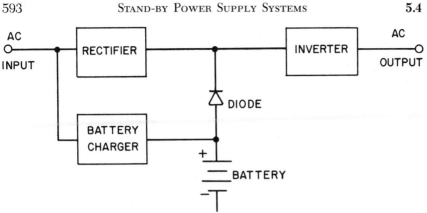

Fig. 5.4h   Stand-by system with battery charger redundancy

of Figure 5.4h, however, should the battery charger fail, the system continues to function so long as the AC input is available. Should the AC input source fail, the system will continue to operate until the energy stored in the battery is consumed. If, prior to this time, the AC input source is restored, the system continues to function properly but the battery charger is not capable of recharging the battery. While the system continues to operate properly, the battery charger may be repaired if it is possible to do so between the time interval defined by that time when the failure of the battery charger is noticed and the next failure of the AC input source. Indeed, if nothing else can be done, it is generally possible to by-pass the diode with some available resistance (even a light bulb) which will restore some energy to the battery. Even a small amount of battery capacity is ample for a number of short transient outages.

In addition to the increased reliability of the two sources noted in Figure 5.4h, a lower cost for this system frequently results. This lower cost is due to the fact that the unregulated rectifier in most cases is providing a larger current than is the battery charger. Thus, the rectifier capacity is greater than that of the battery charger. Since it is less expensive to buy unregulated power than to buy regulated power, it is possible, under many circumstances, to achieve a lower cost. This combination of lower cost and increased reliability is the optimum thought of the system designer.

The system of Figure 5.4h can be extended to include more than one AC source, as illustrated by Figure 5.4i. Alternative sources may include other AC lines and the output from engine-driven alternators, or, indeed, from any alternator regardless of the number of phases, the voltage, the frequency, or the variation in frequency. It may be desirable to "stagger" the input voltage ranges of the sources to favor one or another source. Since this "staggering" of sources results in an increase in input voltage variation over which the inverter must operate, a thyristor has been included to provide a dynamic

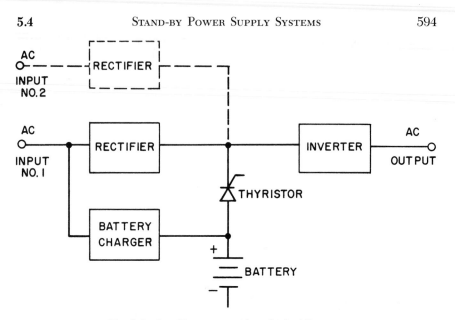

Fig. 5.4i   Stand-by system with multiple AC inputs

switching of sources, minimizing the input voltage variation. The peak value of the alternative sources when rectified must be greater than the battery potential in order to "turn off" the thyristor.

The ultimate in input redundancy occurs with more than one complete power source. Shown in Figure 5.4j is a system with source, battery charger, and battery redundancy. It is also wise to separate the power feeds to inverter. This system can be extended to any desired degree.

The systems of Figures 5.4g, h, i and j offer a continuous source of power to a load without regard to the state of the AC input source so long as the

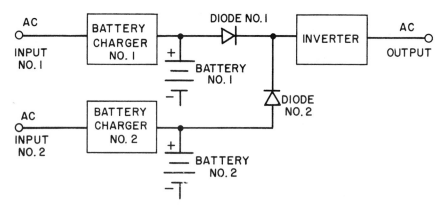

Fig. 5.4j   Stand-by system with multiple input redundancy

stand-by source has sufficient energy to supply the load. The options noted provide any degree of redundancy desired for the stand-by power source. These figures show that this redundancy is adequate to insure the most critical loads. The diagrams also make it clear that, should a failure occur in the inverter, the load source is no longer protected. Thus, our next concern must be the failure of equipment within the inverter block.

Equipment Failure (Inverter)

In many applications of stand-by power, the integrity of the line must be maintained in spite of equipment failure. Failure to preserve this integrity can result in loss of output, scrap material, plant damage, or loss of life. The degree of the protection necessary depends on the damage which can result. Process control computers are particularly important in plant operations because failure of the computer system results in an uncontrolled process. Loss may be sufficiently high to justify greater system redundancy.

The simplest form of output redundancy is illustrated by Figure 5.4k.

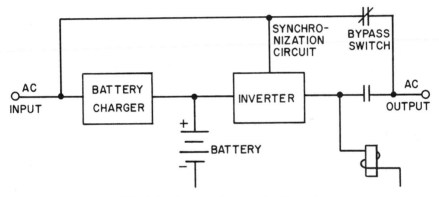

Fig. 5.4k   Output redundancy to AC input

A by-pass switch is provided from the output of the inverter to the AC input line. In this diagram, and in many to follow, the symbol for an electro-mechanical switch (relay) will be employed. This symbol should be construed to include both static and electromechanical devices.

Two items are essential in the operation of the circuit of Figure 5.4k: a synchronization circuit and a means of sensing source failure. A synchronization circuit has been added to insure that both the AC input and the inverter are in phase in order to minimize the switching transient. The switching device or devices have also been added together with appropriate sensing circuitry.

At this point, it is easy to gloss over an essential discussion masked by the obviousness of the preceding remark. Consider for a moment the fact that both voltage waveforms must be of the same frequency and in phase. On failure of the input AC line, no transfer occurs, since the output of the inverter is not impaired. In the event that a loss of output from the inverter occurs, transfer to the AC source takes place. Retransfer to the inverter can occur when the inverter output is re-established and when synchronization of the outputs has been restored. The retransfer may occasion an output voltage transient, as well as the transfer, since stored energy in the filter or the inertia of the rotating alternator used in the inverter requires some time to bring up to full output current.

Note that the addition of the line synchronization capability has increased the number of components in the inverter, thereby decreasing its reliability. The synchronization circuit must be carefully designed to eliminate any AC line transients which may cause a failure in inverter output.

The point of detection of inverter output failure is important. Sensing as early in the circuit as possible provides a better transfer, since energy stored in the filter may be used to reduce the transient. Early detection of thyristor failure or of abnormal vibration are preferable to the simple detection of reduced output voltage.

Difficulties can occur in providing a sensing circuit which operates on an adjustable reduced output voltage level. If no delay is built into the sensing circuit, transfer can occur on line transients, causing frequent operation. If the normal delay is included in the retransfer circuit, no system redundancy occurs in the time interval defined by the time of transfer and the delay before the retransfer occurs.

In order to minimize transfers on simple line transients, which may be due to sudden load demands, an integrating circuit may be inserted in the voltage level sensor. While this eliminates the transfer due to short transients, it causes a greater output transient when a transfer is made because of equipment failure.

The difficulties noted strongly suggest that consideration be given to early failure detection. The importance of this portion of the stand-by power supply system cannot be overemphasized. In many cases it is wise to have a double sensor system which will provide a delay either on failure of a component such as a thyristor or on failure of a bearing resulting in abnormal vibration and also a sensor circuit employing an integrating circuit and detecting the reduction of the output voltage.

Input and output redundancy is shown in Figure 5.41. Any of the input redundant schemes could be employed.

With AC input redundancy, the integrity of the AC line must be considered. Naturally, if frequent line disturbances occur, little is achieved by such an arrangement. The only gain is the possibility of not having a line failure until something can be done to re-establish source redundancy.

The degree of redundancy can be improved by providing a back-up

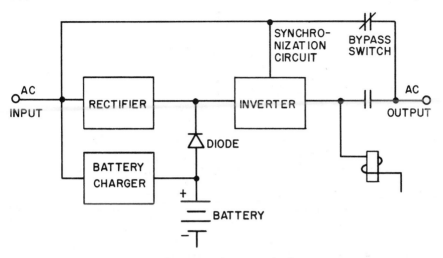

Fig. 5.4l   Input and output redundancy

inverter stand-by system. Figure 5.4m illustrates this emergency power supply backing up an emergency power supply but with dual loads. Here each inverter is capable of handling the full output capacity (both loads). On failure of either, the remaining unit assumes the full load. Momentary paralleling is possible to use the stored energy in the filter of the unit going off. As noted previously, this energy can help to minimize the transfer transient.

Complete input and output redundancy is diagramed in Figure 5.4n.

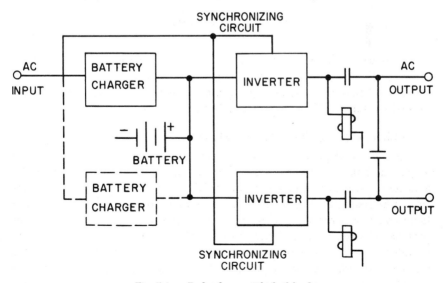

Fig. 5.4m   Redundancy with dual loads

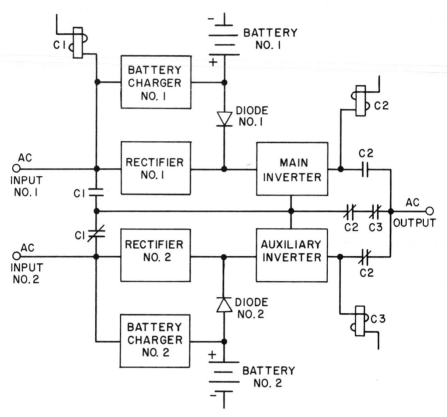

Fig. 5.4n   Three-fold redundancy

Battery, battery charger, rectifier and inverters are repeated. *Each* is sized to handle the full load. Further redundancy is provided by the AC line. If the main set fails, the auxiliary set supplies the load. If it, in turn, fails before the main is repaired, the load is supplied by the line.

Common Bus Branch (Load) Failure

In addition to providing a means of transfer from the output of an inverter which has failed to an alternate power source, the transfer switch provides a means of clearing branch circuit fuses sufficiently fast to protect other loads from a faulted load. The inverter is frequently current-limited to provide a finite overload capability. Thus, on failure of one load, the current limit provides a known amount of current for opening the fuse in the faulted branch. By referring to available fuse characteristics one derives the information that a number of branches are required with even the fastest of available fuses in order to provide load clearing within one-half cycle. By using a transfer switch, which allows the transfer of the load from the inverter output to the AC line, one can, in effect, increase the short-circuit current capability, thus providing a means of rapidly clearing the fuse in the branch circuit. After clearing the faulted branch, one can retransfer back to the inverter stand-by systems.

Ideally, a short-circuit in one fused branch of an n-branch load, such as in Figure 5.4o, would have no effect upon the power supplied to the remaining

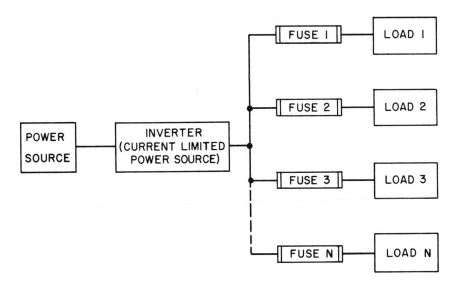

Fig. 5.4o   Inverter system supplying N fused load branches

branches. In reality, however, a fuse requires a finite amount of time to clear. If the power source is current-limited, the supply voltage will drop to a value near zero until the fault is cleared. Unless rapid opening of the fuse occurs, other loads will become inoperable.

Assuming, for any given supply output capacity, that all branches of an n-branch load consume nearly equal parts of the supply power, it follows that the larger the number of load branches, the smaller the average branch fuse rating. A 10-KVA power source, for example, may have five equally loaded (2 KVA) branches. Another 10-KVA supply may have ten equally loaded (1 KVA) branches. In the latter case, the average branch fuse has about one-half the current rating of those in the former case.

In general, the larger the current capacity of a power source, the shorter the time required to clear a fuse in a short-circuited branch. Looking at the same relationship in a slightly different way, we can see that the smaller the fuse rating of a short-circuited branch, the less time required to clear it—assuming the short-circuited supply capacity remains constant.

The preceding two paragraphs lead to the generalization that the smaller the fraction of total supply capacity carried by a fused load branch, the smaller the time required to clear a branch fuse, and the less serious the load power disturbance. Conversely, the larger the fraction of total capacity carried by each branch, the larger the fuses in each, and the longer the time required to clear the fault.

Figure 5.4p demonstrates this generalization. Assuming (1) all branches

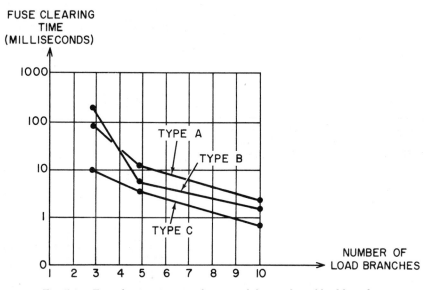

Fig. 5.4p   Fuse clearing time as a function of the number of load branches

consume equal fractions of load power, (2) the source always current-limits at 15 amperes, and (3) the short-circuit load impedance is zero, fuse clearing time is shown as a function of the number of load branches. The graph illustrates this data for three different fuse "speeds."

A well-designed stand-by power supply system requires that consideration be given to all types of failure. The most often neglected area of consideration is the load bus system. Selection of the type of branch circuit protector must be coordinated with the short-circuit characteristic of the inverter as well as the requirements of the loads.

System Components

It is frequently true that the characterization of the system itself is dependent on the components available for use in the system. For each system function there are a number of components from which to choose. Arbitrary selection of any single component without regard to the others can result in an unworkable or, at best, an inefficient system. Thus, the designer's problem

is to choose a compatible set of components which will satisfy his requirements. Specifically, the designer seeks the optimum compatible set of components. In order to select the appropriate component for a given function, it
is necessary to understand the characteristics of the components from which
one must select.

Rotating Equipment

Rotating equipment may be subdivided into two general classes. The first
class includes all devices operating from a source of electric power. This set
includes motors and, because of intimate relationship, alternators and generators. The other general category of rotating equipment includes those devices
driven by engines which have the ability to operate from liquid or gaseous
fuels.

The preceding discussion has not covered the nature of the equipment
in the blocks. As an example, the battery charger could be a motor generator
set with appropriate controls. The inverter could, of course, be a DC motor
driving an alternator. The selection of these components is determined by
economic considerations. The economics involve not only the initial cost,
operating cost, and maintenance cost of the equipment itself but also an
evaluation of the need for reliability based on the criticality of the load. If
load failure results in a vacant lot characterized by a hole or by the need
to repipe a plant because of the solidification of material which under normal
circumstances would have been a usable product, the greatest possible reliability should be built into the stand-by power source. On the other hand, if
the loss of power results in some annoyance but not in the loss of plant capacity
or deterioration of product quality, then the ultimate in redundancy and
reliability is not warranted.

Particular note should be taken of stand-by power supply systems employing engines. The engine has been proven to be a reliable device in many
situations and in very adverse environments. Unhappily, the engine itself is
not the most frequent cause of failure. Most complaints of poor reliability
of engine-driven sets can be traced to unreliable, but necessary, peripheral
equipment such as fuel pumps, cooling systems, and so forth. Care should
therefore be exercised in the specification of all the engine system components.

Two typical stand-by power supply systems employing motors, generators
and alternators are shown in Figures 5.4q and r. In Figure 5.4q, the AC line
provides power until it fails. On failure, the battery supplies power to the
alternator, which, in turn, supplies the output voltage. In Figure 5.4r, the
AC motor has been replaced by a rectifier of the static variety. Now referring
to Figure 5.4r and to Figure 5.4g, it is easy to see that the two are identical,
with the motor-alternator set replacing the block marked "inverter." A system
employing an engine is shown in Figure 5.4s.

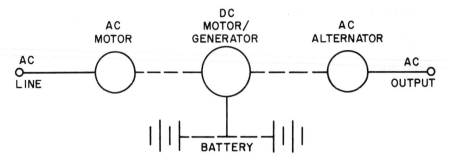

Fig. 5.4q   Stand-by power supply system with motors, generators and alternators

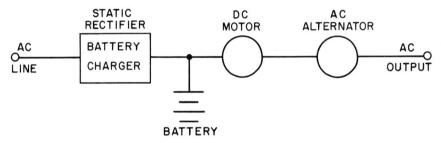

Fig. 5.4r   Stand-by power supply system with static rectifier

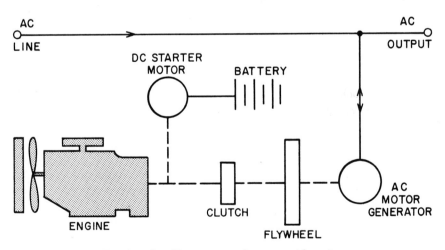

Fig. 5.4s   Stand-by power supply system with engine

Batteries

There are three types of batteries in general use for stand-by systems. They are the lead antimony, lead calcium, and nickel cadmium. The lead antimony and lead calcium are lead acid batteries deriving their name from the hardening material in the lead alloy. Both have approximately the same ampere-hour characteristics on discharge. The lead antimony construction costs less, requires more maintenance due to its higher internal losses, and evolves more hydrogen than the lead calcium. The life expectancies of 14 to 30 years are frequently quoted. The life depends on the construction of the plate and the plate thickness. It also depends in large measure on the care given the batteries in service. Both types can operate over a temperature range of $-10°$ to $110°F$. The lead calcium cell may be floated at 2.25 volts per cell without the necessity for equalizing charge or, at worst, with long periods between equalization.

Nickel cadmium batteries are the alkaline type. They differ from lead acid batteries in having a larger short-time current capability, higher cost, and lower volts per cell. Little hydrogen is generated by this cell and frequent overcharge is recommended. Life expectancy and operating temperature range are similar to the lead acid types.

Battery Chargers

The battery chargers generally used for stand-by systems are "float" chargers. They are characterized by relatively constant output voltage to recharge the battery as their output current varies from almost zero to rated value. Beyond their rated current, output voltage drops rapidly with increased load current. This current limit protects the charger when applied to the battery in the discharge state.

Satisfactory battery life is dependent on the design and operation of the battery charger and so are maintenance costs. The feedback techniques employed in battery chargers to maintain the constant output voltage and current limit are well known and will not be discussed in detail. The rectifiers themselves may take the form of either a polycrystalline cell, such as selenium or copper oxide, or a monocrystalline cell, such as germanium or silicon. The trend is toward the silicon rectifier. Control devices include the magnetic amplifiers and thyristors.

Motor-generator battery chargers are also available for stand-by system recharging service. The use of motor generators in this application predates that of the drive-type rectifiers.

Static Inverters

The static inverter employed in the stand-by system tends to be the most complex piece of equipment in that system. Since static inverters are relatively new, a very brief discussion will be provided to show how they operate.

Transistor inverters are the least expensive of the static inverters. Their principal area of usefulness is the low input voltage (24 direct volts and less) and low output capacity (500 voltamperes and less) range. This type inverter can operate at high frequency and can cease operation under dangerously high output current overloads.

Figure 5.4t shows a typical circuit for the center-tap transistor inverter;

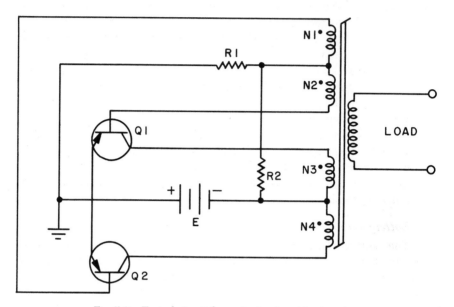

Fig. 5.4t   Typical circuit for center-tap transistor inverter

N_1 and N_2 are feedback windings, R_1 is the feedback resistor, and R_2 the starting resistor. The operating cycle can be traced by assuming Q_1 closed, Q_2 open. Substantially all the supply voltage E appears across N_3, causing a change in flux level by Faraday's law:

$$\Delta\phi = \frac{10^8 Et}{N_3} \qquad 5.4(1)$$

where $\Delta\phi$ = change in flux level,
$\qquad$ N = turns in N_3 and N_4 coil (identical),
$\qquad$ E = supply voltage (appearing across N_3), and
$\qquad$ t = time.

Eventually the core saturates, requiring an increase in exciting current. To supply this increased collector current, the transistor must have an increased base current. This cannot be supplied due to the decreased coupling between N_2 and N_3 because of core saturation. Q_1 begins to open, reducing exciting current. At this point, the change of flux reverses, reversing the coil voltage polarity. Q_2 is then turned on by N_1, Q_1 turned off by N_2. The half cycle begun in this fashion is similar to that described until a reversal occurs again completing the cycle.

Starting resistor R_2 provides enough base bias current to allow exciting current to flow. Natural circuit imbalance insures that only one transistor closes, thus starting the oscillator.

Should the load be short-circuited, no feedback is provided by N_1 or N_2 and oscillations cease. Other modes of operation are also possible. Care must be taken to provide the correct base current (which is dependent on transistor current gain) so that loading does not excessively shift the frequency. The amount of this shift is also dependent on the "rounding" of the B-H loop.

In the center-tap inverter circuit, each transistor must withstand a voltage equal to or greater than twice the supply potential. Transistors having a sufficiently high rating to withstand a 48-volt source are more expensive. Usually, above 24 volts, it is less expensive to go to the bridge circuit shown in Figure 5.4u.

The bridge circuit operates in a manner similar to that described for the center-tap circuit. Two starting resistors R_{s1} and R_{s2} are necessary. The diodes CR 1, 2, 3, 4 provide transient voltage suppression for unsymmetrically wound transformers.

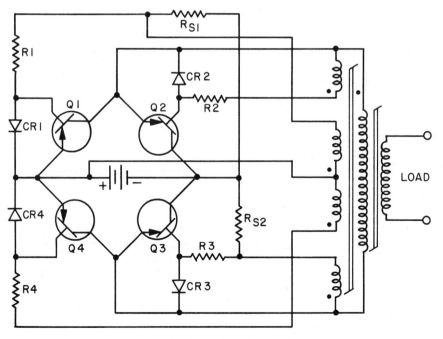

Fig. 5.4u   Bridge transistor inverter

A typical means of stabilizing frequency is shown in Figure 5.4v. Since the saturation flux density and turns are relatively constant, frequency is controlled by the supply voltage E. Use of a "zener" diode stabilizes this voltage and, therefore, frequency.

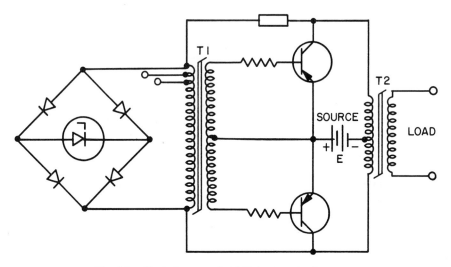

Fig. 5.4v   Typical means of stabilizing inverter frequency

Silicon-controlled rectifier-inverters are the "work horse" of static inverters. They operate efficiently and reliably at high input voltages (130 to 600 direct volts) and high output capacities (500 volt-amperes and larger). Proper specification of the equipment is essential to obtaining reliable operation.

The operation of a static inverter may be simulated by switches, as shown in Figure 5.4w. Switches 1 and 1^1 are operated in unison and switches 2 and

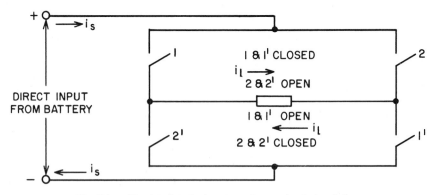

Fig. 5.4w   Simulated static inverter using mechanical switches

2^1 are also operated in unison. When 1 and 1^1 are closed, and 2 and 2^1 are open, load current flows in a direction shown by the arrow in Figure 5.4w. With 2 and 2^1 closed, and 1 and 1^1 open, load current flows in the reverse direction. Thus, while the source current i_s flows in the same direction when

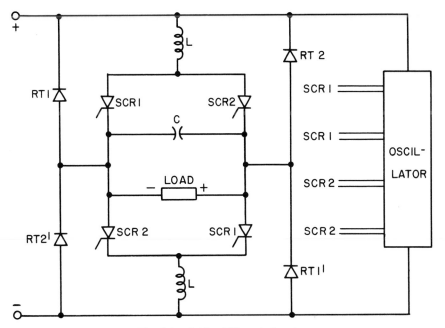

Fig. 5.4x   Bridge SCR static inverter

either set of switches is closed, the load current i_1 reverses polarity as each set is alternately closed and opened. An inversion of current has been performed by the circuit.

The switches used in Figure 5.4w are not static, since a switch contains moving parts. In Figure 5.4x, the mechanical switches have been replaced with electrical switches (silicon-controlled rectifiers). Also shown are the commutating inductors L and commutating capacitor C. These components are necessary to turn the controlled rectifiers off. Turn-on is accomplished by the application of voltage to the gate leads of the controlled rectifiers by the oscillator.

Rectifier diodes RT_1, RT_1^1, RT_2, and RT_2^1 are *not* a part of the basic inverter switching circuit. They serve to clamp the amplitude of the load voltage to a value approximately equal to the magnitude of source voltage.

Figure 5.4x shows the diagram of a bridge-connected static inverter. This arrangement is frequently used for source voltages of 130, 260, and 600 volts. For source potentials of 12, 24, and 48 volts, the circuit of Figure 5.4y is

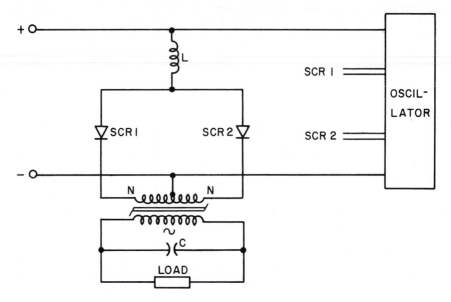

Fig. 5.4y   Center-tap SCR static inverter

frequently employed. Its operation is seen to be similar to that of the bridge circuit. It differs in that half the number of controlled rectifiers are used, and each must hold off a voltage approximately equal to twice the supply voltage.

Various types of output waveforms may be obtained from the square wave, which is the basic output waveform of the static inverter. Sinusoidal waveforms are most common, but triangular, sawtooth, and many rectangular combinations are also possible. Voltage stabilization may be a welcome bonus provided by the output wave-shaping circuitry. Current limiting is also possible.

Bus Transfer Switches

Bus transfer switches have historically been of the electromechanical type. Various techniques have been employed to speed the transfer from one source to the other. These techniques have included a pulsing arrangement on the coil of the electromechanical switch in order to overcome the inherent inertia. "Make-before-break" sequences have also been used.

A newer development has been the use of the static switch for more rapid transfers. Generally, thyristors have been used as the power-handling switching component. Figure 5.4z represents a hybrid transfer switch on the output side of the inverter. Momentary paralleling of sources is achieved by the "drop out" time of the electromechanical contactor. Fast turn-on is achieved by the thyristors. Sensing is done at the output of the inverter switch prior to the filter in order to anticipate output failure.

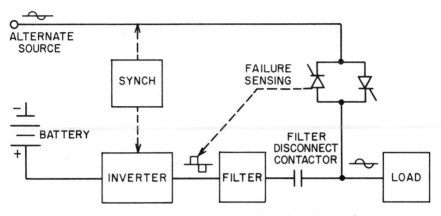

Fig. 5.4z   No-break power system with hybrid transfer switch

A static set which serves the same function is shown in Figure 5.4aa. In this case, momentary paralleling is achieved by the logic circuit which supplies the gating pulses.

Protective Components

It is essential to the proper operation of the emergency power supply system that adequate thought be given to the protective devices. Available as protective devices are fuses of various speeds and circuit breakers. The application of fuses to the load circuit was discussed earlier in this section. While it is not possible to provide a clear definition of the components to be used for the specific applications within a system, it is necessary to urge careful attention to the selection. Inrush current on start-up, transient variation of the input voltage and transient load variation each present their particular problems. A frequent experience is that the protective system may be the least reliable of the system components. This is not intended to mean that the component itself fails, but rather that the system fails through action of

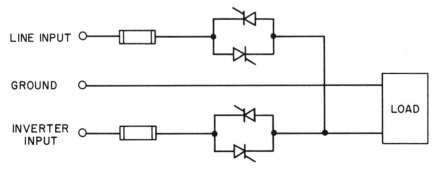

Fig. 5.4aa   Static no-break power system

the protective device under normal operating conditions. Whenever the stand-by power supply system fails to provide output, whether the components are damaged or not, the resulting damage is the same.

Stand-by Power Supply Systems

A number of stand-by power supply systems have been presented in the previous discussion. These have been categorized in terms of the type of power failure, but not by the class of system they represent. It was noted earlier that the systems themselves could be classified by virtue of the time it takes to transfer from the primary to the secondary power source. This grouping is more nearly akin to the thinking of the purchaser than is the classification by power failure noted previously.

Multicycle Transfer System

In general, multicycle transfer systems employ either electromechanical transfer switches or engine- or motor-driven equipment which must start up. Figures 5.4bb and 5.4cc show two composite systems using both rotating and static equipment. In Figure 5.4bb both a battery input and an engine-driven generating input is provided to the inverter. *If* the engine-driven generator is started prior to the time the energy contained in the battery is completely utilized by the inverter and load, *no* interrupt time occurs. Note that the commercial AC power input provides the load normally. On failure of the power lines, the contactor K_1 operates to insert the inverter. Since the inverter is started on the fallout of the contactor K_1, some start-up transient must occur in the inverter requiring some time interval between the failure of the commercial AC input source and the inverter source. Note that the use of a static switch in position K_1 would increase the speed of operation of the electromechanical contactor but would *not* materially affect the start-up of the inverter.

Note that a rearrangement of the components of Figure 5.4bb results in the system of Figure 5.4cc. This system may not have a transfer time, since the battery would normally be chosen to have a capacity sufficient to cover the energy required by the inverter and load during the period of time it takes to start the engine-driven alternator. Should that time interval be long, multicycle start-up will result. Note, however, that in this system there is no protection for inverter failure. The system of Figure 5.4cc does have a by-pass to the AC commercial line providing some protection in the event of component failure in the system.

Subcycle Transfer System

The subcycle transfer systems are generated by the use of static switches on the output side of the inverter. Thus, the circuits of Figures 5.4k, l, m,

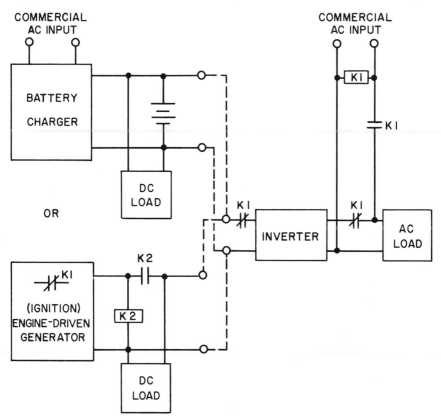

Fig. 5.4bb Composite multi-cycle transfer system using both batteries and engine-driven generator

n, aa and bb may be subcycle transfer systems depending entirely on the arrangement of the switch itself and, in the case of Figure 5.4bb, the start-up time of the inverter.

"No-Break" Transfer System

The so-called "no-break" transfer system may be no-break in the sense that if the AC commercial source fails, no cessation in output power results. No-break may also be applied to those systems having a redundant source on the output side of the inverter. Figures 5.4g, h, i, j and cc are examples of no-break systems with redundant input sources so that no break occurs in the output power should the AC commercial source fail. The switching systems shown in Figures 5.4z and aa can be used in many of the previously defined circuits to provide no-break switching under the right sequence of operations. Figures 5.4k, l, m and n are of this type.

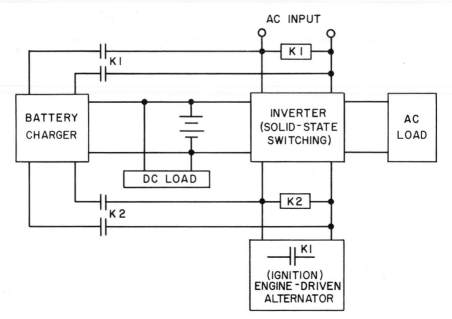

Fig. 5.4cc   Alternative arrangement of a multi-cycle transfer system

System Redundancy

The subject of system redundancy has been frequently mentioned in the previous discussions. Two basic classifications of system redundancy are made on the basis of input and output. Redundancy in the input circuit to the inverter occurs by using multiple battery sources, multiple input lines coming from separate power feeders, or by using various types of rotating alternators and generators with either motor or engine drives. Figures 5.4h, i and j illustrate the various types of input redundancy.

Redundancy in the output circuit of the inverter is obtained by providing a switch which will provide a path to an alternate source. Output redundancy may involve switching from the inverter output to the power line. More complex schemes involve switching from inverter standby system to inverter stand-by system. Figures 5.4k, l, m and n provide examples of output redundancy.

The previous examples have shown the wide number of choices available for component redundancy in stand-by power supply systems. In order to select the best system for a given application, it is necessary first to evaluate the degree of integrity required for the application. It is then essential that the stand-by system be evaluated to determine which component is most likely to fail. A decision can then be made regarding the cost of redundancy in that area, vs the needs of the application.

Specifications

Considerable activity is under way attempting to provide specifications for stand-by power supply systems and system components. It is essential that the user provide in his specifications certain types of information which are important in insuring system reliability. Some of these items will be noted.

Little need be said about the specification of the power handling capability of the stand-by power supply system. Nevertheless, it is suggested that consideration be given, not only to the immediate load requirements, but also to future load requirements. It is generally less expensive to purchase additional capacity when the first system is acquired than to add capacity to that system at a later date.

In many cases, the characteristics of the loads are most important. Power factor as well as transient characteristics should be clearly defined. These definitions are particularly important if static equipment is involved.

Transient data on the input sources are important in proper design of the system. If a battery source is employed, it is desirable to know the transients which can exist on the battery bus. These transients should be specified in terms of their maximum voltage, as well as their energy content. If an existing battery installation is used, any loads which are switched will generally institute a transient voltage due to the inductance of the lines themselves. A knowledge of this transient voltage is particularly important in the proper design of static equipment.

While it is difficult to obtain any meaningful data on the number and duration of outages of the utility power lines, it is necessary to provide information on the length of time the power supply must produce power for the loads without having the primary input source available. This evaluation can best be done on the time required to "shut down" the load system rather than with reference to the input failure.

Particular note should be taken of the characteristics of static inverters. Three overload ratings are important. In order that sufficient commutating capacity can be designed into the inverter, it is necessary to know the maximum instantaneous current. It is also necessary to know the overload current for a one- to two-minute interval in order that ample cooling be provided to the semiconductor devices. The final overload rating of importance is the one- or two-hour overload necessary in order to provide ample thermal capacity in the magnetic components. This third overload rating is also of importance in defining rotating equipment.

In the event that the stand-by system includes a means of transferring the load of the inverter output to an alternate source, the characteristics during transfer and retransfer should be amply defined. Among those characteristics is the time, when switching between the two sources, that zero voltage can be tolerated. Phase shift in voltage from one source to the other should also be stated together with the transient voltage characteristics on transfer or retransfer. These characteristics can be defined by an evaluation of the sensitivity of the loads to varying phase angle, frequency and transient voltage. If dynamic loads are included, the transient response and time constant of these loads should be stated.

Chapter VI

TRANSMITTERS AND TELEMETERING

C. H. Hoeppner, B. G. Lipták,
C. L. Mamzic,
A. F. Marks and J. Valentich

CONTENTS OF CHAPTER VI

INTRODUCTION   618
6.1 PNEUMATIC TRANSMITTERS AND CONVERTERS   619
     Signal Ranges   621
     Baffle-Nozzle Error Detector   621
     Force Balance Transmitters   622
          One-to-One Repeaters   622
          Pressure Transmitter (Force Balance)   625
     Motion-Balance Pressure Transmitter   626
     Transmitters Grouped by Measured Variable   627
          Differential Pressure Transmitter   627
          Square-Root-Extracting d/p Transmitter   628
          Rotameter Transmitter   630
          Temperature Transmitter (Filled)   631
          Buoyancy Transmitter (Level or Density)   633
          Force Transmitter   634
          Motion Transmitter   634
          Speed Transmitter   635
     Current-to-Air Converters   636
     Transmission Lag   638
6.2 ELECTRONIC TRANSMITTERS   640
     Force-Balance Transmitters   641
     Motion-Balance Transmitters   641
          Differential Transformers   643
          Photoelectric Transducers   644
          Capacitance Transmitter   645
          Potentiometric Transmitter   646
          Piezoelectric Transducer   646
     Physical Properties Transducers   647
     Chemical Properties Transducers   648
     Electronic Signal Types   649
          Voltage Signals   649
          Current Signals   649
          Resistance Detection   649

	Frequency Signals	651
	Phase Shift	652
	Digital Signals	652
	On-Off Oscillating Circuits	653
	Desirable Transmitter Features	653
6.3	Signal Converters	655
	Pneumatic-to-Electronic Converters	655
	Millivolt-to-Current Converters	656
	Voltage-to-Current Converters	657
	Current-to-Current Converters	658
	Resistance-to-Current Converters	659
	Other Converters	659
	Signal Conditioners	660
	Pulsation Dampeners and Snubbers	660
	Electronic Noise Rejection	661
6.4	Telemetering Systems	662
	Telemetry Methods	664
	Measuring and Transmitting	668
	Transducers	671
	Strain-Sensing Alloys	673
	Bonding Cements and Carrier Materials	677
	Power Sources	680
	Applications of Telemetry	680
	Chemical Plants	680
	Textile Mills	681
	Conveyor Chains	681
	Gear Train Efficiency	682
	Shaft Horsepower	682
	Power Plants	685
	Underground Cable Tension	686
	Earth-Moving Equipment	687
	Limitations of Telemetry	687
	Choice of Transducers	688
	Strain and Temperature Transmission	689
	Sensing Elements	690
	Transmitters and Batteries	690
	Receivers and Discriminators	691
	Recorders	691
	Antennas and Total System Operation	692
	Calibration	692
	Calibration Procedures	692
	Calibration Results	693

The definition of a transmitter given in the first volume of this Handbook is as follows: "A device that senses a process variable through the medium of a primary element, and that has an output whose steady-state value is a predetermined function of the process variable. The primary element may or may not be integral with the transmitter."

Thus, a transmitter is required neither from a measurement nor from a control point of view, because it serves only as an operating convenience by making the process data available in a centralized location, such as a control room. Because the measurement function is frequently incorporated in transmitters, several types are discussed in Volume I, which concerns process measurement. The coverage of pneumatic and electronic transmitters in the first two sections of this chapter is therefore concerned more with transmission and less with measurement, although some duplication is admittedly present.

The third section of this chapter discusses signal converters. Both transmitters and converters can be referred to as transducers, because a transducer, again as defined in Volume I, is "A general term for a device that receives information in the form of one or more physical quantities; modifies the information or its form or both; and sends out a resultant output signal. Depending on the application, the transducer can be a primary element, a transmitter, a relay, a converter or other device."

The fourth and final section covers the area of telemetering. Although the term telemetering refers only to the practice of transmitting and receiving the measurement of a variable for readout or other uses, it is most commonly applied to electric signal systems (such as radio) involving larger transmission distances.

The discussion presented in this chapter on the subjects of transmitters, converters and telemetering should provide sufficient information to cover the requirements of most chemical plants. The exception to this statement is a project involving computerized process controls. Signal transmission, signal wiring and input-output interfacing requirements on computerized systems are discussed in Sections 8.6 and 8.7 of this volume.

6.1 PNEUMATIC TRANSMITTERS AND CONVERTERS

Flow Sheet Symbols

PRESSURE
TRANSMITTER

INDICATING
TEMPERATURE
TRANSMITTER
(FILLED SYSTEM)

FLOW
TRANSMITTER

THERMOCOUPLE
CONVERTER-
TRANSMITTER

LEVEL
TRANSMITTER

CURRENT TO AIR
CONVERTER

Ranges:

FORCE from 15 lbf to 180,000 lbf.

LEVEL (buoyancy) from $1''$ to $60'$.

MOTION from $\frac{1}{8}''$ to $60''$.

619

	Pressure from 0.2″ H_2O to 80,000 PSIG.
	Pressure differential from 0.2″ H_2O to 1,500 PSID.
	Temperature (filled) spans from 50 to 1,000°F within the range of absolute zero to 1,400°F.
Accuracy:	Generally in the range of $\frac{1}{4}$% to 1% of full scale.
Cost:	Most pneumatic transmitters are priced between $150 and $500.
Partial List of Suppliers:	Bailey Meter Co.; BIF Industries; Black, Sivalls & Bryson, Inc.; Bristol Div., American Chain & Cable Co.; Fischer & Porter Co.; Fisher Controls; Foxboro Co.; Honeywell, Inc.; Kane Air Scale Co.; Kieley & Mueller, Inc.; Magnetrol, Inc.; Masoneilan International, Inc.; Moore Products Co.; Taylor Instrument Cos; Weber Air-Weight Co.; Weighing & Controls Co.

A pneumatic transmitter is a device that senses some process variable and translates the measured value into an air pressure which is transmitted to various receiver devices for indication, recording, alarm and control.

Pneumatic controllers date back to the turn of the century. Pneumatic transmitters, however, did not make their appearance until the late 1930s— some 25 years after electric telemetering had become an established practice. Before pneumatic transmitters were introduced, controllers were all direct-connected, i.e., they contained a measuring element connected to the process. This meant that the controllers and control boards had to be located close to the process.

Pneumatic transmitters were first developed as an alternative to expensive explosion-proof electric transmitters for use in medium-range signal transmission systems in refineries and chemical plants. It was quickly recognized that transmitters offered many advantages over the use of direct-connected controllers, recorders and indicators, such as safety, economy and convenience. They eliminate the need for connecting flammable, corrosive, toxic and pressurized fluids into the control room. Furthermore, since controls can be located remotely, centralized control rooms become practicable and such elements as long, gas-filled temperature-sensing bulbs with expensive armored capillary and with attendant bad ambient temperature errors and sensing lags become unnecessary. As a result, the process variable can be conveniently indicated, recorded and controlled on relatively inexpensive standardized receiver devices. Once introduced, transmitters caught on quickly and when miniature pneumatic controls were introduced in 1948, the concept was based on the use of pneumatic transmitters in all remotely controlled loops.

Signal Ranges

At first, each supplier settled on his own standard range of transmitter output. Generally, the span was selected to be compatible with commonly available pressure sensors and in some cases with the then commonly used operating ranges of pneumatic valve actuators. Too high a pressure would have placed extra demands on the piping and air supply system, whereas too small a span would have meant a sacrifice in resolution or accuracy. Also, the minimum pressure had to be some value above 0 PSIG, since the detecting nozzle back pressure in a transmitter does not (theoretically) drop to atmospheric. In fact, the closer the nozzle back pressure approaches zero, the more critical the seating and the greater the change in baffle clearance per increment of output. This, in turn, results in greater error due to hysteresis and nonlinearity. Using a "live" zero provides an added benefit in that if a transmitter failed, the reading would drop below zero on the scale and give immediate evidence of failure. Originally, ranges such as 2–14 PSIG, 3–18 PSIG and 3–27 PSIG were being used. The benefits from complete standardization became obvious, and by 1950 a 3–15 PSIG range was fast becoming the accepted standard. The 3–27 PSIG range had been used mostly in power plants and combustion-control systems, and it continued as one of the standard ranges. Formal recognition of standard ranges appeared in 1958 with the issuance of SAMA (Scientific Apparatus Makers Association) Standard RC2-1-1958. Three ranges were listed as standard: 3–15 PSIG, 3–27 PSIG and 6–48 PSIG. Today, however, 3–15 PSIG is overwhelmingly the most accepted.

Baffle-Nozzle Error Detector

The heart of practically all pneumatic transmitters and controllers is a baffle-nozzle error detector (Figure 6.1a). The circuit consists of a restriction, detecting nozzle, connecting chamber and baffle. The baffle is effective as long as the clearance is within one-fourth of the inside diameter of the nozzle. Beyond this clearance, the annular escape area is greater than the area of the nozzle itself and the baffle no longer provides a restrictive effect. The restriction must be small enough with respect to the nozzle so that with the baffle wide open the resultant nozzle back pressure is practically atmospheric. The restriction size also determines the continuous air consumption of the circuit, and it should be kept small for this reason. On the other hand, the restriction

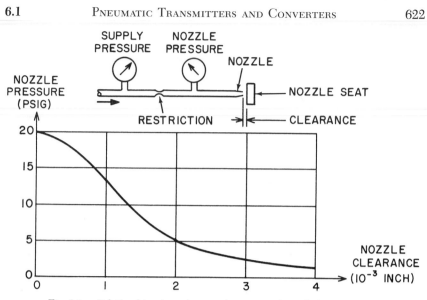

Fig. 6.1a   Relationship of nozzle seat clearance and nozzle back-pressure

cannot be too small, for then it could clog easily with dirt or foreign matter which may be present in the supply air. Conversely, the nozzle should be large enough to give the proper minimum back-pressure, but not so large that the clearance change for full-scale operation is so small as to require near perfect seating. With this in mind, a typical restriction size might be 0.012 in. ID and nozzle size 0.050 in. ID. For such a general configuration, a plot of nozzle back-pressure vs baffle clearance is given in Figure 6.1a.

Pneumatic detector circuits do take some other forms, some of which will be covered in the descriptions of individual transmitter types. The emphasis in this section is on the functioning of the pneumatic transmitter circuitry, since application and measurement details are covered in Volume I. It is not possible to cover all varieties of pneumatic transmitters in the space allotted but the ones selected for discussion are typical and cover the basic types, so that, collectively, they represent a good cross section of the variation in design features.

Force Balance Transmitters

One-to-One Repeaters

The force-balance principle is most commonly used in pneumatic transmitters. A very basic 1:1 force balance transmitter is shown in Figure 6.1b. Other repeater designs are discussed in Sections 1.9, 1.14 and 2.4 of Volume I. Process pressure acts downward on the flexible diaphragm, and the force resulting therefrom is counterbalanced by the force of nozzle back pressure acting upward on the diaphragm. Air, at a supply pressure slightly higher than the maximum process pressure to be measured, flows through the restriction to the underside of the diaphragm and bleeds through the nozzle to atmosphere. At equilibrium, the nozzle seat clearance is such that the flow of air in the nozzle is equal to the continuous flow through the restriction. If the process pressure increases, the diaphragm baffles off the nozzle, causing the back pressure to increase until a new equilibrium is achieved. If the process pressure drops, the diaphragm moves away from the nozzle, causing the back pressure to drop until equlibrium is reestablished. Since nozzle back-pressure is directly related to process pressure, the signal can be used remotely for indication, recording or control.

Though such a transmitter is accurate, it has some limitations. Namely, since all the air must flow through the restriction, the speed of response would be slow, particularly if there were much volume in the output side of the circuit. Also, a leak in the output side would cause an error or even make the unit completely inoperative. Figure 6.1c shows essentially the same repeater, but with refinements. This system employs a volume booster which has considerable air-handling capacity, so that it speeds up the response, minimizes transmission lag and copes with leakage. In addition, it provides a constant pressure-drop across the nozzle regardless of the level of operation. This improves accuracy with regard to linearity and hysteresis as the total diaphragm travel is lessened, and the relationship of nozzle seat position to nozzle back pressure is more linear than that of the direct nozzle circuit in Figure 6.1b.

The booster contains an exhaust diaphragm assembly consisting of two diaphragms which move integrally and which provide a bleed to atmosphere. On the underside of the diaphragm is a differential spring, which at balance exerts a force equivalent to 3 PSID acting on the diaphragm. Therefore, the

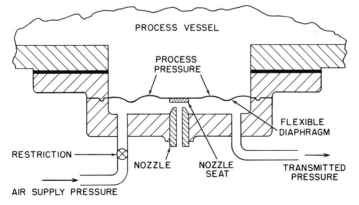

Fig. 6.1b   Simplified 1:1 force-balance pressure transmitter with direct nozzle circuit

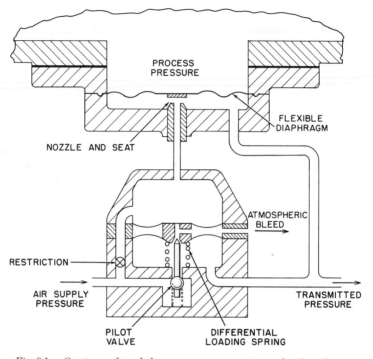

Fig. 6.1c   One-to-one force-balance pressure transmitter with volume booster
and constant differential nozzle circuit

nozzle back pressure which acts above the exhaust diaphragm is always nominally 3 PSI higher than the transmitted pressure. Since the nozzle bleeds
into the transmitted pressure, the pressure-drop across the nozzle is *always
constant* regardless of output pressure. As in Figure 6.1b, the air pressure on
the underside of the transmitter diaphragm counterbalances the process pressure. If the process pressure increases, the diaphragm moves the seat closer
to the nozzle, increasing the nozzle back pressure. This moves the exhaust
diaphragm downward, closing off the exhaust port as it contacts the pilot
valve, and moves the pilot valve downward to open the supply port. The
result is an increase in transmitted pressure until the transmitted pressure,
which is fed back to the underside of the diaphragm, equals the process pressure. A decrease in process pressure causes the nozzle back-pressure to drop,
and the exhaust diaphragm moves upward so that the pilot valve closes off
the supply port while the diaphragm opens the exhaust port, causing the
transmitted pressure to drop until equilibrium is established. At balance,
there is a continuous flow of air passing through the restriction, detection
nozzle and out to atmosphere via the exhaust diaphragm.

Pressure Transmitter (Force Balance)

The applications for pressure repeaters are limited. The more common transmitter is one which converts the measured process variable range into a standard 3–15 PSIG output pressure signal (Figure 6.1d). (Others are dis-

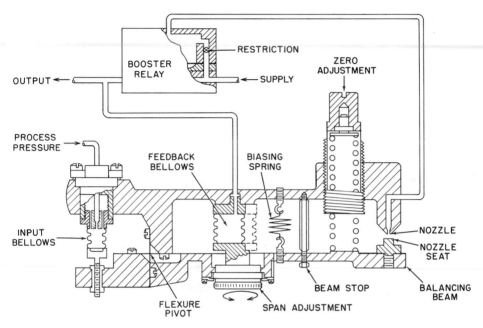

Fig. 6.1d   Force-balance pressure transmitter with variable range

cussed in Chapter II of Volume I.) It operates on a principle of force balance, or more precisely, on a principle of moment balance.

Process pressure acts on the input bellows and applies a force on the balancing beam which rotates on the flexure pivot. A change in input pressure results in a moment which is counterbalanced by an equivalent change in moment as output pressure changes in the feedback bellows. An increase in process pressure rotates the beam counterclockwise and moves the nozzle seat closer to the nozzle, increasing the back-pressure and booster output until moments come into balance. The biasing spring assures sufficient counterclockwise moment on the balancing beam so that, even with zero process pressure, it is possible to set the output at 3 PSIG by adjusting the compression of the zero spring. Added compression of the zero spring gives an elevated zero, or range suppression. The feedback bellows is eccentrically mounted on a rotatable seat. Rotating the assembly changes the moment arm, hence the span. The span is basically a function of the relative ratio of effective areas of the input bellows and the feedback bellows. The booster relay is an amplifying type which minimizes total nozzle clearance and improves accuracy.

Motion-Balance Pressure Transmitter

The pressure transmitter in Figure 6.1e consists of a pressure measuring element and a motion transmitter. Instead of a conventional baffle-nozzle, this transmitter employs an annular orifice with a variable restrictor called the wire pilot. Supply air passes through the fixed restriction into the follow-up

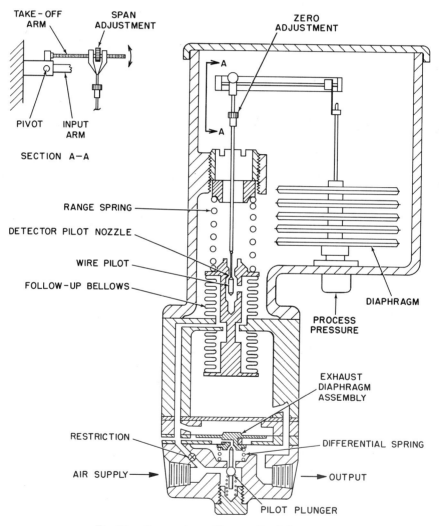

Fig. 6.1e   Pressure transmitter—motion balance type

bellows and out the detector nozzle. The wire pilot throttles the exhaust from the nozzle. It has a sharply tapered step such that when the large-diameter wire restricts the orifice the back-pressure rises to a maximum, and when the small diameter is effective the back pressure drops to atmospheric. At balance, therefore, the follow-up bellows moves to position the detector nozzle in line with the wire pilot's tapered step.

Process pressure acts on the measuring diaphragm. An increase in process pressure moves the diaphragm upward, which, via the U-shaped linkage, moves the wire pilot upward. The wire pilot restricts the annular orifice and the back pressure increases. The two bellows that make up the follow-up bellows system have the same area and are connected rigidly by a center post so that nozzle back pressure has no effect on movement of the bellows assembly. The nozzle back pressure is connected to the top of the exhaust diaphragm assembly. This is an amplifying type diaphragm assembly, since the upper diaphragm has six times the effective area of the lower. Therefore, as the nozzle back pressure increases, the output increases in a 6:1 ratio. The output feeds back to the underside of the follow-up bellows and pushes it upward. Upward motion is resisted by the range spring. The spring constant is such that a 12 PSIG change on the follow-up bellows moves the bellows assembly through the nominal full scale travel of the wire pilot.

Zero is adjusted by setting the initial operating position of the wire pilot. Span is adjusted by varying the radius of the take-off arm shown in section A-A. The total force required to operate such a transmitter is only two grams. Therefore, any type of primary element can be used with it—from low-force draft elements to various types of high-pressure elements. The basic range is determined by the spring rate of the measuring element.

Transmitters Grouped by Measured Variable

Differential Pressure Transmitter

A typical force balance differential pressure transmitter is shown in Figure 6.1f. The high and low pressures act on opposite sides of a diaphragm capsule, and the resulting differential exerts a force on the force bar. The force bar pivots on the diaphragm seal. The external end of the force bar pulls on one end of the range rod. The range rod, with its integral flapper, pivots about the range wheel. A feedback bellows acts on the opposite side of the range rod. A change in differential results in a changed flapper position which alters the nozzle back pressure, relay output and feedback force until all moments come to balance.

Zero is adjusted by adjusting the zero spring tension. Span is adjusted by moving the range wheel, which changes the relative input/output moment arm ratio.

Since the output of the transmitter is linearly proportional to pressure

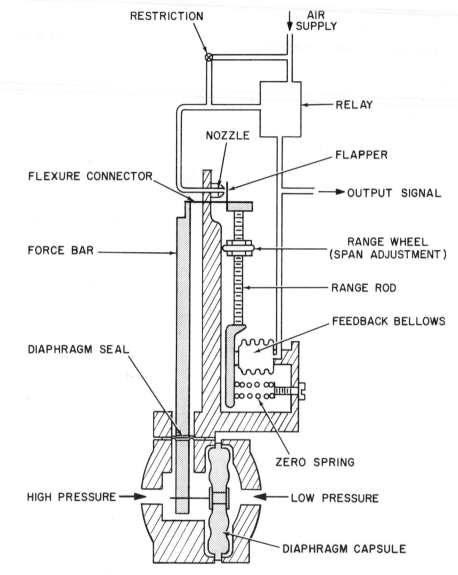

Fig. 6.1f   Differential pressure transmitter—force-balance type

differential, if the unit is used on an orifice type measurement, the flow would
have to be read on a square root calibrated scale.

Square-Root-Extracting d/p Transmitter

There are differential pressure elements such as in Figure 6.1g which
provide a motion output. A motion transmitter as in Figure 6.1e could be

used with such a meter to give an output linearly proportional to the differential. However, the motion transmitter shown in Figure 6.1g converts linear motion of the meter into a square root related output so that orifice type flow can be read on a linear scale. Linear signals are preferred when flows are added, subtracted or averaged and when other analog computing and characterizing requirements exist. They are often specified in order to give better readability and control rangeability.

The meter in Figure 6.1g consists of a high- and a low-pressure bellows joined by a common center shaft. The differential causes the bellows to move

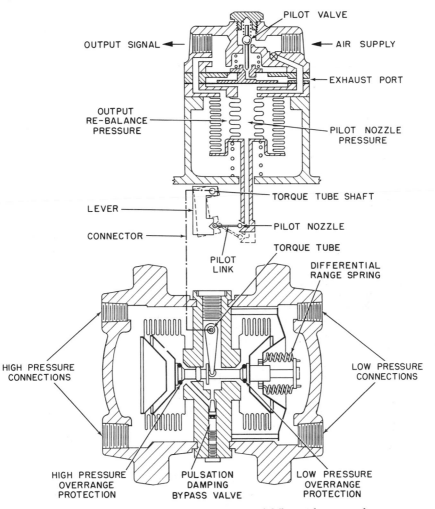

Fig. 6.1g   Linear flowmeter consisting of a combination of differential meter and square-root-extracting motion transmitter

a linearly proportional amount depending upon the total spring rate of the range spring, the bellows and torque tube. Movement of the bellows twists the torque tube causing rotation of the torque tube shaft. It is this motion which actuates the transmitter. The bellows are filled with liquid (usually ethylene glycol), and as the bellows move, liquid is transferred from one bellows to the other via the pulsation dampener needle valve. If the normal differential pressure range is exceeded, the bellows move until an O-ring on the center shaft seals off the liquid in the bellows. The pressure can then build up to the body rating of the meter without damaging the unit.

The torque tube rotates the lever connector of the motion transmitter. A floating pilot link is socketed in the connector at one end and baffles off the detector nozzle at the other. The transmitter is calibrated so that at zero differential the output is at 3 PSIG and the pilot link is horizontal. An increase in differential rotates the connector counterclockwise. This restricts the nozzle and causes a build-up in transmitted pressure, which in turn acts on the large bellows, moving it downward until equilibrium is attained. Since the pilot link has a fixed length and nozzle clearance is for all practical purposes constant, the pilot nozzle must be moved downward according to the cosine law, i.e., at minimum input with link horizontal, there must be considerable motion of the pilot nozzle as compared with the lever connector, and therefore the gain is practically infinite. As the included angle increases, considerably less nozzle travel is required to offset lever connector motion. For small angles, the relationship is almost exactly square root.

Another transmitter having the same function employs a varying spring rate with travel. This is accomplished by picking up added leaf springs as the travel increases.

Rotameter Transmitter

When a rotameter is used as a primary flow-measuring device, a magnetic means of float position detection is utilized. The motion of the magnetic follower mechanism then is converted into a 3–15 PSIG signal. (Section 5.7 of Volume I contains a detailed discussion of rotameters.)

In Figure 6.1h, a permanent magnet is embedded in either the float or in an extension of the float. A magnetic steel helix is supported in an aluminum cylinder mounted between bearings. The leading edge of the helix is constantly attracted to the magnet. The vertical position of the float results in a corresponding radial position of the helix. Attached to the helix follower assembly is a cam. The profile of the cam is sensed by a pneumatic detector circuit consisting of a transmitting and a receiving nozzle. The receiving nozzle pressure serves the same function as nozzle back pressure in a conventional baffle-nozzle circuit. When the flow from the transmitting to receiving nozzle is not interrupted by the cam, the receiver pressure is a maximum. When fully interrupted, receiver pressure is 0 PSIG. At balance, the detector system is throttled by following the cam profile.

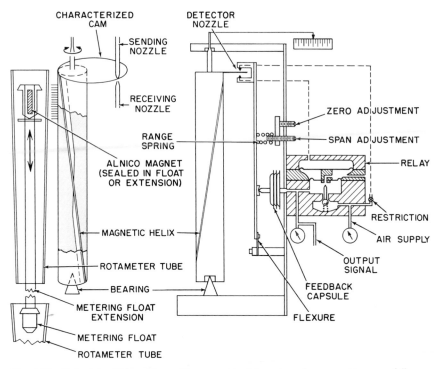

Fig. 6.1h   Rotameter transmitter with magnetic take-out and pneumatic cam follower mechanism

As the float rises, the cam rotates in a direction of decreasing displacement. As the cam edge moves away from the detector nozzles, the increasing receiver nozzle pressure acts on the relay and results in an amplified increase in transmitted pressure. The transmitted pressure acts on the feedback capsule and moves the flexure-mounted detector nozzles toward the cam edge until equilibrium is established. The spring rate of the range spring is set so that a 12 PSI change in output is required to track the full displacement of the cam.

Span is adjusted by turning a screw which takes up coils in the range spring, thus changing its spring rate. Zero is set by adjusting a second screw which sets the initial spring tension.

Temperature Transmitter (Filled)

Pneumatic temperature transmitters are almost exclusively the force-balance types. Figure 6.1i shows a transmitter with a sealed, gas-filled bulb (Class III system). (A detailed discussion of filled thermal systems can be found in Section 4.3 of Volume I.) Except for a negligible volume of gas in the thermal system bellows and connecting capillary, practically all the fill gas

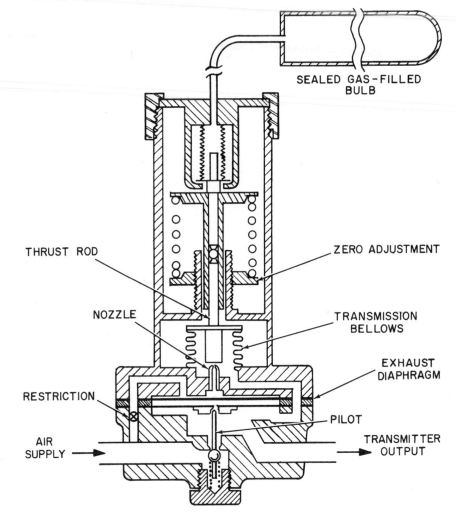

Fig. 6.1i   Force-balance temperature transmitter—filled bulb system

is in the bulb. The volume of the thermal system is constant, and as the bulb senses the process temperature, the pressure of the fill gas varies according to the gas laws. The fill gas pressure creates a force downward on the thermal system bellows. This force acts through the thrust rod and is counterbalanced by the force resulting from transmitted air pressure acting upward on the transmitter bellows. An increase in process temperature increases fill gas pressure, which pushes down on the thrust rod baffling off the nozzle. The consequent increase in nozzle back pressure pushes the exhaust diaphragm and pilot plunger down, closing the exhaust seat and opening the supply port until the increase in transmitted pressure acting on the transmission bellows results in force balance.

The temperature range is a function of the initial fill pressure (the higher the pressure, the narrower the span) and the ratio of effective areas of the thermal system and transmission bellows. The zero spring acts counter to the thermal system and establishes the low end of the measured temperature range.

Buoyancy Transmitter (Level or Density)

Figure 6.1j illustrates a motion balance buoyancy transmitter. (A more detailed discussion of displacement type level instruments is presented in

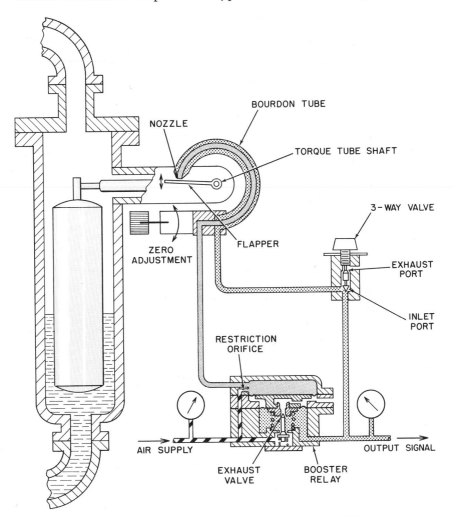

Fig. 6.1j   Level transmitter with torque tube spring and pneumatic follower system

Section 1.15 of Volume I.) Changes in level directly affect the net weight on the float lever as the float displaces the liquid. The float lever is connected to a torque tube. As the torque tube twists, it rotates a center shaft to which the flapper is attached.

Supply air flows through a restriction to the top of the booster relay and through a small tube inside the Bourdon to the detecting nozzle. The booster relay, which is an amplifying type with a 3:1 ratio, provides an output pressure, a portion of which is fed back to the Bourdon via a three-way valve. The three-way valve provides for span adjustment. If its plunger is moved up, closing off the exhaust seat, full transmitted pressure feeds back to the Bourdon. The result is that when the level rises and moves the flapper closer to the nozzle, the consequent increase in transmitted pressure makes the Bourdon move away from the nozzle (negative feedback) so that total flapper travel and hence measuring range is large. Adjusting the three-way valve in its other extreme position where it closes off transmitted pressure results in practically on-off action at the nozzle, representing the narrowest range of measurement. Normally the three-way valve is adjusted somewhere between these limits.

Zero is changed by rotating the Bourdon with respect to the flapper.

Buoyancy transmitters can be used to measure density or specific gravity as well as level and level interface. Force-balance designs are also available.

Force Transmitter

Force transmitters may serve as load cells in weighing applications or in the measurement of variables such as web tension. The transmitter in Figure 6.1k operates on a force-balance principle. (For a more detailed discussion of pneumatic load cells refer to Section 7.3 of Volume I.) Pressure from the tare regulator acts upward on the top diaphragm and is set to counterbalance the fixed weight of a hopper or tank and support structure. Net weight then is counterbalanced by output air pressure acting under the bottom diaphragm. Air flows from supply through a constant differential relay, which maintains a relatively constant flow across a restriction and into the net weight chamber. The net weight chamber is connected to the detector nozzle. The nozzle baffle is attached to the supporting platform. If net weight increases, the platform and flapper move down, baffling off the nozzle causing the back pressure to increase until force-balance is restored.

Smaller pneumatic force transmitters are also available. For stabilizing potentially noisy systems, some incorporate means of hydraulic pulsation dampening.

Motion Transmitter

A variety of approaches are used to detect motion, some of which are described in Section 9.4 of Volume I. A motion transmitter as shown in Figure

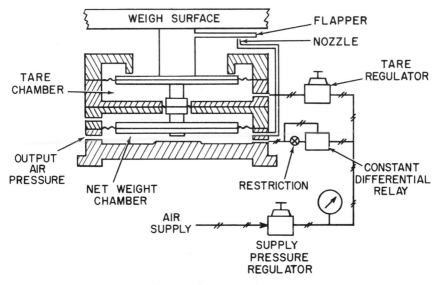

Fig. 6.1k   Force-balance weight transmitter

6.1e can be adapted to measure total motions from $\frac{1}{8}$ in. to approximately 1 in. This type is particularly useful where only a low force is available. Ordinary pressure regulators can be adapted as motion transmitters by substituting a sliding thrust rod for the lead screw which normally adjusts the pressure-setting spring. The most prominent approach, however, is to use a valve positioner and, in effect, reverse its function, i.e., instead of controlling position or motion, have it transmit a signal proportional to position or motion. Figure 6.1l shows a valve positioner connected as a motion transmitter. The pilot valve is reverse-acting, and the output feeds back into what is normally the input bellows.

If the prime mover pulls the connector downward, the parallel lever system pushes down and compresses the range spring. This opens the supply port and closes off the exhaust port of the three-way pilot, causing the output to increase. The output acts upward on the outer bellows to counteract the increase in spring force.

A roll type contact point establishes the gain of the lever system. Hence, its adjustment determines the span of measurement.

Speed Transmitter

In Figure 6.1m, a prime mover drives an input shaft which carries a multi-pole permanent magnet. The combination of magnetomotive pull and rotation of the magnet tends to turn the disc on its flexure pivot. The torque on the flexure is proportional to input shaft speed. Attached to the flexure is a radial force bar which doubles as a flapper and a rebalancing lever. Output

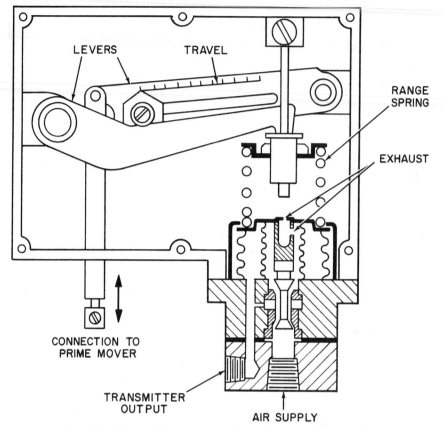

Fig. 6.11   Valve positioner connected for motion transmitter service

pressure feeds back to a ball type piston, and the force derived counterbalances the input torque.

As speed increases, an increase in counterclockwise torque results. This moves the flapper toward the nozzle. The relay amplifies the nozzle back pressure change, which serves as the output and as feedback to the ball piston. The pressure increases until the feedback moment balances the input torque.

Pneumatic speed transmitters are also made up by combining an electrical speed detector-amplifier with an electric-to-pneumatic converter. A typical detector is one which integrates the rate of magnetic pulses generated as gear teeth cut across the field of a magnetic pick-up head. (For more details refer to Section 9.3 of Volume I.)

Current-to-Air Converters

Electro-pneumatic transmitters are also called converters and transducers. They are extremely important, since they form the link between electrical measurements and pneumatic control systems. They also convert electronic controller outputs into air pressures for operation of pneumatic valves. (These devices are also discussed in Section 1.7 and are illustrated in Figures 1.7e and 1.7f.)

Figure 6.1n illustrates one of these converters and also lists the various electric devices with which it is commonly combined. The input is usually a DC current in the range of 1–5, 4–20, or 10–50 milliamperes. An Alnico permanent magnet creates a field which passes through the steel body of the transmitter and across a small air gap to the pole piece. A multi-turn, flexure-mounted voice coil is suspended in the air gap. The input current flows through the coil creating an electro-magnetic force which tends to repel the coil and thus converts the current signal into a mechanical force.

Since the total force obtainable in a typical voice coil motor with such

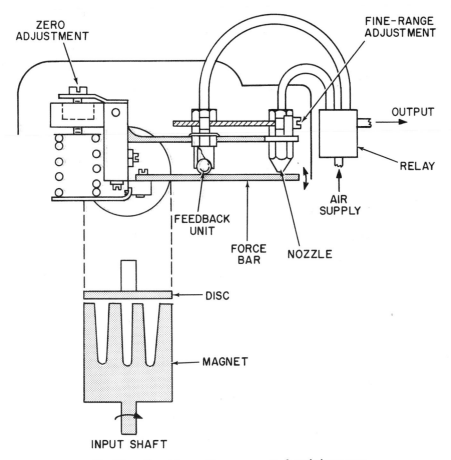

Fig. 6.1m   Speed transmitter—pneumatic force-balance type

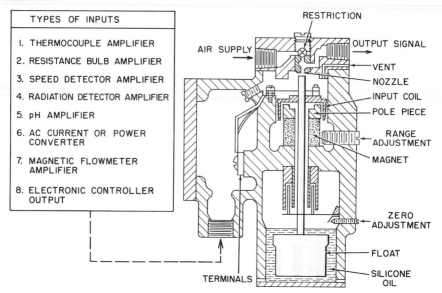

TYPES OF INPUTS
I. THERMOCOUPLE AMPLIFIER
2. RESISTANCE BULB AMPLIFIER
3. SPEED DETECTOR AMPLIFIER
4. RADIATION DETECTOR AMPLIFIER
5. pH AMPLIFIER
6. AC CURRENT OR POWER CONVERTER
7. MAGNETIC FLOWMETER AMPLIFIER
8. ELECTRONIC CONTROLLER OUTPUT

Fig. 6.1n   Electric-to-pneumatic transducer-transmitter with list of typical input sources

small current inputs is only in the order of some ounces, a different approach, namely, the use of a reaction nozzle, is employed to convert the force into a pneumatic output pressure. In this circuit, supply air flows through a restriction and out the detector nozzle. It is the reaction of the air jet as it impinges against the nozzle seat which supplies the counterbalancing force to the voice coil motor. The nozzle back pressure is the transmitted output pressure.

In order to make the transmitter insensitive to vibration, the voice coil is integrally mounted to a float which is submerged in silicone oil. The float is sized so that its buoyant force equals the weight of the assembly, leaving a zero net force.

Zero is adjusted by changing a leaf-spring force. Span is adjusted by turning the range-adjusting screw to change the gap between the screw and the magnet, thus shunting some of the magnetic field away from the pole piece.

Transmission Lag

With pneumatic transmitters, there is a transmission lag which increases with the length of transmission tubing. In order to get some idea of what this lag amounts to, the time for 63.2 percent complete recovery from a step input change is selected as the basis for comparison. The 63.2 percent figure defines the time constant in single-order systems, but since transmission systems

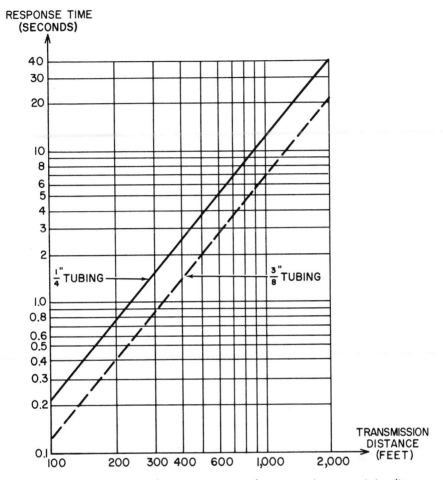

Fig. 6.1o   Response time (time for 63.2 percent complete response) vs transmission distance

do not behave as first-order systems, the figure cannot be so interpreted. Nevertheless, it does serve as an arbitrary benchmark for comparison.

A plot of response time vs transmission distance for $\frac{1}{4}$-in. and $\frac{3}{8}$-in. tubing is given in Figure 6.1o. Response is faster with $\frac{3}{8}$-in. tubing, but $\frac{1}{4}$-in. tubing is much more conventionally used. At 300 ft, for example, the response time for $\frac{1}{4}$-in. tubing is 1.5 seconds and for $\frac{3}{8}$-in. tubing, 0.8 seconds.[*]

The effect of transmission distance on pneumatic control is also discussed in Sections 4.1 and 4.3.

[*] Warnock, J. D., How Pneumatic Tubing Size Influences Controlability, INSTRUMENTATION TECHNOLOGY, February 1967.

PRESSURE
TRANSMITTER

FLOW
TRANSMITTER

LT

LEVEL
TRANSMITTER

TIT

INDICATING
TEMPERATURE
TRANSMITTER
(FILLED SYSTEM)

6.2 ELECTRONIC TRANSMITTERS

Ranges:

Force from 15 lbf to 180,000 lbf.
Level (buoyancy) from 1″ to 60′.

Motion from $\frac{1}{8}$″ to 60″.

Pressure from 0.2″ H_2O to 80,000 PSIG.

Pressure differential from 0.2″ H_2O to 1,500 PSID.

Temperature (filled) spans from 50 to 1,000°F within the range of absolute zero to 1,400°F.

Accuracy:

Generally in the range of $\frac{1}{4}$% to 1% of full scale.

Cost:

Most electronic transmitters are priced between $225 and $700.

Partial List of Suppliers:

Bailey Meter Co.; Baldwin Electronics, Inc.; BIF Industries; BLH Electronics, Inc.; Black, Sivalls &

640

Bryson, Inc.; Bristol Div., American Chain & Cable Co.; Fischer & Porter Co.; Fisher Controls; Foxboro Co.; General Electric Co.; Honeywell, Inc.; Martin-Decker Corp.; Motorola, Inc.; Taylor Instrument Cos.; Weighing & Controls Co.

An example of a simpler "transmitter," or transducer, is a thermocouple for measuring temperature. In this case, the temperature difference between the hot junction and the reference junction creates a DC voltage which is directly proportional to the temperature difference. This principle is explained in Volume I, Section 4.9. Note that in this case, the thermocouple serves two functions: (1) by converting temperature directly to another signal, it makes the primary measurement and (2) it creates an electrical signal that can either be amplified or go directly to the receiving instrument.

The many different kinds of transducers will be considered in several classes—the force-balance transducers, the motion-balance transducers, the physical properties transducers, the chemical properties transducers, and the direct electrical properties transducer/transmitters.

Force-Balance Transmitters

Figure 6.2a illustrates a force-balance differential pressure transmitter, in which the measurement that produces a force tends to move the top of the force bar. This tiny motion, acting through levers, moves the ferrite disc closer to the transformer, changing its output. This changes the amplitude output of the oscillator, which is rectified and then amplified to generate a DC milliampere transmitter signal. This output signal is fed back through the voice coil on the armature of the force motor, which is in series with the output terminals. When this feedback moment is equal to the moment created by the measurement force F_2, the force bar is again in its original position and the amplifier signal stabilizes. The similarity between this unit and its pneumatic equivalent, illustrated in Figure 6.1f, is obvious.

The advantage of force-balance units over motion-balance devices is that by eliminating motion one eliminates flexures, springs, etc. Further, by always returning to the same position, hysteresis is minimized and greater accuracy can be obtained. In general, a force-balance transmitter cannot be used to produce digital signals without the use of a supplementary device external to the transmitter, such as an analog-to-digital converter (see Section 8.7).

Motion-Balance Transmitters

In a motion-balance transmitter (Figure 6.2b) the process measurement produces motion against a calibration spring, resulting in a change of position corresponding to a change in the process variable. This position is detected by a transducer. The output of the transducer is amplified and an electric feedback signal is used to stabilize the amplifier. Depending upon the type of transducer and the signal level it generates, the amplifier may not be required but may be a part of the receiver.

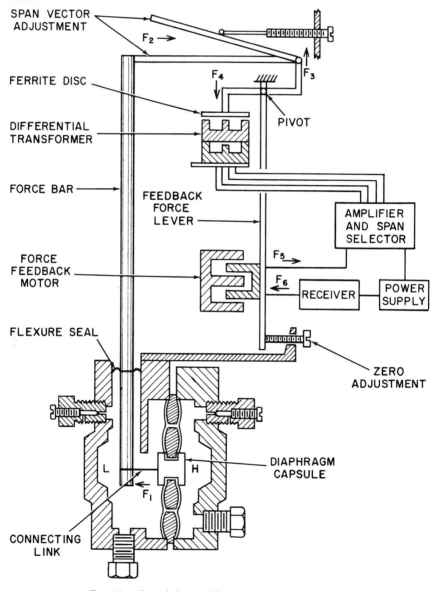

Fig. 6.2a   Force-balance differential pressure transmitter

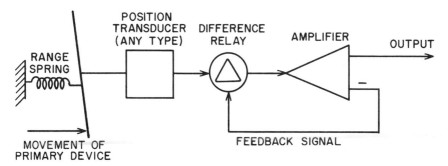

Fig. 6.2b   Motion-balance transmitter

Transducer principles and methods are as numerous as the number of manufacturers.

Differential Transformers

One of the most frequently used transducer principles is known as the differential transformer. Two different designs are shown in Figures 6.2c and 6.2d. Except for the type of motion and the design of the transformer core and windings, these units are very similar.

The linear variable differential transformer (LVDT), illustrated in detail (A) of Figure 6.2c, consists of a transformer coil with a single primary winding and two symmetrically spaced secondary windings. The core or armature is a cylinder of magnetic material, such as ferrite, which can be moved within the "air gap" of the windings.

When AC excitation is applied to the primary winding and the armature slug is centered, or is in the "null" position the induced AC voltage in the two secondary windings is equal and is in the same or opposite direction, depending on the method of winding. If the two secondary windings are connected in series with the voltages opposed, Figure 6.2c, detail (B), they will cancel out and give a "zero" or null reading when the slug is in the null position. As the slug is moved closer to coil A and farther from coil B, as shown in detail (C), the output voltage increases in the direction of the coil A output, and this increase is proportional to the displacement of the slug.

Another typical hookup is shown in detail (D), with a rectifier in the output of each secondary coil and hooked up to a DC zero center meter. The meter will read zero at the null slug position and plus or minus for slug positions displaced from the center.

The DC output of such a differential transformer can be amplified and used directly as a transmitted signal (as in Figure 6.2b), or it can be used as a position detector of other devices as shown in Figure 6.2a. Another differential transformer design is shown in Figure 6.2d, and there are several others.

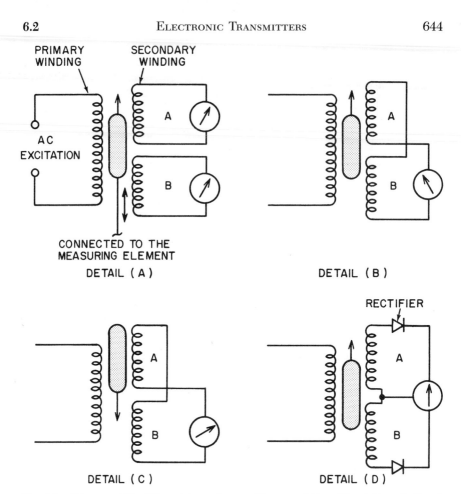

Fig. 6.2c   Linear variable differential transformer. (Courtesy of Consolidated Controls Corp.)

The excitation in a differential transformer must be supplied by an AC circuit. The AC source can be the receiving instrument using an AC transmitted signal or it can be a part of the amplifier in the transmission part of the transmitter, powered by a DC supply taken from the receiver. Alternatively, a power supply can be provided as part of the transmitter, but-this requires additional wiring.

Photoelectric Transducers

Figure 6.2e shows a typical schematic of a photoelectric transducer where the position of the photocoder is proportional to the motion of a primary sensing element. Light from the source shines through perforations in the shutter to energize some photoelectric cells. The output of these cells is

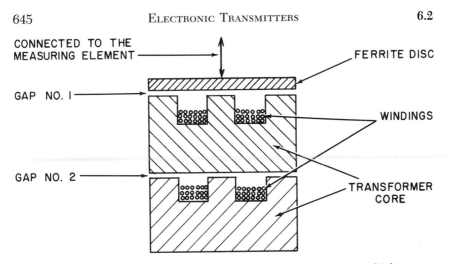

Fig. 6.2d   Differential transformer with ferrite disc. Relative spacing of gaps 1 and 2 determines the output voltage.

scanned, and the pulses are amplified to produce a digital signal or are rectified to produce a DC analog signal.

Capacitance Transmitter

Figure 6.2f shows a capacitance type pressure or differential pressure transducer, which is truly of the motion-balance type. Positioned between two fixed capacitor plates is a highly prestressed, thin metal diaphragm. This forms the separation between two gas-tight enclosures which are connected

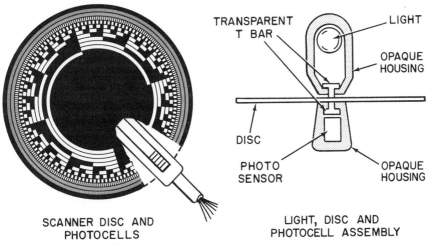

Fig. 6.2e   Photoelectric encoder transducer

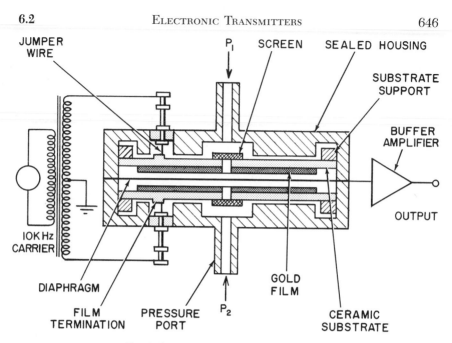

Fig. 6.2f   Capacitance position transmitter

to the process. The difference in pressure between the two chambers produces a force which causes motion of the diaphragm. This moves the diaphragm closer to the fixed capacitor plate of the low-pressure chamber. The transmitter is excited by a 10-KHz AC carrier current, and the unbalance produces a 10-KHz voltage with an amplitude proportional to the difference in pressure.

Potentiometric Transmitter

Figure 6.2g shows a potentiometer (resistance) driven by a Bourdon tube in a manner similar to the movement of a pressure gauge. Rotation due to a pressure change turns the shaft of a precision potentiometer, and the change in resistance is proportional to the process pressure.

Piezoelectric Transducer

Figure 6.2h shows the cross section of a piezoelectric accelerometer transducer in which the change in the amount of strain in the piezoelectric crystal generates a minute voltage. The crystal is sandwiched between a mass and the base. The entire assembly is held together and given an initial strain by the nut on the threaded center bolt. When the transducer is moved suddenly upward, the force on the crystal is increased by a force equal to the mass multiplied by the acceleration. This increased force changes the strain in the crystal and generates a voltage. Since there is no displacement other than

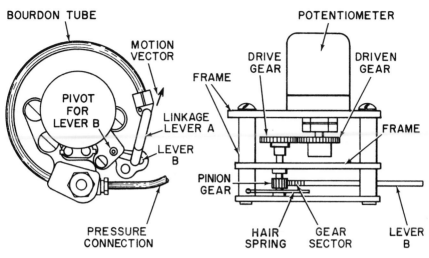

Fig. 6.2g   Potentiometric transmitter

the minute compression of the crystal, this device is a borderline case between a force-balance and a motion-balance device.

Other types of motion transducers include strain gauges, synchrotype servomechanism resolvers, and others.

Physical Properties Transducers

Several types of transducers measure physical properties directly, due to the physical property changing some electrical property. The best known of these are thermocouples, resistance bulbs, and thermistors, in which a

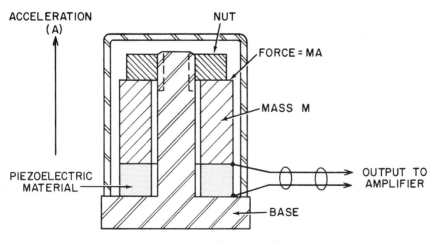

Fig. 6.2h   Piezoelectric transducer

change of electrical properties occurs as temperature changes. Less well known, but similar in character, are piezoelectric crystals sensitive to pressure and/or temperature, photocells sensitive to the intensity of light, capacitance devices sensing the dielectric properties of materials, density instruments, viscosity sensors, thermal conductivity detectors, mass spectrometers, and many others.

Since most of these devices are performing a primary measurement function their description can be found in Volume I.

Chemical Properties Transducers

The pH meter exemplifies the best-known method of direct chemical measurements by electrical means. The potential produced between two electrodes is proportional to the hydrogen ion concentration. There are many other selective ion probes available and more are constantly being developed. A typical ion electrode scheme is shown in Figure 6.2i.

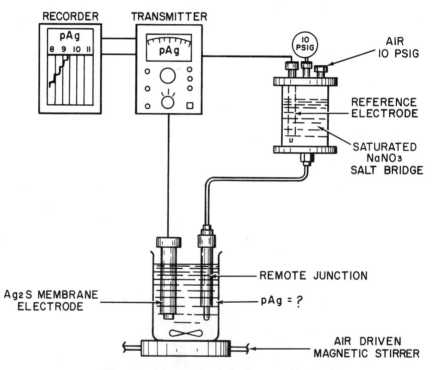

Fig. 6.2i　Selective ion electrode measurement

Electrical conductivity measurements can also be considered as chemical in nature, because the conductivity is proportional to the ion concentration. Conductivity measurement is used to detect ion concentration and is also used to sense the level of conductive liquids. Further details on conductivity type measurements are given in Sections 1.6 and 8.15 of Volume I.

A listing of transducers commercially available in the United States, a compilation of the manufacturers' specifications, has been published by the Instrument Society of America.[1]

Electronic Signal Types

There are many other types of transducers besides those that produce AC or DC output signals. Some have no output at all but are passive, exhibiting electrical properties that have to be measured, such as resistance, capacitance, reluctance, reactance, or even open and closed contacts.

Some of these forms of electrical signals are more convenient than others, and a brief discussion of the advantages and disadvantages of each is presented in the following paragraphs.

Voltage Signals

DC voltage signals cannot be accurately measured in systems in which a current is allowed to flow, but can be measured by potentiometric techniques even when the signal level is as low as 0–1 millivolts. The length and resistance of signal wires are not factors here, because a potentiometer measures at zero current flow by balancing a known voltage against the unknown voltage until current flow stops. DC voltage signals must be made immune to noise by filtering techniques at the receiving instrument and by shielding the leads.

AC voltage signals, on the other hand, are not readily measured at low levels because of the difficulty in removing noise and spurious voltage peaks caused by induction onto the signal leads. Lead lengths are also limited by the inductive and capacitive effects of the wires. Twisting and shielding of the leads and the use of coaxial cables do help to reduce these effects.

Current Signals

DC current signals have found the widest acceptance in industrial electronic control systems in the United States, with current levels varying from 1 to 50 milliamperes. Two ranges are more widely used than others: 4–20 DC milliamperes and 10–50 DC milliamperes. Both of these signal levels are sufficiently high to minimize the need for special wiring, although shielding of signal cables and location away from wiring carrying heavy loads is still advisable when the signals are to be used as computer inputs. Voltage signals can easily be derived from current signals by inserting a series resistor in the wiring and measuring the voltage drop across it.

AC signals are not used for remote transmission.

Resistance Detection

Resistance is not a signal but an electrical property that can be readily measured to very high accuracies by the use of DC bridge circuits, such as the Wheatstone bridge shown in Figure 6.2j.

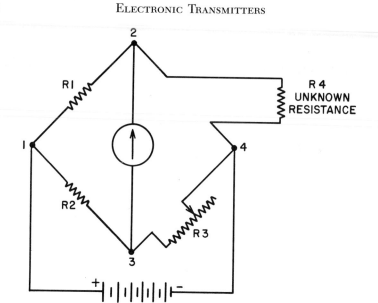

Fig. 6.2j   Resistance detection with Wheatstone bridge

The classical Wheatstone bridge works as follows:

1. The battery causes a voltage difference to be exerted between points 1 and 4, and the resulting current through each branch is proportional to the resistance through that branch. In the center is a galvanometer that can measure very small currents.

2. When the galvanometer registers no current flow, the voltage at points 2 and 3 are equal. This can occur only when,

$$\frac{R_3}{R_1 + R_3} = \frac{R_4}{R_2 + R_4} \qquad 6.2(1)$$

3. R_1, R_2 and R_3 are all known, and R_1 can be made to equal R_2 and be constant. Therefore the unknown resistance R_4 will equal R_3:

$$R_4 = R_3 \qquad 6.2(2)$$

Bridge circuits such as the Wheatstone bridge have been modified in that the galvanometer has been replaced with a differential amplifier. The battery may be replaced by a constant voltage source, and the output of the amplifier may drive a servo motor that operates potentiometer R_3. The principles, however, are still the same. Accuracies of one part in 10,000 are obtainable with laboratory versions of this instrument.

The strain gauge principle of measurement also uses the change of resistances. The resistor is cemented to a support or spring connected to the load so that it is distorted upon straining the material to which it is fastened. Distortion of the resistor wire causes changes in its resistance. This operating principle is also described in Sections 2.8 and 7.4 of Volume I.

Frequency Signals

Because of the difficulties in obtaining accuracy with alternating voltage and current signals, various other methods have been developed for use in alternating current measurements. Alternating current frequency can be accurately measured by several methods; therefore, frequency and frequency shift detection techniques are useable to measure variables such as capacitance, reactance, reluctance, and transformer effects. A typical block diagram for a tuned oscillator circuit is shown in Figure 6.2k.

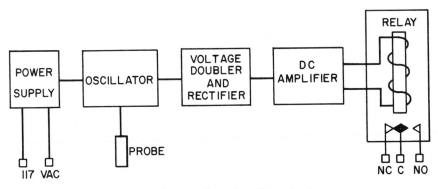

Fig. 6.2k   Typical tuned oscillator circuit

A variable capacitance from the probe, dependent upon the material level in the tank, determines the voltage amplitude of the oscillator section in this circuit. A rising level or capacitance on the probe lowers the amplitude of the oscillator voltage. This voltage is doubled, rectified and amplified in the DC amplifier. Depending on the position of the "fail-safe high or low level" switch, this voltage turns the last transistor on or off. The transistor controls the current through the relay coil, thereby controlling the position of the relay contacts.

An example of variable frequency signals for transmission is the AC tachometer, in which a magnetic slug, gear tooth, key in a shaft, etc. are counted by the pulses created in a coil (Figure 6.2l). When the magnetic slug comes across the face of the coil, a voltage is generated first in one direction as the slug enters, then in the opposite direction as it leaves. These pulses are a direct measurement of the speed of shaft rotation. When counting gear teeth where the space between them is similar to the width of the teeth, a smooth AC without wide gaps between pulses can be generated. The design

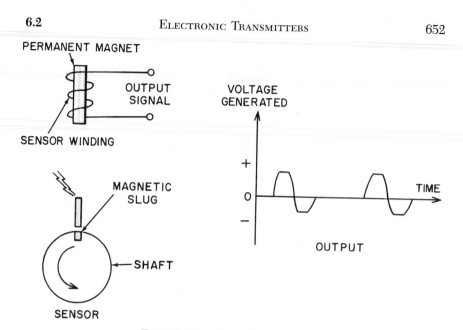

PERMANENT MAGNET

OUTPUT
SIGNAL

SENSOR WINDING

MAGNETIC
SLUG

SHAFT

SENSOR

VOLTAGE
GENERATED

+

O

−

TIME

OUTPUT

Fig. 6.21   Magnetic tachometer sensor

of the sensor probe has to be made with knowledge of the tooth shape, size, etc., so that good waveforms can be generated for easy measurement.

Phase Shift

The introduction of a reactance or a capacitance in an oscillating circuit changes the relationship between the peak voltage and the peak current flowing in the circuit. The measurement of the phase angle is a direct measurement of the variable.

Digital Signals

With the increased use of computing and logging systems, there has been a marked increase in the use of digital transmitters for high-accuracy data transmission. The early applications that led the way in digital transmission were tank farm level transmitters, using coded contact wheels, telephone dials, multiple openings of a single contact, and photo scanning. All of these were developed prior to 1960. Pulse type tachometers were also an early development in digital transmission.

More recent systems, such as the aerospace transmitters using optical encoders, have been developed with greater precision at higher cost. The use of these in the process industries is limited, as they require more maintenance and are not as rugged as other types.

On-Off Oscillating Circuits

One of the developments in using capacitive and reactive transducer methods has been the design of capacitance probes for level measurement. The principle is the same as the frequency shift principle except that the operating frequency of the oscillator is crystal controlled so that it does not oscillate when the capacitance is changed. The output of the oscillation amplifier is sent to the relay coil. When the circuit stops oscillating, the relay opens.

Desirable Transmitter Features

The characteristics most desirable for an electronic transmitter are high resolution, reliability and low cost. In order to meet these requirements, a transmitter should be designed to have the following features:[2]

1. Small size and weight for easy installation and maintenance.
2. Rugged design to withstand industrial environment.
3. Minimum dependence on environmental conditions for accuracy, which necessitates good temperature stability, being unaffected by barometric change, weatherproofing, etc.
4. Elimination of the need for adjustment due to load or line resistance variations.
5. No potential hazard to personnel, equipment or explosive atmospheres. Low-voltage operation with limited current capacity assists in eliminating these problems. By definition, intrinsically safe equipment provides all three kinds of protection. (See also Section 10.11 of Volume I.)
6. Convenient means for accurate field calibration and maintenance. In electrical transmitters this usually means conveniently located test terminals. In explosion-proof designs, calibration should be possible without opening the housing.
7. Capacity to operate during voltage dips and power outages. Generally this is accomplished elsewhere in the control system. (Stand-by power supplies are discussed in Section 5.4.)
8. Minimum number of transmission and power wires. Many systems are now available in which the transmitter is powered from the receiver over the same wires that are used for transmission.
9. Compatible output with both measuring and controlling instruments.
10. Provision for voice communication over the transmission wires between transmitter and receiver locations.
11. Optional local indication of output signal being available.
12. Circuitry designed to make troubleshooting and maintenance easy.

Most systems today are solid state (no tubes) and are mounted on circuit boards or are encapsulated. Plug-in components make fast repairs easy, but encapsulated modules are considered to be throwaway items.

REFERENCES

1. *ISA Transducer Compendium*, E. J. Minar, ed., and P. A. Recchione, asst. ed., Plenum Press, New York, 1963.
2. H. Blake, "Transmitters Transducers With a Purpose," presented at the American Gas Association, Inc., Transmission Conference–Automation Session in Cleveland, Ohio, May 27–28, 1968.

6.3 SIGNAL CONVERTERS

AIR TO CURRENT
CONVERTER
(WHICH BASICALLY
IS A PRESSURE
TRANSMITTER)

OTHER POSSIBLE
LETTER COMBINA-
TIONS TO DESIGNATE
CONVERTERS ARE:

A/D, D/A,
E/D, E/E, E/I, E/P,
I/D, I/E, I/I, I/P,
P/D, P/E, P/I, P/P,
R/D, R/E, R/I, R/P,
etc.,

WHERE THE LETTERS
USED REPRESENT:

A - ANALOG
D - DIGITAL
E - VOLTAGE
I - CURRENT
P - PNEUMATIC
R - RESISTANCE

Signal Ranges:	PNEUMATIC: 3–15, 3–27, 6–30 PSIG.
	VOLTAGE: 0–1 millivolts DC to 1–5 volts DC or 0–5 volts AC.
	CURRENT: 1–5, 4–20, 10–50 millivolts DC.
Accuracy:	Generally in the range of $\frac{1}{10}$% to $\frac{1}{2}$% of full scale.
Cost:	From \$150 to \$600.
Partial List of Suppliers:	Acromag, Inc.; Bailey Meter Co.; Bristol Div., American Chain & Cable Co.; Fischer & Porter Co.; Foxboro Co.; General Electric Co.; Honeywell, Inc.; Motorola, Inc.; Scientific Columbus, Inc.; Taylor Instrument Companies; Voltron Products, Inc.; Westinghouse Electric Corp.

Pneumatic-to-Electronic Converters

The pneumatic-to-electronic transducer is used wherever pneumatic signals must be converted to electronic signals for any one of the following reasons:

    a. Transmission over large distances

    b. Input to an electronic logger or computer

    c. Input to telemetering equipment

    d. Instrument air not available at the receiver controller

In principle, any of the electronic pressure transmitters could be used, but in practice, special devices are utilized to improve accuracy. The air signals are at low pressure levels, (3–15 PSIG), and many of the pressure detectors are not sensitive or not linear enough at these pressures. A P/I transducer should be at least a $\frac{1}{2}$ percent accurate device and preferably a $\frac{1}{4}$ percent to preserve the integrity of the initial signal. Since the total error is the square root of the mean squares of the individual component errors, the greater the precision of the P/I transducer, the better the signal.

Because of this need for accuracy, most P/I transducers use a bellows input and a motion balance sensor. A typical high-quality P/E converter is shown in Figure 6.3a.

Electronic-to-pneumatic converters are discussed in Sections 1.7 and 6.1.

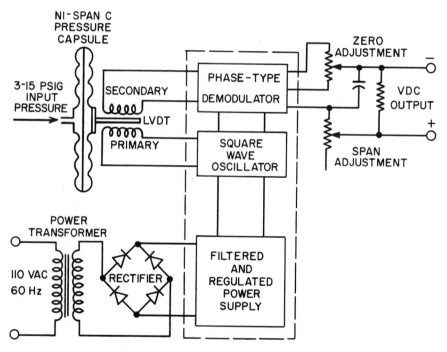

Fig. 6.3a   Pneumatic to electronic converter

Millivolt-to-Current Converters

Millivolt-to-current converters are widely used in the measurement of temperature, using thermocouples or other millivolt-generating sensing elements. They are also utilized in converting the output signals of analyzers into higher level transmission signals. A typical millivolt-to-current converter is illustrated in a block diagram form in Figure 6.3b.

When these devices are used to convert thermocouple outputs, they are also referred to as temperature transmitters. Several companies in the mid-

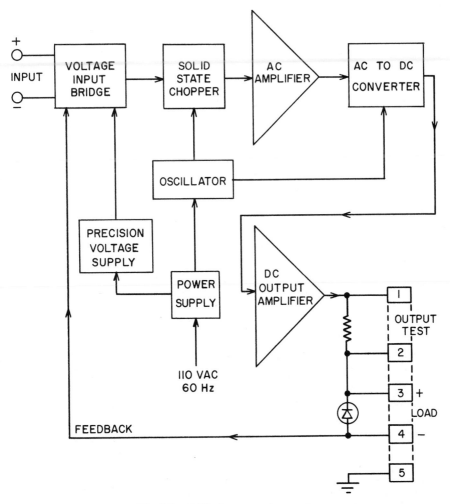

Fig. 6.3b   Millivolt-to-current converter

1960s developed a new, miniaturized converter made to be locally mounted in the thermocouple head directly on the thermowell (Figure 6.3c).

Voltage-to-Current Converters

Converters are also available for the conversion of higher voltages into transmission signals (Figure 6.3d). These usually consist of voltage dividers

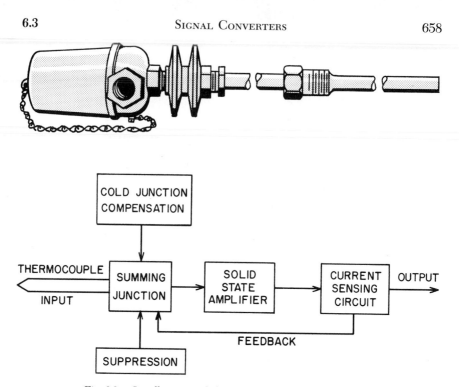

Fig. 6.3c   Locally mounted thermocouple to current converter

(and rectifiers if necessary) to reduce voltages to a level compatible with the receivers.

Current-to-Current Converters

Current-to-current transducers are available to convert AC signals to DC (or DC to AC) and to amplify or reduce their levels as necessary. They are sometimes used in the power industry, but not too widely, because there are other methods of receiving current signals which are capable of handling higher power levels. Levels of alternating current can be changed, when

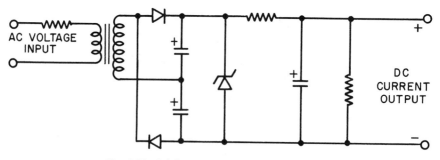

Fig. 6.3d   Solid-state voltage-to-current converter

necessary, by current transformers. A common signal level in the power industry is 0–5 amperes. Direct currents are reranged by putting a series resistor (low ohms) in the circuit and reading the voltage drop across it.

A converter of AC to DC milliamperes is shown in Figure 6.3e. This

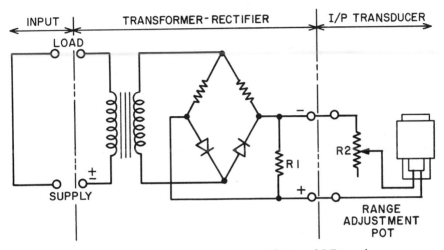

Fig. 6.3e   AC to DC milliampere converter with integral I/P transducer

figure shows three separate devices: a current transformer, an AC to DC milliampere converter and a current-to-pneumatic converter previously discussed. The function of the transformer is to scale down the current to the range normally used for direct AC metering. The AC/DC converter makes this signal compatible with the usual DC milliampere transmission systems. DC to DC "converters" are sometimes used for isolation of electrical circuits, such as with intrinsically safe systems. In this case, it is usually called a current repeater or barrier repeater.

Resistance-to-Current Converters

Resistance measurements are common in temperature measurements and in resistance or strain gauge sensors. The circuits used are similar to those of the millivolt-to-current converters, except that the front end is a resistance bridge instead of a voltage bridge (Figure 6.3f).

The strain gauge bridge, a special form of resistance element, is described in Sections 2.8 and 7.4 of Volume I. The strain gauge elements may take the place of two of the resistors in the resistance bridge shown in Figure 6.3f.

Other Converters

Analog-to-digital and digital-to-analog converters are covered in Section 8.7. Similarly, the analog-to-on-off and digital-to-on-off devices, such as monitor switches, relays, counters etc. are discussed in Chapter IV.

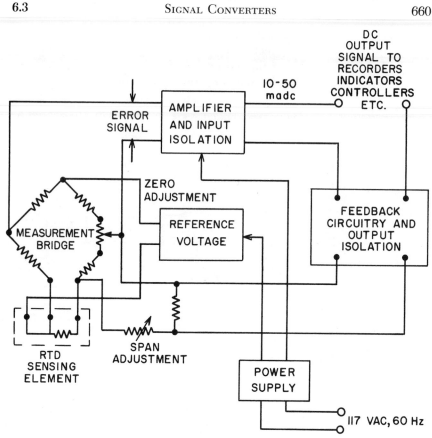

Fig. 6.3f   Resistance-to-current converter

Signal Conditioners

Pulsation Dampeners and Snubbers

Pressure systems, especially liquid-filled ones, transmit process noise very rapidly. An outstanding example is the output from a reciprocating compressor or pump.

In order to obtain precise measurements in systems with pulsation problems, it is necessary to dampen the pressure pulses by the use of restricted flow passages. Some of the various types in use are illustrated in Figure 6.3g.

One design consists of a fitting with a corrosion-resistant porous metal filter disc. By the use of such a device, the equilibrium reading on the indicator is delayed by about 10 seconds. Another snubber design depends for its damping action on a small piston in the inlet fitting, which rises and falls with pressure impulses, and thereby absorbs shock and surge. Still another snubber design uses the adjustable restriction created by a microvalve in the

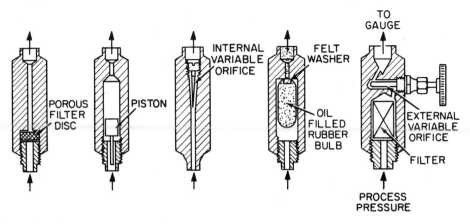

Fig. 6.3g   Variations on snubber designs

inlet fitting, to damp pulsations. This differs from a needle valve in two important ways: the filler consists of stainless steel wool to prevent plugging, and it does not quite shut off. This is to prevent "shutting-in" a false reading.

Other forms of pulsation or noise dampening can be obtained by using dashpots in motion balance devices or by restricting the flow between bellows in fluid-filled devices.

Electronic Noise Rejection

It is a general consensus that the analog electronic signal leads should be twisted pairs, and that they should be run in metallic conduit or be shielded. Less generally agreed upon is a recommendation that these conduits or shielded bundles of wiring be isolated from high-power-level wiring to motors, but this too is followed in many cases.

DC signals of 4–20 or 10–50 milliamperes can be run in the same conduit with telephone wires, low-voltage DC signals and/or thermocouple leads. They should not be run in the same conduit or in the same shielded bundle with alarm signals or power wiring.

AC signals (which are much less common) should be run in twisted pairs, with each pair shielded, and then in conduit or shielded bundles. Some common mode noise rejection is built into the instruments, but this is not sufficient if computers are being used or if higher level spikes from motor loads are experienced.

The various methods of conditioning digital signals, and the subject of computer compatible wiring practices in general, is covered in Section 8.6.

6.4 TELEMETERING SYSTEMS

Features:

Radio telemetry hardware is available that can measure and transmit any physical phenomenon that can be transduced into an electric current, voltage or resistance. These include acceleration, bending moment, current, proximity, resistance, shock, strain, temperature, torque, vibration, and voltage.

Transmitter Types:

BCO (bridge-controlled oscillator) transmitters and VCO (voltage-controlled oscillator) transmitters.

Transmission Signal Band:

88–108 mHz, standard FM (frequency modulated) broadcast band. (mHz = 10^6 cycles per second).

Cost:

Transmitters are available in the price range of $600 to $1,600.

Receivers, depending on their features and accessories, cost from $250 to $2,000. A simple, single-channel system including transmitter, receiver and accessories may cost between $1,000 and $2,000.

Partial List of Suppliers:

Aerotherm Corp.; AMF Alexandria Div.; American Electronic Labs.; Bendix Electrodynamics; Compu-Dyne Corp.; Control Data Ads Div., Electronic Modules Corp.; F & M Systems Co.; Foxboro Co.; Gulton Industrial Data Systems; Industrial Electronetics; Leeds & Northrup Co.; Martin Decker Corp.; Microdot, Inc.; Philco Ford Corp.; Radio Engineering Labs.; Tele Dynamics Div. AMBAC; Vidar Corp.

In this section the principles of telemetry, its uses, advantages, and specific applications are presented. The methods of utilizing telemetry equipment in conjunction with existing transducers and indicators are also described. Following a brief presentation of what radio telemetry is, the section continues by discussing telemetry applications in chemical and other industrial processes. The information provided in some parts of this section (such as the subject of calibration practices) is specific to the products of a particular manufacturer, but the techniques and concepts described are generally applicable to the field of radio telemetry.

Industry is turning to new techniques of measuring physical quantities at relatively inaccessible locations and recording them from a convenient distance. More precise transducers and powerful self-contained radio transmitters are sending strain measurements back from energized high-voltage transmission lines and temperatures from inside chemical process equipment, and they are warning of critical temperatures inside whirling motors. Because transmission distances are relatively short, the equipment is less elaborate and less costly than in space telemetry. But industrial environments and operation by non-technical personnel requires most telemetry system components to be far more rugged, and capable of accurate measurements with only the simplest calibration techniques.

The deeper an instrumented vehicle probes into the remote reaches of outer space, the more technologically spectacular seem the achievements of telemetry. Yet the vast distances spanned by telemetry signals are less challenging technically than the stubborn problems in some industrial applications, of inaccessibility to the quantities being measured.

Signals from a missile-launched space probe soaring toward the sun are often easier to obtain than measurements from inside a solid, earthbound motor only a foot or two away. To find the temperature of the spinning rotor, housed in a steel casing and surrounded by a strong alternating magnetic field, may require more ingenuity to transcend the operating environment than taking measurements from the most distant instrument payload speeding through the unaccommodating environment of space.

The technology that has produced missile and space telemetry is also spawning new forms of industrial radio telemetry, capitalizing on the development of new transducers, powerful miniature radio transmitters, improved self-contained power sources, and better techniques of environmental protection.

Measuring from a distance requires, first, at the remote detection point, a transducer, a device that converts the physical quantity being measured into a signal (usually an electrical one) so that it can be more conveniently transmitted. Then a connecting link is needed between the location where the measurement is being made and the point where the signal may be read or recorded. This link can be either an electrical circuit (there have been *wired* telemetry systems since long before the turn of the century), pneumatic or hydraulic lines, a beam of light, or now more practically, a radio carrier for frequency or pulsed systems of measurement.

In radio telemetry a transmitter generates the carrier, a subcarrier impresses the measurement signal onto the carrier, a radio receiver receives this carrier out of the air and reproduces the measurement signal it has borne, and a meter or recorder displays the measured quantity.

Telemetry Methods

Telemetry is concerned with the transmission of measured physical quantities such as temperature, displacement, velocity, humidity, blood pressure, pressure, acceleration, etc. to a convenient remote location, in a form suitable for display and analysis. The first practical telemetry applications were made by the public utilities prior to World War I. The greatest development took place, however, with the advent of high-speed aircraft, missiles, and satellites. Simultaneously, the continuing development of industrial telemetry applications proceeded at a relatively slow speed until the experience of the military and governmental applications was applied to the industrial telemetry field. Industrial telemetry now takes advantage of new types of transducers for measuring physical quantities to be telemetered, miniature radio transmitters, long-life miniaturized self-contained power supplies and better techniques of environmental protection. Industrial applications cover a broad segment of industry including utility, chemical, transportation, construction and machinery.

Medical science is currently employing telemetry for use in experimental, clinical, and diagnostic applications. Some of the particular body characteristics telemetered include heartbeat, brainwaves, blood pressure, temperature, voice patterns, heart sounds, respiration sounds, and muscle tensions. Similar studies are being pursued in the biological and psychological fields, where more experimental latitude permits embedding of transmitters within living animals.

The basic telemetry system consists of three building blocks. These are (1) the input transducer, (2) the transmitter and (3) the receiving station. Transducers convert the measured physical quantity into a usable form for transmission. The conversion of the desired information into a form capable of being transmitted to the receiver is a function of the type of transducer employed. Transducers convert the physical quantities to be measured into electrical, light, pneumatic, or hydraulic energy. The type of energy conversion is determined by the type of transmission desired.

One of the most common types of transducers generates electrical signals as a function of the changing physical quantity, and one of the most common varieties of this type is the resistance wire strain gauge. In this transducer, the ability of the wire to change its dimension as it is stressed causes a corresponding change in its electrical resistance. A decrease in wire diameter generally results in greater resistance to the flow of electricity. Similarly, temperature-sensitive materials that have electrical characteristics changing with temperature make temperature detection possible.

In most transducers, the electrical output is varied as a function of changes in the physical parameter. These electrical changes can be transmitted by wire direct to a control center, data display area or to a data analysis section for evaluation. However, the difficulties with use of wire in many applications have given rise to wireless telemetry.

In order to transmit the transducer information through the air, it is necessary to apply this information to a high-frequency electrical carrier as is commonly done in radio. Application of the transducer information to a high-frequency carrier is commonly called modulation. High-frequency or rapidly changing electricity has the capability of being propagated through space, whereas low-frequency or battery, non-changing voltage does not possess this ability.

The technique used for applying or modulating the high-frequency carrier by the transducer output involves any one of three different methods. It is possible to modulate a carrier by a change in amplitude, a change in frequency, or a change in the carrier phase. The last technique is similar to the modulation employed in transmitting color by television.

In color TV the brightness signal is transmitted as amplitude modulation (AM), the sound as frequency modulation (FM), and the color as phase modulation (PM), or pulse coding. Pulse coding is used to modulate the radio frequency carrier in either AM, FM, or PM. The various types of modulation that have been used for telemetry are shown in Figure 6.4a.

A common and extremely useful technique for increasing the information carrying capability of a single transmitting telemetry line is called multiplexing. When it is desirable to monitor different physical parameters, such as temperature and pressure, it may be wasteful to have duplicating telemetry transmission lines. Multiplexing techniques can usually be considered to be of two types, frequency division multiplexing and time division multiplexing. In the frequency division multiplexing system, different subcarrier frequencies are modulated by their respective changing physical parameter; these subcarrier frequencies are then used to modulate the carrier frequency enabling the transmission of all desired channels of information, simultaneously by one carrier. At the receiver, these subcarrier frequencies must be individually removed. This is accomplished by filters that allow any one of the respective subcarrier frequencies to pass. Each subcarrier frequency is then converted back to a voltage by the discriminator. The discriminator voltages can be used to actuate recorders and/or similar devices. Time division telemetry systems may employ pulse modulation or pulse code modulation. In these systems the information signal is applied, in time sequence, to modulate the radio carrier. The characteristics of a pulse signal can be affected by modulating its amplitude, frequency, or phase.

Telemetry began as a wire communication technique between two

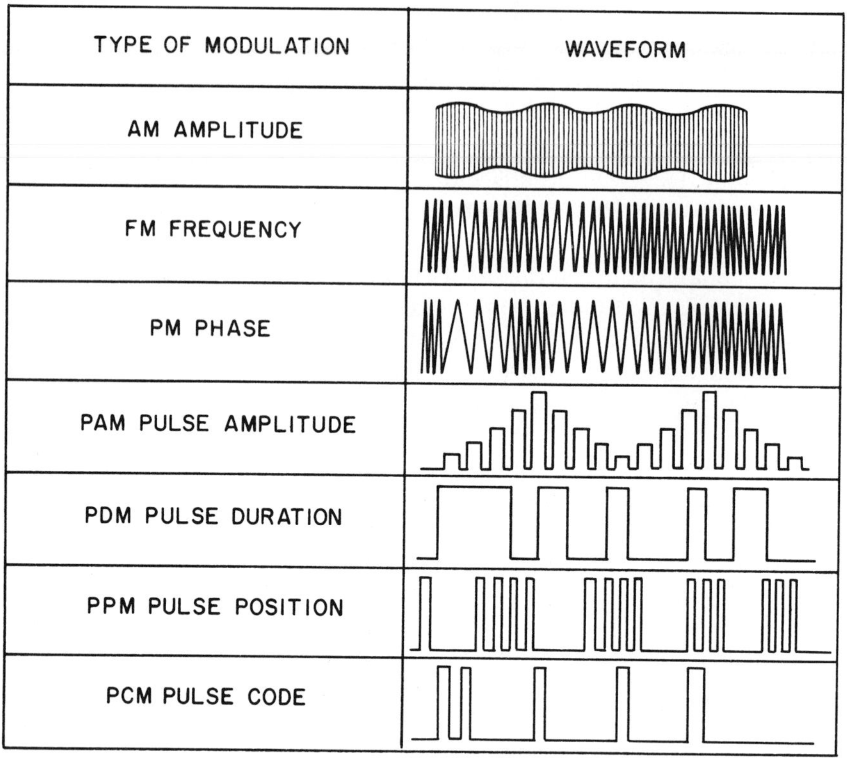

TYPE OF MODULATION	WAVEFORM
AM AMPLITUDE	
FM FREQUENCY	
PM PHASE	
PAM PULSE AMPLITUDE	
PDM PULSE DURATION	
PPM PULSE POSITION	
PCM PULSE CODE	

Fig. 6.4a   Telemetry modulations

remotely located stations. As science extends its domains into the realm of space, telemetry will be the essential communicating link between satellites, space ships, robots and other scientific devices yet to be designed.

The range of a radio link is limited by the strength of the signal radiated by the transmitter toward the receiver and by the sensitivity of that receiver. A 10-microwatt output will transmit data easily one hundred feet with a bandwidth of 100 KHz. The wider the bandwidth, the more the effect from noise, and therefore the more transmitting power required for an acceptable signal.

At the receiving station, there are usually no space restrictions in accommodating large antennas, sensitive radio tuners and recorders, and an ample power supply. But the transmitting station often must be small, possibly doughnut size but sometimes no bigger than a pea, and must be self-sufficient—carrying its own power or perhaps receiving it by radio.

On the surface, industrial radio telemetry seems to be simply a matter of hardware. And it almost is, except that the functional requirements are a lot different from those in missile and space telemetry. Distances are much shorter, a matter of a few feet to a few hundred yards; signal power can be radiated directly from the transmitter circuitry or from an antenna as simple as an inch or two of wire. Most tests are repeatable—no missile blowing up on the pad here, taking with it valuable instruments and invaluable records of the events leading up to that failure.

Quantities can be measured one or two at a time, rather than requiring an enormous amount of information to be transmitted at once. This results in relatively inefficient use of the radio link, but enables simpler circuitry at both the transmitting and receiving ends.

Surprisingly, environment plays the most critical role in industrial telemetry. It makes by far the largest difference between telemetry operations from missiles and spacecraft and those used in industrial remote measurement. While missile telemetry equipment is expected to withstand accelerations of 10 to 20 g, the rotating applications of telemetry in industry, such as the embedding of a transducer in a spinning shaft, require immunity to 10,000 or 20,000 g centrifugal accelerations.

The environmental extremes under which industrial telemeters must work are considered normal operating conditions by their users. Unlike missile telemetry equipment, which is shielded and insulated against extremes of temperature, shock and vibrations, and which is carefully calibrated for weeks before it is used only once in an actual shot, industrial telemeters must operate repeatedly without adjustment and calibration. Used outdoors, they are often subjected to a temperature range of -40 to $+140°F$. They must operate when immersed in hot or cold fluids, and thus it is almost mandatory that they be completely encapsulated to be impervious to not only humidity and water, but to many other chemical fluids and fumes. Many lubricating oils operate at temperatures of 300 or 350°F.

We know that missile telemetry components must be small and light, yet an order of magnitude reduction in size and weight has been necessary to make telemetry suitable for high-speed rotating shafts or for biological implants. They must be so reliable that no maintenance is required, for there are no service centers set up to handle this kind of equipment, and it must work without failure to continue to gain industrial acceptance.

Information theory has been used extensively to develop space telemetry for the most efficient data transmission over a maximum distance with a minimum of transmitted power. Inefficiencies, being of no real consequence in industrial telemetry, make for less elaborate, less costly equipment. Radio channels are used in a relatively inefficient manner, and the distances between transmitter and receiver are usually so short that there are few problems of weak signals. In many cases, measurement and testing via telemetry links takes place in completely shielded buildings or in metal housings.

Though telemetry is usually defined as measurement at a distance, it has gradually begun to embody the concept of control from a distance, too. In telemetry—the transmission of the value of a quantity from a remote point—it may serve merely to communicate the reading on an instrument at a distance. But the output of the instrument can also be fed into a control mechanism, such as a relay or an alarm, so that the telemetered signal can activate, stop, or otherwise regulate a process. Measurement may be taken at one location, indication provided at a second location, and the remote control function initiated at one of those two locations or at a third point.

For example, a motor might be pumping oil from one location while oil pressure is being measured at another. When the pressure reading is telemetered to a control station, a decision can be made there to reduce pump motor speed when the pressure is too high, or a valve can be opened at still another location to direct the oil to flow in another path. The decision-making controller may be an experienced pipeline dispatcher or an automatic device.

Measuring and Transmitting

Telemetry, then, really begins with measurement. A physical quantity is converted to a signal for transmission to another point. The transducer that converts this physical quantity into an electrical signal may be a piezoelectric crystal, a variable resistance, or perhaps an accelerometer.

Telemetered information need be no less accurate than that obtained directly under laboratory conditions. For instance, in telemetering strain measurements, it is possible to achieve accuracies of a few microinches per inch or greater. The only limitation is usually the degree of stability in the bond of the strain gauge to the specimen, and not the strain gauge itself.

If great accuracy in temperature measurement is desired, it can be attained by choosing a transducer which provides a large variation of output signal over a small range of process property variation. The resolution which this provides may be translated into true accuracy by careful transducer calibration. Accuracy is reduced, of course, if a wider range of temperature needs to be detected. Typical single-channel analog telemetry links maintain a measurement accuracy of one to five percent. But this is not a limitation of the total system, since one percent of a 100° temperature change would only be one degree, so several telemetry channels can easily share the total temperature range to be measured—say a 100°F range divided into four 25°F ranges, to produce an accuracy of one-fourth of a degree.

Special temperature probes have been produced for the range of 70 to 400°F, and higher to maximize the stability and accuracy of temperature telemetry. These probes, when used with the proper choice of transmitters and receivers, can provide temperature measurements to closer than 0.05°F.

One of the limitations to accuracy and to repeatability in telemetry is the output level of the transducer. The low electrical levels produced by thermocouples and strain gauges (millivolts) are much more difficult to telemeter than a higher-voltage level of say 5 volts. At low signal levels, extraneous electrical noises produce great degradation. This noise may be thermally generated, caused by atmospheric effects, or generated by nearby electrical equipment. When low-level transducers are used, stable amplifiers are required to raise the signal voltage to useful modulation levels.

There may be great variations in the strength of the radio signal received because of variations in distance between transmitter and receiver, or because of the interposition of metallic objects. In industrial radio telemetry transmission, these effects can be prevented from disturbing the data by resorting to frequency modulation of both the subcarrier and the carrier so that the telemetered signal is unchanged by undesirable amplitude variations. This method is called FM/FM telemetry.

If FM modulation is employed in the subcarrier of the transmitter, the transducer signal modulates the frequency of the subcarrier oscillator. This can be done by a simple resonant circuit which produces a given frequency in the audio range, say 1,000 Hz, which is varied above or below by the signal from the transducer as it responds to the variable it is measuring. If the signal were fed to a loudspeaker, a rising or falling tone could be heard. The subcarrier oscillator then modulates a radio frequency carrier, varying its frequency (FM) or its amplitude (AM) in accordance with the subcarrier signal. The radio frequency in FM industrial radio telemetry links is usually in the 88 to 108 mHz band. At the receiving end of the link, the radio receiver demodulates the signal, removing the carrier and feeding the subcarrier to a special discriminator circuit that removes the modulation, and precisely reproduces the original measurement signal for calibrated indication or recording.

Multiple measurements can also be transmitted over the carrier by sampling the output of several transducers in rapid sequence, a technique called time-division multiplexing. This technique has been employed to handle as many as a million samples per second. It provides for very simple data displays and easier separation of channels for recording or analysis, and it is free of cross talk. If possible, it is advantageous to use no multiplexing at all for concurrent data taking, but to use separate radio carriers for each measurement being transmitted.

A single channel of industrial FM/FM telemetry equipment may cost between $1,000 and $2,000, depending upon the flexibility required and the measurements being made. This buys everything needed for a given remote measurement—transducer, radio link, power supply and simple indicator.

The telemetry data received may be recorded in a number of ways, but such records must preserve the accuracy of the entire system. For example, if a one-percent system is recorded on a graph, and $\frac{1}{64}$ in. is the most that can be distinguished on the graph paper, the minimum size graph for full scale should be approximately 2 in. Similarly, numeric data should be printed to enough decimal places to preserve the accuracy of the system.

Several examples of new industrial telemetry applications in later paragraphs will show how systems are applied to various remote measurement problems. They will also give some idea of the specific equipment requirements for industrial environments.

High-voltage transmission lines are an excellent example of how inaccessible an object of measurement may be. These lines vibrate in the wind, and the stresses and strains require measurement under the dynamic conditions that contribute to fatigue failure. Strain tests to determine fatigue will show quickly whether the endurance limit of the line has been exceeded, and only if it is exceeded need we be concerned about fatigue failure. Therefore it is necessary to measure the number and magnitude of the strain reversals in order to predict the time of failure. Telemetry techniques permit dynamic testing under actual service conditions rather than by simulated laboratory conditions or static tests.

While the transducer that produces an electrical signal proportional to strain may have an output of 0.01 volts, the live transmission line to which it is attached may be at a potential of several hundred thousand volts. The problem is to detect this hundredth of a volt in the presence of a very large signal. In the language of the telemetry engineer, this is rejection of a common mode voltage of the order of 10^8 to 1. Then why not de-energize the line? It's a simple matter of economics—an idle line transmits no power, and the wind forces that cause the line to vibrate are neither predictable nor constant. So, weeks or months may be spent in gathering measurements for a particular set of spans. However, a ratio telemetry link makes it possible to transmit the strain signal even while power is being carried.

A self-contained FM radio transmitter is attached to the transmission line at a point adjacent to a strain gauge. All remain at the same electrical potential as the line, much like a bird sitting safely on the wire, transmitting the strain gauge output to a radio receiver and recorder located at some convenient point on the ground, where vibration analysis can be made. As a result, armor rods may be placed around the line at the vulnerable points, or vibration absorbers of the correct resonant frequency can be installed at the proper points on the line.

More down to earth, but equally inaccessible to measurement, is strain on the chain belt of an earth mover. Too light a chain will quickly fail from fatigue caused by the alternating stresses imposed by the full and empty buckets it transports. Measurements made under actual operating conditions of the earth hoist mean attaching strain gauges to a chain traveling at 500 ft per minute, subjecting them to violent shock and vibration. On this kind of moving equipment, slip rings and wire-link remote measurements will not work.

Here again, radio telemetry is now providing the dynamic measurements needed to test the earth moving equipment at work. A transducer and a small, rugged transmitter are attached to points along the chain—strain varies from link to link depending on the proximity to the bucket—until the most vulnerable part of the chain is found. It is preferable to use several transducers and multiple channel telemetry equipment for such measurements to simplify correlation between load and the resulting strain at various links.

Telemetry can also determine water levels and flow rates of rivers to provide vital data for flood control, or for efficient hydroelectric power generation. Data on the potential watershed into rivers can be obtained by analyzing the water content of snow that would eventually melt and feed them. One requirement is to measure the depth and water content of snow in the mountains, then transmit this data from remote points to a central receiving station. The snow-measuring transducer may consist of a radioactive source atop a tall pole and a radiation intensity meter on the ground beneath the snow. The gamma-ray intensity reaching the meter is a function of the height and water content of the intervening snow. Both the meter and the transmitting equipment can be powered by a storage battery and controlled by a clock timer that sets the time of transmission to a few seconds per day.

Transducers

Improvements in transducers are opening up new measurement possibilities. Typically strain gauges have consisted of a metallic element that was stretched. As its length increased, its cross section decreased (Poisson's ratio), thus increasing its electrical resistance. A metal strain gauge with a gauge factor of two increases resistance twice as fast as it does its length. But new semiconductor materials exhibit gauge factors as high as 100 or 200. Consequently, the output of a bridge of semiconductor materials may be used to modulate telemetry subcarrier oscillators directly without further amplification. There are some drawbacks—semiconductor materials are temperature sensitive and introduce greater drifts than metal foil. This problem is not insurmountable, for the arms of the gauge may be located at a single point for temperature compensation.

Semiconductor piezoresistivity is a property that makes possible simpler and more reliable pressure transducers than the conventional electromechanical pressure cells, which employ an elastic sensor with a deflection proportional to pressure. The ideal pressure transducer must provide a precise, repeatable measure of steady-state pressure. It must also respond linearly to large pressure variations without permitting small pressure changes to be obscured by noise or threshold effects. Furthermore, there must be a minimum of interaction between the transducer and the medium whose pressure is measured. One recent development integrates the pressure sensor and the output devices in a single silicon strip. It has a dynamic response to 6,000 Hz.

Among other promising approaches to pressure transducers is a chemical cell in which liquid displacement is used to unbalance a bridge. In this type of unit, a center electrode divides two sections of liquid which is metered through a small orifice. Electrodes at each end make up a unit which becomes half of the bridge; they are electrically connected by the liquid. The center electrode is displaced by pressure of a fluid, although it also could be moved by mechanical leverage or magnetic energy. A transducer of this type can measure absolute strain or strain rates, producing an output so high that it requires no further amplification before being fed to a telemetry transmitter. In fact, its amplification is limited only by temperature effects, which become increasingly important as the bridge configuration is physically changed to increase amplification.

Semiconductors are also appearing in light-actuated analogs of the mechanically operated potentiometer. Electromechanical potentiometers sense mechanical movement and translate it into voltage by changing the position of a sliding contact on a length of electrically resistive material. They have long been used for position measuring and as pick-offs from accelerometers, gyros, torque angle meters, etc.

In the electro-optical solid-state potentiometers, the conventional wiper arm that makes the sliding contact has been replaced with a tiny light beam. The electrical element consists of a resistance film separated from an adjacent conducting strip by a photoconductive crystal. This photoconductive strip acts as an insulator when it is dark, but where the light beam hits it, an electrical connection is made between the resistive and the conductive strips at that point. The output voltage is a linear function of the light beam displacement on the photoconductive crystal. Such a potentiometer is essentially a friction-free, noise-free device and has a marked advantage over mechanical potentiometers because it can be scanned at speeds up to 7 meters per second. Its resolution is better than 0.0005 in., and its deviation from linearity can be made less than 0.2 percent over a range of 10 to 90 percent of full voltage. While the resistive strip is temperature sensitive, the voltage across its terminals or anywhere along the strip remains invariant. On the other hand, the temperature sensitivity of the light-detecting crystal influences the output voltage and linearity of the potentiometer. The practical temperature limits of a photopotentiometer presently range between −65 and 100°F.

The bonded strain gauge—either wire, foil or semiconductor—is very useful in most industrial measurements. It is applied to the actual machine part in which strain is to be measured. Strain is related to stress by means of the appropriate elasticity modulus, and many operating parameters of the equipment can be obtained. The bonded resistance strain gauge has been used in the field of stress analysis under a variety of environmental conditions from vacuum to very high pressures. Strain has been measured in the very hot environment of jet engine turbines, as well as in superheated steam. The use of these devices in cryogenic fluids at temperatures as low as −452°F and under the nuclear radiation associated with reactors has also been reported.

Most of the adverse effects of the environmental conditions can be minimized by suitable protection of the strain gauge. The major limiting environmental variable is temperature.

Strain-Sensing Alloys

The common strain-sensing alloys undergo changes at high temperatures which alter their resistivity and temperature coefficient of resistance. The changes in the electrical properties are generally time and temperature dependent in the same way that metal physical characteristics are affected by heat treating or cold working. The change in resistivity results in a gauge resistance change with time at constant temperature, and it destroys the initial zero reference which is necessary in determining changes in specimen state of strain with time. This makes it impossible to separate stress-producing thermal or mechanical strains from the error signals produced by unwanted resistance changes due to temperature.

Fortunately, the gauge factor is not drastically changed by metallurgical property changes. Gauge factor vs temperature characteristics for the copper-nickel alloys, nickel-chrome and platinum alloys is shown in Figure 6.4b.

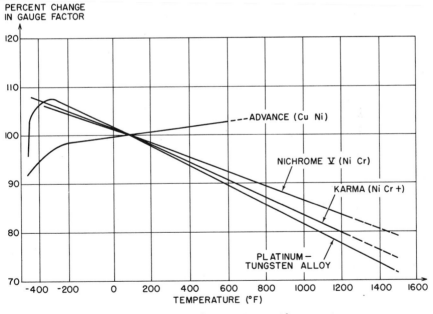

Fig. 6.4b   Typical gauge factor variation with temperature

Copper-nickel alloys such as Advance and Cupron have a nearly uniform gauge factor change with temperature of +0.5 percent per 100°F over the range of −200 to 600°F. This alloy is the most widely used in strain gauge work because of its reasonably high resistivity, low and controllable temperature coefficient of resistance, and uniform gauge factor over an extensive strain range. The nickel-chrome alloys such as Nichrome V, Tophet A, Karma, Evenohm, commonly exhibit a reduction in gauge factor with temperature of -1.5 to -1.8 percent per $100\,°F$ temperature increase, and this decrease is uniform over the temperature range from -452 to $1,200\,°F$. Nickel-chrome alloys are employed for cryogenic testing because of this uniformity of gauge factor and for high-temperature tests because the corrosion resistance is superior to that of the copper-nickel alloys.

The most common platinum alloy used in strain gauges is platinum-tungsten. This material has a high gauge factor (almost twice that of Advance at room temperature) and changes -2 percent per $100\,°F$ temperature increase between -320 and $1,500\,°F$. This alloy, while commonly used for the measurement of dynamic strain at all temperatures, is being used extensively for static strain measurements in the temperature range above $800\,°F$, where nickel-chrome alloys become unstable. Platinum alloys are not considered suitable for use in the cryogenic range below liquid nitrogen temperature. Figure 6.4c shows the typical short-term drift rate vs. tempera-

SHORT TERM (TWO HOUR)
DRIFT RATE
(10^{-6} INCH / INCH / MIN)

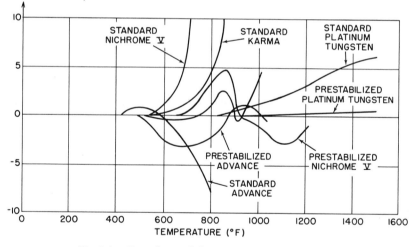

Fig. 6.4c   Typical zero drift at constant temperature

ture characteristics for all of these alloys. Above $450\,°F$ Advance becomes electrically mobile and would be considered unsuitable if absolute stability were required. Intergranular corrosion of this copper base alloy above $600\,°F$ drastically alters the temperature coefficient of resistance.

The temperature coefficient of nickel-chrome alloy and Nichrome V cannot be adjusted for temperature compensation by heat treatment. The material can be stabilized to produce a more drift-free condition. Note that the stable range is increased from 600 to 700°F for material which has been stabilized.

Karma and Evenohm alloys contain small quantities of iron and aluminum in addition to the nickel and chromium. These materials are also unstable above 600°F, but can be heat treated to produce self-temperature compensated strain gauge types. Note that 600°F is a practical operating limit for these alloys, and that 800°F would be considered a maximum for short-term strain measurements where zero stability is required.

The platinum alloys cannot be heat treated to produce self-temperature-compensated strain gauges, although the temperature coefficient is altered by pre-stabilization. Figure 6.4c indicates that reasonably stable operation can be achieved using this alloy to 800°F and, with proper treatment, an additional 400°F can be achieved. Where the reasonably low drift of the platinum-tungsten alloys can be tolerated, 1200°F is a practical limit for short-term measurements.

Figure 6.4d shows the typical change in slope of the apparent strain

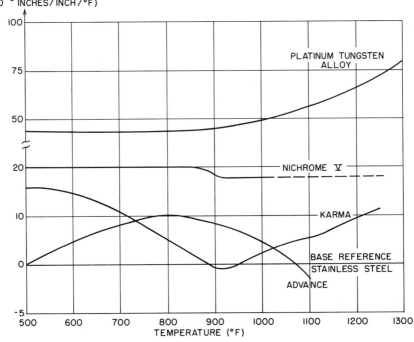

Fig. 6.4d   Typical apparent strain slope change with temperature

characteristic for various strain gauge alloys vs temperature. It can be seen that materials like Advance and Karma can be adjusted to produce zero slope when the strain gauges are bonded to the test material. This is the basis for self-temperature-compensated gauge types as opposed to gauges which can be compensated by circuit techniques. It can also be seen that the platinum alloys cannot be adjusted to produce self-temperature-compensated gauges. Circuit compensation utilizing the resistance thermometer technique as shown in Figure 6.4e has been used to produce temperature-compensated gauges for specific temperature ranges on a particular test material.

Karma, adjusted for a negative temperature coefficient, can be connected in series with a Nichrome V material with a positive temperature coefficient to produce a temperature-compensated gauge for cryogenic range. This composite gauge construction is shown in Figure 6.4e. The most effective

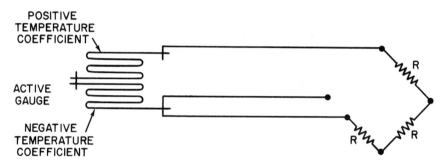

SELF TEMPERATURE COMPENSATED COMPOSITE GAUGE

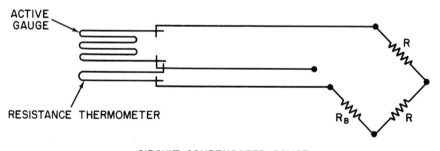

CIRCUIT COMPENSATED GAUGE

Fig. 6.4e   Temperature-compensated gauges

temperature compensation for the cryogenic temperature range is achieved by using Nichrome V and the platinum resistance thermometry circuit compensation technique.

Bonding Cements and Carrier Materials

The strain gauge bonding cements and carrier materials must combine to faithfully transmit the outer fiber strain to the strain-sensing filament, as well as provide electrical insulation between electrically conductive materials and the strain-sensing element. The dimensional stability of these two components will contribute to the electrical stability of the strain gauge. These components should maintain their combined shear strength and electrical insulation properties over the useful operating temperature range of the alloy.

The organic backings and cements are generally limited to the temperature range between -452 and $600°F$ unless nuclear radiation and/or vacuum environments are involved. Organic materials include cellulose fiber and nitrocellulose paper type carriers normally attached with nitrocellulose cements. Paper gauge installations are generally limited to the temperature range from -320 to $180°F$. There is a variety of epoxy, phenolic and modified epoxy-phenolic carrier materials and bonding cements available. The unfilled or non-reinforced epoxy carriers are suitable for testing in the range of -300 to $200°F$. The carriers reinforced with glass cloth, or fibers bonded with filled resin cements, can extend the operational temperature range to -452 and $600°F$. The strain range is somewhat reduced at the cryogenic temperatures, but most materials are capable of measuring at least $\frac{1}{2}$ percent strain over their entire operating temperature range.

Above $600°F$, strain gauges are usually attached with ceramic cements or by the new flame-spray techniques. The limiting temperature for both of these types is related to the electrical characteristics as shown in Figure 6.4f. Foil gauges are supplied either on strippable vinyl carriers, or as "free handling" units for transfer into the ceramic cement. Transfer technique requires a great deal of skill. The commonly used ceramic cements combine metallic oxides with a phosphoric acid base binder and require a $600°F$ cure temperature. These coatings are porous and hygroscopic. The useful strain range of ceramic cements is limited to $\frac{1}{2}$ percent by the brittle nature of the coatings. If a resistance to ground of one million ohms was considered a practical operating limit for electrical and dimensional stability, then foil type strain gauges bonded with ceramic cements would be limited to $1,200°F$ operation. Since the resistance to ground of a strain gauge installation is a function of the distance between gauge elements and ground, the cross-sectional area of the conducting path and the gauge resistance, it can be shown that wire type gauges can achieve a higher temperature limit for a given installation thickness than foil gauges.

Improved electrical properties of the aluminum oxide used in the Rokide process for flame-spray attachment of gauges has contributed greatly to the extension of high temperature limits. The flame-spray coatings exhibit the same problems with porosity and hydroscopic tendencies as ceramic cements, but are capable of 1 percent elongation instead of only ½ percent. Gauge installations require no elevated temperature curing. This is most important in the testing of materials which will alter their basic metallurgical conditions above 600°F cure temperature.

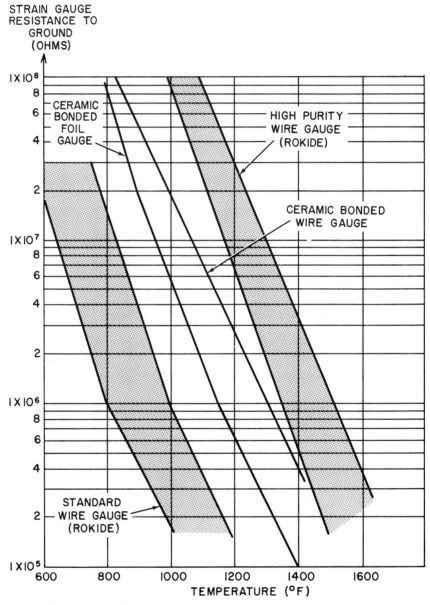

Fig. 6.4f Typical electrical resistance to ground variation with temperature

Lead wires also provide electrical resistance in series with the active sensing element and also have a temperature coefficient of resistance and a strain sensitivity. Selection and installation of lead wires is of extreme importance to the overall strain gauge performance.

Problem areas can be minimized by the use of a three-wire system as shown in Figure 6.4e. The metallic portion of the lead wire system should have a low, stable resistivity and a low temperature coefficient of resistance. For that reason copper wires clad with nickel or stainless steel, or nickel wires clad with silver have been used, depending upon the temperature range. Nickel-chrome alloy wire or lead ribbon is occasionally used at very high temperatures, but extreme care should be exercised, because this material has 40 times the resistivity of copper.

The insulation used on the lead wires is of equal importance since it must withstand the environment and provide an electrical insulation that is at least equivalent to that of the carrier and bonding cements. A variety of plastic and ceramic insulations are available for low and intermediate temperatures. Extreme care should be exercised when selecting a glass type insulation for high-temperature tests. Sodium silicate base glass insulation may have leakage problems as low as 600°F, while high-purity silica materials are capable of operation to 1,500°F. If the lead wires must be bonded to the specimen, they should be carefully installed so as not to be subjected to undue strains and thermal gradients.

In any general discussion of this type, where state of the art is being considered, many special-purpose gauges, designed to solve particular problems, are overlooked. These range from high-elongation measurements on viscoelastic materials to interlaminate strain measurements within filament-wound rocket casings. As testing temperatures are extended, it becomes more difficult to achieve strain gauge temperature compensation, maximum stability and repeatability. Although "active gauge-dummy gauge" or full-bridge circuit compensation techniques have not been discussed here (See Section 2.8 of Volume I), they represent the most accurate form of compensation for all undesirable effects. The full-bridge weldable technique is designed to take advantage of the best qualities of this type of temperature compensation and probably represents the most advanced state of the art for the measurement of static strains over wide temperature ranges. Where weldable techniques or "active gauge-dummy gauge" compensation cannot be used, it becomes necessary to design a gauge to match the specific environmental temperature range, or other test conditions. Techniques and materials are available to solve many problems, but gauges are not, because of the tremendous number of variables to be considered for each test. Bonded resistance strain gauge types can be constructed to meet the requirements if the environmental conditions, the approximate strain range, and metallurgical characteristics of the test materials are known.

Power Sources

Power sources for the transmitter in industrial telemetry applications are seldom a problem. Batteries can be used for temporary applications and at temperatures below 200°F. Small and light, rechargeable and expendable batteries are available solidly encapsulated in epoxy resin to withstand almost as rugged environments as the telemeter itself.

In a moving or rotating application, stationary magnets can be placed so that they generate electricity in a moving coil and are used to provide automatic power generation. If this method is not feasible, a stationary coil can be placed in the vicinity of the transmitter and fed electrical energy at a high frequency, so that its field can easily couple into a moving coil in almost any environment. The stationary coil ring may be large, even encompassing a whole room; usually only one turn of wire is necessary. The stationary coil may also be made extremely small, a quarter to one-half inch in diameter, and coupled to the end of a rotating shaft. These power supplies and coil configurations are standardly available units.

Applications of Telemetry

In the design of machinery, one of the most difficult factors to cope with is alternating fatigue-producing stresses which occur at some parts of the machine. It has long been the custom to measure stress in equipment with bonded strain gauges to predict the failure limits before actual failure occurs. This had only been possible on those portions of the equipment which could be connected by wires. With radio telemetry, it is now possible on all members. Costly fatigue failures are now avoidable through installation of miniature telemetry components that are reliable, rugged, and accurate in heretofore inaccessible locations and environments. Industrial uses are virtually limitless; systems can be built to specifications and encapsulated to withstand the most adverse conditions. Low-cost measurement and telemetry systems have been applied to read internal vibrations and strain in rotating equipment, chains, vehicles, and projectiles—eliminating slip rings and wires. Measurement can be made under operating conditions of vibration, acceleration, strain, temperature, pressure, magnetic fields, electrical current, and voltage, under such adverse conditions as in a field of high electrical potential, in fluids, in steam or in high-velocity gases.

Chemical Plants

In chemical plants it is important to know the exact temperatures at various points in the process. These temperatures are obtained by using thermocouples. An interesting problem has recently occurred in a rubber treatment operation. Here, when the temperature was measured by a fixed thermocouple, the viscosity of the material was so high that a considerable amount of heat was generated by friction with the thermocouple.

It is recognized that a stationary thermocouple will read the average temperature of the fluid passing by it; consequently, if there are hot and cold spots, these may not be detected. A telemeter and its battery can be encapsulated in a small floating ball and made to flow through the process with the fluid, reading the temperature at specific points in the moving flow without causing any stagnation heat. When this is done, it is usually necessary to provide a receiving antenna within the chemical vessel. This may be a simple insulated wire that can be stretched in the space available.

Textile Mills

The increasing applications of telemetry in textile mills include measurement of acceleration forces and shocks on the shuttle of a loom, and of tension and temperature of fibers in treating baths. As fibers are stretched and relaxed in heat-treating ovens, telemeters can measure the uniformity of this process. Telemeters also offer the ability to measure the difference between two temperatures or two strains. These quantities may be matched at the telemeter and only a difference signal transmitted. In this manner, the error involved in taking the difference between two large numbers is usually eliminated.

Conveyor Chains

Transmitters are installed in the form of links on conveyor chains (one transmitter or several per chain) to measure strain and the process speed and temperature. Receivers are located near the chain. The original purpose of telemetry for conveyor chains was to detect link strain leading to breakage. Since the points of greatest strain are at the drive pulleys, strain can be kept within acceptable limits by adjusting the drive torque of each pulley and, if necessary, increasing the number of drive points.

The instrumented link, which contains a strain gauge bridge, is placed on the side of the chain that bears against the pulleys. The lesser strains on the opposite side can be determined by laboratory measurements and extrapolation.

Conveyor chains may be several thousand feet long, with many drive pulleys and supporting blocks, and extend from one floor level to another. In automobile assembly plants the 600°F baking oven is always at the top level, and combination of temperature and height submits the chain and the telemetry system to the greatest strain. Link breakage is generally due to fatigue from cyclic over-stressing.

Ovens, spraying booths and other shielded areas may have an environment not suitable for good antenna reception. By equipping the transmitter link with a tuned dipole antenna in such instances, the signal can be made strong enough to reach the receiver just outside the area.

The measurement system can be portable so that the measuring link can be inserted during a brief down time. The telemetry transmitter, complete with batteries and antenna, is hung on the supports which reach from the conveyor chain to the automobile bodies. Since size restrictions on the telemetry equipment are not severe, sufficient batteries can be used for the desired operating time. With rechargeable batteries, a cycle of ten operating hours during the day, for example, and a 14-hour overnight charge may be provided.

A frequent problem with conveyor chains is lack of tension at points along the line, particularly at the lower end of descending ramps. Here the tension becomes zero and links pile up. This condition can be corrected by changing the drive torques, but the approach of a pileup cannot usually be observed. Without telemetric strain measurements, the pileups may occur regularly and become serious enough to cause the conveyor to jump the track. The presence of loose chain links in the system indicates that the other links are being overstrained. Telemetry permits the speed of the chain and drive torque at any particular point to be observed remotely, so that the operator can adjust torque and prevent pileups.

Gear Train Efficiency

To measure the efficiency of a gear train speed reducer operating between a high-speed power source, such as an electric motor, and a low-speed load, such as a ball crushing mill, radio telemetry transmitters were designed which operate from strain gauges on the input and output shafts simultaneously, and transmit torque readings to stationary indicators. A tachometer on each shaft gives the rpm of the input and output shafts. From these readings, input and output horsepower is calculated and the gear train efficiency is obtained. The telemetry transmitters can be quickly installed and removed and can be used in dirty or corrosive environments.

Shaft Horsepower

Radio telemetry is frequently the most economical method of transmitting strain and temperature signals from rotating parts. In some applications it is the only feasible method. The noise-to-signal ratio is lower than with most other methods, resulting in more accurate data. Radio telemetry represents a significant advance over previously used slip ring methods.

To operate an ocean ship efficiently, it is essential to have an accurate measurement of shaft horsepower. At the boilers and turbines, horsepower measurements are not accurate because of power losses in the speed-reducing drive. Ship speed can be slowed by hull or propeller condition, and therefore an accurate horsepower measurement of the drive shaft, correlated with engine speed measurement, is important information. Radio telemetry systems have been used to measure torque on the drive shaft of a ship. The equipment is attached to strain gauges properly mounted on the shaft, and remotely located receivers display torque and rpm readings. Horsepower is obtained by simple computation.

Telemetry equipment has also been tested and used in generators and commutators to measure shaft torques, winding temperatures, and thermal strains. Shaft torque data can be used to evaluate lubricants for use in transmissions or to measure the tension in a strip being wound into a magnetic core.

A bridge-controlled oscillator telemetry system can be used to transmit the strain readings from gauges mounted on the bars of an experimental commutator, as the temperature and speed change. The transmitters are fastened to the inside surface of the commutator. The data presented in Figure 6.4g show that a linear relationship exists between the stress in the commutator

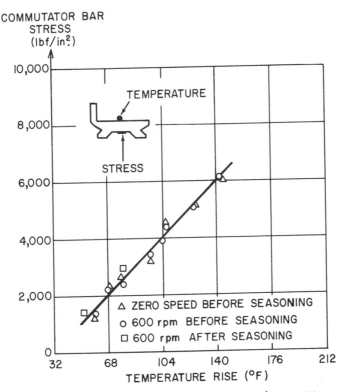

Fig. 6.4g　Stress vs temperature in bars of experimental commutator

bars and the temperature of the bar. The data also show that changes in speed up to 600 rpm do not affect the measured strain. The effect of temperature on the stress was approximately 500 lbf/in.^{2} for every 10°F rise in temperature. A maximum stress of 6,200 lbf/in.^{2} was measured at 144°F.

The bridge-controlled oscillator systems can also be used to transmit the torsional strain from resistance strain gauges mounted on the main shafts of the motors of hot strip mills. Motors of several thousand horsepower, which operate at 150 rpm, have shafts up to 3 ft in diameter. Torque is measured and related to the electrical parameters as hot strip steel is being rolled. The transmitters are strapped to the shaft with large hose clamps. Resistance strain gauges can be used to sense the torsional strain in smaller motor shafts while they are driving a gear train in order to evaluate various dry lubricants. The bridge-controlled oscillator transmitter can be used to transmit the strain value from the motor shaft at speeds up to 2,000 rpm. Since the speed is high, the transmitter is housed inside a hollow steel cylinder, which can be keyed and bolted to the motor shaft. The strain gauges are located behind the drive pulley adjacent to the steel cylinder. The four lead wires from the strain gauge bridge pass through a hole in the pulley and steel cavity to the terminals of the transmitter. From the magnitude of the torsional strains, an evaluation of the various lubricants can be made.

    Another possible application is to monitor the tension load in magnetic steel strips as the strip is wound onto a magnetic core. Resistance strain gauges are mounted on a shaft, and the shaft is calibrated for torsional strain as a function of torsional load. The shaft is then fixed to the base of the shallow steel cylinder and guided by the sleeve bearing in the spider. The mandrel for the magnetic core is fixed to the end of the shaft. As the machine rotates, the magnetic core is wound around the mandrel and the shaft is strained in torsion. From the torsional strain measurements, the tension in the magnetic strip can be calculated. A typical oscillographic trace for this application is shown in Figure 6.4h. From this record, the exact position of the core can be fixed at any time. The change in tension load in the strip, as a function of the induced friction, can also be determined. The friction or drag holding down the supply strip is produced by two pneumatic cylinders acting against a Teflon friction plate.

TENSION

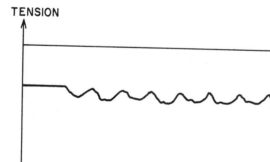

Fig. 6.4h   Typical oscillograms obtained using bridge-controlled transmitters

The voltage-controlled oscillator transmitter can be used to measure temperature in the damper windings of 100,000 HP size motor-generators. The motor-generators can be driven by a water turbine, and the torque in the shaft between the motor-generator and the water turbine can also be measured, using resistance strain gauges. The motor-generators, which operate at about 180 rpm, have stators about 40 ft OD and the shaft is about 3 ft in diameter. Transmitters can be fixed to the shaft and to the rotor. Test data showed that the measured shaft torque agreed well with the calculated torque, but that the damper winding temperatures were considerably lower than expected.

The biggest problems encountered in measuring strains in rotating parts involve the transmitting antenna and shielding against external electrical fields. It is found that the best configuration for the transmitting antenna is a 320 degree ring, with the center screw holding the battery and transmitter together.

A two-element rabbit-ear receiving antenna can be made from $\frac{1}{4}$-in.-diam brass tubing, and can be successfully used in turbulent air when the damper winding temperatures are measured. Two brass tubing antennas hooked in parallel will improve the reception in these measurements.

For best results, all connecting wires should be shielded and grounded. The transmitters should also be shielded to prevent electrical pickup. All wiring must be firmly fastened to prevent any motion. On shafts where torsional strains are measured, the strain gauge bridge circuit need not be wired by a continuous loop around the shaft as is normally done, but can be wired without the loop which caused electrical interference.

Transmitters used in transient temperature conditions must be calibrated under the transient conditions so that the test data can be corrected. Under transient conditions, a temperature gradient will exist over the transmitter, and the correction will depend upon the gradient.

To minimize the transient effect, the transmitter should be enclosed in a metal tube of high thermal conductivity. Although the tube distributes the temperature uniformly over the entire transmitter surface, cooling all parts uniformly, a radial gradient still exists.

The effect of thermocouple length on the accuracy of temperature measurement is negligible when it is from 4 to 12 ft. Generally, copper-constantan thermocouples with a 10-ohm resistance, which is 10 percent of the impedance of the transmitter, can be used for 1 percent accuracy. For a long copper-constantan thermocouple having a resistance greater than 10 ohms, calibration would be necessary. Thermocouples with an initial resistance less than 10 ohms would have an accuracy better than 1 percent.

Power Plants

In power plants, coal is fed in turn to a number of hoppers by conveyor belt. A tripper on the conveyor belt diverts the coal into a particular hopper until it is full. Either an operator or a mechanical sensing device determines when the hopper is full, and a signal is transmitted to the conveyor to move onto the next hopper. Before telemetering equipment was in use, costly accidents could occur if the operator should be away momentarily or if the sensor failed to function. As much as six tons of coal a minute could overflow onto the power station floor.

To prevent this, pressure switches are installed in the tripper chute to activate a radio transmitter if coal backs up into the tripper. The transmitter sends its signal to a receiver located at the conveyor belt and sounds an alarm. This type of control is difficult if not impossible to achieve by wired power connections, because the tripper is moving and because the corrosive coal dust atmosphere attacks the wires. For this reason a radio transmitter equipped with long-life batteries is mounted on the tripper. The receiver at the control end is powered by AC. Subcarrier tone (frequency) coding is used to eliminate effects of interference and noise, giving positive protection at all times.

Underground Cable Tension

In some applications, strain information cannot be transmitted from the measuring transducer to the receiver and recorder by radio. A telemeter developed for measuring tension in underground cables is one such application. When radio cannot be used, it is possible to employ a pulling cable to bring the telemetered information to a location at which it can be recorded. To measure the tension in cables pulled through underground conduit, washer-type load cells can be placed under the heads of each of the three pulling bolts. These load cells are then connected to the telemetry subcarrier oscillators located inside the pulling head. These oscillators include batteries to power the unit for a period of 10 hours. Each subcarrier oscillator is of a different frequency so that it can be multiplexed, with the other subcarriers, on a single line.

The center conductor and the outer braid of all cables are connected electrically, making a complete short circuit. This feature is used to conduct current through the entire length of the electrical cables. A small iron core toroid, with a high impedance winding around it, can be installed around the center conductor and beneath the outer shield. Two small wires are brought out through the insulation and the outer shield, to the telemetry subcarrier oscillators. In this manner currents of several amperes are induced in the large center conductor, circulating through the length of the line, and returning through the outer shield. At the drum end of the cables, another current transformer is used to sense the flowing current and produce a voltage which is then coupled to a radio transmitter.

The radio transmitter is located on the side of the drum. It produces a low-power radio signal to permit the subcarriers, which are coupled from the conductor, to be transmitted a few feet to a receiving antenna without the use of slip rings or rotating joints. A receiving antenna is connected to an ordinary telemetry receiver and recorder. The three signals transmitted simultaneously are separated by filters in the receiver so that they can be discriminated and plotted separately.

The entire system is calibrated electrically by placing a known resistor across one of the arms of the strain gauge resistance bridge in each load cell. The shift in frequency produced by the resistor is referenced directly to the load in pounds, when the cross section (in square inches) of the load cell is known. In the very first tests of this unit, an unexpected phenomenon—large alternating stresses of "violin string effect"—were noted. These stresses varied greatly with the free length of the pulling cable and with the rate of pull.

Earth-Moving Equipment

By attempting to pick up too great a load, a crane hoist operator may damage his hoist or overturn his crane. Measurements along the crane boom or hoist bed are inaccurate at best, and connecting wires are easily broken.

Radio telemetry located in the hook (which picks up the load) transmits the magnitude of the load to the operator in his cab. A red line on a load meter indicates the danger point, and simple scale markings show the effect of the boom angle. The telemetry transmitter in the hook is equipped with rechargeable batteries, which can operate many days between chargings. Provision is made to recharge the batteries overnight from the main battery of the crane. Current requirements are negligible. Zero or tare adjustment is made by simple screwdriver control at the receiver. Recalibration or scale change is also provided as an additional receiver adjustment.

Limitations of Telemetry

The preceding paragraphs describe a number of the requirements placed upon telemetry systems by the transducers and quantities being measured. Unfortunately, the development of telemetry has not been such as to satisfy all requirements, and in many cases the telemetry system seriously limits the measurement. A compromise is therefore required between telemetry capabilities and the requirements of measurement. The shortcomings and limitations of the telemetry system place restrictions upon measurements above and beyond those encountered in the laboratory when the telemeter is not employed. In the first place, an electrical output from the measuring device is required in order that the measurement may be placed on a radio link. Consequently, transducers which produce an electrical output in one form or another are necessary. Also, the telemetry system may not be perfectly stable down to zero frequency (DC), and transducers and methods of measurement must be chosen to minimize the effects of drift. Over-modulating the subcarrier, or the time-division multiplexer, may also affect adjacent channels, as well as produce erroneous data in its own channel. If various measuring devices are switched, the switching transients must be minimized, or the accuracy of the telemetry system may be impaired. When mechanical commutators or time multiplexers are employed, the measurement of the time occurrence of the event, such as the impact of cosmic particles or the receipt of a guidance pulse, is made more difficult and the time ambiguity of the multiplexed system is a serious limitation.

The measurement of a large number of parameters requires extensive and bulky equipment, unless the parameters can be combined in groups of similar inputs to minimize the signal conditioning required. This fact generally dictates a relatively standard transducer rather than an optimum one for each particular measurement.

The bandwidth of the measurement, or the frequency with which the measured quantity changes, is also seriously limited by the telemeter. In the FM/FM telemeter, the permissible bandwidth varies from a relatively low value on the lower frequency subcarriers to a reasonably high value on the high-frequency subcarriers. The bandwidth of the measurement must not exceed the subcarrier bandwidth limitations, or sidebands will be generated in adjacent channels, thereby reducing the accuracy of other measurements (if multiplexed), or interference with adjacent RF signals will be caused.

In a time-multiplexed system, the problem of "folded data" is present whenever the rate of data change is faster than one-half the sampling rate. When this occurs, it is not known whether the measured quantity has reversed itself several times between samples, or if there has been no reversal at all. It is considered desirable to limit the bandwidth of the data so that this ambiguity is not present; however, with refined techniques of analysis, this is not a rigid requirement. The form in which the data is displayed or recorded is also a limitation on measurement. In general, time-history plots of the measured quantity are desired. In this case, the speed at which the recording medium moves is often a severe limitation. If sampling is not regular, demultiplexing difficulties are magnified.

Choice of Transducers

The choice of measuring equipment is often limited by the ability to calibrate it. Calibration is usually made from a graphic plot and is applied to the data, which is then replotted in calibrated form. Calibration corrections, selected by other means, may also be applied before plotting or printing the data. Acquiring the calibration curves in the first place, however, is often a difficult procedure. The transducers are calibrated before they are installed, but the remainder of the data system must be calibrated by substitution methods. This requires accessibility for substitute transducers or signals which may be applied. Also, for this purpose, transducers which have simple simulators, such as resistors for a strain gauge, are desirable.

Within the limitations outlined above, the transducers must be chosen to match the particular telemeter employed. A variable-reluctance transducer may be quite satisfactory in an FM/FM system, but may be very difficult to use with a time-multiplexed system. Analog transducers are chosen to match levels and impedances and are often interleaved with digital transducers, such as shaft encoders and outputs from digital computers. These choices must be made to maximize the utility of the telemeter.

The accuracy of measurement may, in many cases, be limited by the telemeter rather than by the transducer. When this is the case, it is sometimes possible to use several transducers to spread the range of measurement over several telemetry channels. This is done in a manner similar to the display of a watt-hour meter, in which the reading of each dial is transmitted over a separate channel. An example of this technique is the measurement of gaseous pressure by means of a bellows gauge at high pressures, a Pirani gauge at medium vacuum pressures and an ionization gauge at low vacuum pressures. To detect system errors, measurements of the same quantity are made by separate transducers utilizing separate telemetry channels. This form of redundant measurement has unfortunately been used very little in radio telemetry.

Transducers must also be chosen to measure the desired quantity without measuring other effects to which they are subjected. In other words, a pressure transducer should measure pressure and not be affected by temperature changes to which it is subjected. The two principal offenders in this regard are temperature and acceleration. It is a major telemetry problem to select transducers which are free of temperature and acceleration effects, unless they are used to measure those particular quantities. Other parameters which affect transducers to a lesser degree are pressure, humidity, ageing and vibration. Tests of all these quantities can be made before the transducer is mounted, and from the results of these tests, the proper transducer can be chosen. Also, even though transducers are installed in groups with the same kind of measurements handled in the same manner, each transducer in a group may be subjected to different environmental conditions. For this reason, it is not possible to have a single "dummy" transducer to calibrate out the environmental effects on the other live transducers.

Strain and Temperature Transmission

With the development of radio telemetry the cost of strain gauge and thermocouple transmitters to be used on rotating parts has been reduced. Various types of telemetry systems are available, including an FM/FM system which presents a DC voltage to the recording instrumentation. This means that both the subcarrier and the radio carrier are frequency modulated. The FM/FM systems have completely transistorized transmitters and tube type receiving stations. The transmitters are potted in an epoxy resin of low thermal conductivity. The potted transmitter is of rugged construction for environmental extremes and utilizes the 88 mHz to 108 mHz radio band.

An FM/FM telemetry system is superior to one that is amplitude modulated. Electrical signal amplitudes are difficult to maintain with linearity in the presence of environmental changes, whereas radio signal frequencies are not. As a result, the FM/FM principle is widely employed, and the telemetry equipment thereby becomes insensitive to amplitude changes.

Some of the strain gauge transmitters are controlled by the impedance of a Wheatstone bridge and contain a subcarrier oscillator and a radio frequency oscillator. These transmitters can be used for both static and dynamic strain measurements. Another strain gauge transmitter is controlled by the impedance of a single resistance strain gauge and contains only a radio frequency oscillator and no subcarrier oscillator. It can be used only for dynamic strain measurement. The bridge-controlled oscillator transmitters are commonly called BCOs. The thermocouple transmitter is voltage controlled and contains a subcarrier oscillator and a radio frequency oscillator. It is called a VCO.

The FM/FM telemetry system is made up of the following elements: (1) sensing element, (2) battery, (3) transmitter, (4) receiver, (5) discriminator, (6) DC amplifier, (7) recorder and (8) the transmitting and receiving antennas.

Sensing Elements

The sensing element is a transducer which will convert the variable to be measured into a suitable electrical quantity. The sensing element for the BCO is a 120-ohm strain gauge or a resistance type temperature sensor, wired into a Wheatstone bridge type circuit with one, two or four active gauges. When fewer than four gauges are active, the bridge must be completed with dummy gauges or resistors. The bridge resistance can range from 50 to 500 ohms, but to obtain maximum sensitivity and stability, the bridge impedance should be adjusted to 120 ohms.

The sensing element for the VCO is a copper-constantan thermocouple. The VCO has a built-in circuit for cold junction compensation. Other thermocouples can also be used, but they must be calibrated with the VCO, and the cold junction compensation has to be made manually. Other millivolt output devices can also be utilized with the VCO.

Transmitters and Batteries

The BCO transmitter can be used at temperatures from 78 to 140°F, and the VCO from −40 to 258°F. All connections to the transmitters are made by soldered joints to pins protruding through the epoxy compound. Each transmitter has screw adjustments. One rotates a multi-turn potentiometer and adjusts the subcarrier oscillator center frequency. Another moves through the transmitter, and its position sets the transmitting radio frequency. The strain sensitivity or millivolt sensitivity can be changed by the connections to various pins protruding through the epoxy potting compound.

The transmitter is made up of two components: the subcarrier oscillator and the radio frequency oscillator. As was mentioned, the subcarrier can be bridge controlled (BCO) or voltage controlled (VCO). The subcarrier center frequency is 4,000 Hz, which frequency can be modulated ±400 Hz by the strain or voltage being measured. Using BCOs, a strain as large as 2,500 microinches per inch and as small as 2 microinches per inch can be measured and transmitted. The temperature measurement range of the VCO is from −200 to 4,000°F. With a copper-constantan thermocouple, a temperature change as small as 2°F can be sensed and transmitted.

The single-resistance strain gauge transmitter does not have a subcarrier oscillator and can be used from −40 to 212°F. It has only a radio frequency oscillator which is modulated by the sensor signal. For this reason, it is not suitable for static strain measurements and must be used for dynamic strain measurements only. It has a frequency response to 25,000 Hz or greater. A static strain signal transmitted by this device will drift. It is provided with self-contained rechargeable nickel-cadmium batteries. Pins protruding through the epoxy case are used for all electrical connections. Only one screw adjustment is provided, and this is used to set the radio frequency.

Rechargeable nickel-cadmium batteries are used with the BCO and the VCO. The BCO batteries have useful lives of four and nine hours. A VCO battery has 40 hours useful life. The single-resistance strain gauge transmitter has a built-in nickel-cadmium battery with a life of four hours.

Receivers and Discriminators

Receivers are cabinet mounted along with the discriminators and a power supply. The receiver is a vacuum tube FM receiver with a tuning range of 88 to 108 mHz, and the subcarrier discriminator is a vacuum tube phase-locked type with a frequency of 4,000 Hz.

When the transmitter is used in its greatest sensitivity mode, the output of the discriminator is approximately 1 volt for a 25 microinch per inch strain with a single active gauge in the bridge. At the most insensitive mode one volt is obtained for approximately 500 microinches per inch strain. The discriminator can withstand a 500 percent overload, which means that a five-volt signal will be obtained from a strain of 125 microinches per inch at the maximum sensitivity and from 2,500 microinches per inch at the minimum sensitivity.

Recorders

The recorder can be a 12-channel oscillograph using 8-in. recording paper. An accurate paper speed of 0.1 to 80 in. per second with timing lines at 0.01, 0.1, 1.0, and 10-second intervals is a good selection. Galvanometers with a frequency response of 1,000 Hz and a coil resistance of 24 ohms are used in the oscillograph.

Antennas and Total System Operation

A nickel cadmium battery supplies the power to the transmitter. For the BCO, the resistance change of the strain gauge changes the frequency of the subcarrier. In the case of the VCO, the millivolt output of the thermocouple changes the frequency of the subcarrier. This change modulates the radio frequency transmitted by antenna. The receiving antenna picks up the signal and conducts it by wire link to the radio receiver, which is tuned to the transmitting frequency. The radio receiver demodulates the FM carrier to reproduce the subcarrier signal. The subcarrier signal is then fed to the discriminator, which demodulates this signal to obtain a DC voltage, which is then amplified by the DC amplifier and recorded on the oscillograph. The oscillograph record, properly calibrated, is then a display of the strain in microinches per inch for the BCOs, or the temperature in degrees for the VCO. At the same time the DC signal can be read on a VTVM and can be used as a check on the oscillograph.

The transmitter subcarrier oscillators are factory set to operate at a center frequency of about 4,000 Hz. They have a frequency range of ± 400 Hz about the 4,000 Hz center frequency. The center frequency is set with a counter at the time of testing. The change of ± 400 Hz is the information frequency change brought about by the change in strain or temperature measured by the sensor. It is this information frequency change that the discriminators isolate as a DC voltage change, which is proportional to the measured strain and is recorded on the oscillograph.

Calibration

Calibration Procedures

Batteries are calibrated under simulated service conditions for voltage drop vs time.

Bridge-controlled transmitters are calibrated for strain subcarrier frequency change using a cantilever beam instrumented with resistance strain gauges. The beam is calibrated for load vs strain using a strain indicator. It is then used to calibrate the bridge-controlled transmitters statically, by measuring the subcarrier frequency change as a function of strain. A dynamic calibration can also be made using a second cantilever beam driven by a vibration generator. Two resistance strain gauges are mounted back-to-back on the second beam and calibrated. One of the gauges is monitored through the telemetry system and the other by wire link to the oscillograph, and the two signals are then compared. The single-resistance strain gauge transmitter is similarly calibrated, but in this case, the beam is fixed in a fatigue machine operating at 30 Hz. Calibrations are performed at various strain levels. Again two calibrated gauges are monitored and compared, one using the telemetry systems and the other using wire link.

The effect of temperature on a transmitter and battery is measured at temperatures from 65 to 135°F by placing both in an air-circulating oven, with the receiving equipment and the calibration beams at room temperature outside the oven.

The voltage-controlled transmitter is calibrated for temperature subcarrier frequency change from 78 to 640°F. Two calibrated thermocouples, welded next to one another on a piece of stainless steel, are heated simultaneously. After determining by wire link instrumentation that both thermocouples are indicating the same temperature, the millivolt output of one is fed into the transmitter, and the output of the other is fed by wire into a precision potentiometer. The subcarrier frequency change is determined as a function of temperature, and the radio signal is recorded on the calibrated oscillograph, with a galvanometer determining its deflection as a function of temperature. The data obtained by wire link and radio are then compared to establish the calibration. The effect of thermocouple lengths can also be investigated in the same test setup. The receivers and discriminators are calibrated before these tests.

Cold junction compensation may be investigated from −40 to 258°F by cooling the transmitter and battery, with leads shorted and with a 20-millivolt input, in a cold chamber below room temperature and by heating in an air-circulating oven to above room temperature. The 20-millivolt input is imposed with a DC power supply kept outside the temperature chamber.

The discriminators are calibrated with the transmitters. The subcarrier frequency, which is the input to the discriminator, is monitored with a digital counter as the calibration beam is loaded. The voltage output corresponding to the frequency change can be monitored with a vacuum tube voltmeter. The strain, subcarrier frequency change, and the voltage output of the discriminator are then correlated. A digital frequency counter is used to set the transmitter center frequency.

Calibration Results

The circuit shown in Figure 6.4i is used to protect the galvanometers. This is necessary because the DC amplifier, used to amplify the signal from the

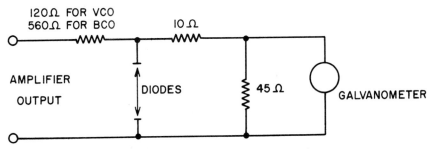

Fig. 6.4i   Galvanometer protection circuit

discriminator, produces voltage transients of 35 volts when the gain is changed, and at this power level the voltage is enough to burn out the galvanometer. The 560-ohm resistor is used to obtain an optimum deflection of the galvanometers with the BCO transmitters. This value is changed to 120 ohms to obtain the optimum deflection when the VCO transmitters are used.

The nickel-cadmium battery life curves in Figures 6.4j, k and l in general show a decrease in voltage during the first hour of discharge. They then retain relatively constant voltage for a period of time, depending on the rated life of the battery. The voltage then drops off quickly, at which time the subcarrier oscillator frequency and the radio frequency change rapidly, and radio emis-

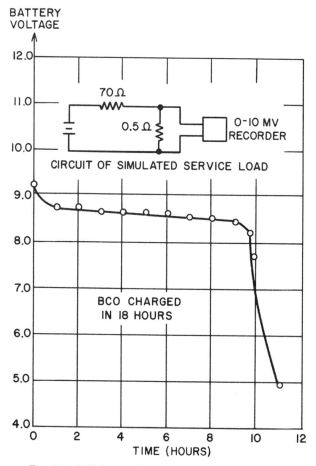

Fig. 6.4j   BCO battery life under simulated service load

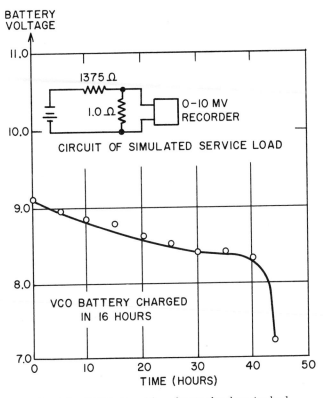

Fig. 6.4k   VCO battery life under simulated service load

sions stop. The circuit diagram shown in Figures 6.4j, k and l is used to check each battery.

The battery provides all DC voltage requirements for the transmitting system and for the sensing transducer. The batteries can be recharged with the battery charger using a 110-volt source.

The calibration curve of discriminator output voltage as a function of subcarrier frequency change for two discriminators is given in Figure 6.4m. This data shows that the center frequency of both discriminators is about 4,200 Hz and that the voltage output of the discriminators is linear for a subcarrier frequency change of ± 400 Hz. This means that the largest strain that the system will measure, with a linear relationship, is one that will change the voltage about ± 3 volts or the information frequency by ± 400 Hz. Strains can be measured beyond the ± 3-volt range, but the relationship between voltage and subcarrier frequency change will not be linear. This data also shows that the output voltage of discriminator #2 is about 25 percent greater

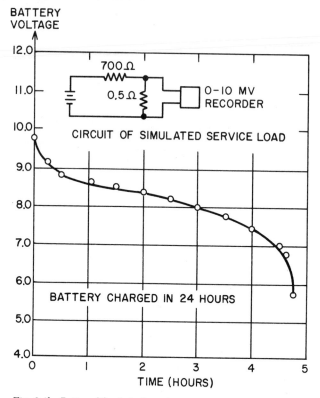

Fig. 6.41   Battery life of single-resistance strain gauge transmitter
under simulated service load

than that of #1. The calibration of the discriminator is independent of the transmitters.

The data in Figures 6.4n, o and p show the static relationship between the strain input to the transmitter and the subcarrier frequency change, at various sensitivity ranges. The data in these Figures can be used with the data in Figure 6.4m to establish the relationship between the change of strain and discriminator voltage output. The sensitivity range shown in Figure 6.4n is the highest that can be obtained with the system, while the data in Figure 6.4o is the lowest sensitivity in the unoverloaded mode. The data in Figure 6.4p are calibration curves of intermediate sensitivities which are attained by using the resistor R shown in the electrical diagram of Figures 6.4q and r.

The data in Figure 6.4q show the attenuation effect of resistor R on the strain gauge bridge and the subcarrier frequency. When R = O, an arbitrary strain of 230 microinches per inch changes the information frequency by 700 Hz, and as R increases, the attenuation also increases. At R = 60, the signal is attenuated 50 percent to a frequency change of 350 Hz. The 60-ohm resistors would change any strain in the linear range by 50 percent. With this curve and the calibration curves in Figures 6.4n, o and p, any strain sensitivity between minimum and maximum can be set.

The data in Figure 6.4r shows the relationship between the strain input to the transmitter and the deflection of a galvanometer, with the amplifier set at a gain of 2. These curves are important in setting the instrumentation when the expected strain can be estimated.

The effect of temperature on the transmitter and battery is shown in Figure 6.4s. Note that a temperature change from 80 to 135°F has less than 1 percent effect on a transmitted signal.

The curves in Figure 6.4t show the dynamic response of the discriminators. The lower curve shows the maximum linear strains that the discriminator will convert to voltage at various input frequencies. The upper curves show the expected variation in strain sensitivity, when the magnitude of the

(Turn to page 705)

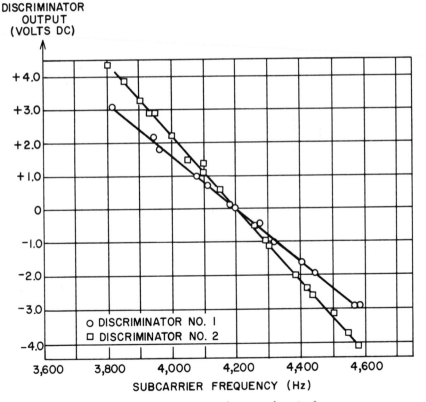

Fig. 6.4m   Discriminator output voltage vs subcarrier frequency

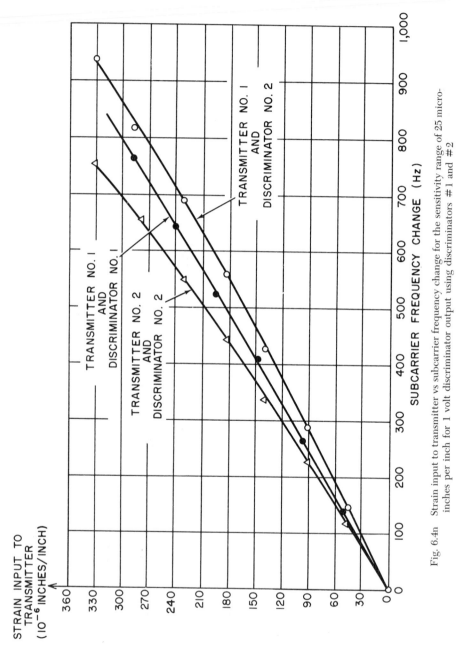

Fig. 6.4n   Strain input to transmitter vs subcarrier frequency change for the sensitivity range of 25 micro-inches per inch for 1 volt discriminator output using discriminators #1 and #2

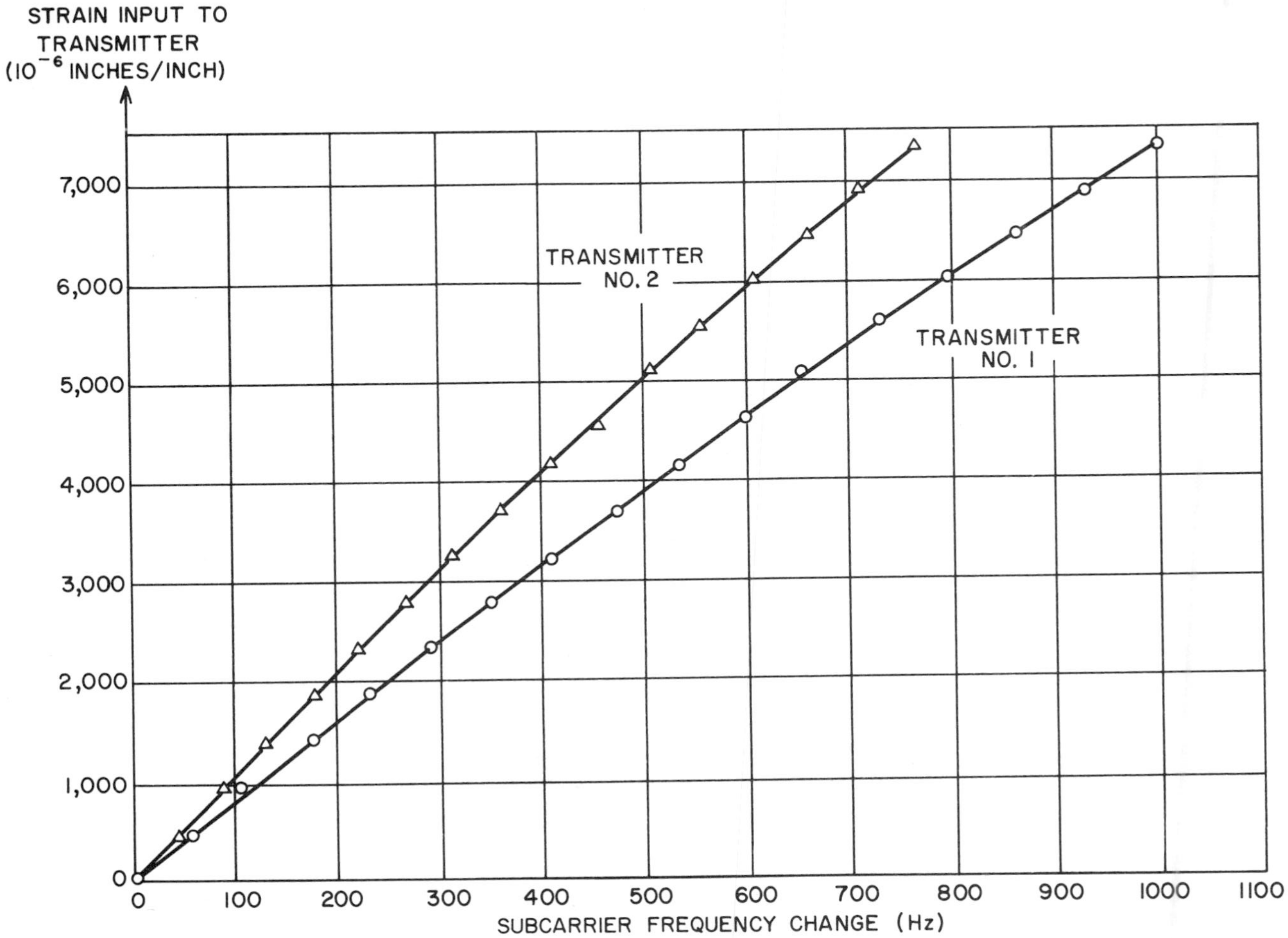

Fig. 6.4o Strain input to transmitter vs subcarrier frequency change for the sensitivity range of 500 microinches per inch strain for nominal 1-volt discriminator output using discriminator #1

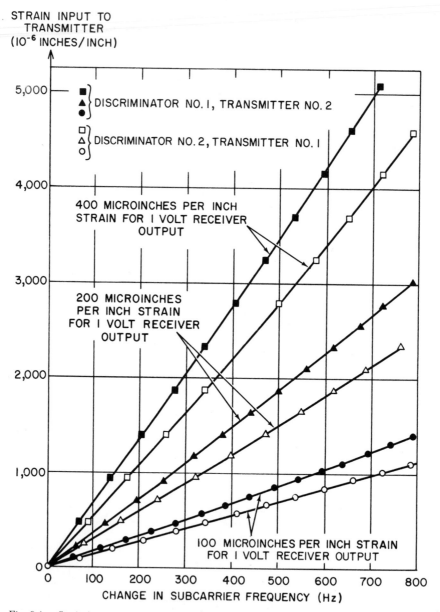

Fig. 6.4p   Strain input to transmitter vs subcarrier frequency change for the strain sensitivity ranges of 100, 200 and 400 microinches per inch for nominal 1-volt discriminator output

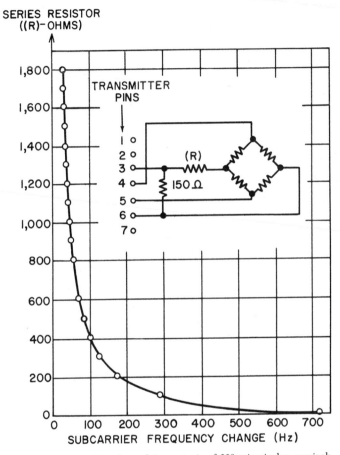

Fig. 6.4q Attenuation of an arbitrary strain of 230 microinches per inch by resistance in series with bridge, using transmitter #1 and discriminator #2

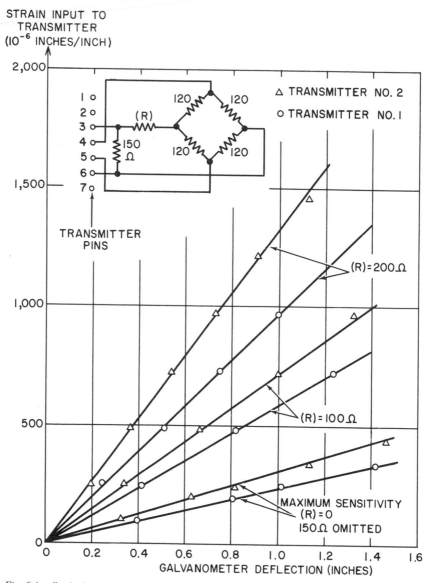

Fig. 6.4r   Strain input to transmitter vs galvanometer deflection for the transmitters #1 and #2 using discriminator #1

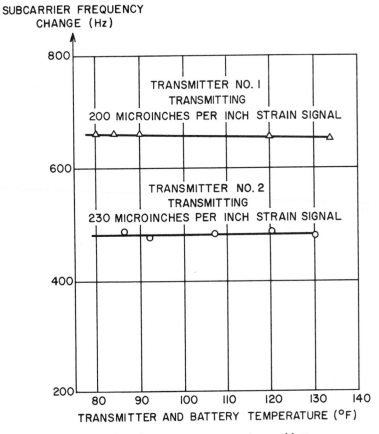

Fig. 6.4s   Effect of temperature on transmitter and battery

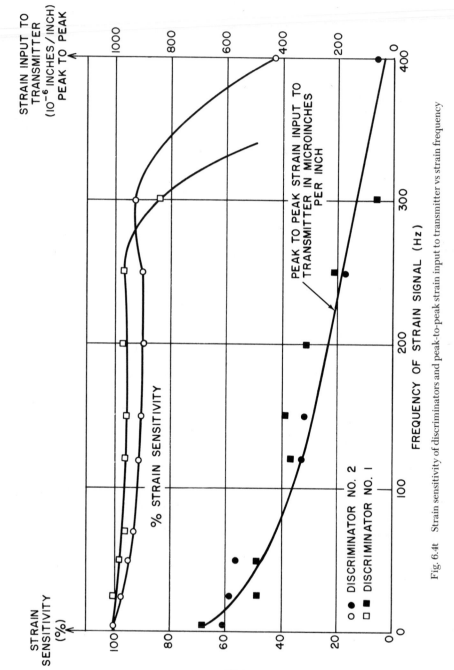

Fig. 6.4t   Strain sensitivity of discriminators and peak-to-peak strain input to transmitter vs strain frequency

704

transmitted strain is at or below the level of the lower curve. The input frequency to the discriminator is a sine wave, and the maximum strain that the discriminator will convert to voltage is that value (peak-to-peak) at which the peaks of the sine waves are a maximum without distortion. The maximum strain that the discriminators will convert to voltage at 5 Hz is about 685 microinches per inch peak-to-peak; it drops to about 40 microinches per inch at 400 Hz. Above these values the peaks of the sine wave are distorted. When the maximum undistorted strains are divided by the corresponding oscillographic deflections, the maximum strain in microinches per inch (per inch of oscillographic chart) is obtained, reducing the data to standard form. If the values at 5 Hz are considered as 100 percent strain sensitivity, the upper two curves in Figure 6.4t are obtained for the two discriminators. These curves show that the strain sensitivity decreases radically above about 300 Hz. This means that the discriminators should not be used above these values unless a special calibration is made beyond that frequency.

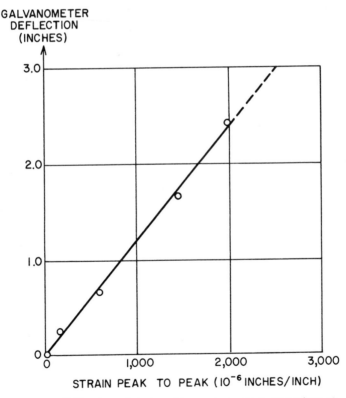

Fig. 6.4u   Calibration of single-resistance strain gauge transmitter at 30 Hz for strain vs galvanometer deflection

The calibration curve for the single-resistance strain gauge transmitter at 30 Hz for strain vs deflection, using a galvanometer and an amplifier set at a gain of 10, is given in Figure 6.4u. These data show that it will transmit a strain of over 2,000 microinches per inch peak-to-peak at 30 Hz. It was found that 5 Hz is about the lowest frequency at which stable strain transmission can be obtained.

The calibration curves for the voltage-controlled transmitter are given in Figures 6.4v and w. The curves in Figure 6.4v show the subcarrier frequency change as a function of the measured temperature using a copper-constantan thermocouple, and the galvanometer deflection using an amplifier gain of 2. Both curves are linear. As in the case of the bridge-controlled transmitters, the galvanometer deflection curves are beneficial to the technician in setting the equipment for the proper sensitivity in a test.

The curve in Figure 6.4w shows that the transmitter automatically compensates for the cold junction temperature changes, because both the compensation curve and the calibration curves show the same change in subcarrier frequency with temperatures from 75 to 140°F.

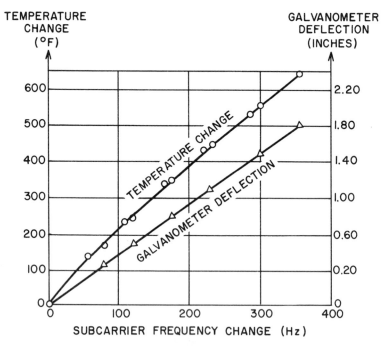

Fig. 6.4v Temperature change and galvanometer deflection vs subcarrier frequency change

SUBCARRIER FREQUENCY
CHANGE
(Hz)

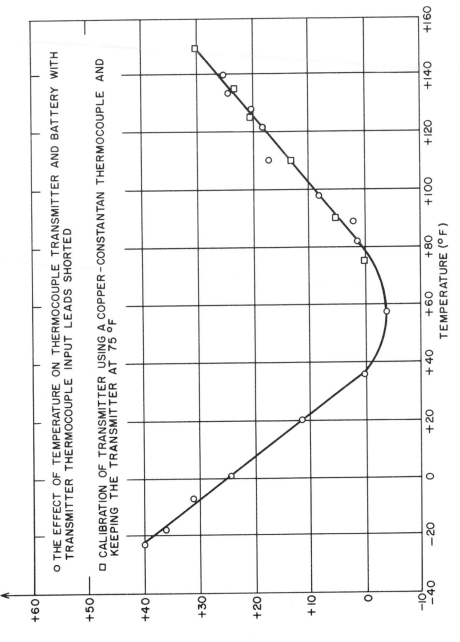

O THE EFFECT OF TEMPERATURE ON THERMOCOUPLE TRANSMITTER AND BATTERY WITH
TRANSMITTER THERMOCOUPLE INPUT LEADS SHORTED

□ CALIBRATION OF TRANSMITTER USING A COPPER-CONSTANTAN THERMOCOUPLE AND
KEEPING THE TRANSMITTER AT 75 °F

TEMPERATURE (°F)

Fig. 6.4w Effect of temperature on subcarrier frequency of transmitter

Chapter VII

CONTROL THEORY

R. M. Bakke, B. G. Lipták,
C. F. Moore, P. W. Murrill,
F. G. Shinskey and C. L. Smith

CONTENTS OF CHAPTER VII

	INTRODUCTION	714
7.1	AUTOMATIC CONTROL TERMINOLOGY AND BASIC CONCEPTS	715
	Why Automatic Control?	715
	Feedback Control	716
	Feedforward Control	717
	Mathematical Representation	719
	Glossary of Analog and Digital Process Control Terms	719
7.2	PROCESS VARIABLES, DYNAMICS, CHARACTERISTICS AND DEGREES OF FREEDOM	736
	Process Variables	736
	Degrees of Freedom	737
	Direct Contact Water Heaters	740
	Controllers	741
	Over-Control	742
	Process Dynamics	743
	Inertia	744
	Resistance and Capacitance	744
	Transportation Time	745
	Differential Equations	746
	Catalyst Preparation Tank	746
	Water Heaters	747
	Chemical Reactors	748
	Mass-Spring Dashpots	749
	Response	750
	First-Order Response	751
	Second-Order Response	752
7.3	CONTROLLERS AND CONTROL MODES	754
	Modes Of Control	754
	Proportional, Integral And Derivative Modes	754
	Inverse Derivative Control Mode	757
	Two-Position (On-Off) Controllers	757
	Single-Speed Floating Control	757

Pneumatic Controller Dynamics   759
    On-Off Controllers   759
    Proportional Controllers   762
    Proportional-Plus-Derivative Controllers   763
    Proportional-Plus-Integral Controllers   766
    PID Controllers   768

7.4   Transfer Functions, Linearization and Stability
Analysis   769
   Laplace Transforms   769
    First Order Lag   772
    Second Order Lag   773
    Partial Fraction Expansion   774
   Transfer Function   776
    First-Order Lag   777
    Second-Order Lag   777
    PID Controllers   778
   Block Diagrams   778
   Linearization   780
   Stability   784
    Descartes' Rule of Signs   784
    Routh's Criterion   785
    Nyquist Criterion   787

7.5   Closed-Loop Response with Various Control Modes   789
   Chemical Reactor Analysis   790
    Proportional Control   793
    Integral Control   795
    Proportional-Plus-Integral Control   796
    PD and PID Control   796
   Selection of Controller Modes   799
    Two-Position Control   799
    Proportional Control   799
    Integral Control   799
    Proportional-Plus-Integral Control   800
    Proportional-Plus-Derivative Control   800
    Inverse Derivative Control   800
    Proportional-Plus-Integral-Plus-Derivative Control   800

7.6   Feedback and Feedforward Control   802
   Feedback Control   802
    Performance of Feedback Control   802
   Feedforward Control   805
    Load Balancing   805
    Steady-State Model   806
    Dynamic Model   807

	Adding Feedback	811
	Performance	812
7.7	ADAPTIVE CONTROL	813
	Adaption of Feedback Parameters	813
	Programmed Adaption	814
	Self-Adaption	816
	Performance	817
	Equipment Limitations	817
	Adaption in Feedforward Systems	818
7.8	CASCADE CONTROL	819
	Primary and Secondary Loops	819
	Types of Secondary Loops	821
	Secondary Control Modes	822
	Instability in Cascade Loops	822
	Saturation in Cascade Loops	824
7.9	RATIO CONTROL	826
	Flow Ratio Control	826
	Ratio Stations	828
	Setting the Ratio Remotely	830
	Cascade Control of Ratio	831
7.10	SELECTIVE CONTROL	832
	Control Within Limits	832
	Limitations on Manipulated Variables	832
	Limits on Controlled Variables	834
	Selection of Extremes	835
	Redundancy	835
	Saturation Problems in Selective Control	836
7.11	OPTIMIZING CONTROL	837
	Feedforward Optimizing Control	837
	Feedback Optimizing Control	839
	Continuous Slope Control	839
	Sampled Data Slope Control	839
	Peak-Seeking Control	841
7.12	TUNING OF CONTROLLERS	842
	Definition of "Good" Control	842
	Closed-Loop Response Methods	843
	Ultimate Method	843
	Damped Oscillation Method	846
	Process Reaction Curve	847
	Open-Loop Tuning	850
	Integral Criteria in Tuning	852
	Digital Control Loops	854
	Summary	854

Example   856
Nomenclature   857
7.13 CONTROLLER TUNING BY COMPUTER   859
     A Practical Process Model   860
     Fitting the Model to the Process   862
     Adjusting the Controller   865
     Sample Problem   867
     Conclusions   871

For many years, process control was approached from a hardware point of view, and little attention was paid to the theoretical considerations contributing to the successful marriage between control instrumentation and controlled process. It is only since about 1950 that instrument engineers started to take the role of the process into consideration, and they began to recognize the importance of the dynamic characteristics of both the control hardware and the controlled system. The purpose of this chapter is to cover the many important aspects of control theory as it applies to process instrumentation.

The first section of this chapter lists the terms found in the jargon of control engineers. The next three sections discuss the response and dynamic characteristics of processes and the control characteristics of available instruments.

The transient behavior of a system (its dynamic characteristics) is usually described by differential equations, and just as the use of logarithms simplifies the handling of multiplications to a requirement for addition, so do Laplace transforms perform a similar function in the solution of differential equations.

Section 7.5 covers the loop response of controlled processes (controller and process acting together).

Starting with Section 7.6, the various basic types of control systems are elaborated on, including feedback and feedforward, cascade and ratio, selective, adaptive and optimizing loops.

In the last two sections of this chapter, the subject of controller tuning is discussed. Following the suggestions and techniques presented in these sections will allow the instrument engineer properly to match the dynamics of the process with that of the controller. In the last section, advanced methods of controller tuning by the use of computers are treated in some detail.

While most other chapters of this Handbook can be used as reference only, it is suggested that the less experienced instrument engineer read this chapter in its entirety.

7.1 AUTOMATIC CONTROL TERMINOLOGY AND BASIC CONCEPTS

The development of automatic control systems in the past 50 years has been equated in importance to the industrial revolution in the nineteenth century. In many respects the introduction of automatic control systems was a second industrial revolution of a sort. While the first was an extension of man's muscle, the second was an extension of his brain. In the nineteenth century we learned to harness and use various forms of natural energy; and in the twentieth century we learned to make devices that could make the decisions necessary to control the various forms of energy.

Principles used in automatic control cut across virtually every scientific field and in the process created a new field of its own. Today the basic principles of automatic control have a wide range of applications and interests, including process control, manufacturing control, aero-space control, and even areas such as traffic control and biomedical control.

Why Automatic Control?

The real need for automatic control in some areas is perhaps more obvious than it is in the process industries. In assembly line oriented manufacturing facilities the need for automation is quite apparent. A machine in many cases is more suitable, both for economic and safety considerations, to perform the numerous tedious and monotonous tasks involved. It is also fairly clear to the casual observer that the control of a supersonic aircraft is much too complicated to be left entirely in the hands of a human pilot. However, in the typical process applications the reasons for control are perhaps a little less apparent.

Since most process equipment operates at a constant load, the tendency might be to suggest that the best solution to the control problem is to set all the variables which affect the process to their proper positions and forget about the process. The difficulty with this reasoning is that seldom can all the inputs to the system be fixed. Most process equipment is subject to many inputs, some of which can be manipulated (or set at a fixed value) and some which will change without regard to the operator's desires. Changes in such variables result in disturbances in the process, unless corrective action of some sort is taken.

Consider the simple direct contact water heater shown in Figure 7.1a.

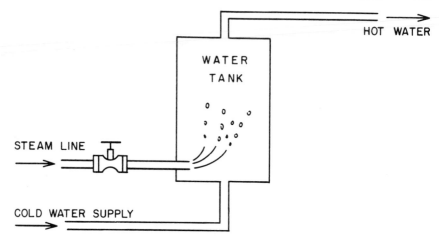

Fig. 7.1a   Direct contact water heater

The heater consists of a tank from which hot water is obtained by bubbling live steam directly into the tank which is full of water. Cool water enters at the bottom of the tank and the hot water leaves the top. A valve is available by which to regulate the flow rate of steam into the heater. In this example, if all other factors were constant the temperature of the outlet could be controlled simply by placing the steam valve at the proper setting. Note, however, that if the temperature of the inlet water changes, the outlet temperature would eventually change by the same amount unless corrective actions were taken. Other variables here besides the inlet water temperature which could disturb the process are the flow rate of the water, the steam supply pressure, the steam quality, and the ambient temperature. A change in any one of these variables would cause a change in the water outlet temperature unless some correction were made.

Feedback Control

Two concepts provide the basis for most automatic control strategies: feedback (closed-loop) control and feedforward (open-loop) control. Feedback control is the more commonly used technique of the two and is the underlying concept on which much of today's automatic control theory is based. Feedback control is a strategy designed to achieve and maintain a desired process condition by measuring the process condition, comparing the measured condition with the desired condition, and initiating corrective action based on the difference between the desired and the actual condition.

The feedback strategy is very similar to the actions of a human operator attempting to control a process manually. Consider the procedure an individual might employ in the control of the direct contact hot water heater described earlier. The operator would read the temperature indicator in the hot water line and compare its value with the temperature he desires (Figure 7.1b). If the temperature were too high, he would reduce the steam flow, and if the temperature were too low, he would increase it. Using this strategy he will manipulate the steam valve until the error is eliminated.

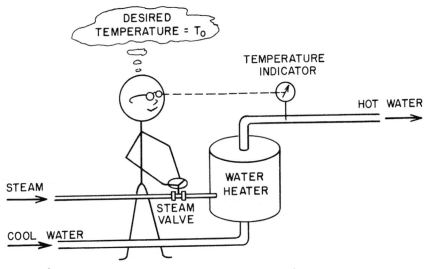

Fig. 7.1b   Manual feedback control

An automatic feedback control system would operate in much the same manner (Figure 7.1c). The temperature of the hot water is measured and a signal is fed back to a device which compares the measured temperature with the desired temperature. If an error exists, a signal is generated to change the valve position in such a manner that the error is eliminated. The only real distinction between the manual and automatic means of controlling the heater is that the automatic controller is more accurate, consistant, and not as likely to become tired or be distracted. Otherwise, both systems contain the essential elements of a feedback control loop (Figures 7.1c and d).

Feedback control has definite advantages over other techniques (such as feedforward control) in relative simplicity and potentially successful operation in the face of unknown contingencies. In general, it works well as a regulator to maintain a desired operating point by compensating for various disturbances which affect the system, and it works equally well as a servo system to initiate and follow changes demanded in the operating point.

Feedforward Control

Feedforward control is another basic technique used to compensate for uncontrolled disturbances entering the system. In this technique the control

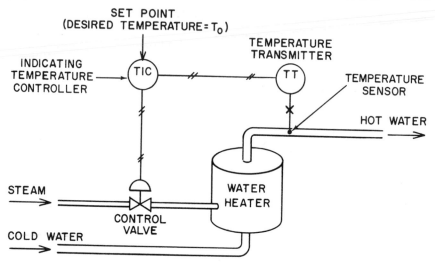

Fig. 7.1c   Automatic feedback control

action is based on the state of a disturbance input without reference to the actual system condition. In concept, feedforward control yields much faster correction than feedback control, and in the ideal case compensation is applied in such a manner that the effect of the disturbance is never seen in the process output.

A skillful operator could use a simple feedforward strategy to compensate for changes in inlet water temperature of a direct-contact water heater. Detecting a change in inlet water temperature, he would increase or decrease the steam rate to counteract the change (Figure 7.1e). The same compensation could be made automatically with an inlet temperature detector designed to initiate the appropriate corrective adjustment in the steam valve opening.

The concept of feedforward control is very powerful, but unfortunately it is difficult to implement in a pure form in most process control applications.

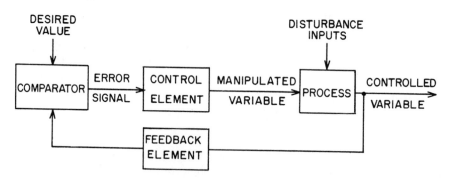

Fig. 7.1d   Basic components of a feedback control loop

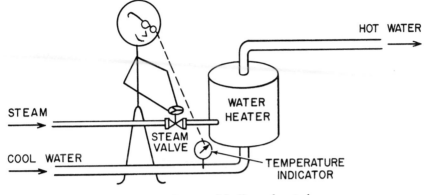

Fig. 7.1e   Concept of feedforward control

In many cases disturbances can not be accurately measured, and therefore feedforward concepts cannot be applied. Even in applications where all the inputs can be either measured or controlled, the "appropriate" action to be taken to compensate for a particular disturbance is not always obvious. In many applications, feedforward control is utilized in conjunction with feedback control in order to handle those unknown contingencies which might otherwise disturb the pure feedforward control system.

Mathematical Representation

A fundamental prerequisite of automatic control theory application is a detailed understanding of the operation of the process under control. While standard equipment and process design requires a detailed knowledge of the equipment operation for constant inputs, automatic control requires a detailed knowledge of the equipment operation when inputs are changing in time. This time-varing behavior is referred to as "process dynamics" and can be conveniently summarized in mathematical terms using differential equations (or the Laplace transfer function representation of differential equations). The application of automatic control theory presumes a knowledge of the entire control system mathematics, and without such knowledge automatic control theory is to a large extent useless.

In many cases the mathematical description (mathematical model) of the various components of the control loop can be obtained analytically based entirely on the physics of the process components. In other cases models can be obtained by experimental testing procedures in which the actual response of the process is analysed in some manner to extract the desired dynamic information.

Glossary of Analog and Digital Process Control Terms

Since the birth of process control a sizable number of terms have become associated with the process control jargon. The following tabulation includes such terms and terminology associated mostly with analog process control. For digital terminology refer to Section 8.1.

Absolute Alarm   An alarm caused by the detection of a variable which has exceeded its high or low limit condition.

Access   The ability to place information into and retrieve information from a storage device.

Actuating Signal   The variable that initiates corrective action in a control system; the same as error signal in feedback control.

Adaptive Control Action   Control action in which the control algorithm or control parameters are changed automatically in such a way as to improve the performance of the control system.

Adder   A device whose output is a representation of the sum of the inputs.

Adiabatic   Occurring without transfer of heat to or from the body or system.

Ambient Pressure   The pressure of the medium surrounding a device.

Ambient Temperature   The temperature of the medium into which the heat of the system is dissipated.

Amplification   The ratio of the signal output amplitude to the signal input amplitude.

Amplifier   A device whose output is an enlarged reproduction of the essential features of an input and which draws power from some external source.

Amplitude   The difference between the average value of a sinusoidal variation and the maximum (or minimum) value.

Amplitude Ratio   A factor expressing the ratio of the output amplitude to the input amplitude when the input is sinusoidal.

Analog (Analogue)   A system whose behavior is mathematically analogous to that of some other system and therefore has the same dynamic equations.

Analog Backup   An alternative means to maintain control over the process in the event of a failure in the primary control system; the backup consists of conventional analog instrumentation.

Analog Controller   A small special-purpose analog computer used to operate on continuous process signals such as voltages, pressures or currents to determine necessary control action. These controllers are distinguished from digital controllers operating on signals with discrete numerical values at discrete intervals of time.

AND   A logic operator having the property that, if P, Q, and R are all logical statements, then the AND of P, Q, and R is true if P, Q, and R are all true, and it is false if any of the three statements are false.

AND Gate   A gate that implements the logic AND operation.

Anticipatory Action   Same as *Rate Action*.

ASCII   American Standard Code for Information Interchange, an eight-level code intended to provide information code compatibility between digital devices of U.S. manufacturers.

Attenuation   A decrease in the strength of a signal between two points or between two frequencies.

Auctioneering Device   A device which automatically selects either the highest or the lowest input signal from among several input signals.

Automatic Control System   Any operable combination of one or more automatic controllers connected in closed loops with one or more processes.

Automatic Controller   A device which measures the value of a variable, quantity, or condition and operates to correct, or limit, deviation of this measured value from a selected reference.

Backlash   A nonlinearity typically associated with the slack in a gear train.

Backup   Provision of alternative means of operation in case of a failure of the primary means of operation.

Bandwidth   The range of frequencies within which performance of a component is accurate, usually extending from zero frequency to some cutoff frequency.

Bang-Bang Control   The same as two-position control.

Boolean Algebra   Pertaining to the operations of formal logic.

Break Point   In frequency response plots the intersection of the asymptotes of the magnitude (or amplitude) ratio plot. Also called break frequency and corner frequency.

Calculating Action   The coupling of primary feedback variables with one another and/or other variables to form a computable function from which the control action is taken.

Capacitance   The amount of energy or material which must be added to a closed system to cause unit change in potential; hence the partial derivative of the content with respect to potential.

Capacity   A measure of the maximum quantity of energy or material which can be stored within a given piece of equipment or system.

Cascade   A series of stages in which the output of one stage is the input of the next.

Cascade Control   Automatic control involving cascading of controllers such that one controller manipulates the set-point input of the other controller instead of manipulating a process variable directly.

Characteristic Function   A polynomial that characterizes the transient response of a system and is the denominator of the system's overall transfer function.

Closed Loop   A signal path which consists of a forward path, a feedback path, and a summing point connected so as to form a closed circuit.

Combination Control   Closed loops connected in combination with the loops being coupled through primary feedback or through any other controller elements.

Command Signal   The set point or the reference input to a control system.

Comparator   The portion of the control elements that determines the feedback error (difference between the reference input and the feedback variable) on which the controller acts.

Compensator   A component or circuit added to a system to improve the characteristics of its response; hence, in many cases the controller.

Constant Value Control   Automatic control of a desired constant value.

Continuous Action   Control action performed continuously (analog control).

Continuous Process   A process in which for extended periods of time uninterrupted flows of fluids enter and products leave a system, as opposed to a batch process, in which the input and output flows are intermittent and periodic.

Control Accuracy   The degree of correspondance between the controlled variable and the desired value of the variable.

Control Action   The response of a control device to an actuating signal for the purpose of decreasing the control error.

Control Algorithm   The mathematical representation of the control action to be performed.

Control Elements   The portion of the control system which relates the error signal to the manipulated variable; the portion of the control system which implements the algorithm.

Control Input   Same as *Set Point.*

Control Mode   A specific type of control action such as proportional, integral, or derivative.

Control Point   The desired value of the variable under control.

Control Ratio   The response of the control variable to a change in set point.

Control System   A system in which deliberate guidance is employed to execute a planned set of control functions.

Controlled System   The body, process, or machine which determines the relationship between the control variable and the manipulated variable.

Controlled Variable   That quantity or condition of the control object which is to be directly measured and controlled.

Controller   A device which operates automatically to regulate a controlled variable.

Corner Frequency   Same as *Break Point.*

Correction   A value to be added to a measured value to compensate for an error.

Correction Time   The time required for a controlled variable to reach and stay within a band about the control point following a change of the set point or operating conditions.

Corrective Action   The variation of the manipulated variable produced by the controller.

Coupled-Control-Element Combination   One in which two or more controller outputs are combined to operate one manipulated variable.

Critical Damping   The smallest degree of damping under which a system can operate without overshooting the desired value after a step change in input.

Critical Gain   A value of system gain beyond which the system is unstable.

Critical Point   In stability analysis by the Nyquist method, the point $s = -1$.

Curve Fitting   The representation of a curve by a mathematical expression. It usually involves determining the "best" optimum expression by some regression technique.

Cutoff Frequency   The frequency of sinusoidal forcing beyond which the amplitude ratio of the response is below some specified lower limit.

Cybernetics   The theory of control and communication in both machines and animals.

Cycling   A periodic change (oscillation) in the controlled variable.

Damping   That property of a system which causes dissipation of energy and hence causes decay in the amplitude of oscillations.

Damping Coefficient   In the characteristic function of a system, the parameter which characterizes the nature of damping of the transient response. Also called damping factor and damping ratio.

Dashpot   A damping device, usually consisting of a cylinder and a piston in which relative motion of either displaces a fluid such as air or oil, resulting in friction.

Dead Band   The range of values through which the measured variable can be changed without initiating a response.

Deadtime   The fixed interval of time between the start of an input to a component and the beginning of response to the input.

Dead Zone   See *Dead Band*.

Decade   Range of frequencies of which the highest is ten times the lowest.

Decibel   In frequency response terminology, a quantitative comparison of the magnitudes of the input and the output sine waves; the number of decibels is 20 times the $\log_{10}$ of the amplitude ratio.

Decimal Numbering System   The numbering system using a base of 10.

Decoder   A circuit which responds to specific coded signals and rejects others.

Delay   Same as *Deadtime*.

Delay Time   The time elapsing from the time input changes to the time the output responds to the input.

Derivative Action   A controller mode in which there is a continuous linear relationship between the controller output and the derivative of the error signal.

Derivative Time   The time difference by which the output of a proportional-derivative controller leads the input when the input changes linearly with time.

Desired Value   The value of the controlled variable which is desired; the same as *Set Point*.

Deviation   The difference between the actual value of the controlled variable and the value of the controlled variable corresponding to the set-point.

Differential Action   Same as *Derivative Action.*

Differential Gap   In a two-position control system an adjustment which determines the smallest range of values through which the controlled variable must pass in order to change the output signal of the controller from maximum to a minimum.

Differentiator   A device, usually of the analog type, whose output is proportional to the derivative of the input signal.

Digit   One of a definite set of characters which are used as coefficients of power of the base in the positional notation of numbers.

Digital   Pertaining to data in the form of digits; discrete data as contrasted to continuous analog data.

Digital Backup   An alternate method of controlling a process which employes a spare digital process control computer.

Direct Acting   Operation of a final control element directly proportional to the control output.

Directly Controlled Variable   That process variable whose value is measured to originate a feedback signal on which the control action will be taken.

Distance-Velocity Lag   Same as *Deadtime.*

Distortion   An undesired change in wave shape.

Disturbance   An input signal other than the set point which directly affects the output of the process.

Downtime   The time interval during which a system is not productive.

Drift   Undesired change in instrument indication with respect to time from an initial value corresponding to a state when the measured variable and ambient conditions are constant.

Driven Response   Same as *Forced Response.*

Droop   An offset, particularly downward (upward is called rise).

Dynamic Error   A measured error caused by variations in time of a quantity being measured, additive to static error.

Dynamic Gain   The amplitude ratio of the steady-state output and input signals of an element or system when the input is sinusoidal. The dynamic gain typically changes with the frequency of the sinusoidal signal. A record of the variations in dynamic gain vs frequency is called the frequency response of the system.

Dynamic Response   The behavior of the output of a device in time with respect to variations in the inputs in time.

Error Rate Control   Same as *Proportional-Plus-Derivative Control.*

Error Ratio   The response of the closed-loop system actuating signal to a change in set point.

Error Signal   The signal resulting from subtracting the feedback signal from the reference signal. The error signal is the input to that part of the controller which contains the algorithm.

Error Squared   The technique of using the square of the error on which to make the control calculation so as to produce a non-linear correction.

Exclusive OR   A logic operator having the property such that, if P and Q are logical statements, then the exclusive OR of P and Q is true if either, but not both, statements are true, and false if both are true or both are false.

Exponential Stage   A system whose transient response to a step input is an exponential decay; hence, a first-order linear system.

Feedback   The signal to the controller representing the condition of the controlled variable.

Feedback Control Action   Control action in which a measured variable is compared with the reference value to produce an actuating error signal which is acted upon to attempt to reduce the error.

Feedback Elements   That portion of the controller which establishes the relationship between the primary feedback and the actual controlled variable.

Feedforward   A control system in which corrective action is based on measurement of disturbance inputs into the process.

Filter   A transducer whose frequency-response characteristics are such that input signals within a certain range are transmitted while other signals are not transmitted.

Final Control Element   That portion of the control loop which directly changes the value of the manipulated variable.

First-Order Lag, or *First-Order Delay*   A system whose dynamic behavior is described by a first-order linear differential equation.

Flip-Flop   A circuit containing active elements capable of assuming either one of two stable positions at a given time.

Floating Control   A mode of control in which the manipulated variable is changed proportional to the integral of the error. A change continuously occurs as long as an error exists.

Floating Rate   Same as *Reset Rate*.

Floating Speed   In single- or multi-speed controller action, the rate of motion of the final control element. It is commonly expressed in percent of full range motion per minute.

Forced Response   That part of a systems output which is a direct result of an input forcing function, and remains after the transient has died out.

Forcing   A change in an input to a system in a specific manner beginning with the system initially at steady state.

Forward Controlling Elements   Those elements in the control system which change a variable in response to the actuating error signal.

Frequency Response   The effect of input frequency on the amplitude ratio and phase shift of a system's or element's output for a sinusoidal input. The frequency response can be directly related to the differential equation which describes the system.

Fundamental Natural Frequency   The lowest of a set of natural frequencies.

Gain Margin   Related to the magnitude of a system response to a sinusoidal input at the frequency for which its phase angle is −180 degrees. A measure of the degree of stability a system will have under feedback control.

Harmonic   A sinusoidal quantity having a frequency which is an integral multiple of some fundamental frequency to which it is related.

Head   Pressure resulting from gravitational forces on liquids; measured in terms of the depth below a surface of the liquid.

High Limiting Control Action   Control action in which the output never exceeds a predetermined high limit.

Higher Order Delay   A system characterized by a high order differential equation.

Hunting   The undesirable motion of an automatic control system in which the controlled variable swings or oscillates about the desired value without seeming to approach it.

Hysteresis   A non-linearity usually attributed to flexibility and loose fits in linkage and to backlash in gear trains.

Idealized System   An imaginary system whose controlled variable has a stipulated relationship to specified set points.

Impulse   A theoretical signal which is a pulse signal of infinite magnitude and infinitesimal duration. In practical applications, a sharp increase or decrease in a variable followed immediately by a return to the original value.

Inclusive OR   Same as *OR*.

Indirectly Controlled Variable   A variable which is not monitored directly by the control system but is related to, and influenced by, the variable which is under direct control.

Initial Error   The transient error appearing immediately after a step function input has been initiated.

Input   A variable that is dependent only on conditions outside the system.

Input Element   The portion of the control system which provides the reference input to the comparator in response to the set point.

Instantaneous Sampling   The process of obtaining a sequence of instantaneous values of a variable continuous in time.

Integral Action   A controller mode in which there is a continuous linear relationship between the integral of the error signal and the output signal of the controller.

Integral Action Limiter   A device which limits the value of the output signal due to integral action to minimize the effect of reset wind-up.

Integral (Reset) Controller   A controller which produces integral control action only.

Integral Time   The proportionality constant in the equation relating the controller output to the error for integral control. It is the time required to produce a change in controller output equal to the change in error input.

Interacting   Two or more consecutive transfer stages whose effective transfer function is not the product of the individual stages.

Interacting Control   Control action produced by an algorithm whose various terms are not independent.

Lag   Any deviation from instantaneously complete response to an input signal. It is usually associated with lags due to resistances and capacitances in the system, however sometimes it is used synonymously with *Delay* and *Dead Time.*

Laplace Transform   A mathematical transformation in which differential equations can be handled much like algebraic equations. The Laplace transform of a variable f(t) is defined as follows:

$$F(s) = \int_0^\infty f(t)e^{-st}\,dt \qquad\qquad 7.1(1)$$

Limit Cycle   A sustained oscillatory response of a feedback system in which the amplitude of oscillation is limited. The cause is usually the presence of some nonlinearity such as saturation or hysteresis.

Limiting   A condition in which the system's response is restricted to a value less than that for a linear response.

Linear Forcing   A forcing function which is linear in time, such as f(t) = A + Bt, where A and B are constants and t is time.

Linear Programming   A procedure for maximizing or minimizing some variable (such as cost or profit). The profit or cost function is written as a linear function of a number of variables which are subject to a number of constraints in the form of linear equalities. The procedure is easily adaptable to computer solution and used quite often as a strategy in supervisory control.

Linearize   To substitute a linear function for a non-linear one, which gives approximately the same relationships over a small range.

Line-Out Time   Same as *Settling Time.*

Live Zone   A zone in the operating cycle of a machine or system during which corrective action can be initiated.

Load, or Load Variable   An outside influence on an automatic control system other than the set point whose effect on the system must be compensated for by the control system.

Load Error   Same as *Offset.*

Log   A periodic summary of process operation data.

Log-Modulus Plot   A rectangular plot of the logarithm of the amplitude ratio vs phase angle of frequency response data; a Nichols plot.

Loop   A series of stages forming a closed path.

Loop Gain (Closed Loop)   The control system gain relating a change in the controlled variable to a change in set point with the feedback element included.

Loop Gain (Open Loop)   The gain of the control system relating a change controlled variable to a change in set point with the feedback element removed.

Low Limiting Control Action   Control action in which the output is never less than a predetermined low limit.

Low-Pass Filter   A wave filter having a transmission band extending from zero frequency up to some cutoff frequency.

Lumping   In the derivation of a model, an assumption that the effects of two or more aspects of the system can be considered as a single quantity.

Magnitude Ratio   In steady-state sinusoidal forcing, the ratio of the amplitude of the output signal to the amplitude of the input signal. The magnitude ratio is usually distinguished from the amplitude ratio in that it is normalized by the system gain such that:

$$\text{magnitude ratio} = \frac{\text{amplitude ratio}}{\text{system gain}} \qquad 7.1(2)$$

Manipulated Variable   The process variable that is changed by the controller to eliminate error.

Manual Backup   An alternative means to maintain process control in the event of a failure in the primary control system which uses manual adjustment of final control elements.

Mathematical Model   A mathematical representation of a process, device, or system derived from either analytical considerations, experimental investigations, or both.

Matrix   A two-dimensional rectangular array of quantities.

Measured Signal   The electrical, mechanical, pneumatic or other variable which is related to some process variable such as flow rate, temperature and level.

Mode   The classification of a controller by the manner in which the manipulated variable responds to the error signal. Some common modes are proportional, integral, and derivative.

Model   A conceptual approximation of a physical element or system of elements; used in the prediction of the behavior of the system; usually mathematical in nature.

Modulation   The process by which some characteristic of one wave is varied in accordance with some characteristic of another wave.

Multi-Element Control System   A control system utilizing input signals derived from two or more process variables which are used jointly in determining the action of the control system.

Multi-velocity Action   Control action in which the velocity of the actuating variable takes one of several predetermined velocities, each corresponding to a definite range of the actuating signal.

Natural Frequency   The frequency of oscillation that a system would have if the damping were reduced to zero.

Natural Response   Same as *Source-Free Response*.

Neutral Zone   A range of error values that gives rise to a value of zero for the controller output.

Noise   An unwanted fluctuation in a variable which tends to obscure its information content.

Non-Interacting Control System   A multi-element control system designed to eliminate effectively the interaction between various process loops so that adjustments can be made in one controlled variable without disturbances being introduced in the other controlled variables.

Non-Self-Operating Control   Control in which energy required to operate the actuating unit is supplied by an external source.

NOR   A logic operator having the property such that, if P, Q, and R are logic statements, then the NOR of P, Q, R is true if all statements are false, and is false if at least one statement is true.

Normalize   To shift the representation of a variable or quantity so that the representation lies in a prescribed range.

NOT   A logic operator having the property such that, if P is a logical statement, then the NOT of P is the opposite statement. If P is true, the NOT of P is false; if P is false, the NOT of P is true.

Octave   A span of frequencies of which the highest is twice the lowest.

Offset   The steady-state deviation of the controlled variables from the set point, usually caused by a disturbance or load change in a system employing a proportional controller.

On-Off Control   A special type two-position control in which the manipulated variable has only one of two possible values: on or off.

Open Loop   Refers to a feedback control system operating with the feedback loop disconnected.

Open-Loop Gain   The ratio of the change in the feedback variable to the change in the set point with the feedback element disconnected.

Open-Loop Transfer Function   The ratio of the transformation of the output to the transformation of the input if the feedback were disconnected.

Operating Conditions   Conditions such as ambient temperature and pressure to which a device or system is subjected besides the measured variable.

Optimization   A procedure whereby the optimum value of a variable, design, program, etc. is found or achieved. The optimum value is determined from a minimization (or maximization) of a criterion function such as cost (or profit).

Optimizing Control Action   Control action that automatically seeks the optimum value of a specific variable or parameter rather than maintaining it at some set value.

OR   A logic operator having the property such that, if P, Q, and R are logical statements, then the OR of P, Q, and R is true if at least one statement is true, and is false is all statements are false.

OR Gate   A gate that implements the OR logical operator.

Oscillation   A period change in the controlled variable.

Output   The variable that is chosen to describe the condition of a system; the dependent variable in a dynamic equation.

Over-Damped   A second-order or higher system which is damped to such a degree that the transient response has no tendency to oscillate or overshoot.

Overshoot   The maximum amount by which a process output exceeds its desired value (or steady-state value) following a step change in input.

Overshoot Time   The time required for a transient error to reach the overshoot point in the response of an automatic control system.

Parallel Cascade Action   The regulation of the set points of two or more automatic controllers using other continuous controllers.

Parameter   A constant coefficient in an equation that is determined by the physical properties of the system.

Peak-to-Peak Amplitude   The difference between the extremes of a quantity.

Performance Operator   Same as *Transfer Function*.

Period   The length of time between consecutively recurring conditions, the reciprocal of frequency.

Perturbation   A disturbance or input forcing, usually of small magnitude, introduced to test a system's response.

Phase   The condition of a sinusoidally varying function at any particular moment in time. It may be expressed in angular form with reference angle taken as zero at the time when the function is at its average value and increasing.

Phase Angle   For two functions varying sinusoidally with the same frequency, that part of the cycle which one signal has reached when the other is at zero phase.

Phase Margin   A measure of the degree of stability a system will have under feedback control computed as the difference between -180 degrees and the phase angle of the system's frequency response when the magnitude ratio is unity.

Phase Shift   The lag or lead that occurs when a sinusoidal signal passes through an element or control system.

Pneumatic Controller   A conventional process controller whose inputs, computations, and output are all pneumatic.

Pole   A real or complex value of the dummy Laplace variable for which the value of the transfer function is infinite; hence, a root of the denominator of the transfer function.

Position Constant   Same as *Steady-State Gain*.

Position Error   Same as *Offset*.

Predictive Control   A control scheme that involves the measurement of changes in load variables and taking corrective action before the system is disturbed. Same as *Feedforward Control*.

Primary Element   That portion of the measuring device which first senses a change in the controlled variable.

Primary Feedback   The signal fed back to the controller which is directly related to the controlled variable.

Process   Those components of a system that are not directly related to control function; hence, the system being controlled.

Process Control System   An automatic control system in which the controlled variable is associated with a process state.

Program Control   A control system in which the set point varies with time according to a predetermined program.

Proportional Action   A control action in which the output of the controller is proportional to the error.

Proportional Band   The proportionality constant in the equation relating the controller output to the error. Physically, it is the error in percent of instrument span required to cause a unit change in controller output. It is the reciprocal of the proportional gain.

Proportional Gain, or *Proportional Sensitivity*   The ratio of the change in output due to proportional control action to the change in error input.

Proportional-Plus-Derivative Control   A control action which is a linear combination of proportional action and derivative (rate) action.

Proportional-Plus-Integral Control   A control action which is a linear combination of proportional and integral (reset) control.

Proportional-Plus-Integral-Plus-Derivative Control   A control action which is a linear combination of proportional, integral, and derivative control.

Proportional-Speed-Floating Control   Same as *Integral Control*.

Pulse   A variation for a short duration of a quantity whose value is normally constant.

Pure Lag   Same as *Deadtime*.

Ramp Forcing   Same as *Linear Forcing*.

Ramp Response   The total time response resulting from a ramp input (an input with a constant rate of change other than zero).

Ramp Response Time   The time interval by which an output lags a ramp input.

Rangeability   The ratio of maximum flow to minimum controllable flow in a final control element.

Rate Action   The controller mode in which there is a linear relationship between the controller output and the derivative of the error signal.

Rate Response   Same as *Derivative Action*.

Rate Time   The proportionality constant in the equation relating the controller output to the error for rate control. Physically, it is the time required for a unit change in the controller output when the derivative of the error with respect to time is unity.

Ratio Controller   A controller that maintains a fixed ratio between two or more variables.

Real-Time Lag   Same as *Lag*.

Reference Input   The variable signal with which the feedback variable is compared in the computation of an error. It is related to the set point usually by a constant.

Regulatory Control   Control with the primary object of maintaining the controlled variable constant in spite of external disturbances.

Relative Damping   For an underdamped system, a number which is the actual damping factor divided by the critical damping factor.

Remote Control   A system for control of remotely located devices.

Reset   An actual or effective change in set point to eliminate an offset or static error. It can be accomplished automatically by the "integral" or "reset" mode.

Reset Rate   Inverse of *Integral Time*.

Reset Wind-up   The undesirable performance of the integral mode in the presence of saturation of the actuating element.

Resonance Peak   A maximum occurring in the output amplitude in frequency response studies.

Response   The effect on a system's output caused by a particular change in an input.

Response Time   The time required for an output to increase from one specified percentage of its final value to another, based on a step input.

Reverse Acting Controller   A controller in which the absolute value of the controller output decreases as the absolute value of the control error increases.

Rise   An offset, particularly upward; opposite of droop.

Rise Time   Same as *Response Time*.

Sampling Action   Process variable sampled and control action taken at intervals.

Sampling Period   The time between the intermittent observations in a sampled data system.

Saturation   A nonlinearity which results from physical limitations on the maximum and minimum value an element will transmit.

Scale   To change a quantity by a factor in order to bring its range within specified limits.

Scan   Sequential interrogation of devices for data from process sensors for the purpose of control or data logging.

Self-Operated Controller   A controller in which the energy necessary to operate the final control element is derived from the controlled process medium.

Self-Regulation   The inherent characteristic of a process which comes to a steady-state value without the aid of an automatic control scheme.

Self-Tuning   The technique of automatically updating the controller tuning parameters based on changing process conditions.

Sensitivity   The ratio of change of output to change of input.

Servocontrol   Control in which the principal objective is to follow a reference value which varies with respect to time.

Servomechanism   A feedback control system in which the controlled variable is a mechanical position.

Servo Operation   A control system operation whose primary objective is to follow a reference value which varies with respect to time.

Set Point   The desired value of the controlled variable.

Settling Time   The time required for the absolute value of the difference between the output of the process and the desired value to become and remain less than a specified amount, following the application of a step input.

Signal   Information being transferred from one device or element to another. It can be accomplished by either mechanical, electrical, pneumatic, hydraulic, or digital means.

Signal Transducer   A device which converts one standardized transmission signal to another.

Signal-to-Noise Ratio   Ratio of the signal amplitude to noise amplitude.

Simulation   Using an analog or digital computer in such a manner as to represent a physical system in which information provided to the computer represents process variables. Information produced by the computer represents the results which would be obtained by the process.

Single-Velocity Action   Control action in which the actuating variable changes with a constant velocity when the actuating signal is within a particular region.

Sinusoidal Change   A signal having sinusoidal or cyclic characteristics.

Source-Free Response   A system's natural response as it relaxes from a state of stored internal energy to an equilibrium condition.

Stability   A property of a physical system in which the natural response is positively damped so that in time the response reaches some finite steady-state value or at least reaches a limit cycle which is bounded.

Stable System   A system whose response to a bounded input is also bounded.

Stage, or *Transfer Stage*   Some part of a larger system which is sufficiently independent of the other part so that a separate transfer function can be written for it.

Static Error   A measurement error effective when the measurement is made at steady-state conditions.

Static Gain   The ratio of an output change to an input change at steady-state conditions.

Steady State   The condition of a system when the transient response has died out. It is implied that all properties are constant with time; however, steady-state conditions can occur even though the output is changing, such as a sine or ramp function.

Steady-State Error   In a control system, the same as *Offset.*

Steady-State Gain   A proportionality constant in a transfer function not containing integration.

Step Change   An instantaneous change, from one value to another value, resembling a step.

Step Response   The time response of an element or system to a step change in input from one operating level to another.

Successive Approximation   An analog-to-digital conversion technique in which increasingly larger or smaller known voltages are compared with the unknown voltages. The logic decision in each comparison generates the binary representation of the voltage.

Superposed Action   Two or more control actions superposed.

Supervisory Control Action   Control action in which the direct control loops operate independently subject to periodic updating of set points of the individual controllers.

System   All the materials and mechanisms contained within arbitrarily defined boundaries.

System Analysis   The definition of a control problem and the development of the solution.

Three-Mode Controller   A controller containing three modes of control, typically, proportional, integral and derivative.

Three-Position Controller   A multi-position controller having three distinct values of output.

Throttling Band   Same as *Proportional Band.*

Time Constant   The time required of the output of a first-order system to reach 63.2 percent of a complete response to a step input.

Time Proportioning Controller   A controller whose output consists of periodic pulses, the duration of which is varied according to the error signal.

Time Schedule Controller   A controller in which the set point is varied automatically according to some predetermined time schedule.

Time Sharing   Pertaining to the interleaved use of the time of a device.

Transducer   An element or device which receives energy (information) from one system and retransmits it, often in a different form, to another. Generally, any device that transmits, amplifies, or changes a signal.

Transfer Function   A mathematical representation of the dynamics of a system using Laplace transform notation. The transfer function is a ratio of the transform of the output of a system to the transform of the input of a system.

Transform   To change, according to some standard formula, a function of a certain variable into a function of another variable, e.g., to take the

Laplace transformation of a function, resulting in the transformation of a time domain to a Laplace domain.

Transient Response   That part of the output of a system which is related to the natural response of a system and eventually disappears if the forcing continues unchanged.

Transmittance   Same as *Transfer Function*.

Transportation Lag   Same as *Deadtime*.

Truth Table   A table that describes a logic function by listing all possible combinations of input values and indicating, for each combination, the true output values.

Tuning   The adjustment of the control parameters (e.g., gain, reset, rate) to give the desired response.

Two-Position Control   A system of regulation in which the manipulated variable has only two discrete values of output.

Type-One Servo   A system under servo control that contains one integration in the control loop.

Undamped   Said of a system capable of oscillator transient response of constant amplitude.

Under-Damped   Said of a system capable of oscillator response which diminishes in time.

Unsteady State   The condition of a system undergoing a state of transient change.

Update   To modify a system, program, strategy, etc. according to current information.

Velocity Constant   The proportionality constant in the transfer function of a system that contains one integration; hence, the gain of a type-one servo.

Velocity-Limiting Control Action   Control action in which the rate of change of a specified variable will not exceed a predetermined limit.

Zero   A real or complex value of an independent variable that makes the function of that variable equal to zero.

7.2 PROCESS VARIABLES, DYNAMICS, CHARACTERISTICS AND DEGREES OF FREEDOM

This section concerns itself with the nature of chemical processes, particularly as far as those requirements and limitations are concerned, which the process places on the associated automatic control systems.

The state of a process may be described by the values of its variables. This state is defined when each of the system's degrees of freedom are specified. The number of degrees of freedom of a process represents the maximum number of independently acting automatic controllers which can be placed on the process.

The transient behavior of a process shows its dynamic characteristics. This involves its stability and speed of response, and it is usually described in terms of inertia, resistance, capacitance and transportation time. Differential equations are utilized to describe these characteristics, and depending on their nature, we distinguish the various process responses by calling them first order or second order.

Process Variables

Many external and internal conditions affect the performance of a process unit. These conditions may be expressed in terms of process variables such as temperature, pressure, flow, concentration, weight, level, etc. The process is usually controlled by measuring one of the variables which represents the state of the system and then by automatically adjusting one of the variables which determines the state of the system. Typically, the variable chosen to represent the state of the system is termed the *controlled variable* and the variable chosen to control the system's state is termed the *manipulated variable*.

The manipulated variable can be any process variable which causes a reasonably fast response and is fairly easy to manipulate. The controlled variable should be the variable which best represents the desired state of the system. Consider the water cooler shown in Figure 7.2a. The purpose of the cooler is to maintain a supply of water at a constant temperature. The variable which best represents this objective is the temperature of the exit water, (T_{wo}), and should be selected as the controlled variable. In other such cases direct control of the variable which best represents the desired condition is not possible. Consider the chemical reactor shown in Figure 7.2b. The variable

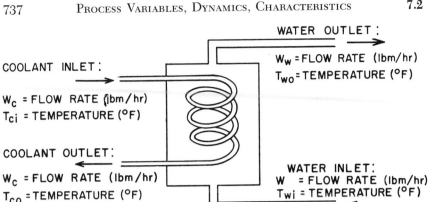

Fig. 7.2a   Process variables in a simple water cooler

which is directly related to the desired state of the product is the composition of the product stream; however, in this case a direct measurement of product composition is not always possible. If product composition is to be controlled, some other process variable must be used which is related to composition. A logical choice for this chemical reactor might be to hold the pressure constant and use reactor temperature as an indication of composition. Such a scheme is often used in the indirect control of composition.

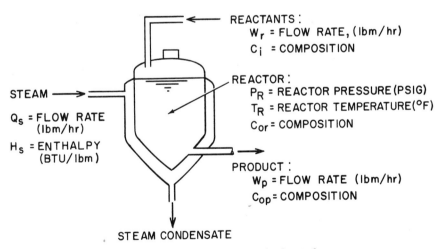

Fig. 7.2b   Process variables in a simple chemical reactor

Degrees of Freedom

Degrees of freedom is a very important concept in process control, but its importance is frequently not appreciated. Mathematically the number of degrees of freedom is defined as

$$df = v - e \qquad\qquad 7.2(1)$$

where df = number of degrees of freedom of a system,

     v = number of variables which describe the system, and

     e = number of independent relationships which exist among the various variables.

A clear understanding of this principle and its application can give a keener insight into the steady state and dynamic behavior of processes and a clearer understanding of the basic framework around which a control system must be designed.

     To illustrate the general principle, consider the degrees-of-freedom equation applied to three common examples: Consider the motion of an airplane, a boat and a train. For each vehicle the variables are the same, latitude, longitude, and altitude. A value for each of the three navigational coordinates specifies the exact position of the vehicle. Therefore, in each case the value of v is 3. The number of independent equations relating the three space coordinates differs in each case. For an airplane (Figure 7.2c) on a casual

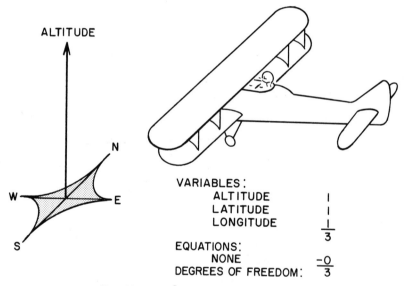

Fig. 7.2c   Degrees of freedom of an airplane

flight (with no specific flight plan), there are no fixed relationships between its latitude, longitude and altitude at any moment of time. Therefore, the number of degrees of freedom of the plane is

$$df = v - e = 3 - 0 = 3$$

This answer is intuitively correct since the plane is free to fly in virtually any direction in the three-dimensional space. The boat (Figure 7.2d) has the

VARIABLES:
    ALTITUDE |
    LATITUDE |
    LONGITUDE |
                           3
EQUATIONS:
    ALTITUDE = SEA LEVEL −|
DEGREES OF FREEDOM   2

Fig. 7.2d   Degrees of freedom of a boat

same three navigational variables as does the plane, but assuming the water is relatively calm the altitude is fixed. Therefore the following equation is established:

$$\text{altitude of boat} = \text{sea level}$$

Hence, the degrees of freedom for the boat are

$$df = v - e = 3 - 1 = 2$$

The boat is free to move in two directions but must remain confined to the water's surface.

The train (Figure 7.2e) also has the same three navigational variables but is confined to a track which has a fixed location described by two equations,

$$\text{altitude of track} = \text{geological contour of earth}$$
$$\text{latitude of track} = \text{specific function of the longitude}$$

The first equation fixes the altitude to be equal to the local ground surface elevation. The second equation describes the exact route of the track in terms of the latitude and longitude. Therefore, the train has the following degrees of freedom:

$$df = v - e = 3 - 2 = 1$$

The train has only one degree of freedom and can only move along the path specified by the position of the track. The exact position of the train can be specified by the relative distance down the track.

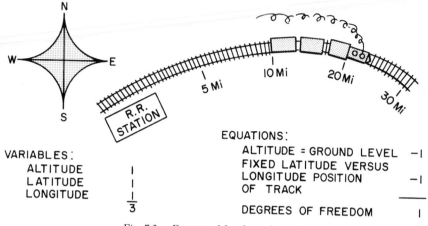

Fig. 7.2e   Degrees of freedom of a train

Direct Contact Water Heaters

As an example of analysis for degrees of freedom inherent in a process unit, consider the direct contact hot water heater shown in Figure 7.2f.

The first step in the analysis is to define the problem and list all the pertinent variables which affect the system. Table 7.2g lists such variables for this particular problem and also notes the corresponding equations which relate these variables. Based on these tabulations the degrees of freedom for the hot water heater is defined as follows:

$$df = v - e = 7 - 5 = 2$$

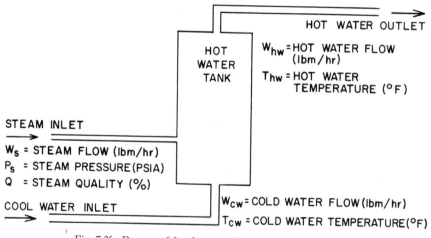

Fig. 7.2f   Degrees of freedom in a direct contact water heater

Table 7.2g

DEGREES-OF-FREEDOM ANALYSIS OF A DIRECT CONTACT WATER HEATER

Variables	Number of Variables
W_s = flow rate of steam	1
W_{cw} = flow rate of cool water	1
W_{hw} = flow rate of hot water	1
Q = quality of steam	1
P_s = supply pressure of steam	1
T_{cw} = temperature of cool water inlet	1
T_{hw} = temperature of hot water outlet	1
Total Number of Variables	7

Equations	Number of Equations
Constant terms P_s, Q, T_{cw}	3
Material balance (Conservation of Mass)	1
Energy balance (Conservation of Energy)	1
Total number of equations	5

Degrees of Freedom = 2

Note that to a large extent the number of variables included in the tabulation is arbitrary. In this particular example we could have included in the list, the enthalpy of the three streams;

H_s, the enthalpy of steam (BTU/lbm)
H_{cw}, the enthalpy of inlet water (BTU/lbm)
H_{hw}, the enthalpy of outlet water (BTU/lbm)

However, in considering these three new variables, *three more* independent *equations* must also be considered, namely,

$$H_{cw} = c_p(T_{cw} - T_r) \qquad\qquad 7.2(2)$$

$$H_{hw} = c_p(T_{hw} - T_r) \qquad\qquad 7.2(3)$$

$$H_s = \text{function of } P_s \text{ and } Q \text{ (found in steam tables)} \qquad 7.2(4)$$

where T_r and c_p are constants.

Hence, the number of degrees of freedom remains unchanged:

$$df = v - e = 10 - 8 = 2$$

Controllers

In Section 7.1 (see Figure 7.1c), the control of a direct contact water heater was implemented by the addition of a feedback control loop which measures the temperature of the outlet water (T_{hw}), compares it with some predetermined value, and adjusts the steam flow according to the error. Therefore, in terms of degrees of freedom the control loop adds another independent equation,

$$F = \text{function of } (T_{hw}\text{-setpoint}) \qquad 7.2(5)$$

The net effect on the overall degrees of freedom is a reduction by one. Therefore, this controlled process has one (1) remaining degree of freedom.

The remaining degree of freedom could be removed by the addition of another controller on the inlet water line such that the water flow is maintained at a constant value.

Over-Control

Perhaps the most beneficial result realized from the understanding of the degrees of freedom concept is a knowledge of what can be controlled and what cannot. According to the degrees-of-freedom rule, *the number of independently acting automatic controllers on a system or on part of a system, may not exceed the number of degrees of freedom.* To illustrate this rule consider the water heater analysed above. The degrees of freedom of the uncontrolled process unit was two. Therefore the maximum number of controllers which can be successfully employed on the tank is also two, as described earlier. More than two would yield an over-specified system resulting in conflicting actions between the controllers.

To illustrate over-control, consider the simple boiler shown in Figure 7.2h, in which water is being boiled. A control valve in the vent line can regulate the flow of water vapor. The heat input can also be regulated. An analysis of the variables and the accompanying equations are given in Table 7.2i. The resulting degrees of freedom of the uncontrolled process is, therefore,

$$df = v - e = 6 - 4 = 2$$

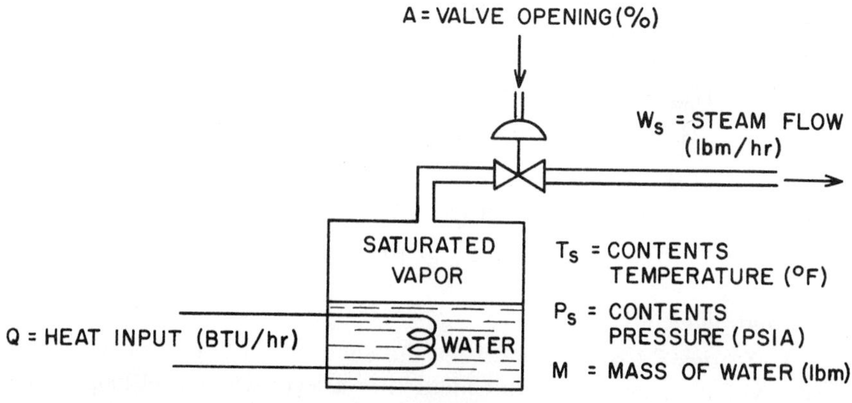

Fig. 7.2h   Degrees of freedom in a simple water boiler

Table 7.2i

DEGREES-OF-FREEDOM ANALYSIS OF A SIMPLE WATER BOILER

Variables	Number of Variables
M = mass of water in tank at any point in time	1
Q = heat input into tank	1
W_s = flow rate of vapor out of tank	1
T_s = temperature of tank contents	1
P_s = pressure in tank	1
A = valve opening	1
Total number of variables	6

Equations	Number of Equations
Material balance	1
Energy balance	1
Valve equation, $V = f(A, P_s)$	1
Vapor pressure vs temperature data (steam tables)	1
Total number of equations	4

Degrees of freedom = 2

The analysis suggests that since the process has only two degrees of freedom, only two independent controllers can be added. In this case there might be a natural tendency to attempt to control directly all of the system outputs such as temperature, pressure, and vapor flow rate; however, the number of degrees of freedom dictates the maximum of two such control loops to be used. The possible combinations to be considered are: pressure and flow rate, temperature and flow rate, or temperature and pressure. A superficial analysis indicates that any two of these combinations would be satisfactory. The individual equations involved point out another basic lesson concerning the degrees of freedom of processes. An analysis of the vapor pressure vs temperature data indicates that there is only one degree of freedom between temperature (T_s) and pressure (P_s). Therefore, an attempt to control both independently would result in an over-controlled system, in spite of the overall degrees of freedom being satisfied. The concept of degrees of freedom should be applied at all levels in the system analysis.

Process Dynamics

Controlling processes would be trivial if all systems responded instantaneously to changes in the process inputs. The difficulty in control lies in the fact that all processes to one degree or another tend to delay and retard the changes in process variables. This time-dependent characteristic of the process is termed process dynamics, and its evaluation is essential to the understanding and application of automatic control.

The dynamic characteristics of all systems, whether mechanical, chemical, thermal, or electrical in nature, can be attributed to one or more of the following effects: (1) inertia, (2) capacitance, (3) resistance, (4) transportation time.

Inertia

Inertia effects pertain to Newton's second law, governing the motion of matter:

$$\Sigma F = (M \times a) \qquad 7.2(6)$$

where ΣF = net force acting on a mass,
  M = total mass, and
  a = acceleration of that mass.

Inertia effects are most commonly associated with mechanical systems which involve moving components and parts, but they are also important in some flow systems in which fluids must be accelerated or decelerated.

Resistance and Capacitance

Resistance and capacitance are perhaps the most important effects in industrial processes involving heat transfer, mass transfer, and fluid flow operations. Those parts of the process which have the ability to store energy or mass are termed capacities, and those parts which resist transfer of energy or mass are termed resistances. The combined effect of supplying a capacity through a resistance is a time retardation, which is very basic to most dynamic systems found in industrial processes. Consider, for example, the water heater system shown in Figure 7.2j, where the capacitance and resistance terms can

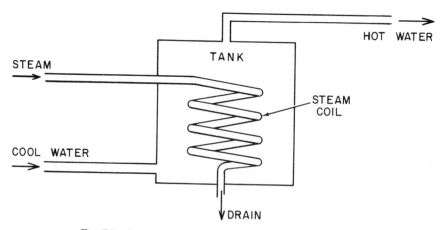

Fig. 7.2j   Resistance and capacitance effects in a water heater

be readily identified. The capacitance is the ability of the tank, and of the water in the tank, to store heat energy. A second capacitance is the ability of the steam coil and its contents to store heat energy. A resistance can be identified with the transfer of energy from the steam coil to the water due to the insulating effect of a stagnant layer of water surrounding the coil. If an instantaneous change is made in the steam flow rate, the temperature of the hot water will also change, but the change will not be instantaneous. It will be sluggish, requiring a finite period of time to reach a new equilibrium. The behavior of the system during this transition period will depend on the amount of material which must be heated in the coil and in the tank (determined by the capacitance) and on the rate at which heat can be transferred to the water (determined by the resistance).

Transportation Time

A contributing factor to the dynamics of many processes involving the movement of mass from one point to another is called the transportation lag, or deadtime. In the simple heater problem discussed earlier consider the effect of piping on the heated water to reach a location some distance away from the heater (Figure 7.2k). The effect of a change in steam rate on the water temperature at the end of the pipe will not only depend on the resistance and capacitance effects in the tank, but will also be influenced by the length

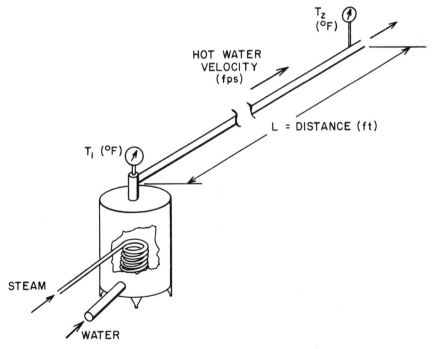

Fig. 7.2k   Transportation time effects on a water heater

of time necessary for the water to be transported through the pipe. All lags associated with the heater system will be seen at the end of the pipe, but they will be delayed. The length of this delay is called the transportation lag, or deadtime. The magnitude is determined as the distance over which the material is transported, divided by the velocity at which the material travels. In the heater example,

$$\theta = v/L \qquad\qquad 7.2(7)$$

Differential Equations

The quantitative effect of resistance, capacitance, transportation lag and inertia on a process can be expressed in terms of the differential equations which describe the process. Such equations can be developed by applying Newton's law, the law of conservation of mass, and the law of conservation of energy along with specific equations related to a particular process (i.e., equations describing heat transfer, fluid flow through valves, chemical kinetics, thermodynamic equilibrium, etc.). To demonstrate the general procedure involved, several simple examples are given of methods for developing differential equations that describe process units.

Catalyst Preparation Tank

Consider a catalyst preparation tank shown in Figure 7.21. The system is used to upgrade a low-concentration catalyst stream by the addition of a

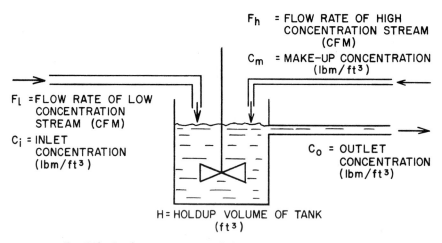

F_h = FLOW RATE OF HIGH CONCENTRATION STREAM (CFM)

C_m = MAKE-UP CONCENTRATION (lbm/ft^3)

F_l = FLOW RATE OF LOW CONCENTRATION STREAM (CFM)

C_i = INLET CONCENTRATION (lbm/ft^3)

C_o = OUTLET CONCENTRATION (lbm/ft^3)

H= HOLDUP VOLUME OF TANK (ft^3)

Fig. 7.21   Catalyst preparation tank described by differential equation

relatively small amount of a highly concentrated stream. The tank is assumed to be well mixed, and the flow rate of highly concentrated material is considered very small in comparison with the flow of material out of the tank.

The only equation needed to describe the system is the material balance on the catalyst. The law of conservation of mass states that:

$$\left(\begin{matrix}\text{Flow of catalyst}\\ \text{into the tank}\end{matrix}\right) - \left(\begin{matrix}\text{Flow of catalyst}\\ \text{from the tank}\end{matrix}\right)$$

$$= \left(\begin{matrix}\text{Accumulation of}\\ \text{catalyst in tank}\end{matrix}\right)$$

$$\left(F_h C_m + C_i F_l\right) - \left(C_o F_l\right) = \left(\frac{d}{dt} HC_o\right) \qquad 7.2(8)$$

Since the holdup, H, can be considered constant, the equation can be written

$$F_h C_m + C_i F_l = C_o F_l + H\frac{d}{dt} C_o \qquad 7.2(9)$$

where F_h = flow rate of highly concentrated makeup stream (CFM),
   F_l = flow rate of low concentrated streams (CFM),
   C_m = concentration of catalyst in the makeup stream (lbm/ft^3),
   C_i = concentration of catalyst in the inlet stream (lbm/ft^3),
   C_o = concentration of catalyst in the outlet stream (lbm/ft^3), and
   H = holdup of tank (ft^3).

Water Heaters

The equations describing the direct contact hot water heater shown in Figure 7.2m are developed as follows. The tank is assumed to be at uniform temperature, and the quantity of water added to the system by steam condensation is assumed to be comparatively small. For this simple example an energy balance is all that is necessary to describe the entire system.

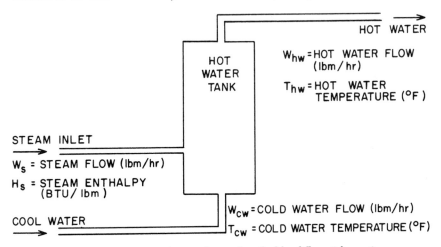

Fig. 7.2m Direct contact water heater described by differential equation

$$\begin{pmatrix} \text{Rate of energy} \\ \text{into heater} \end{pmatrix} - \begin{pmatrix} \text{Rate of energy} \\ \text{out of heater} \end{pmatrix}$$

$$= \begin{pmatrix} \text{Rate of energy} \\ \text{accumulation} \\ \text{in the heater} \end{pmatrix}$$

$$\left(H_s W_s + W_{cw} C_p (T_i - T_r) \right) - \left(W_{hw} C_p (T_{hw} - T_r) \right) =$$
$$\left(\frac{d}{dt} \rho \, H C_p (T_{hw} - T_r) \right) \qquad \text{7.2(10)}$$

Or, since ρ, C_p, and T_r are constants,

$$H_s W_s + W_{cw} C_p T_i = W_{hw} C_p T_{hw} + H\rho C_p \frac{d}{dt} (T_{hw}) \qquad \text{7.2(11)}$$

where H_s = enthalpy of steam (BTU/lbm),
$\quad\quad\quad W_s$ = flow rate of steam (lbm/hr),
$W_{hw} = W_{cw}$ = flow rate of water (lbm/hr),
$\quad\quad\quad T_{cw}$ = inlet water temperature (°F),
$\quad\quad\quad T_{hw}$ = outlet water temperature (°F),
$\quad\quad\quad C_p$ = heat capacity of water (BTU/lbm°F),
$\quad\quad\quad \rho$ = density of water (lbm/ft³),
$\quad\quad\quad H$ = volume of tank (ft³), and
$\quad\quad\quad T_r$ = reference temperature (°F).

Chemical Reactors

Figure 7.2n illustrates a simple back-mix chemical reactor in which component "a" reacts to form component "b" by the following first-order reaction:

$$a \xrightarrow{\ k\ } b$$

The feed is a binary mixture with composition X_{ai}, X_{bi}. The product stream is also a binary mixture containing "b" and unreacted "a". Due to the well-mixed conditions in the reactor, composition of the reactor contents and the product stream composition are identical (X_{ao}, X_{bo}). A material balance of component "a" applied to the reactor is as follows:

$$\begin{pmatrix} \text{Rate of "a"} \\ \text{into reactor} \end{pmatrix} - \begin{pmatrix} \text{Rate of "a"} \\ \text{out of reactor} \end{pmatrix} = \begin{pmatrix} \text{Rate of accumulation} \\ \text{of "a" in reactor} \end{pmatrix}$$

$$(QX_{ai}) - (QX_{ao} + r_a H) = \left(H \frac{d}{dt} (X_{ao}) \right) \qquad \text{7.2(12)}$$

where $r_a H$ represents the amount of component "a" consumed by the reaction. No actual flow is involved, but it does represent the removal of "a" from

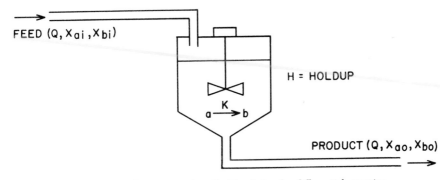

FEED (Q, X_{ai}, X_{bi})

H = HOLDUP

K
a ⟶ b

PRODUCT (Q, X_{ao}, X_{bo})

Fig. 7.2n   Backmix reactor described by first-order differential equation

the system and therefore must be considered. The rate of consumption of "a" is described as

$$r_a = kX_a \qquad\qquad 7.2(13)$$

and can be substituted into equation 7.2(12) to yield the complete differential equation which describes the system:

$$QX_{ai} = (Q + k)X_{ao} + H\frac{d}{dt}X_{ao} \qquad\qquad 7.2(14)$$

Mass-Spring Dashpots

To illustrate a typical mechanical system in which inertia effects are important, consider the mass-spring dashpot of Figure 7.2o. Consider the system to be ideal in that all the mass is located at a point in the center of the block, and both the spring and dashpot are linear. Newton's law of motion states that the sum of the forces on the mass must be equal to the product of the mass times its acceleration:

$$\Sigma F = (M)(a) \qquad\qquad 7.2(15)$$

where

$$a = \frac{d^2}{dt^2}(y) \qquad\qquad 7.2(16)$$

An analysis of the system indicates that the four forces acting on the mass are:

1. Force of gravity (acting downward) $= (M)(g) = $ constant   7.2(17)

2. Force of spring (acting upward) $= (M)(g) - (K)(y)$ 7.2(18)

   When the mass is in the rest position, the upward force of the spring equals the downward force due to gravity.

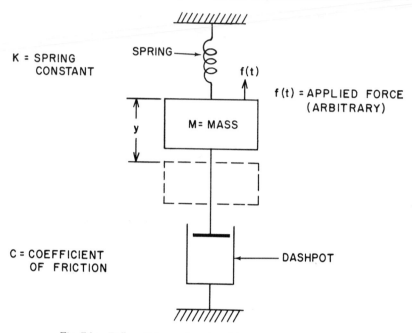

K = SPRING
  CONSTANT

SPRING

f(t)

M = MASS

y

f(t) = APPLIED FORCE
      (ARBITRARY)

C = COEFFICIENT
   OF FRICTION

DASHPOT

Fig. 7.2o   Differential equation describing a mass-spring dashpot

3. Force of dashpot $= -C\dfrac{d}{dt}(y)$ 7.2(19)
   (acting in a direction
   opposite the direction
   of movement)

4. Arbitrary force
   imposed on mass $= f(t)$ 7.2(20)

Substituting the force terms into Newton's equation (considering all forces acting upward to be positive) yields

$$-(M)(g) - (K)(y) + (M)(g) - C\frac{d}{dt}(y) + f(t) = M\frac{d^2}{dt^2}(y) \qquad 7.2(21)$$

or

$$M\frac{d^2}{dt^2}(y) + C\frac{d}{dt}(y) + (K)(y) = f(t) \qquad 7.2(22)$$

Response

The concept of system response is fundamental in automatic control. As inputs to a particular process change, so will the process respond in a certain manner depending on its dynamics. For a particular input the response of the process can be predicted from the solution of the differential equation which describes the process dynamics. The following paragraphs will discuss several differential equations which are typical of dynamics associated with industrial processes.

First-Order Response

The linear, first-order differential equation is typical of a large class of components and control systems. The general form of such equation is

$$\tau \frac{d}{dt} c(t) + c(t) = K r(t) \qquad\qquad 7.2(23)$$

where τ, K = constants of the process, time constant and gain,
       t = time,
       $c(t)$ = process output response, and
       $r(t)$ = process input response.

Process elements of this description are common and are generally referred to as first-order lags. The response of a first-order system is characterized by two constants: a time constant τ and a gain K. The gain is related to the amplification associated with the process and has no effect of the time characteristics of the response. The time characteristics are related entirely to the time constant. The time constant is a measure of the time necessary for the component or system to adjust to an input, and it may be characterized in terms of the capacitance and resistance (or conductance) of the process:

$$\tau = \text{resistance} \times \text{capacitance} = \frac{\text{capacitance}}{\text{conductance}} \qquad\qquad 7.2(24)$$

To illustrate the nature of a first-order system, consider the response which results from an input of the following form:

$$\left. \begin{array}{ll} r(t) = 0.0 & t \leq 0 \\ r(t) = R_0 & t > 0 \end{array} \right\} \qquad 7.2(25)$$

The solution of the first-order differential equation for such an input, considering the initial value of c to be zero, is

$$c(t) = K R_0 (1.0 - e^{-t/\tau}) \qquad \text{for} \qquad t > 0 \qquad\qquad 7.2(26)$$

Details of the solution are given in Section 7.4.

In their responses (Figure 7.2p), two characteristics distinguish the first-order systems. (1) The maximum rate of change of the output occurs immediately following the step input. (Note also that if the initial rate were unchanged the system would reach the final value in a period of time equal to the time constant of the system.) (2) The actual response obtained, when the time lapse

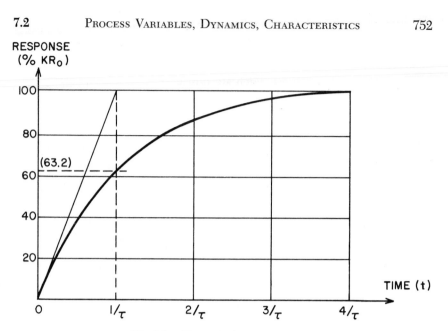

Fig. 7.2p   First-order lag step response

is equal to the time constant of the system, is 63.2 percent of the total response. These two characteristics are common to all first-order processes.

Second-Order Response

Due to inertia effects and various interactions between first-order resistance and capacitance elements, some processes are second order in nature and are described by the following differential equation:

$$\frac{d^2}{dt^2}c(t) + 2\xi\omega_n\frac{d}{dt}c(t) + \omega_n^2 c(t) = K\omega_n^2 r(t) \qquad 7.2(27)$$

where ω_n = the natural frequency of the system,
  ξ = the damping ratio of the system,
  K = the system gain,
  t = time,
  $r(t)$ = input response of system, and
  $c(t)$ = output response of system.

The solution of equation 7.2(27), for a step change in $r(t)$ with all initial conditions zero, can be any one of a family of curves shown in Figure 7.2q. (For details of the solution refer to Section 7.4.)

In the actual solution three possible cases must be considered, depending on the value of the damping ratio,

1. When $\xi < 1.0$, the system is said to be underdamped and will overshoot the final steady-state value. If $\xi < 0.707$, the system

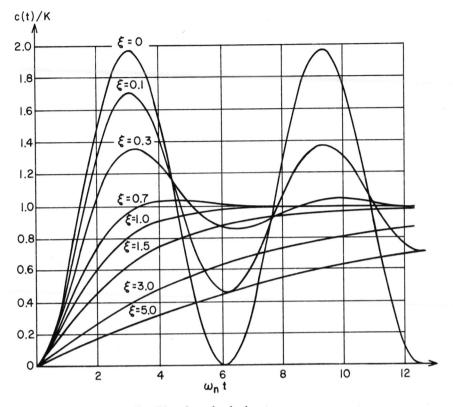

Fig. 7.2q   Second-order lag step response

will not only overshoot but will oscillate about the final steady-state value.

2. When $\xi > 1.0$, the system is said to be "overdamped" and will not oscillate or overshoot the final steady-state position.

3. When $\xi = 1.0$, the system is said to be "critically damped" and yields the fastest response without overshoot or oscillation.

The "natural frequency" term ω_n in the second-order equation is related to the speed of the response for a particular value of ξ. The response illustrated in Figure 7.2q is plotted against a normalized time in which the actual time is divided by the natural frequency, and therefore a large frequency tends to squeeze the response and a small frequency tends to stretch the response. The natural frequency is defined in terms of the "perfect" or "frictionless" situation, where $\xi = 0.0$. In such a situation the response is a sustained sinusoid with a frequency of oscillation equal to ω_n. For the case where ξ is not zero, the actual frequency of an underdamped response is related to the natural frequency by

$$\omega = \omega_n \sqrt{1 - \xi^2} \qquad 7.2(28)$$

7.3 CONTROLLERS AND CONTROL MODES

In automatic control the device used to initiate control action is a special purpose analog (or digital) computer which takes the difference between the desired value and the actual value of a controlled variable and uses this difference so as to eliminate the error in the actual value of the controlled variable. In general, such controllers can be classified in two ways: in terms of the physical mechanism which the controller employs, pneumatic, electronic, hydraulic, digital, etc., or in terms of the manner in which the controller reacts to an error signal. The method by which a controller counteracts a deviation from set point is called the control mode. It is the purpose of this section to emphasize the latter (control mode) distinction between controllers and present the mathematical description of controllers commonly used in industry.

Modes of Control

Proportional, Integral and Derivative Modes

The three most commonly used modes of feedback control are the proportional, integral, and derivative modes, which are defined thus—

Proportional mode:

$$m(t) = K_c[C_R - c(t)] + M_o \qquad 7.3(1)$$

Integral mode:

$$m(t) = \frac{1}{T_i} \int [C_R - c(t)] + M_o \qquad 7.3(2)$$

Derivative mode:

$$m(t) = T_d \frac{d}{dt}[C_R - c(t)] + M_o \qquad 7.3(3)$$

where t = time,

C_R = required variable value (set point),

$m(t)$ = output of controller,

754

$c(t)$ = the signal fed back to the controller representing the controlled variable (measurement signal),

M_o = constant,

K_c = proportional gain,

T_i = integral time, and

T_d = derivative time.

$\left. \right\}$ adjustable controller parameters

Various combinations of these modes comprise most of the controllers found in industry. Some typical combinations are shown in Table 7.3a.

Table 7.3a
DESCRIPTIONS OF CONVENTIONAL CONTROL MODES

Symbol	Description	Mathematic Expression
ONE MODE		
P	Proportional	$m = K_c e$
I	Integral (reset)	$m = \dfrac{1}{T_i} \int e\, dt$
TWO MODE		
PI	Proportional-plus-integral	$m = K_c \left[e + \dfrac{1}{T_i} \int e\, dt \right]$
PD	Proportional-plus-derivative	$m = K_c \left[e + T_d \dfrac{d}{dt} e \right]$
THREE MODE		
PID	Proportional-plus-integral-plus-derivative	$m = K_c \left[e + \dfrac{1}{T_i} \int e\, dt + T_d \dfrac{d}{dt} e \right]$

The proportional mode alone is the simplest of these three basic modes. It is characterized by a continuous linear relationship between the controller input and output. Several synonymous names in common usage are proportional action, correspondence control, droop control and modulating control. The adjustable parameter of the proportional mode, K_c, is called the proportional gain, or proportional sensitivity. It is frequently expressed in terms of percent proportional band, PB, which is related to the proportional gain

$$PB = (1/K_c)100 \qquad\qquad 7.3(4)$$

"Wide bands" (high percentages of PB) correspond to less "sensitive" controller settings, and "narrow bands" (low percentages) correspond to more "sensitive" controller settings.

The integral mode is sometimes used as a single-mode controller but is more commonly found in combination with the proportional mode. The proportional-plus-integral controller is perhaps the most widely used combination. The integral mode is synonymous with the terms reset action and floating control. The adjustable parameter associated with the integral mode is the integral time, T_i, or the reset rate, $1/T_i$.

The derivative mode of control is most commonly referred to as rate control, or preact control, because its output is based on the rate of change of the input variable. Since the output of this mode alone would be zero for a constant value of input, this mode is never used alone and is commonly found in combination with proportional mode.

The response of the individual modes along with the typical combinations of modes is shown in Table 7.3b for several inputs.

Table 7.3b
RESPONSE OF PROPORTIONAL, INTEGRAL AND
DERIVATIVE MODES

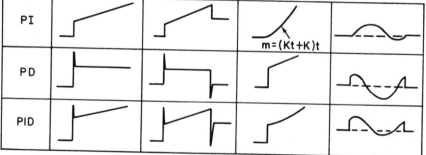

Inverse Derivative Control Mode

A special purpose control action used on extremely fast processes is the so-called inverse-derivative mode. As the name implies, it is the exact opposite of the derivative mode. Where the output of the derivative mode is directly proportional to the rate of change in error, the output of the inverse-derivative mode is inversely proportional to the rate of change in error.

Proportional-plus-derivative:

$$m = K_c\left(e + T_d\frac{d}{dt}e\right) \qquad\qquad 7.3(5)$$

Proportional-plus-inverse-derivative:

$$m = K_c\left(e - T_d\frac{d}{dt}e\right) \qquad\qquad 7.3(6)$$

Two-Position (On-Off) Controllers

The two-position controller is extensively used. In its simplest form, it is the kind of control system used in domestic heating systems, refrigerators and water tanks. A perfect on-off controller is "on" when the measurement is below the set point and the manipulated variable is therefore at its maximum value. When the measured variable is above the set point, the controller is "off" and the manipulated variable is at its minimum value.

$$\left.\begin{array}{ll} e > 0 & m = \text{maximum value} \\ e < 0 & m = \text{minimum value} \end{array}\right\} \quad 7.3(7)$$

In most practical applications, due to mechanical friction or arcing of electrical contacts, there is a narrow band (around zero error), which the error must pass through before a change will occur. This band is known as the differential gap and its presence is sometimes desirable to minimize the cycling tendency of the two-position controller. Figure 7.3c shows the response of a simple on-off controller to a sinusoidal input.

Single-Speed Floating Control

For self-regulating processes with little or no capacitance, the single-speed floating controller is used. The output of this controller is either increasing or decreasing at a certain rate. Such control is commonly associated with systems in which the final control element is a single-speed reversible motor. The controller usually contains a neutral zone for which the output of the controller is zero; otherwise the manipulated variable would be changing continually in one direction or the other. The output of the reversible motor is either forward, reverse, or off. Mathematically the single-speed floating control is expressed as follows:

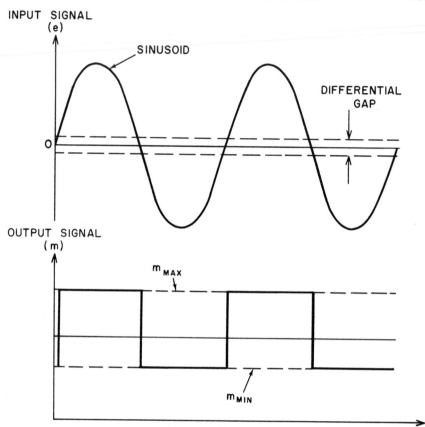

Fig. 7.3c   Response of a two-position controller

$$
\left.
\begin{array}{ll}
e > +\epsilon/2 & m = (T_i \times t) + M_{o_1} \\
e < -\epsilon/2 & m = -(T_i \times t) + M_{o_2} \\
-\epsilon/2 \le e \le +\epsilon/2 & m = 0 + M_{o_3}
\end{array}
\right\}
\qquad 7.3(8)
$$

where t = time,

m = controller output (manipulated variable),

e = controller input (error signal),

ϵ = constant defining neutral zone,

T_i = controller speed constant, and

$M_{o_1}, M_{o_2}, M_{o_3}$ = constants of integration.

The response of a single-speed floating controller to a sinusoidal input is shown in Figure 7.3d.

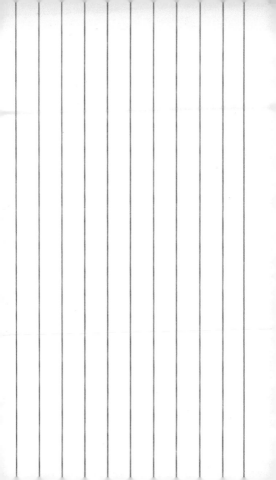

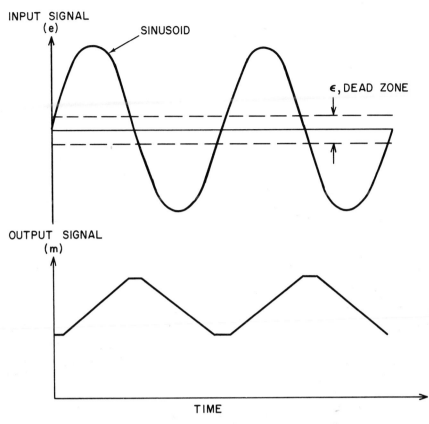

Fig. 7.3d   Response of a single-speed floating controller

Pneumatic Controller Dynamics

On-Off Controllers

The simple device shown in Figure 7.3e illustrates how physical elements are configured to give the desired control modes. It consists of a flapper-nozzle and an "air relay" or power amplifier and is the central component of many pneumatic and hydraulic controllers. The back pressure (P_b) in the chamber of the flapper-nozzle is controlled by the position of the flapper with respect to the nozzle. If the flapper is fully closed (i.e., if $X = 0$), the back pressure will be equal to the supply pressure (P_s). If the flapper is fully open (i.e., if X is very large), the back pressure (P_b) will be approximately equal to the ambient pressure (P_a). The output of the "air relay" (P_o), is a direct function of the back pressure. The relay is essentially a "power amplifier" necessary to supply the air flow required of any pneumatic controller. The relay is

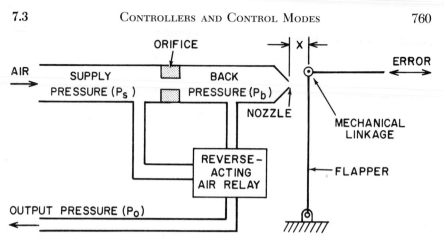

Fig. 7.3e   Pneumatic two-position controller

termed reverse acting because for an increase in P_b there is a corresponding decrease in P_o.

In establishing the equations describing the flapper-nozzle arrangement, its operation will be assumed to be isothermal and the air will be considered as an ideal gas.

A material balance on the nozzle chamber yields

$$w_1 - w_o = \frac{d}{dt} M \qquad\qquad 7.3(9)$$

where w_1 = deviation in weight rate of air flow into chamber, from the normal operating flow rate of W_{1i},

w_o = deviation in weight rate of air flow out of chamber from the normal operating flow rate of W_{oi}, and

M = mass of air in chamber.

The weight rate of flow into the chamber is a function of the pressure drop across the chamber, which can be considered linear over the region of interest.

$$w_1 = K_1 p_b \qquad\qquad 7.3(10)$$

The flow of air out of the nozzle chamber will be a function of the pressure P_b in the chamber and of the position X of the flapper,

$$W_o = f(P_b, X) \qquad\qquad 7.3(11)$$

which in a linearized form can be written as

$$w_o = K_2 p_b + K_3 x \qquad\qquad 7.3(12)$$

where

$$K_2 = \left. \frac{\partial W_o}{\partial P_b} \right|_i \qquad\qquad 7.3(12a)$$

and

$$K_3 = \frac{\partial W_o}{\partial X}\bigg|_i \qquad\qquad 7.3(12b)$$

(For details of the linearization of nonlinear equations see Section 7.4.) The mass M can be defined in terms of the ideal gas law:

$$M = \frac{29P_b V}{RT} \qquad\qquad 7.3(13)$$

or, in terms of deviations from normal operating conditions,

$$m = K_4 p_b \qquad\qquad 7.3(14)$$

where $K_4 = 29V/RT$,
  $V =$ volume of nozzle chamber,
  $T =$ temperature in nozzle chamber,
  $R =$ universal gas constant, and
  $29 =$ molecular weight of air.

Substituting equations 7.3(10), (12) and (14) into the general material balance equation 7.3(9) yields

$$(K_1 + K_2)p_b + K_4\frac{d}{dt}p_b = -K_3 x \qquad\qquad 7.3(15)$$

This can be written as a general first-order lag:

$$p_b + \tau\frac{d}{dt}p_b = Kx \qquad\qquad 7.3(16)$$

where

$$\tau = \frac{K_4}{K_1 + K_2} \qquad\qquad 7.3(16a)$$

and

$$K = \frac{-K_3}{K_1 + K_2} \qquad\qquad 7.3(16b)$$

Considering the relatively small volume of the nozzle, the time constant τ will be negligible with respect to typical process time constants. Therefore, for all practical purposes the equation can be written as

$$-p_b = p_o = -Kx \qquad\qquad 7.3(17)$$

In this equation K is typically in the neighborhood of 5,000 to 8,000 PSI/in. This high sensitivity makes the device undesirable for proportional control, but it can be used as a simple two-position or bang-bang controller because any small positive value of x will result in the maximum value of P_o and any small negative value of x will result in the minimum value of P_o. (The terms negative and positive are used relatively to the normal flapper position.)

Proportional Controllers

The modified configuration shown in Figure 7.3f obtains proportional control from the flapper-nozzle arrangement. A feedback bellows and a spring has been added to the bottom of the flapper. The flapper has also been

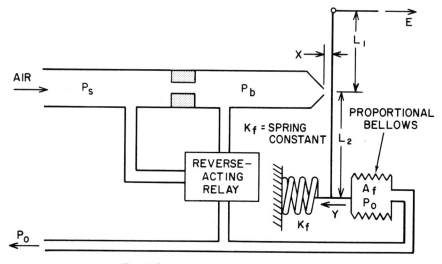

Fig. 7.3f   Pneumatic proportional controller

extended so that the input signal E, which positions the flapper, is mechanically linked to the flapper above the nozzle. The relative position X of the flapper to the nozzle is determined by both the position Y of the feedback bellows and the position E of the input signal.

$$X = f(E, Y) \qquad\qquad 7.3(18)$$

Using simple geometry this can be expressed as a linear function of the deviations in X, E, and Y about normal operating conditions X_i, E_i and Y_i.

$$x = \left(\frac{L_2}{L_1 + L_2}\right)e - \left(\frac{L_1}{L_1 + L_2}\right)y \qquad\qquad 7.3(19)$$

The position of the spring-bellows arrangement for a given pressure, P_o, can be determined by force balance. Since inertia effects are zero, force balance means that the force exerted by the bellows is equal to the force exerted by the spring.

$$A_f P_o = K_f Y \qquad\qquad 7.3(20)$$

In terms of deviations about a normal operating condition, equation 7.3(20) becomes:

$$A_f p_o = K_f y \qquad\qquad 7.3(21)$$

where A_f = cross-sectional area of the bellows,
              K_f = spring constant,
P_o and p_o = absolute output pressure and deviation in output pressure,
 Y and y = absolute position of flapper and deviation in flapper position.

Substituting equations 7.3(19) and (21) into the equation for the flapper-nozzle, 7.3(17), yields:

$$p_o = -K\left(\frac{L_2}{L_1 + L_2}\right)e + \left(\frac{L_1}{L_1 + L_2}\right)\left(\frac{A_f}{K_f}\right)p_o \qquad 7.3(22)$$

or

$$p_o = \frac{-K\left(\dfrac{L_2}{L_1 + L_2}\right)}{1 - K\left(\dfrac{L_1}{L_1 + L_2}\right)\left(\dfrac{A_f}{K_f}\right)}e \qquad 7.3(23)$$

Since K is a very large value, the equation reduces to an expression equivalent to that of a proportional controller.

$$p_o = \frac{L_2 K_f}{L_1 A_f}e \qquad 7.3(24)$$

The addition of the internal feedback mechanism effectively reduced the sensitivity of the device to an acceptable range for proportional control. Note that the proportional gain in the above equation can be easily adjusted by the ratio L_1/L_2, which can be changed by varying the position on the flapper at which the error signal is applied.

Proportional-Plus-Derivative Controllers

With slight modifications additional modes of operation can be incorporated into the proportional controller shown in Figure 7.3f. With the addition of a variable restriction in the line leading to the feedback bellows (Figure 7.3g), there will be a restriction offering resistance to flow into the bellows and therefore creating a time lag effect.

The material balance on the bellows is

$$W_b = \frac{d}{dt}M_b \qquad 7.3(25)$$

or, in terms of deviations about a normal operating point,

$$w_b = \frac{d}{dt}m_b \qquad 7.3(26)$$

where W_b, w_b = the weight rates of air flow into the bellows and
       M_b, m_b = the weights of air in the bellows.

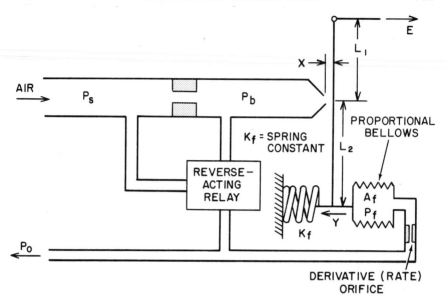

Fig. 7.3g   Pneumatic proportional plus derivative controller

The rate of flow into the bellows will be a function of the pressure drop across the restriction in the line leading to the bellows:

$$W_b = K_5(P_o - P_f) \qquad 7.3(27)$$

Expressing the equation in terms of a linear deviation from a normal operating point yields

$$w_b = K_5(p_o - p_f) \qquad 7.3(28)$$

The mass of air can be expressed in terms of the ideal gas law:

$$M = \frac{29P_fV}{RT} \qquad 7.3(29)$$

where V = volume of bellows and
T = temperature in the bellows.

Linearization of this equation yields

$$m = K_6p_f + K_7v \qquad 7.3(30)$$

The variation in the feedback bellows volume may be expressed in terms of the variation in bellows position

$$v = A_fy \qquad 7.3(31)$$

Equations 7.3(26), (28), (30), and (31) can be combined with the force balance on the bellows expressed by equation 7.3(32),

$$A_f p_f = K_f y \qquad\qquad 7.3(32)$$

to yield equation 7.3(33), which describes the behavior of the bellows:

$$\tau \frac{d}{dt} y + y = \frac{A_f}{K_f} p_o \qquad\qquad 7.3(33)$$

where

$$\tau = \left(\frac{K_6}{K_5} + \frac{K_7 A_f^2}{K_5 K_f} \right) \qquad\qquad 7.3(33a)$$

As before in equation 7.3(19),

$$x = \left(\frac{L_2}{L_1 + L_2} \right) e - \left(\frac{L_1}{L_1 + L_2} \right) y \qquad\qquad 7.3(34)$$

and, in accordance with equation 7.3(17),

$$p_o = -Kx \qquad\qquad 7.3(35)$$

Therefore the equation which describes the entire system is

$$p_o + \tau_s \frac{d}{dt} p_o = K_c \left[e + T_d \frac{d}{dt} e \right] \qquad\qquad 7.3(36)$$

where

$$\tau_s = \frac{T_d}{1 - K \left(\dfrac{L_1}{L_1 + L_2} \right) \left(\dfrac{A_f}{K_f} \right)} \qquad\qquad 7.3(36a)$$

and

$$K_c = \frac{-K \left(\dfrac{L_2}{L_1 + L_2} \right)}{1 - K \left(\dfrac{L_1}{L_1 + L_2} \right) \left(\dfrac{A_f}{K_f} \right)} \qquad\qquad 7.3(36b)$$

But because K is very large, the constants in the equation above can be approximated as

$$\tau_s \cong 0.0 \qquad\qquad 7.3(36c)$$

$$K_c \cong \left(\frac{L_2}{L_1} \right) \left(\frac{K_f}{A_f} \right) \qquad\qquad 7.3(36d)$$

Therefore, the equation for a proportional-plus-derivative controller can be written as

$$p_o = K_c \left[e + T_d \frac{d}{dt} e \right] \qquad\qquad 7.3(37)$$

The proportional gain K_c can be adjusted in the same manner as was the proportional controller, by varying the L_2/L_1 ratio, and the derivative time

T_d, can be adjusted by varying the resistance of the restriction in the air line to the bellows.

Proportional-Plus-Integral Controllers

Another modification of the basic flapper-nozzle device is illustrated in Figure 7.3h, where an additional bellows is included in the feedback loop and is mounted to act opposite the proportional bellows.

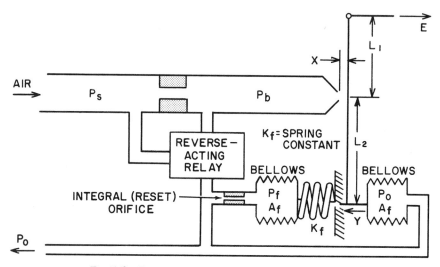

Fig. 7.3h　Pneumatic proportional plus integral controller

Similar to the previous approach, a force balance and a mass balance on this bellows is utilized to determine the position of the flapper. The resulting equation 7.3(38) describes the controller dynamics:

$$y + \tau_i \frac{d}{dt}y = \left(\frac{A_f K_6}{K_f K_5}\right)\frac{d}{dt}p_o \qquad 7.3(38)$$

where τ_i is the time constant of the bellows as defined earlier by equation 7.3(33a);

$$\tau_i = \left(\frac{K_6}{K_5} + \frac{K_7 A_f^2}{K_5 K_f}\right) \qquad 7.3(33a)$$

All other equations remaining unchanged, the entire system equation can be written as

$$G_1 p_o + \frac{d}{dt}p_o = K_c\left(e + T_i\frac{d}{dt}e\right) \qquad 7.3(39)$$

where

$$G_1 = \cfrac{1}{\tau - K\left(\cfrac{L_1}{L_1 + L_2}\right)\left(\cfrac{A_f K_6}{K_f K_5}\right)} \qquad 7.3(39a)$$

$$K_c = G_2 = \cfrac{-K\left(\cfrac{L_2}{L_1 + L_2}\right)}{\tau - K\left(\cfrac{L_1}{L_1 + L_2}\right)\left(\cfrac{A_f K_6}{K_f K_5}\right)} \qquad 7.3(39b)$$

But because K is very large, the constants in equation 7.3(39) can be approximated as

$$G_1 \cong 0.0 \qquad 7.3(39c)$$

$$K_c = G_2 \cong \frac{L_2 K_f K_5}{L_1 A_f K_6} \qquad 7.3(39d)$$

Therefore the simplified equation can be written as

$$\frac{d}{dt} p_o = K_c\left[e + T_i \frac{d}{dt} e\right] \qquad 7.3(40)$$

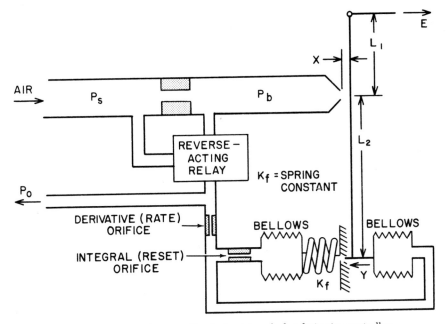

Fig. 7.3i   Pneumatic proportional plus integral plus derivative controller

and integrating both sides yields an equation in the more familiar form of a proportional-plus-integral controller:

$$p_o = K_c \left[e + \frac{1}{T_i} \int e \, dt \right]$$ 7.3(41)

PID Controllers

In a similar manner the basic pneumatic elements can be modified to yield a response which approximates the proportional-plus-integral-plus-derivative action (Figure 7.3i).

The descriptive equation for such a three-mode controller is,

$$p_o = K_c \left[e + \frac{1}{T_i} \int e \, dt + T_d \frac{d}{dt} e \right]$$ 7.3(42)

7.4 TRANSFER FUNCTIONS, LINEARIZATION AND STABILITY ANALYSIS

Most techniques used in the analysis of control problems are dependent on the existence of descriptive mathematical equations. In this section the reader will find some of the mathematical tools needed for the analysis of control problems.

One of the main topics of this section is therefore the subject of transforms, which allows for simple handling of difficult problems. As the use of logarithms simplifies the handling of multiplications to a requirement for addition only, so do Laplace transforms perform a similar function in the solution of differential equations.

Some of the other control system analysis tools discussed here are the block diagrams providing a convenient technique of presentation and linearization, which is a technique to convert non-linear equations into linear form.

One of the main purposes of control system analysis is to guarantee that the particular system will be stable in its operation. Some of the stability criterions discussed in this section include the Descartes, Routh's and Nyquist criterions.

Laplace Transforms

One very useful tool in the analysis of differential equations is the principle of Laplace transforms. The Laplace transform concept is widely used in process control and provides the basic framework on which most automatic control theory is based.

The principle of any transform operation is to transform a difficult problem to a form more convenient to handle. Once the desired manipulations or results have been obtained from the transformed problem, an inverse transformation can be made to determine the solution of the original problem. For example, logarithms are a transform operation by which problems of multiplication and division can be transformed to problems of addition and subtraction. Laplace transforms perform a similar function in the solution of differential equations. The Laplace transform of a linear ordinary differential equation results in a linear algebraic equation. The algebraic problem is usually much simpler to solve than the corresponding differential equation. Once the

Laplace domain solution has been found, the corresponding time domain solution can be determined by an inverse transformation.

The Laplace transform of a time domain function f(t) will be noted by the symbol F(s), defined as follows:

$$F(s) = \mathcal{L}[f(t)] = \int_0^\infty f(t)e^{-st}dt \qquad 7.4(1)$$

where $\mathcal{L}[f(t)]$ is the symbol for indicating the Laplace transformation of the function in brackets. The variable s is a complex variable ($s = a + jb$) introduced by the transformation. All time-dependent functions in the time domain become functions of s in the Laplace domain (s-domain).

For the mathematical concept of the Laplace transformation to be meaningful certain restrictions are placed on the function f(t). However, in most practical control work no such difficulties are encountered and therefore this will not be considered in this treatment. Only concepts normally used in process control will be covered.

A number of theorems exist which facilitate the use of Laplace transform techniques. The following are some of the most useful ones.

1. *Linearity theorem*

$$\mathcal{L}[K\,f(t)] = K\mathcal{L}[f(t)] = K\,F(s) \qquad (K = \text{constant}) \qquad 7.4(2)$$
$$\mathcal{L}[f_1(t) \pm f_2(t)] = F_1(s) \pm F_2(s) \qquad 7.4(3)$$

2. *Real differentiations theorem*
   First-order differential

$$\mathcal{L}\left[\frac{d}{dt}f(t)\right] = sF(s) - f(0) \qquad 7.4(4)$$

General nth-order differential

$$\mathcal{L}\left[\frac{d^n}{dt^n}f(t)\right]$$

$$= s^nF(s) - s^{n-1}f(0) - s^{n-2}\frac{d}{dt}f(0) - \cdots$$

$$- s\frac{d^{n-2}}{dt^{n-2}}f(0) - s\frac{d^{n-1}}{dt^{n-1}}f(0) \qquad 7.4(5)$$

3. *Real integration theorem*

$$\mathcal{L}\left[\int f(t)\,dt\right] = \frac{F(s)}{s} \qquad 7.4(6)$$

In general,

$$\mathcal{L}\left[\int^1\int^2 \cdots \int^n f(t) \, dt^n\right] = \frac{1}{s^n}F(s) \qquad\qquad 7.4(7)$$

4. *Initial value theorem*

$$f(0) = \underset{s\to\infty}{\text{limit}} \, s \, F(s) \qquad\qquad 7.4(8)$$

5. *Final value theorem*

$$f(\infty) = \underset{s\to 0}{\text{limit}} \, s \, F(s) \qquad\qquad 7.4(9)$$

The direct and inverse Laplace transformation of a particular function can be obtained from direct integration of equation 7.4(1) and/or the applications of the above theorems. Extensive tabulations of specific transform pairs are also available in most mathematical tables. A few of the more common transform pairs encountered in control analysis are tabulated in Table 7.4a.

Table 7.4a
LAPLACE TRANSFORM PAIRS

Transform, $F(s)$	Function, $f(t)$	Transform, $F(s)$	Function, $f(t)$
1	$\delta(t)$	$\dfrac{s+a}{(s+a)^2+b^2}$	$e^{-at}\cos bt$
$\dfrac{1}{s}$	$u(t)$	$\dfrac{1}{(s+a)^n}$	$\dfrac{1}{(n-1)!}t^{n-1}e^{-at}$
$\dfrac{1}{s^n}$ $(n=1,2,\ldots)$	$\dfrac{t^{n-1}}{(n-1)!}$	$\dfrac{ab}{(s+a)(s+b)}$	$\dfrac{1}{b-a}(e^{-at}-e^{-bt})$
$\dfrac{1}{s\pm a}$	$e^{\mp at}$	$\dfrac{e^{-as}}{s}$	$u(t-a)$
$\dfrac{1}{s(s\pm a)}$	$\dfrac{1}{\pm a}(1-e^{\mp at})$	$\dfrac{e^{-as}}{s^2}$	$\begin{cases}0 & (0<t<a)\\ t-a & (t>a)\end{cases}$
$\dfrac{s}{s^2+a^2}$	$\cos at$	$\dfrac{1-e^{-as}}{s}$	$\begin{cases}1 & (0<t<a)\\ 0 & (t>a)\end{cases}$
$\dfrac{a}{s^2+a^2}$	$\sin at$		
$\dfrac{s}{s^2-a^2}$	$\cosh at$	$\log\dfrac{s-a}{s-b}$	$\dfrac{1}{t}(e^{bt}-e^{at})$
$\dfrac{a}{s^2-a^2}$	$\sinh at$	$\tan^{-1}\dfrac{a}{s}$	$\dfrac{1}{t}\sin at$
$\dfrac{1}{(s+a)^2+b^2}$	$\dfrac{1}{b}e^{-at}\sin bt$		

First-Order Lag

The following two examples illustrate the solution of differential equations using the underlying principle of the Laplace transform technique. First consider a simple first-order lag described earlier as equation 7.2(23):

$$\tau \frac{d}{dt} c(t) + c(t) = K \, r(t) \qquad\qquad 7.4(10)$$

where
$$c(0) = 0.0$$

and
$$r(t) = \begin{cases} 0 & t < 0 \\ 1 & t > 0 \end{cases} \qquad\qquad 7.4(11)$$

The general procedure in the solution of the above equation will be to

(1) transform the differential equation to the Laplace domain;
(2) solve the resulting algebraic equations for the system output, C(s);
(3) take inverse Laplace transformation of the expression describing C(s) to determine the corresponding time domain solution.

Since Theorem 1 indicates linearity, each term in the differential equation can be transformed individually.

$$\mathcal{L}\left[\tau \frac{d}{dt} c(t)\right] + \mathcal{L}[c(t)] = \mathcal{L}[K \, r(t)] \qquad\qquad 7.4(12)$$

The first term can be determined by the use of Theorems 1 and 2.

$$\mathcal{L}\left[\tau \frac{d}{dt} c(t)\right] = \tau \mathcal{L}\left[\frac{d}{dt} c(t)\right] = \tau(sC(s) = c(0)) \qquad\qquad 7.4(13)$$

The second term by definition is

$$\mathcal{L}[c(t)] = C(s) \qquad\qquad 7.4(14)$$

The third term can be determined from the table of transform pairs in Table 7.4a.

$$\mathcal{L}[K \, r(t)] = \frac{K}{s} \qquad\qquad 7.4(15)$$

The entire Laplace domain equation therefore will read

$$\tau s \, C(s) + C(s) = \frac{K}{s} \qquad\qquad 7.4(16)$$

This can be solved for C(s):

$$C(s) = \frac{K}{s(1 + \tau s)} \qquad\qquad 7.4(17)$$

For the inverse transformation we note from Table 7.4a the following transform pair:

$$\mathcal{L}\left[\frac{1}{\pm a}(1 - e^{\mp at})\right] = \frac{1}{s(s \pm a)} \qquad\qquad 7.4(18)$$

Therefore,

$$c(t) = K(1 - e^{-t/\tau}) \qquad t > 0 \qquad\qquad 7.4(19)$$

Second-Order Lag

The differential equation for a second-order lag is

$$\frac{d^2}{dt^2}c(t) + 4\frac{d}{dt}c(t) + 3\,c(t) = r(t) \qquad\qquad 7.4(20)$$

where

$$c(0) = \frac{d}{dt}c(0) = 1 \qquad\qquad 7.4(20a)$$

and

$$r(t) = e^{-t} \qquad t > 0 \qquad\qquad 7.4(20b)$$

Taking the Laplace transform of each expression,

$$\mathcal{L}\left[\frac{d^2}{dt^2}c(t)\right] = \left[s^2C(s) - sc(0) - \frac{d}{dt}c(0)\right] \qquad\qquad 7.4(21)$$

$$\mathcal{L}\left[4\frac{d}{dt}c(t)\right] = 4[sC(s) - c(0)] \qquad\qquad 7.4(22)$$

$$\mathcal{L}[3c(t)] = 3\,C(s) \qquad\qquad 7.4(23)$$

Using Table 7.4a,

$$r(t) = e^{-t} = \frac{1}{s + 1} \qquad\qquad 7.4(24)$$

The complete transformed expression therefore becomes

$$S^2C(s) + 4s\,C(s) + 3\,C(s) - (s + 5) = \frac{1}{s + 1} \qquad\qquad 7.4(25)$$

By algebra,

$$C(s) = \frac{s^2 + 6s + 6}{(s + 1)^2(s + 3)} \qquad\qquad 7.4(26)$$

The next step is to obtain an inverse transformation of the above expression. But the limited table of transform pairs in Table 7.4a does not contain this function. Therefore, it is necessary to break the function into smaller terms whose inverse transforms are available. Using the rules of partial fraction expansion (the next paragraph in this section),

$$C(s) = \frac{s^2 + 6s + 6}{(s + 1)^2(s + 3)} = \frac{C_1}{s + 1} + \frac{C_2}{(s + 1)^2} + \frac{C_3}{s + 3} \qquad 7.4(27)$$

$$C_3 = \left[\frac{s^2 + 6s + 6}{(s + 1)^2}\right]_{s=-3} = -\frac{3}{4} \qquad 7.4(28)$$

$$C_2 = \left[\frac{s^2 + 6s + 6}{s + 3}\right]_{s=-1} = \frac{1}{2} \qquad 7.4(29)$$

$$C_1 = \left[\frac{1}{1!}\frac{d}{ds}\left(\frac{s^2 + 6s + 6}{s + 3}\right)\right]_{s=-1} = \frac{7}{4} \qquad 7.4(30)$$

(where $1! = 1$, $2! = 1 \cdot 2 = 2$, $3! = 1 \cdot 2 \cdot 3 = 6$ etc.).
Therefore

$$C(s) = \frac{7/4}{(s + 1)} + \frac{1/2}{(s + 1)^2} - \frac{3/4}{(s + 3)} \qquad 7.4(31)$$

Using Table 7.4a to transform each of the terms above,

$$c(t) = \frac{7}{4}e^{-t} + \frac{1}{2}te^{-t} - \frac{3}{4}e^{-3t} \qquad t > 0 \qquad 7.4(32)$$

Partial Fraction Expansion

As demonstrated in the second-order lag example, the necessary inverse transformation may not be directly available in the Laplace transform tables at hand. In such cases the function must be expanded in terms of the roots of the denominator of the Laplace expression, namely,

$$F(s) = \frac{A(s)}{B(s)} = \frac{C_1}{(s + r_1)} + \frac{C_2}{(s + r_2)} + \frac{C_3}{(s + r_3)} + \cdots + \frac{C_n}{(s + r_n)} \qquad 7.4(33)$$

where

$$B(s) = (s - r_1)(s - r_2)(s - r_3) \cdots (s - r_n) \qquad 7.4(34)$$

$r_1, r_2, r_3, \ldots r_n$ = roots of $B(s)$
$C_1, C_2, C_3, \ldots C_n$ = constants in partial fraction

The procedure of evaluating the constants in the expansion depends on the nature of the roots, which can be (1) real and distinct, (2) real and repeated, (3) complex conjugates.

When the roots are real and distinct, the expansion of $F(s)$ is

$$F(s) = \frac{A(s)}{(s - r_1)(s - r_2) \cdots (s - r_n)} \qquad 7.4(35)$$

$$F(s) = \frac{C_1}{(s - r_1)} + \frac{C_2}{(s - r_2)} + \cdots + \frac{C_n}{(s - r_n)} \qquad 7.4(36)$$

and the inverse transformation is

$$f(t) = C_1 e^{r_1 t} + C_2 e^{r_2 t} + \cdots + C_n e^{r_n t} \qquad \text{7.4(37)}$$

where

$$\begin{aligned}
C_1 &= \lim_{s \to r_1} [(s - r_1) F(s)] \\
C_2 &= \lim_{s \to r_2} [(s - r_2) F(s)] \\
&\quad \vdots \\
C_n &= \lim_{s \to r_n} [(s - r_n) F(s)]
\end{aligned} \right\} \qquad \text{7.4(38)}$$

For the case when the roots are real and repeated,

$$F(s) = \frac{A(s)}{(s - r_1) + \cdots + (s - r_j)^q + \cdots + (s - r_n)} \qquad \text{7.4(39)}$$

(The jth root is repeated q times.)

$$F(s) = \frac{C_1}{(s - r_1)} + \cdots + \left[\frac{C_q'}{(s - r_j)^q} + \frac{C_{q-1}'}{(s - r_j)^{q-1}} + \cdots + \frac{C_1'}{(s - r_j)} \right]$$
$$+ \cdots + \frac{C_n}{(s - r_n)} \qquad \text{7.4(40)}$$

The inverse transform is

$$f(t) = C_1 e^{r_1 t} + \cdots + [C_q' t^{q-1} + C_{q-1}' t^{q-2} + \cdots C_1'] e^{r_j t}$$
$$+ \cdots + C_n e^{r_n t} \qquad \text{7.4(41)}$$

where

$$C_q' = \lim_{s \to r_j} \{(s - r_j)^q F(s)\} \qquad \text{7.4(42)}$$

$$C_{q-1}' = \lim_{s \to r_j} \left\{ \frac{1}{1!} \frac{d}{ds} [(s - r_j)^q F(s)] \right\} \qquad \text{7.4(43)}$$

$$C_{q-k} = \lim_{s \to r_j} \left\{ \frac{1}{k!} \frac{d^k}{ds^k} [(s - r_j)^q F(s)] \right\} \qquad \text{7.4(44)}$$

(where $k! = 1 \cdot 2 \cdot 3 \cdot 4 \cdot 5 \cdots k$)

For complex roots,

$$F(s) = \frac{A(s)}{(s - r_1) \cdots (s - a - jb)(s - a + jb) \cdots (s - r_n)} \qquad \text{7.4(45)}$$

$$F(s) = \frac{C_1}{(s - r_1)} + \cdots + \frac{1}{2jb} |\pi_j| \left(\frac{e^{j\alpha}}{s - a - jb} - \frac{e^{-j\alpha}}{s - a + jb} \right)$$
$$+ \cdots + \frac{C_n}{(s - r_n)} \qquad \text{7.4(46)}$$

The inverse transformation results in

$$f(t) = C_1 e^{r_1 t} + \cdots + \frac{1}{b} |\tau_j| e^{at} \sin(bt + \alpha)$$

$$+ \cdots + C_n e^{r_n t} \qquad \qquad 7.4(47)$$

$$\tau_j = \lim_{s \to a + jb} [(s - a - jb)(s - a + jb) F(s)] \qquad 7.4(48)$$

This results in a complex polynomial which can be expressed in terms of a magnitude, $|\tau_j|$, and an angle, α. For example,

$$F(s) = \frac{10}{s(s^2 + 0.5s + 0.4)} = \frac{10}{s(s + 0.25 - j0.58)(s + 0.25 + j0.58)} \qquad 7.4(49)$$

$$F(s) = \frac{C_1}{s} + \frac{1}{2jb} |\tau_j| \left(\frac{e^{j\alpha}}{s - a - jb} - \frac{e^{-j\alpha}}{s - a + jb} \right) \qquad 7.4(50)$$

where $a = -0.25$
$b = 0.58$

$$C_1 = \lim_{s \to 0} \left[s \left(\frac{10}{s(s^2 + 0.5s + 0.4)} \right) \right] = 25 \qquad 7.4(51)$$

$$\tau_j = \lim_{s \to 0.25 + j0.58} \left[(s^2 + 0.5s + 0.4) \left(\frac{10}{s(s^2 + 0.5s + 0.4)} \right) \right] \qquad 7.4(52)$$

$$= \frac{10}{-0.25 + j0.58}$$

$$|\tau_j| = \frac{10}{\sqrt{(0.25)^2 + (0.58)^2}} = \frac{10}{0.63} = 15.9 \qquad 7.4(53)$$

$$\alpha = \text{angle of } \tau_j = 0 - \tan^{-1} \frac{0.58}{-0.25} = -113.5 \qquad 7.4(54)$$

Therefore,

$$c(t) = 25 + \left(\frac{15.9}{0.58} \right) e^{-0.25t} \sin(0.58t - 113.5°) \qquad 7.4(55)$$

Transfer Function

A notation often used to describe the dynamics of a particular process or system is the transfer function. For a system with an input r(t) and an output c(t) the transfer function G(s) is defined as the ratio of the Laplace transform of the output of the system C(s) divided by the Laplace transform of the input to the system R(s).

$$G(s) = \frac{C(s)}{R(s)} \qquad 7.4(56)$$

Inherent in the transfer function concept is an assumption that the process is initially at steady state, meaning that

$$\frac{d}{dt}c(0) = \frac{d^2}{dt^2}c(0) = \cdots = \frac{d^n}{dt^n}c(0) = 0.0 \qquad 7.4(57)$$

$$\frac{d}{dt}r(0) = \frac{d^2}{dt^2}r(0) = \cdots = \frac{d^n}{dt^n}r(0) = 0.0 \qquad 7.4(58)$$

and

$$c(0) = K\,r(0) \qquad 7.4(59)$$

where K is the steady-state gain of the process. Under such conditions the Laplace transform of a particular differential equation can be obtained by substituting

$$s \leftrightarrow d/dt(\ ) \qquad 7.4(60)$$
$$s^2 \leftrightarrow d^2/dt^2(\ ) \qquad 7.4(61)$$
$$s^n \leftrightarrow d^n/dt^n(\ ) \qquad 7.4(62)$$
$$1/s \leftrightarrow \int(\ )\,dt \qquad 7.4(63)$$
$$1/s^2 \leftrightarrow \int\int(\ )\,dt^2 \qquad 7.4(64)$$
$$1/s^n \leftrightarrow \int\int \cdots \int^n(\ )\,dt^n \qquad 7.4(65)$$
$$C(s) \leftrightarrow c(t) \qquad 7.4(66)$$
$$R(s) \leftrightarrow r(t) \qquad 7.4(67)$$

The transfer function can then be determined by solving for $C(s)/R(s)$. Consider the following examples:

First-Order Lag
As explained in connection with equation 7.2(23),

$$c(t) + \tau\frac{d}{dt}c(t) = K\,r(t) \qquad 7.4(68)$$

Using the substitutions

$$C(s) + \tau s\,C(s) = K\,R(s) \qquad 7.4(69)$$

Therefore the transfer function is

$$\frac{C(s)}{R(s)} = \frac{K}{1 + \tau s} \qquad 7.4(70)$$

Second-Order Lag
As described in connection with equation 7.2(27),

$$\frac{d^2}{dt^2}c(t) + 2\xi\omega_n\frac{d}{dt}c(t) + \omega_n^2 c(t) = K\omega_n^2\,r(t) \qquad 7.4(71)$$

After the substitutions:

$$s^2 C(s) + 2\xi\omega_n s C(s) + \omega_n^2 C(s) = K\omega_n^2 R(s)$$ 7.4(72)

Therefore,

$$\frac{C(s)}{R(s)} = \frac{K\omega_n^2}{s^2 + 2\xi\omega_n s + \omega_n^2}$$ 7.4(73)

PID Controllers

In accordance with equation 7.3(42),

$$m(t) = K_c\left(e(t) + \frac{1}{T_i}\int e(t)\,dt + T_d\frac{d}{dt}\,e(t)\right)$$ 7.4(74)

After the substitutions:

$$M(s) = K_c\left(E(s) + \frac{1}{T_i s}E(s) + T_d s\,E(s)\right)$$ 7.4(75)

Therefore,

$$\frac{M(s)}{E(s)} = K_c\left(1 + \frac{1}{T_i s} + T_d s\right)$$ 7.4(76)

Block Diagrams

In the analysis of a control system the use of the various mathematical equations in their conventional form is not generally the most convenient technique to employ. The so-called block diagram representation is more

Table 7.4b
BLOCK DIAGRAM SYMBOLS

Symbol	Interpretation
$X(s) \longrightarrow$	Input or output signal; arrow gives direction.
$X(s)$ $X(s) \longrightarrow$ $\downarrow X(s)$	Branch point. Division of a signal to give two or more paths without modification.
$X(s) \pm$ $Z(s) \longrightarrow$ $\pm$ $\uparrow Y(s)$	Summing point. $Z(s) = \pm X(s) \pm Y(s)$
$X(s) \to \boxed{G(s)} \xrightarrow{Y(s)}$	System element. $Y(s) = G(s)\,X(s)$

desirable. Such diagrams not only present an organized picture of the flow of information and energy, but also, in the framework of the transfer function notation, facilitate the simultaneous solution of the differential equations which describe the system.

The conventions used in the construction of block diagrams are shown in Table 7.4b. The two main symbols are a circle, which indicates summation of two signals, and a rectangle, which indicates multiplication of a signal by a constant K or by a transfer function G(s). It is important to note that both symbols indicate linear operations. In using block diagrams to analyze control problems it is important to be able to convert from one form to another. Such manipulations are termed block diagram algebra. Some examples are shown in Table 7.4c.

Table 7.4c
MANIPULATIONS OF BLOCK DIAGRAMS

In control analysis the most frequent block diagram manipulations involve feedback loops of some sort. The rules for reducing such systems into a single transfer function can be summarized as follows.

$$\text{Output} = \frac{(\text{Input}) \begin{pmatrix} \text{Product of blocks in the forward} \\ \text{path between input signal} \\ \text{and output signal} \end{pmatrix}}{1 + \begin{pmatrix} \text{Product of blocks in the} \\ \text{control loop} \end{pmatrix}} \qquad 7.4(77)$$

For systems with more than one input, equation 7.4(77) becomes

$$\text{Output} = \sum_{i=1}^{N} \frac{(\text{Input-i}) \begin{pmatrix} \text{Product of blocks in the forward} \\ \text{path between input-i and the} \\ \text{output signal} \end{pmatrix}}{1 + \begin{pmatrix} \text{Product of blocks in the} \\ \text{control loop} \end{pmatrix}} \qquad 7.4(78)$$

For example, consider the feedback type control system in Figure 7.4d, where

$$G_c(s) = K_c \text{ (proportional control)} \qquad 7.4(79)$$

$$G_p(s) = \frac{K}{1 + \tau s} \text{ (first-order process)} \qquad 7.4(80)$$

$$H(s) = \frac{K_H}{1 + \tau_H s} \text{ (first-order feedback)} \qquad 7.4(81)$$

Substituting,

$$c(s) = \frac{\dfrac{AK_cK}{1 + \tau s}}{1 + K_c\left(\dfrac{K}{1 + \tau s}\right)\left(\dfrac{K_H}{1 + \tau_H s}\right)} R(s)$$

$$+ \frac{B\left(\dfrac{K}{1 + \tau s}\right)}{1 + K_c\left(\dfrac{K}{1 + \tau s}\right)\left(\dfrac{K_H}{1 + \tau_H s}\right)} D(s) \qquad 7.4(82)$$

or

$$c(s) = \frac{AK_cK(1 + \tau_H s)}{(1 + \tau s)(1 + \tau_H s) + K_cKK_H} R(s)$$

$$+ \frac{BK(1 + \tau_H s)}{(1 + \tau s)(1 + \tau_H s) + K_cKK_H} D(s) \qquad 7.4(83)$$

Linearization

The great majority of techniques used in the analysis of control problems (including block diagrams) are dependent on the existence of mathematical

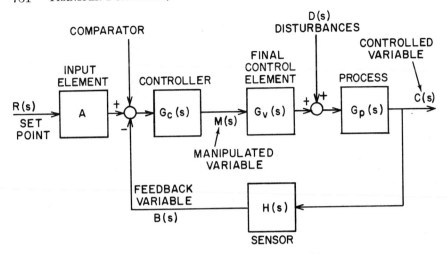

Fig. 7.4d   Block diagram of typical control loop

equations in a linear form. By their very nature most systems are nonlinear to one degree or another and therefore must be approximated by some linear equation. The technique used to linearize a nonlinear function can best be illustrated by considering a nonlinear equation such as;

$$Y = \phi(X_1, X_2, X_3, \ldots) \qquad 7.4(84)$$

where Y = dependent variable,
$\quad\quad\;\; X$ = independent variables, and
$\quad\quad\;\; \phi$ = nonlinear function.

The equation can be expanded about a point $(X_{1i}, X_{2i}, X_{3i}, \ldots)$ using a Taylor's series expansion in which the higher order terms are ignored.

$$Y = Y_i + \left.\frac{\delta Y}{\delta X_1}\right|_i (X_1 - X_{1i}) + \left.\frac{\delta Y}{\delta X_2}\right|_i (X_2 - X_{2i})$$

$$+ \left.\frac{\delta Y}{\delta X_3}\right|_i (X_3 - X_{3i}) + \cdots \qquad 7.4(85)$$

or

$$y = K_1 x_1 + K_2 x_2 + K_3 x_3 + \cdots \qquad 7.4(86)$$

where

$$y = Y - Y_i \qquad\qquad\qquad 7.4(87)$$
$$x = X - X_i \qquad\qquad\qquad 7.4(88)$$

$$K_1 = \left.\frac{\delta Y}{\delta X_1}\right|_i = \text{constant} \qquad 7.4(89)$$

$$K_2 = \left.\frac{\delta Y}{\delta X_2}\right|_i = \text{constant} \qquad 7.4(90)$$

$$K_3 = \left.\frac{\delta Y}{\delta X_3}\right|_i = \text{constant} \qquad 7.4(91)$$

For example, consider the development of the transfer function representation for the water tank shown in Figure 7.4e. A material balance on the tank yields:

$$\rho Q_{in} - \rho Q_{out} = \frac{d}{dt}(\rho H K_s) \qquad 7.4(92)$$

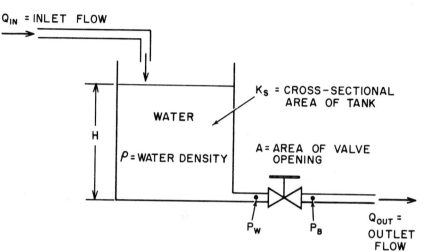

Fig. 7.4e   Nonlinear process represented by a water tank

Flow out of the tank is a function of the area of the valve opening and of the pressure drop across the valve, which can be determined by the following equation:

$$Q_{out} = C_d A \sqrt{(2g_c/\rho)(P_w - P_B)} \qquad 7.4(93)$$

where ρ = water density,
   C_d = orifice discharge coefficient (constant),
   P_B = back pressure (constant),
   P_w = water pressure at valve, and
   A = area of valve opening.

The pressure at the valve can be related to the height of water column in the tank:

$$P_w = \rho \frac{g}{g_c} H \qquad 7.4(94)$$

Therefore the valve equation is

$$Q_{out} = C_d A \sqrt{\frac{2g_c}{\rho}\left(\rho\frac{g}{g_c}H - P_B\right)} \qquad 7.4(95)$$

Substituting into the material balance, yields

$$Q_{in} - C_d A \sqrt{\frac{2g_c}{\rho}\left(\rho\frac{g}{g_c}H - P_B\right)} = \rho K_s \frac{d}{dt}(H) \qquad 7.4(96)$$

which is nonlinear and must be linearized before a Laplace transform can be determined.

Using the linearization technique described above, the nonlinear equation can be expressed as a linear function about a normal operating point $(Q_{out_i}, Q_{in_i}, A_i, H_i)$ as follows:

$$q_{out} = K_1 a + K_2 h \qquad 7.4(97)$$

where

$$7.4(98)$$
$$q_{out} = (Q_{out} - Q_{out_i})$$
$$a = (A - A_i) \qquad 7.4(99)$$
$$h = (H - H_i) \qquad 7.4(100)$$

$$K_1 = \left.\frac{\delta Q_{out}}{\delta A}\right|_i = C_d \sqrt{\frac{2g_c}{\rho}\left(H_i\rho\frac{g}{g_c} - P_B\right)} \qquad 7.4(101)$$

$$K_2 = \left.\frac{\delta Q_{out}}{\delta H}\right|_i = C_d A_i \sqrt{\frac{2g_c/\rho}{H_i\rho\frac{g}{g_c} - P_B}}\left(\rho\frac{g}{g_c}\right) \qquad 7.4(102)$$

Expressing Q_{in} as a deviation from a reference operating condition, Q_{in_i}, the linearized process differential equation is

$$q_{in} - K_1 a - K_2 h = K_3 \frac{dh}{dt} \qquad 7.4(103)$$

The Laplace transformation yields

$$Q_{in}(s) - K_1 A(s) - K_2 H(s) = K_3 s H(s) \qquad 7.4(104)$$

or

$$H(s) = \frac{Q_{in}(s)}{K_2 + K_3 s} - \frac{K_1 A(s)}{K_2 + K_3 s} \qquad 7.4(105)$$

Therefore the two transfer functions which describe the process are

$$\frac{H(s)}{Q_{in}(s)} = \frac{1}{K_2 + K_3 s} \qquad \text{7.4(106)}$$

$$\frac{H(s)}{A(s)} = \frac{K_1}{K_2 + K_3 s} \qquad \text{7.4(107)}$$

Stability

A stable linear system or element is one in which the system response is always bounded for any bounded system input. While most processes encountered in the process industries (with the exception of a few chemical reactors) are inherently stable, a feedback system employed to control the process can lead to a potentially unstable system. Mathematically, the stability of a linear system can be determined by an analysis of the roots of the "characteristic equation" from the differential equation describing the process (which corresponds to the roots of the denominator of the transfer function).

In general, the transfer function describing a system will be of the form

$$G(s) = \frac{B(s)}{A(s)} = \frac{b_m s^m + b_{m-1} s^{m-1} + b_{m-2} s^{m-2} + \cdots + b_0}{a_n s^n + a_{n-1} s^{n-1} + a_{n-2} s^{n-2} + \cdots + a_0} \qquad \text{7.4(108)}$$

The denominator is an nth order polynomial which contains n roots of the general form

$$s = \alpha + j\omega \qquad \text{7.4(109)}$$

where α is the real part and ω is the imaginary part of the root ($\omega = 2\pi f$). In order for a linear system such as $G(s)$ to be stable, all of the roots must lie in the left-hand complex plane (i.e., α must be less than 0.0). If any of the roots lie in the right-hand plane the system will be unstable (Figure 7.4f).

In many cases locating the exact position of each of the n roots is a difficult task. Unfortunately no "quadratic equation" exists for factoring polynomials of a higher order than 2. The techniques used are time consuming and tedious trial-and-error methods. However, in many situations it may not be necessary to actually obtain the exact position of the roots in order to learn the nature of these roots.

Descartes' Rule of Signs

A simple rule-of-thumb analysis which can provide much insight into the nature of roots is Descartes' rule of signs. If the terms in the denominator are arranged in descending powers of s, Descartes' rule of signs states that the number of positive *real* roots cannot exceed the number of variations in sign from term to term. Therefore a simple *necessary* but *not sufficient* condition of stability which can be applied at a glance to any transfer function is that all the coefficients of each term in the denominator must be the same sign. If any variations in sign exist, the system is unstable. If no variations

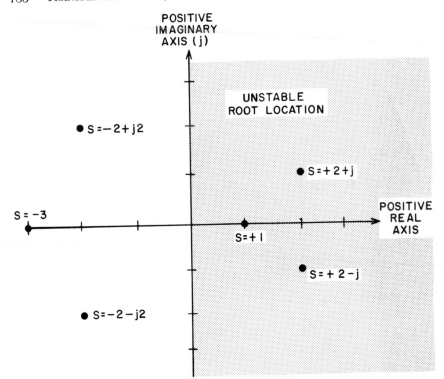

Fig. 7.4f   Stability region of complex plane

in sign exist, the system is potentially stable; however, instabilities could result from complex roots in the right-hand plane which Descartes' rule is unable to predict.

Routh's Criterion

One absolute method to determine if either complex or real roots lie in the right-hand plane is by the use of Routh's criterion. The method consists of a systematic procedure for generating a column of numbers which are analyzed for sign variations. The first step is to arrange the denominator of the transfer function into descending powers of s. All terms including those which are zero should be included.

$$A(s) = a_n s^n + a_{n-1} s^{n-1} + a_{n-2} s^{n-2} + \cdots + a_0 \qquad 7.4(110)$$

Next, arrange the coefficients of s according to the following schedule:

$$\left. \begin{array}{cccccc} a_n & a_{n-2} & a_{n-4} & a_{n-6} & \cdots \\ a_{n-1} & a_{n-3} & a_{n-5} & a_{n-7} & \cdots \end{array} \right\} \qquad 7.4(111)$$

Now, expand according to the following manner:

$$\left. \begin{array}{llll}
a_n & a_{n-2} & a_{n-4} & a_{n-6} \quad \cdots \\
a_{n-1} & a_{n-3} & a_{n-5} & a_{n-7} \quad \cdots \\
b_1 & b_2 & b_3 & b_4 \quad \cdots \\
c_1 & c_2 & c_3 & \cdots \\
\vdots & \vdots & \vdots & \\
d_1 & d_2 & \cdots & \\
e_1 & e_2 & \cdots & \\
f_1 & \cdots & & \\
g_1 & \cdots & &
\end{array} \right\} \quad 7.4(112)$$

where the additional rows are calculated by

$$\left. \begin{aligned}
b_1 &= \frac{a_{n-1}a_{n-2} - a_n a_{n-3}}{a_{n-1}} \\[2mm]
b_2 &= \frac{a_{n-1}a_{n-4} - a_n a_{n-5}}{a_{n-1}} \\[2mm]
b_3 &= \frac{a_{n-1}a_{n-6} - a_n a_{n-7}}{a_{n-1}} \\[2mm]
&\qquad\vdots \\[2mm]
c_1 &= \frac{b_1 a_{n-3} - a_{n-1} b_2}{b_1} \\[2mm]
c_2 &= \frac{b_1 a_{n-5} - a_{n-1} b_3}{b_1} \\[2mm]
&\qquad\vdots
\end{aligned} \right\} \quad 7.4(113)$$

This process is continued until all new terms are zero. Routh's criterion says that the number of roots of the denominator of the transfer function which lie in the right-hand plane is equal to the number of changes of sign in the left-most column of the array shown in 7.4(112).

Example: Consider a system described by the following transfer function:

$$g(s) = \frac{s + 1}{s^3 + 2s^2 + 5s + 24} \qquad 7.4(114)$$

An initial analysis made using Descartes' rule of thumb indicates that, since all the signs of the denominator are the same, the system appears to be stable; however, Routh's criterion must be applied in order to determine this conclusively. The Routh array is

1	5	0
2	24	0
-7	0	
24	0	

An inspection of the left-most column indicates two sign changes (one from $+2$ to -7 and the other from -7 to $+24$); therefore, there are two roots in the right-half plane, resulting in an unstable system.

Nyquist Criterion

In the process industries the Routh criterion has the limitation of not being applicable to systems containing deadtime. Another more general stability criterion is the Nyquist criterion. This is based on the "frequency response" concept of a process and can be easily applied to deadtime or to other distributed parameter effects. To demonstrate the Nyquist approach, consider the simple control loop shown in Figure 7.4g. By block diagram algebra the transfer function describing the system is

$$\frac{C(s)}{R(s)} = \frac{G(s)}{1 + G(s)H(s)} \qquad 7.4(115)$$

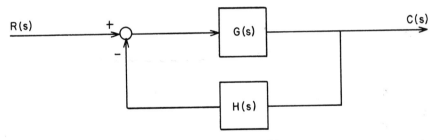

Fig. 7.4g   Feedback control loop

The characteristic equation describing the stability of the system is

$$1 + G(s)H(s) = 0.0 \qquad 7.4(116)$$

The above function can be examined in the frequency domain by substituting $s = -j\omega$. The frequency response of $G(j\omega)H(j\omega)$ can be represented by a polar plot in which the magnitude and phase of the resulting polynomial are plotted in the complex plane for values of ω from 0 to ∞. Consider the polar plot of the frequency response of a second-order system illustrated in Figure 7.4h. From this plot the stability can be determined by an investigation of the $s = -1$ point in the complex plane. The Nyquist stability criterion is

$$N = Z - P \qquad 7.4(117)$$

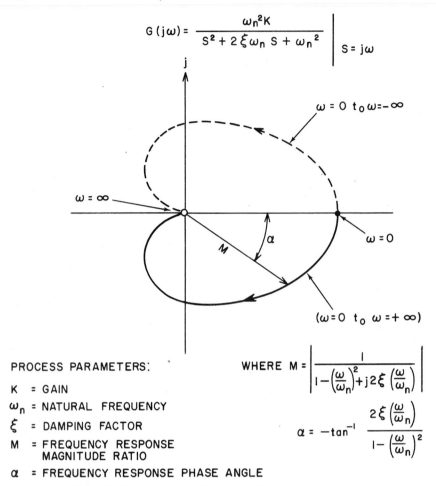

$$G(j\omega) = \left. \frac{\omega_n{}^2 K}{S^2 + 2\xi\omega_n S + \omega_n{}^2} \right|_{S = j\omega}$$

$\omega = 0 \ to \ \omega = -\infty$

$\omega = \infty$

$\omega = 0$

$(\omega = 0 \ to \ \omega = +\infty)$

PROCESS PARAMETERS:

K = GAIN

ω_n = NATURAL FREQUENCY

ξ = DAMPING FACTOR

M = FREQUENCY RESPONSE
     MAGNITUDE RATIO

α = FREQUENCY RESPONSE PHASE ANGLE

WHERE $M = \left| \dfrac{1}{1 - \left(\dfrac{\omega}{\omega_n}\right)^2 + j2\xi\left(\dfrac{\omega}{\omega_n}\right)} \right|$

$\alpha = -\tan^{-1} \dfrac{2\xi\left(\dfrac{\omega}{\omega_n}\right)}{1 - \left(\dfrac{\omega}{\omega_n}\right)^2}$

Fig. 7.4h   Polar plot of transfer function for second-order system

where N = net number of encirclements of
           point s = −1 in a clockwise direction,
      Z = number of zeros of G(s)H(s)
           that lie in the right-hand plane, and
      P = number of poles of G(s)H(s)
           that lie in the right-hand plane.

7.5 CLOSED-LOOP RESPONSE WITH VARIOUS CONTROL MODES

In the performance of a control system the interest is in the overall response of the control loop and process acting together. Response means the dynamic behavior of the total system after an upset, which may be caused by a process disturbance, a load change or by a set-point adjustment. The response quality is evaluated by the speed with which the controlled variable returns to set point, by the amount of over-correction (over-shoot) that occurs and by the stability of the system during this upset condition.

Depending on the nature of the process involved, different control modes are required for optimum performance. The purpose of this section is to describe the closed-loop response using various control modes and to analyze the system response utilizing transfer functions to characterize both the controller and the process it controls.

Also in this section is a set of recommendations on when to use which control mode as a function of the type of process involved.

In the evaluation or design of a control system one is generally interested in the "closed-loop" response of the system to changes in the set point or to changes in load or disturbance variables. For a particular system this response can be characterized by a differential equation (or transfer function) which describes the system. Consider the block diagram of a general control system in Figure 7.5a. Block diagram algebra indicates that the closed-loop transfer function which describes the system is as follows:

$$C(s) = \frac{AG_c(s)G_v(s)G_p(s)}{1 + G_c(s)G_v(s)G_p(s)G_h(s)} R(s)$$

$$+ \frac{G_p(s)G_d(s)}{1 + G_c(s)G_v(s)G_p(s)G_h(s)} D(s) \qquad 7.5(1)$$

where $C(s)$ = Laplace transform of the system output,
   $R(s)$ = Laplace transform of the system set point,
   $D(s)$ = Laplace transform of the disturbance input,
   $G_c(s)$ = transfer function of the controller,
   $G_v(s)$ = transfer function of the final control element,
   $G_p(s)$ = transfer function of the process,

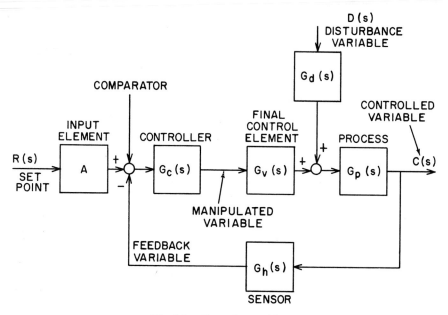

Fig. 7.5a   General control loop

$G_h(s)$ = transfer function of the feedback sensor device, and
$G_d(s)$ = transfer function of the disturbance variable.

It is important to note that in general the differential equation which describes the behavior of a control system is a function not only of the controller but of the process dynamics and of the control hardware as well. Each element in the control loop contributes significantly to the overall performance of the system and should therefore be considered.

Chemical Reactor Analysis

To illustrate the overall analysis of a control system consider the reactor shown in Figure 7.5b. The reactor is a simple backmix reactor heated by a steam jacket. The reactor temperature is controlled by a feedback controller which regulates the steam flow. As indicated above, the first step in the analysis of the system is to describe each element in the system, which are the reactor $G_p(s)$, the valve $G_v(s)$, the sensor $G_h(s)$, and the controller $G_c(s)$.

The valve and sensor for this example will be assumed to be instantaneous.

$$G_v(s) = K_v \qquad\qquad 7.5(2)$$

$$G_h(s) = K_h \qquad\qquad 7.5(3)$$

In general such an assumption may not be accurate. In practice, no valve or sensor acts instantaneously. Both introduce some lag into the loop, but for

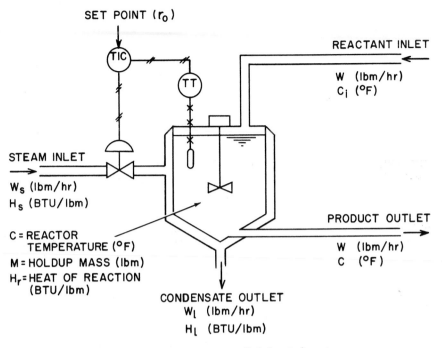

Fig. 7.5b   Temperature-controlled chemical reactor

the highly over-damped, slow-responding processes these hardware lags may be insignificant. However, for the relatively fast processes, such lags will have considerable effect on the overall system performance and should always be considered.

The description of the process can be determined by an analysis of the reactor. An energy balance yields

$$(W_sH_s + WK(C_i - C_r) - (WK(C - C_r) + MH_r + W_lH_l)$$
$$= \left(MK\frac{d}{dt}(C - C_r)\right) \qquad 7.5(4)$$

or, in terms of deviations from normal operating conditions,

$$(H_s - H_l)w_s + (WC_p)c_i - (WC_p)c = (MC_p)\frac{d}{dt}c \qquad 7.5(5)$$

where W_s = flow rate of steam, variable (lbm/hr),
    W = flow rate of product, constant (lbm/hr),
    K = heat capacity of product and reactant stream, constant (BTU/lbm),
    M = mass in reactor, constant (lbm),

C_i = temperature of reactant inlet stream, variable (°F),
C = temperature of the reactor, variable (°F),
H_s = enthalpy of steam, constant (BTU/lbm),
H_l = enthalpy of condensate, constant (BTU/lbm),
H_r = heat of reaction, constant (BTU/lbm),
C_r = reference temperature, constant (°F),
w_s = deviation in flow rate of steam,
c_i = deviation in temperature of inlet stream, and
c = deviation in temperature of reactor.

Assuming the system to be initially at steady state, the Laplace transformation yields

$$(H_s - H_l)W_s(s) + (WC_p)C_i(s) - (WC_p)C(s) = (MC_p)sC(s) \qquad 7.5(6)$$

Solving for $C(s)$:

$$C(s) = \frac{K_p}{1 + \tau_p s}W_s(s) + \frac{1.0}{1 + \tau_p s}C_i(s) \qquad 7.5(7)$$

where

$$K_p = \frac{H_s - H_l}{WC_p} \text{ (process gain)} \qquad 7.5(7a)$$

$$\tau_p = M/W \text{ (process time constant)} \qquad 7.5(7b)$$

Therefore in terms of the transfer function for the general block diagram,

$$G_p(s) = \frac{K_p}{1 + \tau_p s} \qquad 7.5(8)$$

$$G_d(s) = \frac{1}{K_p} \qquad 7.5(9)$$

The resulting block diagram for the system is shown in Figure 7.5c.

The controller is yet to be specified. It can be any one of the controllers discussed in Section 7.3:

$$(P): G_c(s) = K_c \qquad 7.5(10)$$

$$(I): G_c(s) = \frac{1}{T_i s} \qquad 7.5(11)$$

$$(PI): G_c(s) = K_c\left(1 + \frac{1}{T_i s}\right) \qquad 7.5(12)$$

$$(PD): G_c(s) = K_c(1 + T_d s) \qquad 7.5(13)$$

$$(PID): G_c(s) = K_c\left(1 + \frac{1}{T_i s} + T_d s\right) \qquad 7.5(14)$$

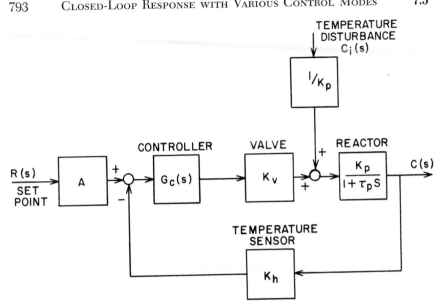

Fig. 7.5c   Reactor system

The closed-loop transfer function can be determined for any of the above controllers by substituting equations 7.5(2), 7.5(3), 7.5(8), 7.5(9) and the appropriate controller equation into the general closed-loop transfer function, equation 7.5(1).

Proportional Control

Consider the response of the reactor system when a proportional controller is used, equation 7.5(10). Substituting into equation 7.5(1) yields

$$C(s) = \frac{AK_cK_v\dfrac{K_p}{1 + \tau_p s}}{1 + \dfrac{K_cK_vK_pK_h}{1 + \tau_p s}} R(s)$$

$$+ \frac{\dfrac{1}{1 + \tau_p s}}{1 + \dfrac{K_cK_vK_pK_h}{1 + \tau_p s}} C_i(s) \qquad 7.5(15)$$

or,

$$C(s) = \frac{K_{s1}}{1 + \tau_s s} R(s) + \frac{K_{s2}}{1 + \tau_s s} C_i(s) \qquad 7.5(16)$$

where

$$K_{s1} = \frac{AK_cK_vK_p}{1 + K_cK_vK_pK_h} \qquad \text{7.5(16a)}$$

$$K_{s2} = \frac{1}{1 + K_cK_vK_pK_h} \qquad \text{7.5(16b)}$$

$$\tau_s = \frac{\tau_p}{1 + K_cK_vK_pK_h} \qquad \text{7.5(16c)}$$

Therefore the closed-loop response is described by a first-order transfer function. The rapidity of response will depend on the order of magnitude of the system time constant, τ_s. Note the process is also first order, but its time constant is larger than the time constant of the overall system; see equations 7.5(7b) and 7.5(16c).

One undesirable feature of a simple proportional controller is its steady-state response. Consider the response of the system to a step change in disturbance.

For $t < 0$,

$$c_i(t) = 0.0$$
$$r(t) = 0.0$$

and for $t \geq 0$,

$$c_i(t) = d_0$$
$$r(t) = 0.0 \qquad \text{(set-point constant)}$$

Therefore,

$$c(t) = K_{s1}(1 - e^{-t/\tau_s})0.0 + K_{s2}(1 + e^{-t/\tau_s})d_0 \qquad \text{7.5(17)}$$

Under steady-state conditions $(t \rightarrow \infty)$,

$$c(t) = C_{ss} = K_{s2}d_0 \qquad \text{7.5(18)}$$

or

$$C_{ss} = \frac{1}{1 + K_cK_vK_pK_h}d_0 \qquad \text{7.5(19)}$$

The result is a steady-state deviation from the desired set point. Such steady-state error is referred to as "offset" and is a characteristic of all pure proportional controllers.

Offset is generally associated only with the response of the system due to the disturbance variable. Similar errors due to set-point changes can be eliminated by calibration. Consider the steady-state response of the system to a step change in set point.

$$C_{ss} = \lim_{t \rightarrow \infty} K_{s1}(1 - e^{-t/\tau_s})R_0 \qquad \text{7.5(20)}$$

$$C_{ss} = K_{s1}R_0 \qquad 7.5(21)$$

Note that the steady-state response is equal to the set point ($C_{ss} = R_0$) if K_{s1} is calibrated to be equal to 1.0, which is achieved by setting A in equation 7.5(16a) to equal

$$A = \frac{1 + K_cK_vK_pK_h}{K_cK_vK_p} \qquad 7.5(22)$$

Integral Control

Consider the response of the reactor system if an integral controller is used, equation 7.5(11). The closed-loop transfer function is

$$C(s) = \frac{A\dfrac{K_vK_p}{T_is(1 + \tau_ps)}}{1 + \dfrac{K_vK_pK_h}{T_is(1 + \tau_ps)}}R(s) + \frac{\dfrac{1}{(1 + \tau_ps)}}{1 + \dfrac{K_vK_pK_h}{T_is(1 + \tau_ps)}}C_i(s) \qquad 7.5(23)$$

or, rearranging,

$$C(s) = \frac{A\dfrac{K_vK_p}{T_i\tau_p}}{s^2 + \dfrac{1}{\tau_p}s + \dfrac{K_vK_pK_h}{T_i\tau_p}}R(s) + \frac{\dfrac{s}{\tau_p}}{s^2 + \dfrac{1}{\tau_p}s + \dfrac{K_vK_pK_h}{T_i\tau_p}}C_i(s) \qquad 7.5(24)$$

One notable difference between the performance of the system with integral control and that with proportional control is the steady-state behavior. For an arbitrary step change in both the set point and disturbance,

$$R(s) = R_0/s \qquad 7.5(25)$$
$$C_i(s) = D_0/s \qquad 7.5(26)$$

the steady-state behavior can be determined using the final value theorem; see equation 7.4(9).

$$C_{ss} = \lim_{t \to 0}\{sC(s)\} = \frac{A}{K_h}R_0 + 0.0D_0 \qquad 7.5(27)$$

$$= \frac{A}{K_h}R_0 \qquad 7.5(28)$$

The disturbance does not ultimately affect the steady-state output of the control system. Indeed, if the controller is calibrated such that $A = K_h$, the final value of the response will always equal the desired value. One serious drawback to the integral controller, however, is the relatively slow response time. For the proportional control system the response time was fast, and indeed extremely fast for large values of gain (equation 7.5(16c)). The integral controller on the other hand results in a second-order response which is more sluggish and requires much longer to reach the final steady-state value.

Proportional-Plus-Integral Control

In many cases both proportional and integral modes are combined to take advantage of the fast transient behavior of proportional control and of the offset-free steady-state behavior of integral control (equation 7.5(12)). Applied to the reactor example, the closed-loop transfer function is as follows:

$$C(s) = \frac{A\left(\dfrac{K_c K_v K_p}{\tau_p T_i}\right)(T_i s + 1)}{s^2 + \left(\dfrac{1 + K_h K_c K_v K_p}{\tau_p}\right)s + \dfrac{K_h K_c K_v K_p}{\tau_p T_i}} \cdot R(s)$$

$$+ \frac{\dfrac{s}{\tau_p}}{s^2 + \left(\dfrac{1 + K_h K_c K_v K_p}{\tau_p}\right)s + \dfrac{K_h K_c K_v K_p}{\tau_p T_i}} \cdot C_i(s) \qquad 7.5(29)$$

At steady state,

$$C_{ss} = \frac{A}{K_h} \cdot R_0 + 0.\cancel{0}D_0 \qquad 7.5(30)$$

The steady-state performance is identical to a pure integral controller, but the transient response is much more rapid. Figures 7.5d and e compare the response of these controllers to both a change in set point and a change in disturbance. For the disturbance case, note that the steady-state offset of proportional control is eliminated by both the integral and proportional-plus-integral controllers. The combined modes, however, react much faster than does the integral mode alone. For the set-point change (Figure 7.5e), the best response occurs with the pure proportional control. Steady-state offset in this case can be effectively eliminated by calibration. On the other hand, the integral controllers must eliminate the steady-state error by integration. Of the two integral controllers, the combined controller responds faster than the pure integral one.

PD and PID Control

A similar analysis can be made of the reactor example for algorithms containing derivative action. For example, the closed-loop transfer function resulting from a proportional-plus-derivative controller is as follows:

$$C(s) = \frac{K_{s1}(1 + T_d s)}{1 + \tau_s' s} R(s) + \frac{K_{s2}}{1 + \tau_s' s} C_i(s) \qquad 7.5(31)$$

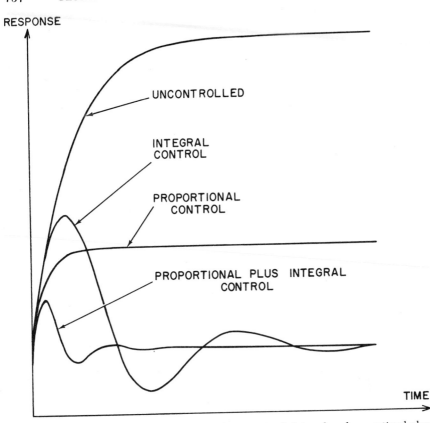

Fig. 7.5d   Response to a disturbance input with proportional, integral, and proportional plus integral controllers

where, similarly to equations 7.5(16a, b and c),

$$K_{s1} = \frac{AK_cK_vK_p}{1 + K_cK_vK_pK_h} \qquad \text{7.5(31a)}$$

$$K_{s2} = \frac{1}{1 + K_cK_vK_pK_h} \qquad \text{7.5(31b)}$$

$$\tau'_s = \frac{\tau_p + K_cK_vK_pK_hT_d}{1 + K_cK_vK_pK_h} \qquad \text{7.5(31c)}$$

The closed-loop equation is the same order as the process (first order). In general the system is fast responding, but it does not compensate for steady-state offsets. Figure 7.5f illustrates the effect of the derivative mode addition to a proportional-plus-integral controller.

RESPONSE

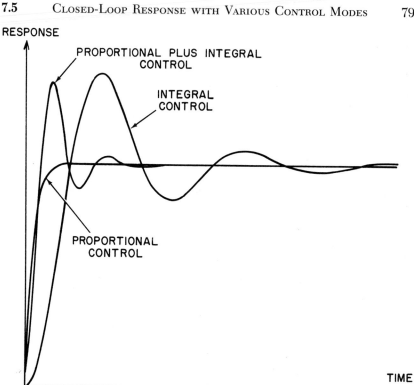

Fig. 7.5e   Response to a step change in set point with proportional, integral, and proportional plus integral controllers

RESPONSE

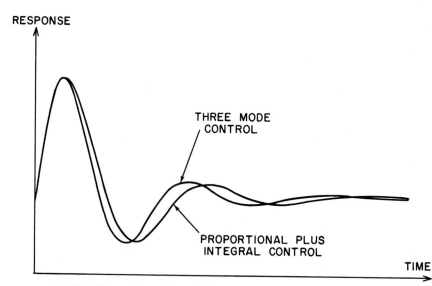

Fig. 7.5f   Effect of the addition of derivative action to a two-mode controller

Selection of Controller Modes

Selecting the proper mode of control is an important step in the design of a control loop. Unfortunately there is no universally perfect controller that can be applied to all processes. In general the factors which the application engineer must consider are (1) control quality, (2) costs and (3) ease of operation. For a particular controller the cost and the relative ease of operation are generally fixed, but the quality of control realized varies from process to process.

A few general comments to aid in the selection of the appropriate control modes follow.

Two-Position Control

Two-position control is the least expensive and simplest controller. It is sometimes used with relatively slow-responding, low-order systems. It cannot be used effectively on high-order processes which do not contain a dominant time constant or on systems containing sizable deadtimes. The areas of application for on-off controllers in the process industry are few and are fast disappearing.

Proportional Control

The pure proportional controller is the simplest regulating controller and can be applied with some degree of success to most processes in the process industry. It is particularly useful for process transfer functions which contain a $1/s$ term and have a single dominant time constant, as illustrated by equation 7.5(32).

$$G_p(s) = \frac{k}{(1 + \tau_1 s)(1 + \tau_2 s)(1 + \tau_3 s) \cdots} \qquad 7.5(32)$$

where $\tau_1 \gg \tau_2$, $\tau_1 \gg \tau_3$, $\ldots$

The pure proportional controller responds rapidly to both set-point and disturbance changes but it possesses the undesirable characteristic of steady-state error (offset). Tuning is relatively easy and is accomplished by the adjustment of a single parameter. Their use in the process industry is generally limited to local regulators, where offsets can be tolerated.

Integral Control

Integral (floating) control is another relatively simple one-parameter controller. It is particularly effective for (1) very fast processes, in particular fast processes which contain noise, (2) processes dominated by deadtime and (3) high-order processes in which all of the time constants are roughly the same order of magnitude. Integral control does not exhibit steady-state error, but it is relatively slow responding. It decreases the stability of a system and therefore should not be used for process transfer functions which contain $1/s$

terms. Only a few applications exist in the process industry for this single-mode controller.

Proportional-Plus-Integral Control

The proportional-plus-integral controller is perhaps the best conventional controller in use today. It does not have the offset associated with proportional control, yet it yields a much faster dynamic response than does integral action alone. Due to the presence of the integral mode the stability of the control loop is decreased. Extreme caution should be used in applying this controller to process transfer functions containing $1/s$ terms.

In situations where control signal saturation may occur, saturating the final control element, the integral mode should be used with care, because of the possibility for reset "wind-up." This is a condition in which the controller continues to integrate the error signal even though no further corrective action can be realized. The result is a confused controller which must "unwind" before effective control can be realized, resulting in a very poor transient behavior of the control system.

This two mode controller is widely used in the process industry, controlling level, pressure, flow and all types of other variables which do not have large time lags.

Proportional-Plus-Derivative Control

The proportional-plus-derivative (rate) controller is effective for systems containing a number of time constants. It results in more rapid response and less offset than is possible by pure proportional control. In general it increases the overall stability of the control loop. Care should be employed in using the derivative mode in the control of very fast processes or if the measurement signal is noisy. Applications in the process industry tend to favor the PID controllers against the PD controllers.

Inverse Derivative Control

The inverse derivative controller is a special two-mode controller which uses the derivative concept. It is used primarily to introduce a lag into the control loop when it is attempted to control very fast or noisy processes.

Proportional-Plus-Integral-Plus-Derivative Control

The proportional-plus-integral-plus-derivative controller is the most complex of the conventional control mode combinations. In theory, the PID controller results in better control than any of the one- or two-mode controllers mentioned earlier. In practice, however, the control advantage is difficult to realize due to the difficulty of selecting the proper tuning parameters. The addition of the derivative mode to the PI controller is often specified to compensate not for process lags but for hardware lags which might best be corrected at the source. The addition of the derivative mode is sometimes specified due to the mistaken belief that derivative action is helpful in overcoming process deadtimes.

PID controllers are utilized in the process industry to control such slow variables as most temperature, pH and other analytical variables are likely to be.

7.6 FEEDBACK AND FEEDFORWARD CONTROL

Sections 7.1 through 7.5 have been in large part dedicated to the theory of feedback control. The purpose of this section is to discuss the concept of feedforward control, its theory and application. As the name implies, feed forward control complements feedback and can either substitute for, or work in conjunction with, feedback control. For these reasons, it is important to identify the principal features and limitations of feedback as a basis for the development of the feedforward concept.

The purpose of any form of control is to maintain the controlled variable at a desired value known as the set point. The term regulation is used to identify control in the face of disturbing forces. A control system can only achieve regulation by opposing the disturbing forces—called the load—with equivalent changes in one or more *manipulated* variables. For the controlled variable to remain stationary, the controlled process must be in a state of balance. One of the means by which this balance is brought about is feedforward control, which provides corrective action before the effect of a disturbance is seen as an error in the controlled variable.

Feedback Control

Regulation through feedback control is achieved by acting on the change in the controlled variable induced by a change in load. Deviations in the controlled variable are converted into proportional changes in the manipulated variable and "fed back" into the process to restore its balance. Figure 7.6a points out the backward flow of information from the output of the process back to its manipulated input. The load is shown to be divided into components such as feed rate, feed composition and ambient temperature. All components of the load, however, can be balanced by a single manipulated variable. There must always be at least one manipulated variable for each independently controlled variable.

Performance of Feedback Control

Feedback, by its very nature, is incapable of correcting a deviation in the controlled variable at the time of its detection. In any process, no matter how simple, a finite delay exists between action of the manipulated variable and its subsequent effect on the controlled variable. Where this delay is

802

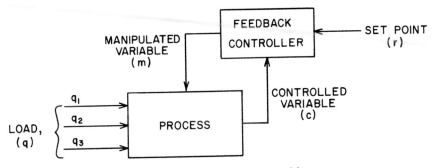

Fig. 7.6a   Simplified feedback control loop

substantial and the process is subject to frequent disturbances, considerable difficulty is encountered in maintaining control. Moreover, perfect control is not theoretically obtainable, even in the simplest process, because a deviation in the controlled variable must appear *before* any corrective action can begin. In addition, the value of the manipulated variable needed to balance the load must be sought by trial and error, with the feedback controller observing the effect of its output on the controlled variable.

The effectiveness of feedback control depends on the dynamic gain of the controller in relation to the frequency and magnitude of the disturbances encountered. Although a high controller gain is obviously desirable, the dynamic gain of the closed loop at its natural period of oscillation must be less than unity if the loop is to remain stable. So if the process has a high dynamic gain, the controller gain must be correspondingly low—in essence, the process dictates how well it can be controlled.

The job required of a feedback controller can best be appreciated by defining how it operates. The output m of an ideal three-mode controller is related to the deviation e between the controlled variable and its set point:

$$m = \frac{100}{PB}\left(e + \frac{1}{T_i}\int e\, dt + T_d \frac{d}{dt}\, e\right) \qquad 7.6(1)$$

where 100/PB is equivalent to the gain K_c.

The settings of the three modes are identified as the percent proportional band PB, reset time-constant T_i, and derivative time-constant T_d, both in the same units as time t. Consider the process to be in the steady state at time t_1, so that the deviation is zero. Solving equation 7.6(1) at time t_1 yields the output m_1:

$$m_1 = \frac{100}{PBT_i}\int_{t_0}^{t_1} e\, dt \qquad 7.6(2)$$

The output is related to the history of the deviation e from the time control began (t_0).

Let the process encounter a sustained change in load, requiring a controller output m_2 to return the system to balance. At some time, t_2, the process will have returned to the steady state, with e again zero, so that

$$m_2 = \frac{100}{\text{PBT}_i} \int_{t_0}^{t_2} e \, dt \qquad\qquad 7.6(3)$$

In order to change its output from m_1 to m_2, the controller had to sustain an integrated error:

$$m_2 - m_1 = \frac{100}{\text{PBT}_i} \int_{t_1}^{t_2} e \, dt \qquad\qquad 7.6(4)$$

This integrated error appears as the area between the controlled variable and the set point in Figure 7.6b. For convenience in notation, let the integrated

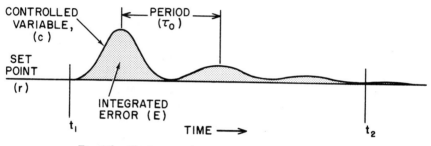

Fig. 7.6b   The integrated error resulting from a load change

error be denoted by E and the change in output by Δm. Then, the integrated error caused by a load change is

$$E = \Delta m \frac{\text{PBT}_i}{100} \qquad\qquad 7.6(5)$$

The point remains, however, that optimum values of PB and T_i exist for each control loop, which therefore determine the minimum value of E developed by a given load change. Some processes are rapid in response, such that T_i may be only a few seconds. Unless load changes are unusually severe or frequent, feedback control is ordinarily quite acceptable. A flow control loop is a good example. Other processes, while slow to respond, have a low dynamic gain, thereby accommodating a narrow proportional band. Control of temperature in a simple stirred tank is an example of this type of process, where feedback is acceptable. Still other processes are characterized by long delays and high sensitivity, such as the control of the composition of a product leaving a distillation tower. Here a wide proportional band and long reset time may result in off-specification product, if load changes are frequent or severe. This type of process can benefit from feedforward control.

Feedforward Control

Feedforward provides a more direct solution to control problems than finding the correct value of the manipulated variable by trial and error. In a feedforward system, the major components of load are sensed and used to calculate the value of the manipulated variable required to maintain control at the set point. Figure 7.6c shows how information flows forward from the

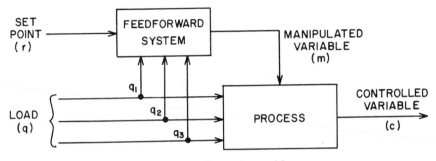

Fig. 7.6c   Feedforward control loop

load to the manipulated input of the process. If the controlled variable were used in the calculation, a positive feedback loop would be formed. The set point is necessary, however, to give the system a command.

The term "system" is used rather than "controller" when discussing feedforward, because it is not possible conveniently to provide the computing functions required by the forward loop with a single controlling device. Instead, the feedforward system consists of several devices whose related functions comprise a mathematical model of the process, including both its steady-state and dynamic characteristics.

Load Balancing

A dynamic balance is achieved over the process by solving its material and/or energy balance equations continuously. Whenever a change in load is sensed, the manipulated variable is automatically adjusted to the correct value at a rate which keeps the process continually in balance. It is theoretically possible to achieve perfect control, although in practice the computing system cannot be made to duplicate the process exactly.

Fortunately, the material and energy balance equations are the easiest to write for a given process. Consequently variations in non-stationary parameters like heat-transfer coefficients and the efficiency of mass-transfer units do not ordinarily affect the performance of a feedforward system. Load components are usually feed flow and feed composition, where product composition is to be controlled, or feed flow and temperature where product temperature is to be controlled. Feed flow is the primary component of load in virtually every application, because it can change widely and rapidly. Feed composition and temperature are less likely to exhibit wide excursions, and their rate of change is always limited by upstream capacity. In some feedforward systems these secondary load components are left out of the design.

The output of a feedforward system should be an accurately controlled flow rate. This controlled flow rate cannot be obtained by manipulating a control valve directly, because valve characteristics are non-linear and changeable, and their flow is subject to external influences. Therefore most feedforward systems depend on measurement and feedback control of flow to obtain accurate manipulation of the flow rate. Only when the rangeability of the manipulated variable exceeds that which is available in flowmeters should one consider having the valves positioned directly, and in such cases great care must be taken to obtain reproducible response.

Steady-State Model

The first step in designing a feedforward control system is to formulate the steady-state mathematical model of the process. The resulting equation is solved for the manipulated variable, which is to be the output of the control system. Then the set point is substituted where the controlled variable appears in the model.

Instead of further generalization with abstract symbols, the procedure will be demonstrated by using the example of temperature control in a liquid heat exchanger.

A liquid flowing at rate W is to be heated from temperature T_1 to T_2 by steam at rate W_s. The energy balance, excluding losses, is

$$WC(T_2 - T_1) = \lambda W_s \qquad\qquad 7.6(6)$$

Here coefficient C is the heat capacity of the liquid in BTU/lbm°F and λ is the heat in BTU/lbm given up by the steam in condensing, while the flow rates are in lbm/hr and the temperatures in degrees Fahrenheit. Solving for W_s yields

$$W_s = W\frac{C}{\lambda}(T_2 - T_1) \qquad\qquad 7.6(7)$$

Now T_2 becomes the set point, and W and T_1 are components of load to which the control system must respond.

An implementation of equation 7.6(7) is given in Figure 7.6d. A hand-control station introduces the set point for T_2 into the summing relay where input T_1 is subtracted. The gain of the summing relay is adjusted to implement the constant ratio C/λ. Next, the linear liquid flow signal is multiplied by the $(C/\lambda)(T_2 - T_1)$ term to produce the required setting of steam flow. Both flow signals must be linear, because the temperature difference signal is linear. (For a detailed presentation on the function and scaling of these analog computing devices, refer to Section 9.1 on pneumatic analog computers.)

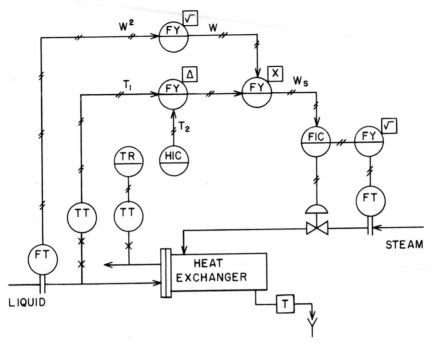

Fig. 7.6d   Feedforward control with the steady-state model alone

If, upon startup, the actual value of the controlled variable does not equal the set point, an adjustment should be made to the gain of the summing amplifier. Then, after making this adjustment, if the controlled variable does not return to the set point following a change in load, the presence of an error somewhere in the system is indicated. The error could be in one of the computing instruments, the flowmeter, or some factor affecting the heat balance which was not included in the feedforward system. Heat losses and variations in steam pressure are two possible sources of error in the mathematical model. Since any error will cause a proportional offset, the system designer should weigh the various sources of error, compensating for the largest or most changeable components where practicable. For example, if steam pressure variations were a source of error, the steam flowmeter could easily be pressure compensated.

Dynamic Model

In addition to the possibility of offset following a change in load, a transient error may also appear. A typical dynamic response of the controlled variable is shown in Figure 7.6e. Following a step increase in liquid flow, which is accompanied by a simultaneous and proportional step increase in steam flow as dictated by the feedforward system, the exit temperature will

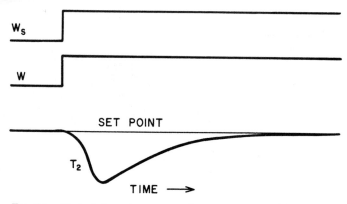

Fig. 7.6e Typical dynamic response of an uncompensated feedforward system

still fall temporarily. This transient error reveals a dynamic imbalance in the system, because for a time, the exchanger needs more heat than the steam flow controller is allowing. The reason for this is that the energy level of the process must be increased before an increase in heat transfer level can take place. Some of the added flow of heat from the increased steam flow is diverted in this example to increase the process energy level. Since the balance of the added heat flow is less than that calculated as needed to maintain the exit temperature of the increased liquid flow at steady state, this exit temperature will dip until the process energy level has been increased adequately. In order to correct the temperature error, an additional amount of energy must be added to the exchanger over what is required for the steady-state balance.

    An equally sound and more general explanation can be provided with the use of Figure 7.6f. This figure shows the load and the manipulated variable entering the process at different points, where they encounter different dynamic elements. In the heat exchanger example, the liquid enters the tubes

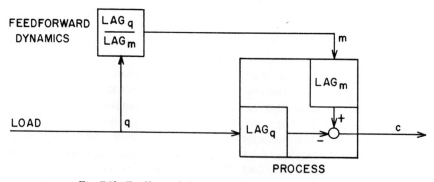

Fig. 7.6f Feedforward dynamics model for a general process

while the steam enters the shell. The heat capacities of the two locations are different. As a result, the controlled variable (the liquid temperature) responds more rapidly to a change in liquid flow than to a change in steam flow. In other words, the lag characteristic of the load input is less than that of the manipulated variable.

The objective of the feedforward system is to balance the manipulated variable against the load, providing the forward loop with compensating dynamic elements. Neglecting steady-state considerations, Figure 7.6f shows what the dynamic model must contain to balance a process with two first-order lags. The lag characteristic of the load input must be duplicated in the forward loop, and the lag in the manipulated input of the process must be cancelled. Thus the forward loop should contain a lag divided by a lag. Since the inverse of a lag is a lead, the dynamic compensating function is a "lead-lag." The lead time constant should be equal to the time constant of the controlled variable in response to the manipulated variable, and the lag time constant should equal the time constant of the controlled variable in response to the load. In the case of the heat exchanger, the fact that lag m is longer than lag q causes the temperature decrease on a load increase. This is the direction in which the load change would drive the process in the absence of control.

In transfer function form, the response of a lead-lag unit, as derived in equation 7.4(70) is

$$G(s) = \frac{1 + \tau_1 s}{1 + \tau_2 s} \qquad\qquad 7.6(8)$$

where τ_1 = lead time constant,
τ_2 = lag time constant and
s = the Laplace operator.

To speak of frequency response in reference to feedforward control is meaningless, since forward loops cannot oscillate. The most severe test that can be applied to a forward loop is a step change in load. Therefore the step response of the lead-lag unit is important. The gain of the unit to a step input in accordance with equation 7.4(19) is

$$c(t) = 1 + \frac{\tau_1 - \tau_2}{\tau_2} e^{-t/\tau_2} \qquad\qquad 7.6(9)$$

where e is the base of natural logarithms. The maximum gain (at $t = 0$) is the lead/lag ratio τ_1/τ_2. The response curve decays from this maximum exponentially at the rate of the lag time constant τ_2. Note how the 63 percent recovery point is reached in Figure 7.6g at $t = \tau_2$.

When properly adjusted, the dynamic compensation afforded by the lead-lag unit can produce the controlled variable response appearing in Figure 7.6h. This is the "signature" of a feedforward-controlled process. Because real processes do not consist of simple first-order lags, a first-order lead-lag unit

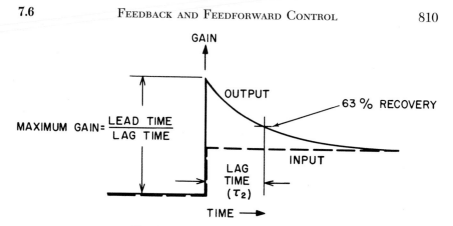

Fig. 7.6g Step response of a lead-lag unit

cannot produce perfect dynamic balance. Yet a marked improvement over the uncompensated response is attainable, and, as shown, the error can be equally distributed on both sides of the set point.

Adjusting the lead-lag unit requires some judgment. First, a load change should be introduced without dynamic compensation to observe the direction of the error. (Compensation is inhibited by setting lead and lag time constants at equal values.) If the resulting response is in the direction of the load response, lead time should exceed lag time; if not, lag time should be greater. Next, observe the time required for the controlled variable to reach its maximum or minimum value. This time should be introduced into the lead-lag unit as the lesser time constant. In other words, if lead should dominate, this would be the lag setting. Set the greater time constant at twice this value and repeat the load change. If the error curve is still not equally distributed

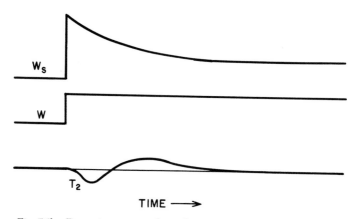

Fig. 7.6h Dynamic response of an adequately compensated feedforward system

across the set point, increase the greater time constant and repeat the load change.

One outstanding characteristic of the signature curve is that its area will be equally distributed about the set point if the difference between the lead and lag settings is correct. So once area equalization has been obtained, both settings should be increased or decreased with their difference constant until a minimum error amplitude is reached.

Adding Feedback

The offset resulting from steady-state errors mentioned earlier can be eliminated by adding feedback. This is most expeditiously done by replacing the feedforward set-point station with a controller, as shown in Figure 7.6i. The feedback controller adjusts the set point of the feedforward system in cascade, as the feedforward system adjusts the set point of the manipulated flow controller in cascade.

The feedback controller should have the same control modes as it would without feedforward control, but the settings should not be as tight. A feedback controller reacts to a disturbance by creating another disturbance in the opposite direction one-half cycle later. The feedforward system in the mean-

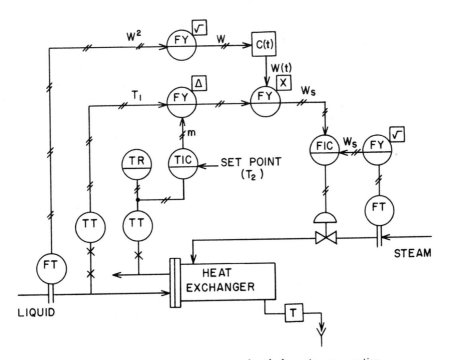

Fig. 7.6i   Feedforward-feedback control with dynamic compensation

time has already positioned the manipulated variable properly so that the error in the controlled variable will disappear. But if acted upon by tightly set feedback, the correct position will be altered, producing another disturbance. This prolongs the settling time of the system.

Note the location of the lead-lag unit, designated c(t) in Figure 7.6i. Its settings are particular to one load input, namely, liquid flow; therefore it must be placed where it can modify that signal and no other. It should not be placed anywhere in the feedback loop, as it would hinder the feedback function, whether performed automatically or manually.

Performance

The use of feedback in a feedforward system does *not* detract from the performance improvement gained by feedforward. Without feedforward, the feedback controller was required to change its output to follow all changes in load. With feedforward, the feedback controller must only change its output by an amount equal to what the feedforward system fails to correct. A feedforward system applied to a heat exchanger can control the steam flow to within 2 percent of what is required by the load. Thus, the feedback is only required to compensate 2 percent of a load change, rather than the full amount. This reduction of Δm in equation 7.6(5) by 50/1 results in the reduction of E by the same ratio. Reduction by 10/1 in errors resulting from load changes are relatively easy to obtain, and improvements of 100/1 have been reported.

Since a feedforward system is both more costly and requires more engineering effort than a feedback system, the control improvement it brings must be worthwhile. Consequently, feedforward systems have been applied chiefly to processes which are very sensitive to disturbances and slow to respond to corrective action, and whose product flow or value is relatively high. Distillation columns of 50 trays or more have been the principal beneficiaries of this technology. Boilers, multiple-effect evaporators and direct-fired heaters have also been controlled in this way, as well as waste neutralization plants, solids dryers, turbo-compressors and a scattering of other hard-to-control processes. Reference (1) contains a more detailed presentation on both feedforward and feedback control, along with many representative applications. Chapter X of this volume is devoted to unit operations and covers many examples of feedforward applications.

REFERENCE

1. Shinskey, F. G.: "Process-Control Systems", McGraw-Hill Book Company, New York, 1967.

7.7 ADAPTIVE CONTROL

An adaptive control system is one whose parameters are automatically adjusted to meet corresponding variations in the parameters of the process being controlled, in order to maximize the response of the loop. The significant point is the adjustment of *parameters*, which are normally fixed, as opposed to the outputs of a system, which are expected to vary.

The parameters set into controllers and control systems naturally reflect the characteristics of the processes they control. To maintain optimum performance, these parameters should change as the associated process characteristics change. If these changing characteristics can be directly related to the magnitude of the controller input or output, it is possible to compensate for their variation by the introduction of suitable non-linear functions at the input or output of the controller. Examples of this type of compensation would be the introduction of a square-root converter in a differential pressure flow-control loop to linearize the measurement, or the selection of a particular valve characteristic to offset the effect of line resistance on flow. This is not considered to be adaptation, in that these functions are fixed. Whenever the factors affecting the response of the process are independent of controller input and output, however, the compensating adjustment must be made either from a knowledge of the disturbing factor, or on the basis of loop response. A system which adapts on the basis of a measurement of the disturbing factor is said to be programmed and one which uses a measurement of its own performance is called self-adaptive. Programmed adaption of feedback control systems is being used to a limited extent, while techniques of self-adaption are not yet fully developed, and, although desirable, suffer from some serious disadvantages.

The name adaptive is occasionally applied to control systems designed to maintain a certain steady-state gain between a controlled and a manipulated variable. A most common example of this is the control at the maximum of a curve representing process efficiency as a function of the manipulated variable. Optimizing systems of this sort are covered in Section 7.11.

Adaption of Feedback Parameters

When a feedback control loop is disturbed by a change in load or set point, a deviation will appear in the controlled variable for a certain period of time. The function of the controller is to return this deviation to zero. The size of the deviation (relative to the upset) and the path it takes in returning to zero reveal how well the controller is doing its job. This response should be well damped, yet the time required for the controlled variable to return to the set point should be minimized. Ordinarily, the responsibility for achieving a desirable response rests with the engineer who selects and adjusts the controller. Since the process characteristics may change, however, the control loop response may deteriorate unless the engineer readjusts the controller.

It would be desirable, therefore, to have the controller settings adjusted automatically to compensate for variations in process parameters as they are detected.

Programmed Adaption

Where a measurable process variable produces a predictable effect on the gain of the control loop, compensation for its effect can be programmed into the control system. The most notable example in process plants is the variation in dynamic gain with flow in longitudinal equipment where no backmixing takes place. It is commonly seen in heat exchangers, but causes a particularly severe problem in once-through boilers.

In a once-through boiler, feedwater enters the economizer tubing section, passes directly into the evaporative tubing, and thence into the super-heater tubing, leaving as super-heated steam whose temperature must be accurately controlled. No mixing takes place as in the drum boiler, so a sizable amount of deadtime exists in the temperature control loop, particularly at low flow rates. Figure 7.7a shows how the step response of steam temperature to

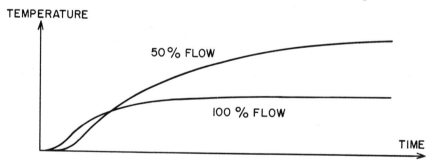

Fig. 7.7a   Step response of steam temperature to firing rate in a once-through boiler

changes in firing rate varies with feedwater flow. At 50 percent flow, the steady-state gain is twice as high as at 100 percent flow, because only half as much water is available to absorb the same increase in heat input. The deadtime and the dominant time constant are also twice as great at 50 percent flow.

The effect of these variable properties on the dynamic gain of the process is evidenced in Figure 7.7b. The same size load upset produces a larger

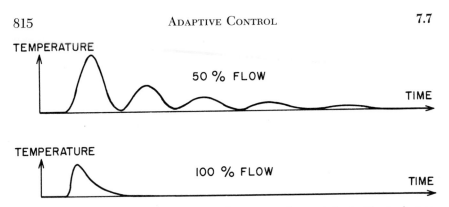

Fig. 7.7b   Response of the steam temperature loop to step changes in load without adaption

excursion in temperature at 50 percent flow, indicating a higher dynamic gain. The difference in damping between the two conditions also reveals the change in dynamic gain and so does the response, which is twice as fast at 100 percent flow. If oscillations were evident in the 100 percent flow response curve, their period would be much shorter than for 50 percent flow, due to the difference in deadtime. At 25 percent flow, oscillations would be still slower and damping would disappear altogether.

If proportional were the only feedback mode used in this application, the period of oscillation would vary inversely with flow, as would the dynamic gain at that period. The only adjustment that could be made would be that of the proportional band, which should vary inversely with flow. This would make the damping uniform, but nothing can change the increased sensitivity of the process to upsets at low flow rates.

Since reset and derivative modes are normally used to control temperature, some consideration should also be given to their adaption to changes in flow. The process deadtime—hence its period of oscillation under proportional control—varies inversely with flow. Therefore, the reset and derivative time constants also should vary inversely with flow for the adaption to be complete.

The equation for the flow-adapted three-mode controller is

$$m = \frac{100w}{PB}\left(e + \frac{w}{T_i}\int e\, dt + \frac{T_d}{w}\frac{d}{dt}e\right) \qquad 7.7(1)$$

where w is the fraction of full-scale flow, and PB, T_i and T_d are the proportional, reset and derivative settings at full-scale flow. Equation 7.7(1) can be rewritten to reduce the adaptive terms to two:

$$m = \frac{100}{PB}\left(we + \frac{w^2}{T_i}\int e\, dt + T_d\frac{d}{dt}e\right) \qquad 7.7(2)$$

Other parameters can be substituted for flow in instances where they apply.

Self-Adaption

Where the cause of changes in control loop response is unknown or unmeasurable, programmed adaption cannot be used. If adaption is to be applied, it must be based on the response of the loop itself, i.e., self-adaption. Self-adaption is a much more difficult problem than programming, because it requires an accurate evaluation of loop responses, ideally without a forced input.

Evaluation of set-point response is not especially difficult, because the magnitude and rate of the set-point change are available. This is of little merit in process control, however, where most of the troublesome systems have constant set points but are subject to load changes. These load changes are generally of unpredictable magnitude and rate, so that the loop response must be evaluated *without* foreknowledge of the nature of the disturbance. Some typical responses to both types of upsets are presented in Figures 7.5d and e.

The block diagram of a self-adaptive system is given in Figure 7.7c. This

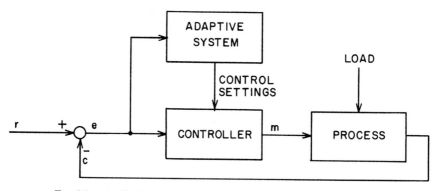

Fig. 7.7c   A self-adaptive system is a control loop around a control loop

system has all the problems of implementing the programmed adaption *plus* the problems of evaluating response and making a decision on the correct adjustment. Adaption of the proportional control mode is the first step. This can be done from a calculation of loop damping. The calculation requires a comparison of the amplitude of successive cycles on a damped oscillation, which is not easy to mechanize, since it must also recognize over-damped or non-oscillatory response.

If adaption of all three control modes were easy, it would be useful not only for compensating variable process parameters but also for controller adjustment on startup of stationary processes. By the same token, the best procedure for adapting the controller settings may well be to imitate the engineer who is skillful in the art of controller adjustment. The fact that proper controller adjustment is a skill practiced efficiently by few engineers suggests that reliable self-adaption of all modes may not be available for some time.

Performance

The performance difference between programmed and self-adaptive systems is the same difference that exists between feedforward and feedback control systems. A self-adaptive controller cannot make an adjustment to correct its settings until an unsatisfactory response is encountered. Two or more cycles must pass before an evaluation can be made to base an adjustment upon. Therefore, the cycle time of the adaptive loop must be much longer than the natural period of the control loop itself. Consequently, the self-adaptive system cannot correct for the present poor response but can only prepare for the response to the next upset, assuming that the control settings presently generated will also be valid then. On the other hand, the programmed system should always have the correct settings.

Another problem with the self-adaptive system is undetected drift, either of the process parameters or of the controller settings to the point where dangerous instability could result due to the slightest upset. This would particularly occur following a shutdown condition or similar instance when the control loop is open. Not only would the controller have to be transferred to manual for startup (as is the present practice) but its settings should also be initialized. If not carefully coordinated, this could place more burden on the operator than was removed by the adaptive functions. The programmed system presents no such hazards or operational problems.

Equipment Limitations

Adapting the proportional band of a controller is practical by the use of an analog multiplier. Such a scheme is shown in Figure 7.7d. Although

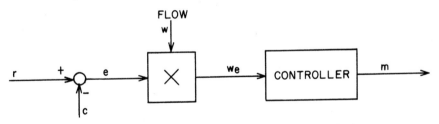

Fig. 7.7d   Adaption of the proportional mode using a multiplier

the adaptive term is shown to be flow, the same configuration could be used for self-adaption, where w would be the output of the adaptive system. The multiplier is special in that it must operate in two quadrants—its output must be negative when the error is negative. The controller itself may have reset and derivative modes, but, if so, they must be set to insure stability for both extremes of flow.

To adjust reset and derivative terms automatically is much more complicated. They must be adapted separately, and a once-simple device then becomes a multi-unit system. The error signal must be split into three components, operated upon dynamically, each component adapted and recombined into the system. Another method would be to adjust variable resistors either mechanically or through solid-state circuits. Although more direct, this method is not altogether practical at present. Extreme reliability, instead of high accuracy, is required. Complete programmed adaptation is therefore generally restricted to digital computer control systems, where it presents no special problems.

Adaption in Feedforward Systems

In Section 7.6, feedback was added to a feedforward control system to eliminate offset resulting from errors in the load balancing calculation. The errors could be in the computing system, in the flowmeters, or in the representation of the process. If an error is constant, it can be calibrated out by appropriate adjustments. But if the error is variable, feedback is necessary if offset is to be eliminated. In essence, the feedback controller is adjusting or adapting the feedforward system for variations in the characteristics of the controlled process.

For maximum effectiveness, the feedback should be applied where it can adjust for the parameter likely to introduce the greatest change. In the system shown in Figure 7.6i, the feedback controller can correct the model for changes in set point T_2, for heat losses that are proportional to T_2 and W, and for changes in C, λ, or the calibration of either flowmeter, since its output is multiplied by W. It is also possible to let feedback compensate for variations in T_1, if these variations are limited in magnitude and rate, by removing the liquid inlet temperature transmitter and the subtracting amplifier from the system.

The feedforward part of the system also provides proportional adaption for variations in feedback loop gain. Whereas the dynamic gain of the process varies inversely with flow, the multiplier in the forward loop makes the gain of the feedback loop vary directly with flow. This is the same type of feedback gain adaption as is shown in Figure 7.7d, but with three distinct advantages:

1. This mutual adaption is inherent in feedforward-feedback systems; therefore no additional cost is incurred.
2. A standard single-quadrant multiplier can be used since the controller output is always positive.
3. The action of the forward loop is so powerful in opposing disturbances, that adaptive derivative and reset time constants are not required.

This last form of adaption is the most effective and easiest to implement of all the forms described here. As a result, it is the only form being used extensively in high-performance process-control systems of variable dynamic gain of which the once-through boiler system is the most notable example.

7.8 CASCADE CONTROL

An intermediate process variable that responds both to the manipulated variable and to some disturbances can be used to achieve more effective control over the primary process variable. This technique, called "cascade control," is shown in block diagram form in Figure 7.8a. Two controllers are

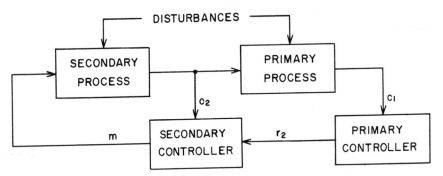

Fig. 7.8a   Process divided into two parts by the cascade control system

used, but only one process variable (m) is manipulated. The primary controller maintains the primary variable c_1 at its set point by adjusting the set point r_2 of the secondary controller. The secondary controller, in turn, responds both to the output of the primary controller, and to the secondary controlled variable c_2.

There are two distinct advantages gained with cascade control:

1. Disturbances affecting the secondary variable can be corrected by the secondary controller before a pronounced influence is felt by the primary variable.
2. Closing the control loop around the secondary part of the process reduces the phase lag seen by the primary controller, resulting in increased speed of response.

Primary and Secondary Loops

Figure 7.8b shows how closing the loop around the secondary part of the process can reduce its time lag. Here the response of the secondary closed

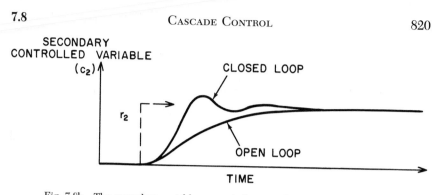

Fig. 7.8b   The secondary variable responds faster with its control loop closed

loop to a step change in its set point r_2 is compared with the response of the secondary variable to an equivalent step in the manipulated variable. If one refers to Figure 7.8c the secondary variable there is the jacket temperature and the manipulated variables are the steam and cold water flows. The secondary variable will always come to rest sooner under control, because initially the controller will demand a greater quantity of the manipulated variable, than what is represented by the "equivalent" step change in the manipulated variable. This is particularly true when the secondary part of the process contains a dominant lag (as opposed to deadtime). Because the gain of the secondary controller can be high with a dominant lag, closing

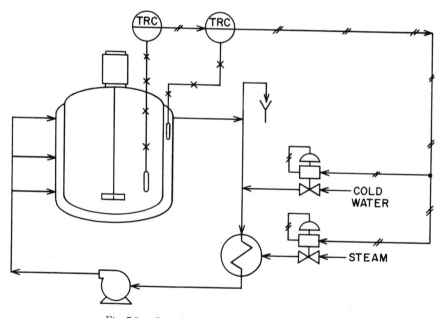

Fig. 7.8c   Cascade control of a stirred-tank reactor

the loop is particularly effective. The dominant lag in the secondary loop of Figure 7.8c is the time lag associated with heat transfer.

One very important factor in designing a cascade system is the proper choice of the secondary variable. Ideally, the process should be split in half by the cascade loop, i.e., the secondary loop should be closed around half of the time lags in the process. To demonstrate, consider the two extremes:

1. If the secondary variable were to respond instantly to the manipulated variable, the secondary controller would accomplish nothing.
2. If the secondary loop were closed around the entire process, the primary controller would have no function.

For optimum performance, the dynamic elements in the process should also be distributed as equitably as possible between the two controllers.

An additional problem appears when most of the process dynamics are enclosed in the secondary loop. Although the response of this loop is faster than the process alone, its dynamic gain is also higher, as indicated by the damped oscillation in Figure 7.8b. This means that if stability is to be retained, the proportional band of the primary controller must be wider than it would be without a secondary loop. Proper choice of the secondary variable will allow a reduction in the proportional band of the primary controller, because the high-gain region of the secondary loop lies beyond the natural frequency of the primary loop. In essence, reducing the response time of the secondary loop moves it out of resonance with the primary loop.

Types of Secondary Loops

The most common forms of secondary loops are listed in order of their frequency of application.

1. *Valve Position:* The position assumed by the plug of a control valve is affected by forces other than the control signal, principally friction and line pressure (Section 1.8). Change in line pressure can cause a change in position and thereby upset a primary variable, and stem friction has an even more pronounced effect. Friction produces a square-loop hysteresis between the action of the control signal and its effect on the valve position. Hysteresis is a non-linear dynamic element whose phase and gain vary with the amplitude of the control signal. Hysteresis always degrades performance, particularly where liquid level or gas pressure are being controlled with reset action in the controller. The combination of the natural integration of the process, reset integration in the controller, and hysteresis, causes a "limit cycle" which is a constant-amplitude oscillation. Adjusting the controller settings will not dampen this limit cycle but just change its amplitude and period. The only way of overcoming a limit cycle is to close the loop around the valve motor (Section 1.7, Positioners).

2. *Flow Control:* A cascade flow loop can overcome the effects of valve hysteresis as well as a positioner can. It also insures that line pressure variations or undesirable valve characteristics will not affect the primary loop. For these reasons, in composition control systems, flow is usually set in cascade. Cascade flow loops are also used where accurate manipulation of flow is mandatory, as in the feedforward systems shown in Figures 7.6d and i.

3. *Temperature Control:* Chemical reactions are so sensitive to temperature that special consideration must be given to controlling the rate of heat transfer. The most commonly accepted configuration has the reactor temperature controlled by manipulating the coolant temperature in cascade. A typical system for a stirred tank reactor is shown in Figure 7.8c. Cascade control of coolant temperature at the exit of the jacket is much more effective than at the inlet, because the dynamics of the jacket are thereby transferred from the primary to the secondary loop. Adding cascade control to this system can lower both the proportional band and reset time of the primary controller by a factor of 2 or more. Since exothermic reactors require heating for startup as well as cooling during the reaction, heating and cooling valves must be operated in split range. The sequencing of the valves is ordinarily done with positioners, resulting in two additional cascade sub-loops in the system.

Secondary Control Modes

Valve positioners are proportional controllers, usually with a fixed band of about 5 percent. Flow controllers invariably have both proportional and reset modes. In temperature-on-temperature systems, such as shown in Figure 7.8c, the secondary controller should not have reset. Reset is used to eliminate proportional offset, and in this situation a small amount of offset between the coolant temperature and its set point is inconsequential. Furthermore, reset adds the penalty of slowing the response of the secondary loop. The proportional band of the secondary temperature controller is usually as narrow as 10 to 15 percent. A secondary flow controller, however, with its proportional band exceeding 100 percent does definitely require reset.

Derivative cannot be used in the secondary controller if it acts on set-point changes. Derivative action is designed to overcome some of the lag inside the control loop, and if applied to the set-point changes it results in excessive valve motion and over-shoot. Some controllers are now available with derivative action on the measurement input only, and these can be used effectively in secondary loops where the measurement is sufficiently free of noise.

Instability in Cascade Loops

Adding cascade control to a system can destabilize the primary loop if most of the process dynamics are within the secondary loop. The most common example of this is the practice of using a valve positioner in a flow-control loop. Closing the loop around the valve increases its dynamic gain so much that the proportional band of the flow controller may have to be increased by a factor of 4 to maintain stability. The resulting wide proportional band means slower set-point response and deficient recovery from upsets. If a large valve motor or long pneumatic transmission lines cause problems, a volume booster should be used to load the valve motor, rather than a positioner. (See Section 1.7 on boosters and positioners.)

    Instability can also appear in a composition or temperature-control system where flow is set in cascade. These variables ordinarily respond linearly with flow, but the non-linear characteristic of a head flowmeter can produce a variable gain in the primary loop. Figure 7.8d compares the manipulated flow

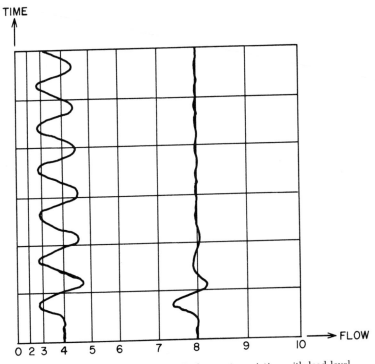

TIME

0 2 3   4   5   6   7   8   9   10   → FLOW

Fig. 7.8d   Manipulated flow records show gain variation with load level

record following a load change at 40 percent flow with a similar upset at 80 percent flow. Differential pressure h is proportional to flow squared:

$$h = kF^2 \qquad\qquad 7.8(1)$$

    If the process is linear with flow, its loop gain will vary with flow when h is the manipulated variable, because flow is not linear with h:

$$\frac{dF}{dh} = \frac{1}{2kF} \qquad\qquad 7.8(2)$$

Thus, if the primary controller is adjusted for $\frac{1}{4}$-amplitude damping at 80 percent flow, the primary loop will be undamped at 40 percent flow, and entirely unstable at lower rates.

Whenever a head flowmeter provides the secondary measurement in a cascade system, a square-root extractor should be used to linearize the flow signal, unless flow will always be above 50 percent of scale.

Saturation in Cascade Loops

When both the primary and secondary controllers have automatic reset, a saturation problem can develop. Should the primary controller saturate, limits can be placed on its reset mode, or logic can be used to inhibit automatic reset as is done for controllers on batch processes. Saturation of the secondary controller poses another problem, however, because once the secondary loop is opened due to saturation, the primary controller will also saturate. A method for inhibiting reset action in the primary controller when the secondary loop is open for any reason is shown in Figure 7.8e.

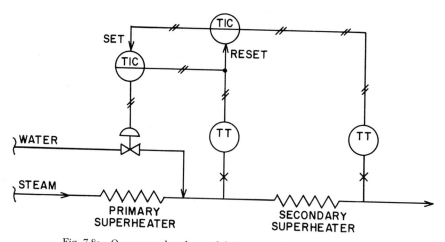

Fig. 7.8e   Open secondary loop inhibits reset in the primary controller

In many pneumatic controllers, the output is fed back through a restrictor to a positive input bellows to develop automatic reset (Figure 7.3h). If the secondary controller has reset, its set point and measurement will be equal in the steady state, so the primary controller can be effectively reset by feeding back either signal. But if the secondary loop is open, so that its measurement no longer responds to its set point, the positive feedback loop to the primary controller will also open, inhibiting reset action.

Placing the dynamics of the secondary loop in the primary reset circuit is no detriment to control. In fact it tends to stabilize the primary loop by retarding the reset action. Figure 7.8e shows the first known application of this technique. The primary measurement in this example is at the outlet of a steam superheater, whereas the secondary measurement is at its inlet, with the valve delivering water. At low load, no water is necessary to keep the secondary temperature at its set point, but the controllers must be prepared to act should the load suddenly increase. In this example the proportional band of the secondary controller is generally wide enough to require reset action.

When putting a cascade system into automatic operation, the secondary controller must first be transferred to automatic. The same is true in adjusting the control modes, insofar as the secondary should always be adjusted first, with the primary in manual.

7.9 RATIO CONTROL

Ratio control systems maintain a relationship between two variables to provide regulation of a third variable. Ratio systems are used primarily for blending ingredients into a product, or as feed controls to a chemical reactor. An example would be the addition of tetraethyl lead to a motor gasoline. A proper lead-to-gasoline ratio must be maintained to produce the desired octane number, which may or may not be measured.

Ratio systems actually portray the most elementary form of feedforward control (Section 7.6). The load input to the system (gasoline flow), if changed, would cause a variation in the controlled variable (octane number), which can be avoided by proper adjustment of the manipulated variable (additive flow). The load, or wild flow, as it is called, may be uncontrolled, controlled independently or manipulated by another controller which responds to the variables of pressure, level, etc.

Flow Ratio Control

Ratio control is applied almost exclusively to flows. Consider maintaining a certain ratio R of ingredient B to ingredient A:

$$R = B/A \qquad 7.9(1)$$

There are two ways to accomplish this. The more common method manipulates a flow loop whose set point is calculated:

$$B = RA \qquad 7.9(2)$$

This system is shown in Figure 7.9a. The set point for the flow controller is generated by an adjustable-gain device known as a ratio station. Since the calculation is made outside of the control loop, it does not interfere with loop response.

The second method is to calculate R from the individual measurements of flows A and B, using equation 7.9(1). The calculated ratio would then be the controlled-variable input to a manually set controller. Changes in the set point would change the ratio. Such a scheme is shown in Figure 7.9b.

The principal disadvantage of this system is that it places a divider inside a closed loop. If flow B responds linearly with valve B, the gain of the loop

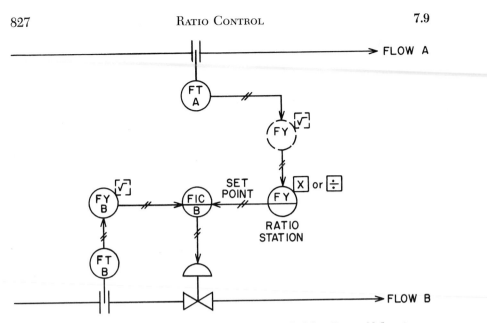

Fig. 7.9a   System maintaining a constant ratio of controlled flow B to wild flow A

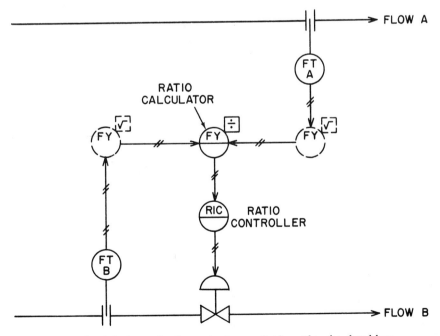

Fig. 7.9b   Calculating the flow ratio places a divider within the closed loop

will vary because of the divider. Differentiation of equation 7.9(1) explains why this is true:

$$\frac{dR}{dB} = \frac{1}{A} = \frac{R}{B} \qquad 7.9(3)$$

The gain varies both with the ratio and with flow B. In most instances, the ratio would not be subject to change, but gain varying inversely with flow can cause instability at low rates. An equal-percentage valve must be used to overcome this danger. If the ratio were inverted,

$$R = A/B \qquad 7.9(4)$$

then

$$\frac{dR}{dB} = -\frac{A}{B^2} = -\frac{R}{B} \qquad 7.9(5)$$

and the results are essentially the same.

The square-root extractors in Figures 7.9a and b are shown in broken lines to indicate that the systems can operate without them using flow-squared signals. In this case, the controlled variable is

$$R^2 = B^2/A^2 \qquad 7.9(6)$$

The scale of the ratio controller or ratio station has to be non-linear as a result. Differentiating,

$$\frac{d(R^2)}{dB} = \frac{2B}{A^2} = \frac{2R}{A} = \frac{2R^2}{B} \qquad 7.9(7)$$

Again, for the system in Figure 7.9b, the gain varies inversely with flow.

The principal advantage of using the ratio computing system is that the controlled variable—flow ratio—is constant and can be recorded to verify control. Using the system of Figure 7.9a, two records would have to be compared for verification.

Ratio Stations

No matter how ratio control is brought about, a computing device must be used whose scaling requires some consideration. The ratio station of Figure 7.9a normally has a gain range of about 0.3 to 3.0. Flow signal A, in percent of scale, is multiplied by the gain setting to produce a set point for flow controller B, in percent of scale. The true flow ratio must take into account the scales of the two flowmeters. The setting of the ratio station R is related to the true flow ratio by

$$R = \text{true flow ratio} \frac{\text{Flow A Scale}}{\text{Flow B Scale}} \qquad 7.9(8)$$

For example, the true flow ratio of the additive to gasoline is to be 2.0 cc/gal. If the additive flow scale is 0–1,200 cc/min, and the gasoline flow scale is 0–500 gal/min, then

$$R = 2.0 \text{ cc/gal} \frac{500 \text{ gal/min}}{1,200 \text{ cc/min}} = 0.833 \qquad 7.9(9)$$

When head flow signals are used, R should appear with a square root scale in order to be meaningful. Table 7.9c compares the gain of a ratio station (corresponding to linear flow ratio) with the ratio setting for head flowmeters. As shown by Table 7.9c, the available rangeability of ratio settings is seriously limited when using squared flow signals.

Table 7.9c
RATIO SETTINGS FOR HEAD FLOWMETERS

Flow Ratio Desired	Actual Ratio (Gain) Setting Required to Achieve Desired Ratio	
	If Signals Are Linear	If Square Root Signals Are Used and the Scale Is Linear
0.6	0.6	0.36
0.8	0.8	0.64
1.0	1.0	1.0
1.2	1.2	1.44
1.4	1.4	1.96
1.6	1.6	2.56

A device known as a ratio controller combines the ratio function and the controller in one unit. This is economical, saving not only cost, but panel space as well.

Since ratio stations are used with remote-set controllers, some means must be available for setting flow locally, for startup, or during abnormal operation. An auto-manual station is sometimes provided for this purpose. With the ratio controller, this feature requires two scales on the set-point mechanism—one reading in ratio for remote-set operation, the other reading in flow units for local-set operation.

When using a divider as a ratio computer, with linear flow signals, the scale factor for the divider should be $\frac{1}{2}$. This places a ratio of 1.0 at midscale:

$$R = \frac{1}{2} \frac{B}{A} \qquad 7.9(10)$$

Equal flow signals will produce a 50 percent output from the divider, and the full ratio range is then 0–2.0, linear. If flow-squared signals are used, the divider should have a scale factor of $\frac{1}{3}$, to provide a full ratio range of 0–1.73, with a square root scale. This places a ratio of 1.0 (A = B) at 0.58 on the square root scale.

$$R^2 = \frac{1}{3}\frac{B^2}{A^2} \qquad\qquad 7.9(11)$$

If a scale factor of $\frac{1}{2}$ were used, the ratio range would be restricted to 0–1.41.

Setting the Ratio Remotely

When a divider is used to calculate the flow ratio, then this ratio may be set with a remote-set controller. In order to set the ratio remotely when the controlled flow set point is calculated, a multiplier must replace the ratio station. With linear flowmeters, the usual choice of scaling factor for the multiplier is 2.0:

$$B = 2RA \qquad\qquad 7.9(12)$$

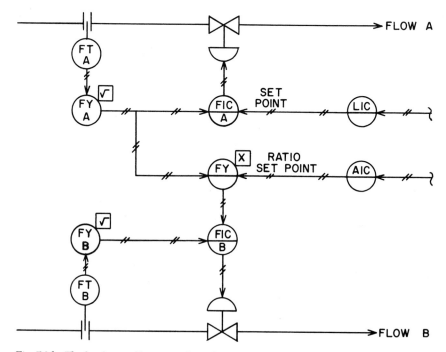

Fig. 7.9d   The level controller manipulates flow rate, and the composition controller manipulates flow ratio

In this way, a ratio of 1.0 appears at midscale of the ratio input, i.e., where A = B, R = 50 percent of scale. If squared flow signals are used, a scaling factor of 3.0 provides a ratio range of 0–1.73:

$$B^2 = 3R^2A^2 \qquad\qquad 7.9(13)$$

One reason for using a multiplier to set the flow ratio is the availability of very narrow ratio ranges for those applications in which the need for precision—not rangeability—is paramount. Ratio ranges of 0.9–1.1, 0.8–1.2, etc. are possible. A typical application would be the accurate proportioning of ammonia and air to an oxidation reactor for the production of nitric acid. If the compositions of the individual feeds are constant, only a very fine adjustment of the ratio is required.

Cascade Control of Ratio

Figure 7.9d shows two combinations of cascade and ratio control. Liquid level is affected by total flow, hence the liquid-level controller sets flow A, which in turn sets flow B proportionately. (To prevent the instability possible at low flow, when setting a head flowmeter in cascade, linear flow signals are used.)

Conversely, composition is not affected by the absolute value of either flow, but only by their ratio. Therefore, to make a change in composition, the controller must adjust the ratio set point of the multiplier. To minimize the effect of the composition controller on liquid level (through its manipulation of flow B), flow B should be the smaller of the two streams.

7.10 SELECTIVE CONTROL

For each controlled variable there must be a manipulated variable. In a given process, however, the number of controlled variables and the number of manipulated variables may not necessarily be equal. When this situation develops, a logical means of sharing the various control loops must be devised. Switching a manipulated variable from one controller to another, or switching a controller from one controlled variable to another, can be done smoothly and efficiently with selective devices. These devices receive two or more inputs and produce a single output which may be the highest, lowest, or median of the inputs, depending on the function desired. Using these devices effectively requires a thorough understanding of the process application.

Control Within Limits

No matter what type of process is to be controlled, constraints are always involved—flows are limited, vessel capacities and stress ratings cannot be exceeded and dangerous operating conditions must be avoided. When hazards to life and equipment exist, these constraints must be enforced by automatic means. Some of these constraints are on controlled variables such as temperature, pressure and liquid level. In other cases, manipulated variables—principally flow rates—must be limited. Both types of constraints use the same selecting devices, but the remainder of their control systems differ.

Limitations on Manipulated Variables

In a situation of constrained control, there may be one controlled variable, with a choice of manipulated variables. A typical example is the firing of a process heater with either of two fuels. The choice between the fuels is usually made on an availability basis. Fuel A may be burned to the limit of its availability and then it is supplemented with Fuel B. This limit may be set manually, or by a controller responding to the storage capacity of fuel A.

A successful control system incorporating this limitation must have the following features: (1) Capability of manipulating the limited variable on the allowable side of its limit. (2) Smooth transition from one manipulated variable to the other without adversely affecting the controlled variable.

832

Accommodating these features requires coordination of the manipulated variables and proper weighing of their effects on the process.

A system wherein the temperature of the process is controlled by manipulating limited and supplemental fuels is shown in Figure 7.10a. Fuel A is

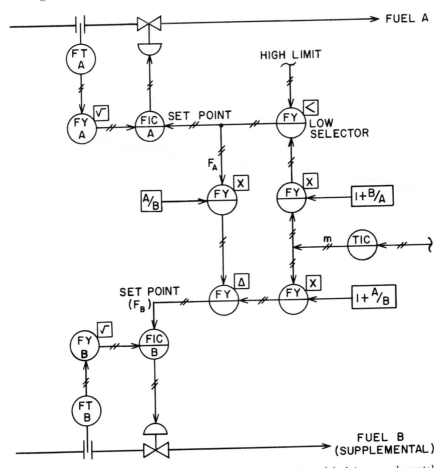

Fig. 7.10a   Temperature is controlled by manipulating either limited fuel A or supplemental fuel B

limited by a second (high limit) input to the low selector, on its set-point signal. The limit could be set manually or could come from the output of a pressure controller on the fuel header. The objective is that fuel A, being less expensive, should be used to the limit of its availability before admitting fuel B.

The smooth transition through the limit can only be brought about by using the computing devices as shown. Any difference between the output of the temperature controller and the equivalent flow of fuel A is converted into a set-point signal for the supplemental fuel flow controller. The temperature controller output represents the total heat input to the system by the combined fuels. This is converted into the set point for fuel A by a multiplying factor of $1 + B/A$, where B/A is the ratio of full-scale heat input of fuel B to that of fuel A. This equates the heat-flow values of controller output and fuel A set point.

As long as fuel A is at or below its high limit, the two inputs to the subtracting relay will cancel one another:

$$m(1 + A/B) - m(1 + B/A)(A/B) = 0 \qquad 7.10(1)$$

When the high limit is exceeded, F_A becomes constant and F_B starts increasing smoothly from zero without affecting the gain of the temperature-control loop. The system will also work with the fuel A controller in manual, as long as the heat load is above what fuel A can supply.

Limits on Controlled Variables

When limits must be placed on one or more controlled variables with a single manipulated variable, selection is made at the output of the controllers. An example is the pump-control system shown in Figure 7.10b. It is

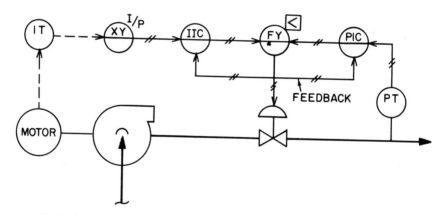

Fig. 7.10b   Pressure can be controlled only if motor current is below its set point

desired to control pressure in the process header by throttling the pump discharge, but only when motor current is below its rated limit. If an abnormal situation should cause motor current to reach its limit, the current controller would take over manipulation of the valve from the pressure controller by calling for the valve to close further. Pressure would remain below set point if this occurred.

Low-signal selectors are used on controller outputs more often than high-signal selectors, because the intent of this type of system is protection, and most final operators are designed to fail safely upon loss of signal. Other typical applications of controller output selection include: temperature/pressure control of a reactor by manipulating feed rate; pressure control of a compressor by manipulating a bypass valve with flow override for surge protection; and flow control of steam to a reboiler with override by level to avoid uncovering tubes in an abnormal situation.

Selection of Extremes

Another use for selective devices is the control of an extreme of several similar variables. For example, the hottest temperature in a fixed-bed chemical reactor should be controlled to avoid damage—yet its location may change with flow and time. So temperatures at several locations may be measured and sent to a high-selector, whose output becomes the controlled variable.

Similarly, in a nuclear reactor, two or more coolant streams are used to remove heat from the core. More than one coolant temperatures need to be controlled, but control-rod position is the only available manipulated variable. One solution is to control off the highest coolant temperature by manipulating the rod position.

Redundancy

The same concept can be employed to protect the plant from instrument or equipment failure. If one of the coolant temperature measurements in the preceding example is low due to loss of load, instrument failure or open circuit, it will not affect control of the reactor because it is disregarded by the control system. If it is higher than the other measurements for any reason, however, it would become the controlled variable, and reactor power would be reduced or shut down. Thus the high selector can protect the plant from an unsafe condition, regardless of the source of the failure.

Where multiple sensors are used on a single variable due to questionable reliability, a high selector can protect the plant against unsafe operation due to sensor failure. This would be the case where the sensors are located in a usually hostile environment—high temperature or corrosive, dirty or vibrating surroundings. If the sensors were able to fail in either direction, high failures will result in safe but costly shutdowns. Maximum reliability is obtained with a median selector, which rejects the highest and lowest of three or more input signals. The circuit is shown in Figure 7.10c. Individual or simultaneous high and low failures will not result in loss of control, whereas two signals of the same value are recognized as a true measurement, not a failure.

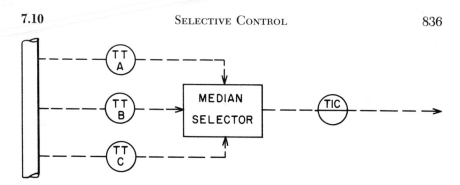

Fig. 7.10c   Maximum effectiveness in a redundant system obtained by using a median selector

Saturation Problems in Selective Control

Controller saturation problems arise when a controller with reset action has its loop opened. This can take place in selective control only when controllers share a single manipulated variable, as in Figure 7.10b. In this case, if the process load is low, pressure can be controlled with motor current below its set point. Reset action in the current controller will therefore continue to raise its output to full scale. Upon an increase in load, considerable over-shoot could develop before the current controller began throttling the valve. To avoid this reset saturation, the output of the selector is used for reset feedback to both controllers. When the pressure controller is selected, its feedback and output are equal, closing the reset loop and thereby allowing reset action. Meanwhile, the current controller is being reset by a signal other than its own output; therefore its reset feedback loop is open, and it acts like a proportional controller whose output is biased by the pressure controller.

When both measurements are at their respective set points, the two controller outputs will be equal. This is where switching occurs. The controller whose measurement is increasing will take control smoothly from the other, at this point.

7.11 OPTIMIZING CONTROL

The objective of an optimizing control system is to maintain a process in its most efficient state within the prevailing constraints. "Most efficient" could hold several meanings, depending on the intent of the plant manager: maximum yield, profit, production, product concentration, etc., or minimum losses, operating costs, power consumption, etc. The controlled variable in each case is usually an economic function. Product concentration and some other controlled variables are exceptions. Economic functions are not often directly measurable.

For example, assume that it is desired to minimize power consumed in a given electrochemical reaction. Obviously the power consumption would be zero if the reactor were shut down; so the objective, more properly stated, must be to minimize the power consumed per pound of product made. Therefore, even in this case where the significant terms—power consumption and product flow—may be measurable, the controlled variable must be computed. In other cases, the controlled variable cannot even be computed, because it requires off-line information such as laboratory analyses. In addition, the quality of a product may be the result of variables manipulated several hours earlier.

Consequently, in most processes the variable to be optimized is not available for feedback control. But the engineer who knows his process can still achieve efficient operation by intelligent coordination of variables, such as manipulating air in the proper ratio to fuel flow for efficient combustion.

Feedforward Optimizing Control

Fuel-air ratio control is an example of feedforward control applied to a combustion process. Where greater precision is required than is provided by this intuitive approach, or where the process is more complex or less easily assimilated, a profit-loss function should be prepared. This, in combination with the process model, can yield a control equation optimizing the objective function.

A case in point is the minimum-cost operation of a column separating components A and B by distillation. Let A be the more volatile component whose concentration in the feed is z and in the distillate is y. The distillate product is more valuable than the bottom product by Δv dollars per gallon, if its purity is controlled at the required specification. The loss function includes the amount of component A in the feed which is not recovered as distillate, and the cost of heating and cooling the column:

$$L = \Delta v(Fz - D) + cV \qquad 7.11(1)$$

where   c = cost of boiling and condensing one unit of vapor,
       V = vapor flow,
       F = rate of feed, and
       D = rate of distillate.

The loss function can be placed on the basis of feed rate by dividing all the terms in equation 7.11(1) by F:

$$\frac{L}{F} = \Delta v\left(z - \frac{D}{F}\right) + c\frac{V}{F} \qquad 7.11(2)$$

Next, Equation 7.11(2) can be solved for several values of the manipulated variable V/F and the independent variable z, with the value of D/F required for control of the distillate concentration (y) having been obtained from a mathematical model of the column. A tabulation of the loss function for typical values of z and V/F might appear as in Table 7.11a.

Table 7.11a
LOSS AS A FUNCTION OF
V/F AND Z

z	V/F	D/F	L/F
0.6	4.8	0.545	0.590
↓	5.0	0.556	0.588
↓	5.2	0.564	0.592
0.5	5.0	0.435	0.630
↓	5.2	0.448	0.624
↓	5.4	0.458	0.648

From the table, the optimum value of V/F changes from 5.0 when z = 0.6 to 5.2 when z = 0.5. A feedforward system for maintaining optimum performance with varying feed composition would adjust V/F from a feed analysis using a simple linear equation connecting the above points:

$$V/F = 3.8 + 2z \qquad 7.11(3)$$

or

$$V = F(3.8 + 2z) \qquad 7.11(4)$$

This feedforward system is readily implemented with available computing devices, and, following installation, the coefficients can be adjusted for a close fit or to allow for cost variations.

Feedback Optimizing Control

The control problem for a feedback optimizer is to find by trial and error the correct value of the manipulated variable that will produce a maximum or minimum value of the controlled variable. Compared with feedforward optimizing, this requires relatively little knowledge about the behavior of the process, and any number of load variables can be counter-acted by the controller. The principal drawbacks are the need for a measurable or calculable controlled variable and the relatively poor dynamic response obtainable from a feedback optimizer. The fact that the steady-state gain of the process is variable causes a stability problem. This variable process gain is the reason why some experts consider feedback optimizing to be an adaptive function (Section 7.7).

Continuous Slope Control

The problem faced by an optimizing system is that of controlling the steady-state gain of the process at some desired value K:

$$dc/dm = K \qquad 7.11(5)$$

Here c is the controlled variable and m the manipulated variable. K could be set at any desirable value, but a value of zero representing minimum or maximum operation is most common.

Before any control action can be taken, the process must be disturbed by dm to determine its present gain. Then the transient following the disturbance must be allowed to die before the steady-state gain can be determined. A continuous optimizing controller changes m at a rate dm/dt and observes the resulting effect on the process, dc/dt. The calculation is then made for gain:

$$K = \frac{dc}{dm} = \frac{dc/dt}{dm/dt} \qquad 7.11(6)$$

The system may use differentiators to calculate dc/dt and dm/dt and a divider to solve equation 7.11(6). Another solution is to manipulate m with an integrating controller whose error signal is then inserted into the denominator of equation 7.11(6) as being proportional to dm/dt. In either scheme, when dm/dt approaches zero, the gain of the process becomes indeterminate. The difficulty in making the calculations accurately and reliably has prevented extensive application of this system.

Sampled Data Slope Control

This concept has the same basis as continuous slope control but operates incrementally:

$$\frac{\Delta c}{\Delta m} = K \qquad 7.11(7)$$

The manipulated variable is stepped at time intervals Δt, which are sufficiently long to allow response transients to disappear. The observed process changes by Δc over this interval and allows for a reasonably accurate calculation of gain to be made, if no other upset has been sustained during the interval. The controller compares the calculated $\Delta c/\Delta m$ with K and adjusts m correspondingly.

$$\frac{\Delta m}{\Delta t} = \frac{1}{T_i}\left(K - \frac{\Delta c}{\Delta m}\right)$$ 7.11(8)

A block diagram of the system is given in Figure 7.11b.

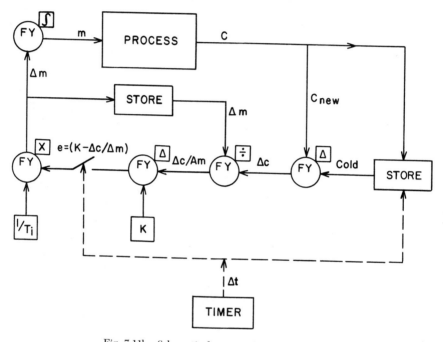

Fig. 7.11b   Schematic for a sampled data optimizer

Equation 7.11(8) is an integrating control function whose reset time constant is adjustable as T_i. A motor may be used to provide the integrating function appearing in Figure 7.11b. Typical of integrating controllers, it has poor dynamic response and cannot be used at all on non-self-regulating (integrating) processes, as undamped oscillations will result. Similar to the continuous controller, it is not self-starting but requires a manual increment Δm or a process upset to make the first calculation.

Sometimes several manipulated variables affect the optimum, such as the yield from a reactor varying with pressure, temperature, feed rate, feed composition, etc. In this case, the optimizing calculation must be shared among the manipulated variables, each with its own control setting. The controlled variable is then optimized independently for each manipulated variable in succession.

Peak-Seeking Control

While the slope controllers come to rest when the optimum is reached, the peak-seeking controller must hunt about the optimum. It works by continually comparing the present value of the controlled variable against the last. As long as the difference is positive, the manipulated variable continues to be driven in the same direction. When the controlled variable starts to decrease, the sign reversal causes m to reverse direction. Thus the peak is approached from the other side, until the controlled variable again decreases, reversing the action once more. A continuous oscillation results, whose amplitude varies with the rate of change of the manipulated variable and with the sensitivity of the reversal detector. While simpler than the slope controllers, its hunting characteristic is undesirable.

None of these feedback optimizers are used to any great extent, principally because measurable controlled variables which pass through a minimum or maximum are rarely encountered in process control.

7.12 TUNING OF CONTROLLERS[1]

The adjustment or tuning of controllers is one of the least understood, poorly practiced, yet extremely important aspects of the application of automatic control theory. In this section the objective is to present several procedures for estimating the optimum settings for a controller. Superficially, the best way to present this subject would seem to be to discuss only the best method of tuning controllers in more detail, but there is no general agreement as to which method is the best. Some methods lean heavily on experience while others rely more on mathematical considerations. Although the methods discussed in this section attempt to yield optimum settings, the only criterion stated is that the response have a decay ratio of $\frac{1}{4}$. This has been shown to be insufficient to obtain a unique combination of settings for a controller with more than one mode. When tuning a controller, one should be aware of this possibility, and remember that although the response has a decay ratio of $\frac{1}{4}$, the controller may still not be at its optimum settings.

Definition of "Good" Control

The first difficult problem encountered in tuning controllers is to define what is "good" control, and this unfortunately differs from process to process. The adjustment of process controllers is usually based on time domain criteria as opposed to frequency domain criteria,[2] such as gain margin or phase margin. Table 7.12a gives the four most commonly used criteria. The first item, the decay ratio, has the advantage of being readily measured, as it is based on only two points on the step response. The latter three, the integral criteria, have the advantage of being more precise, that is, more than one combination of controller settings will usually give a $\frac{1}{4}$ decay ratio, but only one combination will minimize the respective integral criteria.

The desired decay ratio is usually $\frac{1}{4}$, which is a good compromise between a rapid rise time and a short line-out time.

Although the shape of the response that minimizes the respective integral criteria differs from process to process, some general relative characteristics may be noted. ISE penalizes large errors whenever they occur; minimizing ISE will thus favor responses with short rise times (consequently being less damped). ITAE penalizes even small errors occurring late in time; minimizing

Table 7.12a
CRITERIA FOR CONTROLLER TUNING

1. Specified Decay Ratio, Usually $\frac{1}{4}$	Decay ratio $= \dfrac{\text{second peak overshoot}}{\text{first peak overshoot}}$ (see Figure 7.12b)
2. Minimum Integral of Square Error (ISE)	$ISE = \int_0^\infty [e(t)]^2 dt$ where $e(t) = $ (set point $-$ process output)
3. Minimum Integral of Absolute Error (IAE)	$IAE = \int_0^\infty |e(t)|\, dt$
4. Minimum Integral of Time and Absolute Error (ITAE)	$ITAE = \int_0^\infty |e(t)|\, t\, dt$

ITAE will thus favor responses with short line-out times (highly damped). IAE is intermediate, and the corresponding response frequently has a decay ratio near $\frac{1}{4}$.

Closed-Loop Response Methods

Techniques for adjusting controllers generally fall into one of two classes. First, there are a few methods based upon parameters determined from the closed-loop response of the system, i.e., with the controller on "automatic." Second, some methods are based upon parameters determined from the open-loop response curve, commonly called the process reaction curve. To use the open-loop methods the controller does not even have to be installed before the settings can be determined. Here we are concerned with closed-loop methods, of which the two most common are the ultimate method and the damped oscillation method.

Ultimate Method

One of the first methods proposed for tuning controllers was the ultimate method, reported by Ziegler and Nichols[3] in 1942. The term "ultimate" was attached to this method because its use requires the determination of the ultimate gain (sensitivity) and ultimate period. The ultimate sensitivity K_u is the maximum allowable value of gain (for a controller with only a proportional mode) for which the system is stable. The ultimate period is the period of the response with the gain set at its ultimate value (Figure 7.12c).

To determine the ultimate gain and the ultimate period, the gain of the controller (with all reset and derivative action turned off) is gradually adjusted

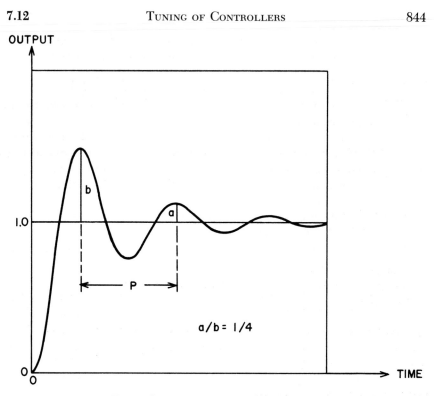

Fig. 7.12b   Response curve for ¼ decay ratio

until the process cycles continuously. To do this, the following steps are recommended.

1. Switch the controller on automatic.
2. Tune all reset and derivative action out of the controller, leaving only the proportional mode, i.e., set $T_d = 0$ and $T_i = \infty$.
3. With the gain at some arbitrary value, impose an upset on the process and observe the response. One easy method for imposing the upset is to move the set point for a few seconds and then return it to its original value.
4. If the resulting response curve does not damp out (as in curve A in Figure 7.12c), the gain is too high (proportional band setting too low). Therefore the proportional band setting is increased and step 3 repeated.
5. If the response curve in step 3 damps out (as in curve C in Figure 7.12c), the gain is too low (proportional band is too high). The proportional band setting is therefore decreased and step 3 repeated.
6. When a response similar to curve B in Figure 7.12c is obtained,

the values of the proportional band setting and the period of the response are noted. (This must not be a limit cycle.)

There are a few exceptions to steps 4 and 5, because in some cases decreasing the gain makes the process more unstable. In these cases the "ultimate" method will not give good settings. Usually in cases of this type

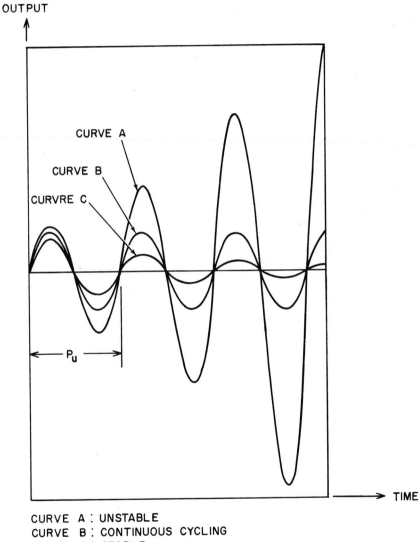

CURVE A : UNSTABLE
CURVE B : CONTINUOUS CYCLING
CURVE C : STABLE

Fig. 7.12c   Typical responses obtained when determining ultimate gain and ultimate period

the system is stable at high and low values of gain but unstable at intermediate values. Thus, the ultimate gain for systems of this type has a different meaning. To use the ultimate method for these cases, the lower value of the ultimate gain is sought.

To use the ultimate gain and the ultimate period to obtain controller settings, Ziegler and Nichols correlated, in the case of the proportional controllers, the decay ratio vs gain in the controller expressed as a fraction of the ultimate gain for several systems. From the results they concluded that a value of gain equal to one-half the ultimate gain would often give a decay ratio of $\frac{1}{4}$. i.e.,

$$K_c = 0.5K_u \qquad (PB = 2PBu) \qquad\qquad 7.12(1)$$

By analogous reasoning and testing, the following equations were found to give reasonably good settings for more complex controllers:

Proportional-plus-reset:

$$K_c = 0.45K_u \qquad (PB = 2.2PBu) \qquad\qquad 7.12(2)$$
$$T_i = P_u/1.2 \qquad\qquad 7.12(3)$$

Proportional-plus-derivative:

$$K_c = 0.6K_u \qquad (PB = 1.65PBu) \qquad\qquad 7.12(4)$$
$$T_d = P_u/8 \qquad\qquad 7.12(5)$$

Three mode (proportional-plus-reset-plus-derivative):

$$K_c = 0.6K_u \qquad (PB = 1.65PBu) \qquad\qquad 7.12(6)$$
$$T_i = 0.5P_u \qquad\qquad 7.12(7)$$
$$T_d = P_u/8 \qquad\qquad 7.12(8)$$

Again it should be noted that the above equations are empirical and exceptions are inherent.

Damped Oscillation Method

A slight modification of the previous procedure has also been proposed by Harriott.[4] For most processes it is not feasible to allow sustained oscillations, and the ultimate method cannot be used. In this modification of the ultimate method the gain (proportional control only) is adjusted, using steps analogous to those used in the ultimate method, until a response curve with a decay ratio of $\frac{1}{4}$ is obtained. However, it is necessary to note only the period P of the response. With this value P, the reset and derivative modes are set as

$$T_d = P/6 \qquad\qquad 7.12(9)$$
$$T_i = P/1.5 \qquad\qquad 7.12(10)$$

After these modes are set, the sensitivity is again adjusted until a response curve with a decay ratio of $\frac{1}{4}$ is obtained. This method usually requires about the same amount of work as the ultimate method, since it is often necessary to experimentally adjust the value of the gain determined from the ultimate method to obtain a decay ratio of $\frac{1}{4}$.

In general, there are two obvious disadvantages to these methods. First, both are essentially trial and error, since several values of gain must be tested before the ultimate gain, or the gain to give a $\frac{1}{4}$ decay ratio, is determined. To make one test, especially at values of gain near the desired, it is often necessary to wait for the completion of several oscillations before it can be determined if the trial value of gain is the desired one. Second, while one loop is being tested in this manner its output may affect several other loops, thus possibly upsetting an entire unit. While all tuning methods require that some change be made in the control loop, other techniques require only one test and not several as in the closed-loop method.

Process Reaction Curve

In contrast to the closed-loop methods, the open-loop technique necessitates only one upset to be imposed on the process. Actually, the controller is not in the loop when the process is tested. Thus, these methods seek to characterize the process, and then determine controller settings from the parameters used to characterize the process. In general, it is not possible to completely characterize a process; hence, approximation techniques are employed. Most of these techniques are based on the process reaction curve which is the response of the process to a unit step change in the manipulated variable, i.e., the output of the controller. To determine the process reaction curve, the following steps are recommended:

1. Let the system come to steady state at the normal load level.
2. Place the controller on manual.
3. Manually set the output of the controller at the value at which it was operating in the automatic mode.
4. Allow the system to reach steady state.
5. With the controller still on manual, impose a step change in the output of the controller, which is the air signal to the valve.
6. Record the response of the controlled variable. Although the response is usually recorded by the controller itself, it is often desirable to have a supplementary recorder or a faster chart drive for the existing controller to insure greater accuracy.
7. Return the controller output to its previous value and return the controller to automatic operation.

It is undoubtedly easier to obtain the process reaction curve than to obtain the ultimate gain.

Most open-loop methods are based on approximating the process reaction curve by a simpler system, and several techniques are available for doing this. By far the most common approximation is that of a pure time delay plus a first-order lag. One reason for the popularity of this approximation is that a real time delay of any duration can only be represented by a pure time delay, because there is no other simple yet adequate approximation. Although it is theoretically possible to use systems higher than first order in conjunction with a pure time delay, the approximations are difficult to obtain accurately. Thus, the real system is usually approximated by a pure time delay plus a first-order lag. This approximation is easy to obtain, and it is sufficiently accurate for most purposes.

Figure 7.12d shows one popular procedure of approximating the process

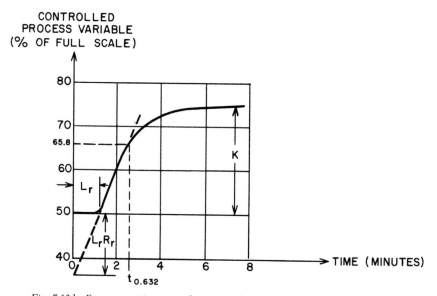

Fig. 7.12d   Process reaction curve for a step change of 1 PSI in controller output

reaction curve by a first-order lag plus time delay. The first step is to draw a straight line tangent to the process reaction curve at its point of maximum rate of ascent (point of inflection). Although this is easy to visualize, it is quite difficult to do in practice. This is one of the main difficulties in this procedure, and a considerable amount of error can be introduced at this point. The slope of this line is termed the reaction rate R_r, and the time at which this line intersects the value of the initial condition from which the process reaction curve originated is the deadtime, or time delay L_r. In Figure 7.12d the determination of these values for a 1-PSI step change in the controller output to a process, is illustrated. If a different step change in controller output were used, the value of L_r would not change significantly. However, the value of R_r is essentially directly proportional to the magnitude of the change in controller output, and therefore if a 2-PSI change in output were used instead of 1 PSI, the value of R_r would be approximately twice as large. For this reason the value of R_r used in the equations to be presented later must be the value that would be obtained for a 1-PSI change in controller output. In addition, the value of the process gain K must be determined as follows:

$$K = \frac{\text{final steady-state change in controlled variable (\%)}}{\text{change in controller output (PSI)}} \qquad 7.12(11)$$

The determination of this value is also illustrated for the process reaction curve in Figure 7.12d.

There is a second good method for determining the pure time delay plus first-order lag approximation. In order to distinguish between these two methods, they will be called Fit 1 (described previously) and Fit 2. The only difference between the two methods is in the first-order lag time constant which is obtained. The pure time delay for both fits is the same as described before and is given by

$$\text{Fit 1 and Fit 2: } t_0 = L_r \qquad 7.12(12)$$

The first-order lag time constants are given by

$$\text{Fit 1: } \tau_{F_1} = K/R_r \qquad 7.12(13)$$

$$\text{Fit 2: } \tau_{F_2} = t_{0.632} - t_0 \qquad 7.12(14)$$

where $t_{0.632}$ is the time necessary to reach 63.2 percent of the final value. Note that the parameters for Fit 1 are based on a single point on the response curve, which is the point of maximum rate of ascent. However, the parameters obtained with Fit 2 are based on two separate points. Studies[5] indicate that the open-loop response based on Fit 2 always provides an approximation to the actual response that is as good or better than the Fit 1 approximation. A typical curve resulting from the above procedure is shown in Figure 7.12d. From the graph, the following parameters are determined directly (response shown is from 1 PSI change in controller output; for different step changes, K and R_r must be adjusted accordingly):

$$L_r = 1.19 \text{ minutes}$$
$$L_r R_r = 14.9\%/\text{PSI}$$
$$K = 25\%/\text{PSI}$$
$$t_{0.632} = 2.58 \text{ minutes}$$

The following parameters can be calculated from those above:

From equation 7.12(12):

$$t_0 = L_r = 1.19 \text{ minutes.}$$

From equation 7.12(13):

$$\tau_{F_1} = KL_r/L_r R_r = 2.00.$$
$$t_0/\tau_{F_1} = \mu = L_r R_r/K = 0.595.$$

From equation 7.12(14):

$$\tau_{F_2} = t_{0.632} - t_0 = 1.39.$$
$$t_0/\tau_{F_2} = 0.857.$$

Open-Loop Tuning

One of the earliest methods using the process reaction curve was that proposed by Ziegler and Nichols[3] in the same article that presented the ultimate method. To use their process reaction curve method only R_r and L_r must be determined. Using these parameters, the empirical equations used to predict controller settings for a decay ratio of $\frac{1}{4}$ are given in Table 7.12e in terms of L_r and R_r and in Table 7.12f in terms of t_0 and τ.

Table 7.12e
EQUATIONS FOR ZIEGLER-NICHOLS AND COHEN-COON

Controller	Ziegler-Nichols	Cohen-Coon
Proportional	$K_c = \dfrac{1}{L_r R_r}$	$K_c = \dfrac{1 + \mu/3}{R_r L_r}$
Proportional + Reset	$K_c = \dfrac{0.9}{L_r R_r}$ $T_i = 3.33 L_r$	$K_c = 0.9\dfrac{1 + \mu/11}{R_r L_r}$ $T_i = 3.33 L_r \dfrac{1 + \mu/11}{1 + 11\mu/5}$
Proportional + Reset + Rate	$K_c = \dfrac{1.2}{L_r R_r}$ $T_i = 2.0 L_r$ $T_d = 0.5 L_r$	$K_c = 1.35\dfrac{1 + \mu/5}{R_r L_r}$ $T_i = 2.5 L_r \dfrac{1 + \mu/5}{1 + 3\mu/5}$ $T_d = \dfrac{0.37 L_r}{1 + \mu/5}$

In developing their equations Ziegler and Nichols considered processes that were not "self-regulating." To illustrate, consider the level control of a tank with a constant rate of liquid removal. Assume that the tank is initially operating so that the level is constant. If a step change is made in the inlet liquid flow, the level in the tank would rise until it overflows. This process is not "self-regulating." On the other hand, if the outlet valve opening and outlet back pressure are constant, the rate of liquid removal increases as the

Table 7.12f
COMPARISON OF ZIEGLER-NICHOLS, COHEN-COON, AND 3C EQUATIONS

Controller	Ziegler-Nichols	Cohen-Coon	3C
Proportional	$KK_c = (t_0/\tau)^{-1.0}$	$KK_c = (t_0/\tau)^{-1.0} + 0.333$	$KK_c = 1.208(t_0/\tau)^{-0.956}$
Proportional + Reset	$KK_c = 0.9(t_0/\tau)^{-1.0}$ $\dfrac{T_i}{\tau} = 3.33(t_0/\tau)$	$KK_c = 0.9(t_0/\tau)^{-1.0} + 0.082$ $\dfrac{T_i}{\tau} = \dfrac{3.33(t_0/\tau)[1 + (t_0/\tau)/11.0]}{1.0 + 2.2(t_0/\tau)}$	$KK_c = 0.928(t_0/\tau)^{-0.946}$ $\dfrac{T_i}{\tau} = 0.928(t_0/\tau)^{0.583}$
Proportional + Reset + Rate	$KK_c = 1.2(t_0/\tau)^{-1.0}$ $\dfrac{T_i}{\tau} = 2.0(t_0/\tau)$ $\dfrac{T_d}{\tau} = 0.5(t_0/\tau)$	$KK_c = 1.35(t_0/\tau)^{-1.0} + 0.270$ $\dfrac{T_i}{\tau} = \dfrac{2.5(t_0/\tau)[1.0 + (t_0/\tau)/5.0]}{1.0 + 0.6(t_0/\tau)}$ $\dfrac{T_d}{\tau} = \dfrac{0.37(t_0/\tau)}{1.0 + 0.2(t_0/\tau)}$	$KK_c = 1.370(t_0/\tau)^{-0.950}$ $\dfrac{T_i}{\tau} = 0.740(t_0/\tau)^{0.738}$ $\dfrac{T_d}{\tau} = 0.365(t_0/\tau)^{0.950}$

liquid level increases. Hence, in this latter case the level in the tank will rise to some new position but would not increase indefinitely, and the system is self-regulating. To account for self-regulation, Cohen and Coon[6] introduced an index of self-regulation μ defined as

$$\mu = R_r L_r / K \qquad\qquad 7.12(15)$$

Note that this term can also be determined from the process reaction curve. For processes originally considered by Ziegler and Nichols, μ equals zero, and therefore there is no self-regulation. To account for variations in μ, Cohen and Coon suggested the equations given in Table 7.12e in terms of L_r and R_r and in Table 7.12f in terms of t_0 and τ.

For the case of proportional control, the requirement that the decay ratio be $\frac{1}{4}$ is sufficient to insure a unique solution, but for the case of proportional-plus-reset control, this restraint is not sufficient to insure a unique solution. Another constraint, in addition to the $\frac{1}{4}$ decay ratio, can be placed on the response to determine unique values of K_c and T_i. Requiring that the control area of the response be a minimum, meaning the area between the response curve and the set point be a minimum, is a possible second constraint. This area is called the error integral or the integral of the error with respect to time.

With the proportional-plus-reset-plus-rate controller the same problem of not having a unique solution exists even when the $\frac{1}{4}$ decay ratio and minimum error integral constraints are applied. Therefore, a third constraint must be chosen to determine a unique solution. A value 0.5 for the dimensionless group $K_c K T_d / \tau$ is one such constraint (based on the work of Cohen and Coon.)[6] The tuning relations which will result from applying these three constraints are given in Table 7.12f. This method has been referred to as the 3C method.[7,8,9]

Integral Criteria in Tuning[10]

Tables 7.12g and h relate the controller settings that minimize the respective integral criteria to the ratio t_0/τ. The settings differ if tuning is based on load (disturbance) changes as opposed to set-point changes. Settings based on load changes will generally be much *tighter* than those based on set-point changes. When loops tuned to load changes are subjected to a set-point change, a more oscillatory response is observed.

The relationship between integral criteria controller settings and the ratio t_0/τ is expressed by the tuning relation given in equation 7.12(16).

$$Y = A \left(\frac{t_0}{\tau} \right)^B \qquad\qquad 7.12(16)$$

where $Y = KK_c$ for proportional mode, τ/T_i for reset mode, T_d/τ for rate mode,

Table 7.12g

TUNING RELATIONS BASED ON INTEGRAL CRITERIA AND
LOAD DISTURBANCE[5,10,11]

Criterion	Controller	Mode	A	B
IAE	Proportional	Proportional	0.902	−0.985
ISE	Proportional	Proportional	1.411	−0.917
ITAE	Proportional	Proportional	0.490	−1.084
IAE	Prop. + Reset	Proportional	0.984	−0.986
		Reset	0.608	−0.707
ISE	Prop. + Reset	Proportional	1.305	−0.959
		Reset	0.492	−0.739
ITAE	Prop. + Reset	Proportional	0.859	−0.977
		Reset	0.674	−0.680
IAE	Prop. + Reset + Rate	Proportional	1.435	−0.921
		Reset	0.878	−0.749
		Rate	0.482	1.137
ISE	Prop. + Reset + Rate	Proportional	1.495	−0.945
		Reset	1.101	−0.771
		Rate	0.560	1.006
ITAE	Prop. + Reset + Rate	Proportional	1.357	−0.947
		Reset	0.842	−0.738
		Rate	0.381	0.995

Table 7.12h

TUNING RELATIONS BASED ON INTEGRAL CRITERIA AND
SET-POINT DISTURBANCE[12]

Criterion	Controller	Mode	A	B
IAE	Prop. + Reset	Proportional	0.758	−0.861
		Reset°	1.02	−0.323
ITAE	Prop. + Reset	Proportional	0.586	−0.916
		Reset°	1.03	−0.165
IAE	Prop. + Reset + Rate	Proportional	1.086	−0.869
		Reset°	0.740	−0.130
		Rate	0.348	0.914
ITAE	Prop. + Reset + Rate	Proportional	0.965	−0.855
		Reset°	0.796	−0.147
		Rate	0.308	0.929

° For this mode, equation 7.12(16) should be of the form $Y = A + B (t_0/\tau)$.

A, B = constant for given controller and mode, and

t_0, τ = pure delay time and first-order lag time constant based on the process reaction curve $t_0 = L_r$.

Digital Control Loops

Digital control loops differ from continuous control loops in that the continuous controller is replaced by a sampler, a discrete control algorithm calculated by the computer, and by a hold device (usually a zero-order hold). In such cases, Moore et al.[11] have shown that the open-loop tuning methods presented previously may be used, provided the deadtime used is the sum of the true process deadtime and one-half the sampling time, as expressed by equation 7.12(17);

$$t_0' = t_0 + T/2 \qquad\qquad 7.12(17)$$

where T is the sampling time and t_0' is used in the tuning relationships instead of t_0. (Section 7.13 is devoted to the subject of controller tuning by computer.)

Summary

In this section several techniques for estimating controller settings have been presented. Superficially, the best way to present this subject would seem to be to discuss only the best method of tuning controllers in more detail, but there is no general agreement as to which method is best. Some methods lean heavily on experience while others rely more on mathematical considerations. Although the methods presented first attempt to yield optimum settings, the only criterion stated is that the response have a decay ratio of $\frac{1}{4}$. This has been shown to be insufficient to obtain a unique combination of settings for a controller with more than one mode. When tuning a controller, one should be aware of this possibility, and remember that although the response has a decay ratio of $\frac{1}{4}$, the controller may still not be at its optimum settings.

Although some difficulties encountered in presenting methods for tuning controllers have been mentioned, a few others will be discussed here. First, industrial controllers are not ideal. Furthermore, the non-idealities differ from one commercial model to another, and this makes empirical compensation for these effects difficult. Although it has not been mentioned, the process has been assumed to be linear. This assumption is reasonable over limited ranges, but usually not for the entire span of the controller. Thus, the process-reaction curve or the ultimate gain should be determined at or near the operating level at which the system is to perform.

In a few processes the non-linear effects are more pronounced. For example, the controller may be adjusted to give a $\frac{1}{4}$ decay ratio for an increase in the set point, but when the set point is decreased, the response may look entirely different. This phenomenon is a consequence of the non-linearities, and it may be impossible to adjust the controller to obtain the desired response in both directions. In these cases, the engineer must compromise between the optimum settings for each direction of change.

This general question of non-linearities, or "non-symmetrical" systems, is perhaps one of the most frustrating aspects of controller tuning. It is manifest in many day-to-day situations in which a process is operating satisfactorily with all loops well tuned, and then a "load change" is experienced and all the controllers must be retuned. The extent to which this problem is present can be determined in advance and experimentally by running *two* process reaction curves—one for a "plus" step input and the other for a "minus" step input. If the two process reaction curves are mirror images of one another, the process is symmetrical and no variation in tuning parameters should be expected in and about the operating level from which the step inputs were made. If the two process reaction curves are non-symmetrical, meaning that they are not mirror images, the process is non-linear, and the tuning parameters cannot be expected to be valid over a wide range. With conventional control hardware the only alternative to retuning is to use very conservative tuning parameters. This implies taking the "most sluggish" process reaction curve and using it to determine the tuning parameters.

When a process is non-linear or non-symmetrical, what is really needed is automatic adaptive tuning (Section 7.7). This implies that the control system will "sense its state" and automatically retune itself. There are automatic hardware systems for doing this, particularly for the controller gain, but they are not justified in the great majority of cases. With the increased usage of digital control there is a trend toward software techniques for adaptive tuning.

Due to the assumptions required to develop the various techniques for tuning controllers, the results using any of these methods are approximations of the optimum combination. In every case, the results are much better than the average engineer can estimate without these techniques. In most cases the results obtained with these methods are very satisfactory, but in a few cases subsequent adjustments are necessary. At this stage the question of how to define "good" control often becomes the main issue, and the decision is usually a qualitative judgment made by the instrument engineer.

In making subsequent adjustments, one must understand the effect of each mode upon the overall response. Thus, the following generalities about the effects of adjustments for each mode are pertinent.

Adjustment of the proportional band: Decreasing the proportional band (increasing the gain) increases the decay ratio, thus making the system less stable. However, the frequency of the response is also increased, which is usually desirable. Increasing the proportional band has an opposite effect.

Adjustment of the reset mode: When the reset time is increased, the decay ratio is decreased, thus making the system more stable. Simultaneously, the frequency increases. Decreasing the reset time has an opposite effect. One should recall that when the reset time is at its maximum value, this mode has been tuned out of the controller.

Adjustment of the derivative mode: Of all the modes, the effect of this mode is the most difficult to predict. Starting at a derivative time of zero, increasing the derivative time usually is beneficial, but not always. In almost all practical cases there is a point beyond which increasing the derivative time will prove detrimental. Thus, about all that one can do is try a change in the derivative time and see what happens.

In this section several methods have been presented for tuning automatic controllers. Although each attempts to give the optimum combination, the answers from each method are different, and the question arises as to which method is best. Since fewer assumptions are involved and the basic mathematics is more sound, the settings determined from the integral criteria are closer to the optimum combination. No matter which method is used, the controller will invariably be tuned closer to its optimum combination of settings in much less time than by the "guessing" which is often done.

Example

For the example shown in Figure 7.12d, the model parameters are

$$L_r = 1.19 \text{ minutes} \qquad K = 25\%/\text{PSI}$$
$$R_r = 12.5\%/\text{PSI minute} \qquad \tau = 1.39 \text{ minute (using Fit 2)}$$
$$\mu = 0.595 \qquad t_0 = 1.19 \text{ minutes}$$
$$t_0/\tau = 0.857$$

For a PI controller, the settings predicted by the various tuning techniques are

Ziegler-Nichols (Table 7.12e)

$$K_c = 0.9/(14.9\%/\text{PSI}) = 0.0605 \text{ PSI}/\%$$
$$T_i = (3.33)(1.19 \text{ min}) = 3.96 \text{ minutes/repeat}$$

Cohen-Coon (Table 7.12e)

$$K_c = 0.9(1 + 0.595/11)/14.9\%/\text{PSI} = 0.0638 \text{ PSI}/\%$$
$$T_i = \frac{3.33(1.19 \text{ min})(1 + 0.595/11)}{1 + (11)(0.595)/5} = 1.81 \text{ minutes/repeat}$$

3C (Table 7.12f)

$$K_c = \frac{0.928(0.857)^{-0.946}}{25\%/\text{PSI}} = 0.0430 \text{ PSI}/\%$$
$$T_i = (1.39 \text{ minutes})(0.928)(0.857)^{0.583} = 1.18 \text{ minutes/repeat}$$

IAE criterion and load disturbance (Table 7.12g)

$$K_c = \frac{(0.984)(0.857)^{-0.986}}{25\%/PSI} = 0.0459 \text{ PSI}/\%$$

$$T_i = \frac{1.39 \text{ minutes}}{(0.608)(0.857)^{-0.707}} = 2.05 \text{ minutes/repeat}$$

IAE criterion and set-point disturbance (Table 7.12h)

$$K_c = \frac{(0.758)(0.857)^{-0.861}}{25\%/PSI} = 0.0347 \text{ PSI}/\%$$

$$T_i = \frac{1.39 \text{ minutes}}{1.02 - (0.323)(0.857)} = 1.88 \text{ minutes/repeat}$$

Note: On some controllers the reset mode is calibrated in repeats/minute units. In those cases use the $1/T_i$ values for tuning. The relationship between K_c and proportional band is

$$PB = 100/K_c \qquad\qquad 7.12(18)$$

Nomenclature

    (Units are given for example in Figure 7.12d)

$e(t)$	Error signal (percent of scale reading)
K_c	Proportional gain of the controller (PSI/% of scale)
K	Process gain (percent of scale/PSI)
L_r Deadtime,	defined in Figure 7.12d (minutes); $(L_r = t_0)$
P	Period (minutes)
PB	Proportional band $(100/K_c)$
PB_u	Ultimate proportional band $(1/K_u)$
P_u	Ultimate period (minutes)
R_r Reaction rate,	defined in Figure 7.12d (percent of scale/PSI minute)
K_u	Ultimate sensitivity (PSI/percent of scale)
τ First-order lag	time constant (minutes)
T	Sampling time (minutes)
T_i	Reset time (minutes/repeat)
T_d	Derivative time (minutes)
t	Time (minutes)
μ Index of self regulation	defined in Figure 7.12d, dimensionless
t_0 Deadtime, or pure	delay time (minutes); $t_0 = L_r$
t_0'	$t_0 + T/2$ (minutes)
$t_{0.632}$	Time to achieve 63.2 percent of response (minutes)

REFERENCES

1. Murrill, P. W., *Automatic Control of Processes*, Ch. 17, International Textbook Co., Scranton, Pennsylvania, 1967.
2. Caldwell, W. I., G. A. Coon, and L. M. Zoss, *Frequency Response for Process Control*, McGraw-Hill Book Company, New York, 1959.
3. Ziegler, J. G., and N. B. Nichols, "Optimum Settings for Automatic Controllers," Trans. ASME, Vol. 64 (1942), pp. 759–765.
4. Harriott, P., *Process Control*, McGraw-Hill Book Company, New York, 1964.
5. Miller, J. A., A. M. Lopez, C. L. Smith, and P. W. Murrill, "A Comparison of Open-Loop Techniques for Tuning Controllers," Control Engineering, Vol. 14, No. 12 (December 1967), pp. 72–76.
6. Cohen, G. H., and G. A. Coon, "Theoretical Considerations of Retarded Control," Taylor Instrument Companies Bulletin #TDS-10A102.
7. Smith, C. L., and P. W. Murrill, "A More Precise Method for the Tuning of Controllers," ISA J. (May 1966).
8. Smith, C. L., and P. W. Murrill, "An Analytic Technique for Tuning Underdamped Control Systems," ISA J. (September 1966).
9. Smith, C. L. and P. W. Murrill, "Controllers—Set Them Right," Hydrocarbon Processing and Petroleum Refiner, Vol. 45, No. 2 (February 1966), p. 105.
10. Lopez, A. M., J. A. Miller, C. L. Smith, and P. W. Murrill, "Controller Tuning Relationships Based on Integral Performance Criteria," Instrumentation Technology, Vol. 14, No. 11 (November 1967), pp. 57–62.
11. Moore, C. F., et al., "Simplification of Digital Control Dynamics for Tuning and Hardware Lag Effects," Instrument Practice, January 1969, pp. 45–49.
12. Rovira, A. A., Ph.D. dissertation, Louisiana State University, Baton Rouge, Louisiana (1969).

ACKNOWLEDGMENT

The authors wish to express their appreciation for the support of this work by a Project THEMIS contract administered by the U.S. Air Force Office of Scientific Research for the U.S. Department of Defense.

7.13 CONTROLLER TUNING BY COMPUTER

Controller tuning is a job that industrial management usually places in the hands of instrument engineers, process engineers or instrument technicians. Experience has shown that an understanding of feedback control practice and principles leads to the successful adjustment of these industrial instruments.

In practice, the application of feedback control to industrial processes has also required the participation of process operators. In some industries, company policy imposes "hands off" rules on process engineers, and any manipulation of the process required for instrumentation selection and adjustment must be performed only by qualified process operators. In other industries (usually involving potentially explosive units such as boilers), legal codes prohibit anyone but a specially licensed operator from manipulating the process. Generally, the only men able to exercise on-the-spot judgment for maintaining process operation are the operators.

When process variables are under feedback control, process operation is physically coupled to the design and adjustment of the controllers. Systems with continuous process instrumentation or under direct digital control are presently designed to provide process operators with only limited access to control functions. The operator may alternate between manual and automatic control; in the manual mode he has command of the manipulated variable, and in the automatic mode he has command of the set point.

The division of labor between operators and instrument engineers is reflected in the design of many instruments. For example, the operator functions (set point, mode transfer, output adjustments) are physically separated from the tuning adjustments (proportional band, reset time, derivative time). Present computer systems generally provide for this same division between operator and tuning adjustments. Since only thousands of ddc loops are now implemented as opposed to millions of continuous loops, it is too early to know if these similarities will persist.

An inherent difficulty in this division of labor is that two men must do what one could if it were not for a gap in skills. A reasonable way to reduce this gap is to provide the operators with the tools required for adjusting controllers, rather than training and licensing instrument men to operate processes. Functionally, this is a practical approach because the operators maintain 24-hour surveillance of the process, while instrument men usually monitor the unit periodically on the day shift only.

The skills gap can be overcome if the operator is allowed access to scientific computing functions by means of a time-shared control computer, scientific computer, or a terminal. The instrumentation and engineering departments can define the control strategies for any particular set of process dynamics and performance requirements for storage in the computer. The operators can then approach the computer with data and, in return, receive appropriate adjustments.

The intention is not to reduce the instrument engineer's participation in defining control strategy, but rather to relieve him of a tedious adjustment task. In current industrial practice, qualified instrument people initially tune controllers when a plant goes on line. As time progresses and the plant experiences operation changes, these controllers are either detuned to prevent cycling, or merely switched to a manual operating mode until an instrument engineer can be called. Providing operators with simple testing techniques and with access to the computer can avoid these remedial solutions and prevent the inherent operating losses that result from having detuned or open-loop controllers.

A few brief words are given here about the practical details of preparing for the tuning procedure. Care should be taken to insure that unit conversions from electrical to computing to engineering units are correct. When closed-loop cycling tests are run, the process operator should pulse the controller set point to avoid hanging the loop in hysteresis and not displaying its linear damping. If these precautions are exercised and the recommended procedure is followed, the technique is self-checking.

A Practical Process Model

The effectiveness of any control system depends largely on how well the process under control has been mathematically modelled. Some control designs are less sensitive to model error than others. However, as control performance requirements increase, so does the requirement for model accuracy.

Mathematical modeling of a process is generally a two-step operation: defining a model structure and finding the parameters in the model. How can process operators experimentally perform this second step with minimal procedural requirements and, at the same time, still be in control of the test and unit safety? The technique that follows assumes that the operator has access to a scientific computer through a terminal or operator's console. Solutions are performed in a manner that automatically checks the validity of the process model and the accuracy of the operator's data.

Most continuous industrial processes, when viewed as a single loop from manipulated to controlled variable, consist of a series of time lags associated with loop elements such as valves, heaters, terminal capacitance and instrument delay. Occasionally, these processes also involve transport delay (pure deadtime), resulting, for example, from plug flow in pipes. Unlike nuclear reactors and modern missiles, these processes are usually stable without feedback (open loop).

A general mathematical structure for these processes that is difficult to improve upon is the first-order model with deadtime given in equation 7.13(1). This still popular and traditional model, which has been used since at least the early 1940s, has three parameters: gain, time constant and deadtime. (For a detailed discussion of this relationship refer to equation 7.2(23)).

$$\frac{dc(t)}{dt} = \frac{1}{\tau}[Km(t - t_0) - c(t)] \qquad\qquad 7.13(1)$$

where t = variable time,

$\dfrac{dc(t)}{dt}$ = derivative with respect to time
    of the controlled variable,

c(t) = controlled or measured process variable,
    process output response,

m = manipulated process variable,
    a function of time and deadtime,

K = process gain,

τ = process time constant, and

$t_0 = L_r$ = apparent process deadtime.

This mathematical structure has the advantage of constrained dimensionality. Deadtime in the model conveniently lumps the higher order process delays and the actual process deadtime into one parameter.

Why not try discrete model structures? The main reason, in this case, is that engineering judgment is required for successful implementation. The problem with the discrete approach centers on the questions: how often and for how long should data be taken? No clearly defined rules are available which would allow the application of these approaches by people with skill levels lower than that of an engineer.

Also, in the discrete approach, the complexity of the model increases with the ratio of apparent deadtime to sample rate. If a 10-percent resolution is desired in the accuracy defining the deadtime, then the number of parameters in the model will be greater than 10. Requirements on time resolution accuracy for continuous models do not impose this increase in model dimensionality. Model parameters all require storage, and programming to service this storage, in the computer. Constraining dimensions therefore saves money.

Fitting the Model to the Process

The two most popular tests for analytically adjusting controllers are the step response test and the ultimate sensitivity test. (For their detailed description refer to Section 7.12.) The ultimate sensitivity test provides a closed-loop estimate of process performance. However, it has limitations when fitting the classical three-parameter process model because only two parameters can be found, ultimate gain and period. This approach is widely used because the knowledge of closed-loop ultimate gain in the testing mode is usually adequate to arrive at stable closed-loop settings for the controller.

The conventional step response test, Figure 7.13a, though less frequently used in industrial practice, is currently the most popular method for explicitly fitting the three-term model. For process loops that will readily achieve new

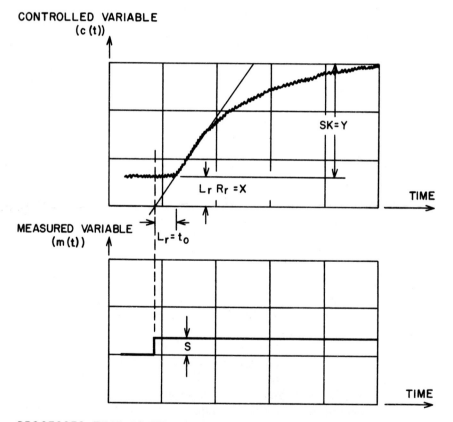

PROCESSES THAT DRIFT DURING THE CONVENTIONAL STEP RESPONSE TEST WILL INVALIDATE THE MEASURED PARAMETERS X, Y AND L_r.

Fig. 7.13a   First-order response

final values (their time constant τ is small), and for processes that do not appreciably drift during the test, this method works well. However, if gain K is not known from experience (it usually is not), this technique does have the disadvantage that step size for a satisfactory implementation must be selected by trial and error.

To fit the three-parameter model using this test, the following measurements should manually or automatically be obtained: manipulated variable step size (S), apparent deadtime ($L_r = t_0$), total computed change in the controlled variable ($X = L_r R_r$) at the maximum rate of change (R_r) over the period of the apparent deadtime ($L_r = t_0$), and the final value of the controlled variable's response ($Y = K$). Then

$$R_r = \frac{X}{SL_r} \qquad \qquad 7.13(2)$$

$$K = \frac{Y}{S} \qquad \qquad 7.13(3)$$

$$\tau = \frac{K}{R_r} \qquad \qquad 7.13(4)$$

Ordinarily, the conventional step response test is not practical in loops with extremely large time constants. Furthermore, it does not provide a check on the testing accuracy itself unless it is repeated. Even if the test is repeated, errors in unit conversion are not always uncovered.

A combination of a modified step response and an ultimate sensitivity test, overcomes the above limitations (Figure 7.13b). When performing the modified step response, the operator retains the step just long enough to be certain that the process response reaches the maximum rate of change. At this time he returns the manipulated variable to its initial condition. Because the controlled variable does not have to achieve a new steady state, selection of step size S is not critical. This technique should not be confused with pulse testing, which requires analysis of the total pulse response curve. The maximum rate R_r still yields to equation 7.13(2).

The ultimate sensitivity test calls for increasing the gain of a proportional controller to K_u until the process cycles without convergence or divergence. The following equations govern the stability of the system:

$$\frac{K_u K}{\sqrt{\left(\dfrac{2\pi\tau}{P_u}\right)^2 + 1}} = 1 \qquad \qquad 7.13(5)$$

$$\tan^{-1}\left(\frac{2\pi\tau}{P_u}\right) + \frac{2\pi L_r}{P_u} = \pi \qquad \qquad 7.13(6)$$

where K_u is the ultimate proportional gain and P_u is the cycle period. These equations assume continuous proportional control, or sampling at such a rate

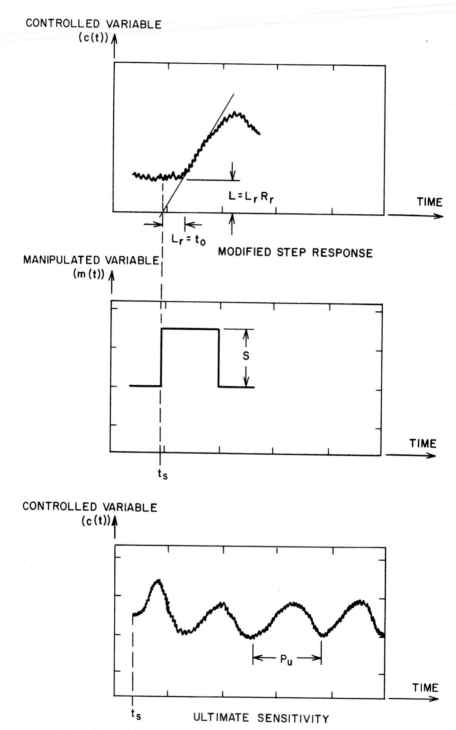

Fig. 7.13b   Modified step response test combined with ultimate sensitivity test

that it is apparently continuous with respect to either L_r or τ. Reduction of the equations to determine the process time constant and gain gives

$$\tau = \frac{P_u}{2\pi}\left[\tan\left(\pi - \frac{2\pi L_r}{P_u}\right)\right] \qquad 7.13(7)$$

$$K = \frac{1}{K_u}\sqrt{\left(\frac{2\pi\tau}{P_u}\right)^2 + 1} \qquad 7.13(8)$$

With τ calculated from equation 7.13(7) and gain from equation 7.13(8) the equation 7.13(4) gives a cross check on the accuracy of the test and on the validity of the model structure.

Another equation describing process gain may be readily obtained by discrete cycling of the loop with sample rate equal to the deadtime ($L_r = t_0$). If the gain required for discrete ultimate sensitivity is K_{ud}, then, in accordance with equation 7.4(19), the conditions of stability are defined by

$$K_{ud}K[1 - e^{-t_0/\tau}] = 1 \qquad 7.13(9)$$

Process gain can be computed from this test by

$$K = \frac{1}{K_{ud}}\frac{e^{t_0/\tau}}{e^{t_0/\tau} - 1} \qquad 7.13(10)$$

The amplitude of cycling for the discrete test will be different from that of the continuous test, giving an additional check for linearity of gain.

To implement these modeling procedures, the operator follows the testing sequence shown in Figure 7.13c. The degree of automatic operation possible in this sequence is strictly a function of available hardware and software.

If a valid model cannot be demonstrated, the instrument engineer should tune a controller by trial and error, or, if a more scientific approach is desired, then more general model structures and identification approaches should be tried. If gain is quite non-linear with operating conditions, frequent adjustment or on-line adaptation should be considered.

Adjusting the Controller

For those processes that fit the first-order-plus-deadtime model, a variety of approaches are described in the literature for computing adjustment coefficients. The techniques developed have been found to work well in practice with conventional controllers. A set of tables is provided for tuning controllers that relate manipulated variables M(s), with s being the Laplace operator, to process error E(s), by the following transfer functions (as derived in equations 7.4(76), 7.5(10), 7.5(12) and 7.5(14)):

Model(s)	Transfer Function	
P	$M(s) = K_c E(s)$	7.13(11)
PI	$M(s) = K_c(1 + 1/T_i s)E(s)$	7.13(12)
PID	$M(s) = K_c(1 + 1/T_i s + T_d s)E(s)$	7.13(13)

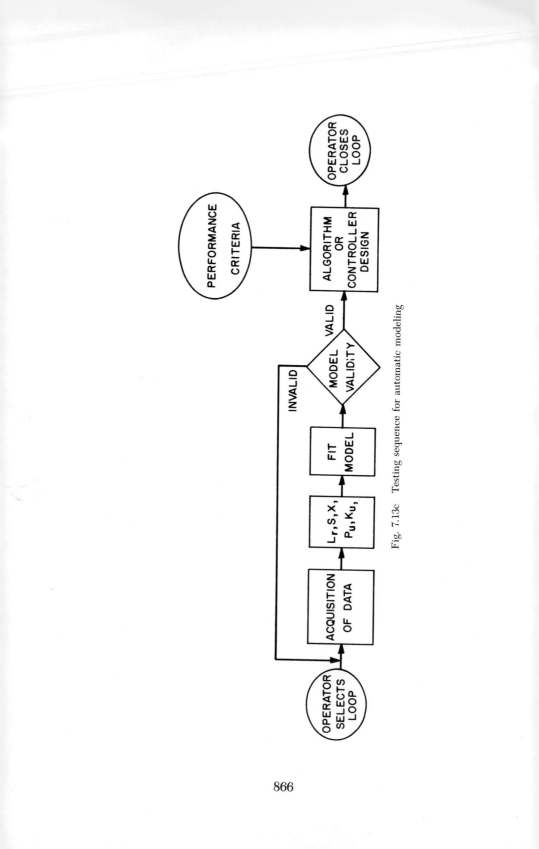

Fig. 7.13c   Testing sequence for automatic modeling

The tables provide tuning for a variety of tuning objectives that are summarized in Table 7.13d. This summary is an elaboration of the contents of Table 7.12a presented earlier.

Tuning that will minimize the performance object we selected from Table 7.13d may be obtained for any standard one- through three-mode controller by consulting (or pre-programming) the formulas given in Tables 7.13e, f and g, which in their content are similar to Tables 7.12g and h.

Table 7.13d

CONTROLLER TUNING CRITERIA

FOR PROPORTIONAL CONTROL SYSTEMS[*]

$$\text{ISE} - 1 = \int_0^\infty [c(t) - c(\infty)]^2 \, dt$$

$$\text{ISE} - 2 = \int_{\theta_0}^\infty [c(t) - c(\infty)]^2 \, dt$$

$$\text{ISE} - 3 = \int_0^\infty \left[\frac{c(t) - c(\infty)}{c(\infty)} \right]^2 \, dt$$

$$\text{IAE} - 1 = \int_0^\infty |c(t) - c(\infty)| \, dt$$

$$\text{IAE} - 2 = \int_{\theta_0}^\infty |c(t) - c(\infty)| \, dt$$

$$\text{IAE} - 3 = \int_0^\infty \left| \frac{c(t) - c(\infty)}{c(\infty)} \right| \, dt$$

$$\text{ITAE} - 1 = \int_0^\infty |c(t) - c(\infty)| \, t \, dt$$

$$\text{ITAE} - 2 = \int_{\theta_0}^\infty |c(t) - c(\infty)| \, t \, dt$$

$$\text{ITAE} - 3 = \int_0^\infty \left| \frac{c(t) - c(\infty)}{c(t)} \right| \, t \, dt$$

[*] Table 7.13d is provided by A. M. Lopez, J. A. Miller, C. L. Smith and P. W. Murrill from their paper on page 57 of the November 1967 issue of Instrument Technology.

Sample Problem

The previously outlined tuning method is applied here, as a sample problem, to a 1,000 lbm/hr pilot plant distillation unit under the direct digital control of a time-shared computer. A process operator in cooperation with a computer tunes the main temperature control loop of this unit (Figure 7.13h).

Figure 7.13i shows the modified step response and the ultimate sensitivity test data. In this example, process gain computed from equation 7.13(8) is 74 percent of that computed from equation 7.13(4), which is considered to be adequate verification.

Table 7.13e
TUNING EQUATIONS FOR PROPORTIONAL CONTROL BASED ON LOAD DISTURBANCE

$$K_c = \frac{A}{K_u}\left(\frac{t_0}{\tau}\right)^B \qquad\qquad 7.13(14)$$

Criterion	Constants	
	A	B
Ultimate	2.133	-0.877
1/4 Decay	1.235	-0.924
ISE—1	1.411	-0.917
ISE—2	0.9889	-0.993
ISE—3	0.6659	-1.027
IAE—1	0.9023	-0.985
IAE—2	0.6191	-1.067
IAE—3	0.4373	-1.098
ITAE—1	0.4897	-1.085
ITAE—2	0.4420	-1.108
ITAE—3	0.3620	-1.119

Table 7.13f
TUNING EQUATIONS FOR PI CONTROLLERS BASED ON LOAD DISTURBANCE

$$K_c = \frac{A}{K_u}\left(\frac{t_0}{\tau}\right)^B \qquad\qquad 7.13(14)$$

$$\frac{1}{T_i} = \frac{A}{\tau}\left(\frac{t_0}{\tau}\right)^B \qquad\qquad 7.13(15)$$

Criterion	Controller Mode	Constants	
		A	B
ISE	Proportional	1.305	-0.960
	Reset	0.492	-0.739
IAE	Proportional	0.984	-0.986
	Reset	0.608	-0.707
ITAE	Proportional	0.859	-0.977
	Reset	0.674	-0.680

Table 7.13g
TUNING EQUATIONS FOR PID CONTROLLERS BASED ON LOAD DISTURBANCE

$$K_c = \frac{A}{K_u}\left(\frac{t_0}{\tau}\right)^B \qquad\qquad 7.13(14)$$

$$\frac{1}{T_i} = \frac{A}{\tau}\left(\frac{t_0}{\tau}\right)^B \qquad\qquad 7.13(15)$$

$$T_d = \tau A\left(\frac{t_0}{\tau}\right)^B \qquad\qquad 7.13(16)$$

Criterion	Controller Mode	Constants	
		A	B
ISE	Proportional	1.495	−0.945
	Reset	1.101	−0.771
	Rate	0.560	1.006
IAE	Proportional	1.435	−0.921
	Reset	0.878	−0.749
	Rate	0.482	1.137
ITAE	Proportional	1.357	−0.947
	Reset	0.842	−0.738
	Rate	0.381	0.995

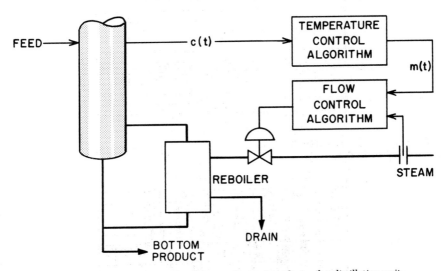

Fig. 7.13h   Operator tuning techniques demonstrated on pilot distillation unit

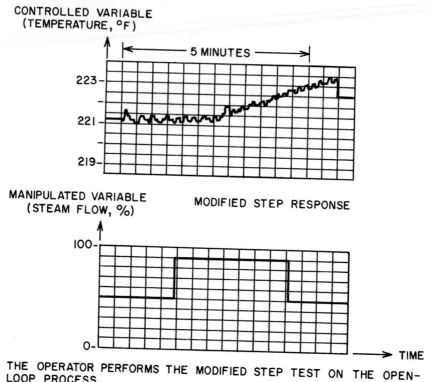

CONTROLLED VARIABLE
(TEMPERATURE, °F)

5 MINUTES

223 —

221 —

219 —

MANIPULATED VARIABLE
(STEAM FLOW, %)

MODIFIED STEP RESPONSE

100 —

0 —

TIME

THE OPERATOR PERFORMS THE MODIFIED STEP TEST ON THE OPEN-
LOOP PROCESS.

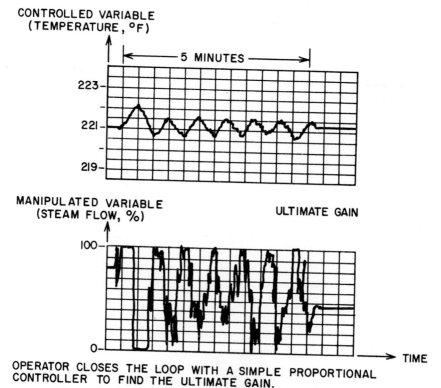

CONTROLLED VARIABLE
(TEMPERATURE, °F)

5 MINUTES

223 —

221 —

219 —

MANIPULATED VARIABLE
(STEAM FLOW, %)

ULTIMATE GAIN

100 —

0 —

TIME

OPERATOR CLOSES THE LOOP WITH A SIMPLE PROPORTIONAL
CONTROLLER TO FIND THE ULTIMATE GAIN.

Fig. 7.13i Controller tuning performed by process operator in cooperation with a computer

In Figure 7.13j a PI ddc algorithm closes the loop, using the higher gain value for the process model. The sampling period is one second, essentially removing this parameter from the design. Controller settings, computed from Table 7.13f, are chosen to minimize the integral time multiplied by absolute error (ITAE criterion).

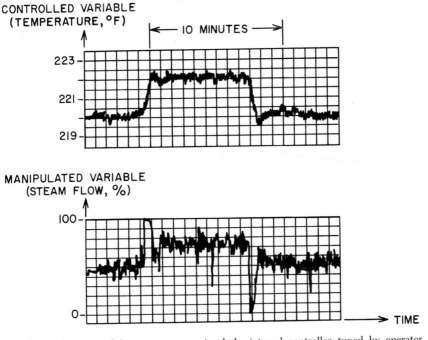

Fig. 7.13j   Response of loop to a proportional-plus-integral controller tuned by operator and computer

The operator is able to tune the loop with relative ease, since all he has to do is gather data. The computer calculates the model parameters, verifies the operator's tests and supplies tuning adjustments for the controller.

Conclusions

Fundamental commodities traded in the selection and adjustment of control modes are performance, safety and cost. Controllers and ddc systems designed to satisfy sophisticated mathematical criteria are generally complex. Instrument engineers must provide reasonable stability margins with these systems to insure process safety. Both time and complexity are directly convertible to cost.

Relatively simple control structures can be implemented by downgrading performance objectives. Also, operating these simple controllers at low sensitivity gives wide margins of safety at the expense of performance. The general practice of feedback regulation favors the conservative approach by accepting performance losses in trade for lower instrument cost and for a wider safety margin.

With the technique described above, the operator can check model accuracy periodically or when operating conditions change. Up-to-date modeling and tuning should greatly improve process performance. The technique maintains process safety by allowing the operators to remain in the loop during the critical testing phase.

Cost reductions are realized by reducing the amount of trial-and-error tuning. The operator need not bring the process to a final value in the step response test. This saves time when dealing with slow processes. Test signal size is not critical, reducing trial-and-error selection.

Time-shared systems with ddc offer a convenient vehicle for implementation of operator tuning. The ddc computer has greater scaling and sensitivity capabilities than conventional industrial instruments. Also, it can perform unit conversion, and can be easily programmed to abort the tests automatically and set off alarms if necessary.

ACKNOWLEDGMENT

The material contained in this section has been presented by the author at the IFAC/IFIP Symposium on Digital Control of Large Industrial Systems, June 17–19, 1968, Toronto, Canada and later published by the ISA in the September 1968 issue of the magazine Instrument Technology.

Chapter VIII

PROCESS COMPUTERS

R. M. BAKKE, J. W. BERNARD,
R. F. JAKUBIK, V. A. KAISER,
C. H. KIM, B. G. LIPTÁK,
W. L. SKAGGS and M. D. WEISS

CONTENTS OF CHAPTER VIII

	INTRODUCTION	879
8.1	COMPUTER TERMINOLOGY	882
8.2	PROCESS COMPUTER INSTALLATIONS	900
	Chemical	900
	Petrochemical	901
	Petroleum	902
	Glass	904
	Rubber	905
	Manufacturing	906
8.3	PLANNING OF COMPUTER PROJECTS	907
	Feasibility and Scope Study	907
	Detailed Process Analysis	909
	Programming	910
	Software Checkout	911
	Installation and Evaluation	911
	Operator Training	912
	Documentation	913
8.4	COMPUTER LANGUAGES	915
	Hardware vs Software	915
	Machine Language	917
	Binary Numbers	917
	Bits, Bytes and Words	918
	Machine Language Instructions	919
	Assembly Language	922
	Higher Level Languages	923
	FORTRAN	925
	Control Languages	926
	Control Software	926
8.5	CONTROL ALGORITHMS	931
	Position Algorithms	931
	Velocity Algorithms	932
	Feedforward Algorithms	933

	Series Algorithms	934
	Optimization of Algorithm Use	935
8.6	Signal Conditioning and Desirable Wiring Practices	937
	Types and Sources of Noise	938
	Reducing Electrical Noise Interference	941
	Circuit Arrangement	941
	Impedance Level	943
	Ground Systems	945
	Wiring	948
	Filtering	950
	Amplifier Guard	952
	Power Transformer Guard	953
	Low-Level Signal Multiplexing	953
	Noise Rejection in A/D Converters	956
	Common-Mode Rejection Measurement	957
	Conclusions	960
8.7	Computer Interface Hardware	961
	Digital Input	962
	Digital Outputs	963
	Analog Outputs	964
	Analog Inputs	967
	Multiplexing of Analog Signals	970
	Signal Conditioning	975
	Summary	977
8.8	Set-Point Stations	978
	Features	979
	Fail-Safe	979
	Stepping and Synchronous Motors	982
	Integrating Amplifier	982
	Speed, Resolution and Feedback	982
	Summary	984
8.9	Memory Devices	985
	Main Memories	986
	Ferrite Core Memory, Operating Principle	986
	Ferrite Core Memory, Organization	988
	Ferrite Core Memory, Cost	993
	Planar Film	993
	Cylindrical Film	995
	Integrated Circuit Memories	996
	Auxiliary Memory Devices	997
	Drum Memories	997
	Disc Memories	998

	Bulk Memories	1000
	Removable Discs	1000
	Magnetic Tape Memories	1001
	Read-Only Memories	1004
	Other Memory Techniques	1005
8.10	Data Logging and Supervisory Program	1006
	System Identification Numbers	1007
	Entry Types and Uses	1008
	Variable Information Sheet	1009
	Purpose	1009
	Variable Processing	1009
	Variable Identification Number	1011
	Variable Values	1011
	Variable Information Sheet Organization	1012
	General Action Sheet	1012
	Purpose	1012
	General Action	1012
	General Equation Sheet	1014
	Purpose	1014
	General Equation	1015
	Adjustment Information Sheet	1016
	Purpose	1016
	Adjustment Equation	1018
	Adjustment Requests	1018
	Adjustment Stack Servicing	1018
	Example of Stack Servicing	1020
	Request Table Processing	1021
	Simultaneous Equations	1022
	Adjustment Information Sheet Organization	1023
	Sample Problems	1023
	Mass Flow Calculation Problem	1024
	Logic Problem	1024
	Furnace Control Problem	1027
	Measurement Information	1027
	Alarm Information	1030
	Control Information	1030
	Data Sheets	1031
8.11	Optimizing Program	1041
	Program Description	1042
	Solution Method	1042
	Detection of an Optimal Solution	1044
	Primary Convergence Technique	1045

Secondary Convergence Technique   1046
Selection of Convergence Technique   1046
Methods of Resolving Infeasibilities   1047
Program Features   1048
    Multiunit Capability   1048
    Multiple Sets of Limits   1048
    Adaptive Linearity Limits   1049
    Move Penalties   1050
    Target Constraints   1050
    Dead Zones   1051
    Curtaining Feature   1051
    Linear Programming Algorithm and Error Check   1052
Using the Program   1052
    Model   1053
    Input   1054
    Output   1055
Mathematical Methods   1055
    Problem Statement   1055
    Problem Solution   1056
    Partial Derivatives   1057
    Adaptive Linearity Limits   1057
    Most Constraining Limits and Move Penalties   1058
Sample Problem   1058
8.12   DIRECT DIGITAL CONTROL   1069
Definitions   1069
Operation of ddc   1070
System Components   1072
    Instrument Hardware   1072
    Computer Hardware   1073
    Auxiliary Hardware   1077
    Input-Output Hardware   1077
    Backup Hardware   1078
Analysis and Programming   1078
Problems With ddc   1079
Benefits of ddc   1079
Implementation of ddc   1081
Conclusions   1084
8.13   OPERATOR COMMUNICATION   1086
Computer-Manual (C/M) Stations   1086
Computer-Manual-Automatic (C/M/A) Stations   1088
Operator's Console   1091
    Keys and Displays   1091

	Operating Sequence	1094
	Variations, Modifications	1095
8.14	HIERARCHICAL COMPUTERS	1097
	The Hierarchical Concept	1097
	The Hierarchical Structure	1098
	Parallel Cascade Processing	1104
	Conclusions	1107
8.15	APPLICATION OF PROCESS COMPUTERS	1109
	System Design	1109
	Decomposing the Process	1112
	Process Control System Levels	1116
	Analog vs Digital	1120
	System Availability	1121
	Control Strategies	1123
	Man-Process Communication	1124
	Measurement Methods	1126
	Implementation	1127
	Operation	1131
	Management	1132
	Conclusions	1136

This chapter is devoted to the subject of digital process computers and the next one discusses the analog computers. Figure 8.0 illustrates the organization of the family of computers. The following sections provide an in-depth coverage of hardware, software, application and installation aspects of process computers.

Some of the frequently asked basic questions raised in connection with computer applications are—

When should computers be installed?
What are the cost aspects?
What are the potential benefits?
What are their limitations?

When should computers be installed? Two types of processes justify the installation of a control computer. One involves those processes that can not be optimized without the high-speed solution of complex mathematical equations, and the other involves those that require high-speed handling and reorganizing of great quantities of data, such as in multi-reactor batch processes, where one might be faced with hundreds of ingredients, recipes and destinations. There also are two reasons which do not justify the installation of process computers. The first is the prestige motivation, where the computer is installed more as a status symbol than a necessary tool serving improved plant operation. The second wrong reason for computer installation is for replacing conventional instrumentation without improving the operation.

Conversion to computer control involves more than the acquisition and plugging in of a piece of equipment. It involves a survey of the plant and its operation and a study of the amount of redesigning required for accommodation of the new system.

What are the cost aspects? It might seem reasonable to assume that if the number of conventional control loops to be placed under computer control is great enough, an economic break-even point exists which, if exceeded, will make the computer installation economically more attractive. This assumption has been found to be without foundation and should be discounted as a motivation for considering computers. Simply stated, computerized plants cost

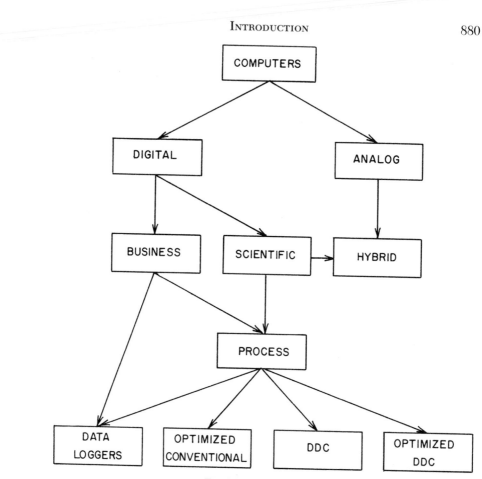

Fig. 8.0 Computer types

more than conventionally controlled ones, the cost margin being a function of the backup instrumentation used to prevent plant shutdown in case of computer failure. The cost of installation should not be evaluated in hardware terms, but in terms of the benefits to be gained.

What are the potential benefits? It is difficult to generalize as to which are likely to be the greater or the lesser benefits of a computer installation, and therefore only a brief listing of potential areas is furnished here. Optimization aimed at operating costs, production rate or product quality can be performed on a unit, plant or company-wide level. Savings can be achieved in raw materials, fuels, operating manpower and space. In addition, it becomes possible to adjust for uncontrollable variables such as feed compositions or ambient conditions, to obtain a better overall understanding of the process, to improve the quality of operating records and to better plant safety. On a lower priority basis, the process computer can also handle such tasks as accounting, inventory control, scheduled maintenance, purchasing or production planning.

What are their limitations? The justification for the installation should be established together with the process interrelationships (mathematical model) which are needed for the computer to be able to manipulate and optimize the process. The first computer installation on a new process is likely to be an expensive and time-consuming project. It is usually wise ro proceed gradually with the installation. This means, for example, that the installation of a data logging system is the usual first step in data gathering for a better understanding of the process. Most experienced engineers recommend using a supervisory computer—as the second step—which only adjusts the set points of the conventional instruments; in case of difficulty the plant can then be operated by the conventional controllers as it was before the computer was installed.

Another limitation is the availability of adequately trained personnel for operation and maintenance. Similarly, the number of computer engineers familiar with process, and the number of process engineers that have computer experience, is very limited.

Probably the most important limitation involves the realization that the computer's performance is directly related to the information it is given to work with.

8.1 COMPUTER TERMINOLOGY

Absolute Coding   Coding written in language acceptable to a computer without further modification. Same as *Machine Language.*

Access Time   (1) The time interval between the instant at which data are called for from a storage device and the instant delivery is completed, i.e., the read time. (2) The time interval between the instant at which data are requested to be stored and the instant at which storage is completed, i.e., the write time.

Accumulator   A register in the arithmetic unit which stores operands and in which arithmetical results are formed. See also *Register.*

Accuracy   The degree of freedom from error; the degree of conformity to truth or to a rule. Accuracy is contrasted with precision, e.g., four-place numerals are less precise than six-place numerals, but a properly computed four-place numeral might be more accurate than an improperly computed six-place numeral.

Adaptive Control Action   Control action whereby automatic means are used to change the type and/or influence of control parameters in such a way as to improve the performance of the control system.

Address   A label, such as an integer or other set of characters, which identifies a register, location, or device in which data are stored.

   Absolute Address   Actual location in storage of a particular unit of data; address that the control unit can interpret directly. Also, the label assigned by the engineer to a particular storage location in the computer.

   Relative Address   A label used to identify a word in a routine or subroutine with respect to its relative position in that routine or subroutine. A relative address is translated into an absolute address by addition of some specific starting address for the subroutine within the main routine.

   Relativization   A means by which the next instruction address and the operand address are given relative addresses when written. The relative addresses are translated automatically to absolute addresses during execution of the program.

   Symbolic Address   A label assigned to a selected word in a routine for the convenience of the programmer. The symbol used is independent of the location of a word within a routine. It identifies the field of data to be operated on or the operation to be used rather than its storage location.

Alarm   An audible or visible signal that indicates an abnormal or out-of-limits condition in the plant or control system.

ALGOL   ALGOrithm Language

Allocation   The assignment of blocks of data to specified blocks of storage.

Analog   The representation of numerical quantities by means of physical variables such as voltage, current, resistance, rotation, etc. Contrasted with *Digital*.

Analog Backup   An alternate method of process control by conventional analog instrumentation in the event of a failure in the computer system.

Analog Computer   See items under *Computer*.

Analog Input Module   A device which converts analog input signals from process instrumentation into a digital code for transmission to the computer.

Annotation   An added descriptive comment or explanatory note.

Argument   The known reference factor necessary to find the desired item (functional) in a table.

Arithmetic Unit   That portion of the hardware of an automatic digital computer in which arithmetical and logical operations are performed.

ASCII   American Standard Code for Information Interchange. An eight-level code intended to provide information code compatibility between digital devices of U.S. manufacture.

Assembler   A program which converts symbolic language to machine language by substitution of absolute operation codes for symbolic operation codes and absolute or relocatable addresses for symbolic addresses.

Asynchronous Computer   A computer in which performance of the next command is started by a signal that the previous command has been completed. Contrasted with *Synchronous Computer*, which is characterized by a fixed time cycle for the execution of operations.

Attenuation   (1) A decrease in signal magnitude between two points, or between two frequencies. (2) The reciprocal of gain, when the gain is less than one.

Auctioneering Device   A device which automatically selects either the highest or the lowest input signal from among two or more input signals.

Automatic Data-Processing System   A system that uses minimum manual operations in processing data.

Automatic Programming   The process of using a computer to perform some stages of the work involved in preparing a program.

Availability   The total amount of time that a computer is properly operating.

Background Program   A program of no particular urgency with regard to time that may be preempted by a program of greater urgency and priority. Contrasted with *Foreground Program.*

Backup   Provision of alternative means of operation in case of a failure of the primary means of operation.

Batch Processing   Collection of data over a period of time to be sorted and processed as a group during a particular machine run.

Baud   A unit of signaling speed equal to the number of discrete conditions or signal events per second. For example, one baud equals one half-dot cycle per second in Morse code, one bit per second in a train of binary signals, and one three-bit value per second in a train of signals each of which can assume one of eight different states.

Benchmark Problem   A sample problem used to evaluate the performance of computers relative to each other.

Binary   (1) Pertaining to a characteristic or property involving a selection, choice, or condition in which there are two possibilities. (2) Pertaining to the numeration system with a radix of two.

Binary Coded Decimal (BCD)   Pertaining to a decimal notation in which the individual decimal digits are each represented by a group of binary digits. In the 8-4-2-1 binary coded decimal notation, the number 23 is represented as 0010 0011 whereas in binary notation 23 is represented as 10111.

Bit   A binary digit; hence, a unit of data in binary notation. In the binary numbering system, only two marks (0 and 1) are used. The number 10111 contains five bits.

Block   A group of consecutive machine words considered or transferred as a unit, particularly with reference to input and output.

Bootstrap   A technique or device designed to bring itself into a desired state by means of its own action, e.g., a machine routine whose first few instructions are sufficient to bring the rest of itself into the computer from an input device.

Breakpoint   A point in a program at which a computer may be made to stop automatically for a check on the progress of the routine.

     Conditional Breakpoint   A breakpoint at which the routine may be continued as coded if desired conditions are satisfied.

Buffer   A storage device used to compensate for a difference in rate of flow of data, or time of occurrence of events, when transmitting data from one device to another.

Bug   An error or malfunction in a program or hardware.

Bulk Memory   An auxiliary memory device with storage capacity greatly in excess of working (core) memory; e.g., disc file, drum.

Bus   One or more conductors used for transmitting signals or power.

Byte   A sequence of adjacent binary digits operated upon as a unit and usually shorter than a word.

Calibrate   (1) To ascertain, usually by comparison with a standard, the locations at which scale or chart graduations should be placed to correspond to a series of values of the quantity which the instrument is to measure, receive, or transmit. (2) To adjust the output of a device, to bring it to a desired value, within a specified tolerance, for a particular value of the input. (3) To ascertain the error in the output of a device by checking it against a standard.

Call   To transfer control to a specified subroutine.

Calling Sequence   A specified arrangement of instructions and data necessary to set up a given subroutine.

Capacity   In computer terminology, the quantity of information that can be contained in a storage device, defined in terms of the basic information size, such as words or characters.

Central Processor   That portion of any computer system that performs the actual computation. It normally consists of the arithmetic and control units and working memory.

Channel   A path along which data, particularly a series of digits or characters, may flow or be stored.

Check   A means of verifying the accuracy of data transmitted, manipulated, or stored by any unit or device in a computer.

     Built-in Check (Automatic Check)   Any check constructed in hardware.

     Duplication Check   A check that requires identical results of two independent performances of the same operation.

     Mathematical Check   A check making use of mathematical identities or other properties, frequently with some degree of discrepancy being acceptable; e.g., checking multiplication by verifying that $A \cdot B = B \cdot A$, checking a tabulated function by the difference method.

     Parity Check   A summation check in which the binary digits, in a character or word, are added (modulo 2) and the sum checked against a single, previously computed parity digit; e.g., a check that tests whether the number of ones is odd or even.

     Programmed Check   A mathematical check inserted into the operating program.

     Redundant Check   A check that attaches one or more extra digits to a word according to rules, so that if any digit changes, the malfunction or mistake can be detected.

     Summation Check   A redundant check in which groups of digits are summed, usually without regard for overflow, and that sum checked against a previously computed sum to verify accuracy.

     Transfer Check   A check on transmitted data by temporarily storing, retransmitting, and comparing.

     Twin Check   A continuous duplication check achieved by duplication of hardware.

Clear   To replace information in a storage device by the character zero.

Clock Frequency   Master frequency of periodic pulses which schedule the operation of the computer.

Clock Rate   The rate at which a word or characters of a word (bits) are transferred from one internal computer element to another. Clock rate is expressed in cycles (if a parallel-operation machine—words; if a serial-operation machine—bits) per second.

Closed Loop   A signal path that includes a forward path, a feedback path, and a summing point and forms a closed circuit.

Closed Subroutine   A limited-use subroutine that can be stored at one place and can be connected to a routine by linkages at one or more locations. Contrasted with *Open Subroutine*.

Code (noun)   A system of rules for using a set of symbols to represent data.

> *Computer Code; Machine Code*   The code that the computer hardware was built to interpret and execute.

> *Instruction Code*   The symbols, names, and operation descriptions for all instructions represented by computer code.

> *Numerical Code*   A code in which the symbols used are all numerals.

> *Pseudo Code*   An arbitrary code, independent of the hardware of a computer and designed for convenience in programming, that must be translated into computer code if it is to direct the computer.

Code (verb)   To express a program in a code that a specific computer was built or programmed to interpret and execute.

Coding   The act of preparing in code or pseudo code a list of the successive computer operations required to solve a specific problem. Also, the list itself.

Command   A pulse, signal, or set of signals initiating one step in the performance of a computer operation. A command is one part of an instruction.

Common Mode Interference   A form of interference which appears between any measuring circuit terminals and ground.

Common Mode Rejection   The ability of a circuit to discriminate against common mode voltage, usually expressed as a ratio or in decibels.

Comparator   A device for comparing two different transcriptions of the same information to verify agreement or determine disagreement. A circuit that compares two signals and indicates agreement or disagreement; a signal may be given, indicating whether they are equal or unequal.

Compiler   A program which translates a problem-oriented language to a machine-oriented language, e.g., *Fortran, Algol*. A compiler, as contrasted with an assembler, can substitute subroutines as well as single machine instructions for certain symbolic inputs. A compiler which translates directly from source language to machine language is known as a single-pass compiler. A compiler that generates an interim object language which requires further translation or modification is known as a multiple-pass compiler.

Computer   Any device capable of accepting data, applying prescribed processes to them, and supplying the results of these processes. The word "computer" in this glossary usually refers to a stored-program digital computer.

    Analog Computer   A computer which calculates by using physical analogs of the variables. Usually a one-to-one correspondence exists between each numerical variable occurring in the problem and a varying physical measurement in the analog computer. The physical quantities in an analog computer are varied continuously instead of in discrete steps as in the digital computer.

    Digital Computer   A computer capable of accepting and operating on only the representations of real numbers, or other characters coded numerically.

    Stored Program Computer   A digital computer capable of performing sequences of internally stored instructions, as opposed to calculators on which the sequence is impressed manually. Such computers usually possess the further ability to operate upon the instructions themselves, and to alter the sequence of instructions in accordance with results already calculated.

Contact Sense Module   A device which monitors and converts program-specified groups of field switch contacts into digital codes for input to the computer. Inputs are scanned by the computer at programmed intervals.

Control Action   Of a controller or a controlling system, the nature of the change of the output effected by the input. The output may be a signal or the value of a manipulated variable. The input may be the control loop feedback signal when the set point is constant, an actuating error signal, or the output of another controller.

Control Algorithm   A mathematical representation of the control action to be performed.

Control Computer   A computer which, by means of inputs from and outputs to a process, directly controls the operation of elements in that process.

Control Counter; Program Counter; Instruction Counter; Control Register   A counter built into the control unit and used for sequencing instructions to be executed. It normally contains the address of the next instruction to be performed.

Control Logic   The sequence of steps or events necessary to perform a particular function. Each step or event is defined to be either a single arithmetic expression or a single Boolean expression.

Control Output Module   A device which stores computer commands and translates them into signals which can be used for control purposes. It can generate digital outputs to control on-off devices or to pulse set-point stations, or it can generate analog outputs—voltage or current—to operate valves and other process control devices.

Control Panel (Automatic)   A panel of indicator lights and switches on which are displayed a particular sequence of routines, and from which an operator can control the operation of these routines.

Control Panel (Maintenance)   A panel of indicator lights and switches on which are displayed a particular sequence of routines, and from which an operator can control the operation of these routines.

Control Panel (Operator's Request)   A panel consisting of indicator lights and switches by which an operator can request the computer to perform particular functions.

Control Panel (Programming)   A panel consisting of indicator lights and switches by which a programmer can enter or change routines in the computer.

Control System   A system in which deliberate guidance or manipulation is used to achieve a prescribed value of a variable.

Control Unit   Portion of the hardware of an automatic digital computer that directs sequence of operations, interprets coded instructions, and initiates proper commands to computer circuits to execute instructions.

Controller   A device which operates automatically to regulate a controlled variable.

Converter   A device for transferring data from one storage medium to another, as from punched cards to magnetic tape.

Core Memory   A high-speed random-access storage device utilizing matrix arrays of ferrite cores, usually used as the computer's working memory.

Core Resident   A term pertaining to certain pivotal programs permanently stored in core memory for frequent execution.

Counter   A device or memory location whose contents can be successively incremented or decremented.

Cycle Time   The basic unit of computer speed, usually the time required for a read and write operation in core memory.

Data Break   An automatic input/output channel which provides external equipment with direct access to core memory.

Data Display Module   A device which stores computer output and translates this output into signals which are distributed to a program-determined group of lights, annunciators, and numerical indicators in operator consoles and remote stations.

ddc   Same as *Direct Digital Control.*

Dead Band   The range through which an input can be varied without initiating response.

Deadtime   The interval of time between initiation of an input change or stimulus and the start of the resulting response.

Debug   To test a computer program to find whether it works properly and to trace and correct any errors.

Debug On-Line   To *Debug* a computer performing on-line functions and utilizing another routine which has previously been checked out.

Diagnostic Routine   A program designed to locate malfunctions in computer hardware or software.

Digital   Pertaining to data in the form of digits. Contrasted with *Analog.*

Digital Backup   An alternative method of digital process control initiated by use of special purpose digital logic in the event of a failure in the computer system.

Digital Computer   See items under *Computer.*

Direct Digital Control (ddc)   Control performed by a digital device which establishes the signal to the final controlling element. Examples of possible digital (D) and analog (A) combinations for this definition are:

	Feedback Elements $\rightarrow$	Controller $\rightarrow$	Final Controlling Element
1.	D	D	D
2.	A	D	D
3.	A	D	A
4.	D	D	A

Disc   A flat circular plate with a magnetic surface on which data can be stored by selective magnetization of portions of the flat surface.

Disturbance   An undesired change in a variable applied to a system which tends to affect adversely the value of a controlled variable.

Downtime   Time when a computer is not operating.

Driver   A small program or routine that handles the control of an external peripheral device or executes other programs.

Drum   A right circular cylinder with a magnetic surface on which data can be stored by selective magnetization of portions of the curved surface.

Dummy   An artificial address, instruction, or other unit of information inserted solely to fulfill prescribed conditions (such as word length or block length) without affecting operations.

Dump   To copy the contents of all or part of a storage, usually from an internal storage into an external storage.

Error   The difference between the indication and the true value of the measured signal. A positive error denotes that the indication of the instrument is greater than the true value.

$$\text{Error} = \text{Indication} - \text{True Value}$$

Error Signal   In a closed loop, the signal resulting from subtracting a particular return signal from its corresponding input signal.

Error Squared   The technique of introducing the square of the error in the error term of a linear algorithm so as to produce a non-linear correction.

Exception-Principle System   An information system or data-processing system which reports on situations only when actual results differ from planned results. When results occur within a "normal range," they are not reported.

Executive Program   A program that controls the execution of all other programs in the computer based on established hardware and software priorities and real time or demand requirements.

Extract   To obtain certain digits from a machine word as may be specified. Or, to replace contents of specific columns of another machine word, depending on the instruction. Or, to remove from a set of items of information all those items that meet some arbitrary condition.

Field   A set of one or more characters constituting a unit of data. Compare *Word*. (A field need not correspond in length to a word.)

Final Controlling Element   That forward controlling element which directly changes the value of the manipulated variable.

Fixed Heads   Pertaining to the use of stationary, rigidly mounted, reading and writing heads on bulk memory devices.

Flip-Flop   A circuit or device containing active elements, capable of assuming either one of two stable states at a given time.

Foreground Program   A time-dependent program initiated via an outside request whose urgency preempts operation of a background program.

FORTRAN (*FORmula TRANslating system*)   A procedure-oriented language designed for solution of arithmetic and logical programs.

Gate Circuit   An electronic circuit with one or more inputs and one output, with the property that a pulse goes to the output line only if some specified combination of pulses occurs on the input lines. Gate circuits constitute much of the hardware by means of which logical operations are built into a computer.

Guard Bit   A bit contained in each word or groups of words of memory which indicates to computer hardware or software whether the content of that memory location can be altered by a program.

Half Duplex   In communications, pertaining to an alternative, one way at a time, independent transmission.

Hardware   Physical equipment, e.g., mechanical, magnetic, electrical or electronic devices. Contrast with Software.

Hardware Priority Interrupt   The hardware implementation of priority interrupt functions.

Head   A device that reads, records, or erases data on a storage medium; e.g., a small electromagnet used to read, write, or erase data on a magnetic drum or tape, or the set of perforating, reading, or marking devices used for punching, reading, or printing on paper tape.

Index Register   A register to which an arbitrary integer, usually one, is added or subtracted upon the execution of each machine instruction. The register may be reset to zero or an arbitrary number. Used with indexable instructions to get "effective" instruction addresses during execution. Also called "cycle counter" and "B-box."

Indirect Address   An address that specifies a storage location that contains either a direct address or another indirect address.

Initialize   To set counters, switches, and addresses to zero or other starting values at the beginning of, or at prescribed points in, a computer routine.

In-Line Processing   The processing of data without sorting or any other prior treatment other than storage.

Input   (1) The data to be processed. (2) The state or sequence of states occurring on a specified input channel. (3) The device or collective set of devices used for bringing data into another device. (4) A channel for impressing a state on a device or logic element. (5) The process of transferring data from an external storage to an internal storage.

Instruction Counter   Same as *Control Counter.*

Interacting Control   Control action produced by an algorithm whose various terms are interdependent.

Interface   Logic necessary to provide electrical and communication compatibility between two devices.

Interrupt   See *Priority Interrupt, Hardware Priority Interrupt,* and *Software Priority Interrupt.*

Key Punch   A typewriter-like machine for recording data by cutting holes or notches in cards.

Linear Programming   The analysis or solution of problems in which the linear function of a number of variables is to be maximized or minimized when those variables are subject to a number of constraints in the form of linear inequalities.

Linkage   In programming, coding that connects two separately coded routines.

Logical Operation   An operation in which a decision affecting the future sequence of instructions is automatically made by the computer. The decision is based upon comparisons between all or some of the characters in an arithmetic register and their counterparts in any other register on a less than, equal to, or greater than basis; or between certain characters in arithmetic registers and built-in standards.

Machine Language   Same as *Absolute Coding.*

Magnetic Core Storage   A storage device consisting of magnetically permeable binary cells arrayed in a two-dimensional matrix. (A large storage unit contains many such matrices.) Each cell (core) is wire wound and can be polarized in either of two directions for the storage of one binary digit. The direction of polarization can be sensed by a wire running through the core.

Magnetic Disc Storage   A storage device consisting of magnetically coated disks accessible to a reading and writing arm in much the manner of an automatic record player. Binary data are stored on the surface of each disk as small, magnetized spots arranged in circular tracks around the disk. The arm is moved mechanically to the desired disc and then to the desired track on that disc. Data from a given track are read or written sequentially as the disc rotates.

Magnetic Drum Storage   A storage device consisting of a rotating cylinder surfaced with a magnetic coating. Binary data are stored as small, magnetized spots arranged in closed tracks around the surface. A magnetic reading and writing head is associated with each track so that the desired track can be selected by electric switching. Data from a given track are read or written sequentially as the drum rotates.

Manipulated Variable   A quantity or condition which is varied as a function of the actuating signal so as to change the value of the directly controlled variable. In any practical control system, there may be more than one manipulated variable. Accordingly, when using the term, it is necessary to state which manipulated variable is being discussed. In process control work, the one immediately preceding the directly controlled system is usually intended.

Masking   An operation that replaces characters in the accumulator with characters from a specified storage location that correspond to the "ones" in the mask which is in a specified storage location or register.

Measured Signal   The electrical, mechanical, pneumatic, or other variable applied to the input of a device. It is the analog of the measured variable produced by a transducer (when such is used). In a thermocouple-thermometer system, for example, the measured signal is an emf which is the electrical analog of the temperature applied to the thermocouple. In a flowmeter, the measured signal may be a differential pressure which is the analog of the rate of flow through the orifice. In an electric tachometer system, the measured signal may be a voltage which is the electrical analog of the speed of rotation of the part coupled to the tachometer generator.

Measured Variable   The physical quantity, property, or condition which is to be measured. It is sometimes referred to as the measurand.

Memory   The capacity of a computer to receive and store data subject to recall. Loosely, any device that can store data.

Memory Protect   A technique of protecting the contents of sections of memory from alteration by inhibiting the execution of any memory modification instruction upon detection of the presence of a guard bit associated with the accessed memory location. Memory modification instructions accessing protected memory are usually executed as a no-operation, and a memory protect violation program interrupt is generated.

Merge   To produce a single sequence of items, ordered according to some rule (that is, arranged in some orderly sequence), from two or more sequences previously ordered according to the same rule, without changing the items in size, structure, or total number. Merging is a special kind of collating.

Microprogramming A programming capability wherein several instruction operations can be combined in one instruction for greater speed and efficient use of memory.

Mnemonic An alphanumeric designation, easy to remember and commonly used to designate a memory location or computer operation; e.g., START might represent the location of the first instruction in a routine.

Multiplex The process of transferring data from several storage devices operating at relatively low transfer rates to one storage device operating at a high transfer rate in such a manner that the high-speed device is not obliged to "wait" for the low-speed units.

Multiprogramming The interleaved or time-shared execution of two or more programs by a computer.

Noise An unwanted component of a signal or variable which obscures its information content.

Normalize To shift the representation of a quantity so that the representation lies in a prescribed range.

Object Program The coding which is the output of an automatic code translation program such as an assembler or compiler.

Off Line (1) Pertaining to equipment or programs not under the direct control of the central processor. (2) Pertaining to a computer that is not actively monitoring or controlling a process or operation, or pertaining to a computer operation performed while the computer is not monitoring or controlling a process or operation.

On Line (1) Pertaining to equipment or programs under direct control of a central processor. (2) Pertaining to a computer that is actively monitoring or controlling a process or operation, or pertaining to a computer operation performed while the computer is monitoring or controlling a process or operation.

On-Line Equipment Equipment for which the transfer of data to or from the unit is under direction of the control unit of the computer.

Open Subroutine A general-use subroutine that must be relocated and inserted into a routine at each place it is used. Contrasted with *Closed Subroutine.*

Operand That which is operated upon. An operand is usually identified by an address part of an instruction.

Operating Conditions Conditions (such as ambient temperature, ambient pressure, vibration, etc.) to which a device is subjected, but not including the variable measured by the device.

Operation Code The part of a computer instruction which specifies the operation to be performed.

Operator's Console Equipment which provides for manual intervention and for monitoring computer operation.

Optimize To establish control parameters so as to make control as effective as possible.

Output   Process of transferring data from internal storage of a computer to some other storage device.

Pack   (1) To include several discrete items of information in one unit of information. (2) To relocate programs and data to make efficient use of available storage capacity.

Page Addressing   A memory addressing technique utilized with certain computers whose addressing capability is limited to less than the total memory capacity available. Using page addressing, memory is divided into segments (pages), each of which can be addressed by the available addressing capability.

Parameter   A controllable or variable characteristic of a system or device, temporarily regarded as a constant, the respective values of which serve to distinguish the various specific states of the system or device.

Parity Check   See items under *Check.*

Patch   A section of coding inserted into a routine (usually by explicitly transferring control from the routine to the patch and back again) to correct a mistake or alter the routine.

Peripheral Equipment   Equipment used for entering data into or receiving data from a computer.

Power Consumption   The maximum wattage used by a device within its operating range during steady-state signal condition. For a power factor other than unity, power consumption is the maximum volt-amperes used under the condition stated above.

Priority Interrupt   The temporary suspension of a program currently being executed in order to execute a program of higher priority. Priority interrupt functions usually include distinguishing the highest priority interrupt active, remembering lower priority interrupts which are active, selectively enabling or disabling priority interrupts, executing a jump instruction to a specific memory location(s), and storing the program counter register in a specific location(s). See *Hardware Priority Interrupt* and *Software Priority Interrupt.*

Priority Interrupt Module   A device which monitors a number of priority-designated field contacts and immediately notifies the computer when any of these external priority requests have been generated. It assures servicing of urgent interrupt requests on the basis of programmer-assigned priorities when requests occur simultaneously.

Process   The collective functions performed in and by the equipment in which a variable or variables is or are controlled. "Equipment" in this definition does not include automatic control equipment. The process may also be referred to as the controlled system.

Process Control Loop   A system of control devices linked together to control one phase of a process.

Program   A plan for the automatic solution of a problem. A complete program includes plans for the transcription of data, coding for the computer, and plans for the absorption of the result into the system. The list of coded instructions is called a "routine."

Program Counter   Same as *Control Counter.*

Programmer   A person who prepares computer operation procedures by means of flow charts and coding.

Proportional Control Action   Control action in which there is a continuous linear relation between the output and the input. This condition applies when both the output and input are within their normal operating ranges and when operation is at a frequency below a limiting value.

Pulse   A significant and sudden change of short duration in the level of an electrical variable, usually voltage.

Pulse-Counting Module   A device which counts and stores a number of high- or low-speed pulse channels and transmits their status to the computer upon command.

Random Access   (1) Pertaining to the process of obtaining data from, or placing data into, storage where the time required for such access is independent of the location of the data most recently obtained or placed in storage. (2) Pertaining to a storage device in which the access time is effectively independent of the location of the data.

Ratio Controller   A controller that maintains a predetermined ratio between two or more variables.

Read   (1) To copy, usually from one form of storage to another, particularly from external or secondary storage to internal storage. (2) To sense the meaning of arrangements of hardware.

Real Time Operation   Processing data in synchronism with a physical process so that results of data processing are useful to the physical operation.

Recursive   Pertaining to the use of a subroutine iteratively in the solution of a problem.

Register   A device for the temporary storage of one or more words to facilitate arithmetical, logical, or transferral operations. Examples are the accumulator and the address, index, instruction, and M-Q registers.

Relative Address   A label used to identify the location of data in a program by reference to its position with respect to some other location in that program. Relative addresses are translated into absolute addresses by the addition of the reference address.

Relocate   In programming, to move a routine from one portion of storage to another and to adjust the necessary address references so that the routine, in its new location, can be executed.

Resolution   The least interval between two adjacent discrete details which can be distinguished one from the other.

Routine   A set of coded instructions arranged in proper sequence to direct the computer to perform a desired operation or series of operations. See also *Subroutine*.

Diagnostic Routine   A specific routine designed to locate either a malfunction in the computer or a mistake in coding.

Executive Routine; Master Routine   A routine designed to process and control other routines. A routine used in realizing *Automatic Programming*.

General Routine   A routine expressed in computer coding designed to solve a class of problems, specializing to a specific problem when appropriate parametric values are supplied.

Generator   A general routine that accepts a set of parameters and causes the computer to compute a specific routine for further use. Among other things, the parameters may specify the input-output devices to use, designate subroutines, or describe the form of a record.

Interpretive Routine; Interpretation   An executive routine which, during the course of data-handling operations, translates a stored pseudocode program into a machine code and at once performs the indicated operations by means of subroutines.

Postmortem Routine   A routine which, either automatically or on demand, prints data concerning contents of registers and storage locations when the routine is stopped in order to assist in locating a mistake in coding.

Rerun Routine; Rollback Routine   A routine designed to be used in the wake of a computer malfunction or a coding or operating mistake to reconstitute a routine from the last previous rerun point.

Sampling, Analog   The process by which the computer selects individual analog input signals from the process, converts them to an equivalent binary form, and stores the data in memory.

Scale Factor   A number used as a multiplier, so chosen that it causes a set of quantities to fall within a given range of values. To scale the values 856, 432, -95, and -182 between $-1 +1$, a scale factor of $\frac{1}{1000}$ would be suitable.

Scanning   The action of comparing input variables to determine a particular action.

Scanning Limits   The action of comparing input variables against either prestored or calculated high and/or low limits to determine if an alarm condition is present.

Search, Binary   A technique for finding a particular item in an ordered set of items by repeatedly dividing in half the portion of the ordered set containing the sought-for item until only the sought-for item remains.

Self-Tuning   The technique of automatic modification of control algorithm constants based upon process conditions.

Sequence Monitor   Computer monitoring of the step-by-step actions that should be taken by the operator during a startup and/or shutdown of a power unit. As a minimum, the computer would check that certain milestones had been reached in the operation of the unit. The maximum coverage would have the computer check that each required step is performed, that the correct sequence is followed, and that every checked point falls within its prescribed limits. Should an incorrect action or result occur, the computer would record the fault and signal the operator.

Sequential Control   The manner of control of a computer in which instructions to it are set up in a sequence and are fed in that sequence to the computer during solution of a problem

Service Routine; Utility Routine   A routine in general support of the operation of a computer, e.g., an input-output, diagnostic, tracing, or monitoring routine.

Servomechanism   (1) A feedback control system in which at least one of the system signals represents mechanical motion. (2) Any feedback control system.

Set Point (Command)   An input variable which sets the desired value of the controlled variable. The input variable may be manually set, automatically set or programmed. It is expressed in the same units as the controlled variable.

Set-Point Control   A control technique in which the computer supplies a calculated set point to a conventional analog instrumentation control loop.

Simulation   A pseudo-experimental analysis of an operating system by means of mathematical or physical models which operate in a time-sequential manner similar to the system itself.

Smooth   To apply procedures that decrease or eliminate rapid fluctuations in data.

Software   (1) The collection of programs and routines associated with a computer, e.g., compilers; library routines. (2) All the documents associated with a computer, such as manuals and circuit diagrams. Contrasted with *Hardware*.

Software Priority Interrupt   The programmed implementation of priority interrupt functions. See *Priority Interrupt* and *Hardware Priority Interrupt*.

Source Language   A program language used as an input to a translation program such as an assembler or compiler.

Steady State   A condition, such as value, rate, periodicity, or amplitude, exhibiting only negligible change over an arbitrarily long period of time.

Storage   Same as *Memory*.

Suboptimization   The process of fulfilling or optimizing a chosen objective which is an integral part of a broader objective. Usually the broad objective and lower-level objective are different.

Subroutine   A series of computer instructions which perform a specific task for another routine. It is distinguishable from a routine in that it requires, as one of its parameters, a location specifying where in the main program to return to after its function has been accomplished.

Successive Approximation   An analog-to-digital conversion technique in which increasingly larger or smaller known voltages are compared with the unknown voltage. The equality decision made in each iteration ultimately forms the binary representation of the analog value.

Supervisory   A process computer application wherein the computer performs higher-level process calculations but does not actuate final elements, such as valves. Contrasted with *Direct Digital Control*. For example, the computer may handle mathematical models of the process, or may perform process calculations and relay the results to controllers for valve actuation.

Supervisory Control   Control action in which the control loops operate independently, subject to intermittent corrective action, e.g., set-point changes from an external source.

Symbolic Coding   Any coding in which symbols other than actual binary machine language are used.

Synchronous Computer   A computer in which each event, or the performance of each operation, starts as a result of a signal generated by a clock.

System   A collection of hardware and software organized in such a way as to achieve an operational objective.

Systems Analysis   The definition of a control problem and the development of a solution to the control problem.

Systems Engineering   The implementation of a hardware and software system resulting from analysis of a control problem.

Table   A block of information in memory which is used as data by a program.

Termination Rack   An equipment rack containing field wiring terminals and associated signal conditioning equipment. It provides the termination interface between a computer control system and field-mounted instrumentation.

Three-Position Controller   A multi-position controller having three discrete values of output.

Time Sharing   Pertaining to the interleaved use of the time of a device.

Track   The portion of a moving storage medium, such as a drum, tape, or disc, that is accessible to a given reading head position.

Transducer   An element or device which receives information in the form of one physical quantity and converts it to information in the form of the same or other physical quantity.

Transmitter   A transducer which responds to a measured variable by means of a sensing element and converts it to a standardized transmission signal which is a function only of the measurement.

Tuning   The adjustment of control constants in algorithms or analog controllers to produce the desired control effect.

Utility Routine   Same as *Service Routine.*

Valve Output Module   A device that translates the computer's output data into analog signals suitable to position control valves or other devices.

Velocity Limit   A limit which the rate of change of a specified variable cannot exceed.

Verify   (1) To check, usually with an automatic machine, one typing or recording of data against another in order to minimize the number of human errors in the data transcription. (2) In preparing data for a computer, to make certain that data prepared are correct.

Watchdog Timer   An electronic interval timer that generates a priority interrupt unless periodically recycled by a computer. It is used to detect program stall or hardware failure conditions.

Word   A sequence of bits or characters treated as a unit and capable of being stored in one computer location.

Word Time   The data transfer rate (words per second) between a device and the computer.

Write   (1) To copy information usually from internal to external storage. (2) To transfer information to an output medium. (3) To record information in a register, location, or other storage device or medium.

8.2 PROCESS COMPUTER INSTALLATIONS

The digital computer was invented during World War II. It was originally designed for engineering and scientific calculations and later applied to accounting and business in general.

The first application of the digital computer to manufacturing and processing plants was in early 1959 at the Texaco refinery in Port Arthur, Texas. A decade later, the number of on-line digital computers used in the process industry was estimated at 3,000 to 4,000.

In the early applications, the primary function of the computer was to calculate the best operating conditions in the manufacturing process and then adjust the set point of conventional controllers, which "moved" the process to the desired state. As computer speeds and reliabilities increased and computer costs correspondingly decreased, the digital computer was employed for direct digital control (ddc), wherein the computer substituted for normal analog feedback controllers. The first ddc installation in the process industry occurred in 1962. Since then, significant technical advances in the design and manufacture of internal computer circuitry have resulted in the development of the minicomputer. This is a small, compact and low-cost general purpose computer. The minicomputer has made it possible to undertake applications that heretofore could not be justified on larger computers.

Process control computers are used in just about every industry including those of petroleum, chemical, metals, power, paper, cement, discrete manufacturing, glass and rubber and in research. The functions performed by the computer vary depending on the industry but usually include one or more of the following: data acquisition, logging and operator display, performance and "state-of-unit" calculations, supervisory control of set point, ddc and logic operations.

The status of this field is examined by describing a few of the specific installations. A detailed description of the applications is beyond the scope of this section, but enough information is provided to give a fairly complete picture of techniques, functions and objectives.

Chemical

Texaco refinery at Port Arthur, Texas The application is a catalytic polymerization unit where molecules of light hydrocarbon gas are joined together to make a motor-gasoline blending stock.

The computer performs calculations and initiates several independent control actions, all of which are designed to increase the conversion of feed material into useful product. The computer functions include (1) determination of the maximum reactor inlet pressure consistent with compressor and vessel ratings, (2) distribution of the available feed material to five parallel pairs of reactors to take advantage of any differences in catalyst activity which may exist, (3) determination of the amount of water to be injected for catalyst activation, and (4) when needed, determination of the amount of unreacted material that should be returned to the reactor inlet.

The calculations performed are based on 104 plant inputs, and the computer provides supervisory control of 15 set points.

Monsanto at Luling, Louisiana   The computer is applied to an ammonia plant. Natural gas is reacted with steam and air to provide hydrogen and nitrogen in the proper ratio for ammonia synthesis. The reaction of hydrogen and nitrogen is carried out at high pressure.

The computer system is programmed to keep the gas compressors fully loaded by reacting promptly to changes in atmospheric conditions which affect compressor capacity and by adjusting flow rates as changes occur in the number of available compressors.

The type of control is supervisory.

Unnamed company   Batch polymerization application. Monomers are converted to the polymer by free radical batch polymerization. Ingredients in the batch include monomer, water, catalyst, etc. The contents of the reactor are heated to the desired reaction temperature and catalyst is added to start the polymerization. An emulsifier is continually added to the reaction vessel, the rate of addition being dependent on the extent of reaction. When the desired conversion level is reached and other operating conditions are met, the batch is discharged into a recovery and processing system.

The computer reads in several measurements for each polymerization reactor and uses the data to calculate a heat balance. The information is used to check system performance, to calculate conversion level and to determine rate of conversion. These results together with data contained in a stored "recipe" are used to calculate the optimum emulsifier addition rate and the proper time to terminate this addition.

The supervisory control system receives 96 plant measurements and provides 12 pulse outputs.

Petrochemical

Staatsmijnen's Geleen facility in The Netherlands   Ethylene production is the application. Naptha feedstock is heated to high temperatures in thermal cracking furnaces which produces a hydrocarbon effluent material ranging from H_2 to butanes, a gasoline fraction and a heavy oil fraction. The gaseous material is then compressed to a high pressure prior to fractionation of the numerous components. Fractionation takes place in two parts, a so-called low-temperature separation for the recovery of high-purity ethylene product and a so-called high-temperature separation for the recovery of propane, butane, and heavier materials.

The computer controls the operation of the cracking furnaces and the low-temperature distillation columns with the overall objective being to maximize profit. The computer calculates the "best" operating conditions by periodically executing an optimization program. The computer's optimizer uses a model of the furnaces to evaluate the effect of changes to furnace variables on profit. By use of an overall plant model, the computer calculates the effect of the "proposed" furnace operation on compressor capacities and downstream column constraints. The modified profit function takes into account the distance a constrained variable is away from its constraint value and insures that the proposed new operation, when implemented, will not exceed physical plant limitations or constraints. The computer system outputs 66 supervisory control set points and accepts 500 analog inputs.

Union Carbide, Seadrift Plant   A hierarchy of computers govern a variety of process units. Satellite computers control individual processes; a central computer supervises and optimizes overall operation. Off-line activities may be time-shared with the central computer. A schematic of the system is shown as Figure 8.2a.

The satellite computers are programmed to meet specific needs of each process unit and perform basic control functions; including scan, alarm, setpoint drive or ddc, log, and operator demand functions. The central computer transmits information to the satellite computers for implementing advanced control via a data link.

The central computer receives converted process information from the satellite computers via the data link and uses the information for advanced concepts of control such as supervisory calculations, operating guides, and optimization. The central computer uses its spare time for compilations and assemblies of new programs and execution of off-line programs.

The intercommunication system provides the means for communicating between central computer and up to eight satellite computers. The communication system operates in a central/peripheral relationship such that only the central computer can initiate data transfer. A satellite may request a data transfer through a priority interrupt, but no transfer is made until the central computer can act.

The design of the system is such that the satellites will continue to monitor, log, and control the process unit operations even if the central computer or data link should fail.

Petroleum

Shell Oil's Cardón Refinery in Venezuela   The application is a refinery management information system that reports on the movement of oil through the refinery.

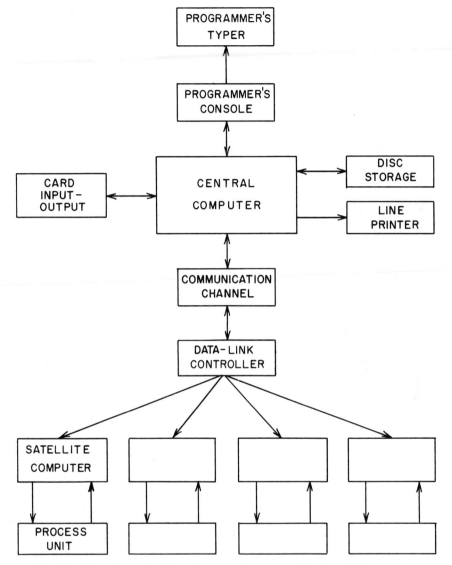

Fig. 8.2a   Computer control hierarchy; central computer linked to satellite computers

The computer automatically collects plant data, integrates flows, processes digital information from the tank-gauging system, accepts operator data regarding changes in refinery equipment configuration, and prepares reports for various supervisors.

The main reports produced by the system are (1) 24-hour plant balances, (2) spot balances (snapshot of the plant at a certain moment), (3) 24-hour tank balance, (4) blending and shipping reports, and (5) summary of product analysis data. The reports permit better operation of the refinery.

The operator communicates changes via a keyboard typer to insure that the reports are realistic. The communication is carried out in a conversational (question/answer) mode, with operator responses keyed to specific codes to guard against operator entry of invalid information.

Sinclair Refining Company at Houston, Texas This computer application involves the monitoring and peak integration of in-line process gas chromatographs. The chromatographs analyze the composition of materials flowing through the plant. The chromatograph's detector cell output is normally recorded for operator use in the form of chromatograms which require interpretation before they can be used in plant control work.

The computer reads the detector cell outputs directly, calculates peak areas, identifies components and prints analyses reports. Since the chromatographs are in-line, the computer supervises all gas chromatograph activity, including the opening and closing of solenoid valves for sample injection, column backflushing and injecting standard calibration samples.

The computer system provides more accurate analytical results and eliminates the need for certain chromatograph equipment, such as the "black box" programmers.

Glass

Owens-Corning Fiberglass Corporation at Newark, Ohio Application involves a recuperative-fired glass furnace. The furnace operates at 2,500 to 6,200°F. The usual analog control system for such a furnace is shown in Figure 8.2b. Cool air at constant temperature enters the bottom of the recuperator. The air is heated by the recuperator and some of the air is exhausted (or "spilled") to atmosphere to control the air temperature. The combustion air temperature cascades into spill air flow control to control the air temperature. The hot air not spilled is burned in the furnace. Gas flow and air flow vary according to the oxygen content of the burned fuel and of the melter temperature. The furnace pressure control maintains pressure by restricting the flow of burned fuel with an air damper. The glass level control adjusts the speed of the batch charger (raw material feeder). Other temperature controls adjust full flow to maintain temperature.

The computer is programmed to perform all of the functions of the conventional analog control system. The computer reads in analog inputs associated with each loop at a predesignated frequency (poll time), converts the analog signal to a digital value, compares the digital value with the stored set point, and calculates the new valve position based on a 3-mode (PID) control. The computer then outputs analog signals to set the valves at the new positions.

The operator/engineer communicates with the computer via a process

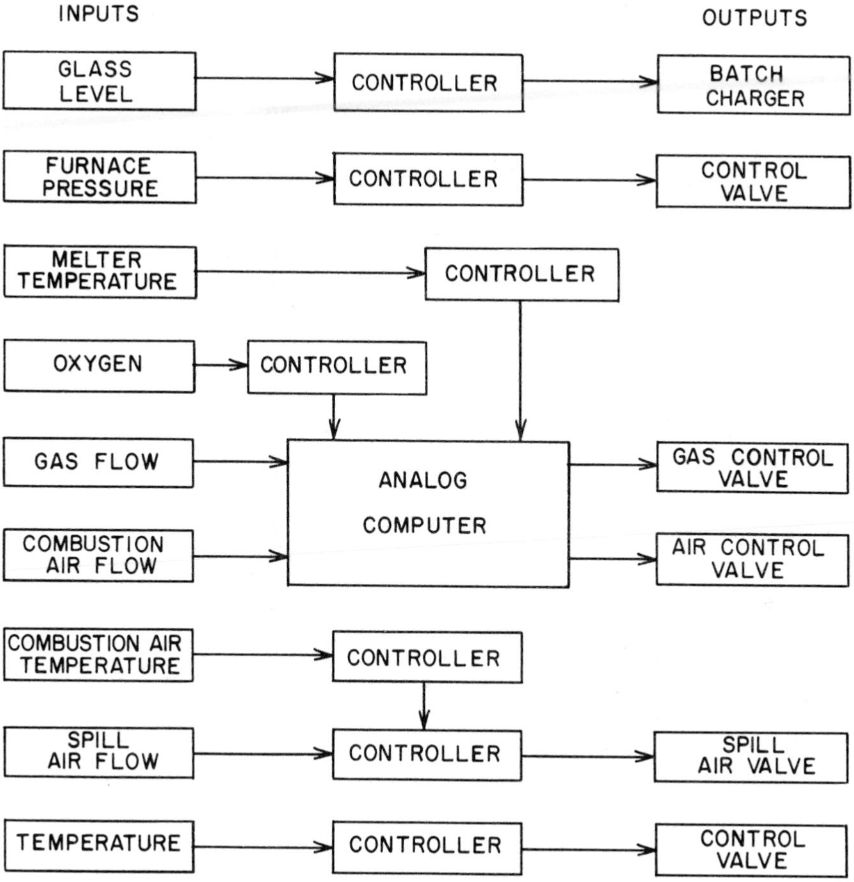

INPUTS   OUTPUTS

| GLASS LEVEL | → | CONTROLLER | → | BATCH CHARGER |

| FURNACE PRESSURE | → | CONTROLLER | → | CONTROL VALVE |

| MELTER TEMPERATURE | → | CONTROLLER |

| OXYGEN | → | CONTROLLER |

| GAS FLOW | → | ANALOG COMPUTER | → | GAS CONTROL VALVE |

| COMBUSTION AIR FLOW | → | | → | AIR CONTROL VALVE |

| COMBUSTION AIR TEMPERATURE | → | CONTROLLER |

| SPILL AIR FLOW | → | CONTROLLER | → | SPILL AIR VALVE |

| TEMPERATURE | → | CONTROLLER | → | CONTROL VALVE |

Fig. 8.2b   Glass furnace analog control system

operator's console. For any loop, the computer program allows the operator to display set point, input, error, output and alarm conditions. The engineer can display and change all the other information in that computer for any loop, such as limits, gains, and poll times. Each change is recorded by the computer on an output typer.

The ddc system has manual station backup. It reads nine analog inputs and outputs six analog signals.

Rubber

Unnamed Application is styrene-butadiene rubber (SBR) reactor charging and control. Styrene and butadiene are converted to the polymer by free radical, exothermal polymerization. The polymerization recipe is made by continuously metering individually controlled charge streams into parallel reactors. After reaction is completed, the reactor contents are transferred to strippers where the unreacted styrene and butadiene are recovered.

The computer is involved in preparing the reactor charge and then in monitoring the reaction itself. The computer prepares the reactor batch by controlling the flow of "recipe" ingredients from charge tanks into the pre-designated reactors. The charging operation is initiated and monitored by the operator via the process operator's console. After the recipe catalyst is added, the computer maintains "reaction" at its maximum rate, subject to rec-ipe-defined chemical restraints and equipment limitations.

The mode of control is ddc, but full analog backup is provided. The computer accepts 75 analog inputs and 50 digital inputs, and it outputs 100 digital and 35 analog signals.

Manufacturing

Rochester Products, Division of General Motors at Rochester, New York The application is testing and adjusting automobile carburetors. The carbu-retors are placed on test stands where their ability to mix air and fuel at desired rates under all anticipated road conditions is tested. Each test stand is equipped with several precision orifices and nozzles to meter the desired fuel/air mixtures.

The computer, upon operator initiation, conducts 14 tests at each stand. Each test includes a sequence of commands in accordance with specifications for the particular carburetor on test. The computer outputs control signals to the test stand to select the specified nozzles and orifices in order to provide the proper amount of air and fuel flow to the carburetor and moves the carburetor adjustment screws to the test position. The computer then measures the air and fuel flow passing through the carburetor and compares the flows with specified values. A reject or accept code is stored for the completed test, and the next test is started. After the final test is conducted, the operator is alerted.

The computer system accepts 362 analog inputs, 1,200 digital inputs and 64 priority interrupts, and it outputs 106 analog, 4,272 digital and 248 pulse signals to the test stands.

8.3 PLANNING OF COMPUTER PROJECTS

In order to successfully implement a computer control project, one must plan a significant number of tasks to be performed in a logical sequence. In this section a discussion is provided of the more important tasks in this plan.

The logical first step involves a feasibility evaluation which should not be limited to cost considerations, but should also establish if other than economic prerequisites are available. When the proposed installation has been proved feasible, the next logical step is to define the approximate scope of the planned installation. Once the feasibility is established and the scope is defined, attention should next be focused on analyzing the process to establish its steady state and dynamic model and to determine the instrumentation needs, in terms of both sensing and control elements.

The next major task involves the preparation of the various programs and the checking out of the software. Simultaneously with this activity, procurement of the various hardware components takes place together with the preparation of design documents for their installation in the field. After the physical installation is completed, the system is tested, evaluated and if necessary redesigned. While these activities are in progress the training of operators and the assembling of system documents can also be completed.

Figure 8.3a shows a typical critical path schedule for a process computer project.

Feasibility and Scope Study

The purpose of the initial study is to select the processing units most likely to benefit from computer control and determine the associated project costs.

One or more of the following are considered to be "accepted criteria" for a computer control application.

Value of raw material is substantially upgraded in the process. Process is subject to disturbances by uncontrollable variables (feed rate composition, ambient conditions, etc.). Therefore, the process is normally operated below ultimate capacity to preclude unsafe operating conditions or the making of an off-specification product.

Process is adequately instrumented so that the measurement data needed to define its state is available.

The process is amenable to analyses; hence theoretical and engineering concepts can be meaningfully applied to it.

After the process is selected, the study team (1) defines the functions to be performed by the computer, (2) evaluates the type of control for each loop (supervisory, ddc, operator guide), (3) identifies the computer/process interface

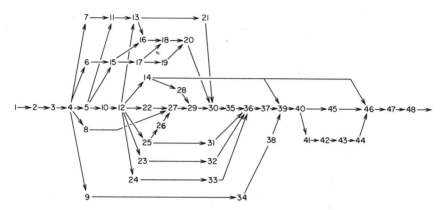

Fig. 8.3a   Typical tasks and their sequence in a computer project

1 Evaluate feasibility	24 Specify and order new sensors
2 Write and send out specification	25 Specify cables and other wiring
3 Evaluate bids and select vendor	26 Manufacture cables
4 Form project team	27 Assemble hardware
5 Review system functions	28 Build or procure test equipment
6 Take system design course	29 Check equipment in vendor's shop
7 Take programing course	30 Check system in vendor's shop
8 Take maintenance course	31 Do on-site wiring
9 Start management and operator familiarization	32 Prepare site and modify plant
	33 Install new sensors
10 Review equipment requirements	34 Prepare operator manuals
11 Make general plans for programming	35 Ship hardware
12 Specify input-output terminal locations	36 Connect hardware at site
13 Assign input-output memory locations	37 Check noncontrol functions
14 Specify factory and field test procedures	38 Train operators
15 Develop first control function(s)	39 Perform equipment acceptance tests
16 Program first control function(s)	40 Check first control function(s)
17 Develop second control function(s)	41 Collect data
18 Program second control function(s)	42 Develop new control functions
19 Develop third control function(s)	43 Program new control functions
20 Program third control function(s)	44 Check new control functions
21 Program executive and noncontrol functions	45 Check other control functions
	46 Perform functional acceptance test
22 Manufacture or procure special equipment	47 Evaluate system performance
	48 Improve and expand system
23 Specify site details and plant modifications	

characteristics (signal type and range, number of signals, etc.) and (4) prepares an initial design of the computer control system which includes hardware configuration, type of programs and (approximate) size, and manpower requirements.

The results of the study form the basis for a cost analysis and computer justification report to management.

Detailed Process Analysis

The purpose of this phase of the project is to (1) write the equations describing the process, (2) develop a (programmable) procedure for selecting the best operating conditions, (3) investigate instrumentation required for computer control, (4) examine the effect of process dynamics, and (5) carry out an evaluation of the probable payout which might result from computer control.

Three process equations are usually written. The first equation expresses the "operating profit" as a function of the way the process is operated. It is often referred to as the "objective equation." A second equation relates the process dependent variables to the process independent variables. This equation is referred to as the "process model." Third, a set of equations is written that describe limitations on process operation (e.g., reactor temperature cannot exceed 750°F). These equations are called "constraint equations."

Next a procedure or technique is specified that uses the process equations as a "model" of the process and solves them periodically to determine if better operation can be achieved by manipulating any of the controlled process variables through optimization.

All information needed to solve the process equations must reside in the computer's memory for use as required. The information is obtained directly from plant instruments, inferred from other measurements if direct detection is impossible, or entered manually by the operator via an input device. In some instances, additional instrumentation may be required to provide data needed by the computer that the conventional control system does without.

A computer system scans (reads in) process inputs and performs calculations to ascertain the need for changes and implements the changes by outputting control signals to analog controllers or directly to valves. The frequency of sampling and the size of corrections must be consistent with process dynamics and control instrument characteristics. In a complicated process it is often necessary to measure the dynamic characteristics of the process, analyze them and perhaps even simulate the proposed sampled-data system on a computer. The information obtained from this activity is then made a part of the computer's stored program.

At this point in the project, equations developed for the process are calibrated against typical plant data. Another set of plant measurements are then substituted into the equations which are solved for the dependent variables. The predicted values of the dependent variables are compared with the actual values to evaluate the accuracy of the process equation.

Next, the performance of the equations for control is evaluated by substituting data from typical operating periods and solving the equations (in manner similar to that done by the computer) to determine how the on-line computer would have controlled the plant. The targets, conversions, yields, quality, etc. calculated by the computer are then compared with actual plant values, and an economic improvement is calculated.

Programming

The design of an on-line computer control system can be very complex. It involves integrating three types of programs. These are supervisory programs, application programs, and support programs.

Supervisory programs provide the basic framework upon which the user designs and builds the remainder of the system. The supervisory programs direct the handling of interrupts, supervise the execution of programs dictated by the process, service all error conditions, maintain timers and counters and handle input/output operations.

The heart of the supervisory programs is the "system director," or "scheduler." When an application program or other system routine has completed its calculations, it returns control to the system director, which decides what is to be done next.

The supervisory programs are normally provided by the computer hardware vendor.

Application programs carry out specific engineering type functions for the system, such as (1) making periodic logs of all important variables within the system, (2) establishing closed-loop control over specified process variables, (3) responding to external interrupts (a supervisory program recognizes the occurrence of the interrupt), and (4) displaying up-to-the-minute information about the process.

The major subsystems are the input conversion subsystem, information subsystem, operator/engineer communication subsystem, and control subsystem.

The *input conversion subsystem* converts process measurements to engineering units, checks them against high/low limits and provides digital filtering. The *information subsystem* maintains a data base from time-averages of process and calculated variable data for use by other programs in the system. The *operator/engineer communication subsystem* allows the operator/engineer to access and/or change process information contained in the computer's memory. The *control subsystem* is a sequence of operations that calculates new set points and/or operating guides, performs validity checks on the resulting values and outputs the changes to the process. The subsystem utilizes the process equations, procedures, and logic developed by the process analysts.

The support programs are needed to insure that the supervisory and application programs are working satisfactorily. They analyze for hardware and software errors in the system and execute standard recovery procedures when an error is detected. The hardware checks include testing the analog-to-digital and digital-to-analog converters with standard signals. The software checks verify that software linkages between the various subsystems are functioning properly, insure that process data is checked, converted, and accumulated within a specified update period, and protect the data base against major system disturbances such as disc failures.

Although the support programs are normally provided by the computer vendor, the systems programmer often makes modifications and/or additions to the support programs in order to make them consistent with overall system objectives. This is particularly true with regard to the recovery routines provided.

Once program design has been completed, every effort should be made to freeze the design at an early date. Additions to the programs can always be made after startup at plant site.

Software Checkout

A computer control software system consists of one large program that has been highly segmented. The number of program segments is typically 50 to 100.

The software effort is usually performed by a team of programmers. Each programmer is given a specification for his routines and the responsibility to code, debug, test, and integrate the programs into the system. Because there is a tendency on the part of programmers to make some changes during the cycle of code, debug, etc., it is often found that the individual programs do not interface properly. At this point the lead programmer must decide which program is to be changed to make all the links fit properly.

System testing and integration must be carried out prior to computer delivery to the plant site. This will minimize problems in the field. Most of the computer vendors provide testing facilities that include computer and simulation equipment in or near the large metropolitan areas.

Installation and Evaluation

Following delivery of the computer hardware, the process instruments are connected to computer termination points. A special application program is used to check out all input and output connections. A number of additional programs must be written during system testing to facilitate tying together the hardware/software system. The need for some of the programs is obvious prior to installation and can be written before field testing starts, while the need for others is not apparent until testing is well underway.

Following connection of all inputs and outputs, the supervisory routine is entered into the computer and checked out using one or more application programs. Special programs are written to pass data between system programs as required. The other application programs are added one by one. Many of the programs will still require revision because of mating problems with other programs. After the addition of major program segments, test problems are run. When the control subsystem is to be added, the program is modified in such a way that output is to an output typer rather than to the plant. Once the engineer gains confidence in the results, output directly to the plant is permitted.

Once the computer system is up and running, the system evaluation should be started. By collecting and analyzing plant data, it will be possible to compare expected benefits from computer control against the actual benefits. The task will not be easy, however, since the improvements by percentage are small, and it may be difficult to duplicate operating conditions and plant objectives (for which process and economic data are available) before computer control, with plant operation under computer control.

As plant management and operators become more familiar with the system and with the reports generated by the computer, they will probably ask for more information or want the information presented in a different manner. This will require changes in some of the programs. Also, the process analysts may find it desirable to place some changes into the application programs after studying the results of the process calculations. The systems programmer is likely to initiate changes and modifications to make the system appear more elegant. In any case, changes to existing programs and/or addition of new programs should be carefully checked to avoid damaging other programs.

Operator Training

The process operators should be trained in the use of the computer system. Training should commence shortly after the computer hardware is installed and checked out. The operator should learn the functions of the computer system and the ways he can use it to perform his work more easily and efficiently. He should be told that it is just a machine and is at his disposal, just like the analog controllers.

The procedures the operator must follow in communicating with the computer should be carefully reviewed. Several planned sessions of from 2 to 4 hours each should be scheduled and devoted to actual "hands-on" experience with computer equipment and particularly with the process operator's console.

The operator should also know when and how to take control loops off computer control. In addition, the operator and instrument engineer should be aware of the implications that instrument or process equipment failure have on the computer control scheme, logs, displays, etc. and be aware of what action is required.

The operator should learn the backup or fail-safe provisions that have been provided in case the computer or major peripherals fail, and what role, if any, he plays in such situation.

A computer log book should be made available to the operators where they can record system malfunctions or happenings that are unusual or unexpected.

Finally, an operator's manual is prepared containing all the information described above. Periodically, the systems programmer should review pertinent changes to the program with the operators and see that the operator's manual is updated accordingly.

Documentation

Although documentation is discussed last, its importance ranks close to the top. System documentation should not be an afterthought and should not be prepared after the computer system is in operation. The procedures for documenting the system should be layed out and followed from the very beginning.

It was pointed out earlier that there are 50 to 100 program segments that must be fitted together. This requires that detailed specifications be written for each program and strictly adhered to.

Although changes to program specifications should be avoided, they will happen. By providing good documentation along with proper updates, programmers affected by the changes will be alerted. Interfacing and testing of the system programs are made easier by good documentation. Program bugs and their cause can be pinpointed sooner and corrective action taken.

There are basically four kinds of software documentation: a functional description, data specification, program specification, and a flow chart.

The *functional description* is a narrative of what the system is programmed to do. The inputs (direct and manual) to the system, calculations performed, operator logs, management reports, control outputs, etc. are explained in detail. This document includes a description of the computer modules, their characteristics, and how they are used in the computer system. It also includes an overall flow chart to show the flow of information through the system.

The *data specification* contains a layout of core and bulk memory, describes all file records, tables, and the format and meaning of system messages.

The *program specification* describes each of the system programs. It indicates the purpose of the program, the coding language, inputs to the routine, outputs and calculated quantities, linkages with other programs, calling sequence, variables and constants used, the program size and its normal storage location (core or bulk memory).

A *flow chart of the program* is included, if warranted. When the programs are completed, the flow chart is updated, and written comments are inserted next to it to make it easier to understand. Cross reference to the coding listing is made on the flow chart by an identification number, letter, or symbol.

The hardware system documentation includes an operational description of the computer equipment and a block diagram of major equipment modules. The description covers the computer's physical and electrical specifications.

At an early stage of project planning a site plan is prepared. It shows the computer room, the location of computing equipment, power requirements, the layout of computer conduit and signal troughs, and the grounding scheme.

A detailed tabulation of analog/digital inputs and analog/digital outputs is prepared showing multiplexer addresses, computer I/O terminal designations and corresponding plant identification. The information is used by the instrument engineer during tie-in of the plant with the computer system and by the systems programmer when he builds tables that contain the input/output information.

The documentation also includes a description of hardware acceptance tests and includes sign-off sheets used in accepting the equipment.

8.4 COMPUTER LANGUAGES

Accurate and efficient communication between humans of the same intelligence level is a difficult enough task. Communication between intelligent programmers and flip-flop elements should represent an even greater challenge. This section is devoted to the subject of computer languages, serving as communication tools in this field.

The internal workings of a digital computer must necessarily be confined to the manipulation of binary bits. This provides the maximum efficient use of the components of the computer and the best possible speed of operation. Several stages of complexity are available to the computer designer to interface this binary machine with the digitally oriented human. These levels include machine language, assembly language, FORTRAN, other higher level languages, and the *control language*. The subsequent discussion of computer languages will be divided into these logical stages, to parallel the stages of complexity from the machine to the human programmer, engineer and operator. This order is also roughly parallel to the historical development of computer languages. Early computers, programmed in machine language, were difficult to program because the idiosyncracies of each machine were built into the language. The *assembly language* provided a more generalized approach, since the idiosyncracies of the machine could be built into the translator program which converted a more generalized assembly language to machine language. FORTRAN was planned to be a universal language, written the same by all programmers and translated into machine language by the machine itself. But FORTRAN has been presented in versions I, II, III, IV, each of which has specific rules, for specific classes of machines. (Only FORTRAN II and FORTRAN IV have survived). Now that control languages are available, they are as diversified as the machines. With each user and each manufacturer having his own version of a control language, the permutations and combinations have become large enough to warrant a full-scale effort by several academic groups to try to develop a unified control language.

Hardware vs Software

In implementing a computer project, policy decisions must be made as to which part of the project should be implemented in hardware and which

915

part should be implemented in software. The hardware portion of the computer comprises the electronic and mechanical devices in the computer. The software portion includes the program itself and the manner in which it is written. Certain portions of every calculation, such as multiplication, can be implemented by hardware using a wired program instead of the usual software sequence of instructions. A hardware implementation of an arithmetic operation reduces its operation time considerably. For example, a 4 microsecond machine might be able to do a multiplication in 4 microseconds using a wired program, while the software version using a sequence of instructions would depend upon the length of each of the numbers, and might take 200 to 400 microseconds.

The distinctive useful feature of the digital computer is its use of the stored program. It can reduce a sequence of operations to a sequence of symbols in its memory which are called upon when needed to go into operation. It can select any part of the program stored to carry out operations based on conditional instructions and logical decisions. The nature and character of the computer's software determines its capability for rapid and efficient programming. Deep in the machine, software for the control computer is contained in machine language instructions, i.e., it is written in the language of the machine. But the machines are equipped with translating capabilities so that at the input-output interface, the language that the programmer must use may be oriented toward mnemonic symbols, or even algebraic or control symbols. Each machine language instruction requires the machine to perform one operation. The mnemonic equivalent of this step-by-step instruction procedure is called *assembly language*. The computer assembles the program by translating each instruction symbol from its alphabetical shorthand comprehensible to man, to a numeric shorthand comprehensible to the machine. Even more compatible with man's normal language is FORTRAN (for FORmula TRANslation). This category of language, now available in many versions, is written, and punched on cards in algebraic-type symbols. The computer compiles the program from FORTRAN to a computer-oriented machine language, and acts upon the machine language instructions to carry out the program.

Control oriented languages similar to FORTRAN have been written for specific control computers. These contain instructions like LOG and SCAN, which are difficult to implement in lower level languages by inexperienced programmers. This evolution in control languages has culminated in a process engineers' oriented "language" in which the engineer need only fill in the blanks of a standard form. The standard form is converted into punched cards, which the computer is programmed to accept, and is translated by a suitable compiler to a machine-oriented language. In some control computers, the data on this form are submitted by the operator's console. On others, the console is designed to modify some of the data, such as the control set points. The use of the console is a hardware implementation in contrast to the software implementation available from the punched cards.

The software program of a control computer must function in real time as each point of the process is scanned at an appropriate rate corresponding to the speed of response required. Thus, a flow loop would be scanned more rapidly than a temperature loop, since temperature is usually associated with a large thermal lag, which allows only slow changes in this variable. The cycle time of a computer scanning mechanism (multiplexer) is usually in the range of 100–250 points per second. The software must advise the scanner how often to look at each point. Most flow loops are scanned once every second, pressure or level loops once every 5 seconds, and temperature measurements once every 30 seconds. High-speed aerospace data scanners can utilize speeds as fast as 1,000 to 4,000 points per second, but this kind of speed is not usually required in industrial applications.

Machine Language

To provide ease in reading by the programmer or operator, machines programmed in machine language translate their binary numbers into a more intelligible mode before displaying them to their human mentors. Thus though all machines operate in simple binary, the machines may display the contents of their registers in binary-code-decimal, in octal, or in hexadecimal form. The Greek-Latin subscripts used for naming the radix (or base) of these number systems are octal for base 8, and hexadecimal for base 16. Both the octal and hexadecimal systems are combinations of binary systems, and hence can readily be explained in terms of the properties of simple binary numbers.

Binary Numbers

The binary number system is most efficient for use by the electrical relays, switches, tubes, transistors, or other bistable devices available in the digital machines. The magnetic core memory, and the codes on punched tape and magnetic tape also are most amenable to this sort of information system. The system requires only two symbols, which can be called ON and OFF, HIGH and LOW, or just 0 and 1. Use of the 0 and 1 symbols relates the system to the field of binary numbers and permits representation of any integer, fraction, decimal number, positive or negative number by such a symbol. Letters of the alphabet can be represented by a binary code, as can instructions to the computer such as ADD, or SKIP, or READ.

The binary number, therefore, meets the basic requirements of a digital system because it can represent both *data,* in the form of numbers and letters, and *instructions,* which are part of a program to be acted upon by the computer. The two basic electronic components used to store binary numbers are the flip-flop and the magnetic toroid. The flip-flop is used to store numbers and instructions in transfer, while the toroid is used to store numbers in the core memory, awaiting use by the rest of the computer. The bulk memory contains the binary digits in storage, as does the magnetic tape, which stores the data as the presence or absence of a magnetic pulse. The punched tape and the punched cards provide binary information in the form of the presence or absence of a hole in the card or tape (See Table 8.4a).

Table 8.4a

NUMBER SYSTEMS:

BINARY, DECIMAL, OCTAL AND

HEXADECIMAL

Binary	Decimal	3-Bit Octal	4-Bit Hexadecimal	Hex Symbol
0	0	000	0000	0
1	1	001	0001	1
10	2	010	0010	2
11	3	011	0011	3
100	4	100	0100	4
101	5	101	0101	5
110	6	110	0110	6
111	7	111	0111	7
1000	8	1 000	1000	8
1001	9	1 001	1001	9
1010	10	1 010	1010	A
1011	11	1 011	1011	B
1100	12	1 100	1100	C
1101	13	1 101	1101	D
1110	14	1 110	1110	E
1111	15	1 111	1111	F

Just as a decimal number represents information as a sum of multiples of powers of ten, a binary number represents the information as a sum of multiples of powers of two. The decimal number 453 represents $4 \times 10^2 + 5 \times 10^1 + 3 \times 10^0$. In general a number to a base r is represented by a sum $= a_0 r^0 + a_1 r^1 + a_2 r^2 + \cdots$ plus as many terms as are required to represent the number. Thus, if we had a duodecimal number system (base 12)

$$(453)_{12} = 4 \times 12^2 + 5 \times 12^1 + 3 \times 12^0 = (639)_{10} \qquad 8.4(1)$$

453 to the base 12 would be equal to 639 to the base 10. If we were looking for a representation of 453 in the binary system, we would note that

$$(453)_{10} = 1 \times 2^8 + 1 \times 2^7 + 1 \times 2^6 + 1 \times 2^2 + 1 \times 2^0 \qquad 8.4(2)$$

Bits, Bytes and Words

To the computer, the binary digit is a bit, represented in transition by a pulse in a computer word. At rest, the bit becomes a position of a flip-flop, or the magnetic direction of a ferrite core, or a mark on tape, or disc, or the presence or absence of a hole on a punched tape or card. Computer *words* contain 10 to 36 bits, or pulses. A *byte* is a collection of bits, shorter than a word, but long enough to have some decimal or alpha-numeric meaning. The EBCDIC (Extended Binary Coded Decimal Interchange Code) can be represented by an 8-bit *byte*, which in turn can be represented by two hexadecimal symbols.

A 4-bit byte can represent the decimal numbers from zero to fifteen as shown in Table 8.4a. Note the octal and hexadecimal representations of these decimal numbers. To represent symbols for which there are no names (i.e., 10, 11, 12, 13, 14 and 15) in base 16, the letters A, B, C, D, E and F are used. Obviously,

$$(10)_{16} = (16)_{10} \qquad\qquad 8.4(3)$$

that is, the symbol 10 in base 16 represents the decimal number 16, in base ten.

Table 8.4b illustrates the relationship between the EBCDIC code and several other common codes used in computer interfacing. The Hollerith code is used for punched cards, with the rows numbered from the top down, 12, 11, 0–1 to 9, as shown in Figure 8.4c. The ASCII (American Standard Code for Information Interchange) used for punched tape, teletype transmission, etc. requires a 7-bit byte. One hole of the 8-hole tape is used for feed and reference. The magnetic tape code also used a 7-bit byte but is not identical to the ASCII code for punched tape.

Machine Language Instructions

The computer word required to make the computer perform an operation must be comprehensible to the machine and hence must be based upon the inner structure of the machine. Multiple address machines require not only the instruction, but the address of each of the operands, the address of the result and the address of the next instruction. This results in a four-address machine. These functions in modern machines are fulfilled by hardware. Thus, the address of the next instruction is obtainable from an instruction counter. The use of the instruction counter also facilitates the use of "indexing." If a number other than 1 is added to the instruction counter it will SKIP, or BRANCH, to that instruction.

A three-address machine might have the instruction

$$345 \quad 456 \quad 01 \quad 567$$

which means, "take the number contained in cell 345," "add it to" "the number contained in cell 456" and "place the result in cell 567," with 01 being the code for ADD. An accumulator, a register connected to an adding circuit, which accumulates a sum like an adding machine register, reduces

Table 8.4b

ALPHANUMERIC COMPUTER CODES INCLUDING EBCDIC, HOLLERITH AND ASCII

Character	EBCDIC Code – Binary Code	EBCDIC Code – Hexadecimal Code	Hollerith Card Code	ASCII Code – Punched Tape Code	ASCII Code – Magnetic Tape Code
A	1100 0001	C1	12–1	0 · 0 0	1 · 1 1 1
B	1100 0010	C2	12–2	0 · 0 0	1 · 1 1 1
C	1100 0011	C3	12–3	0 0 · 0 0 0	1 1 · 1 1
D	1100 0100	C4	12–4	0 · 0 0	1 · 1 1 1
E	1100 0101	C5	12–5	0 0 · 0 0 0	1 1 · 1 1
F	1100 0110	C6	12–6	0 0 · 0 0 0	1 1 · 1 1
G	1100 0111	C7	12–7	0 0 0 · 0 0	1 1 1 · 1 1 1
H	1100 1000	C8	12–8	· 0 0 0	1 · 1 1 1
I	1100 1001	C9	12–9	0 · 0 0 0 0	1 1 · 1 1
J	1101 0001	D1	11–1	0 · 0 0	1 · 1
K	1101 0010	D2	11–2	0 · 0 0	1 · 1
L	1101 0011	D3	11–3	0 0 · 0	1 1 · 1 1
M	1101 0100	D4	11–4	0 · 0 0	1 · 1
N	1101 0101	D5	11–5	0 0 · 0	1 1 · 1 1
O	1101 0110	D6	11–6	0 0 · 0	1 1 · 1 1
P	1101 0111	D7	11–7	0 0 0 · 0 0	1 1 1 · 1
Q	1101 1000	D8	11–8	· 0 0 0	1 · 1
R	1101 1001	D9	11–9	0 · 0 0	1 1 · 1 1
S	1110 0010	E2	0–2	0 · 0 0	1 · 1
T	1110 0011	E3	0–3	0 0 · 0	1 1 · 1 1
U	1110 0100	E4	0–4	0 · 0 0	1 · 1
V	1110 0101	E5	0–5	0 0 · 0	1 1 · 1 1
W	1110 0110	E6	0–6	0 0 · 0	1 1 · 1 1
X	1110 0111	E7	0–7	0 0 0 · 0 0	1 1 1 · 1
Y	1110 1000	E8	0–8	· 0 0 0	1 · 1
Z	1110 1001	E9	0–9	0 · 0 0	1 1 · 1 1
				1 2 4 · 8 C 0 X (check)	1 2 4 8 · A B C (check)

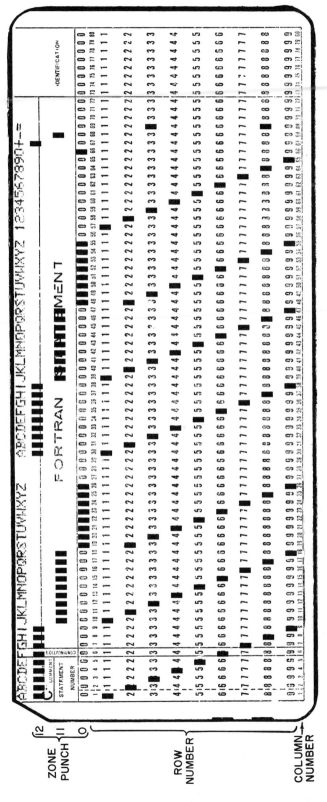

Fig. 8.4c   The Hollerith punched card code

the requirement to a single address per instruction. In a machine with an accumulator the operation above would be carried out in three steps:

	Operation Code	Address
1. Load contents of cell 345 into accumulator.	1100	345
2. Add contents of cell 456.	1000	456
3. Store sum at cell 567.	1101	567

where the symbols 1100, 1000 and 1101 are recognized by the computer as commands to LOAD, ADD and STORE. A list of typical operation codes in machine language is contained in Table 8.4d.

Table 8.4d
TYPICAL MACHINE LANGUAGE OPERATION CODES

	Type of Operation	Equivalent in Various Codes		
		Binary	Hex	Decimal
Load and	Load Accumulator	11000	18	10
Store	Store from Accumulator	11010	1A	11
Arithmetic	Add	10000	10	21
	Subtract	10010	12	22
	Multiply	10100	14	23
	Divide	10101	15	24
Input and	Read	00001	01	36
Output	Write	00010	02	38
Branch or	Unconditional	01000	08	49
Skip	Conditional	01001	09	—
	Wait	00110	05	41
	Halt	—	—	48

Assembly Language

Assembly language is a symbolic representation of the actual machine language of the computer. It provides direct access to the full power of the computer yet is readily understandable, and it requires no knowledge of binary systems. Instead it requires a knowledge of a symbolic language which indicates in alphabetical terms the function that each operation is required to perform. With assembly language one can maximize computer speed and minimize computer use of core memory. Table 8.4e illustrates a typical assembly language mnemonic code. Let us use that code to perform the typical PID algorithm given in equation 8.4(4) and control the temperature of a fluid emerging from a steam-heated heat exchanger (Table 8.4f).

Table 8.4e

TYPICAL ASSEMBLY LANGUAGE MNEMONIC OPERATION CODE

	Type of Operation	Code
Load and Store	Load Accumulator	LDA
	Load Index	LDX
	Load Status	LDS
	Store from Accumulator	STA
Arithmetic	Add	ADD
	Subtract	SUB
	Multiply	MUL
	Divide	DIV
Shift (Accumulator)	Shift Left	SLA
	Shift Right	SLR
Branch or Skip	Branch or Skip on Condition	BSC
	Modify Index	MDX
Input and Output	Read	RED
	Write	PNT

$$P = K_P e_n + K_I \left(e_n + \sum_0^{n-1} e_i \right) + K_D \frac{e_n - e_{n-1}}{\text{time}} \qquad 8.4(4)$$

First we must assume that another program is scanning the input sensors and is depositing the temperatures (and other variables) in proper positions in the computer memory. Our assembly language can use symbols both for the operations we ask the computer to perform and for the locations in memory.

The source program in assembly language consists of a series of statements, each one of which corresponds to a single computer operation. Each statement contains a label, an operation and an address. In addition, a tag index advises whether an index register is used and, if so, which one. The contents of the index register may be used in addition to an address to provide for locating information (a varying number of increments from a fixed index). If the fixed index refers to a variable, the displacement may refer to a function of that variable, such as its set point, proportional band, reset rate, derivative constant, minimum or maximum alarm level, etc. (see Table 8.4g).

Higher Level Languages

Of the many higher level languages that have been devised to make the computer available to engineers, FORTRAN has become the most popular in the United States. In Europe, ALGOL (a universal ALGOrithm Language) has become accepted. For business applications COBOL (COmmon Business

Table 8.4f

ASSEMBLY PROGRAM TO CALCULATE REQUIRED
OUTPUT SIGNAL FROM TEMPERATURE DEVIATION

Operation Tag	Operand	Comments
LDA	TEM5	Load accumulator with temperature 5
SUB	SPT5	Subtract set point of temperature 5
STA	ERT5	Store error signal of temperature 5
SUB	PET5	Subtract previous error of temperature 5
DIV	DLT5	Divide by time interval of sampling #5
MUL	KDT5	Multiply by derivative mode constant
STA	DET5	*Store derivative correction term*
LDA	ERT5	Recall error signal of temperature 5
MUL	KPT5	Multiply by proportional mode constant
STA	PRT5	*Store proportional correction term*
LDA	ERT5	Recall error signal of temperature 5
MUL	KIT5	Multiply by integral mode constant
ADD	INT5	Add previous accumulated error integral
STA	INT5	Store new accumulated error integral
ADD	PRT5	Add proportional correction term
ADD	DET5	Add derivative correction term
STA	VOT5	*Store result as valve output for #5*
LDA	ERT5	Recall error signal of temperature 5
STA	PET5	Store as next previous error signal

Oriented Language) is preferred. The triumvirate ALGOL, COBOL, and FORTRAN constitute the major man-oriented languages for use with computers. Another language, PL/1 (Programming Language—one) has been devised to combine the advantages of COBOL and FORTRAN. A discussion of FORTRAN will serve to indicate how such languages are devised, how they provide an improved

Table 8.4g

OVERALL PID CONTROL PROGRAM FOR ENTIRE PLANT

Label	Operation	Tag	Operand	Comments
	LDX	X1	1	Load index with number 1
START	LDA	X1	VAR	Load accumulator with variable located at VAR +1
	SUB	X1	SPT	Subtract set point of variable 1
				For specific steps to follow, see Table 8.4f.
	STA	X1	PET	Store error signal as next previous error
	BSC	X1	MAX	Compare X1 with maximum number of points
	MDX	X1	1	Increase index by 1 if X1 < MAX
	LDX	X1	1	Return to point 1 if X1 = MAX
	MDX	1C	START	Modify instruction counter to go to start

technological interface with the computer, and how the computer is programmed to use them.

FORTRAN

Of the two surviving FORTRAN systems, FORTRAN IV has been designated "standard FORTRAN" by the United States of America Standards Institute (USASI), and FORTRAN II USASI has designated "basic FORTRAN."

In writing a FORTRAN program one must observe a series of rules for variable, constant, and statement. A FORTRAN variable is represented by one to five letters or numbers, but the first symbol must be a letter. If the first letter is I, J, K, L, M or N, the variable is an integer. If it is another letter of the alphabet, the variable is a decimal number. This makes it possible to use a symbol like TEMP5 for temperature-five. Symbols like VEL, DISP, RATE, TIME can be used for velocity, displacement, rate and time.

Constants are symbolized by the presence of the decimal point to indicate a decimal number. Hence 1 represents an integer, and 1.0 a decimal number.

Arithmetic operations, such as multiply, divide, add and subtract, are represented as in algebra, except that the asterisk (*) represents multiply instead of the ambiguous ×, and the slant (/) represents divide. To raise to a power, a double asterisk is used (**). The equal sign is utilized to complete an equation, but its effect on the computer is significantly different. The equal sign advises the computer that the result of a calculation should be placed in a memory position. Consequently, only a single variable can appear to the left of an equal sign. When this variable is called upon in later calculations, the contents of the memory cell specified by that variable is fetched and used. Only numerical calculations can be performed in FORTRAN, although it often gives the illusion that actual algebraic equations are being solved.

The general second-degree quadratic equation

$$y = ax^2 + bx + c \qquad\qquad 8.4(5)$$

is written in FORTRAN as

$$Y = A*X**2 + B*X + C \qquad\qquad 8.4(6)$$

For the computer to perform this calculation it must have values of A, B, C and X. The A, B and C may be fixed and X varied until Y becomes zero. When this is done, the procedure illustrates the FORTRAN equivalent of solving an equation.

Other FORTRAN expressions include READ, WRITE, GO TO, IF, and DO. The DO loop is the most powerful tool of FORTRAN, since it permits iterative repetition of the same calculation, with different numbers, for a finite number of times. Thus one can obtain a solution to an equation, or to a series of equations, by iterative procedures. Specific rules for use of these statements are contained in the FORTRAN manual that accompanies each machine. The general nature of the technique illustrates how FORTRAN can be used for process control and how it can lead to other higher level control languages (see Table 8.4h).

FORTRAN can handle arrays of numbers, work with matrices, use a list of supplied functions such as logarithm and arc tangent, or use a list of functions prepared by the programmer, for a specific type of problem. For example, each of the control algorithms can be called up as a specific FORTRAN subfunction if it has been written into the memory of the computer.

Table 8.4h

FORTRAN PROGRAM FOR PID ALGORITHM GIVEN IN
EQUATION 8.4(4)

	DO 30 N	= 1,K
	PRE(N)	= ERR(N)
	ERR(N)	= TEMP(N) − SPT(N)
	PSI(N)	= KP*ERR(N) + KI*ERR(N) +
		+ SUM(N) + KD*(ERR(N) − PRE(N))/TIME
	SUM(N)	= SUM(N-1) + KI*ERR(N)
30	CONTINUE	

Control Languages

Higher level languages for use in control are not like FORTRAN, which can be arranged in sentences to perform a computation. Higher level control languages are implemented by filling in boxes in a form sheet corresponding to parameters of equations. The operational equations are written and supplied by the manufacturer as part of the software of the computer. In computer systems terminology, the control used by a program operates in the "interpretive mode," rather than in the "compiler mode" typical of FORTRAN.

The engineer will make his first contact with higher level languages when evaluating them for tasks of the purchasing decision type. Later he will have occasion to use such a language in filling out the forms supplied to him by a vendor, to implement vendor-supplied software. A typical "fill-in-the-blanks" form is idealized in Table 8.4i. In this form the engineer supplies details for each loop of his process, such as identification of the process, the unit, and the loop; engineering units to be used in console display; alarm conditions for maximum and minimum values; set points; maximum rate of error change; proportional, reset, and rate constants, etc. Provisions are made for standards, for special control algorithms, for selection of timing intervals and for calling special programs under alarm action, should an emergency develop.

Control Software

Let us now follow through a complete software implementation program, starting with a process-oriented control drawing and ending with the machine

Table 8.4i

IDEALIZED FILL-IN-THE-BLANKS COMPUTER FORM

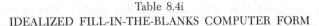

	Primary	Secondary
Sensor Type	TC–IC	TC–IC
Signal Range	0–6 mv	0–6 mv
Engineering Units	0–200°F	0–200°F
Control Algorithm	PID	P + batch type I
Sampling Interval	60 sec	60 sec
Maximum Measurement	180°F	200°F
Minimum Measurement	50°F	50°F
High Alarm	175°F	200°F
Low Alarm	125°F (during reaction)	—
Maximum Rate of Change	1°F/interval	1°F/interval
Set Point	150°F	160°F (initial)
Proportional Constant	0.50	0.50
Reset Constant	0.42	0.42
Derivative Constant	0.21	—
Control Action	Direct	Direct
Valve Action	Direct	Reverse
Valve Characteristic	Equal %	Equal %
Output Signal Range	4–20 madc	4–20 madc

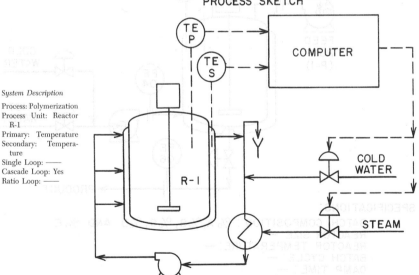

PROCESS SKETCH

System Description

Process: Polymerization
Process Unit: Reactor
 R-1
Primary: Temperature
Secondary: Tempera-
 ture
Single Loop: ——
Cascade Loop: Yes
Ratio Loop: ——

TE P

TE S

COMPUTER

R-1

COLD WATER

STEAM

language program that will carry out this control. The process-oriented performance drawing prepared by an instrument or process engineer is shown in Figure 8.4j. A similar drawing is prepared for each functioning piece of equipment or unit in the plant. An instrumentation or control engineer next prepares a logic diagram such as in Figure 8.4k from this sketch. This logic flow chart provides the program analyst with sufficient information to write the control program given in Table 8.4l. Additional control-oriented operations, available in the particular control computer used, would also be included in the control program.

The program is then debugged by the analyst-programmer and converted to a machine language program by a computer compiler. Debugging of the total computer package is performed by a computer systems engineer on the actual computer to be used, while the computer is being tested at the manufacturer's facility. The previous debugging of the program has been performed

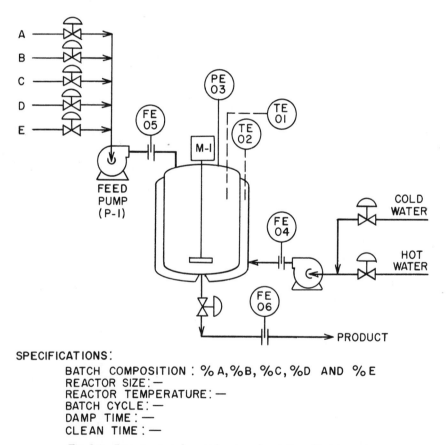

SPECIFICATIONS:

      BATCH COMPOSITION : % A, % B, % C, % D AND % E
      REACTOR SIZE: —
      REACTOR TEMPERATURE: —
      BATCH CYCLE: —
      DAMP TIME : —
      CLEAN TIME : —

Fig. 8.4j   Process-oriented control system description of batch reactor

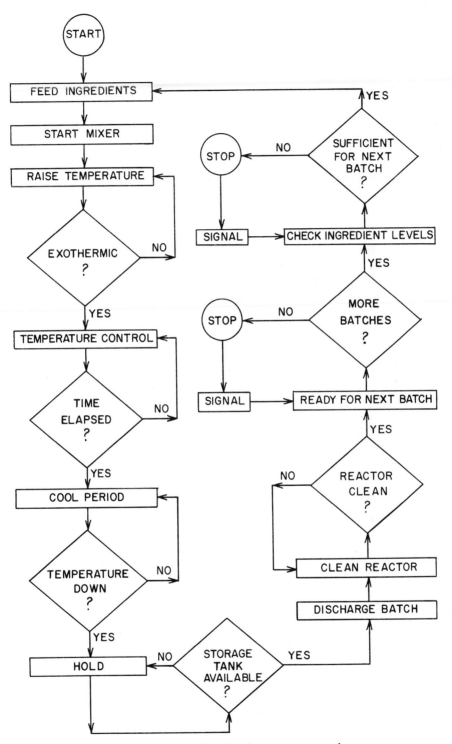

Fig. 8.4k   Logic flow chart for computer control

Table 8.4l

IDEALIZED COMPUTER CONTROL PROGRAM

Statement Number	Idealized Control Language
0001	START PROGRAM
0002	START PUMP P-1
0003	OPEN VALVE A
0004	CALL FLOW 05A
0005	CALL FLOW 05B
0006	CALL FLOW 05C
0007	CALL FLOW 05D
0008	CALL FLOW 05E
0009	START MIXER M-1
0010	RAISE TEMPERATURE
0011	IF (TEO2—TEO1) 12, 10, 10
0012	CALL TEMPERATURE CONTROL
0013	IF (TIME—BATCH TIME) 12, 14, 14
0014	CALL COOL
0015	IF (TE01–100) 16, 16, 14
0016	IF (PE03–20) 14, 17, 17
0017	CALL STORAGE
0018	IF (VESSEL) 19, 19, 20
0019	CALL HOLD
0020	CALL DUMP
0021	IF (FE06 TOTAL—BATCH) 20, 22, 22
0022	START WASH
0023	CALL WASH PROGRAM
0024	PRINT (3, 25)
0025	FORMAT ("READY FOR NEXT BATCH")
0026	SENSE SWITCH, 27, 28
0027	STOP
0028	CALL INGREDIENT LEVEL CHECK
0029	GO TO 02
0030	PRINT (3, 31)
0031	FORMAT ("INGREDIENT" 3X "LOW")
0032	STOP

in a simulated environment, an off-line facility available to user, vendor, engineer or consultant specifically for the purpose of program debugging. This procedure of debugging the program itself, evaluating interaction of the program with the computer, etc. (discussed in some detail in Section 8.3) will ensure a smooth startup of the computer in the process plant.

8.5 CONTROL ALGORITHMS

The pay-out from the control capability of the digital computer comes from special process control techniques available and defined as "control algorithms." In addition to the proportional, proportional-integral, proportional-derivative, and three-mode control techniques used in analog control, the digital computer is capable of providing a variety of control modes, ranging from the simple two-position (bang-bang) control, to a complex control mode based upon the solution of a state space matrix. The success of the digital control system is determined by the capability of the control engineer to select the optimum control algorithm for the control of each loop in the process. Features like adaptive control are readily realizable in a digital control package and prove most effective in operation under changing load conditions. An algorithm is defined as a sequence of calculations performed to obtain a given result. Control algorithms have been developed for either fixed or adaptive control modes, and they may use a variety of control logic, including equations, tables, and matrix algebra.

Position Algorithms

The position algorithm is the digital equivalent of the predominantly used analog control scheme. The output of the controller is a valve position, scaled 0 to 100 percent open. In reverse acting controllers, as the deviation increases, the valve closes. Equation 8.5(1) describes the analog equation for a proportional controller,

$$P = M_0 + KP(\text{set point} - \text{measurement}) \qquad 8.5(1)$$

where P = required valve position in PSIG;
   M_0 = normal valve position, when the system is on set point; assumed to be 9 PSIG;
   KP = proportional gain.

Assuming that temperature is the measured variable and that the gain KP is 0.22, if the process temperature is 10°F below the set point of 100°F, the resulting valve position is 11.2 PSIG. On the other hand, if the detected temperature is 110°F, the resulting valve position signal is 6.8 PSIG.

931

Equation 8.5(1) can be expressed as a digital algorithm:

$$PSI = MID + KP[SP - TEMP(N)] \qquad 8.5(2)$$

since the temperature at time N is a discrete number, unlike the continuously varying number we have in the analog case.

Since we shall be scanning all points with the same digital device, we need a tag for the particular temperature loop considered in this calculation. Using K for this symbol, and again using the FORTRAN notation as more appropriate for a digital calculation, we obtain

$$PSI(K) = MID(K) + KP(K)*[SP(K) - TEMP(K, N)] \qquad 8.5(3)$$

where a dual index array is required for the temperature, the Nth value of the Kth loop.

To expand this notation to proportional integral control, note that digital (discrete) integration is merely the sum of the errors, and therefore

$$PSI(K) = MID(K) + KP(K)*ERR(K, N) + KI(K)*SUM(K, N) \qquad 8.5(4)$$

which is the digital algorithm for PI control modes, where

$$ERR(K, N) = SP(K) - TEMP(K, N) \qquad 8.5(5)$$

and

$$SUM(K, N) = SUM(K, N - 1) + ERR(K, N) \qquad 8.5(6)$$

This notation can also be applied to a 3-mode controller to arrive at the corresponding PID algorithm:

$$\begin{aligned} PSI(K) = MID(K) &+ KP(K)*ERR(K, N) \\ &+ KI(K)*SUM(K, N) + KD(K)*DEL(K, N) \end{aligned} \qquad 8.5(7)$$

For a PD controller, the algorithm is

$$PSI(K) = MID(K) + KP(K)*ERR(K, N) + KD(K)*DEL(K, N) \qquad 8.5(8)$$

In equations 8.5(7) and 8.5(8),

$$DEL(K, N) = [ERR(K, N) - ERR(K, N - 1)]/TIME(K) \qquad 8.5(9)$$

and TIME(K) is the sampling time for loop K.

Velocity Algorithms

An interesting observation made early in digital control history, was that the valve can be used as an integrator. Since the valve maintains its previous position we need only inform the valve, at any given sampling interval N, how far it must move from its previous position. This information is available from the rate of error change in the last sampling period, i.e., the velocity of its change. The resulting velocity algorithm can be obtained by numerical differentiation of the previously derived position algorithm. For the PID case, we have

$$\frac{DPSI(K)}{TIME} = KP(K)*[ERR(K, N) - ERR(K, N - 1)]$$

$$+ KI(K)*ERR(K, N) + KD(K)*\frac{DEDEL(K, N)}{TIME**2} \qquad 8.5(10)$$

The problem of how to determine the second derivative of the error DEDEL has been disposed of by various approximations for this term. One method might be to note that

$$DEDEL = [ERR(K, N) - ERR(K, N - 1)]/TIME**2$$
$$-[ERR(K, N - 1) - ERR(K, N - 2)]/TIME**2 \qquad 8.5(11)$$

which becomes

$$DEDEL = [ERR(K, N) - 2.0*(ERR(K, N - 1)$$
$$+ ERR(K, N - 2)]/TIME**2 \qquad 8.5(12)$$

Fortunately, the digital computer can easily store the errors:

$$ERR(K, N), \qquad ERR(K, N - 1) \qquad and \qquad ERR(K, N - 2)$$

A problem occurs when the sign of the error changes during a sampling period, causing an incorrect directional signal to the valve. This is corrected by an additional logic check, which is made to guarantee that the sign of the error signal and the direction of valve motion are in agreement.

Feedforward Algorithms

Control of many processes with long deadtimes, or with long time lags, can be improved by feedforward control, and digital control provides an easy means to accomplish this improvement. Figure 8.5a illustrates a much simplified feedforward control diagram. The purpose of the feedforward loop in this computer is to adjust the set points of the product flow controllers as a function of feed flow and composition.

$$SPX = X*F/NUM + SPX \qquad 8.5(13)$$
$$SPY = Y*F/NUM + SPY \qquad 8.5(14)$$

where

$$NUM = (1 + INDEX)/TIME \qquad 8.5(15)$$

and

$$INDEX = ITER + 1. \qquad 8.5(16)$$

With this program, the feedforward control moves the set points by one increment toward their final position.

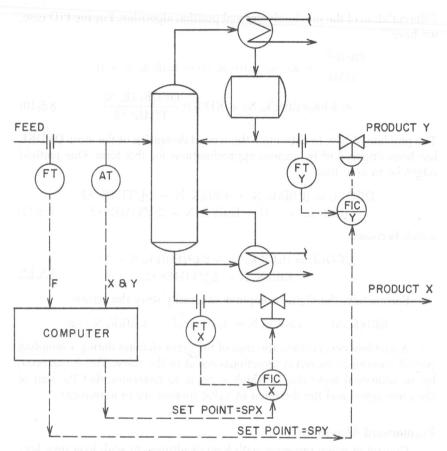

Fig. 8.5a   Feedforward control diagram for a distillation column

Series Algorithms

The use of control schemes similar to that employed in analog technology restricts the total power of digital control. A generalized control scheme can be derived involving a function of both the error signals and the measured quantities, each term appropriately weighted to obtain optimal control. A series equation has been used to express this concept as follows:

$$PSI = WK(1)*ERR(N) + WK(2)*ERR(N - 1) + \cdots$$
$$+ WM(1)*VAR(N) + WM(2)*VAR(N - 1) + \cdots \qquad 8.5(17)$$

where $WK(1)$ = weighing error constant 1 [for error (N)] and

           $WM(1)$ = weighing measurement constant 1 [for measurement (N)].

Rather than being oriented in time, another series algorithm emphasizes the deviation from a *target* value. It calculates the velocity signal from the target error, the controlled variable error, and the uncontrolled variable deviation, as follows:

$$DELPSI = WTF(1)*DELT + WTF(2)*DELM$$
$$+ WTF(3)*DELU \qquad 8.5(18)$$

where DELPSI = velocity signal to control element,

    WTF(k) = k weighing function,

    DELT = deviation in target value (e.g., efficiency of unit heater),

    DELM = deviation in controlled variable (e.g., process temperature), and

    DELU = deviation in uncontrolled variable (e.g., environment temperature).

The weighing constants are obtained empirically using an analog simulation of the process. If the process is not readily simulated, the constants can be obtained by adaptive development, with the unit on-line, using the ability of the computer to find optimum values. The incremental control feature prevents gross instabilities or overall errors, since correction is taking place periodically, and repeatedly over small increments of time.

Optimization of Algorithm Use

Optimization of algorithm use involves, first, selection of the optimum algorithm, and second, optimization of the algorithm parameters. Various techniques have been suggested in Section 7.12 to arrive at the optimum proportional band, reset rate and derivative time settings for PI, PD or PID controllers. However, the digital computer can do better than that because it can tune itself to the prevailing conditions of the process. (Also refer to Section 7.7 on adaptive control.)

A process unit for which self-adjusting control is required, is one with variable deadtime. A heat exchanger, for example, will vary its deadtime as a function of the process fluid flow rate. Some computer manufacturers provide a self-tuning algorithm in their software package. The algorithm approximates the time delay of the process by comparing the present measured variable with a variable stored for N periods.

$$BETA = ETD - SMV*N*TIME \qquad 8.5(19)$$

$$EMV(K) = PEMV(K) - SMV(K) - CORR \qquad 8.5(20)$$

$$CORR = ((PEMV(K + N) - PEMV(K - N))$$
$$*BETA/2*N*TIME)$$
$$+ ((PEMV(K + N) + PEMV(K - N) - 2PEMV(K))$$
$$*BETA**2)/(2*(N*TIME)**2 \qquad 8.5(21)$$

where   BETA = correction factor,

                 ETD = estimated time delay,

SMV = stored measured variable from N periods age,

TIME = sampling time,

EMV(K) = estimated measured variable at sampling time K,

PEMV(K) = previously estimated measured variable at sampling time K − 1,

SMV(K) = stored measured variable at sampling time K, and

CORR = correction to measured variable.

At any given time the estimated error EE(I) is obtained from a series of previously stored EMV values as follows:

$$EE(I) = ALPA(1)*EMV(1) + ALPA(2)*EMV(I - 1) + \cdots \quad 8.5(22)$$

where ALPA(K) are estimated weighing factors, based on actual or estimated instrument and process dynamics.

8.6 SIGNAL CONDITIONING AND DESIRABLE WIRING PRACTICES

In process control, the accuracy, sensitivity and response time are all critically dependent on how transducers are connected to the controller or computer. The measurement input to a control system contains two components: the signal and noise. The latter incorporates all other fluctuations in the input signal that do not contain useful control information. The useful information is limited to those fluctuations which the control action can reduce by appropriate manipulation of process parameters. Very often the signal is deteriorated by noise interference which, in many cases, can be more significant than the signal itself.

Generally speaking, noise includes three main sources:

1. Uncontrollable process disturbances that are too rapid to be reduced by control action.
2. Measurement noise resulting from such factors as turbulence around flow sensors, instrument noise, etc.
3. Stray electrical pickup such as from AC power lines, pulses from power switching, RF interference, etc.

Noise adversely affects both analog and digital electronic equipment, but in digital systems the noise effects are lasting, resulting in miscounts and loss of data.

Typical ranges for these noise components are shown in Figure 8.6a. Stray electrical pickup can be minimized by such good engineering and installation practices as proper shielding, screening, grounding, and routing of wires. Even so, noise still exists at the input, and the effect of noise on controlled and manipulated variables can be significantly reduced through judicious selection of filtering. In most cases, the relatively high-frequency noise (i.e., above approximately 5 Hz) is removed by a conventional analog filter, and the remaining noise occupying the range of 0.002 to 5 Hz, may require either special analog filtering and/or digital filtering.

It is good practice to follow these three steps to reduce the effects of noise on control quality:

1. Estimate the dynamic characteristics (amplitude and bandwidth) of the noise.

937

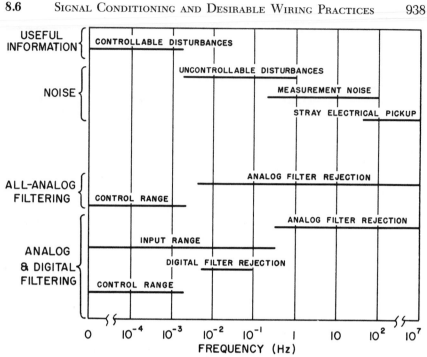

Fig. 8.6a   Frequency range of input signals and filtering functions

2. Determine its effect on the controlled and manipulated variables.
3. Select the best choice of analog and/or digital filtering.

This section is devoted to a close examination of noise suppression techniques. Noise suppression is a system problem, and discussion of noise therefore should not be limited to the signal transmission medium. The noise suppression methods must be applied consistently from the transducer, through the signal transmission line to the signal conditioner, multiplexer and/or analog to digital converters.

Types and Sources of Noise

As shown in Figure 8.6b, signal leads can pick up two types of external noises: normal mode and common mode interferences.

Interference which enters the signal path as a differential voltage across the two wires is referred to as normal mode interference. The normal mode noise cannot be distinguished from the signal coming from the transducer.

Interference which appears between ground and both signal leads, as an identical voltage, is referred to as common mode interference.

There are many sources of noise interference. The following are the most common of these sources:

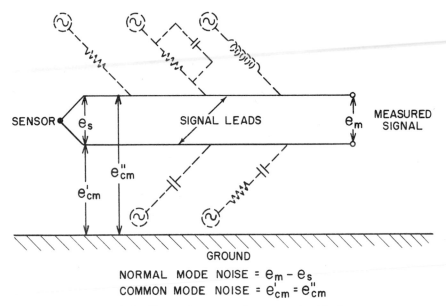

$$\text{NORMAL MODE NOISE} = e_m - e_s$$
$$\text{COMMON MODE NOISE} = e'_{cm} = e''_{cm}$$

Fig. 8.6b   Electrical noise interferences on signal leads

Inductive pickup from power sources. This includes 60-Hz noise from power lines, 120-Hz noise from fluorescent lighting, as well as high-frequency noise generated by electric arcs and pulse transmitting devices. The worst man-made electrical noise results from the opening or closing of electrical circuits containing inductance. The amplitude of transients generated in an inductive circuit is given by equation 8.6(1):

$$\epsilon_i = L(d_i/d_t) \qquad\qquad 8.6(1)$$

This amplitude is calculated as the inductance times the rate of change in the switched current. Closing an inductive circuit can cause a transient of twice the input voltage, and opening such a circuit can cause transients as high as ten times the supply voltage. The noise transients are proportional to the instantaneous value of the AC supply when switching inductive loads from an AC supply.

Common impedance coupling "ground loops." Placing more than one ground on a signal circuit produces a ground loop which is a very good antenna, and noise signals induced in the loop are easily coupled into the signal lines. This can generate noise that will completely obscure the useful signal.

Electrostatic coupling to AC signals. The distributed capacity between signal conductors and from signal conductors to ground provides a low-impedance path for cross talk and for signal contamination from external sources.

Ineffective temperature compensation in transducers, lead wire systems, amplifiers and measuring instruments can create changes in system sensitivity and drift of "zero" line. This is especially important in strain gauge transducers, resistance type temperature sensors and other balanced-bridge devices.

Loading the signal source. When a transducer or other signal source is connected in parallel with an amplifier or measuring device with a low input impedance, the signal voltage is attenuated because of the shunting effect. This can considerably decrease the sensitivity of the system.

Variable contact resistance. All resistance type transducers, as well as bridge type circuits, are susceptible to changes in contact resistance. The measuring instrument is unable to distinguish between a resistance change in the sensor and a resistance change in the external wiring.

Conducted AC line transients. Large voltage fluctuations or other forms of severe electrical transients in AC power lines (such as are caused by lightning) are frequently conducted into the electronic systems by AC power supply cords.

Conduction pickup refers to interfering signals which appear across the receiver input terminals because of leakage paths caused by moisture, poor insulation, etc.

Thermoelectric drift is created by the junction of two dissimilar metal wires, and it varies with changes in temperature. This is more critical in DC circuits in the microvolt level.

Electrochemically generated corrosion potentials can often occur, especially at carelessly soldered junctions.

Failure to distinguish between a ground and a common line. The ground currents that exist in a "daisy-chain" equipment arrangement are primarily noise signals induced into the equipment, or perhaps generated by some equipment, and they find their way back to earth. The resulting voltage drops in the ground wire place each piece of equipment at a different potential with respect to earth. This potential difference is easily coupled into the signal lines between equipment.

Use of the same common line for both power and signal circuits or between two different signal lines can cause the appearance of transients on the signal and can also cause cross talk between the two signals.

Inadequate common mode rejection, due to line unbalance, will convert a common mode noise interference to a normal mode noise signal.

Table 8.6c shows the average noise conditions that exist in various industries. Some chemical plants have experienced as high as 60 volts common mode noise interference. Electrical noise interference can be a severe problem in industries such as steel, power and petroleum, where power consumption is high and complex electrical networks exist.

Table 8.6c
AVERAGE NOISE CONDITIONS EXPERIENCED IN VARIOUS INDUSTRIES

Description		Industry		
		Chemical	Steel	Aerospace
Normal Signal Levels (mv)		10–100	10–100	10–1,000
Possible or Expected Noise Level	Normal mode (mv)	1–10	2–7	1–10
	Common mode (volts)	4–5	4–5	4–5
Desired Noise Rejection	Normal mode	$10^3:1$	$10^3:1$	$10^3:1$
	Common mode	$10^6:1$	$10^6:1$	$10^6:1$

Reducing Electrical Noise Interference

There are several ways of reducing the effects of noise on control systems, such as line filtering, integrating, digital filtering, and improving the signal-to-noise ratio, but none of these will completely solve the problems without proper cable selection and installation.

Careful examination of the various sources of noise shows that some can be eliminated by proper wiring practice. This implies good temperature compensation, perfect contacts and insulation, carefully soldered joints, use of good quality solder flux, and use of the same wire material. However additional precautions are needed to eliminate pickup noise entirely.

Circuit Arrangement

One of the important practices in eliminating noise in low-level signal systems is the arrangement of external circuits so that noise pickup will appear equally on both sides of the signal pair, remain equal, and appear simultaneously at both input terminals of the differential amplifier (or measuring device). In this case it can be rejected without affecting the frequency response or the accuracy of the system by means of common mode rejection.

If a noise signal is allowed to combine with the useful signal, it can be separated only by selective filtering.

A special discussion of thermocouple signal conditioning is warranted, since it is widely used in various industries. Thermocouple signals are liable to serious noise problems because they are often in contact with the device or structure under test. Additional thermoelectric potentials developed in the bimetallic leads should be accounted for when measuring a signal from a thermocouple. The best thermocouple materials are not the best electrical conductors, yet lead wires must be of these same materials all the way back to the reference junction. (See Volume I, Section 4.9, for more details.)

Standard practice to avoid the problem of accounting for each bimetallic junction is to compare the output of the thermocouple with the output of an identical thermocouple in a controlled temperature environment. The balance of the circuit in that case is represented by copper wires creating an identical couple in each lead and producing a zero net effect when both couples are at the same temperature.

The constant-temperature reference junction is the most accurate method, but it is not the most convenient. A simulated constant-temperature reference can be had with devices containing compensating junctions, millivolt sources and temperature-sensitive resistors. They permit a transition from thermocouple material leads to copper lead wires.

A more complex conditioning circuitry is shown in Figure 8.6d, which performs the functions of thermocouple signal conditioning, noise discrimination, balancing, ranging and standardizing. This circuit must be tied to a calibration scheme for the system output. The balance circuit shown accomplishes both variable offset and calibration. The signal-conditioning circuitry makes it possible to convert the transducer output to make best use of the optimum range of the measuring system for best linearity, accuracy and resolution.

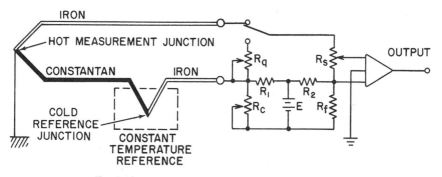

Fig. 8.6d   A typical (complex) thermocouple conditioner

E = A precision isolated power supply
R_1 and R_2 = Precision bridge resistors, low values (5 or 10 ohms)
R_c = Balance control variable resistor, large resistance value
R_f = Reference resistor, large resistance value
R_q = Equivalent line resistor for use during standardizing and calibrating
R_s = Span adjustment potentiometer

Because the thermocouples are often in electrical contact with the device or structure of which the temperature is being measured in addition to an intentional *ground* at the sensor, differential input amplifiers with very good common mode rejection are required to effectively discriminate against noise in the structure or grounding system.

Impedance Level

The common mode rejection of a good low-level signal amplifier is in the order of 10^6 to 1. This value is always decreased when the signal source and input signal leads are connected to the amplifier. Signal source impedance and signal lead impedance have a shunting effect on the signal input to the amplifier. From a point of view of maximum power transfer, equal impedance on both sides of a circuit is an ideal termination. In the transmission of voltage signals, however, current flow has to be reduced to as close to zero as possible. This is achieved by selecting amplifiers with high input impedance and using transducers with low output impedance. Good instrumentation practice dictates that the input impedance of the amplifier be at least ten times the output impedance of the signal source.

However, low signal source impedance has an adverse effect on pickup signals, because these will be shunted by the low impedance of the transducer. An additional step is to ground the transducer whenever possible, and to run a balanced signal line to the amplifier input.

There are three basic techniques to condition transducers with high output impedance:

> High-impedance voltage amplifiers
> Charge amplifiers
> Transducer-integrated unloading amplifiers

Voltage amplifiers must have an amplifier input impedance that is very high compared with the source impedance. In this way, the amplifier's effect on the phase and amplitude characteristics of the system is minimized. A typical transducer with output amplifier is shown in Figure 8.6e. High impedance is practical for an amplifier, but an equally significant portion of the

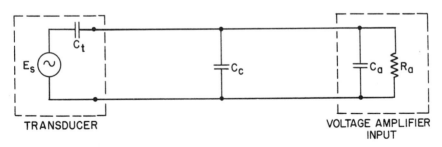

Fig. 8.6e   A basic voltage equivalent transducer with output amplifier

E_s = Transducer voltage source
C_t = Transducer capacitance
C_c = Signal cable shunt capacitance
C_a = Amplifier input capacitance
R_a = Amplifier input resistance

load is in the interconnecting cable itself, and it does not take much cable to substantially lower the available voltage.

In applying a voltage amplifier, low-frequency system response must be considered. The voltage amplifier input resistance (R_a) in combination with the total shunt capacitance forms a high-pass first-order filter with a time constant (T), defined by

$$T = R_a(C_t + C_c + C_a) \qquad 8.6(2)$$

Cutoff frequency can become a problem when it approaches information frequency at low source capacitances (short cable or transducers with very low capacitance) or at lower amplifier-input resistances.

Charge amplifiers have been widely used in recent years. This approach avoids the cable capacitance effects on system gain and frequency response. The typical charge amplifier shown in Figure 8.6f is essentially an operational

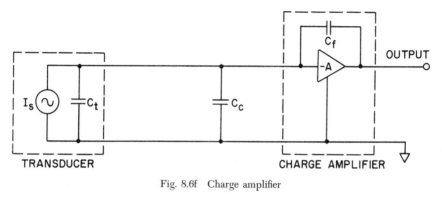

Fig. 8.6f   Charge amplifier

I_s = Transducer current source
C_f = Feedback capacitance
C_t = Transducer capacitance
C_c = Signal cable shunt capacitance

amplifier with integrating feedback. A charge amplifier is a device with a complex input impedance which includes a dynamic capacitive component so large that the effect of varying input shunt capacitance is swamped, and the output is the integral of the input current. Filtering of the resultant signal on both the low and high ends of the information band is desirable at times so that it is possible to get a rather high order of rejection without affecting information by using bandpass filters. The addition of a resistor in parallel with the feedback capacitor in a charge amplifier will decrease the closed-loop gain at low frequencies, resulting in the desired high-pass filter characteristics.

Unloading amplifiers integrally mounted in a transducer housing (Figure 8.6g) have become available from transducer manufacturers, because neither voltage amplifiers nor charge amplifiers offer a very satisfactory solution to

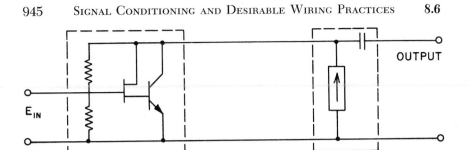

Fig. 8.6g   The unloading amplifier, located at the transducer, reduces the input capacitance relative to voltage or charge amplifiers

the conditioning problem for systems with very high input capacitance (usually a result of very long lines). With the voltage amplifier, signal-to-noise ratio suffers, because capacitance loading decreases the available signal. In a charge amplifier, the signal-to-noise ratio suffers because of the increased noise level (input noise is a direct function of input capacitance).

Thus, remote signal conditioning appears to offer a satisfactory solution to the accommodation of long data lines. If closely located to the transducer, the voltage-responding or charge-responding amplifiers are equally effective. However, these techniques decrease the dynamic range capability (changing input amplifier gain to accomplish a range change is not possible) and restrict the high-frequency response signal amplitudes due to the limited current capability of the remote amplifier. When these two limitations are overcome, almost all of the signal-conditioning equipment will be at the remote location.

In summary the following steps are recommended:

1. Select a signal source with low-impedance output.
2. Select an amplifier (or measuring device) with high-impedance output.
3. Use a balanced line from signal source to amplifier input (maximum allowable unbalance is 100 ohms/1,000 ft).
4. Keep signal cables as short as possible.
5. Use remotely located amplifiers, when long signal cables are required (except for thermocouple and RTD signals).
6. Select a signal source that can be grounded (thermocouples, center-tapped sensors, etc.).

It is evident that common-mode rejection must be maintained at a high level in order to attain noise-free results from low-level signal sources.

Ground Systems

Good grounding is essential for normal operation of any measurement system. The term "grounding" is generally defined as a low-impedance metallic connection to a properly designed ground grid, located in the earth. In large equipment, it is very difficult to identify where the ground is. On standard all-steel racks, less than 6 ft in length, differences in potential of up to 15 volts peak-to-peak have been measured. Stable, low-impedance grounding is necessary to attain effective shielding of low-level circuits, to provide a stable reference for making voltage measurements and to establish a solid base for the rejection of unwanted common-mode signals.

In a relatively small installation, two basic grounding systems should be provided. First, all low-level measurements and recording systems should be provided with a stable *system ground*. Its primary function is to assure that electronic enclosures and chassis are maintained at zero potential. As previously mentioned, a stable system ground is not easy to accomplish. However, in most cases, a third copper conductor in all electrical circuits, which is *not* a current carrying conductor, but is firmly tied to the electric power ground or to the building ground on the cold water system, would provide a satisfactory system ground. In the event that such a ground is unavailable or unstable, a satisfactory system ground can usually be established by running one or more heavy copper conductors to copper water lines or to properly designed ground grids.

Signals can be measured with respect to the system reference ground only if the input signals are fully floating with respect to ground. In this case, the stable system ground fulfills the task of providing a base for common-mode noise rejection.

The other important ground is the *signal ground*. This system is necessary to ensure a low-noise signal reference to ground. This ground should be a low-impedance circuit providing a solid reference to all low-level signal sources and thus minimizing the introduction of interference voltages into the signal circuit. The signal ground should be insulated from other grounding systems, and it is generally undesirable to connect it to the system ground at any point (Figure 8.6h). In a single-point grounding system, no current flows in the ground reference, and if signal cable is properly selected, noise due to large and hard-to-handle low-frequency magnetic fields will not exist. It should be emphasized that a signal circuit should be grounded at one point and at one point only preferably at the signal source (Figure 8.6i).

By connecting more than one ground to a single signal circuit, as shown in Figure 8.6h, a ground loop is created, because two separate grounds are seldom, if ever, at the same potential, this will generate a current flow which is in series with the signal leads. Thus the noise signal is combined with the useful signal. These ground loops are capable of generating noise signals that can be 100 times larger than the typical low-level signal.

In off-ground measurements and recording, the cable shield is not grounded, but it is stabilized with respect to the useful signal through a

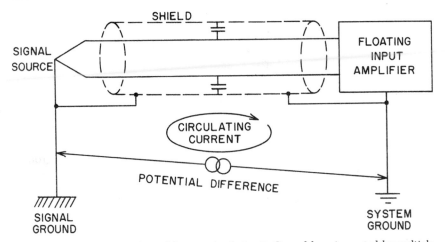

Fig. 8.6h   Incorrect grounding of floating signal circuit. Ground loop is created by multiple grounds in a circuit, by grounding the shield at both ends

connection to either the center tap or the low side of the signal source. Since the shield is driven by an off-ground voltage, appropriate insulation is needed between the shield and the outside of the cable.

It is important that electric racks and cabinets be connected to a proper system ground and not allowed to contact any other grounded element in the building.

Guidelines on grounding can be summarized as follows:

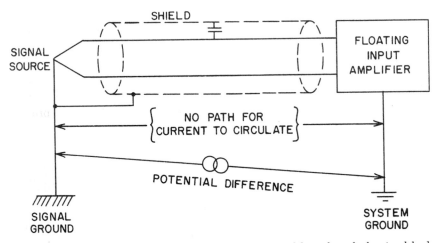

Fig. 8.6i   Correct grounding of floating signal circuit. Ground loop through the signal lead is eliminated by grounding the shield at signal end only

1. Intentional or accidental ground loops in either the signal circuit or the signal cable shield will produce excessive electrical noise in all low-level circuits and will destroy the useful signal.
2. Every low-level data system should have a stable system ground and a good signal ground.
3. A signal circuit should be grounded at only one point.
4. The signal cable shield should not be attached to more than one grounding system.
5. Always ground a floating signal circuit and its signal cable shield at the signal source only.

Wiring

Another important aspect of reducing noise pickup involves the wiring system used to transmit the signal from its source to the measuring device or computer.

In less demanding low-frequency systems, where the signal bandwidth is virtually steady state and system accuracy requirements are not very high, two-wire signal leads will normally suffice. A third wire, or shield, becomes necessary when any of the above parameters are exceeded.

Where top performance is required, the shield is run all the way from the signal source to the receiving device. As already mentioned, the shield should be grounded at the signal source and not at the recorder, because this arrangement provides maximum rejection of common-mode noise. The cable shield reduces electrostatic noise pickup in the signal cable, improves system accuracy and is indispensable in low-level signal applications where high source impedance, good accuracy or high-frequency response is involved. As the signal frequency approaches that of the noise, which is usually at 60 Hz, filtering can no longer be used to separate noise from the useful signal. Therefore the only practical solution is to protect the signal lines and prevent their noise pickup in the first place.

Elimination of noise interference due to magnetic fields, can also be accomplished by wire-twisting (transpositions). If a signal line consisting of two parallel leads is run along with a third wire, carrying an alternating voltage and an alternating current, the magnetic field surrounding the disturbing line will be intercepted by both wires of the signal circuit. Since these two wires are at different distances from the disturbing line, a differential voltage will be developed across them. If the signal wires are twisted (Figure 8.6j), the induced disturbing voltage will have the same magnitude and cancel out.

To prevent noise pickup from electrostatic fields, low-level signal conductors must be surrounded by an effective shield. One type of shield consists of a woven metal braid around the signal pair, which is placed under an outside layer of insulation. This type of shield gives only 85 percent coverage of the

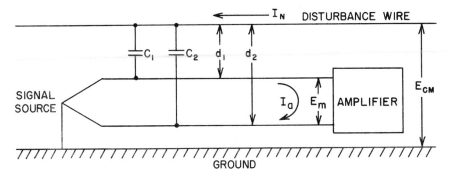

WHERE $d_1 \neq d_2$. THUS, $C_1 \neq C_2$, ∴ DIFFERENCE IN DISTANCE FROM
DISTURBANCE WIRE CREATES A DIFFERENTIAL VOLTAGE, AND A
NOISE SIGNAL WILL BE INDUCED INTO THE SIGNAL LEADS.

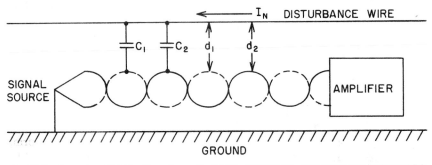

WHERE $d_1 = d_2$. THUS $C_1 = C_2$, ∴ THE INDUCED DISTURBING VOLTAGES
WILL HAVE THE SAME MAGNITUDE AND CANCEL OUT WHEN
TWISTED WIRES ARE USED.

Fig. 8.6j   Twisted wire eliminates the noise interference due to magnetic field from disturbing
wires

signal line and is adequate for some applications if the signal conductors have
at least ten twists per foot. Its leakage capacity is about 0.1 picofarad per
foot. At the microvolt signal levels this kind of shielding is not satisfactory.
Another type of signal cable is shielded with lapped foil (usually aluminum-
Mylar tape) shields, plus a low-resistance drain wire. This type of shielding
provides 100 percent coverage of the signal line, reducing the leakage capacity
to 0.01 picofarads per foot. To be most effective, the wires should have six
twists per foot. The lapped foil shield is covered with an insulating jacket
to prevent accidental grounding of the shield. Care should be exercised to
prevent the foil from opening at bends.

This type of low-level signal cable provides the following benefits:

1. An almost perfect shield between the signal leads and ground.
2. Magnetic pickup is very low because the signal leads are twisted.

3. Shield resistance is very low because of the low-resistance drain wire.

The use of ordinary conduit is a questionable means of reducing noise pickup. Its serious limitation is in obtaining or selecting a single ground point. If the conduit could be insulated from its supports, it would provide a far better electrostatic shield.

The following general wiring rules should be observed in installing low-level signal circuits:

1. Never use the signal cable shield as a signal conductor, and never splice a low-level circuit.
2. The signal cable shield must be maintained at a fixed potential with respect to the circuit being protected.
3. The minimum signal interconnection must be a pair of uniform, twisted wires, and all return current paths must be confined to the same signal cable.
4. Low-level signal cables should be terminated with short, untwisted lengths of wire, which expose a minimum area to inductive pickup.
5. Reduce exposed circuit area by connecting all signal pairs to adjacent pins in connector.
6. Cable shields must be carried through the connector on pins adjacent to the signal pairs.
7. Use extra pins in connector as a shield around signal pairs by shorting pins together at both ends and by connecting to signal cable shield.
8. Separate low-level circuits from noisy circuits and power cables by maximum physical distance of up to 3 ft and definitely not less than 1 ft.
9. Cross low-level circuits and noisy circuits at right angles and at maximum practical distance.
10. Use individual twisted shielded pairs for each transducer. Thermocouple transducers may be used with a common shield when the physical layout allows multiple pair extension leads.
11. Unused shielded conductors in a low-level signal cable should be single-end grounded with the shield grounded at the opposite end.
12. High standards of workmanship must be rigidly enforced at all times.

Filtering

Filtering is required to stabilize the input signal and to remove AC noise components (particularly 60-Hz noise) resulting from direct connection of normal mode AC signals, from normal mode noise pick-up or from conversion of common-mode noise to normal-mode noise due to line unbalance. A reliable, low-cost and effective filter for analog inputs is the balanced resistance-capacitance filter, shown in Figure 8.6k. Its ability to eliminate AC

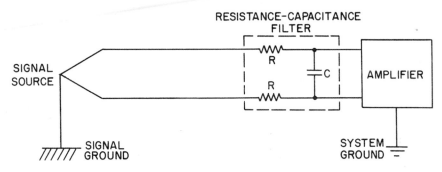

Fig. 8.6k   A basic balanced resistance-capacitance filter

components increases exponentially with the frequency of noise signal components. Common-mode noise rejection of about 40 decibels is possible with this type of filter, with decibel defined as

$$\text{decibel} = 20 \log\frac{\text{inlet noise amplitude}}{\text{outlet noise amplitude}} \qquad 8.6(3)$$

Filtering action causes a time delay (time constant, T = RC) between a signal change at the transducer and the time of recognition of this change by the measuring device.

As the time delay may be in the order of one second or more, there are situations in which this has to be reduced (systems with high-frequency response). In order not to decrease the filtering efficiency but only the time delay, inductance-capacitance filters may be used (Figure 8.6l). This increases the filter cost.

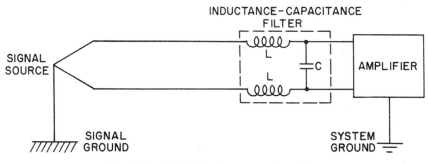

Fig. 8.6l   A basic inductance-capacitance filter

Since a filter limits the bandwidth of the transmitted signal, it might be desirable to use more complicated and expensive filters when higher-frequency AC transducer signals are involved. Fortunately, this type of situation seldom occurs.

Amplifier Guard

In a normal measuring device, a signal line is connected to a differential DC amplifier with floating inputs. Generally, these amplifiers are provided with an internal floating shield which surrounds the entire input section, as shown in Figure 8.6m. This floating internal shield is called a "guard shield" or simply a "guard."

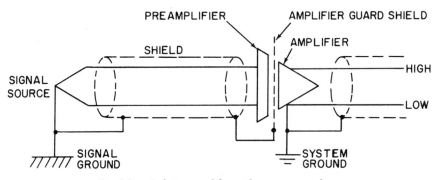

Fig. 8.6m   Isolating amplifier with proper grounding

The guard principle requires that the amplifier guard be driven at the common-mode voltage appearing at the amplifier inputs. The most effective way to do this is as follows:

1. Connect amplifier guard to the signal cable shield and make sure that the cable shield is insulated from chassis ground or from any extension of the system ground.
2. Connect the ground of the signal source to the signal cable shield. In this way, the amplifier guard and signal cable shield are stabilized with respect to the signal from the source.
3. Connect the signal cable shield and its tap to the signal source and also to the signal ground, which should be as close as possible to the signal source. This will limit the maximum common-mode voltage. The signal pair must not be connected to ground at any other point.
4. Connect the amplifier chassis, equipment enclosure, the low side of amplifier output and output cable shield to the system ground.

Power Transformer Guard

It is important to avoid strong magnetic fields emanating from the power supply transformer. To avoid capacitive coupling, the transformer should be provided with at least two or three shielding systems. The third or final shield should be connected to the power supply output common. The inner shield should be connected to the signal ground.

Low-Level Signal Multiplexing

Conventional noise rejection techniques such as twisting leads, shielding, and single-point grounding, have been discussed. It was pointed out that input guarding involves isolating the transducer input signal from the common-mode voltage between signal leads and ground. This change from differential to single-ended signal can occur any place in the system after the multiplexer. Early versions of multiplexing systems employed crossbar switches, relays, and similar electromechanical devices for low-level input signal multiplexing.

A simple passive filter (RC) circuit is shown in Figure 8.6n, which can

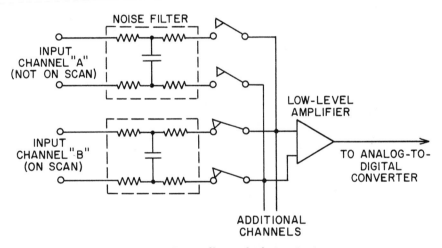

Fig. 8.6n   Passive filter multiplexing circuit

be designed to reject common-mode noise from about 40 decibels, at the selected frequency. More sophisticated passive networks (such as parallel T or notch filters) improve noise rejection, but it is hard to obtain 60-decibel noise rejection with passive circuits.

Because of deficiencies in the noise rejection capabilities of the earlier approaches, the limitations in scan rates, and with the ever-increasing use of data acquisition systems in many fields of application, a myriad of devices have been developed with extended capability to cover the spectrum of present-day requirements. Each general category contains subsets of devices using the same basic switching element, but offering application-dependent variations. The most important variations are the programmable range (i.e. gas chromatograph signal) switching and input grounding.

There are three commonly used multiplexing techniques: (1) capacitive-transfer-multiplexer, (2) three-wire multiplexer, and (3) solid-state multiplexer.

The capacitive-transfer (flying capacitor) switching arrangement, shown in Figure 8.6o is simple, economical and capable of great noise rejection

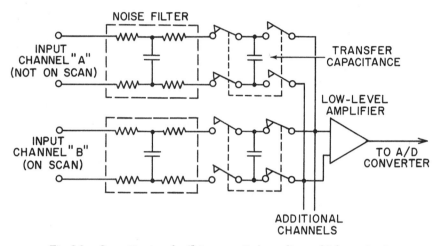

Fig. 8.6o   Capacitive transfer (flying capacitor) coupling multiplexer circuit

(including random noise spikes). This system is limited to applications where signal bandwidth requirements are narrow (on the order of 0.5 to 1 Hz) due to a necessarily large transfer capacitance to minimize the effect on system resolution of charge losses and of delays during charging, settling and digitizing.

In the capacitive transfer circuit, between scans, a set of normally closed contacts connect the low-leakage transfer capacitor across the input signal. Common practice is to short the amplifier input during this between-scan period to avoid stray pickup (due to high-impedance open circuit). When the multiplexer selects the input for scan, the amplifier input short is removed, and the contacts switch to connect the transfer capacitor to the amplifier input.

This transfer circuit introduces input attenuation and phase lag, which must be considered in circuit design. An additional RC filter is usually necessary to achieve acceptable common mode rejection of 100 to 120 decibels.

The three-wire multiplexing system requires the transducer lead wires to be shielded with the shield terminated at the signal ground, and this guard shield must be carried through the multiplexer, at each point, up where the differential signal is transferred to a ground-referenced signal.

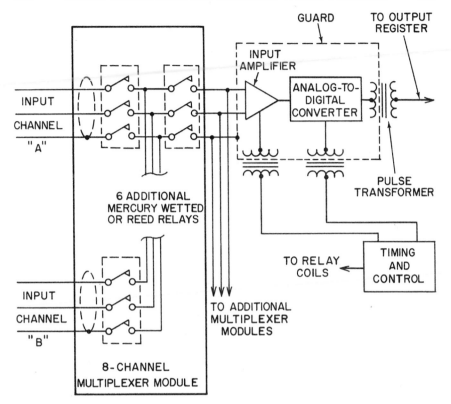

Fig. 8.6p   Circuit diagram of electromechanical, high common-mode voltage rejection, three-wire multiplexing system

In Figure 8.6p the input amplifier and analog-to-digital converter are enclosed within the guard. The serial digital data is transmitted through a shielded pulse transformer to an output register for presentation in parallel form. Relay coils are matrixed and controlled by an address logic. This system is used when input signal bandwidths of several hertz are required, since filtering is not essential to obtain good common-mode rejection. Common-mode rejection from DC to 60 Hz is about 100 decibels at reduced input bandwidth.

A solid-state multiplexing system is shown in Figure 8.6q. In a typical high-performance solid-state multiplexer, each input has a matched pair of transistor switches terminating at the primary of shielded input transformers, which are driven through an isolation transformer. One cycle of a square wave of peak amplitude equal to the input signal level is transferred across the transformer by alternately pulsing the switches. This signal is synchronously rectified to preserve original input polarity integrated over the cycle period, and it is amplified and held for digitizing by an integrator and hold amplifier.

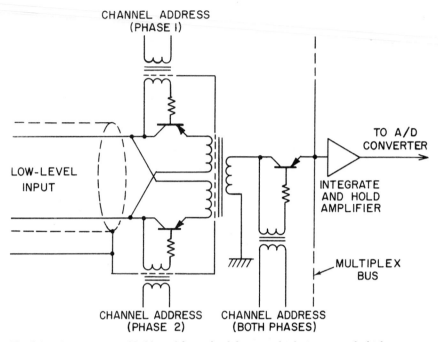

Fig. 8.6q   An integrate-and-hold amplifier with solid-state multiplexing system for high common-mode rejection

The cost of the system is relatively high, making the application economically impractical unless the high common-mode tolerance is required. Common-mode rejection from DC to 60 Hz of 120 decibels is easily obtained.

Selection of any multiplexing system should be based on the performance, reliability and cost of a particular application. Table 8.6r summarizes the features of the systems discussed.

Noise Rejection in A/D Converters

The dominant noise in A/D converters is line frequency noise. One approach toward reducing this noise is to integrate the input signal. The integrating technique relies on A/D converter hardware. The operation of the A/D converter is such that it converts the continuously monitored measurement into a pulse train and totals the number of pulses it receives. If the measurement time is equal to the period of the line frequency, integration yields the true value of signal level, and the line frequency noise effect becomes zero, as shown in Figure 8.6s.

This method is usually applied to slow multiplexers (e.g., 40 points per second scan rate) and is suitable for applications in process monitoring and control where high sampling frequencies are not required. In appropriate situations, it provides noise rejection of about 1,000 to 1 (or 60 decibels) at 60 Hz, and offers good rejection of other frequency noises.

Table 8.6r

FEATURE SUMMARY OF COMMONLY USED MULTIPLEXERS

Features	Capacitive-Transfer	Three-Wire	Solid-State
Cost	Low	Low	High
Input Signal Ranges	±50 to ±500mv	±5mv to ±5 volts	From 5mv full scale to 100mv full scale
Scan Rates (points per second)	Up to 200	Up to 200	Up to 20,000[1]
Operable Common-Mode Environment (volts)	Up to 200 to 300	Up to 200 to 300	Up to 500
Common-Mode Rejection from DC to 60 Hz (decibels)	100 to 120[2]	100 to 120	120
Accuracy	$\pm0.1\%$ full scale at 1 to 10 samples per second scan rate. $\pm0.25\%$ full scale at 10 to 50 samples per second	$\pm0.1\%$ full scale for all scan rates	$\pm0.1\%$ full scale[3]

[1] 10-Microvolt resolution in high common-mode environments.
[2] With a two-section filter.
[3] Overall accuracy.

Common-Mode Rejection Measurement

A typical configuration of a guard-shielded measuring system is shown in Figure 8.6t. The basic definition of common-mode rejection is the ratio of the common-mode voltage (E_{cm}) between points C and D, to the portion of E_{cm} which appears at the amplifier inputs A and B. Almost all of the remainder of the voltage E_{cm} will appear across C_g and R_g, in which case the common-mode rejection (CMR) is approximately

$$CMR_{(AC)} = \frac{1}{2\pi f C_g R_c} \qquad 8.6(4)$$

$$CMR_{(DC)} = \frac{R_g}{R_c} \qquad 8.6(5)$$

It can be seen from these relations that the common-mode rejection is dependent on the value of R_c, which is the unbalance of the transducer and

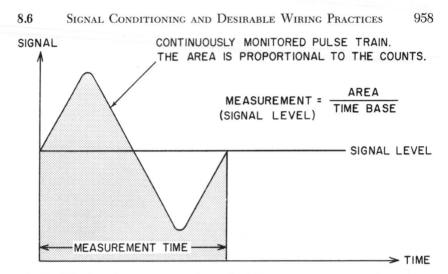

Fig. 8.6s   Line frequency noise reduction by A/D converter integration technique

signal lines. In an ideal case, when R_c is zero, the common-mode rejection will be infinite.

Measuring that portion of E_{cm} which appears across R_c by direct methods is not possible, since it is unlikely that any instrument could be connected to the source to measure the voltage across R_c without also changing the current through it. Thus, when testing the guarding of a system, the component due to E_{cm} at the amplifier output is measured. This value is divided by the amplifier gain, and it is assumed that the quotient is a good measure of the voltage across R_c.

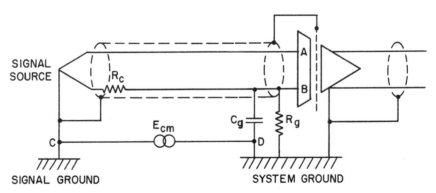

Fig. 8.6t   A typical configuration of a guard-shielded measuring system

R_c = Cable unbalance resistance
C_g = Measuring circuits/system ground capacitance
R_g = Measuring circuits/system ground resistance
E_{cm} = Common mode voltage

A practical setup for measuring rejection in a common-mode system is shown in Figure 8.6u. With the signal source disconnected from the signal ground, a 100-volt, 60-Hz signal from a signal generator is applied between the system ground and signal source. The change in the digital voltmeter

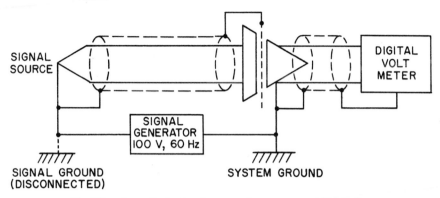

Fig. 8.6u   A practical setup for measuring common-mode rejection

(DVM) reading will show the effect of this common-mode signal on the system. To obtain the effective voltage across the amplifier inputs, this change is divided by the amplifier gain. The common-mode rejection is then obtained by dividing the signal generator voltage with the assumed voltage at the amplifier input. For example,

$$\text{Amplifier gain } (G) = 1{,}000$$
$$\text{Signal generator voltage } (E_{cm}) = 100 \text{ volts}$$
$$\text{DVM reading before applying } E_{cm} \ (e'_{cm}) = 2.0 \text{ volts}$$
$$\text{DVM reading after applying } E_{cm} \ (e''_{cm}) = 2.1 \text{ volts}$$

The increase in voltage at the amplifier output (Δe_{cm}) due to E_{cm} is

$$\Delta e_{cm} = e''_{cm} - e'_{cm} = 2.1 - 2.0 = 0.1 \text{ volts} \qquad 8.6(6)$$

The voltage at the amplifier input $(\Delta e'_{cm})$ due to E_{cm} is

$$\Delta e'_{cm} = \frac{\Delta e_{cm}}{G} = \frac{0.1}{1{,}000} = 0.0001 \text{ volts} \qquad 8.6(7)$$

Common-mode rejection of the system is

$$CMR = \frac{E_{cm}}{\Delta e'_{cm}} = \frac{100}{0.0001} = \frac{1{,}000{,}000}{1} = 10^6{:}1 \qquad 8.6(8)$$

When dealing with AC common-mode signals applied to DC measuring instruments, it is important to consider the effect of inherent noise rejection in the measuring instrument (or amplifier). Many DC instruments have an input filter which allows the undisturbed measurement of DC signals in the presence of AC noise. Such instruments are said to have normal-mode interference rejection. Thus, if common-mode rejection is measured by the indirect method just described, the apparent common-mode rejection will be the product of the actual common-mode rejection and of the filter rejection. In the earlier example, if a filter with 10:1 normal-mode rejection were inserted in front of the amplifier input, the apparent common-mode rejection would increase to 10^7:1.

Conclusions

Electrical noise is not unique to new installations. Even systems that are in satisfactory operation can develop noise as a result of burned or worn electrical contacts, defective suppressors, loose connections and the like. Any equipment used to identify the noise source must have the frequency response and rise time capability to display noise signals. Minimum requirements would be 10 MHz bandwidth and 35 nanosecond rise time. Once the noise signal is identified, its elimination involves applying the basic rules discussed earlier. The most important ones are listed below:

1. Select signal sources with low output impedance and with grounding capability.
2. Use only top quality signal cable, in which the signal pair is twisted and protected with a lapped foil shield, plus a low-resistance drain wire.
3. Provide a good low-resistance system ground and a stable signal ground located near the signal source.
4. Ground the signal cable shield and signal cable circuit at the source only, and keep them from contacting ground at any other point.
5. Select appropriate multiplexing and amplifier circuits having input guarding systems of at least 10^6:1 common-mode rejection at 60 Hz.
6. Provide triple-shielded power supplies for measuring devices.
7. Measure the common-mode rejection of the system by an indirect method whenever in doubt of the rejection capability.
8. Pay particular attention to the quality of the workmanship and system wiring practices.

8.7 COMPUTER INTERFACE HARDWARE

For a computer system with digital process control to be in direct communication with process instrumentation and equipment, connections between the two are required. Types of signals that can be sent and received by the computer and the process will usually be different, and conversions of signals to the appropriate type between the sending and receiving devices are required. The I/O (input-output) interface hardware handles the necessary signal conversions so that direct communication can take place.

In a process control computer system (Figure 8.7a), information concerning process operating conditions is made available through process instrumentation, either as continuous (analog) electrical or pneumatic signals or as

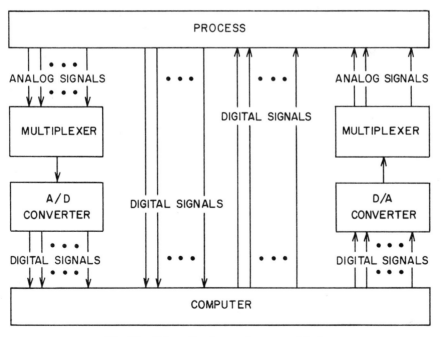

Fig. 8.7a Schematic of computer-process interface

discrete (digital) on-off type signals. To adjust process operating conditions, the process instrumentation may require either type of signal. The digital computer itself, on the other hand, can only accept or transmit digital signals having certain characteristics. Techniques and equipment for getting digital information to (digital input) and from (digital output) the computer, and for getting analog information to (analog input) and from (analog output) the computer are described in this section.

Digital Input

Digital information is generally electrical in nature and appears in the process as the status (open/closed) of relay contacts, as alternative voltage levels, or as a pulse or sequence of pulses. These digital signals are used to represent the status of valves (open/closed), switches (on/off), alarms (high/low), and the output of digital sensors such as turbine flowmeters and digital tachometers. The information can be scanned by the computer to determine the status of the process or to interrupt the computer in case of specific changes in status.

The most common form of digital input is the contact closure type, where an isolated set of contacts operates together with a process-related switch. The computer determines the status of these contacts by applying an electrical signal on one line and detecting with circuitry on the other line, whether an open or closed circuit exists. Typically, a current of 10 ma flows when the contacts are closed, and transistor switching at the computer interface indicates "1" for closed circuit and "0" for open circuit. Restrictions are placed on total loop resistance (maximum for closed circuit, minimum for open circuit), and maximum levels of shunt capacitance and series inductance. For voltage level inputs a similar scheme is used, except that the sensing voltage is that generated in the field, and an additional voltage supply at the computer interface is unnecessary (Figure 8.7b).

Digital inputs are usually treated at the computer end in groups corre-

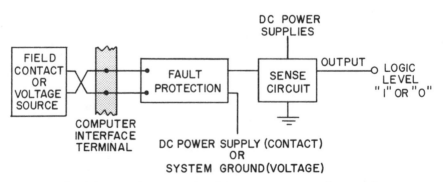

Fig. 8.7b   Digital input interface for sensing contact or voltage signals

sponding to the word length of the computer. A 16-bit machine, for example, would read digital input data on 16 lines simultaneously. The data, as a series of ones and zeros, would be stored internally as one word. The correspondence between bit position in the word and field function is maintained by software. If the number of digital inputs, exceeds the length of one computer word, multiplexing can be used to share one group of computer input lines among the several groups of digital input lines from the field.

Maximum scan rates depend on such factors as line and filter capacitances as well as on computer speed and interface design. Computer instructions can be used to address one word, and subroutines can be written to address, and store in memory tables, a group of words. Alternatively, hard-wired logic can be used to scan and store a group of words concurrently with the execution of other computer instructions. With short signal lines, no filtering, and direct-to-memory input channels, burst-mode scan rates on the order of 500,000 words per second are possible with modern process control computer systems.

Pulse inputs are handled in much the same way as logic voltage level inputs, except that the output of the sensing circuit is normally connected to a hardware counter rather than to a specific bit position in the input register. Each successive pulse increments or decrements the previous count by one, and computer instructions can be used to read and reset the counter register. Pulse rates are established by comparing contents of the counter register with the elapsed time between successive readings. Specifications on pulse inputs generally include voltage level ranges for both states and information on minimum pulse widths. Maximum pulse rates are limited by counter circuit requirements. If, for example, the electronics requires pulse widths of at least 100 microseconds (200 microseconds between successive pulses), the maximum pulse rate would be 5,000 pulses/second.

Change-of-state information, used to interrupt the computer, is sensed in the same way as other digital inputs. The output of the sense circuitry is, however, connected directly to the central processor control unit for interrupt signals. The occurrence of an external interrupt suspends the normal sequence of instruction execution and allows special interrupt response software routines to be executed.

Digital Outputs

Digital outputs enable the computer to provide on-off control to external devices such as relays, indicator lamps, small DC motors, etc., to generate pulse trains for stepping motor devices, and to transmit data to other computers. Digital outputs can also be used to generate analog outputs through digital-to-analog converters. In general, the computer interface provides solid-state or relay switching for a group of signal pairs which are connected to external devices. As with digital inputs, the size of a group depends on the computer word length, and several groups of field lines can share one group of computer lines.

Upon computer instruction, a word contained in an internal register or in a memory location is transferred to a digital output register.

For latching outputs, the "ones" in the word typically cause the switch or relay associated with the bit position to latch, while the "zeros" cause the switch to unlatch. In turn, external devices assigned to switches (bit positions) can be turned on or off, depending on whether the external contacts are normally open or normally closed.

In pulse outputs, the presence of a 1 usually causes a pulse (momentary closure) to be generated on the line corresponding to the bit position, while a 0 causes no pulse. In so-called register outputs, the computer supplies power to lines corresponding to bit positions containing ones, and equipment similar to the digital input equipment at the other end of the lines reads the data to effect the transfer of information (Figure 8.7c).

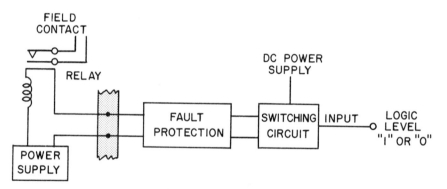

Fig. 8.7c　Digital output control

Because the switching of external power is performed at the computer interface, restrictions are placed on the voltage and current corresponding to the rating of the switching device. Maximum voltages are normally in the range of 40 to 50 volts DC, and maximum currents are on the order of 0.4 to 0.5 amperes for solid-state switches. Ratings on the order of 100 to 200 VA (volt-amperes) for relays are common.

Analog Outputs

Continuous electrical signals are generated from digital data using digital-to-analog (D/A) converters. If a large number of analog outputs are required, it is usually economical to share a common D/A converter among several analog channels by multiplexing the D/A output. Computer interface hardware thus consists of the logic circuitry necessary to obtain a binary digital value in a register ready for conversion, the D/A converter itself and an analog multiplexer.

The circuitry for analog output logic accepts a digital word from an internal register or computer memory, either in serial or parallel form. For signed outputs, one bit in the word indicates whether the analog output should be positive or negative. The logic circuitry recognizes this "sign bit" and switches the polarity on the reference supply accordingly. The remaining magnitude bits can be in natural binary form or in a number of other binary codes (binary coded decimal, excess three, etc.). (See Section 8.4 for details.) Unless a digital code with multiple-valued bit weighing (e.g., the Gray code) is used, no code conversion is normally required prior to D/A conversion. For example, code conversion is not normally performed when magnitude bits are in complementary form for negative numbers, because the D/A converter can be designed to accept this single-valued digital code.

D/A converters either supply a voltage output proportional to the magnitude represented by the digital value or a current output corresponding to the digital value. For *voltage outputs*, the voltage drop, between the reference voltage and the output, is adjusted by switching resistors in and out of the circuit, with each magnitude bit either opening or closing one switch depending on whether the bit is "1" or "0." For *current outputs*, either summing of constant-current source outputs can be used, or switching in resistors to cause current branching, can be utilized.

The resolution of the D/A conversion is dependent on the number of magnitude bits employed in the digital representation. For example, 10 magnitude bits can represent $2^{10} = 1,024$ possible states, or 1,023 possible changes in magnitude, between full scale and zero. For a full-scale output of 5 volts and a zero output of 0 volts, the smallest change which could be made in the analog output signal would be 5/1,023 volt, or 4.887 millivolts. Higher resolutions require more magnitude bits, and each additional bit approximately halves the smallest possible change in analog output signal. It is approximate only because of the $2^N - 1$ possible steps, where N is the number of bits converted. The maximum decoding error due to the converter resolution is $\pm\frac{1}{2}$ of the value of the least significant bit, or, for the 10-bit, 5-volt full-scale converter, ±2.444 millivolts.

D/A voltage conversion can be performed with resistor ladder networks made up of either single- or weighted-value resistors (Figures 8.7d and e). For the former, all resistors used can be identical, making it fairly easy to obtain matched temperature coefficients for accuracy considerations. The weighed-resistor network can provide some savings in power consumption and reference supply costs, but for high-resolution conversion, the resistor values can become very high and therefore difficult to obtain. In addition, temperature coefficients are more difficult to match if resistance values differ over a large range.

Errors associated with converting digital values to analog signals accumulate from several sources. In D/A converters which use relays for the analog signal switching, most errors are due to imperfectly matched resistors. Because

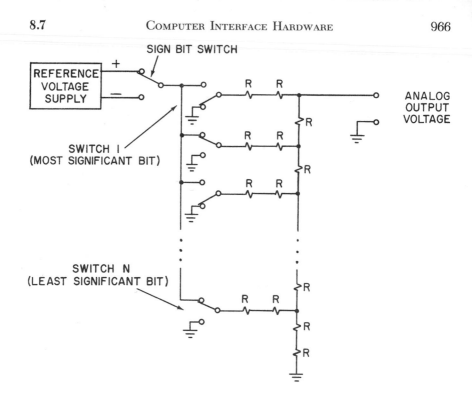

Fig. 8.7d   Single-valued resistor network for D/A conversion

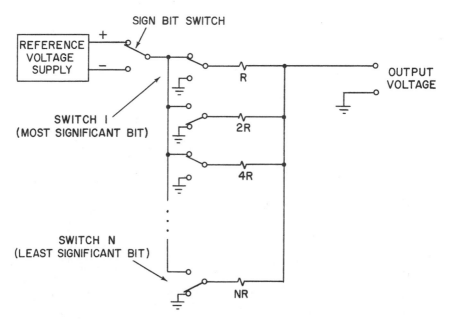

Fig. 8.7e   Weighted resistor network for D/A conversion

outputs are based on ratios of resistances, individual resistors must be matched as they age and as temperature changes. Metal film and wire-wound resistors can be manufactured with tighter and more easily matched tolerances than carbon composition or deposited carbon resistors, and are favored for medium and high-accuracy converters.

Since a relay is a nearly perfect analog switch, the use of relays for analog switching in the D/A converter is preferred for high-accuracy requirements. (For details on relays, refer to Section 4.5.) Relays are, however, larger, require more operating power, have more limited switching speeds, and are less reliable for high-speed operation than solid-state switches. (For a detailed discussion of solid-state switches see Section 4.6.) Most solid-state switches are not perfect in that they exhibit a finite resistance when closed and leakage currents when open. In addition they allow some feedthrough of the switch-driving signal. For most industrial applications, where size and power consumption are not particularly important and where reliable operation at moderate conversion speeds is important, relay-resistor D/A converters are preferred.

The accuracy and stability of the reference voltage is also of prime importance in overall D/A converter accuracy. The most commonly used source for a reference voltage is the silicon Zener diode, reverse-biased at some current value where the dynamic resistance and the voltage-temperature relationship are known. If the operating temperature range is large, temperature-compensated Zener reference diodes are favored. Changes in the voltage, applied to the Zener circuit can also affect overall D/A conversion accuracy, but these errors can be reduced with additional circuitry.

Analog Inputs

Continuous electrical signals are converted to digital values using analog-to-digital (A/D) converters. Multiplexing of analog signals into shared A/D converters is common because of the relative costs of multiplexers and converters. Depending on the multiplexer-converter system used, some signal conditioning may be required for amplification or impedance matching. Computer interface hardware therefore includes the A/D converter and the logic circuitry necessary to control the multiplexer and to supply the digital representation to computer memory or to an internal register.

The heart of the A/D converter is the comparator, which is an electronic circuit that compares unknown voltages and makes a binary decision as to which is larger. The comparator can be used in a number of ways in A/D conversion.

The simultaneous method of conversion uses $2^N - 1$ comparators to convert the analog signal to N bits of data (Figure 8.7f). It is fast, since voltage comparisons for each bit value are done simultaneously, but is expensive for high-resolution conversion because of the large number of comparators required.

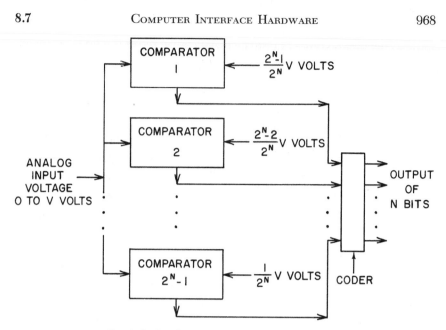

Fig. 8.7f Simultaneous method of A/D conversion

Slower, but less expensive for high-resolution conversion, are the A/D converters which use one comparator circuit and a digital-to-analog (D/A) converter. As illustrated in Figure 8.7g, the converter operates with pulses gated to a counter, while, with each new count, the D/A unit converts the counter contents to an analog voltage which is compared with the unknown

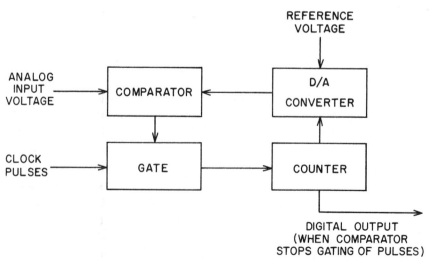

Fig. 8.7g Counter method of A/D conversion

voltage. When the comparator indicates agreement (within the one-bit resolution), counting ceases and the counter contains the digital value of the
unknown analog signal. Conversion time depends on the size of the counter,
which also determines resolution, but the converter cost is not particularly
sensitive to resolution. The counter method is therefore favored when high
resolution and low speed are required.

A variation on the counter method of A/D conversion is the ramp method,
in which a ramp generator (e.g., an integrating operational amplifier) rather
than the D/A converter supplies the signal to the comparator circuit. Because
the ramp does not have to be generated by the D/A converter logic, this
technique is somewhat faster than the counter method.

The most popular A/D converter design is based on the so-called successive approximation method. Since each bit position in a natural binary code
differs from its adjacent bits by a factor of two in weighting, the reference
voltage can be repeatedly divided by two and compared with the unknown
voltage, to sequentially determine the bit values. The successive-approximation
A/D converter handles large and small resolution conversions at moderate
speed and moderate cost (Figure 8.7h).

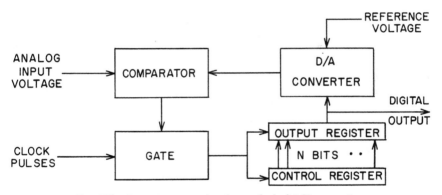

Fig. 8.7h   Successive approximation method of A/D conversion

Conversion speeds for the successive approximation converter depend
on the resolution, not only because of the number of sequential logic decisions
required but also because of settling time considerations. Each step must allow
time for settling to within total system accuracy. A redundancy technique
allows a fast initial approximate conversion, followed by a correction step
which adjusts the least significant bit after allowing sufficient settling time.
The conversion is therefore completed faster at the expense of additional
hardware. Redundancy is useful when both high speed and high resolution
are required.

There are many variations on these more common methods of A/D
conversion. Subranging, for example, combines techniques that are part

parallel (e.g., the simultaneous method) and part sequential (e.g., successive approximation). The entire analog input voltage range is divided into subranges using simultaneous comparators to determine the value of the most significant bits, and the remaining bits are determined sequentially. Speed requirements and hardware costs dictate the proper division between the simultaneous and sequential operations.

The merits of A/D converters include resolution, which is determined by the number of bits; accuracy, which is a function of the comparator performance and of the reference voltage stability; speed, which depends on the conversion technique, logic speed, and settling times; and cost (Table 8.7i).

Table 8.7i
CHARACTERISTICS OF A/D CONVERTER METHODS

Conversion Method	Time for 10 Bits		Relative Cost
	Aperture	Conversion	
Simultaneous	100 nsec	100 nsec	High for high resolution
Counter	0.003 msec°	1.8 msec°°	Low
Successive Approximation	0.03 msec	0.03 msec	Medium
Ramp	0.001 msec°	0.5 msec°°	Low
Successive Approximation With Redundancy	0.003 msec	0.03 msec	High

° Not constant.
°° Average.

Aperture time, or the time that the converter must "see" the analog voltage in order to complete a conversion, may also be an important consideration if the analog voltage can change rapidly. Aperture time can be decreased by using a sample-and-hold circuit which, in effect, charges a capacitor while sampling and holds the charge during conversion. (Sample-and-hold circuits are also used in multiplexed D/A conversion to "hold" the analog voltages on channels which share a common D/A converter.) In the successive approximation converter, aperture time can be decreased using the redundancy technique described above, where the final correction step compensates for changes in analog input voltage which occur during the initial approximate conversion.

Multiplexing of Analog Signals

Where many analog signals must be converted to or from digital values, the costs of using separate D/A or A/D converters for each analog signal are usually prohibitive. Analog multiplexers are used to time-share D/A and A/D equipment among many analog signals. The multiplexer selects and switches one analog signal at a time. The selection may be performed either sequentially or randomly, and the switching can be done with either electromechanical or solid-state devices (Figure 8.7j).

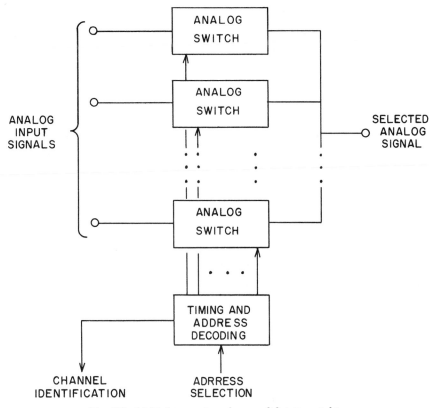

Fig. 8.7j   Multiplexer, using relay or solid-state switching

Electromechanical switches use metallic contacts that are closed when the switch is on and open when the switch is off. Because of the very low resistance of a metal-to-metal contact (typically less than one ohm), and the very high (for most practical purposes, infinite) resistance when the contacts are separated, electromechanical switches permit very accurate multiplexing.

Solid-state switches, on the other hand, can exhibit higher resistances when closed and, when open, leakage and driving signal feedthrough, and therefore permit less accurate multiplexing. (For a discussion of the relative merits of mechanical and solid-state switches, see also Sections 4.5 and 4.6). Solid-state switches, besides being smaller and consuming less power, can be operated at much higher speeds than electromechanical devices. When these attributes are important and only moderate accuracy is required, solid-state multiplexers are used.

One of the earlier electromechanical switches developed is the so-called crossbar switch, which is a three-dimensional array of contacts. One of several x-dimension and one of several y-dimension actuators are selected to close a vertical column of contacts. Further electromechanical selection chooses one of the several vertical contacts. Crossbar switches can operate at speeds of nominally 100 points/second and have small contact resistances and cross talk (Figure 8.7k).

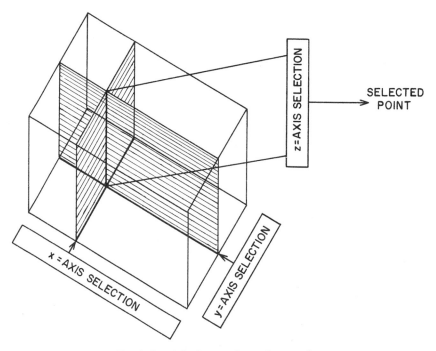

Fig. 8.7k　Multiplexer with crossbar switch

Motor-driven rotary multiplexers are used for both pneumatic and electrical signals. In pneumatic applications, the multiplexer connects one pressure signal at a time to a pressure transducer which supplies a single electrical signal to the computer for A/D conversion (Figure 8.7l). Typically, 64 pressure signals share one high-quality transducer and one input line on an electrical signal multiplexer at the computer interface. Savings in transducer and multiplexer costs are obtained at the expense of lower scanning rates, which are in the order of 5 to 10 ports per second.

Electrical signal switching with motor-driven rotary multiplexers can be done with brushes or wipers, or with rotating magnets which cause reed

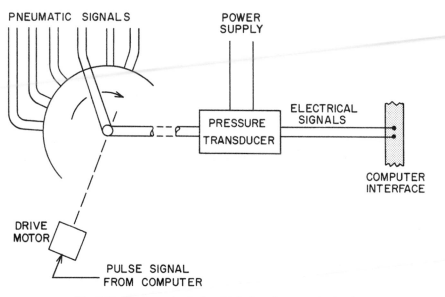

Fig. 8.71   Electromechanical multiplexing of pneumatic signals

switches to close. Analog input signals are wired to contacts circularly located on a round plate. The other side of the contacts is common and can be wired directly to the computer input terminals or to intermediate signal conditioning equipment (Figure 8.7m). With thermocouple signals that are multiplexed at some distance from the computer, for example, it may be preferable to convert

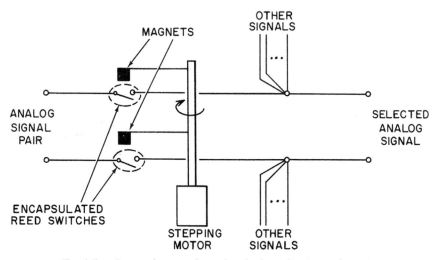

Fig. 8.7m   Rotary electromechanical multiplexer for electrical signals

the thermocouple emf to a high-level voltage or current signal prior to transmission to the computer.

Except for the fact that relay switches involve mechanical motion and solid-state analog switches do not, the basic principles of these two methods of switching are the same: each analog input signal is connected to a separate analog switch. The opposite side of the switch is common to all analog switches and carries the analog signal for which the switch has been closed. Timing and address decoding logic is used to close one analog switch at a time, and either random or sequential operation can be used.

As described previously, analog switching with relays is usually preferred for accurate but moderate-speed applications when size and power requirements are not stringent. Solid-state switching circuits include those using silicon junction diodes, bipolar transistors, and semiconductor junction and metalized-oxide semiconductor (MOS) field effect transistors (FET). MOSFET is, because of relative accuracy and circuit simplicity, often favored for multiplexing low-current analog voltages.

Random selection of multiplexer channels is performed by decoding a digital word containing the "address" of the desired channel, and closing the analog switch for the selected channel. The D/A converter output, or the A/D converter input, is then connected to the desired analog signal line. An address N bits in length can contain one of 2^N different binary addresses and therefore can be used to select one of 2^N multiplexer channels (Figure 8.7n).

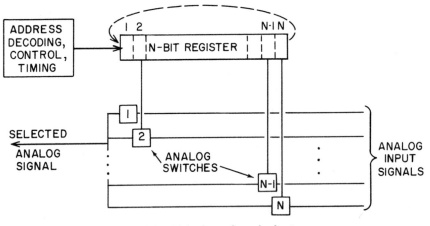

Fig. 8.7n   Multiplexer channel selection

Sequential scanning can be performed with either software or hardware. A computer program which continually increments or decrements the multiplexer address can effect a sequential D/A or A/D converter scan. Faster scan rates are possible using separate hardware counters, which count pulses at the desired channel selection rate, or with shift registers or ring counters. Shift registers are a series of connected flip-flops which transfer bits in one direction through the chain. In multiplexing applications a "1" bit in a flip-flop closes an analog switch and a "0" opens the switch. An N-bit shift register controls an N-channel multiplexer. By inserting zeros in all flip-flops but the first, and shifting this "1" bit through all flip-flops at the desired scan rate, sequential scanning is performed. In ring counters, the final-stage flip-flop is connected back to the first-stage flip-flop, thus effecting a continuous scan controlled by a digital clock.

Signal Conditioning

Signal conditioning at the I/O interface may be required to make process and computer signals compatible. In pulse counter operation, for example, pulses from the process may require reshaping in order for the counter to work properly. Digital inputs and outputs may require signal level changes so that contact ratings are not exceeded and on-off logical voltage levels are sufficient to cause switching. Analog signals may require circuitry for imped-ance matching, amplification of low-level signals or attenuation of high-level signals, and filtering to reject noise.

In process control computer installations, signal conditioning require-ments are most often associated with analog inputs. Multiplexer-converters that use solid-state switching may require high-level analog signals. For low-level signals (e.g., from thermocouples) amplification is required. With a relay multiplexer and a solid-state A/D converter, amplification is usually performed on low-level signals *after* multiplexing, since low-level signals can be accurately switched with relays and the amplifier can therefore be shared along with the A/D converter. In some cases, gain-selectable amplifiers are used and the computer program can control the gain as necessary while selecting multiplexer channels for conversion.

Some solid-state multiplexers require amplification of low-level signals prior to switching. For thermocouples, emf-to-current conversion, with voltage output across a resistor in the current loop, is used to supply the high-level voltage signal to the computer interface. If conditions warrant, it may be attractive to use remote electromechanical multiplexing of the low-level signals to share a common emf-to-current converter and one channel on the solid-state multiplexer at the I/O interface. In an application where most analog input signals are from pneumatic instrumentation and thermocouples, and high scan rates are not required, separate electromechanical multiplexing is often favored over installing individual pressure-to-current and emf-to-current transducers for each signal.

Noise considerations also influence the selection of I/O interface hardware and signal conditioning equipment. Electrical analog input signals are received from the process as a current loop or as a potential (voltage) difference on a pair of signal lines. As described previously, typical high-level current signals can be easily converted into high-level voltage signals of the proper range by the insertion of a resistor in the current loop. In some cases, all electrical signals can share a common ground potential, and single-ended multiplexing can then be used. If it is necessary to use both signal lines (e.g., thermocouples), then both lines must be switched together by the multiplexer and differential amplifiers are required to obtain the proper voltage range for conversion.

Common-mode noise and normal-mode noise are important considerations in both system design and cabling practice (see Section 8.6). Common-mode noise does not affect the system performance unless it is converted to normal-mode. Normal-mode noise, principally from common-mode conversion resulting from unbalanced line and leakage impedances, and from inductive coupling between adjacent conductors, can seriously affect performance. The use of twisted signal pairs with shielding is usually recommended to reduce

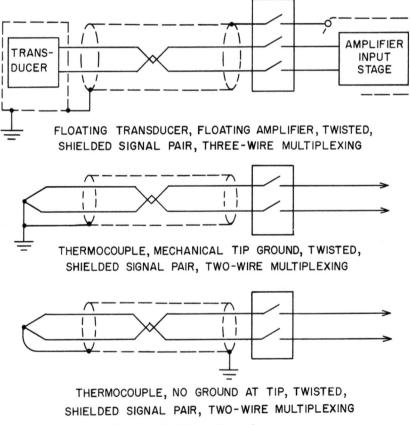

FLOATING TRANSDUCER, FLOATING AMPLIFIER, TWISTED, SHIELDED SIGNAL PAIR, THREE-WIRE MULTIPLEXING

THERMOCOUPLE, MECHANICAL TIP GROUND, TWISTED, SHIELDED SIGNAL PAIR, TWO-WIRE MULTIPLEXING

THERMOCOUPLE, NO GROUND AT TIP, TWISTED, SHIELDED SIGNAL PAIR, TWO-WIRE MULTIPLEXING

Fig. 8.7o   Shielding and grounding practices

electromagnetic interference. Methods for the reduction of common-mode interference include single-point grounding and isolation. The use of a differential amplifier guarded with a floating input and with a common ground for transducer (or, if floating, transducer shield) and cable and amplifier guard shields, helps isolate the computer system from the effects of common-mode interference. However, since the shield must be continuous through the input section of the amplifier, expensive three-wire multiplexing is required. Because of the expense, three-wire multiplexing is avoided where possible, and attention is concentrated on cable selection, routing and good grounding practices. Some two- and three-wire multiplexers are illustrated in Figure 8.7o.

The higher-frequency alternating current components of the normal-mode noise that do appear at the computer I/O interface can usually be inexpensively filtered out. The most common 60-Hz noise component is normally attenuated with simple, first-order, RC circuits (Figure 8.6k) in process control applications. If the process changes at frequencies which are affected by the RC noise filter, more expensive LRC filters can be used to get more selective filtering. A more detailed discussion of this subject can be found in the preceding section.

Summary

Connecting computers to external devices can involve many considerations and tradeoffs in hardware costs, effective operating speeds, flexibility, resulting signal fidelity, and reliability. The many parameters in I/O interface design are, however, much reduced with the selection of a particular control computer system. In general, most control computer manufacturers and suppliers offer a configuration of computer and interface equipment design to be mutually compatible and which will interface with the normal complement of conventional process instrumentation and control devices. With the selection of a computer system, many of the design decisions (for example, in A/D conversion method) are already made, and the major attention can be directed toward communication with devices having unique requirements. Most suppliers of computer systems publish manuals which describe the characteristics of the I/O interface system in detail and give advice on connecting process signals to the interface. These manuals should be consulted prior to connecting any signals from external devices to the computer system.

8.8   SET-POINT STATIONS

XIC SP

Available Designs:	(a) Electronic output,
	(b) Pneumatic output,
	(c) Stepping motor,
	(d) Synchronous motor,
	(e) Integrating amplifier.
	Note: In the feature summary below, the letters "a" to "e" refer to the designs listed above.
Output Signal Ranges:	3–15 PSIG, 1–5 madc, 4–20 madc and 10–50 madc.
Speed of Full-Scale Travel:	(c) 9 to 33 seconds,
	(d) 5 to 1,000 seconds,
	(e) under 1 second.
Resolution (minimum step in set-point change):	(c) 0.05% to 0.1%,
	(d) 0.1 to 0.5%.
Other Features:	Limit switches are available (c and d), available separately (c), available combined with controller (c, d and e), available combined with recorder controller (bc and bd).
Sizes:	3″ × 6″ (ac and ad), 2″ × 6″ (e, contains two units), 2″ × 6″ or 6″ × 6″ (bc and bd).
Cost:	$300 to $500.
Partial List of Suppliers:	Foxboro Co. (ac), General Electric Co. (ac and ad), Moore Products Co. (bc and bd), Motorola Inc. (ae), Taylor Instrument Companies (ac and ad).

Features

Set-point stations are used to interface the digital control computer with plant controllers and actuators, and they enable the computer to make changes in plant operating conditions. In supervisory control applications, the computer adjusts the set point of an analog controller, while in direct digital control, the computer may adjust the actuator directly. In either case, the types of signals available from the computer must be translated into the signal required by the receiving device. In addition to signal translation, the set-point station has several other important features that are necessary to provide efficient operator communication. These features, and the usual methods of their implementation, are listed in Table 8.8a. An illustrative set-point station design incorporating these features is shown in Figure 8.8b.

Set-point stations are usually packaged for panel mounting, using the common 2 × 6, 3 × 6, or 6 × 6-in. nominal cutouts. Where separate controllers are used in supervisory applications, the set-point station is preferably mounted adjacent to the associated controller for ease of operator association. In new installations, it is sometimes more economical to use packages which combine the set-point station with an indicating or recording controller and require less panel space than the separately mounted units. In direct digital control applications, all set-point stations may be mounted on a separate operator's console to create a "cockpit" type plant control station.

Conventional controllers typically obtain their set-point signals from manually adjusted knobs, in which case the mechanical motion must be translated into an electrical or pneumatic signal internally, or, in case of the cascade arrangement, the electrical or pneumatic output of other controllers has to be translated. Set-point stations can be designed to interface with the electronic or pneumatic controller through either input. Several methods have been used, but the most commonly accepted one is to use the set-point station to generate, from computer output signals, a set point that can be directly used in the controller, through, for example, a remote or cascade input connection. The set-point station must therefore supply a standard electronic or pneumatic controller signal.

One earlier method of interfacing the computer with conventional controllers, though no longer acceptable, illustrates some of the functional design requirements for this interface. For electronic controllers, a voltage output was generated by the computer (via a digital-to-analog converter) proportional to the desired set point. A switch on the controller allowed the operator to select the effective set point as either the computer output signal, for computer control, or the signal from the operator-adjusted set-point potentiometer.

Fail-Safe

This computer-manual switch installation was a relatively simple modification to an existing electronic controller, and additional panel space for a

Table 8.8a
SET-POINT STATION FEATURES

Function	Method of Implementation
A means for the operator to observe the value of the computer-generated set point	Vertical, horizontal, or circular scales, usually with 0 to 100% range and with pointer indicating set-point position
A means for the operator to switch from computer to manual control	Selector switch or back lighted pushbutton
A means for the operator to detect, at a glance, if the station is on computer control	Switch position or indicating lamp, which is on when the computer controls set point
A means for the operator to manually adjust, or to over-ride, the computer's set point	Knob or push-bar for manual adjusting. Pointer on scale indicates set-point value.
A means to limit the range over which the computer can adjust the set point	Mechanical stops, with separate lamp to indicate when set point engages upper or lower limit
A means to remove the limits when the set point is under manual control	Automatic, with switching from computer to manual or a separate limit-no-limit switch
A means for the operator to observe the set-point station output signal	A scale, in output units with pointer indicating output value
A means for the station to accept electrical signals from the computer, indicating change in set point	Stepping or synchronous motors or integrating amplifiers
A means for the computer to determine status of set-point station	Contacts, indicating position of computer/manual and limit/no-limit switches and information on whether set point is at limit

separate set-point station was not required. However, this method has serious drawbacks and is not normally recommended. This controller modification scheme is not "fail-safe," in the sense that a computer system failure and loss of the computer-supplied set-point voltage could cause a "zero" set point to the controller. This sudden change on one or more controllers could cause dangerous plant conditions or initiate plant shutdown procedures. Means can and should be provided to generate automatic switching to the manually adjusted set-point potentiometer in the event of computer failure. In this case the operator is required to continually adjust the manual set point to keep it up to date with the computer's, so that bumpless transfer is guaranteed. In addition, when the transition from operator control to computer control is made, the computer must supply a signal identical to that supplied manually, in order to achieve bumpless transfer. The computer therefore should read the operator's set-point voltage through the analog-to-digital converter and then generate an identical voltage, using the digital-to-analog converter in order to "line-up" with the existing set point. This procedure is difficult to perform without a resulting offset and set-point "bump."

Modern set-point stations are designed to provide fail-safe operation, in case of computer failure, and furnish bumpless transfer during computer-to-manual and manual-to-computer changes in set-point source. These features are provided by designing the set-point station so that the computer must supply a live signal only while making changes in the set point, and

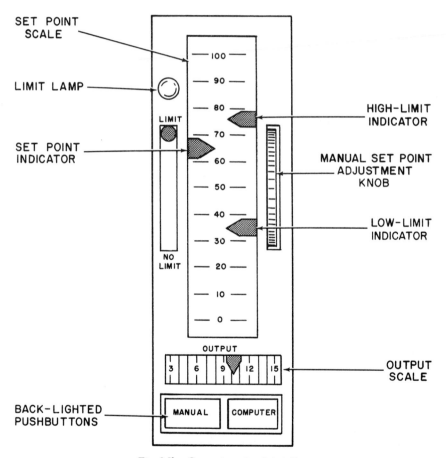

Fig. 8.8b　Computer set-point station

by providing means to automatically maintain the manual set point at the computer-set value while under computer control.

Stepping and Synchronous Motors

The most common technique used is that of operating a stepping or synchronous motor with computer digital output signals. A stepping motor receives a pulse train from the computer, with the number of pulses proportional to the change in set point to be made. Failure of the computer results in no more pulses and leaves the set point in its last position. A synchronous motor receives a timed contact closure signal from the computer, and the change made in the set point is proportional to the duration of the contact closure signal. Unless the computer system fails while changing the set point in such a way that the motor is driven all the way up or down scale, fail-safe operation is achieved. To guard against this latter type of failure, operator-adjustable mechanical stops, to keep the computer set point within preselected limits, are recommended, particularly with synchronous motor type set-point generators.

For electronic controllers, the motor in the set-point station drives a potentiometer to supply the appropriate DC milliampere signal to the controller set point. In pneumatic set-point stations, the motor adjusts a pressure regulator to provide the required 3–15 PSIG pressure signal. In either case, the mechanical movement provided by the motor is identical to the mechanical adjustment made when the operator manually adjusts the set point on the set-point station. Bumpless transfer is therefore inherent in this design. (In this arrangement, manual set-point adjustments are made at the set-point station rather than at the conventional controller.)

Integrating Amplifier

Another method of generating continuous set-point signals from discontinuous computer signals, indicating desired changes in set points, is that of using an integrating electronic amplifier. Computer voltage pulses are integrated, and the integral voltage, representing the sum of all previous changes in set point, is continuously supplied to the controller set point. In this scheme, the amplifier drift rate may be an important consideration when the computer is not adjusting the set point often. This of course is not a consideration when stepping or synchronous motors are used.

Speed, Resolution and Feedback

Maximum operating speed and resolution of set-point position can be major considerations in the selection of set-point stations. For a stepping motor station, resolution is defined by the number of steps required to drive the set point from zero to full scale. A full-scale travel of 1,000 steps, for example, gives a resolution of 0.10 percent. The maximum pulse rate is defined by the sum of the minimum "on" pulse duration and the minimum allowable time between pulses. If, for example, the sum is 30 milliseconds, the maximum pulse rate is 33 pulses/second, and assuming 0.1 percent resolution, 30 seconds would be required for a full-scale change (1,000 pulses) in set point. Depending on the application, faster changes in set points may be required.

In general, higher operating speeds and lower resolutions are characteristic of synchronous motor-operated set-point positioners. Full-scale travel times of 5 seconds are available with 0.5 percent resolution and with correspondingly higher resolutions for slower speeds (e.g., 10 seconds, 0.25 percent resolution), depending on motor speeds and gearing. Anti-backlash gearing usually provided for stepping motors is not available with synchronous motors, and gear train backlash of nominally 0.2 percent is typical.

In supervisory control, the computer program must be initialized prior to taking over control of the set points from the operator, and therefore the computer must be able to read the value of manually set set points. Also, during operation, in either supervisory or direct digital control mode, the computer should check the results of every set-point adjustment to insure that the requested change was effected accurately. Set-point feedback signals are therefore required from the set-point stations to the computer. These feedback signals can be obtained from a retransmitting slidewire to insure proper motor positioning in electronic set-point stations. For pneumatic set-point stations, the motor position provides only the input to the pressure regulator and therefore to further check that the pressure regulator responded properly, the pneumatic output signal, through appropriate transducers, should be fed back to the computer. Computer-set-point station interface signals are shown in Figure 8.8c.

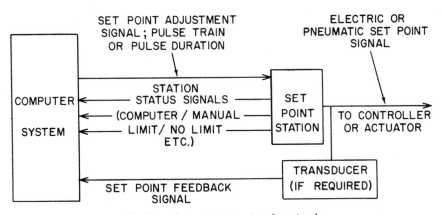

Fig. 8.8c   Set-point station interface signals

Summary

Major features of some available set-point stations are listed at the beginning of this section. Indicating electronic controllers, with integral motor-driven potentiometers, are available from a number of suppliers, and the selection of a particular model will be influenced by the type of plant instrumentation currently in use and the particular features of the set-point station and controller required for the application. Separate set-point stations, particularly for pneumatic instrumentation, and set-point stations combined with recording controllers, are supplied by fewer instrument manufacturers, sometimes leaving little choice for selection. Costs for set-point stations generally range from about $300 for stripped-down separate stations, to about $1,000 for set-point stations combined with strip-chart recording controllers and with a full complement of switches, indicators, etc. for complete operator and computer communication.

8.9 MEMORY DEVICES

Available Types:

(a) Internal registers, with solid-state logic,

(b) Main memories, with magnetic core elements,

(c) Auxiliary memories with
  (c1) Fixed head drum,
  (c2) Fixed head disc,
  (c3) Small moving head disc designs.

(d) Bulk memories with
  (d0) Removable disc packs only,
  (d1) Removable disc drives,
  (d2) Magnetic tape storage systems.
  Note: In the feature summary below the letters "a" to "d2" refer to the above listed designs.

Capacity Range in Bits:

(a) $<10^3$, (b) 10^4–10^7, (c1 and c2) 10^5–5×10^7, (c3) 7×10^6–3×10^7, (d1) 10^7–2×10^9, (d2) 10^8–2×10^8 per reel.

Access Times in Milliseconds:

(a) $<10^{-2}$, (b) 0.1–5, (c1 and c2) 8–18, (c3) 25–350, (d1) 50–100, (d2) serial type.

Cost per Bit:

(a) \$1–\$10, (b) 2–8¢, (c1) 0.1–1.2¢, (c2) 0.1–0.9¢, (c3) 0.05–0.12¢, (d1) 10^{-4}–10^{-3}¢, (d2) well below 0.01¢ as function of size.

Unit Cost Range:

(c1) \$4,000–\$40,000, (c2) \$1,000–\$40,000, (c3) \$8,000–\$16,000, (d0) \$100–\$700 or \$30–\$50 per recording surface, (d1) \$10,000–\$200,000.

Monthly Rental Cost:

(d1) \$300–\$5,000, (d2) \$300–\$1,200.

　　　　　　　Ampex Corp. (b, d2), Athana Corp. (d0), Bryant Computer Products, Inc. (c1, c3), Burroughs Corp. (b, d2), Caelus Memories, Inc. (d0), Computer Peripherials, Inc. (c2, c3), Control Data Corp. (d1, d2), Datacraft Co. (b), Data Disc, Inc. (c2, c3), Digital Development Corp. (c2), Electronic Memories, Inc. (b), Fabri-Tek, Inc. (b), Fairchild Camera & Instrument Corp. (b), Ferrox Cube Corp. (b), General Electric Co. (d1), General Instrument Co. (c1), Honeywell, Inc. (b, c3, d0, d1, d2), IBM Corp. (d0, d1, d2), Information Storage, Inc. (c2), Interdata (b), Lockheed Electronics Co. (b), Magnafile, Inc. (c1), Memorex (d0), NRC Corp. (d2), RCA Corp. (b, d2), Raytheon Computer (b), Scientific Data Systems, Inc. (c2), Univac Div., Sperry Rand Corp. (d2), Vermont Research Corp. (c1).

Memory requirements for computers include (1) those registers necessary to store internal arithmetic and control information, (2) main working memories, to store programs and data which need rapid and direct access to the central processor, (3) high-speed on-line auxiliary memories, to store programs and data needed less often, and (4) large-capacity bulk memories for filing large volumes of seldom-used information (Table 8.9a). Internal storage devices generally operate at the speed of the computer logic and are small, on the order of 1,000 binary digits (bits) or less in size. The other memories become, in the order listed, both slower and larger in capacity, because of the sensitivity of memory costs to operating speeds.

Internal registers are generally an integral part of the computer central processing unit, and independent selection of alternative devices is usually not practical. Consequently the following discussion is limited to those devices used for the main, auxiliary and bulk memories.

Main Memories

Ferrite Core Memory, Operating Principle

The hysteresis loop, or the relationship between the magnetization state and the applied magnetic field, is the fundamental characteristic of any

Table 8.9a
CHARACTERISTICS OF MEMORY DEVICES

Function of Computer Memory	Suitable Storage Devices	Access Time (ms)	Memory Capacity (bits)	Cost per Bit (¢)
Internal Registers	Solid-state logic systems	<0.01	$<1,000$	100–1,000
Main Working Memories	Ferrite core, film or integrated circuit memories	0.1–5	10^5–10^7	1–50
Auxiliary Memories	Drums or discs	15–150	10^6–10^9	0.01–1
Bulk Storage	Discs or tapes	Serial	$>10^7$	<0.01

magnetic memory element. The hysteresis loop for a ferrite memory core is shown in Figure 8.9b. When no magnetic field is being applied, the magnetization state can be either $+B_r$ or $-B_r$, depending on previously applied magnetic fields. By assigning logic "1" and logic "0" to these two states, binary data can be stored into or read from the element by applying magnetic fields.

Ferrite core memory elements are ceramic materials in the shape of a toroid (Figure 8.9c) with two or more wires passing through their center, which are used to apply and sense changes in magnetic fields. Cores are "read out" by applying on one of the wires a full read current $(-I_f)$ which is

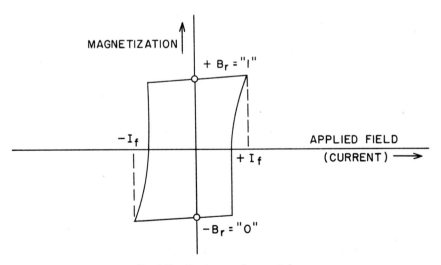

Fig. 8.9b　Ferrite core hysteresis loop

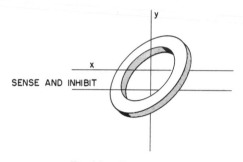

Fig. 8.9c　Ferrite core

sufficient to switch the core to the $(-B_r)$ state. If the core had been in the $(+B_r)$ state, a large change in flux would induce a voltage on the wires. If the core had been in the $(-B_r)$ state, a smaller change in flux, and a smaller induced voltage, would result. After rejecting the lower voltages with a threshold stage in the sense amplifier, the presence or absence of an induced voltage indicates whether the core had been in the "1" or "0" state.

The earliest core memories were 0.1 in. or more in outside diameter and their switching times, for changes in magnetization state, were several micro-seconds. Since smaller cores can be switched faster, continuing efforts have been made to reduce the core size. Diameters in the range of 18 to 22 mils are widely used today, having switching times as low as 140 nanoseconds. Smaller cores and faster switching times enable faster memory read-write cycle times, a continuing objective of computer manufacturers and users (Table 8.9d). The reduction in core size is, however, limited by core manufacturing and testing problems and by the difficulty of threading wires through arrays of the individual cores in order to assemble a working core memory.

Ferrite Core Memory, Organization

Several wiring configurations can be used to make a computer memory from arrays of individual cores. For moderate storage capacities, the most economical system configuration is the coincident-current, or three-dimen-sional (3-D), organization (Figure 8.9e). This configuration allows the core itself to assist in address selection and therefore requires a minimum number of drivers and decoders. Two wires are used to induce the magnetic field. The hysteresis loop is nearly square, so that the current must be traveling in the same direction in both wires in order to switch the magnetization state of the core, which is not disturbed by either one of the currents alone. Individual cores are arranged in a three-dimensional configuration, and they are threaded with x- and y-direction addressing windings and with sense and inhibit wind-ings. The sense and inhibit windings are used to sense a core's switching (when reading a "1") and to inhibit its switching (when writing a "0"). Since the sense and inhibit operations need not be performed at the same time, these

Table 8.9d

FEATURES OF FERRITE CORE MEMORY SYSTEMS

Manufacturer	Model	Cycle Time (μsec)	Capacity (words)	Word Length (bits)	Comments
Ampex Corp.	RF 4	1.0	512–4k	4–20	3D
	RG	0.9	4k–16k	8–80	3D
	RM	2.7	65k	72	2-½D, 2W
Burroughs Corp.	BFC 600	0.6	4k–8k	10–20	2-½D, 3W
	BFC 1000	1.0	4k–16k	10–16	2-½D, 3W
Datacraft	DC–31	1.0	32–1k	4–8	2D
	DC–30	1.0	32–4k	4–64	3D
	DC–32	1.5	4k–8k	4–72	3D
Electronic Memories, Inc.	NANO 2900	0.9	4k–16k	9–36	2-½D, 3W
	NANO 2650	0.65	4k–16k	9–36	2-½D, 3W
Fabri-Tek, Inc.	MSA 2	2.0	4k–16k	16–40	3D
	MFA 1	1.0	4k–16k	Up to 25	3D
	MD	0.65	4k–16k	Up to 28	2-½D, 3W
	MT	4.0	65k–262k	20–80	2-½D, 2W
Fairchild Camera & Instrument Co.	Miser	1.6	1k–4k	8–24	3D
	Pacer	0.65	2k–8k	16–30	2-½D, 3W
Ferroxcube Corp.	FX22	4.0	128–512	Up to 8	3D
	FC18	8.0	8k–16k	16–32	3D
Honeywell, Inc.	TCM–32	5.0	128–4k	8–36	3D
	ICM–47	0.75	4k–8k	4–28	2-½D, 3W
	ICM–40	1.0	4k–16k	4–26	2-½D, 3W
Interdata		1.4	1k–4k	8–16	2-½D, 3W
Lockheed Electronics Co.	CI–200	2.0	4k	4–36	3D
	CC–100	1.0	4k	4–84	3D
	CD–65	0.65	8k–32k	8–32	2-½D, 3W
RCA Corp.	EANDRO	1.0	4k	4–36	3D, 3W
		5.0	1k–8k	4–36	3D
	MS–3250	0.5	8k–16k	Up to 36	2-½D, 3W
	MS–3300	1.0	Up to 32k	Up to 72	3D, 3W
Raytheon Computer	300	0.9	8k–16k	4–28	2-½D, 3W

functions can be combined in a single wire. Because of the additional circuitry costs, the three-wire version is somewhat more expensive than the four-wire arrangement, but this additional cost is partially offset by the reduced threading costs. With the trend toward the use of smaller cores, the four-wire version is becoming impractical, and most high-speed 3-D core memories are of the three-wire type.

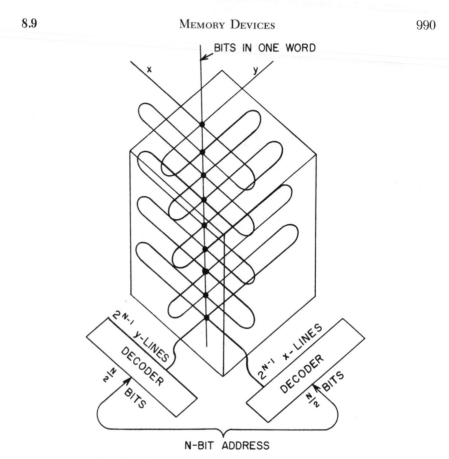

Fig. 8.9e   Coincident-current (3-D) memory organization

    In the 3-D memory, the x and y windings are used to address a vertical column of cores, and the particular column which receives a coincidence of currents is the address selected. The vertical length of the column, representing the number of bits in a word or character, depends on the currents in the sense and inhibit winding(s), each of which is threaded through the cores on one horizontal plane. A typical stack has 4,096 18-bit words, arranged as 18 horizontal planes of 4,096 cores each. The cores in each plane are arranged in a 64 × 64 array, and they are threaded by 64 address selection wires in each (x and y) direction, plus by the sense and inhibit winding(s). Corresponding selection wires in all planes are connected and are driven by a single driver circuit.

    The simplest core memory organization is the linear-select (2-D) form (Figure 8.9f), with cores arranged in a single plane and with word wires and

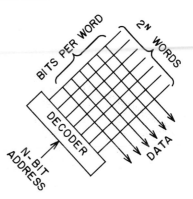

Fig. 8.9f   Linear-select (2-D) memory
organization

bit wires at right angles. Addressing is done along one side of the plane, and data is read along the other side. For storage, currents on one word wire and all bit wires combine at the selected intersection to either switch to "1" or prevent switching to store "0" in the cores. For readout, a larger single current on a word line is used to switch all cores on the wire. Currents greater than the normal full switching current can be used to cause faster switching than would be possible using the coincident-current principle of the 3-D type of organization. However, in the 2-D arrangement, all address decoding must be done by external electronic circuitry, and the 2-D organization is therefore more expensive than the 3-D type.

A compromise between the economy of the 3-D memory and the speed of the 2-D memory is the so-called $2\frac{1}{2}$-D memory. The $2\frac{1}{2}$-D arrangement, like the 2-D memory, has one wire threading the cores of a particular word and a bit wire through each core at right angles to the word wire (Figure 8.9g). The same word wire, however, threads the cores for two words, and each bit wire loops to thread the corresponding cores in both words. Coincident currents in word and bit lines are used to switch the cores, with the

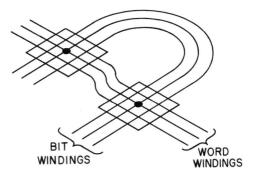

Fig. 8.9g The $2\frac{1}{2}$-D arrangement
(bit selection depends on current
direction)

BIT
WINDINGS

WORD
WINDINGS

bit current direction determining the cores that switch. When the two currents oppose one another, no switching takes place. A sense wire threads all bits, common to one doubled bit line. Inhibit windings are not needed because the bit current drivers control whether or not a particular core switches. For large bulk core memories, where the costs of the core array predominate over the cost of the electronics, the bit and sense wires are common, and only two wires need be threaded through each core.

Since readout operations in core memories set all cores in a particular word to the "0" magnetization state, the information in the word is not preserved during the readout operation. In main computer memories, the contents of any word, whether it contains instruction or data, should only be changed during a write operation. When reading the contents of the memory location, the contents of the location are therefore immediately rewritten into that location. The principal difference between a memory-read operation and a clear-and-write operation is whether the memory output is sensed in the interval between the read and write cycles. The distinction is therefore made in the logic circuitry and is not related to the memory magnetics. Memory operating speeds are usually measured by the time required for a total read-

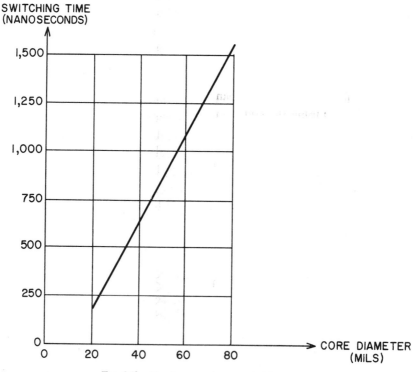

Fig. 8.9h   Ferrite core size vs switching time

write cycle, which includes two core switching times plus allowances for noise problems, current rise times, and delays in peripheral circuits (Figure 8.9h).

Ferrite Core Memory, Cost

Core memory production costs include those for manufacturing the individual ferrite cores, the labor for stringing the wires through the cores, the labor for testing, and for manufacturing the associated core memory electronic circuits. Memory prices are therefore a function of speeds, because smaller cores are more expensive to manufacture and string than larger cores, and a function of the electronics required, which in turn are a function of memory organization and total size. A $2\frac{1}{2}$-D core organization reaches its lowest cost with short memory words, and the 3-D memory cost is almost unaffected by word length. The 2-D organization using ferrite cores generally has no cost advantage over $2\frac{1}{2}$-D or 3-D. The cost of the magnetic memory elements tends to be a linear function of total capacity, while the cost of electronics is an exponential function. Therefore in smaller memories the cost is dominated by the electronics, and in larger sizes, by the memory stack.

For a 64,000-bit memory of 16–24 bit words, costs range from 6¢/bit (3-D) to 7–8¢/bit ($2\frac{1}{2}$-D). Relative costs for a 512,000-bit memory would be 4¢/bit (3-D) to $3\frac{1}{2}$¢/bit ($2\frac{1}{2}$-D). Since computer main memories are usually expandable in modules of 4,096 or 8,192 words, the higher prices for the smaller memories generally prevail where such optional expansion capability is provided. Large bulk memories of 20 to 40 million bits capacity, usually in $2\frac{1}{2}$-D, 2-wire organization, can be obtained for as low as $1\frac{1}{2}$ to 2¢/bit.

Planar Film

A very thin, flat layer of nickel-iron alloy, with the property of magnetic anisotropy, is the basis of thin-film memories. The magnetic properties of the film element along one axis are similar to those for the ferrite core (Figure 8.9b), but along the orthogonal axis, the hysteresis loop is as shown in Figure 8.9i. Along the axis with the rectangular hysteresis loop, the "easy" axis, two stable storage states of magnetization exist as in the ferrite core. Along the other, or "hard" axis, no magnetization can exist in the absence of an applied magnetic field. These properties are used to operate the film element in a different way than ferrite cores.

When subjected to a magnetic field along the hard axis, the magnetization along the easy axis is reduced to zero. Therefore the net change in flux is negative if a "1" had been stored and positive if a "0" had been stored. The flux change is sensed as a positive or negative voltage pulse. A small magnetic field is then applied along the easy axis, and the hard axis field is removed. The resulting stable magnetization state along the easy axis depends on the polarity of the applied easy axis field; thus either a "1" or "0" can be stored.

The basic thin-film element is made into a memory array by arranging

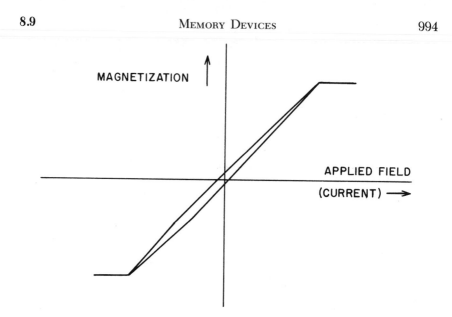

Fig. 8.9i   Magnetic film "hard" axis hysteresis loop

a number of small film spots (Figure 8.9j). A series of flat conductors, placed adjacent to the spots and parallel to the film easy axis, are the word wires used to generate the hard axis field for reading. At right angles, magnetically coupled to the easy axis, are the sense and digit conductors. The sense line is used to pick up the readout pulse, and the digit line is used to generate the easy-axis field for writing. These two lines can be combined into a common sense-digit line.

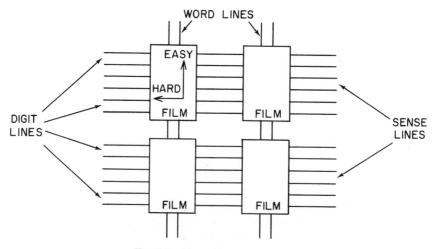

Fig. 8.9j   Planar film memory array

The planar film has an open flux path along both axes, and the magnetic flux from one end of the bit must return through the air to the other end. Read currents which would rotate the field less than 90 degrees would therefore result in slow demagnetization and an eventual loss of information. Non-destructive readout (NDRO) is therefore not possible with planar film memories.

Thin-film memories offer more speed and shorter access times than ferrite cores, and they dissipate less power than 2-D or $2\frac{1}{2}$-D ferrite core arrays of comparable speed. Power dissipation can be as low as 1 milliwatt per bit vs 1.4 milliwatts per bit for a comparable $2\frac{1}{2}$-D core array. Films are capable of switching in 5 nanoseconds or less vs the 100 nanoseconds or more for state-of-the-art ferrite cores. These switching speeds are possible because the film elements switch by magnetic domain rotation rather than domain wall motion as in solid ferrite cores. Film memories are less expensive than core arrays because the memory planes are batch fabricated while core arrays must be manually strung, and because the driving circuitry is less complex. The cost of thin-film memories can be expected to drop as designs are refined and as integrated circuit techniques are applied.

Cylindrical Film

Cylindrical film, also referred to as plated wire or woven wire, is essentially the same as the planar film described above, except that the film is deposited around a wire (often of beryllium-copper) along the easy axis, and this wire is used as the sense-digit line (Figure 8.9k). The word line is at right

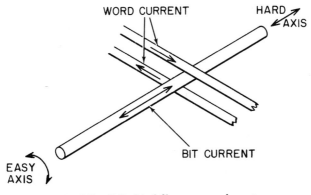

Fig. 8.9k Cylindrical film memory element

angles to the digit wire, and in a hairpin type loop. Basic operation of the cylindrical film element is the same as of the planar film. A current in the word line generates a magnetic field along the hard axis, and the readout signal is sensed on the sense line (the plated wire). A current in the wire generates the easy-axis field for write operations.

Cylindrical film arrays are usually constructed in either of two forms. In the "plated-wire" form, the word line is a series of flat strips attached to the outside of a sandwich containing the plated wires. In the "woven-wire" form, the word lines are formed by weaving a series of round copper wires into a fabric with, and at right angles to, the plated wire. Neither form is clearly superior to the other.

Cylindrical films have a closed flux path along the easy axis to prevent demagnetization, and are therefore amenable to non-destructive readout operation. The non-destructive readout properties of cylindrical film can be utilized to make a $2\frac{1}{2}$-D organization memory unit. The costs are significantly reduced by simplifying the necessary electronic circuitry.

Integrated Circuit Memories

Large-scale integrated (LSI) circuit memories are based on one of two approaches: the bipolar transistor or P-N junction technology and on the insulated gate field-effect transistor (FET) technology. Both approaches use silicon planar processing, and many of the manufacturing processes are common to both. Section 4.6 provides further details on integrated circuits.

In bipolar technology, common transistors with emitter, base, and collector regions are integrated with resistors and diodes. These transistors depend upon the diffusion of holes and electrons for power gain. Isolation between circuits is achieved through reverse-biased P-N junctions. Insulated gate FET technology, in which power gain depends on the propagation of electric fields across an insulating layer, is also an attractive route to LSI memories, primarily because circuits implemented with either P-channel or N-channel, metal, oxide, silicon (MOS) FET's alone, require no additional isolation between circuits. MOS technology allows order-of-magnitude higher circuit densities per chip, but bipolar devices are 5–10 times faster. The processing of MOS devices is more critical and sensitive to spectrographic levels of impurity than are bipolar devices, but fewer process steps are required.

Whereas magnetic memories (cores, film, wire) relay on the interpretation of millivolt analog signals, integrated circuit memories use only digital logic levels. LSI memories can therefore be expected to be less susceptible to the effects of extraneous noise and to environmental changes. LSI chips can be organized with internal address decoding to reduce the number of connections to external electronics, to increase the reliability, and to reduce total memory costs. Integrated circuit memories are, however, volatile, stored information is lost in the event of power failure, and provisions must be made to either dump key locations to a non-volatile memory or to provide a separate sustaining power source for the memory.

The prime costs in LSI memory manufacture are incurred in the basic masking and diffusion processes. Since these costs are relatively insensitive to circuit complexity, LSI memory costs are somewhat independent of memory size. For this reason, as manufacturing costs are reduced and yields improved, LSI memories will be attractive for all computers.

Auxiliary Memory Devices

Below 50×10^6 bits, most auxiliary memory systems are based on either rotating magnetic drums or discs. Below 10×10^6 bits, most of these devices are fixed head-per-track memories because of the high costs of moving head mechanisms. In some small, low-cost computer systems, the main memory is a rotating drum, but the use of core for the main memory and disc or drum storage for auxiliary memory is the common practice in medium-scale general purpose computer systems such as those used for process control.

Drum Memories

A magnetic drum, Figure 8.9l, consists basically of a rotating cylinder coated with a thin layer of magnetic material having a hysteresis loop similar

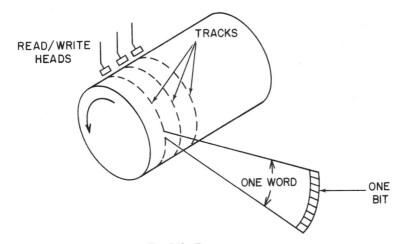

Fig. 8.9l   Drum memory

to that of the ferrite core shown in Figure 8.9b. A number of recording heads are mounted along the drum surface and these heads are used to read and write information by sensing or changing the magnetization of the areas passing under the heads. The area of the drum passing under any one head while the drum rotates is known as a track. Each track is divided into words, located sequentially around the track, and each word area is further divided into bits.

Some drums use separate read-write heads, and others have single read-write heads for each track. Writing is accomplished by polarizing the magnetic surface as it passes under the head. Opposite polarities are used to store "1's" and "0's" serially on the drum surface. Signals stored on the drum are read by coupling the changes in flux into the read head to induce signals that are amplified and interpreted.

Most drums are operated in serial mode, meaning that only one track is read from or written into at any one time. Addressing is performed in both space (selection of a track) and time (position of the drum). Access times therefore depend primarily on drum rotation speeds, and can range up to the time required for one complete drum revolution. A drum rotating at 3,600 rpm, for example, would have a maximum rotational delay (before the addressed word was under the head) of 16.7 milliseconds. In practice, average access times can be expected to be approximately half the maximum delay, and usually range from 8 to 17 milliseconds in most modern magnetic drum storage systems.

Information transfer to or from the drum is normally performed at drum speed and the bit transfer rate therefore depends on the bit packing density along the track. If 128 20-bit words, or a total of 2,560 bits, are stored in one track, for example, these bits would be transferred at a rate of

$$\frac{2,560}{16.7 \times 10^{-3}} = 153.6 \times 10^3 \text{ bits/second}$$

for a 3,600 rpm drum rotational speed.

Most drum memories use hydrodynamically balanced flying heads to read and write data. These heads are spring loaded toward the drum surface and depend on drum rotation to maintain the proper gap between head and drum surface. In some systems, mechanical schemes are used to retract the heads if drum rotational speed is reduced (e.g., during power failures) so the drum surface will not be damaged. Alternatively, the drum surface can be moved away from the head by, for example, using a tapered drum which moves axially in the event of reduced drum speed. Modern bit packing densities (to 1,200 bits/in.) and track widths (5 to 10 thousandths of an inch) preclude the use of fixed position heads or heads hydrostatically balanced with external air supplies.

Rotating drum memories, sealed in a pressurized inert gas atmosphere, can operate over wide temperature and vibration extremes, and are probably the most reliable mechanical, large-capacity computer storage devices available. Some of the features of drum type memory units are summarized in Table 8.9m.

Disc Memories

The magnetic disc (or disk) memory is based on one or more flat, rotating discs coated with magnetic material. The principles of reading and writing

Table 8.9m
FIXED-HEAD DRUM MEMORIES

Capacity Range	0.3 to 50 million bits
Average Access Time	Most common 8 to 9 milli-seconds (for one-half revolution at 3,600 rpm), some 16–18 milliseconds
Number of Recording Tracks	16 to 1,024

information on a disc are the same as those described for the magnetic drum memory device.

Information is stored on a disc in circular bands, called tracks, that are analogous to the tracks of drum memories (Figure 8.9n). Each track is divided, by imaginary disc radii, into sectors. Each sector contains a number of words or bytes of binary information. The sector is usually the smallest addressable unit in a disc storage system where addressing is performed by specifying track and sector numbers. One or both sides of a disc may be used for recording. Discs may be stacked and driven together with a common drive. These systems are termed disc files. Removable and replaceable disc files are called disc packs.

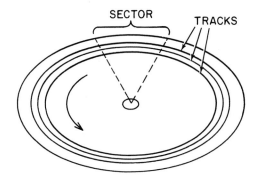

Fig. 8.9n   Recording surface of disc

As with the drum system, a read-write head is required to transfer information to or from a particular track while the device is rotating. Disc memories may either have one assigned head per track, one or more heads for each surface or disc, or one head for a stacked group of discs. Tradeoffs between performance (primarily access times) and costs (large number of heads vs moving head mechanism) determine the choice for a particular application. Below 10 million bits, economics generally favor head-per-track (fixed-head) memories (Table 8.9o). Most disc file systems with capacities of greater than 10 million bits use movable heads. The most common arrangement is to have one movable read-write head for each disc surface (Table 8.9p).

Table 8.9o
HEAD-PER-TRACK (FIXED-HEAD) DISC MEMORIES

Capacity Range	0.1 to 50 million bits
Average Access Time	8.3 or 16.7 milliseconds (3,600 or 1,800 rpm)
Number of Recording Tracks	8 to 1,024

Table 8.9p
SMALL MOVING-HEAD DISC MEMORIES

Capacity Range	7 to 30 million bits
Average Rotational Delay	8 to 50 milliseconds
Average Access Time	25 to 350 milliseconds
Number of Recording Tracks	100 to 200

Bulk Memories

Removable Discs

Disc access times depend largely on rotational delay, or latency, for head-per-track units. Fixed-head disc units are therefore very similar in operating characteristics to magnetic drums. For moving-head units, access times also depend on the time required to position the head over the track addressed. This positioning time can vary widely, depending on the distance that the head must travel. Disc memories therefore fall into two main classifications, fixed-head and moving-head types, and the choice for any application is a compromise between cost and access time. Moving-head memories can be further classified as having either fixed or removable storage media. Most small computers used in dedicated systems for process control, message switching, etc., do not require removable storage media. Data processing and other more general purpose applications of computers may require the use of removable storage units (Table 8.9q).

A disc memory system using removable discs has virtually unlimited storage capacity, because the file medium is interchangeable. Disc packs, or

Table 8.9q
REMOVABLE DISC DRIVES

Capacity Range	10 to 2,000 million bits
Average Rotational Delay	12 to 20 milliseconds
Average Access Time	50 to 100 milliseconds
Number of Recording Surfaces	2 to 160

removable groups of discs that share a common spindle, are a popular form of storage for up to billions of bits at access times between those of fixed-head discs or drums and magnetic tape (Table 8.9r). The most popular type of pack consists of six 14-in. diam discs mounted on a common spindle, 0.35 in. apart. The ten inside disc surfaces are used for recording, and the outer two surfaces are not used. The discs are nominally 200 μin. thick, and data are recorded on 200 tracks, at bit densities varying from 765 to 1,105 bits/in., on a 2-in.-wide band on each surface. For the head positioning system to be accurate, disc packs must be at nearly the same temperature as the disc drive unit.

Table 8.9r
DISC PACKS

Capacity Range	10 to 2,000 million bits
Number of Discs	1 to 11
Number of Recording Surfaces	2 to 20
Number of Tracks per Surface	200/100
Weight	4 to 16 pounds

Magnetic Tape Memories

Magnetic tape has long been a popular medium for storing very large quantities of information. The tape itself is generally a flexible plastic tape with a thin coating of ferromagnetic material either on the surface or sandwiched between two layers of plastic. Information is accessed serially by moving the tape past the reading and writing systems. Reel-to-reel tape transports (Table 8.9s) are the most common devices for handling magnetic recording tape while endless-loop cartridge devices are newer and less standardized.

Table 8.9s
REEL-TO-REEL MAGNETIC TAPE
STORAGE SYSTEMS

Range of Transfer Rates	2,000 to 150,000 characters/ second
Speed Range	30 to 150 in./second
Recording Density	200; 400; 556; 800; 1,600 bits/in.
Number of Recording Tracks	Usually either 8 or 9; some 10 or 16
Tape Width	Usually 0.5 in.; some 0.75 or 1.0 in.

A standard 10½-in. reel of tape contains 2,400 ft of ½-in.-wide tape. The typical tape can store 800 bits of information on each of the 7 or 9 channels (800 characters). One such reel can therefore store some 160 to 200 million bits. Most modern tapes use a base, or substrate, of polyethylene terphthalate polyester film 1.42 mils thick and a magnetic oxide coating (usually gamma ferric oxide) about 400 μin. thick.

A tape transport mechanism is required to move the tape past the recording heads. The transport must have the ability to start and stop quickly, and to cause the tape to pass the recording heads at high speed. Supply and take-up reels, which have a high inertia, are generally isolated from the mechanism which drives the tape past the heads so that the tape can be quickly accelerated or decelerated. The tape is usually either laced around sets of tension arms (Figure 8.9t) or passed through a vacuum column (Figure 8.9u)

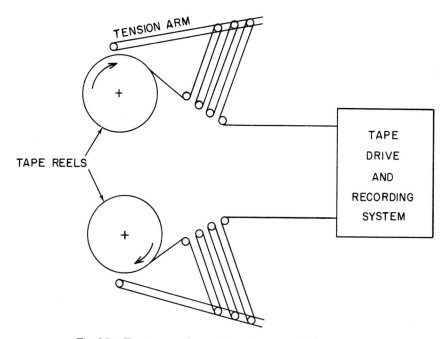

Fig. 8.9t Tension arms for starting and stopping high-speed tape

between the reels and continuously rotating capstans. Pressure rolls press the tape against the capstans when activated. Mechanisms with start/stop times of the order of 1–5 milliseconds, and tape speeds of up to 200 in./second, are available.

Mechanical buffers, tape reels, and multiple drives are not needed in a more recent tape drive design (Figure 8.9v) based on the so-called Newell principle. A single motor drives a large capstan against which the tape supply and take-up rolls are forced. The center of each tape roll moves toward or

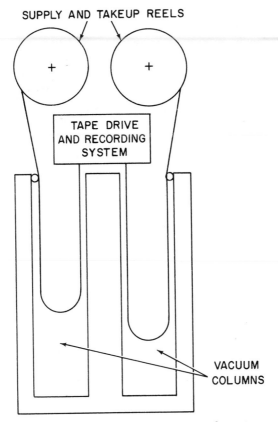

Fig. 8.9u   Vacuum columns for starting and stopping high-speed tape

away from the center capstan as the tape is removed or added to the roll. Tape accelerations on the order of 1,000 in./sec^2 can be obtained.

A large variety of other systems for handling magnetic tape are being introduced, most using some form of endless-loop tape cartridges. Tape lengths are usually shorter than those used with reel-to-reel systems, and therefore average random accessing times are less for smaller quantities of data. A

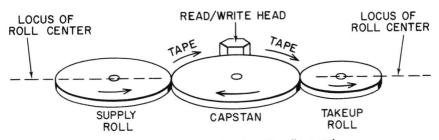

Fig. 8.9v   Tape transport based on Newell principle

typical cartridge system might use endless 100-ft, $\frac{1}{4}$-in. tapes, operate at a speed of 10 in./second, and have a 7- to 9-bit character transfer rate of about 500 characters/second. Capacities vary widely, but are generally one or two orders of magnitude less than that of the standard $10\frac{1}{2}$-in. reel-to-reel devices.

Magnetic tape recording systems are best suited for sequentially organized data. Other forms of processing, where either the data are randomly organized or the processing is done randomly, are inefficiently handled with magnetic tape, because the serial nature of the storage medium means long average access times (in seconds). For large-capacity on-line storage requirements the magnetic tape is the lowest-cost storage medium. Because of the smaller average access times, removable disc pack storage is better suited than magnetic tape for medium-sized storage requirements.

Read-Only Memories

Read-only memory (ROM) can be read from but not changed or erased on-line and therefore provides permanent storage in a computer system. ROM units occupy a position in computer architecture intermediate between hardware and software. They are used for hard-wired storage of programs and data (e.g., arithmetic subroutines, function tables, test routines) or for micro-programming computer instruction logic so that individual logic elements can be connected in a variety of ways. ROM allows more efficient use of computing hardware, enables faster overall operation of the computer, and allows flexibility in basic computer design. A general purpose computer with ROM can be made into a special-purpose device by hard-wiring the read-only memory to best suit the particular application.

Two types of read-only memories are those based on capacitors and transformers. In capacitor based ROMs, word lines and sense (bit) lines are orthogonal to each other in a planar array. The presence or absence of a capacitor at each intersection is used to indicate whether a "1" or "0" is stored. Such memories that can be read out in fractions of a microsecond can be batch fabricated by either first forming capacitors at each intersection and etching away plates at particular intersections to form zeros, or by punching holes in a conductor placed between all plates to form ones at the intersections desired. When a pulse is applied to a selected word line, it is coupled by the capacitor to the sense line at intersections that contain capacitors, and the output signals indicate the bit pattern of ones and zeros stored in the word.

Transformer-based ROMs can use E-, U- or toroid-shaped elements. The elements are usually placed in a line, and each element has a sense (bit) winding. A particular word wire is threaded through some of the transformer elements (secondary winding) to store a "1," and it by-passes other elements to store a "0." (In an E-shaped element, the side of the center leg on which the word line passes determines whether "1" or "0" is stored.)

A current pulse on the word wire will induce currents in the sense windings on elements that the word wire passes through, and will not induce currents in sense windings of elements by-passed. The stored bit pattern of ones and zeros is then read on the sense windings. A number of word wires can be used with a given set of transformers, each word wire having a distinct thread/by-pass pattern.

The significant use of read-only memories in computers is a recent development, and the type of ROM which will predominate will depend on such factors as access times and cost of fabrication. The latter consideration will be largely dependent on the degree of automation that can be introduced in the manufacturing process.

Other Memory Techniques

In addition to those described earlier, many other computer memory techniques and devices have been used and are under development. Mercury and magnetostrictive delay lines have characteristics like the magnetic drum, for example, but are seldom used in newer computer models. Relays, vacuum tubes, and cathode-ray tubes are additional examples. None of these devices is in significant use today, and the potential for future developments using these techniques appears to be limited.

Optical memory systems, based on the use of laser beams, are the subject of much current interest. These optical systems hold much promise for the storage of vast amounts of information in small quantities of recording media. One system on the market, for example, utilizes a laser beam to burn tiny (1 micron) holes in the surface of a coated tape, and a less powerful beam to detect whether a hole ("1" or "0") is present. Capacities of up to a trillion bits, at densities of 30 million bits/in.^{2} and average access times of 200 milliseconds, are possible, but, once stored, the information is unalterable. Other laser-based systems under development include those based on holography (10 million bits/in.^{2}), the use of selected-wavelength lasers and a photochromic medium (100 million bits/in.^{2}) and thermomagnetic recording, with which the laser beam is used to cause a local temperature increase which, in the presence of a biasing magnetic field, locally switches the magnetization of the material.

Regardless of the memory devices available at any time, three major criteria will determine if, when, and how they are used in computer systems: (1) the speed at which a change of state can be made (and reset, if destructive readout), (2) cost per bit, and (3) physical size and operating environmental constraints. As new techniques and devices are developed, they will gradually replace the present hierarchy of core, drum, disc, and tape. In particular, large-scale integrated circuits may well replace all upper-level (higher speed, moderate size) memories, and optical systems all lower-level (lower speed, large size, low cost) memories, and the two may replace all devices that currently predominate.

8.10 DATA LOGGING AND SUPERVISORY PROGRAM

In the chemical industry the most frequent installation of a computer is either for logging process data or to furnish set points to conventional controllers. This latter mode of operation is also referred to as supervisory computer control. In Section 8.4 the various basic computer languages are discussed and the existence of "fill-in-the-blanks" type programs is also mentioned.

This section is devoted to a more detailed coverage of this "fill-in-the-blanks" language as it applies to data logging and supervisory tasks. Unfortunately, the nature of this language—the use of fill-in forms—makes it specific to the particular computer manufacturer's system. Therefore in order to provide a detailed and specific description of this program, the coverage must be limited to one manufacturer's language. The reader is reminded that similar programs are available from other suppliers.

The first part of this section is devoted to the explanation of the program and of the forms used, and the second part contains process-oriented examples of how this language is used.

The dedication of a digital computer to supervisory control generally indicates that the computer is capable of translating the solutions of general and optimizing equations into controller set-point adjustments. In addition, supervisory control computers are generally capable of performing other control room functions such as data logging.

PROSPRO° (PROcess Scheduler PROgram) is an application program for supervisory control of an industrial process. The measuring instruments and control strategy are described to the program by completing entries on standard forms. The use of those standard forms, required to perform data logging and set-point adjustments, is explained in this section. Overall control of the computer is handled by the *processing scheduler program*, which selects the major programs to be executed on the basis of their priorities. The major programs all return to the *processing scheduler* so it can select the next task. The data from the *variable information sheets* are used by a program called the *variable processor*. The program called *adjustment processor* uses the

---

° IBM Language Specification Manual. H20-0473-0. International Business Machines Corporation, Data Processing Division, 112 East Post Road, White Plains, N.Y. 10601.

adjustment information. The *general action and general equation* sheets are used by a program called the *general processor.*

Data logging with PROSPRO is performed by a logging call that is initiated from either intermediate or final special action entries on the variable information form.

System Identification Numbers

Five types of identification numbers are recognized by the PROSPRO. They are identified and grouped by the first or left-most digit or character and are followed by a unique four-digit number. Below is a brief description of each type.

"0nnnn"   Identifies a *process control variable.* When such a number is used as an entry it signifies that the identified process variable is to be processed "special" by the *variable processor* at the first available opportunity. The variable identification number is placed in the "variables to be processed special" list. Upon completion of the current variable or adjustment processing, the *variable processor* goes to this "special" list and processes all variables identified in it. This temporarily defers variables being processed from the "routine" list, which is generated from routine processing interval information.

"1nnnn"   Specifies that (1) a *general equation* is to be solved to develop a required value before processing can continue or (2) a *general action* is to be executed before processing can continue. Normal processing of the variable or adjustment is temporarily suspended while the designated *general equation* solution or *general action* execution is carried out. Upon completion of the general processing, the variable or adjustment processing proceeds normally unless notified to terminate.

"2nnnn"   Specifies that a request is to be made to the *adjustment processor* to adjust for a detected unusual change in a variable or to correct for a deviation between a variable's target and current values. A request is placed in an "adjustments to be made" stack for the *adjustment processor.* Upon completion of all variable processing, the *adjustment processor* removes these requests from the stack and processes them.

"3nnnn"   Specifies that a special program is to be executed on a "core-exchange" basis to (1) develop a value required for continuation of processing or (2) perform some type of special action. The entire processor (variable or adjustment) in control of computer is saved; the special program is then located, placed in core, and executed until an effective return of control is made. The saved processor is returned to core, and processing continues from the point following the call of the special program.

"–nnnn" Specifies that a value from miscellaneous data storage is to be obtained for the desired value.

Entry Types and Uses

There are three types of entries on the sheets: alphameric, numeric, and numeric with decimal. Whenever an entry is made, certain rules and restrictions must be applied.

The alphameric entries may contain any combination of letters, numbers, and special characters. The information may be placed anywhere within the available spaces, although entries are normally made from the left. Blanks (spaces) may be included to improve readability.

The numeric entries may contain only numbers except when a leading minus sign is required. The minus sign, when used, must be in the first or left-most space. Entries may be filled with zeros or left blank to indicate "not applicable." When a numeric entry is used, all spaces including insignificant zeros must be filled in.

The numeric with decimal entries must have an embedded decimal point within the numerical portion. A number is preceded by a minus sign to indicate a negative value. The position of the significant characters within the available spaces is immaterial except that there must not be any embedded blanks (spaces), and all characters must be adjacent. The range of the value may be extended by the use of the conventional exponential format. To ease the transcribing of information to cards and improve readability, these entries should be left justified.

Shown below are some typical entries of each type.

Alphameric

F R C – 1 0 1   F D   R A T E

M B / D

8.10(1)

Numeric

1

0 0 0 6

– 1 2 3 4 5

8.10(2)

Numeric with Decimal

4 2 . 0

– 0 . 0 3

0 . 0 0 0 1 2 3 4 5 6 7

– 1 2 3 4 5 6 7 0 . 0 0 0

1 . 2 3 E 4

1 2 . 3 4 E 0 3

8.10(3)

Variable Information Sheet

Purpose

The purpose of the Variable Information Sheet (Table 8.10a) is to identify uniquely all the variables used and to collect the necessary information required to process these variables. The process control variables are all the measured variables (electrical signal inputs) read from the process units and the calculated variables developed from other measured and calculated variables. The completed forms are (1) reviewed for completeness, validity, and consistency of information, (2) transcribed into 80-column data processing cards and (3) placed in a loose-leaf notebook as a permanent reference.

Variable Processing

When the variable processor acquires control of the computer, it analyzes the "variables-to-be-processed" lists. Entries to these lists are made from various other sections of the system. Each list has an assigned processing priority. Entries in a list are processed on a "first-in, first-out" basis. After selection of the variable to process, its status is checked to insure that it is in service and should be processed. If not in service, the selected variable is ignored and another variable is selected for processing.

The first task of the *variable processor* is to determine a new current value in engineering units for the variable. From information specified, the new current value is developed either by converting a measured electrical signal from a process unit or from other stored values and constants. After the new current value is developed, any specified intermediate action is executed. Also, if specified, a new average value is developed for every routine processing pass of the variable.

The *variable processor* then checks the new current value against any specified control limits or points to determine if operator notification is required. Also, any specified special action is executed the first time a control limit violation or control point passage is detected. Whenever the new current value exceeds the variable's assigned maximum or minimum limit, the developed value is stored with an "out-of-limits" indicator. This is to permit other sections of the control system to reject it when necessary.

If the variable being processed does not have a target value to be maintained, then any specified final action is executed and the processing of the variable is considered complete for this pass.

For those variables with target values to be maintained, the variable processor determines if it is time to calculate a new target value and/or make adjustments to correct for the deviation between the variable's target and current values. The operator is always kept informed of the deviation status relative to the specified deviation limit. If the deviation is significant enough to warrant adjustment action, then the action is initiated or executed. This action is usually either the submission to a request of the *adjustment processor*

Table 8.10a

VARIABLE INFORMATION SHEET

Description _____

_____

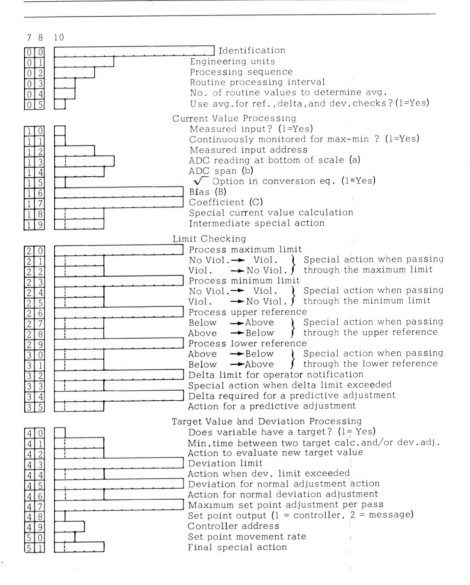

7 8 10

0 0 ⎤ Identification
0 1   Engineering units
0 2   Processing sequence
0 3   Routine processing interval
0 4   No. of routine values to determine avg.
0 5   Use avg.for ref., delta, and dev.checks? (1=Yes)

Current Value Processing
1 0   Measured input? (1=Yes)
1 1   Continuously monitored for max-min ? (1=Yes)
1 2   Measured input address
1 3   ADC reading at bottom of scale (a)
1 4   ADC span (b)
1 5   √ Option in conversion eq. (1=Yes)
1 6   Bias (B)
1 7   Coefficient (C)
1 8   Special current value calculation
1 9   Intermediate special action

Limit Checking
2 0   Process maximum limit
2 1   No Viol.→ Viol.   ⎱ Special action when passing
2 2   Viol.   → No Viol. ⎰ through the maximum limit
2 3   Process minimum limit
2 4   No Viol.→ Viol.   ⎱ Special action when passing
2 5   Viol.   → No Viol. ⎰ through the minimum limit
2 6   Process upper reference
2 7   Below →Above ⎱ Special action when passing
2 8   Above →Below ⎰ through the upper reference
2 9   Process lower reference
3 0   Above →Below ⎱ Special action when passing
3 1   Below →Above ⎰ through the lower reference
3 2   Delta limit for operator notification
3 3   Special action when delta limit exceeded
3 4   Delta required for a predictive adjustment
3 5   Action for a predictive adjustment

Target Value and Deviation Processing
4 0   Does variable have a target? (1= Yes)
4 1   Min.time between two target calc.and/or dev.adj.
4 2   Action to evaluate new target value
4 3   Deviation limit
4 4   Action when dev. limit exceeded
4 5   Deviation for normal adjustment action
4 6   Action for normal deviation adjustment
4 7   Maximum set point adjustment per pass
4 8   Set point output (1 = controller, 2 = message)
4 9   Controller address
5 0   Set point movement rate
5 1   Final special action

or the changing of a controller set point. After the execution of any specified final action, the processing of this variable is considered complete for this pass.

If no more variables remain to be processed, a permanent release of the computer is made to the *processing scheduler*. If, however, other variables are waiting, a check of the processing scheduler is made to determine if any higher-priority program is waiting for control of the computer. If no program is waiting, then another variable is selected and processing of it commences as described above. If, however, something more important requires attention, a release of computer control is made to the processing scheduler with the notification that variable processing is not complete and must be restarted at the first available opportunity.

Variable Identification Number

Each process control variable is uniquely identified. This identification is an assigned number greater than zero but less than 3275. The number is placed in the four spaces provided in the upper right box on the form (Table 8.10a). Although the system permits variable identification numbers to range to 3275, for efficiency they should be numbered from 0001 to the approximate total number of variables in use.

Variable Values

In addition to the current value of a variable which is developed and maintained by the *variable processor*, there are nine other identifiable values attributed to it. Each of these ten values is designated by an appended units digit on the variable identification number. These various values are normally referenced when using the *general action* and *general equation forms* or when required in special programs. The appended digit and its meaning are as follows:

Digit	Meaning
0	Current or last developed value
1	Target value
2	Average value
3	Last base value
4	Assigned MAXIMUM value
5	Assigned MINIMUM value
6	Assigned upper action reference point value
7	Assigned lower action reference point value
8	Assigned delta-limit value
9	Assigned deviation-limit value

Whenever any of these values are not specified or cannot be calculated or developed, a system-recognizable "unavailable" value is stored for it. Any attempt to use an "unavailable" value during processing will usually result in process termination.

Variable Information Sheet Organization

The information entries on the sheet are grouped into six major categories:

Entries	*Categories*
00–05	General information
10–18	Input or current value development information
19	Intermediate special action
20–35	Limit-checking and special-action information
40–50	Control output information
51	Final special action

General Action Sheet

Purpose

The purpose of the General Action Sheet (Table 8.10b) is to identify uniquely all the *general actions* and to collect the information necessary to define the specific action steps. A *general action* is a specified sequence of standard action steps which are interpreted and executed "in line" during generalized process control. The completed sheets are (1) reviewed for completeness, validity, and consistency of information, (2) transcribed into 80-column data processing cards, and (3) placed in a loose-leaf notebook as a permanent reference.

General Action

Where the variable and adjustment processors fail to provide for a particular action, situation, or calculation, a need exists for a specific identified subprogram to perform the action or cope with the circumstance. This exception could be accomplished by the creation of a special FORTRAN language subprogram such as an optimizing equation (Section 8.11) which is compiled, tested, and stored in a program library within the computer and is available for execution when required. However, in many instances the special processing required is so trivial that it does not warrant the creation (and associated maintenance problems) of such a program.

The *general action* is a small program which is defined in a specific generalized system language which permits the execution of a nominal amount of calculation and decision making without resorting to a FORTRAN program. The language consists of a few standard action codes which are listed as action steps and define the special processing desired. These action steps are interpreted and executed sequentially, starting with the first step and continuing until end of action or until directed to another action step by a branch step. Blank or undefined action steps are ignored.

Table 8.10b

GENERAL ACTION SHEET

```
                                                  1        5
                                                 ┌─┬────────┐
                                                 │1│        │
                                                 └─┴────────┘
```

Description _____

_____

_____

```
7 8  10 11  13        18      22
┌─┬─┐ ┌─┬─┐ ┌─┐       ┌─┬─────┐
│0│0│ │G│A│ │ │       │1│     │
└─┴─┘ └─┴─┘ └─┘       └─┴─────┘
```

13:       Are values beyond MAX-MIN limits acceptable? (0=No, 1=Yes)
18-22:   Next General Action to be executed. (Chaining)

Step	Code	A	B	
7 8	10	12 14	16 20 22	26 Remarks

Step	Code
0 1	S
0 2	S
0 3	S
0 4	S
0 5	S
0 6	S

Step	Code
0 7	S
0 8	S
0 9	S
1 0	S
1 1	S
1 2	S
1 3	S
1 4	S
1 5	S
1 6	S

RTN	– Return to processing	VTC	– Variable Times Constant
END	– End Variable Processing	VDV	– Variable Divided by Var.
CON	– Constant	VPV	– Variable Plus Variable
MSG	– Type Message No./Value	VPC	– Variable Plus Constant
PVS	– Process Variable Special	VMV	– Variable Minus Variable
NOP	– No Operation	ABS	– Absolute Result
FBA	– Feedback Adjustment	CVV	– Compare Var. to Variable
FFA	– Feedforward Adjustment	CVC	– Compare Var. to Constant
QVS	– Query Variable Status	BRM	– Branch Result Minus
SRT	– Save Real Time	BRZ	– Branch Result Zero
CDT	– Calculate Delta Time	BRP	– Branch Result Plus
VEE	– Variable Equals Equation	BCL	– Branch Compare Low
VEV	– Variable Equals Variable	BCE	– Branch Compare Equal
VEC	– Variable Equals Constant	BCH	– Branch Compare High
VTV	– Variable Times Variable	BRA	– Branch unconditionally

General Equation Sheet

Purpose

The purpose of the General Equation Sheet (Table 8.10c) is to identify uniquely all the *general equations* and to collect the information necessary

Table 8.10c

GENERAL EQUATION SHEET

$$\begin{array}{cc} 1 & 5 \\ \boxed{1} \end{array}$$

Description
_____

_____

_____

General Equation form: $X = A + B \left[C + (F/D) \right]^{E}$

7 8	10 11	13	16	18	22
0 0	G E	☐	-☐☐	1	

13: Are values beyond MAX-MIN limits acceptable? (0=No, 1=Yes)
16: Temp. Stg. Loc. of X value when chaining Eqs. (-1 thru -7)
18-22: Next General Equation to be executed. (Chaining)

T

7 8	10	12 V'	16	18 V"	22	24 Constant Coefficient 36	Remarks
0 1							
0 2							
0 3							
0 4							
0 5							
0 6							
0 7							
0 8							
0 9							

$\text{Term}_{T} = 0.0 + \sum \left[(V')_{T} \times (V'')_{T} \times (\text{Constant Coefficient})_{T} \right]$

Where T is A, B, C, D, E or F
(V) is the value of an identified variable V' or V" as follows:

V = 0 (or omitted); Value assumed to be 1.0
V > 0; Contents of Identified Process Variable Value (iiiij)
$-7 \leqslant V < 0$; Contents of Identified Temp. Stg. Loc. (Chaining)
$-9999 \leqslant V \leqslant -10$; Contents of Identified Miscellaneous Data Storage

Constant Coefficient.

If omitted, the coefficient is assumed to be 1.0.
When specified, a decimal point (.) must be placed within the field; all other significant characters must be a digit (0-9) except the first which may be a minus sign (-) to indicate a negative value; and all characters from the first to the last must be adjacent. The numerical range of the Constant Field (24-36) can be extended through use of the conventional exponential (E) format.

to define the form and content of each of these equations. The *general equation* is used to develop a single solution value from a combination of other variable and constant values. The completed sheets are (1) reviewed for completeness, validity, and consistency of information, (2) transcribed into 80-column data processing cards, and (3) placed in a loose-leaf notebook as a permanent reference.

General Equation

The *general equation* is an arithmetic expression whose general form is

$$X = A + B[C + (F/D)]^E \qquad 8.10(4)$$

Rarely, however, is the equation used in its most general form. Instead, by the judicious selection of general terms (A, B, C, D, E, and F), it is possible to tailor the equation to a specific subform. Since the equation is solved mathematically with only the specified terms included, any term may normally be omitted. Therefore, the following are assumed when a general term is not specified:

$$A = O, B = 1, C = 0, D = 1, E = 1, \text{ and } F = 0.$$

General Equation Terms   The value of each *general equation* term specified (A, B, C, D, E, or F) is developed as the summation of one to a maximum of nine multiplication products (VVC). Each product (VVC) is equal to the multiplication of the value of one identified variable (V′) times the value of another identified variable (V″) times the value of an indicated constant (C). Whenever any item in the product development is not specified, a value equal to unity is assumed. Therefore, as with the form of the *general equation*, selective omission of one or more product items makes it possible to develop several forms of the product used for general term development.

General Equation Chaining   It may be impossible to arrive at the desired calculated "X" by using only a single equation. However, the *general equation processor* has the ability to chain or join together two or more *general equations* to form a series of equations with intermediate "X's" or answers. The next or any subsequent equation in the chain retrieves the intermediate "X's" from the temporary storage locations as variables for that equation. The final equation indicates no equation to be solved next. This breaks the chain, leaves the final "X" in the *result register*, and returns to processing at the point following the call to the first equation in the chain.

General Equation Identification Number   Each general equation is uniquely identified. This identification is an assigned number greater than 10000 but less than 20000. (Note: General actions are identified similarly.) The number is placed in the four spaces provided in the upper right box (1xxxx) on Table 8.10c. Although the system permits equation (or action) identification numbers to range from 10001 through 19999, for efficiency they are numbered from 10001 to the approximate total number of equations (and actions) in use.

For additional system efficiency and to conserve mass storage requirements, the general equation sheet is partitioned into two groups of entries—basic (entries 01 through 04) and expanded (entries 05 through 09). Whenever only the basic entries are required, the assigned identification number is optional (even or odd). However, when the expanded entries are used, the assigned identification number must be *even* and the next sequential number (i.e., the assigned number plus one) must *not be assigned* to an equation (or action). The uniting of the two groups of entries, basic and expanded, is automatic.

General Equation and Action Grouping   One of the several uses of the general equation is by a general action. Although the system will function with randomly assigned equation and action identification numbers, there is an incentive to group them together.

Whenever an equation (or action) is referenced, the system first determines if it is already in core storage. If so, processing proceeds immediately. If not, it must go to mass storage and transfer it to core storage. When it does this, it not only transfers the required equation (or action) but 15 other identified equations (or actions). Therefore if related equations are grouped, the number of mass storage accesses can be reduced. Efficiency is increased because core storage accessing is measured in millionths of seconds while mass storage accesses are measured in seconds or tenths of seconds.

To determine the equations (and actions) which are grouped and accessed together, the following formula can be used:

$$I.d = \frac{\text{Identification Number} - 1000}{16} \qquad 8.10(5)$$

Ignoring the decimal fraction portion (d) of the quotient, the equations (or actions) are grouped together when the integer portions (I) are the same. Example: Identification numbers 10160 through 10175 are grouped together; likewise 10256 through 10271.

In addition to increased efficiency, grouping of related equations (and actions) is beneficial for simplifying future editing and maintenance of documentation references.

Adjustment Information Sheet

Purpose

The purpose of the Adjustment Information Sheet (Table 8.10d) is to identify uniquely all the *adjustment equations* used and to collect the information necessary to (1) define the form and content of the adjustment equation and (2) control and communicate the output of the equation. The adjustment

Table 8.10d

1   5
2 [   ]

ADJUSTMENT INFORMATION SHEET

Description _____

_____

_____

Adjustment Equation: $0 = F_1 \Delta T + F_2 \Delta M + F_3 \Delta U + F_4 \Delta V_1 + F_5 \Delta V_2 + F_6 \Delta V_3$
Simultaneous Equation Information.   (Omit if single equation solution)

7 8   10   14   16   20   22   26   28   32 34
[0][0]   [2 [   ]]   [2 [   ]]   [2 [   ]]   [2 [   ]]   □ Re-solution acceptable?
                                                              (1=Yes)

Definition and Development of Terms in Adjustment Equation.

7 8   10   14   16 Const. F_n Coeff.   28   30   33
[0][1]   [:]   [   ]   [   ]   T-Targeted Variable
[0][2]   [:]   [   ]   [   ]   M-Manipulated Variable
[0][3]   [:]   [   ]   [   ]   U-Uncontrolled Variable
[0][4]   [:]   [   ]   [   ] V_1- } Other
[0][5]   [:]   [   ]   [   ] V_2- } Variables
[0][6]   [:]   [   ]   [   ] V_3- }

Adjustment Equation Output Restrictions.

[1][0]   [:]   Action taken when Variable M Out of Service.
[1][1]   [:]   Action taken when Variable M Off Computer (On Operator).
[1][2]   [   ]   Maximum allowable adjustment to M per pass.
[1][3]   [:]   Action taken when Target of M already at MAX-MIN limit.
[1][4]   [:]   Adjustment Reference List for Partial Loop Test. (2nnnn or Blank)
[1][5]   [   ]   If Entry 14 specifies a Reference List, Entries 15-18 specify
[1][6]   [   ]   sublists (-1, -2, -3, -4) referenced for all subsequent M's in
[1][7]   [   ]   control loop. If Entry 14 is Blank, Entries 15-18 specify all of
[1][8]   [   ]   the subsequent M Variables in control loop. Is Target of M set
[1][9]   [ ]   to Average instead of Current for Partial Loop? (1=Yes)

Manipulated Variable Adjustment (ΔM) Output Control.

[2][1]   [   ]   TIME in minutes (bXXXX) or as developed (-nnnn or 1nnnn).
7 8   10 12   14   17
[2][2]   [| |]   10-12: %T = Cumulative elapsed time from present time
[2][3]   [| |]   expressed as a percent of TIME.
[2][4]   [| |]   14-17: %ΔM = Percent of ΔM change made after %T time.
[2][5]   [| |]

Special Action Initiated Upon Significant Change in Manipulated Variable (M).

7 8   10 Initiating M Change 22   24 Action
[3][1]   [   ]   [   ]   Action may be specified as follows:
[3][2]   [   ]   [   ]   (a) Process Variable Special (0nnnn).
[3][3]   [   ]   [   ]   (b) Execute General Action (1nnnn).
[3][4]   [   ]   [   ]   (c) Feedforward Request Call of
[3][5]   [   ]   [   ]   Adjustment Equation (2nnnn).
[3][6]   [   ]   [   ]   (d) Execute Special Program (3nnnn).

equation is part of the control system by which the implemented control schemes normally make adjustments to manipulatable variables. The completed forms are (1) reviewed for completeness, validity, and consistency of information, (2) transcribed into 80-column data processing cards, and (3) placed in a loose-leaf notebook as a permanent reference.

Adjustment Equation

The adjustment equation is used to calculate changes to manipulated process variables which compensate for (1) a noted difference between a process variable's target and current values (feedback) or (2) an anticipated difference between a variable's target and current values due to a detected or anticipated change in other process variables (feedforward). Equations may be solved singly, or several may be solved simultaneously.

The general form of the adjustment equation is

$$0 = F_1 \Delta T + F_2 \Delta M + F_3 \Delta U + F_4 \Delta V_1 + F_5 \Delta V_2 + F_6 \Delta V_3 \quad 8.10(6)$$

The symbols in the equation are defined as follows: ΔT represents the noted difference between a process variable's target and current values when the equation is solved for a feedback adjustment; ΔM represents the solution answer of the equation or the necessary change to a manipulated process variable's target to adjust for the known ΔT difference or ΔU change; ΔU represents a detected change in an uncontrolled process variable or an anticipated change in a manipulated process variable when the equation is solved for a feedforward adjustment; ΔV's are optional variables available for defining other ΔU's or for defining interacting control variables when adjustment equations are solved simultaneously. The F's are the term coefficients in the solution which can either be specified as constant values or defined elsewhere by the system.

Adjustment Requests

All requests to the *adjustment processor* which processes the adjustment equations must furnish the following information: (1) mode of request (feedback or feedforward), (2) identification of the adjustment equation (2nnnn), (3) identification of the process variable in the equation for which the request was initiated (0nnnn), and (4) the specified difference in the identified process variable for which the equation is to make an adjustment. The adjustment request information is added to an adjustment stack.

Adjustment Stack Servicing

When a request is placed in the adjustment stack, the *processing scheduler* is signaled to execute the *adjustment processor* at the first available opportunity. When the *adjustment processor* acquires control of the computer, its task is to select and extract requests from the stack, to process the extracted requests using the identified adjustment equations, and to transmit the output of the adjustment equation solution to other parts of the system.

The selection of the next or primary request to process is made on a "first-in, first-out" basis except that all waiting predictive or feedforward requests are processed before any waiting feedback requests. An analysis of the stack determines whether the primary request is of a feedforward or feedback mode. This request is moved from the stack to an internal request table. Whenever a request is transferred to the request table, a check of the stack is made to determine if there are any similar requests later in the stack. A similar request is defined as one of the same mode, same identified adjustment equation, and same identified process variable. When a similar request is found, the adjustment processor moves the request out of the stack, and cancels the entry and disposes of the similar request in one of two ways. If the primary request is a feedforward request, then the subsequent similar request amount is added to the primary request amount. If the primary request is a feedback request, then the subsequent similar request amount replaces the primary request amount.

From the primary request in the request table, the primary adjustment equation is identified. Using the primary adjustment equation's identification, the designated information record is transferred from mass storage to core storage. During this transfer, the primary manipulated variable is identified and its status is checked.

From other primary adjustment equation information, a determination is made as to whether the adjustment is to be made alone, or simultaneously with other adjustments. For a single solution, processing proceeds as described in the next paragraph. For simultaneous solution, the other adjustment equation identifiers are listed as secondary equations. As with the primary information, each identified secondary adjustment equation information record is transferred from mass storage to core storage. Also during this transfer, each secondary manipulated variable is identified and its status is checked and indicators are set accordingly. Further, as each secondary adjustment equation information record is placed in working storage, a check is made to ensure that a consistent set of information is specified. A consistent set means that each adjustment equation information record must specify the others. If an inconsistent set is detected, the primary request is cancelled and another primary request is chosen from the stacker and processing proceeds as described above.

The primary manipulated variable status indicator is checked to insure that the variable is in service. If it is out of service, processing of the primary request is terminated except that any alternative action specified is executed before termination. If the primary manipulated variable is in service, a status indicator is checked to insure that it is on computer control (i.e., any calculated adjustment can be made by the computer). If on operator control, processing of the primary request is terminated except that, if alternative action is specified, this action is executed before termination.

If the manipulated variables are all available for adjustment, then the stack is checked to determine if there are any other requests waiting for any of the identified primary and secondary adjustment equations. If any of these additional requests are found, they are transferred from the stack to the request table to become secondary requests and the stack entry is cancelled. Any located secondary requests, which are similar requests, are handled in the same manner as were the similar primary requests (i.e., feedforward requests amounts are added together, whereas the subsequent feedback amounts replace the preceding amounts in the request table).

All requests that are to be processed at this time have been extracted from the stack and tabulated in the request table. After compacting the remaining request entries, the request stack is returned to mass storage for next pass processing. If processing is terminated beyond this point for any reason, the requests in the table are effectively cancelled.

Example of Stack Servicing

To illustrate the above description of the adjustment stack servicing, an example request stack is shown in Table 8.10e and the resultant request table and adjustment stack after servicing are shown in Tables 8.10f and g.

The *adjustment processor* selects the fourth request in the stack as the primary request since it is the first feedforward request found in the stack. It then adds the amount of the eighth request in stack to the primary request, since this is a similar request. It then searches the stack for any secondary requests it can process at this time and finds the second request in the stack which becomes a secondary request. This secondary request amount is replaced by the sixth request's amount (9.0) as it is a similar request. As no more requests

Table 8.10e
ADJUSTMENT STACK BEFORE SERVICING

Adjustment Identification	Variable	Amount	Mode
20001	●	●	FB
20002	T	6.0	FB
20003	●	●	FB
20002	U	− 10.0	FF
20004	●	●	FB
20002	T	9.0	FB
20004	●	●	FF
20002	U	5.0	FF
20005	●	●	FB

Table 8.10f
ADJUSTMENT STACK AFTER SERVICING

Adjustment Identification	Variable	Amount	Type
20001	●	●	FB
20003	●	●	FB
20004	●	●	FB
20004	●	●	FF
20005	●	●	FB

Table 8.10g
REQUEST TABLE FOR PROCESSING

Adjustment Identification	Variable	Amount	Type
20002	U	−5.0	FF
20002	T	9.0	FB

are to be processed at this time, it compacts the stack and saves it for later processing and proceeds to process the requests in the request table (Table 8.10g).

Request Table Processing

The request table is then combined with the adjustment equation information in working storage to set up the equation (or equations) to solve for the calculated changes to the manipulated variable (or variables). For a single equation, this is a rather simple and straightforward task. For simultaneous equations, the procedure is somewhat more involved and is illustrated by the discussion on Simultaneous Equations.

The calculated adjustments to the manipulated variables (ΔM) are checked against restrictions to determine if the total amount of calculated adjustment can be made. The first restriction is an allowable step size which the *adjustment processor* is permitted to make. The second restriction is to ensure that the calculated adjustment, when added to the manipulated variable's present target value, does not exceed its assigned maximum and minimum limits. Failure to pass either or both of the above restrictions causes the calculated adjustment to be reduced to the actual calculated adjustment for the remainder of the adjustment request processing.

Prior to disposing of the calculated adjustment, a check is made to insure that all subsequent manipulated variables, if any, are available for computer adjustment (i.e., in service and on computer). If any subsequent specified manipulated variable is not available, then the control loop is considered only partially controlled by the computer. Therefore, to protect against target "windups," the present value of the manipulated variable's target is set equal to its current (or average) value.

The dynamic distribution of the calculated adjustment is determined from specified information. The manipulated variable's target value is adjusted at this time only for the amount indicated at this time. The remaining amount, if any, will be distributed as specified by the *dynamic control program.*

After disposing of the calculated adjustment, a check of specified information is made to determine whether or not this adjustment or an anticipated adjustment is significant enough to cause a need for additional actions. If found significant, the additional actions are initiated or executed.

This completes the processing of the tabulated requests in the request table. If the adjustment stack contains no more requests to process, a permanent release of processing control is made to the *processing scheduler.* If, however, other requests are waiting, a check is made of the processing scheduler, to determine if a higher-priority program is waiting. If none are waiting, the *adjustment processor* returns to the stack servicing operation to select the next request to process. If, however, something more important requires attention, a release of control is made to the processing scheduler with the notification that adjustment processor has not completed its task and must be restarted at the first available opportunity.

Simultaneous Equations

When more than one equation is to be solved simultaneously, the processor incorporates the calling process differences into the solution of the equations and executes the calculated control adjustments. This can best be illustrated by a simple example:

Given: The following entries in the *adjustment request stack.*

STACK REQUESTS

Equation	Variable	Differential
1	T_1	ΔT_1
3	T_3	ΔT_3
1	U_1	ΔU_1

The adjustment processor selects equation 1 as the next equation to be processed and the U_1 entry as the primary request (first-come, first-served except that feedforward requests take precedence). It determines from the adjustment information of Equation (1) that the equation must be solved simultaneously with Equations (2) and (3). The processor collects the additional information and assembles the equations in question.

Simultaneous Equations

(1) $0 = F_{11}\Delta T_1 + F_{12}\Delta M_1 + F_{13}\Delta M_2 + F_{14}\Delta U_1$

(2) $0 = F_{21}\Delta T_2 + F_{22}\Delta M_2 + F_{23}\Delta M_1 + F_{24}\Delta T_3 + F_{25}\Delta U_2$

(3) $0 = F_{31}\Delta T_3 + F_{32}\Delta M_3 + F_{33}\Delta M_2$

The three simultaneous equations appear in matrix notation:

$$
\begin{array}{ccc}
 & \text{Calculated} & \\
\text{Coefficient Matrix} & \text{Matrix} & \text{Constant Matrix}
\end{array}
$$

$$
\begin{array}{ccc}
\Delta M_1 & \Delta M_2 & \Delta M_3
\end{array}
$$

$$
\begin{array}{c}
(1) \\
(2) \\
(3)
\end{array}
\begin{vmatrix}
F_{12} & F_{13} & 0 \\
F_{23} & F_{22} & 0 \\
0 & F_{33} & F_{32}
\end{vmatrix}
\times
\begin{vmatrix}
\Delta M_1 \\
\Delta M_2 \\
\Delta M_3
\end{vmatrix}
=
\begin{vmatrix}
-F_{11}\Delta T_1 - F_{14}\Delta U_1 \\
-F_{21}\Delta T_2 - F_{24}\Delta T_3 - F_{25}\Delta U_2 \\
-F_{31}\Delta T_3
\end{vmatrix}
$$

The adjustment processor solves the equations through use of matrix algebra. It sets up the required tables and matrices and performs the necessary operations.

Adjustment Information Sheet Organization

The information entries on the sheet are grouped into five major categories.

Entries	Categories
00	Simultaneous equation information
01–06	Adjustment equation definition
10–19	Adjustment equation output control restrictions
21–25	Calculation adjustment output control
31–36	Additional action which is executed or initiated on significant calculated adjustments (ΔM) to the manipulated variable

A concerted effort has been made to present a sheet which is as readable and self-explanatory as possible and yet fit on a single page (except for the extended reference lists). Although the number of entries is quite large, the majority of the adjustment equations do not require all of the entries. It is not necessary to fill in an entry which is not applicable or needed.

Sample Problems

Three sample problems together with the completed sheets which solve them, are included. The first two illustrate the use of *general action* and *general equation* sheets to calculate the current value of a variable. The third illustrates the use of the variable information and adjustment information sheets to implement a control strategy.

Mass Flow Calculation Problem

Variable 00400 is the mass flow of a liquid stream. There is no direct measurement of mass flow, so its current value must be calculated from a combination of other measurements.

Variable 00350 is the pressure drop across an orifice in the stream. The temperature at the orifice is 00351. The calibration constant for the pressure measurement is stored as *miscellaneous data* -0052.

The mass flow is to be computed by the formula

$$\text{Mass Flow} = (\text{Calibration Constant}) \sqrt{\frac{\text{Differential Pressure}}{\text{Absolute Temperature}}} \qquad 8.10(7)$$

Tables 8.10h and i show how this calculation could be performed by PROSPRO. Entry 18 of the variable information sheet (Table 8.10a) for the mass flow would contain 10400 to call for the general action of Table 8.10h.

Logic Problem

This example illustrates the use of the general action sheets to solve a problem involving logical decisions. The differential pressure of the preceding problem varies over such a wide range that a single differential pressure transmitter is not sufficiently accurate, so two are installed. The wide-range instrument has four times the range of the second instrument.

For high values of differential pressure, transmitter #1 is used. For low values the reading of the second instrument is multiplied by 4. To ensure a smooth transition between instruments, a combination of readings is used over the interval 20 to 25 percent of the range of transmitter #1. Variable number 00348 is the wide range pressure and variable number 00349 is the narrow range.

The current value of variable 00350, the effective differential pressure, is calculated by equations 8.10(8) through 8.10(11) (V03480 etc. represent the current value of variable 00348):

If

$$V03480 > 25\%$$

then

$$V03500 = V03480 \qquad 8.10(8)$$

If

$$V03480 < 20\%$$

then

$$V03500 = 4.0*V03490 \qquad 8.10(9)$$

If

$$20\% < V03480 < 25\%$$

then

$$V03500 = A*V03480 + (1 - A)*4.0*V03490$$
$$= A*(V03480 - 4.0*V03490) + 4.0*V03490 \qquad 8.10(10)$$

Table 8.10h

GENERAL ACTION SHEET

1		5		
1	0	4	0	0

Description Calculate the Current Value of Mass Flow

Identified by the Variable V00400

ΔP is V00350 and T (°F) is V00351

7 8	10 11	13		18	22
0 0	G A	☐		1	

13:        Are values beyond MAX–MIN limits acceptable? (0=No, 1=Yes)
18–22:   Next General Action to be executed. (Chaining)

Step	Code	A	B	Remarks

7 8	10	12 14	16 20	22 26	Remarks
0 1	S	V P C	0 3 5 / 0	0 2	Determine absolute temperature
0 2	S	C O N	4 5 9 · 6 9		°F + 459.69 = °R
0 3	S	V D V	0 3 5 0 0		ΔP / Tabs
0 4	S	V E V	– 0 0 0 3		Save result for general equation
0 5	S	V E E		1 0 4 0 1	Result from general equation
0 6	S				

0 7	S				
0 8	S				
0 9	S				
1 0	S				
1 1	S				
1 2	S				
1 3	S				
1 4	S				
1 5	S				
1 6	S				

RTN	– Return to processing	VTC	– Variable Times Constant
END	– End Variable Processing	VDV	– Variable Divided by Var.
CON	– Constant	VPV	– Variable Plus Variable
MSG	– Type Message No./Value	VPC	– Variable Plus Constant
PVS	– Process Variable Special	VMV	– Variable Minus Variable
NOP	– No Operation	ABS	– Absolute Result
FBA	– Feedback Adjustment	CVV	– Compare Var. to Variable
FFA	– Feedforward Adjustment	CVC	– Compare Var. to Constant
QVS	– Query Variable Status	BRM	– Branch Result Minus
SRT	– Save Real Time	BRZ	– Branch Result Zero
CDT	– Calculate Delta Time	BRP	– Branch Result Plus
VEE	– Variable Equals Equation	BCL	– Branch Compare Low
VEV	– Variable Equals Variable	BCE	– Branch Compare Equal
VEC	– Variable Equals Constant	BCH	– Branch Compare High
VTV	– Variable Times Variable	BRA	– Branch unconditionally

Table 8.10i

GENERAL EQUATION SHEET

1			5
1	0,4,0,1		

Description *Square root and Calibrate for Mass Flow*

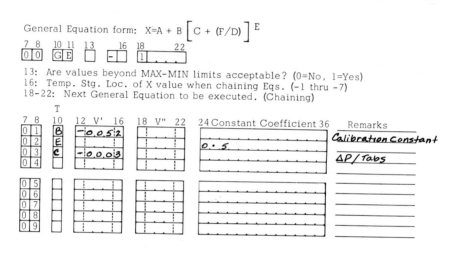

General Equation form: $X = A + B \left[C + (F/D) \right]^E$

7 8	10 11	13	16	18	22
0 0	G E		-	1	

13: Are values beyond MAX-MIN limits acceptable? (0=No, 1=Yes)
16: Temp. Stg. Loc. of X value when chaining Eqs. (-1 thru -7)
18-22: Next General Equation to be executed. (Chaining)

7 8	10 T	12 V' 16	18 V" 22	24 Constant Coefficient 36	Remarks
0 1	B	-0.052			*Calibration Constant*
0 2	E			0.5	
0 3	C	-0.003			*ΔP/Tabs*
0 4					
0 5					
0 6					
0 7					
0 8					
0 9					

$$\text{Term}_T = 0.0 + \sum \left[(V')_T \times (V'')_T \times (\text{Constant Coefficient})_T \right]$$

Where T is A, B, C, D, E or F
(V) is the value of an identified variable V' or V" as follows:

$V = 0$ (or omitted); Value assumed to be 1.0
$V > 0$; Contents of Identified Process Variable Value (iiiij)
$-7 \leq V < 0$; Contents of Identified Temp. Stg. Loc. (Chaining)
$-9999 \leq V \leq -10$; Contents of Identified Miscellaneous Data Storage

Constant Coefficient.

If omitted, the coefficient is assumed to be 1.0.
When specified, a decimal point (.) must be placed within the field;
all other significant characters must be a digit (0-9) except the first
which may be a minus sign (-) to indicate a negative value; and all
characters from the first to the last must be adjacent. The numerical
range of the Constant Field (24-36) can be extended through use of the
conventional exponential (E) format.

where

$$A = \frac{V03480 - 20}{25 - 20} \qquad 8.10(11)$$

Tables 8.10j and k show completed sheets for this example. Note that the first general action uses expanded entries so it must be assigned an even number (10302) and that 10303 cannot be used for another general action or general equation.

Furnace Control Problem

As an example of the manner in which the data forms might be completed, consider the process shown in Figure 8.10l. A furnace is heated by a combination of oil and gas. Raw material is heated in the furnace and then enters a catalytic reactor. In the reactor a portion of the raw material is converted to products.

The following strategy is to be implemented:

1. Use all the gas feed available.
2. Use 3 barrels of oil for every 1,000 standard cubic feet of gas.
3. Control the temperature at the furnace outlet by adjusting the flow rate of the raw material.
4. Maintain 18-minute average conversion at 75 percent by adjusting the furnace outlet temperature.
5. Use feedforward correction from gas feed to raw material flow.

Measurement Information

Oil feed (variable 00001) is measured by a flow transmitter with a range of 0 to 30,000 barrels per day connected to input address 11. Routine calculations should be made every 3 minutes.

Gas feed (variable 00002) is measured by a flow transmitter with a range of 0 to 90,000,000 standard cubic feet per day (0–90,000 MSCF/D) connected to input address 12. Routine calculations should be made every 6 minutes.

Temperature (variable 00003) is measured by an iron-constantan thermocouple connected to input address 13. Routine calculations should be made every 2 minutes.

Conversion (variable 00004) is measured by a transmitter with a range of 40–100 percent connected to input address 14. Routine calculations should be made every 6 minutes.

Raw material flow (variable 00005) is measured by a flow transmitter with a range of 0 to 20,000 barrels per day connected to input address 15. Routine calculations should be made every 1 minute.

All transmitters give an ADC reading of 6552 at the low end of the scale and 32766 at the high end. Note that the code for iron-constantan subroutines is 31 in entry 18 for the variable information sheet.

Table 8.10j

GENERAL ACTION SHEET

1		5
1	0.3.0.2	

Description   *Compute Effective Differential Pressure, VO3500*

7 8	10 11	13	18	22
0 0	G A	1		1 0.3.0.4

13:       Are values beyond MAX–MIN limits acceptable? (0=No, 1=Yes)

18-22:   Next General Action to be executed. (Chaining)

Step		Code		A		B		Remarks
7 8	10	12 14	16	20	22		26	
0 1	S	C.V.C	0.3.4.8.0				0 2	*Compare VO3480 to 25%*
0 2	S	C.O.N	2.5.•.0					
0 3	S	B.C.L					0 6	
0 4	S	V.E.V				0.3.4.8.0		*Result is VO3480*
0 5	S	R.T.N						
0 6	S	V.T.C	0.3.4.9.0				0 7	
0 7	S	C.O.N		4.•.0				
0 8	S	V.E.V	-.0.0.0.5					*Save 4 times VO3490*
0 9	S	C.V.C	0.3.4.8.0				1 0	
1 0	S	C.O.N	2.0.•.0					
1 1	S	B.C.H					1 4	
1 2	S	V.E.V				-.0.0.0.5		*Result is 4 times VO3490*
1 3	S	R.T.N						
1 4	S	V.P.C	0.3.4.8.0				1 5	
1 5	S	C.O.N	-.2.0.•.0					
1 6	S	N.O.P						*Complete link for chaining*

RTN	– Return to processing		VTC	– Variable Times Constant
END	– End Variable Processing		VDV	– Variable Divided by Var.
CON	– Constant		VPV	– Variable Plus Variable
MSG	– Type Message No./Value		VPC	– Variable Plus Constant
PVS	– Process Variable Special		VMV	– Variable Minus Variable
NOP	– No Operation		ABS	– Absolute Result
FBA	– Feedback Adjustment		CVV	– Compare Var. to Variable
FFA	– Feedforward Adjustment		CVC	– Compare Var. to Constant
QVS	– Query Variable Status		BRM	– Branch Result Minus
SRT	– Save Real Time		BRZ	– Branch Result Zero
CDT	– Calculate Delta Time		BRP	– Branch Result Plus
VEE	– Variable Equals Equation		BCL	– Branch Compare Low
VEV	– Variable Equals Variable		BCE	– Branch Compare Equal
VEC	– Variable Equals Constant		BCH	– Branch Compare High
VTV	– Variable Times Variable		BRA	– Branch unconditionally

Table 8.10k

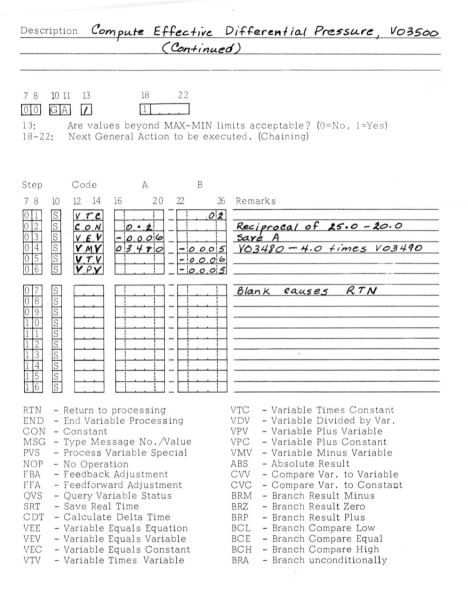

GENERAL ACTION SHEET

```
        1        5
       |1|0 3 0 4|
```

Description *Compute Effective Differential Pressure, VO3500*
                        (Continued)

```
7 8  10 11  13        18       22
|0|0| |G|A|  |/|      |1 . . .|
```
13:     Are values beyond MAX-MIN limits acceptable? (0=No, 1=Yes)
18-22: Next General Action to be executed. (Chaining)

Step	Code	A	B	Remarks
7 8	10 12 14	16 20	22 26	
0 1	S	V T C	. 0 2	
0 2	S	C O N	0 . 2	*Reciprocal of 25.0 - 20.0*
0 3	S	V E V	- 0 . 0 0 6	*Save A*
0 4	S	V M Y	0 3 4 8 0	- 0 . 0 0 5	*VO3480 - 4.0 times VO3490*
0 5	S	V T V		- 0 . 0 0 6	
0 6	S	V P Y		- 0 . 0 0 5	
0 7	S			*Blank causes RTN*
0 8	S			
0 9	S			
1 0	S			
1 1	S			
1 2	S			
1 3	S			
1 4	S			
1 5	S			
1 6	S			

RTN	– Return to processing	VTC	– Variable Times Constant
END	– End Variable Processing	VDV	– Variable Divided by Var.
CON	– Constant	VPV	– Variable Plus Variable
MSG	– Type Message No./Value	VPC	– Variable Plus Constant
PVS	– Process Variable Special	VMV	– Variable Minus Variable
NOP	– No Operation	ABS	– Absolute Result
FBA	– Feedback Adjustment	CVV	– Compare Var. to Variable
FFA	– Feedforward Adjustment	CVC	– Compare Var. to Constant
QVS	– Query Variable Status	BRM	– Branch Result Minus
SRT	– Save Real Time	BRZ	– Branch Result Zero
CDT	– Calculate Delta Time	BRP	– Branch Result Plus
VEE	– Variable Equals Equation	BCL	– Branch Compare Low
VEV	– Variable Equals Variable	BCE	– Branch Compare Equal
VEC	– Variable Equals Constant	BCH	– Branch Compare High
VTV	– Variable Times Variable	BRA	– Branch unconditionally

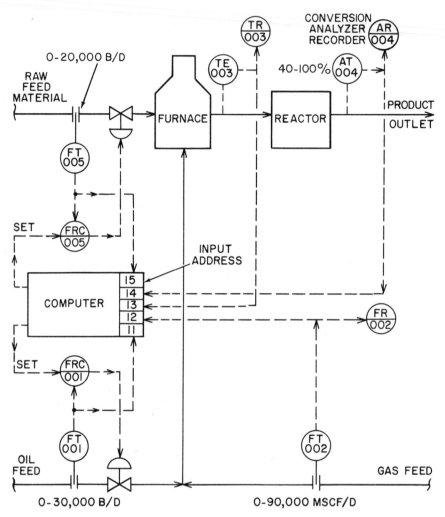

Fig. 8.101 Furance control utilizing a supervisory computer

Alarm Information   The alarm conditions are listed in Table 8.10m.

Control Information   If conversion is higher than its target by 1 percent furnace outlet temperature should be reduced 2.5°F. This feedback adjustment is to be made every 18 minutes. If a change in the target for furnace outlet temperature is computed, 40 percent of the change is to be applied immediately, a further 80 percent 2 minutes later with a subsequent −20 percent applied 4 minutes after the computation.

If the furnace outlet temperature is higher than its target 1°F, increase

Table 8.10m

Variable	Maximum	Minimum	Deviation
Oil Feed (B/D)	19,000	14,000	1,000
Gas Feed (MSCF/D)	57,000	42,000	None
Furnace Temperature (°F)	800	700	10
Conversion (%)	80	70	3
Raw Material Flow (B/D)	18,000	10,000	4,000

raw material flow by 1,000 B/D. This feedback adjustment is to be made every 2 minutes.

If the gas feed increases by 1 MSCF/D, the raw material flow is to be increased by 0.3 B/D. This feedforward correction is to be applied whenever two successive readings of gas feed differ by more than 1,500 MSCF/D.

The maximum acceptable adjustments of manipulated variables are listed in Table 8.10n.

Table 8.10n

Manipulated Variable	Maximum Adjustment
Furnace Outlet Temperature (°F)	5
Raw Material Flow (B/D)	1,000
Oil Feed (B/D)	1,200

Data Sheets   Tables 8.10o through 8.10w are the completed data sheets, which solve the furnace control problem.

Table 8.10o

VARIABLE INFORMATION SHEET

Description   *Oil Feed*

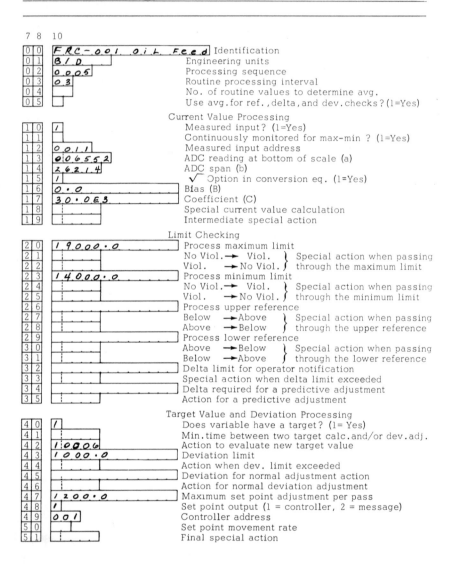

Table 8.10p

VARIABLE INFORMATION SHEET

1   5
| 0 | O.O.O.2 |

Description *Gas Feed*

7 8 10

0	0	F.R.-O.O2 G.A.S. .F.E.E.D.	Identification
0	1	M.3.C.F/D	Engineering units
0	2	0.0.0.4	Processing sequence
0	3	0.6	Routine processing interval
0	4		No. of routine values to determine avg.
0	5		Use avg. for ref., delta, and dev. checks? (1=Yes)

Current Value Processing

1	0	1	Measured input? (1=Yes)
1	1		Continuously monitored for max–min? (1=Yes)
1	2	0.0.1.2	Measured input address
1	3	0.0.6.5.5.2	ADC reading at bottom of scale (a)
1	4	2.6.2.1.4	ADC span (b)
1	5	√	Option in conversion eq. (1=Yes)
1	6	0.0	Bias (B)
1	7	9.0.0.E.3	Coefficient (C)
1	8		Special current value calculation
1	9		Intermediate special action

Limit Checking

2	0	5.7.0.0.0.0	Process maximum limit
2	1		No Viol. → Viol. ⎫ Special action when passing
2	2		Viol. → No Viol. ⎬ through the maximum limit
2	3	4.2.0.0.0.0	Process minimum limit
2	4		No Viol. → Viol. ⎫ Special action when passing
2	5		Viol. → No Viol. ⎬ through the minimum limit
2	6		Process upper reference
2	7		Below → Above ⎫ Special action when passing
2	8		Above → Below ⎬ through the upper reference
2	9		Process lower reference
3	0		Above → Below ⎫ Special action when passing
3	1		Below → Above ⎬ through the lower reference
3	2		Delta limit for operator notification
3	3		Special action when delta limit exceeded
3	4	1.5.0.0	Delta required for a predictive adjustment
3	5	2.0.0.0.1	Action for a predictive adjustment

Target Value and Deviation Processing

4	0		Does variable have a target? (1= Yes)
4	1		Min. time between two target calc. and/or dev. adj.
4	2		Action to evaluate new target value
4	3		Deviation limit
4	4		Action when dev. limit exceeded
4	5		Deviation for normal adjustment action
4	6		Action for normal deviation adjustment
4	7		Maximum set point adjustment per pass
4	8		Set point output (1 = controller, 2 = message)
4	9		Controller address
5	0		Set point movement rate
5	1		Final special action

Table 8.10q

VARIABLE INFORMATION SHEET

Description   *Furnace Temperature*

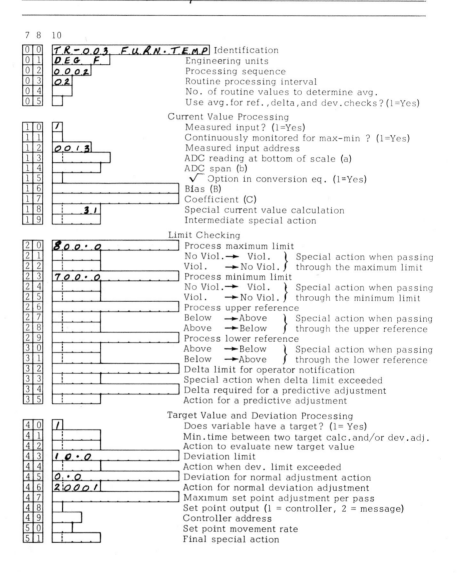

7 8 10

0 0	TR-003 FURN.TEMP	Identification
0 1	DEG F.	Engineering units
0 2	0.002	Processing sequence
0 3	02	Routine processing interval
0 4		No. of routine values to determine avg.
0 5		Use avg. for ref., delta, and dev. checks? (1=Yes)

Current Value Processing

1 0	1	Measured input? (1=Yes)
1 1		Continuously monitored for max-min ? (1=Yes)
1 2	00.13	Measured input address
1 3		ADC reading at bottom of scale (a)
1 4		ADC span (b)
1 5		✓ Option in conversion eq. (1=Yes)
1 6		Bias (B)
1 7		Coefficient (C)
1 8	3.1	Special current value calculation
1 9		Intermediate special action

Limit Checking

2 0	800.0	Process maximum limit
2 1		No Viol.→ Viol. } Special action when passing
2 2		Viol. →No Viol. } through the maximum limit
2 3	700.0	Process minimum limit
2 4		No Viol.→ Viol. } Special action when passing
2 5		Viol. →No Viol. } through the minimum limit
2 6		Process upper reference
2 7		Below →Above } Special action when passing
2 8		Above →Below } through the upper reference
2 9		Process lower reference
3 0		Above →Below } Special action when passing
3 1		Below →Above } through the lower reference
3 2		Delta limit for operator notification
3 3		Special action when delta limit exceeded
3 4		Delta required for a predictive adjustment
3 5		Action for a predictive adjustment

Target Value and Deviation Processing

4 0	1	Does variable have a target? (1= Yes)
4 1		Min. time between two target calc. and/or dev. adj.
4 2		Action to evaluate new target value
4 3	10.0	Deviation limit
4 4		Action when dev. limit exceeded
4 5	0.0	Deviation for normal adjustment action
4 6	20001	Action for normal deviation adjustment
4 7		Maximum set point adjustment per pass
4 8		Set point output (1 = controller, 2 = message)
4 9		Controller address
5 0		Set point movement rate
5 1		Final special action

Table 8.10r

VARIABLE INFORMATION SHEET

	1	5
0	0.0.0.4	

Description　*Conversion in Reactor*

7 8　10

0 0	AR-0.0.4 CONV	Identification
0 1	P.C.T.	Engineering units
0 2	0.0.0.1	Processing sequence
0 3	0.6	Routine processing interval
0 4	0.3	No. of routine values to determine avg.
0 5	1	Use avg. for ref., delta, and dev. checks? (1=Yes)

Current Value Processing

1 0	1	Measured input? (1=Yes)
1 1		Continuously monitored for max-min? (1=Yes)
1 2	0.0.1.4	Measured input address
1 3	0.0.6.5.5.2	ADC reading at bottom of scale (a)
1 4	2.6.2.1.4	ADC span (b)
1 5		√ Option in conversion eq. (1=Yes)
1 6	40.0	Bias (B)
1 7	60.0	Coefficient (C)
1 8		Special current value calculation
1 9		Intermediate special action

Limit Checking

2 0	80.0	Process maximum limit
2 1		No Viol. → Viol. ⎱ Special action when passing
2 2		Viol. → No Viol. ⎰ through the maximum limit
2 3	70.0	Process minimum limit
2 4		No Viol. → Viol. ⎱ Special action when passing
2 5		Viol. → No Viol. ⎰ through the minimum limit
2 6		Process upper reference
2 7		Below → Above ⎱ Special action when passing
2 8		Above → Below ⎰ through the upper reference
2 9		Process lower reference
3 0		Above → Below ⎱ Special action when passing
3 1		Below → Above ⎰ through the lower reference
3 2		Delta limit for operator notification
3 3		Special action when delta limit exceeded
3 4		Delta required for a predictive adjustment
3 5		Action for a predictive adjustment

Target Value and Deviation Processing

4 0	1	Does variable have a target? (1= Yes)
4 1	0.0.0.1.8	Min. time between two target calc. and/or dev. adj.
4 2	1.0.0.0.7	Action to evaluate new target value
4 3	3.0	Deviation limit
4 4		Action when dev. limit exceeded
4 5	0.0	Deviation for normal adjustment action
4 6	2.0.0.0.2	Action for normal deviation adjustment
4 7		Maximum set point adjustment per pass
4 8		Set point output (1 = controller, 2 = message)
4 9		Controller address
5 0		Set point movement rate
5 1		Final special action

Table 8.10s

VARIABLE INFORMATION SHEET

Description   *Raw Material Flow*

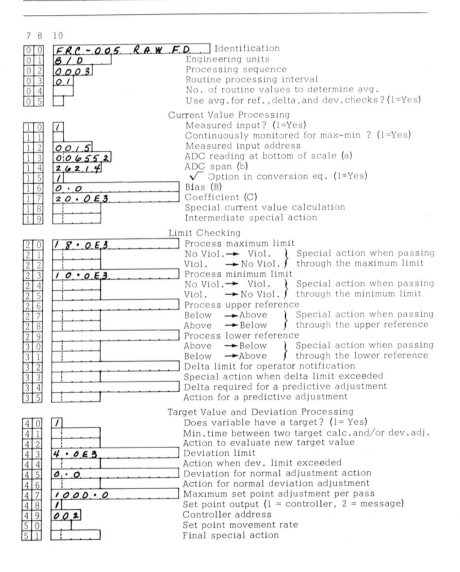

7 8 10

0	0	FRC.-0.05 R.A.W. F.D.	Identification
0	1	B/D	Engineering units
0	2	0003	Processing sequence
0	3	0.1	Routine processing interval
0	4		No. of routine values to determine avg.
0	5		Use avg. for ref., delta, and dev. checks? (1=Yes)

Current Value Processing

1	0	1	Measured input? (1=Yes)
1	1		Continuously monitored for max-min ? (1=Yes)
1	2	0015	Measured input address
1	3	006552	ADC reading at bottom of scale (a)
1	4	26214	ADC span (b)
1	5	1	√ Option in conversion eq. (1=Yes)
1	6	0.0	Bias (B)
1	7	20.0E3	Coefficient (C)
1	8		Special current value calculation
1	9		Intermediate special action

Limit Checking

2	0	18.0E3	Process maximum limit
2	1		No Viol.→ Viol. ⎫ Special action when passing
2	2		Viol. → No Viol. ⎭ through the maximum limit
2	3	10.0E3	Process minimum limit
2	4		No Viol.→ Viol. ⎫ Special action when passing
2	5		Viol. → No Viol. ⎭ through the minimum limit
2	6		Process upper reference
2	7		Below → Above ⎫ Special action when passing
2	8		Above → Below ⎭ through the upper reference
2	9		Process lower reference
3	0		Above → Below ⎫ Special action when passing
3	1		Below → Above ⎭ through the lower reference
3	2		Delta limit for operator notification
3	3		Special action when delta limit exceeded
3	4		Delta required for a predictive adjustment
3	5		Action for a predictive adjustment

Target Value and Deviation Processing

4	0	1	Does variable have a target? (1= Yes)
4	1		Min. time between two target calc. and/or dev. adj.
4	2		Action to evaluate new target value
4	3	4.0E3	Deviation limit
4	4		Action when dev. limit exceeded
4	5	0.0	Deviation for normal adjustment action
4	6		Action for normal deviation adjustment
4	7	1000.0	Maximum set point adjustment per pass
4	8	1	Set point output (1 = controller, 2 = message)
4	9	002	Controller address
5	0		Set point movement rate
5	1		Final special action

Table 8.10t

GENERAL EQUATION SHEET

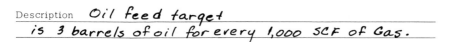

1		5
1	0,0,0,6

Description   *Oil feed target*

is 3 barrels of oil for every 1,000 SCF of Gas.

General Equation form: $X = A + B\left[C + (F/D)\right]^E$

7 8	10 11	13	16	18	22
0 0	G E	0	-	1	

13: Are values beyond MAX–MIN limits acceptable? (0=No, 1=Yes)
16: Temp. Stg. Loc. of X value when chaining Eqs. (-1 thru -7)
18-22: Next General Equation to be executed. (Chaining)

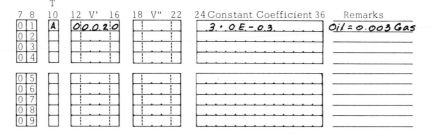

7 8	T 10	12 V' 16	18 V" 22	24 Constant Coefficient 36	Remarks
0 1	A	0,0.0.2,0		3.•.0.E.-.0.3	Oil = 0.003 Gas
0 2					
0 3					
0 4					
0 5					
0 6					
0 7					
0 8					
0 9					

$$\text{Term}_T = 0.0 + \sum\left[(V')_T \times (V'')_T \times (\text{Constant Coefficient})_T\right]$$

Where T is A, B, C, D, E or F
(V) is the value of an identified variable V' or V" as follows:

$V = 0$ (or omitted); Value assumed to be 1.0
$V > 0$; Contents of Identified Process Variable Value (iiiij)
$-7 \leqslant V < 0$; Contents of Identified Temp. Stg. Loc. (Chaining)
$-9999 \leqslant V \leqslant -10$; Contents of Identified Miscellaneous Data Storage

Constant Coefficient.

If omitted, the coefficient is assumed to be 1.0.
When specified, a decimal point (.) must be placed within the field;
all other significant characters must be a digit (0-9) except the first
which may be a minus sign (-) to indicate a negative value; and all
characters from the first to the last must be adjacent. The numerical
range of the Constant Field (24-36) can be extended through use of the
conventional exponential (E) format.

Table 8.10u

GENERAL EQUATION SHEET

Description   *Conversion target = 75 %*

General Equation form: $X = A + B \left[C + (F/D) \right]^E$

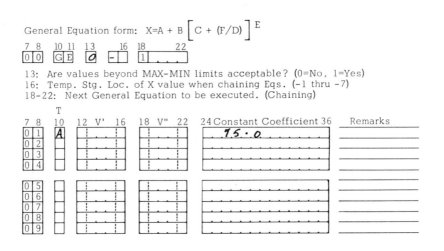

7 8   10 11 13   16 18   22
| 0 0 | G E | 0 | - | 1 . . . |

13: Are values beyond MAX-MIN limits acceptable? (0=No, 1=Yes)
16: Temp. Stg. Loc. of X value when chaining Eqs. (-1 thru -7)
18-22: Next General Equation to be executed. (Chaining)

T				24 Constant Coefficient 36	Remarks
7 8	10	12 V' 16	18 V" 22		
0 1	A			75.0	
0 2					
0 3					
0 4					
0 5					
0 6					
0 7					
0 8					
0 9					

$$\text{Term}_T = 0.0 + \sum \left[(V')_T \times (V'')_T \times (\text{Constant Coefficient})_T \right]$$

Where T is A, B, C, D, E or F
(V) is the value of an identified variable V' or V" as follows:

V = 0 (or omitted); Value assumed to be 1.0
V > 0; Contents of Identified Process Variable Value (iiiij)
$-7 \leq V < 0$; Contents of Identified Temp. Stg. Loc. (Chaining)
$-9999 \leq V \leq -10$; Contents of Identified Miscellaneous Data Storage

Constant Coefficient.

If omitted, the coefficient is assumed to be 1.0.
When specified, a decimal point (.) must be placed within the field;
all other significant characters must be a digit (0-9) except the first
which may be a minus sign (-) to indicate a negative value; and all
characters from the first to the last must be adjacent. The numerical
range of the Constant Field (24-36) can be extended through use of the
conventional exponential (E) format.

Table 8.10v

ADJUSTMENT INFORMATION SHEET

$$\begin{array}{cc} 1 & 5 \\ \hline 2 & 0\ 0\ 0\ 1 \end{array}$$

Description *Adjust raw material flow feedforward from gas feed, feedback from temperature*

Adjustment Equation: $0 = F_1 \Delta T + F_2 \Delta M + F_3 \Delta U + F_4 \Delta V_1 + F_5 \Delta V_2 + F_6 \Delta V_3$

Simultaneous Equation Information. (Omit if single equation solution)

```
7 8   10    14  16    20  22    26  28    32  34
[0 0] [2      ] [2      ] [2      ] [2      ]  ☐ Re-solution acceptable?
                                                 (1=Yes)
```

Definition and Development of Terms in Adjustment Equation.

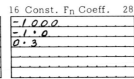

7 8	10	14	16 Const. F_n Coeff.	28	30 33	
0 1	:		-1.000		0003	T-Targeted Variable
0 2	:		-1.0		0005	M-Manipulated Variable
0 3	:		0.3		0002	U- Uncontrolled Variable
0 4	:					V_1- ⎫ Other
0 5	:					V_2- ⎬ Variables
0 6	:					V_3- ⎭

Adjustment Equation Output Restrictions.

1 0	:	Action taken when Variable M Out of Service.
1 1	:	Action taken when Variable M Off Computer (On Operator).
1 2	1000.0	Maximum allowable adjustment to M per pass.
1 3	:	Action taken when Target of M already at MAX-MIN limit.
1 4	:	Adjustment Reference List for Partial Loop Test. (2nnnn or Blank)
1 5	0.005 ⎫	If Entry 14 specifies a Reference List, Entries 15-18 specify
1 6	⎪	sublists (-1, -2, -3, -4) referenced for all subsequent M's in
1 7	⎬	control loop. If Entry 14 is Blank, Entries 15-18 specify all of
1 8	⎪	the subsequent M Variables in control loop. Is Target of M set
1 9	⎭	to Average instead of Current for Partial Loop? (1=Yes)

Manipulated Variable Adjustment (Δ M) Output Control.

```
[2 1] [          ]  TIME in minutes (bXXXX) or as developed (-nnnn or 1nnnn).
7 8   10 12 14    17
```

7 8	10 12	14	17	
2 2				10-12: %T = Cumulative elapsed time from present time
2 3				expressed as a percent of TIME.
2 4				14-17: % Δ M = Percent of Δ M change made after %T time.
2 5				

Special Action Initiated Upon Significant Change in Manipulated Variable (M).

7 8	10 Initiating M Change 22	24 Action	
3 1		:	Action may be specified as follows:
3 2		:	(a) Process Variable Special (0nnnn).
3 3		:	(b) Execute General Action (1nnnn).
3 4		:	(c) Feedforward Request Call of
3 5		:	Adjustment Equation (2nnnn).
3 6		:	(d) Execute Special Program (3nnnn).

Table 8.10w

ADJUSTMENT INFORMATION SHEET

$$\overset{1}{\boxed{2}}\overset{\qquad 5}{\boxed{0.0.02}}$$

Description *Adjust furnace temperature to maintain conversion*

Adjustment Equation: $0 = F_1 \Delta T + F_2 \Delta M + F_3 \Delta U + F_4 \Delta V_1 + F_5 \Delta V_2 + F_6 \Delta V_3$

Simultaneous Equation Information. (Omit if single equation solution)

7 8	10	14	16	20	22	26	28	32	34

$\boxed{0\,0}$ $\boxed{2}$ $\boxed{2}$ $\boxed{2}$ $\boxed{2}$ $\square$ Re-solution acceptable?
(1=Yes)

Definition and Development of Terms in Adjustment Equation.

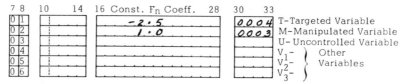

7 8   10   14   16 Const. Fn Coeff.   28   30   33

0 1		
0 2		
0 3		
0 4		
0 5		
0 6		

-2.5
1.0

0004 T-Targeted Variable
0003 M-Manipulated Variable
       U- Uncontrolled Variable
       V_1- ⎫
       V_2- ⎬ Other Variables
       V_3- ⎭

Adjustment Equation Output Restrictions.

1 0	Action taken when Variable M Out of Service.
1 1	Action taken when Variable M Off Computer (On Operator).
1 2	5.0	Maximum allowable adjustment to M per pass.
1 3	Action taken when Target of M already at MAX-MIN limit.
1 4	Adjustment Reference List for Partial Loop Test. (2nnnn or Blank)
1 5	0003	If Entry 14 specifies a Reference List, Entries 15-18 specify
1 6	0005	sublists (-1, -2, -3, -4) referenced for all subsequent M's in
1 7		control loop. If Entry 14 is Blank, Entries 15-18 specify all of
1 8		the subsequent M Variables in control loop. Is Target of M set
1 9		to Average instead of Current for Partial Loop? (1=Yes)

Manipulated Variable Adjustment (Δ M) Output Control.

$\boxed{2\,1}$ $\boxed{0004}$ TIME in minutes (bXXXX) or as developed (-nnnn or 1nnnn).

7 8   10 12   14   17

2 2	0 0 0	0040	10-12: %T = Cumulative elapsed time from present time
2 3	050	0080	expressed as a percent of TIME.
2 4	100	-020	14-17: % Δ M = Percent of Δ M change made after %T time.
2 5			

Special Action Initiated Upon Significant Change in Manipulated Variable (M).

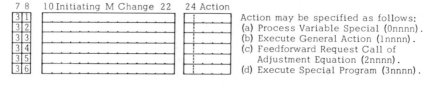

7 8   10 Initiating M Change 22   24 Action

3 1			Action may be specified as follows:
3 2			(a) Process Variable Special (0nnnn).
3 3			(b) Execute General Action (1nnnn).
3 4			(c) Feedforward Request Call of
3 5			Adjustment Equation (2nnnn).
3 6			(d) Execute Special Program (3nnnn).

ACKNOWLEDGMENT

This section contains material which is the property of and is copyrighted by the International Business Machines Corporation and is included with their permission.

8.11 OPTIMIZING PROGRAM

In addition to data logging and supervisory control, computers are frequently applied in the chemical industry to provide unit-, plant- or company-wide optimization. Section 8.4 covered the basic computer languages, and Section 8.10 discussed the "fill-in-the-blanks" approach to supervisory computer control.

This section is devoted to the description of an optimizing program which can be implemented through such supervisory systems as described in Section 8.10. Optimization is frequently done either "off-line," where the output is a typewriter output containing suggestions to the operator, or "on-line," where the optimization program is allowed to adjust computer-generated set points.

Because the approaches to optimization are somewhat specific to the computer manufacturer's systems, the optimization program outlined here is one that is utilized by the same supplier and therefore can be implemented with the supervisory system described in Section 8.10. Although this section outlines the details of one manufacturer's optimization program, sufficient attention is given to problem formulation and to underlying mathematics so that the reader will be informed as to the general approach taken to optimization in both dedicated and general purpose computing systems.

The many abbreviations that appear in the text have been left in by the editor, not because this is considered to be desirable, but because it illustrates the present trend in computer engineering jargon.

The optimization program described in this section uses a method of "sectional linear programming" to optimize a non-linear mathematical model. The model and objective function are linearized about a starting point (initial position) by calculating the incremental rate of change (partial derivative) of each constraint or dependent variable (Y) with respect to each independent variable (X). Linear programming is applied to find the optimum solution within the range for which the linearized model equations are reasonable approximations of the true non-linear equations. The model and objective function are linearized about this solution point (present position), and another local optimum is found. Alternate relinearizations and optimizations are performed until a solution is found that closely approximates an optimal solution to the non-linear problem.

The design of the program is particularly directed toward its use as an efficient non-linear strategy for the on-line control of a plant. It is expected that the routine will be frequently entered to detect any movement of an optimum away from the specified constraints. It can also be used as an off-line program for investigation of non-linear optimizing problems.

Program Description

The Control Optimization Program° (COP) is a general purpose, non-linear optimization program. The program provides all input and output, data handling, optimization, and logical control functions required to solve a non-linear optimization problem.

The user must define the problem to be solved by means of a mathematical model of the system or process to be optimized and provide a set of data defining the starting values and limits (bounds and constraints) of the problem variables. The various items of data and the formats in which they are presented are covered in the following paragraphs.

The mathematical model required is basically a program consisting of a series of mathematical equations describing the relationships that exist between the dependent or constraint (Y) variables and the independent or decision (X) variables. The model also includes the objective function that is to be maximized or minimized. The objective function and constraint equations may be non-linear and/or nonseparable; that is, a Y variable may be expressed as a function of X variables and other Y variables.

Solution Method

This program uses a method of "sectional linear programming" to optimize the system or process described by the model. The constraint and objective functions are linearized about the starting point (initial position) by calculating the incremental rate of change (partial derivative) of each Y variable with respect to each X variable. The linearized constraint and objective functions have the general form

$$\Delta Y = f(\Delta X)$$

and may be considered valid approximations to the non-linear constraint and objective functions over some limited range in each X, defined by a linearity limit (LLIM) which is provided by the user as input data. The bounds (upper and lower limits) on each X variable and constraints (upper and lower limits) on each Y variable are also specified in the input data.

The COP formulates and solves an incremental linear programming (LP)

° IBM Application Manual #20-0208-2, International Business Machines Corporation, Data Processing Division, 112 East Post Road, White Plains, N.Y. 10601.

problem that represents the linearized subset of the non-linear problem. The term "incremental" is used because the LP problem formulated is in terms of the changes in X and Y variables (ΔX and ΔY) rather than the values of the variables. The elements of the LP matrix are the partial derivatives of Y's with respect to X's evaluated at the initial position (IPSN). The bounds and constraints used in the LP are the distances of each variable from its upper and lower limits. Since X variables have linearity limits imposed, as well as absolute upper and lower bounds, each ΔX in the incremental LP problem is bounded by the lesser of either the linearity limit or the distance of the X variable from its upper or lower bound.

The solution to the incremental LP problem is a set of values of ΔY and ΔX. These changes are added to the starting values of X and Y to give an improved (more nearly optimal) set of X and Y values called the "approximate present position" (APPN). The adjective "approximate" is used because the Y values may be somewhat in error due to the assumption of linearity. The X values of APPN are presented to the model, which calculates the corresponding true Y values. The resulting set of X and Y values is termed the "present position" (PPSN) and represents a new point on the model surface that is nearer to the optimum solution.

Figure 8.11a depicts graphically the sequence of operations described above. For simplicity, only two variables are shown: the objective function

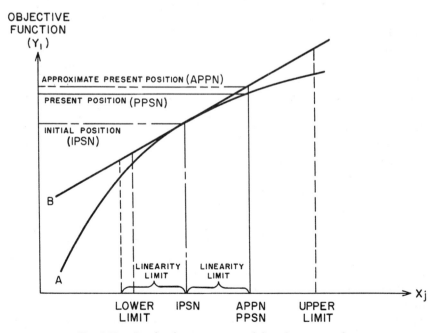

Fig. 8.11a   Graphical representation of the solution procedure

Y_1, and some X variable X_j. The curve, A, is the true non-linear relationship of Y_1 to X_j. The straight line B is the linearized relationship

$$\Delta Y_1 = \frac{\partial Y_1}{\partial X_j} \Delta X_j$$

where the partial derivative is evaluated at the point defined by IPSN. The upper and lower limits (bounds) and the linearity limit of X_j are shown as vertical lines intersecting both the non-linear and the linearized functions. The lower bound on ΔX_j in the incremental LP problem is the distance from IPSN to the lower limit, while the linearity limit becomes the upper bound on ΔX_j. The solution of the incremental LP problem (ΔX_j and ΔY_1) when added to the initial values of X_j and Y_1 yields the point APPN (X_j), APPN (Y_1). Because the relationship was assumed to be linear, the value of Y_1 obtained is greater than the true value of Y_1 corresponding to the new value of X_j. The model calculates the correct (PPSN) value of Y_1 and corrects for the "linearity error." Note that the value of X_j does not change during linearity error correction, since X is the independent variable.

The operations described above (linearization, formulation and solution of the incremental LP problem, calculation of APPN, and linearity error correction to determine PPSN) constitute one step, or loop, in the optimization. The program continues by relinearizing about the current solution point, PPSN, formulating a new LP problem, and finding another APPN and PPSN. The loop is repeated until a solution is found that closely approximates an optimal solution to the non-linear problem, or until the maximum loop count (LMAX) is reached.

The optimal solution found may not be the global optimum but represents the maximum (or minimum) value of the objective function that can be reached through a series of stepwise moves, always in the direction of improved objective function value, without violating any of the bounds or constraints. If alternative optima exist or are suspected, the problem should be rerun, using various diverse starting points, in an attempt to verify the first solution or to discover better solutions.

Detection of an Optimal Solution

The optimum point may lie outside the area enclosed by the bounds and constraints (exterior optimum) or within the bounded and constrained area (interior optimum). An exterior optimum is detected when, on any loop, no X variable moves to its linearity limit. This solution represents the feasible point closest to the infeasible exterior optimum.

The presence of an interior optimum is indicated when the objective function value, after linearity error correction, shows a decrease when compared with the value at the start of a loop, as illustrated in Figure 8.11b. The values of X_j and Y_1 at the start of the loop are represented by point

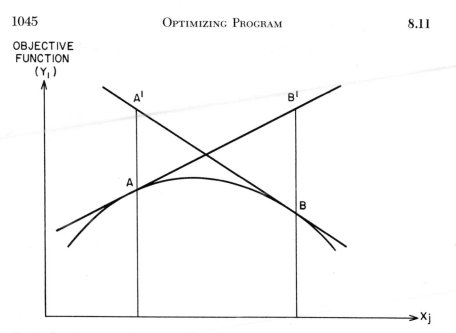

Fig. 8.11b   Decrease in objective function (Y_1) after linearity error correction, when optimum point is overrun

A. The LP solution, APPN, and the corrected solution, PPSN, are represented by B' and B respectively. The decrease in the objective function indicates the COP has overrun the optimum point, which lies somewhere between A and B. On the succeeding loops, the overrun will be repeated in going from point B back to point A and returning again to B. This oscillation will continue until the maximum loop count is reached, unless some technique is employed to converge on the optimal point known to lie somewhere between A and B.

There is a choice of two techniques for converging on the interior optimum. Both convergence techniques make use of the fact that the linearity limits on the X variables generally control the length of the "step" taken on each loop. If the linearity limits are reduced each time the optimum point is overrun, it is possible to determine with increasing precision the location of the optimum point.

Primary Convergence Technique

The primary convergence technique (CVT1) employs a simultaneous reduction in all linearity limits whenever the optimum point is overrun. A linearity limit divisor (LDVR), which has an initial value of 1, is maintained by the convergence subroutine CVT1. On each loop, linearity limits are divided by LDVR to obtain the values used in formulation of the incremental LP problem. CVT1 can, therefore, control the size of each optimization step by varying the value of LDVR. On each loop, CVT1 compares the corrected objective function value with the value at the start of the loop. If a decrease is noted, LDVR is multiplied by a user-specified factor DFACT.

If the new value of LDVR is greater than the user-specified factor DMAX, the starting point of the current loop is taken to be the optimum solution. If the new value of LDVR is less than or equal to DMAX, the program continues by one of two procedures at the option of the user:

The previous loop is repeated with reduced linearity limits, using the same partials matrix and starting values for each variable.

PPSN is updated with the current solution, the model is relinearized, and optimization continues with reduced linearity limits from the point of decreased objective function value.

It is possible that bumps or ridges on the objective function surface may at times give a false indication of overrunning the optimum solution on one loop, while steady improvement is shown in objective value on succeeding loops. If LDVR is not reduced under such circumstances, the number of loops taken to reach the optimum solution may be excessive. A count (NMOV) is kept of the number of successive loops showing improved objective function value since LDVR was last modified. When NMOV reaches a maximum value (NMAX) specified by the user, LDVR is divided by DFACT to permit larger steps to be taken on each loop.

Secondary Convergence Technique

The secondary convergence technique (CVT2) employs a selective reduction in linearity limits. A separate linearity limit divisor (LDIVR) is maintained for each X variable. When the partial derivative of Y_1 with respect to any X changes in sign from one loop to the next, the LDIVR for that X is multiplied by DFACT. An optimum solution is assumed when the values of LDIVR for all X variables that are bounded by their linearity limits have reached the maximum value, DMAX. As in CVT1, provision is made for decreasing LDIVR when the partial derivative has not changed sign in NMAX consecutive loops.

Selection of Convergence Technique

The convergence techniques described are designed to handle various objective function surface configurations. Unfortunately, the nature of the objective function surface is seldom apparent from an inspection of the model equations. Therefore, it is generally not possible to predict which convergence method will be most effective and most efficient for a given model.

CVT1, using the basic option of repeating the current loop and reusing the partials matrix when limits are reduced, is particularly suited to a model with a relatively smooth convex (concave if minimizing) objective function surface. It is computationally the most efficient method, since the time-consuming calculation of partial derivatives is by-passed when a loop is repeated. Since this method always returns to the better solution when a decrease in objective function is detected, the solution which is finally selected as the optimum is always the best solution.

When the objective surface features a gently sloping ridge with relatively steep sides, frequent decreases in objective function value will be noted as the solution path continually recrosses the ridge line. CVT1 exhibits a pronounced tendency to shut off prematurely on such a surface. The alternative option of proceeding from the point of decreased objective function value reduces the tendency of premature shutoff and yields satisfactory results in most cases.

CVT2 is primarily intended for use with models having severe ridges on the objective function surface. Since linearity limits are reduced selectively, the effect of those X variables inducing oscillation across the ridge is reduced, while those X variables promoting progress along the crest of the ridge are not affected.

CVT2 is the only convergence method which can be used with a response model that does not have the capability of correcting for linearity error on each loop. Generally, CVT2 should be selected over CVT1 only when the model cannot correct for linearity error, or when the nature of the objective surface is such that CVT1 does not yield satisfactory results. CVT2 may take from 10 to 50 percent more loops than CVT1 to arrive at a solution.

Note that there is one form of objective function ridge that none of the convergence methods is capable of handling. This is the situation in which the partial derivatives of objective function with respect to every X variable change sign when the ridge is crossed, and all of the convergence methods will converge upon the point at which the ridge is first crossed.

Methods of Resolving Infeasibilities

An infeasible point is one at which one or more variables have values outside their respective limits. COP will accept an infeasible initial position and will employ the methods described below to resolve the infeasibility.

If an infeasibility occurs in an X variable, COP makes an "arbitrary move" (ARBM) to bring the X variable within its limits. The arbitrary move is made without regard to the linearity limit, if any, imposed on the variable. Thus an arbitrary move may cause violation of the linearity limit. An arbitrary move may also cause a related Y variable to become infeasible.

If an infeasibility occurs in a Y variable, the dual LP algorithm is employed to find a feasible solution. If the dual algorithm is unable to find a feasible solution, the X variable bounds are examined. If no X variable is bounded by its linearity limit, the problem is infeasible and execution is terminated. If any X variable is bounded by its linearity limit, COP may, at the option of the user, relax the linearity limits and/or relinearize about the infeasible dual solution point to attain feasibility.

The control vector entry RXMAX specifies the maximum number of times linearity limits may be relaxed on any one loop to attain feasibility. A count of the number of relaxations that have taken place (RXNO) is maintained by COP. In formulating the LP problem, the effective value of the linearity limit is taken to the (RXNO + 1) times the input limit value, so that on consecutive relaxations the effective limit value is 2, 3, 4, 5, etc., times the input value. If RXMAX is specified as zero, no relaxation is permitted unless linearity limits have already been reduced by the convergence subroutine CVT1, in which case one relaxation is permitted. When limits are both reduced by CVT1 and relaxed, the effective linearity limit value is (RXNO + 1)/LDVR times the input value. RXNO is set to zero initially and reset to zero when a feasible solution is found.

If the infeasibility persists after RXMAX relaxations of linearity limits, COP will consider the possibility of relinearizing the model about the infeasible LP solution point in a further attempt to get a feasible solution. The control vector entry RLMAX specifies the maximum number of times the model may be relinearized about an infeasible point in seeking to attain feasibility. If RLMAX is set to zero, no relinearization is permitted. RXNO is set to zero prior to relinearization.

Program Features

Multiunit Capability

The program operates on-line in a multiunit control system in which one computer is controlling two or more process units. When COP is called, the input data specifies size (number of X and Y variables) and identification (MDLNO) of the model to be used. The program adjusts to problem size and calls the appropriate model for calculation of partial derivatives and dependent variable values. The multiunit capability may also be used when COP is utilized off-line and two or more models are used on a more or less regular basis.

The multiunit capability of COP is initially set at five models. By very simple program modifications the user may vary the capability to suit his requirements. The theoretical upper limit on multiunit capability is 32,767 models, but a more realistic limit is imposed by the amount of disc storage available for storing the models.

Multiple Sets of Limits

Generally, the limits imposed upon process variables are dependent on three factors: (1) unit design characteristics, (2) management decisions and (3) operating environment.

While unit design factors are usually constant, management objectives may vary periodically with changes in economic conditions and control strategy, and the operating environment (ambient air temperature, relative humidity, feed stream composition, etc.) may also be changing continually. At a given point in time, any one of these three factors may be the controlling influence on process variable limits. It is desirable, therefore, to maintain a separate set of values for each limiting factor.

In addition to setting an upper and lower limit on the value of a variable, it may be desirable to limit the amount by which a variable may be changed during the optimization.

COP provides for up to five limits on each independent (X) variable and four limits on each dependent (Y) variable. The program selects the most constraining limits for each variable in formulating the optimization problem. The various limits are summarized below:

Extreme limits—usually correspond to the design or safety limits of the process unit.

Management limits—reflect management objectives and control strategy.

Variable limits—generally calculated and updated in response to changes in operating environment.

Change limits—specify the maximum permissible change for each variable.

Linearity limits—define the range for which the linearized constraint equations may be considered valid. Linearity limits apply only to independent (X) variables.

All of the above limits are available when COP is used on-line. When it is off-line, only extreme limits, change limits, and linearity limits may be used. Extreme limits are always required. The other limits may be used at the option of the user.

Adaptive Linearity Limits

It is frequently impossible to determine with any precision the range in each X variable for which the linearized constraint and objective functions are valid. Since the linearity limits also define the size of each optimization step, the use of excessively small linearity limits will increase the number of steps or loops required to reach the optimum solution, with a resultant increase in execution time. The ideal value of a linearity limit is that which permits the largest step size for each loop without incurring excessive linearity error.

When the adaptive linearity limit feature is used, COP calculates the maximum change in each X variable that may be made without incurring an error in any Y variable greater than a maximum value specified by the user. This procedure allows the linearity limit (and step size) to vary during the course of an optimization as the degree of non-linearity of the model surface varies.

The adaptive linearity limit feature may also be used as an aid in selecting linearity limit values as input to the program.

Move Penalties

It may be desirable to prevent independent variable moves that result in relatively small objective function improvements. Imposing a move penalty on an independent variable inhibits changes that would cause an objective function increase of less than the amount of the move penalty.

Target Constraints

A dependent variable is normally constrained by an upper and lower limit, with the program free to seek an optimum anywhere within these constraints. For some variables, typically product specifications, deviation from some "target" value should be minimized. Deviation from the target does not make the product worthless but does lower its value because additional treatment or processing will be required. There may also be upper and lower limits that must not be violated, but these are generally too wide to be the only constraints on the variable. When a target constraint is employed, a penalty is associated with deviation of the variable from its target value, and the program is encouraged to find a solution at or near the target.

The Y-variable constraints may be viewed as cost functions, where the "constraint cost" is an artificial cost concerned only with imposing limits on Y variables and is not related to the objective function. Figure 8.11c shows conventional upper and lower constraints represented as cost functions. The

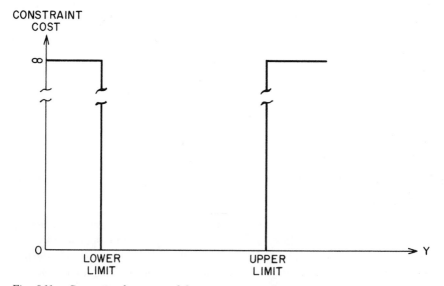

Fig. 8.11c   Conventional upper and lower constraints shown as constraint cost functions

cost associated with any value of Y less than the lower limit or greater than the upper limit is infinite. Zero cost is associated with Y values between the limits.

Figure 8.11d illustrates the target constraint. The cost of deviating from the target value is shown by the slanted lines converging at the target. The slope of the lines is determined by the slack cost (SLKC) value specified by the user. The target constraint may be used independently or in combination with conventional constraints, as shown.

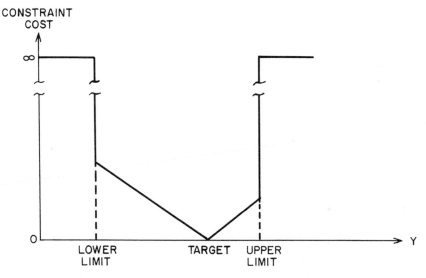

Fig. 8.11d   Target constraint with conventional upper and lower constraints

A special case of the target constraint is the one-sided target or "soft" constraint. Figure 8.11e depicts a Y variable with a conventional or "hard" lower constraint and a soft upper constraint. Either the upper or lower constraint, but not both, may be soft. The soft constraint may be imposed independently or in combination with an opposing conventional constraint.

Dead Zones

It is often possible to tolerate minor constraint violations, provided they do not persist for too long. The dead zone (DZON) defines the tolerable amount of constraint violation for each Y variable.

Curtaining Feature

COP may be directed to use the LP solution on each loop as a starting LP basis for the following loop. This procedure may significantly reduce the number of LP iterations taken on each loop, provided a majority of the variables move in the same direction on each loop. When a substantial number

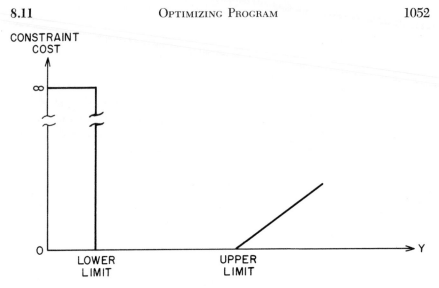

Fig. 8.11e   Soft upper constraint with conventional lower constraint

of variables move in alternating directions from loop to loop, the curtaining feature may cause an increase in the number of LP iterations required.

It is also possible to direct COP to use the final solution (PPSN) of the previous optimization as a starting point (IPSN) for the current problem. When this option is taken, COP will ignore the initial position in the input data.

Linear Programming Algorithm and Error Check

COP uses a linear programming (LP) algorithm (based on the simplex procedure) that has proved very effective for process control applications in which execution time and core storage requirements are critical factors. Special features of the LP algorithm permit handling move penalties, target constraints, upper and lower variable bounding, and two-sided or "free" variables without adding rows or columns to the LP matrix.

COP can be directed to check the numerical error introduced during LP iterations. The reduced cost row and/or right-hand-side values are re-computed and compared to the corresponding values from the LP solution. An indicator is set (on-line) or a message is printed (off-line) if the relative error in any element exceeds the user-specified tolerance NETOL.

Using the Program

COP provides all input and output functions, interface with the model, and all linearization and optimization calculations. Two items are required of the user for off-line operation:

    1. A properly formulated mathematical model and the objective function to be optimized, or a programmed routine to calculate partial derivatives, or a matrix of partial derivatives.

    2. A set of data defining the initial values of model variables, limits on model variables, and the required logical control parameters.

When this program is to be used on-line in a process control system, the user must also provide programs to perform the necessary instrument scans, averaging and digital filtering of process data, and routines to transform optimizer solutions into appropriate signals to effect changes in control variables.

Model

The model is a user-written program that describes the system or process to be optimized. COP uses the model to correct the Y variable values of the current solution (APPN) for linearity error and to calculate partial derivatives for the next optimization step or loop.

In the context of the COP program, two general types of models are recognized:

A simulation model is a program which, given values for all X variables, can calculate corresponding values for all Y variables.

A response model is a program which, given values for all X variables, can determine the change in each Y variable that will result from a unit change in each X variable (the partial derivative of each Y with respect to each X) and can write a matrix of partial derivatives. A response model may also be capable of calculating Y-variable values corresponding to the given X variable values. This capability is highly desirable but not essential.

In general, the formulation and coding of the model are left completely to the discretion of the user. Any program that accurately performs the required functions is acceptable as a model. The single area in which COP imposes rigid specifications upon model coding is that of communication and interface between COP and the model. All models communicate with COP through an in-core interface. Response models also communicate via a partials matrix, which they write to the disc.

With a simulation model, the following procedure is used: At the start of each loop, COP calls the model to correct Y-variable values for any linearity error incurred on the previous loop or for any inconsistencies in input (IPSN) values. After Y-error correction, COP uses the model to compute partial derivatives and, if required, adaptive linearity limits.

The model is expected to provide for *accurate* calculation of Y-variable values corresponding to a given set of X-variable values. The model should not concern itself with constraint violations, as this is a function performed by the optimizing program.

The method used in calculating partial derivatives permits simulation model equations to be both non-linear and nonseparable; that is, a Y may be expressed as a function of Xs and other Ys. Model equations must, however, be continuous functions, at least over the region to be optimized. Discontinuities in model equations can cause very erratic results during the optimization.

Note that on each optimization step or loop the simulation model is loaded and executed once for Y error correction and either once or twice for each X variable during calculation of partial derivatives. Thus, a model with 20 X variables is loaded and executed either 21 or 41 times (depending upon whether X's are perturbed in one or both directions during partial calculation) on each loop. The time required to perform the optimization is therefore greatly affected by the computational efficiency of the model.

A response model is assumed to have the capability of calculating Y variable values as well as calculating the partials matrix. COP calls the model twice on each loop, once for Y error correction and once for calculation of partial derivatives.

A response model that does not have the capability of correcting linearity error in Y values may be used, but some loss of accuracy must be expected. Linearity error incurred in the Y variable values is accumulated from one loop to the next and reflected in the final results. The total cumulative error in the final Y values generally depends upon the number of loops to final solution, the accuracy of partial derivatives calculated at each loop, and the selection of linearity limit values. Such a model may be entirely satisfactory for an on-line system in which the optimization is restricted to one or two loops, but it is generally poorly suited to off-line studies in which the optimization may run many loops. Note also that the primary convergence technique (CVT1) cannot be used effectively when the model cannot correct linearity error.

Input

Input data for on-line operation is usually on disc storage in FORTRAN-format arrays. Floating-point data may be in either normal precision (two-word) or extended precision (three-word) format. At the beginning of an optimization, COP copies the input from the user's disc area to the COP data area and converts from normal to extended precision format if necessary. This ensures a consistent set of data during the optimization, while the user's data area may be updated by various interrupt routines.

For off-line operation, input data may originate on punched cards. The card input routine stores the data arrays directly in the data area. Change cards are accepted to modify any input data value in the data area.

The amount of input required varies considerably according to the program features to be used. Four items of input are always required: the control vector (CTLV), the starting value of each X and Y value (XLIM), and the table of bounds and constraints (TBC), which specifies the type of bounds or constraints to be applied to each X and Y variable. Many COP features require an additional set of input data. For example, if the target constraint feature is used, a set of data specifying the target value of each Y variable must be entered.

The control vector (CTLV) is a group of 21 one-word entries which specify the program features and options to be used, the dimensions (number of X and Y variables and model constants) of the current problem, identification of the model to be used for partials calculation, various user-controlled factors for use in the optimization calculations, and the type of printed output desired.

Output

Output is printed and/or written in FORTRAN-format arrays. The principal output is the value of each variable at the conclusion of the optimization. This is the final present position (PPSN) and represents either an optimal solution to the non-linear problem or the most nearly optimal solution that can be found in the number of loops COP was permitted to run (LMAX).

An alternative method of expressing the final solution, and one that may be of primary importance in some on-line systems, is the total change in value of each variable during the course of the optimization. This is computed by subtracting the initial (IPSN) value of each variable from its final (PPSN) value.

Additional output available in both on-line and off-line systems includes the most constraining limit values, excluding linearity limits (MCXL) used by COP in formulating the optimization problem; the arbitrary X variable moves (ARBM) made by COP to bring all X variables within their upper and lower limits; the final solution obtained from the LP before correcting Y values for linearity error (APPN); and the adaptive linearity limit values (ALIM). In on-line output, ARBM, APPN, and ALIM values are available for the final loop only. In off-line systems and those on-line systems using printed output, ARBM, APPN and ALIM are available on each loop if desired.

Additional output available only in printed form includes the matrix of partial derivatives, the LP tableau, the basis and non-basis vector IDs used by the LP, and the upper and lower bound tables used in the LP. All of the above output is available for each loop. The frequency with which additional output is printed is controlled by the user.

Mathematical Methods

Problem Statement

Given: N independent variables X_j, M dependent variables Y_i, an objective function of the general form

$$P = f(X_j, Y_i)$$

and a mathematical model of the form

$$Y_i = f(X_j, Y_i)$$

where the objective function and each of the constraint equations may be non-linear and/or non-separable.

Maximize the objective function $P = f(X_j, Y_i)$ subject to:

$$MCLL_i \leq Y_i \leq MCUL_i$$
$$MCLL_j \leq X_j \leq MCUL_j$$

where $MCLL_i$ = most constraining lower limit on Y_i,
$\quad\quad MCUL_i$ = most constraining upper limit on Y_i,
$\quad\quad MCLL_j$ = most constraining lower limit on X_j, and
$\quad\quad MCUL_j$ = most constraining upper limit on X_j.

Problem Solution

A starting point or initial position of the independent variables is supplied as input, and corresponding Y values are calculated. The initial position (PSN) is then represented as a vector of initial values of X_j and Y_i:

$$\begin{vmatrix} X_0 \\ Y_0 \end{vmatrix}$$

The constraints and objective function are linearized about the initial position by using the linear terms of a Taylor series:

$$\Delta P = \frac{\partial P}{\partial X_1}\Delta X_1 + \frac{\partial P}{\partial X_2}\Delta X_2 + \cdots + \frac{\partial P}{\partial X_n}\Delta X_n$$

$$\Delta Y_i = \frac{\partial Y_i}{\partial X_1}\Delta X_1 + \frac{\partial Y_i}{\partial X_2}\Delta X_2 + \cdots + \frac{\partial Y_i}{\partial X_n}\Delta X_n$$

where all partial derivatives are evaluated at the point $\begin{vmatrix} X_0 \\ Y_0 \end{vmatrix}$.

Let

$$C_j = \partial P/\partial X_j$$
$$A_{ij} = \partial Y_i/\partial X_j$$

Then the linearized problem is as follows: Maximize

$$\Delta P = \sum_j C_j \Delta X_j$$

subject to:

$$\Delta Y_i = \sum_j A_{ij} \Delta X_j$$

$$YLC_i \leq \Delta Y_i \leq YUC_i$$
$$XLBND_j \leq \Delta X_j \leq XUBND_j$$

where $YLC_i = MCLL_i - (Y_0)_i$,
$\quad\quad YUC_i = MCUL_i - (Y_0)_i$,

$\text{XLBND}_j = \text{MCLL}_j - (X_0)_j$, and
$\text{XUBND}_j = \text{MCUL}_j - (X_0)_j$.

The solution to the linearized problem is $\begin{vmatrix} \Delta X \\ \Delta Y \end{vmatrix}$.

The local optimum point $\begin{vmatrix} X_1 \\ Y_1 \end{vmatrix}$ is

$$\begin{vmatrix} X_1 \\ Y_1 \end{vmatrix} = \begin{vmatrix} X_0 \\ Y_0 \end{vmatrix} + \begin{vmatrix} \Delta X \\ \Delta Y \end{vmatrix}$$

The next step is to relinearize about point 1 and repeat the process until the optimum solution to the nonlinear problem is reached.

Partial Derivatives

When a simulation model is used, partial derivatives are calculated by incrementing and decrementing each X in turn by an amount DELX, using the model to calculate the resulting Y values. Thus,

$$\frac{\partial Y_i}{\partial X_j} \approx \frac{Y_p - Y_m}{2\text{DELX}_j}$$

where Y_p = the value of Y_i at $|X_0| + \text{DELX}_j$ and
$\qquad Y_m$ = the value of Y_i at $|X_0| - \text{DELX}_j$.

DELX_j is a user-specified increment for perturbing X_j. At the option of the user, partial derivatives may be calculated by perturbing the X variables in the positive direction only.

Adaptive Linearity Limits

Adaptive linearity limits are determined during the calculation of partial derivatives. Let

$\qquad Y_0$ = the value of Y_i at $|X_0|$
$\qquad Y_p$ = the value of Y_i at $|X_0| + \text{DELX}_j$
$\qquad Y_m$ = the value of Y_i at $|X_0| - \text{DELX}_j$

The linearity error in each Y_i over the range $|X_0| \pm \text{DELX}_j$ is calculated as

$$\text{LINERR}_i = Y_0 - \frac{Y_p + Y_m}{2}$$

Given a maximum permissible linearity error (YERR_i) for each Y variable, DELX_j is quadratically scaled to determine the linearity limit in X_j with respect to each Y_i:

$$LINLIM_{ij} = DELX_j \sqrt{\frac{YERR_i}{LINERR_i}}$$

The adaptive linearity limit for each X_j is then taken to be the minimum of the m values of $LINLIM_{ij}$ and AFACT times the linearity limit ($LLIM_j$) supplied as input. AFACT is a user-specified integer variable.

Note that the method used in calculating linearity limits assumes a constant second derivative and ignores interaction between X variables.

Most Constraining Limits and Move Penalties

For each variable, the most constraining upper limit (MCUL) is selected as the *minimum* of upper extreme limit, upper management limit, upper variable limit, initial position plus change limit, and present position plus linearity limit. Similarly, the most constraining lower limit (MCLL) is the *maximum* of lower extreme limit, lower management limit, lower variable limit, initial position minus change limit, and present position minus linearity limit.

Where a move penalty is specified, the objective function coefficient C_j is modified so that

$$C_j = \frac{\partial P}{\partial X_j} - (MP)_j$$

where $(MP)_j$ is the move penalty specified for X_j.

Sample Problem

The sample problem outlined here is a very simplified simulation model of an alkylation unit in a petroleum refinery. The model contains three independent variables and eight dependent variables. Model constants are used for objective function coefficients to afford flexibility in varying economic incentives in off-line process studies. The model is shown in Table 8.11f.

Table 8.11g shows a listing of the sample problem data check. The output of the sample problem is shown in Table 8.11h.

ACKNOWLEDGMENT

This section contains material which is the property of and is copyrighted by the International Business Machines Corporation and is included with their permission.

Table 8.11f

SAMPLE PROBLEM MODEL

```
// JOB                                                                      SMPL001
// FOR MODLO                                                                SMPL002
*NONPROCESS PROGRAM                                                         SMPL003
*ONE WORD INTEGERS                                                          SMPL004
**COP/1800 SAMPLE PROBLEM - ALKYLATION UNIT MODEL                           SMPL005
      EXTERNAL COPB                                                         SMPL006
      COMMON ICTLV(54),X(3),Y(8),CONST(6)                                   SMPL007
C************************************************************************    SMPL008
C     X(1) IS OLEFIN FEED RATE, B/D                                         SMPL009
C     X(2) IS ISOBUTANE RECYCLE RATE, B/D                                   SMPL010
C     X(3) IS FRESH ACID ADDITION RATE, MLB/D                               SMPL011
C                                                                           SMPL012
C                                                                           SMPL013
C     Y(1) IS PROFIT, $/D                                                   SMPL014
C     Y(2) IS ALKYLATE YIELD, B/D                                           SMPL015
C     Y(3) IS ISOBUTANE MAKE-UP, B/D                                        SMPL016
C     Y(4) IS SPENT ACID STRENGTH, WT.PCT                                   SMPL017
C     Y(5) IS OCTANE NUMBER                                                 SMPL018
C     Y(6) IS ISOBUTANE/OLEFIN RATIO                                        SMPL019
C     Y(7) IS ACID DILUTION FACTOR                                          SMPL020
C     Y(8) IS F-4 PERFORMANCE NUMBER                                        SMPL021
C                                                                           SMPL022
C     CONST(1) IS ALKYLATE PRODUCT VALUE, $/OCTANE-BBL                      SMPL023
C     CONST(2) IS OLEFIN COST, $/BBL                                        SMPL024
C     CONST(3) IS ISOBUTANE COST, $/BBL                                     SMPL025
C     CONST(4) IS RECYCLE FRACTIONATION COST, $/BBL                         SMPL026
C     CONST(5) IS ACID COST, $/TON                                          SMPL027
C     CONST(6) IS CONVERGENCE TOLERANCE
```

1059

Table 8.11f (Continued)

```
C************************************************************SMPL028
C      ASSUME Y(2) THEN CALC. Y(2), Y(3) AND Y(6) BY ITERATIVE PROCEDURE SMPL029
C************************************************************SMPL030
       Y(2)=1.5*X(1)                                        SMPL031
   1   Y(3)=1.22*Y(2)-X(1)                                  SMPL032
       Y(6)=(X(2)+Y(3))/X(1)                                SMPL033
       Y2CAL=.01*X(1)*(112.0+13.167*Y(6)-.667*Y(6)**2)      SMPL034
       ERROR=Y2CAL-Y(2)                                     SMPL035
       IF (ABS(ERROR)-CONST(6)) 3,3,2                       SMPL036
   2   Y(2)=Y2CAL                                           SMPL037
       GO TO 1                                              SMPL038
C************************************************************SMPL039
C      ASSUME Y(4), THEN CALC. Y(5),Y(7) AND Y(8) BY ITERATIVE PROCEDURE SMPL040
C************************************************************SMPL041
   3   Y(4)=90.0                                            SMPL042
   4   Y(5)=86.35+1.098*Y(6)-.038*Y(6)**2+.325*(Y(4)-89.0)  SMPL043
       Y(8)=3.0*Y(5)-133.0                                  SMPL044
       Y(7)=35.82-.222*Y(8)                                 SMPL045
       Y4CAL=(98000.0*X(3))/(Y(2)*Y(7)+1000.0*X(3))         SMPL046
       ERROR=Y4CAL-Y(4)                                     SMPL047
       IF (ABS(ERROR)-CONST(6)) 6,6,5                       SMPL048
   5   Y(4)=Y4CAL                                           SMPL049
       GO TO 4                                              SMPL050
C************************************************************SMPL051
C      CALC. OBJECTIVE FUNCTION Y(1)                        SMPL052
C************************************************************SMPL053
   6   Y(1)=Y(2)*Y(5)*CONST(1)-X(1)*CONST(2)-Y(3)*CONST(3)-X(2)*CONST(4) SMPL054
      *        -(X(3)/2.0)*CONST(5)                         SMPL055
       CALL LINK(COPB)                                      SMPL056
       END                                                  SMPL057
```

1060

Table 8.11g
LISTING OF SAMPLE PROBLEM DATA

```
**HEADCOP/1800 SAMPLE PROBLEM          RUN NO. 1
**CTLV100180008010 3  8  6  1
**DATA
OLEFIN RATE  B/D0100      1600      2500       1
I-C4 RECYCLE B/D0100      1.2E4    15000       5
FRESH ACID MLB/D0100       110       120       .2
PROFIT ****  $/D0000
ALKY YIELD   B/D0011      2000      4000
I-C4 MAKE-UP B/D0011                4000
SPENT ACID WTPCT0011        85        97
OCTANE NUMBER  0011         90        98
ISO/OLEFIN RATIO0011         3        14
ACID DIL. FACTOR0011       .01         5
F-4 PERFORM. NO.0011       140       165
**CNST        .062      3.36      5.04      .0267      20      .001

**CPUT
**HEADCOP/1800 SAMPLE PROBLEM          RUN NO. 2
**CMNTCHANGE UPPER LIMIT ON X(2) TO 20000
**CMNTSTART AT SOLUTION TO PREVIOUS RUN
**CTLV100184008010  3  8  6  1
***CNGEEXULX  2   20000
**CPUT
**FINI
```

1061

Table 8.11h
SAMPLE PROBLEM OUTPUT

INPUT SUMMARY

VARIABLE NAME	TBC	IPSN	LOWER LIMIT	UPPER LIMIT	LIN. LIMIT	MOVE PENALTY	DELTA-X
X 1 OLEFIN RATE B/D	0100	1600.0000	0.0000	2500.0000	50.0000	0.0000	1.0000
X 2 I-C4 RECYCLE B/D	0100	12000.0000	0.0000	15000.0000	500.0000	0.0000	5.0000
X 3 FRESH ACID MLB/D	0100	110.0000	0.0000	120.0000	5.0000	0.0000	0.2000

VARIABLE NAME	TBC	IPSN	LOWER LIMIT	UPPER LIMIT	TARGET	SLACK COST	YERRMAX
Y 1 PROFIT **** $/D	0000	0.0000	0.0000	0.0000	0.0000	0.0000	0.0000
Y 2 ALKY YIELD B/D	0011	0.0000	2000.0000	4000.0000	0.0000	0.0000	0.0000
Y 3 I-C4 MAKE-UP B/D	0011	0.0000	0.0000	4000.0000	0.0000	0.0000	0.0000
Y 4 SPENT ACID WTPCT	0011	0.0000	85.0000	97.0000	0.0000	0.0000	0.0000
Y 5 OCTANE NUMBER	0011	0.0000	90.0000	98.0000	0.0000	0.0000	0.0000
Y 6 ISO/OLEFIN RATIO	0011	0.0000	3.0000	14.0000	0.0000	0.0000	3.0000
Y 7 ACID DIL. FACTOR	0011	0.0000	0.0100	5.0000	0.0000	0.0000	0.0000
Y 8 F-4 PERFORM. NO.	0011	0.0000	140.0000	165.0000	0.0000	0.0000	0.0000

CONTROL VECTOR

CNTL1	CNTL2	OUTPT	XNO	YNO	CNO	LFREQ	DFREQ	LMAX	DFACT	DMAX
1001	8000	8010	3	8	6	1	0	50	2	8

NMAX	MDLNO	AFACT	RXMAX	RLMAX	DLIM	PLIM	RDOFF	PTOL	NETOL
3	0	2	1	1	25	50	28	100	122

CONSTANTS

1)	0.0620	2)	5.0400	3)	3.3600	4)	0.0267	5)	20.0000	5)	0.0010

Table 8.11h (Continued)

LOOP SUMMARY

VARIABLE NAME	LOOP 0		LOOP 1		LOOP 2		LOOP 3		LOOP 4		LOOP 5	
X 1 OLEFIN RATE B/D	1693.7500		1743.7500	*	1693.7500	*	1743.7500	*	1793.7500	*	1843.7500	*
X 2 I-C4 RECYCLE B/D	15000.0000		15500.0000	*	16000.0000	*	16500.0000	*	17000.0000	*	17500.0000	*
X 3 FRESH ACID MLB/D	95.6250		90.6250	*	95.6250	*	90.6250	*	95.6250	*	100.6250	*
Y 1 PROFIT **** $/D	953.7114		967.1977		997.5786		1016.4343		1053.3508		1087.1645	
Y 2 ALKY YIELD B/D	2997.3818		3085.7421		2991.5820		3079.6245		3167.6625		3259.6953	
Y 3 I-C4 MAKE-UP B/D	1963.0556		2020.8552		1955.9799		2013.3918		2070.7983		2128.1982	
Y 4 SPENT ACID WTPCT	90.0015		88.5017		90.7703		89.3443		89.8475		90.2763	
Y 5 OCTANE NUMBER	93.8605		93.3840		94.2948		93.8359		94.0037		94.1472	
Y 6 ISO/OLEFIN RATIO	10.0150		10.0478		10.6013		10.6169		10.6318		10.6458	
Y 7 ACID DIL. FACTOR	2.8348		3.1521		2.5456		2.8512		2.7394		2.6439	
Y 8 F-4 PERFORM. NO.	148.5816		147.1522		149.8844		148.5079		149.0113		149.4416	

	LOOP 0	LOOP 1	LOOP 2	LOOP 3	LOOP 4	LOOP 5
PRIMAL ITER CNT	0	3	3	3	3	3
DUAL ITER COUNT	0	0	0	0	0	0
LIN-LMT DIVISOR	1	1	1	1	1	1
CONDITION						

LOOP SUMMARY

VARIABLE NAME	LOOP 6		LOOP 7		LOOP 8		LOOP 9		LOOP 10		LOOP 11	
X 1 OLEFIN RATE B/D	1893.7500	*	1943.7500	*	1993.7500	*	2043.7500	*	2093.7500	*	2143.7500	*
X 2 I-C4 RECYCLE B/D	18000.0000	*	18500.0000	*	19000.0000	*	19500.0000	*	20000.0000	U	20000.0000	U
X 3 FRESH ACID MLB/D	105.6250	*	110.6250	*	105.6250	*	110.6250	*	115.6250	*	120.0000	U
Y 1 PROFIT **** $/D	1118.9687		1148.7832		1172.9541		1207.3657		1239.7670		1253.6704	
Y 2 ALKY YIELD B/D	3343.7236		3431.7475		3519.7675		3607.7832		3695.7973		3788.6220	
Y 3 I-C4 MAKE-UP B/D	2185.5927		2242.9819		2300.3662		2357.7451		2415.1225		2478.3686	
Y 4 SPENT ACID WTPCT	90.6530		90.9817		89.8045		90.2013		90.5523		90.4850	
Y 5 OCTANE NUMBER	94.2734		94.3838		94.0047		94.1369		94.2540		94.1677	
Y 6 ISO/OLEFIN RATIO	10.6590		10.6716		10.6835		10.6949		10.7057		10.4855	
Y 7 ACID DIL. FACTOR	2.5598		2.4863		2.7388		2.6508		2.5727		2.6302	
Y 8 F-4 PERFORM. NO.	149.8204		150.1516		149.0141		149.4107		149.7622		149.5032	

Table 8.11h (Continued)

PRIMAL ITER CNT	3	3	3	3
DUAL ITER COUNT	0	0	0	0
LIN-LMIT DIVISOR	1	1	1	2
CONDITION				

LOOP SUMMARY

VARIABLE NAME	LOOP 12		LOOP 13		LOOP 14	
X 1 OLEFIN RATE B/D	1687.5000	*	1693.7500	*	1700.0000	*
X 2 I-C4 RECYCLE B/D	15000.0000	U	15000.0000	U	15000.0000	U
X 3 FRESH ACID MLB/D	96.2500	*	95.6250	*	95.0000	*
Y 1 PROFIT **** $/D	953.4882		953.7114		953.6646	
Y 2 ALKY YIELD B/D	2986.2021		2997.3818		3008.5375	
Y 3 I-C4 MAKE-UP B/D	1955.6665		1963.0556		1970.4157	
Y 4 SPENT ACID WTPCT	90.2131		90.0015		89.7877	
Y 5 OCTANE NUMBER	93.9402		93.8605		93.7801	
Y 6 ISO/OLEFIN RATIO	10.0478		10.0150		9.9825	
Y 7 ACID DIL. FACTOR	2.7817		2.8348		2.8884	
Y 8 F=4 PERFORM. NO.	148.8208		148.5816		148.3403	

PRIMAL ITER CNT	3	3
DUAL ITER COUNT	0	0
LIN-LMIT DIVISOR	4	8
CONDITION		

VARIABLE NAME	TBC	IPSN	PPSN	CHANGE	LOWER LIMIT	UPPER LIMIT	LIN. LIMIT
X 1 OLEFIN RATE B/D	0100	1600.0000	1693.7500	93.7500	0.0000	2500.0000	3.1250
X 2 I-C4 RECYCLE B/D	0100	12000.0000	15000.0000	3000.0000	0.0000	15000.0000	31.2500
X 3 FRESH ACID MLB/D	0100	110.0000	95.6250	-14.3750	0.0000	120.0000	0.3125

VARIABLE NAME	TBC	IPSN	PPSN	CHANGE	LOWER LIMIT	UPPER LIMIT	TARGET
Y 1 PROFIT **** $/D	0000	706.2121	953.7114	247.4992	0.0000	0.0000	0.0000
Y 2 ALKY YIELD B/D	0011	2815.7275	2997.3818	181.6542	2000.0000	4000.0000	0.0000
Y 3 I-C4 MAKE-UP B/D	0011	1835.1875	1963.0556	127.8681	0.0000	4000.0000	0.0000
Y 4 SPENT ACID WTPCT	0011	91.1294	90.0015	-1.1278	85.0000	97.0000	0.0000
Y 5 OCTANE NUMBER	0011	93.6951	93.8605	0.1654	90.0000	98.0000	0.0000
Y 6 ISO/OLEFIN RATIO	0011	8.6469	10.0150	1.3680	3.0000	14.0000	0.0000
Y 7 ACID DIL. FACTOR	0011	2.9450	2.8348	-0.1101	3.0000	5.0000	0.0000
Y 8 F=4 PERFORM. NO.	0011	148.0854	148.5816	0.4962	140.0000	165.0000	0.0000

OPTIMUM SOLUTION AT LOOP 14

Table 8.11h (Continued)

```
CHANGE UPPER LIMIT ON X(2) TO 20000
START AT SOLUTION TO PREVIOUS RUN
**CNGEEXULX 2    20000
```

INPUT SUMMARY

VARIABLE NAME	TBC	IPSN	LOWER LIMIT	UPPER LIMIT	LIN. LIMIT	MOVE PENALTY	DELTA-X
X 1 OLEFIN RATE B/D	0100	1600.0000	0.0000	2500.0000	50.0000	0.0000	1.0000
X 2 I-C4 RECYCLE R/D	0100	12000.0000	0.0000	20000.0000	500.0000	0.0000	5.0000
X 3 FRFSH ACID MLR/D	0100	110.0000	0.0000	120.0000	5.0000	0.0000	0.2000

VARIABLE NAME	TBC	IPSN	LOWER LIMIT	UPPER LIMIT	TARGET	SLACK COST	YERRMAX
Y 1 PROFIT **** $/D	0000	706.2121	0.0000	0.0000	0.0000	0.0000	0.0000
Y 2 ALKY YIELD B/D	0011	2815.7275	2000.0000	4000.0000	0.0000	0.0000	0.0000
Y 3 I-C4 MAKE-UP B/D	0011	1835.1875	0.0000	4000.0000	0.0000	0.0000	0.0000
Y 4 SPENT ACID WTPCT	0011	91.1294	85.0000	97.0000	0.0000	0.0000	0.0000
Y 5 OCTANE NUMBER	0011	93.6951	90.0000	98.0000	0.0000	0.0000	0.0000
Y 6 ISO/OLEFIN RATIO	0011	8.6469	3.0000	14.0000	0.0000	0.0000	0.0000
Y 7 ACID DIL. FACTOR	0011	2.9450	0.0100	5.0000	0.0000	0.0000	0.0000
Y 8 F-4 PERFORM. NO.	0011	148.0854	140.0000	165.0000	0.0000	0.0000	0.0000

CONTROL VECTOR

CNTL1	CNTL2	OUTPT	XNO	YNO	CNO	LFREQ	DFREQ	LMAX	DFACT	DMAX
1001	8400	9010	3	8	6	1	0	50	2	8

NMAX	MDLNO	AFACT	RXMAX	RLMAX	DLIM	PLIM	RDOFF	PTOL	NETOL
3	0	2	1	1	25	50	28	100	122

CONSTANTS

1) 0.0620	2) 5.0400	3) 3.3600	4) 0.0267	5) 20.0000	6) 0.0010

Table 8.11h (Continued)

LOOP SUMMARY

VARIABLE NAME	LOOP 0	LOOP 1	LOOP 2	LOOP 3	LOOP 4	LOOP 5
X 1 OLEFIN RATE B/D	1600.0000 *	1650.0000 *	1600.0000 *	1650.0000 *	1600.0000 *	1650.0000 *
X 2 I-C4 RECYCLE B/D	12000.0000 U	12500.0000 U	13000.0000 U	13500.0000 U	14000.0000 U	14500.0000 U
X 3 FRESH ACID MLR/D	110.0000 *	105.0000 *	100.0000 *	95.0000 *	100.0000 *	95.0000 *
Y 1 PROFIT **** $/D	706.2121	76.6445	817.8574	857.1645	880.7004	923.2912
Y 2 ALKY YIELD B/D	2815.7275	2905.7329	2827.9965	2917.0781	2831.6835	2920.1250
Y 3 I-C4 MAKE-UP B/D	1835.1875	1894.9941	1850.1557	1908.8352	1854.6538	1912.5524
Y 4 SPENT ACID WTPCT	91.1294	90.2035	90.6208	89.4255	91.3669	90.2319
Y 5 OCTANE NUMBER	93.6951	93.4280	93.7942	93.4281	94.2682	93.9123
Y 6 ISO/OLEFIN RATIO	8.6469	8.7242	9.2813	9.3386	9.9091	9.9470
Y 7 ACID DIL. FACTOR	2.9450	3.1229	2.8790	3.1228	2.5633	2.8003
Y 8 F-4 PERFORM. NO.	148.0854	147.2842	148.3826	147.2843	149.8045	148.7369
PRIMAL ITER CNT	0	3	3	3	3	3
DUAL ITER COUNT	0	0	0	0	0	0
LIN-LMIT DIVISOR	1	1	1	1	1	1
CONDITION						

LOOP SUMMARY

VARIABLE NAME	LOOP 6	LOOP 7	LOOP 8	LOOP 9	LOOP 10	LOOP 11
X 1 OLEFIN RATE B/D	1700.0000 *	1650.0000 *	1700.0000 *	1675.0000 *	1700.0000 *	1675.0000 *
X 2 I-C4 RECYCLE B/D	15000.0000 U	15000.0000 U	15000.0000 U	15000.0000 U	15000.0000 U	15000.0000 U
X 3 FRESH ACID MLR/D	90.0000 *	95.0000 *	90.0000 *	92.5000 *	95.0000 *	97.5000 *
Y 1 PROFIT **** $/D	942.7119	948.4062	942.7119	950.3115	953.6416	951.5419
Y 2 ALKY YIELD B/D	3008.5375	2918.6118	3008.5375	2963.7714	3008.5375	2963.7714
Y 3 I-C4 MAKE-UP B/D	1970.4157	1910.7062	1970.4157	1940.8010	1970.4157	1940.8010
Y 4 SPENT ACID WTPCT	88.7827	90.6233	88.7827	89.7607	89.7877	90.6185
Y 5 OCTANE NUMBER	93.4534	94.1393	93.4534	93.8152	93.7801	94.0940
Y 6 ISO/OLEFIN RATIO	9.9825	10.2489	9.9825	10.1139	9.9825	10.1139
Y 7 ACID DIL. FACTOR	3.1059	2.6492	3.1059	2.8650	2.8884	2.6793
Y 8 F-4 PERFORM. NO.	147.3603	149.4179	147.3603	148.4456	148.3403	149.2820

1066

Table 8.11h (Continued)

PRIMAL ITER CNT	3	3	3	3	3
DUAL ITER COUNT	0	0	0	0	0
LIN-LMIT DIVISOR	1	1	1	1	1
CONDITION					* U U

LOOP SUMMARY

VARIABLE NAME	LOOP 12	LOOP 13	LOOP 14	LOOP 15	LOOP 16	LOOP 17
X 1 OLEFIN RATE B/D	2193.7500	2243.7500	2218.7500	2193.7500	2206.2500	2212.5000
X 2 I—C4 RECYCLE B/D	20000.0000	20000.0000	20000.0000	20000.0000	20000.0000	20000.0000
X 3 FRESH ACID MLB/D	120.0000	120.0000	120.0000	120.0000	120.0000	120.0000
Y 1 PROFIT **** $/D	1260.7817	1259.8881	1261.2675	1260.7817	1261.2114	1261.2978
Y 2 ALKY YIELD B/D	3880.1323	3970.4033	3925.4179	3880.1323	3902.6129	3914.1250
Y 3 I—C4 MAKE—UP B/D	2540.0112	2600.1416	2570.2597	2540.0112	2555.1816	2562.7324
Y 4 SPENT ACID WTPCT	89.8152	89.1134	89.4676	89.8152	89.6416	89.5548
Y 5 OCTANE NUMBER	93.8849	93.5911	93.7391	93.8849	93.8120	93.7756
Y 6 ISO/OLEFIN RATIO	10.2746	10.0724	10.1725	10.2746	10.2233	10.1978
Y 7 ACID DIL. FACTOR	2.8186	3.0142	2.9157	2.8186	2.8671	2.8914
Y 8 F—4 PERFORM. NO.	148.6546	147.7734	148.2173	148.6546	148.4362	148.3269

PRIMAL ITER CNT	3	3	3	3	3
DUAL ITER COUNT	0	0	0	0	0
LIN-LMIT DIVISOR	1	1	2	4	8
CONDITION					

LOOP SUMMARY

VARIABLE NAME	LOOP 18
X 1 OLEFIN RATE B/D	2218.7500
X 2 I—C4 RECYCLE B/D	20000.0000
X 3 FRESH ACID MLB/D	120.0000

Table 8.11h (Continued)

```
Y 1 PROFIT ****    $/D    1261.2675
Y 2 ALKY YIELD     B/D    3925.4179
Y 3 I-C4 MAKE-UP   B/D    2570.2597
Y 4 SPENT ACID WTPCT       89.4676
Y 5 OCTANE NUMBER            93.7391
Y 6 ISO/OLEFIN RATIO        10.1725
Y 7 ACID DIL. FACTOR         2.9157
Y 8 F-4 PERFORM. NO.       148.2173

PRIMAL ITER CNT     3
DUAL ITER COUNT     0
LIN-LMIT DIVISOR    8
CONDITION
```

VARIABLE NAME	TBC	IPSN	PPSN	CHANGE	LOWER LIMIT	UPPER LIMIT	LIN. LIMIT
X 1 OLEFIN RATE B/D	0100	1692.7500	2212.5000	518.7500	0.0000	2500.0000	3.1250
X 2 I-C4 RECYCLE B/D	0100	15000.0000	20000.0000	5000.0000	0.0000	20000.0000	31.2500
X 3 FRESH ACID MLB/D	0100	95.6250	120.0000	24.3750	0.0000	120.0000	0.3125

VARIABLE NAME	TBC	IPSN	PPSN	CHANGE	LOWER LIMIT	UPPER LIMIT	TARGET
Y 1 PROFIT **** $/D	0000	953.7114	1261.2978	307.5864	0.0000	0.0000	0.0000
Y 2 ALKY YIELD B/D	0011	2997.3818	3914.1250	916.7431	2000.0000	4000.0000	0.0000
Y 3 I-C4 MAKE-UP B/D	0011	1963.0556	2562.7324	599.6767	0.0000	4000.0000	0.0000
Y 4 SPENT ACID WTPCT	0011	90.0015	89.5548	-0.4467	85.0000	97.0000	0.0000
Y 5 OCTANE NUMBER	0011	93.8605	93.7756	-0.0848	90.0000	98.0000	0.0000
Y 6 ISO/OLEFIN RATIO	0011	10.0150	10.1978	0.1827	3.0000	14.0000	0.0000
Y 7 ACID DIL. FACTOR	0011	2.8348	2.8914	0.0565	0.0100	5.0000	0.0000
Y 8 F-4 PERFORM. NO.	0011	148.5816	148.3269	-0.2546	140.0000	165.0000	0.0000

OPTIMUM SOLUTION AT LOOP 18

8.12 DIRECT DIGITAL CONTROL[1]

The term direct digital control (ddc) as used in this section refers to the direct computer adjustment of control valve openings, without utilizing conventional controllers. This technique is basically different from supervisory or digital set-point control, which is discussed in previous sections. The benefits and drawbacks of ddc are covered here together with the description of its operating principles, hardware and software requirements, reliability, accuracy and application considerations.

Some of the topics treated have also been covered in other sections of this chapter, but the discussion here emphasizes the ddc aspects of these subjects and as such the dual coverage serves a useful function.

Definitions

The term *computer* as used in this section refers to a stored program digital computer.

Direct digital control (ddc) is a term that has widely differing connotations for different people. In general, ddc is considered to be the regulatory control of a process loop by a computer without the use of a conventional analog controller in the loop. Feedback control, feedforward control, ratio control and cascade control are techniques that may also be implemented by ddc. On-off control, such as the control of a sequencing operation, is not usually considered to be ddc, even though it is the control of a loop by a computer without recourse to a controller.

Digital set-point control, also called *supervisory control*, differs from ddc in that an analog controller is in the loop. In digital set-point control, the following actions take place:

(1) The digital computer calculates a new set point. (2) A signal is generated in the computer for transmission to the controller. The signal may be either an electrical analog signal that would be received by the cascade mechanism of the controller for changing the set point, or it may be a pulse signal. (3) Control action of the loop is performed by the analog controller. Digital set-point control is described in Section 8.8.

An *algorithm* is a calculational technique composed of computer instructions stored in the computer for the solution of a specific problem. As an

1069

example, all computers have the capability of adding numbers through wired-in instructions in the computer. Very few computers have the capability of extracting a square root. In order to extract a square root, known mathematical techniques are used to compute the square root, and this technique is programmed into the computer using the instructions of the computer. Such a program would be known as a square root algorithm. As there are a variety of techniques of calculating a square root, such as the Newton method and the Newton-Raphson method, there could be different types of square root algorithms. Algorithms are described in more detail in Section 8.5.

Two main types of algorithms are used in the ddc systems: the position algorithm and the velocity algorithm. Both algorithms are described in Section 8.5.

The *positional algorithm* calculates the output to the valve and transmits that signal to a holding amplifier and then to the valve.

The *velocity algorithm* calculates the difference in the present signal and the previous signal, manipulates the data, and transmits the differential signal to the holding amplifier and then to the valve.

Either a position or a velocity algorithm is used on a single project, but rarely both. A more complete treatment of computer terminology is given in Section 8.1.

Operation of ddc

Direct digital control is a feasible technique because of the slow dynamics of the control loops compared to the calculational speed of the computer. The response time of loops is such that "continual" control with an analog controller cannot be distinguished from "shared time" control with a ddc system, if the updating time of the loop is carefully chosen. A curve indicating a generalized relationship between square root of mean square (rms) error and update time is shown in Figure 8.12a. As long as the loop is updated more frequently than the time corresponding to the minimum error desired, control will be satisfactory and, in fact, indistinguishable from analog control. The update time is a function of the type of loop and the dynamics associated with the loop. As a general rule, *flow* loops have to be updated every second, pressure and level loops every five seconds, and temperature loops every thirty seconds. Because the signal level is held until the loop is updated, one computer may service many loops.

The information required in a ddc system is identical to the information required in a conventional analog system: tuning constants for proportional, integral (reset time), derivative (rate), set point, direct/reverse action, etc., must be provided in ddc as in analog control. This information is entered into the memory of the computer by means of an operator's console. Also stored in the computer memory are the algorithms for the type of control desired, such as feedback, feedforward, cascade, etc. All data stored in the computer are available for change through the keyboard by qualified personnel.

**rms ERROR
FROM SET POINT**

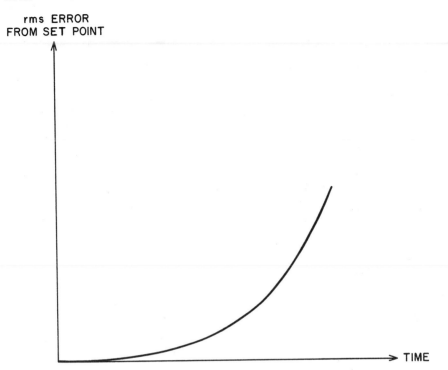

Fig. 8.12a   Error-to-time relationship in a ddc loop

When assembling a ddc system, one must classify loops by the speed of updating required, and the loops must be scanned at the required frequency. On scanning an input, the computer executes the following sequence of steps:

1. The predefined scan program selects the point to be scanned.

2. The input relay in the input multiplexer associated with that point is closed.

3. The analog signal is brought into the computer after being filtered, amplified and converted to a binary representation.

4. The input is usually converted to engineering units and compared with alarm limits. If the point is outside of alarm limits, a message is typed on an alarm typer.

5. The analog input is used in the control equation pertaining to that point, along with the requisite tuning constants to calculate the output signal.

6. The scan program selects the correct output line associated with the input line.

7. The output relay in the output multiplexer is closed.

8. In the case of a system utilizing a position algorithm, the desired output is converted from digital form to analog form by a digital-to-analog converter.

The output signal is transmitted to a holding amplifier or ladder network which holds the last signal until the loop is again updated.

9. In the case of a system utilizing a velocity algorithm, a differential signal is calculated, i.e., the difference between the new signal and the last signal. This signal may be converted to a timed voltage, often of 10 milliseconds duration, and this voltage applied to the appropriate holding amplifier. Thus, the total charge on the holding amplifier may be increased or decreased.

10. In the case of both the velocity algorithm and the positional algorithm, the output of the holding amplifier is converted to a pneumatic signal and applied to the valve diaphragm, thus completing the control loop.

Communication with the system is through the operator's console for changing data, such as set-point and tuning constants. Manual startup of a system is implemented by the use of computer-manual or computer-manual-automatic stations.

Thus careful planning must go into the timing of updating a loop. If a loop is not updated frequently enough, control will be impaired. If it is updated too frequently, control will not be improved and computer time will be wasted.

The number of loops that a ddc system can control is a function of the system speed, the organization, and the number and types of loops to be controlled. Operational speed of ddc systems is therefore expressed in the number of loops per second that the unit can handle, rather than in terms of a total number of loops. Most ddc computer vendors have the capability of handling 150–200 loops per second, largely due to input/output data-handling limitations. If a "typical plant" has 50 percent flow loops, 25 percent pressure and level loops, and 25 percent temperature loops that are sampled at the rates of 60, 12, 12 and 2 times per minute, respectively, then a ddc system operating at 200 points per second could theoretically handle 358 loops.

System Components

A ddc control loop has many of the same elements as a conventional analog control loop. In fact, the loops are identical, with the exception that the controller has been replaced with a computer system and its attendant peripheral equipment, such as typers, computer-manual stations and consoles.

Instrument Hardware

1. *Sensor*   A selection of sensors can be made based on the various sections of Volume I.

2. *Transmitter*   A transmitter is not essential for the operation of a ddc control loop. Some signals can be taken directly into the computer without amplification or conversion. However, this procedure is not usually followed, because it would be impossible to display process variables without strong

transmitted signals. Different types of transmitters are discussed in Sections 6.1 and 6.2.

3. *Current-to-air converter*   Since most valves are air-actuated, a current-to-air converter must be installed. Sections 1.7 and 6.3 describe this component.

4. *Valve*   The control valve, as the most common final control element, is discussed in the first chapter of this volume.

Computer Hardware

1. An *analog input system* is required to receive the signals from the transmitters or sensors and convert them into a form suitable for handling by the computer. A typical low-level analog input system is illustrated in Figure 8.12b. A detailed discussion on computer input-output interfaces is given in Section 8.7. Typical components and features of an analog input system include the following:

Terminal connectors are used to connect the inputs from the signal source to the computer.

A *dropping resistor* develops the correct voltage required by the amplifier. If, for example, the amplifier has a fixed gain of 200, and if the input to the analog-to-digital converter must be a maximum of 10 volts, it follows that the input signal to the amplifier must be a maximum of 50 millivolts. If a transmitter supplies a 4–20 madc signal, then a $2\frac{1}{2}$-ohm resistor can be used to produce the 50 millivolts, and if the transmitter supplies a 10–50 madc signal, then a 1-ohm resistor is required. If a voltage signal is supplied, no dropping resistor is required.

A *fusible resistor* may be included as a safety precaution to prevent damage to the computer if an over-voltage is inadvertently applied to the signal leads. In such a case, the fusible resistor will be destroyed, thus preventing an overload on the computer.

Resistive-capacitive (RC) filter networks are provided to filter the incoming signal. For a discussion on filtering techniques refer to Section 8.6. The value of the resistor and the capacitor may be changed to provide proper filtering for different types of input signals.

Relays are used to switch in the input signal. The relay is closed on command from the computer. Most computers have a random addressing structure so that any relay in any order may be addressed; they do not need to be addressed sequentially. The relays and associated equipment comprise the *multiplexer*. The purpose of this equipment is to channel many inputs through single items, such as the amplifier, analog-to-digital converter, and buffer.

Relays in the multiplexer may handle low-level (less than 50-millivolt) signals or high-level (greater than 1-volt) signals. One standard type of low-

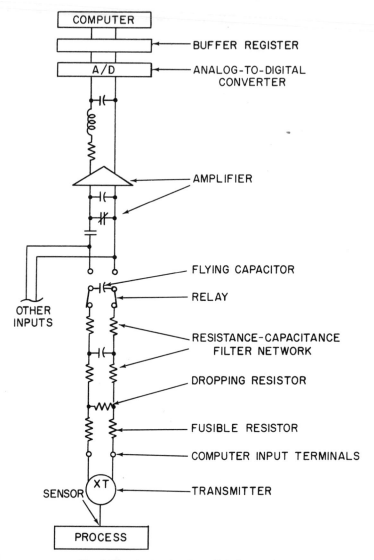

Fig. 8.12b　Analog inputs to digital computers

level relay used is the mercury-wetted relay. It can handle either low- or high-level signals at switching rates of up to 200 points per second.

If a faster switching rate is required, solid-state switches may be used that provide switching speeds up to 40,000 points per second. These switches usually require a signal of 1 volt or higher and therefore, low-level signals cannot be directly switched by a solid-state multiplexer but must be pre-amplified. Fast switching speeds are not required in direct digital control. They are often used to enhance the reliability of the system, however, since the solid-state switch has no moving parts and thus is less subject to failure. Other considerations and more details on multiplexers are given in Sections 8.6 and 8.7.

A *flying capacitor* may be included in an analog input system to improve the input sample. The capacitor averages the signal, provides an additional filtering effect and holds the input value while the signal is being switched into the amplifier. This isolated switching provides excellent common mode noise rejection of 120 decibels at 60 Hz. Equation 8.6(3) provides the definition.

A *preamplifier* is used to raise the voltage of the incoming signal to the voltage level required by the analog-to-digital converter. Input systems may be designed to handle only one level of low-voltage signals, in which case one preamplifier is used. If more than one low-level signal is used, the signals may be segregated and several preamplifiers of different gains may be used, or a single preamplifier with a variable gain may be utilized. Settling time on the preamplifier is about 2 milliseconds.

High-level signals are attenuated to the level required by the analog-to-digital converter and do not require a preamplifier.

Associated with the preamplifier may be choppers on the input side and a RLC (resistance-inductance-capacitance) filter on the output side.

An *analog-to-digital converter* is used to convert the voltage level from the preamplifier into a binary representation of that voltage. The output may be of any precision (resolution) desired. A commonly used output is 12 bits, with an extra bit for the sign position. For operating speeds and conversion times refer to Table 8.7i.

A *buffer* is used to hold the output of the analog-to-digital converter and transfer it into the computer.

The *computer* is the final element in the analog input system, and it is the device that initiates action. A *real time clock* is part of the computer system and a stored program in the computer receives timing pulses from it. At specified times, programs are actuated. A scan program is one that would close a specific relay, and if an amplifier with variable gain is used, set the gain of the amplifier to the correct value.

2. *Digital input system*   Digital inputs are two-state contact closures or voltages which serve to tell the status of a particular device. Some of the more important digital inputs are the status of the computer-manual, or of the computer-manual-automatic, transfer switch on the backup stations or interrupts from the operator's console, etc.

Digital inputs are *not* read into the computer by a multiplexer. These contact closures are wired into a buffer register, and the register is read by the computer as frequently as desired, which is usually once a second in ddc systems. The size of the buffer register is the same as the word size of the computer. It usually takes two cycle times of the computer to read in one word. As an example, if a computer has a cycle time of 1.0 microsecond and a word size of 16 bits, and if there are 240 digital inputs to be read, they could be entered into the computer in 30 microseconds. Further information on digital inputs is provided in Section 8.7.

3. *Analog output system*   The analog output system in direct digital control varies with the supplier of the equipment and with the type of algorithm used. A comprehensive system using the position algorithm is described below.

Two outputs are produced from the computer: An analog signal denoting the value of the signal and a code designating the station involved. The following computer hardware is used:

Digital-to-analog converter   The output generated by the computer is converted from a binary representation to an analog voltage by a digital-to-analog converter. This device usually employs 10 magnitude bits and is described in more detail in Section 8.7. The output voltage generated is placed on a bus bar and is made available to every output station. This type of action is possible only in a system using a position algorithm where the position of the valve is calculated.

Digital outputs are put into a code. These outputs are for identification of the proper loop to be updated.

The analog signal and the digital signals are transmitted to the station logic interface, where the following actions occur:

(1) The digital signals are decoded and the correct station is selected.
(2) When the station is selected, the analog signal is gated into a memory-holding amplifier.
(3) An output signal to the process (of 4–20 or 10–50 madc) is generated.
(4) The memory-holding amplifier holds the signal until the next updating time.

A validity system may be used to check the data being generated. The following items are checked:

(1) The case of the stations is checked. The stations are grouped in lots of 72, each group being called a case. The system is checked to see if the correct case of stations has been identified.
(2) The station number within the case is checked to see if the number is valid. Both the case identification and the station identification is coded with a simplified parity scheme. Control bits are also provided to indicate whether the station is to be connected for updating or whether it is to be read for feedback.
(3) The complementary value of the analog signal is put in an adder, and added to the value of the signal produced from the station. If the sum does not equal zero, it is obvious that the output analog signal has become garbled and is incorrect.

The holding amplifier is usually mounted in the housing for the computer-manual station or the housing for the computer-manual-automatic backup control station. These devices are described in more detail in Sections 8.8 and 8.13.

If a position algorithm is used, a feedback signal is required to provide the computer algorithm with the value of the signal being sent to the loop, but if a velocity algorithm is used, this feedback signal is not required. This subject is more thoroughly covered in Section 8.5. Therefore, if a position algorithm is used, one must have an additional analog input for every loop that is on control. This feedback signal may be handled either as a separate input to the computer or as a signal that is multiplexed into a single input position, thus requiring less hardware.

4. *Digital output system* Digital outputs are two-state output signals from a computer that serve to open or close contact devices, such as relays and solenoids. They are handled in a similar but opposite manner to digital inputs. The speed of operation with digital outputs is the same as with digital inputs. The computer usually supplies the power (24 or 48 volts DC) to actuate the relay.

Commonly used digital outputs in a ddc computer system are the contact closures to switch a station from computer to manual, to turn on lights on the console, to start or stop devices, etc. Further information on digital outputs is found in Section 8.7.

Auxiliary Hardware

Each ddc loop must have certain hardware associated with it in order to perform routine operations such as manual control. Among these items are the holding amplifiers, meters to display the measured variable signal, meters to display the output signal to the valve, and automatic-manual devices, such as a computer-manual station or a backup controller.

Input-Output Hardware

Input-output (I/O) devices for items other than process signals must also be provided with a system. There are several categories of I/O devices:

1. *Information*—for display of process data.
   Logging typers are used to produce log sheets on a periodic or demand basis.
   Message typers are used to make a typewritten copy of all messages associated with the system.
   Cathode ray tube may be used to display information such as data trends or points in alarm status.
2. *Program*—for entry of computer program.
   Paper tape reader and paper tape punch may be used for program entry.

Card reader and punch may be used as another method of program entry. It is considerably more flexible than the paper tape system, but it is also more expensive.

A card or tape preparation unit should be available for program preparation with the system.

3. *Console*—for operator-computer communication.

Operator functions should be made readily accessible to the operator.

Instrument engineer functions, such as entry devices for tuning constants, must also be available from the console. Consoles are discussed in more detail in Section 8.13.

Backup Hardware

Backup devices are used in most systems in order to allow for controlling the loops, when the computer fails. Two kinds of backup devices are used. One approach serves to keep the computer system running by providing spare components, and the other is used when the computer system fails. Among the former type are spare hardware items, such as a redundant amplifier, and even a redundant computer. Among the latter type are computer-manual stations and backup controllers.

Analysis and Programming

Analysis refers to the selection of the control scheme for each loop, establishing the scan rate, establishing the alarm limits, determining what to do in case of failure of a portion of the control loop, determining the algorithm to use, etc. In cases where there is optimization to perform in addition to ddc, the mathematical model of the process must be established together with optimizing procedures.

Programming may be defined as the approach employed to enter the calculational techniques determined by the analysis of the problem into the computer. Programming may involve only the filling in of blanks in program forms developed by vendors (Section 8.10). It may also mean the implementation of elaborate control schemes without the benefit of the simplifying program supplied by the vendor, or it may involve the construction of the simplifying routine itself.

In ddc projects in particular, extreme care should be exercised in the startup of a system, particularly if the backup devices are manual stations and not controllers. The programs and the computer equipment must be carefully and rigorously tested, often with the use of simulated inputs. Thorough checking in the factory and in the field before commissioning the equipment will eliminate many problems.

It should be kept in mind at all times that analysis and programming is a significant part of every computer project. In small, well-defined ddc systems, the analysis and programming cost may only be 10 percent of the hardware. In large jobs with a large amount of calculations, the analysis and programming costs can be several times the equipment cost.

Problems with ddc

The various problem areas are summarized below:

Cost—It is obvious that ddc costs more than conventional analog control on a strictly replacement basis. Direct digital control allows the analog controller in the loop to be replaced with an analog and digital input system, a computer, an analog and digital output system, auxiliary hardware, various input-output and backup devices (including even the analog controller that was originally replaced in some cases) and a large amount of software. Obviously, functions other than original cost must be considered before reaching a decision to install ddc.

Reliability The fear of computer failure, shutting down the operation of an entire plant, is a very real concern in the evaluation of ddc units. With analog controllers, failures occur in individual units, and there is no likelihood of all controllers failing at once unless there is a general power failure. However, ddc places the control for all loops into one device. Although safety procedures are applied to offset this problem—as described in Sections 8.13 and 8.15—it remains a serious impediment to the use of ddc. Many ddc users go to the expense of installing an analog backup controller for the more important loops to be controlled or for every loop in the system.

Delivery Process control computer systems are chronically late in delivery and startup, usually due to analysis and programming problems. Inadequacy of backup devices may mean that it is difficult, if not impossible, to start up a new unit. As the delay due to long computer delivery can be quite costly, it often works against the use of a ddc system.

Operator acceptance It is felt by some companies that ddc operation is too complicated for the operator to understand and therefore he might not accept the system. This fear has no real basis, as almost all ddc systems installed have had good operator acceptance. Operation of operator consoles is described in the next section.

Training Personnel must be trained to operate and maintain this new type of equipment. This work must be carried out on a continuing basis.

Benefits of ddc

Among the advantages of the ddc technique are the following:

Improved control ddc affords better control than is available with conventional controllers. Reset time, for example, can be made longer than is available with conventional analog controls. Special algorithms may be written and used to improve control. An example of that is an error squared algorithm which might be used in level control of a horizontal vessel.

Advanced control   Special techniques may be used for performing advanced control with a ddc computer system. Cascade control, ratio control, feedforward control, and non-interacting control are all commonly available. These techniques are also available in conventional analog control equipment, but they are much more easily implemented with ddc.

Flexibility   Loops may be rearranged by ddc without having to install additional equipment or without the need for shutdown. The inherent flexibility of ddc is exceedingly valuable during plant startup. Loops may be reconfigured and different control schemes implemented and examined within the span of a few hours.

Self-tuning   Two types of self-tuning may be used with ddc. One variety involves the computer (on program control) to replace the present tuning constants with preselected tuning constants. This technique is most widely utilized in batch processing. The other technique is to use a true self-tuning algorithm, which calculates new tuning constants based on the detected dynamics of the control loop. See Section 7.13 for more details.

Control enforcement   This item is considered by many to be one of the most important benefits of ddc. It means that the operator is being monitored by the computer in three important areas: set-point change, computer-manual change, and tuning constant change. The operator is permitted to change the set point or the computer-manual status whenever he desires, but when he does, a message indicating this action and the time is printed on the alarm typer. Close monitoring of this information allows plant supervisors to improve the operation of the process. Tuning constants are made unavailable to unauthorized personnel by means of a key lock switch on the operator's console.

Cost   As previously mentioned, a ddc system is more expensive on a first-cost basis than is conventional analog control. Once a system is installed, however, it is often quite inexpensive to add loops, particularly if this was allowed for in the original design of the system.

Process information fallout   Considerable process information is generated in the normal course of ddc operations, such as scan and alarm. Logs are easily generated at little cost if a computer is used. This advantage is not particular to ddc but is true for any computer control system.

Other benefits   While the computer may be installed for the purpose of performing ddc, it may also perform calculational or sequential operations, such as making weight balances and heat balances and performing optimization.

Direct digital control has not been patented. The use of a computer to change the set point of an analog controller is patented in the United States by the Exner patent granted to Hughes Aircraft. The use of ddc allows the computer to control the loop without violating the patent.

On-stream factors of over 99 percent may be expected, with a mean time between failure of the main frame approaching 4,500 hours.

The computer has many checks built into it, and, through programming, an exceptionally reliable system can be provided. Table 8.12c summarizes some of the protection techniques applied.

The accuracy of the computer system is a function of many factors, among which are:

> The noise level of the signal.
> The temperature drift of the thermocouple isothermal reference plane.
> The temperature drift of the preamplifier.
> The stability of the preamplifier.
> The resolution of the analog-to-digital converter.
> The round-off error in the conversion equation.
> The round-off error in the calculations.
> The resolution of the digital-to-analog converter.

Overall accuracy of a signal from input to output is usually ±0.1 percent of full scale. This figure is usually better than the accuracy of the transmitted signal.

Implementation of ddc

The first item in implementing a ddc system is to determine which loops are to be ddc controlled. Each loop should be closely examined to determine if the benefits of ddc outweigh the disadvantages for that particular loop. It is difficult to improve upon an analog flow control loop. As flow loops must be sampled far more frequently than other loops, they will have a much greater effect on the amount of computer time consumed than any other loop. The results of a careful consideration of loops to be controlled by ddc will provide increased availability of computer time. More computer time means more time available for other ddc loops and more time available for process optimization.

From a knowledge of the loops to be controlled, scan cycles can be set. The loops must be scanned frequently enough to provide adequate control as shown in Figure 8.12a. The dynamics of each loop will be different, but they can be grouped in different categories to simplify programming. As a general rule, most flow loops can be adequately handled by sampling once every second, although there may be some flow loops that should be sampled twice every second. Pressure and level loops are usually sampled once every five seconds, and slow temperature loops may be sampled once every 30 seconds.

In order to simplify programming, the *number of scan cycles* should be kept as small as possible concommittant with good control. Usually three to six scan cycles are employed.

After the loops are categorized into scan cycle groups, an input point designation is assigned. The format of the point designation is specified by

Table 8.12c
FAILURE DETECTION AND PREVENTION IN ddc SYSTEMS

Component That Might Fail	Failure Detection Device	Failure Protection Technique
Power to System	Power failure monitor on computer	Auxiliary power source such as static inverter and battery (Section 5.4)
Transmitter	Program comparison with floating zero, limits, or last value	Outside of system—may use second transmitter as standby
Analog Input Multiplexer	Same as above	Locate critical inputs on different cards in multiplexer; have multiple inputs
Amplifier or Analog-to-Digital Converter	Two check signals of known voltage	Through program control, calculate slope and intercept of amplifier curve. Compare known inputs with calculated value
Computer Hardware	Use a check problem exercising all commands to produce a known answer	Go to backup device: manual station, backup controller, backup computer
Computer Software	Use a rundown timer (watchdog timer) that must be reset by the computer program	Same as above
Output Multiplexer for ddc	By ddc validity system	Locate critical outputs on different cards; otherwise same as above
Analog Output for ddc, Station Selection, Data Transmitted	By ddc validity system	See computer hardware above
Analog Output for ddc value	Comparison with previous value; rate of change of value	By program; if the value is excessive, program could print alarm. If it happens too frequently, use backup devices
Analog Output for ddc, Updating	Update timer in automatic-manual station	See computer hardware above
Holding Amplifier Power Failure	Loss of power	Battery backup (Section 5.4)

the vendors of the ddc software package, as are the details of implementing the system. The loops are identified by name and are grouped into the scan tables mentioned above. When implementing the scan tables, one should build in flexibility by incorporating non-existent points. In this manner points can be easily added or changed from one table to another without significantly increasing the reprogramming time.

From the knowledge of the signal, the input resistor can be sized and the amplifier gain can be computed to provide a signal corresponding to the maximum voltage of the analog-to-digital converter. This operation is a function of the computer hardware used. If several fixed-gain amplifiers are used, the inputs must be physically wired to specific terminal connections which the computer vendor has wired to the correct amplifier. If a variable-gain amplifier is used, a computer instruction must be coded in order to set the amplifier to the correct gain when that point is scanned. In order to improve the accuracy of the system, the maximum anticipated signal level when amplified should closely approach the maximum input signal level of the analog-to-digital converter. If the system is not designed to operate in this manner, the accuracy can be improved by increasing the number of bits in the analog-to-digital converter.

Conversion equations must be determined for each analog input to the computer and programmed into the computer. The conversion equation is a function of the type of device used and the range of the transmitter. Conversion equations can be calculated as precisely as desired. For instance, curves for different types of thermocouples may be expressed as linear equations over a narrow range, quadratic or cubic equations may be used to represent the entire range, or a series of linear equations can be utilized, with each line representing a small segment of the range. Flow rates can be calculated using a variety of compensating terms, such as changes in temperature, viscosity and specific gravity. For each analog input, there always is a conversion equation, but a number of different inputs may share a common conversion equation.

Points that are to be alarmed must be incorporated into the system. Usually all points on ddc control must be alarmed, and often other points not on ddc control but incorporated into the computer system are also alarmed. The types of abnormal conditions which usually result in an alarm include:

1. *Zero voltage alarm*—on loops with a transmitter. This alarm indicates that the transmitter has failed.
2. *High-low* alarm—indicates that the process variable is outside of normal limits.
3. *Extreme* high-low alarm—indicates that the process is at a critical point.

A standard message format is usually available from the vendor for outputting the alarm message on a typewriter or on a cathode ray tube display.

Additional information that must be furnished for each analog point is virtually identical with the information that must be furnished for a control system using analog components. A summary of the most important data is given in Table 8.12d.

Table 8.12d
INFORMATION REQUIRED IN A COMPUTER-CONTROLLED SYSTEM

Process Variable	Loop Summary
Process Variable High Limit	Computer Station Status
Process Variable Low Limit	Auto-Manual Status
Scan Status	Feedforward Address
Sample Time	Auto Reset Status
Automatic Update	Direct-Reverse Status
Deviation	Gain
High Deviation Limit	Reset Rate
Low Deviation Limit	Rate Time
Ratio	Bias
Supervisory	Deadtime Status
Set Point	Batch Constant
ddc Calculated Output	Deadtime Gain
Calculated Output High Limit	Deadtime Lag
Calculated Output Low Limit	Deadtime
Feedforward Status	Octal Loop Address
Cascade Status	Output to Stations
Ratio Status	Adaptive Status

After all requisite information has been gathered it must be programmed. Often the programming is merely the completion of forms for inserting data into existing programs supplied by the computer vendor (Section 8.10). The finished programs are carefully debugged before operation is initiated on an existing plant. Sometimes, errors will be found in the computer software such as an incorrect entry of the requisite information. The user will usually test the unit with simulated analog signals to prove out the control system before it is installed.

Section 8.13 discusses the backup system and the operator's console and briefly describes their use.

Conclusions

Direct digital control received its initial impetus in the early 1960s. The peak of activity in the field probably occurred in 1966 with the purchase of about 15 ddc systems, several of which were designed to control 600 loops or more. In these systems, the computer was used to control all the loops in the plant. In these early ddc systems, analog controller backup to insure reliability of control was used on all critical loops and often on the non-critical loops also.

By 1969, the pace in the use of ddc had slowed considerably. It was found to provide control superior to analog control, but the advantages did not always balance the economic considerations. Most installations in which ddc is being implemented today are those with 50 loops or less to control.

There now appears to be a tendency to use ddc on loops that will particularly benefit by this type of control, such as those that are hard to control or that require advanced control techniques. On this basis, ddc loops may represent about 20 percent of the loops on a "typical" continuous process.

At the present time, ddc is not the governing reason for a computer to be installed on a process. It is usually due to requirements for process optimization or process sequencing, with ddc being used only to augment those concepts. A large investment in computer hardware and software is required by ddc. However, digital computers and their peripherals continue to drop in price as the technology changes, and the economics of the use of direct digital control should be continually re-examined.

ACKNOWLEDGMENT

(1) By permission of Honeywell, Inc., portions of this section have been prepared from the Honeywell *DDC System Operator's Guide:* Series 16 Technical Bulletin, "Process Interface Controller," and "Operator's Console" (each Honeywell, Inc., CCD, Copyright 1968).

8.13 OPERATOR COMMUNICATION[1]

This section is devoted to the interface requirements between the computer and the human operator, in terms of both information display and data entry. From a hardware point of view this would include operator's consoles, computer-manual stations, backup controllers, logging and alarm typewriters and cathode ray tube displays. The discussion of the features of these devices presented in this section is oriented toward the thinking of one manufacturer, yet it is broad enough is scope to be of general value.

If the conventional analog controls were to be replaced with a computer having no operator's communication hardware, two serious problems would arise: (1) The human operator would have no information as to the status of the process and (2) The system would have no alternative mode of operation in case of computer failure. Overcoming these problems is the main motivation in the design of the operator's interface hardware.

Some of the information that is continuously displayed by an analog controller, such as the tuning constants, need only be provided when the operator asks for it. Such data is therefore usually provided on the operator's console, where any of the loops can be selected for display. Other information, such as the value of the process variable, should be continuously displayed. This type of indication is furnished on the backup stations, which are specifically assigned to the particular loops and which provide the capability for alternative (other than computer) modes of operation in case of computer failure. Depending on the plant requirements, these stations will either provide a choice between computer and manual modes of operation or will allow for three operating modes, the third being analog control.

Some of the discussion presented in connection with set-point stations in Section 8.8 also applies here.

Computer-Manual (C/M) Stations

The least expensive device for displaying a large amount of process information is a computer-manual station. One station is used on each loop.

The C/M station shown in Figure 8.13a contains the following components:

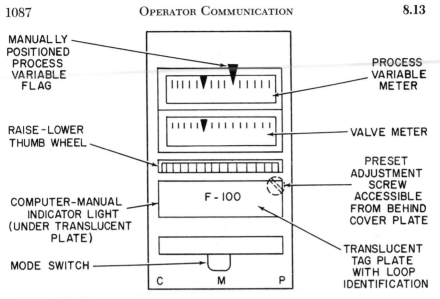

Fig. 8.13a   Computer-manual station

1. Valve meter   This meter is usually calibrated in units of 0 to 100 percent. It always indicates the current to the valve regardless of whether the station is on or off computer control.

2. Process variable meter   This meter is usually calibrated in units of 0 to 100 percent. It is driven by a current signal proportional to the process variable value in that loop.

3. Process variable flag   Located on the process variable meter is a manually positioned flag which can be used by the operator as a reference. For example, he can use the flag to indicate where the computer set point is or where the computer high or low alarm is.

4. Mode switch   A switch is provided with the following three positions:

    a. *Computer*   In this position, control of the station is by the computer. Should the computer fail to update the station, it maintains the output signal to the valve at the last value requested by the computer.

    b. *Manual*   In this position, the output signal to the valve is manually set by the operator. The station maintains this value at a drift rate of usually no greater than 1 percent per hour.

    c. *Preset*   In this position, the signal to the valve is maintained at a *predetermined value* which is set by a screwdriver-adjusted potentiometer accessible from behind the cover plate. This potentiometer can be set at any time, regardless of the mode switch position. By depressing the potentiometer with a screwdriver, the valve meter is switched to indicate the preset value, and the shaft can then be turned to the desired preset value. This adjustment in no way affects the value of the signal to the valve, when the mode switch is in computer or manual position. The preset option can also be remotely initiated by the computer. There is no drift in the preset mode. All switching transitions from one state to another are accomplished bumplessly except the transition to and from preset. This transition causes the output signal to go to the preset value regardless where it was prior to the transition. The amount of bump when switching from preset to computer depends on how long the station has been in the preset mode. If it remains in preset for more than five minutes, the switching will be bumpless.

5. Raise-lower thumbwheel   When the station mode switch is in the manual position, this thumbwheel allows the operator to reposition the valve. This is done on an incremental basis, with the rate of change in either direction being determined by how far the spring-return thumbwheel is rotated from its neutral position. When the mode switch is in the preset position, the operation of this thumbwheel is overridden.

6. Tag plate   A translucent plate is provided on which the loop identification may be engraved.

7. Indicator lamps   Two indicator lamps may be provided. One operates in conjunction with a timer circuit and automatically lights when the computer fails to update. The time-out interval on this lamp can be set at either 6–12 seconds or 45–90 seconds. This setting must be longer than the time interval used by the computer to update. The other lamp is available for unspecified use via terminals in the rear of the case. Both lamps are mounted behind the translucent tag plate.

This C/M station can be quite compact, with a front dimension of 2 by 3 in. However, this device is often mounted in a housing of the same size as the analog controller housing. The C/M station usually also houses the holding amplifier.

The system is so designed that when the computer fails, the C/M stations will automatically go either to manual control at the present valve position or to preset control at a predesignated valve position.

The C/M station is designed for those loops that require manual control and display of variables only. It should be satisfactory for these loops to operate in the manual mode for extended periods while the computer is undergoing maintenance.

Computer-Manual-Automatic (C/M/A) Stations

The computer-manual-automatic station is specifically designed for those situations where the loop cannot be satisfactorily operated in the manual mode for long periods and requires an analog controller for backup. It is an analog backup controller.

The C/M/A station closely resembles a conventional analog controller and is also similar to computer set-point stations (Figure 8.8b). It actually is an analog controller with an extra switch for computer control position, in addition to the conventional manual and analog automatic positions. The C/M/A station contains the holding amplifier as an integral part of its circuitry. Signals from the computer update the holding amplifier. In case of computer failure, the station automatically switches to the analog automatic position, and continuous process control is thereby assured.

When the loop is on computer control, the analog controller is by-passed and control action is handled by computer algorithms. On computer control, the set point of the controller is provided by the computer. In order to prevent bumping the loop when it is switched from computer to analog control, the operator must either balance the loop manually, or the controller must be equipped with a servo motor so that the analog controller can track the computer set point. There is no problem in switching from automatic or manual to the computer mode, because, the computer is programmed to use the existing value of the process variable as the initial set point, and therefore bumpless transfer is achieved.

The functions of the C/M/A station include the following:

1. It affords backup on computer failure.
2. It provides the same information as the C/M station.
3. It is easily understood by the operators.
4. It allows great flexibility in having loops included or excluded from computer control.
5. A plant may be started up either on manual, automatic or ddc control.
6. Use of C/M/A stations frees the plant from startup delays due to computer delivery problems. If the computer is not delivered on time, the plant can run on analog control.

The C/M/A station shown in Figure 8.13b contains the following displays and components:

1. Process variable meter   A vertical scale meter which indicates deviation from set point. The value of the process variable is indicated on the tape scale by the meter pointer. This arrangement allows the operator to determine the deviation from set point as well as to read the variable. The meter accuracy is usually ±0.5 percent of full scale.

2. Set-point adjustment   A thumbwheel is provided beside the meter face to adjust the station set point. Set-point indication is provided by the tape scale on the meter face. Changing set point moves a calibrated tape scale which underlies a fixed set-point indicator. On the non-tracking version of these stations, the set point is positioned only by this adjustment. However, when the tracking version is used and the unit is in the computer mode, as

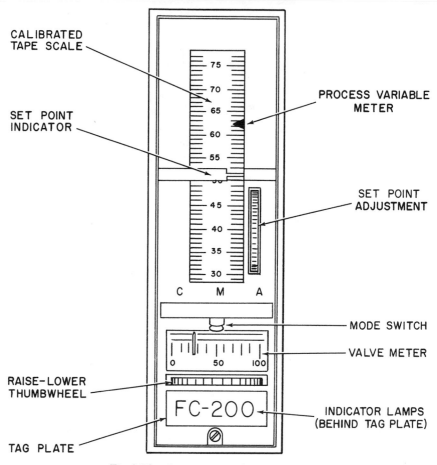

CALIBRATED
TAPE SCALE

SET POINT
INDICATOR

PROCESS VARIABLE
METER

SET POINT
ADJUSTMENT

MODE SWITCH

VALVE METER

RAISE–LOWER
THUMBWHEEL

FC-200

INDICATOR LAMPS
(BEHIND TAG PLATE)

TAG PLATE

Fig. 8.13b   Computer-manual-automatic station

it is being updated by the computer, its set-point value automatically tracks the process variable by means of a servo system.

    3. *Mode switch*   A mode switch is provided, having the following three positions:

       a. *Computer*   In this position, the computer controls the valve output signal directly. When the computer fails to update the station, one option allows the stations to revert to local automatic control, and the second option allows the station to hold the last value outputted by the computer. Additionally, the local set point on the tracking type station follows the process variable during operation in the computer mode.

       b. *Manual*   In this position, the valve output signal is set by the operator.

c. *Automatic*   In this position, the station operates as an analog controller. Set point is locally set by the operator.

4. Valve meter   This meter is usually calibrated in units of 0 to 100 percent. It always indicates the signal to the valve regardless of the mode of operation.

5. Raise-lower thumbwheel   When the station is in the manual position, the raise-lower thumbwheel allows the operator to reposition the valve. This is done on an incremental basis with the rate of change being determined by how far the spring return thumbwheel is rotated from its neutral position.

6. Tag plate   A translucent plate is provided on which the loop identification is engraved.

7. Indicator lamps   Two indicator lamps are provided. One operates in conjunction with a timer circuit (optional on the non-tracking, standard on the tracking design), and the lamp automatically lights when the computer fails to update. The time-out interval can be set at either 6–12 seconds or 45–90 seconds. This setting must be longer than the time interval used by the computer to update. The other lamp is available for unspecified use via terminals in the rear of the case. Both lamps are mounted behind the translucent tag plate.

In addition to the functions just described, other adjustments are available on these controllers when the chassis is partially withdrawn from the case. These include the control mode settings and the controller action (direct or reverse) jumper.

Operator's Console

A typical system might have 75 percent of the loops backed up by C/M/A stations and 25 percent by C/M stations. These stations indicate the process variable and the signal to the valve, but there is no indication of the set point, tuning constants, direct-reverse acting status, etc. This information is displayed by means of the operator's console.

There are as many designs of operator's consoles as there are computer suppliers. These consoles are supported by supplier developed programs.

Keys and Displays

The operator's console is shown in Figure 8.13c and a detailed keyboard in Figure 8.13d. The console contains the following components:

1. Function keys   These keys are used to designate the variable to be displayed or entered. Thirty keys are shown. All functions are associated with the loop identification appearing in the identification register.

2. Action keys   These keys are used to indicate what type of action the operator is requesting, such as enter or display. They also cause the execution

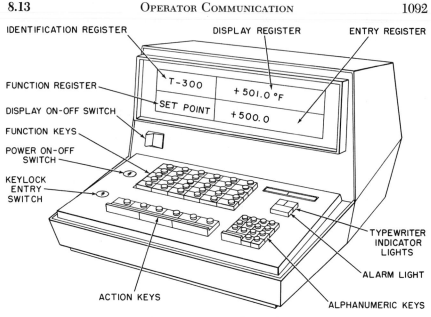

IDENTIFICATION REGISTER   DISPLAY REGISTER   ENTRY REGISTER

FUNCTION REGISTER

DISPLAY ON-OFF SWITCH

FUNCTION KEYS

POWER ON-OFF SWITCH

KEYLOCK ENTRY SWITCH

T-300

SET POINT

+ 50I.0 °F

+ 500. 0

TYPEWRITER INDICATOR LIGHTS

ALARM LIGHT

ACTION KEYS

ALPHANUMERIC KEYS

Fig. 8.13c   Operator's console

of system operations which are not associated with a specific control point. Examples of this are demand, log, or octal display-enter requests.

3. Alphanumeric keys   These keys are used to key alphanumeric information into the display registers. Some keys have both an alphabetic character and a numeric character, and the system determines which of the two is indicated by the sequence in which the keys are depressed. The keys are always interpreted as numerics, except for the first key depressed after the loop identification/clear key, which is interpreted as an alphabetic character.

4. Power on-off switch   This key-operated switch allows for disabling the operation of the console keyboard.

5. Keylock entry switch   This key-operated switch is used to establish the priority of the operator. In the *restricted* position, only a limited number of entries can be executed. In the unrestricted position, all operator console entries present in the system may be used.

6. Alarm light   This light is on whenever an alarm condition occurs within the areas of responsibility of the console. It remains on until the operator manually depresses the light itself.

7. Typewriter indicator lights   These lights indicate whether the alarm and logging typewriters associated with the console's operating areas are enabled. An illuminated light indicates that the device is disabled.

8. Display on-off switch   This switch is used to disconnect the power to the register displays.

9. Identification register   This is often a four-window register used by

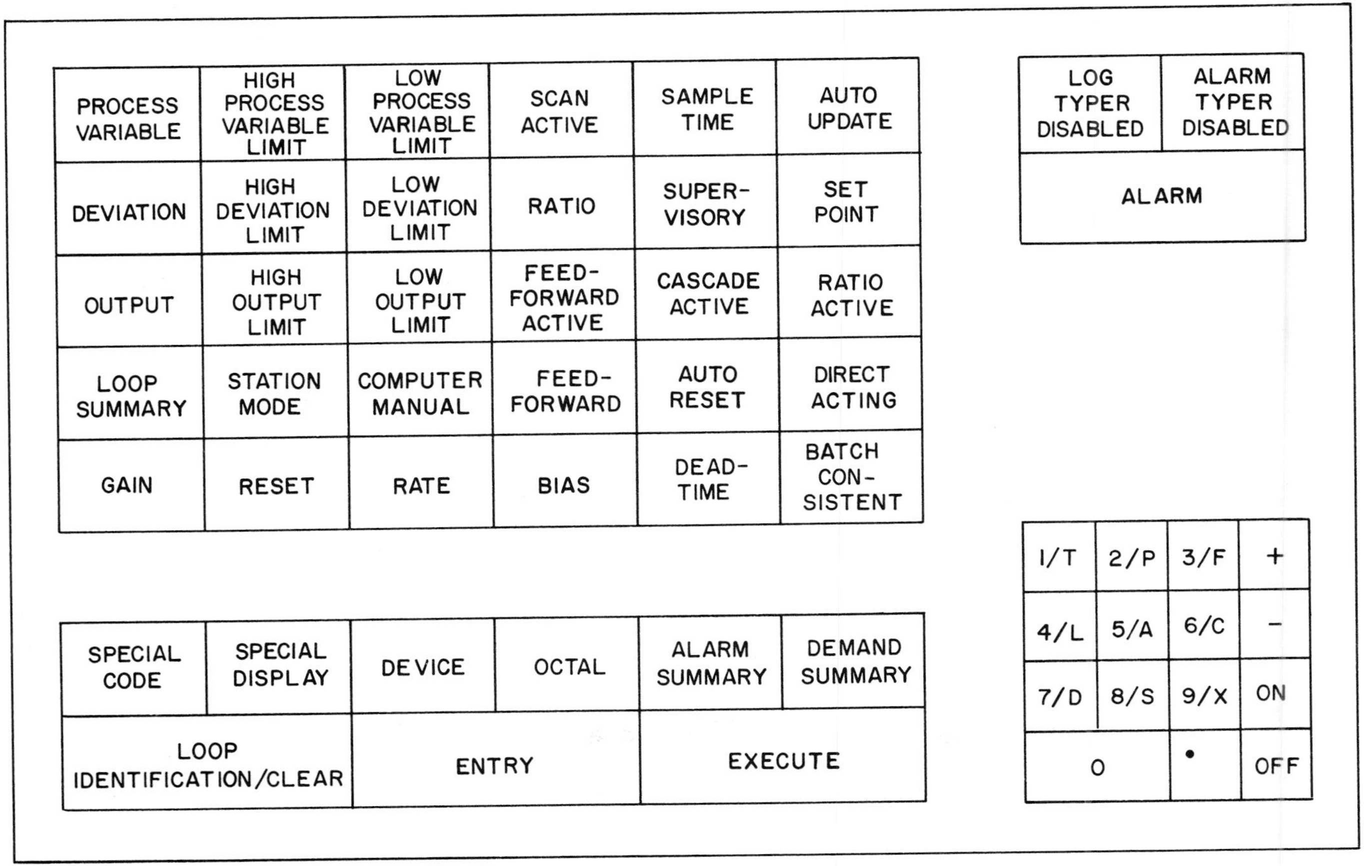

LOG TYPER DISABLED	ALARM TYPER DISABLED
ALARM	

1/T	2/P	3/F	+
4/L	5/A	6/C	−
7/D	8/S	9/X	ON
	O	•	OFF

PROCESS VARIABLE	HIGH PROCESS VARIABLE LIMIT	LOW PROCESS VARIABLE LIMIT	SCAN ACTIVE	SAMPLE TIME	AUTO UPDATE
DEVIATION	HIGH DEVIATION LIMIT	LOW DEVIATION LIMIT	RATIO	SUPER- VISORY	SET POINT
OUTPUT	HIGH OUTPUT LIMIT	LOW OUTPUT LIMIT	FEED- FORWARD ACTIVE	CASCADE ACTIVE	RATIO ACTIVE
LOOP SUMMARY	STATION MODE	COMPUTER MANUAL	FEED- FORWARD	AUTO RESET	DIRECT ACTING
GAIN	RESET	RATE	BIAS	DEAD- TIME	BATCH CON- SISTENT

SPECIAL CODE	SPECIAL DISPLAY	DEVICE	OCTAL	ALARM SUMMARY	DEMAND SUMMARY
LOOP IDENTIFICATION/CLEAR		ENTRY		EXECUTE	

Fig. 8.13d Operator's console keyboard

the operator to designate the loop he is addressing. This register is activated by depressing the loop identification/clear action key and entering the desired loop identification characters on the alphanumeric key module.

10. Function register   This is often a three-window register which is used to indicate the function requested by the function key module.

11. Display register   This is usually a seven-window register which is used to display the value or status of the function requested by the operator in its associated engineering units.

12. Entry register   This may be a six-window register used to display new values to be entered into the computer. This register is activated by the entry action key. The desired value is keyed into the register with the alphanumeric keys and the value is entered by depressing the *execute* action key.

Operating Sequence

A typical operating sequence for the use of the console is described here, with the aim of displaying the present value of a set point and then entering a new value.

1. Depress the loop identification/clear action key. This clears all display windows.

2. Sequentially depress the four alphanumeric keys which correspond to the point identification. As each key is depressed, the corresponding character is displayed in the identification register. As soon as the fourth alphanumeric key is depressed, an automatic display of the point process variable appears in the entry register, and the words PROCESS VARIABLE appear in the function register.

3. Depress the dedicated function key labeled SET POINT. This produces an automatic display of the set-point value in the data register, and the words SET POINT appear in the function register. At this point, any other dedicated function key may be depressed, and that function value or state will be displayed in the data register with its associated name in the function register. To enter a new value of any system variable, the old value must first be displayed. Thus, to enter a new set point, the previous steps must first be executed. The sequence for entering a new value of set point follows.

4. Depress the ENTRY action key, which clears the process variable from the entry register.

5. Sequentially depress the six alphanumeric keys for the new value. The first key depressed must be either the + or − key, and one of the following five entries must be the decimal key. The remaining four entries must be numeric keys (0 to 9). As each alphanumeric key is depressed, the corresponding character is displayed in the entry register. At the completion of the sixth entry, the value of the desired set point appears in the entry register.

However, it is not yet entered into the point record. Notice that the old and new values of the set point are simultaneously displayed at this point. If the operator has made a mistake, at this point, he may abort by depressing the ENTRY action key and re-entering a new value.

6. When the operator is satisfied that the value displayed in the entry register is correct, he depresses the EXECUTE action key. This causes the new value to be entered into the point record and the new set point to overlay the old one in the data register.

The "display and enter new set point" operation described here is typical of the console operation. Some operations are not concerned with the value of the function but with its status. The operation of these functions is similar to that of the value function, except that an on-off status is displayed in the data register, and corresponding on-off entries are made in the work register to change status.

A feature for checking reasonability of entries is built into the software. Whenever a reasonability check reveals a violation, an error message is displayed in the console function window. Some of the available reasonability checks include:

> Calling for a point identification which does not exist.
> Entering octal information with the characters 8 and 9 in them.
> Entering a value which is beyond the allowable range. For example, attempting to enter a set point of $-100°F$ when the defined range is 0 to $500°F$.
> Key lock violation.
> Requesting a status or variable which does not exist on that point.
> Attempting to enter functions in an area not assigned to the requesting console.

Variations, Modifications

Most suppliers provide one console to be used by both the plant operator and by the instrument engineer for tuning the loops. Others keep the two functions separate so that the instrument engineer will not be in the way of the operator when tuning a loop. Consoles are usually table top mounted, as shown in Figure 8.13c, but they may also be rack mounted.

There has been considerable difference of opinion as to how many loops an operator can handle through an operator's console. Knowledgeable users have suggested anywhere from between 50 to 220 loops! It may be necessary to have multiple consoles, since there may be too many loops for one operator to handle or too large an area for one man to operate. A method of using multiple consoles is called the area concept. The plant may be divided into a number of distinct areas. One of the area codes is assigned to each point in the system. This area code does three things:

It is used on console entries to determine if the console requesting the entry is allowed to modify data in that location.

It is used to route messages to the proper typewriter for that area.

It is used to route messages to a predetermined backup typewriter in case the primary typewriter is unavailable.

This arrangement allows for segregating points among operators so that the actions of one operator cannot interfere with those of another operator. However, capability is included to allow any given area to be transferred from one console to another. This transfer capability is normally under key lock and cannot be accomplished without proper authority.

The area concept also allows for automatic routing of messages in case of a temporary disablement of a typewriter. For instance, if paper or ribbon is being changed on a typewriter, a "device disable message" is entered via the console specifying which typewriter is disabled. Any alarms or system messages which would normally be transmitted to that typewriter are now channeled to a preselected backup typewriter during the disablement.

There may be up to seven distinct areas in a system. Each of these areas is assigned to a console, an alarm and message typewriter, and backup typewriter for alarms and messages. The data can be modified only by that console which is assigned the area code corresponding to that location. In the case of point 2F101, for example, only the console assigned to area 2 can modify an alarm limit or other data related to variable F101. Alarm messages concerning point 2F101 are normally printed on the typewriter assigned to area 2. However, if this typewriter is disabled, the message is routed to an alternative typewriter which is the backup typewriter for that area.

ACKNOWLEDGMENT

[1] By permission of Honeywell, Inc., portions of this section have been prepared from the Honeywell *DDC System Operator's Guide:* Series 16 Technical Bulletin, "Process Interface Controller," and "Operator's Console" (each Honeywell, Inc., CCD, Copyright 1968).

8.14 HIERARCHICAL COMPUTERS

As computer systems increase in complexity, including supervisory and management functions as well as on-line process control, multi-layer structures offer a practical solution to satisfy the system requirements. Different computer systems, designed for widely different types of functions, can be interconnected in a hierarchical fashion to meet the overall needs more efficiently. Many higher-level management functions may well be performed by a large time-sharing system, interconnected to smaller dedicated computers for immediate process supervisory and direct control functions.

The hierarchical structuring of computer systems for supervisory and direct process control is discussed in this section. The logical development of this concept is reviewed. Various forms of system structure are examined. The parallel cascade processing system, available commercially, is summarized, and the pros and cons of hierarchical ddc systems are discussed.

One major consideration in ddc computer installations is reliability, and the hierarchical concept represents one method of failure protection by having a second computer available to take over the critical functions from a defective unit.

The Hierarchical Concept

The hierarchical concept for digital process control systems evolved from earlier experience with data acquisition systems, supervisory computer control, and direct digital control (ddc) systems. The major technical problems solved by data acquisition systems were the process-to-computer data links and supporting software required to convert process measurements into digital data. Methods were developed to handle measurement scaling, non-linearities, and other routine engineering unit conversions required to produce accurate data. However, it was soon learned that data is not information and that usable information required much more specific computation for a particular process.

Supervisory computer control systems were developed to use measurement information directly on-line to control the process. Computer-to-process data links were developed to adjust the set points of conventional analog controllers. This experience showed that the necessary process control models were difficult to derive, because considerably less was known about the process

1097

interrelationships than had been supposed. Programming systems for on-line operation created many complications, due primarily to their complexity and sometimes to the slow speeds of available computers. This necessitated the use of machine language programming to provide the speed required with reasonable hardware efficiency.

Direct digital control systems took an additional step in the total development of digital process control by incorporating the analog control task of directly manipulating the process valves. It was quickly learned that proper execution required the development of computer-to-man and man-to-computer links that provided rapid operator communication with the process. This led to the development of consoles serviced by high-speed interrupt programs to give adequate performance (see Section 8.13).

The ddc experience demonstrated that digital computers could easily replace the analog controllers with equivalent or better technical performance, but the control programming, relatively simple for these well-known algorithms, requires a high operating frequency to meet the needs of fast control loops.

The factor of control system reliability became very apparent, particularly during the development of ddc systems. This was solved by providing appropriate backup for the direct control functions and operator communication to operate the plant safely during computer outages. Reliability of hardware and software is of primary importance to enable the system to operate a process in real time. It often turns out that programming failures are much more frequent than hardware failures.

As shown on Figure 8.14a, requirements for digital process control systems tend toward a natural hierarchical structure. The tasks vary from high frequency with low complexity to low frequency with high complexity.

These functions might be implemented better by incorporating digital hardware and software components which best meet the task requirements in each area, into the complete process control system.

Flexibility is required at all levels of the hierarchy shown in Figure 8.14a. At the lower complexity levels it is needed to change the systems' inputs, direct control loops and outputs to meet changing plant conditions. In operator communications, flexibility of information presentation is required to best meet changing operating needs. Maximum flexibility at the supervisory and optimizing control level is needed to develop and incorporate new control models, advanced control schemes, methods of data reduction and reporting procedures. This will enhance the operation as more is learned about the process.

The Hierarchical Structure

The general objectives for a digital process control system are listed in Table 8.14b. Various hardware and software structures can be devised to meet these objectives. However, methods for efficient programming consistent with

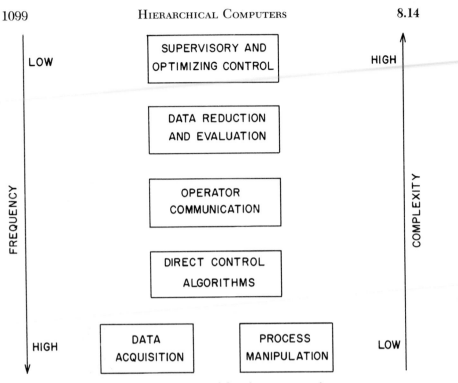

Fig. 8.14a   Functions of digital process control

system reliability must be included, because software frequently represents the greatest engineering effort in installing a system.

One structure that does achieve this is shown in Figure 8.14c. The supervisory computer uses higher-level languages to give maximum flexibility for on-line programming of the more complex functions. The second computer (the ddc machine) handles the high-frequency input-output tasks for all process and operator communication. Calculations in this computer are generally limited to engineering units' conversion and standard control algorithms. A bidirectional data link is provided between the computers so that the supervisory computer can obtain current process data for its calculations and can change the control action in the ddc computer.

Table 8.14b
GENERAL SYSTEM OBJECTIVES

1. Reliability of Hardware and Software
2. Flexibility of All System Levels
3. Efficient Operation for
      —High-Frequency, Low-Complexity Functions
      —Low-Frequency, High-Complexity Functions

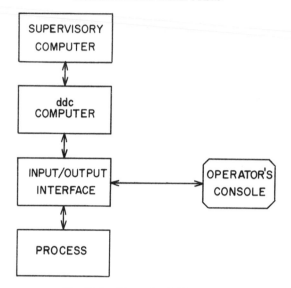

Fig. 8.14c   Hierarchical ddc system

The basic functions in the ddc computer are standardized and will be repeated from system to system. This permits the use of standard machine language programs for efficient operation of these high-frequency functions. Special software in the supervisory computer can be used to modify or add new control loops to give the flexibility required. This can be performed in a fashion that prevents mistakes in on-line programming of the ddc computer and gives maximum programming reliability.

Flexible custom programming is the rule rather than the exception in the supervisory computer, so a higher-level language such as FORTRAN should be provided. This would have to operate in the presence of real-time interrupts for both compiling and running programs. Some special machine language subroutines can be incorporated into the compiler to maximize computing efficiency while retaining many of the compiler's inherent programming advantages. As most mistakes occur in the application programs, features can be included in the supervisory software to check out new programs thoroughly before allowing commands to be sent to the ddc computer.

The availability of two separate but linked computers allows one computer to back up the other and maintain the more critical ddc loops. The supervisory computer can include duplicate software to take over these functions in an emergency, while suspending the supervisory programs (Figure 8.14d).

This form of digital backup requires the capability to switch I/O interface equipment to the second computer, and the maintenance of ddc programs and a current data base in the backup computer. If complete system backup

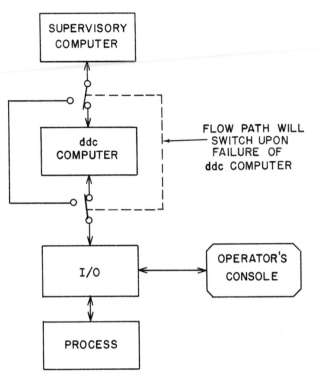

Fig. 8.14d   Backup for control computer

is desired, the common portion of the I/O equipment, such as the A/D converter, can be switched over on a failure (Figure 8.14e).

Many other backup structures can be arranged. In some cases, it is only necessary to back up the critical loops, in lieu of complete I/O redundancy. This can be accomplished by partial switching of the I/O channels as illustrated in the system in Figure 8.14e, or by a three-computer structure (Figure 8.14f). In the latter case, the ddc computers must have spare capacity so that all critical loops can be handled by either computer.

Necessary data may also be updated by sharing the memory between the computers, as shown in Figure 8.14g. This can be arranged so that either computer can address a common memory section. However, total system reliability is then dependent on the availability of the common memory unit.

Multiple computers, memory units, common I/O equipment and consoles can be linked in various fashions to provide complete redundancy, as illustrated in Figure 8.14h. Generally such systems are structured so that all multiple components can perform the same functions, with the computing load continuously divided among them. When one component fails, the total system functionally performs the same tasks, but degrades to a lower system loading.

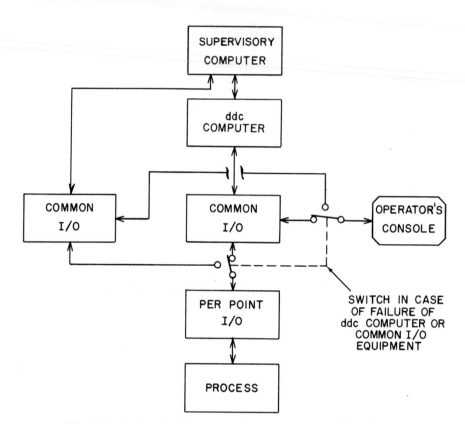

Fig. 8.14e   Backup for both the control computer and the I/O hardware

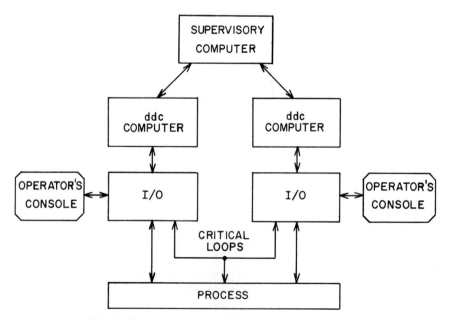

Fig. 8.14f   Critical loop backup utilizing three computers

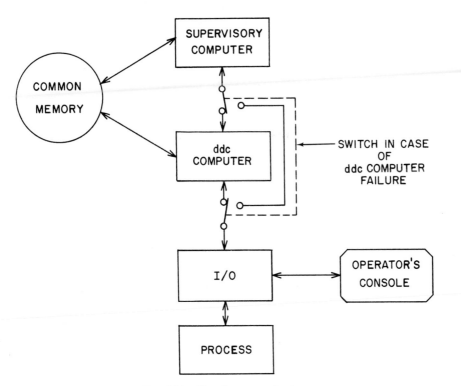

Fig. 8.14g   Shared memory structure

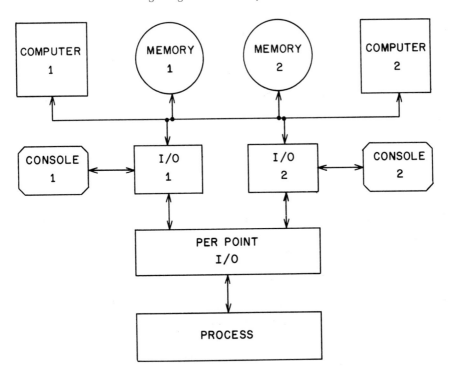

Fig. 8.14h   Multiple component structure

This approach does not allow separation of the different types of programming unless more computer units are incorporated into the system.

Parallel Cascade Processing

The parallel cascade processor is a commercial system, specifically designed for process control, which is based on a hierarchical concept. This system accomplishes *parallel* computing in that the ddc processor calculates the feedback control equations and operates the process input-output equipment while the supervisory processor calculates the correct values of the ddc set points. These set points are then *cascaded* into the correct location in the control processor memory in a fashion similar to analog cascade control.

A typical structure for this system is shown in Figure 8.14i. The structure is based on the use of two identical computers with very fast access auxiliary memory. Some of the system characteristics are listed in Table 8.14j. Complete input-output interface for supervisory and ddc control is available.

Very high availability of the control computations is assured because a completely updated copy of the control program is maintained in the supervisory bulk memory. In case of failure of the control processor, the input-output interface is automatically switched to the supervisory processor. This processor drops its supervisory functions and takes over the control functions within one second. When the control processor is again operational, the latest control program can be transferred back to the control processor before it again takes over control. This assures continuous direct process control with bumpless transfer despite a control computer failure. In addition, partial or complete control loop backup can be provided with a modified redundant I/O structure and/or with analog backup equipment, as was discussed in Section 8.13.

An important feature of this system structure is the use of fast access bulk memory (fixed-head disc) with minimum core memory. In the control processor, this allows a large process data base for high-frequency alarm and control operations without excess use of core memory. This data base is frequently updated by cycling sections of the bulk memory through the processor core memory on a regular basis.

The fast access bulk memory used in the supervisory computer allows the swapping of lower priority computation in an incomplete state when it is necessary to respond to other real-time requests. The computation is then recalled as often as necessary for completion in its order of priority. Various computations can be efficiently handled in this multi-programming environment regardless of execution time.

The major software packages provided for this system are listed in Table 8.14k. In the control processor, the core/drum package incorporates all the facilities for a real-time control system including ddc, set-point control, monitoring, alarming, logging, and communication with operating personnel.

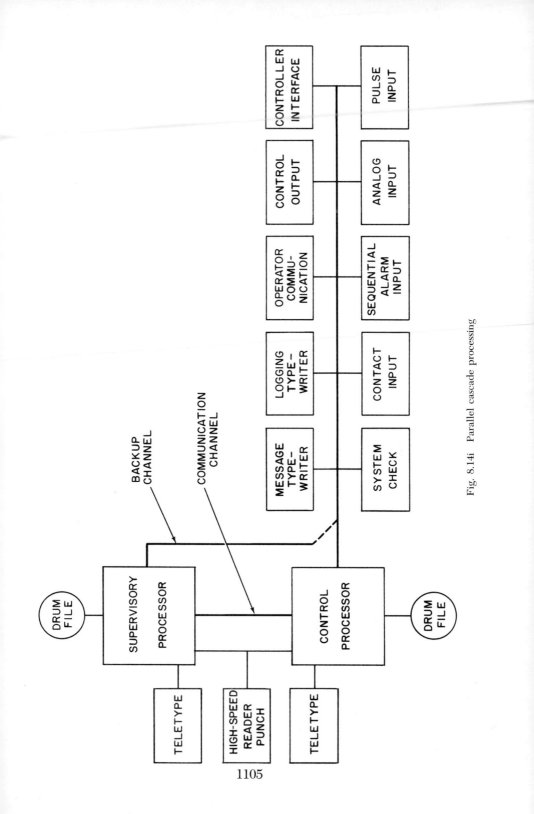

Fig. 8.14i Parallel cascade processing

Table 8.14j
CENTRAL PROCESSOR CHARACTERISTICS°

Organization	Fully parallel, single address, binary
Circuitry	Solid-state hybrid
Word Size	12 data bits (optional 13th parity bit)
Operating Times	1.6 μsec cycle time
	3.2 μsec add or subtract time
	21 μsec (max) multiply time
	37 μsec (max) divide time
Instructions	260 implemented
Main Memory	Ferrite core: maximum of 32,768 words in increments of 4,096, minimum of 12,278 words in the control processor and 16,374 words in the supervisory processor.
Auxiliary Memory	Magnetic drum file: 8.7 msec average access time, 92,000 words per second effective transfer rate, internal parity checking, lockout write-over protection; 204,800 words in the control processor, 409,600 words in the supervisory processor.
Addressing	Direct and indirect, auto-indexing in 8 core registers per each 4K core bank.
Data Channels	Programmed I/O bus, 250K words per second
	Direct memory access, 633K words per second, true cycle stealing
	Process data bus, 96K words per second
	Peripheral data bus, 113K words per second
Peripheral I/O Controllers	Drum file
	Teletype
	Paper tape reader
	Paper tape punch

° The features in this table are specific to the PCP-88 computer system of the Foxboro Co.

The batch process control software operates in conjunction with this package for simplified batch process programming and operation.

The supervisory operating system contains core and bulk memory portions to allocate the supervisory processor time efficiently among various requests for real-time programs and for time-shared background programs. A modified FORTRAN package provides on-line program development, compilation and testing in a protected environment. Various subroutines use real-time or dummy data for testing, provide access to data files, and link various programs together.

Table 8.14k
SOFTWARE CAPABILITIES

Supervisory Processor	Control Processor
Supervisory Operating System	Core/Drum Real-Time Direct
On-Line FORTRAN	Control Package
On-Line Assembler	Batch Process Control
Drum Librarian	
File Handler	
Loop Adder and Modifier	
Display	
Calculator	
Timer Setup	

A highly efficient symbolic assembly language is provided with various aids to simplify custom programming in assembly language. An on-line drum librarian is furnished to assume the clerical work involved in allocating bulk memory to library entries and in securing linkage between such entries when required.

Several conversational mode programs are provided for easy use by less experienced personnel. The loop adder and modifier provides on-line access to the data base in the control processor in the form of multiple choice questions from the computer. Highly restrictive safeguards prevent accidental changes to the data base.

Similarly, the display routine serves as a debugging tool and data logger by displaying information contained in bulk or core memory. The calculator program provides access to mathematical subroutines in the program library and allows process engineers to use the computer for immediate problem solving. The timer setup permits almost any program in the drum library to be automatically called and executed at a specified future time.

Conclusions

Hierarchical ddc systems have the principal advantages and disadvantages listed in Table 8.14l. The use of separate computers for supervisory and direct control functions with separate software systems permits simpler on-line program development. Operating programs can be protected by restricting the data transmitted between the computers to specific types and locations. More elaborate safeguards for program modification permit changes only in particular computers in a specified fashion. For example, the control programming can be modified in the supervisory computer and extensive checks applied before allowing the change to be communicated to the control computer.

The supervisory computer can automatically assume the tasks of the

Table 8.14l
HIERARCHICAL ddc SYSTEMS

Advantages	Disadvantages
Easier On-Line Program Development	Redundant Hardware and Software
Inherent Program Protection	Maintaining Two Identical Data Bases
Safe On-Line Program Modification	Additional Computer Linkage Channels
Digital Backup Of Control Functions	
Faster Process Control	
Higher Overall System Efficiency	

control computer upon failure with exactly the same form of control. Any degree of hardware and software redundancy can be provided to give significantly higher system availability over that of a single computer system.

The separation of the control and supervisory tasks allows faster control action as the control computer responds immediately to real-time process demands. Overall system efficiency is greater because both computers operate asynchronously and in parallel with each other to meet the real-time process demands as well as the background demands of operating personnel.

The major disadvantage is the need for redundant hardware and software which generally raises the system cost. This requires additional system checking and switchover equipment for automatic digital backup. Keeping redundant software updated requires the maintaining of identical data bases for each computer so either can assume the critical process functions. Additional computer linkage channels must also be provided for high-speed data transmission between the computers and the input-output channels to the process.

8.15 APPLICATION OF PROCESS COMPUTERS

Many factors affect the design of a modern process control system, as Figure 8.15a shows. The process is the dominant influence, because it defines the actions that must be taken for proper operation. The three major parts—measurements, control, and communication—are implemented in an equipment assemblage to form the overall process control system. This system is the intelligence network of the process and can vary from brilliant to completely idiotic in its operation. It must be designed so that the "proper" intelligence is available at the "proper" place in the process control system to meet the process needs.

This section reviews the major application factors to consider in defining the functions required in a control system for a process application. These factors, shown in Figure 8.15a, are individually discussed, starting with better methods of understanding the process, extending through methods of defining the control system and its various elements, and concluding with the operating and management considerations.

Table 8.15b contains useful definitions of control systems' terminology. The meaning of the word "system" is evasive. This definition emphasizes the material nature of process systems. It excludes the concepts of activities, sets of procedures or flow paths which others have found useful in different contexts. For a complete set of definitions relating to computer terminology refer to Section 8.1.

The concept of functions includes many actions, such as measurement, control, communication, evaluation, and management, needed to satisfy the operating requirements of a process system. For example, a particular operating function may be process startup coordination. This requires information from other parts of the system, evaluation and decisions based on the information, and subsequent communication of desired actions.

System Design

There are five major steps in determining the structure of a process control system to meet the application requirements of a specific process (Table 8.15c). Perhaps the most difficult step is the first, since the remainder of the design depends on the quality of problem definition. All process requirements must

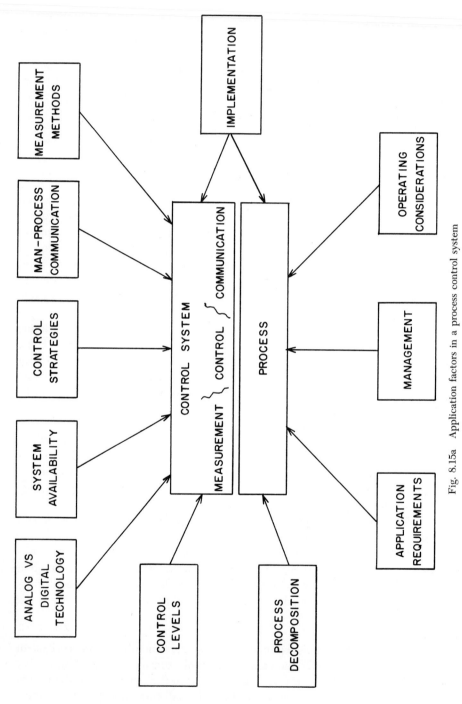

Fig. 8.15a Application factors in a process control system

Table 8.15b
WORKING DEFINITIONS OF LARGE INDUSTRIAL SYSTEMS

System: A collection of people and/or objects united by some form of interaction or interdependence. A system may consist of several subsystems and may, in turn, be part of a larger multi-level system.

Purpose: The reason for the establishment or evolution of a system. A system may have several purposes.

Goal or Objective: The standards of performance which are set to indicate how well the system is fulfilling its purpose.

Function: The acts or operations required to satisfy the goals of a particular system.

Control: In the broadest sense, obtaining system behavior in accordance with some predetermined plan to satisfy its goals. In the narrow sense, maintaining a system variable at a specific value or sequence of values.

Process Control System: An assemblage of equipment to provide measurement, communication and control functions required to operate a process.

Operating Strategy: The chosen independent manipulated variables, the dependent variables used to indicate the process status, and the control functions required in the process control system.

Multi-level System: A collection of subsystems in which the functions of one include setting the goals and monitoring and coordinating the functions of others in a hierarchical fashion.

Table 8.15c
PROCESS CONTROL SYSTEM DESIGN

Problem Definition
—Process Understanding
—Methods of Operation
—Process Disturbances
—Supporting Organization

Analysis of Process Economic Sensitivity

Solution of Technical Problems
—Measurement
—Control
—Communication

Cost Analysis
—Alternate Systems Implementation
—Process Benefits

Structure Process Control System

be understood, including methods of operation, possible process disturbances, and supporting organization of personnel and service.

Several techniques are helpful in understanding the process in an organized manner. The first is the use of material and energy balances as detective tools to determine major process flow and energy requirements. These can be only approximate balances, because their main purpose is better to understand the many process interactions. Sometimes, computer simulations can be used effectively.

Another helpful technique is to decompose the process vertically and horizontally into subsystems which can be defined and understood more easily than the total process complex. A useful approach is outlined in the following paragraph.

With a good understanding of the process operation and of major disturbances, it is much easier to determine the relative economic sensitivity of the various subsystems. This is a vital step in planning the detailed project studies, as both the technical problems and the economics need to be identified. Solving interesting technical problems without economic significance is a common cause of poor project results.

The technical problem areas in measurement, control, and communication must be studied in relation to process design and operation, and practical solutions proposed. These detailed studies usually involve computer simulation studies, plant studies in existing or similar plants, and functional engineering studies. It is important to plan the complete study program before individual studies are started, so that the best techniques will be used to solve each specific problem.

The next step is a cost analysis of alternate system solutions and their process design or operating benefits. Today, control system structures can be designed with analog and/or digital techniques to accomplish almost any system objective. The real problem is the determination of what is worth doing and what makes practical economic sense.

Finally, based on all this information, the process control system can be structured to accomplish the system objectives in a practical manner. Assuming good economic incentives, the project can then start into the actual implementation, as discussed in Section 8.3. However, it is essential to treat the process system as a dynamic entity with change as a way of extending its life, rather than as a static unit that will remain as it is at the beginning of the project or at startup. Improvements learned in operating the process and changing operating requirements will continually require the updating of the control schemes.

Decomposing the Process

To help determine a useful vertical structure of a process, Table 8.15d suggests hierarchy names for industrial multi-level systems. A system structure should use a minimum number of hierarchy names to define the required functions. Most chemical processes, for example, would probably not require section and subsection levels.

A useful horizontal breakdown can be based on the concept of subsystems, as defined in Table 8.15e. There should be minimum interaction between these

Table 8.15d
HIERARCHY LEVEL NAMES FOR
INDUSTRIAL MULTI-LEVEL SYSTEMS

1. Corporation	5. Building	9. Systems
2. Company	6. Process	10. Subsystem
3. Division	7. Section	11. Unit
4. Plant	8. Subsection	12. Component

Table 8.15e
SUBSYSTEMS

Operational Subsystems—formed by units that are closely linked by operating conditions and common purposes.

Sequential Subsystems—formed by units that have a similar purpose and function and are sequentially connected.

Functional Subsystems—formed by units that have similar purposes and functions but are not connected in the process.

Stand-Alone Subsystems—units or components that are totally independent of all others.

subsystems on a given level while maintaining the multi-level hierarchical concept.

The process operating structure is developed by an iterative procedure as outlined in Table 8.15f. Carefully defined purposes, goals, and broad functional requirements are basic to a system structure. The key to a good analysis is the willingness to list purposes and functions of all sections of the system even if they seem obvious.

Table 8.15f
STEPS IN ORGANIZING A MULTI-LEVEL PROCESS SYSTEM

Study the complete process.
Develop a hierarchy of process system purposes.
Develop corresponding operating goals and functions.
List process components.
Combine components into units.
Combine units into subsystems.
Formalize subsystem purposes, goals, and functions.
Analyze subsystem operation.
Develop subsystem operating strategy.
Iterate previous steps.
Develop system level operating strategy.
Iterate previous steps.
Finish development of operating strategies for higher levels.
Continued reevaluation of operating strategies and systems structure in light of changing process performance, goals, or components.

Avoid using simplified process diagrams in process analysis, because many pertinent details are usually omitted. The function to be performed and its implementation must be defined separately. Frequently, one man carries out functions at several different levels. The functions could then be associated with the office rather than the correct position in the hierarchical structure.

The specification of operating functions, rather than control functions, provides the means for comparing analog, digital, and human implementation in terms of performance and reliability requirements. The depth of technical investigation in specific areas can be related to the potential return without fear of upsetting the whole control system structure. The resulting control system will probably be simpler and will have a higher probability of good performance.

This method of decomposing a process forces systematic organization and insures that the implementation of various functions will be compatible. This gives a much clearer process understanding. The operating functions are well documented and can become a significant portion of the process operating manual. Changing operation requirements can be met much more easily because the specific function which must be changed can be pinpointed.

An example of a system structure based on purpose and function is shown in Figure 8.15g. This was put together following the steps outlined in Table 8.15f. The figure can show only a simple descriptive name for each block which is actually characterized by detailed lists of purposes and functions.

The building level purpose is to make several different kinds of chemicals. Its functions include: production scheduling, setting production rate and product distribution, startup and shutdown decisions, process and service coordination, economic optimization and management functions. The process or system level makes products C, D, and E. The operating functions at this level include the monitoring of the process operation, process equipment, subsystem coordination, maintaining product quality, and setting feed rates to meet production goals and operating conditions.

There are seven subsystems in the illustrated process. Their purposes include such tasks as completing reaction of A with B and C to make C, D, and E; or providing aqueous B to the reactor and removing unreacted B from the process stream. Some required subsystem operating functions are maintaining B feed flow, maintaining ratio A/B, maintaining reactor temperature distribution, supplying aqueous B at desired concentration and coordinating units. These are exactly the functions required to operate the process. They permit the direct definition of the operating strategy and of the necessary control system.

All the different types of subsystems are present in the structure. The operational subsystems are the reactor, absorber and evaporator. The evaporator could be considered partly sequential in that a dehydrator column follows the more closely connected evaporator effects. The distillation columns form

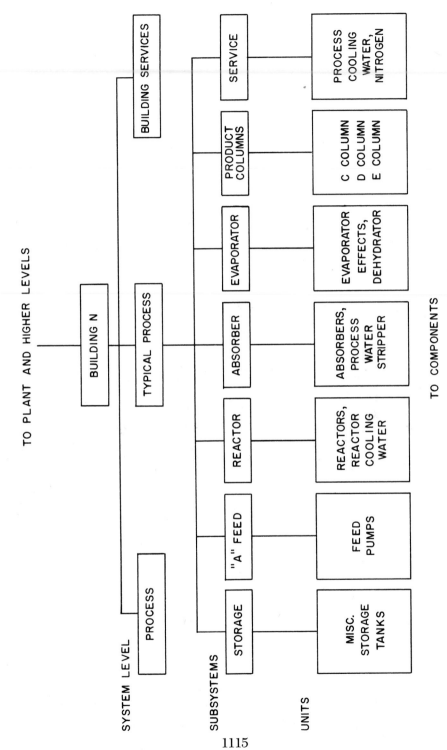

Fig. 8.15g Process decomposition based on purpose and function

a true sequential subsystem, while the service and storage units form functional subsystems. Finally, the feed subsystem is of the stand-alone type and consists of only pumps and high-pressure protective devices. In order to keep the diagram reasonably clear, the component level has not been shown, but there are 99 components in this system.

Process Control System Levels

In most cases, the types of functions and their different uses in fulfilling plant operating requirements tend to lead to a natural vertical structure for a control system, as shown in Figure 8.15h. The process operating functions are shown on the right side of the figure.

Emergency plant operation is provided in level 0. True process safety systems must be considered separate and independent of any other control functions: i.e., the high temperature or pressure shutdown of a reactor. Under

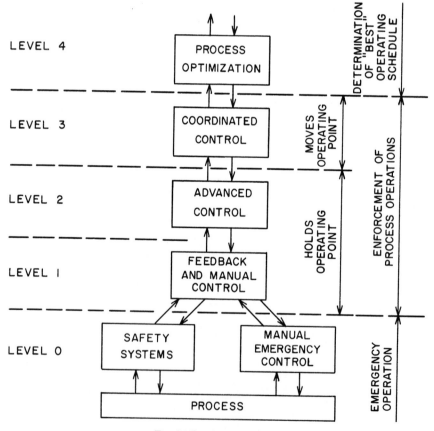

Fig. 8.15h   Control levels

control system failure, emergency manual control provides the means of manually manipulating critical valves to move the process to a safe operating point, or to shutdown. With proper system design, this mode should be rarely used and could be separate from the normal process-operating console.

Process operations enforcement is provided by levels 1 through 3. Levels 1 and 2 hold a given operating point while level 3 is used to move the plant from one operating point to another.

The first level of control contains the conventional feedback control loops, such as flow, temperature, pressure and level, that are normally employed today. They can include a wide variety of specific functions, such as cascade control or ratio control, but all are concerned with enforcing specific operating set points that were determined from some other source. This level also includes the normal manual function required to manipulate valves directly during startup, shutdowns, and maintenance.

Advanced control, level 2, includes such techniques as simple feed-forward-feedback control schemes based on material or heat balance that compensate for process disturbances which can be measured in advance. Multi-variable control schemes that eliminate interaction of one control loop with another are also included here. Unconventional feedback schemes employing *nonlinear, logic,* or *sampled* data concepts also fall into this level.

The third level, coordinated control, includes functions concerned with changing the operating level of the plant. This level contains two distinct classes of control. The first are functions, such as sequencing control and automatic startup or shutdown, that change operating points in a *predetermined* fashion. Plant performance can be evaluated under predetermined circumstances and conditions can be automatically changed to suit the operating rate. The plant can be shut down if safety is impaired. The second class includes schemes that require *calculation,* such as bang-bang or adaptive control. This level includes automatic compensation for shifting conditions such as aging catalyst or methods for improving the efficiency of batch operations.

Process optimization, level 4, is used to determine the best operating schedule for the process. This may involve direct optimizing schemes, calculations based on economic scheduling, and linear or dynamic programming. This level also includes the development of process models for other control schemes. Even higher levels of control can be added to this hierarchy to tie into divisional and company level management systems.

Generally, as control levels rise, the complexity increases, but the frequency at which the task has to be carried out decreases. This is a nonlinear effect, as shown in Figure 8.15i, with the complexity directly related to memory requirements and engineering effort, increasing rapidly with the control levels. On the other hand, cumulative computer loading rises rapidly with the high frequency of operation at the lower control levels. The rise

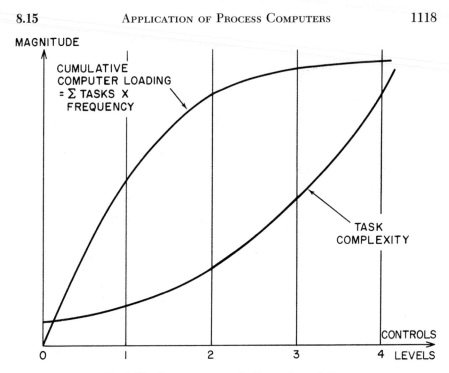

Fig. 8.15i   Process computer loading and complexity

is much slower at higher, more complex control levels, because the frequency of operation is significantly less.

The control system levels can be expanded to include measurement and man-process communication functions that may be required in a complete process control system (Figure 8.15j). The obvious interactions between measurement, control, and communication are not shown individually. It is assumed that each higher-level function has access to all information in lower-level functions.

The emergency environmental measurements at level 0 are those required to move the plant to safe operating condition, or to shutdown. Normally, this should be only a small percentage of the total plant measurements. The first-level environmental, inferential, and simple composition measurements are those required for use directly in control and process operating communications.

Level 2 includes indirect measurements determined by calculation from level 1 measurements such as temperature or pressure-compensated gas flows. These are used for supervisory, accounting, communication, and some control functions. More complex composition measurements such as infrared, mass, and/or X-ray spectrometry are included in level 3. These are seldom, if ever, used to manipulate the process directly. The level 4 laboratory quality meas-

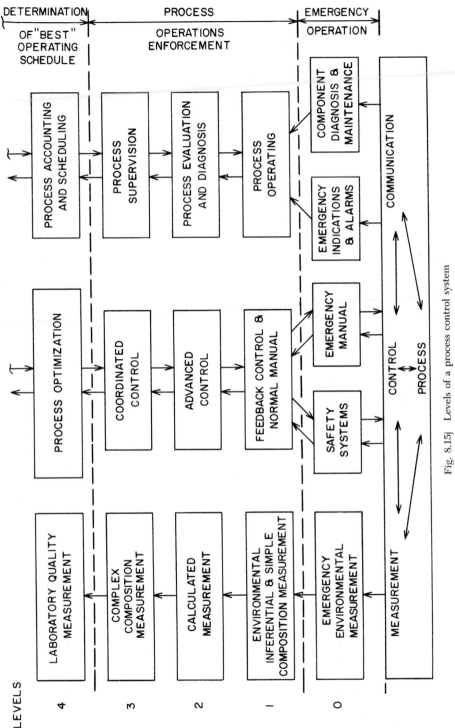

Fig. 8.15j   Levels of a process control system

1119

urements are seldom used directly in control functions, but are involved in process optimizing or higher-level communication functions.

The man-process communication functions also form a level structure in the process control system. The zero level emergency indication and alarms are needed in moving the process to a safe operating point, or to shutdown. Information required to diagnose component failures and to make other system maintenance checks is also required at this level.

The first level communication function includes all the information needed to keep the process operating in a safe condition. Process evaluation and diagnosis, level 2, includes communication to determine how well the process is operating, to ascertain potential problems, and to rapidly diagnose process failures when they occur. The communication for process supervision at the third level includes current schedules, feedstock availability, utilities, and best operating policies required to make day-by-day adjustments of operating conditions. Level 4 communicates all information on quantities of production, feedstock, supplies, shipping, labor, etc., process accounting and determination of production schedules. The results of off-line or on-line process optimizing calculations can also be used at this level.

By breaking up the process control system as in Figure 8.15j, one may show the various desirable functions in more detail. It is not implied that all of these functions can or should be handled automatically, although in some fashion, all must be carried out in industrial processing plants.

Analog vs Digital

The possibility of using digital techniques opened a completely new dimension of achievement in process control systems. It is claimed that anything can be done, but the real problem is to take practical advantage of digital techniques, possibly in combination with analog techniques, to arrive at the best overall process control system.

Basic technological advantages for analog and digital techniques are listed in Table 8.15k. Estimates of the level where each advantage would be useful for the functions in Figure 8.15j are given. The method of implementation (i.e., general purpose or special purpose digital computers, or digital logic elements) is not meant to be part of this table and depends on many other factors.

It is apparent from this table that digital advantages outweigh analog techniques in most functions above level 1. In level 1, they have advantages in the communication function. Most tasks can be performed either way in level 1 control functions, but combined digital-analog methods will probably provide the best practical solutions.

Analog techniques will probably have advantages for level 0 emergency functions for some time, because these are usually implemented in small independent units for each safety system.

Table 8.15k

ADVANTAGES OF ANALOG AND DIGITAL TECHNIQUES

Advantages	Comments	Levels Where Advantage Is Useful		
		Measurement	Control	Communication
DIGITAL TECHNIQUES				
Memory	Necessary in higher-level control functions; useful at most levels for improved communication functions	2 to 4	2 to 4	1 to 4
Complex Calculations	Almost any desired accuracy feasible with drift-free calibration	2 to 4	2 to 4	2 to 4
Logic Operations	Easier to implement sequencing, startup, alarming and diagnostic functions	3 & 4	0 to 4	0 to 4
Flexibility	System may be evolved as plant design and operations are improved	2 to 4	1 to 4	1 to 4
ANALOG TECHNIQUES				
Speed	Frequency of digital sampling and calculation practically limited to about once per second	0	0 & 1	0
Differential Equations	Can be solved directly with analog methods, where digital approximations require tradeoffs of time against accuracy.	3 & 4	0 to 4	2
Simplicity and Cost	Independent lower level functions implemented in a simpler manner at lower cost.	0 & 1	0 & 1	0

System Availability

System availability is the total time in a given period that all functions are operational, in contrast to system reliability, which is given only in terms of frequency of component failure. It is possible to have a system with lower reliability, but with a higher system availability for its more important functions.

Total system availability is a function of both frequency of failure and time to repair. If the frequency of failure is often enough, and the time to repair short enough, then the only requirement may be to shut down the process in a safe manner (Figure 8.151). This backup for process shutdown must be considered whenever a computer is used in direct process control. This may be completely manual, but it should not be confused with the manual operation requirements of the system when it is normally running. The needed backup is usually simpler and may be handled more effectively in another way.

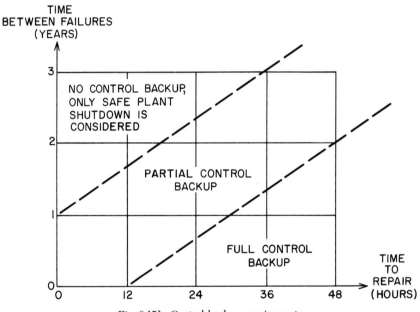

Fig. 8.151 Control backup requirements

However, in many cases partial control backup is desired to keep the process running, even if at reduced efficiency, until repairs can be made. This normally requires a mixture of backup control and manual operations. In most cases the backup control can be relatively simple and usually amounts to about 10–20 percent of the total direct control loops in the system. Full backup control is not required with present control computer systems unless the cost of process downtime is high.

These backup requirements can be provided by separate analog units or by a second digital computer, where several computers are combined in a system (see Section 8.14). The need for manual or automatic switchover to backup control is a function of the response time of a particular loop and the safety of the process. Normally, only a few control loops in a system require automatic switchover. Manual switchover is preferable with analog backup in order to further reduce the system complexity.

Obviously, maximum reliability of the system functions is desirable to all concerned, but minimum acceptable reliability gives a better base from which to design a system. In Table 8.15m an estimate of this factor is shown for each function in Figure 8.15j.

Table 8.15m
MINIMUM ACCEPTABLE RELIABILITY
OF PLANT FUNCTIONS IN UNITS OF
MONTHS BETWEEN FAILURES

Levels	Measurement	Control	Communication
4	1–3	1–3	3–6
3	3–6	6–9	6–9
2	9–12	9–12	6–9
1	12–18	12–18	12–18
0°	24–48	24–48	12–18

° Level 0 is in terms of per point failure, all higher levels are in terms of a system failure.

This table is based on the assumption that a few hours are required to get the proper serviceman to the site and to repair the system. If time to repair a given function could be cut to a few minutes, perhaps by a simple substitution of a complete functional unit by operating personnel, then these minimum acceptable times could be reduced.

Control Strategies

There are two basic methods of developing the control strategies of a process control system to meet the operating objectives. One is to develop some overall "optimizing strategy" that attempts to encompass all of the control problems in one general scheme. This is generally the view observed by looking "down" on the whole process from above and considering it as one problem complex. The second method is to begin with the process and look "up" at the various improvements, starting with the process subsystem designs, and then applying the various levels of control in increasing complexity until the process objectives are met. It is this latter approach that is the most practical to use in developing control strategies.

This method of breaking up the development of the control strategy into various levels (Figure 8.15h) allows a proper distribution of the control tasks and can reduce the control system complexity. Each added level of control can be evaluated on its merits of meeting the process objectives. Each problem is considered at its level and solved in the simplest possible manner. It does no good to devise complex control schemes if the methods used to manipulate the process are inadequate. On the other hand, if the various process disturbances are eliminated by the lower levels of control, the higher levels can often be simplified, and they may only have to change commands to the lower level control schemes for their implementation.

It is essential to permit flexibility and to expect that the process will be improved not only after startup but possibly throughout its working life. As more experience is gained and operating requirements change, control schemes must be updated. This is particularly true for the higher levels of control, but it is also a factor at lower levels where control loops may have to be added or modified.

Control enforcement is essential at the first three levels. This function assures that all control schemes are operational and properly tuned for the particular operating conditions. Digital computers are very advantageous for this function as they can store, and log, if desired, all changes of control parameters. It is also possible to repeat the exact parameters used previously for a particular type of plant operation, and to restrict control tuning, alarm limit setting, etc., to the plant engineering or instrument personnel. This assures that all control loops are used properly and that difficulties are noted. This function of control enforcement can have high economic significance. This is impossible with analog control, and it is common to find plants with 25–50 percent of the control loops detuned or on manual control.

Process models must be developed for some of the higher control levels. However, there should be careful distinction between the process control and process design models. Control models are usually much simpler except when they are used for dynamic optimizing. The model must either be solved directly or by an iterative method to manipulate the process. Models developed from static mass or energy balances, with a separate dynamic correction term, are often very effective. These models are usually used with some type of on-line correction for frequent updating.

Man-Process Communication

System communication requirements must be designed to be effective not only during normal operations, but also during the other conditions shown in Figure 8.15n. The system assembly and checkout performed by the supplier would require those listed to communicate with the system. Commissioning and installation requires both supplier and user communication. Customer programmer, maintenance and plant operating personnel must be involved.

During normal operation only the plant supervisor and operators should have to communicate directly with the process. First-level maintenance and

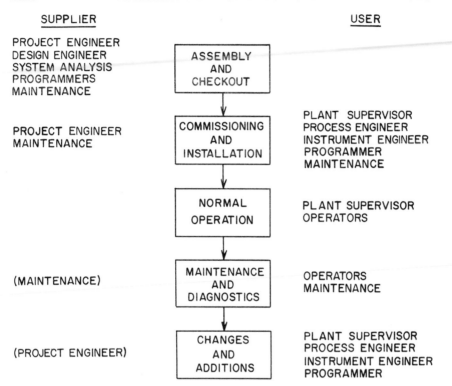

Fig. 8.15n   Typical communication requirements involving supplier and user.

system diagnostics should be performed by customer personnel with backup from the supplier. The user should handle application changes in the system, with advice from the supplier as required.

Consideration must be given to the abilities of all people communicating with the system during these periods. Normal plant operating tasks should be implemented as simply as possible. Tasks performed rarely, and then only by technical people, can be handled in a more complicated manner.

One important concept possible with computer systems is *operation by exception.* This method of man-process communication alerts operating personnel when measurements or calculations go outside predetermined limits. Yellow deviation alarms, for example, warn when measurements deviate from their set points, and red absolute alarms indicate serious process problems.

This concept also permits the use of shared devices for information presentation, as the monitoring function is taken care of automatically. This results in significantly smaller operating consoles with more convenient access by the operator. As plants become larger and more complex, these factors are not only desirable, but necessary to put large volumes of data into meaningful context for proper interaction between the process and operating personnel.

Audible and/or visual communication is an important operating factor in some processes and must be considered in the design and location of the communication consoles. Plant radio or telephone systems and closed-circuit TV have been used successfully to meet particular operating problems.

Communication consoles normally involved in these systems are listed in Table 8.15o. Various communication devices may be considered for specific console requirements. These may include a graphic panel for point indication and alarm, analog and/or digital indication, cathode ray tube displays, selectable strip chart or X-Y recorders, printers, typewriters, and tape reader or punch. Reliability of some of these devices in industrial environments may pose serious restrictions.

Table 8.15o
NORMAL COMMUNICATION CONSOLE FUNCTIONS
FOR PROCESS CONTROL SYSTEMS

Operations Console	Engineers Console	Programmers Console
Measurements and Set Points	Loop Tuning	Initial Program Setup
Valve Position	Alarm Limits	New Program Development
Alarm Indication	Changing Control Loops and Alarms	System Changes or Additions
Data Trending	Maintenance Functions	Program Debugging
Manipulation of Set Points, Valves, etc.		
Special Data Entries		

It is advisable to use standard modules as much as possible in any console design, as these normally involve considerable supplier software support, and changes could involve extensive custom programming.

Consideration must be given to communication with other digital systems for higher-level control tasks or for accounting information and order requests of various types. In most cases, this should not be raw data, but "information" resulting from some predigestion by the process computer. The concepts of reporting only exception data are important here.

Measurement Methods

It is often said that before you can control something, you have to measure it. With modern digital techniques, this should be extended to: measure *or calculate it*. Although it is usually easier to measure something directly for control purposes, in some instances, calculation has proven more reliable and less expensive than a complex analytical device. In other control instances, it is possible to use approximate calculations which are periodically updated by off-line lab analysis.

Measurements should be placed where they are most sensitive to process variations so they will give maximum control leverage. The on-line location, and in some cases, even the type of measurement, is usually difficult to determine in a complex process because of variations which make comparative testing difficult. However, in many cases, this process sensitivity can be determined experimentally by process simulations or estimated by calculation.

Measurement enforcement can be simpler in a digital system, since irregularities can be checked against normal high and low limits to warn of problems. It is possible to cross check results of some calculations, such as material balances, to detect important measurement variations. Even though the measurement problem may not be pinpointed, it is usually possible to suspend the use of the measurement to prevent operating problems which could be hazardous or costly. This is done by using the latest acceptable measurement values in a calculation or control scheme until the live measurement is again operative.

Digital techniques make practical the use of various types of statistical techniques to enhance the data collected from difficult measurements. For example, repeated measurement samples can be statistically combined to determine the true information content. However, some weighing factor for the value of past information, which for control purposes acts like a measurement lag, must usually be applied. This can produce poorer control in faster loops.

Many process measurements are inferential rather than absolute, and their calibration can be affected by other measurable factors. Where this is true, it is possible to compensate these measurements to give a higher accuracy over the process operating range. In other cases, it is possible to include on-line measurement calibration relative to other process measurements or to standards introduced periodically.

Implementation

When implementing a process control system, it is essential to establish a consistent design and operating philosophy so the many separate tasks will be properly coordinated. This will assure a system which uniformly executes all functions for easier process operation.

Before hardware design starts, all system functions should be defined, including the extent of flexibility needed to make changes and additions after installation. It is important not to become enamored with all the complex tasks that can be done with a computer, but rather to determine the simplest method that will meet the process objectives adequately. The first imperative is to start the system smoothly on worthwhile tasks. Operating personnel will gain confidence and encouragement will be generated to add more complex functions if needed.

All the various system tasks can be classified to determine a more realistic equipment selection, as shown in Figure 8.15p. Factors to be considered, in addition to duty cycle, memory requirements, and randomness, include the ratio of input/output operations to internal calculations, the amount and location of core memory, and the type and complexity of programming.

Higher-level tasks such as plant optimization can involve very complex calculations. However, these calculations are normally required at a relatively low frequency (i.e., hours or days) compared with the lower-level control functions. Because of this difference in complexity and frequency of calculation, the best computer system for this function would probably be somewhat different from one used for lower-level control and communication functions. Depending on the particular situation, these higher-level tasks might be more economically performed off-line, or shared on a computer doing these functions for several plants.

After the project functions have been defined, it is possible to start specifying the measurements needed, the methods of process manipulation, and the computing and communication equipment system. The selection and location of the proper type of process measurement transducers must be developed in conjunction with the process design. Necessary analytical instrument installations should be carefully engineered to assure proper sampling and range of operations. Frequency of manual insertion of all off-line information must be specified.

The necessary algorithms for each control loop must be designated at all levels of control, along with their input information requirements and the dispatching of their outputs. Also, the calculations needed for other plant uses must be designated and data inputs and outputs specified. Examples of these algorithms are shown in Figure 8.15q.

After the control and calculation requirements for the system are determined, the design of the communication system with the various consoles and readout devices can be finalized. The operator's consoles may require some special engineering to give the most effective operator interface, but the balance of the system usually can use standardized units.

When the entire project is reviewed at this point, it may have to be modified somewhat to match that which can be practically accomplished with available equipment. However, this approach of starting with a thorough understanding of the system functions before specifying equipment gives a better process control system than trying to make the process fit a particular equipment configuration.

In comparing various computers for a particular process control system, it is essential to evaluate *both* the hardware and software for the *planned*

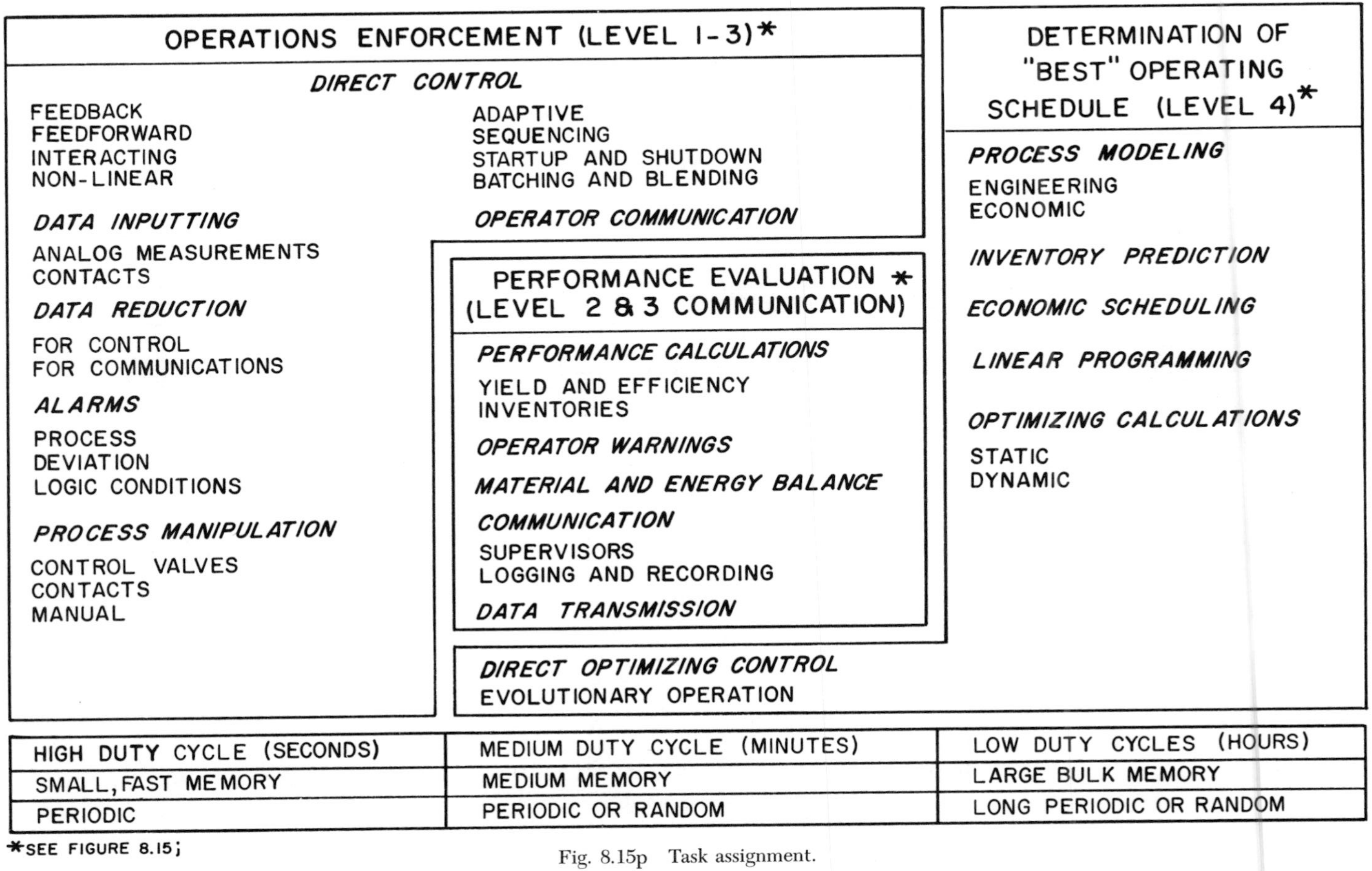

OPERATIONS ENFORCEMENT (LEVEL 1-3)*

DIRECT CONTROL

ADAPTIVE
SEQUENCING
STARTUP AND SHUTDOWN
BATCHING AND BLENDING

OPERATOR COMMUNICATION

DATA INPUTTING

ANALOG MEASUREMENTS
CONTACTS

DATA REDUCTION

FOR CONTROL
FOR COMMUNICATIONS

ALARMS

PROCESS
DEVIATION
LOGIC CONDITIONS

PROCESS MANIPULATION

CONTROL VALVES
CONTACTS
MANUAL

FEEDBACK
FEEDFORWARD
INTERACTING
NON-LINEAR

PERFORMANCE EVALUATION *
(LEVEL 2 & 3 COMMUNICATION)

PERFORMANCE CALCULATIONS

YIELD AND EFFICIENCY
INVENTORIES

OPERATOR WARNINGS

MATERIAL AND ENERGY BALANCE

COMMUNICATION

SUPERVISORS
LOGGING AND RECORDING

DATA TRANSMISSION

DIRECT OPTIMIZING CONTROL
EVOLUTIONARY OPERATION

DETERMINATION OF "BEST" OPERATING SCHEDULE (LEVEL 4)*

PROCESS MODELING

ENGINEERING
ECONOMIC

INVENTORY PREDICTION

ECONOMIC SCHEDULING

LINEAR PROGRAMMING

OPTIMIZING CALCULATIONS

STATIC
DYNAMIC

HIGH DUTY CYCLE (SECONDS)	MEDIUM DUTY CYCLE (MINUTES)	LOW DUTY CYCLES (HOURS)
SMALL, FAST MEMORY	MEDIUM MEMORY	LARGE BULK MEMORY
PERIODIC	PERIODIC OR RANDOM	LONG PERIODIC OR RANDOM

*SEE FIGURE 8.15j

Fig. 8.15p   Task assignment.

1129

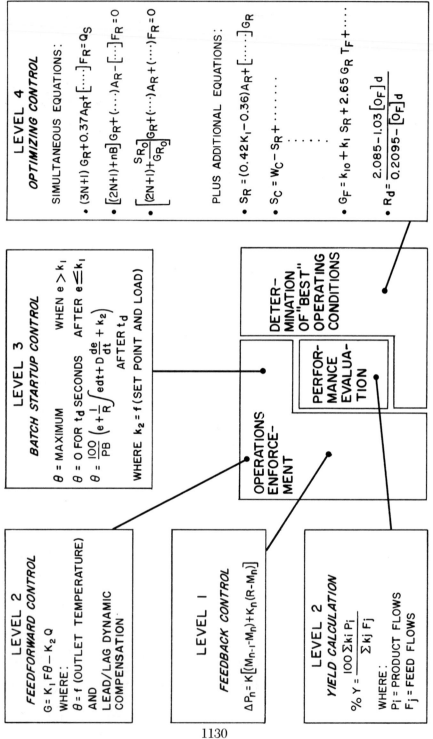

LEVEL 4
OPTIMIZING CONTROL

SIMULTANEOUS EQUATIONS:

- $(3N+1)\,G_R + 0.37\,A_R + [\dots]\,F_R = Q_S$

- $[(2N+1)+nB]\,G_R + (\dots)\,A_R - [\dots]\,F_R = 0$

- $\left[(2N+1) + \dfrac{S_{R_0}}{G_{R_0}}\right]\,G_R + (\dots)\,A_R + (\dots)\,F_R = 0$

PLUS ADDITIONAL EQUATIONS:

- $S_R = (0.42\,K_1 - 0.36)\,A_R + [\dots]\,G_R$

- $S_C = W_C - S_R + \dots$

- $G_F = k_{10} + k_1\,S_R + 2.65\,G_R\,T_F + \dots$

- $R_d = \dfrac{2.085 - 1.03\,[O_F]\,d}{0.2095 - [O_F]\,d}$

LEVEL 3
BATCH STARTUP CONTROL

θ = MAXIMUM WHEN $e > k_1$

θ = 0 FOR t_d SECONDS AFTER $e \le k_1$

$\theta = \dfrac{100}{PB}\left(e + \dfrac{1}{R}\displaystyle\int e\,dt + D\dfrac{de}{dt} + k_2\right)$ AFTER t_d

WHERE $k_2 = f$ (SET POINT AND LOAD)

DETER-MINATION OF "BEST" OPERATING CONDITIONS

PERFOR-MANCE EVALUA-TION

OPERATIONS ENFORCE-MENT

LEVEL 2
FEEDFORWARD CONTROL

$G = K_1\,F\theta - K_2\,Q$

WHERE:

$\theta = f$ (OUTLET TEMPERATURE)

AND

LEAD/LAG DYNAMIC COMPENSATION

LEVEL I
FEEDBACK CONTROL

$\Delta P_n = K\left[(M_{n-1} - M_n) + K_n(R - M_n)\right]$

LEVEL 2
YIELD CALCULATION

$\%\,Y = \dfrac{100\,\Sigma k_i\,P_i}{\Sigma k_j\,F_j}$

WHERE:

P_i = PRODUCT FLOWS

F_j = FEED FLOWS

Fig. 8.15q Control algorithm

1130

tasks thoroughly. Frequently, the hardware is completely specified with almost no consideration being given to the related software. In other cases, the evaluation is based on benchmark programs unrelated to the functions planned for the system.

If at all possible, standard hardware and software should be used to reduce expensive custom engineering and programming. Realistic environmental and reliability needs should be specified to obtain the best match of computer system to application.

The supplier should carefully check out the complete system before shipment to the plant site, where only limited test facilities are usually available.

Operation

Operations enforcement is one of the most beneficial factors of a well-designed process control system. Usually, all variations from the specified plant operation are stored and read out to operating and supervisory personnel. This insures that all process equipment and controls are either operating properly at desired conditions or that supervision is alerted to problems. A review of operating experience of the preceding shift is very useful to operating personnel.

The improved regulation offered by higher-level control techniques normally provides significant operating benefits. However, in many cases this is even further enhanced by an ability to change process operating conditions in a coordinated way. This frequently shortens the time to make changes, provides for safer operation, and, in some cases, permits the process to be moved to operating conditions which are difficult or impossible to achieve manually. This is usually caused by process interactions which require tight timing of changes to prevent instabilities or the exceeding of safety limits.

Training the personnel using the system is always a major factor in a successful installation. The experience of the operators has to be considered, as they often do not fully understand the elaborate control schemes. It is necessary to determine the exact tasks that supervision desires to delegate to the operators. In many situations it may be better to have the operators request assistance from their supervisors if it becomes necessary to change the method of operation. However, it is essential to provide the operators with a simple method to assume control of the process in emergency situations until needed assistance can be obtained. Methods that enhance the operator's abilities should be included. Warnings and alarms that alert the operator to potential process difficulties will allow an expansion of his job function. An operator can then operate larger plant segments or take over more functions in a given size operation.

Training of maintenance and programming personnel is essential. This is best done during system assembly with these personnel taking an active part in the factory checkout and plant startup.

With better software systems, the instrument engineer can easily learn enough programming to make all changes and additions which are required after installation. This ability is very important in upgrading the process operations as conditions change or as more is learned about the plant. As most processes are continually improved over their life-span, this factor should not be overlooked in a system specification.

Management

The functions and scope of an individual process control system must be related to the plant management organization. For example, as shown in Figure 8.15r, a given management level has the primary responsibility for

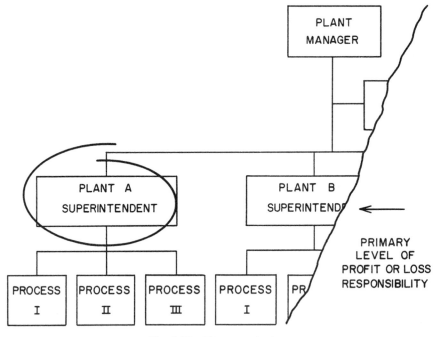

Fig. 8.15r   Plant organization

profit and loss. This level needs complete information and control of its operations in order properly to execute the assigned responsibility. If more computers are combined into a larger system, their organization should follow the plant organization and supervisory staffing to keep the system manageable.

There has been much discussion about controlling processes with a remote time-sharing computer. This is not suitable for the lower level control and for communication functions which are essential to process safety and basic regulatory functions. As shown in Figures 8.15i and 8.15p, these lower levels require a high frequency of operation with relatively simple calculations done in a small committed computer. Depending on their complexity, time-sharing higher level functions may be worthwhile. However, a local checking method and "switch" for the plant personnel must be provided to override command signals from the remote computer system. Otherwise, any misoperation or failure in the remote system could potentially cause serious process operating problems.

To achieve maximum advantage, a modern process control system should be structured as an integral part of the process design. With traditional methods, the process concept, design, and economic evaluation are usually completed before the control system is considered, as shown in Figure 8.15s.

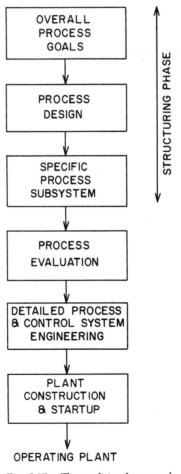

Fig. 8.15s The traditional approach
to process design

All the instrument engineer can do at this point is to design conventional control schemes, usually based on the individual plant unit operations. These are then hurriedly combined into some form of process control system to meet plant construction and startup schedules. In most cases, these time schedules do not allow for tailoring or innovation to better meet the process operating needs.

The ideal approach to developing a process design incorporates the control system as part of the structuring phase, as shown in Figure 8.15t. The process goals and plant operating goals are considered together by the design and control engineers in an iterative fashion. Thus, specific process subsystems can be developed to meet both goals together and to achieve the best tradeoffs between the process design and the process control system. Significant process

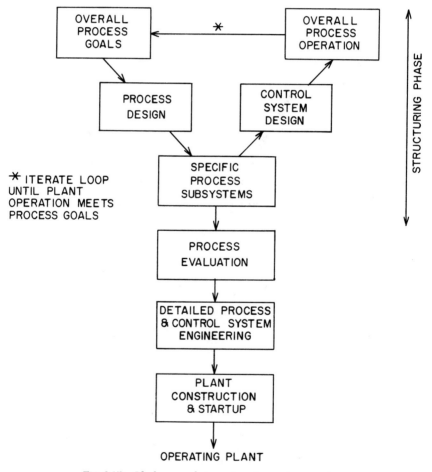

Fig. 8.15t   Ideal approach to engineering a process system

equipment can be saved, in some cases, by reduced holdup, better scheduling, and tighter process control.

The proper incorporation of digital control systems usually makes it easier to attract top level people to the process design and operation, which in itself can be a distinct advantage. In most cases, the detailed effort required to employ a digital control system gives substantial payoff in improved process design and operation, whether the computer is actually installed or not. If this effect is the primary goal of a process computer installation, higher management should be well aware of it.

The economic return from a process control system can be affected by many factors. Many persons tend to study and consider automating, only the higher economic levels of plant decisionmaking. They assume that the rest of the process control system will be satisfactory. All control schemes must eventually manipulate the process. Without proper structuring of all functions, higher-level systems can be unduly complicated, or in some cases, totally unworkable. The way to insure the best control system design is to specify all required functions before determining the hardware implementation.

Different types of processes have significantly different economic incentives. In continuous processes, the real gains are normally from increased throughput and/or higher quality, sometimes with less usage of utilities. In batch processes, savings in operating time and personnel are very significant. Newer types of higher level control schemes can add substantially to the operating economics in some discontinuous processes and in those involving long process delays. In some cases, process computers can also decrease plant startup time.

Operating manpower savings can usually be achieved in the form of less personnel or with the same personnel effectively operating larger plant areas. However, personnel must usually be added to maintain a digital system and

Table 8.15u
MAJOR APPLICATION FACTORS

Technical and Economic Understanding	Structuring of a Process Control System
—Problem Definition	—Consistent Philosophy
—Process Decomposition	—Task Assignment
—Process Economics	—Operating Considerations
	—Plant Organization
Levels of Process Control Systems	—Methods of Design
—Complexity Versus Frequency	—Personnel
—Analog Versus Digital Techniques	—Control System Economics
—System Availability	
—Control Factors	
—Man-Process Communication	
—Measurement Methods	

to change software as more is learned about the process operation. Additional personnel are needed to design and implement a digital system and properly manage the combination of engineering disciplines required for a successful installation.

Conclusions

This section discussed the major factors that have to be considered for the successful planning of a modern process control system. These were explored separately, but they all interrelate in achieving the best system design.

The importance of defining the objectives of the system and understanding the process was covered with methods outlined to do this more effectively. The levels of a process control system were explored with detailed discussion of the important factors. Finally, an approach to structuring the system was outlined and related operating and management factors were discussed. Table 8.15u reviews the major application factors as a simplified checklist.

Chapter IX

ANALOG AND HYBRID COMPUTERS

A. B. Corripio, P. A. Holst,
A. E. Nisenfeld and F. G. Shinskey

CONTENTS OF CHAPTER IX

9.1	PNEUMATIC ANALOG COMPUTERS	1140
	On-Line Computation	1140
	Pneumatic Computing Functions	1141
	Adding and Subtracting	1141
	Multiplying and Dividing	1142
	Arbitrary Functions	1142
	Dynamic Functions	1144
	Scaling Procedures	1145
	Accuracy Estimation	1148
	Calibration	1149
9.2	ELECTRONIC ANALOG AND HYBRID COMPUTERS	1151
	Electronic Analog Computers	1152
	Principles	1152
	Operational Amplifier	1154
	Computing Units	1158
	Coefficient Units	1167
	Control System	1170
	Design	1171
	Hybrid Computers	1172
	Concept	1172
	Hardware	1172
	Operation	1175
	Software	1176
	Literature	1177
9.3	ANALOG AND HYBRID COMPUTER APPLICATIONS	1179
	Programming	1180
	Circuit Diagram	1180
	Scaling	1183
	Time Scale Selection	1185
	Testing	1186
	Documentation	1187
	Hybrid Programs	1188
	Computations (Analog)	1189

Differential Equations   1189
Polynomials   1193
Algebraic Equations   1194
Repetitive Computations   1196
Simulation (Analog)   1199
Transfer Functions   1201
Non-Linear Functions   1202
Process Simulation   1210
Control System Simulation   1213
Simulation With ddc   1215
Control Valve Simulation   1216
On-Line Applications   1218
Signal Generation   1218
Signal Conditioning   1219
Data Processing   1222
Hybrid Computer Applications   1224
Hybrid Analog Computers   1225
Hybrid Digital Computers   1225
Balanced Hybrid Computers   1227
Partial Differential Equations   1230
9.4   ANALOG COMPUTERS IN DISTILLATION COLUMN CONTROLS   1232
Process Model and Control System   1232
Computing Functions   1234
Scaling   1236
Internal Reflux Computer   1238
Flow Control of Distillate (Fast Response)   1241
Constant Separation (Feedforward)   1244
Maximum Recovery (Feedforward)   1246
Composition Control of Two Products (Feedforward)   1248
Control of Two Products With Interaction   1251
Control of Two Products With Sidedraw   1252
Feed Composition Compensation   1254
The Total Model   1256
Batch Distillation   1257
9.5   PROCESS REACTOR MODELS AND SIMULATION   1259
Reaction Kinetics   1259
Perfectly Mixed Reactor   1261
Plug-Flow Reactor   1264
Non-Ideal Flow   1267
Heterogeneous Reactions   1267
Catalytic Reactions   1268
Conclusions   1268

9.1 PNEUMATIC ANALOG COMPUTERS

Pneumatic analog computers have been solving process operation problems on-line since about 1960, when accurate multipliers and dividers became available. They are used primarily for special purpose calculations of process parameters using transmitted measurements, according to a fixed program. Typical applications include mass flow calculation of gases from orifice differential pressure, temperature and static pressure, heat flow from liquid flow rate and temperature difference, and feedforward calculations such as those discussed in Sections 7.6 and 9.4.

The accuracy of pneumatic computing devices is comparable to that of the transmitters which supply their inputs. Consequently they are acceptable for all but the most complex or demanding calculations. Component prices lie in the $150–$250 range, which is their main advantage over comparable electronic devices. For a comprehensive tabulation of the pneumatic computer component characteristics, refer to the feature summary at the beginning of Section 4.4.

On-Line Computation

All transmitted signals look alike to analog computing devices, varying from 0 to 100 percent of the standard 3–15 PSIG range. Whatever information they contain can be operated upon properly only if the computer is designed by taking into account the range and significance of each input and output. For example, a certain calculation requires the division of a flow signal by absolute temperature. The signal from the temperature transmitter may have a range of 0–100°F, however. The divider must be "scaled" to recognize that 0 percent scale is not absolute zero temperature but 460°R on a scale of 460–560°R. The denominator of the divider should vary between 460/560 and 1.00 as the temperature signal moves from 0–100 percent of scale; otherwise the calculation will be in error.

Programming analog computers is simplified by considering each term in a given calculation as varying between 0 and 1.00 instead of 0–100 percent or 3–15 PSIG. This is most easily recognized in the operation of a square-root extractor. The input and output of a square-root extractor are equal only at

3 and 15 PSIG (0 and 100 percent) and nowhere in between. Since the square root of 15 is not 15, and the square root of 100 is not 100, this leads to much confusion. The explanation is that $\sqrt{0} = 0$ and $\sqrt{1} = 1$. At 50 percent input, the output is $\sqrt{0.5} = 0.707$, or 70.7 percent.

Many applications require compensating process measurement for variations in environmental conditions. An example would be converting a measurement of product density at a variable temperature to specific gravity at a reference temperature. The first step is to model the process from known data, to determine how density varies with temperature, i.e., whether multiplication, division, addition, or subtraction is required. Then the computer must be scaled so that no compensation is applied at reference temperature. In the case of addition or subtraction, the compensating term is zero at reference conditions and when multiplying or dividing is required, the compensating term is 1.00 at reference conditions.

Since most pneumatic computing devices will perform only one operation, even fairly simple calculations may require several devices. In addition to arranging and properly scaling the devices for the given inputs and outputs, the designer must also see that the intermediate signals remain on-scale for all reasonable input combinations. It is possible for all inputs and outputs to be on-scale, but the calculation can still be in error because one of the devices has saturated.

Pneumatic Computing Functions

The devices performing these functions are all described in Section 4.4, but the functions themselves are elaborated on in detail here as a guide to their proper application.

Adding and Subtracting

A device capable of summing is usually also capable of subtraction and averaging, depending on the arrangement of inputs. A general formula for a summing relay would be

$$D = K(A - C) + B \qquad 9.1(1)$$

where A, B, and C are input signals, D is the output, and K is an adjustable gain. Alternatively, one or two of the inputs may be replaced by a spring to apply a constant bias force. To generate a linear model:

$$D = KA + b \qquad 9.1(2)$$

Input C would be replaced with a zero spring and input B replaced with a bias spring b.

Averaging is accomplished by feeding back output D into output C:

$$D = K(A - D) + B \qquad 9.1(3)$$

Solving for D:

$$D = \frac{KA + B}{K + 1} \qquad\qquad 9.1(4)$$

Summing relays are often used for adding and subtracting flow signals. For this application, attention must be given to the ranges of inputs and outputs. If adding flows of 0–100 and 0–200 GPM, for example, should the output scale be 0–200 or 0–300 GPM? This depends on whether the sum *can* exceed 200 GPM in actual practice. If not, greater accuracy will be achieved using the smaller scale, by applying equation 9.1(1) with input C as zero and input B as the 0–200 GPM signal. Otherwise the averaging equation 9.1(4) should be used, whose output scale is the sum of the input scales, or 0–300 GPM.

A similar problem may develop when subtracting flow signals. Assume the 0–200 GPM flow is to be subtracted from the 0–300 GPM flow. An output scale of 0–300 GPM is possible using equation 9.1(1) with A as zero. However, if the difference never exceeds 100 GPM, greater accuracy will be achieved by applying the 0–300 GPM signal to both A and B. Then,

$$D = (1 + K)B - KC \qquad\qquad 9.1(5)$$

The output scale is then the difference between the input scales. This arrangement was used in the selective control system shown in Figure 7.10a.

Multiplying and Dividing

Pneumatic multipliers typically have only two inputs, but several adjustable coefficients are available to facilitate scaling. The general formula for a multiplier is

$$A = a + f(B - b)[C(1 - c) + c] \qquad\qquad 9.1(6)$$

Coefficients a, b, and c are bias or zero adjustments for the three signals, f is the gain of the device with both inputs at 100 percent, and $1 - c$ is the span of the C input. (For calibrating these devices refer to the last paragraph of this section.) The formula for a divider is found by solving equation 9.1(6) for signal B:

$$B = b + \frac{A - a}{f[C(1 - c) + c]} \qquad\qquad 9.1(7)$$

The linear relationship between absolute and Fahrenheit temperatures from the example cited earlier may be solved prior to dividing, by selection of coefficient c as 460/560.

Arbitrary Functions

An arbitrary function is a particular nonlinear relationship between two terms. The most common nonlinear function is the square root. It can be approximated by the motion of an angular mechanical linkage, or by feeding back the output of a divider into the denominator. If B is substituted for C in equation 9.1(7) and a, b, and c are zero and f is 1.0, then

$$B = A/B \qquad\qquad 9.1(8)$$

or

$$B = \sqrt{A} \qquad\qquad 9.1(9)$$

Similarly, a square function can be made by applying the input to both B and C of a multiplier:

$$A = B^2 \qquad\qquad 9.1(10)$$

Curves, which do not pass through points (0,0) and (1,1) can also be modeled by multipliers and dividers by using coefficients a, b, c, and f in equations 9.1(6) and 9.1(7). Three types of curves that can be modeled in this way are shown in Figure 9.1a:

Curve 1 is a polynomial with positive coefficients, i.e., a positive slope increasing with the input signal. Using equation 9.1(6) for a multiplier and substituting B for C,

$$A = a + f(B - b)[B(1 - c) + c] \qquad\qquad 9.1(11)$$

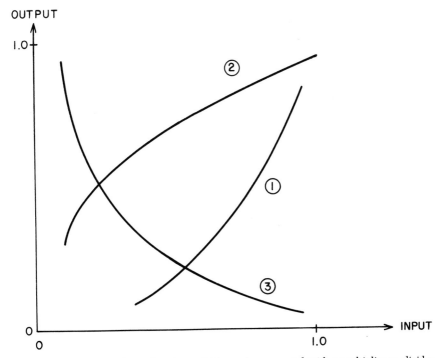

Fig. 9.1a   Typical non-linear functions which can be generated with a multiplier or divider

Let b = 0 and expand into the familiar polynomial.

$$A = a + fcB + f(1 - c)B^2 \qquad 9.1(12)$$

Curve 2 has a positive but decreasing slope and can be modeled by substituting B for C in divider equation 9.1(7). Its formula is identical to Equation 9.1(12) except that A is the input and B is the output.

Curve 3 is a hyperbola, described by divider equation 9.1(7), whose input is C, with numerator A fixed.

Cams, or four-bar linkages can also be used to generate arbitrary functions, but they lack the flexibility obtainable with multipliers and dividers.

Nonlinear functions modeled by connecting segments of lines can be generated using the selective devices described in Section 7.10. One or more linear functions can be generated with summing devices and the selectors are then used to switch from one summer to another at the point where these lines intersect. Horizontal segments of the function can be developed by limiters, which are selective relays with one fixed input.

Dynamic Functions

Dynamic functions commonly applied to pneumatic signals include integrating and differentiating, lag and lead-lag. The lead-lag function was described in detail in Section 7.6, since it is used extensively in feedforward control. The lag function is actually a special case of lead-lag with zero lead time. It, too, is used in feedforward control, and also to filter or dampen fluctuating signals.

Integration is the reset function of a controller. In fact, a pneumatic controller can be used to perform the integrating function by using the pressure in the reset bellows as the output, as shown in Figure 9.1b.

$$D = \frac{100}{(PB)T_i} \int_{t_0} (A - C)dt + B \qquad 9.1(13)$$

Here, input B is the "initial condition" of the integrator, i.e., its output D at time zero. When the block valve admitting signal B is closed, integration

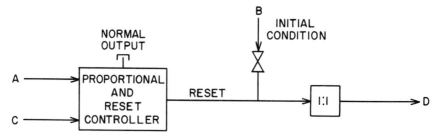

Fig. 9.1b   A pneumatic integrator made from a controller

begins at a rate determined by the proportional band PB and reset time T_i of the controller. The proportional band should be fixed at a high value (500 percent) to avoid saturating the normal output; inaccuracy will otherwise result.

The output of a pneumatic integrator deviates from linearity by about 5 percent due to the *change* in airflow through the reset restrictor with density (output pressure). The 1:1 relay is required at the output because flow cannot be withdrawn from the reset bellows without causing an error. By the same token, the initial-condition valve and associated piping must be free of leaks.

Pneumatic components have also been used for differentiation, as shown in Figure 9.1c. The most serious obstacle to obtaining accurate time-

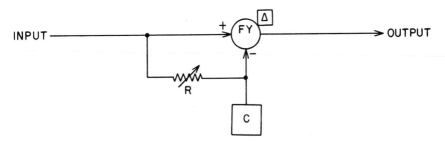

Fig. 9.1c   Pneumatic differentiator made from a subtractor, a restrictor and a capacity tank

differentials is that a differentiator has a very high sensitivity to noise. In fact, differentiators tend to be unstable unless their maximum gain is limited to 20 or less. The effect of this limitation is that a differentiator typically responds with a first-order lag whose time constant is the derivative time constant divided by the maximum gain. This is best expressed using Laplace transforms:

$$G(s) = \frac{\tau s}{1 + \tau_0 s/\alpha} \qquad 9.1(14)$$

The transfer function is $G(s)$, where s is the Laplace operator, τ is the time constant (τ_0 at time t_0) of the resistance-capacity combination (RC), and α is the gain of the unit.

The response of a differentiator to a ramp input is shown in Figure 9.1d. The output is superimposed on the input to demonstrate the role of the time constant τ. In actual operation, the output would be zero or some value fixed by a bias spring when the input is steady.

Scaling Procedures

Every analog computer requires scaling to insure compatibility with the input and output signal ranges. Coefficients a, b, c, f, K, etc., appearing in the equations of this section must be evaluated according to a carefully

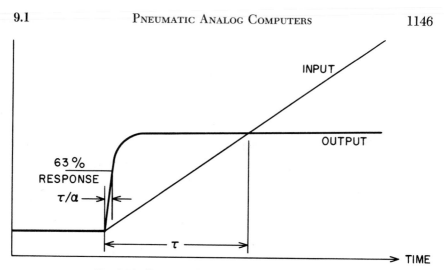

Fig. 9.1d   Response of a differentiator to a ramp input

organized procedure, or the wrong answers will result. A simple, yet highly effective, procedure is:

1. Write the equation to be solved, including all conversion factors, with all signals given in the units in which they are to be measured or displayed.
2. Relate each input and output signal (having a range of 0–1.0) to the range of each variable, by a set of "normalizing" equations.
3. Substitute the normalizing equations into the original equation and solve for the output signal.

As an example, follow the application of the foregoing procedure to the calculation of the rate of heat transfer, Q, to a liquid cooling medium. The liquid is flowing at rate F, with an inlet temperature T_1, and a higher outlet temperature T_2. The conditions are given in Table 9.1e:

The equation to be solved is

$$Q = Fk(T_2 - T_1) \qquad\qquad 9.1(15)$$

Table 9.1e
HEAT EXCHANGER OPERATING CONDITIONS

Signal	Variable	Range	Normal Value
B	F	0–100 GPM	60 GPM
A	T_2	25–75°F	50°F
C	T_1	0–50°F	30°F
E	Q	0–60,000 BTU/hr	48,000 BTU/hr

Coefficient k includes liquid density and specific heat, but it can be calculated from the normal operating conditions given in Table 9.1e.

$$k = \frac{48,000}{60(50 - 30)} = 40 \qquad 9.1(16)$$

Next the normalizing equations are written, relating signals A, B, C, and E to the variables in equation 9.1(15):

$$Q = 60,000E \qquad 9.1(17)$$
$$F = 100B \qquad 9.1(18)$$
$$T_2 = 25 + 50A \qquad 9.1(19)$$
$$T_1 = 50C \qquad 9.1(20)$$

Note that when signal A is zero, T_2 is actually 25°F, and when signal A is 100 percent (1.0), T_2 is 75°F.

Substituting the normalized equation into equation 9.1(15) yields

$$60,000E = (100B)40(25 + 50A - 50C) \qquad 9.1(21)$$

Solving for E,

$$E = 3.33B(0.5 + A - C) \qquad 9.1(22)$$

Equation 9.1(22) must be solved by a subtractor and multiplier in combination, as shown in Figure 9.1f. Since it is solved in two operations, equation 9.1(22) must be separated into two pieces. Let the intermediate variable be identified as D:

$$D = 0.5 + A - C \qquad 9.1(23)$$
$$E = 3.33BD \qquad 9.1(24)$$

Fig. 9.1f   Computer for pneumatic heat transfer calculation

The entire factor 3.33 in the equation is shown applied to the multiplier. This is not altogether necessary—for example, the subtractor could have a gain of 2.0 and the multiplier a gain of 1.67:

$$D = 1.0 + 2.0(A - C) \qquad 9.1(25)$$
$$E = 1.67BD \qquad 9.1(26)$$

This tends to improve the accuracy of the calculation but increases the danger of saturating the subtractor. Whenever more than one device is used in this way, each operation should be tested for saturation with reasonable combinations of inputs. In this example, a combination of 75°F for T_2 and 0°F for T_1 would not be reasonable.

A shortcut method of scaling can be used for compensating computers. Consider the example where a gas flowmeter requires compensation for absolute pressure:

$$W = k\sqrt{hP} \qquad\qquad 9.1(27)$$

In equation 9.1(27), W is the mass flow, h is the orifice differential pressure, and P is the absolute pressure. The orifice and differential range are selected so that 100 percent differential equals 100 percent flow at the normal operating pressure. With this in mind, scaling requires knowing only the normal pressure and the range of the pressure transmitter.

Let the normal pressure be 64.7 PSIA, and the transmitter range be 0–75 PSIG with a base of 14.7 PSIA. Compensation requires that the multiplier exhibit a gain of 1.00 at the normal pressure. The absolute pressure range is 14.7 to 89.7 PSIA; therefore,

$$P = 14.7 + 75C \qquad\qquad 9.1(28)$$

Here signal C represents the pressure input to the multiplier. Then the compensating factor is:

$$\frac{P}{64.7} = 0.227 + 1.160C \qquad\qquad 9.1(29)$$

To conform with equation 9.1(6), the maximum value of the compensating factor is extracted as $f = 89.7/64.7 = 1.387$

$$\frac{P}{64.7} = 1.387(0.836C + 0.164) \qquad\qquad 9.1(30)$$

Then the scaled equation for the multiplier is

$$A = 1.387B(0.836C + 0.164) \qquad\qquad 9.1(31)$$

Where A is the multiplier output and B and C are the orifice differential and pressure inputs, respectively. A square-root extractor following the multiplier completes the calculation.

Accuracy Estimation

Three factors must be considered before an estimate of accuracy can be made for a given computer.

1. The sensitivity of the system equation to errors in the input signals.
2. The statistical combination of individual errors.
3. The contribution of errors in the computing components.

The first consideration requires differentiating the output of the system equation with respect to each input, with all scaling factors included. Equation 9.1(31) can be differentiated as an example:

$$dA = 1.387dB(0.836C + 0.164) + 1.387B(0.836dC) \qquad 9.1(32)$$

Differential dA is the maximum output error which would result from error dB in signal B and error dC in signal C. Note that the output error depends on signal levels as well as scaling factors.

Equation 9.1(32) ignores probability considerations, however. If both dB and dC are random errors, the resulting error dA with the same probability of occurrence would be related to them by a root-mean-square combination:

$$(dA)^2 = [1.387dB(0.836C + 0.164)]^2 + [1.387B(0.836dC)]^2 \quad 9.1(33)$$

The error e contributed by the computing device—a multiplier in the example—is similarly added in rms fashion to equation 9.1(33):

$$(dA)^2 = [1.387dB(0.836C + 0.164)]^2 + [1.387B(0.836dC)]^2 + e^2 \quad 9.1(34)$$

Because accuracy changes with the magnitude of the input signals, an estimate of error is valid for only one set of conditions. This is particularly true where a square root is taken:

$$A = \sqrt{B} \qquad\qquad 9.1(35)$$

$$dA = \frac{dB}{2\sqrt{B}} = \frac{dB}{2A} \qquad\qquad 9.1(36)$$

An error of 0.5 percent in differential pressure B would represent an error in flow A of 0.5 percent at 50 percent flow, 1 percent at 25 percent flow, and 2.5 percent at 10 percent flow.

When flow signals are added, they must, of course, be linear. If one of the streams is shut off, its flow signal will be zero, representing the possibility of a sizeable error. To prevent a large negative error from affecting the summation, high-select relays may be inserted downstream of each square-root extractor as shown in Figure 9.1g. These selectors prefer a small negative error in differential to a large negative error in flow.

Calibration

Each instrument in a computing system, including indicators, recorders and transmitters, should be calibrated against a common standard where possible. The most reliable standard to use is a mercury column, accurate to ±0.1 percent of the 3–15 PSIG range.

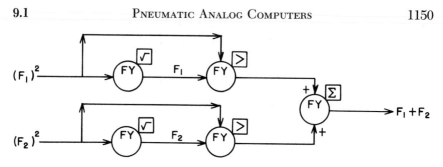

Fig. 9.1g   The use of high selectors will eliminate large negative errors when flow signals are added

Calibrating multipliers and dividers is particularly painstaking because the adjustments must be made in a specific order. Referring to Equation 9.1(6) for a multiplier:

1. Zero adjustment a must be made with signals B and C at zero.
2. Zero adjustment b must be made with signal B at zero and C at 100 percent.
3. Zero adjustment c must be made with signal B at 100 percent and signal C at zero.
4. Span adjustment $1 - c$ must be made with B at some specified intermediate value and C at 100 percent.

A similar procedure must be followed with dividers.

After all components are calibrated individually, the system must be calibrated as a whole, to offset systematic (nonstatistical) errors. Almost any of the adjustments can be used to calibrate the system at a single operating point, but the wrong choice may cause a greater error at some other point. So the accuracy of the system should be evaluated at several sets of conditions to determine which of the available adjustments would minimize the average error for all sets. Often more than one coefficient may require adjustment.

In summary, pneumatic analog computers are simple, reliable, flexible, and sufficiently accurate for most on-line computing applications.

9.2 ELECTRONIC ANALOG AND HYBRID COMPUTERS

Feature Summary: For a tabulation of electronic computer component characteristics and symbols refer to the feature summary at the beginning of Section 4.4.

Size: Small—under 50 amplifiers.
Medium—from 50 to 150 amplifiers.
Large—over 150 amplifiers.

Components: Integrators, adders, multipliers, function generators, resolvers, analog switches, signal relays, digital attenuators, logic gates, flip-flops, counters, monostables, differentiators, pulse generators and coefficient potentiometers.

Interface Components: Analog-to-digital and digital-to-analog converters, multiplexers, track-store amplifiers and logic signal registers.

Cost: For computer with 20 to 100 amplifiers—$5,000 to $15,000. For computer with over 100 amplifiers—$15,000 to $60,000. For hybrid interface to cover 10 to 50 channels—$12,000 to $35,000.

Accuracy: For static, linear analog units, ±0.005% to ±0.1%.
For static, non-linear analog units, ±0.01% to ±2%.

Frequency: Clock cycle frequency, 100 KHz to 2 MHz. Conversion rate, 10 KHz to 1 MHz. Full power signal bandwidth, from DC to 25 to 500 KHz.

Partial List of Suppliers: Applied Dynamics; Astrodata; Electronic Associates, Inc.; Hitachi, Ltd.; Hybrid Systems, Inc.; Simulators, Inc.; Solartron Enterprises; Systron Donner Corp.; Telefunken GmbH; Zeltex, Inc.

Electronic Analog Computer

Principles

The basic building block of an analog computer is the operational amplifier. All the fundamental functions of an analog computer, summation, integration, multiplication, and function generation, can be carried out with operational amplifiers. The operational amplifier is single ended (all signals are referenced to a common ground), DC coupled wide bandwidth, and has a very high static gain (often 10^5 to 10^8). Special precautions are taken to reduce or eliminate offsets, drift with temperature changes or time, and to avoid electronic noise in the circuitry. The computer is designed to minimize cross talk (electromagnetic and static coupling) between different computing units, and other disturbing or impairing electronic effects due to the presence of a large number of components and signal sources within one system or console.

Figure 9.2a shows the primary analog computing units or functions obtained with an operational amplifier. Note that the amplifier in addition to the high gain, also inverts the polarity of the resultant input signal combination. In case of the integrator, the actual input to the operational amplifier is switched between two points: the initial condition (e_1) input network, and the integrand (e_2) input network. In the initial condition position shown, the capacitor C charges up to a signal value e_0 (at time t = 0) with an electronic time constant determined by Z_0C. When the switch is in the lower (operate) position, the output signal e_0 will correspond to the time integral of the integrand input e_2 with a speed or time scale determined by $1/Z_2C$.

The squarer and the multiplier are based on nonlinear input networks that cause the output signal to follow a square law relationship to the input. In the multiplier this is utilized in the quarter-share principle, which states that

$$XY = \tfrac{1}{4}[(X + Y)^2 - (X - Y)^2] \qquad 9.2(1)$$

Similarly, the function-generating unit employs an input (or feedback) network which is programmed to change impedance according to a specified function, the program. This input network program may be either fixed, as in the case of mathematical functions, such as $\sin(x)$, $\cos(x)$, $\log_e(x)$, or variable, to allow empirical and other functions to be developed.

CIRCUIT   **SYMBOL**

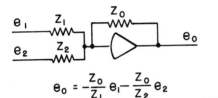

$$e_0 = -\frac{Z_0}{Z_1}\, e_1$$

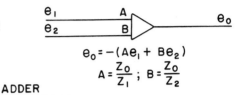

$$e_0 = -Ae \; ; \; A = \frac{Z_0}{Z_1}$$

INVERTER

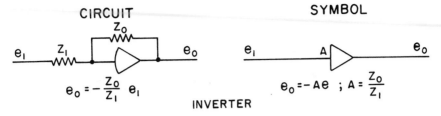

$$e_0 = -\frac{Z_0}{Z_1}\, e_1 - \frac{Z_0}{Z_2}\, e_2$$

$$e_0 = -(Ae_1 + Be_2)$$

$$A = \frac{Z_0}{Z_1} \; ; \; B = \frac{Z_0}{Z_2}$$

ADDER

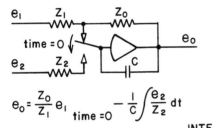

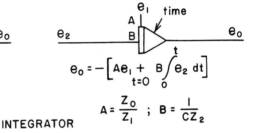

$$e_0 = \frac{Z_0}{Z_1}\, e_1 \Big|_{time=0} - \frac{1}{C}\int \frac{e_2}{Z_2}\, dt$$

$$e_0 = -\left[Ae_1 + B \int_{0}^{t} e_2\, dt \right]_{t=0}$$

$$A = \frac{Z_0}{Z_1} \; ; \; B = \frac{1}{CZ_2}$$

INTEGRATOR

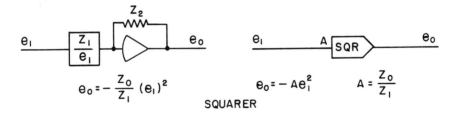

$$e_0 = -\frac{Z_0}{Z_1}\, (e_1)^2$$

$$e_0 = -Ae_1^2 \qquad A = \frac{Z_0}{Z_1}$$

SQUARER

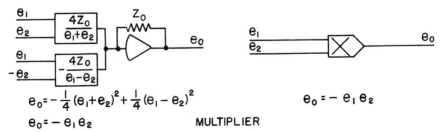

$$e_0 = -\frac{1}{4}(e_1 + e_2)^2 + \frac{1}{4}(e_1 - e_2)^2$$

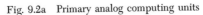

$$e_0 = -e_1 e_2 \qquad\qquad e_0 = -e_1 e_2$$

MULTIPLIER

Fig. 9.2a   Primary analog computing units

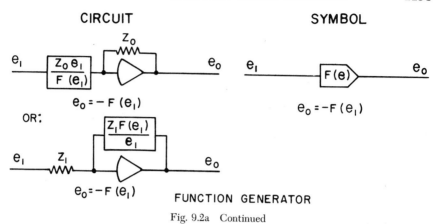

CIRCUIT

$$e_o = -F(e_1)$$

OR:

$$e_o = -F(e_1)$$

SYMBOL

$$e_o = -F(e_1)$$

FUNCTION GENERATOR

Fig. 9.2a   Continued

Coefficient and parameter values are introduced in the analog computation and simulation circuits using potentiometers, which may be adjusted manually or by a servo motor. The potentiometer serves as a signal (voltage) divider (Figure 9.2b) in different circuits, enabling a wide variety of input or feedback gains to be established. Note that in all these circuits the signal-division ratio α is defined by the potentiometer setting and the actual input network load effects due to the finite input impedances used.

Other analog computing units (often called hybrid units) are the comparator, the analog switch, and the signal relay shown in Figure 9.2c. These units tie the analog signal variables to logic signals (discrete, binary variables) to detect relative signal values and to introduce changes in the analog circuits.

Operational Amplifier

Operational amplifiers are used for three basic purposes in the analog computer:

1. to generate the necessary computing functions,
2. to amplify in signal level and power the analog variables,
3. to provide isolation and unloading between the different input and output signals within the computing units.

In many analog computers all three purposes are served by the use of one amplifier type. This gives commonality in the design of the computing units, allows easy exchange of components and parts, and simplifies service and maintenance. In larger and more specialized analog computers, however, it is generally better to utilize a number of amplifier types, each with some desired characteristics for particular purposes. Examples of this are *chopper-stabilized amplifiers* for low-drift integrators, *wide-bandwidth amplifiers* in inverters and output stages, and *high-gain amplifiers* in adders.

The general features of an operational analog computer amplifier include:

Single-ended input and output

Very high gain, especially at DC

Linear, inverting output-input relationship

Low electronic noise offsets, and drift characteristics

Stable and rugged in use, allowing extensive cable capacitances and
 long wiring in the console, with little performance deterioration

Resistance to overrange and short-circuit abuse without being impaired

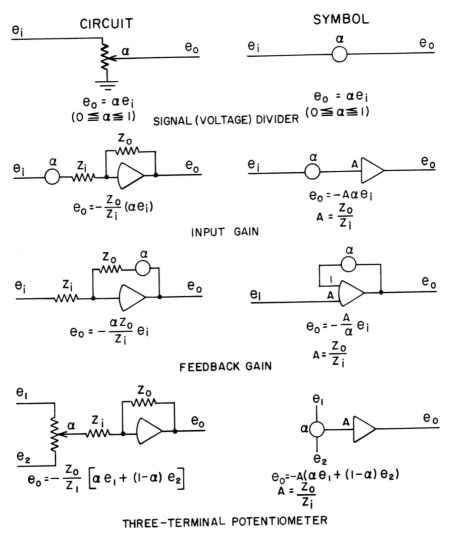

Fig. 9.2b   Potentiometer circuits

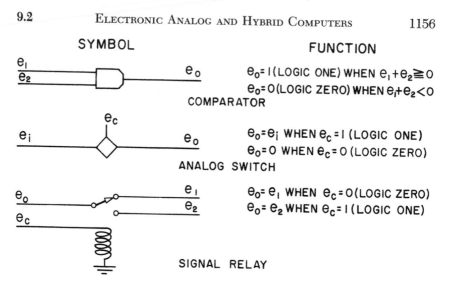

SYMBOL   FUNCTION

$e_0 = I$ (LOGIC ONE) WHEN $e_1 + e_2 \geqq 0$
$e_0 = 0$ (LOGIC ZERO) WHEN $e_1 + e_2 < 0$

COMPARATOR

$e_0 = e_i$ WHEN $e_c = I$ (LOGIC ONE)
$e_0 = 0$ WHEN $e_c = 0$ (LOGIC ZERO)

ANALOG SWITCH

$e_0 = e_1$ WHEN $e_c = 0$ (LOGIC ZERO)
$e_0 = e_2$ WHEN $e_c = I$ (LOGIC ONE)

SIGNAL RELAY

Fig. 9.2c   Hybrid computing units

Figure 9.2d shows the use of an operational amplifier as an adder. Since the amplification μ is normally in the order of 10^6 or more, the electrical potential e_g at the summing junction is virtually zero ($e_g = -e_0/\mu$). Disregarding the input current i_g, the balanced condition of the amplifier gives

$$i_0 = -(i_1 + i_2 + i_3 + \cdots) \qquad 9.2(2)$$

or

$$e_0 = -\left(\frac{z_0}{z_1}e_1 + \frac{z_0}{z_2}e_2 + \frac{z_0}{z_3}e_3 + \cdots\right) \qquad 9.2(3)$$

Introducing relative input gains $g_i = z^*/z_i$ where z^* is a conveniently selected reference value, the equation states

$$e_0 = -\frac{1}{g_0}(g_1e_1 + g_2e_2 + g_3e_3 + \cdots) \qquad 9.2(4)$$

BASIC CIRCUIT   SYMBOL

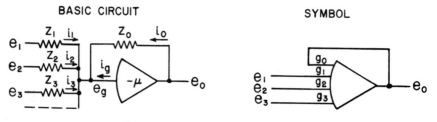

Fig. 9.2d   Operational amplifier as adder

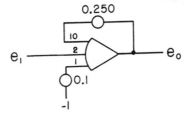

Fig. 9.2e   Example of adder

This corresponds to a weighted summation of the input variables, where the weighing factors are the input and feedback gains. The circuit shown in Figure 9.2e thus corresponds to the summation

$$e_0 = -\frac{1}{10 \times 0.250}(2e_1 - 0.1) \qquad\qquad 9.2(5)$$

or

$$e_0 = -0.8e_1 + 0.04 \qquad\qquad 9.2(6)$$

If the amplification μ is low, the summing junction potential e_g must be considered in equation 9.2(3), resulting in a correcting factor f_μ

$$e_0 = -\frac{f_\mu}{g_0}(g_1e_1 + g_2e_2 + g_3e_3 + \cdots) \qquad\qquad 9.2(7)$$

where

$$f_\mu = \frac{\mu}{\mu + \dfrac{1 + g_1 + g_2 + g_3 + \cdots}{g_0}} \qquad\qquad 9.2(8)$$

Equation 9.2(8) expresses the dependence of the accurate summation of the input variables on the amplification and the input gains used in the actual analog computer circuit. Since the amplification (the open-loop gain) of the operational amplifier falls off with increasing signal frequencies, the analog computations must be carried out within the corresponding frequency limits.

An example of open-loop gain characteristic is shown in Figure 9.2f, for an analog computer amplifier. If this amplifier is used as a summer with input gains of 10, 10, 1, and 1 (total of 22), for a summation accuracy of 0.01 percent, the open-loop gain must be at least

$$\mu = \frac{0.9999(1 + 22)}{0.0001} = 230,000 \qquad\qquad 9.2(9)$$

or corresponding to signal frequencies below 120 Hz. If the same amplifier is used in a unity gain inverter configuration, the necessary open-loop gain

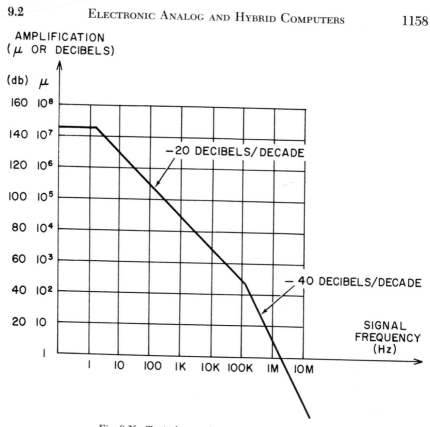

Fig. 9.2f   Typical gain characteristics of open loop

is only 20,000 and the signal frequency limit extends to 1.6 kHz. Practical considerations, such as amplifier input current, accuracy of input and feedback networks, etc., further reduce these limits.

Computing Units

Typical analog computing units are listed in the feature summary at the beginning of this section and Section 4.4. Outmoded or specialized units, such as servomultipliers, tapped potentiometers, and CRT function generators, fall outside the scope of this section.

The key analog computing unit is the integrator, which distinguishes the analog computer from an assembly of operational amplifiers. The integrator carries out integration with respect to time, which is the only independent variable in the analog computation or simulation. All other variables appear as functions of time. The integrator is controlled by *mode commands* which specify the computational (operational) mode, such as initial condition (IC), operate (OP), or hold (HD). Other modes, such as potentiometer set (PS) and static testing (ST), are also used. Mode switching is accomplished with electronic or electromechanical switches, as shown in Figure 9.2g. The closure of the switches IC, OP, and PS gives the basic operational modes, as listed in Table 9.2h. Also included is the time scale (TS) switch which selects one of two or more time scales to be used in the integration.

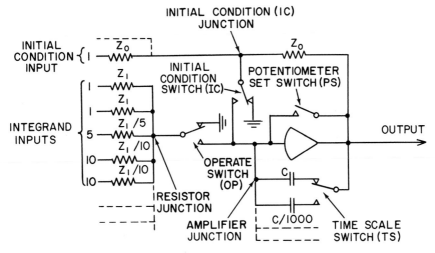

Fig. 9.2g Integrator circuit

Table 9.2h
INTEGRATOR MODE SWITCHING

Switches Are in Noted Condition°				Operating Mode Required	Time Scale Required
(IC)	(OP)	(PS)	(TS)		
0	0	0	0	Hold (HD)	Normal (X1)
0	0	0	1		Fast (X1,000)
0	0	1	0	Potentiometer	
1	0	1	0	Set (PS)	
0	1	1	0		
0	0	1	1		
0	1	0	0	Operate	Normal (X1)
0	1	0	1	(OP)	Fast (X1,000)
1	0	0	0	Initial Condition	Normal (X1)
1	0	0	1	(IC)	Fast (X1,000)

° "0" means that switch is in condition noted in Figure 9.2g. "1" means contact has been transferred.

The PS mode is the neutral or passive mode which effectively cancels out any inputs to the integrator. This mode should be used whenever the integrator (or the computer) is inactive. In other modes, notably in OP and HD, the integrators may drift off scale, causing overloads.

The mode switches and the time scale switch may be commanded either by the common computer control system (for all the integrators that operate in synchronism), or individually, by manual or logic signal, or by commands originating in a hybrid computer. Particular computer designs offer other modes, such as extended *hold* (HD) for long time constants, priority of IC over OP modes, and reversed mode control functions. Usually only the IC and the OP modes are critical for the proper use of the integrator.

The circuit in Figure 9.2g shows that both the initial condition junction and the resistor junction are connected to ground (reference zero) when not in the active modes. This ensures proper loading of the input networks under all conditions. There is no resistor in the circuit that shorts the PS from the amplifier output to the input (amplifier junction). If the shorting is carried out with a non-zero impedance, improper and error-causing output signals may result in the PS mode.

In Figure 9.2i are given four alternate applications of the integrator, employing the mode switches for functional and operational signal switching. Of these applications, the track-store circuit, and the analog switch are the most useful. For practical reasons only one mode switch is used in each circuit. The remaining modifications are done by regular program (patching) connections.

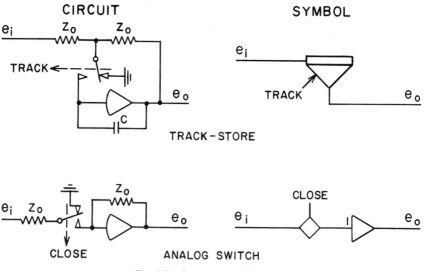

Fig. 9.2i Integrator applications

CIRCUIT   SYMBOL

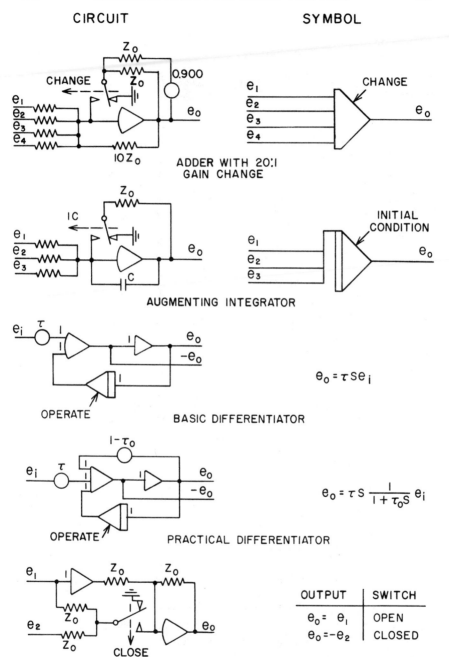

ADDER WITH 20:1
GAIN CHANGE

AUGMENTING INTEGRATOR

$$e_0 = \tau s e_i$$

BASIC DIFFERENTIATOR

$$e_0 = \tau S \frac{1}{1 + \tau_0 S} e_i$$

PRACTICAL DIFFERENTIATOR

OUTPUT	SWITCH
$e_0 = e_1$	OPEN
$e_0 = -e_2$	CLOSED

SINGLE-POLE DOUBLE-THROW ANALOG SWITCH

Fig. 9.2i   Continued

The circuit for the adder, or inverter, is shown in Figure 9.2j. The PS switch is used to clamp the output to zero during the potentiometer set mode. Various input gain configurations are given in Figure 9.2k for the most common adder amplifier circuits. The normal (standard) input gains are given in parentheses.

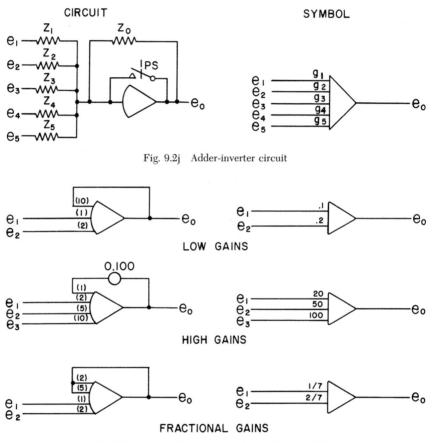

Fig. 9.2j   Adder-inverter circuit

LOW GAINS

HIGH GAINS

FRACTIONAL GAINS

Fig. 9.2k   Input gain arrangements in adder amplifiers

The electronic multiplier is outlined in Figure 9.2l. Four square-law networks require the inputs $+X$, $-X$, $+Y$ and $-Y$ for four quadrant operations. Some analog computer multipliers require only $+X$ and $+Y$ inputs when the necessary inverters are built in. Other multiplier types, such as those based on the trans-conductance principle, logarithmic functions, or time division, may require alternate inputs. In the following application circuits only the $+X$ and the $+Y$ inputs are shown when no ambiguity exists.

Figure 9.2m shows some of the most common applications of the multi-

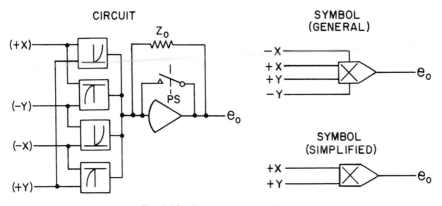

Fig. 9.2 1   Quarter square multiplier

plier in analog computations and simulations. In order to preserve stability the denominator input (e_2) must be of one polarity only and equal to or greater in value than the numerator (e_1). Similarly, the square rooter argument (e_1) must be of only one polarity.

Of special interest in process control and instrumentation applications are the sign-preserving squaring and square-root circuits shown in Figure 9.2m. In some types of computing units these functions can be obtained directly by rearranging the connections to the square-law networks shown in Figure 9.2l.

The function generator is outlined in Figure 9.2n. Two fixed or variable program networks are associated with an output (operational) amplifier circuit. The function generator unit comprises 10 to 20 segments or elements arranged to additively contribute to the output (e_0), depending on the value of the input variable (e_1) and the program. In this way a broad range of functions can be approximated by straight line segments of appropriate lengths and slopes (gain effects). The programming thus consists of selecting the necessary number of segments, and defining and adjusting the intercept points between adjoining segments over the required functional range. The example in Figure 9.2n shows the approximation of the function $e_0 = \sin(2e_1)$ with a total of 14 segments, 7 for each polarity of the input.

A variable function generator is programmed by manual adjustments of the segment intercept coordinates, using some program setup units or control systems within the analog or hybrid computer. Automated programming systems are also available by which the adjustments can be made according to information read in from punched cards, tapes or from the hybrid computer, using either incrementally defined function values, or servo-assisted, continuously adjusted program points.

Commonly available fixed program function generators include $\sin(x)$ and $\cos(x)$ functions, $\log_e(x)$ and $\exp_e(x)$ functions, as well as squaring and square-

CIRCUIT   SYMBOL

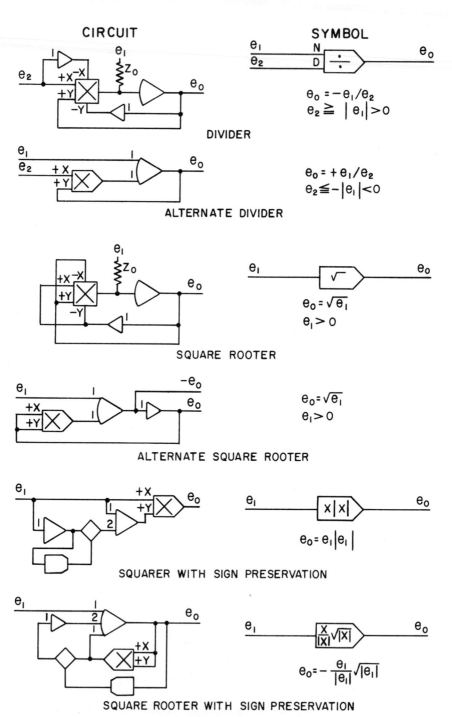

$$e_0 = -e_1/e_2$$
$$e_2 \gtreqqless |e_1| > 0$$

DIVIDER

$$e_0 = +e_1/e_2$$
$$e_2 \lesseqqgtr -|e_1| < 0$$

ALTERNATE DIVIDER

$$e_0 = \sqrt{e_1}$$
$$e_1 > 0$$

SQUARE ROOTER

$$e_0 = \sqrt{e_1}$$
$$e_1 > 0$$

ALTERNATE SQUARE ROOTER

$$e_0 = e_1 |e_1|$$

SQUARER WITH SIGN PRESERVATION

$$e_0 = -\frac{e_1}{|e_1|}\sqrt{|e_1|}$$

SQUARE ROOTER WITH SIGN PRESERVATION

Fig. 9.2m   Multiplier applications

1164

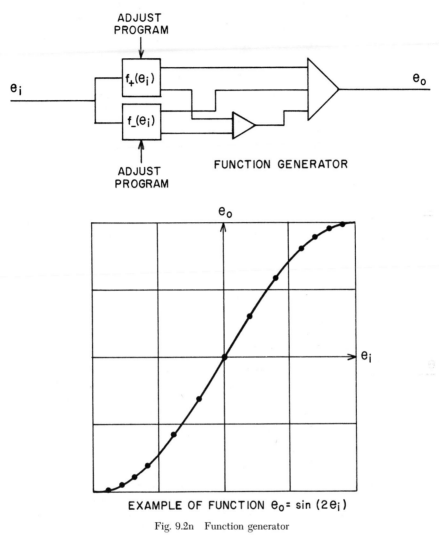

EXAMPLE OF FUNCTION $e_0 = \sin(2e_i)$

Fig. 9.2n   Function generator

root functions. Widely used in process control and instrumentation applications are the $\log_e(x)$ function generators, with the input-output relationship

$$e_0 = -\tfrac{1}{10}\log_e(e_1) \qquad\qquad 9.2(10)$$

and the inverse,

$$e_0 = \exp_e(-10e_1) \qquad\qquad 9.2(11)$$

These functions are generated by input networks which will accept input
signals of only positive or negative polarity, as shown in Figure 9.2o, with

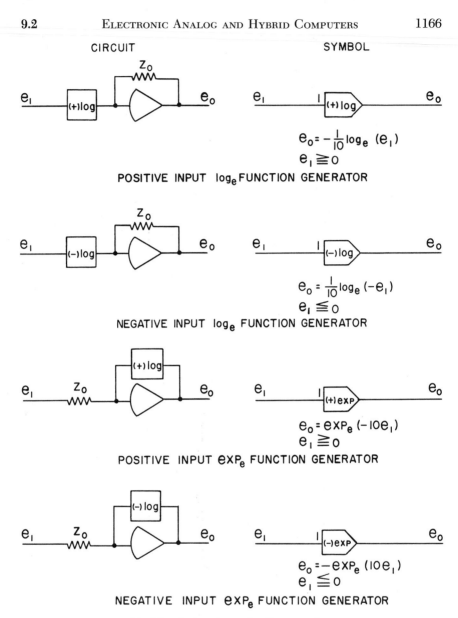

Fig. 9.2o   Log$_e$ and exp$_e$ function generators

the functions indicated in Figure 9.2p. Typical applications of the log functions are shown in Figure 9.2q, consisting of log and exp circuits. In general the circuits can be simplified and considerably improved by combining the function-generating networks in single-amplifier circuits, as is the case in the four-quadrant multiplication illustration.

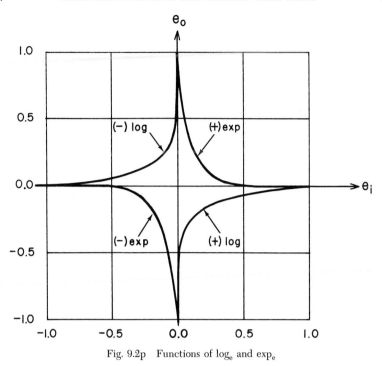

Fig. 9.2p   Functions of $\log_e$ and $\exp_e$

Coefficient Units

The main units for establishing parametric values in the analog computing circuits are the coefficient potentiometers and the digital attenuators. The potentiometers are generally wire-wound, multi-turn linear voltage dividers, as shown in Figure 9.2r, with the low terminal connected to analog signal ground. Ungrounded potentiometers with three terminals available for signal connections are often used in voltage division circuits where the reference is non-zero, as indicated by the limiter circuit in Figure 9.2r. The potentiometers may have one or more taps on the winding to provide terminals for compensation networks, which decrease the signal phase shift effects introduced by the winding construction.

Since the potentiometers are passive units with no power or signal amplification, it is important to adjust them with the desired load (input terminals) connected to the wiper, and make sure the load does not change during actual signal attenuation applications. This means the load impedance must remain constant during signal variations. Non-constant input impedances are usually associated with multipliers, function generators, and signal switches or relays actuated during the computations. The adjustments are made by applying a fixed voltage to the high terminal of the potentiometer, with the low terminal connected to reference ground. The output voltage (e_0) is the

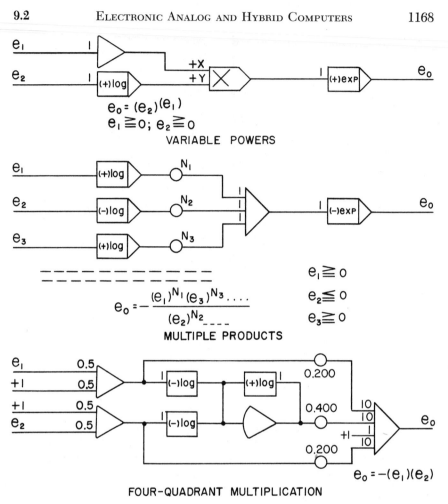

Fig. 9.2q   Applications of the log$_e$ function generator

ratio of the applied fixed voltage (e_1), measuring the actual signal voltage division (e_0/e_1) as the potentiometer setting. The control system of the analog computer provides automatic signal and ground connections to the potentiometers in certain modes, such as potentiometer set or potentiometer read. Normally the computer reference voltage will be applied as the fixed voltage, and the potentiometer setting read as a decimal fraction $0 \leq p \leq 1$. The limits 0 and 1 may be difficult to achieve in practice, and the potentiometer should not be used near these values. Best results are usually obtained with the potentiometer set between the limits of 0.100 and 0.950, or further restricted to 0.500–0.900. Some analog computers do not automatically provide one or both of the required connections for potentiometer setting of the three-

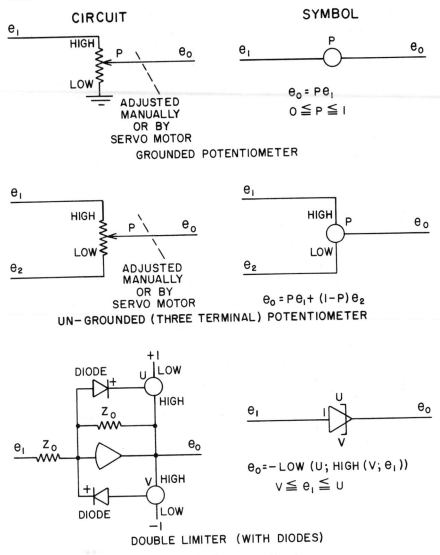

CIRCUIT

SYMBOL

$e_o = P e_1$

$0 \leqq P \leqq 1$

GROUNDED POTENTIOMETER

$e_o = P e_1 + (1-P) e_2$

UN-GROUNDED (THREE TERMINAL) POTENTIOMETER

$e_o = -\text{LOW} (U; \text{HIGH} (V; e_1))$

$V \leqq e_1 \leqq U$

DOUBLE LIMITER (WITH DIODES)

Fig. 9.2r   Coefficient potentiometer

terminal type potentiometers. Also in some designs all the potentiometer terminals are switched at the same time in the PS mode, which means two potentiometers in cascade (series connected) cannot be properly adjusted when one loads the other.

Digital attenuators take the place of coefficient potentiometers in many hybrid computers, primarily for functional and efficiency reasons. These attenuators consist of digitally actuated switch networks which alter the attenuation (or signal conductance) in fixed and standard amounts. The digital attenuators also provide signal level and power amplification, at least in terms of unloading the effects of the input impedances that may be connected to the attenuator. Thus, they may be used with less restrictions than conventional potentiometers. Some manufacturers offer special input networks as a permanent part of the attenuators, by which input signal gains ranging from 0 to 1.6 or more may be obtained.

Control System

The control system has two purposes:

> To provide for easy setup and checkout of the
> analog and hybrid programs.
> To support and assist the operator in carrying out
> the analog and hybrid computations and simulations.

Due to the nature of analog and hybrid programs, the setup and checkout phase accounts for the greater part of the control system uses.

Typical applications of the control system functions include:

> Inserting the analog and hybrid program signal connections
> Establishing the appropriate modes of operation of the computing
> units
> Adjusting the coefficient values of the program
> Selecting the initial states of bistable units
> Setting up the values of programmable function generations
> Initializing the program control units within the control system
> Verifying the signal connections by measuring initial computer
> variables
> Comparing integrand values with precalculated reference values
> Checking interface channel functioning
> Stepping through logic unit sequences and iterative routines (slow
> motion)
> Connecting the desired display and recording devices, and setting
> the necessary scales and modes of operations
> Energizing the appropriate operational modes of the computing units
> Controlling and commanding the execution of particular computa-
> tional sequences
> Monitoring and reacting to the overload and underranging of the
> computational variables
> Introducing variations in coefficient values in the analog and hybrid
> program
> Reading and printing out the values associated with the analog and
> hybrid program variables, coefficients, interface channels, and
> program conditions

Analog computer control systems vary greatly in the degree these functions are supported by automated and power assisted routines within the system. Present computers will permit the operator to carry out most of these functions by centralized pushbutton and switch manipulations, with relatively few advanced and extended auxiliary systems provided. Few of the functions are operated (activated or controlled) through the hybrid interface.

Design

Most analog and hybrid computers are designed in the form of a console, arranged modularly with computing units, control systems, display and recording devices as plug-in components. The computer consoles are constructed of rigid, steel frames with removable side panels for easy access. The front sides are usually furnished with screw-in standard sized panels that mount the monitoring and control elements in an assembled and convenient arrangement. Components and modules are connected by extensive and bulky electrical cables, often furnished as elaborate wire harnesses, with individually color-coded signal and power conductors.

The computing units are plug-in individual or dual/quadruple trays, which are inserted into assigned positions or slots in the computer. All the slots are normally prewired, allowing easy and flexible expansion and modification of computing units. A disadvantage is the high initial price of an incomplete computer. Often field conversions or adaptations turn out to be impractical and expensive.

Critical to precise computations are stable and accurate power supply units in the computer, with good voltage regulation characteristics, ample power reserves, and excellent isolation from line voltage and frequency transients. Also of vital importance are signal and power ground systems, which serve as reference for all analog signal measurements and drain all electrical operating and energizing currents.

Ground systems are divided into four categories:

Reference ground (no current drain allowed)
Analog signal return (grounding of potentiometers)
Electrical power return
Operating and energizing signal returns (relays, switches)

For proper operation a computer must be installed with adequate electrical power and ground connections, to permit as noise-free and stable operations as possible, and to take full advantage of the computer capabilities and signal ranges.

Similarly, to ensure reliable, high-precision operation, the environment of the computer may require air conditioning or closed system circulation of air. Temperature-insensitized or compensated solid-state components have greatly alleviated some of the heat problems and have eliminated the needs for special temperature-controlled chambers (ovens) within the consoles. Electrical contact contaminations, however, are still a major concern in the maintenance of electronic computers, since the designs usually include numerous multi-pin connectors for low-level signals, and large numbers of electromechanical relay contacts activate computational modes, address computing units, and implement signal connections.

Hybrid Computers

Concept

A hybrid computer is a combination in hardware and software of one or more analog and digital computers. Such a combination aims at providing faster, more efficient and economical computational power than is otherwise available by computers of either type alone. The results depend to a large extent on the exchange of information between the analog and the digital computers, and on the compatibility in operations and mutual interactions between the two parts. (See Section 9.3.)

The basic concept of a hybrid computer is a device for intimate, relevant and rapid exchange of information between the parallel and simultaneous computations and simulations within the analog computer, with the serial and sequential operations of the digital computer. This information exchange links the two computational domains for combined advantages of the fast and flexible analog computer with the precise and logic controllable digital computer.

The extent of the information exchange between the two parts and the sophistication of the control structures and instruction repertoires determine the capability and the capacity of the hybrid computer. Best results are obtained when both computers are designed and developed with hybrid applications as the major purpose. If a hybrid computer is made up of general purpose analog and digital computers, with an interface tailored to these, the resulting hybrid computer often poses limitations in equipment complement and operational features. Hybrid computers of this kind may represent a restricting compromise between available designs, hardware and the desired characteristics of an efficient hybrid computer.

Hardware

The characteristic part of a hybrid computer is the interface which connects the analog and the digital computers. The interface consists of data communication channels in which information is passed between the two computational parts. The interface does not carry out computations, but it may contain equipment that incorporates computational units, such as multiplying digital-to-analog converters.

The interface contains a number of conversion channels in which information is converted between an analog signal (voltage) and an encoded numerical (discrete) digital computer representation, according to programmed instructions and hardware executions. The number of conversion channels states the total parallel capacity of the interface, for conversions in both directions. In practice, the number of A/D (analog-to-digital) channels may differ from the number of D/A channels, depending on applications and implementations. Since the conversion channels link parallel and concurrent analog computer variables with serial and sequential program steps in the digital computer, the interface must provide storage facilities, by which information (variables) can be stored (held) while all channels are being prepared (loaded), so all conversions can take place simultaneously, in terms of analog variables, and sequentially, in terms of digital variables.

If the converted information is not buffered, it reflects computational variables or conditions that are not concurrent or co-referenced in time. Such a skewness is indicated in Figure 9.2s, showing the effect of a sequential conversion capability (among several time-shared units). To a degree, and at the cost of computer time, the skewness effects may be reduced by a common hold mode for all the analog computing units, during which the information

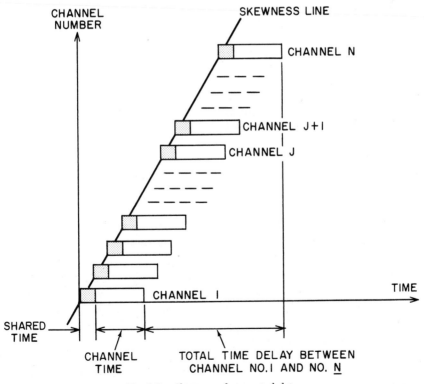

Fig. 9.2s Skewness of converted data

conversion may take place. Fast, accurate mode control facilities which allow rapid computer interruption and resumption of the analog program are necessary for this.

Figure 9.2t shows an example of an A/D conversion channel in which the analog signal is tracked and stored in analog form. The conversion channel utilizes a multiplexer and converter which is shared between a number of channels, typically 24 to 36, and it reads the converted information into a programmed (controlled) memory location in the digital computer. From this memory location the digital program can then obtain the converted information.

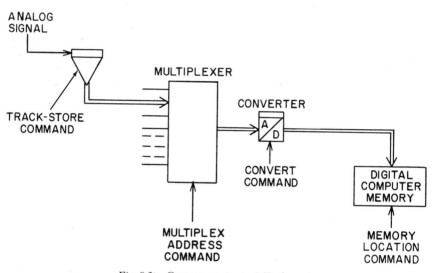

Fig. 9.2t   Components in an A/D channel

An example of a D/A conversion channel is shown in Figure 9.2u, consisting of a buffer register for the digital information and a converter unit. The buffer register holds the digital information until the moment of conversion, when it is loaded into the converter with other D/A channels. For single-channel or continuous D/A conversion, the buffer register is sometimes by-passed, and the digital information read directly (jammed) into the converter. The analog conversion output is a direct signal output or, as in a *multiplying D/A*, the product of an analog signal input and the converted D/A information. The conversion moment is only triggered by digital computer instructions, or indirectly, by interrupts from the analog computer through the digital computer program.

A number of *status channels* in the interface provide binary information exchange between the analog and the digital computers. This information relates to the status of particular computational variables or computing units,

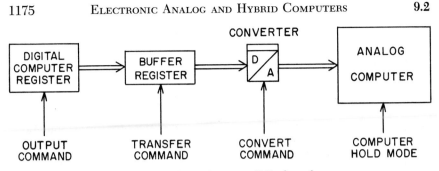

Fig. 9.2u   Components in a D/A channel

and to the condition and progress of the programmed functions. Status channels are used to ensure proper relationships in terms of operation sequences, timing of program events, steps, and coordination of the two hybrid computer parts.

There are three main types of status channels, depending on the importance or immediacy of the reported conditions:

> *Status indicating*, which at all times corresponds in status to the condition of its input variable.
>
> *Status retaining*, which will retain the predetermined status of the input variable from the first time it is set or activated and until the status channel is interrogated by the receiving computer and the channel is reset or deactivated.
>
> *Status responding*, which will respond immediately to the condition of the input variable when this is set or activated, and will cause the receiving computer to interrupt its tasks or change its modes or functions according to the programmed actions.

Status channels are generally furnished in the interface as one-way signal lines which must be assigned to the particular conditions or computing units which are to be reported during the programming.

The interaction and control of operations between the two computers are handled by a number of binary *command channels*. The command channels represent direct, fixed interactive links between the control systems or command sources of the two computers. They control the executions of programs, such as start and stop computations and iterations, and the initialization or reset of programmed functions and routines.

Operation

The operational efficiency of a hybrid computer depends on the command and control (instruction) structures of the two computer parts. Analog and digital computers are sufficiently different in organization and operation to present profound problems in terms of command characteristics and functional orientation. In a hybrid computer the analog and digital computers are linked together on the fundamental control level, providing facilities for mutual interruption and interaction of tasks.

In terms of analog computer control, the hybrid computer permits the digital computer program to carry out the control system functions previously outlined. These functions span the setup and checkout of analog and hybrid programs, the initialization and presetting of conditions for the computations and simulations, and the measurement, recording, and monitoring of the analog computer variables and functions during productive runs. Of prime importance in a hybrid computer is the ability of the computer programs to govern the progress of the computations, and take the appropriate control actions depending on the obtained results and responses.

The digital computer instruction repertoire normally includes many bit handling instructions designed to facilitate the exchange of information through the status channels, command channels, and the direct mode control and command channels. These instructions permit multiple level priority assignments and convenient program handling of the exchanged information, either in converted data format or in single bit format. The fast access to information in the digital computer memory through direct access or cycle stealing is important for high-speed conversion channel utilization and efficient hybrid computations.

Software

Proper and complete software is required to attain the optimum operational efficiency and utilization of the hybrid computer. This is especially important in hybrid computer applications, since the complexity and extent of many hybrid programs and the sophistication of the instruction repertoire and control routines in the computers otherwise limit the usefulness and understanding of the computer capabilities.

The software is designed with practical problem-solving objectives in mind, such as handling the systems functions and presenting conditions and variables in concise and efficient formats. For best results the software is written for a particular hybrid computer and is defined and specified according to the characteristics of the hardware configuration. The software consists of three types of system programs:

> *Batch-oriented* hybrid computer operation programs; organized and utilized for large, complex hybrid problems, with extensive program setup and checkout demands.
>
> *Conversational* computer operation programs; designed for experimental and development-oriented type hybrid computations and simulations.
>
> *Utility routines;* for efficient and convenient setup, checkout and

documentation of intermediate or limited hybrid computer programs, such as test programs, experimental circuit evaluations, and hybrid program developments.

The software enables the operator to carry out a hybrid computation or simulation of his specific problem with a minimum of effort. In a hybrid application this includes determining the signal connections and tie-lines, calculating the scale factors and coefficients of analog variables, adjusting coefficient units to the appropriate values prior to, or during, computations, and selecting the states or modes of the analog computing units. An important aspect of the software is the readout and documentation capabilities in terms of obtaining the values within the hybrid program, decoding or interpreting these in the language of the problem (such as in engineering units or mathematical terms), and making this information available to the operator, either by typewriter output, graphic displays, storage on punched cards, paper tape, or recorded in other appropriate forms.

LITERATURE

1. J. J. Blum, *Introduction to Analog Computation*, Harcourt, Brace & World, Inc., New York, 1969.
2. J. McLeod, edit., *Simulation—The Dynamic Modeling of Ideas and Systems With Computers*, Simulation Councils, Inc., La Jolla, California, 1968.
3. G. R. Peterson, *Basic Analog Computation*, The Macmillan Company, New York/London, 1967.
4. M. L. James, G. M. Smith, J. C. Wolford, *Analog Computer Simulation of Engineering Systems*, International Textbook Co., Scranton, Pennsylvania, 1966.
5. G. A. Korn, *Random Process Simulation and Measurement*, McGraw-Hill Book Co., New York, 1966.
6. R. R. Jenness, *Analog Computation and Simulation: Laboratory Approach*, Allyn and Bacon, Inc., Boston, 1965.
7. G. A. Korn, T. M. Korn, *Electronic Analog and Hybrid Computers*, McGraw-Hill Book Co., Inc., New York, 1964.
8. J. R. Ashley, *Introduction to Analog Computation*, John Wiley & Sons, Inc., New York, 1963.
9. W. Giloi, R. Lauber, *Analogrechnen* (in German). Springer-Verlag Berlin/Göttingen/Heidelberg, 1963.
10. D. M. MacKay, M. E. Fisher, *Analog Computing at Ultra-High Speed*, John Wiley & Sons, Inc., New York, 1962.
11. R. J. Tomovic, W. J. Karplus, *High Speed Analog Computers*, John Wiley & Sons, Inc., New York, 1962.
12. H. D. Huskey, G. A. Korn, *Computer Handbook*, McGraw-Hill Book Co., New York, 1962.
13. Anon., *Computer Basics*, Vol. I, II, and V (U.S. Navy Course) H. W. Sams & Co., Inc., The Bobbs-Merrill Co., Inc., Indianapolis/New York, 1961.
14. S. Fifer, *Analogue Computation*, I–IV, McGraw-Hill Book Co., New York, 1961.
15. R. M. Howe, *Design Fundamentals of Analog Computer Components*, Van Nostrand Book Co., New York, 1961.
16. A. S. Jackson, *Analog Computation*, McGraw-Hill Book Co., New York, 1960.
17. A. E. Rogers, T. W. Connolly, *Analog Computation in Engineering Design*, McGraw-Hill Book Co., New York, 1960.
18. N. J. Scott, *Analog and Digital Computer Technology*, McGraw-Hill Book Co., New York, 1960.

19. G. W. Smith, R. C. Wood, *Principles of Analog Computation*, McGraw-Hill Book Co., New York, 1959.
20. T. D. Truitt, A. E. Rogers, *Basics of Analog Computers*, J. F. Rider Publisher, Inc., New York, 1960.
21. W. J. Karplus, *Analog Simulation* (Solution of Field Problems), McGraw-Hill Book Co., New York, 1958.
22. C. L. Johnson, *Analog Computer Techniques*, McGraw-Hill Book Co., New York, 1956.
23. J. N. Warfield, *Introduction to Electronic Analog Computers*, Prentice-Hall, Inc., New York, 1956.
24. Edit., *A Palimpsest on the Electronic Analog Art*, G. A. Philbrick Researches, Inc., Boston, 1955.
25. C. A. A. Wass, *Introduction to Electronic Analogue Computers*, Pergamon Press Ltd., London, 1955.
26. W. W. Soroka, *Analog Methods in Computations and Simulation*, McGraw-Hill Book Co., New York, 1954.
27. G. A. Korn, T. M. Korn, *Electronic Analog Computers*, McGraw-Hill Book Co., New York, 1952.

9.3 ANALOG AND HYBRID COMPUTER APPLICATIONS

Analog and hybrid computers are important tools for investigating, evaluating, experimenting, and designing dynamic systems and components. Traditional analog computations aim at solving sets of differential equations (often non-linear equations with time-varying coefficients). Hybrid (analog-digital) computers combine this differential equation capability with general logic and decision-making functions, memory, and efficient and convenient control facilities. Thus, the hybrid system offers computational and simulational powers in a wider application area, and allows more extensive and complex problems to be solved.

The complexity and size of problems that can be solved by an analog computer depend basically on the number of computing units. More computing units provide a greater computer capacity, while the problem solution time remains the same. In contrast, the more complex or extensive a digital program becomes, the longer the digital computation time will be independent of computer size. Hybrid computers attempt to combine the solution speed of the analog computer with the program control of the digital computer for a better system with high speed.

From an application point of view, the analog computer is a continuous, parallel, fixed-point computer. Its fundamental feature is the simultaneous and interactive operation of the computing units, by which the solution to the programmed problem, or the response of the simulated system, is generated. The hybrid computer represents a combination in hardware and software of analog and digital computers. The characteristics of a hybrid computer depend strongly on the interface between the analog and the digital computers. Interface includes rates of information conversions, mutual interaction and coordination in computer operations.

The hybrid computer can solve problems of greater complexity than either type alone. Its overall resolution corresponds to that of the analog computer. Added uncertainties are due to the information conversions and to the time differences incurred in each exchange. The hybrid computer program must be composed and scaled in the same manner as an analog program as far as the analog part is concerned. The digital part of the hybrid program is usually composed in a hybrid language developed particularly for the available hardware system, or in an adapted version of a standard programming language such as FORTRAN.

Analog and hybrid computations may be categorized as the solutions of mathematically expressed problems. Such applications include well defined and parameterized input-output relationships, where the objectives are to find a particular solution or function within a family of solutions, or to arrive at some collective characteristics of a range of solutions, or for a variety of parametric values.

Analog and hybrid simulations consist of applications where the emphasis is on the functional representation of the system or component to be "modeled," or on the composite responses of a number of "black boxes." Simulations often aim at the overall effects of a number of subsystems or units interacting with each other. Some, or all such subsystems may be mathematically expressed, without the whole system being formulated in analytical terms. Usually the simulations imply experimentations and wide-range variations with little or no prior knowledge of the final results or of the circuit values and configurations.

Programming

Electronic analog and hybrid computations and simulations are based on the parallel operations of a number of computing units. Each unit represents a specific mathematical or functional relationship between a set of input variables and the unit's output variable, the output signal. This relationship applies only to a defined signal (voltage) range and within a limited spectrum of signal frequencies (rates of change) of the output signal. Analog and hybrid applications are generally based on similar circuits and employ identical techniques, as in the following examples.

Circuit Diagram

Analog and hybrid programs are defined by a circuit diagram showing the interconnections of analog computing units and channels of information or control signals through the interface to the digital computer. The circuit diagram is a block diagram representation of the signal flows by which the interactions of the computing units is visualized. The circuit diagram usually includes only those computing units and control signal lines which are specifically related to the computations or simulations within the program.

The circuit diagram is developed according to the mathematical equations to be solved, the systems or components to be simulated, or the functions to be generated. It is based on the graphic symbols given in Table 9.3a. The reader may also refer to Sections 4.4, 4.6, A.1 and A.2 for standard symbols. These symbols state the basic functions to be carried out by the units, and do not include specific hardware design features or particular applicational aspects. Other symbols are available for more detailed functional descriptions or specialized operations.

Table 9.3a
GRAPHIC SYMBOLS FOR ANALOG COMPUTERS

SYMBOL	DESCRIPTION

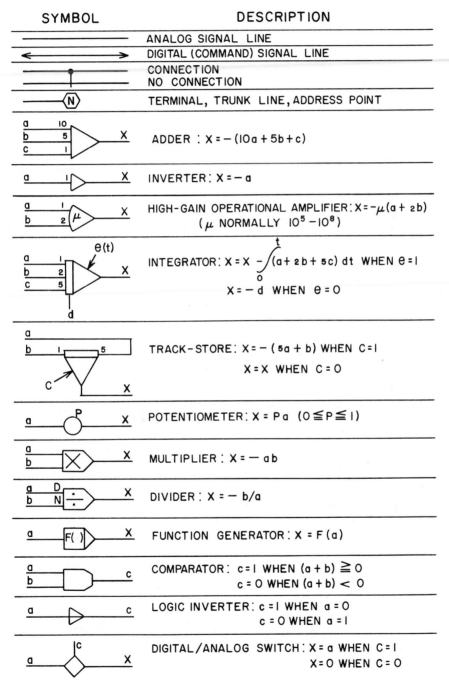

ANALOG SIGNAL LINE

DIGITAL (COMMAND) SIGNAL LINE

CONNECTION
NO CONNECTION

TERMINAL, TRUNK LINE, ADDRESS POINT

ADDER : $X = -(10a + 5b + c)$

INVERTER: $X = -a$

HIGH-GAIN OPERATIONAL AMPLIFIER: $X = -\mu(a + 2b)$
(μ NORMALLY $10^5 - 10^8$)

INTEGRATOR: $X = X - \int_0^t (a + 2b + 5c)\, dt$ WHEN $e = 1$
$X = -d$ WHEN $e = 0$

TRACK-STORE: $X = -(5a + b)$ WHEN $C = 1$
$X = X$ WHEN $C = 0$

POTENTIOMETER: $X = Pa$ $(0 \leqq P \leqq 1)$

MULTIPLIER : $X = -ab$

DIVIDER : $X = -b/a$

FUNCTION GENERATOR : $X = F(a)$

COMPARATOR : $c = 1$ WHEN $(a + b) \geqq 0$
$c = 0$ WHEN $(a + b) < 0$

LOGIC INVERTER: $c = 1$ WHEN $a = 0$
$c = 0$ WHEN $a = 1$

DIGITAL/ANALOG SWITCH: $X = a$ WHEN $C = 1$
$X = 0$ WHEN $C = 0$

SYMBOL	DESCRIPTION

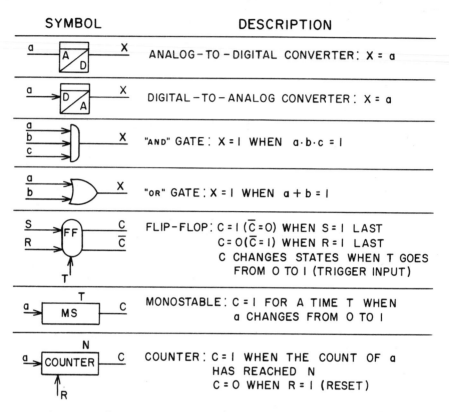

	ANALOG – TO – DIGITAL CONVERTER : $X = a$
	DIGITAL – TO – ANALOG CONVERTER : $X = a$
	"AND" GATE : $X = 1$ WHEN $a \cdot b \cdot c = 1$
	"OR" GATE : $X = 1$ WHEN $a + b = 1$
	FLIP-FLOP : $C = 1$ $(\bar{C} = 0)$ WHEN $S = 1$ LAST $C = 0 (\bar{C} = 1)$ WHEN $R = 1$ LAST C CHANGES STATES WHEN T GOES FROM 0 TO 1 (TRIGGER INPUT)
	MONOSTABLE : $C = 1$ FOR A TIME T WHEN a CHANGES FROM 0 TO 1
	COUNTER : $C = 1$ WHEN THE COUNT OF a HAS REACHED N $C = 0$ WHEN $R = 1$ (RESET)

The circuit diagram should be as simple as possible, including only those details that are necessary for the proper interpretation. Additional information of operation or application should be given in appended or referenced notes. The diagram is usually drawn with the main flow of information or succession of operations from left to right. Minor loops or feedback signals are in vertical or right-to-left direction.

The circuit diagrams included in this and the preceding section are arranged and scaled for the unity time scale or speed of integrations, with a few exceptions. For other time scales or speeds, the circuits can be easily adapted by changing the speed of the integrators according to the applicational requirements. Integrators are shown with only *one* mode control signal (OPERATE), implying that when this control signal is zero (low) the integrator is in the initial condition mode (RESET). Other modes, such as HOLD, STATIC TEST, or POTENTIOMETER SET, may be used as common computer modes, without affecting the circuits. Where the circuit integrators are in the standard OPERATE mode, the circuits can be used in repetitive operations as well as in single runs with proper resetting and initialization.

Scaling

For proper operation, the analog computer variables (analog signals) must be scaled in magnitude (amplitude) and speed (rate-of-change, or frequency) to fit the computer requirements and to get the most efficient and accurate performance from the computing units. Normally, the computer reference (voltage) is used as the full range (unity) measure of magnitude, although many computing units will allow signals up to 20 to 50 percent above this without deterioration in linearity or functional integrity.

Scaling in magnitude means that the analog variable at its maximum or minimum point should be as near unity as practical or possible. Exceeding the unity value may cause overload (overvoltage or excessive current), which will impair the functioning of the affected units, and may cause the analog computer to be out of balance. Similarly, an analog variable that never approaches unity is under-ranged and should be rescaled to take full advantage of the computing range (better signal-to-noise ratio).

The magnitude scale is defined by the relation

$$\text{Analog Variable} = \frac{\text{Problem Variable}}{\text{Top of Scale of Problem Variable}} \qquad 9.3(1)$$

With this definition, the analog variables will be dimensionless and will range between a high and a low value depending on the selected *full-scale value.* This top of scale (full scale) is also called the *scale factor,* since the value of the problem variable at all times may be determined according to the "scale,"

$$\text{Problem Variable} = (\text{Analog Variable}) \times (\text{Scale Factor}) \qquad 9.3(2)$$

Initial scaling is carried out using expected or calculated maximum (minimum) values of the problem variables as the top of scale values. If, for instance, an automobile speed v is anticipated to range from 24 to 55 mph during a simulation, the full scale may be chosen at 60 mph. With this value, the analog variable will range between 0.400 and 0.9166 as a representation of v. If the full scale had been set at 100 mph, the analog variations would only be 0.240 to 0.550, or under-ranged.

Figure 9.3b shows an example of a scaled analog circuit in which a variable v enters from the left. At point A, the scaled variable (the product of v and y) is

$$A = \left(\frac{v}{60}\right)(-5y)(-1) = \frac{vy}{12} \qquad 9.3(3)$$

The summation produces an analog variable at B which is

$$B = -[(0.600)(10)(A) - (0.250)(0.2u)] \qquad 9.3(4)$$

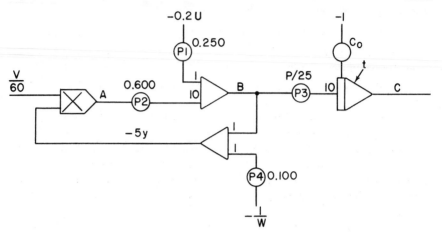

Fig. 9.3b   Example of scaled analog circuit

or

$$B = -\frac{vy - u}{20} \qquad\qquad 9.3(5)$$

The integrand, accordingly, will be equal to

$$\frac{dC}{dt} = -\left[\left(\frac{P}{25}\right)(10)(B)\right] \qquad\qquad 9.3(6)$$

If the variable B is rescaled (after a test run or an evaluation of the circuit) to represent

$$B = -\frac{vy - u}{10} = -2\left[\frac{vy - u}{20}\right] \qquad\qquad 9.3(7)$$

then the inputs to it must be raised in proportion to this rescaling, ×2, while the outputs from it must be attentuated in a similar ratio, ×0.500.

Figure 9.3c shows the rescaled version of the circuit, with the new potentiometer settings for P1, P2 and P3. Since the output B is now twice what it was before, the input to y must be reduced to 0.500. The potentiometer P5 takes care of this. Since the original connection from A to B did not allow an increase by a factor of 2, an additional input with a gain of 5 was introduced:

$$5 + (0.700)(10) = 12 = (2)(6) \qquad\qquad 9.3(8)$$

The same effect could also have been achieved by rescaling v, or y, or both, together with readjusting P2. In general, several ways must be evaluated to find the simplest circuit with *standard* input gains and *practical* potentiometer settings.

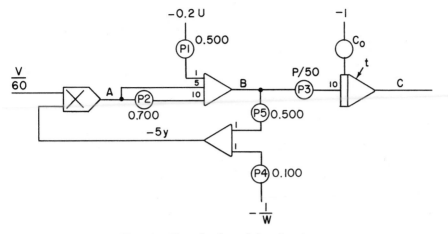

Fig. 9.3c　Example of rescaled analog circuit

Time Scale Selection

The rate-of-change, or speed, of the computational variables is directly related to the time scale or integration speed used in the program. Unity time scale, or speed, corresponds to real (clock) time, with a one-to-one time lapse correspondence between problem variables and computer variables. If the computer is operated at a higher speed k, the computations or simulations will occur k times faster than the real time version. Consequently they must be measured in a time scale which is 1/k of real time scale. This is the basis for using speeded-up repetitive computations with CRT display for trial-and-error type programs, and real time single run recording for hard copy documentations. Within such time scale changes the analog program remains the same and no special rescaling is required.

In Laplace transform notation (Section 7.4), the integrator in the unity gain-speed configuration is described by the transfer function

$$H(s) = -1/s \qquad\qquad 9.3(9)$$

If this is defined instead to be $H(s) = -k_p/s$ without altering the time scale definition, an increase in speed with a factor of k_p is introduced. Thus, all integrators appear with k_p higher input (integrand) gains when the analog problem is programmed, making higher gains available. However, the problem is "slowed down" by the same factor, since the time scale (which remains the same) measures k_p times shorter.

An example of this is given in Figure 9.3d which shows the circuit for solving the equation:

$$\frac{d^2y}{dt^2} + Ay = 0 \qquad\qquad 9.3(10)$$

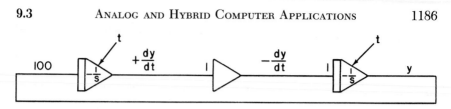

Fig. 9.3d   Circuit to solve equation 9.3(10), with A = 100

Here, A is first given as 100, which leads to the circuit shown, with integrator input gains of 1 and 100. If the integrator transfer function is defined as $-10/s$, the circuit appears as in Figure 9.3e with only unity gain inputs. Now the problem is solved 10 times slower, which means analog events observed m seconds apart really correspond to problem events which are m/10 seconds apart.

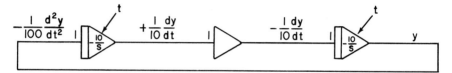

Fig. 9.3e   Redefined integrator transfer function to solve equation 9.3(10)

If $A = 0.0001$, the factor $k_p = 10^{-2}$ may be used, giving an integrator transfer function of $-1/100s$ with unity integrator inputs. In this case the measured m seconds of the analog computation must be interpreted as 10^2m seconds of problem time. In other words the problem has been "speeded up" by a factor of 100 due to this definition.

Testing

The purpose of testing the analog computer circuits is to verify that the desired signal connections have been properly established, and that the selected computing units operate correctly and within satisfactorily scaled ranges of variation. Using the circuit shown in Figure 9.3c as an example, for given initial values of the variables, such as $v_0 = 30$, $u_0 = -4$, $W_0 = 1.0$, $C_0 = 0.25$, and so on, the appropriate outputs of all units may be calculated. In this case the variables in the *initial condition mode* are

$$\left. \begin{array}{c} (v_0/60) = 0.500 \\ -0.2u_0 = -0.2(-4) = 0.8 \\ -(1/W_0) = -1.0 \\ C_0 = 0.25 \end{array} \right\} \qquad 9.3(11)$$

From these values the remaining variables in Figure 9.3c may be found:

$$-5y_0 = -[0.1 + 0.5(-1)(0.4 + 12(-1)(0.5)(-5y))] \qquad 9.3(12)$$

or

$$-5y_0 = 0.075 \quad \text{(thus } y_0 = -0.015) \qquad 9.3(13)$$

Also,

$$\left.\begin{array}{l} A_0 = -0.0375 \\ B_0 = 0.05 \\ (dC/dt)_0 = 0.02P \end{array}\right\} \qquad 9.3(14)$$

The initial value of the integrand of C is tested in the STATIC TEST mode, which allows the value to be read as an analog signal similar to the other variables. The test values may be calculated by the digital computer in a hybrid system, and may be tested as a part of the initialization of a hybrid program.

Documentation

For identification and interpretation of variables, the analog and hybrid computer program must be documented. In addition to the circuit diagram with the scaled variables, it is desirable to list all variables with the appropriate full-scale dimensions, and the values of the coefficients introduced in the program, such as scaling values or initial conditions.

Table 9.3f is a list of program variables for a sample problem. The table contains entries for computing unit number, the function of this unit (if several alternatives exist), the output signal representation, initial or test value, and the problem variable dimension (for reference only). From this listing a measured value of the analog signal, for instance, $(v/60) = 0.880$, at the computing unit 004, can be evaluated to give the problem variable $v = (60)(0.88) = 52.8$ in miles per hour.

Table 9.3f
A LISTING OF PROGRAM VARIABLES (EXAMPLE)

Computing Unit Number	Function	Program Variable	Test (Initial) Value	Full Scale		Comments
001	Add	$-5y$	0.0750	0.2	GPM	
002	Invert	$-0.2u$	0.8000	5	GPH	$u_0 = -4$
003	Add	B	0.0500	10	GPH	
004	Integrate	$v/60$	0.5000	60	mph	Car speed
005	Multiply	$vy/12$	-0.0375	12	GPH	
006	Integrate	C	0.2500	1	gallon	
007						
008						

Table 9.3g gives the list of coefficient values to be entered in the program. The list contains information pertaining to the purpose of the coefficient (scaling, parameter or initial value), the representation of the coefficient value

Table 9.3g
A LISTING OF COEFFICIENTS (EXAMPLE)

Coefficient Number	Function	Program Representation	Test Value	Input Variable	Comments
010					
011	Scale	0.500	0.500	0.2u	
012	Parameter	P/25	0.100	B	To be varied 0.1–0.8
013	Scale	0.700	0.700	vy/12	Part of $\times 12$ input
014	Initial Value	C_0	0.500	−1	
015	Scale	0.100	0.100	1/W	
016					

and the numerical value to be used in tests, or in running the program. Additional information on the analog variable to which the coefficient is applied and possible comments as to variation, scale changes, and reference units are also given.

Hybrid Programs

Hybrid computer programs consist of three parts:

1. Analog circuit diagram (see previous paragraphs)
2. Digital computer flow chart (see Section 8.4 for details)
3. Hybrid interface program

Characteristic of hybrid computer applications are the interactions and coordinated operations of analog and digital computers. These tie together the sequential (step-by-step) program actions in the digital computer with the parallel (simultaneous) computations in the analog computer. The purpose of the hybrid program and documentation is to define the cause-and-event relationships and to identify the conditions and priorities under which these relationships are valid.

In general, the transmittal of information between the analog and the digital computers takes place in a number of *interface channels,* each with its typical use and function. These channels are of the following types:

Analog-to-digital conversion channels (high or low speed).
Digital-to-analog conversion channels (high or low speed).
Status reporting lines (bits) for present events or conditions.
Status retaining lines (bits) for past (previous) events or conditions.
Status commanding lines (bits) for event or condition commanding.
Program control lines for computer operations and synchronizations.

Depending on the particular hybrid computer arrangement and the qualities of the interface channels, the hybrid program and documentation may take various forms. A list of interface variables is given in Table 9.3h,

Table 9.3h

A LISTING OF INTERFACE VARIABLES (EXAMPLE)

Analog Variable	Computing Unit Number	Check Value	Interface Channel	Program Step Number	Digital Instruction	Comments
vy/12	005	0.025	A/D-1	2.06		Start point
v/60	004	0.500	A/D-2	2.15		Read v
C	006	0.250	D/A-4	2.32		Set IC
P/25	P012	0.100	P.S.	4.40		New parameter
−0.2u	002	0.800	A/D-5	4.42		
x + y	014	0.200	SL-2	4.60		
g	044	—	SC-1	4.82		Close switch

exemplifying the ties between the analog and the digital programs. Key entries are the analog variables (computing units) and the digital program steps.

Of special concern in most hybrid computer applications are the utilization of time in the two computers, notably, the apparent reduction in efficiency due to waiting, information conversion, and loading of buffering units. In hybrid programming, care must be taken to mesh and sequence properly the analog and digital operations. "Time-event" diagrams and hybrid computer flow charts may be used for this purpose.

Computations (Analog)

Analog computations were explained as the use of the analog computer in the solution of mathematically expressed problems, such as an analytical function or a set of equations. This was based on the observation that within the operational ranges of its computing units, the analog computer functions in a precise and well-defined manner, with mathematical expressions governing the relationships between input and output variables. In this sense the analog computer is a mathematical instrument, or an analytical tool with which certain mathematical problems or equations may be solved.

Differential Equations

A major application area for analog computations is the solution of sets of differential equations. Such equations are usually related to each other through some common variable or function, and the sets of equations express and define a system behavior or the relationships governing a higher entity. Of particular importance are sets of ordinary differential equations, which are related through one independent variable, such as time.

The solution of ordinary differential equations may be carried out in three different ways:

1. Direct instrumentation.
2. Indirect, closed-loop approach.
3. Implicit function generation.

The direct instrumentation method, as the name implies, is based on a straightforward use of analog computing units according to the equations to be solved, with little consideration to the functional or operational characteristics of the analog computer. The direct instrumentation method may also be called "feedforward" or "open-loop" instrumentation, because there is only one-way action (or computation) from input to output, such as from given problem variables to displayed or recorded output variables.

An example of such an instrumentation is shown in Figure 9.3i corresponding to the equation

$$dy/dt = ax^2 - b(\sin x) \qquad 9.3(15)$$

where both x and y are functions of t. In this circuit the output is dy/dt and depends only on the input x.

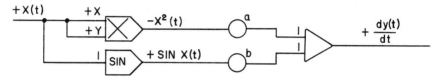

Fig. 9.3i   Direct instrumentation method for solving differential equations

In the example the information flows from left to right, or from independent to the dependent variable, with no "feedback" or "closed loops" in the circuit. In general the direct instrumentation method applies to equations in which a complete separation of variables is possible. Their main use is in connecting otherwise computed or generated variables or functions.

The indirect method of solving differential equations is based on the successive integration of a computer variable from its highest to lower order terms (with respect to time). This is expressed mathematically by

$$\frac{d^{(n-1)}x}{dt^{(n-1)}} = C_0 + \int \frac{d^{(n)}x}{dt^{(n)}} \, dt \qquad 9.3(16)$$

where C_0 is the integration constant. For a typical equation, such as

$$\frac{d^3y}{dt^3} + a\frac{d^2y}{dt^2} + b\frac{dy}{dt} + cy + d = 0 \qquad 9.3(17)$$

the method states

$$\frac{d^3y}{dt^3} = -a\frac{d^2y}{dt^2} - b\frac{dy}{dt} - cy - d \qquad 9.3(18)$$

where the left side is the integrand input to an integrator which produces as its output the variable d^2y/dt^2. This integrand consists of the terms on the right-hand side of equation 9.3(18) and is thus made up of several input and feedback variables. In other words, the right-hand side of the expression states what lower order terms are used in making up the highest order term of the original equation, and what other input functions (if any) affect the variable. The analog computer circuit for this example is given in Figure 9.3j.

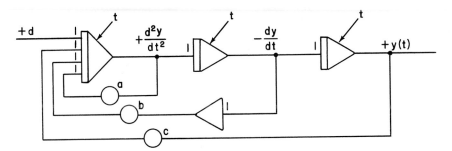

Fig. 9.3j   Indirect method of solving differential equations

The implicit method of solving differential equations is based on the principle of finding an implied (derived, or related) variable or set of variables that may be more easily determined than the given equations. The implicit method is primarily useful in the solution of equations with analytical functions, such as $\sin \omega t$ or $e^{-\alpha t}$, where time t is the only independent variable. Other applications are in generating inverse functions, for instance, division instead of multiplication, or differentiation instead of integration.

An example of the use of implicit variables is given in the generation of the function:

$$y(t) = A(b + t)^n \qquad 9.3(19)$$

where n is a variable. Solving this equation by direct instrumentation or by the indirect method would require a number of function generating units. By differentiation with respect to time, an implicit equation may be found

$$\frac{dy}{dt} = nA(b + t)^{n-1} = \frac{ny}{(b + t)} \qquad 9.3(20)$$

or more specifically,

$$(b + t)\frac{dy}{dt} - ny = 0 \qquad 9.3(21)$$

This may be interpreted as the "balanced" condition at the input of a high-gain operational amplifier. This condition is used in the computer diagram shown in Figure 9.3k. The operational amplifier in the circuit seeks to balance its inputs according to equation 9.3(21) by finding the appropriate output $-dy/dt$ which will satisfy this by changing the feedback signal from the

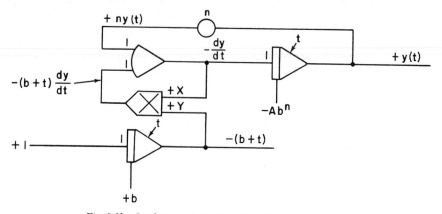

Fig. 9.3k   Implicit method of solving differential equations

multiplier. In this example the time t is the independent variable. Note that the output function y(t) has the initial value of Ab^n.

Figure 9.3l illustrates the general use of the implicit method with the function $F(x)$ in the feedback path of the operational amplifier. In the "balanced" condition the function output equals the input y with opposite polarity:

$$F(x) = -y \qquad 9.3(22)$$

The amplifier output is thus the inverse function

$$x = F^{-1}(y) \qquad 9.3(23)$$

If, for example, the function $F(x) = \sin(x)$ the output will correspond to

$$F^{-1}(y) = \arcsin(y) \qquad 9.3(24)$$

The circuit shown in Figure 9.3l will remain stable only if the feedback path through the function generator does not change the polarity of the feedback signal. Since most operational amplifiers are limited in velocity, it may sometimes be necessary to restrain the use of the implicit method to functions that are continuous, simple, and well within the performance limits of the units. In other applications the operational amplifier may have to be "slowed down" by a small capacitive feedback.

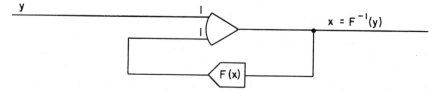

Fig. 9.3l   The general use of implicit function generation

Polynomials

Another area of analog computations is solving, or finding the roots of, polynomials. Root locus plots are often used in stability analyses and design evaluations. Two categories of methods are available for solving polynomials: the first is based on a systematic variation (sweeping or scanning) of the independent variable over its full range while the corresponding functional value of the polynomial is observed until it becomes zero or minimum. In the second category the methods are based on some strategy for ascending or descending a "terrain" (hill-climbing techniques). These methods use series of minor variations (perturbations) of the variables to find how the resulting polynomial value varies locally, and thereby in what direction to search.

Finding the polynomial roots by scanning is often the most immediate and easily applied method, with small equipment requirements. If the polynomial is known to have only real roots, these may be determined by successive integrations:

$$P(z) = az^3 + bz^2 + cz + d \qquad 9.3(25)$$

with the circuit shown in Figure 9.3m. In this case time t represents the independent variable z, and the roots are found by slowly letting t assume all values between -1 and $+1$ and observing the output function $P(z) = P(t)$.

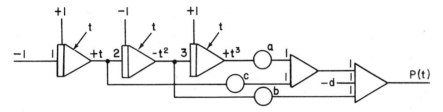

Fig. 9.3m   Successive integration method for finding roots of polynomials

If the polynomial has complex roots (or complex coefficients) a method of harmonic synthesis may be used, in which the variable z is replaced by a vector

$$z = re^{j\alpha} \qquad 9.3(26)$$

Higher order terms of z correspond to $r^2e^{j2\alpha}$, $r^3e^{j3\alpha}$ and so on. Substituting time ωt for the angle α (measured counterclockwise) the real parts and the imaginary parts may be collected in two polynomials:

$$\mathrm{Re}P(z) = ar^3\cos 3\,\omega t + br^2\cos 2\,\omega t + cr\cos\omega t + d \qquad 9.3(27)$$

$$\mathrm{Im}P(z) = ar^3\sin 3\,\omega t + br^2\sin 2\,\omega t + cr\sin\omega t \qquad 9.3(28)$$

The roots are found as the values of r and ωt (or α) when both polynomials $\mathrm{Re}P(z)$ and $\mathrm{Im}P(z)$ are zero *at the same time*. This is practically done by detecting when the sum of the absolute values of $\mathrm{Re}P(z)$ and $\mathrm{Im}P(z)$ is zero.

For finding the roots of a third order polynomial, such as given in equation 9.3(25), the "vector length" r is varied from zero to unity at a slow pace, while the scanning is done repetitively. The computer is programmed to stop or hold at the moment the sum of the absolute values becomes zero. The roots of the polynomial are then determined from

$$z_0 = r_0 \cos \omega t_0 + jr_0 \sin \omega t_0 \qquad\qquad 9.3(29)$$

where the subscript "0" identifies the polynomial zero conditions. The sum of absolute values is recorded in Figure 9.3n, indicating the location of the roots as minimum (zero) points.

There are two groups of strategies used to descend a "hill," those that for each step in the search determine which way is the best to go (the steepest descent for instance), and those that rely on pursuing one direction until it definitely bears uphill—and then changing direction at a right angle (resulting in a criss-cross search pattern). Characteristic of both groups is the need to carry out several search missions from different starting points in order to find all the minima (or zeros) of the polynomial.

Algebraic Equations

Algebraic equations are independent of time and thus have less applicational immediacy than differential equations. When algebraic equations form a part of a set of equations, or are inherently defined by the common variables, certain precautions are necessary. Since an algebraic equation (or algebraic loop) consists of time-independent functions, the analog computing units provide the solution practically instantaneously, which may cause instability and inaccuracy, especially if high loop gains and positive (regenerative) feedback effects are present. Negative feedback loops (an odd number of polarity inversions) do not normally pose any problems, whereas positive loops with gains equalling unity or greater may cause error buildup and instability. For such algebraic loops, the loop gains should be kept at less than unity, i.e., less than 0.99 or 0.98, to ensure proper performance.

The solution of a set of algebraic equations can be carried out in basically two ways: by purposely slowing down the algebraic loops through the introduction of time constants, and by carrying out the computations in a step-by-step manner, using track-store techniques and numerical (sequential) approximation methods. Figure 9.3o shows an example of an algebraic loop (heavy broken line) which may be stabilized by integrators. The penalty for this is the time needed for the integrator-filters to settle down, while the remaining parts of the computations must be in HOLD. Note that the loop gain actually remains the same.

Figure 9.3p shows the numerical approximation method of solving a set of algebraic equations in which the three equations

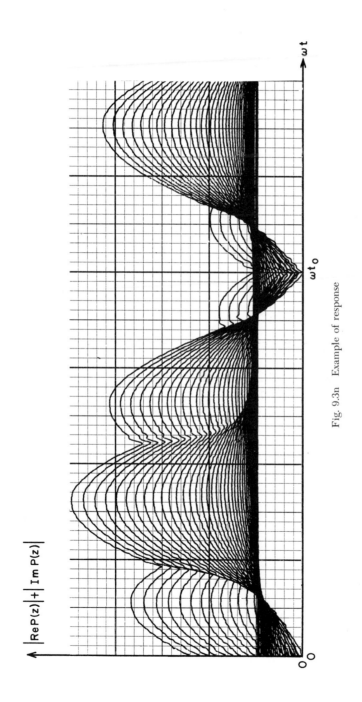

Fig. 9.3n Example of response

$$Ax - y + 2z = B$$
$$x - y - z = B/10$$
$$2x - y - 2z = C$$
$$9.3(30)$$

form the basis of the circuit. The variables x, y and z appear at the outputs of three track-store units which track by three sequential pulses T_1, T_2, and T_3. The circuit takes a few "go-arounds" of the command cycles before the variables settle down at the appropriate solutions, as indicated by Figure 9.3q. For proper convergence, the set of equations should be arranged to get the larger coefficients along the main diagonal (in the example, A, -1 and -2 for the variables x, y and z respectively).

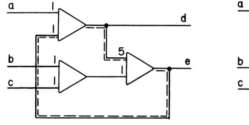 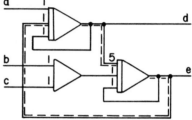

Fig. 9.3o   Algebraic loops

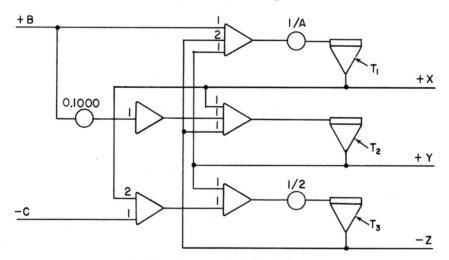

Fig. 9.3p   Successive approximations circuit

Repetitive Computations

In many analog computations, the solution may not be in the form of functions of the independent variable, but rather the values of coefficients and parameters in the problem which satisfy certain boundary requirements,

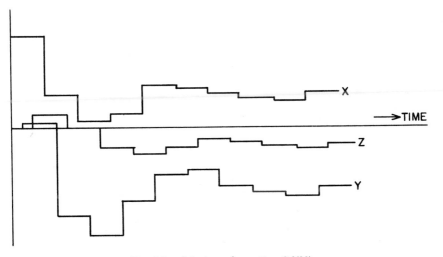

Fig. 9.3q   Solutions of equations 9.3(30)

limit the range of variations, or otherwise characterize the dependent varia-bles. Such computations are generally done repetitively, at a high speed, making it possible to display the variables or functions on a CRO (cathode ray oscilloscope). During such repetitive computations the operator may vary the problem coefficients and almost instantaneously see the effects on the computer solutions. He may also find the particular combinations (or charac-teristic functions) he is searching for.

A form of repetitive computations which makes use of the information storage capabilities of the track-store units is called *iterative computations*. By sampling (tracking) variables in one solution and carrying these into the succeeding ones, it is possible to let the computer find desired values, or special conditions, speedily and efficiently. Such iterative computations are useful when a large number of computer solutions must be examined, or when a multiple of combinations must be scanned in order to determine the required parameters or functions.

A simple example of an iterative analog computation is given in Figure 9.3r, in which the computer "finds" the value of the problem parameter "a" so that boundary requirements are satisfied. The computer solves repetitively the equation

$$d^2U/dt^2 = aU^2 + P \qquad\qquad 9.3(31)$$

and determines whether U_T (at time $t = T$) equals the required U_1. If not, a corrective feedback action occurs through the slow integrator A which is in continuous operation, as shown in Figure 9.3s.

Another form of repetitive computations that makes use of information

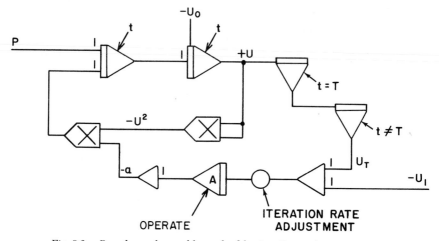

Fig. 9.3r   Boundary value problem solved by iterative analog computation

storage is *statistical analog computations,* in which a large number of computations are carried out to achieve a satisfactory statistical confidence level. The purpose of the computations may range from statistical parameters, such as variance and standard deviations, to probability functions and regression analysis. Often random signals (from noise generators) are used to "disturb"

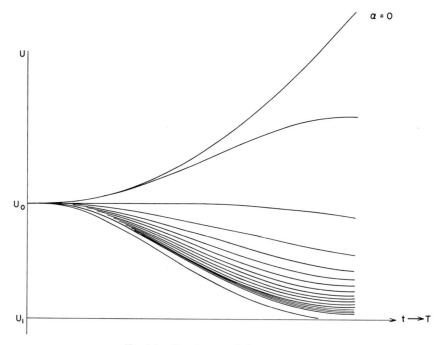

Fig. 9.3s   Iterative search for parameter α

or force the values of the computational variables, or to provide different initial conditions for multiple trials. The variables are usually sampled at some characteristic moments, and the values stored by track-store units, until the final statistical computations are completed.

Figure 9.3t shows an example of the computation of the probability function of the variable $y(t) = x + \phi(t)$, where $\phi(t)$ is a random variation within the limits ± 1. The circuit works on the basis of a fixed number of samples N in an ensemble, from which a number N_L is determined that corresponds to the probability p(L) of the function y(t) being equal or less than the reference level L. The reference is scanned over the range from -1 to $+1$ by increments of ΔL, each time the count N is "full" and a new counting starts. The probability p(L) is represented by the integrated pulse inputs to the unit A, which is determined by the comparison between the reference level L and the function y(t) at sample times C_T. The timing of the incrementing pulses ΔN and ΔL is governed by two monostables, MS-1 and MS-2 respectively.

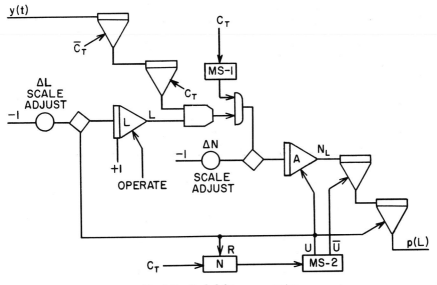

Fig. 9.3t   Probability computation

The probability function p(L) of the level L is shown in Figure 9.3u, as recorded by an XY-plotter.

Simulation (Analog)

Simulation is the study of processes, mechanisms, and devices by manipulation and observation of an analog computer representation (model) which has the same, or adequately similar, response characteristics for the particular

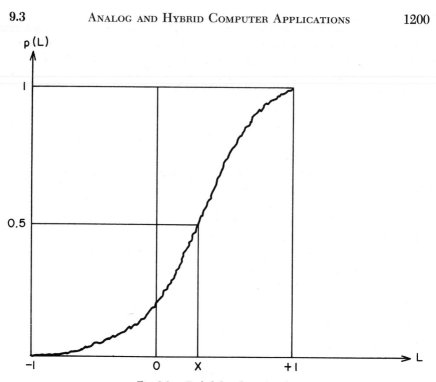

Fig. 9.3u   Probability function plot

aspects being studied, and which is convenient, economical, and efficient for the purpose. Such an analog representation is generally developed by systematic substitutions of computing units and circuits for some or all parts of the object of investigation, without emphasis on inherent mathematical relationships or overall systems aspects.

Analog simulations may involve only computing units, in which case the speed of simulated responses may be chosen for reasons of convenience or efficiency, and repetitive or multi-run operations may be utilized as desired. When special components, hardware units, or "live" processes are tied to the computer, the speed of response must be in *real time* and the simulation carried out in a CONTINUOUS RUN mode. Real time on-line simulations often span long time intervals, and may have many stability and reliability requirements.

Analog computer simulations normally include two varieties; simulations based on transfer function representations or similar input-output relationship definitions, or simulations based on a physical representation by knowledge and experience of the response characteristics involved. The understanding of the processes or mechanisms to be simulated, coupled with the imaginative utilization of analog computing units, makes simulation a valuable and far-reaching tool for systems analysis and synthesis.

Transfer Functions

Transfer functions are widely used in the study of dynamic systems. Most transfer functions are based on the Laplace transformation principle, which applies to linear systems (Section 7.4). In short, the transformation changes a time-domain (time-related) relationship into one of frequency-domain, where the characteristic dependence is on the s variable (frequency), instead of time t. If, for example, the transfer function is given as $H(s) = Y(s)/X(s)$, with

$$H(s) = \frac{k}{s(as^2 + bs + c)} \qquad 9.3(32)$$

then the corresponding relationship in the time-domain may be found by inverse Laplace transformation, or found in a table of transform pairs, as $h(t) = y(t)/x(t)$, or

$$h(t) = k[1 + C_1 e^{-bt/2} \sin(C_2 t + C_3)] \qquad 9.3(33)$$

Here $x(t)$ and $y(t)$ are input and output variables, respectively, and the constants are $C_1 = 2\sqrt{a}/\sqrt{4a - b^2}$, $C_2 = 1/\sqrt{a}\, C_1$, and $C_3 = \tan^{-1} 2C_2/bC_1^2$. This unique transformation facilitates the simulation of dynamic systems.

For a given transfer function $H(s)$ the corresponding analog computer circuit may be determined, as shown in the following example:

$$H(s) = \frac{Y(s)}{X(s)} = k\frac{s^3 + as^2 + bs + c}{s^3 + ps^2 + qs + r} \qquad 9.3(34)$$

A rearrangement gives

$$Y(s) = kX(s) + \frac{1}{s}\left[kaX(s) - pY(s) + \frac{1}{s}\left(kbX(s) - qY(s) \right. \right.$$
$$\left. \left. + \frac{1}{s}\left(kcX(s) - rY(s) \right) \right) \right] \qquad 9.3(35)$$

From this the analog computer circuit is derived, as shown in Figure 9.3v, with the successive integrations corresponding to the successive multiplications with $1/s$. Note that the actual transfer function for an integrator is $-1/s$, which calls for inversion of some of the terms in equation 9.3(35).

Figure 9.3w shows the generalized transfer function circuit of the first order (one integration), with two outputs or variables y_1 and y_2:

$$H_1(s) = \frac{Y_1(s)}{X(s)} = -\frac{as + (ad - bc)}{s + (d - ce)} \qquad 9.3(36)$$

$$H_2(s) = \frac{Y_2(s)}{X(s)} = \frac{(ae - b)}{s + (d - ce)} \qquad 9.3(37)$$

The generalized second-order transfer functions circuit is shown in Figure 9.3x, for which three equations can be formed:

$$H_1(s) = \frac{Y_1(s)}{X(s)}$$

$$= \frac{-as^2 + (bc + fg - a(d + m))s + [am(k - d) + b(cm - gh) + f(dg - ck)]}{s^2 + (d + m - ce - gn)s + [dm - hk + c(kn - em) + g(eh - dn)]}$$

$$9.3(38)$$

$$H_2(s) = \frac{Y_2(s)}{X(s)}$$

$$= \frac{(ae - b)s + [fk - bm + a(em - kn) + g(bn - ef)]}{s^2 + (d + m - ce - gn)s + [dm - hk + c(kn - em) + g(eh - dn)]}$$

$$9.3(39)$$

$$H_3(s) = \frac{Y_3(s)}{X(s)}$$

$$= \frac{(an - f)s + bh - df + a(dn - eh) + c(ef - bn)}{s^2 + (d + m - ce - gn)s + [dm - hk + c(kn - em) + g(eh - dn)]}$$

$$9.3(40)$$

The transfer function of a given analog computer circuit can be derived by conventional block diagram rules. For instance, an integrator circuit, as shown in Figure 9.3y, has the equivalent block diagram as shown on the right side of the illustration.

Non-Linear Functions

Often used in analog computer simulations are the non-linear functions, such as limiting, deadband, and backlash effects. General purpose function generating units may be programmed for many nonlinearities, although effects which call for "memory" of past responses, or sensitivity to more than one

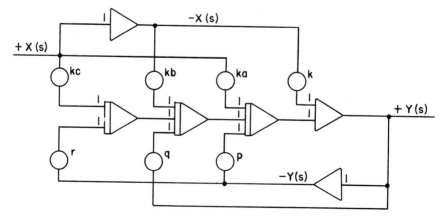

Fig. 9.3v   Analog circuit for the transfer function in equation 9.3(34)

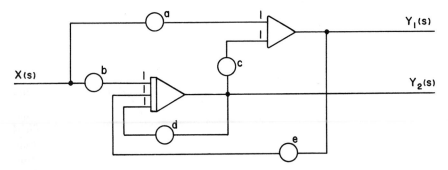

Fig. 9.3w   Generalized first-order transfer function circuit for equations 10.3(36) and 10.3(37)

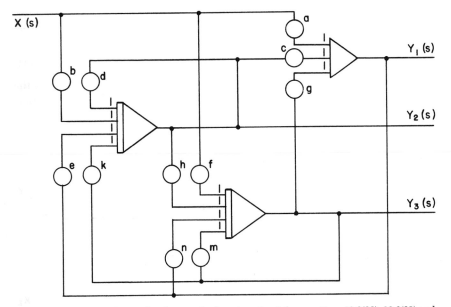

Fig. 9.3x   Generalized second-order transfer function circuit for equations 10.3(38), 10.3(39) and 10.3(40)

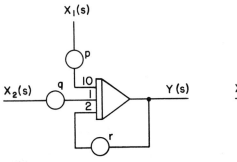

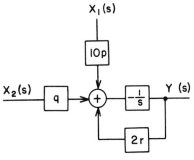

ANALOG COMPUTER CIRCUIT

EQUIVALENT BLOCK DIAGRAM

Fig. 9.3y   Deriving the Laplace transfer function

input usually require special circuits. The traditional network of operational amplifiers and diodes has been superceded by more precise and well defined actions achieved by signal comparators, analog switches and relays. One important aspect of this is the ability of the nonlinear function generating circuit to respond to parametric input signals, by which the nonlinearity characteristics become "variable" during simulations.

One of the basic non-linear functions is the limiter, with the circuit shown in Figure 9.3z. This function expresses a "lower" selection by the output, from two input signals X_1 and X_2 which may be stated:

$$Y = \text{LOW}(X_1; X_2) \qquad\qquad 9.3(41)$$

The normal input-output relationship is shown in Figure 9.3z, with the "limiting" value determined by the input X_2. Interchanging the two inputs reverses the limiting function and makes the circuit inverting.

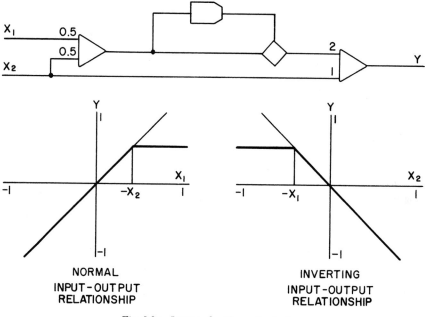

NORMAL
INPUT-OUTPUT
RELATIONSHIP

INVERTING
INPUT-OUTPUT
RELATIONSHIP

Fig. 9.3z   Limiter for "lower" selection

The "higher" selection is expressed by:

$$Y = \text{HIGH}(X_1; X_2) \qquad\qquad 9.3(42)$$

Two simple adaptions of the limiter are shown in Figure 9.3aa and Figure 9.3bb of the *absolute value function* and the *complementary value function*. They are expressed by the equations

$$Y = |X| = X - 2\,\text{LOW}(0;X) \qquad\qquad 9.3(43)$$

and

$$Y = 1 - |X| = 1 - X + 2 \, \text{LOW}(0;X) \qquad 9.3(44)$$

With the *double limiter* function, two parametric input signals X_1 and X_2 determine the limits of Y (and thereby of X). This function may be expressed

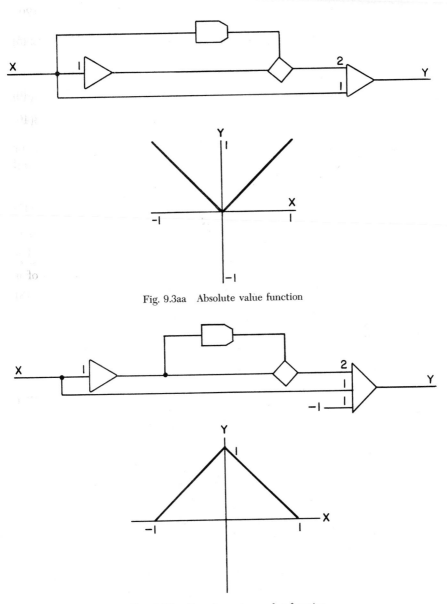

Fig. 9.3aa   Absolute value function

Fig. 9.3bb   Complementary value function

as

$$Y = \text{LOW}(X_1; \text{HIGH}(X; -X_2)) \qquad 9.3(45)$$

or

$$Y = \text{HIGH}(-X_2; \text{LOW}(X; X_1)) \qquad 9.3(46)$$

where $X_1 \geq 0$ and $X_2 \geq 0$.

A non-linear function resembling the double limiter is the *deadzone*, or *deadband*. This function is generated with the two parametric inputs X_1 and X_2 according to the equation

$$Y = \text{HIGH}(X - X_1; \text{LOW}(X + X_2; 0)) \qquad 9.3(47)$$

for $X_1 \geq 0$ and $X_2 \geq 0$. Another version of the deadzone type function is shown in Figure 9.3cc, which resembles the input-output relationship of a relay. This function may be expressed as

$$Y = X[1 + \tfrac{1}{2}\text{SGN}(X - X_1) - \tfrac{1}{2}\text{SGN}(X + X_2)] \qquad 9.3(48)$$

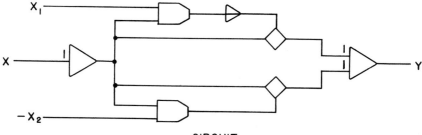

CIRCUIT

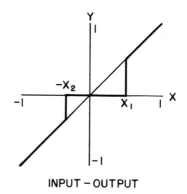

INPUT – OUTPUT

Fig. 9.3cc   Relay function

where the *signum function* is defined as

$$\text{SGN}(X) = \frac{X}{|X|} \qquad 9.3(49)$$

For many applications, an on-off type relay function is required, as shown in Figure 9.3dd, where the output is

$$Y = 1 + \tfrac{1}{2}\text{SGN}(X - X_1) - \tfrac{1}{2}\text{SGN}(X + X_2) \qquad 9.3(50)$$

with $X_1 \geq 0$ and $X_2 \geq 0$.

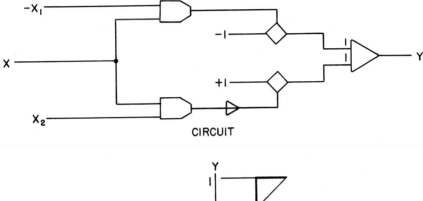

CIRCUIT

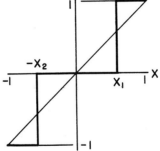

INPUT-OUTPUT

Fig. 9.3dd   On-off relay function

A second category of non-linear functions is based on the detection and retention of the highest or lowest value assumed by a variable or function during a preceding interval of time T. Such a "peak" detection during the interval T may be stated as

$$Y = \text{MAX}(X; T) \qquad 9.3(51)$$

for the highest assumed value of the variable X, and

$$Y = \text{MIN}(X; T) \qquad 9.3(52)$$

for the lowest value of X during the time period. In the corresponding circuits the integrator input gain A (or the input gain - integration speed product) must be large enough to give the output variable Y a velocity (slewing rate or speed) at least as high as that of the input variable X, in order for the circuit to "catch up" to the peaks. Also, the initial value Y_0 must be low enough in the range within which the "peak" value will occur.

An important application of the peak value detection circuit is in the *backlash function* generation, which is shown in Figure 9.3ee. For proper operation the constraints are

$$X_2 \leq 0 \leq X_1 \qquad\qquad 9.3(53)$$

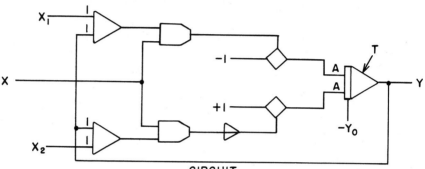

CIRCUIT

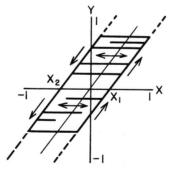

INPUT-OUTPUT

Fig. 9.3ee   Backlash function

In this case the integrator may be "initialized" with Y_0, since the output value must be obtained correctly for any given initial X_0 if

$$X_2 \leq X_0 \leq X_1 \qquad\qquad 9.3(54)$$

Another non-linearity with stringent memory requirements is the *hysteresis function*, in which the output variable Y depends not only on the present input value X, but also on the immediately preceding "history" of X and thereby of Y. The circuits for the hysteresis function and the input-output relationship are indicated in Figure 9.3ff. Note that the initial value of the output Y_0 may be zero, which will produce the correct "virgin" characteristic in response to X, within the time interval T after resetting the integrator. The two track-store units carry out the memorizing of the output Y for opposite polarities of the input signal.

A simple and hysteresis-like non-linearity is the *square-loop hysteresis effect.* Here the output variables are Y_1 and Y_2, corresponding to the para-

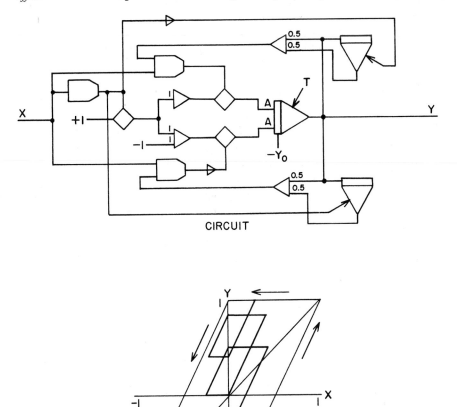

Fig. 9.3ff   Hysteresis function

metric input values (X_1 and X_2) and the full range output ± 1, respectively. For proper operation of this circuit, it is required that

$$X_2 \leq X_1 \qquad\qquad 9.3(55)$$

Process Simulation

The simulation of most processes, reactions, and plants may be carried out by developing a representation or building a model of common elements, where each element is adapted to the particular functions it must simulate. When all the parts have been defined and assembled, the overall effects and constraints may be imposed, such as material and energy balances, operating modes, limitations, instrumentation and control system characteristics.

A common element in the process and plant simulations is the lag, defined by the transfer function (in Laplace notation)

$$H(s) = \frac{Y(s)}{X(s)} = \frac{K}{1 + \tau s} \qquad\qquad 9.3(56)$$

where K is the steady-state gain, and τ is the time constant. In general, the gain and time constant will vary with other simulation variables, such as flows, temperatures and pressures. A generalized circuit as shown in Figure 9.3gg may have to be used. Other versions of the lag are also shown with different aspects and advantages, such as independent adjustment of gain and time constant, bipolar output, and simplicity and independent adjustment (but poorly scaled in many cases) in the illustration at the bottom of this figure.

Another commonly used element is the *lead-lag*, for which the transfer function is

$$H(s) = K\frac{s}{1 + \tau s} \qquad\qquad 9.3(57)$$

or

$$H(s) = K\frac{1 + \tau_1 s}{1 + \tau_2 s} \qquad\qquad 9.3(58)$$

Figure 9.3hh shows a generalized circuit for the type of transfer function in equation 9.3(57), with variable gain K and time constant τ.

The class of lead-lag transfer functions stated in equation 9.3(58) may be simulated with circuits of the type shown in Figure 9.3ii. The generalized arrangement used for individual time constants and gain variations by signal inputs often results in poor integrator scaling, especially for large τ_2. The common circuit permits the ratio between τ_1 and τ_2 to be adjusted over a wide range.

In many analog computer simulations a *signal delay* effect, or simulation of a *transport lag*, such as the flow of fluids through pipes or the movement of material on a conveyor, is desired. Delays or deadtimes like this often

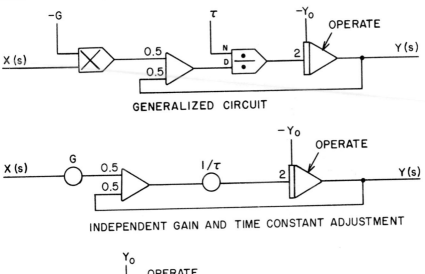

GENERALIZED CIRCUIT

INDEPENDENT GAIN AND TIME CONSTANT ADJUSTMENT

DUAL POLARITY OUTPUT

SIMPLE ALTERNATE CIRCUIT

Fig. 9.3gg   The lag function

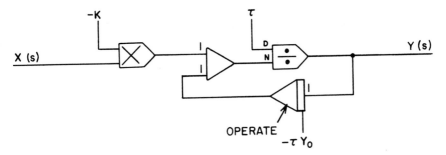

Fig. 9.3hh   The lead-lag function for equation 9.3(57)

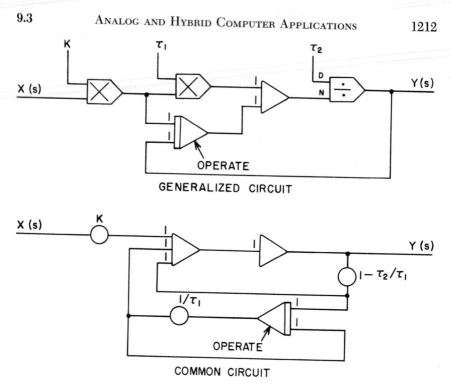

Fig. 9.3ii   The lead-lag function for equation 9.3(58)

represent important dynamic factors and influence the stability and operation of the plant or process. In general there is no true analog computer "model" of a transportation delay, and for most applications some form of approximation of the delay function must suffice.

The most commonly known approximations are the Padé functions. In process and plant simulations these often do not satisfy the needs, and other empirically determined functions are used. Figure 9.3jj shows an example of delay approximation, for which the time domain responses are improved above those of the Padé functions, to give more stable, damped, and consistent

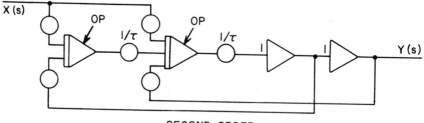

SECOND ORDER

Fig. 9.3jj   Transport lag approximation

performance. The delay approximation is the low-pass type, with the transfer function

$$H_2(s) = \frac{3.45 - 0.345(\tau s)}{3.45 + 3.18(\tau s) + (\tau s)^2}$$
9.3(59)

The transportation delay time τ may be varied by adjusting the coefficient potentiometers in the circuits in Figure 9.3jj. If the delay time is a function of a simulation variable, such as the flow rate in a pipe, the potentiometers must be replaced by multipliers. The range of variation is generally limited to less than 100:1 for the type of circuit given, without rescaling or rearrangement of the gain distributions. The shortest delay is determined by the highest input gains available (or gain-integrator speed combinations).

Control System Simulation

The simulation of instrumentation and control systems includes measuring or detecting devices, the controllers, and the actuating or manipulating output elements. Most measuring devices may be simulated by one or more simple lags (see *Process Simulation*) for dynamic representation, with added non-linearities for device characteristics, sensitivity, or operating point modeling (see *Non-linear Functions*). For special functions, such as logarithmic input-output relationships (pH measurements), the general purpose function generating units may be used, or the functions generated by implicit techniques (see *Differential Equations*). When the measuring device puts out a discontinuous (pulse type) signal, the signal-generating circuits may be used (see *Signal Generation*), as discussed later in this section.

A direct, three-mode (PID) controller has the transfer function

$$M(s) = \frac{100}{PB}\left[1 + \frac{1}{T_i s}\right]\left[1 + \frac{T_d s}{1 + T_0 s}\right][R(s) - C(s)]$$
9.3(60)

where $M(s)$ = controller output in Laplace representation,
$\quad C(s)$ = measured (controlled) input variable,
$\quad R(s)$ = reference (set point),
$\quad PB$ = proportional band (in percent),
$\quad T_i$ = reset (integration) time,
$\quad T_d$ = derivative (rate) time, and
$\quad T_0$ = stabilizing (filtering) time constant.

The filter time constant is usually made as small as possible and is often a fraction of the derivative time T_d (such as $T_d/16$).

Figure 9.3kk shows a simulation circuit for the three-mode controller, with the initial value M_0 corresponding to the controller output in "manual."

The controller transfer function in equation 9.3(60) is of the interacting

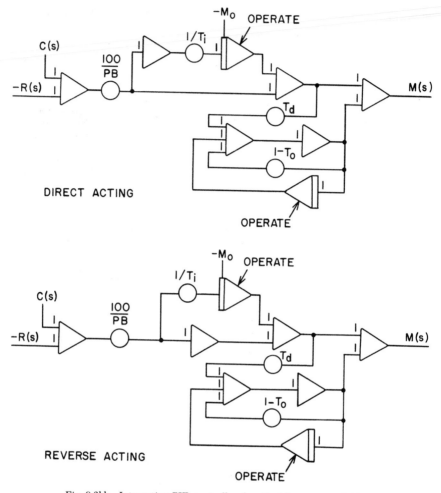

Fig. 9.3kk   Interacting PID controller described by equation 9.3(60)

type, which is the common case for most industrial applications. A mathematically non-interacting controller is expressed by

$$M(s) = \left[\frac{100}{PB} + \frac{1}{T_i s} + \frac{T_d s}{1 + T_0 s} \right] [R(s) - C(s)] \qquad 9.3(61)$$

in which all the three modes can be adjusted independently.

In general, the proportional band should be an *overall* adjusting effect, which leads to the practical non-interacting controller with the transfer function

$$M(s) = \frac{100}{PB} \left[1 + \frac{1}{T_i s} + \frac{T_d s}{1 + T_0 s} \right] [R(s) - C(s)] \qquad 9.3(62)$$

with the circuit arrangement shown in Figure 9.3ll. Also see equation 7.4(76).

Modern controllers often limit the derivative action to respond only to measured variable changes, and not to rapid set-point changes, which could upset the process. This is stated in the modified transfer function

$$M(s) = \frac{100}{PB}\left[1 + \frac{1}{T_i s}\right]\left[R(s) - \left(1 + \frac{T_d s}{1 + T_0 s}\right)C(s)\right] \qquad 9.3(63)$$

which is of the interacting kind.

Simulation with ddc

Direct digital control may be simulated with a controller circuit as shown in Figure 9.3mm, where the measured variable C(s) and the set point R(s)

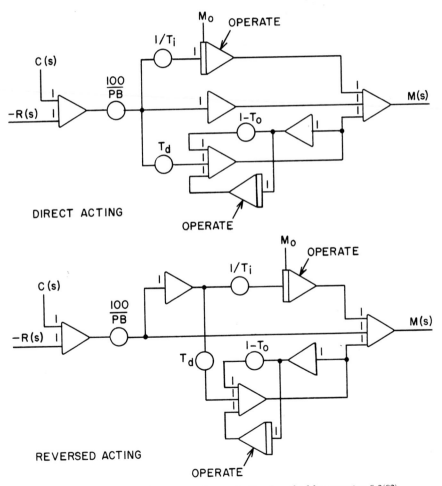

Fig. 9.3 ll   Practical non-interacting controller described by equation 9.3(62)

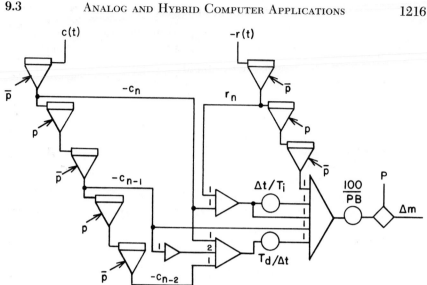

Fig. 9.3mm   Simulation of ddc based on equation 9.3(64)

are sampled at fixed intervals Δt and held in track-store units. The controller algorithm in this case is

$$\Delta m = \frac{100}{PB}\left[\left(1 + \frac{\Delta t}{T_i}\right)(r_n - c_n) - r_{n-1} + c_{n-1} - \frac{T_d}{\Delta t}(c_n - 2c_{n-1} + c_{n-2})\right]$$

$$9.3(64)$$

with Δm the change of the output computed in the time interval. This control corresponds to the one given in equation 9.3(63) with derivative action responding to the measured variable only. The track-store operation is controlled by the pulse P occurring once during each time interval Δt to produce an output pulse Δm of amplitude (height) corresponding to the desired change in the control variable.

Control Valve Simulation

The most important actuating or manipulating output element is the control valve, which may have several distinct functional aspects in an analog computer simulation. The flow characteristics can generally be linear, equal percentage, quick opening, or butterfly type. Except for the linear valve, the flow characteristics must be generated by a special analog computer circuit (by implicit techniques), or programmed in a function-generating unit, for a true representation (see Chapter I for details). Additional effects such as limiting or backlash in valve stem movements must be included (see *Non-linear Functions*).

The dynamic performance of the control valve may be represented by a time lag of first or second order, with a limited velocity in stem movement. The time lags are simulated by circuits which were described under transfer functions, lags, and lead-lags. Velocity limiting may be expressed by the equation

$$\frac{dm_0}{dt} = \text{LOW}\left(m_L; \frac{dm_i}{dt}\right) \qquad\qquad 9.3(65)$$

where m_0 is the stem position, m_L the velocity limit, and m_i the input stem position of an ideal, unconstrained valve. The simulation circuit for velocity limiting is shown in Figure 9.3nn.

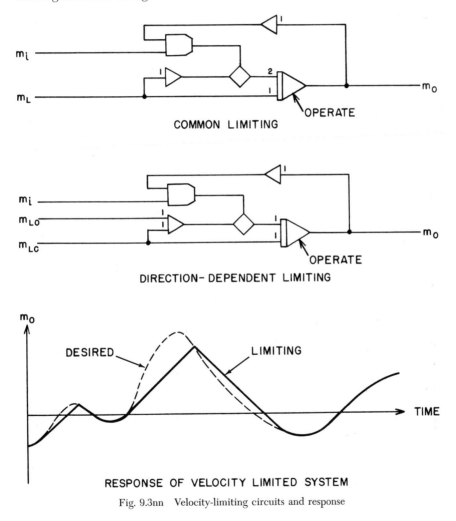

COMMON LIMITING

DIRECTION- DEPENDENT LIMITING

RESPONSE OF VELOCITY LIMITED SYSTEM

Fig. 9.3nn   Velocity-limiting circuits and response

On-Line Applications

On-line analog and hybrid computers offer many advantages in terms of signal conversion, analysis and accumulation, as well as in carrying out specific control tasks, running experiments, collecting data, training operating personnel, and in evaluating equipment and concepts. Such computers combine the simulation and computation capabilities of the parallel analog domain, with the logic, memory, and communication facilities of the digital computer, in a coordinated and consistent operating system.

Signal Generation

One prime function in an on-line application is the generation or conversion of analog and digital information into signals that are used as inputs to the process or plant. Such signals include forcing functions, perturbations, for instance, and control signals to keep the variables within range, as well as command signals for sequencing and communicating with the reactions or mechanisms involved in the application.

The harmonic oscillator is used for generating sinusoid signals by solving the equation

$$\frac{d^2y}{dt^2} + y = 0 \qquad\qquad 9.3(66)$$

with the initial conditions $Y_{t=0} = -A \sin \phi$ and $dy/dt_{t=0} = -A \cos \phi$. This gives the output

$$y(t) = A \sin (\omega t + \phi) \qquad\qquad 9.3(67)$$

where ω is the angular frequency and ϕ is the phase angle with respect to t.

Amplitude stabilization (counteracting divergence or convergence of the generated sinusoid), using the expression

$$\sin^2(x) + \cos^2(x) = 1 \qquad\qquad 9.3(68)$$

is valuable in long continuous runs, where integrator offsets or drift may otherwise influence the amplitude of the sinusoids, or inadequate dynamic performance of the computing units otherwise makes the solution unstable.

In Figure 9.3oo three circuits are shown for modulation of the generated sinusoidal signals: *amplitude modulation, frequency modulation,* and *phase modulation,* based on the harmonic generation. In these circuits the modulation variable is an input signal. The phase modulation is based on the relation

$$\sin(x + y) = \sin(x)\cos(y) + \cos(x)\sin(y) \qquad\qquad 9.3(69)$$

For continuously running signals or simulations, the stabilized harmonic generator circuit should be used.

Discontinuous signals, such as triangular wave and pulse trains, may be

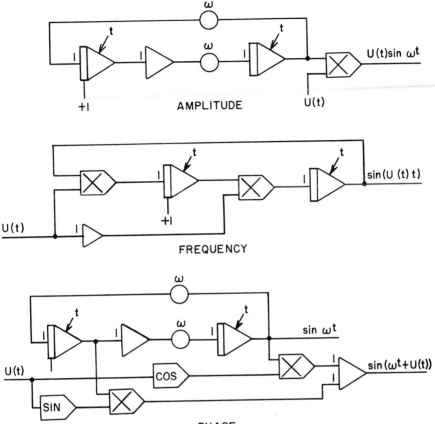

Fig. 9.3oo　Sinusoidal signal modulation

generated by the type of circuit shown in Figure 9.3pp. For an input variable $u(t) \geq 0$, the frequency is

$$f = u(t)/4 \qquad 9.3(70)$$

Signal Conditioning

Signal-conditioning circuits are used on many input variables from a process by filtering, range adjusting and gain variations. Filtering is commonly done by the type of circuits shown in Figure 9.3qq. Three types of filters, with the transfer functions all having real poles only, are used:

$$H_L(s) = \frac{1}{1 + \tau s} \qquad \text{(Low-pass filter)} \qquad 9.3(71)$$

$$H_H(s) = \frac{\tau s}{1 + \tau s} \qquad \text{(High-pass filter)} \qquad 9.3(72)$$

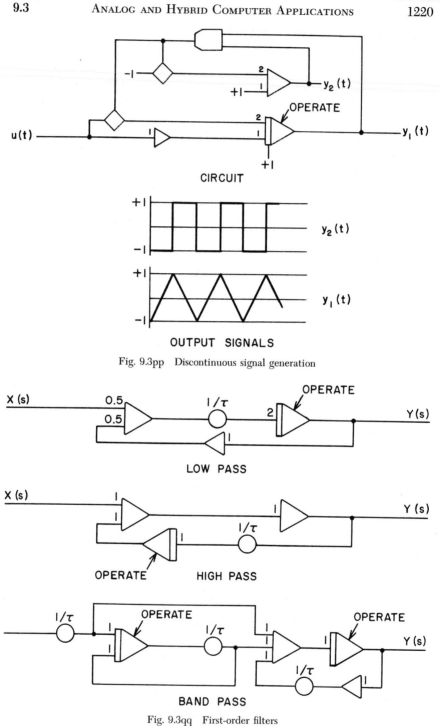

Fig. 9.3pp   Discontinuous signal generation

Fig. 9.3qq   First-order filters

$$H_B(s) = \frac{\tau s}{(1 + \tau s)^2} \qquad \text{(Bandpass filter)} \qquad 9.3(73)$$

These filters all have roll-offs of 6 decibels per octave (20 decibels per decade) and unity gain at the passing frequency band, with convenient adjustment of the frequency $1/t$.

Figure 9.3rr shows an example of a second-order filter with complex poles.

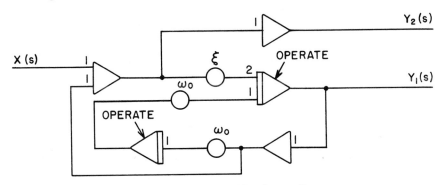

Fig. 9.3rr   Band-pass and band-reject filters

It may be used for bandpass or band-reject filtering. The transfer functions are

$$H_P(s) = \frac{Y_1(s)}{X(s)} = \frac{2\xi s}{s^2 + 2\xi s + \omega_0{}^2} \qquad 9.3(74)$$

and since

$$H_R(s) = 1 - H_P(s) \qquad 9.3(75)$$

$$H_R(s) = \frac{Y_2(s)}{X(s)} = \frac{s^2 + \omega_0{}^2}{s^2 + 2\xi s + \omega_0{}^2} \qquad 9.3(76)$$

where ω is the filter center frequency and ξ the damping factor, with its Q value according to the relation

$$Q = \frac{1}{2\xi} \qquad 9.3(77)$$

The filter characteristic for $2\xi = 0.2$ or $Q = 5$ is given in Figure 9.3ss, indicating a base roll-off of 6 db per octave (20 db per decade) for frequencies considerably different from ω_0. The *passband* width (for which $|H_P(s)| \geq 0.5$) is approximately

$$f \cong 0.343\xi = 0.1715/Q \qquad \text{(relative to } \omega_0) \qquad 9.3(78)$$

Range adjustment is based on detection of the amplitudes (the peak values) of the input signal and varying a bias value so that the output signal will be symmetrical with respect to zero.

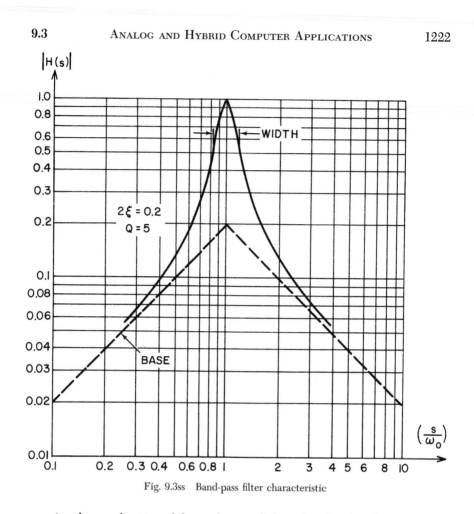

Fig. 9.3ss   Band-pass filter characteristic

Another application of the *peak-to-peak detection* circuit is the *automatic gain control* circuit. In this case the sampled and held values of the input signal are divided into the input signal $Y_I(n)$ to produce a gain-adjusted output signal according to

$$Y_G(n) = \frac{2Y_I(n)}{Y_{MAX}(n-2) - Y_{MIN}(n-1)} \qquad 9.3(79)$$

Both circuits work with essentially one period's delay in adjusting the signal and are thus faster than usual control circuits based on filter-averaging (long filter time constants). For proper operation, the integrator input gains (A) must be chosen according to the highest rate of change of the input signal.

Data Processing

The need for *data reduction* and analysis is often present in process industries and instrumentation facilities. Measurements of process variables or system parameters are often subject to disturbances or fluctuations (random or deterministic), or in other cases the desired information is implicit only in the combination of several variables.

Analog and hybrid computers may be applied with little difficulty in this type of data analysis. The process signals are filtered and adjusted in the analog computer and are presented in a format suitable for further reduction. Such data processing may take place on-line (in real time) or be applied to recorded variables off-line (in a practical time scale). Often, the end result is stored in the digital computer until printing or dumping out, or for further analysis by repetitive play-backs etc.

An example of data processing is shown in Figure 9.3tt, where the cross correlation of two input variables is computed. In this case, the two input signals are filtered in the analog computer, converted into digital format, and stored (tabulated) and played back by the digital computer. The cross correlation function is determined by the analog computer, and the obtained results displayed and recorded graphically, enabling the operator or experimentor to evaluate the correlation between the two input variables.

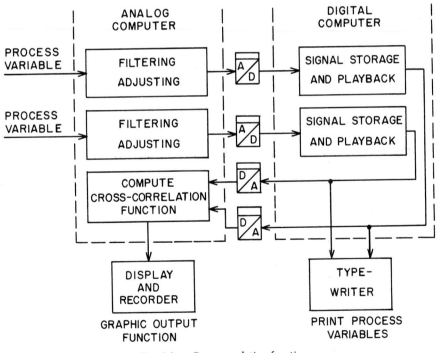

Fig. 9.3tt   Cross-correlation function

A second example is given in Figure 9.3uu of a test facility application. It is desired to detect (and record) a characteristic performance variable during the short transient of the test. The computation of the performance variable

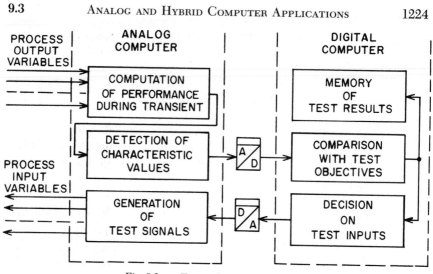

Fig. 9.3uu   Test evaluation and control

is carried out in the analog computer because of speed requirements, with the detected value transmitted to the digital computer. Based on the performance obtained during the test, the digital computer program reorients the tests and calculates new inputs to the process.

Similar on-line applications of analog and hybrid computers may be found in:

> Statistical analysis of variations
> Probability functions
> Regression analysis
> Detection of characteristic values and functions
> Power spectrum analysis
> Curve or function fitting procedures
> Logging of bivariable or multivariable functions
> Extensive and complex process monitoring
> Behavioral or diagnostic analysis

Hybrid Computer Applications

Hybrid computers may be divided into three types according to predominant computational capacity:

1. *Hybrid analog computer,* in which a digital computer serves mainly to control the operations of the analog computer.
2. *Hybrid digital computer,* in which an analog computer serves mainly as an extended interface of the digital computer.
3. *Balanced hybrid computer,* in which the computational capacities of both computers are significant to the purposes of the hybrid system.

This division generally reflects the relative speeds of the analog and digital computers, since a functionally unbalanced system is oriented toward the faster, and thus more productive part, leaving the other part to do the minor, or less pressing tasks.

Hybrid Analog Computers

Hybrid analog computers range from the analog computer with some logic facilities (limited number of logic units) to hybrid systems where the digital computer represents an extensive control unit for the analog computer. The digital part of an analog hybrid computer carries out only a few very simple computations as an active contribution to the solution of the programmed problem. The amount of information converted between analog and digital representations is small compared with the status and command type of information exchange.

Figure 9.3vv shows an example of the use of a hybrid analog computer in engineering design. A model of the design unit is programmed on the analog computer to be run at high speed and in a repetitive mode, with the relevant operating characteristics of the unit displayed on a CRT oscilloscope. The model simulates the design unit for selected values of design parameters, such as dimensions, speeds, or mass distributions. The operator evaluates the performance of the unit, enters alternate parameter values in engineering units at the digital computer keyboard, and requests data and status information. The digital computer calculates the corresponding analog scale factors and coefficient values of the entered parameters, and adjusts the potentiometers and computing units of the analog computer accordingly. In this example the hybrid computer leaves the operator free to evaluate the design unit in engineering terms.

Another application of hybrid analog computers is shown in Figure 9.3ww for the simulation of a chemical reactor. The objective of the simulation is to determine optimal operating conditions of the reactor for given production requirements and raw material. The reactor and control system is programmed on the analog computer to run in real time and continuously. Measurements of reactor efficiency or economics are derived from and processed by the analog computer program, which provides inputs to the digital computer for data collection and logic decisions as to the appropriate control actions. Simulation in this case is all analog. The digital computer provides memory for the generated data of the reactor and its performance, and logic decisions for control of the analog computer and modes of the simulated control system. Appropriate optimization programs are used to find the optimal reactor control configuration and operating points.

Hybrid Digital Computers

In a hybrid digital computer, the analog computer generally provides extended interface and display facilities for a larger and faster digital com-

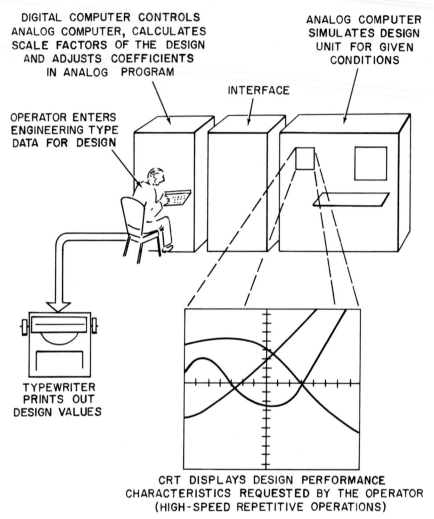

DIGITAL COMPUTER CONTROLS
ANALOG COMPUTER, CALCULATES
SCALE FACTORS OF THE DESIGN
AND ADJUSTS COEFFICIENTS
IN ANALOG PROGRAM

ANALOG COMPUTER
SIMULATES DESIGN
UNIT FOR GIVEN
CONDITIONS

INTERFACE

OPERATOR ENTERS
ENGINEERING TYPE
DATA FOR DESIGN

TYPEWRITER
PRINTS OUT
DESIGN VALUES

CRT DISPLAYS DESIGN PERFORMANCE
CHARACTERISTICS REQUESTED BY THE OPERATOR
(HIGH-SPEED REPETITIVE OPERATIONS)

Fig. 9.3vv   Example of hybrid analog computer used in engineering design

puter. This includes *signal conditioning and buffering* (see *On-Line Applications*), presentation of hybrid program status and computational results for display, or recording of variables, and limited computations, such as slow integrations or solutions of differential equations (see *Computations (Analog)*).

Figure 9.3xx is an example of a hybrid digital computer application. A large, fast digital computer is programmed to simulate a nuclear reactor with its associated steam generator, turbine, and power plant. The simulation runs in real time and is used to train operating personnel for the nuclear power plant, and to investigate characteristic operations. A slow and limited capa-

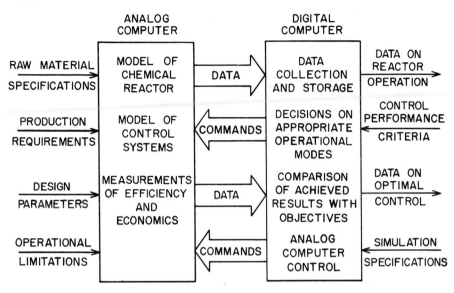

Fig. 9.3ww   Optimal reactor control simulation

bility analog computer is used to tie the operator or trainee console to the simulated plant, and to provide realistic plant environmental effects such as noisy signals, interactive and non-linear equipment, and long time constants (see *Simulations (Analog)*). In addition to the simulation program, the digital computer also contains the training program, or the operational investigation program.

Figure 9.3yy shows a hybrid digital computer application in data reduction and analysis. Many electrical-electronic signals from a pilot plant, or experimental setup, are filtered and adjusted in the analog computer for conversion to digital formats. The digital computer stores the data in its memory, computes auto- and cross-correlations of the measurements, statistical parameters such as average and median values, and variance and standard deviations and produces the results of the experiments in the appropriate form. The analog computer displays the results and records the values graphically, for later evaluation by the investigators. The digital computer is programmed and instructed to carry out the experiments or tests, monitor the results, inform the operator (investigator) of status and alarm conditions, and suggest possible diagnosis of reported conditions or performance characteristics.

Balanced Hybrid Computers

The most powerful, efficient, and economical hybrid computer is balanced in terms of analog and digital computational capabilities. In such a computer there are no speed or bandwidth discrepancies between the two parts, and

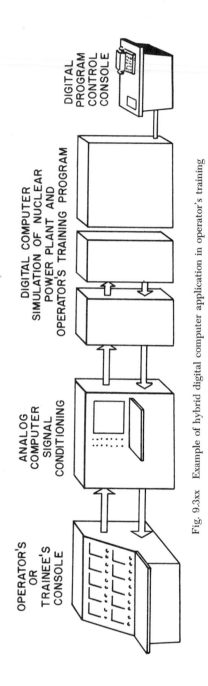

OPERATOR'S
OR
TRAINEE'S
CONSOLE

ANALOG
COMPUTER
SIGNAL
CONDITIONING

DIGITAL COMPUTER
SIMULATION OF NUCLEAR
POWER PLANT AND
OPERATOR'S TRAINING PROGRAM

DIGITAL
PROGRAM
CONTROL
CONSOLE

Fig. 9.3xx   Example of hybrid digital computer application in operator's training

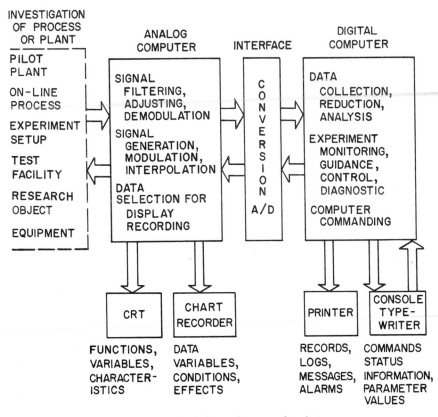

Fig. 9.3yy   Hybrid data reduction and analysis

a consistent and coordinated performance may be obtained. A balanced hybrid computer offers the fastest solution; often one or more orders of magnitude faster than other hybrid or stand-alone computers, of comparable types.

The balanced hybrid computer is typically applied to large sets of complex differential equations, partial differential equations, iterative and statistical problems including optimization and search type problems, and in simulating extensive and complicated systems such as plants, processes, or apparatus. An example of such an application is the simulation shown in Figure 9.3zz. This is a hybrid simulation of a power plant with control system. The plant consists of a steam generator fired with pulverized coal, a double-flow turbine and a reheater system, with associated subsystems and equipment. A general model of the plant, representing deviations from steady-state conditions only, consists of more than 180 equations describing the relationship between the variables in the plant, the operating conditions and the plant design parameters.

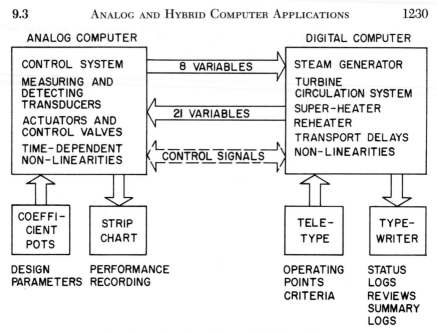

Fig. 9.3zz   Example of balanced hybrid simulation

The hybrid simulation of the plant model is carried out in digital computers. For practical and economic purposes, the simulation is reduced to 27 equations, comprising eight input variables from the control system, and producing 21 output variables (state variables for monitoring and control). The control system is modeled on the analog computer with more than 30 integrators, 60 adders and 75 coefficient potentiometers.

This hybrid simulation allows the plant model to be run three times faster than the real-time plant, which substantially shortens the design and study time. Designers claim that the most important advantage of the hybrid simulation (over any other form of simulation) is the ability to observe and evalute the effects of design parameter changes in the plant model.

Partial Differential Equations

Another important application area of the balanced hybrid computer is in solving partial differential equations. Conventional analog computer solutions are limited to a few independent variables and to relatively simple types of equations, since the solutions are generated by finite difference approximations in terms of analog representations. The coarseness of such an approximation "grid" directly depends on the available analog computing units, thereby prohibiting increased solution accuracy by using finer grid spacing.

Hybrid computer solution of partial differential equations makes use of the memory capabilities in the digital computer by which an analog sub-solution in terms of one independent variable can be stored and reproduced in forming or approximating another independent variable by increments. In solving elliptic partial differential equations, such incremental approximations may be recursive, and the final solutions or set of solutions arrived at after a number of "go-arounds." It is difficult or impossible to design an iterative - recursive solution method for hyperbolic, and especially parabolic, types of partial differential equations. The solutions can only be arrived at in a step-by-step manner. For these types of equations the incremental step (from one continuous analog solution to the next) must be very fine, and special measures must be introduced in the programs to prevent or reduce errors.

The hybrid computer applications in solving partial differential equations allow the errors implicit in the analog representation to be reduced, mainly by correcting the finite approximation method. Another improvement may be obtained in the smoothing of derivatives when truncated Fourier series are used in this correction of first-order finite differences. When boundary conditions are imposed on the partial differential equations, the digital computer may be used to derive and define more accurate values for the analog sub-solutions by extrapolation methods, by axis transformations, or by transient error corrections.

9.4 ANALOG COMPUTERS IN DISTILLATION COLUMN CONTROLS

Analog computers are used to solve equations necessary for the proper control of processing units. Their prime application in distillation is to minimize upsets to the unit, caused by a change in process inputs. The computer calculates the effects of the changes and determines the corrective action needed to counteract them. The control actions are implemented by directly manipulating the final control elements or by changing the set points of controllers.

The benefits derived from better control are reduced energy consumption, reduced operating manpower, increased throughput, reduced disturbances to other processing units, increased plant flexibility and increased product recovery.

Analog computers—properly applied—will improve the operation of a distillation column, but their use is justified only when the resultant system provides a good payout or when automatic control of the column was not possible without them.

The first step in the design of a control system using analog computers is the derivation of a process model. Knowing these equations, the manipulated variables can be selected and the operating equations for the control system are developed. The computer components are then selected for the correct solution of these equations. The components are then scaled, because analog computers act on normalized numbers (0 to 100 percent) rather than on the actual process values.

The final control system may contain a single computing relay or be a complex, multicomponent analog computer. In this section a general discussion of the procedures for designing distillation controls is followed by examples of the common applications for analog computers in distillation column control. For a more detailed discussion of conventional and optimized distillation column controls refer to Sections 10.3 and 10.4.

Process Model and Control System

The first step in the design of a control system must be the development of a process model. Frequently omitted in simple distillation columns, this step is essential if there is any thought of using an analog computer.

The model defines the process with equations developed from the unit's material and energy balances. The model is kept simple by the use of one basic rule:

> The degrees of freedom for control limit the controlled variables (product compositions) specified in the equations. (See Section 7.2 for a detailed discussion.)

For example, for a given feed rate only one degree of freedom exists for material balance control. If overhead product is a manipulated variable (controlled directly to maintain composition), then bottoms product cannot be independent but must be manipulated to close the overall material balance according to the following equations:

$$F = D + B \qquad\qquad 9.4(1)$$

$$\text{Accumulation} = \text{Inflow} - \text{Outflow} \qquad\qquad 9.4(2)$$

$$= F - (D + B) \qquad\qquad 9.4(3)$$

If accumulation must be zero, then B is dependent upon F and D, as expressed by equation 9.4(4).

$$B = F - D \qquad\qquad 9.4(4)$$

where F = feed rate (the inflow), D = overhead rate (an outflow) and B = bottoms rate (an outflow).

Similarly, the criterion for separation is the ratio of reflux (L) to boilup (V). Manipulating reflux has the same effect (though opposite) as manipulating boilup. Consequently only one degree of freedom exists to control separation.

For a column of the type shown in Figure 9.4a, two equations define the process: an equation for separation and an equation for material balance control.

The load and the controlled and manipulated variables are identified first. Load and controlled variables are usually obvious, but identification of manipulated variables can be difficult. (The selection of manipulated variables is discussed in greater detail in Sections 10.3 and 10.4.)

The general guidelines are:

1. Manipulate the stream which has the greatest influence on the associated controlled variable.
2. Manipulate the smaller stream if two streams have the same effect on the controlled variable.
3. Manipulate the stream which has the most nearly linear correlation with the controlled variable.
4. Manipulate the stream which is least sensitive to ambient conditions.
5. Manipulate the stream least likely to cause interaction problems.

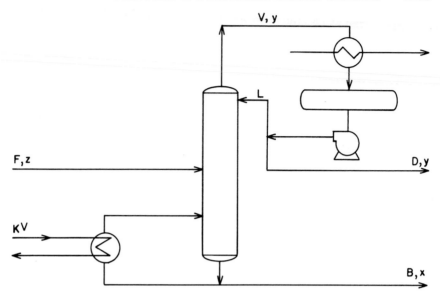

Fig. 9.4a   Typical two-product distillation column

The equations are then solved for the manipulated variables in terms of the controlled and load variables. In that form the equations are the mathematical representation of the control systems.

Computing Functions

The form of the control system equations essentially defines the computing functions required. Economies can often be made by combining functions. For example, the solution of the equation

$$C = (A + B)(D) \qquad\qquad 9.4(5)$$

for C requires a summing and a multiplying operation. If the computing relays are pneumatic, then two separate instruments must be used. If electronic, only one instrument is required, because an electronic multiplier (divider) algebraically adds several signals and multiplies (divides) the sum by another signal which can also be a summation. Sections 4.4, 9.1 and 9.2 give a listing of the operations and combinations of operations that can be performed by the available devices.

The prime selection factor, however, is the integrity of the equation, and therefore a *different arithmetic operation cannot be substituted to enable the use of less expensive instruments.* A frequent abuse of this rule is found in electronic systems where a ratio station with external bias is substituted for a multiplier in a ratio system with feedback. Figure 9.4b illustrates the correct way to instrument such a system. The equation would be

$$M = (L)(R) \qquad 9.4(6)$$

where M is the manipulated variable, L is the load variable and R desired ratio. The analyzer controller (ARC) adjusts the ratio to maintain the controlled variable at the desired value. If the system involved two-component blending, with the ARC controlling the composition of the blend, the difference between a multiplier and a ratio station with bias is easily demonstrated.

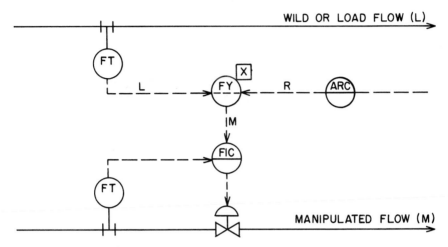

Fig. 9.4b   Ratio control system with feedback

Line Q in Figure 9.4c represents the relationship between M and L for a given set of conditions using a multiplier. If the load changes from L_0 to L_1, equation 9.4(6) requires that M be changed from M_1 to M_2. The multiplier output follows line Q to the intersection with L_1, and the process is satisfied without changing R_1, that is, without upsetting the product composition and causing the ARC to change its output. This situation will also prevail if load is brought back to L_0.

Line T_0 in Figure 9.4d is the initial operating line of the system using a ratio station with external bias. The equation solved by this instrument is

$$M = (L)(R_a) + B \qquad 9.4(7)$$

where R_a is the manual setting on the ratio station and B is the output of the ARC (bias). Note that the ratio R_3 is not equal to R_1 in Figure 9.4c. The process is satisfied at initial load condition (L_0 and M_1), but if the load changes to L_1, the ratio station calls not for M_2 to be delivered, but for M_5, and the process is upset until the ARC changes the bias of line T_0 from B_1 to B_2. When the load is restored to L_0, the system will be in error once more.

If a change in the composition of M or L should occur, the ARC will change its output, adjusting R in equation 9.4(6), to restore the product

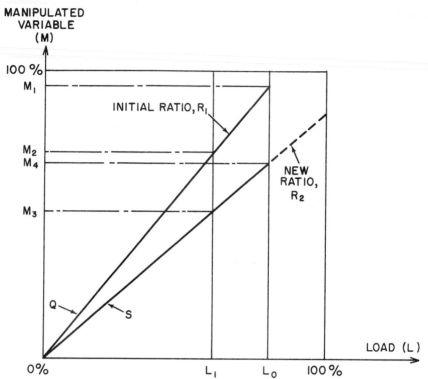

Fig. 9.4c   Multiplier in ratio with feedback system solves control equation exactly and prevents upsets

composition to specification. The new operating line is S in Figure 9.4c and U in Figure 9.4d. The performance differences with respect to load changes will still be there ($M_3 \neq M_6$).

Any deviation in arithmetic operations from that required by the process equations will have similarly undesirable results.

Scaling

An analog computer operates on normalized values of the process variables, as discussed in Sections 9.1 and 9.2. That is, the analog signal will vary from 0–1 (0 to 100 percent) as the process variable varies from zero to its maximum value. Figure 9.4e illustrates the relationship between the various forms of analog signals and some typical process measurements.

The actual value of a process measurement is found by multiplying the analog signal by the calibrated full-scale value (*meter factor*) of the process variable.

In the examples of Figure 9.4e the temperature, represented by a 75 percent analog signal, is 320°F, the linear flow is 775 GPH, the output of

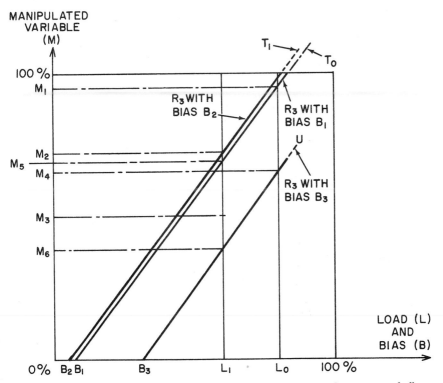

Fig. 9.4d   Ratio station with bias provides incorrect solution to control equation and allows upsets to occur

the differential pressure transmitter (flow squared) is 779 GPH and the composition is 3.75 percent.

As an example, let us review a flow ratio system in which the load stream L has the range of 0–1,000 GPM, the manipulated stream M has a range of 0–700 GPM and the ratio range R is 0–0.8. (R = M/L.)

Let L', R' and M' represent the normalized values of flow L, ratio R and stream M, respectively. The scaled equation is

$$700M' = (1000L')(0.80R') \qquad 9.4(8)$$

Reducing to the lowest form,

$$M' = 1.143(L')(R') \qquad 9.4(9)$$

The number 1.143 is the *scaling factor*. M' is plotted as a function of L' and R' in Figure 9.4f.

In applications such as the constant separation system, exact scaling is not critical. The flexible scaling cannot be used (1) when compensation for feed composition is part of the model, (2) when narrow spans must be used

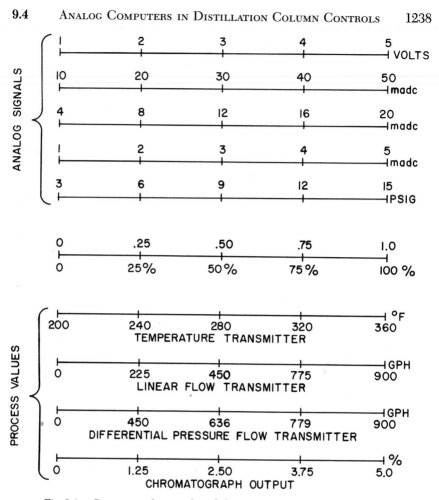

Fig. 9.4e   Common analog signals and their relationship to process variables

for reasons of stability, and (3) when transmitter calibrations are inconsistent with material balance ratios. Exact scaling techniques must be used for these cases.

Internal Reflux Computer

A major heat load in some columns is the energy required to heat the reflux returning to the column up to its boiling point. This heat—and the heat required to vaporize this *external* reflux—is obtained from the rising vapors which condense to become *internal* reflux. The latent heat of vaporization is essentially constant, and the ratio of internal reflux rate to external reflux is a function of the amount of subcooling of the external reflux.

Subcooling can be expressed as the difference between the overhead vapor temperature and the external reflux temperature. The equation solved by an internal reflux[1] system is

$$L_i = L\left[1 + \frac{C_p}{H}(T_v - T_l)\right] \qquad 9.4(10)$$

where L_i = internal reflux rate,
  L = external reflux rate (GPM),
  C_p = specific heat of external reflux,
  T_v = vapor temperature,
  T_l = reflux temperature,
  ρ = density of the external reflux liquid, and
  H = heat of vaporization of the external reflux.

Figure 9.4g is a pneumatic computing system for solving equation 9.4(10). The temperatures must be subtracted in a separate device before the multiplication. In the electronic version, the subtraction is performed in the multiplier and one device is thereby eliminated.

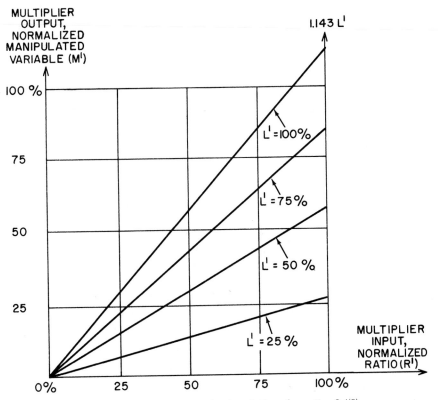

Fig. 9.4f   Multiplier output for the solution of equation 9.4(9)

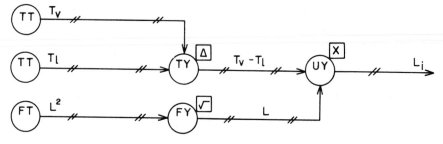

Fig. 9.4g Pneumatic version of the internal reflux computer

The subtracter is scaled first. Assuming a ΔT_{max} of 50°F, the span of T_v between 150–250°F and the span of T_l between 125–225°F, the equation for the subtracter is written

$$\Delta T' = \frac{(150 + 100T'_v) - (125 + 100T'_l)}{50} \qquad 9.4(11)$$

This reduces to the scaled equation

$$\Delta T' = 2(T'_v - T'_l + 0.25) \qquad 9.4(12)$$

If the following assumptions are made,

$$L_{i\,max} = 15{,}000 \text{ GPM} \qquad L_{max} = 10{,}000 \text{ GPM}$$
$$C_p = 0.65 \qquad H = 250$$

The equation for the multiplier then becomes:

$$L'_i = \frac{10{,}000L'}{15{,}000}\left[1 + \left(\frac{0.65}{250}\right)50\Delta T'\right] \qquad 9.4(13)$$

then equation 9.4(13) reduces to

$$L'_i = 0.667L'(1 + 0.13\,\Delta T') \qquad 9.4(14)$$

When $\Delta T'$ is zero, the internal reflux equals 0.667 times the external reflux. The number one (1) within the parentheses therefore sets the minimum internal reflux. When $\Delta T'$ is 100 percent, the ratio of internal reflux to external reflux is at a maximum.

The expression within the parentheses must be normalized. This is done by dividing both terms by the total numerical value, that is, 1.13. To preserve the equality, the coefficient of L is multiplied by 1.13. The scaled equation becomes

$$L'_i = 0.754L'(0.885 + 0.115\,\Delta T') \qquad 9.4(14a)$$

Internal reflux systems are designed to compensate for changes outside the column, and it should be understood that a change within the column can introduce a positive feedback. Figure 9.4h shows a typical internal reflux application and its response to an upset *within* the column. The computer reacts the same to an increase in vapor temperature as it does to a decrease in liquid temperature, but the required control actions are in the opposite direction.

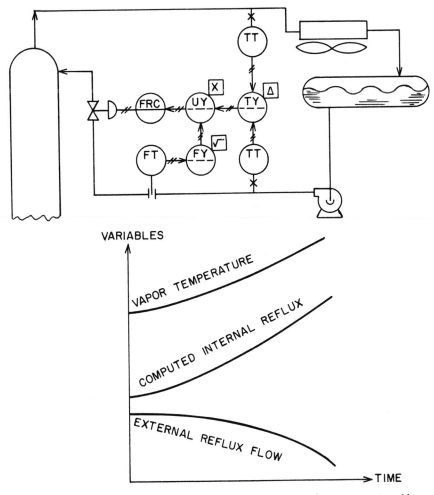

Fig. 9.4h   Response of internal reflux computer to an increase in the concentration of heavy components in the overhead vapors

Flow Control of Distillate (Fast Response)

The column interactions which otherwise may necessitate the use of internal reflux computers, can be eliminated in some cases by controlling the flow of distillate product draw-off and putting reflux under accumulator level control. This is a slower system than one which flow controls the reflux, but its response is usually adequate. If necessary the response can be sped up by reducing the accumulator lag[2].

The material balance around the accumulator (Figure 9.4i) is expressed by

$$V = L + D \qquad\qquad 9.4(15)$$

where V is boilup (vapor rate), L is reflux rate and D is distillate rate. To overcome the accumulator lag, L must be manipulated in direct response

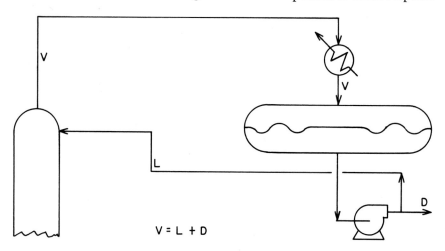

Fig. 9.4i   Reflux accumulator material balance

to a change in D rather than by a level controller. If V is constant (k), equation 9.4(15) is solved for L, which is the manipulated variable in this part of the system:

$$L = k - D \qquad\qquad 9.4(16)$$

For this equation to be satisfied, L must be decreased one unit for every unit D is increased and vice versa.

If V is indeed constant and the computations and flow manipulations are perfectly accurate, no level controller is needed. These conditions cannot be met and a trimming function is usually introduced. The system equation becomes

$$L = m - KD \qquad\qquad 9.4(17)$$

where m is the output of the level controller and K is an adjustable coefficient. The resulting control system is shown in Figure 9.4j.

The range of coefficient K should be broad enough to allow scaling and adjustment to be done during commissioning. Because the level controller trims

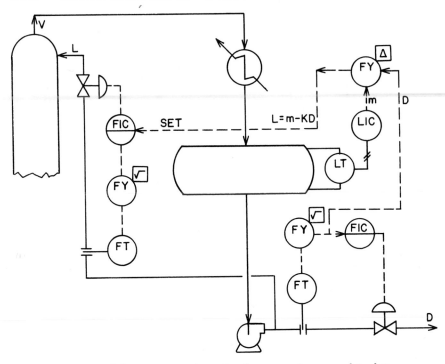

Fig. 9.4j   Reflux rate computing system for overcoming accumulator lag

the computation, the scaling and the value of K do not alter the steady-state value of reflux L, since these factors affect the transient response only. The response of the reflux flow to changes in distillate for several values of K are given in Figure 9.4k. The full-scale values of reflux and distillate flows in this case are 1,000 GPM and 500 GPM, respectively.

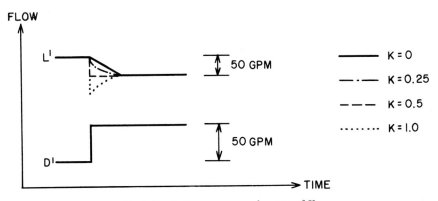

Fig. 9.4k   Reflux response as function of K

When K = 0, the reflux is adjusted by the level controller. In other cases the reflux flow is immediately changed by some percentage for a change in distillate, and the level controller forces the balance of the change.

If K = 0.5, the reflux flow is changed to the exact new steady-state value, because K equals the ratio of D_{max}/L_{max}, and therefore the computation is exact. If K = 1.0, the initial response of reflux is greater than required for the new steady state, and the level controller corrects the flow.

The value of K does not change the steady-state flow. It affects the transient response only, and therefore it can be used to adjust the dynamics of the loop. The greater the value of K, the faster the response.

Care must be taken to prevent increasing the response to the point of instability. A rule of thumb to follow is

$$K_{max} = 1.5(D_{max}/L_{max}) \qquad 9.4(18)$$

In some instruments the range of adjustability of K is limited and scaling is necessary. For the values used in the illustration, equation 9.4(17) becomes

$$1,000L' = 1,000m' - 500KD' \qquad 9.4(19)$$

where L' and D' are the normalized values of L and D. The maximum value of m is equal to the maximum value of L, since the level controller by itself can cause the level control valve to open fully.

The scaled equation is

$$L' = m' - 0.5KD' \qquad 9.4(20)$$

K must be adjustable over a range of ±10 percent for satisfactory tuning flexibility.

A similar system can be used on column bottoms where the bottoms product is flow controlled and the bottoms level is maintained by manipulating the heat input or boilup (V). The equation for that system is

$$V = m - KB \qquad 9.4(21)$$

where V is the boilup and B is the bottoms flow.

Constant Separation (Feedforward)

A distillation column operating under constant separation conditions has one fewer degrees of freedom than others, because the energy-to-feed ratio is constant. At a given separation, for each concentration of the key component in the distillate, there is a corresponding concentration in the bottoms.

In other words, holding the concentration of a component constant in one product stream fixes it in the other.[3] Figure 9.4l is an example of a constant separation feedforward system where distillate is the manipulated variable. If the flow measurements are of the differential pressure type, and realizing that $Fz = Dy + Bx = Dy + (F - D)x$, the equations are

$$D^2 = F^2 \left(\frac{z - x}{y - x}\right)^2 = F^2 \left(\frac{D}{F}\right)^2 \qquad 9.4(22)$$

$$Q^2 = F^2[Q/F]^2 \qquad 9.4(23)$$

where z, y, x = concentration of the key light component in feed, overheads
and bottoms, respectively,

D/F = required distillate-to-feed ratio and

Q/F = required energy-to-feed ratio.

No scaling is required of this equation, because the manually adjusted ratio allows signal weighting to be done in the field.

Normal design practice calls for the output of the trim controller ARC to be at 50 percent when the design or normal distillate-to-feed ratio is

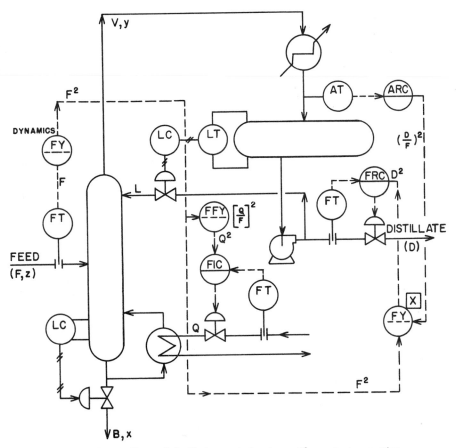

Fig. 9.4 l   Feedforward distillation control system with constant separation

required. If the gain of the multiplier is set at 2, the output tracks the load when this normal distillate-to-feed ratio is in force.

In a linear system the gain of the multiplier equals the scaling factor. In this system, however, the gain of the multiplier equals the square root of the scaling factor. Applying this rule to the example, the scaled form of equation 9.4(22) is

$$D^{2\prime} = 4.0(F^2)'[(D/F)^2]' \qquad\qquad 9.4(24)$$

where $D^{2\prime}$, $F^{2\prime}$ and $[(D/F)^2]'$ are the normalized values of the respective terms in 9.4(22).

The instrument labeled "dynamics" is a special analog module designed to influence the transient response. In the steady state its output equals its input, so no scaling is needed. See Section 7.6 for a more complete discussion on dynamic compensation elements, such as the one shown in Figure 7.6i.

Maximum Recovery (Feedforward)

In many distillations, one product is worth much more than the other, and the control system is designed to maximize the more valuable stream. The most common equation for this type system is[3]

$$D = m(KF + K_2F^2) \qquad\qquad 9.4(25)$$

where D = distillate rate,
    F = feed rate,
    K = adjustable coefficient,
    K_2 = 1 − K and
    m = feedback trim.

The control diagram for a maximum recovery system is shown in Figure 9.4m. Note that the distillate-to-reflux loop for accelerated response is used also. The summing amplifier utilized to compute $(KF + K_2F^2)$ needs no special scaling. Although the coefficients can be calculated in advance with reasonable accuracy, on-line adjustment is quite easy, and the rigour of the calculations can be avoided. The multiplier may require exact scaling, however.

In a typical system the maximum value of the parenthetical term is equal to the maximum feed flow, say 150 GPM. The values of m are computed from the feed composition. A typical range for m is 0.35 to 0.65. The distillate flow transmitter is calibrated for 90 GPM maximum flow.

The scaled equation is

$$90D' = (150S')(0.35 + 0.3m') \qquad\qquad 9.4(26)$$

where $S' = (KF + K_2F^2)$. Note that in this equation the minimum value of D is (0.35)(S). The number 0.35 is the zero value for m', and 0.30 is the span value.

$S^I = (KF + K_2 F^2)$

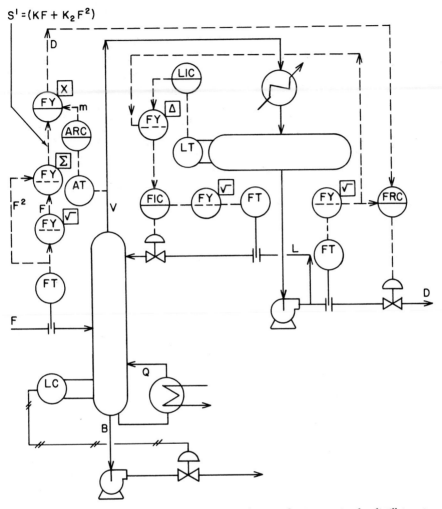

Fig. 9.4m   Maximum recovery system: computer solves quadratic equation for distillate rate

Reducing 9.4(26) to its normalized form gives

$$D' = \frac{150}{90}(S')(0.35 + 0.30m') \qquad 9.4(27)$$

The zero and span values must be included in the normalized procedure, and the right-hand side of 9.4(27) is therefore multiplied by and divided by the zero plus the span:

$$D' = (0.65)(1.667)(S')\left(\frac{0.35 + 0.30m'}{0.65}\right) \qquad 9.4(28)$$

Reducing the remaining fraction to its simplest form, the scaled equation becomes

$$D' = 1.084(S')(0.538 + 0.462m') \qquad 9.4(29)$$

The response curves of the scaled multiplier are shown in Figure 9.4n.

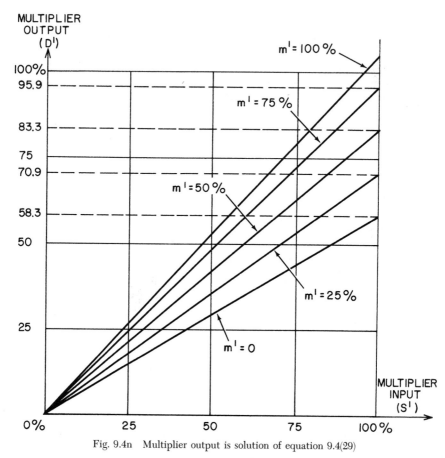

Fig. 9.4n   Multiplier output is solution of equation 9.4(29)

Composition Control of Two Products (Feedforward)

Some columns require tighter control of both product streams than can be had from the constant separation system described above. These columns require closed-loop composition control of distillate and bottoms.

The control of distillate composition is still done by manipulating distillate flow as required by

$$D = F\left(\frac{z - x}{y - x}\right) \qquad 9.4(30)$$

However, also to enforce composition control of the bottoms product, an additional manipulated variable is needed. Another product stream cannot be independently manipulated without changing the accumulation in the column, and this is not practical. The energy balance must therefore be adjusted to control bottoms composition x.

The relationship between x and the energy balance is developed by Shinskey[3] as a function of separation S:

$$S = \frac{y(1 - x)}{x(1 - y)} \qquad 9.4(31)$$

The relationship between separation S and the boilup-to-feed ratio V/F over a reasonable operating range is

$$V/F = a + bS \qquad 9.4(32)$$

where a and b are functions of the relative volatility, number of trays, feed composition and the minimum V/F. The control system therefore computes V based on the equation

$$V = F\left[a + b\left(\frac{y(1 - x)}{x(1 - y)}\right)\right] \qquad 9.4(33)$$

Since y is held constant, the bottoms composition controller adjusts the value of the parenthetical expression if an error should appear in x. Let $V/F = y(1 - x)/x(1 - y)$, and the control equation becomes:

$$V = F(a + b[V/F]) \qquad 9.4(34)$$

where [V/F] = the desired boilup-to-feed ratio.

The system implementing equations 9.4(30) and 9.4(34) is shown in Figure 9.4o. The two instruments respectively labeled "FY − 1" and "FY − 2" must be multipliers. The instrument labeled "dynamics" is a special analog computer for dynamic compensation and is discussed in Section 7.6. The scaling of FY − 1 may be done as described in connection with Figures 9.4l or m. (The illustration is based on the first.) Multiplier FY − 2 must be scaled using the latter technique, outlined in connection with Figure 9.4m.

Included in a and b is the relationship between boilup (vapor rate) and energy flow Q, and the minimum boilup-to-feed ratio. Equation 9.4(34) can therefore be written

$$Q = kF([V/F]_{min} + [V/F]) \qquad 9.4(35)$$

where k represents the proportionality constant.

$$\text{Let } Q = 10,000 \text{ lb/hr,}$$
$$F = 60 \text{ GPM,}$$
$$[V/F]_{min} = 1.0,$$

$$[V/F]_{max} = 3.0 \text{ and}$$
$$k = 0.600.$$

The values of $[V/F]_{min}$ and $[V/F]_{max}$ are utilized to establish zero and span values for the scaled equation, as explained in connection with equation 9.4(14). The scaled equation is

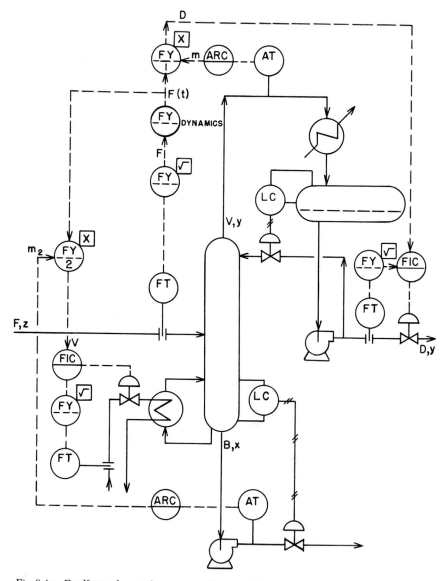

Fig. 9.4o   Feedforward control system provides closed-loop composition control of two product streams

$$10,000Q' = (0.60)(60F')(1.0 + 2.0[V/F]') \quad\quad 9.4(36)$$

$$Q' = 0.0036F'(1.0 + 2.0[V/F]') \quad\quad 9.4(37)$$

$$Q' = 0.01080F'(0.333 + 0.667[V/F]') \quad\quad 9.4(38)$$

Control of Two Products with Interaction

There is always interaction between the material and energy balances in a distillation column. In some columns it is not severe enough to impede closed-loop composition control of two product streams, but in others it is. The severity is a function of feed composition, product specifications and manipulated and controlled variable pairing. These subjects are covered in greater detail in Section 10.4.

The severe interactions frequently occur when the energy balance is manipulated by two independent composition controllers. A column in which reflux flow and steam flow are the manipulated variables is an example of a severely interacting column. The interaction index[4] for such a column will usually be on the order of -0.9[1], as opposed to a typical interaction index of $+0.4$ for the system illustrated in the preceding section. The significance of these indices is that a column with the -0.9 index is about 40 times more difficult to control than one with the $+0.4$.

The control system equations are

$$L = F([L/F]) \quad\quad 9.4(39)$$

$$Q = kF([V/F]_{min} + [V/F]) \quad\quad 9.4(35)$$

where L is the reflux rate and [L/F] is the desired reflux-to-feed ratio.

Note that in the control system described by these two equations, the rate of products leaving the column is dependent on two energy balance terms. Increasing heat input forces the composition controller resetting reflux flow to increase heat withdrawal, and the top and bottom composition controllers therefore fight each other. The only way to avoid this "fighting" is by preventing a change at one end of the column from upsetting the other end.

The heat input is changed when the bottom composition controller is upset. If the upset is because of a high concentration of light ends in the bottom product, heat is increased to adjust the separation being performed *and to drive the extra light ends out the top*. The top composition controller does not know how to split the increased vapor load, but it sees a measurement

---

[1] For a column where two manipulated variables are paired with two controlled variables, the interaction index of -0.9 means that a $10°F$ increase in bottoms temperature (due to an increase in steam flow) will increase the top temperature by $9°F$; i.e., $10°F$ multiplied by the absolute value of the index. Then, if reflux flow is increased to lower the top temperature by $9°F$, the bottom temperature will fall by $8.1°F$; i.e., $9°F$ multiplied by the index, etc. If the manipulated and controlled variables are properly paired, the index is less than one and the interaction-induced upset decreases with each cycle. If the index is greater than one, the system will never come to equilibrium, but will run away due to incorrect pairing.

indicating an upset and responds to an increase in heat input by increasing the reflux flow. If the reflux rate is compensated for the change in heat input, the top composition controller upset can be avoided.

The relationship between reflux L and heat input Q can be found by solving equations 9.4(39) and 9.4(35) for L in terms of Q. The resultant equation is of the form

$$L = k_1 Q - k_2 F \qquad\qquad 9.4(40)$$

The values of k_1 and k_2 are found by deriving equation 9.4(40) graphically, using actual process values of $[L/F]$, $[V/F]_{min}$ and $[V/F]$.

The decoupling equation 9.4(40) replaces 9.4(39) in the control model. The resulting system is shown in Figure 9.4p.

The system is now half-decoupled: a change in heat input will not upset the top temperature because the decoupling loop adjusts the reflux independently of the top temperature controller. However, the heat input is still coupled to reflux, because a change in reflux will still cause the bottom temperature controller to adjust steam flow. This degree of decoupling is enough to reduce the interaction index to about -0.60, a 20-fold decrease in severity. The two multipliers are scaled as described previously, while the adder is tuned on line.

Control of Two Products with Sidedraw

The presence of a sidestream product in addition to an overhead and bottom product *adds a degree of freedom* to a control system. This extra degree of freedom can be seen from the overall material balance equation:

$$F = D + C + B \qquad\qquad 9.4(41)$$

where C is the sidestream rate. Two of the product streams are available for manipulation, and the material balance can still be closed by the third product stream.

The added degree of freedom makes careful analysis of the process even more essential to avoid mismatching of manipulated and controlled variables. As in previous systems, the analysis involves developing the process model and determining the relationship among the several controlled and manipulated variables. There are, however, several possible combinations of variables which must be explored.

The possible combinations of manipulated variables for the column where the bottoms composition and the sidestream composition must be controlled are

> Distillate and sidestream flows
> Distillate and bottom flows
> Distillate flow and heat input

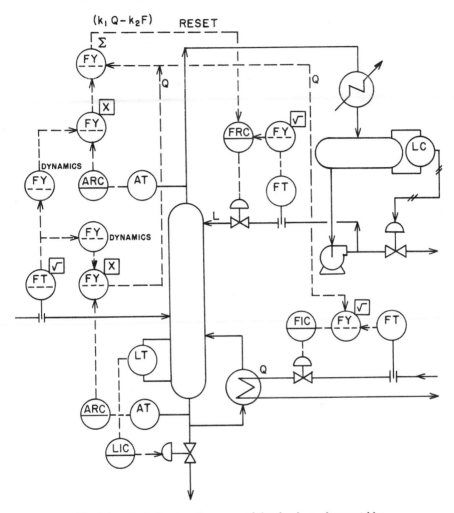

Fig. 9.4p   Control system improves stability by decoupling variables

Sidestream and bottom flows
Sidestream flow and heat input
Bottom flow and heat input

The most frequently successful pairing is to control sidestream composition
by manipulating distillate flow, and bottoms composition by manipulating
sidestream flow. The equations are

$$D = F\left(\frac{z_1 - c_1}{y_1 - c_1}\right) \qquad 9.4(42)$$

$$C = F\left(\frac{z_2 - x_2}{c_2 - x_2}\right) \qquad\qquad 9.4(43)$$

The symbols z_1, y_1, c_1 are the concentrations in the feed, distillate, and sidestream of the component under control in the sidestream. The concentrations of the key component in the bottom are respectively expressed by z_2, x_2 and c_2 for the feed, the bottoms and the sidestream.

The control system is shown in Figure 9.4q. Note that the ratio of heat input to feed (and therefore boilup to feed) is held constant. Separate dynamic elements are used for the distillate loop and for the heat input and sidestream loops.

The computing relays may be scaled by either of the methods discussed previously.

Feed Composition Compensation

Occasionally, changes in feed composition occur too fast for feedback control, and compensation for these changes is necessary (Figure 9.4r).

The basic equation 9.4(30) already has a term z representing concentration of the key component in the feed:

$$D = F\left(\frac{z - x}{y - x}\right) \qquad\qquad 9.4(30)$$

When z is measured, the equation for distillate can be simplified to

$$D = zF/m \qquad\qquad 9.4(44)$$

where m is the output of the feedback trim controller.

In an electronic system, equation 9.4(44) can be solved with one instrument. To scale the computer, assume the following range of values:

$$D = 0\text{--}100 \text{ GPM}$$
$$F = 0\text{--}500 \text{ GPM}$$
$$z = 0\text{--}0.3$$

A value for m will be assigned after the numerator is scaled.

$$100D' = (0.3z')(500F') \qquad\qquad 9.4(45)$$
$$D' = (1.50)(z')(F') \qquad\qquad 9.4(46)$$

Under normal conditions the value of the denominator is to be one and the output of the feedback controller will be 50 percent. Based on this, and assuming a reasonable span (20 percent) for the feedback trim, the complete equation is written

$$D' = \frac{(1.50)(z')(F')}{(0.9 + 0.2m')} \qquad\qquad 9.4(47)$$

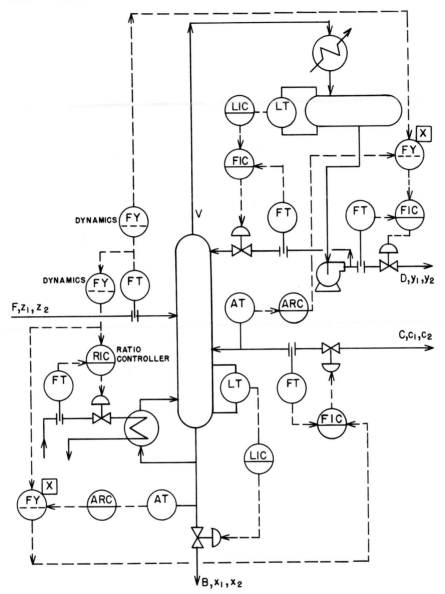

Fig. 9.4q   Control of composition in two product streams with a sidedraw

To prevent the denominator from exceeding one, divide the numerator and the denominator by the maximum value of the denominator (1.1). The result is the scaled equation for the system:

$$D' = \frac{1.363(z')(F')}{(0.82 + 0.18m')} \qquad 9.4(48)$$

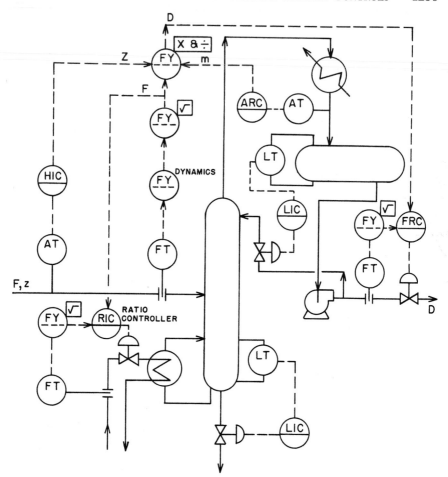

Fig. 9.4r   Feed composition measured and used to compute distillate flow

The auto manual station (HIC) is utilized in the event of analyzer failure. Dynamic compensation is placed on the feed signal only.

The Total Model

It is possible to design a system to compensate for all load variables: feed rate, composition, enthalpy, reflux and bottoms enthalpy. The goal of these systems is to overcome the problems associated with unfavorable interactions and to isolate the column from changes in ambient conditions. These problems can usually be solved by careful system analysis and variable pairing, thus avoiding complicated total energy and material balance control systems.

The complexity of the total material and energy balance systems is made apparent by the list of equations required in the model:

Feed enthalpy balance
Bottoms enthalpy balance
Internal reflux computation
Reboiler heat balance
Overall material balance

Examples of such systems can be found in References 5 through 8.

Batch Distillation

The goal of a batch distillation is to produce a product of specified composition at minimum cost. This means that operating time must be reduced to some minimum while maintaining product recovery.

If product removal is too fast, separation and the quantity of product recovered is reduced. Conversely, if the product is withdrawn so as to maintain separation, its withdrawal rate is reduced and operating time is increased. However, the set point to a composition controller can be programmed so that the average composition of the product is within specifications and thereby withdrawal rate is maximized.[9]

Figure 9.4s shows a system for doing this when a constant vapor rate is maintained from the reactor. The equation is

$$y = mD + y_i \qquad\qquad 9.4(49)$$

where y is the fraction of key component in the product, m is the rate of change of y with respect to the distillate (D) and y_i is the initial concentration of the product.

The only adjustment required is in setting m. The higher its value, the faster y will change and the smaller will be the quantity of material recovered.

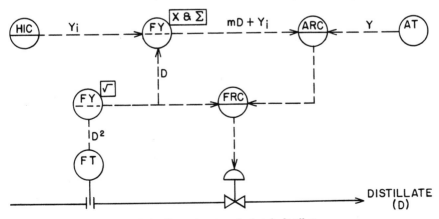

Fig. 9.4s   Control system for batch distillation

REFERENCES

1. Lupfer, D. E., and M. L. Johnson, "Automatic Control of Distillation Columns to Achieve Optimum Operations," ISA Trans. (April 1964).
2. Van Kampen, J. A., "Automatic Control by Chromatograph of a Distillation Column," Convention on Advances in Automatic Control, Nottingham, England (April 1961).
3. Shinskey, F. G., *Process Control Systems*, McGraw-Hill Book Co., New York, 1967.
4. Nisenfeld, A. Eli, "Feedforward Control for Azeotropic Distillation," Chemical Engineering, September 23, 1968.
5. Lupfer, D. E., and M. W. Oglesby, "Automatic Control of Distillation Column Heat Inputs," Industrial and Engineering Chemistry, December 1961.
6. Lupfer, D. E., and M. W. Oglesby, "Feed Enthalpy Computer Control of a Distillation Column," Control Engineering, February 1962.
7. Lupfer, D. E., and D. E. Berger, "Computer Control of Distillation Reflux," ISA Journal, Vol. 6, No. 6 (June 1959).
8. Williams, T. J.; Harnett, R. T.; Rose, Arthur; Industrial and Engineering Chemistry, Vol. 48, (1956), pp. 1008–1019.
9. Converse, A. O., and G. D. Gross, "Optimal Distillate Rate Policy in Batch Distillation," Ind. Eng. Chem. (August 1963).

9.5 PROCESS REACTOR MODELS AND SIMULATION

The reactor is the heart of the chemical process. The rest of the process consists of the reactant preparation hardware and of the devices required for the separation and purification of the products or materials produced in the reactor. Thus, the plant efficiency hinges on the efficient operation of the reactor. Process reactors are known by different names in different processes: electrolytic cells, reformers, cracking furnaces, and catalytic crackers are some examples.

A mathematical model of the process reactor is the name we give to the group of equations that describe the operation of the reactor. To write these equations the control engineer must use his knowledge of thermo-dynamics, fluid dynamics, heat transmission, molecular diffusion, reaction kinetics, control theory and economics. With the aid of modern digital and analog computers he will need only to refresh his calculus and command of differential equations to use these equations in simulation.[1]

If writing the equations that describe the reactor is a science, using these equations to simulate the reactor on a computer is an art. As any art, it must be learned through practice. The detailed mathematical model of a process reactor will often consist of a large number of non-linear equations, which cannot be practically solved in a computer. The art of simulation consists of picking out those equations which are relevant to the objectives of the simulation and of dropping those terms that contribute little to the answers. In addition, the engineer must determine what variables are to be solved for from what equations.

The type of process reactor determines the type of mathematical model. The models describing stirred tank reactors and plug-flow reactors will be presented in this section after a brief introduction to reaction kinetics. Following this discussion the modifications necessary to represent non-ideal flow, heterogeneous reactions and catalytic reactors will be presented.

Reaction Kinetics

Reaction kinetics deals with the rate of reaction with respect to time. This rate depends on the concentration of the reactants and on temperature. For gas phase reactions the concentrations of the reactants are proportional to the pressure, and therefore the pressure also affects the rate of reaction. Consider the reaction

$$aA + bB \rightleftharpoons qQ + sS \qquad 9.5(1)$$

which means that a moles of A react with b moles of B to produce q moles of Q and s moles of S. A and B are therefore the reactants and Q and S are the products. The arrow pointing to the left indicates that the reaction is reversible, and if left reacting long enough, it will reach equilibrium when the rate of reaction to the right equals the rate of reaction to the left. If the reaction is irreversible, the reaction will stop when either reactant is totally consumed.

For homogeneous reactions the rate of reaction is expressed in terms of the rate of appearance of one of the reactants per unit volume of reacting mixture.

$$r_A = \left(\frac{1}{V}\right)\frac{dN_A}{dt} \qquad 9.5(2)$$

A homogeneous reaction is one that takes place in a single phase, whether liquid, solid or gas. In equation 9.5(2), r_A is the rate of appearance of reactant A per unit volume, and N_A is the total moles of reactant A present. When A is disappearing, r_A must be a negative quantity. The rates of appearance of the other reactants and products can be expressed in terms of r_A since, from equation 9.5(1), for each mole of A that disappears b/a moles of B will also disappear, and q/a moles of Q will appear, etc.

The equations that express the rate of reaction r_A in terms of the concentrations of the reactants and products are called the *kinetic model* of the reaction. This model is obtained by proposing different models for the reaction and then choosing the one that can best predict the reaction rate data taken in the laboratory or pilot plant. Simulation with the laboratory apparatus is an excellent tool for testing the different models. A model for the reaction represented by equation 9.5(1) could be

$$r_A = -k\left(C_A^\alpha C_B^\beta - \frac{1}{K}C_Q^\gamma C_S^\sigma\right) \qquad 9.5(3)$$

where C_A and C_B are the concentrations of the reactants in moles per unit volume, and C_Q and C_S are the concentrations of the products. The reaction coefficient k is a function of temperature and must be determined experimentally. The equilibrium constant K is also a function of temperature and can be determined from thermodynamic data. The second term inside the parentheses in equation 9.5(3) is the rate of the reverse reaction and is zero for irreversible reactions. The exponents α, β, γ and σ are part of the kinetic model. When they are equal to a, b, q and s, respectively, the reaction is said to be elementary. If the exponents differ from the coefficients of equation

9.5(1), the reaction is not elementary, meaning that the actual mechanism of the reaction consists of a series of intermediate elementary reactions that are summarized by equations 9.5(1) and 9.5(3).

The reaction rate coefficient k is usually an exponential function of the absolute temperature called the Arrhenius equation:

$$k = k_0 e^{-(E/RT)} \qquad\qquad 9.5(4)$$

where E is called the activation energy of the reaction, R is the ideal gas law constant (1.98 BTU/lb-mole °R) and T is the absolute temperature of the reacting mixture. The constants E and k_0 must be determined experimentally by conducting the reaction at different temperatures. A kinetic model is essential for the simulation of a process reactor and without it the simulation is worthless. However, the kinetic model[2] does not have to be as accurate for control simulation as it would have to be for the design of the reactor.

Perfectly Mixed Reactor

A flow diagram for a stirred tank (perfectly mixed reactor) is shown in Figure 9.5a. The reactor shown is a continuous one, since the reactants flow continuously into it and the product stream is continuously withdrawn. The mathematical model is essentially the same for a batch reactor in which the reactants are added at the beginning of the batch, "cooked" for a certain period of time and then discharged.

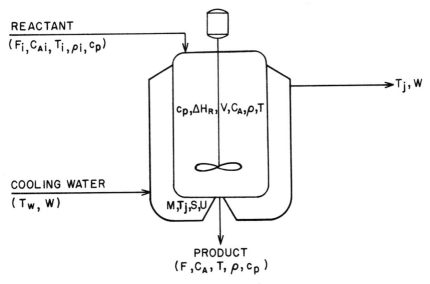

Fig. 9.5a   Perfectly mixed reactor

As we develop the mathematical model, we shall make repeated use of the accounting equation:

$$\text{Rate of accumulation} = (\text{rate in}) - (\text{rate out}) \qquad 9.5(5)$$

This equation is applied to any quantity that is conserved, such as total mass, energy, momentum, mass of components that do not participate in the reaction (i.e., solvent), etc. If we include the rate of appearance by chemical reaction as one of the "rate in" terms, we can also write a balance on mass or moles of each of the reactants and reaction products. When applied to mass, equation 9.5(5) is called the *mass, or material balance;* if to energy, then *energy balance,* and so on.

A total mass balance on the reactor in Figure 9.5a is

$$\frac{d}{dt}(V\rho) = F_i\rho_i - F\rho \qquad 9.5(6)$$

where d/dt represents the derivative, or instantaneous rate of change, with respect to time of the mass of reacting mixture, which is the product of its volume V and its density ρ. F_i and ρ_i are the volumetric rate of flow and the density, respectively, of material fed into the reactor, and F is the volumetric rate of flow of material out of the reactor. If the density of the reacting mixture is not changed appreciably by the reaction ($\rho = \rho_i$), then equation 9.5(6) can be further simplified:

$$\frac{dV}{dt} = F_i - F \qquad 9.5(7)$$

A material balance on reactant A is given by

$$\frac{d}{dt}(VC_A) = F_iC_{Ai} + r_AV - FC_A \qquad 9.5(8)$$

where C_{Ai} is the concentration of reactant A in the input stream and C_A is the concentration of A in the reactor and, since it is a perfectly mixed reactor, in the output stream. Similar balances can be made on each of the reactants and reaction products. Differentiation by parts of the left-hand side of equation 9.5(8) gives

$$\frac{d}{dt}(VC_A) = V\frac{dC_A}{dt} + C_A\frac{dV}{dt} \qquad 9.5(9)$$

and from equation 9.5(7),

$$\frac{d}{dt}(VC_A) = V\frac{dC_A}{dt} + C_A(F_i - F) \qquad 9.5(10)$$

Substituting into equation 9.5(8) and rearranging,

$$\frac{dC_A}{dt} = r_A + \frac{F_i}{V}(C_{Ai} - C_A) \qquad 9.5(11)$$

The ratio F_i/V is the reciprocal of the residence time of the reactor, which is also the reactor time constant.

If, as illustrated by the upper portion of Figure 9.5b, there were two inlet streams (F_{iA} and F_{iB}), with corresponding concentrations (C_{iA} and C_{iB}), the flow and concentration equations can be written as

$$dV/dt = F_{iA} + F_{iB} - F \qquad 9.5(12)$$

$$V\frac{dC}{dt} = F_{iA}C_{iA} + F_{iB}C_{iB} - FC \qquad 9.5(13)$$

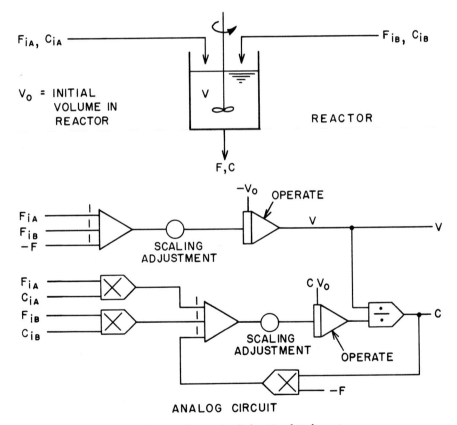

Fig. 9.5b   Simulation circuit for stirred tank reactor

Then the analog circuit to simulate equations 9.5(12) and 9.5(13) is shown in Figure 9.5b.

An energy balance on the reactor, illustrated in Figure 9.5a gives,

$$\frac{d}{dt}(V\rho C_p T) = F_i \rho C_p T_i + V r_A(-\Delta H_R) - US(T - T_j) - F\rho C_p T$$

$$9.5(14)$$

where T is the temperature of the reacting mixture, C_p is its specific heat or amount of energy necessary to raise a unit mass by one degree of temperature, T_i is the temperature of the input stream, $-\Delta H_R$ is the heat liberated in the reaction per unit mass of A reacted, T_j is the temperature of the cooling water in the jacket, S is the effective heat transfer surface area of the jacket and U is the heat transfer coefficient or rate of heat transfer to the jacket per unit area per degree of temperature difference. U is either experimentally determined or estimated from heat transfer correlations. Another simple application of differential calculus, with the help of equation 9.5(6), reduces equation 9.5(14) to

$$\frac{dT}{dt} = \frac{F_0}{V}(T_i - T) + \frac{-\Delta H_R}{\rho C_p} r_A - \frac{US}{V\rho C_p}(T - T_j) \qquad 9.5(15)$$

An energy balance on the jacket gives us an equation for T_j:

$$\frac{dT_j}{dt} = \frac{W}{M}(T_w - T_j) + \frac{US}{M}(T - T_j) \qquad 9.5(16)$$

where W is the mass flow rate of cooling water, M is the heat capacity of the water in the jacket and the walls of the reactor, and T_w is the cooling water inlet temperature. It has been assumed that the jacket is also perfectly mixed and that the specific heat of water is unity.

The pressure in the reactor can be calculated, if necessary, from thermodynamic equilibrium relationships, as a function of temperature and concentration. Equations 9.5(7), (11), (15) and (16), coupled with the reaction kinetic equations similar to 9.5(3) and (4), constitute the mathematical model of the system. The volume V, concentrations C_A, C_B, etc, and temperatures T and T_j are the dependent variables, since they depend on time, which is the independent variable. At steady state the dependent variables do not change with respect to time, and therefore the derivatives are equal to zero. In addition to the equations shown, the initial values of each of the dependent variables are necessary to complete the model (Figure 9.5b). The equations listed are ready to be programmed on an analog computer or to be solved digitally by either numerical integration techniques or any of the so-called "simulation languages," without any further manipulation.

Plug-Flow Reactor

The flow diagram of a plug-flow reactor is shown in Figure 9.5c. The term plug-flow arises from the assumption that each element of fluid flows through the reactor as a small "plug," without mixing with the fluid behind

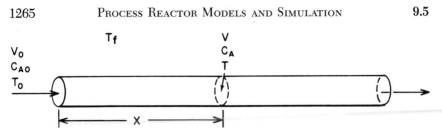

Fig. 9.5c   Plug-flow reactor

or ahead of it. It is also assumed that the concentrations and temperatures are uniform across the cross-sectional area of flow. Cracking furnaces, with very high velocities inside the tubes, approach this condition of plug-flow.

Whereas compositions and temperatures are functions of time in the perfectly mixed reactor, in the plug-flow reactor they are functions of time and distance X from the entrance. Thus there will be derivatives with respect to time at each point in the reactor, and derivatives with respect to distance at each instant of time. These are partial derivatives and the equations they are part of are partial differential equations. To solve a partial differential equation on the analog computer, it must be approximated by a number of ordinary differential equations. It is easier to derive these equations directly by the method that will be shown here.

The reactor is divided into a number of small "pools" that are assumed to be perfectly mixed. As shown in Figure 9.5d, the length of each pool is

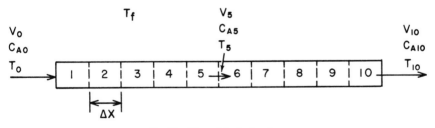

Fig. 9.5d   "Pool" model of plug-flow reactor

ΔX and its volume $A\ \Delta X$, where A is the cross-sectional area of flow. The volumetric rate of flow is equal to the velocity v times the cross-sectional area of A. A magnified sketch of pool number i is shown in Figure 9.5e. Assuming a gas phase reaction, a material balance can be written around each of the pools to give

$$\frac{d}{dt}(A\ \Delta X\ \rho_i) = A(v_{i-1}\rho_{i-1}) - A(v_i\rho_i) \qquad 9.5(17)$$

where ρ_i is the density of the reacting mixture, which can be expressed in terms of the temperature T, pressure P and average molecular weight M_{av} by the ideal gas law:

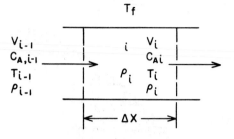

Fig. 9.5e   Pool Number i

$$\rho_i = M_{avi}P_i/RT_i \qquad 9.5(18)$$

or in terms of any other equation of state. R is the ideal gas constant and M_{avi} can be calculated as a function of the composition of the mixture in the pool. Equation 9.5(17) can be simplified, since A and ΔX are constants, to give

$$\frac{d\rho_i}{dt} = \frac{1}{\Delta X}(v_{i-1}\rho_{i-1} - v_i\rho_i) \qquad 9.5(19)$$

A material balance on reactant A will give us

$$\frac{d}{dt}(A\ \Delta X\ C_{Ai}) = A(v_{i-1}C_{A,i-1}) + A(\Delta X\ r_A) - A(v_iC_{Ai}) \qquad 9.5(20)$$

where C_{Ai} is the concentration of reactant A in the ith pool. Again we can simplify the equation by dividing $A\ \Delta X$ out:

$$\frac{dC_{Ai}}{dt} = r_A + \frac{1}{\Delta X}(v_{i-1}C_{A,i-1} - v_iC_{Ai}) \qquad 9.5(21)$$

Similar equations can be written for each of the other reactants and reaction products.

An energy balance on the ith pool gives us

$$\frac{d}{dt}(A\ \Delta X\rho_iC_pT_i) = A\ \Delta Xr_A(-\Delta H_R) + A(v_{i-1}\rho_{i-1}C_pT_{i-1})$$

$$+\ Up\ \Delta X(T_f - T_i) - A(v_i\rho_iC_pT_i) \qquad 9.5(22)$$

where p is the perimeter of the pipe, so that $p\ \Delta X$ is the heat transfer area of one pool and T_f is the firing box temperature. With the help of equation 9.5(17) and some calculus, this last equation becomes

$$\frac{dT_i}{dt} = \frac{-\Delta H_R}{\rho_iC_p}r_A + \frac{Up}{A\rho_iC_p}(T_f - T_i) + \frac{v_{i-1}\rho_{i-1}}{\rho_i\Delta X}(T_{i-1} - T_i) \qquad 9.5(23)$$

The velocity from each pool to the next can be calculated with a pressure balance:

$$P_i - P_{i+1} = k_f v_i^2 \qquad 9.5(24)$$

where the constant k_f is proportional to the friction factor which can be calculated with fluid dynamics correlations. The pressure can be calculated from equation 9.5(18).

Equations 9.5(18), (19), (21), (23) and (24) must be solved for each one of the pools. Thus an increase in the number of pools requires a greater number of analog computer components or an increase in the time required to simulate the reactor on the digital computer. The greater the number of pools, the closer we approach the partial differential equations that describe the plug-flow reactor.

Non-Ideal Flow

Process reactors deviate somewhat from the ideal flow conditions of perfectly mixed and plug-flow. A stirred tank reactor may contain "pockets" of fluid in which the concentration of reactants is different from that in other parts of the reactor. The way to model this type of non-ideality is to divide the reactor into two or more "pools" and apply the material balance to each of the pools. The sum of the volumes of the pools must be equal to the volume of the reactor. If backflow is allowed for from a pool to the one behind it, the rate of recirculation between the pools must be assumed; as this rate is increased, the results of the divided reactor model approach those of the perfectly mixed reactor model. If no backflow is allowed for, as the number of pools is increased, the model approaches that of the plug-flow reactor.

Another type of non-ideal flow is created by a tubular reactor in which the velocity, temperature and concentrations vary across the radius of the tube. If axial symmetry is present, each of the pools representing the reactor can be divided into a number of "rings" and the accounting equations can be written for each of the rings. It is easy to see how the equipment and time requirements for the simulation increase geometrically with the number of pools and rings used. For the transfer of heat and mass by conduction, diffusion and eddy currents between two adjacent rings, appropriate flow models and heat and mass transfer equations must be used.

Heterogeneous Reactions

When a reaction takes place in more than one phase, it is called heterogeneous. This occurs when the reactants cannot mix completely to form a single phase, that is, when they are immiscible or partially miscible. In heterogeneous reactions, the concentration of reactants and products depends not only on the rate of reaction but also on the rate at which they diffuse from one phase to the other. For fast reactions and slow diffusion rates, the rate of diffusion will have greater influence on the rate at which the reaction occurs than the kinetics of the reaction. In general, material and energy balance equations must be written for each phase and the appropriate diffusion equations used for the transfer of reactants and products between the phases. In addition, equilibrium solubilities and phase equilibrium relationships must be considered.

Catalytic Reactions

A catalytic reaction is one conducted in the presence of a catalyst. A catalyst is a material that speeds up the reaction without being consumed or produced by it. The reaction rate coefficient k is then a function of the temperature, catalyst concentration and "age." This age is the accumulated time since the catalyst was replaced or regenerated. These functions must be determined experimentally and incorporated into the model. When the catalyst is a solid, the additional complications of heterogeneity that were discussed in the previous paragraph, are also present. Sometimes the equations can be simplified considerably by expressing the rate of reaction on the basis of unit mass of catalyst instead of on the basis of unit volume. This is true for fixed catalyst bed reactors.

Conclusions

Presented in this section are methods for developing the equations that describe process reactors. Since the reactor is only a part of the control loop, for the simulation to be complete, it must include the equations that represent the temperature, level, pressure, and concentration sensors, transmission lines, controllers and control valves (Section 9.3). One of the advantages of simulating the reactor on an analog rather than a digital computer is that actual process controllers can be used and tuned in the simulation. This advantage reverts back to the digital computer if a digital computer is to be used to control the reactor.

A reactor control simulation allows the control engineer to tune the controllers without any loss of production or danger of blowing up the plant. It also provides him with the perfect tool to train plant operators for smoother and safer startups, it serves as a "live" model with which to try new control ideas that will result in safer and more efficient operation of the reactor, and it gives him an insight of the behavior of the reactor equivalent to several years of reactor operation. This latter advantage derives from the ability to look at variables in the simulation which are impossible or impractical to measure in the process reactor.

REFERENCES

1. Murrill, Pike, and Smith, *Formulation and Utilization of Mathematical Models,* International Textbook Co.
2. O. Levenspiel, *Chemical Reaction Engineering: An Introduction to the Design of Chemical Reactors,* John Wiley & Sons, Inc., New York, 1962.

Chapter X

PROCESS CONTROL SYSTEMS

R. J. Baker, B. Block,
A. M. Calabrese, S. G. Dukelow,
F. B. Horowitz, H. L. Hoffman,
D. C. Kendall, C. H. Kim,
B. G. Lipták, D. E. Lupfer,
C. J. Santhanam and W. F. Schlegel

CONTENTS OF CHAPTER X

	INTRODUCTION	1276
10.1	CONTROL OF CHEMICAL REACTORS	1277
	Temperature Control	1277
	Pressure Control	1282
	Optimization	1284
10.2	COMPUTER CONTROL OF BATCH REACTORS	1287
	Batch Reactor	1287
	Computer Control Functions	1289
	Measurement and Output Signals	1290
	Control Requirements	1291
	Sequence Logic Generation	1294
	Sequence Logic Structure	1294
	Process Logic Flow Charts	1296
	Reactor System Flow Charts	1299
	Programming Control Logic	1307
	The Operating System	1311
	Simulation and Checkout	1314
	Conclusions	1316
10.3	CONTROL OF DISTILLATION TOWERS	1319
	Distillation Equipment	1319
	Variables and Degrees of Freedom	1320
	Pressure Control	1322
	Liquid Distillate and Inerts	1322
	Vapor Distillate and Inerts	1323
	Liquid Distillate and Negligible Inerts	1324
	Condenser Below Receiver	1326
	Vacuum Systems	1326
	Feed Control	1329
	Reboiler Controls	1331
	Reflux Controls	1332
	Variable Column Feed	1333
	Superdistillation	1336

Analyzers 1336
    Analyzers as Operator Guides 1338
    Sampling 1340
Computers 1340
Feedforward Control 1342
Conclusions 1342

10.4   Optimizing Control of Distillation Columns 1343
Suboptimization 1343
    Control Equations 1344
    Reflux Equation 1347
    Dynamics 1348
Optimization 1349
    Product Prices Unknown 1349
    Product Prices Known 1350
    Optimizing Policies 1350
    Column Operating Constraints 1350
        Condenser Constraint 1352
        Reboiler Constraint 1352
        Column Constraint 1353
    Limited Market and Feedstock 1354
    Control Equations 1355
        Top Product More Expensive 1355
        Bottom Product More Expensive 1357
    Unlimited Market and Feedstock 1360
        Optimum Concentrations 1362
        Loading Constraint 1362
        Optimum Feed Flow Rate 1362
        Reboiler Limiting 1364
Conclusions 1368

10.5   Control of Refrigeration Units 1369
Refrigeration Units; Heat Pumps 1369
Refrigerants 1370
Circulated Fluids 1373
Household Refrigerators—Capillary Type 1373
Household Refrigerators—Float Type 1374
Small Industrial Refrigerators 1375
Expansion Valves 1376
    Evaporator Pressure Drop Low 1376
    Evaporator Pressure Drop High 1377
    Specialized Installations 1377
On-Off Control: Small Industrial Units 1378
Multi-Stage Refrigeration Units 1380

Large Industrial Refrigerators With High
    Turn-Down Ratio   1381
Conclusions   1384
10.6   Control of Steam Boilers   1385
Performance   1386
Safety Interlocks   1387
Combustion Control   1387
    Fuel Controls (Measurable Fuels)   1390
    Unmeasured Fuels   1394
    Air Control and Measurement   1396
    Damper and Fan Controls   1399
    Furnace Draft Control   1400
    Fuel-Air Ratio   1403
    Fuel and Air Limiting   1408
Feedwater Control   1411
    Single Element   1412
    Two Elements   1412
    Three Elements   1414
    Valve Sizing   1414
    Pump Speed Control   1416
Steam Temperature Control   1416
    Desuperheater Spray Controls   1416
    Variable Excess Air   1417
Integration of Loops   1419
10.7   Control of Furnaces   1420
Control System Functions   1420
Combustion Air Requirements   1422
Control Systems   1422
Startup Heaters   1423
    Process Considerations   1423
    Process Controls   1424
    Firing Controls   1424
    Safety Controls   1425
Fired Reboilers   1425
    Process Considerations   1425
    Process Controls   1425
    Firing Controls   1426
    Safety Controls   1426
Process Heaters and Vaporizers   1427
    Process Controls   1427
    Fuel Gas Firing Controls   1429
    Fuel Oil Firing Controls   1430

Multiple Fuel Firing Controls   1431
Safety Controls   1432
Reformer Furnaces   1433
     Process Controls   1433
     Firing Controls   1436
     Safety Controls   1437
Cracking (Pyrolysis) Furnaces   1437
     Process Controls   1438
     Firing Controls   1439
     Safety Controls   1440
Analytical Instruments   1441
     Oxygen Analyzers   1441
     Combustible Analyzers   1442
     Calorimetric Analyzers   1442
Advanced Controls   1442
     Feedforward Control   1443
     Analog Computer Control   1444
     Digital Computer Control   1445
Conclusions   1448
10.8   CONTROL OF DRYERS   1449
Principles   1449
Control   1452
Batch Dryers   1453
     Atmospheric   1453
     Vacuum   1455
     Special   1456
Continuous Dryers   1457
     Heated Cylinder   1457
     Rotary   1458
     Turbo   1459
     Spray   1460
     Fluid Bed   1461
     Feedforward System   1463
Conclusions   1465
10.9   CONTROL OF CRYSTALLIZERS   1466
Control Basis   1466
Cooling Crystallizers   1468
     Controlled Growth Magma Crystallizers   1468
     Classifying Crystallizers   1470
     Direct Contact Crystallizers   1471
Evaporator Crystallizers   1471
     Submerged Combustion Crystallizers   1471

Indirect Heating 1471
Multiple Effect Operation 1472
Vacuum Crystallizers 1473
Reaction Crystallizers 1475
Auxiliary Equipment 1475
Conclusions 1475
10.10 Control of Centrifuges 1477
Centrifuge Selection 1477
Sedimentation Centrifuges 1479
Batch 1479
Semi-Continuous 1480
Continuous 1481
Filtration Centrifuges 1482
Automatic Batch 1483
Continuous 1484
Conclusions 1485
10.11 Control of Heat Exchangers 1486
Variables and Degrees of Freedom 1486
Liquid-to-Liquid Heat Exchangers 1487
Component Selection 1488
Three-Way Valves 1490
Cooling Water Conservation 1492
Balancing the Three-Way Valve 1492
Two Two-Way Valves 1494
Steam Heaters 1495
Minimum Condensing Pressure 1497
Control Valve in the Condensate Line 1497
Pumping Traps 1498
Level Controllers 1499
By-pass Control 1500
Condensers 1502
Reboilers and Vaporizers 1504
Cascade Control 1505
Feedforward Control 1507
Multipurpose Systems 1509
Conclusions 1511
10.12 Control of Pumps 1513
Centrifugal Pumps 1513
On-Off Level Control 1514
On-Off Flow Control 1516
On-Off Pressure Control 1517
Throttling Control 1518
Speed Variation 1519

	Rotary Pumps	1520
	On-Off Control	1521
	Safety and Throttling Controls	1521
	Reciprocating Pumps	1522
	On-Off Control	1523
	Throttling Control	1523
10.13	Control of Compressors	1525
	Centrifugal Compressors	1526
	Suction Throttling	1526
	Discharge Throttling	1528
	Inlet Guide Vane	1528
	Variable Speed	1529
	Surge Control, Fixed Set Point	1530
	Anti-Surge Controller	1531
	Rotary Compressors	1532
	Variable Speed	1532
	By-pass and Suction Control	1533
	Suction Pressure Control	1534
	Reciprocating Compressors	1534
	On-Off Control	1535
	Constant Speed Unloading	1535
	Variable Speed	1537
	Lube and Seal Systems	1537
10.14	Effluent and Water Treatment Controls	1540
	Chemical Oxidation	1541
	Chemical Reduction	1545
	Neutralization	1547
	Precipitation	1552
	Biological Control	1555
	Conclusions	1558
10.15	Analog and Digital Blending Systems	1559
	Blending Methods	1560
	Rate Blending	1562
	Totalizing Blending	1562
	Optimizing Blending	1564
	Analog Blending	1564
	Mechanical Ratio Control	1565
	Pneumatic Ratio Control	1568
	Electronic Ratio Control	1570
	Ratio Dial Setting	1573
	Ratio Controller Tuning	1574
	Digital Blending System	1575
	Conclusions	1578

There is practically no limit to the topic of process control systems or unit operations. For the purposes of this Handbook, the scope of coverage has been limited to the most frequently utilized processing units in the chemical industry.

In the following sections, the measurement and control practices for chemical reactors, distillation columns, refrigeration units, boilers, furnaces, dryers, crystallizers, centrifuges, heat exchangers, pumps, compressors, effluent treatment systems and digital blending systems will be discussed. The coverage of these topics is *system oriented,* meaning that on hardware aspects or on control theory considerations, the reader should review the corresponding sections of these volumes.

In case of the most important unit operations (chemical reactors and distillation columns), in addition to the treatment of measurement and control practices, the subjects of computer control and optimization are also covered.

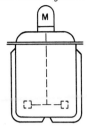

10.1 CONTROL OF CHEMICAL REACTORS

The reactor is probably the most important piece of equipment in a chemical plant. The performance of this unit usually determines the product quality, and its efficiency is a major contributing factor to total plant production.

This section is devoted to the subject of conventional reactor control, including the regulation of reactor temperatures and pressures together with some optimization techniques to maximize production as a function of available heat removal capacity.

In connection with other aspects of reactor control, the reader is referred to Section 10.2 for a discussion of computerized batch reactor control and to Section 9.5 for the treatment of reactor modeling and simulation.

Temperature Control

Reaction temperature is frequently selected as the controlled variable in reactor control. It may be necessary to control reaction rate, side reactions, distribution of side products, or polymer molecular weight and molecular weight distribution. All of these are sensitive to temperature. It is frequently necessary to control reaction temperature to within $\frac{1}{2}°F$. Many reactions are exothermic. In order to control reaction temperature, the released heat must be removed from the system as it is liberated by the reactants. A simple temperature control scheme is depicted in Figure 10.1a. The reaction temperature is sensed, and flow of heat transfer medium to the reactor jacket is manipulated.

In case of a large number of installations this scheme is considered to be unsatisfactory because, under throttled conditions, the flow of heat transfer medium can be inadequate to maintain a good heat transfer coefficient, and also because the temperature gradient in the heat transfer medium across the jacket may be large enough to keep different areas of the jacket heat transfer surface at different temperatures. This can result in localized temperature differences within the reactor (hot and cold spots) which are both uncontrollable and undesirable.

A more desirable arrangement is shown in Figure 10.1b. In this scheme, the liquid heat transfer medium is recirculated at a high rate through the

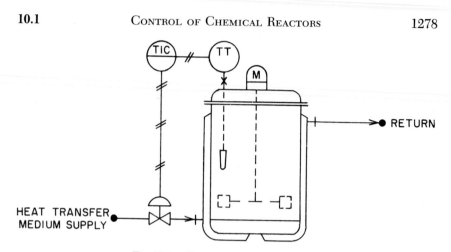

Fig. 10.1a   Reactor temperature control

jacket by way of an external pumping loop. The fluid velocity in the reactor jacket is maintained high enough to produce satisfactory film coefficients for heat transfer. In addition, a sufficient volume of liquid is circulated to keep the temperature gradient in the heat transfer medium as it passes through the jacket at a low enough level to maintain the jacket wall temperatures uniform throughout the reactor.

Both of these temperature control systems suffer from deficiencies which relate to time lags inherent in most reactor systems. First, there is a time lag in the response of the loop of the heat transfer medium in adjusting the

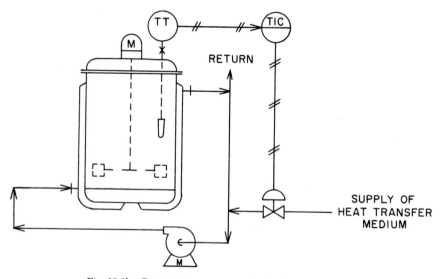

Fig. 10.1b   Reactor temperature control with recirculation

temperature of the cooling (or heating) medium. Second, there is a time lag due to the physical mass of the reactor itself and to the heat load imposed on the cooling system to readjust the reactor temperature. Third, a very significant time lag is caused by the reactant mass and by the relatively large quantity of heat which must be removed to bring about a small change in reactant temperature level. Due to these process lags, a simple temperature control system tends to over-compensate for system disturbances. Each time an upset occurs, there can be cycling and poor control before the controller compensates for the system disturbance. Usually the period of this oscillation is several times the reactor heat transfer time lag. When the controller is properly adjusted, there still may be three cycles of oscillation before the product temperature returns to its control point. Periods of cycling up to one hour can exist, resulting in poor reactor temperature control for extended periods. A superior method of reactor temperature control, a cascade loop, is depicted in Figure 10.1c. Here the controlled process variable (reactor batch

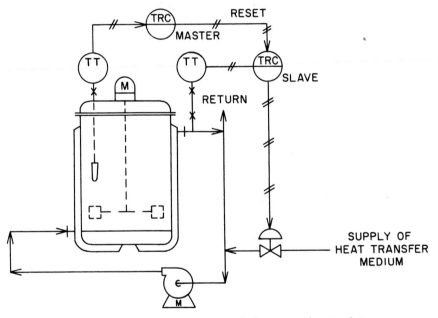

Fig. 10.1c   Cascade temperature control of reactor with recirculation

temperature), whose response is slow to changes in the heat transfer medium flow (manipulated variable), is allowed to adjust the set point of a secondary loop, whose response to changes is rapid. In this case, the reactor batch temperature controller varies the set point of the jacket temperature control loop.

An essential feature of a successful cascade control system installation is that the secondary control loop should be able to correct for disturbances in the heat transfer medium source without allowing its effects to be felt by the master controller. For example, a change in cooling water supply temperature is corrected for in the slave loop and is not allowed to upset the master controller. As pointed out in the detailed discussion of cascade systems in Section 7.8, the process lags should be distributed between master and slave loops.

In most processes, a certain temperature has to be reached in order to initiate the reaction. In the case of a steam-cooling water system, the steam may be directly injected into the cooling water circulating loop by way of a ring heater or a steam-water eductor. Where other heat transfer fluids are involved, indirect means of heating the circulating heat transfer medium are employed. A heat exchanger in the circulating loop is used to indirectly heat the fluid.

Figure 10.1d depicts a cascade temperature control system with provisions for batch heat-up. The heating and cooling medium control valves are

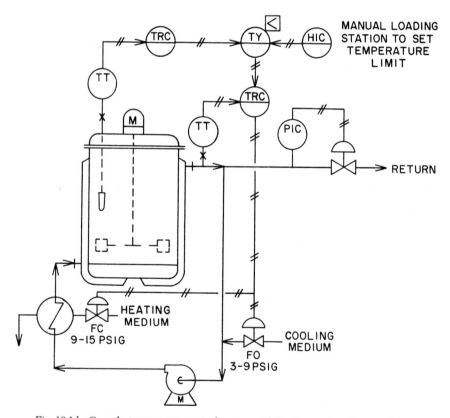

Fig. 10.1d   Cascade temperature control system with heating and cooling capability

split-range controlled, such that the heating medium control valve operates between the air signal values of 9 and 15 PSIG, and the cooling medium control valve operates between 3 and 9 PSIG. This is a fail-safe arrangement. In the event of instrument air failure, the heating medium control valve closes and the cooling medium control valve opens to provide emergency cooling for the batch.

Figure 10.1d also shows an arrangement whereby an upper temperature limit is set on the recirculating heat transfer medium stream. This is an important consideration if the product is temperature sensitive, or if the reaction is adversely affected by high reactor wall temperature. In this particular case, the set point to the slave controller is prevented from exceeding a preset high temperature limit. Another feature shown is a back-pressure control loop in the heat transfer medium return line. This may be needed to impose an artificial system back pressure, so that during the heat-up cycle no water leaves the recirculation loop and therefore the pump does not experience cavitation problems.

The complexity and details of control for all types of heat exchangers are discussed in Section 10.11.

If the temperature control loops are tuned to give optimal control during the reaction phase (cooling cycle), then the control loop will not be sufficiently damped to prevent overshoot of set point during the heat-up period. If the loop is damped to minimize overshoot during heat-up, then control during reaction will suffer. Frequently, operators will manually control the approach to the temperature set point in order to prevent overshoot, which otherwise can result in undesirable product properties or in uncontrollable reaction rates. Instrumentation is available to permit rapid automatic heat-up without temperature overshoot, in conjunction with optimal reaction control.

One such system operates as an on-off controller during the heat-up cycle. When the set point has been approached within some small margin, the control is first reversed to temporary cooling to remove the thermal inertia from the system. After a brief period of full cooling, the loop is switched from on-off to PID control, where the three modes of the controller have already been tuned for the dynamics of the cooling cycle. This method of temperature overshoot prevention is referred to as *dual mode control*. Other techniques for the prevention of reset windup are discussed in detail in Sections 4.2 and 4.3.

Sometimes it is necessary to control the reaction at several temperatures or control the rate of temperature rise during a reaction. For this purpose, clock-actuated cams or other types of function generators are used to regulate the controller set point as a function of time. For a more detailed discussion of this hardware refer to Section 4.4.

Occasionally the design engineer must use his imagination in developing indirect temperature control systems. As an example of this, high-pressure polyethylene processes operate at pressures sufficiently high (e.g., 20,000–50,000 PSIG) that the resultant wall thickness of the tubular reactor is too thick to permit good temperature control due to the poor heat transfer through the reactor wall. In such a case, temperature control can be obtained by having reaction temperature control the catalyst flow to various points in the reactor which, in turn, controls reaction rate.

In a process where the reactor pressure is a function of temperature (e.g., the reactor pressure is essentially the vapor pressure of one of the major components in the reaction), this pressure may be sensed and used to control temperature.

Pressure Control

Certain types of chemical reactions require, in addition to temperature control, some form of pressure control. Typical of these reactions are oxidation and hydrogenation reactions, in which the concentration of oxygen and hydrogen in the liquid reactants and the consequent reaction rate is a function of pressure. The reaction rate is also a function of pressure in gas phase reactions. In high-pressure polyethylene polymerization, both the reaction rate and the resultant polymer properties are sensitive to reaction pressure.

Figure 10.1e depicts a batch reaction in which the process gas is wholly absorbed in the course of the reaction. Here, the concentration of process gas in the reactants is related to the partial pressure of the process gas over the reactants. Pressure may be sensed and controlled, thereby controlling the concentration of process gas in the reactants and the resultant reaction rate. In this mode of control, system pressure responds quickly to changes in controller output, and therefore a fairly narrow proportional band may be utilized. Derivative response is not required. In addition to pressure control,

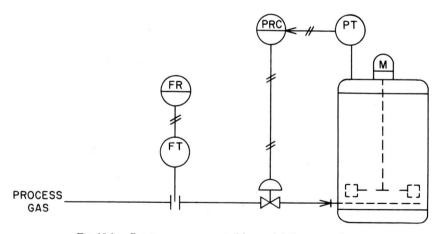

Fig. 10.1e   Reactor pressure control by modulating gas makeup

the reactor will also require one of the previously discussed temperature control systems. Figure 10.1e is simplified in this regard.

Certain reactions not only absorb the process gas feed, but generate by-product gases. Such a process might involve the formation of carbon dioxide in an oxidation reaction. Figure 10.1f illustrates the corresponding pressure control system. Here, the process gas feed to the reactor is on flow control, and reactor pressure is maintained by throttling a gas vent line. This particular illustration also shows a vent condenser, which is used to minimize the loss of reactor products through the vent.

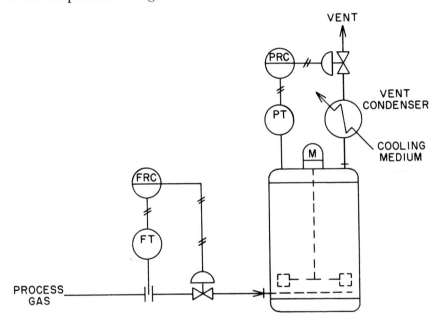

Fig. 10.1f   Reactor pressure control by throttling flow of vent gas

In the case of a continuous reactor, a system such as is shown in Figure 10.1g is often employed. Here the reactor is liquid full, and both the reactor liquid and any unreacted or resultant by-product gases are relieved through the same outlet line. Reactor pressure is sensed, and the overflow from the reactor is throttled to maintain the desired operating pressure. Process gas feed and process liquid feed streams are on flow control.

All of these illustrations are simplified. For instance, it may be desirable to place one of the flow controllers on ratio in order to maintain a constant relationship between feed streams. If the reaction is hazardous, and there is the possibility of an explosion in the reactor (e.g., oxidation of hydrocarbons), it may be desirable to also add safety devices such as a high-pressure switch to automatically stop the feed to the reactor.

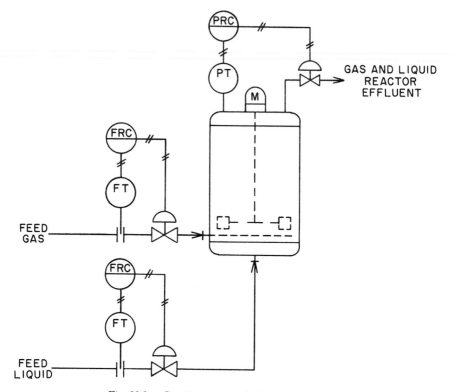

Fig. 10.1g   Continuous control of reactor pressure

Optimization

It is sometimes desirable to optimize the performance of the reactor. For example, it may be desirable to maintain the reactor throughput at as high a rate as the heat removal capabilities of the system will permit. This is a broad topic, and, within the scope of this discussion, it is difficult to do more than give a few specific examples.

One form of optimization is frequently applied in the batch polymerization of copolymers. In copolymer polymerization, reactivity ratios are generally such that one of the monomers is depleted at a faster rate than the other. Product requirements sometimes demand that the ratio of monomer to comonomer in the final product be uniform. That is, as the polymerization progresses, the ratio of unreacted monomer and comonomer must be kept constant. In such cases, the more reactive monomer or a mixture of monomers is fed continuously to the reactor to maintain the desired ratio of monomers in the reactor. In order to do this properly it is necessary to know how far polymerization has advanced. This can be determined by measuring the total heat released by the system and relating this heat release to the degree of polymerization. If the flow of coolant makeup is multiplied by its temperature rise, this will give the instantaneous rate of heat removal from the system.

This rate of heat removal can be used to reset a flow controller maintaining the monomer feed rate to the reactor. This rate of heat evolution need only be integrated by time to give the total heat released by the batch reaction at any instant in time. This total is related to the total conversion at a given time, and therefore its measurement can be useful for the batch addition of certain ingredients. For example, modifiers must be added at certain percent

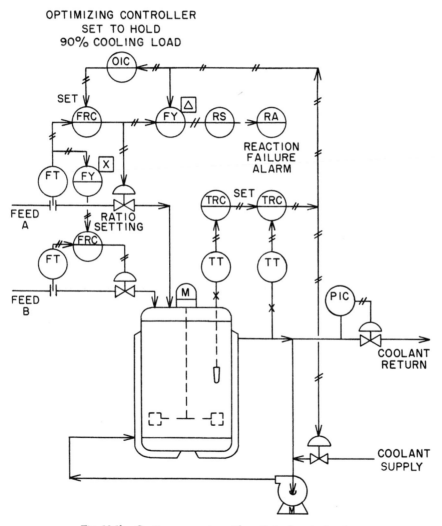

Fig. 10.1h   Continuous reactor with optimized control system

monomer conversions to polymer during the reaction. This approach is applicable to many rubber polymerization processes.

Another type of reactor optimization may involve maintaining the reaction at as high a rate as the system will tolerate. Figure 10.1h depicts such a system. Here a monomer is polymerized essentially at the same rate at which it is fed to the reactor. The reactor is controlled to maintain a feed rate high enough that polymerization will occur as rapidly as the cooling system can remove the reaction heat. The cooling system capacity will, of course, vary with ambient temperature conditions as they affect the corresponding cooling water temperature. There may also be a variation in the heat transfer coefficient as the concentration of product solids and its viscosity increases.

The system shown in Figure 10.1h utilizes a conventional cascade temperature control loop. The two feed streams are on ratio control. Reaction optimization is achieved by charging feed A at a high enough rate that the coolant makeup to the circulating cooling loop is maintained at near its maximum capacity (90 percent open valve). This is achieved by sensing the air signal to the coolant control valve and using this information to reset the feed flow set point on the master feed control loop (in this case, feed A). By maintaining the coolant makeup at as high a rate as is practical, the jacket temperature is maintained at as low a temperature as the coolant source permits, thereby assuring maximum permissible heat removal at any point in the reaction. Coolant supply temperature levels may fluctuate. This is the case with tower water which depends on ambient humidity and temperature conditions and on cooling tower heat load to establish the cooling water supply temperature. The performance of the illustrated control system is unaffected by these fluctuations. In order to insure that charging is interrupted if for any reason the reaction ceases, the air signals to the control valve on feed A and to the coolant makeup are compared. When the reaction is proceeding normally, there is a predictable relationship between these air signals. In the event that the reaction ceases, the generated reaction heat drops, and the control signal to the monomer feed control valve will change to admit more and more feed, such that the valve would approach a fully open condition. The difference between the two air signals can be used to actuate an audible alarm or to initiate a safe and automatic shutdown procedure.

10.2 COMPUTER CONTROL OF BATCH REACTORS

Batch processing is employed in the production of many important materials. Major industrial applications lie in chemical polymers—resins, fibers, and elastomers. The key unit operation in these processes is the batch reactor. It presents a variety of control problems stemming from the need for precise regulation of the reaction conditions in order to stay within close product specifications. Product quality and uniformity between batches is essential, since products are blended in storage containers with previous batches and must not adulterate the bulk quantities. A lost batch, or a number of lost batches due to poor control, can severely affect plant production costs, because the raw materials are usually expensive chemicals.

For these reasons, computer control of batch reactors is becoming increasingly important. It is a proved method of implementing regulatory tasks in complex plants containing multiple reactor units with the capability of processing many different recipes.

This section deals with the techniques for computer control of batch reactors. It is intended to serve as a guide for the instrument engineer confronted with the task of generating the sequential control programs and integrating them into a software environment designed to permit sequential control as well as ddc or set-point control of process loops. Recognizing that there are a definite series of steps involved in going from operating instructions and control diagrams to working control programs, the material is presented in the natural order of analysis, programming and checkout.

Batch Reactor

In a batch reactor, measured quantities of reactants are charged and allowed to react for a given time under carefully controlled conditions. In a continuous reactor, material constantly flows through the reactor. Batch reactors have separate phases of operation, such as charge, react, and discharge.

Chemical polymerization is the most common type of batch reaction process. The reactions involved may be highly exothermic, releasing an amount of heat proportional to the quantity and nature of the material reacting. The equipment must therefore be designed to provide adequate mixing and heat removal.

For the purpose of studying the specific control requirements, a problem which exemplifies the most common elements of reactor operation and control is used here. Figure 10.2a shows a process and instrument diagram of the

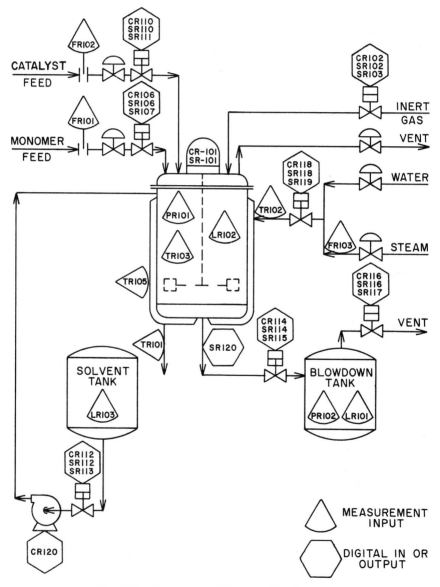

Fig. 10.2a   Process control diagram of batch reactor

example reactor system. The reactor is charged with catalyst, solvents (to maintain a fluid viscosity and serve as a heat sink) and the monomer reactant. The polymerization takes place in a stirred tank, isothermal reactor that can be heated or cooled by water flowing through the jacket. The temperature of the water is governed by the amount of steam injected into the line entering the jacket. The pressure in the reactor is maintained by venting the vapor to a recovery system (not shown) which condenses the unreacted monomer. The product from the reactor goes to a holding tank prior to transfer to another process unit.

Computer Control Functions

The control system for the reactor provides automation of all operations from the beginning to the end of the batch. This reactor is one of many identical or nearly identical reactors in a large hypothetical processing complex. Such plants contain as many as 80 reactors, all requiring the same kind of attention and regulation.

When batch processes are designed for computer control, the measurement and control hardware and the mode of operation are geared to the needs and capabilities of the computer system. The concept of manual backup commonly implemented in computer-controlled continuous processes becomes less practical to apply to the batch reactors. In many applications, processing must be suspended or drastically curtailed upon computer control system failure.

The computer has the built-in capability to respond to process failures in such a way that downtimes are minimized and maximum process safety is provided. It can determine the corrective action needed from the type of failure. Such decisions might determine whether to cause a partial or complete shutdown, and whether it is necessary to shut down all reactors or only those immediately affected by an abnormal condition.

The first level of regulatory actions required to fulfill the proposed control requirements is individual loop control. That is, driving a process variable to its target value, or set point, and maintaining it at that value.

When the computer directly manipulates the control valve, it is said that the variable is under direct digital control (ddc) (Sections 8.12 and 8.14). The measurement transmitter, control algorithm, and control valve constitute the control loop. If the computer transmits a set point to analog control hardware, which in turn determines and maintains the valve position, it is performing set-point control. Although set-point control has been widely applied to continuous processes, ddc is still preferred for some batch applications. This is because the cyclic nature of batch operations creates a need to activate and deactivate loops and to change the loop configuration (input or algorithm type) during a batch. It is easier to do this with software than with hardware.

The next control level consists of executing all of the steps required to operate the reactor and its related equipment. These include such sequential operations as starting and stopping pumps, operating on/off valves, and activating timers and controllers. The procedure for operating the reactor is called *sequence control logic.* Sequential and ddc control will be covered in more detail later in this section.

A still higher level of control, which begins to take on supervisory aspects, contains scheduling and optimization functions. Production scheduling makes batch assignments to processing units in response to production requests and inventory requirements. Optimization programs determine a set of operating conditions that will maximize plant output, or minimize a variable such as steam consumption or elapsed time. Some aspects of optimizing control concepts are covered in Sections 8.11 and 10.1.

Measurement and Output Signals

The reactor and supporting equipment described above and illustrated in Figure 10.2a constitute a process system. For simplicity, batch processes are subdivided into functional groups of equipment called systems which perform essentially independent tasks. For instance, a process may consist of three types of systems: blending, reaction, and purification. Each type of system has an operating procedure and control requirements. Only one operation may occur within a system at any time.

Once the boundaries of the reactor system have been established, the first step is to list the process input/output (I/O) requirements. This type of specification should include the following information:

1. *Point/loop number identification*

   Example:

| Point Type (Level) | System Type (Reaction) | System Number | Level Transmitter Number in System |

   Point Types

A = Analyzer	T = Temp
L = Level	C = Digital output
F = Flow	(contact)
P = Pressure	S = Digital input

   Process System Types

   Feed preparation, blending, reaction and purification.

2. *Point description*

   Example: Reactor level   (LR102) Opened feed valve (SR106)

   Start solvent pump (CR120) Closed feed valve (SR107)

For digital outputs and inputs, the description gives the condition that exists when the contact is closed.

3. *Conversion type* (Analog inputs only): linear, square root, quadratic, etc.
4. *Signal range* (Analog inputs only): volts, millivolts, milliamperes, etc.
5. *Measurement range:* engineering units, etc.
6. *Filtering required,* if any.
7. *Input service:* Alarm scan only or scan and control.

Control Requirements

The following list is prepared for all ddc loops.

1. *Identifying label:* FR101 Label denotes analog input number of controlled variable.
2. *Algorithm:* I, PI, PID, ratio, etc. (Section 8.5).
3. *Secondary loop number,* if this loop is the primary of a cascade system.
4. *Primary loop number* if this loop is the secondary of a cascade system.
5. *Valve action:* air to close, air to open.
6. *Control action:* direct acting or reverse acting.

The controller types (feedback, ratio, cascade, etc.) are discussed in Chapter VII and the ddc algorithms are derived from the corresponding equations.

In batch computer control, especially reactor control, the need for special-purpose control loops is often created by the process characteristics. For instance, auto-selector control is used to select automatically one of two related variables to control a final control element (Section 7.10). Bang-bang (on-off) control is a technique for utilizing the full range of the manipulated variable. The control scheme in this example utilizes an adaptation of these two techniques to control the temperature of the reactor contents. Figure 10.2b shows a schematic of the control configuration. With this scheme, there are four levels of control:

> Feedback—TR102 and FR103
> Cascade—TR103
> Bang-bang—SR101
> Auto-select between FR103 and TR102 outputs

When the output to the split-range-operated valves is zero, the steam valve is closed and the water valve is full open. The control scheme is used to (1) heat up the reactor after it has been charged, (2) maintain reactor temperature during the run, and (3) maintain or adjust temperature after the run, prior to discharge.

During reactor heatup, the bang-bang algorithm supplies a set point to

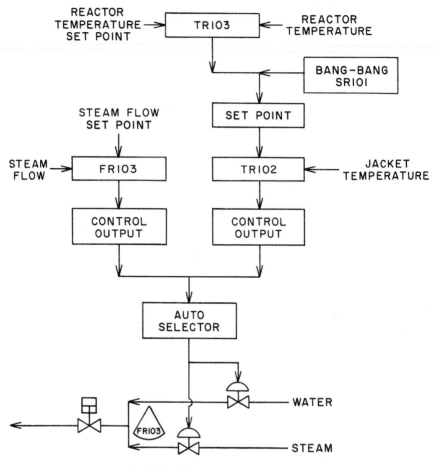

Fig. 10.2b   Temperature control system

the jacket temperature control loop. The steam flow control is inactive, the reactor temperature cascade is open, and the auto/select algorithm is not in service. The bang-bang algorithm compares the reactor temperature with the desired value. If it is less than the set point minus a constant (A + B), the algorithm sends a maximum set point to TR102, causing the valve to allow full steam flow to the jacket. When the jacket temperature enters the hatched area in Figure 10.2c, the bang-bang algorithm supplies a minimum set point to allow full cooling. With proper adjustment of the switching point and bandwidths, overshoot will drive the jacket temperature to between T_a and the initiation target temperature (T_s). At this point, the bang-bang algorithm shuts itself off, after changing the set point to the target value.

The reactor temperature cascade then closes, and the steam flow controller and auto-selector are activated. The reactor temperature control adjusts the jacket temperature as necessary to maintain the temperature during the reaction. The steam flow controller has a fixed set point which is determined as the maximum desired heating rate during a run. The auto-selector algorithm chooses the lowest valve output signal, which means that the steam flow can never exceed the set point of FR103. This provides an override feature which acts as a safeguard against runaway temperature, due to a sudden rise in the amount of material reacting. This could otherwise occur after a dip in temperature caused by the introduction of cold reactant at the beginning of the run, when the reactor contents are not yet an adequate heat sink.

This control scheme can be easily implemented in a computer system, because all of the controllers are combinations of basic algorithms, and the only hardware devices required are the split range control valves and the three inputs, TR103, TR102 and FR103.

Another important element in reactor control schemes is set-point ramping. The discontinuous nature of batch processes necessitates large set-point changes which must be accomplished at a rate compatible with process dynamics. A set-point ramping algorithm increases and decreases the set point by a constant amount until the measurement equals the target set point, or until the actual set point equals the target set point, whichever occurs first. The set-point increments and target set points are supplied by the sequence logic.

A metering algorithm is used to integrate a flow and cause cutoff when the accumulating integral reaches the desired integral. The computer system integrates the flow signal by counting pulses from a digital flowmeter, or by reading the analog flow input and calculating the amount fed over the sample interval and adding it to the accumulated quantity. When the integral approaches the final value (within a preset limit) the computer reduces the flow rate set point to a minimal value, and when the desired total is satisfied, the computer changes the set point to zero. An additional feature of this algorithm

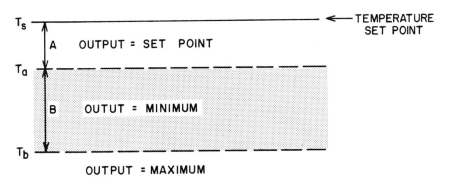

Fig. 10.2c   Bang-bang algorithm switching points

is the ability to *"set/clear"* a bit in core memory upon reaching the final total. This bit can be a request to close an on-off valve, and/or can indicate to the sequence program that the integration is completed.

A variation of the metering algorithm can be used to read a load cell signal, compare this measurement with the desired weight of the vessel being charged, and perform the throttling, cutoff and signaling functions described earlier.

Sequence Logic Generation

The recipes, together with the I/O information, control requirements and operating description provide the basis for constructing the sequence control logic.

The recipes are intended to specify all control parameters which are determined by the desired product characteristics. For example, when the computer encounters an instruction to supply a temperature set point to the reactor temperature controller, it gets the value from a recipe item.

There is one recipe for each type of product. A recipe, as described above, is not a set of instructions, but a list of parameters to be referenced by the control logic. The following items can appear in the recipe:

1. *Operating parameters*
   a. Temperatures, pressures, flows, etc.
   b. Batch total
   c. Charge quantities
   d. Feed ratios
   e. Feed components
2. *Material routing parameters*
   a. Task valve numbers
   b. Valve header configurations
3. *Procedural indicators*
   a. Stop for analysis after run
   b. Wait for a manual operation such as reactor cleanup after a batch

The source material used in developing the sequence logic for the reactor system is the detailed operating description—a procedure which states all of the process steps from the beginning to the end of the batch. This information should be at a level of detail sufficient to specify all actions an operator would make in operating the batch, including those steps required to correct a process problem or abnormal condition.

Sequence Logic Structure

An operating procedure for a process system can be divided into control states which define the particular mode of operation of the system.

Typical control states include:

> *Normal*—Process operating according to prescribed procedure
> *Hold*—Partial shutdown, maintaining conditions near the operating level
> *Emergency shutdown*—Complete, sudden shutdown
> *Return*—Transitional logic to go from a hold or shutdown state back to the normal state

Figure 10.2d depicts types of transfers between control states. The computer transfers in and out of these control states as a function of system performance. For example, if a pump or valve failed, the computer would cause transfer to hold or to emergency shutdown. The system might be programmed to go automatically into hold at a certain point to allow sampling and analysis of the reactor product. These transfers among control states can also occur upon operator request.

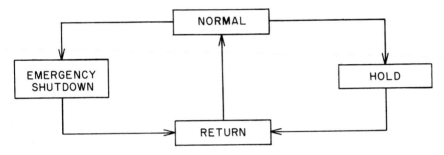

Fig. 10.2d   Reactor control states

The normal control state is further subdivided into process states, as shown in Figure 10.2e. A process state is a distinct phase of operations in a batch cycle, such as charge, reaction initiation or run. When a system is returning to the normal control state, the reentry logic determines what process state the system was in at the time it went into hold or shutdown. It may be necessary to return to a preceding process state to reestablish the conditions that existed at the time of interruption. For instance, if the system was in

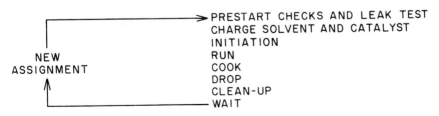

Fig. 10.2e   Reactor process states

hold and the reactor contents were allowed to cool below the run temperature, the system must return to the heat-up state and repeat a series of steps designed to achieve run conditions before reinitializing the reaction.

The sequence logic can be reduced to a time-ordered combination of a few basic actions, which are the components of the process and control states described above. These actions are—

1. Operate on-off device (pumps, valves, etc.)
2. Activate or deactivate ddc loops
3. Open and close cascade loops
4. Supply ddc loop parameters such as set point, ramp rate, alarm limits, tuning parameters, flow integrals
5. Integrate flows
6. Initiate times or delays between processing steps
7. Perform calculations
8. Compare values, measurements, test indicators and branch according to the results of the comparison
9. Initiate alarm status, or operator messages
10. Release control if nothing more can be done until the next sequencing interval.

It is important to note that the data acquisition and loop control functions are not included as part of the sequence logic. Rather, the sequence logic serves in a supervisory role, communicating changes to the control loops as required, in the same way that an operator would adjust set points and alarm limits during the course of a batch.

Process Logic Flow Charts

The most common approach to representing a logic network is the flow chart. There are a number of other formats, such as decision and action tables, which some engineers prefer to use. The flow chart will be used here as the medium for presenting the set of sequence logic developed for the example reactor system.

The flow chart must be self-sufficient in defining all aspects of the process sequencing and equipment control. It must be organized and labeled in a manner that permits both process engineer and programmer easily to interpret the intent of the logic. Input-output information, control loops, process parameters and indicator bits are referenced by preassigned symbolic tags wherever they are used in the flow chart.

The flow chart block symbols used in the example flow charts are defined in Figure 10.2f. A flow chart block is a step, or function, to be performed by the computer at execution time. Each step is related to the process and must be completed before the next operation can begin.

Some flow chart blocks imply a delay or action confirmation. For instance,

TYPE OF BLOCK	EXPLANATION	FLOW CHART SYMBOL
ACTION	STARTING PUMPS, OPENING AND CLOSING VALVES, ARITHMETIC OPERATIONS DELAYS, PARAMETER AND BIT MANIPULATION AND COMMANDS TO CONTROLLERS.	
DECISION	COMPARE DATA AND PROCESS INPUTS AND PERFORM LOGICAL DECISIONS SUCH AS "GREATER THAN, LESS THAN, OR EQUAL TO".	
PREDEFINED FUNCTION	REPRESENTS A FUNCTION CONSISTING OF A NUMBER OF STEPS DEFINED ELSEWHERE AS A LOGICAL UNIT OR SUBROUTINE.	
MESSAGE OUTPUT	ALARMS, STATUS INFORMATION AND OPERATOR GUIDES.	
CONNECTOR	JUNCTION, ENTRY OR EXIT POINT IN A LINE OF LOGIC. ALSO USED AS A CONNECTOR.	

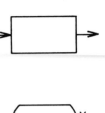

Fig. 10.2f   Flow chart symbols

block 3 in the reactor system flow charts, Figure 10.2g, will cause the following operations:

1. Output of digital signals to the contacts listed above the block to set them in the specified state (all contacts are opened in the example).
2. Delay for a time equal to the processing interval of the system to allow the valve or pump to reach its desired state.
3. Check the contact sensors associated with each valve or pump operated to confirm that it is in the desired state.
4. If a failure occurred, print a message identifying the failure, and transfer to an alternate control state, such as hold, to allow the operator to correct the fault.
5. If the equipment operated satisfactorily, proceed to the next step (block 5 in the example).

Block 15 means that the next step, block 16, will not be executed until 5 minutes has elapsed. An exit statement, as shown in block 13, implies that control will be returned to an executive program, postponing any further execution of control logic until the next processing interval. A processing interval may be from 1 to 30 seconds and will be discussed in more detail later in this section. The return point after the delay caused by an exit statement is indicated in the exit symbol.

The flow charts in Figure 10.2g describe all operations performed by the sequential control program from the beginning (prestart checks) until the end of the batch. The logic for the alternate control states, i.e., hold, shutdown, and return, are also flow charted separately. Each of the process states in Figure 10.2e has its operations defined in a branch from the level-seeking trunk path described by blocks 1, 19, 51, 61, 89, 103, 130, and 134 in Figure 10.2g.

Level seeking means that the recurrent execution of the programs at regular intervals always begins at the first statement and proceeds through the logic until it reaches either a function requiring process response or an exit point. This type of logic causes repeated execution of the same sequencing statements, which may be necessary in situations where regular checks are required. However, it also creates the necessity of adding logic to skip operations, which should only be performed once, such as a delay.

The alternative to the level-seeking approach is "returning" logic. In other words, an instruction causing an exit sets up the block number to return to the next sequencing interval, which can be the next statement or a preceding statement. This technique provides the flexibility needed in most sequential programs, since it can allow varying degrees of level seeking as appropriate to operations in different parts of the program.

The reactor flow charts illustrate an organization which, although primarily "returning" in nature, can permit level seeking if necessary at some point(s) in the program.

Some statements, such as blocks 43, 56, and 93, should be executed only once during a batch. This is indicated by the "FO" (first time only) designation under the block in the flow charts. This implies that the coding will call for a skip around the statement on all subsequent passes through that path of logic for the remainder of the batch.

The sequence control program operates contacts which turn status lights on and off, indicating the current process and control state of the reactor system. These lights are labeled and arranged on a panel in the control room.

Bits (two-state indicators), variables and recipe items are labeled in the flow charts according to the following conventions:

$$YZNNN$$

where Y = Class of data:
B = Bit
R = Recipe item
V = Variable

   Z = Local/common indicator:
      R = Local (used only by one system program)
      C = Common (for communicating with other programs)
NNN = Sequentially assigned item number

Reactor System Flow Charts

The first process state in the reactor control program, the prestart state, is concerned with setting up initial conditions in the reactor system. All on-off valves are closed in block 3, Figure 10.2g, and blocks 5–10 communicate information to the control loops, to activate them and supply initial set points. Each block deals with one control loop, and it can supply as many parameters as required to specify the desired control action. Blocks 11–16 pressure the reactor and perform a leak test.

Charging the reactor is a two-stage operation: (1) feeding a calculated quantity of solvent based on batch size and (2) adding the amount of catalyst required by the recipe. The level change in the solvent tank is used by the sequence logic to confirm the pump operation and determine when the charge is complete. The level monitor subroutine (blocks 36–41) is a general set of logic for this purpose.

The catalyst is metered into the reactor by an integration/cutoff technique described in the control requirements. When the sequence logic finds that the catalyst set point has been changed to zero (block 45), it assumes that the catalyst charge is complete and closes the catalyst feed valve (block 47).

The initiation state logic activates the bang-bang control loop to bring the reactor temperature to the first level and begins feeding reactant. Initiation criteria (blocks 57–60) are reactor temperature and steam flow. Once the reactor temperature reaches the set point (within tolerances) and the controllers are no longer calling for steam, the initiation is complete.

At the start of the "run" state, the feed rate is increased and the temperature control configuration is modified (blocks 64–67) to provide an override limit to the steam flow. Reactor pressure is maintained during the run by venting overhead vapors to a recovery unit.

The logic in blocks 72–88 adjusts the reactor feed rate as necessary to maintain a balance between the heat released and the cooling capability of the system. If the temperature exceeds the control band, the set point is reduced incrementally, unless the temperature exceeds the maximum, which causes feed cutoff. If the recipe feed rate cannot be maintained, as evidenced by repeated temperature excursions beyond the control limit, it is reduced for the remainder of the run. This type of logic is designed to compensate for variations in the reactor heat transfer capability caused by fouling or limited flow of cooling water.

The feed cutoff at the desired integral is sensed by the program as a

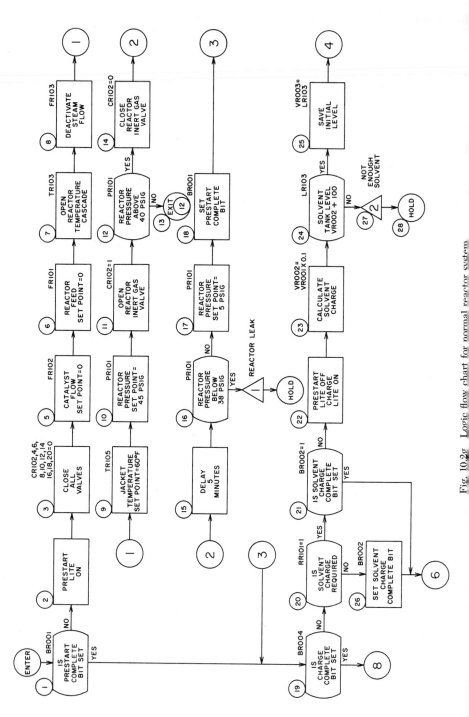

Fig. 10.2g Logic flow chart for normal reactor system

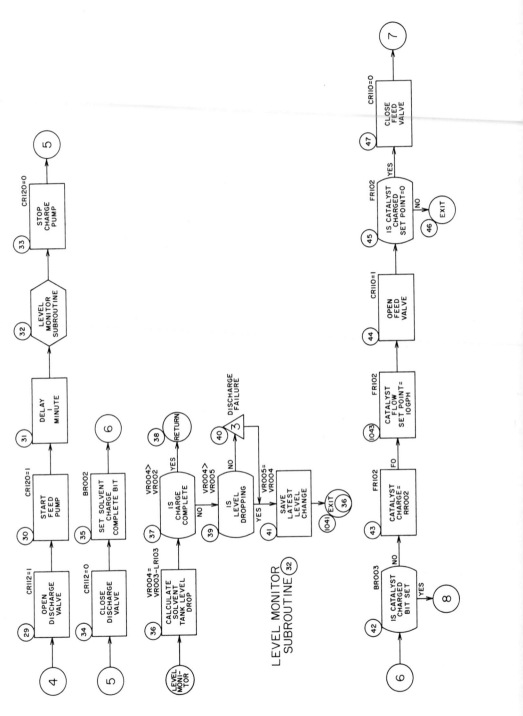

LEVEL MONITOR
SUBROUTINE ③②

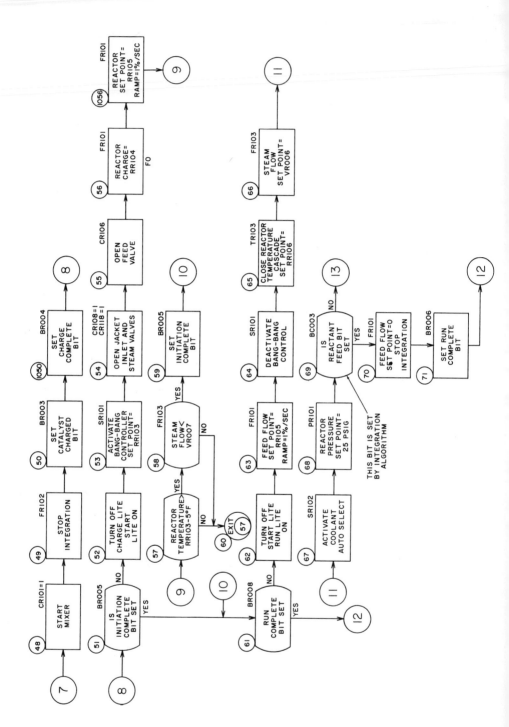

1302

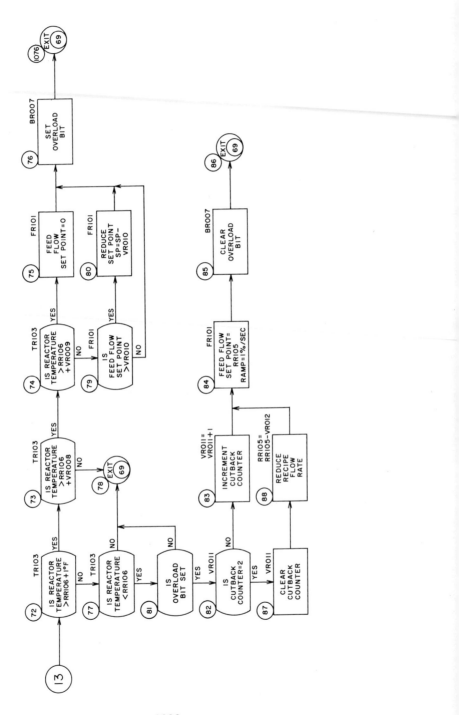

1303

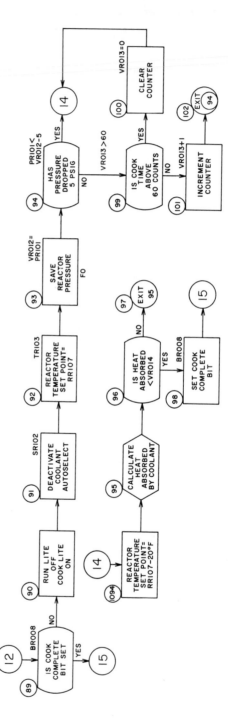

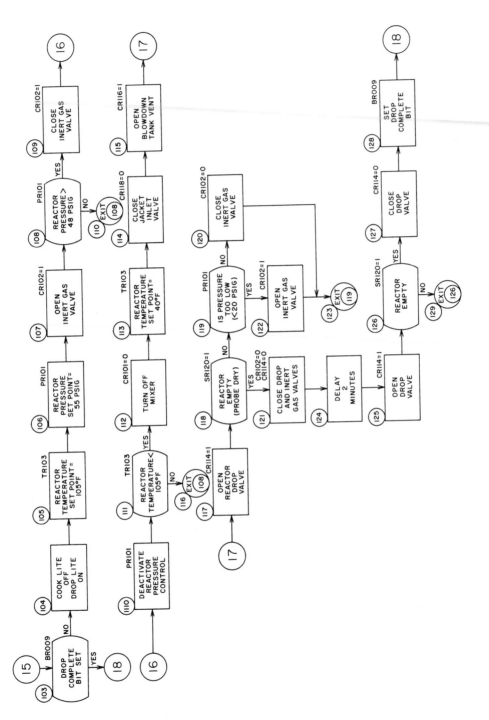

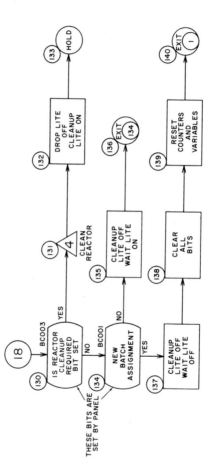

change of state of a bit, labeled the "reactant fed" bit (block 69). It is not sufficient to use feed set point as an end-point indicator, as in the catalyst charge, since the feed rate adjustment logic could temporarily change the set point to zero before the recipe quantity has been fed.

After all the reactant has been fed, a "cook" state allows time for monomer still present to react at an elevated temperature. The following criteria (blocks 93–102) must be satisfied to terminate the cook state.

> A reactor pressure drop or an elapsed time, whichever occurs first. The rate of heat being released falls below a minimal value.

The reactor discharge, or drop, is accomplished by pressuring the contents into a blowdown tank as controlled in steps 106–127. After discharge, if a manual cleanup is required, the system goes into hold. Otherwise, the reactor can begin another batch as soon as it receives the next assignment from a scheduling routine. The "hold" logic is described by Figure 10.2h.

The operational logic for the hold state is defined by blocks 220–223. It determines the process state that the system is currently in, and performs the necessary actions to arrest the progress of the batch. For the initiation and run state, the hold logic stops the feed and reduces the reaction temperature. If the reactor were in the cook state, the program would continue to increment a counter as long as the reactor temperature was within $10°F$ of the cook temperature.

Emergency shutdown logic (blocks 230–237 of Figure 10.2i) performs shutdown action without regard for the process state, as it closes feed and discharge valves, vents the pressure and cools the reactor.

The purpose of the return logic (blocks 250–273 in Figure 10.2i), as stated earlier, is to permit the operations to resume at the proper point after a hold or emergency shutdown. This can usually be accomplished by a transfer to the beginning of the logic, and allowing the self-seeking nature of the program to return to the beginning of the unfinished process state, as listed in Figure 10.2e. In some instances, however, such as the "solvent charge" state, actions already performed must not be repeated. If the solvent charge had begun before the transfer to hold, then reentering the program at the beginning would cause an excess solvent charge, since no account would be taken of the solvent already fed. In this event, the return logic directs control to a point inside the solvent charge state, after the previously executed steps.

Programming Control Logic

Translating the reactor system flow charts into a format acceptable by the computer is the next step in the generation of the control system. A variety of program languages are available for this purpose. These languages can be classified into the following general categories, as discussed in more detail in Section 8.4.

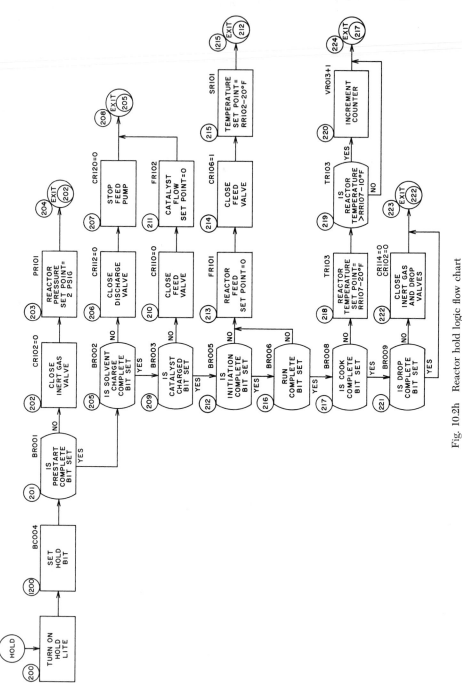

Fig. 10.2h Reactor hold logic flow chart

1308

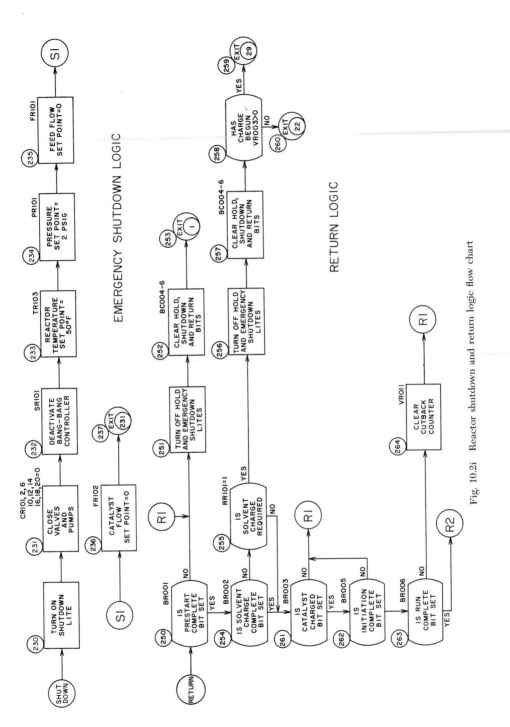

Fig. 10.2i Reactor shutdown and return logic flow chart

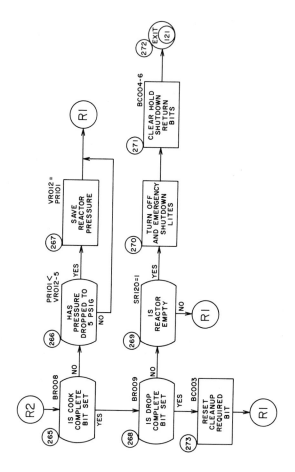

1310

1. *Assembly language*   Codified instructions which perform basic computer operations such as addition, logical operations, data, storage and program transfers.
2. *Macro assembly language*   Subroutines consisting of assembly language statements. The subroutines can be written to correspond to the flow chart blocks.
3. *Compiler language*   "Problem-oriented" set of instructions (FORTRAN, ALGOL, etc.).

The choice of a programming language depends upon (1) the function of the program, (2) who will do the programming, (3) the possibility of extensive program modifications and (4) the desired efficiency of the program in terms of execution time and memory requirements. A sequence control program should be coded in a language that is easily modified, debugged and understood by an engineer. These considerations are usually more important than the execution time and memory requirements and therefore, higher-level languages, such as compiler or macro instructions are preferred over the computer-oriented assembly languages.

The ddc algorithms can be coded in assembly language, since they are not as likely to be changed and require fewer instructions. However, if experienced programmers are not available, or are needed for coding the operating system programs, the ddc routines can be written in FORTRAN, or in an equivalent arithmetic type language with process input-output capability. The I/O instructions are needed to obtain process data and output to final control devices. Most computer system vendors supply "canned" scan and control software. These packages permit the user to generate the control loops by filling out questionnaires rather than by writing control programs (Sections 8.10 and 8.11). The forms request such information as the type of control algorithms, engineering units, range of measurement, input and output wiring designations, sampling and control interval, and control parameters.

The Operating System

The sequence control and ddc programs must be executed at regular intervals, and called into and out of core memory if they normally reside in bulk memory (disc or drum). Programs which coordinate all of these activities comprise the operating system. In order to better understand the structure and functions of the operating system, it can be separated into two areas: the process executive and the system executive. Figure 10.2j is a diagram of the function and data in the two domains.

The process executive serves as the coordinator for the sequence control programs. A typical batch control system would regulate more than one reactor as well as other process systems such as feed preparation or blending, solvent purification, and product separation. Each type of system has its own

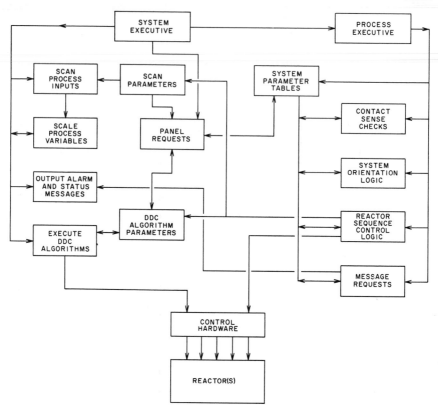

Fig. 10.2j   Functional organization of system software

sequential control program. If multiple systems (many units of the same type) exist, then one program would be used for all these identical systems. The batch (sequential) program for each system is serviced by the process executive at a frequency in accordance with process requirements. For example, a reactor whose total batch cycle takes 8 hours can be serviced every 30 seconds. A system with a turn-around time of 1 or 2 hours may require 5- or 10-second intervals for adequate response. (The ddc loops are handled separately, so that their dynamics do not normally affect this sequencing period.)

The process executive program calls on each frequency level, servicing all batch programs on that frequency before proceeding to the next level. When it recognizes that a system is to be serviced, it calls the appropriate system data, or system parameter table. This table contains the following parameters:

1. Timer status (timers are automatically incremented by a clock program)

2. Contact sense input values
3. Indicator bits
4. Variables
5. Recipe data
6. Message request information
7. Addresses of related information elsewhere in memory
8. Address of next instruction to be executed in the sequence logic
9. Task assignment data (batch size, recipe number, etc.)

The contact sense inputs for the system are read regularly and updated in a core table. The process executive calls a program which compares actual and desired status to confirm that all pumps and valves being monitored are in the desired state. If they are not, the program calls for a hold or emergency shutdown by setting the appropriate bit.

Before the process executive transfers control to the batch program, it must determine whether the system is still in the same control state it was in during the last sequencing interval. In the meantime, for example, an operator hold request or a valve failure may have occurred requiring a transfer into the hold state. An orientation program, common to all systems, determines this and sets up the block number to begin execution of the alternative control state logic. The flow chart for the orientation logic is shown in Figure 10.2k.

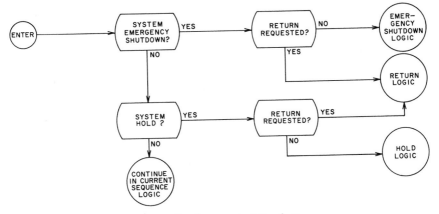

Fig. 10.2k   System orientation logic

The batch program is executed until an exit instruction is encountered or until a valve instruction requires waiting for confirmation of a position change (see *Process Logic Flow Charts*). At this point, the program returns control to the executive which then requests the outputting of messages originating in the program or in the contact sense checks for that system. The system will need no further attention until the next scan interval; hence, the process executive can now call in the next system.

At any time while the process is servicing a batch system, control may be temporarily transferred to a higher priority program under the system executive. The system executive controls programs which perform the following kinds of functions:

1. Scan and convert process variables
2. Execute ddc algorithms and output results to position the final control elements
3. Print alarms, status messages, and logs (requests can come from ddc or batch programs)
4. Service operator panel requests
5. Output to contact closures operating plant equipment (requests can come from ddc or batch programs)
6. Respond to interrupt from the computer clock or process I/O
7. Transfer control among programs on different priority levels
8. Handle core, drum or disc transfers

By partitioning the working areas in core, the system executive can transfer control to and from the process executive without having to restore any data. In other words, the activities occurring within the region controlled by the process executive can take place in parallel with activities on other priority levels that have their own working areas. This implies a time-sharing concept, with program priority levels governing the order of transfer among the partitioned working areas.

The process control panel(s) service the following functions:

1. Display and change control loop parameters—set points, alarm limits, tuning constants, etc.
2. Open/close cascade loops, remove points from scan, add and delete control loops.
3. Request special batch operations, transfer process systems into hold or shutdown, or resume processing.
4. Change recipe data or system parameters.
5. Request logic and summaries.

Programs to service the panel requests are written modularly, so that when a panel interrupt is recognized by the system executive, it identifies the function requested and calls the appropriate panel handler program.

Simulation and Checkout

Once all of the system programs have been coded and loaded into the computer, they must be "debugged" and tested under simulated plant conditions before connecting the control systems to the process.

Before any meaningful performance evaluations can be made, the system must be debugged, or cleared of errors and inconsistencies which keep the program from operating as intended. Incorrect memory assignments or erroneous instructions can hang up the computer or cause it to "run amok," destroying information or improperly executing parts of the programs. The operating system must be checked to assure that it is correctly performing the basic functions such as panel servicing, message output, I/O signal processing, and transferring control to and from ddc and batch programs.

The next phase of checkout involves connecting simulated process measurement inputs to the computer and checking the ddc programs under open and closed loop conditions. During these tests, the reactor system control logic is deactivated so that only continuous loop control is taking place. The control variable set points that would normally come from the reactor control logic can be inserted through the operator's console.

For closed-loop tests, an analog computer, or equivalent electrical circuitry, is used to simulate the process dynamics. The process variable input from the analog computer is monitored to determine the response and stability of the ddc algorithms to set-point changes and simulated process load changes.

Open-loop testing without simulating process dynamics is adequate where the speed of response of the algorithm is more important than loop stability. The only test hardware required is a potentiometer input to the computer and a current meter or recorder connected to the control output.

Where many loops containing the same combination of control algorithms exist, only one loop need be tested, assuming the data base, or loop parameters of all other loops of the same type, have been checked during the debugging phase.

After the performance of the ddc algorithms has been proved, the checkout of the reactor sequence control logic begins. One reactor system is activated and a number of test runs are made under various sets of normal and abnormal conditions. The design of test runs deserves careful planning by engineers with a thorough understanding of the process.

Since there are many possible paths of execution in any chain of decision-making and branching logic, only those paths that seem probable from the standpoint of the process behavior should be checked with computer runs. A desk check, or dry run, can reveal a large percent of the errors that would be discovered later during computer runs. Moreover, a desk check might suffice for those paths which do not seem likely to occur during actual operation. The path of execution of the logic for each run should be laid out on the flow charts, and a table should be prepared to indicate the desired equipment status at the end of each predefined section of the logic. This table should also include the desired results of any calculations made in the program.

Diagnostic aids, such as the program trace and break point, greatly simplify the batch program checkout. The trace is a computer-generated record of each statement number executed during the run. Often during checkout unforeseen situations cause the computer to wander out of the expected path of execution, which becomes more obvious from the trace than from the status of the simulation hardware.

Break points interrupt and stop the computer from executing any further sequence logic until a request to continue is issued via the panel. A break point does not inhibit the computer system from continuing to perform all its other activities outside the system program being checked.

A simulation run should resemble the normal course of events in the actual reactor process. This means that the action of valves, pumps, and sense indicators must be simulated as well as all process inputs referenced by the reactor program. The process variables can be represented by potentiometer regulated voltage inputs. The contact outputs controlling pumps and valves should be wired to lights on a simulation panel. Contact-operated process equipment which has sensors to indicate its status will have the sense inputs wired through relays to the associated light, or contact output. Then, when a light goes on or off, the limit switch(es) will change automatically to their desired state. A valve failure can be simulated by operating a sense relay manually, overriding the signal from the contact output.

Both analog and digital I/O can, of course, be simulated, in the computer or via inputs from cards, or other peripheral devices, without having to connect wires to external hardware (pots, lights, relays, switches).

After each sequential control program has been tested individually, the next and final phase of the system checkout is a loading test, with all system programs running without break points or special halts. These tests are conducted under simulated conditions for an extended period of at least three days. All panel functions should be exercised regularly. An error-free run of this type constitutes reasonable evidence that the system is ready to be hooked up to the process for plant startup.

When the control system is being applied to a new plant, there is a shakedown period during which minor modifications to the process equipment must be made. These changes almost inevitably reflect upon and require changes in the system parameters or batch programs.

Conclusions

This section presented a description of the tasks involved in generating a computer control system for one or a group of batch reactors. To summarize, the major steps in the system development are:

1. Define the scope of the control system in terms of the number and type of process systems under computer control.
2. Define the boundaries of each process system and the interaction with other systems.

3. Identify and list system I/O, analog inputs and outputs, and digital inputs and outputs.
4. Specify the control loops in terms of manipulated variable, ddc algorithm(s), and final control element. List control frequencies, alarm action, engineering units, etc.
5. Generate scan and control data base from information in steps 3 and 4.
6. Prepare or obtain an operating description of each type of system, specifying normal procedures and required response to abnormal situations.
7. Generate sequence logic flow charts from the operating description, keeping cumulative lists of assigned variables, recipe items, indicator bits, timers, contact sensors, and contact outputs.
8. Code and compile, or assemble the batch programs in the selected computer language.
9. Prepare system parameter tables and recipes (batch data base).
10. Load the ddc and batch programs and data base into the computer together with the process and system executives, including all utility programs such as I/O and panel service routines.
11. Hook up computer I/O cables to panels and simulation hardware.
12. Debug system.
13. Simulate and checkout ddc loops.
14. Simulate and check each batch program individually.
15. Test integrated system by running all programs simultaneously.
16. Document all system programs, memory assignments and data base. Compile manuals describing system startup, operations, troubleshooting, and maintenance procedures.

Documentation is not intended to be started after the system is completely operational. Rather, it is an on-going task to maintain correct, well-organized, up-to-date flow charts, program listings, memory assignments, etc. The success of the project is in no small way dependent upon the quality of the system documentation and manuals.

A properly organized computer system for reactor control can bring about substantial economic benefits: (1) uniform product quality and more precise, repeatable control, (2) increased production as a result of reduced process operations time, (3) fewer lost batches due to operator errors and (4) increased safety resulting from regular monitoring and fast response to critical conditions.

Greater incentives exist for automating batch than continuous processes because of the nature of batch processes. Given the fact that operating conditions vary depending upon what process state the system is in, more kinds of regulatory actions are required. Moreover, factors such as decision points, or the variety of possible situations that can develop, each demanding an established sequence of actions, create more control problems and increase the probability of human error.

Comparing the cost of a computer control installation with that of a conventional control system is always difficult. If no backup instrumentation is provided, then for plants with more than 25 reactors, the initial outlay for a ddc computer system can be less than for a conventional control system. Also, instrument maintenance costs are usually reduced.

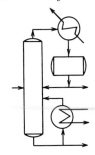

10.3 CONTROL OF DISTILLATION TOWERS

Distillation separates a mixture on the basis of a difference in composition between a liquid and the vapor formed from the liquid. In the process industry, distillation is widely used to isolate and purify volatile materials. Thus, proper instrumentation of the distillation operation is vital to achieve maximum product of satisfactory purity.

This section is devoted to a discussion of measurement and control techniques common in distillation. Some of the more specialized or advanced concepts of column control are covered in other sections. For example, Section 9.4 is devoted to the use of analog computers in tower control, while Section 10.4 discusses optimization of distillation units.

Although one speaks of controlling a distillation tower, many of the instruments actually are associated with equipment other than the tower. For this reason, it might be well to review the equipment used in distillation.

Distillation Equipment

The tower or column has two purposes: First, it separates a feed into a vapor portion which ascends the column and a liquid which descends the column. Second, it achieves intimate mixing between the two counter-current flowing phases. The purpose of the mixing is to get an effective transfer of the more volatile components into the ascending vapor and a corresponding transfer of the less volatile components into the descending liquid.

In continuous distillation, the feed is introduced continuously into the side of the distillation column. If the feed is all liquid, the temperature at which it first starts to boil is called the *bubble point*. If the feed is all vapor, the temperature at which it first starts to condense is called the *dew point*. The column is operated in a temperature range that usually is intermediate to the two extremes of dew point and bubble point. For effective separation of the feed, it is important that both vapor and liquid phases exist throughout the column.

The separation of phases is accomplished by gravity, with the lighter vapor rising to the top of the column, while the heavier liquid flows to the bottom. The intimate mixing is obtained by one or more of several ways. A simple method is to fill the column with lumps of an inert material, or *packing*,

1319

that will provide surface for the contacting of vapor and liquid. A more effective way is to use a number of horizontal plates, or trays, which cause the ascending vapor to be bubbled through the descending liquid.

The other equipment associated with the column is shown schematically in Figure 10.3a. The overhead vapor leaving the column is sent to a cooler, or condenser, and collected as a liquid in a receiver, or accumulator. A part of the accumulated liquid is returned to the column as reflux. The remainder is withdrawn as overhead product or distillate.

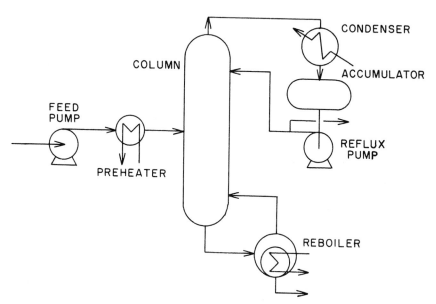

Fig. 10.3a Distillation equipment

The bottom liquid leaving the column is heated in a reboiler. Part of this liquid is vaporized and injected into the column as boil-up. The remaining liquid is withdrawn as a bottom product, or residue.

In some cases, additional vapor or liquid is withdrawn from the column at points above or below the point at which the feed enters. All or a portion of this sidestream can be used as intermediate product. Sometimes, economical column design dictates that the sidestream be cooled and returned to the column to furnish localized reflux. The equipment that does this is called a sidestream cooler. At other times, localized heat is required. Then some of the liquid in the column is removed and passed through a sidestream reboiler before being returned to the column.

Variables and Degrees of Freedom

To set all of the variables for distillation, only a certain number of them can be assigned independent values. The other variables automatically will have fixed values dependent upon their relation to the operation. It is possible that even the independent variables will be subject to certain physical limitations. Nevertheless, only a fixed number of independent variables exist for the system. (See Section 7.2 for more details.)

The discussion of control systems given here will be simplified by assuming that the column already exists. Then many of the variables will be fixed by the engineering design of the column. It also will be assumed that the column is a reboiled one, having a fixed feed point and making only two products—a distillate product and a bottom product. Furthermore, the thermal conditions of the two products will be dictated by the economies of heat recovery. The independent variables which then remain to be used for control of the column's operation can be shown to be limited to six.

Some of the variables which can be manipulated when controlling a column (Figure 10.3b) are as follows:

Column pressure

Feed flow rate

Feed composition

Feed temperature
(or feed quality)
Heat added (boil-up)

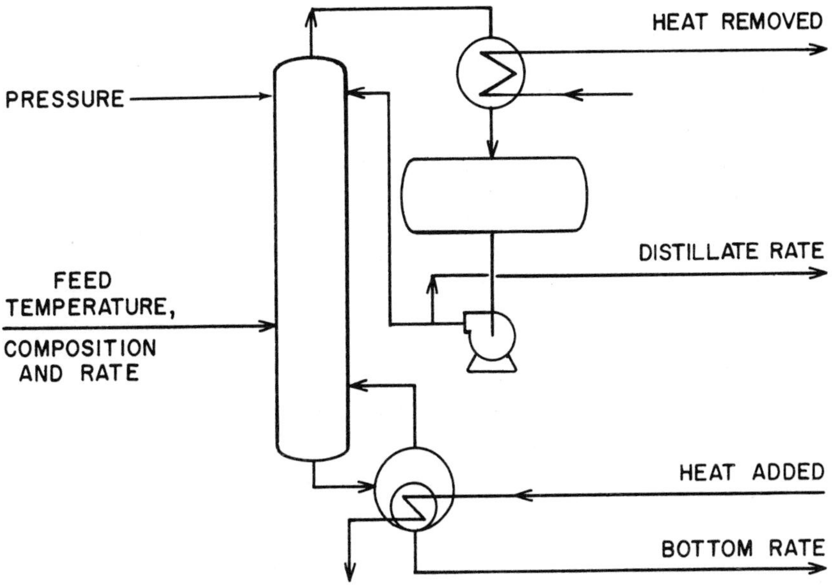

Bottom product
flow rate

Heat removed
(reflux)

Distillate product
flow rate

Fig. 10.3b   Variables that fix the distillation operation

Only six of these variables at one time can be varied independently and at will. The others will necessarily seek values dependent upon their relation to the six independent variables.

Compositions for distillate and bottom products are omitted purposely from the foregoing list—even though these two variables frequently are the only ones which dictate the operation of a column. Being dependent variables, they are not controlled directly. Instead, discrete manipulation of the other variables is required.

Pressure Control

Most distillation control systems are based upon maintaining the column pressure at some constant value. Any variation of the pressure will upset the control system by changing the equilibrium conditions of the material in the column. Therefore, the first systems to be discussed are those which maintain constant pressure.

The set point for pressure is a compromise between two extremes. The pressure must be high enough to cause condensation of the overhead vapor by heat exchange with the cooling medium—usually cooling water. On the other hand, the pressure must be low enough to permit vaporization of the bottom liquid by heat exchange with the heating medium—usually steam.

The specific set point for pressure is determined from economic considerations. For example, if operating pressure increases, the column temperature increases. Thus, the driving force for transferring heat in the condenser is greater. On the other hand, the driving force for the reboiler is lower. Therefore, when pressure goes up, the size of the condenser is reduced and the size of the reboiler is increased.

The optimum pressure is determined by considering the cost for heat transfer equipment, plus costs for utilities and construction of a column to withstand the pressure. It usually is more economical to select the lowest pressure which will permit satisfactory condensation of the distillate product at normal cooling-water temperatures. Once this pressure is set, there is very little incentive to stabilize column operation by varying pressure.

The best way to maintain a column at some pressure depends upon the amount of uncondensables which enter with the feed. Control systems are discussed for the following situations: (1) liquid distillate withdrawn, uncondensables present, (2) vapor distillate withdrawn, uncondensables present and (3) liquid distillate withdrawn, negligible uncondensables.

Liquid Distillate and Inerts

The problem of pressure control is complicated by the presence of large percentages of inert gases. The uncondensables must be removed or they will accumulate and blanket off the condensing surface, thereby causing loss of column pressure control.

The simplest method of handling this problem is to bleed off a fixed amount of gases and vapors to a lower pressure unit, such as to an absorption tower, if such is present in the system. If an absorber is not present, it is possible to install a vent condenser to recover the condensable vapors from this purge stream.

It is recommended that the fixed continuous purge be used wherever economically possible; however, when this is not permitted, it is possible to modulate the purge stream. This might be desirable when the amount of inerts is subject to wide variations over a period of time.

As the uncondensables build up in the condenser, the pressure controller will tend to open the control valve to maintain the proper rate of condensation. This is done by a change of air-loading pressure on the diaphragm control valve. This air-loading pressure could also be used to operate a purge control valve, as the pressure passed a certain operating point. This could be done by means of a calibrated valve positioner or a second pressure controller (Figure 10.3c).

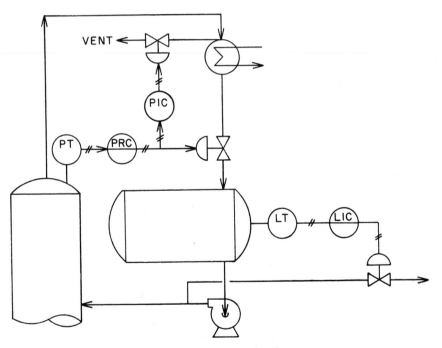

Fig. 10.3c   Column pressure control with inerts present

Vapor Distillate and Inerts

In this case the overhead product is removed from the system as a vapor and, consequently, the pressure controller can be used to modulate this flow as shown in Figure 10.3d. The system pressure will quickly respond to changes in this flow.

A level controller is installed on the overhead receiver to control the cooling water to the condenser. It will condense only enough condensate to provide the column with reflux.

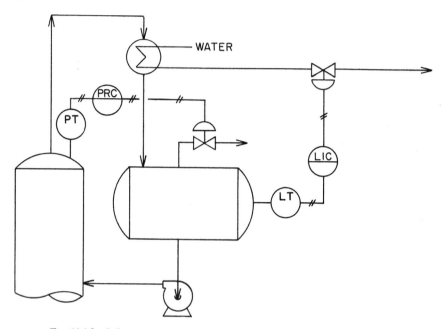

Fig. 10.3d   Column pressure control with vapor distillate and inerts present

This control system depends upon having a properly designed condenser in order to operate satisfactorily. The condenser requires a short residence time for the water to minimize the level control time lag.

If the condenser is improperly designed for cooling water control, it is recommended that the cooling water flow be maintained at a constant rate and the level controller control a stream of condensate through a small vaporizer and mix it with the vapor from the pressure-control valve. If the cooling water has bad fouling tendencies, it would be preferable to use a control system similar to that in Figure 10.3f, using the level controller to control a vapor by-pass around the condenser.

Liquid Distillate and Negligible Inerts

This situation is the most common and is usually controlled by adjusting the rate of condensation in the condenser. The method of controlling the rate of condensation will depend upon the mechanical construction of the condensing equipment.

One method of control is to place the control valve on the cooling water from the condenser (Figure 10.3e). This system is recommended only where the cooling water contains chemicals to prevent fouling of the tubes in the event of high temperature rises encountered in the condenser tubes. The maintenance costs are low because the valve is on the water line and gives satisfactory service provided the condenser is properly designed.

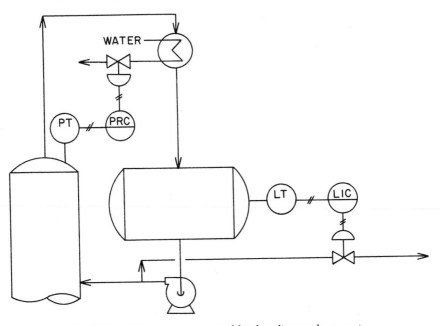

Fig. 10.3e   Column pressure control by throttling condenser water

The best condenser for this service is a bundle type with the cooling water flowing through the tubes. This water should be flowing at a rate of more than $4\frac{1}{2}$ feet per second, and the water should have a residence time of less than 45 seconds. The shorter the residence time for the water, the better will be the control obtained, owing to the decrease in deadtime or lag in the system.

With a properly designed condenser, the pressure controller need have only proportional control, as a narrow throttling range is sufficient. However, as the residence time of the water increases, it will increase the time lag of the system, and consequently the controller will require a wider throttling range and will need automatic reset to compensate for the load changes. The control obtained by using a wide proportional band is not satisfactory for precision-distillation columns because of the length of time required for the system to recover from an upset.

It would be impossible to use this control system, for instance, on a condenser box with submerged tube sections, because there would be a large time lag in the system due to the large volume of water in the box. It would take quite a while for a change in water-flow rate to change the temperature of the water in the box and finally affect the condensing rate.

In the presence of such unfavorable time lags, it becomes necessary to use a different type of control system, one which permits the water rate to remain constant and controls the condensing rate by controlling the amount of surface exposed to the vapors. This is done by placing a control valve in the condensate line and modulating the flow of condensate from the condenser. When the pressure is dropping, the valve cuts back on the condensate flow, causing it to flood more tube surface and, consequently, reducing the surface exposed to the vapors. The condensing rate is reduced and the pressure tends to rise. It is suggested that a vent valve be installed to purge the uncondensables from the top of the condenser, if it is thought that there is a possibility of their building up and blanketing the condensing surface.

Condenser Below Receiver

A third possibility for this type of service is used when the condenser is located below the receiver. This is frequently done to make the condenser available for servicing and to save on steel work. It is the usual practice to elevate the bottom of the accumulator 10 to 15 feet above the suction of the pump in order to provide a positive suction head on the pump.

In this type of installation the control valve is placed in a by-pass from the vapor line to the accumulator (Figure 10.3f). When this valve is open, it equalizes the pressure between the vapor line and the receiver. This causes the condensing surface to become flooded with condensate because of the 10 to 15 feet of head that exist in the condensate line from the condenser to the receiver. The flooding of the condensing surface causes the pressure to build up because of the decrease in the condensing rate. Under normal operating conditions the sub-cooling that the condensate receives in the condenser is sufficient to reduce the vapor pressure in the receiver. The difference in pressure permits the condensate to flow up the 10 to 15 feet of pipe that exist between the condenser and accumulator.

A modification of this latter system controls the pressure in the accumulator by throttling the condenser by-pass flow (Figure 10.3g). The column pressure is maintained by throttling the flow of vapor through the condenser. Controlling the rate of flow through the condenser gives faster pressure regulation for the column.

Vacuum Systems

For some liquid mixtures, the temperature required to vaporize the feed would need to be so high that decomposition would result. To avoid this, it is necessary to operate the column at pressures below atmospheric.

The common means for creating a vacuum in distillation is to use steam

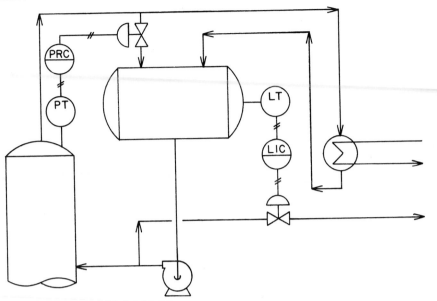

Fig. 10.3f   Column pressure control utilizing lowered condenser

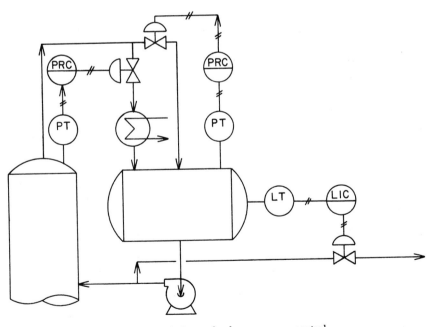

Fig. 10.3g   High-speed column pressure control

jet ejectors. They can be employed singly or in stages to create a wide range of vacuum conditions. Their wide acceptance is based upon their having no moving parts and requiring very little maintenance.

Most ejectors are designed for a fixed capacity and work best at one steam condition. Increasing the steam pressure above the design point will not usually increase the capacity of the ejector; as a matter of fact, it will sometimes decrease the capacity because of the choking effect of the excess steam in the diffuser throat.

Steam pressure below a critical value for a jet will cause the ejector operation to be unstable. Therefore, it is recommended to install a pressure controller on the steam to keep it at the optimum pressure required by the ejector.

The recommended control system for vacuum distillation is shown in Figure 10.3h. Air or gas is bled into the vacuum line just ahead of the ejector. This makes the maximum capacity of the ejector available to handle any surges or upsets.

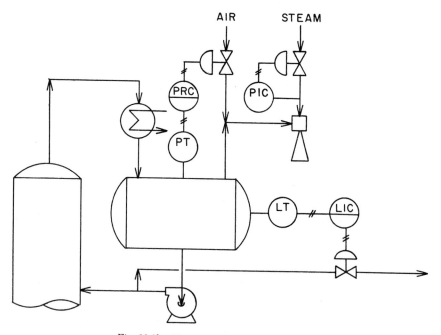

Fig. 10.3h   Vacuum column pressure control

A control valve regulates the amount of bleed air used to maintain the pressure on the reflux accumulator. Using the pressure of the accumulator for control involves less time lag than if the column pressure were utilized as the control variable.

Feed Control

Only six variables need to be fixed in order to predetermine all conditions in a distillation column. The column pressure can be the first controlled variable. If fixed feed conditions are assumed, this will necessitate that three more independent variables be determined—feed rate, feed composition and feed temperature.

One of the best means for obtaining stability of operation in almost any continuous-flow unit, including distillation, is by holding the flow rates constant. Therefore, a controller is used on the feed to maintain a constant rate of flow.

Using a narrow proportional band setting, the controller will cause the control valve to take large corrective action for small changes in flow rate. Therefore, the feed rate will be maintained constant by this high-gain controller.

In rare instances, the feed pump of a distillation unit is a steam-driven pump instead of an electrically driven one. In this case the feed flow control valve can be placed on the steam to the driver.

Feed composition has a great influence upon the operation of a distillation unit. Unfortunately, though, feed composition is seldom subject to adjustment. For this reason, it is necessary to make changes elsewhere to the operation of the column in order to compensate for variations in feed composition. The corrective steps which can be taken are discussed later in this section and also in Sections 9.4 and 10.4. For the time being, it is assumed that a constant feed composition exists.

The thermal condition of the feed determines how much additional heat must be added to the column by the reboiler. For efficient separation, it usually is desirable to have the feed at its bubble point when it enters the column. Unless the feed comes directly from some preceding distillation step, an outside source of heat is required.

Steam may be used to heat the feed, and the sensing device may be a thermocouple contained in a thermowell placed inside the feed line. Any change in the temperature of the feed leaving the exchanger will cause a corrective adjustment to the supply of steam into the exchanger. To maintain feed temperature, usually a three-mode controller is used. On startup, the initially large correction provided by rate action also helps to get the unit lined-out faster.

As discussed in Section 7.8, the use of a cascade loop (Figure 10.3i) can provide superior temperature control.

Constant temperature feed does not necessarily mean constant feed quality. If feed composition varies, its bubble point also varies. It is common practice to set the temperature control at a point which is equivalent to the bubble point of the heaviest feed. As the feed becomes lighter, some of it will evaporate, but this variation can be handled by subsequent controls.

When a furnace is used to add heat to a distillation unit, the temperature controller is used to regulate the amount of fuel to the furnace. For greater stability, a flow controller can be used on the fuel line, and its control setting can be determined by the temperature controller in a manner analogous to the cascade loop in Figure 10.3i.

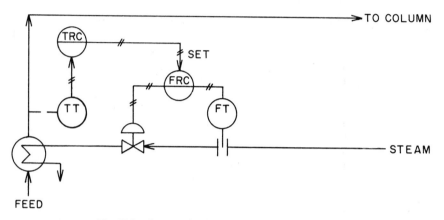

Fig. 10.3i   Improved column feed temperature control

The problem in furnace control arises in trying to maintain efficient combustion of the fuel. Combustion depends upon the air-fuel ratio. Insufficient air causes only a portion of the fuel to be consumed. With fixed air flow, a control system which would increase the flow of fuel to the furnace would not necessarily increase the heat transfer.

At one time it was necessary to have process furnaces use large amounts of excess air to avoid overheating the tubes in the radiant section. Since then designers have learned that it is practical to design and operate a heater with small amounts of excess air. Nevertheless, it is not uncommon to find a furnace which is being operated with more excess air than is required. With sufficient excess air, an adjustment of the fuel rate results in immediate changes in heat supplied to the process stream.

Excess air also contributes to loss of heat, because it reduces the overall furnace temperature by absorbing some of the heat and carrying it away in the flue gas.

Control of air is maintained by controlling the draft in the furnace. However, if the draft is allowed to vary as the volume of fuel varies, variations in secondary air rate would be too great and would cause less efficient combustion. It is also possible to have the furnace draft become positive so that hot gases might leak from the furnace into operating areas. Therefore, it has been proposed to use a motor-operated damper in the exhaust stack line to insure constant draft at all times.

The time-lag of most furnaces is usually between 3 and 12 minutes. The time lag is reduced significantly if a cascade controller is used. The master controller senses the temperature of the process feed stream while the slave controller senses the furnace temperature. For more details on furnace controls refer to Section 10.7.

Reboiler Controls

Since fixing the feed conditions and column pressure determines four of the variables listed earlier, only two other variables remain to be controlled in order to fix the operation of the distillation column. Frequently, the boil-up rate is chosen to be one of the two remaining independent variables.

Boil-up rate is controlled by setting the flow of heat to the reboiler. A flow controller is placed in the line carrying the heating medium to the reboiler (Figure 10.3j).

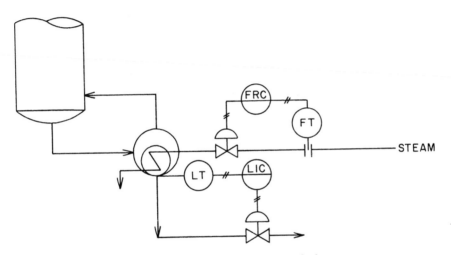

Fig. 10.3j   Reboiler control by maintaining boilup rate

The amount of lighter boiling material in the bottom product is determined by the set point of the steam rate controller. A setting which permits a greater amount of steam into the reboiler will cause more of the lighter material to be driven back into the column as vapors.

Although the rate of bottom product withdrawal shown in Figure 10.3j is controlled by the level in the reboiler, this does not constitute an independent control of bottom product flow rate. The liquid level reflects the difference between what flows into the reboiler from the column and what goes back into the column as boil-up vapor.

Figure 10.3j depicts a kettle type reboiler. Other types include: thermosyphon reboilers and forced-circulation reboilers. For them, the bottom product is withdrawn from the column (Figure 10.3k).

Another method of bottoms control is to have the steam to the reboiler manipulated by the reboiler liquid level. In this case the bottom product withdrawal can be set by a flow controller for which the adjustments are established from measured or computed conditions. This arrangement is better because it gives fast response to disturbances in steam quality or pressure and prevents these disturbances from entering the column and affecting the separation.

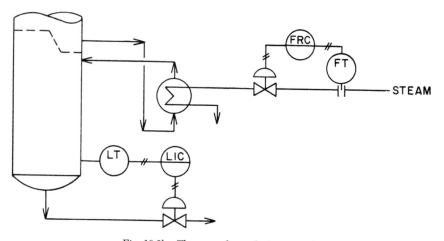

Fig. 10.3k   Thermosyphon reboiler controls

Reflux Controls

For the total control of the distillation operation, reflux rate represents the sixth and last independent variable to be controlled. Since distillation achieves separation of materials by countercurrently contacting vapor and liquid, reflux furnishes the continuous supply of liquid to the top of the column, just as the reboiler furnishes the continuous supply of vapor to the bottom of the column. Reflux rate and boil-up then go hand in hand to determine the quantity and composition of products which are made from a given feed. The rate of reflux is regulated by a flow controller on the reflux line (Figure 10.3l). See Figures 9.4h, j, l, m, o, p, q, and r for other reflux control methods.

The rate of distillate product withdrawal is controlled by the liquid level in the accumulator. This controller is also shown in Figure 10.3l. This rate of withdrawal does not represent another independent variable, since it is forced to be the difference between what comes overhead from the column and what is returned as reflux. This is analogous to the situation described earlier for the bottom withdrawal rate.

Other methods of reflux control are discussed in Sections 9.4 and 10.4.

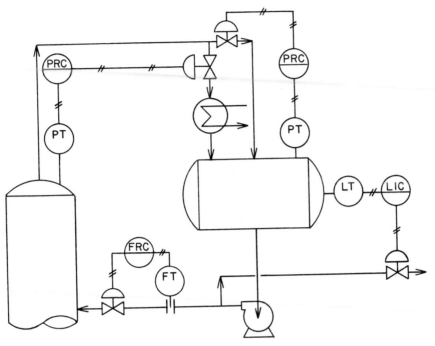

Fig. 10.3 1   Column reflux on flow control

Variable Column Feed

Having constant feed conditions certainly simplifies the amount of control required to achieve stable operation. However, suppose the distillate product is to be fed into a second column. Then any inadvertent changes which occur in the first column would be reflected in the quantity and composition of the feed to the second. If the flow of products from the previous column are controlled by liquid level controllers, these controllers can have a wide proportional band so that changes in the reboiler or accumulator level can swing over a wide range without drastically upsetting the flow of products. Nevertheless, a second column will receive a varying flow of feed if it is linked to the first column.

One way to iron out temporary variations caused by liquid level changes is to add flow controllers to the product lines.

With variable feed rates and variable feed compositions, cascade controls are justified. If the feed rate and composition are relatively constant, hand reset of the major control loop is sometimes adequate, in other cases the flow set point is continuously adjusted by the level controller in a cascade arrangement (Figure 10.3m). A distillation unit operates between two extremes: In one case, insufficient separation is reflected in unacceptable product purity.

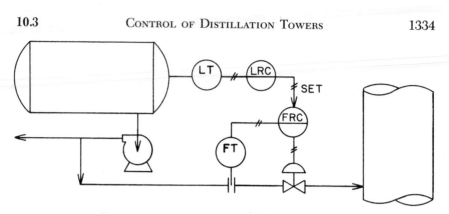

Fig. 10.3m   Cascade control of feed to second column

On the other hand, separation can be far in excess of what is demanded so that utilities and unit capacity are wasted. The goal of distillation then is to achieve specified product purity without causing waste. To obtain this goal, some measure of product composition is needed to modulate the control devices discussed earlier.

Since distillation separates materials according to their difference in vapor pressures, and since vapor pressure is a temperature controlled function, temperature measurement can be used to indicate composition. This presumes that the column pressure remains constant. Then any change in composition within a column will be detected as a temperature change.

Many suggestions have been made in the published literature regarding the best location for a temperature control point. One of these asserts that the control point should be located on the tray where a maximum slope occurs in the curve of temperature vs tray location. However, calculations based upon the equilibrium concentrations for each tray reveal no apparent justification for this plausible sounding method for control point location.

The best point to locate the temperature sensor cannot be established from generalizations. The important consideration is to measure temperature on a tray which most reflects changes in composition.

When composition of the bottom product is the important consideration, it is desirable to maintain a constant temperature in the lower section. This can be done by letting the temperature measurement set the control point of the reboiler steam supply (Figure 10.3n).

When composition of the distillation product is the important consideration it is desirable to maintain a constant temperature in the upper section as in Figure 10.3o. Notice that the point of column pressure control is near the temperature control point. This arrangement helps to fix the relation between temperature and composition at this particular point.

Measuring temperature in a column usually requires that the sensing device be in the liquid on the tray. Heat transfer from a liquid medium to

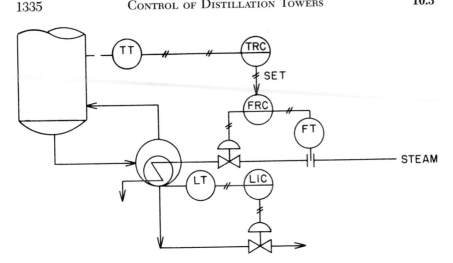

Fig. 10.3n   Temperature cascaded heat addition to the reboiler

the sensing device is much greater than the heat transfer from a gas medium.

In any thermal system there is an error caused by transfer of heat along the neck of the bulb to the surroundings. This neck conduction error can be minimized by increasing the velocity past the bulb, by increasing the length of the neck, by decreasing the cross-sectional area of the neck and by building the neck from materials of low thermal conductivity.

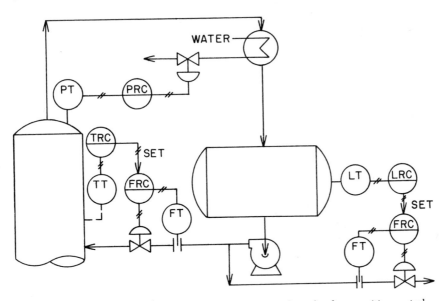

Fig. 10.3o   Temperature cascaded reflux flow for improved overhead composition control

Superdistillation

Distillation temperature is an indication of composition only when column pressure remains constant. When separation by distillation is sought between two compounds having vapor pressures close together, temperature measurement as an indication of composition is not satisfactory.

Sometimes, a change in temperature with composition is far less than, even overshadowed completely by, a change in column pressure. Stated another way, changes in feed rate and column pressure can cause a temperature change which bears little relation to composition.

Fixing two temperatures in a column is equivalent to fixing one temperature and the pressure. Thus, by controlling two temperatures, or a temperature difference, the effect of pressure variations can be nullified. Actually, the assumption used here is that the vapor pressure curves for two components have constant slopes.

Controlling two temperatures is not equivalent to controlling a temperature difference. A plot of temperature difference vs bottom product composition exhibits a maximum. Thus for some temperature differences below the maximum it is possible to get two different product compositions.

Separation of normal butane and isobutane, in the absence of other components such as pentanes and heavier, can be accomplished very well by using temperature difference control. A schematic for one commercial installation is shown in Figure 10.3p.

Analyzers

Analytical or specific composition control is another way to sidestep the problems of temperature control. Although additional investment is needed for the analytical equipment, a savings from improved operation usually results.

Several types of instruments are available for composition analysis. Of these, the chromatograph is the most versatile. (For details refer to Chapter VIII of Volume I.) Once, the time required for a chromatographic analysis (approximately 15 minutes) was a great barrier to its use for automatic control. Since then the equipment has been modified so that analyses can now be made in less than five minutes. With careful handling, this sampling rate will permit closed-loop distillate control.

Light ends distillation have been satisfactorily controlled by the use of chromatography. The diagram for a superfractionator designed to separate isobutane and normal butane is shown in Figure 10.3q.

The chromatograph continuously analyzes a sample of vapor from one of the intermediate trays. The chromatographic output is used to modulate the product drawoff valve. This directly varies the column pressure by adjusting the liquid level in the flooded overhead condenser. The absolute pressure variations induced by this control system are less than $\pm\frac{1}{2}$ PSI from the average value.

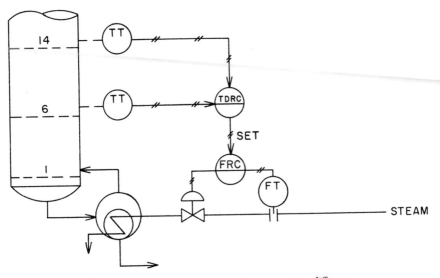

Fig. 10.3p   Heat input controlled by temperature difference

The choice of analyzer control depends upon the analytical equipment available and on the type of separation desired. Each type of separation requires a compromise between the controllability and the time delay of the control system. For example, a three-column system (Figure 10.3r) was explored to determine the best analyzer system to use.

In the depropanizer, where isobutane was to be measured in the presence of ethane, propane, and normal butane, and in the deisobutanizer where

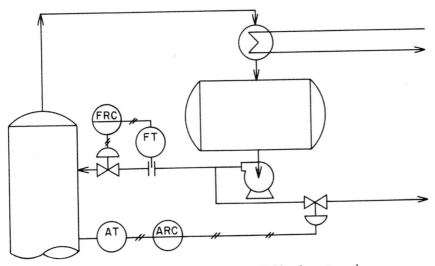

Fig. 10.3q   Distillate withdrawal controlled by chromatograph

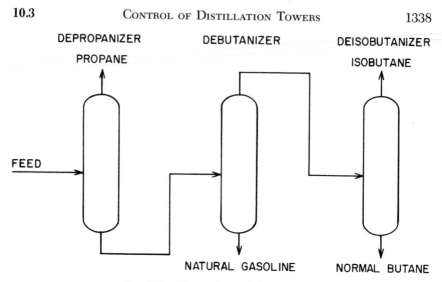

Fig. 10.3r   Three columns linked together

isobutane was to be measured in the presence of normal butane and isopentane, an infrared analysis was to be preferred.

However, in the debutanizer the aim was to measure the combined isopentane plus normal pentane concentrations in the presence of isobutane and normal butane effectively to control the butane-pentane separation. Here, investigation revealed that measurement of the refractive index of the sample in the gaseous state would give in effect a molecular weight analysis and would not differentiate among the isomers. Uniquely, the non-specificity of refractive index was found to be desirable.

Analyzers as Operator Guides

In some cases, the additional cost for connecting an on-stream analyzer directly to an automatic control system is not justified by the improvement in the column operation. In these cases the on-stream analyzers furnish information which helps the operator decide on the set-point changes required.

Chromatography is used by many processors to determine the effectiveness of a distillation operation. One installation uses a single chromatograph to analyze alternately five streams for eight common components. This system pays out in four months on an alkylation unit—based on savings in acid, better isobutane utilization, ease of making octane specifications and savings in laboratory expense.

The five streams which are analyzed are shown in Figure 10.3s.

Devices for measuring boiling point are reliable enough to give continuous guidance to operation. One installation uses six such analyzers on a combination distillation unit. The analyzers follow the 50 percent boiling point

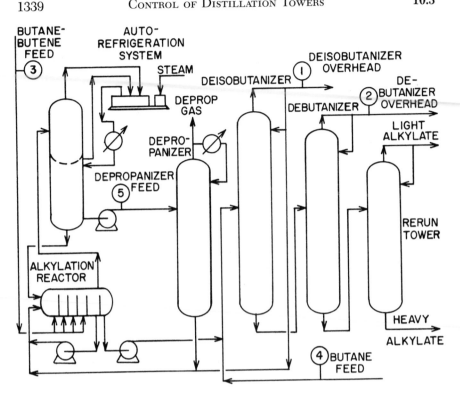

Fig. 10.3s   Single chromatograph alternately analyzes many streams

of the straight-run gasoline cut, the 90 percent boiling point of the catalytic gasoline, the 95 percent boiling point of the light naphtha cut, the 10 percent boiling point and end point of the furnace oil cut, and the fraction of light gas oil cut which boils below 440°F.

The analyzers were tested over a two-month period by substituting standard samples in place of the plant streams normally monitored. It was determined that three analyzers were outstanding, with 90 percent of their readings deviating from the average by not over 2°F. With only one exception, all of the analyzers gave more consistent results than the laboratory distillation apparatus.

Viscosity is another property which is measured continuously to give speedier corrections to existing controls. In vacuum distillation the continuous viscosimeter monitors each of the streams for which viscosity is a specification. Any deviation from the desired viscosity is corrected by changing the set point of the control loop involved. For a discussion of viscosity sensors, refer to Chapter VI of Volume I.

Many other analytical instruments are being moved out of the laboratory and into the processing area. Mobile units containing several different kinds of analyzers are now in use to learn the best place to locate onstream analyzers. In cases where permanent analyzers can not be justified, the mobile unit is connected to the process long enough to find the best operating conditions. Then the mobile unit can be moved elsewhere.

Sampling

Proper sampling of material in a column is necessary if in-plant analyzers are to control effectively. The problem of proper sampling has been the subject of several symposia and certainly is worthy of special attention. A poor sampling system often is responsible for the unsatisfactory performance of plant analyzers. See Chapter VIII of Volume I for added details.

The factors favored by sampling at, or very near, the column terminals are (1) freedom from ambiguity in the correlation of sample composition with terminal composition, and (2) improved control loop behavior due to reduction of transport lag (deadtime) and of the time constants (lags) describing the sampling point's compositional behavior. This assumes that the manipulation for control is applied at the same terminal (steam or reflux as the manipulated variable).

The factors favored by sampling nearer the feed entry are (1) improved terminal composition behavior due to earlier recognition of composition transients as they proceed from the feed entry toward the column terminals, and (2) less stringent analytical requirements due to (a) analyzing the control component at a higher concentration and over a wider range, (b) simplifying the multicomponent mixture, since non-key components tend to exhibit constant composition zones in the column. These latter points often make "impossible" analyses possible.

Computers

As automatic control technology grew, it became more difficult to determine the composite effect of several controllers making simultaneous changes. To offset this growing complexity of control systems, methods were developed for studying the dynamics of processes and to model control systems by the use of mathematical equations. However, the computations required were too cumbersome or complicated to handle by conventional means.

Now electronic computers are available to make control computations. Computers at first were used only to simulate processes and controls. They later were used to reduce plant instrumentation data to more readily understandable operating guides. Now the computer is taking over actual plant control by automatically making control settings.

Since on-line computers are expensive, improved profits resulting from the installation must be substantial in order to justify it. The only units capable of justifying large-size on-line computers are units with high throughputs or units that produce very valuable products.

First installations of computer control were made on existing units. The computer was small and usually performed a single control function. One of the first applications of computers to distillation was one which computed the internal reflux of a column, as discussed in detail in connection with Figure 9.4h of Chapter IX.

The inputs to the computer are (1) the overhead temperature, (2) the reflux temperature and (3) the reflux flow rate. From these, the computer determines the resulting internal reflux. Control is achieved by comparing the output from the computer with a pre-established magnitude for the internal reflux. Any difference between the two, causes the control valve on the external reflux line to be changed so as to reduce the difference.

Another application uses measurements around the column to determine the overall efficiency of the separation from a material balance. Here the actual yield is compared with the theoretical yield which would be computed from feed analysis and flow rate. The ratio of the two yields is an indication of the distillation column's performance.

On-line, digital computer control was included in the original design of the three-stage crude oil distillation unit put on-stream at the Tulsa, Oklahoma, refinery of Sunray DX Oil Company. Having a capacity of 85,000 barrels per day, it was considered large enough to justify the additional expenses of computer control.

The primary role of the computer control system is to minimize the operating cost for utilities, consistent with good operation. It is estimated that the cost of steam, electric power, fuel gas, and water is approximately $2,500 per day. At the same time it is estimated that an increase in gross profit of approximately $250 per day is required for justification of an on-line computer system.

Some of the control systems for which the digital computer generates the optimum set points are as follows:

Main crude oil heater. An oxygen analyzer measures the oxygen content of the flue gas from the main crude oil heater. The result of this analysis along with a measure of the temperature of the flue gas is used by the computer to calculate the excess air admitted to the heater. The computer then causes the heater dampers to be positioned to maintain the best possible efficiency for the heater. It is estimated that the computer control saves almost $100 of the approximately $1,300 daily fuel cost for this heater.

Heat Recovery. The exchange of heat from products to crude oil is achieved in two parallel circuits. The distribution of crude oil between these two circuits is determined by computer control. The computer usually can improve heat recovery by approximately 2,000,000 BTU per hour compared with operation under average manual settings.

Distillation Efficiency. Tray loadings are calculated at several points within the main crude oil column and the lubricating oil fractionating column. Loadings outside of the acceptable range are corrected by adjusting heater outlet temperature or flows through pump-around heat exchanger streams. Thus distillation efficiency is maintained at minimum operating cost.

Another distillation unit which was designed with computer control in mind was the 140,000 barrels per day pipestill at the Whiting refinery of American Oil Company. This is another example of a unit being so big that even slight improvements in operation can give the incentive for computer control. Over half of the control loops on this distillation column were controlled by the computer. The others were set manually on the bases of operating guides calculated by the computer. Since then, complete processing plants have been designed to operate under the guidance and surveillance of large computer systems.

Feedforward Control

An advanced control technique has been devised to handle the many disturbances which can occur during a distillation operation. This technique is called feedforward control and is explained in detail in Section 7.6. The concept of feedforward control is usually implemented by the use of analog or digital computers. Section 9.4 is devoted to the subject of analog computers in distillation control and therefore the treatment here will be brief.

Some of the computerized feedforward control loops might include the control of external reflux by using an internal reflux computer in which the feed rate alters the computed set point. Also the heat to the feed preheater may be regulated by a feed enthalpy computer which makes adjustments on the basis of both feed rate and data from a feed analyzer. A bottom product flow computer can regulate the bottom product withdrawal rate as a function of feed composition, feed flow and product specifications. When feed composition or feed flow changes, each computer accounts for the dynamic response of the particular system and computes the action it should take to counteract the effects of a disturbance.

Conclusions

Since distillation is applied to so many types of operations, each type should be examined separately. The nature of the feed material, the desired product purity and the origin of the most likely disturbances are all factors influencing the proper choice from among the control schemes discussed in this section. In general, the processing industry finds it more economical to spend money for more and better controls in order to avoid product wastage during system upsets.

10.4 OPTIMIZING CONTROL OF DISTILLATION COLUMNS

Optimization of a single distillation column normally implies a maximum profit operation, but, to achieve maximum profit, the price of the column's products must be known. It is therefore impossible to carry control to this extent for every column in a system, because price of products for many columns is unknown. Such prices are unknown because the products are feedstreams to other units whose operations would also need to be taken into account to establish the column's product prices.

When product prices are unknown, it is possible to carry optimization only to the stage where specified products can be produced for the least operating cost. This can be called an optimum with respect to the column involved, but only a suboptimum with respect to the system of which the column is a part.

When column product prices are known, complete economic optimization can be achieved. However, a number of different situations may exist. If there is a limited market for the products, then the control problem will be to establish the separation that results in the maximum profit rate. Such an optimum separation will be a function of all independent inputs to the column involved.

When an unlimited market exists for the products, and sufficient feedstock is available, the optimization problem becomes more difficult. Not only must the optimum separation be established, but the value of feed must also be determined. Optimization for this case will result in operating the column at maximum loading. One of three possible constraints will be involved. Throughput will be limited either by the overhead vapor condenser, the reboiler, or the column itself. In some cases the constraint will change depending upon product prices and other independent variables of the system.

Suboptimization

Application of automatic control systems to single columns should follow three logical steps in the overall hierarchy toward the goal of optimization.

1. Application of basic controls to regulate the most basic functions such as pressures, temperatures, levels and flows.

2. Application of controls to regulate the main source of heat inputs, including regulation of internal reflux flow rate, feed enthalpy and reboiler heat flow rate.

3. Application of controls to regulate the specified separation.

A single column automated through these three stages will have a sub-optimized operation. This suboptimum is defined as an operation that will produce close to the specified separation, whether or not that separation is right or wrong in regard to the total system of which the column is a part. If product purities are better than specified, the operation does not approach a suboptimum.

When a single column is automated through the suboptimum operation stage, it will still exhibit up to 5 degrees of freedom (Sections 7.2 and 10.3). As a basis for proceeding into optimization, Figure 10.4a is presented as one example of a column automated through the suboptimization stage. Figure 10.4b is another example of suboptimization. Both figures achieve the goal of a suboptimum and the difference mainly involves the basic controls.

As shown in Figure 10.4a, the system used to regulate the separation is a *predictive control system*, similar to that described in Section 9.4. Function of the predictive control system is to manipulate the energy balance (reflux flow rate) and the material balance (bottom product flow rate) to give the separation specified. The equations derived for these manipulations are called the *operating control equations*.

Control Equations

The equation used to predict bottom product flow rate is derived from the four material balance equations that follow:

$$1 = (LLT) + (LT) + (\overline{HT}) \qquad 10.4(1)$$

$$(LLF)(F) = (LLT)(T) \qquad 10.4(2)$$

$$(LF)(F) = (LT)(T) + (\overline{LB})(B) \qquad 10.4(3)$$

$$T = F - B \qquad 10.4(4)$$

where LLF = concentration of components lighter than the light key component in the feedstream,

LLT = concentration of components lighter than the light key component in the top product,

LF = concentration of light key component in the feed,

LT = concentration of light key component in top product,

F = feed flow rate,

T = top product flow rate,

B = bottom product flow rate,

$(\overline{HT})$ = specified concentration of heavy key component in the top product, and

$(\overline{LB})$ = specified value of light key component in the bottom product.

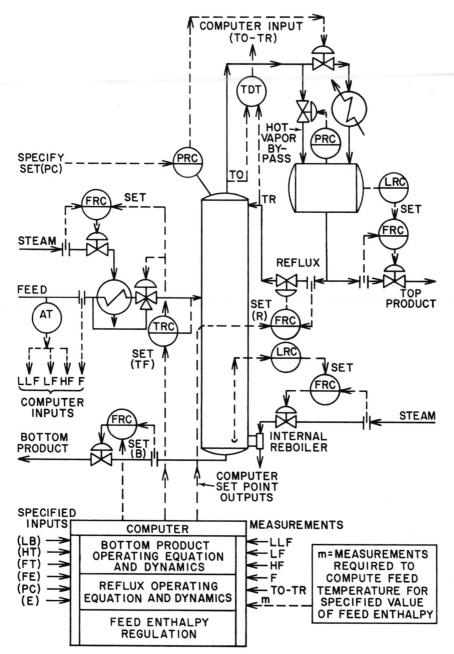

Fig. 10.4a   Distillation column automated through the suboptimization stage (to produce close to the specified separation)

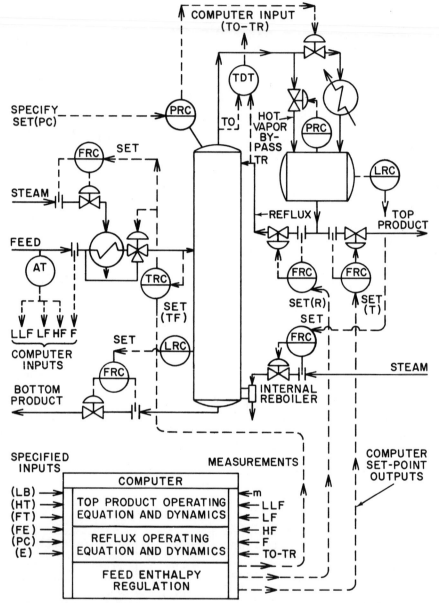

Fig. 10.4b   Alternative method of column automation through the suboptimization control stage

LLT is eliminated from equation 10.4(1) by equation 10.4(2) and LT is eliminated from equation 10.4(1) by equation 10.4(3). T is then eliminated by equation 10.4(4). The solution for the ratio of bottom product flow rate to feed flow rate will be

$$(B/F) = [1 - (\overline{HT}) - (LLF) - (LF)]/[1 - (HT) - (\overline{LB})] \quad 10.4(5)$$

Derivation of the internal reflux operating equation is more difficult. Typically this equation is developed in two parts:

$$(RI/F) = (RI/F)t + (RI/F)e \qquad\qquad 10.4(6)$$

where (RI/F) = internal reflux to feed flow rate ratio required to give a specified separation,

$(RI/F)t$ = theoretical part of reflux operating equation, and

$(RI/F)e$ = experimental part of reflux operating equation.

The experimental part of this equation is necessary because the effect of loading on overall separating efficiency (E) is normally unpredictable. Both parts of the reflux operating equation are functions of all independent inputs to the system. However, simplifications are normally considered for the experimental part as shown below.

$$(RI/F)t = f_1[(LLF), (LF), (HF), (\overline{E}), (\overline{FT}), (\overline{FE}) \ldots$$

$$(\overline{PC}), (\overline{LB}), (\overline{HT})] \qquad\qquad 10.4(7)$$

$$(RI/F)e = f_2(RI) \qquad\qquad 10.4(8)$$

The theoretical part as given in functional form is normally developed by tray-to-tray runs on an off-line digital computer. A statistically designed set of runs is made, and the information thus obtained is curve-fitted to an assumed equation form. Once the steady-state theoretical equation is developed and placed in service, the experimental part is determined by on-line tests. These tests involve operating the column at different loads to determine the correction required to $(RI/F)t$ for the separation to be equal to that specified. Average overall efficiency $(\overline{E})$ is set to give $(RI/F)t$ required to equal the actual (RI/F) existing. The loading tests are carried out under this condition.

Reflux Equation

Internal reflux flow rate of a distillation column can be obtained by making a heat balance around the top tray. The following equation is obtained:

$$(RI) = R[1 + K(TO - TR - dtr)] \qquad\qquad 10.4(9)$$

Substitute this equation into 10.4(6) to eliminate RI:

$$(R/F) = \frac{[(RI/F)t + (RI/F)e]}{[1 + K(TO - TR - dtr)]} \qquad\qquad 10.4(10)$$

where R = external reflux flow rate,

K = ratio of specific heat to heat of vaporization of the reflux,

TO = overhead vapor temperature,

TR = external reflux temperature, and

dtr = difference in temperature between dew point of overhead vapor and boiling point of external reflux.

Dynamics

Equations 10.4(6) and 10.4(5) are steady-state equations and therefore if applied without alteration, undesirable column response will result, especially for sudden feed flow rate changes. Feed composition changes are less severe than are feed flow rate changes and seldom require dynamic compensation.

One useful criterion for the design of dynamic elements to compensate for feed flow rate changes requires the following: When feed flow rate changes, the column's terminal stream flows should respond *fast*, in the correct direction and without overshoot. The simplest form of dynamics to achieve this criterion involves deadtime plus a second-order exponential response. The feed flow rate signal is brought through this dynamic element before being used in the operating equations to obtain bottom product flow rate (B) and reflux flow rate (R) set points. The transfer function for the dynamic element is

$$\frac{(FL)}{(FM)} = \frac{e^{-pt}}{(T1)p + 1} \qquad 10.4(11)$$

where (FL) = feed flow rate lagged,
    (FM) = feed flow rate measured,
    e = e of log to the base e,
    t = deadtime,
    T1 = time constant, and
    p = differential operator.

Using equation 10.4(11), F is eliminated from the left side of equations 10.4(5) and 10.4(6) to obtain the complete set of operating equations as used in Figure 10.4a and also shown below.

$$B = (FM)\left[\frac{e^{-pt}}{(T1)p + 1}\right]\left[\frac{1 - (\overline{HT}) - (LLF) - (LF)}{1 - (\overline{HT}) - (\overline{LB})}\right] \qquad 10.4(12)$$

$$R = (FM)\left[\frac{e^{-pt}}{(T1)p + 1}\right]\left[\frac{(RI/F)t + (RI/F)e}{1 + K(TO - TR - dtr)}\right] \qquad 10.4(13)$$

where, in functional form,

$$(RI/F)t + (RI/F)e = f_1[(LLF), (LF), (HF), (\overline{E}), (\overline{FE}), (\overline{PC}) \ldots$$
$$(\overline{FT}), (\overline{HT}), (\overline{LB})] + f_2(RI) \qquad 10.4(14)$$

where $(\overline{E})$ = specified constant average efficiency,
    $(\overline{FT})$ = specified value of feed tray location,
    $(\overline{FE})$ = specified value of feed enthalpy, and
    $(\overline{PC})$ = specified value of column pressure, and
  LLF, LF, HF = measured values of concentration of feed components.

Application of equations 10.4(12), (13), and (14) will result in a sub-optimized operation. This is an operation producing a performance close to the specified one.

Inspection of equations 10.4(12) and (14) shows that the system still has five degrees of freedom. Therefore, feed tray location $(\overline{FT})$, feed enthalpy $(\overline{FE})$, column pressure $(\overline{PC})$, concentration of heavy key component in the top product $(\overline{HT})$, and concentration of light key component in the bottom product $(\overline{LB})$ must all be specified. Although tray efficiency is specified, it remains to be a fixed value as explained earlier.

Optimization

Product Prices Unknown

Optimization implies maximum profit rate. Unless the terminal products of a distillation column have known prices, however, it is impossible to maximize profit rate for that column without taking into account all other parts of the process of which the column is a part. Thus optimization for a single distillation column whose terminal products prices are unknown is defined as an operation producing specification products for the least operating cost. The problem of determining the required separation for the column is a problem in optimizing the overall system of which the column is a part.

Optimization of a single column whose product prices are unknown involves determining the values for $(\overline{FT})$, $(\overline{FE})$ and $(\overline{PC})$ that result in minimum operating costs for whatever separation is specified. Any mathematical approach applicable can be used to establish values for $(\overline{FT})$, $(\overline{FE})$ and $(\overline{PC})$ that will result in minimum operating costs. Assuming that equations 10.4(12), (13) and (14) are available, it is a relatively simple matter to establish optimum values for these three variables that result in minimum operating costs. Since these variables have specific constraint values, one method involves a search technique. It is usually difficult to justify the search technique for on-line computer control. Therefore, a statistical design study can be made off-line on another computer that allows correlation of the variables with each of the three optimizing variables.

Theoretically, three equations in functional form describe the optimum for $(\overline{FT})$, $(\overline{FE})$ and $(\overline{PC})$:

$$(\overline{FT})_o = f_3(LLF), (LF), (HF), (F), (\overline{HT}), (\overline{LB}) \qquad 10.4(15)$$

$$(\overline{FE})_o = f_4(LLF), (LF), (HF), (F), (\overline{HT}), (\overline{LB}) \qquad 10.4(16)$$

$$(\overline{PC})_o = f_5(LLF), (LF), (HF), (F), (\overline{HT}), (\overline{LB}) \qquad 10.4(17)$$

In a practical sense, the optimum for $(\overline{PC})$, column pressure, will be the minimum value within constraints of the system. Normally $(\overline{PC})$ can be lowered until the condenser capacity is reached, or until entrainment of liquid in the vapor on the trays is initiated.

One way to maintain minimum column pressure involves measuring stem position of the hot vapor by-pass valve around the condenser. As the condenser capacity is approached, the hot vapor by-pass valve will close. A controller called a *valve position controller* (VPC) can be used to manipulate the column pressure controller set point to maintain the by-pass valve near its closed position. This controller must be tuned very slow to decrease the possibilities of interactions caused by varying column pressure.

Table 10.4c illustrates all the control equations involved to optimize the operation of a column when product prices are unknown. Determination of column pressure is handled by the predictive control equation, 10.4(17).

Product Prices Known

When terminal product prices for a single column are known, the design to optimize the operation will start at the level of control just described. In other words, the column is automated to obtain the specified separation for the least operating cost. The problem now is to determine values for the separation that will maximize profit rate. There are, however, a number of different situations that may exist, which will be covered soon. Several general optimizing policies can be stated first. The purpose of these general policies is to reduce the number of variables involved in design of the optimizing control system.

Optimizing Policies

1. The optimum separation to specify for a single distillation column can be determined independent of feed cost.

2. One condition resulting in an optimized operation for a single distillation column is to produce that product with the highest unit price at minimum specified purity.

3. The optimum separation to specify for a single distillation column is not a function of each terminal product price, but is a function of the price difference between products.

4. When the individual components in the products of a single column have separate assigned prices, the optimum separation is not a function of all component prices, but is a function of the price difference between the heavy key component in the top and bottom (PHT—PHB), and the price difference between the light key component in the top and bottom products, (PLT—PLB).

Proof of these policies can be obtained by evaluation of the partial differential equations that describe the profit rate for a single column with respect to the specified separation, $(\overline{LT})$, $(\overline{HB})$.

Column Operating Constraints

When product prices for a distillation column are known, complete economic optimization almost always requires the operation to be against a

Table 10.4c

OPTIMIZING CONTROL FOR SINGLE COLUMN WHEN
PRODUCT PRICES ARE UNKNOWN.

*(Criterion is to Produce Specification Products for the
Least Operating Cost)*

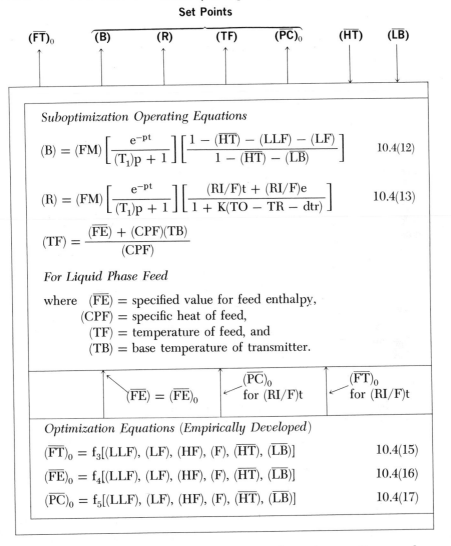

Set Points

$(\overline{FT})_0$ (B) (R) (TF) $(\overline{PC})_0$ $(\overline{HT})$ $(\overline{LB})$

Suboptimization Operating Equations

$$(B) = (FM)\left[\frac{e^{-pt}}{(T_1)p + 1}\right]\left[\frac{1 - (\overline{HT}) - (LLF) - (LF)}{1 - (\overline{HT}) - (\overline{LB})}\right] \qquad 10.4(12)$$

$$(R) = (FM)\left[\frac{e^{-pt}}{(T_1)p + 1}\right]\left[\frac{(RI/F)t + (RI/F)e}{1 + K(TO - TR - dtr)}\right] \qquad 10.4(13)$$

$$(TF) = \frac{(\overline{FE}) + (CPF)(TB)}{(CPF)}$$

For Liquid Phase Feed

where $(\overline{FE})$ = specified value for feed enthalpy,
      (CPF) = specific heat of feed,
      (TF) = temperature of feed, and
      (TB) = base temperature of transmitter.

$(\overline{FE}) = (\overline{FE})_0$ $(\overline{PC})_0$ for $(RI/F)t$ $(\overline{FT})_0$ for $(RI/F)t$

Optimization Equations (Empirically Developed)

$(\overline{FT})_0 = f_3[(LLF), (LF), (HF), (F), (\overline{HT}), (\overline{LB})]$ 10.4(15)

$(\overline{FE})_0 = f_4[(LLF), (LF), (HF), (F), (\overline{HT}), (\overline{LB})]$ 10.4(16)

$(\overline{PC})_0 = f_5[(LLF), (LF), (HF), (F), (\overline{HT}), (\overline{LB})]$ 10.4(17)

constraint. If not against an operating constraint, the optimum will occur when the specified separation $(\overline{LB})$, $(\overline{HT})$ is of such a value that incremental gain in product worth is equal to incremental gain in operating cost. Since the majority of cases will involve operating constraints, it is important to understand the principles involved.

Loading of a distillation column is affected by the specified separation and by the existing feed rate. Loading is increased by specifying a better separation or by increasing the feed rate at a constant separation. In general, both the feed rate and separation are involved as optimizing variables when an unlimited market exists for the products and when sufficient feed stock is available.

The operating constraints normally involve the respective capacities of (1) the condenser, (2) the reboiler and (3) the column. As feed rate is increased, or a better and better separation specified, one of these three constraints will be approached.

Condenser Constraint   Capacity of a given condenser at maximum coolant flow rate is a function of the differential temperature between the overhead vapor and the coolant media. One useful approach to operating a column against the condenser constraint requires correlation of maximum vapor flow rate with this temperature difference. Such a correlation can be obtained by column testing. The information obtained by on-line tests is curve-fitted to some general form such as the one in equation 10.4(18):

$$(V_{oh})_{max} = a_1 + a_2(\Delta T) + a_3(\Delta T)^2 \qquad 10.4(18)$$

where $(V_{oh})_{max}$ = maximum overhead vapor flow rate that will load the condenser maximally,

ΔT = temperature difference between overhead vapor and coolant to the condenser, and

a_1, a_2, a_3 = coefficients.

Values for feed rate and for the separation can be determined that will result in $(V_{oh})_{max}$ to load the condenser.

Accuracy of the condenser-loading equation can be carried as far as desired. For example, an equation can be developed for temperature of the overhead vapor as a function of all independent variables of the system for use in the ΔT determination. This would result in a completely predictive system for loading the condenser.

If the condenser becomes fouled, new coefficients must be established for equation 10.4(18).

Reboiler Constraint   Capacity of the reboiler at maximum flow rate of the heating media is a function of the temperature difference between the heating media in the reboiler tubes and the liquid being reboiled. Just as with the condenser, tests can be conducted to correlate maximum vapor flow rate out of the reboiler with temperature difference across the reboiler tubes. Also, temperature of the reboiler can be expressed as a function of the column's independent inputs. Column pressure will be one of the major variables that will affect temperature of the reboiler liquid. Also, this will be one of the major variables affecting overhead vapor temperature which is used in the condenser loading equation.

Column Constraint Capacity of a given column will be a function of liquid and vapor flow rates within the column as well as of the column pressure. In general, capacity is limited by entrainment of liquid by the vapor. At low internal liquid flow rates a higher vapor flow rate can be used. Also, column capacity will be higher at higher pressures. However, if column capacity is limited by the tray downcomers, internal liquid flow rate can be increased by lowering pressures. If capacity is limited by entrainment, then loading can be increased at higher pressures. Therefore, the capacity limiting parameter must be known.

Over a limited range one can assume a linear relationship between column pressure, liquid flow rate (L), and the maximum vapor rate (V_{max}) that will initiate entrainment. Data such as is found in Figure 10.4d is obtained by

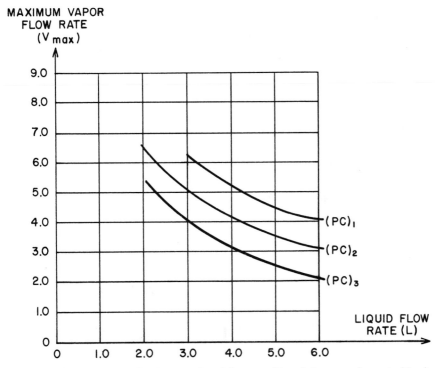

Fig. 10.4d   Typical relationship between liquid flow rate (L) and the vapor flow rate (V_{max}) that will initiate entrainment

column testing, since technology is not sufficiently advanced to predict these effects. An equation can be developed from the test data to cover a limited range of liquid and vapor flow rates, and pressure. The relationship can usually be considered linear. Equation 10.4(19)

$$V_{max} = a_1 + a_2(L) + a_3(PC) \qquad 10.4(19)$$

is useful in predicting the values of feed flow rate and of the separation that will cause maximum vapor flow rate to exist. The use of these loading functions to optimize the operation of a column will be covered soon.

Limited Market and Feedstock

Assuming that the column is already equipped with the operating control functions given in Table 10.4c, the overall situation is illustrated in Figure 10.4e. Basic controls are not shown for purpose of clarity, although the basic controls can be assumed to be the same as in Figure 10.4a.

INPUTS AND OUTPUTS FOR COMPUTER CONTROL

Measured Inputs	Specified Inputs	Constants	Outputs
(LLF)	$(\overline{HT})$	t, $(LT)_{SS}$	$(\overline{FT})_0$
(LF)		T_1, $(HB)_{SS}$	$(PC)_0$
(HF)	$(\overline{LB})$	K	(TF) from $(FT)_0$
(FM)		dtr	(R)
(TO-TR)		CPF	(B)
		TB	
		PF	
		PT	
		PB	

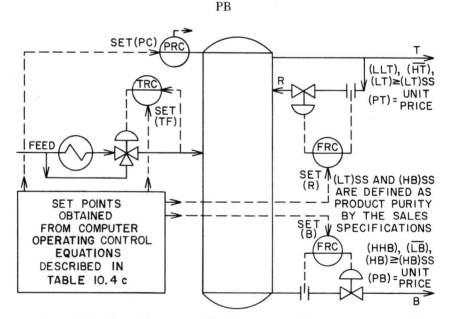

Fig. 10.4e   With prices of the products (PT, PB) known, the problem is to find the separation (HT, LB) that will maximize profit rate

This column has four feed components having concentrations of LLF, LF, HF and HHF. Three of these components appear in each product. The separation for the column is fixed by specifying the concentration of heavy key component in the top product $(\overline{HT})$, and concentration of light key component in the bottom product $(\overline{LB})$. The top and bottom products have unit prices of PT and PB respectively. All the component in the feed that is lighter than the light key component (LLF) will go to the top product, while all the component that is heavier than the heavy key component (HHF) will go to the bottom product. Therefore the operation of the column can do nothing about the distribution of these two components between the top and bottom products. $(\overline{HT})$ then will be the impurity component in the top, and $(\overline{LB})$ the impurity component in the bottom. The concentrations of these two components in the products can be fixed by specifying their values as inputs to the predictive controller. (LT) then is the purity component in the top product, and (HB) the purity component in the bottom. Sales specifications for the two products usually state the minimum acceptable purities.

$$(HB) \geq (HB)_{SS} \qquad\qquad 10.4(20)$$

$$(LT) \geq (LT)_{SS} \qquad\qquad 10.4(21)$$

where $(HB)_{SS}$ and $(LT)_{SS}$ are minimum purity sales specifications.

Control Equations

Top Product More Expensive   As stated earlier, the general optimizing policy is that the product with the highest unit price must be produced at minimum specified purity. The heavy key component in the top product $(\overline{HT})$ is the control component. Its value can be calculated as a function of feed component concentrations, minimum sales purity specification, $(LT)_{SS}$, and concentration of light key component in the bottom product $(\overline{LB})$.

Optimum Concentration of Heavy Key in Top Product

The following material balance equations are involved in determining the optimum value for the concentration of the heavy key component in the top product, $(\overline{HT})_0$.

$$(\overline{HT})_0 = 1.00 - (LT)_{SS} - (LLT) \qquad\qquad 10.4(22)$$

$$(LLF)(F) = (LLT)(T) \qquad\qquad 10.4(23)$$

$$T = (F - B) \qquad\qquad 10.4(24)$$

$$B = (F)\left[\frac{1 - (\overline{HT})_0 - (LLF + LF)}{1 - (\overline{HT})_0 - (\overline{LB})}\right] \qquad\qquad 10.4(25)$$

In equation 10.4(22), (LLT) is eliminated by use of equation 10.4(23). T is eliminated by equation 10.4(24), and B is eliminated by 10.4(25). The following is obtained:

$$(\overline{HT})_0 = \frac{(LF) + (\overline{LB})[(LLF - 1)] - (LT)_{SS}[(LLF) + (LF) - (\overline{LB})]}{(LF) - (\overline{LB})} \quad 10.4(26)$$

Optimum Concentration of Light Key in Bottom Product

Having obtained the value for optimum concentration of heavy key in the top product, the problem reduces to finding the value for optimum concentration of light key component in the bottom product, $(\overline{LB})_0$.

The maximum value that can be specified for $(\overline{LB})$ to meet the minimum sales specifications $(HB)_{SS}$ will be as follows:

$$(\overline{LB})_{max} = \frac{[1-(\overline{HT})_0][(LLF)+(LF)+(HF)-(HB)_{SS}]-(LLF+LF)[1-(HB)_{SS}]}{LF-(\overline{HT})_0}$$

$$10.4(27)$$

Derivation of this equation is made by the same procedure used for equation 10.4(26). As $(\overline{LB})$ is lowered from its maximum value, the flow rate of bottom product will decrease and the top product flow rate will increase. Since the top product is the highest unit price, profit will increase as $(\overline{LB})$ is lowered. Also, operating costs will increase since a better and better separation is being specified.

As $(\overline{LB})$ is lowered further, one of two things will occur to establish $(\overline{LB})_0$ (the lowest value for LB). Either the change in operating cost will approach the change in profit rate, or the column will be loaded against an operating constraint. This constraint will be a condenser limit, a reboiler limit, or a column limit. Whichever occurs first will establish $(\overline{LB})_0$.

Assume for example that no constraints are involved. Worth of the products will be given by

$$PW = (PT)(T) + (PB)(B) \quad\quad 10.4(28)$$

where PW is the total product worth rate, dollars/unit time. Eliminate B by

$$B = F - T \quad\quad 10.4(29)$$

to obtain

$$PW = (PT - PB)(T) + (PB)(F) \quad\quad 10.4(30)$$

Taking the partial derivative of this equation with respect to $(\overline{LB})$, the following is obtained:

$$\frac{\partial(PW)}{\partial(\overline{LB})} = (PT - PB)\frac{\partial(T)}{\partial(\overline{LB})} \quad\quad 10.4(31)$$

Operating costs can be approximated closely by

$$(OC) = (CRI)(RI) \quad\quad 10.4(32)$$

where CRI is the unit cost of operation per unit of internal reflux.

Taking the partial derivative of this equation with respect to $(\overline{LB})$, the following is obtained:

$$\frac{\partial(OC)}{\partial(\overline{LB})} = (CRI)\frac{\partial(RI)}{\partial(\overline{LB})} \qquad\qquad 10.4(33)$$

When the change in operating cost is equal to the change in product worth, the value for $(\overline{LB})_0$ is determined.

$$(PT - PB)\left[\frac{\partial(T)}{\partial(\overline{LB})}\right] = (CRI)\left[\frac{\partial(RI)}{\partial(\overline{LB})}\right] \qquad\qquad 10.4(34)$$

Evaluation of this equation will yield a value for $(\overline{LB})_0$. (T) must be expressed in terms of the specified separation, the feed composition and flow rate. (RI) must be expressed in terms of the independent inputs to the system as used in the internal reflux control equation. The simultaneous solution of equations 10.4(26) and 10.4(34) will yield optimum values for LB and HT: $(\overline{LB})_0$, $(\overline{HT})_0$. The total solution is indicated in Table 10.4f.

Bottom Product More Expensive   The general optimizing policy requires that the product with the highest unit price be produced at minimum specified purity.

Optimum Concentration of Light Key in Bottom Product

Since the bottom product must be produced at minimum purity, equation 10.4(27) will give the optimum value to specify for $(\overline{LB})$. The optimum $(\overline{LB})$ will be equal to $(\overline{LB})_{max}$. This in turn will give minimum specified sales purity $(HB)_{ss}$.

Optimum Concentration of Heavy Key in Top Product

The maximum value for heavy key in the top product is given by equation 10.4(26). As $(\overline{HT})$ is reduced from its maximum allowable value, flow rate of the top product will decrease and flow rate of the bottom product will increase. Therefore, total worth of the products will also increase together with operating costs. The optimum value for $(\overline{HT})$ will occur when an operating constraint is encountered or when the incremental gain in product worth is equal to incremental gain in operating cost.

For this case assume that the column approaches an operating constraint as $(\overline{HT})$ is lowered. Let this constraint be flooding above the feed tray due to excessive entrainment of liquid in the vapor.

First, an equation needs to be developed for vapor flow rate above the feed tray in terms of variables that are contained in the bottom product and reflux operating equations. The loading equation is developed as follows:

$$RI + T = V_t \qquad\qquad 10.4(35)$$

where RI = internal reflux flow rate,
T = top product flow rate (distillate), and
V_t = vapor flow rate above feed tray.

Table 10.4f
CONTROL HIERARCHY INVOLVED FOR OPTIMIZATION
TO MAXIMIZE PROFIT RATE WHEN NO OPERATING
CONSTRAINTS ARE INVOLVED

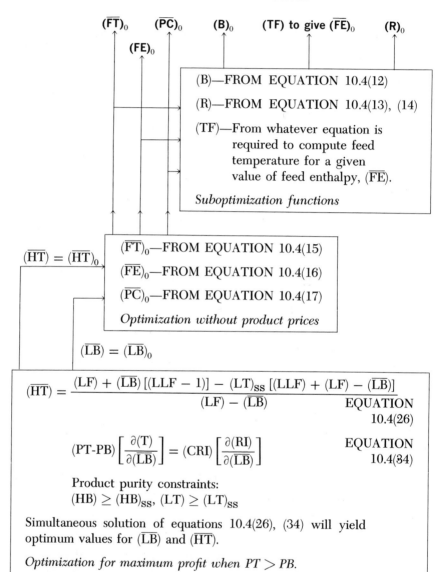

$(\overline{FT})_0$ $(\overline{PC})_0$ $(B)_0$ (TF) to give $(\overline{FE})_0$ $(R)_0$

$(FE)_0$

(B)—FROM EQUATION 10.4(12)

(R)—FROM EQUATION 10.4(13), (14)

(TF)—From whatever equation is
required to compute feed
temperature for a given
value of feed enthalpy, $(\overline{FE})$.

Suboptimization functions

$(\overline{HT}) = (\overline{HT})_0$

$(\overline{FT})_0$—FROM EQUATION 10.4(15)

$(\overline{FE})_0$—FROM EQUATION 10.4(16)

$(\overline{PC})_0$—FROM EQUATION 10.4(17)

Optimization without product prices

$(\overline{LB}) = (\overline{LB})_0$

$$(\overline{HT}) = \frac{(LF) + (\overline{LB})\,[(LLF - 1)] - (LT)_{SS}\,[(LLF) + (LF) - (\overline{LB})]}{(LF) - (\overline{LB})} \qquad \text{EQUATION} \atop 10.4(26)$$

$$(PT\text{-}PB)\left[\frac{\partial(T)}{\partial(\overline{LB})}\right] = (CRI)\left[\frac{\partial(RI)}{\partial(\overline{LB})}\right] \qquad \text{EQUATION} \atop 10.4(34)$$

Product purity constraints:
$(HB) \geq (HB)_{SS}$, $(LT) \geq (LT)_{SS}$

Simultaneous solution of equations 10.4(26), (34) will yield
optimum values for $(\overline{LB})$ and $(\overline{HT})$.

Optimization for maximum profit when PT > PB.

V_t can have a maximum value as given by equation 10.4(19) that is developed specifically for the column involved. Therefore, equate the right side of equation 10.4(19) with the left side of 10.4(35) to obtain

$$RI + T = a_1 + a_2(L) + a_3(\overline{PC}) \qquad 10.4(36)$$

where a_1, a_2, a_3 = coefficients of experimental loading equation,
$ PC$ = column pressure, and
$ L$ = liquid flow rate in column at the point where flooding occurs.

For this case L will equal (RI). Also, eliminate (T) by substituting $(F - B)$. Substitute

$$RI = \left[\frac{RI}{F}\right]F$$

$$B = \left[\frac{B}{F}\right]F$$

The following is obtained:

$$a_1 + a_3(\overline{PC}) + \left[(a_2 - 1)\frac{RI}{F} + \frac{B}{F} - 1\right]F = 0 \qquad 10.4(37)$$

In this equation a_1, a_2 and a_3 are known from the experimental loading equation. (PC) and F are measured, (RI/F) and (B/F) are obtained from the operating control equations for reflux and bottom product flow rate.

It is now possible to find the optimum value for $(\overline{HT})$. As $(\overline{HT})$ is lowered from its maximum allowable value as given by equation 10.4(26), the values for (RI/F) and (B/F) will change. $(\overline{HT})$ can be lowered until loading equation 10.4(37) is equal to zero. Also column pressure $(\overline{PC})$ can be raised to allow a greater loading that will result from a lower $(\overline{HT})$. A point will be found where maximum profit rate will exist.

The maximum limit for $(\overline{PC})$ will be determined by several factors. As $(\overline{PC})$ is increased, operating costs will increase, due to the resulting smaller differential temperature at the reboiler. Therefore, one possible limit would be a reboiler limit. Another possibility will be set by the column pressure rating. Another limit for maximum $(\overline{PC})$ may be determined by requirements of upstream processing equipment. Yet another limit for maximum $(\overline{PC})$ could be loading of the downcomers between each tray. This comes about because at higher vapor densities, disengagement of vapor from the liquid becomes more difficult. Therefore, at some maximum pressure, density of the vapor can approach the point where vapor will not have sufficient time to disengage from the liquid in the downcomers, and a condition known as *downcomer flooding* will occur.

Assume for purpose of illustration here that the maximum for $(\overline{PC})$ is set by pressure rating of the column. Therefore, $\overline{PC}$ will be set and left at this value.

Figure 10.4g shows the optimizing control system connected to a typical distillation column. Only part of the basic controls are shown for purpose of clarity. Assume, however, that all basic controls are the same as shown in Figure 10.4a.

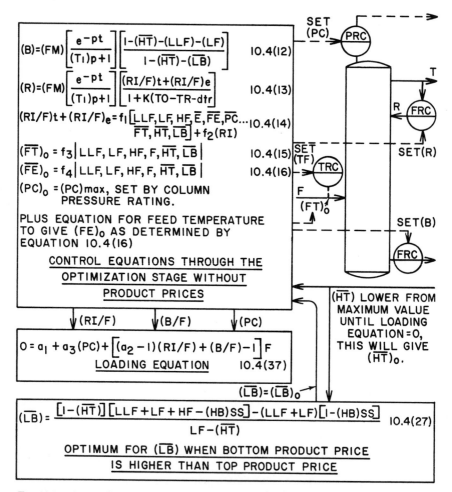

$$(B)=(FM)\left[\frac{e^{-pt}}{(T_1)p+1}\right]\left[\frac{1-(\overline{HT})-(LLF)-(LF)}{1-(\overline{HT})-(\overline{LB})}\right] \quad 10.4(12)$$

$$(R)=(FM)\left[\frac{e^{-pt}}{(T_1)p+1}\right]\left[\frac{(RI/F)t+(RI/F)e}{1+K(TO-TR-dtr}\right] \quad 10.4(13)$$

$$(RI/F)t+(RI/F)_e=f_1\left[\frac{LLF,LF,HF,\overline{E},\overline{FE},\overline{PC}...}{\overline{FT},\overline{HT},\overline{LB}}\right]+f_2(RI) \quad 10.4(14)$$

$$(\overline{FT})_0=f_3\left|LLF,LF,HF,F,\overline{HT},\overline{LB}\right| \quad 10.4(15)$$

$$(\overline{FE})_0=f_4\left|LLF,LF,HF,F,\overline{HT},\overline{LB}\right| \quad 10.4(16)$$

$(PC)_0 =(PC)$max, SET BY COLUMN PRESSURE RATING.

PLUS EQUATION FOR FEED TEMPERATURE TO GIVE $(FE)_0$ AS DETERMINED BY EQUATION 10.4(16)

CONTROL EQUATIONS THROUGH THE
OPTIMIZATION STAGE WITHOUT
PRODUCT PRICES

(RI/F) (B/F) (PC)

$$0=a_1+a_3(PC)+\left[(a_2-1)(RI/F)+(B/F)-1\right]F$$

LOADING EQUATION   10.4(37)

$(\overline{LB})=(\overline{LB})_0$

$$(\overline{LB})=\frac{[1-(\overline{HT})][LLF+LF+HF-(HB)SS]-(LLF+LF)[1-(HB)SS]}{LF-(\overline{HT})} \quad 10.4(27)$$

OPTIMUM FOR $(\overline{LB})$ WHEN BOTTOM PRODUCT PRICE
IS HIGHER THAN TOP PRODUCT PRICE

SET (PC)   PRC
T
R   FRC
SET(R)
SET (TF)   TRC
F
$(FT)_0^{\prime}$
SET(B)
FRC

$(\overline{HT})$ LOWER FROM MAXIMUM VALUE UNTIL LOADING EQUATION =0, THIS WILL GIVE $(\overline{HT})_0$.

Fig. 10.4g System for maximizing profit rate when bottom product price is higher than top product price and when entrainment above the feed tray limits loading

Unlimited Market and Feedstock

When an unlimited market exists for the products, the feed flow rate and separation resulting in maximum profit rate must be determined. Operation for a column under this condition will always be against an operating constraint. Assume for the purpose of the example to be given that the operating constraint involved will be the overhead vapor condenser capacity.

In general, the overall optimization problem for this case is illustrated in Figure 10.4h. Values must be determined for the separation, $(\overline{HT})_0$, $(\overline{LB})_0$, for feed flow rate, $(F)_0$, and column pressure, $(\overline{PC})_0$, that will give maximum profit rate. One of the key component specifications, $(\overline{HT})$ or $(\overline{LB})$, can be easily determined from the general optimizing policies. In general, the incremental gain in operating costs will never exceed the incremental gain realized by increasing feed rate unless the cost of operating utilities is very high. Also, the incremental gain in recovery of the most valuable product will not exceed the incremental gain resulting from increasing feed flow rate. This then means that both products operated at minimum purity will allow the largest quantity of feed to be charged for maximum profit rate.

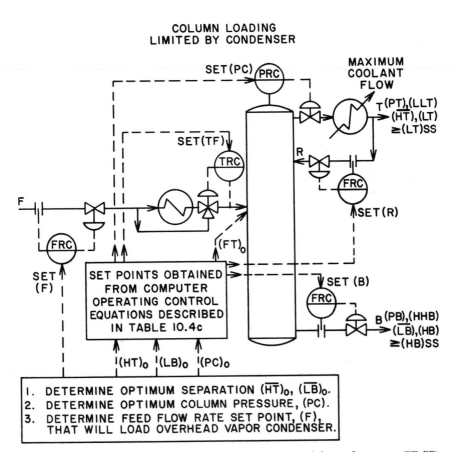

Fig. 10.4h   When an unlimited market exists for the products and the product prices (PT, PB) are known, the problem requires finding the optimum separation, optimum column pressure, and value for feed flow rate that will give maximum loading

Optimum Concentrations   Both products must be produced at minimum purity to achieve the maximum profit operation for this case. Therefore,

$$\overline{LT} = (\overline{LT})_{SS} \tag{10.4(38)}$$

$$\overline{HB} = (\overline{HB})_{SS} \tag{10.4(39)}$$

The concentration of each control component, $\overline{LB}$ and $\overline{HT}$, must be as follows to satisfy 10.4(38) and (39):

$$(\overline{LB})_0 = (\overline{LB})_{max} \tag{10.4(40)}$$

$$(\overline{HT})_0 = (\overline{HT})_{max} \tag{10.4(41)}$$

$(\overline{HT})_{max}$ and $(\overline{LB})_{max}$ are given by equations 10.4(26) and (27). The separation is optimized independent of column pressure $(\overline{PC})$ and feed rate (F).

Loading Constraint   As stated before, the overhead vapor condenser limits loading for this example. Column pressure must therefore be operated at maximum to obtain maximum condensing capacity. (ΔT across the condenser tubes will be the largest at maximum column pressure, and maximum condenser capacity will result.) The maximum pressure that can be used will be determined by one of the following five constraints.

1. Column downcomer capacity.
2. Reboiler capacity.
3. Upstream equipment pressure requirements.
4. Pressure rating of the column shell.
5. Fouling of the reboiler or condenser tubes.

As column pressure is increased, capacity of the reboiler and tray downcomers will be approached. Also, pressure rating of the shell and of other processing equipment will be approached. The one of the five constraints which is approached first, as column pressure is raised, will set the maximum pressure. Assume for purpose of illustration that the shell pressure rating is the limit on column pressure. Pressure can therefore be set constant at the rating value.

Optimum Feed Flow Rate   The optimum separation $(\overline{LB})_0$ and $(\overline{HT})_0$ having been determined, as well as the optimum column-operating pressure $(\overline{PC})_0$, the feed rate can now be increased until the condenser capacity is approached. There are several ways that feed rate can be manipulated to maintain condenser loading. One very useful method involves a predictive control technique.

Overhead vapor flow rate can be expressed in terms of various independent inputs and terms obtained from the operating control equations. Development of the predictive control equation proceeds as follows:

$$V_{oh} = T + R \tag{10.4(42)}$$

where V_{oh} = vapor flow rate overhead,
$\quad\quad$ T = top product flow rate, and
$\quad\quad$ R = external reflux flow rate.

First, eliminate T by $T = F - B$. Now, the maximum overhead vapor flow rate will be given by equation 10.4(18). Therefore, equations 10.4(18) and (42) can be set equal.

$$a_1 + a_2(\Delta T) + a_3(\Delta T)^2 = R + F - B \quad\quad\quad 10.4(43)$$

or

$$F = a_1 + a_2(T_{oh} - T_c) + a_3(T_{oh} - T_c)^2 + B - R \quad\quad 10.4(44)$$

where a_1, a_2, a_3 = coefficients for condenser-loading equation that are determined by column tests,
$\quad\quad$ F = feed flow rate,
$\quad\quad$ T_{oh} = overhead vapor temperature,
$\quad\quad$ T_c = temperature of coolant to overhead vapor condenser, and
$\quad$ B and R = bottom product and reflux flow rates from outputs of operating control equations.

Temperature of the overhead vapor will be a function of all independent inputs to the system. However, column pressure is usually the main variable of concern. For this example let

$$T_{oh} = fn(\overline{PC}) \quad\quad\quad 10.4(45)$$

where $fn(\overline{PC})$ is some function of column pressure.

In many cases $fn(\overline{PC})$ can be considered a linear function such as

$$fn(\overline{PC}) = d_1 + d_2(\overline{PC}) \quad\quad\quad 10.4(46)$$

This equation can be determined off-line from correlation of data obtained from flash calculations at the average composition of the overhead vapor existing. If changes in composition of the overhead vapor affect temperature of the overhead vapor a significant amount, then composition would also have to be taken into account. Composition for the overhead vapor can be easily approximated from feed composition analysis. If equation 10.4(44) is carried to this extent, then the feed flow rate can be predicted to be of such a value to keep the condenser against its maximum capacity. For the purpose of illustration here, T_{oh} is assumed to be a function of column pressure only.

Eliminate T_{oh} from 10.4(44) by equations 10.4(45) and (46) to obtain

$$a_1 + a_2[d_1 + d_2(\overline{PC}) - T_c] + a_3[d_1 + d_2(\overline{PC}) - T_c]^2 + B - R = F_{max}$$
$$10.4(47)$$

$(\overline{PC})$ and (T_c) are measured, B and R are obtained from the operating equations setpoint calculations. F_{max} will be the feed rate required to load the condenser for the particular values of $(\overline{HT})_0$, and $(\overline{LB})_0$ determined.

Figure 10.4i shows the overall optimizing control system applied. Only the necessary basic controls are shown. The other controls can be assumed to be as those shown in Figure 10.4a.

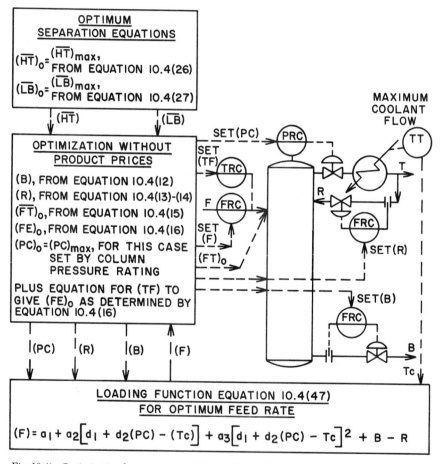

Fig. 10.4i Optimization for maximum profit rate when unlimited market exists for the products and loading is limited by condenser capacity

Reboiler Limiting Assume now that loading is limited by the reboiler instead of the condenser as just covered. Optimum separation remains the same. However, column pressure must be operated at a minimum to gain maximum reboiler capacity. For this example assume that minimum column pressure is set by pressure requirements of downstream equipment. Therefore, column pressure is set at a constant value and will not be changed unless pressure requirements of downstream equipment are changed.

Having achieved the optimum separation (minimum purity of products), and optimum column pressure, the feed rate can now be increased until the reboiler limits. This then will represent the maximum profit operation.

Again, manipulation of the feed flow rate can be handled by a predictive control technique. Liquid flow rate below the feed tray (LR) is given by

$$(LR) = RI + FI \qquad\qquad 10.4(48)$$

where FI is the internal feed flow rate.

$$FI = F[1 + (KF)(T_v - T_f)] \qquad\qquad 10.4(49)$$

where (KF) = a constant equal to the specific heat of the feed divided by the heat of vaporization,

T_v = temperature of vapor above the feed tray, and

T_f = temperature of feed at column entry.

The vapor flow rate out of the reboiler is given by

$$(VB) = (LR) - B \qquad\qquad 10.4(50)$$

Now substitute equation 10.4(49) into (48) to eliminate FI. Then substitute 10.4(48) into 10.4(50) to eliminate (LR). The following is obtained:

$$(VB) = RI + F[1 + (KF)(T_v - T_f)] - B \qquad\qquad 10.4(51)$$

Next substitute (RI/F)F for RI, and (B/F)F for B. Then solve for F to obtain

$$(F) = \frac{(VB)_{max}}{(RI/F) + 1 + (KF)(T_v - T_f) - (B/F)} \qquad\qquad 10.4(52)$$

where (F) = set point of feed flow controller and

$(VB)_{max}$ = maximum reboiler heat input rate.

Equation 10.4(52) calculates that feed flow rate which will cause the vapor rate $(VB)_{max}$ to exist for all separations specified. (RI/F) and (B/F) are obtained from the operating control equations used to achieve a suboptimum operation. T_v and T_f are measured. Equation 10.4(52) is used by specifying $(VB)_{max}$ and then evaluating the column operation. After sufficient time for the column to stabilize, the reboiler valve position (output of reboiler heat flow controller) is observed. Say for example, that the reboiler valve is 85 percent open. $(VB)_{max}$ can then be increased until the reboiler heat control valve is near its maximum opening, say 95 percent open. Enough room must be left to maintain control. $(VB)_{max}$ can be adjusted by the plant operator to maintain the reboiler valve near open or can be handled automatically by logic type feedback. Once $(VB)_{max}$ is established by experience, few adjustments will be required to maintain the column loaded. Adjustments to

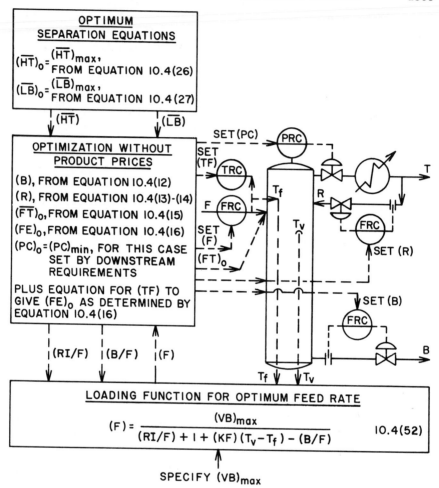

OPTIMUM SEPARATION EQUATIONS

$(\overline{HT})_0 = (\overline{HT})_{max}$, FROM EQUATION 10.4(26)

$(\overline{LB})_0 = (\overline{LB})_{max}$, FROM EQUATION 10.4(27)

OPTIMIZATION WITHOUT PRODUCT PRICES

(B), FROM EQUATION 10.4(12)

(R), FROM EQUATION 10.4(13)-(14)

$(\overline{FT})_0$, FROM EQUATION 10.4(15)

$(FE)_0$, FROM EQUATION 10.4(16)

$(PC)_0 = (PC)_{min}$, FOR THIS CASE SET BY DOWNSTREAM REQUIREMENTS

PLUS EQUATION FOR (TF) TO GIVE $(FE)_0$ AS DETERMINED BY EQUATION 10.4(16)

LOADING FUNCTION FOR OPTIMUM FEED RATE

$$(F) = \frac{(VB)_{max}}{(RI/F) + I + (KF)(T_v - T_f) - (B/F)} \qquad 10.4(52)$$

SPECIFY $(VB)_{max}$

Fig. 10.4j   Optimization of a distillation column when unlimited market exists for the products, prices of the products are known, and loading is limited by the reboiler

$(VB)_{max}$ will only be required as the maximum reboiler duty available varies. The control scheme is illustrated in Figure 10.4j.

Alternative Method for Maintaining Maximum Feed Rate
When a Reboiler or Condenser Constraint Exists

Another approach that can be considered for maintaining maximum feed flow rate involves straight feedback control. For example, the reboiler valve stem position (output of heat flow controller) can be measured and feed flow rate manipulated to maintain the reboiler valve near open.

Likewise, if a condenser limit is involved, stem position of the column pressure control valve can be measured and the feed rate manipulated by a conventional controller to maintain the back-pressure valve near open. Also, if a hot vapor by-pass around the condenser exists, stem position of the by-pass valve will also indicate the state of condensing. These procedures are illustrated in Figure 10.4k.

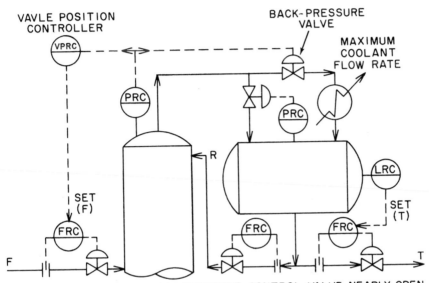

VPC MAINTAINS COLUMN BACK-PRESSURE CONTROL VALVE NEARLY OPEN
BY MANIPULATING FEED FLOW RATE TO KEEP CONDENSER LOADED

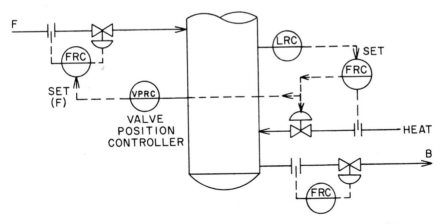

VPC MAINTAINS REBOILER HEAT VALVE NEARLY OPEN BY
MANIPULATING FEED FLOW RATE TO KEEP REBOILER LOADED

Fig. 10.4k   Feedback control method for maintaining condenser or reboiler loaded

Conclusions

Example solutions to some of the common distillation column optimizing problems have been given. Although many different situations can exist, they usually are combinations of those presented.

It is important to realize that optimization by feedback control methods can not approach the quality of control possible by predictive (feedforward) techniques. This is true even though the predictive control equations may be required to be updated by feedback. In effect, predictive optimization control greatly attenuates any error that must be handled by feedback (updating).

The application of feedforward optimizing control forces development of mathematical models of the component parts of a process. The mathematical models developed for optimizing unit operations will eventually be required to extend optimization to include an entire plant complex.

10.5 CONTROL OF REFRIGERATION UNITS

Refrigeration systems are seldom designed by the average application engineer, because the usual industrial practice is to select one of the pre-designed packages on the market. The intent of this section is to point out some of the basic system variations and to note the merits or drawbacks of several control schemes. A brief discussion is first given of the thermodynamic aspects and refrigerant characteristics. This is followed by a description of some typical household refrigerators and a listing of the main types of expansion valves. Then control systems for small, on-off and for large, continuous refrigeration units are described, including some discussion on compound and cascade units.

Refrigeration Units; Heat Pumps

In the sense that refrigerators remove heat from a low-temperature body and transfer it to one with a higher temperature, it is consistent to refer to them as *heat pumps*. Similarly to hydraulic pumps, these too require the input of work in fulfilling their function. This operation is schematically illustrated on the left side of Figure 10.5a. This shows the refrigeration unit as it removes Q_1 amount of heat from a low-temperature reservoir (the process being cooled), with the investment of W amount of work, and transfers a Q_h quantity of heat to a high-temperature reservoir (usually cooling water).

On the right side of this same illustration, the idealized temperature-entropy cycle is shown for the refrigerator, consisting of two isothermal and two isentropic (adiabatic) processes:

1-2; adiabatic process through expansion valve.
2-3; isothermal process through evaporator.
3-4; adiabatic process through compressor.
4-1; isothermal process through condenser.

The isothermal processes in this cycle are also isobaric (constant pressure). The efficiency of a refrigerator is defined as the ratio between the heat removed from the process (Q_1) and the work required (W) to achieve this heat removal.

$$\beta = \frac{Q_1}{W} = \frac{T_1}{T_h - T_1} \qquad 10.5(1)$$

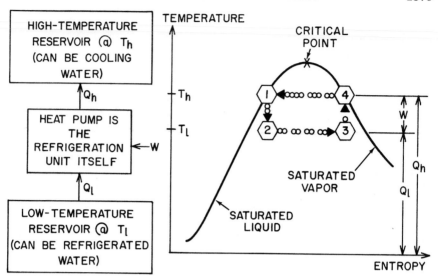

Fig. 10.5a   Thermodynamic characteristics of the refrigeration units

From this idealized definition of efficiency, it is seen that an increase in T_1 or a decrease in T_h will both contribute to increased refrigeration efficiency.

Figure 10.5b shows the four principal pieces of equipment involved. In path 1-2, through the *expansion valve* the high-pressure subcooled refrigerant liquid becomes a low-pressure, liquid-vapor mixture. In path 2-3, through the *evaporator* this becomes a superheated low-pressure vapor stream, while in the *compressor* (path 3-4) the pressure of the refrigerant vapor is increased. Finally, in path 4-1 this vapor is *condensed* at constant pressure. It should be emphasized that the liquid leaving the condenser is usually subcooled, whereas the vapors leaving the evaporator are usually superheated by a controlled amount.

The unit most frequently used in describing refrigeration loads is the ton. Because several tons are referred to in the literature, it is important to distinguish between them:

Standard ton:	200 BTU/min
British ton:	237.6 BTU/min
European ton (Frigorie):	50 BTU/min

Refrigerants

A full understanding of the control requirements for refrigeration units can be gained only by being familiar with both the equipment and the "working fluid" involved. The fluid that carries the heat from a low to a high temperature level is referred to as the refrigerant. Table 10.5c provides a summary of the more frequently used refrigerants.

The data is presented on the assumption that the evaporator will operate at 5°F and the temperature of the cooling water supply for the condenser will allow it to maintain 86°F. Other temperature levels would have illustrated the relative characteristics of the various refrigerants equally well. It is generally desirable to avoid operating under vacuum in any parts of the cycle because of sealing problems, and at the same time, very high condensing pressures are also undesirable because of the resulting structural strength requirements. From this point of view, the refrigerants between propane and methyl chloride in the tabulation display favorable characteristics. An exception to this reasoning is when very low temperatures are required. For such service ethane can be the proper selection in spite of the resulting high system design pressure.

Another consideration is the latent heat of the refrigerant. The higher it is the more heat can be carried by the same amount of working fluid, and therefore the corresponding equipment size can be reduced. This feature has caused many users in the past to compromise with the undesirable characteristics of ammonia.

One of the most important considerations is safety. In industrial installations, the desirability of non-toxic, non-irritating, non-flammable refrigerants cannot be overemphasized. It is similarly important that the working fluid be compatible with the compressor lubricating oil. The refrigerants which

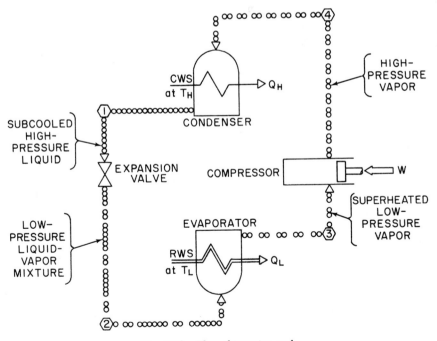

Fig. 10.5b   The refrigeration cycle

Table 10.5c
REFRIGERANT CHARACTERISTICS

Refrigerant		Applicable Compressor (R = Reciprocating, RO = Rotary, C = Centrifugal)	Boiling Point in °F at Atmospheric Pressure	Evaporator Pressure in PSIA, If Operating Temperature is 5°F	Condenser Pressure in PSIA, If Operating Temperature is 86°F	Latent Heat in BTU/lbm at 18°F	Toxic (T), Flammable (F), Irritating (I)	Mixes and/or Compatible with the Lubricating Oil	Chemically Inert and Non-Corrosive	Remarks
ETHANE	C_2H_6	R	−127	236	675	148	T&F	NO	YES	For low-temperature service
Carbon Dioxide	CO_2	R	−108	334	1039	116	NO	YES	YES	Low-efficiency refrigerant
PROPANE	C_3H_8	R	−48	42	155	132	T&F	NO	YES	For low-temperature service
FREON-22	$CHClF_2$	R	−41	43	175	92	NO	(1)	YES	High-efficiency refrigerant
AMMONIA	NH_3	R	−28	34	169	555	T&F	NO	(2)	High-efficiency refrigerant
FREON-12	CCl_2F_2	R	−22	26	108	67	NO	YES	YES	Most recommended
Methyl Chloride	CH_3Cl	R	−11	21	95	178	(3)	YES	(4)	Expansion valve may freeze if water is present
Sulphur Dioxide	SO_2	R	+14	12	66	166	T&I	NO	(4)	Common to these refrigerants:
FREON-21	$CHCl_2F$	RO	+48	5	31	108	NO	YES	YES	a. Evaporator under vacuum
Ethyl Chloride	C_2H_5Cl	RO	+54	5	27	175	F&I	NO	(5)	b. Low compressor discharge-pressure.
FREON-11	CCl_3F	C	+75	3	18	83	NO	YES	YES	c. High volume-to-mass ratio across compressor
Dichlor Methane	CH_2Cl_2	C	+105	1	10	155	NO	YES	YES	

(1) Oil floats on it at low temperatures.
(2) Corrosive to copper-bearing alloys.
(3) Anesthetic.
(4) Corrosive in the presence of water.
(5) Attacks rubber compounds.

are corrosive are undesirable for the obvious reasons of higher first cost and maintenance.

Most working fluids listed are compatible with reciprocating compressors. Only the last four fluids in the tabulation having high volume-to-mass ratios and low compressor discharge pressures can justify the consideration of rotary or centrifugal machines.

Considering all the advantages and drawbacks of the many refrigerants, overall it is Freon-12 that can be recommended for most installations.

Circulated Fluids

In the majority of industrial installations, the refrigerant evaporator is not used directly to cool the process. More frequently the evaporator cools a circulated fluid which is then piped to cool the process.

For temperatures below the point where water can be used for a coolant, brine is frequently used. It is important to remember that weak brines may freeze, and that strong brines, if they are not true solutions, may plug the evaporator tubes. For operation around 0°F, the sodium brines (NaCl) are recommended, and for services down to −45°F, the calcium brines (CaCl) are best.

Care must be exercised in handling brines because they are corrosive if not at the pH of seven or if oxygen is present. In addition, brine will initiate galvanic corrosion between dissimilar metals.

Household Refrigerators—Capillary Type

Small refrigerators for the household are usually either of the capillary or float type design. The capillary system (Figure 10.5d) is one of the oldest designs. A small-diameter tube is introduced between the condenser and the evaporator as a pressure drop source. The friction through the capillary is responsible for changing the high-pressure refrigerant liquid into a low-pressure liquid-vapor mixture.

The advantages of such a simple design include its low cost and the fact

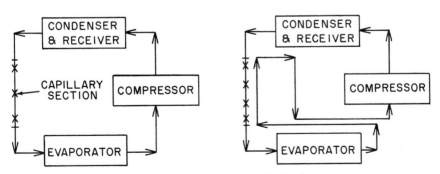

Fig. 10.5d   Fixed capillary type household refrigerators

that it contains no moving parts other than the compressor. Its efficiency is fairly high and the required refrigerant charge is small.

One of the drawbacks of the capillary design is the problem of plugging. Another disadvantage is that the condenser and evaporator pressures will equalize during shutdown, resulting in a low startup torque condition for the compressor.

If it is desired to operate the evaporator at or below 0°F, it is necessary to introduce the economizer heat-exchanger, shown on the right side of the illustration. Due to this heat transfer, additional liquid is vaporized in the capillary while the vapor line to the compressor suction is cooled. This reduces the load on the compressor. Features such as the economizing exchanger can be applied to all refrigeration units, and if such is not shown on the following pages, the reason for this omission is only to minimize the complexity of the illustrations.

Household Refrigerators—Float Type

Figure 10.5e shows the two basic variations of the float type systems, distinguished only by the float being located either on the high- or the low-pressure side.

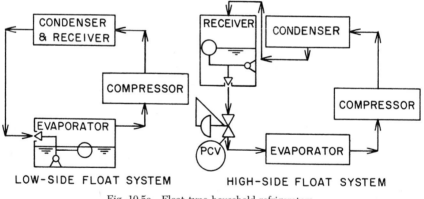

LOW-SIDE FLOAT SYSTEM   HIGH-SIDE FLOAT SYSTEM

Fig. 10.5e   Float type household refrigerators

The "low-side float system" is the earlier design in household refrigerators. It maintains constant refrigerant level in the evaporator by throttling the refrigerant flow from the condenser. Since the float is downstream to the let-down valve, it is on the low-pressure or compressor suction side. The float size and the resulting buoyant force is specific to a particular refrigerant, and therefore changing the refrigerant usually requires a new float size.

On the right side of the illustration the "high-side float system" is shown, representing some of the newer household refrigerator designs. Here the float is located in the condensate receiver, where it maintains constant level. If all of the pressure difference between the condenser and evaporator is taken across this let-down valve, the piping to the evaporator will be substantially subcooled, resulting in sweating or frosting. To prevent this, either the pipe should be insulated or, as shown on the sketch, a back-pressure regulator should be introduced to share the pressure-drop load with the float valve.

Small Industrial Refrigerators

For small industrial refrigerators, throttling by level float is replaced by some simple expansion valve. On the left side of Figure 10.5f, the direct expansion type control is shown. Here a pressure-reducing valve maintains a constant evaporator pressure. The pressure setting is a function of load, and therefore these controls are recommended for constant load installations only. The proper setting is found by adjusting the pressure-control valve until the frost stops just at the end of the evaporator, indicating the presence of liquid refrigerant up to that point.

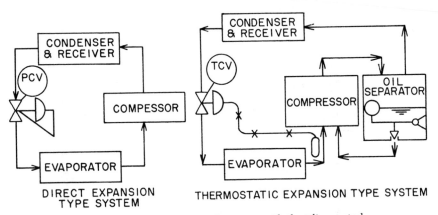

Fig. 10.5f   Small industrial refrigerators with throttling control

If the load increases, all the refrigerant will vaporize before the end of the evaporator, causing low efficiency as the unit is "starved." This condition will be relieved only by changing the pressure setting. When the unit is down, the pressure-control valve closes, isolating the high- and low-pressure sides of the system. This guarantees the desirable high startup torque.

On the right side of the same illustration the thermostatic expansion type control is shown. This system, instead of maintaining evaporator pressure, controls the superheat of the evaporated vapors. This design is therefore not limited to constant loads, because it guarantees the presence of liquid refrigerant at the end of the evaporator under all load conditions.

Figure 10.5f also shows a typical oil separator. As noted earlier in connection with the economizer exchangers, components such as oil separators may also be used on any one of the systems described, but in order to minimize the details, it has been shown only in Figure 10.5f.

Expansion Valves

Evaporator Pressure Drop Low

On the left side of Figure 10.5g a fairly standard superheat control valve is shown. It detects the pressure into and the temperature out of the evaporator. If the evaporator pressure drop is low, these measurements (the saturation pressure and the temperature of the refrigerant) are an indication of superheat. The desired superheat is set by the spring in the valve operator, which together with the saturation pressure in the evaporator opposes the opening of the valve. The "superheat feeler bulb" pressure balances these forces when the unit is in equilibrium, operating at the desired (usually 9°F) superheat.

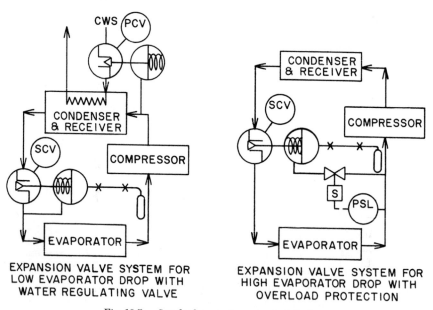

EXPANSION VALVE SYSTEM FOR
LOW EVAPORATOR DROP WITH
WATER REGULATING VALVE

EXPANSION VALVE SYSTEM FOR
HIGH EVAPORATOR DROP WITH
OVERLOAD PROTECTION

Fig. 10.5g   Standard expansion valve installations

If the process load increases, it causes an increase in the evaporator outlet temperature. An increase in this temperature results in raising the "feeler bulb" pressure, which in turn further opens the superheat control valve. This greater flow from the condenser to the evaporator increases the saturating pressure and temperature, and the increased saturating pressure balances against the increased feeler bulb pressure at a new (greater) valve opening in a new equilibrium. To adjust to an increased load condition, the evaporator pressure had been increased, but the amount of superheat (set by the valve spring) is kept constant.

In this same sketch, the operation of the cooling water regulating valve is also illustrated. This valve maintains the condenser pressure constant and at the same time conserves cooling water. At low condenser pressure, such as when the compressor is down, the water valve closes. It starts to open when the compressor is restarted and its discharge pressure reaches the setting of the valve. The water valve opening follows the load, further opening at higher loads to maintain the condenser pressure constant. This feature too can be incorporated into any one of the refrigeration units.

Evaporator Pressure Drop High

As illustrated on the right side of Figure 10.5g, the pressure-sensing tap of the superheat control valve has to be moved downstream of the evaporator if there is a substantial drop across the evaporator.

In order fully to understand the operation of the superheat control valves, it is necessary to elaborate on the types of fillings available for the "feeler bulbs."

If the bulb is filled with a refrigerant *vapor*, it is necessary to locate the valve in a warmer location than the bulb, so that the small amount of filling liquid will stay in the bulb. This type of filling is pressure-limiting, because above a certain temperature the filling refrigerant is fully vaporized and a further rise in temperature will not increase the bulb pressure.

A refrigerant *liquid* filling is illustrated in Figure 10.5g. This filling too will maintain the desired superheat, but it requires special attention during startup because without additional instrumentation it will open when the compressor is started, thereby equalizing the suction and discharge pressures and overloading the motor during pull-down. The solution shown in Figure 10.5g requires a low-pressure switch and a solenoid pilot in the pressure-sensing line. At higher than normal pressures this solenoid is closed, trapping the high pressure and keeping the expansion valve closed until the evaporator is pulled down by the compressor to a sufficiently low pressure. The pressure switch then opens the solenoid pilot, lowering the pressure in the sensing line and thereby reducing the force which kept the expansion valve closed. From this point on, the opening of the expansion valve is automatically adjusted to maintain constant superheat.

Specialized Installations

At very low temperatures, a small change in refrigerant vapor pressure is accompanied by a fairly large change in temperature. For example, Freon-12 at the temperature level of $-100°F$ will show a 5°F temperature change, corresponding to a 0.3 PSIG variation of saturation pressure. Therefore, the use of a thermal bulb to detect indirectly the saturation pressure in the evaporator will result in a more sensitive measurement. A differential temperature expansion valve, taking advantage of this phenomenon, is illustrated on the left side of Figure 10.5h.

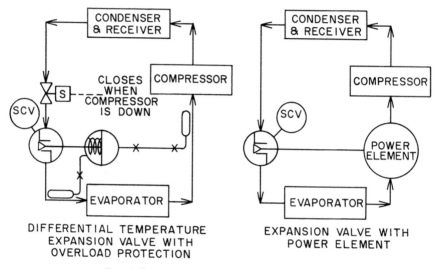

DIFFERENTIAL TEMPERATURE
EXPANSION VALVE WITH
OVERLOAD PROTECTION

EXPANSION VALVE WITH
POWER ELEMENT

Fig. 10.5h   Specialized expansion valve installation

One limitation of this differential temperature superheat control valve is that it does not isolate the high- and low-pressure sides when the system is idle. Thus in order to prevent the equalization of compressor suction and discharge pressures when the compressor is down, a separate solenoid is necessary. This valve is wired to the motor starter and opens only when the compressor is running.

At the right side of the same illustration a more recent development in expansion valve actuator design, the *power element* is shown. This element requires no external connections because it is designed to take the total flow from the evaporator to the compressor, and therefore it tends to require less maintenance. Another advantage of the flow through power element in comparison to other superheat control valve actuators is its almost instant response. One limitation of this design is the physical proximity required between valve actuator and valve body, placing certain restrictions on the routing of refrigerant piping.

On-Off Control: Small Industrial Units

Figure 10.5i shows the controls for a fairly simple and small refrigeration unit. This system includes a conventional superheat control valve, a low-

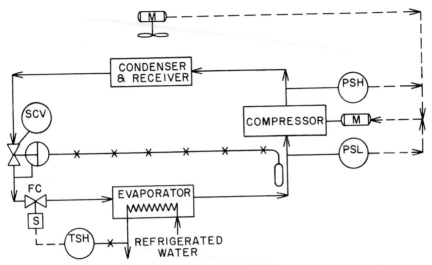

Fig. 10.5i   Small industrial refrigeration unit with electric on-off control

pressure-drop evaporator, a reciprocating on-off compressor and an air-cooled condenser. The purpose of this refrigeration package is to maintain a refrigerated water supply to the plant within some set limits.

The high-temperature switch (TSH) shown on the illustration is the main control device. Whenever the temperature of the refrigerated water drops below a preset value (say 38°F), the refrigeration unit is turned off, and when it rises to some other level (say 42°F), it is restarted. This on-off cycling control is accomplished by the temperature switch closing the solenoid valve when the water temperature is low enough. The closing of this valve causes the compressor suction pressure to drop until it reaches the set point of the low-pressure switch, which in turn stops the compressor.

While the unit is running, the expansion valve maintains the refrigerant superheat constant and the safety interlocks protect the equipment. These interlocks include such features as turning off the compressor if the fan motor stops or if the compressor discharge pressure becomes too high for some other reason.

This unit can operate only at full compressor capacity or not at all. Such machine is referred to as one with *two-stage unloading*. When it is desired to vary the cooling capacity of the unit instead of turning it on and off, two possible control techniques are available. One approach involves the multi-step unloading of reciprocating compressors. In a three-step system the available operating loads are 100%, 50% and 0%, while with five-step unloading, 100%, 75%, 50%, 25% and 0% loads can be handled. For the details of step-wise unloading controls, refer to Section 10.13. The other possible approach is to use continuous throttling, such as the type illustrated in Figure 10.5l.

Multi-Stage Refrigeration Units

It is not practical to obtain a compression ratio outside the range of 3:1 to 8:1 with the compressors used in the process industry. This places a limitation on the minimum temperature that a single stage refrigeration unit can achieve.

For example, in order to maintain the evaporator at −80°F and the condenser at 86°F (compatible with standard supplies of cooling water), the required compression ratio would be as follows, if Freon-12 were the refrigerant:

$$\text{Compression ratio} = \frac{\text{refrigerant pressure at } 86°F}{\text{refrigerant pressure at } -80°F}$$

$$= \frac{108}{2.9} = 37 \qquad\qquad 10.5(2)$$

Such compression ratio is obviously not practical, and therefore a multi-stage system is required. Figure 10.5j illustrates a compound system which is applicable to ultra-low temperature service due to the two compressor stages in series. The *inter-cooler* prevents excessive temperatures from developing as the result of the high overall compression ratio.

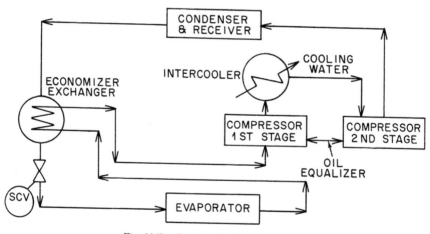

Fig. 10.5j   Compound refrigeration unit

Another standard component in ultra-low temperature refrigeration units is the economizer exchanger, which is useful in any cooling system. Its value is increasing as the evaporator temperature decreases. The most important contribution of this heat transfer unit is to cool the otherwise superheated suction to the first stage of the compressor, thereby reducing its load.

Another approach to multi-stage ultra-low temperature operation is illustrated in Figure 10.5k. Here a cascaded refrigeration unit is shown with two complete refrigeration units working in series. With the terminology used at the beginning of this section, it could be said that one "heat pump" transfers heat from an ultra-low to a low temperature level and the second pumps this heat from the low temperature level to the higher level of the cooling water. In case of a cascaded multi-stage refrigeration unit, the condenser of the first stage is the evaporator of the second, and therefore the two stages are coupled through the refrigerant fluids. The instrumentation used must reflect this fact by recognizing that a load change in the first stage will be reflected in the second stage, or that a change in cooling water temperature will eventually affect both stages.

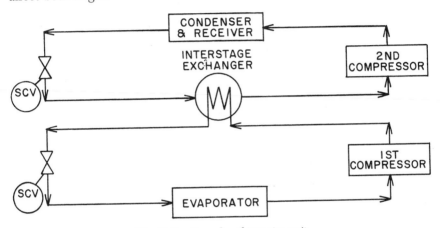

Fig. 10.5k   Cascade refrigeration unit

Large Industrial Refrigerators With High Turn-Down Ratio

The refrigeration unit shown in Figure 10.5l, although being far from a "standard" system, does contain some of the features typical to industrial units in the 500-ton and larger sizes. Some of these features include: the capability for continuous load adjustment as contrasted with step-wise unloading, the application of the economizer expansion valve system and the use of hot gas by-pass to increase rangeability.

The unit illustrated provides refrigerated water at 40°F through the circulating header system of an industrial plant. The flow rate is fairly constant, and therefore process load changes are reflected by the temperature of the returning refrigerated water. Under normal load conditions this return water temperature is 51°F. As process load decreases, the return water temperature drops correspondingly. With the reduced load on the evaporator, TIC-1 gradually closes the suction vane to the compressor. By throttling the suction vane, a 10:1 turn-down ratio can be accomplished. If the load drops below this ratio, the hot gas by-pass system is automatically activated.

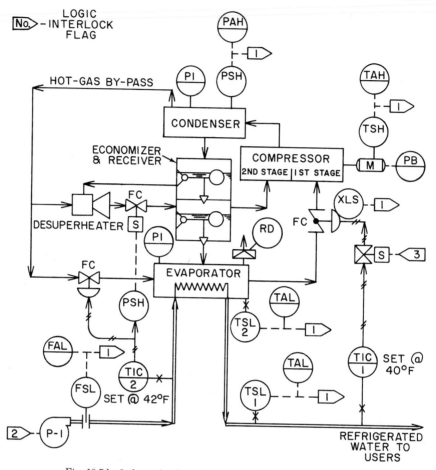

Fig. 10.51   Industrial refrigeration unit with high turn-down ratio

Generally the suction vanes do not close further than the 10 percent load level, thereby protecting the compressor against surges. The setting of TIC-2 corresponds to a value slightly above surge level. If, with the constant refrigerated water flow rate, 100 percent load is represented by a temperature change of 51°F to 40°F or by an 11°F change, then the temperature difference of 1.1°F would correspond to 10 percent load which is the surge level. This would suggest that when the return water temperature dropped to 41.1°F, the hot-gas by-pass should be activated to prevent surging. Usually it is preferred to activate the hot-gas by-pass at a slightly higher load level, which in this case is represented by the setting of 42°F on TIC-2.

When the return water temperature is above 42°F, the output of TIC-2 is zero, and therefore the hot-gas by-pass valve is closed. When the temperature drops to 42°F, this valve starts to open, and its opening is in proportion to the load detected. This means that the valve is fully closed at 42°F, fully open at 40°F and throttled in-between. Therefore, it is theoretically possible to achieve a very high turn-down ratio by temporarily running the machine on close to zero process load. This can be visualized as a heat pump, transferring heat energy from the refrigerant itself to the cooling water, and in the process of so doing, some of the refrigerant vapors are condensed, resulting in an overall lowering of operating pressures in the system.

The economizer shown in this system can increase the efficiency of operation by 5 to 10 percent. This is achieved by the reduction of space requirements, savings on compressor power consumption, reduction of condenser and evaporator surfaces, etc. The economizer shown in Figure 10.51 is a two-stage expansion valve with condensate collection chambers. When the load is above 10 percent, the hot-gas by-pass system is inactive. Condensate is collected in the upper chamber of the economizer, and it is drained under float level control, driven by the condenser pressure. The pressure in the lower chamber floats off the second stage of the compressor, and it too is drained into the evaporator under float level control, driven by the pressure of the compressor second stage. Economy is achieved due to the vaporization in the lower chamber by precooling the liquid which enters the evaporator and at the same time by desuperheating the vapors that are sent to the compressor second stage.

When the load is below 10 percent the hot-gas by-pass is in operation, and the solenoid valve, (Figure 10.51), which is actuated by the high-pressure switch (PSH), opens. Some of the hot gas goes through the evaporator and is cooled by contact with the liquid refrigerant, and some of the hot gas flows through the open solenoid. This second portion is desuperheated by the injection of liquid refrigerant upstream of the solenoid, which protects against overheating the compressor.

In addition to the load controls described, it is also important to consider the safety instrumentation required to protect the unit. (Detailed instrumentation requirements for compressor seal and lube oil controls are covered in Section 10.13.) Usually there are three safety interlock systems protecting a refrigeration unit of this type.

The first logic interlock system prevents the compressor motor from being started if one or more of the following conditions exist, and it also stops the compressor if any except the first condition listed occurs while it is running:

1. Suction vane is open, detected by limit switch (XLS).
2. Refrigerated water temperature is dangerously low, sensed by (TSL-1).
3. Refrigerated water flow is low, measured by (FSL).
4. Evaporator temperature is dangerously low, signaled by (TSL-2).

5. Pressure in the condenser is high, indicated by (PSH).
6. Temperature of motor bearing and/or winding is high, detected by (TSH).
7. Lubricating oil pressure is low (not shown).

The second logic interlock system guarantees that the following pieces of equipment are started or are already running upon starting the compressor:

1. Refrigerated water pump (P-1).
2. Lubricating oil pump (not shown).
3. Water to lubricating oil cooler, if such exists (not shown).

The third interlock usually assures that the suction vane is completely closed when the compressor is stopped.

In addition to these safety interlocks, other instruments are also desirable (Figure 10.51). This would include a small local panel board accommodating the controllers, a few dial gauges and alarm lights, as shown on the illustration. Various local gauges and pressure relief devices should also be utilized.

Conclusions

The intent of this presentation was not to give an all inclusive treatment of the multitude of control considerations applicable to refrigeration units, but to make the reader aware of the most important choices and alternatives.

The features of the various refrigerants (Table 10.5c) and the characteristics of refrigerated liquids must be fully understood, because the control hardware has to be selected for compatibility with these fluids. Generally, it is suggested that a refrigerant be selected which is non-corrosive, safe and compatible with the lubricating oil, and which allows the evaporator to work under positive pressures while it does not require the condenser to operate at very high pressures.

After the selection of the refrigerant, the next most important decision involves the nature of the process load and the range within which control is desired. The merits and drawbacks of two-stage, multi-stage, continuous and hot-gas by-pass load control should be carefully considered, with economic and maintenance aspects balanced against operating efficiency and control quality considerations.

Once the method of load control is decided on, the other remaining decisions are fairly easy to make. For example, it is desirable to throttle the expansion valve so that it maintains the evaporator superheat constant and to provide controls which will protect against pressure equalization during compressor shutdown.

The desirable extent of safety interlocks, the possible need for an oil separator or the installation of an economizer exchanger should all be taken into consideration.

The controls shown in Figure 10.51 reflect some of the noted features.

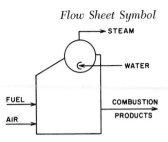

10.6 CONTROL OF STEAM BOILERS

Steam boilers are used industrially both as a power source and in processing. They consist of a furnace where air and fuel are combined and burned to produce combustion gases and of a water tube system, the contents of which are heated by these gases. The tubes are connected to the steam drum, where the generated water vapor is withdrawn. If superheated steam is to be generated, the steam from the drum is passed through the superheater tubes, which are exposed to the combustion gases.

This section provides a detailed treatment of steam generator controls. The control systems discussed include those for the combustion, the feedwater and the steam temperature, in addition to safety instrumentation.

If one defines high pressure as exceeding 100 PSIG, then this section is devoted to the general discussion of high-pressure steam boilers.

Figure 10.6a shows a typical boiler arrangement for gas or oil fuel with measurement points for control indicated.

Steam boilers as referred to in this section are drum type boilers. Very large, supercritical pressure boilers are the "once through" type and are found only in the largest electric generating plants.

A steam boiler steam outlet may be connected to a header in parallel with other boilers or directly to a single steam user. In either case the master controller may be controlled by pressure or flow, depending upon the process requirements. In either case pressure needs to be essentially constant for proper boiler operation. The "load" on a steam boiler refers to the amount of steam demanded by the steam users.

The boiler is a utility supplier of the process, and as such, must follow the needs of the plant in its demand for steam. The boiler and its control system must therefore be capable of satisfying rapid changes in load. Load changes can be due to rapidly changing process requirements or to cycling control equipment. While load may be constant and steady over prolonged periods, the utility system must have sufficient "turn down" to stay in operation at reduced capacities as portions of the plant may be shut down. This consideration usually leads to a greater "turn down" requirement for the boilers than for any other portion of the plant.

Generally speaking, computer use is presently limited in boiler plants

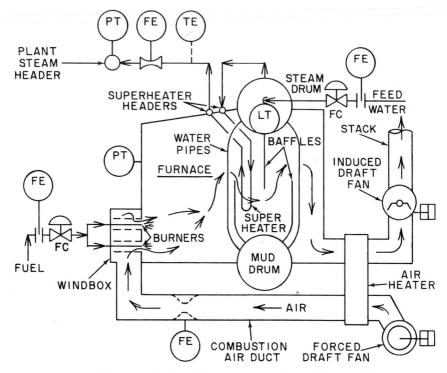

Fig. 10.6a   Drum type boiler with measurement and control points indicated

to information gathering, such as logging, alarm monitoring and display. Computers are not required to supervise and/or modify analog control action, since a properly designed and tuned analog system can provide adequate control quality. In the future, boiler plants are expected to be computerized to a greater extent. The justification in these cases will be optimization, safety and the savings that may result by replacing the analog control hardware with computer software.

Performance

The normal "on-line" requirements for steam boilers are to control steam pressure within ±1% to ±2% of desired pressure, fuel-air ratio within ±5% of excess air based on a desired "load" vs "excess air" curve, steam drum water level within ±1 in. of desired level and steam temperature (where provision is made for its control) within ±10°F of desired temperature. Control of this sort is normally possible provided that the load does not change more than 20% of full scale per minute and there are no boiler design problems limiting this ability. The various loops tend to interact so that integration into an overall system is necessary both during design and when the loops are being field "tuned."

Safety Interlocks

Operations relating to safety, startup, shutdown, and burner sequencing are basically digital in nature and operate from digital binary inputs such as contact closures.

The basic safety interlocks are as follows:

Purge interlock—Prevents fuel from being admitted to an unfired furnace until the furnace has been thoroughly air purged.

Low air flow interlock and/or fan interlock—Fuel is shut off upon loss of air flow and/or combustion air fan or blower. (Fan rotor differential is an accepted measurement.)

Low fuel supply interlock—Fuel is shut off upon loss of fuel supply that would otherwise result in unstable flame conditions.

Loss of flame interlock—All fuel is shut off upon loss of flame in the furnace and/or fuel to an individual burner is shut off upon loss of flame to that burner.

Fan interlock—Stop forced draft fan upon loss of induced draft fan.

Low water interlock (optional)—Shut off fuel on low water level in boiler drum.

High combustibles interlock (optional)—Shut off fuel on highly combustible content in combustion gases.

Where fans are operated in parallel, an additional interlock is required to close the shutoff dampers of either fan when it is not in operation. This is necessary to prevent air recirculation around the operating fan.

Automatic startup sequencing for lighting the burners and for sequencing them in and out of operation is common. This is accomplished at the present time by either relays or solid-state hard-wired logic systems. In the future these functions will probably be accomplished by computers or through fluidic systems.

Design and specification of the various safety interlocks is somewhat specialized due to involvement of insurance company regulations, NFPA (National Fire Protection Association) and of state regulations.

Combustion Control

A combustion control system can be broken down for ease of examination into the *fuel control* and *combustion air control* subsystems. The interrelationships between these two subsystems necessitate the use of the *fuel-air ratio controls*. For safety purposes, fuel addition should be limited by the amount of available combustion air, and combustion air may need minimum limiting for flame stability.

An initial consideration is that there be a master controller for the fuel and air flow control. Various forms of the master controller are shown in Figures 10.6b through 10.6f.

When more than one boiler is operated from the same master controller, a panel station should be provided for *each* boiler, for the purpose of balancing or intentionally unbalancing the loads.

This station provides for hand operation of the entire combustion control system of the associated boiler in addition to the ability to bias the master controller signal up or down when on automatic (Figure 10.6b).

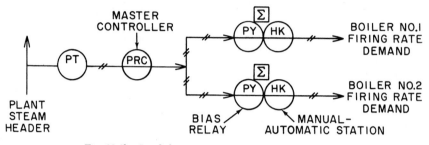

Fig. 10.6b   Load sharing controls between several boilers

When steam pressure is controlled by other means, steam flow can be the master controller.

If variations in fuel heating value are minor, the master flow controller shown might be eliminated and the master load signal generated by a manual loading station.

Situations may arise when it is desirable to have either flow or pressure control. In these cases a master control arrangement as shown in Figure 10.6c can be used. Though it may appear simpler to switch transmitters, it is desirable to transfer the controller outputs so that the controller does not have to be returned each time the measurement is switched.

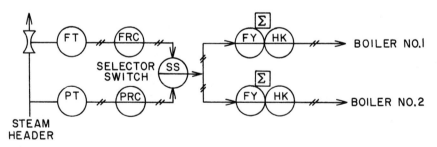

Fig. 10.6c   Boiler control with alternative pressure or flow master

The average setting for a pressure controlled "master" is 16 percent for proportional band and 0.25 repeats per minute for reset, while for a flow control "master," the comparable settings might be 100 percent proportional band and 3.0 repeats per minute for reset. (See Section 7.12 for details on controller tuning.)

In single boiler installations where steam lines are small, steam line velocities high and the relative capacity in the steam system is low, it is sometimes necessary for the sake of control stability to correct the pressure "master" in a feedforward manner as shown in Figures 10.6d and 10.6e.

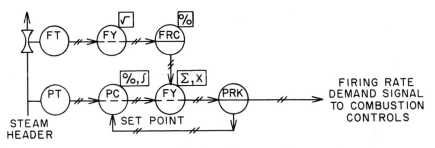

Fig. 10.6d   Boiler control utilizing flow-corrected pressure signals

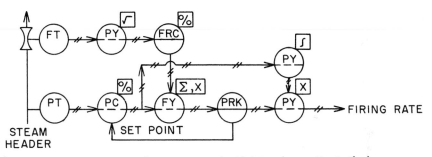

Fig. 10.6e   Flow-corrected pressure control with integral correction in the by-pass

These systems allow the use of wider proportional bands and lower reset settings on the pressure controller than would otherwise be possible for desirable plant operation. Of these two, Figure 10.6e is the more sophisticated, because the integral mode is in the by-pass, and therefore it can continuously calibrate the fuel and air flow controls to match steam flow, thus providing for quick, stable response to major and rapid load changes.

Neither "master controller" loop in Figures 10.6d or 10.6e can be used with more than one boiler, unless the steam flow measurement is a total from all boilers or a high-flow selector is employed, selecting the highest of the steam flows. This is due to the positive feedback from steam flow to fuel feed rate. In the case of a single boiler, steam flow is only a function of plant demand. In the case of multiple boilers on a header, total steam flow is determined by the user, but individual boiler steam flows are determined primarily by the firing rates.

Figure 10.6f illustrates a feedforward concept of determining firing rate demand.

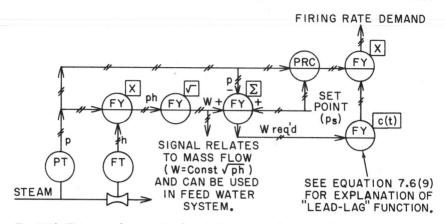

Fig. 10.6f   Firing rate determination by calculating steam flow required as the measured mass flow rate less the rate of loss of boiler contents. (Approximated as const $(p_s - p)$.)

Fuel Controls (Measurable Fuels)

The primary boiler fuels are coal, oil, and gas, but there are a large variety of auxiliary fuels such as waste gases, waste sludges, waste wood products such as bark, sawdust, hogged fuel and coffee grounds. In many cases these auxiliary fuels are dumped to the boiler plant on an uncontrolled basis for immediate burning. There are myriads of these combinations, and only the more common fuel control problems will be covered in the discussion to follow.

Gas and oil involve the simplest controls, since they are easily measured and require little more than a control valve in the fuel line. A valve positioner capable of providing a linear relationship between flow and control signal is desirable. A flow controller is often needed as a means of more precise linearization. Examples of fuel controls are shown in Figures 10.6g and h. (See Section 1.7 for details on valve positioners.)

When gas or oil or a combination of both are used, several alternative methods of fuel control are possible as illustrated in Figures 10.6i, j and k.

The instruments shown ("panel stations") provide the means of manual control plus the ability of automatic control, with bias of one fuel with respect to the other. When it is desired to fire the fuels in a predetermined ratio to each other regardless of load, a manually adjustable signal splitter can be used as shown. The most precise and complex method of ratioing fuel (not shown)

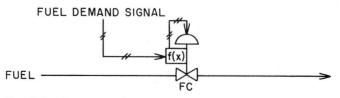
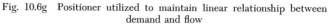

Fig. 10.6g   Positioner utilized to maintain linear relationship between demand and flow

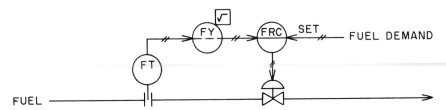

Fig. 10.6h　Fuel controller used to keep demand and flow in linear relationship

is to split the demand signal and send that to individual flow control loops.

Since one of the requirements ultimately is to have fuel ratioed to combustion air, any totalization of fuel for control purposes should be on an "air required for combustion" basis. If totalization is needed on any other basis, such as BTU for other purposes, a separate totalizer should be used.

When auxiliary fuel is burned on an uncontrolled availability basis, the fuel and air control system needs to be able to accommodate sudden changes in auxiliary flow without upsetting the master controller. The master controller should be designed and used to respond to total load demands only and not to correct for fuel upsets. A typical fuel control system for accommodating variations in auxiliary fuel without upsetting the master is shown in Figure 10.6l.

NOTE: ON THIS AND FIGURES TO FOLLOW, THE SYMBOL

REPLACES THE PREVIOUSLY USED REPRESENTATION OF

REPRESENTING BIAS AND MANUAL – AUTOMATIC FUNCTIONS

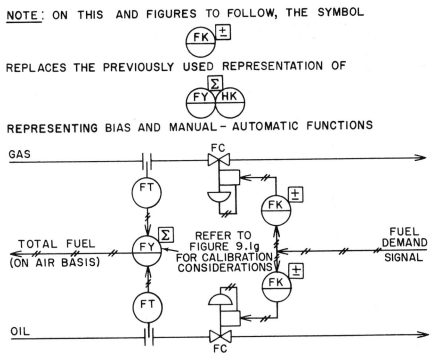

Fig. 10.6i　Fuel demand is manually split between the two fuels on an open-loop basis

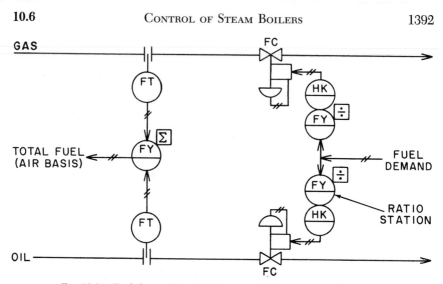

Fig. 10.6j   Fuel demand is ratioed between fuels on an open-loop basis

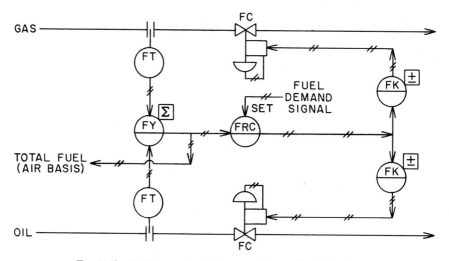

Fig. 10.6k   Fuel demand split between fuels on closed-loop basis

In the basic system without auxiliary fuel, the signal is relayed directly to the control valve. Addition of auxiliary fuel shifts the primary fuel control valve opening to prevent fuel variations affecting overall boiler performance. A more precise system is shown in Figure 10.6m. Here the flow controller adjusts the primary fuel control valve to satisfy total fuel demand and prevents auxiliary fuel variations from upsetting the master controller. It may appear that both greater precision and linearization can always be obtained by using closed-loop flow controls on the streams.

A word of caution is in order because the closed-loop systems are more limited in their "turn-down" capability due to the transmitter rangeabilities.

Signal transportation lags in long pneumatic lines can also be a problem.

Other considerations in the choice between the open and closed loop are in the arrangement of the integrated system. If the integrated system does not compare fuel flow and air flow directly in a ratio or difference controller, the closed-loop system should be used.

Whatever type of fuel control is used, the maximum flexibility in design will be present if all flow signals are linear and all control valve characteristics are also linear. In this manner the various flows and signals can be combined, subtracted, multiplied or divided to produce the desired control. One desired

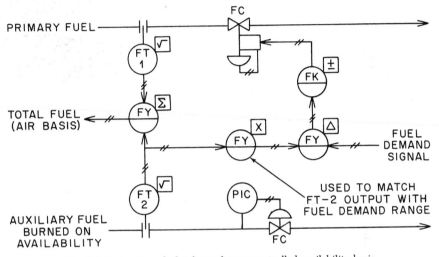

Fig. 10.6 l   Auxiliary fuel is burned on uncontrolled availability basis

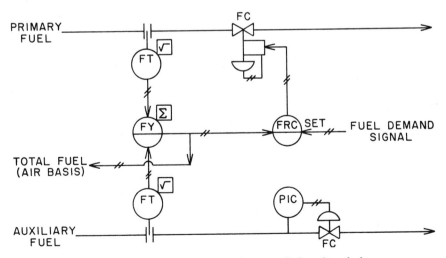

Fig. 10.6m   Closed-loop control of uncontrolled auxiliary fuel

condition is to have the total fuel demand signal linear, with fuel totalized on a basis of combustion air required. The other desired end condition is to have the total fuel control capacity maximum at a value approximately 10 percent greater than required for maximum boiler capacity. This excess is necessary for control flexibility at maximum boiler load. Additional excess capacity should not be considered, since it reduces turn-down capability.

Unmeasured Fuels

Coal is typical of normally unmeasured fuels. Because of the lack of measurement, most coal control systems are open loop, wherein a control signal positions a coal-feeding device directly. This is the case with a spreader stoker or cyclone furnace or indirectly with pulverized coal. There is little choice of fuel control methods with coal, since stoker and pulverizer manufacturers prescribe how these devices are to be controlled.

A spreader stoker consists of a coal hopper on the boiler front with air jets or rotating paddles that flip the coal into the furnace, where a portion burns in suspension and the rest drops to a grate. Combustion air is admitted under the grate. There is no way to control fuel to a spreader stoker except in an open-loop manner by positioning a feeder lever that regulates coal to the paddles.

In pulverized coal fired boilers the coal is ground to a fine powder and is carried into the furnace by an air stream. There are normally two or more pulverizers (in parallel) per boiler. Pulverized coal flow is regulated at the pulverizer, and each manufacturer has a different design requiring different controls. One control arrangement is shown in Figure 10.6n. Here the pulverizer primary air comes from a pressure fan that blows through the pulverizer, picking up the coal and transporting it to the furnace.

In addition to the controls shown, an air temperature control is also required. In this loop, cold and hot combustion air is mixed ahead of the primary air fan to control the temperature of the coal air mixture in the pulverizer. This control is necessary to maintain a maximum safe operating temperature in the pulverizer. This is a simple feedback loop, usually utilizing proportional control only.

A control arrangement for a bowl type pulverizer is shown in Figure 10.6o. In this type of pulverizer the air fan sucks air through the pulverizers with the fan (called an exhauster fan) located between the pulverizer and the burners. The coal-air temperature loop is similar to the one described in Figure 10.6n.

The control of a ball mill type pulverizer is again different from a control standpoint. This is shown in Figure 10.6p, including the application of manual compensation for the number of pulverizers in service.

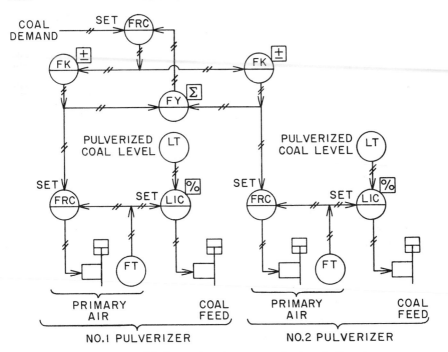

Fig. 10.6n   Coal fuel controls utilizing two pulverizers (ball type)

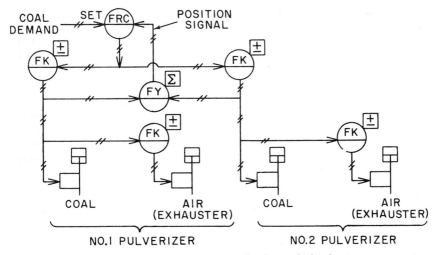

Fig. 10.6o   Coal fuel control utilizing bowl type dual pulverizers

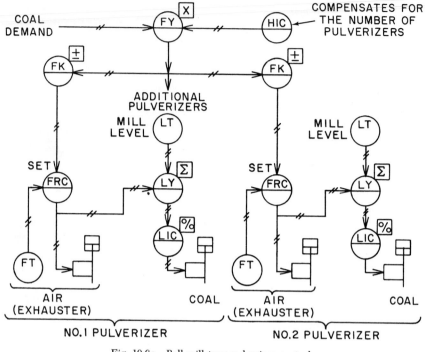

Fig. 10.6p   Ball mill type pulverizer controls

Air Control and Measurement

Combustion air for steam boilers may be supplied by induced draft (suction fan at boiler outlet or stack draft), forced draft (pressure fan at inlet) or a combination of forced and induced draft known as balanced draft fans. With balanced draft boilers, a slight negative pressure is maintained in the furnace.

In the control of combustion air (if there are both forced and induced draft fans), one fan should be selected for basic control of air flow and the other assigned to maintaining the draft pressure in the furnace. The following discussion is based on a single air flow source (fan) per boiler. Balanced draft and its effects on air flow control will be covered later in this section.

For successful control of the air-fuel ratio, combustion air flow measurement is important. From a practical point of view it is impossible to obtain ideal flow detection conditions. Therefore one must provide some device in the flow path of combustion air or combustion gases and field calibrate it, by running combustion tests on the boiler.

These field tests, carried out at various boiler loads, use fuel flow measurements (direct or inferred from steam flow), measurements of percent excess air by gas analysis, and they also utilize the combustion equations to determine air flow. Since we are concerned with a relative measurement with respect to fuel flow, the air flow measurement is normally calibrated and presented on a relative basis. Flow vs differential pressure characteristics, compensations for normal variations in temperature, and variations in desired excess air as a function of load—all are included in the calibration. The desired end result is to have the air flow signal match the steam or fuel flow signals when combustion conditions are as desired.

The following sources of pressure differential can be considered:

> Burner differential
>> Windbox pressure minus furnace pressure
> Boiler differential
>> Differential across baffle in combustion gas stream
> Air heater differential
>> Gas side differential on air heater
> Air heater differential
>> Air side differential on air heater
> Venturi section or flow tube
>> Installed in stack
> Piezometer ring
>> At forced draft fan inlet
> Venturi section
>> Section of forced draft duct
> Orifice segment
>> Section of forced draft duct
> Air foil segments
>> Section of forced draft duct

Of these, the most desirable are the last four, because they use a primary element designed for the purpose of flow detection and measure flow on the clean air side. Some typical installations are shown in Figure 10.6q. For a detailed discussion of flow measurement, refer to Section V in the first volume.

It is often necessary to obtain a total measurement signal, which totalizes two or more flows, because of the boiler and duct work configurations.

Regardless of the type of primary element used, the signal obtained will be noisy due to pulsations from the pumping action of the fans or from the combustion process. Provision should be made for dampening the flow signals, for otherwise the controller can not be tuned for sensitive response. Normal differential pressure ranges for these measurements are between 1 and 6 in. of water, with one in. usually considered as minimum.

Control devices for boiler air flow control on pneumatic installations are double-acting pistons, but in some cases electric motors are also used. In either case linear relationship is required between control signal and combustion air flow. Characteristics of most constant-speed fans or dampers approximate

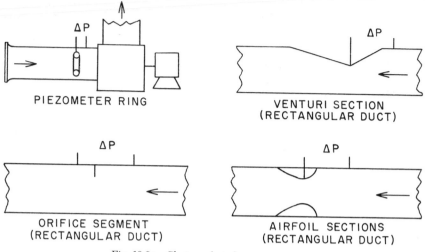

PIEZOMETER RING

VENTURI SECTION
(RECTANGULAR DUCT)

ORIFICE SEGMENT
(RECTANGULAR DUCT)

AIRFOIL SECTIONS
(RECTANGULAR DUCT)

Fig. 10.6q   Choices of air flow measurement

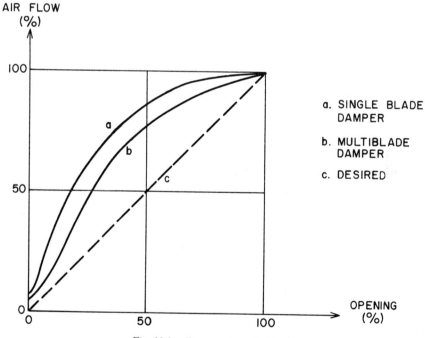

AIR FLOW
(%)

a. SINGLE BLADE
   DAMPER

b. MULTIBLADE
   DAMPER

c. DESIRED

OPENING
(%)

Fig. 10.6r   Damper characteristics

those given in Figure 10.6r. (For more details on dampers, refer to Section 2.4).

Damper and Fan Controls

The relationships between open and closed-loop control that were noted in connection with fuel control, also exist in air flow control. Closed-loop air control may be more precise, and self-linearizing, if the integrated system does not compare air flow with fuel flow (or as inferred from steam flow) directly in a ratio or difference controller.

Open-loop air control variations that may be used depending on the arrangement of fans are as shown in Figure 10.6s through w. Figure 10.6s illustrates a single damper-controlled open loop. Figure 10.6t shows a combination of damper and speed control. A system of this sort is often necessary to increase turn-down (rangeability), where fan speed is variable. Good response of air flow based on fan speed adjustment alone is normally not attainable below about $\frac{1}{3}$ maximum speed. Depending on fan design, this may correspond to 50 percent of boiler capacity. Use of a damper in combination with speed adjustment allows further turndown, since fan speed is blocked at approximately $\frac{1}{3}$ speed. Split ranging, as shown in Figure 10.6t, conserves steam or fan power.

Due to inlet damper leakage that is normally present, it may be necessary

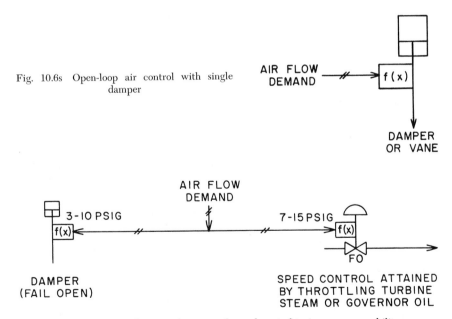

Fig. 10.6s   Open-loop air control with single damper

Fig. 10.6t   Combination damper and speed control to increase rangeability

for wide-range low-load or startup control to parallel inlet and outlet dampers. To save fan power the inlet damper may be operated over the full 3 to 15 PSIG range, while the discharge damper can be fully open at 3 PSIG and closed at 9 PSIG.

As shown in Figure 10.6u, when operating two fans in parallel or on single-fan operation, the idle fan should have its damper closed to prevent recirculation from the operating fan.

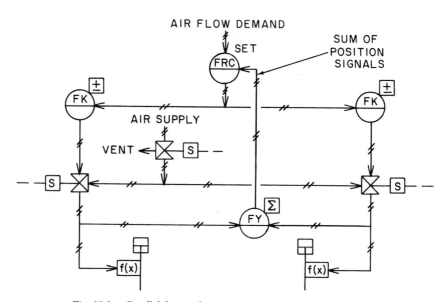

Fig. 10.6u   Parallel fans with automatic air flow balancing controls

When one fan of a two-fan system is switched on or off, considerable operator manual operation is required to prevent serious air flow upsets.

The system shown in Figure 10.6u eliminates this problem by automatically compensating the operating fan damper position as the parallel fan is started up or shut down. Separate discharge dampers may be used for shutoff purposes, supplementing the interlocks shown.

Closed-loop versions of the loops illustrated in Figures 10.6s, t and u would consist of flow controllers with air flow feedback superimposed on the components shown. For example, Figure 10.6v shows the closed-loop version of Figure 10.6u.

Furnace Draft Control

Whenever both forced draft and induced draft are used together, at some point in the system the pressure will be the same as that of the atmosphere. Balanced-draft boilers are not normally designed for positive furnace pressure.

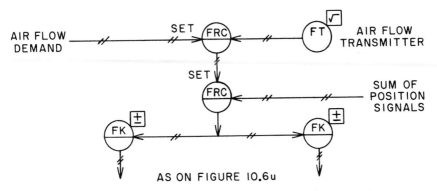

AIR FLOW DEMAND

Fig. 10.6v   Closed-loop control of parallel fans with balancing controls

Therefore the furnace pressure must be negative to prevent hot gas leakage. Excessive vacuum in the furnace, however, produces heat losses through air infiltration. The most desirable condition is thus one where there is a very slight (about 0.1 in. H_2O) negative pressure at the top of the furnace.

Pressure taps for measuring furnace pressure may be located some distance below the top. Due to the chimney effect of the hot furnace gases, pressures measured below the top of the furnace will be lower by approximately 0.01 in. of H_2O per foot of elevation. Thus if the pressure tap is 20 ft below the top of the furnace, the desirable pressure to maintain is approximately 0.3 in. H_2O vacuum.

In the case of a balanced draft boiler the maintenance of constant furnace pressure or draft keeps the forced and induced draft in balance. The purpose of this balance is to share properly the duty of providing combustion air, and to protect furnaces not designed for positive pressure operation.

The measurement of furnace draft produces a noisy signal, limiting the loop gain to relatively low values. In order to provide control without undue noise effects, it is desirable to use a full span of approximately 4 to 5 in. of water. This is normally a compound range such as $+1$ to -4, or $+2$ to -3, in. of water. Even with this span and with a set point of -0.1 to -0.3 in. of water, the controller gain can still not exceed 1.0. In some cases it may be necessary to use only the integral control mode to stabilize the loop.

Additionally stability problems and interactions may occur in the overall system due to measurement lags. It is recommended that the pressure transmitter connections to the boiler furnace be made with at least 1-in.-diam pipe, due to the very low pressures involved. If the distance is less than 25 ft, $\frac{3}{4}$-in. pipe may also be used.

Either the forced or induced draft fan can be used to control the furnace draft, with the other fan performing the basic air flow control function. Interaction cannot be completely eliminated between these two loops, but it can be minimized by system design, as shown in Figure 10.6w.

The common rule is that air flow should be measured and controlled on the same side (air or combustion gas) of the furnace to minimize interaction between the flow and pressure loops.

FURNACE
DRAFT

Signal to Forced Draft Fan
If air flow is detected by
   (a) boiler differential,
   (b) gas side ΔP of air heater,
   (c) venturi in stack,
and if air flow is controlled by
throttling the induced draft fan.

Signal to Induced Draft Fan
If air flow is detected by
   (a) air duct venturi,
   (b) air duct orifice,
   (c) air side ΔP of air heater,
   (d) piezometer ring in forced
      draft fan inlet,
and if air flow is controlled by
throttling the forced draft fan.

Fig. 10.6w   Choices of furnace draft control

Another design reduces interaction by connecting the two fans in parallel and using the furnace pressure as a trimming signal. (This can also be used to overcome problems resulting from (1) noisy furnace pressure signals and (2) slow response due to the series relationship between flow and pressure loops). This control system is illustrated in Figure 10.6x.

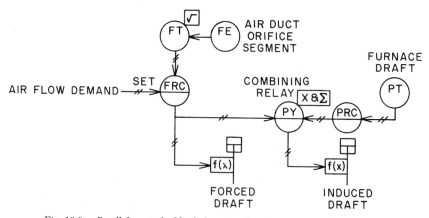

Fig. 10.6x   Parallel control of both fans to reduce interaction, noise and lags

This simple illustration may become more complex, due to fan and damper complexities—in the same manner that the system in Figure 10.6s evolved into Figure 10.6u and for the same reason.

Fuel-Air Ratio

In considering the controls for fuel-air ratio, one consideration is very important. Due to the combustion gas velocity through the boiler, for safety reasons the fuel-air ratio should be maintained on an instant-by-instant rather than a time-averaged basis.

As a general rule (except for very slow-changing boiler loads), fuel and air should be controlled in parallel rather than series. This is necessary because a lag of only one or two seconds in measurement or transmission will seriously upset combustion conditions in a series system. This can result in alternating periods of excess and deficient combustion air. Consequently the discussion here will be limited to parallel air and fuel control systems.

The simplest control of fuel-air ratio is with a system calibrated in parallel, with provision for the operator to make manual corrections (Figure 10.6y). In this system the operator uses the bias provision of the panel station (FK) to compensate for variations in fuel pressure, temperature, heating value or for air temperature, humidity, etc. A system of this sort should be commissioned with detailed testing at various loads for characterizing and matching fuel and air control devices. In addition, simple systems of this type should be adjusted for higher excess air, since they have no means of automatically compensating for the fuel and air variations.

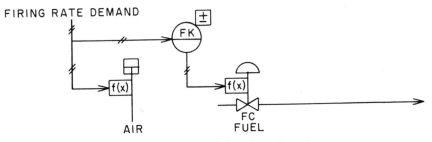

Fig. 10.6y   Simple parallel air-fuel ratio system

The next higher degree of sophistication is a system with simple proportional compensation. This can be done by balancing fuel burner pressure to the differential produced between windbox and furnace pressures.

In the system shown in Figure 10.6z the burner fuel apertures are used to measure fuel flow, and the burner air throat is used as a primary element to detect air flow. There is no square root extraction (such extraction would not show true flows, due to the nature of the primary elements), and therefore

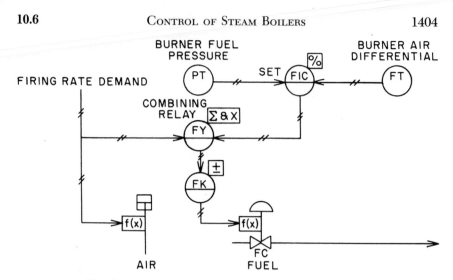

Fig. 10.6z   Proportional compensation in air-fuel ratio control

the actual controller loop gain changes with load (capacity). Rangeability is limited unless there are multiple burners which can be put into or taken out of service.

The arrangement in this control system can also be reversed, with firing rate demand directly adjusting the fuel, and the correction control being on air flow. Consideration of this choice will be taken up later in this section.

As boilers become larger, the need for precision control becomes greater, together with the potential for savings. The following series of diagrams represent further degrees of system sophistication.

The system in Figure 10.6aa is quite similar to that shown in Figure 10.6z,

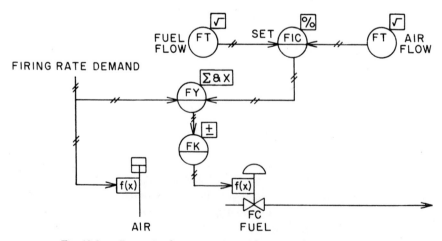

Fig. 10.6aa   Proportional compensation with accurately measured flows

except that here flows are measured as accurately as possible and are used in a flow controller to readjust the primary loop through the combining relay (FY).

An advantage of this system when used with long transmission lines or with maximum turndown is that fuel and air flow are open-loop controlled, and only secondary use is made of their measurements. A disadvantage is that fuel disturbances need master controller action for correction, and the fuel and air loop can interact with each other. The effect of interaction and disturbances in the fuel and air control loops can be minimized by using closed-loop fuel and air control. Except when long pneumatic transmission lines or wide turndown ratios exist, the system shown in Figure 10.6bb (which uses essentially the same equipment) is more desirable, from the non-interacting, self-linearizing standpoint.

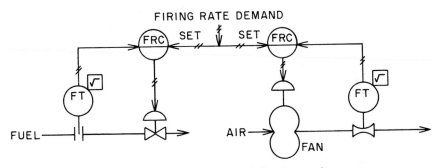

Fig. 10.6bb   Closed-loop air-fuel ratio control

In the system shown in Figures 10.6aa and 10.6bb, fuel flow and air flow signals for proper fuel-air ratio are matched. This is done by matching air to fuel in the field combustion test calibration of the air flow measurement. Field testing is less stringent with the system shown in Figure 10.6bb, since the self-linearizing feature of the closed-loop system reduces the work to characterize the fuel and air-control devices.

The systems shown here are for measurable fuels. In burning coal, fuel flow is often inferred from steam flow, and the steam flow–air flow relationship is used as a control index. Steam flow, however, is not a good fuel flow index to use in closing a fuel flow control loop. Therefore when an unmeasurable fuel is burned, the system using the combining relay in Figure 10.6aa should be used.

A manual adjustment of the fuel-air ratio can be provided with either of the systems illustrated in Figures 10.6aa and bb. This adjustment can be inserted as shown in Figure 10.6cc to keep from changing the gain of the loop as adjustments are made. If ratio adjustment is manual, flue gas analysis systems such as oxygen and combustibles analyzers should be provided for

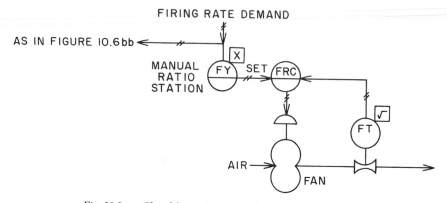

Fig. 10.6cc   Closed loop with manually adjusted ratio station

the operator's guidance or as feedback control of the ratio desired. A measurement of percent oxygen in the combustion gases is most necessary for operator guidance when multiple fuels are being burned or when fuel properties vary. (A detailed discussion on oxygen analyzers is given in Section 8.7 of Volume I.)

Although fuel is measurable and can be totalized (on an air-required basis), variations in fuel BTU content, specific gravity, temperature, and other physical properties of either fuel or air will disrupt the fuel-air ratio. This results in safety and fuel economy problems. Where the boiler is large, the fuel savings are greater, and therefore automatic ratio adjustment based on oxygen analysis should be provided. If this correction is required, it should be provided with means of programming the percent oxygen set point as a function of boiler load. In practically all boilers, desired excess air is not constant, but decreases as load is increased. An example of this relationship is shown in Figure 10.6dd.

In designing a control system to follow the desired curve, provision should be made for shifting the curve to the right or left. Figure 10.6ee shows an example of automatic fuel-air ratio correction based on load and excess air indicated by percent oxygen.

To obtain some of the advantages of the closed fuel loop system, noninteracting oxygen analysis may be used to calibrate continuously the inherently poor fuel flow signal, if it could not otherwise be used with accuracy. An example showing how a satisfactory coal flow signal can be obtained by continuously calibrating a summation signal of pulverizer feeder speeds is shown in Figure 10.6ff. (See Section 2.3 for pump and feeder speed controls.)

Maintaining the correct fuel-air ratio also contributes to limiting the fuel rate to available air, limiting air minimum and maximum to fuel flow, air leading fuel on load increase and air lagging fuel on load decrease. These subjects are covered in the succeeding paragraphs.

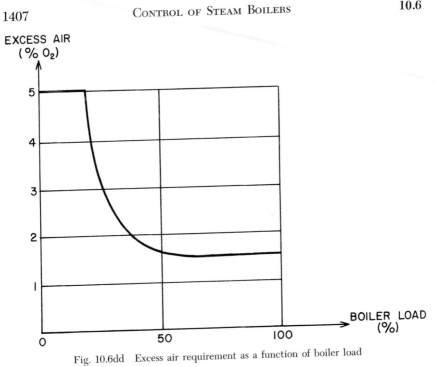

Fig. 10.6dd   Excess air requirement as a function of boiler load

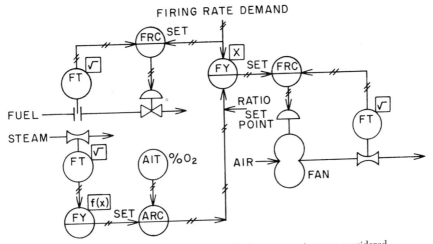

Fig. 10.6ee   Air-fuel ratio control, with load vs excess air curve considered

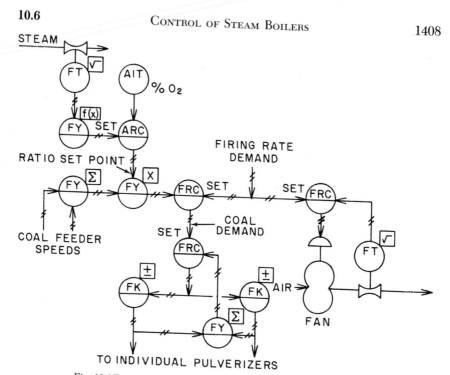

Fig. 10.6ff   Load vs excess air correction on boiler with coal fuel

Fuel and Air Limiting

Though not normally furnished on smaller boilers, the following limiting actions are desirable for safety purposes:

1. Limiting fuel to available air flow
2. Minimum limiting of air flow to match minimum fuel flow or to other safe minimum limit.
3. Limiting minimum fuel flow to maintain stable flame.

These limiting features are simple to apply with the basic non-interacting self-linearizing system in Figure 10.6bb. Figure 10.6gg shows the necessary modifications that can provide these features at low additional cost and without upsetting the set point of the fuel-air ratio. The following is accomplished by the illustrated system:

1. If actual air flow decreases below firing rate demand, then the actual air flow signal is selected to become the fuel demand, by low selector (FY-1).
2. If fuel flow is at minimum and firing rate demand further decreases, actual fuel flow becomes the air flow demand, because FY-2 will select the fuel signal if it is greater than firing rate

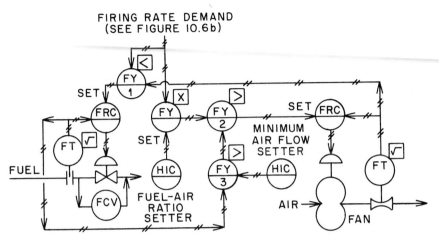

Fig. 10.6gg   Parallel, closed-loop control with air and fuel limiting

demand signal. A manual air flow minimum is also available to come into use through FY-3, such that if fuel flow signal drops below the HIC setting, this manual setting will become the air flow set point.

3. Fuel is minimum limited by separate direct-acting pressure or flow regulator (FCV).

In open-loop systems these limits are more difficult to apply. The application of these limits to an open-loop system is shown in Figure 10.6hh. Here the fuel set point is determined (limited) by actual air flow, and a "fuel cutback" due to reduced firing rate demand is accomplished at the expense of a temporary fuel-air ratio offset.

Limiting combustion air flow to a minimum or to the rate at which fuel is being burned creates special problems, because when the limit is in force, provision must be made to block the integral action in the air flow controller.

It may seem that a better way to limit fuel would be to have the firing rate demand directly set the air flow, with fuel being controlled through the combining relay (Figure 10.6ii). In this arrangement final fuel-air ratio correction is through the integral mode correction of the fuel-air ratio controller. This system is only partially effective, however, because on a sudden decrease of firing rate demand, the resulting reduction in fuel flow will occur only after the air flow has already been reduced.

A further consideration in setting fuel rather than air directly by the firing rate demand is that the parallel boilers can be more easily kept in balance, because balancing fuel directly balances the heat input without consideration of excess air between boilers.

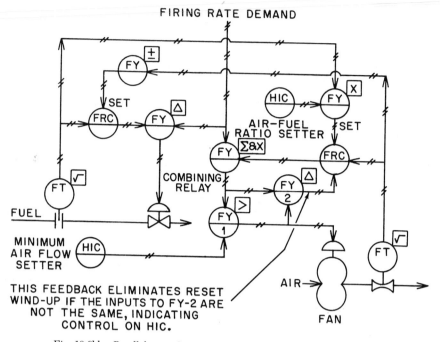

Fig. 10.6hh   Parallel, open-loop control with fuel and air limiting applied

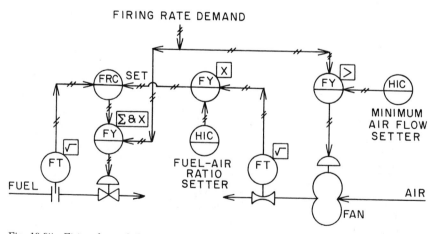

Fig. 10.6ii   Firing demand determines air flow while fuel set point is adjusted by air flow

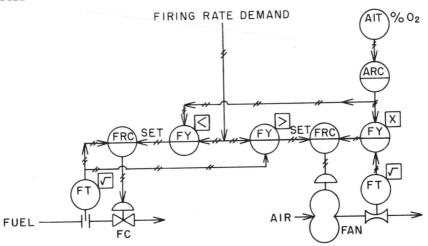

Fig. 10.6jj Feedforward control system which automatically maintains excess air during upsets

Figure 10.6jj illustrates a closed-loop control system corrected by oxygen analysis and provided with safety limits to protect against air deficiency.

Feedwater Control

Feedwater control is the regulation of water to the boiler drum. This water is admitted to the steam drum and after absorbing the heat from the furnace generates the steam produced by the boiler.

Proper boiler operation requires that the level of water in the steam drum be maintained within a certain band. A decrease in this level may uncover boiler tubes, allowing them to become overheated. An increase in this level may interfere with the operation of the internal devices in the drum which separate the moisture from the steam.

The water level in the steam drum is related to, but is not a direct indicator of, the quantity of water in the drum. At each boiler load there is a different volume in the water that is occupied by steam bubbles. Thus as load is increased there are more steam bubbles, and this causes the water to "swell," or rise, rather than fall, due to the added water usage. Therefore if the drum volume is kept constant, the corresponding mass of water is minimum at high boiler loads and maximum at low boiler loads. The control of feedwater therefore needs to respond to load changes and maintain water by constantly adjusting the mass of water stored in the system.

Control of feedwater addition based on drum level alone tends to be self-defeating since, on a load increase, it tends to decrease water feed, when it should be increasing. Figure 10.6kk shows the response relationship between steam flow, water flow and drum level that a properly designed system should accomplish if constant level under variable load is desired. For special reasons

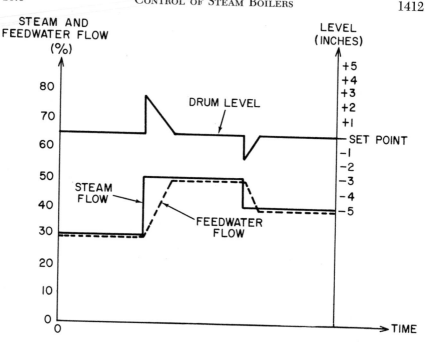

Fig. 10.6kk   Response relationship between steam flow (load), feedwater flow (manipulated variable) and level (controlled variable)

one may wish to increase level with load. Boilers are designed, however, for constant level operation.

Single Element

For small boilers having relatively high storage volumes and slow-changing loads, a simple proportional control may suffice, imprecise as it is. Integral action should not be used, because of resulting instability, which is due to integration of the swell on load changes which must later be removed. Control of this type therefore involves the addition of feedwater on straight proportional level control.

Two Elements

For larger boilers and particularly where there is a consistent relationship between valve position and flow, a two-element system (Figure 10.6ll) can do a very adequate job under most operating conditions. Two-element control is primarily used on intermediate size boilers, where volumes and capacities of the steam and water system would make the simple level control inadequate because of "swell." Smaller boilers, where load changes may be rapid, frequent or of large magnitude, will also require the two-element system.

Field testing, characterization and adjustment of the control valve are

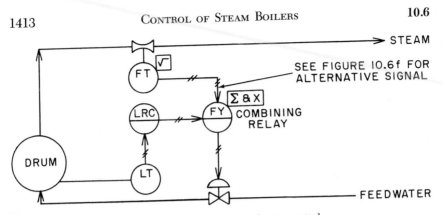

Fig. 10.6 ll   Two-element feedwater control

required so that the relationship of *control signal to feedwater valve flow matches that of the steam flow to the flow transmitter output.*

Any deviations in this matching will cause a permanent level offset at the particular capacity and less than optimum control (Figure 10.6mm). The level controller gain should be such that, on a load change, the level controller output step will match the change in the steam flow transmitter signal.

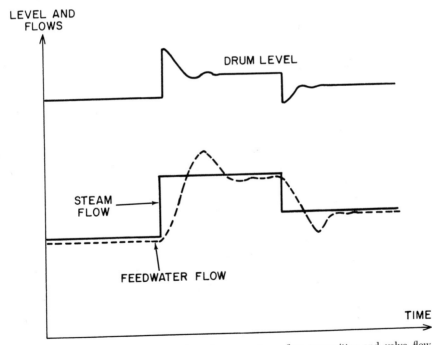

Fig. 10.6mm   The effect of mismatching between steam flow transmitter and valve flow characteristics

Three Elements

As boilers become greater in capacity, economic considerations make it highly desirable to reduce drum sizes and increase velocities in the water and steam systems. Under these conditions the boiler is less able to act as an integrator to absorb the results of incorrect or insufficient control. A three-element system is used on the largest and most sophisticated drum type boilers.

Three-element control is similar to the two-element system except that the water flow loop is closed rather than open. There are several ways of connecting a three-element feedwater system, each of which can produce the results shown in Figure 10.6kk. Figure 10.6nn illustrates various ways of connecting this system.

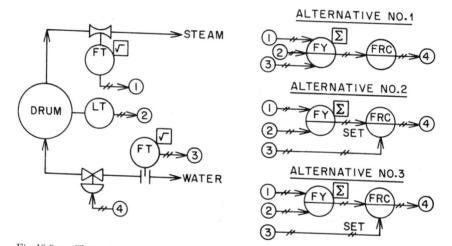

Fig. 10.6nn   Three-element feedwater system. FY is provided with independent gain adjustment capability for level and flow.

In making gain adjustments on a three-element feedwater system, the first step is to determine the relative gains between level and flow loops. By observing a change in boiler load the particular boiler "swell" characteristics can be observed. Maximum system stability results when the negative effect of swell equals the positive effect of flow. For example: If a 20 percent of maximum flow change produces a 2.4 PSI change in flow transmitter output, and this flow change also produces a 3-in. swell on a 30-in. range transmitter or a 1.2 PSI transmitter output change, then the gain of the level loop should be double the gain on the flow loop.

Valve Sizing

Square-root-extracted flow signals are required for calculation flexibility. To keep the controller gain independent of load variations the feedwater valve should have linear characteristics. The simplest way to assure this is by using a characterizing positioner on the control valve. It may also be required to dampen the control valve, to make it less susceptible to noisy control signals.

For control valve sizing a system "head" curve showing the relationship between system pressures and capacities should be developed. A typical head curve is shown in Figure 10.600.

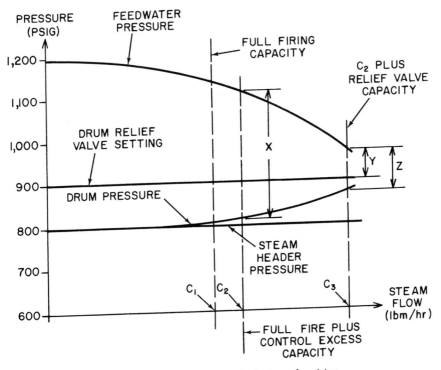

Fig. 10.600　Head curve for feedwater valve sizing

The "head" curve demonstrates a basic question in feedwater control valve sizing. What capacity and pressure drop should be the sizing basis? Capacity C_2 and differential X are the most desirable from a control standpoint. Capacity C_3 and differential Y or Z are often used in an attempt to satisfy a boiler code requirement. The code requires that sufficient water be furnished to the drum with safety valves blowing.

This code requirement is interpreted as full capacity of boiler plus safety valve capacity even though maximum firing rate could not hold the pressure. Safety valves will not blow unless the firing rate is higher than capacity. The code does not require all the capacity to be provided in the feedwater control valve; therefore there is no need to degrade the control quality by furnishing an oversized valve.

It is not uncommon to see a valve that was designed for more than required capacity, and for a 30 PSI differential, operating at a 500 PSI differential and at a fraction of its design capacity. The duty required of a feedwater control valve is quite heavy because of the large energy dissipation as the water passes through the valve. Very pure water as normally used in boilers tends to be "metal hungry." This, combined with water velocity, produces a corrosion-erosion (or cavitation) effect that calls for a chrome-moly steel valve body when velocities are in excess of 10 fps. (See Section 1.2 for details on valve sizing.)

Pump Speed Control

Control of pump speed to regulate feedwater flow can be accomplished if the pump is driven by a steam turbine or is furnished with a scoop tube hydraulic coupling. This can be used in place of a control valve to save pump power on single boiler systems. In systems where several boilers are operating in parallel, the speed control can be used to save pump power by controlling the discharge pressure at a constant differential pressure above boiler pressure.

When pump speed is being used in place of a feedwater valve on a single boiler system, a large part of the speed control range is used in developing pump head at low flow. Characterization of the signal is thus necessary for good operation and constant gain throughout the operating range.

This is demonstrated in Figure 10.6pp, showing the pump characteristics. The reader will note the control ranges used for 0 load and from 0 to full load. (Further discussions are presented in Sections 2.3 and 10.12.)

Steam Temperature Control

The purpose of steam temperature control is to improve the thermal efficiency of steam turbines. Its most common application is for steam turbine electric power generation. The factors affecting steam temperature in a convection type superheater are superheater area, flue gas flow pattern across the superheater, flue gas mass flow, temperature of flue gasses leaving the furnace, and steam flow through the superheater. Additionally, furnace temperature affects a radiant superheater. Some superheaters may be designed for a flat curve combining radiant and convection surface, but most superheaters are the convection type.

Desuperheater Spray Controls

To use a desuperheater spray for steam temperature control, the boiler would normally be provided with added superheater area. Figure 10.6qq demonstrates the effect of the water spray (which is usually between a primary and secondary superheater section) for temperature control.

Provision must be made to prevent reset wind up when in the uncontrolled load range. Control of this system is shown in Figure 10.6rr. Control by desuperheat controls is used by several boiler manufacturers.

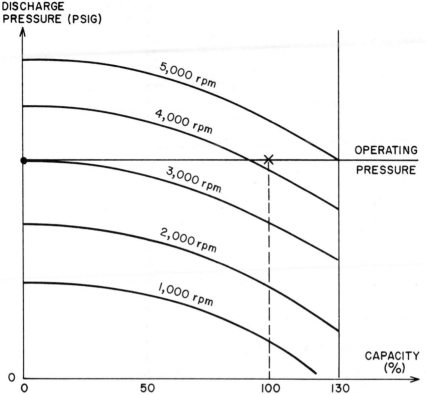

Fig. 10.6pp   Feedwater pump on speed control. Speed range at zero capacity to reach operating pressure is 0–3000 rpm. Speed range for zero to 100% capacity is 3000–4200 rpm.

Large boilers may have burners that tilt up and down approximately 30 degrees for steam temperature control. This effectively changes the furnace heat transfer area, resulting in temperature changes of flue gases leaving the boiler. Spray is frequently used with these systems as an override control. These systems are used on large power plant boilers and normally require combinations of feedforward and cascade control.

Variable Excess Air

Since mass flow of flue gas affects steam temperature, variation in excess combustion air can be used to regulate steam temperature. This method may be utilized on a boiler which was not specifically designed for temperature control. Though increasing excess air flow increases boiler stack heat losses, turbine thermal efficiency will also increase, and maintaining steam temperature provides greater economic benefits. The control arrangement in Figure 10.6ss implements this method.

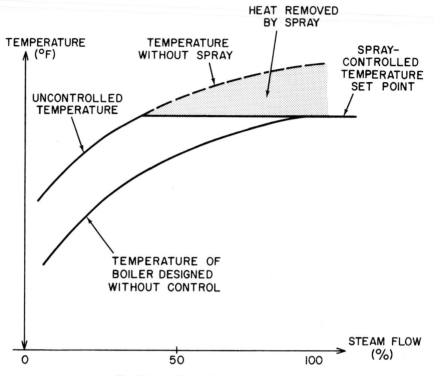

Fig. 10.6qq   Desuperheater characteristics

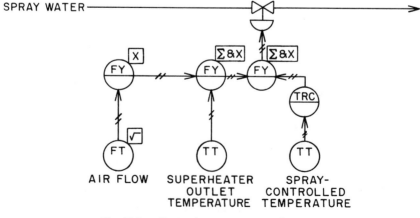

Fig. 10.6rr   Desuperheater spray control system

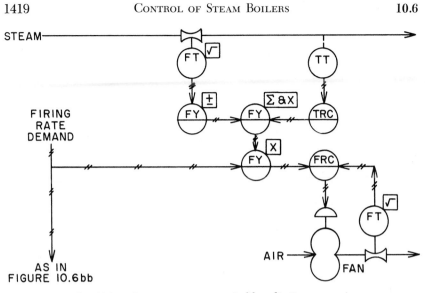

Fig. 10.6ss   Steam temperature control by adjusting excess air

Integration of Loops

By combining the loops and subsystems shown in Figures 10.6b, gg, nn and rr, an integrated control system is achieved. Other combinations of the subsystems can similarly be put together to form a coordinated system. When designing, it is well to break the overall system down into these subsystems and examine them individually. Only then should the subsystems be put together into the total system.

Major design and operation problems in complex systems are created by "tie-backs."

In a complex system, due to adding, subtracting, multiplying, dividing and comparing control signals, transmitter signals, etc., it is a major problem to get the system on "automatic" control easily and quickly.

The chief points to remember are these:

1. The systems often interact, e.g., air flow affects steam temperature, feedwater flow affects steam pressure, and fuel flow affects drum level and furnace draft. Designs with the minimum of interaction are the most desirable.
2. For flexibility and rangeability linear flow signals are necessary. Control valves and piston operators need linearizing positioners.
3. Fuels should be totalized on an air-required basis.
4. Tie-back arrangements, which simplify the task of getting quickly on automatic control, are very important in complex systems.
5. The flows of fuel and air should be controlled such that the flow rates reaching the burner always represent a safe combination.

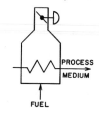

10.7 CONTROL OF FURNACES

Furnaces and heaters are devices in which heat energy is transferred to a charge or feed in a controlled manner. The typical furnace, or heater, (the terms will be used interchangeably here) usually takes the form of a metal housing lined with a heat-conserving refractory. The charge can enter as a solid, liquid or gas, and may or may not be transformed to a different state by the energy supplied. The charge can be carried through the furnace or heater continuously, through metal tubes or troughs, or may be batch-heated by remaining stationary after entering. The functions of furnaces can be broken down into three categories:

1. To heat and/or vaporize the charge
2. To provide heat of reaction to reacting feeds
3. To provide an elevated and controlled temperature for the physical change of charge materials (i.e., change of grain structure).

An example of a heating and vaporizing furnace is a refinery crude oil heater, where crude oil is heated and partially vaporized preparatory to distillation. Typical of a furnace in which reaction heat is supplied is an ammonia reformer furnace, where hydrocarbon vapors and steam are reacted to form hydrogen and carbon monoxide. An annealing furnace is an example of a furnace where physical change takes place. A metal charge is conveyed through or placed in the furnace for a fixed time at various temperatures, resulting in the desired final grain structure. In many instances a combination of these functions are carried out in the same furnace. For example, a reformer furnace first brings the charge up to reacting temperature and then supplies the reaction heat.

Control System Functions

The primary functions of furnace control systems are;

1. To maintain the desired rate of energy transfer to the charge
2. To maintain a controlled and efficient combustion of fuel
3. To maintain safe conditions in all phases of furnace operation.

The control system must ensure that the charge receives the heat energy at the proper rate. Most commonly, the temperature of the charge is the index used as the measure of the heat transferred. Where vaporization occurs along with the transfer of sensible heat, the charge temperature may not be a good index. In this case the total final heat content of the charge is important, but this is a more difficult condition to measure.

Proper combustion of fuel, the second purpose of the control system, involves many factors, such as the regulation of combustion air, the preferential burning of one fuel over another, and the control of atomizing steam when fuel oil is burned. Such factors will be discussed more fully under the description of furnace types.

Safety is always an important consideration in any process, but it takes on added importance in furnaces where combustion takes place. There always exists the possibility of an explosive mixture being formed in the furnace, which can have disastrous effects. Where a combustible charge is being handled, the danger is even greater. Although such a condition can develop during normal operations, its likelihood is greatest during furnace startup and shutdown. To minimize these hazards, industry and local government agencies have developed various codes and practices. These tend to set minimum standards in control equipment and in acceptable startup and operating practices. Strong reliance is placed on instrumentation to warn and/or shut down the unit if a dangerous condition arises. Local codes and industry practices, usually set by the insurers of the plant, should be consulted before firing controls are designed and installed. Safety controls are also discussed in Section 10.6.

One danger that exists in any control system using air-operated control valves is the loss of motivating instrument air. If this occurs, the valves will usually go completely open or completely closed. The failure positions are shown on the control system sketches next to each valve symbol. FC indicates that the valve fails closed and FO indicates it fails open. The air failure actions are chosen by the instrument engineer to give as safe an emergency operation as possible. The choice is dependent on the valve function in the process stream.

The majority of furnaces in operation today, burn some type of fossil fuel (such as fuel gas, fuel oil and coal), and the selection of the particular fuel or combination of fuels is generally a matter of overall plant economics. Some of the factors to be considered in selecting a fuel are the lower thermal efficiency of natural gas vs the inefficiency in burning fuel oil due to poor atomization and need for more excess air.

Many installations burn gas and oil simultaneously or with one or the other as a stand-by. The more common approach is to burn fuel gas preferentially because of the difficulty in storing it, and to use fuel oil as the stand-by. For these reasons, the systems described in this section will cover arrangements for both fuel gas and fuel oil firing as well as combination firing (i.e., firing both fuels simultaneously). Coal firing is rarely used in process furnaces but is commonly used in electric power generation stations and metallurgical furnaces. (For a discussion on coal firing, refer to Section 10.6.)

Combustion Air Requirements

To completely burn the combustible portion of any given fuel, there is an ideal quantity of oxygen, and therefore an ideal quantity of air, required. Under real conditions, the inefficiencies in combustion require that some additional air be added to ensure complete combustion. The quantity over and above the ideal amount of air is called the "excess air," and it is usual to talk about "percent excess air" which is the ratio of excess air to ideal combustion air times one hundred.

Each fuel has a practical minimum limit on percent excess air, and probably the most important function of the firing control system is to maintain the excess air percentage above that minimum. The excess air requirements are usually satisfied by maintaining a suitable furnace draft (i.e., a suitable low pressure at the stack) to ensure that sufficient air is drawn into the combustion chamber. For more accurate measurement of excess air, the measurement of the oxygen content of the flue gases is required (usually by a continuous automatic analyzer) since oxygen content of the flue gas is directly related to percent excess air.

Control Systems

In discussing controls for the more common types of process furnaces, this section is subdivided by type of furnace rather than by the type of control system. Each type of furnace will be described from a process standpoint first, since the process function of the heater naturally determines the furnace operation and therefore the control requirements. Next, the important process variables will be noted and the methods used to measure and control them outlined. The various types of firing controls applicable to the particular type of furnace will then be described. Finally, specific safety considerations and instrumentation will be presented.

The control systems to be outlined have been empirically developed, and most of them are tried and reliable methods. The question of control dynamics is a necessary consideration in most industrial processes, but for the most part, these problems are solved in practice by the use of versatile instrument hardware.

The analog process controller has a wide range of gain adjustment and at least two modes of dynamic compensation (integral and derivative) which also are adjustable over a wide range. Transmitters are available with suppressed ranges and with noise-damping and range-change adjustments that are either built in or provided by component substitution. Final control elements, usually control valves or dampers, may be selected with a wide variety of characteristics, and in many cases these characteristics are field modified by the appropriate cams on the valve operator. These components are fully described in Chapter I.

Normally the instrument engineer can select the proper components to cover the range of system dynamics and final "tuning" of the control loop is done in the field under actual operating conditions. For the great majority of the cases, this approach, although not the most systematic, is most efficient in that little time is expended in lengthy analytical calculations based on questionable assumptions.

For the special cases where it is recognized that long deadtimes or highly interacting process gains will cause instabilities, the use of mathematical analysis and process simulation becomes a necessity.

The intent of this section is to emphasize the process considerations and related control concepts in furnace operation. The instrumentation is shown in symbolic form in the various illustrations. References to actual instrument component types is avoided unless a particular type of instrument is vital to the operation of the system. The various types of instruments for measuring and controlling flow, pressure and temperature utilized in the systems to be outlined are described in detail in Volume I of this Handbook.

Startup Heaters

Startup heaters are treated first because of their simplicity. These units are usually required, as their name implies, at the start up of a process unit, and their span of use is usually from a few hours to at most a few weeks. They are usually vertical, cylindrical units with vertical process tubes along the inner walls and with a single burner centered in the floor. Draft is by natural convection induced by a stack mounted on the top of the heater. Some process systems that require startup heaters are catalytic cracking units to heat up fluidized catalyst beds, ammonia units to heat up the ammonia converter catalysts, and fixed-bed gas-drying units to regenerate the dryer beds.

Process Considerations

The startup heater usually heats an intermediate stream such as air or natural gas, which in turn heats another fluid or solid such as a reactor catalyst. As an example, an ammonia unit startup heater will be described. This unit is used to heat "synthesis gas" (primarily a mixture of hydrogen and nitrogen) which in turn is used to heat the ammonia converter catalyst bed to a temperature of 700°F by burning natural gas. The synthesis gas is recirculated through the catalyst bed gradually bringing it up to operating temperature. After normal operation has begun, the exothermic nature of the reaction keeps the bed at temperature and the startup heater can be shut down.

Process Controls

The important variables are "synthesis gas" flow and temperature. Figure 10.7a shows the necessary controls for the operation of the unit. The flow of synthesis gas is measured by the flow transmitter (FT-1) and is indicated on a flow indicator (FI-1A) mounted on a panel board. The desired gas flow is manually adjusted in the field by operating the hand valve (HV-1) and observing the local flow indicator (FI-1B). The effluent synthesis gas temperature is maintained by the temperature recorder controller (TRC-1), which measures the gas temperature by means of a thermocouple and controls the quantity of fuel gas burned by modulating the control valve (TV-1).

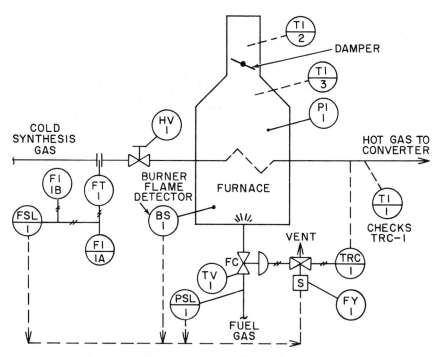

Fig. 10.7a   Startup heater controls

Firing Controls

The fuel gas firing is set by the process temperature controller (TRC-1). The heater draft (i.e., negative pressure in the fire box) is produced by the stack and is set by observing the draft gauge (PI-1), usually an inclined manometer, and by manually adjusting the position of the stack damper. Once initially set, the damper is rarely adjusted again, unless furnace conditions or loads change drastically.

Safety Controls

The major hazards involved with these heaters are (1) the interruption of process gas flow and the resulting possible tube rupture and (2) flame failure and the generation of dangerous fuel-air mixtures in the fire box.

The loss of process gas flow is detected by the low-flow alarm switch (FSL-1), which trips the solenoid (FY-1) to vent the diaphragm on control valve (TV-1). This immediately cuts off fuel and thus terminates firing. Two other devices may cut off the fuel supply in the same manner: the flame detector switch (BS-1), which activates on loss of flame, and the low-pressure switch (PSL-1), which activates on low fuel gas pressure. The loss of flame or the loss of fuel pressure requires that the fuel gas be cut off to prevent the formation of a dangerous air-fuel gas mixture in the fire box, should the fuel gas pressure be restored. The solenoid valve (FY-1) should be the manually reset type (requiring manual opening after trip) to ensure that the control valve stays shut until the operator desires to reopen it. The control valve (TV-1) should be a single-seated valve for tight shutoff, to prevent leakage of fuel into the fire box when it is closed. Failure of instrument air will result in the control valve (TV-1) failing closed (FC) as indicated next to the valve symbol. This is the safest possible action, since it terminates the furnace firing.

Fired Reboilers

The fired reboiler provides heat input to a distillation tower by heating the tower bottoms and vaporizing a portion of it. Normally a tower reboiler uses steam or another hot fluid as a source of heat, but where heat duties are great or where tower bottom temperatures are high, the fired reboiler is used. In construction, depending on its size, the reboiler may be of the vertical, cylindrical type or the larger, conventional horizontal type furnace.

Process Considerations

The fired reboiler heats and vaporizes the tower bottoms as this liquid circulates by natural convection through the heater tubes. The coils are generously sized to ensure adequate circulation of the bottoms liquid. Temperature of the reboiler return fluid is generally used as the means of controlling the heat input to the tower. Overheating the process fluid is a contingency which must be guarded against, since most tower bottoms will coke or polymerize if under excessive temperatures for some length of time.

Process Controls

A common control scheme is shown in Figure 10.7b, which depicts the tower bottom along with the fired reboiler. It is not usually practical to measure the flow of tower bottoms to the reboiler, first, because the liquid is near equilibrium (near the flash point), and second, because it is usually of a fouling nature, tending to plug most flow elements. Proper circulation

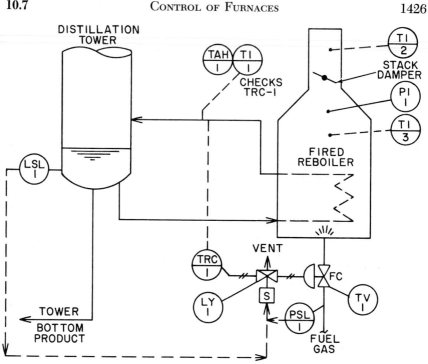

Fig. 10.7b   Fired reboiler controls

of the fluid is provided for in the careful hydraulic design of the interconnecting piping. The other important variable is the reboiler return temperature, which is controlled by TRC-1 throttling the fuel gas control valve (TV-1). The high-temperature alarm, (TAH-1) is provided to warn the operator that the process fluid has suddenly reached an excessive temperature, indicating that manual adjustments are needed to cut back on the firing.

Firing Controls

The firing controls are relatively simple and are similar to those of the startup heater described earlier. The process temperature controller (TRC-1) actuates the fuel valve (TV-1) to satisfy process requirements. The fuel in this case is gas, but it could just as well be fuel oil. The furnace draft is set by means of the stack damper observing the draft gauge (PI-1). The stack temperature (TI-2) and the firebox temperature (TI-3) are detected as a check for excessive temperatures that may develop during periods of heavy firing.

Safety Controls

The major dangers in this furnace type are the interruption of process fluid flow or the stoppage of fuel. The loss of process fluid can occur if the liquid level in the tower bottom is lost. If this occurs, flow will stop in the reboiler tubes and a dangerous overheating of tubes may result. To protect against this, the low-level switch (LSL-1) is wired to trip the solenoid valve (LY-1), which vents the diaphragm of the control valve (TV-1). This causes the fuel valve to close.

A momentary loss of fuel can be dangerous because the flames can be extinguished, and on resumption of fuel flow a dangerous air-fuel mixture will develop in the fire box. To prevent this occurrence, the low-pressure switch (PSL-1) trips the solenoid (LY-1) on low fuel pressure, thereby closing the fuel valve (TV-1). The solenoid valve (LY-1) should be the manual reset type which remains vented, and therefore the valve (TV-1) remains closed even on return of fuel gas pressure.

The air failure action of the control valve (TV-1) is closed (FC) on loss of motivating instrument air. This action on fuel valves is the safe mode, since firing is discontinued during the emergency.

Process Heaters and Vaporizers

The unit feed heater of a crude oil refinery is representative of this class of furnaces. Crude oil, prior to distillation in the "crude tower" into the various petroleum fractions (gasoline, naphtha, gas oil, heavy fuel oil, etc.) must be heated and partially vaporized. The heating and vaporization is done in the crude heater furnace, which consists of a fire box with preheat coils and vaporizing coils. The heating is usually done in coils in the convection section of the furnace, which is the portion that does not see the flame, but is exposed to the hot flue gases on their way to the stack. The vaporizing takes place at the end of each pass in the radiant section of the furnace (where the coils see the flame and the luminous walls of the fire box). The partially vaporized effluent then enters the crude tower where it flashes and is distilled into the desired "cuts." Other process heaters that fall into this category are refinery vacuum tower preheaters, reformer heaters, hydrocracker heaters and de-waxing unit furnaces. The control systems presented will also be applicable to these types of process heaters.

Process Controls

The prime variables with regard to the process are

1. Flow control of feed to the unit
2. Proper splitting of flow in the parallel paths through the furnace (to prevent overheating of any one stream and its resultant coking)
3. Correct amount of heat supplied to the crude tower

Figure 10.7c shows the typical process controls for this type of furnace. The crude feed rate to the unit is set by the flow controller (FRC-2). The flow is split through the parallel paths of the furnace by remote manual adjustment of the control valves (HV-1 through HV-4) via the manual stations

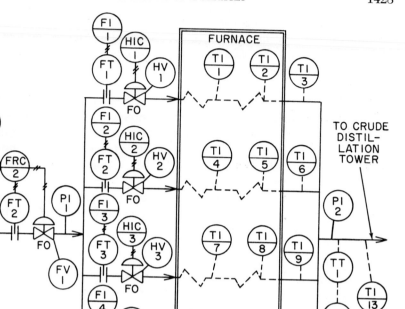

Fig. 10.7c   Crude heater-vaporizer process controls

(HIC-1 through HIC-4). The proper settings are determined by equalizing the flow indications on FI-1 through FI-4. The temperature indicators TI-1 through TI-12 are periodically observed to determine if there are any rising or falling trends in any one of the passes. If such trend develops, the flow through that pass is altered slightly to drive the temperature back toward the norm.

The desired heat input into the feed stream is more difficult to control, because the effluent of the furnace is partially vaporized and the feed stock varies in composition depending on its source. If the feed was only heated, with no vaporization taking place, the control would require only that an effluent temperature be maintained. If complete vaporization and superheating occurred, this too could be well handled by straight temperature control. But in the case of partial vaporization with a variable feed composition, effluent temperature control alone is not a reliable approach. The composition and hence the boiling point curve of the feed is not constant with time, and the required control temperature itself varies. Additional information is thus required, and it is obtained from the distillation downstream of the furnace.

By observation of the product distribution from the fractionation, the need for a change in heat input can be determined. This approach is slow and not precise, but to do a more accurate job would require a great deal of sophisticated instrumentation, involving special analyzers to measure feed composition, and a computer to optimize the mathematical model, and thereby determine the required heat input to the feed. Current practice is to achieve approximate control with a temperature controller (TRC-1), whose set point the unit operator periodically changes to account for feed variations. He depends on his experience and on the results in the fractionator to determine the proper temperature setting.

Fuel Gas Firing Controls

Figure 10.7d shows the firing controls for the furnace using fuel gas only. The local pressure controller (PIC-1) maintains the burner header pressure. The fuel gas headers serve many burners spaced equally along the floor of the fire box. The burners, being essentially fixed-diameter orifices, will pass more or less fuel depending on the header pressure. In order to change the

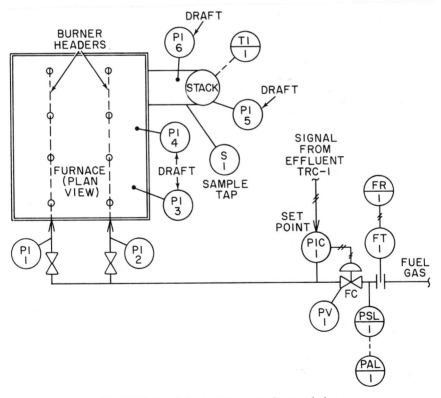

Fig. 10.7d Crude heater firing controls using fuel gas

rate of firing (furnace heat input), the burner header pressure must be altered. This is accomplished by the effluent temperature controller (TRC-1) resetting the set point on PIC-1 as it attempts to maintain a constant furnace effluent temperature. Cascading the TRC-1 output to PIC-1 is beneficial in that the pressure controller compensates for fast local disturbances (i.e., changes in fuel supply pressure) without allowing them to upset the effluent temperature, while the TRC-1 provides the slower correction required to correct for process and ambient disturbances.

The flow recorder (FR-1) is necessary to make efficiency checks on the furnace and also to determine the heat input.

Fuel Oil Firing Controls

Figure 10.7e shows the firing controls for fuel oil. The process effluent temperature control, cascaded to the burner pressure controller (PIC-1), operates as described under fuel gas firing. The major difference with oil firing is the introduction of atomizing steam, which is required to disperse the fuel oil sufficiently to produce efficient burning. The pressure differential controller (PDIC-1) maintains the atomizing steam header at a fixed pressure above the fuel oil header to ensure that it will always have enough atomizing capability. The differential pressure controller (PDIC-1) is necessary because the pressure

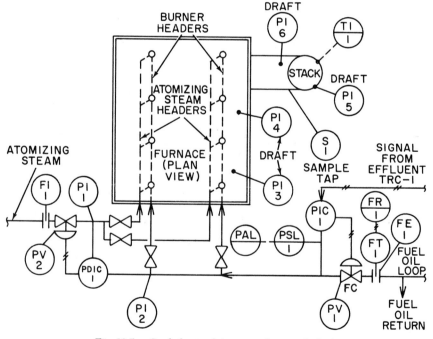

Fig. 10.7e   Crude heater firing controls using fuel oil

in the fuel oil header varies due to the automatic control by TRC-1. PI-1 and PI-2 are required to set PDIC-1 initially and to check its accuracy during operation.

Multiple Fuel Firing Controls

Figure 10.7f shows the more common arrangement for mixed fuel firing with the fuel oil being the make-up fuel and fuel gas being burned preferentially. This situation is common where variable but insufficient amounts of fuel gas are available from a process unit such as a petroleum refinery. The fuel gas pressure is reduced and the burner header pressure is maintained constant by PIC-1. In effect, the amount of fuel gas burned is held constant as long as the fuel gas availability is constant. The set point of PIC-1 is adjusted to change the rate of gas firing. If the pressure in the gas header falls below a predetermined minimum limit for stable burner operation, the fuel gas supply is shut off by the low-pressure switch (PSL-1), tripping the solenoid valve (PY-1) and causing the control valve (PV-1) to close. When this occurs, the full load is taken up by the fuel oil system.

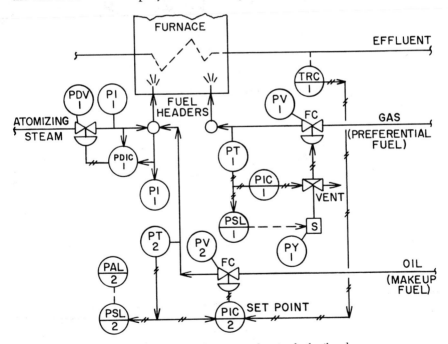

Fig. 10.7f   Crude heater firing controls using both oil and gas

Although the fuel oil is the trim medium, it must be available in sufficient quantity to take the whole load in the event of a fuel gas interruption. The temperature controller (TRC-1) varies the set point of the fuel oil header pressure controller (PIC-2) to satisfy process requirements. On any burner there is a "turn-down" limitation, which is the ratio of minimum to maximum fuel burning capacity (a common ratio is 3:1). Low pressure in the fuel oil header is indicative of the approach of a minimum firing condition; therefore the low-pressure alarm (PAL-2), is used to warn the unit operator when this situation arises. The operator then has the option of going out to the furnace and manually turning off the individual fuel oil burners or reducing the fuel gas firing rate by lowering the set point of PIC-1. Either change will increase the fuel oil demand and thus bring the system into a stable operating zone.

Furnace draft (Figure 10.7e) is normally maintained by free convection in this type of furnace (i.e., the stack produces a negative pressure in the fire box), and air is drawn in through louvers along the sides of the furnace. Furnace draft is set at crucial points in the furnace by means of the draft points PI-3 through PI-6 and by manually setting the louvers on the furnace. These louver adjustments are made initially at startup and are changed only occasionally when firing is being optimized. PI-3 through PI-6 are switchable points on an inclined manometer mounted on the outside wall of the furnace.

The sample tap (S-1) is used to take samples of flue gas to determine its oxygen and combustibles content. This serves to indicate the efficiency of the furnace and to warn of dangerous conditions developing. The samples may be taken periodically and analyzed in the laboratory, or they may be taken continuously and analyzed by an on-stream process analyzer. The analysis of the gas usually is the easiest part of the operation. The delivery of a sample for analysis, properly conditioned (i.e., free of soot and water), is the difficult job. Section 8.7 of Volume I of this Handbook gives a detailed description of this type of analysis.

Safety Controls

The primary hazards in process heaters are

1. The interruption of charge flow
2. The interruption of fuel flow and loss of flame

The interruption of crude charge below a minimum flow will result in over-heating and possible rupturing of the tubes. Resumption of flow may cause hydrocarbon charge leakage into the fire box, with catastrophic results. The low flow alarm (FAL-2 in Figure 10.7c) alerts the operator to this impending condition with the option of correcting the fault or terminating firing. The control valves (FV-1 and HV-1 through HV-4) on instrument air failure will fail open, maintaining the flow through the furnace coils.

Interruption of fuel can cause burner flameout, and the resumption of fuel flow can cause an explosive mixture in the fire box. Flame failure can be detected by sensors (described in Section 10.9 of Volume I), and the shutdown of fuel firing can be automated. This approach is not generally followed in process heaters because of the expense (there can be scores of burners to scan) and because operation is relatively steady (startups and shutdowns occur a year or two apart). Usually a low-pressure alarm (PAL-1, Figure 10.7d) on the fuel header is relied upon to warn the operator of fuel loss, leaving the shutdown decision to him.

On instrument air failure, the fuel firing valve PV-1 closes, shutting down the furnace firing.

Reformer Furnaces

The purpose of a typical reforming furnace, the primary reformer in an ammonia plant is to produce hydrogen for combination with nitrogen in the synthesis of ammonia (NH_3). Hydrogen is produced by reforming hydrocarbon feed (usually methane or naphtha) using high-pressure steam, as shown in the reaction

$$CH_4 + H_2O = 3H_2 + CO \qquad 10.7(1)$$
$$\text{(Methane)} \quad \text{(Steam)} \quad \text{(Hydrogen)} \quad \text{(Carbon monoxide)}$$

General: $\qquad C_nH_{(2n+2)} + nH_2O = (2n + 1)H_2 + nCO \qquad 10.7(2)$

The carbon monoxide is removed by further reaction later in the process. The reaction is endothermic (absorbs heat) and takes place at pressures of about 450 PSIG and at temperatures of around 1,500°F. The reaction takes place as the feed gas and steam pass through tubes filled with a nickel catalyst which are heated in the radiant section of the reformer furnace. Steam must be provided in excess of the reaction requirements to prevent the side reaction of coke formation on the catalyst. The coking of the catalyst deactivates it, resulting in expensive replacements. To minimize the coking, steam is usually supplied in a ratio of 3.5:1 by weight, relative to feed gas. Special precaution must be taken to maintain the excess steam at all times, since even a few seconds' interruption, while feed gas continues, can completely ruin the whole catalyst charge.

Process Controls

As indicated, the major process variables to be held are

1. The feed gas flow
2. The reforming steam flow
3. The effluent temperature and composition

As illustrated in Figure 10.7g, the feed gas flow is maintained by means of the flow controller, (FRC-1), which is pressure compensated via PT-1 and FY-1. Pressure compensation of flow corrects the measurement for fluctuations in feed gas pressure. The steam rate is maintained by means of FRC-2, and the ratio of steam to feed gas flow is continually monitored through FY-2.

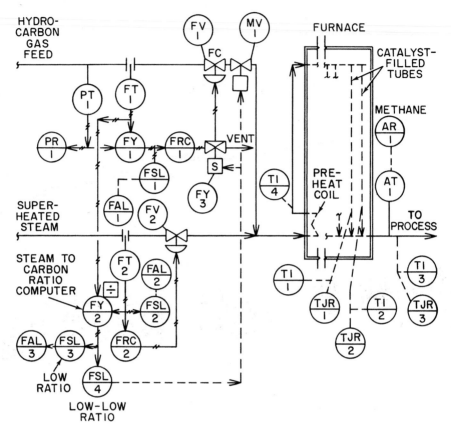

Fig. 10.7g   Reforming furnace process controls

The ratio computer FY-2 accepts a gas flow signal from FT-1 and a steam flow signal from FT-2, and its output is scaled in the ratio of steam to carbon. If this ratio falls below the limit of about 3:1, the low flow ratio alarm FAL-3 is sounded in the control house. If the ratio continues to fall below about 2.7:1, the feed gas is shut off via FSL-4, which trips the solenoid FY-3 and causes the air diaphragm operator on the feed control valve (FV-1), to be vented, thereby closing the valve. The valve (FV-1) is a quick-closing valve (4–5 seconds for full closure) so that the flow of gas can be almost instantly stopped on reforming steam failure, thus protecting the reformer catalyst. It must also be a single-seated tight shutoff valve, to prevent leakage during the time it takes to close the electric operated shut-off valve MV-1. The operation of MV-1 is relatively slow due to the electric motor and speed reduction gearing used. An alternative approach to this ratio control arrangement is shown in Figure 10.7h. Here the feed gas flow is measured by flow transmitter (FT-1), as is the steam by FT-2, and a constant ratio of steam

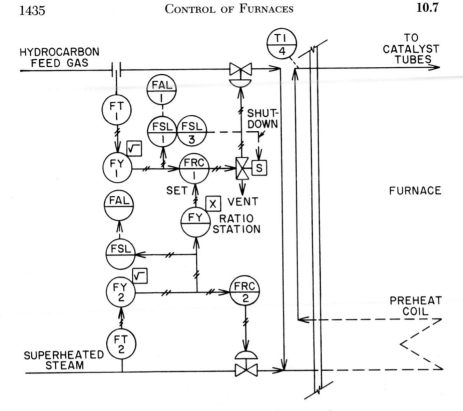

Fig. 10.7h   Reformer furnace control of steam-gas flow ratio

to gas is maintained. The steam is on straight flow control and the gas flow tracks the steam flow. In principle this appears to be an improvement over the previous control system, but in practice the dynamics of measurement and the process makes it less stable. The steam flow measurement noise and lags may result in cycling the gas flow signal, such that the integrated error of gas flow in the latter system is considerably greater than in the former.

The effluent analyzer AT-1 (Figure 10.7g) is used to determine the extent of reaction completion by measuring the methane (CH_4) content of the stream. An optimum furnace temperature profile can be arrived at by manipulating furnace temperatures to achieve the desired degree of conversion. This analyzer can be either an infrared or chromatographic analyzer. The analysis is relatively simple, involving the measurement of 0 to 10 percent methane in a background of hydrogen and carbon monoxide. The only difficulty is the high water content in the stream due to excess steam used in the reaction, therefore, water separating components are required in the analyzer sampling system.

Firing Controls

Firing of a reformer furnace (Figure 10.7i) because of its massive design and great heat inertia, is essentially manual. The furnace has approximately a dozen fuel headers with about 20 individual burners per header. Process temperature indicators TI-1 and TI-2 (Figure 10.7g) are provided at the exit of the reaction tubes. These are constantly monitored and periodic adjustment of firing is made by manipulating the fuel header control valves (HV-1, HV-2, HV-3) via the remote manual stations (HIC-1, HIC-2, HIC-3). The pressure upstream of the header control valves is controlled, so that once the valve stroke is set, fuel flow remains constant.

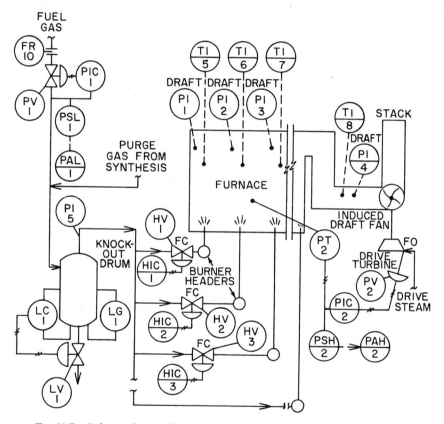

Fig. 10.7i   Reformer furnace firing controls. (PV-2 is the speed governor operator.)

The fuel generally used is natural gas, with a small amount of purge gas from the NH_3 synthesis loop. The fuel gas is pressure-controlled by PIC-1 before it enters the knockout drum. The knockout drum removes the small trace of water that is carried by the fuel gas. The level controller (LC-1), is a snap-acting on-off controller which periodically expels the water accumulated through its valve (LV-1). The flow recorder (FR-10) measures the fuel consumption and is used for efficiency checks and energy balances. Draft is usually maintained in these furnaces by an induced draft fan, which is usually steam driven and the speed of the fan sets the furnace draft (negative pressure) in the fire box. Pressure indicator controller (PIC-2), measures the pressure in the furnace (via PT-2) and controls the speed of the fan by adjusting the operator of the fan turbine governor to hold the desired furnace pressure. The draft points PI-1 through PI-4 are used to manually set the air inlet louvers, on the side of the furnace, to balance the drafts at various points in the furnace.

Safety Controls

The main safety items involve

1. Loss of furnace draft
2. Interruption of fuel supply

The loss of furnace draft may result from a malfunction of the induced draft fan or from loss of steam to the fan drive turbine. The effect is drastically to restrict combustion air, causing hazardous fuel air ratios and in addition the fire-box pressure will most likely rise above atmospheric, forcing flames out into the area around the furnace housing. The high-pressure alarm (PAH-2 in Figure 10.7i) signals the operator that fire-box pressure is rising. The operator at this point begins to cut firing to minimum levels by means of the hand control stations (HIC-1, HIC-2, HIC-3).

Interruption of fuel supply can result in the possibility of a burner flameout, and the resumption of fuel flow may cause a dangerous air-fuel mixture. Flame detection on this type of furnace is impractical, mainly due to the number required (there are in most cases in the neighborhood of 100 burners). Therefore, fuel pressure below a minimum (as indicated by PAL-1) implies a flameout and firing should be cut. The manual procedure, using HIC-1, HIC-2 and HIC-3 to cut back on fuel to the burner headers, is normally employed.

On instrument air failure (Figures 10.7g and i) the gas feed valve (FV-1) closes to prevent coking the reformer catalyst. The fuel valves (PV-1, HV-1, HV-2 and HV-3) close to stop firing during emergency. The steam valve (FV-2) fails open to maintain the flow of cooling steam through the furnace coils.

Cracking (Pyrolysis) Furnaces

An example of a cracking furnace is an ethylene pyrolysis unit. Feed stock, which can vary from heavy gas oil to ethane, is preheated and vaporized in a preheat coil in the furnace; then it is mixed with steam and cracked. Steam is added to the feed in a fixed ratio to hydrocarbon to reduce the partial pressure of the hydrocarbon feed. This tends to maximize the amount of olefins produced and to minimize the coke build-up in the coils. The feed is heated to 1,500°F in the pyrolysis coil, which causes the cracking of the long chain hydrocarbons into shorter chain molecules and initiates the forming of unsaturated (olefin) molecules. The severity of cracking is dependent on the temperature achieved and on the residence time in the pyrolysis coils. Therefore the distribution of furnace products is dependent on the degree of firing and on the temperature profile in the furnace. Effluent from the furnace is quickly quenched to prevent recombination of products into undesirable polymers.

Process Controls

The important variables to be controlled are hydrocarbon feed flow, steam flow, coil temperatures and fire box temperature, which results in the desired effluent product distribution. The process control system (Figure 10.7j) is typical of cracking furnace instrumentation. The charge to the unit is fixed by the flow controllers (FRC-1 through FRC-3) on the individual passes and the total flow is recorded on FR-4. It is important to keep the flow through the coils constant, since coking gradually builds up the pressure drop in the coils. If distribution were left purely to hydraulic splitting, the coking would start in one coil and reduce the flow through that coil, which in turn would over-heat, producing more coke. Eventually the flow would reduce to such a low rate that over-heat of the coil will cause its rupture. The individual flow control valves introduce the variable pressure drop that allows all coils to coke at an almost equal rate.

The steam flow controllers (FIC-5 through FIC-7) maintain the appropriate amount of steam to match the flow of hydrocarbon feed. The temperature controller (TRC-1) sets the total heat input to the process by controlling the effluent temperature to the desired value. The temperature indicator points (TI-1 through TI-6 as well as TI-7 through TI-11), are monitored by the process operator to maintain a certain furnace temperature profile, and hence a certain product distribution in the furnace effluent. The temperature relationships are accomplished by manually trimming burners at the required place in the fire box.

In order to determine whether the desired product specifications are being achieved, an analyzer is commonly used at the exit of the quench system (AT-1). This analyzer is usually a chromatograph, measuring most of the components in the stream that are lighter than butane. This analysis is a difficult one, primarily because of the sample handling requirements. The sample has a very high water content which must be condensed and removed prior to entering the analyzer. It also has a large amount of entrained coke and tars which likewise must be scrubbed out. For a detailed description of chromatographs see Section 8.1 in Volume I.

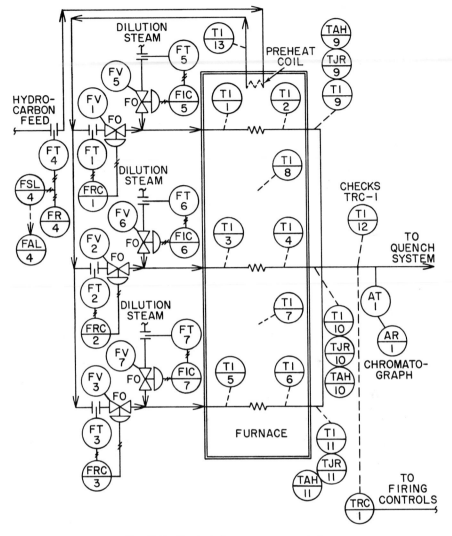

Fig. 10.7j   Pyrolysis furnace process controls

Firing Controls

Figure 10.7k shows the firing controls with gas as fuel. The temperature controller (TRC-1) maintains the process gas effluent temperature at the desired value by resetting the set points of the local pressure controllers (PIC-1 and PIC-2). These pressure controllers maintain the desired burner header pressure to satisfy the heat liberation required to sustain the degree of cracking being achieved. The details of the controls are the same as described under

Firing Controls for *Process Heaters and Vaporizers,* if the fuel is gas only. Fuel oil and multiple fuel cases are also applicable. (See Figures 10.7d, 10.7e, 10.7f).

Safety Controls

The major hazards in operating cracking furnaces are

1. interruption of feed flow
2. interruption of fuel flow
3. coking of individual coils
4. instrument air supply failure.

Interruption of feed flow can result in a dangerous situation if firing is maintained at the normal rate, because the tubes are not designed for the excessive temperatures that result if charge is stopped or is drastically reduced, and the danger of tube rupture is pronounced. The low-flow alarm (FAL-4, Figure 10.7k) is provided in the feed stream to warn the operator of impending danger. Once he verifies that the danger is real he can trip the vent solenoid valve (HY-1) by operating the pushbutton (HS-1), which vents the diaphragm on the emergency valve (HV-1), causing the complete stoppage of fuel gas. This process can be automated, so that on continued fall of feed flow, a

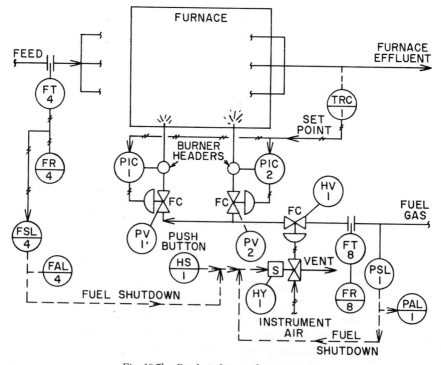

Fig. 10.7k   Pyrolysis furnace firing controls

low-flow switch (FSL-4) automatically actuates the trip solenoid (HY-1), which cuts off the fuel flow.

Interruption of fuel flow will cause burner flameout and on resumption of fuel flow may result in a dangerous fuel-air mixture. To prevent this occurrence, the low-pressure switch (PSL-1) is installed (Figure 10.7k), which trips the emergency shutoff valve (HV-1) via solenoid valve (HY-1). The solenoid valve is the manual reset type; therefore once tripped, it must be manually reopened.

Excessive coking of an individual furnace coil can occur due to the restriction of flow through it. This is a self-worsening effect, and tends to cause dangerous overheats. The prevention of this situation is of prime concern to the operator. Should such a condition occur, it will be detected by the high-temperature alarms (TAH-9 through TAH-11, Figure 10.7j) which warn the operator. The operator can increase the flow setting on the appropriate feed flow controller, thus forcing more fluid through the hot tube hoping to bring the temperature down. If this does not alleviate the condition, he has no alternative but to shut down the firing via HS-1 in Figure 10.7k.

Failure of instrument air will result in the control valve actions shown under the individual valves in Figures 10.7j and 10.7k. In Figure 10.7j the feed valves and the steam valves open on air failure to continue flow through the furnace coils and prevent over-heating and possible rupture of the coils. In Figure 10.7k the fuel control valves close, extinguishing the fire in the furnace, which results in the safest condition in an emergency.

Analytical Instruments

There are two important analyzer measurements used in furnace controls, oxygen analysis and combustibles detection. A third, less commonly used measurement is a calorific analysis of the fuel gas being burned. Details of these analytical instruments and their applications will be found in the Sections 8.7, and 10.8 in Volume I. Here the intent is briefly to indicate how and when these analyzers are tied into combustion control systems.

Oxygen Analyzers

The oxygen analyzer is a common instrument for furnaces consuming large amounts of fuel. Its main function is to measure the O_2 content of the flue gas from which the actual excess air being delivered to the furnace may be computed. Keeping the excess air down to the desired minimum results in the conservation of heat, since the sensible heat of excess air is an outright loss. By trimming down on the air intake into the furnace with louvers or dampers, while observing the O_2 content of flue gas, the efficiency of combustion can be optimized with respect to air consumption. The analyzer has a secondary function in that it can be tied to a low O_2 alarm and warn the operator of an impending hazardous furnace atmosphere. Whether to acquire an oxygen analyzer is generally decided on economic grounds. The cost of the analyzer and its daily upkeep must be saved, as a minimum, for the installation to be justified. For large fuel consumers, the time required to pay for the analyzer installation can be a matter of months. Where furnace operation is such that there are frequent changes in firing loads and/or frequent changes in types of fuel, the analyzer could be considered a necessity from a safety standpoint alone.

Combustible Analyzers

Combustible analyzers, which measure the unburned fuel in the furnace flue gas, are essentially safety devices. If rapid changes in firing and frequently varying fuels are used a combustible analyzer may be a necessity. The analyzer is usually connected to an alarm device to warn the operator of a dangerous situation. It may also be used to shut down the furnace firing if the combustibles exceed some safe limit.

Calorimetric Analyzers

Calorimetric analyzers are used to measure the heating value of the fuel being fed to the furnace. When gas streams of varying heating values are combined, the resulting mixture will have a varying heating value. For stable furnace firing, the heating value of the fuel should be kept as constant as possible. The usual method of accomplishing this is to have a supplemental fuel with high and constant heating value as a control stream. If the heating value of the total stream drops, more of the supplemental fuel is blended in. Like the other analyzers mentioned above, this instrument too can serve as a safety device by giving an alarm when the heating value of the fuel reaches such a low point that it will only marginally support combustion. Both fuel blending and emergency shutdown of firing due to low heating values can be automated (Figure 10.7l). Liquid propane is the supplemental fuel under flow control by FRC-1, being reset by the BTU analyzer controller ARC-1. The level controller LC-1 on the vaporizer adds just the right amount of steam required to vaporize the supplemental fuel. The knockout drum is required to keep any unvaporized fuel from reaching the burners, which are designed for gas firing only. If the analyzer detects a dangerously low heating value, the fuel supply is shut down by ASL-1 actuating the shutoff valve (AV-1).

Advanced Controls

Typical advanced control systems discussed here include feedforward control, analog computer control and digital computer control as applied to furnaces. An example of each type of control will be presented to give the reader an idea of the possibilities of this approach. The divisions made are arbitrary in that the control systems can be designed to include two or all three techniques in the overall system. Thus, a digital computer control system

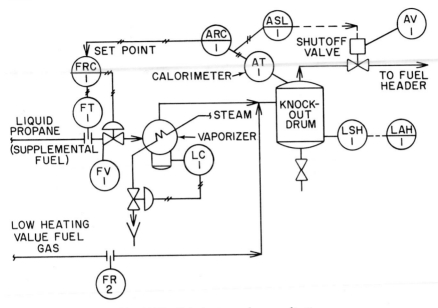

Fig. 10.71   Calorimeter analyzer application

could incorporate the feedforward technique with an analog computer to obtain input data.

Feedforward Control

In feedforward control, known relationships between variables are used to adjust for disturbances before they enter the process. This is contrary to feedback control, which must first detect an error in the process variable before corrective action is initiated.

A simple example of feedforward control is given in Figure 10.7m. The normal control is a feedback, with the temperature controller (TRC-1) providing the set point for the fuel pressure controller (PIC-1). On change of feed flow to the furnace, which is an independent variable in this system, there is a definite need to change the rate of fuel firing. As shown, the flow transmitter (FT-1) detects the change in feed flow and sends an altered signal to the multiplying relay (FY-1). The multiplying relay sets the relationship between a change in feed flow and the required change in fuel header pressure. This is an empirically determined value and can be field-adjusted. The modified signal then enters the summing relay (FY-2), which adds the signal to the temperature controller output, thereby setting a new fuel gas header pressure via PIC-1. In effect, advance information is fed forward to the firing controller (PIC-1), indicating that a change in process load is taking place and that firing conditions should begin to change. Without the feedforward leg of the loop

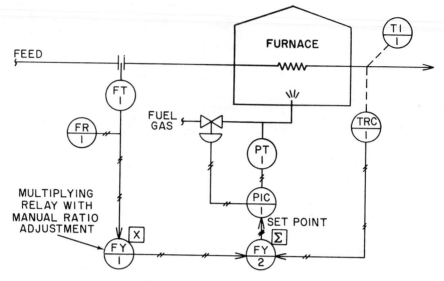

Fig. 10.7m Feedforward control

(FY-1 and FY-2), the required change in firing conditions would take place much later, after the temperature controller (TRC-1) detected a change in the controlled variable (the effluent temperature of the furnace). If a constant ratio existed between feed flow and header pressure, the temperature controller would not be necessary, but in practice, with changing ambient and process conditions, this ratio changes with time. The TRC-1 thus acts as a slow trim-controller, keeping the controlled variable at the desired value.

Analog Computer Control

There are certain variables which are impractical to measure yet are important parameters in a control system. These may be obtained through computation, using one or more easily measured variables. Relatively inexpensive computing components are available to do such computations. They are available in both pneumatic and electronic versions, and some instrument suppliers offer these as assembled systems (Sections 9.1 and 9.2).

An example of the use of an analog computer is in the calculation of the heating value of a fuel gas mixture, blended from varying quantities of two known fuel gases. Figure 10.7n depicts this computation. The heating values of fuel gases #1 and #2 are known, but the percentage of each component in the final mixture is variable, resulting in a varying heating value of the total mixture. The computation shown is quite straightforward, using two multiplying relays FY-1 and FY-2, two summing relays FY-3 and FY-4 and dividing relay FY-5. The results are continually recorded on the BTU recorder XR-1. The above system may be combined into the fuel-firing system

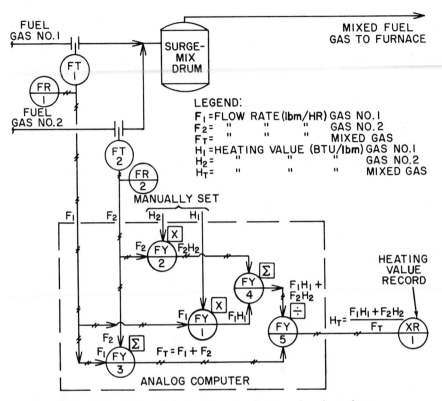

Fig. 10.7n　Analog computer determines heating value of mixed gas

shown in Figure 10.7o. The fuel gas header pressure for stable burner operation, as controlled by PIC-1, is dependent on the heating values of the fuels
consumed. This is so because the two fuels differ greatly in heating value and
further, the amounts of each gas consumed can vary considerably with time.
The combined heating value is therefore computed using the same analog
computer described earlier, and its signal provides the set point of the fuel
gas pressure controller PIC-1. The signal-limiting relays FY-6 and FY-7 are
installed to maintain the header pressure within safe operating limits even
if the computer calls for extreme pressure settings.

Digital Computer Control

Digital computers have been used successfully in process control in a
wide range of applications. Its potential in future installations is enormous.
The degree of complexity of hardware and software can vary widely depending on its functions. Chapter VIII provides a detailed treatment of the subject.
Here the attempt is to give an example of digital computer control for the
purpose of indicating the potential of this tool in furnace operations. Figure

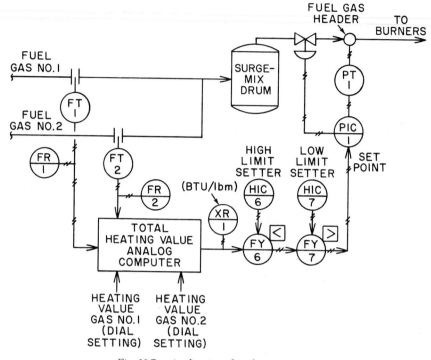

Fig. 10.7o   Application of analog computer

10.7p represents a computer control system for an ethylene cracking furnace. The process control considerations for this type of furnace have been described earlier. One of the main considerations was the proper splitting of the feed among the parallel passes through the furnace. In a large ethylene plant there may be from six to eight identical cracking furnaces with from 10 to 20 parallel coils per furnace, and maintaining proper flow splits in these coils is a formidable task. This task is well suited to computer control because of the repetitious nature of the computations involved. In Figure 10.7p the signals which are inputs to the computer are shown by the symbol A/D (analog to digital), and the computer outputs to the process by D/A (digital to analog). The total feed flow A/D-1 and the individual coil flows (A/D-2, -3, -4, etc.), are taken in by the computer from which it calculates the proper valve settings. The proper valve adjustments are sent out (through signals D/A-2, -3, -4, etc.), which causes the valves to assume their required position to divide the total feed flow. At the start, the total flow will be divided equally among the passes, but as time progresses, coke will gradually build up in the individual coils, causing outlet temperatures to vary from coil to coil. The computer is programmed to keep the effluent temperatures the same (within some tolerance) by altering the flow of the feeds through each coil. It does this by taking

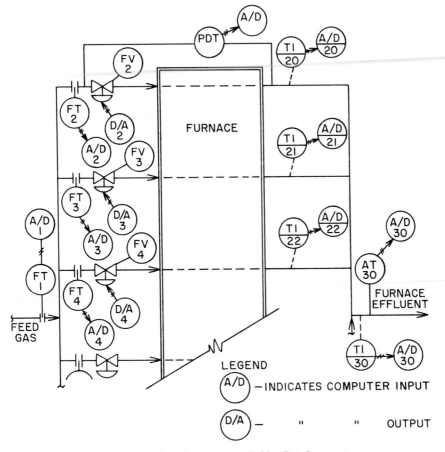

Fig. 10.7p   Pyrolysis furnace controlled by digital computer

in the effluent coil temperature information (A/D-20, -21, -22, etc.), comparing them and modifying the "feed-splitting" computation. The results of the computation are again sent out to reposition the feed valves (FV-2, FV-3, FV-4, etc.). The analyzer, measuring the effluent stream composition (A/D-30), also sends its signal to the computer, where it is compared with the desired values. Based on the differences, the degree of fuel firing at various zones in the furnace is computed. These may be printed out as instructions to the operator or sent out as outputs to the fuel valves, changing the heat flow pattern throughout the furnace and thus optimizing the cracking operation.

Many additional functions can be performed using much of the same input data. Some of these are "off-normal" alarms on feed flow, "off-normal" alarms on effluent temperatures, alarms to signal excessive pressure drop across any coil, high coil-metal temperatures and other scanning tasks. Both the control and scanning functions are done on a periodic basis, with the cycle period determined by the nature of the measurement. Flows are usually handled at much higher frequencies than temperature or pressure inputs. Other computations, such as optimization and furnace efficiency calculations, can be performed at a much slower rate.

The question of when to use a computer is a difficult one to answer, but when a high-quantity throughput or an expensive product is involved, there is a good possibility that justification exists.

Conclusions

The examples given in this section are representative of a wide spectrum of furnaces throughout the process industry. Important aspects of furnace control were covered, with stress being placed on the process requirements.

Safety of operation is a requirement which cannot be compromised. Although many risks are involved in furnace operation, many beyond the control of the instrument engineer, those which can be controlled should obviously be. This does not always mean the use of additional instrument hardware, but rather the effective use of minimum amounts of components to do the practical job. The use of excessive interlocks has, in many instances, made a unit so cumbersome to operate that all interlocks were deactivated, leaving the unit virtually unprotected. There is no substitute for careful planning in the design of furnace safety controls and their mode of operation.

Finally, with the expansion of industrial facilities all over the world and with the use of larger furnace units, thought should be given to the use of the more advanced forms of controls. Proper use of feedforward, analog and digital computer controls can offer more stable, more efficient and therefore a more profitable furnace operation.

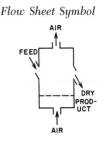

10.8   CONTROL OF DRYERS

The subject of dryers is so extensive that some limitations must be imposed in order to present a worthwhile discussion. In general, this presentation will be based on these principles:

1. The dryers and drying principles discussed will relate to process requirements, that is, to the removal of a volatile solvent from a solid material. Such processes as the removal of a solvent from an air stream (air drying), or the removal of one solvent from another (drying of an organic solvent), will be considered beyond the scope of this section.

2. The solvent to be removed will be taken as water, although an occasional reference may be made to other solvents.

3. The heating medium used will be steam, except in cases where a different heat source is customarily used.

4. Safety controls such as flame burnout indicators will not be included in the discussion except where they specifically relate to the process.

Principles

A drying curve of a typical material is shown in Figure 10.8a. It consists of four zones. Section A-B represents the period of product entry into the dryer. Since some heat is necessary to bring the material to the initial drying temperature, evaporation is slow. The next section, B-C, is a period of evaporation of surface moisture or moisture that migrates readily. The temperature during this time is the wet-bulb temperature of the air directly in contact with the product. After the surface moisture has evaporated, the rate of evaporation drops off (section C-D). This phenomenon is due in part to a "case-hardening" of the surface and in part to the long path necessary for the water to migrate to the surface. If the solid is one which has water-of-crystallization, or bound water, the rate will then drop off even more, as shown in section D-E of the curve.

Two other curves are useful for an understanding of dryer control requirements. The first of these, Figure 10.8b, represents the derivative of Figure 10.8a, that is, the variation in rate of drying of the material as a function of time. The other curve, Figure 10.8c, depicts the temperature levels during drying. In each case the identifying letters correspond to those in Figure 10.8a.

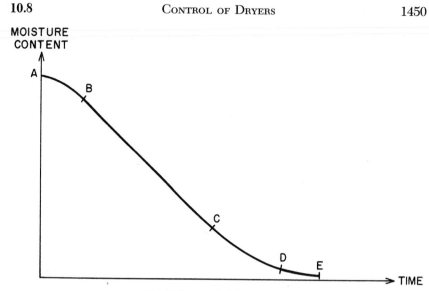

Fig. 10.8a   Typical drying curve

Section B-C is thus a period of *constant rate* and the temperature is also constant, as already mentioned. The remainder of the curve is known as the period of *falling rate*. The temperature of the product will rise rather rapidly to some point close to the temperature of the surroundings within the dryer. It will then asymptotically approach a final value.

     Each material to be dried will have a unique drying characteristic depending upon its substance, the solvent, the affinity of one of these for the other, and the surroundings characteristic of the particular dryer.

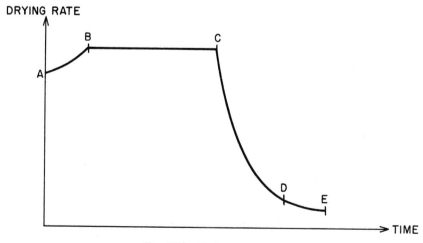

Fig. 10.8b   Drying rate curve

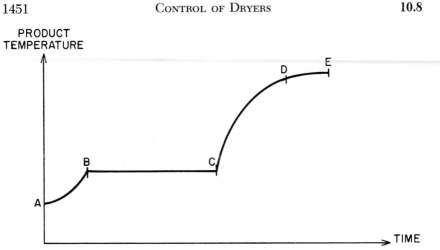

Fig. 10.8c   Variation in product temperature

The theoretical drying rate depends upon feed rate and driving force. The latter is a combination of temperature difference between the product and the surroundings, and the moisture condition prevailing in the dryer atmosphere. Other factors of equal importance are the intimacy of material surface contact with the heating medium and the degree of agitation or surface-renewal.

The three essential operating factors necessary to any drying operation are a source of heat, a means of removing the solvent from the environment of the product and a mechanism to provide agitation or surface-renewal. The extensive assortment of dryer types available serves as testimony to the fact that drying is among the most difficult of process problems. A few representative dryer types are presented in Tables 10.8d (batch dryers) and 10.8e (continuous dryers) by way of general illustration. Further elaboration of control schemes for them will be presented in the following paragraphs.

Table 10.8d
TYPICAL BATCH DRYER CHARACTERISTICS

Type	Examples	Heat Source	Method of Moisture Removal	Method of Agitation	Typical Feed
Atmospheric	Tray	Hot air	Air flow	Manual	Granular or powder
Vacuum	Blender-tumbler	Hot surface	Vacuum	Tumbling	Granular or powder
	Tray	Hot surface	Vacuum	None	Solid or liquid
	Rotating blade	Hot surface	Vacuum	Mixing blade	Solid or liquid
Specialty	Fluid bed	Hot air	Air flow	Air	Granular or powder

Table 10.8e
TYPICAL CONTINUOUS DRYER CHARACTERISTICS

Type	Examples	Heat Source	Method of Moisture Removal	Method of Agitation	Typical Feed
Heated cylinder	Double-drum	Hot surface	Air flow	None	Liquid or slurry
Tumbling	Rotary	Hot air	Air flow	Tumbling	Granular
	Turbo	Hot air	Air flow	Wipe to successive shelves	Granular
Air stream	Spray	Hot air or combustion gas	Air flow	Not required	Liquid or slurry
	Flash	Hot air	Air flow	Air	Granular or powder
	Fluid bed	Hot air	Air flow	Air	Granular

Control

The specific property of interest in dryer operation is moisture content of the final product. This property is difficult to measure directly, particularly for continuous systems.

In batch systems, an empirical cycle is set up by taking samples for moisture analysis at various time periods. Once approximate drying conditions have been established, checking of moisture is required only near the estimated end point.

A similar procedure is followed for continuous systems. Empirical runs are made to establish dryness of a product with given feed and dryer characteristics so that operating curves can be developed. The control scheme is designed to maintain these conditions with occasional feedback and with corrections from grab samples. There are a few analyzers available for moisture analysis in flow systems that can be adapted for automatic, closed-loop dryer control. The most successful of these units rely on infrared or capacitance detection of material flowing through a sample chamber. Care must be exercised in the selection to allow for changes in bulk density or for void spaces, which will introduce an error. These sensors are sometimes ineffective with relation to bound moisture. The discussion of moisture analyzers in Volume I, Section 8.11, gives further elaboration.

Although spray and flash dryers have a retention period of less than a second, holdup in the majority of continuous dryers ranges from 30 minutes to an hour. This fact makes automatic feed control so difficult that it is rarely used. It is usually not feasible to measure the variation in moisture content of the feed. The feed rate is metered by a screw conveyor or other similar device. The variation in entering moisture is accounted for by the dryer control system or by periodic analysis of the dried product followed by suitable controller adjustments.

Dryer controllers usually include proportional and reset modes, only because dryer dynamics do not generally warrant the use of the rate response.

Most dryer control systems are still relatively unsophisticated. The drying rate curve levels off in the area where most product specifications lie, so control of final conditions is not critical. Most moisture specifications are somewhat liberal in recognition of the difficulties present in moist and dry material handling. They also allow for a possible change in moisture content due to ambient air conditions after drying, a characteristic of most dried materials.

To get a general view of the use of controls as applied to dryers, it is desirable to examine and discuss each of the basic units listed in Tables 10.8d and e. This is far from a complete listing but it covers the field thoroughly enough to obtain a good background.

Batch Dryers

Although continuous processing infers a more modern approach, batch dryers are still installed in many plants. These are particularly well adapted for drying relatively small quantities, especially where batch identity might be of value, and for processes where a variety of products are manufactured. The selection and operation of controls for batch units is complicated by the fact that the state of the product changes with time.

Atmospheric

The term atmospheric dryer is applied here to designate batch dryers which operate at close to atmospheric pressure and with the heat for drying supplied by the air within the cabinet. The same medium is relied upon to remove the generated moisture. The two most common forms are the tray dryers, where trays covered with material to be dried are loaded into racks, and the truck dryers, which are similar except that the racks of trays are mounted on trucks.

The available control parameters are the air velocity and distribution, temperature and humidity. The velocity and distribution of the air are usually set manually by dampers and are not changed. Humidity is also a manual setting by virtue of adjusting the damper in the recirculation line. If recirculation is not used and if product requirements warrant, dehumidification is used on the inlet air. Both of these provisions are shown dotted in Figure 10.8f.

Control of the atmospheric temperature within the dryer is accomplished by regulation of the steam to the heating coil. The thermal bulb can be placed in the dryer or in the outlet air. The sizing of the steam valve is determined by the dryer air flow and temperature requirements as analyzed on psychometric charts (see Figure 8.9a in Volume I) and by the pressure of steam used.

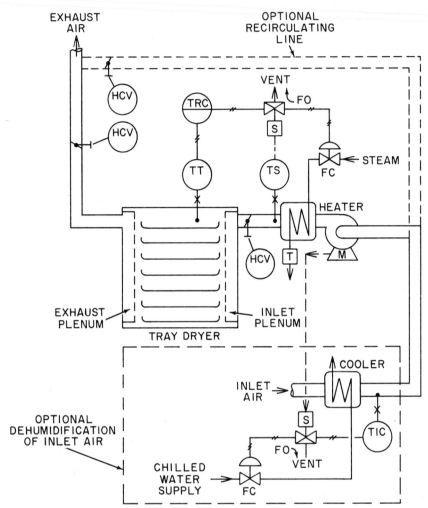

Fig. 10.8f　Atmospheric tray dryer

It is necessary during the latter portion of the drying cycle to reduce steam flow to a level sufficient only to heat the air to the dryer temperature. The load at this time is limited to the dryer heat loss. A temperature switch is frequently included in the heater discharge to limit this maximum inlet temperature.

It is not uncommon to use a program controller (see Section 4.4 for details) for temperature control of a batch tray dryer, particularly on solvent service. In this way the rate of evaporation is controlled and the concentration of solvent in the air *is held below the explosive range.*

Vacuum

The operation of a vacuum dryer relies on the principle of heat transfer by conduction to the product while it is contained in a vessel under vacuum. In its simplest form this dryer is a tray unit in a vacuum chamber with hollow shelves, through which a heat transfer medium is circulated. Such dryers usually start operation at a very low temperature and are known as *freeze dryers*. Other types are the double-cone *blender-dryer* and the vacuum *rotary dryer*. The blender-dryer tumbles the product in a jacketed vessel. The shell of the rotary dryer is also jacketed, but it is fixed, with material agitation being supplied by a hollow mixing blade (for the passage of heat-transfer fluid).

Parameters which can be controlled in batch vacuum dryers are absolute pressure (vacuum), rate of tumbling, or agitation, and temperature of the heat-transfer medium. The general rule is to control the first two manually and concentrate on the transfer fluid temperature for automatic control. A typical scheme for the blender-dryer is shown in Figure 10.8g. The requirements for control of the freeze dryer are somewhat more complex. Refrigeration is required for the heat-transfer fluid in the first part of the cycle to freeze the dryer and its contents. The circulating fluid is then gradually warmed to provide heat for evaporation.

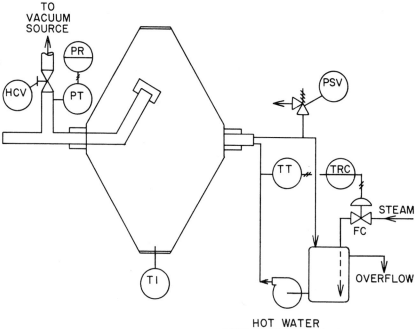

Fig. 10.8g   Vacuum blender-dryer

Special

There are a number of alternative dryers used for batch work that do not fall into the discussed categories. A unit that has become popular in the past few years is the batch fluid-bed dryer.

In a fluid-bed dryer the product is held in a portable cart with a perforated bottom plate. Heated air, blown (pressure) or sucked (vacuum) through the plate, fluidizes the product to cause drying. The air flow is manually adjusted to get proper bed fluidization, and a timer is set for the drying period. Control is restricted to temperature control of the inlet air, as indicated in Figure 10.8h.

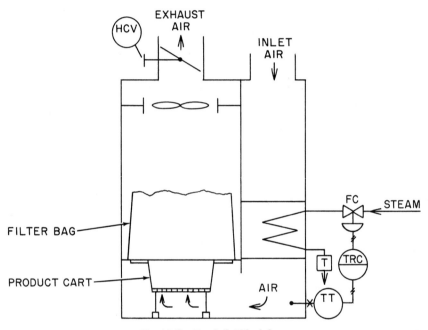

Fig. 10.8h   Batch fluid-bed dryer

A duplex control scheme is sometimes utilized to take advantage of the high heat transfer properties of the fluid bed. The bed is sufficiently well mixed so that a high inlet temperature can be used at the beginning of the cycle without fear of damaging the product as long as the temperature is reduced during the latter stages of drying. This is accomplished by the use of a controller on the outlet air which pneumatically adjusts the set point on the inlet air controller. As the bed temperature rises during drying, the set point of the inlet air temperature controller is reduced (Figure 10.8i).

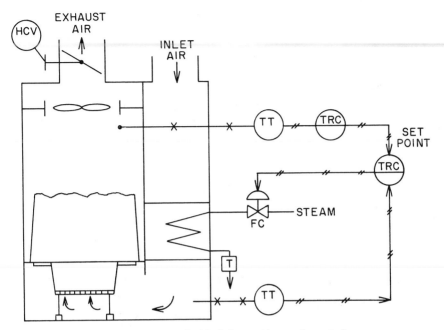

Fig. 10.8i   Batch fluid-bed dryer with cascade control

Continuous Dryers

Most large-scale industrial processes producing dry solid materials employ continuous dryers. The control of a dryer is similar to that of other process equipment with a few important exceptions. First, as stated above, actual dryness (moisture content) of the product is difficult to measure and therefore control generally depends on secondary variables. The majority of continuous dryers require control of the flow and temperature of an air stream. Temperature detection of an air stream is inherently sluggish in response. On the more positive side, the holdup in the majority of dryers is on the order of 30 minutes to an hour, and effects of input changes are blended and fast response is therefore not required. Notable exceptions to the last items are flash and spray dryers with very short holdup periods.

Dryer controllers are generally two-mode units with proportional and reset modes. Setting them in the field requires a good deal of patience, since the full effect of a change in a variable may take half an hour or more. Some variables, notably humidity of the inlet air, may defy regulation.

Heated Cylinder

Many years ago, when the concept of continuous drying was first developed, the heated-cylinder dryer was invented. The best known of these units is the double-drum dryer depicted in Figure 10.8j. Liquid is fed into the "valley" between the heated cylinders. The drums, rotating downward at the center, receive a coating of the liquid with the thickness depending upon the spacing between the rolls. The material must be dry by the time it rotates to the doctor knife, where it is cut off the roll.

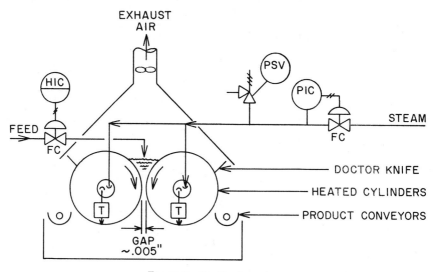

Fig. 10.8j   Double drum dryer

The variables available for control are the speed of the cylinders, the spacing between them, the liquid level in the valley, and the steam pressure in the cylinders. The first two (speed and spacing) are usually adjusted manually. The liquid level is maintained by throttling the feed stream. Some attempts have been made to control this level automatically, but they have been thwarted by three basic difficulties: the height of the level is only 6 to 9 in., the liquid is constantly in a state of extreme agitation, bubbling and boiling, and the liquid is highly concentrated and tends to plug those level sensors which depend on physical contact for measurement. (See Chapter I in Volume I for non-contact level detectors.) Frequently the control of feed is manual, indicated in the diagram by a manual loading station (HIC). With all of the other controls on manual, only the steam pressure is controlled automatically.

Rotary

The term "tumbling dryers" is used here to designate equipment in which material is tumbled, or mechanically turned over, during the drying process. Two examples in this class are the rotary dryer and the turbo dryer.

The cross section of a rotary dryer is shown in Figure 10.8k. As the shell

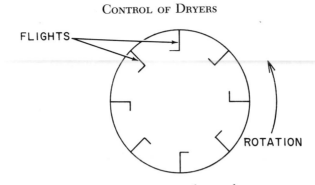

Fig. 10.8k   Cross section of rotary dryer

is rotated, the material is lifted by the flights and then dropped through an air stream. The speed of rotation, the angle of elevation and the air velocity determine the material holdup time. Variable-speed drives are sometimes incorporated to change the rotation rate, but they are usually manually controlled. Air flow is not varied as a control parameter.

A typical control scheme for a rotary, counterflow dryer (material direction opposite to air direction) is shown in Figure 10.8l. Primary controls maintain the air flow and inlet air temperature. Secondary considerations are pressure control on the outlet air to maintain pressure within the dryer, and temperature and level alarms at the dry product outlet. A temperature switch is also provided at the air outlet to prevent dryer over-heating when feed is stopped. Direct setting of the air inlet temperature is based on the supposition that the product approaches this value before it reaches the discharge end. Better control of moisture is obtained if the control is based directly on the temperature of the product, but there are practical difficulties in providing a material holdup for measurement with consistent renewal. The feed rate is also manually controlled, since the long holdup makes automatic adjustments impractical.

Turbo

In a turbo dryer the material is dried on rotating horizontal shelves (Figure 10.8m). They are arranged in a vertical stack, and the product is wiped from each shelf through a slot after a little less than one revolution. A leveler bar then spreads it evenly on the shelf below. In addition to the general countercurrent of hot air, fan blades on a central shaft impart a horizontal velocity pattern.

Since the internal fan provides consistent air circulation, it is feasible to control the air throughout. Motorized dampers are provided for the lower air input and for the combined flow to the upper sections. The division between the upper inlets is adjusted by manual dampers. The inlet air temperature is controlled by throttling the steam valve and the motorized dampers

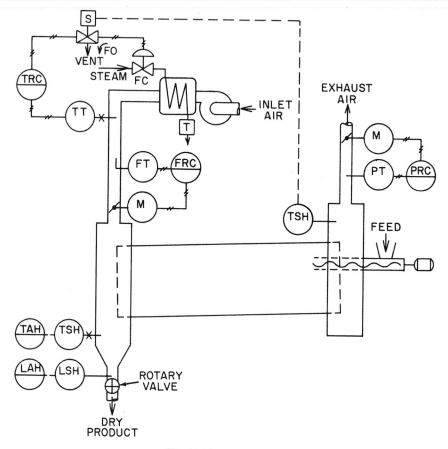

Fig. 10.8 l   Rotary dryer

are adjusted by the corresponding temperatures. Circulation rates of the center fan and the shelves and the feed rate of material are manually controlled.

Spray

The spray dryer is an exception to the generalization of large sojourn times in dryers, with holdup normally on the order of tenths of a second. A liquid or thin slurry feed is atomized into a chamber in which there is a large flow of hot air (Figure 10.8n). The inlet air temperature must be quite high, and therefore direct-fired heaters are used whenever possible.

The feed is introduced by a high-pressure manually controlled pump. Air flow is also manually regulated and balanced. Process conditions are maintained by temperature control near the outlet end. The temperature controller regulates the firing rate of the fuel-air mixture through a gas control

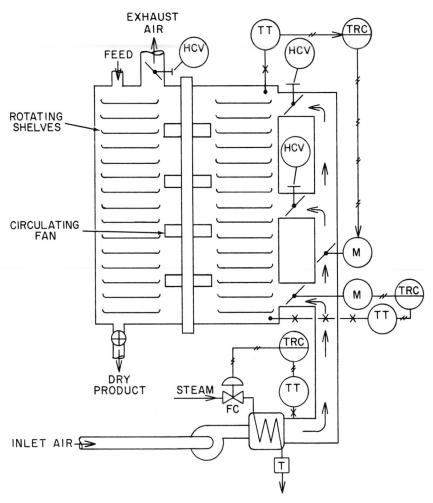

Fig. 10.8m   Turbo dryer

unit. (See Section 10.7 for details). A temperature switch is provided to shut off both the feed and fuel in case of fire or other abnormally high temperature conditions.

Fluid Bed

The fluid-bed dryer has a fluidized bed of material maintained by an airflow upward through a perforated plate (Figure 10.8o). Feed is controlled by a variable-speed screw and discharge is by overflow of the bed through a side arm.

The bed is maintained at the desired product moisture level by a temper-

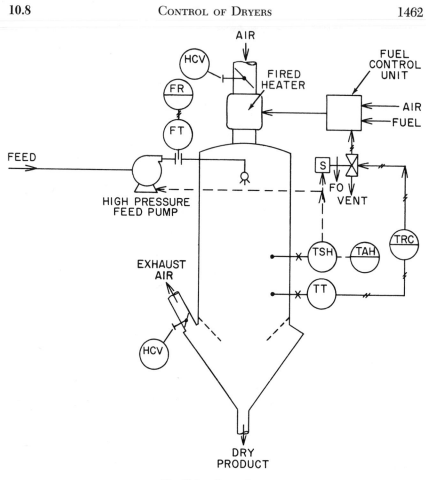

Fig. 10.8n Spray dryer

ature controller with its bulb within the fluidized product or in the air space above.

An improved but rather more sophisticated control scheme has been suggested recently. Since the system of Figure 10.8o is sensitive to the temperature but not to the absolute humidity of the air stream, an increase in humidity of the entering air will cause a reduction in inlet air temperature instead of an increase. The effect of an increased humidity is to reduce the drying rate, which represents less heat loss from the air, and to start to raise the outlet temperature. The controller compensates by lowering the inlet temperature, which further reduces the drying capacity of the air.

Two alternative control schemes are available which do not exhibit this effect. One is an adaptation of the system of Figure 10.8l, where air flow rate

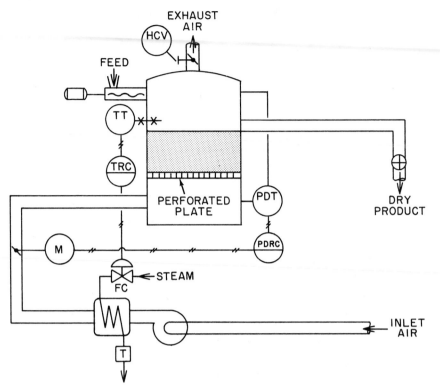

Fig. 10.8o　Fluid-bed dryer

and inlet temperature are directly controlled. The other is the feedforward cascade control scheme illustrated in Figure 10.8p.

Feedforward System

For a condition where the product moisture content required is in the "falling rate region," ($X_p < X_c$, Figure 10.8q) the product moisture can be expressed as

$$X_p = \text{constant} \ln \frac{T_i - T_w}{T_o - T_w} \qquad 10.8(1)$$

where X_p is product wetness and T_i, T_o, T_w respectively are air inlet, outlet and wet bulb temperatures. Note the absence of the variables: feed rate, air rate, feed moisture, air humidity. Thus any one of these variables can be changed without affecting product dryness if the relationship between these three temperatures is maintained. Equation 10.8(1) has three temperature variables and therefore gives specific solutions only after incorporation of the

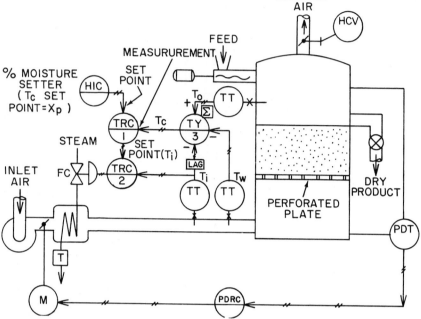

Fig. 10.8p   Humidity corrected feedforward control of fluid-bed dryer

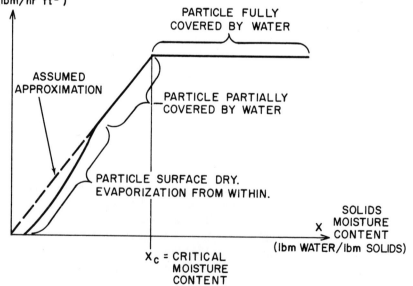

Fig. 10.8q   Drying rate vs. moisture content

dry bulb–wet bulb relationships and of the conservation of energy equation. For a given dryer and a given product moisture content,

$$\frac{T_i - T_w}{T_o - T_w} = K \qquad\qquad 10.8(2)$$

is constant. With this definition of K, the outlet air temperature required to compensate for load changes is given by

$$T_c = T_{on} - \overbrace{K(T_i - T_{in})}^{a} - \overbrace{(1 - K)(T_w - T_{wn})}^{b} \qquad 10.8(3)$$

In this equation the second subscript, n, denotes normal load conditions. As shown in Figure 10.8p, the correcting terms (a and b in equation 10.8(3)) are subtracted from the detected value of T_o in TY-3, and the resulting T_c (corrected) signal becomes the measurement to TRC-1. The set point of TRC-1 is manually set to allow adjustment for new product dryness requirements, and therefore TRC-1 in effect is a moisture controller.

The loop operation is as follows: If feed rate or feed moisture content increases, it reduces T_o and therefore T_c drops. This causes an increase in TRC-1 output, which raises the set point of TRC-2. If air humidity is increased, this increases T_w, which also reduces T_c and therefore raises the set point of TRC-2. (To stabilize the system, a lag is shown, the function of which is elaborated on in Section 7.6.)

To control moisture, temperatures at *both* ends of the dryer *must* change, and the difference between T_i and T_o is a measure of dryer load.

If the product is heat sensitive, then a high limit is introduced. If the inlet temperature (T_i) is too high for convenient wet bulb (T_w) measurement, this signal can also be obtained by measuring air humidity *upstream* to the heater and by use of a function generator. The wet bulb temperature is then obtained from this data.

Conclusions

The control systems which have been described represent actual current practice. The scheme shown for any particular dryer is not a unique solution. Each application could be handled in several different ways. They are intended not to innovate, but rather to illustrate, combinations taken from existing installations.

Improved analysis of dryer operation is now available through utilization of computer techniques. The feedforward cascade scheme for fluid-bed dryers presented earlier is one example of the review of basic drying equations which can lead to better control. With the improved methods available, the control engineer will face the new challenge of optimizing dryer operations by the use of automatic controls.

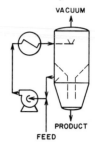

10.9 CONTROL OF CRYSTALLIZERS

Crystallization is one of the cheapest methods of purification and is well suited to the recovery of dissolved substances from a solution. Further, the product has many desirable properties, such as good flow characteristics, handling ease, suitability for packaging and pleasant appearance.

Crystallization can be achieved by any one of the following methods: cooling, evaporation, vacuum, reaction (direct combination). The choice of method depends on the composition of feed liquor, product specifications, solubility curve of the product and other engineering considerations.

Most crystallizers in industrial use operate on a continuous basis, but batch crystallizers are also used, in certain instances, for fine chemicals and pharmaceuticals.

This section is devoted to the subject of crystallizer controls. The discussion of measurement and control practices is presented here; together with the description of the basic crystallization techniques.

Control Basis

Crystallization can occur only if some degree of supersaturation or supercooling has been achieved. Any crystallization operation has three basic steps: (1) achievement of supersaturation, (2) formation of nuclei, and (3) growth of nuclei into crystals.

The definition of supersaturation is given by

$$S = c/c' \qquad 10.9(1)$$

where S = degree of supersaturation,

$\quad c$ = actual concentration of the substance in the solution, and

$\quad c'$ = normal equilibrium concentration of the substance in pure solvent.

Nucleation and growth rates are controlled by supersaturation (or supercooling). The ideal process would be a step-wise procedure, but nucleation cannot be eliminated in a growing mass of crystals. The influence of supersaturation on nucleation and growth rates is shown in Figure 10.9a. While growth rate increases linearly with supersaturation, nucleation rate increases exponentially. Figure 10.9a shows the three regions of supersaturation: (1) metastable (nucleation very low and growth predominates), (2) intermediate

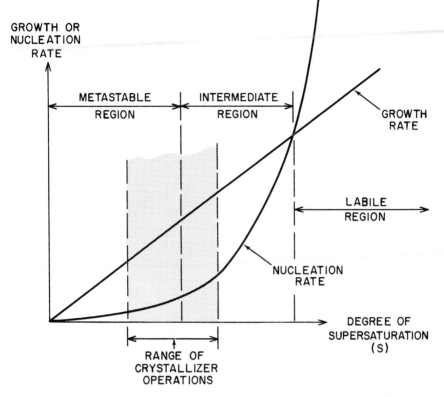

Fig. 10.9a   Influence of degree of supersaturation on nucleation and growth. $S = C/C'$, where

   S = degree of supersaturation,
   C = actual concentration of substance in the solution (parts/100 parts of solvent), and
   C' = normal equilibrium concentration of substance in pure solvent.

(nucleation becoming larger but growth still significant), and (3) labile (nucleation predominates).

     Since very small crystals are difficult to dewater, dry or handle, crystallizer design and control is directed towards reasonably large crystals, by minimizing nucleation. Usual industrial practice involves sufficiently low supersaturation to minimize nucleation but still be adequate for reasonable growth.

     Thus, control of supersaturation becomes the basic element of crystallizer system control. Unfortunately, the workable degree of supersaturation is usually 0.5 to 1 percent, and the degree of supercooling associated with it is so small that it cannot be directly measured. Hence the vast majority of crystallizers use indirect means of control.

     The one exception to this rule is sugar, where degree of supersaturation can be greater and correlations have been developed to directly measure supersaturation. Sugar supersaturation recorders that automatically compute supersaturation are based on these empirical correlations.

To define the maximum number of controllers, let us consider the degrees of freedom for a crystallization system. The variables are

> Temperature and flow of the process fluid
> Temperature and flow of coolant
> Degree of supersaturation
> Ratio of mother liquor to crystals. This can be changed by varying recycles of mother liquor to feed stream.

The limitations are

> Enthalpy balance based on the first law of thermodynamics.
> Stoichiometric balance based on solubility curves.

Since the system has six variables and two equations, it has four degrees of freedom. Thus the maximum number of automatic controllers permissible in a crystallizer system (except reaction crystallizers), without overdefining it, is four.

Cooling Crystallizers

Cooling crystallizers operate at substantially atmospheric pressure, and their heat is transferred to air or to a cooling medium by direct or indirect contact.

There are many types of cooling crystallizers, but the majority fall into three types: (1) controlled growth magma crystallizers, (2) classifying crystallizers, and (3) direct contact crystallizers.

Controlled Growth Magma Crystallizers

The various cradle crystallizers and scraped surface units belong in this category. The cradle types are used in small applications and involve little instrumental control. The various scraped surface crystallizers are used in:

> Crystallization from high-viscosity liquors.
> Open-tank crystallizers as coolers to induce nucleation.

Two versions of this type and the associated control instrumentation are shown in Figures 10.9b and 10.9c.

In Figure 10.9b, an evaporating refrigerant in the shell cools and crystallizes the product from the liquor flowing in the tubes. Usually several scraped surface tubes are used in series.

The control scheme is as follows:

> Feed liquor is under flow control. (Feedback control from product outlet rate is not practical due to high system lag.)

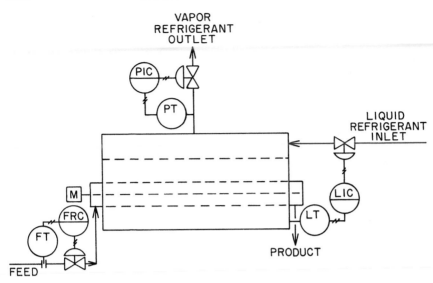

Fig. 10.9b Control of a cooling crystallizer

Liquid refrigerant enters under level control and leaves as a vapor under pressure control. In many instances flow control of the feed liquor is accomplished by a metering pump. Two-, and three-mode controllers are used on pressure control.

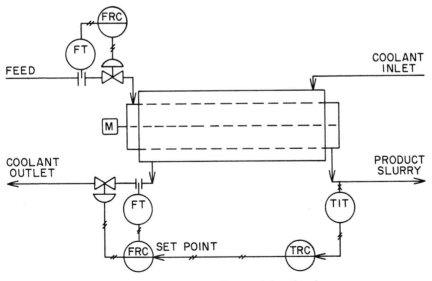

Fig. 10.9c Cooling crystallizer with liquid coolant

A more conventional system uses a heat transfer fluid for cooling and can be controlled as shown in Figure 10.9c, where the flow of cooling fluid is controlled by outlet temperature of the process slurry. One refinement to this scheme is to cascade the outlet temperature of process slurry to the flow loop on the coolant.

Classifying Crystallizers

There are various types of classifying crystallizers where supersaturation is produced entirely by cooling. One of these is the Oslo or Krystal type cooling crystallizer.

The control scheme for a classifying crystallizer is shown in Figure 10.9d. For a given quantity of feed, the ratio of crystals to liquor is fixed by stoichiometry, but frequently a ratio of crystals to mother liquor other than the stoichiometric one is required for good operation. If excess mother liquor is required, the controls can be as shown in Figure 10.9d:

> Flow control of feed liquor is provided to maintain constant throughput.
> Flow control on coolant is based on process fluid outlet temperature

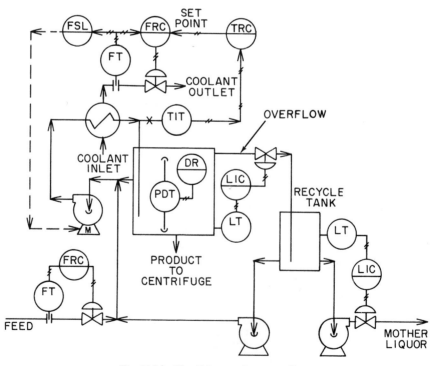

Fig. 10.9d   Classifying, cooling crystallizer

at the heat exchanger. One refinement is cascade control, and another is to interlock the coolant flow with the circulation pump motor.

Control of overflow from the crystallizer is on level control.

This is a case where differential-pressure type sensors can be utilized.

Mother liquor outflow from the recycle vessel is on level control.

In crystallization operations, one important instrument is the density sensor (differential pressure or other type, see Chapter III of Volume I), which gives an indication of the amount of crystals in suspension. This measurement is based on the fact that the density of the circulating stream is constant. Hence the measurement of differential pressure between two points in the vessel is a measure of the amount of crystals in suspension, because any change in density is caused by a change in the amount of crystals in suspension.

Direct Contact Crystallizers

In this unit the coolant is an evaporating refrigerant (or brine) and is in direct contact with the process slurry. Basic control methods are similar to those in controlled growth crystallizers. These involve constant feed rate and evaporation or refrigerant flow controls, based on process fluid outlet temperature or on tank level.

Evaporator Crystallizers

This type of crystallization takes place by evaporation of the solvent. Evaporation can be effected by (1) spray evaporation, using hot gases or air, (2) submerged combustion, and (3) indirect heating.

Of these three, indirect heating is the predominant method.

Submerged Combustion Crystallizers

Submerged combustion crystallizers are used for corrosive materials and for salts having inverted solubility. One version of a control scheme is shown in Figure 10.9e, where feed liquor to the system is on flow control, and burner fuel gas may be under flow and pressure control, while a by-pass controls the combustion flow of air to the burners.

Only the basic elements of control are shown here. Safety interlocks for combustion equipment per FIA or other requirements and auxiliaries are also to be considered in developing a control scheme. The reader is therefore referred to Section 10.7 for a review of direct-fired heater controls.

Indirect Heating

The majority of crystallizers in operation today belong in this category. Two types are important for the production of large crystals: (1) draft tube baffle crystallizers and (2) classifying crystallizers.

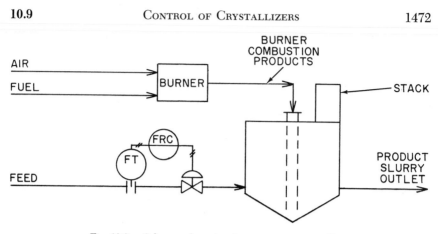

Fig. 10.9e   Submerged combustion evaporator crystallizer

These are different in design, but their approach to process control is similar. One version of control is shown in Figure 10.9f, where four control loops are involved:

1. Feed liquor is fed on level control to the feed tank. Here it is mixed with mother liquor from the centrifuge.
2. The mixed liquor is fed to the suction side of the crystallizer recirculation pump. This feed is adjusted by a level controller. (For methods of protecting the level sensor from plugging and material build-up, refer to Chapter I of Volume I.)
3. Steam flow to the heat exchanger is on flow control. Once the steam rate is fixed, production rate is also fixed, provided that the feed composition does not change.
4. Temperature control in the vessel may be achieved by controlling the evaporator chamber pressure by an air bleed.

Many refinements to this basic system are possible, such as an interlock between steam and circulating pump or the addition of a density recorder as described earlier.

Multiple Effect Operation

Evaporator crystallizers are often employed as multiple effect systems. Instrumentation on such a system is more involved and depends on the type of process flow employed. As an example, a simplified control scheme on a triple effect unit is shown in Figure 10.9g. The unit has these features:

Level in each unit is an important process variable, because it determines the residence time. This is usually controlled by throttling the mixed liquor makeup.
Steam flow to the first unit is usually on flow control.

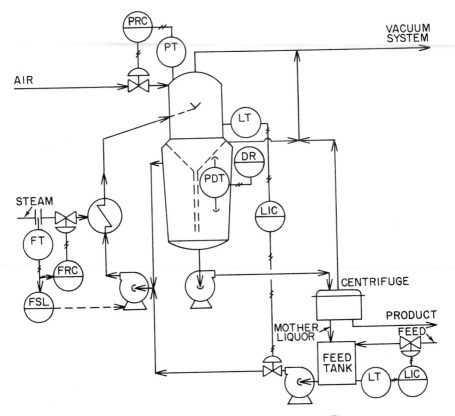

Fig. 10.9f   Indirect heating type evaporator crystallizer

Feed enters the recirculation vessel on (feed tank) level control. Temperature control in the last unit is obtained by absolute pressure control of the air bleed.

In this system, density recorders can also be used. If boiling point elevation is sufficiently large, these may be used to control the effluent liquor concentration directly.

Vacuum Crystallizers

In vacuum crystallizers, heat input to effect evaporation comes entirely from the sensible heat of the feed liquor and from the heat of crystallization of the product. Thus supersaturation is produced by a combination of cooling and concentration effects.

Two basic types are important:

       Draft tube baffle type
       Classification type

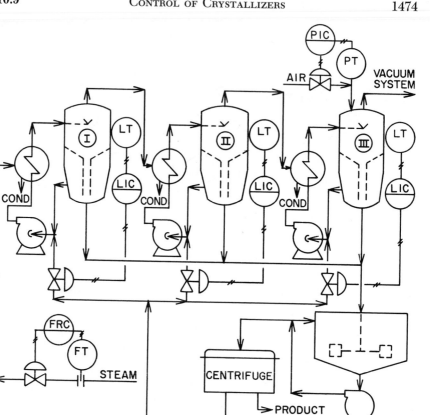

Fig. 10.9g   Multiple effect crystallizer

One control scheme of a vacuum crystallizer is shown in Figure 10.9h. The system shown achieves pressure control and hence maintains process temperature by throttling a coolant in the crystallizer jacket. A more conventional and faster method of pressure control is an air bleed, while the illustrated technique is probably more suitable for high-vacuum applications.

The control system consists of (1) feed on flow control, (2) process slurry leaving the system on level control (the level transmitter generally requires protection against plugging, see Chapter I of Volume I), and (3) temperature control indirectly by manipulating jacket coolant. Use of suspension density recorder and other refinements can also be considered. The control scheme for classifying crystallizers is similar to the one illustrated in Figure 10.9h.

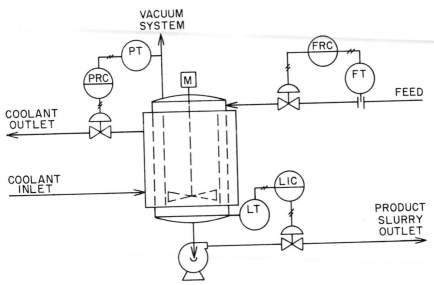

Fig. 10.9h   Draft tube baffle type vacuum crystallizer

Reaction Crystallizers

In reaction crystallizers, crystallization is associated with a chemical reaction, which frequently provides the heat required for evaporation and crystallization.

One design used in ammonium sulfate processes is shown in Figure 10.9i. With the associated control scheme, the acid-to-feed tank is on flow control, total feed is on level control, and gaseous ammonia is added on pH-cascaded flow control. The pH metering lines should be continuously flushed.

Auxiliary Equipment

Control of a crystallizer is affected by the control of auxiliary equipment associated with it. There are three systems whose control should be considered in particular:

> The feed system, including feed liquor, recycle and wash streams.
> Vacuum control to maintain predetermined pressure in the system as a common means of crystallizer temperature control (discussed earlier).
> Dewatering system control. This includes filters or centrifuges (considered in Section 10.10).

Conclusions

In this section some of the common methods of crystallizer control are outlined, but it should be noted that crystallization is still more an art than

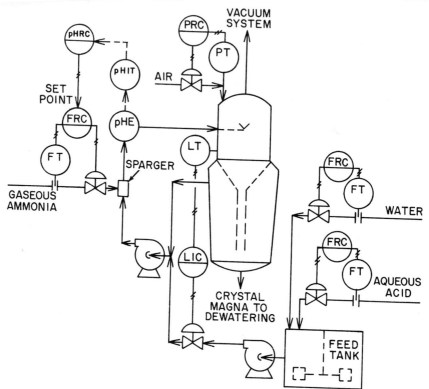

Fig. 10.9i   Reaction crystallizer on ammonium sulfate

a science. The system for any particular application should be chosen after a step-by-step analysis of the complete process scheme. Even in well-known systems, manufacturers often recommend pilot plant study due to the tremendous influence of minor constituents in the feed liquor upon crystal growth. The control system should be developed by taking these factors into consideration.

10.10 CONTROL OF CENTRIFUGES

Centrifugal machines are employed for liquid-liquid or liquid-solid separation. The machines in use can be classified into two groups:

Sedimentation centrifuges: These have solid walls and separation is by sedimentation. The operating principle is shown in Figure 10.10a. The feed enters a solid-walled bowl rotating about a vertical axis. The solid and liquid phases are acted upon by centrifugal force and gravity. The former is preponderant and separation of solid and liquid phases takes place as shown.

Filtering centrifuges: These machines have perforated walls which retain the solids on a permeable surface and through which the liquid can escape. This design is also shown in Figure 10.10a. The action is similar to a filter but with a much higher "g force" than possible in gravity or pressure filtration. Nearly all the liquid is removed, leaving behind an almost dry cake.

The centrifugal force obtained in industrial machines is several times the force of gravity. For a particle rotating with an angular velocity ω at radius r from the axis of rotation, the centrifugal separating effect is given by

$$G = \frac{\omega r}{g} \qquad 10.10(1)$$

Filtering centrifuges operate at a g range of 400 to 1,500 while the g forces for sedimentation units range from 3,000 to over 60,000 in laboratory machines (ultracentrifuges).

As with any high-speed machine, critical speed phenomenon must be considered in the design and operation of centrifuges. At critical speed, frequency of rotation matches natural frequency of the rotating member. At this speed, even the minute vibrations induced by slight imbalances are strongly reinforced. Centrifuges operate well above the critical speed and pass through it during acceleration and deceleration. Except in case of major bowl imbalance, this does not cause problems.

Centrifuge Selection

The selection of a centrifuge involves a balance between several factors: (1) the machines' ability to process the given feed slurry or emulsion at the

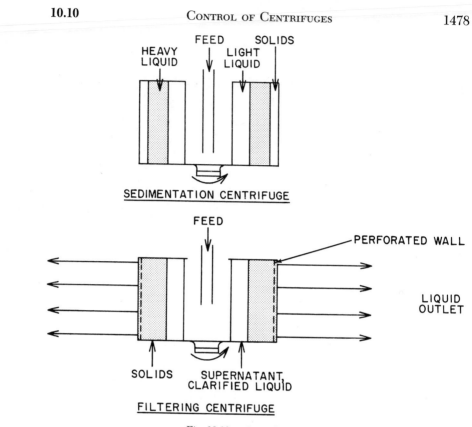

Fig. 10.10a Centrifuge types

desired degree of separation, (2) the reliability of the machine, (3) the operating and maintenance requirements and (4) the investment.

In the majority of cases, equipment manufacturers have standard machines which are adapted to the specific applications. Sedimentation machines are usually chosen on the basis of small-scale tests in laboratory centrifuges. Filtering centrifuges are chosen on the basis of tests in batch machines. Based on such tests, the manufacturers will offer specific machines and will outline the anticipated performance.

The instrument engineer should concentrate on two aspects of the process centrifuge installations:

Feed slurry control—Regulation of feed slurry at the correct continuous rate or in the right batch sizes is of utmost importance, since the machine cannot usually tolerate major variations in feed rate or composition. Control of wash liquor feed is equally important.

Sequencing operations—All batch machines are sequentially operated, and the related interlock design is one of the important steps in engineering a system.

The approach to centrifuge control is outlined in this section. Also to be considered are instruments on drive and discharge mechanisms, auxiliary devices and safety interlocks.

Sedimentation Centrifuges

Sedimentation units are used as clarifiers, desludgers and liquid-liquid phase separators. Particle size is such that separation usually obeys Stokes law. Sedimentation units are generally of small diameter and run at high speed. These can be classified into the following types: (1) tubular, (2) disc and (3) solid bowl.

The characteristics of these centrifuge designs are shown in Table 10.10b.

Table 10.10b
SEDIMENTATION CENTRIFUGE CHARACTERISTICS

Rotor Type	Range of Bowl Diameters (Inches)	Maximum Centrifugal Force (G)	Method of Solids Discharge	Method of Liquid Discharge	Maximum Capacity
Batch	2	60,000	Manual	Manual	1.5 GPH
Tubular	2–5	60,000	Manual (batch)	Continuous	2,500 GPH
Disc	7–24	7,000	Batch or semi-continuous	Continuous	1,200–18,000 GPH
Solid Bowl Constant Speed, Horizontal	6–50	3,000	Automatic, batch	Continuous overflow	60 ft^3/batch
Variable Speed, Vertical	12–84	3,200	Automatic, batch	Continuous	15 ft^3/batch
Continuous Decanting	to 50	3,200	Continuous	Continuous	5,000 GPH, or 65 tons/hr of solids

The sigma concept is often employed to compare the performance of solid wall centrifuges. This equivalence converts the geometry, size and speed of the bowl to the area of a settling tank theoretically capable of the same separation in unit gravity field.

The control concepts applicable to these designs are best discussed in three categories: batch, semi-continuous and continuous.

Batch

All laboratory centrifuges and some of the small sedimentation machines in industry are operated as batch units with manual control of feed and manual discharge of solids. In such instances, little automatic control is required, since all sequencing is being done manually.

Semi-Continuous

In this category of automatic batch type centrifuges, one finds the fall disc type machines, such as the desludging centrifuges and also some solid bowl types.

One process control scheme for a desludging or clarifier type centrifuge is shown in Figure 10.10c. The feed to the machine is introduced from a sludge or magma feed tank by gravity. It has been found that if a minimum head of 10 ft is available, gravity flow is adequate for control. Feed tank location and size is determined along with pipe size for the specific application.

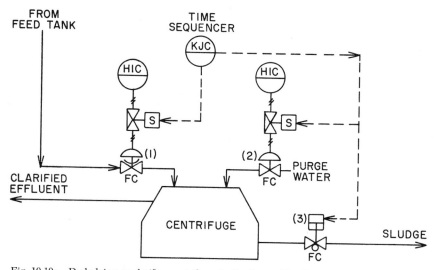

Fig. 10.10c   Desludging or clarifier centrifuge. In the figure, (1) valve is normally open, except during flushing; (2) valve normally closed, except during flushing; (3) valve operated intermittently to discharge solids.

The feed enters this unit at a predetermined rate and is continuously clarified. The clarified effluent is then discharged. The sludge accumulates in the system and is periodically ejected. The unit may or may not run at full speed while sludge is ejected.

The interlock system needs at least one sequence timer and three or more working contacts to control valves on the following lines: feed, sludge and purge water.

Usually added are partial desludging and other refinements, such as water spray in specific parts of the unit.

In these units the choice of control valves is of major importance. Properly chosen ball or plug type valves work satisfactorily, as do diaphragm types on low pressure, on-off service. For throttling valve designs applicable to slurry service, refer to Section 1.21.

In one modification of this machine, the sludge accumulates in an outer chamber. When the chamber is full of sludge, a sensing device activates the desludging operation. Such a unit can tolerate significant variations in the proportion of sludge in feed liquor.

Continuous

Among the versatile units employed in petrochemical processing, polymerization and waste treatment systems are the fully continuous solid bowl sedimentation centrifuges.

One example of a solid bowl machine is shown in Figure 10.10d. The

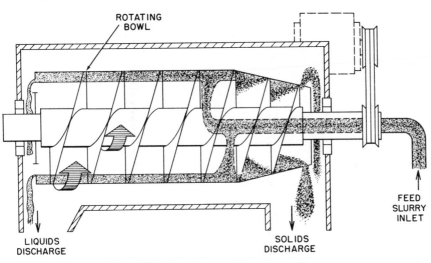

Fig. 10.10d   Continuous, solid-bowl centrifuge

slurry is introduced in the revolving bowl of the machine through a stationary feed tube at the center. It is acted upon by centrifugal force, and the solids (denser phase) are thrown against the wall. Inside the rotating bowl is a screw conveyor with a slight speed differential with respect to the bowl rotation, and it moves the solids up the beach and out of the liquid layer. Solids and clarified liquid are continuously discharged. Control of these solid bowl machines requires consideration of the torque developed by the unloading plow and other internal design considerations. One control scheme shown in Figure 10.10e automatically adjusts the feed rate to maintain the torque. Load cells are frequently applied in such installations as torque detectors.

With interlock in effect (not shown in figure), feed is closed, flush is opened and unit is shut down. This is brought about by (1) low coolant or

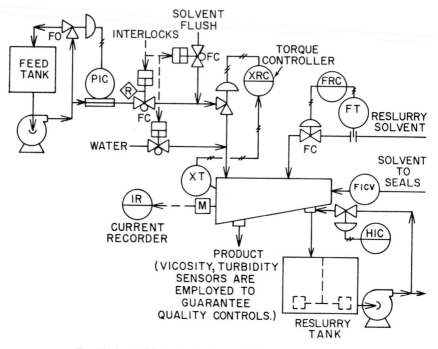

Fig. 10.10e   Solid-bowl centrifuge with feed on torque control

flush flows to seals, (2) low flow or high temperature in lubricating oil system, (3) high motor current and (4) high torque. Other, less severe conditions, such as low reslurry solvent flow, might temporarily stop feed and open flush without shutdown.

It is important that a reasonably uniform slurry be supplied to the machine. For this purpose, a circulating pump with a recycle line is usually installed to keep the slurry in motion and thus preventing settling out of crystals in the tank or in the pipelines. In case the slurry does not settle quickly, magma extraction can be done by a siphon.

Another method of controlling a solid bowl machine is to adjust the relative speed of the conveyor and the bowl to balance the unloading requirements. In such cases, the process slurry feed to the machine can be on flow control. Balancing of the machine is done by measuring the torque and adjusting the differential on the scroll (conveyor).

Filtration Centrifuges

Filtration centrifuges are also of three types: batch, automatic-batch and fully continuous. In all of them a cake is deposited on a filter medium held in a rotating basket, which is then washed and spun dry. The method of solids discharge distinguishes the type. Some of the more common designs are summarized in Table 10.10f. The important control aspects are illustrated by considering the control scheme around an automatic batch machine and on a fully continuous one.

Table 10.10f

CHARACTERISTICS OF FILTERING CENTRIFUGES

Rotor Type	Machine Type	Maximum Centrifugal Force (g)	Method of Solids Discharge	Method of Liquid Discharge	Maximum Capacity
Conical Screen	Wide Angle	1,400	Continuous	Continuous	15,000 GPM
	Differential Scroll	1,800	Continuous	Continuous	70 tons/hr of solids
	Vibrating Screen	500	Continuous	Continuous	100 tons/hr of solids
	Pusher	1,800	Batch	Continuous	10 tons/hr of solids
Cylindrical Screen	Pusher	1,500	Continuous	Continuous	40 tons/hr of solids
	Differential Scroll	1,500	Continuous	Continuous	40 tons/hr of solids
	Horizontal	1,250	Batch	Intermittent	25 tons/hr of solids
	Vertical	900	Batch	Intermittent	10 tons/hr of solids

Automatic Batch

Two versions of the control scheme around a horizontal cylindrical basket machine (also called a "peeler") on automatic-batch sequence control are shown in Figure 10.10g. In one case the circulating feed pump feeds one or more centrifuges against a head tank, and the overflow from this tank is returned to the crystallizer. In the second scheme the feed is by gravity alone.

A typical automatic batch machine will operate on the following sequence cycle:

1. *Screen rinse*   Residual layer of crystals is rinsed prior to loading.
2. *Loading*   Feed is admitted through the feed valves.
3. *Cake rinse*   After the crystal layer is established, feed is stopped and rinsing is started.
4. *Drying*   The cake (after washing) is spun dry.
5. *Unloading*   The unloading knife is cut in to discharge solids.

In these machines, the feed slurry and wash liquor line pressure are assumed to be reasonably steady, and therefore feed flow controllers are not

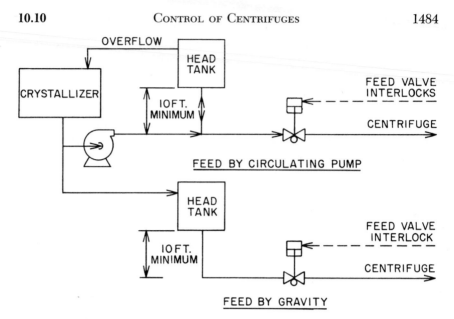

Fig. 10.10g   Intermittent feeding of automatic batch centrifuges

employed. The sequence just discussed is that of a constant-speed machine. A variable-speed machine has a similar sequence but with time periods provided for acceleration and deceleration at appropriate points.

The entire sequence of the batch cycle is governed by one or more sequence timers with contacts to actuate the valves.

Continuous

In a fully continuous machine of the reciprocating pusher type or the differential scroll type, the only control required is that of constant slurry and wash-water feed rate.

When a magma is to be extracted continuously, there is the need to control the quantity of solids. This involves measuring the volumetric feed flow rate, detecting the solids concentration and throttling a valve to maintain the solids charge rate constant. (Sections 5.21 and 3.4 of Volume I and Section 1.21 of Volume II cover the hardware required for such a system.) In many cases the expense is not justified to install such systems.

There are two simple and inexpensive methods by which a crystal magma can be continuously withdrawn at controlled rates:

Removal is in small batches but at such high frequency as to make the flow virtually continuous. The air-operated pulsing valves with adjustable stroke and frequency may be used in this application. The volumetric feeders of the ball check type (Section 2.3) are also suitable for these services.

An elutriation leg attached to a crystallizer with a laundering fluid inflow can also be used to control magma density. One version of this is shown in

Figure 10.10h. The laundering fluid is on flow control with its set point adjusted to maintain density as a measure of solids concentration. Measurement of the differential pressure between two points in the crystallizer can be a method of density detection.

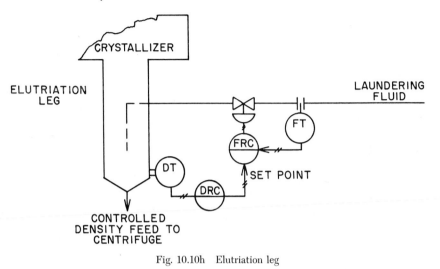

Fig. 10.10h   Elutriation leg

Occasionally raw feed may be used as laundering fluid, but filtrate is more commonly employed.

Washing in these machines is also fully continuous, and wash fluid flow is on automatic control.

A special application is that of a scraped surface crystallizer, producing the slurry for separation in a continuous centrifuge. In such case the liquor can be fed by a metering pump, with the centrifuge mounted in series with the crystallizer.

Conclusions

Some of the more common methods of controlling a centrifuge were outlined in this section. It should be noted that the actual centrifuge operation includes many auxiliaries, such as wash lines for preventing solid build-up in various parts of the unit, lube oil pumps and coolers, and discharge hoppers, with devices to prevent bridging.

As with other high-speed machinery, manufacturers have standard units that are adapted to the application. Operating characteristics vary significantly, while sludge and solid handling problems may add to the overall complexity.

ACKNOWLEDGMENT

F. T. Costigan of the Sharples Corporation assisted in the preparation of this material.

10.11 CONTROL OF HEAT EXCHANGERS

Practically all chemical plants include some form of heat transfer equipment. The control of these devices will be covered in this section. Both the nature of heat exchange and the quality of control desired play a part in determining the proper instrumentation for the unit. As far as the nature of heat transfer is concerned, the following paragraphs will discuss liquid-to-liquid exchangers, steam heaters, condensers and reboilers as the basic forms of heat exchange. Next the more sophisticated control systems—such as cascade or feedforward—are covered, and the section ends with a brief evaluation of multipurpose heat transfer systems.

Variables and Degrees of Freedom

All control loops function on the basis of controlling one variable by manipulating the same or some other process variable(s). It is important to be able to determine the maximum number of independently acting automatic controllers that can be placed on a process. By definition, this is called the number of degrees of freedom. (See Section 7.2 for more details.)

Figure 10.11a shows a steam heater with its variables and defining parameters. The temperatures and flows are variables, while the specific and latent heats are parameters. The available degrees of freedom are arrived at by subtracting the number of system-defining equations from the number of variables.

Degrees of freedom = (number of variables) − (number of equations)

$$10.11(1)$$

In this case, there are four variables and one equation, which is obtained from the first law of thermodynamics, stating the conservation of energy:

$$\lambda W_s = C_p W(T_2 - T_1) \qquad 10.11(2)$$

Therefore, this system has three degrees of freedom which at the same time is the maximum number of automatic controllers that can be employed.

In a liquid to liquid heat exchanger there are four temperature and two flow variables with still only one (conservation of energy) defining equation, (10.11(3)) resulting in five degrees of freedom.

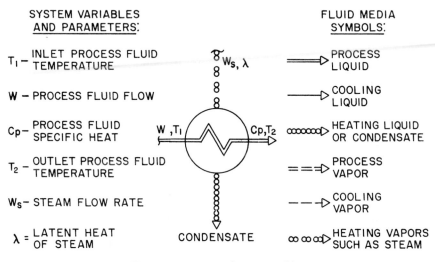

Fig. 10.11a   Heat exchanger variables

$$C_pF(T_2 - T_1) = C_{pc}F_c(T_{1c} - T_{2c}) \qquad 10.11(3)$$

In a steam-heated reboiler or in a condenser cooled by a vaporizing refrigerant (assuming no superheating or supercooling) there are only two flow variables and one defining equation (10.11(4)) allowing for only a single degree of freedom or one automatic controller.

$$\lambda W_s = \lambda_1 W_1 \qquad 10.11(4)$$

In the majority of installations fewer controllers are utilized than there are degrees of freedom available, but every once in a while problems associated with overdefining a system are also likely to arise.

Liquid-to-Liquid Heat Exchangers

In Figure 10.11b the pneumatic transmitters are of the indicating, filled-bulb type, and the indicating controller is panel mounted. These features have

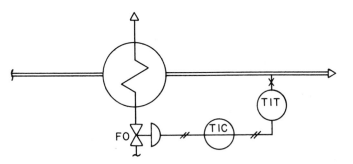

Fig. 10.11b   Feedback control by throttling coolant inlet

been uniformly used on all illustrations in this section for reasons of consistency only.

The purpose of this section is to evaluate control techniques and not to recommend hardware selection. Therefore, it should be clearly understood that it is not the writer's intent to express a preference for filled thermal bulbs over thermocouples or for pneumatic indicating instruments over electronic recording ones. In fact, the repeatability, sensitivity and speed of response of electronic systems with thermocouple elements is considered to be superior. For more details on temperature detectors, see Chapter IV in Volume I.

Figures 10.11b and c illustrate cooler and heater installations with the control valve mounted on the exchanger inlet and outlet respectively. From a control quality point of view, it makes little difference whether the control valve is upstream or downstream to the heater. The location is normally based on the desirability of operating the heat-transfer-medium side of the exchanger under supply or return header pressures.

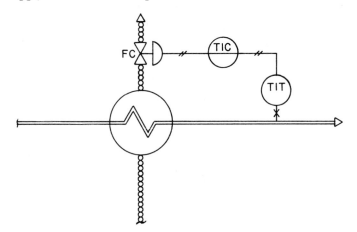

Fig. 10.11c   Feedback control by throttling heating media outlet

Component Selection

It is generally recommended to provide positioners for these valves to minimize the valve friction effects. The use of equal percentage valve trims (Section 1.5) is also recommended, because it usually contributes to maintaining the control system gain constant, under changing throughput conditions. This is due to the equal percent trim, which keeps the relationship between valve opening and temperature change (reflecting load variations) constant.

In the majority of installations, a three-mode controller would be used for heat exchanger service. The derivative or rate action becomes essential in long time-lag systems or where sudden changes in heat exchanger throughput are expected. Due to the relatively slow nature of these control loops, the proportional band setting must be wide to maintain stability. This means that the valve will be fully stroked only due to a substantial deviation from desired temperature set point. (For controller tuning, see Section 7.12.) The reset or integral control mode is required to correct for temperature offsets due to process load changes. Besides the changes in process fluid flow rate, other variables can also give the appearance of load changes, such as inlet temperature or header pressure changes of the heat transfer medium.

The selection and location of the thermal element is also important. It must be placed in a representative location, without increasing measurement time lag. In reference to Figure 10.11b, this would mean that the bulb should be located far enough from the exchanger for adequate mixing of the process fluid, but close enough so that the introduced time delay will not be substantial. If the process fluid velocity is 3 ft/second, then a one-second distance-velocity lag is introduced for each 3 ft of pipe between the exchanger and the bulb. This lag can be one of the factors which will limit the dynamic performance of the system, but it should also be clearly understood that all thermal elements are dynamically imperfect to start with. This inherent limitation results from the fact that in order to change the temperature of a sensor, heat must be introduced, and it has to enter the bulb through a fixed area.

Let us calculate the dynamic lag of a typical filled bulb having the area of 0.02 sq ft and the heat capacity of 0.005 BTU/°F. If this bulb is immersed in a fluid with a heat transfer coefficient (based on flow velocity) of 60 BTU/hr/°F/ft^2, and then the process temperature is changed at a rate of 25°F/minute, the dynamic lag can be calculated. First the amount of heat flowing into the element under these conditions will be determined:

$$q = \text{(rate of temperature change)(bulb heat capacity)}$$
$$= (25)(60)(0.005) = 7.5 \text{ BTU/hr} \qquad 10.11(5)$$

The dynamic measurement error is calculated by determining that temperature differential across the fluid film surrounding the bulb which is required to produce a heat flow of 7.5 BTU/hr.

$$q = Ah \, \Delta T \qquad 10.11(6)$$

and therefore,

$$\Delta T = q/Ah = 7.5/(0.02)(60)$$
$$= 6.25°F \qquad 10.11(7)$$

If the rate of process temperature change is 25°F/minute and the dynamic error based on that rate is 6.25°F, then the dynamic time lag is

$$t_0 = 6.25/25 = 0.25 \text{ minutes}$$
$$= 15 \text{ seconds} \qquad 10.11(8)$$

This lag can also be calculated as

$$t_0 = \frac{\text{(bulb heat capacity)}}{\text{(bulb area)(heat transfer coefficient)}}$$

$$= \frac{60 \times 0.005}{0.02 \times 60} = 0.25 \text{ minutes} \qquad\qquad 10.11(9)$$

Bulb time lags vary from a few seconds to minutes, depending on the nature and velocity of the process fluid being detected. Measurement of gases at low velocity involves the longest time lags, and water (or dilute solutions) at high velocity the shortest. From equation 10.11(9), it is clear that one method of reducing time lag is by miniaturizing the sensing element.

The above numbers were based on a bare bulb diameter of $\frac{3}{8}$ in.

The addition of a thermal well will further increase the lag time, but in most industrial installations, their use is necessary for reasons of safety and maintenance. When thermowells are used, it is important to eliminate any air gaps between the bulb and the socket.

The measurement lag is only part of the total time lag of the control loop. For example, an air heater might have a total lag of 15 minutes, of which 14 minutes is the *process lag*, 50 seconds is the bulb lag, and 10 seconds is the time lag in the control valve.

Three-Way Valves

The limits within which process temperature can be controlled are a function of the nature of load changes expected and of the speed of response for the whole unit. In many installations, the process time lag in the heat exchanger is too great to allow for effective control during load changes. In such cases, it is possible to circumvent the dynamic characteristics of the exchanger by partially bypassing it and blending the warm process liquid with the cooled process fluid, as shown in Figure 10.11d. The resulting increased

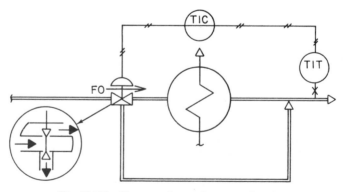

Fig. 10.11d   Diverter valve used to control cooler

system speed of response, together with some cost savings are the main motivations for considering 3-way valves in such services. The "bulb time lag" discussed in the previous paragraph has an increased importance in these systems, because it represents a much greater *percentage* of total loop lag time, than in the previously discussed installations.

As illustrated in Figures 10.11d and e either a diverter or a mixing valve can be used for this purpose. Stable operation of these valves is achieved by *flow* tending *to open* the plugs in both cases. If a mixing valve is used for diverting service or a diverting one for mixing, the operation becomes unstable due to the "bathtub effect" resulting from the flow to close path across the valve. Therefore, it is not good enough to just install a 3-way valve, it has to be either the mixing or the diverting design, to match the particular service.

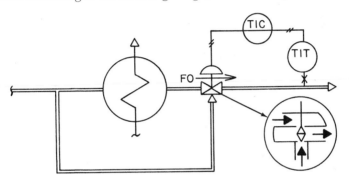

Fig. 10.11e   Mixing valve used to control cooler

Three-way valves are unbalanced designs and are normally provided with linear ports. Its unbalanced nature places a limitation on allowable shutoff pressure difference across the valve, and its availability with linear ports only prevents the relationship between valve movement and temperature change from being constant. (Equal percentage characteristics are required for that, because a change in detected temperature usually requires an exponential change for the flow rate of the heat transfer fluid to reach a new equilibrium at the changed load level.)

Misalignment or distortion in a control valve installation can cause binding, leakage at the seats, high deadband and packing friction. Such conditions commonly arise as a result of high-temperature service on three-way valves. The valve, having been installed at ambient conditions and rigidly connected at three flanges, cannot accommodate pipe line expansion due to high process temperature and therefore distortion results. Similarly, on mixing applications, when the temperature difference is substantial between the two ports, the resulting differential expansion can also cause distortion. For these reasons, the use of 3-way valves at temperatures above 500°F or at differential temperatures exceeding 300°F is not recommended.

The choice of three-way valve location relative to the exchanger (Figures 10.11d and e) is normally based on pressure and temperature considerations, with the upstream location (Figure 10.11d) being usually favored for reasons of uniformity of valve temperature. Where the overriding consideration is to operate the exchanger at a high pressure, the downstream location might be selected.

Cooling Water Conservation

Figure 10.11f modifies the previous sketch to include the additional feature of cooling water conservation. This system tends to maximize the outlet cooling water temperature and thereby minimize the rate of water usage. The application engineer, when using this concept, should be careful in evaluating the temperature levels involved and to make sure that the water contains chemicals to prevent tube fouling, if high temperature operation is planned. If the cooling water supply is sufficient this system will not only conserve water, but will also protect against excessively high outlet water temperatures.

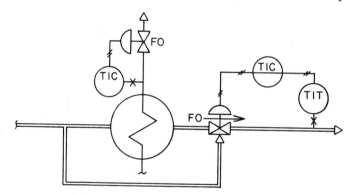

Fig. 10.11f   Conservation of cooling fluid

Balancing the Three-Way Valve

It is recommended to install a manual balancing valve in the exchanger bypass as shown in Figure 10.11g. This valve is so adjusted that its resistance to flow equals that of the exchanger.

When sizing the pump it should be kept in mind that the *resistance* to flow in such installation will be *maximum, when one of the paths is closed and the other is fully open*, while minimum resistance will be experienced when the valve equally divides the flow between the two paths. This is not necessarily self-evident and therefore it will be illustrated on an example.

It is assumed that the flow rate of process fluid is 100 GPM, the pressure drop at full flow through either the exchanger or the balancing valve is 9 PSI and that the diverting valve has a valve coefficient of Cv = 100. The equivalent coefficient for the exchanger (or balancing valve) is calculated as:

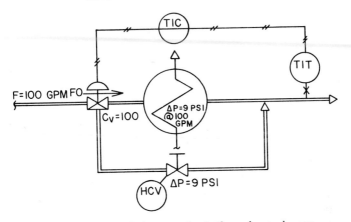

Fig. 10.11g   Use of balancing valve in the exchanger by-pass

$$C_e = \frac{\text{flow}}{\sqrt{\text{pressure drop}}} = \frac{100}{\sqrt{9}}$$

$$= 33.3 \qquad\qquad 10.11(10)$$

Therefore in either extreme position (closed or full bypass) the total system resistance expressed in valve coefficient units, is;

$$\frac{1}{(C_t)^2} = \frac{1}{(C_v)^2} + \frac{1}{(C_e)^2} = \frac{1}{(100)^2} + \frac{1}{(33.3)^2} \qquad 10.11(11)$$

$$\therefore C_t = 31.7$$

When the valve divides the flow equally between the two paths, due to the linear characteristics of all three-way valves, its coefficient at each port will be $C_v = 50$. The equivalent coefficient, $C_e = 33.3$ of the exchanger and balancing valve being unaffected, the total system resistance in valve coefficient units, is $2C_t$, where;

$$\frac{1}{(C_t)^2} = \frac{1}{(C_v)^2} + \frac{1}{(C_e)^2} = \frac{1}{(50)^2} + \frac{1}{(33.3)^2} \qquad 10.11(12)$$

$$\therefore 2C_t = 55.6$$

If we calculate the total pressure drop through the system when the valve is in its extreme and when it is in its middle position, handling the same 100 GPM flow;

$$\Delta P \text{ extreme} = \left(\frac{\text{flow}}{C_t}\right)^2 = \left(\frac{100}{31.7}\right)^2 = 10 \text{ PSI} \qquad 10.11(13)$$

$$\Delta P \text{ middle} = \left(\frac{100}{55.6}\right)^2 = 3.25 \text{ PSI} \qquad 10.11(14)$$

the results indicate that the system drop in one of the extreme positions is more than three times that of the middle position.

Two Two-Way Valves

When for reasons of temperature or because of other considerations, three-way valves cannot be used, but it is desired to improve the system response speed by the use of exchanger by-pass control, the installation of two two-way valves is the logical solution.

As illustrated in Figure 10.11h the two valves should have opposite failure positions. Therefore, when one is open the other is closed and at a 9 PSI signal both are half-way open. In order for these valves to give the same control as a three-way valve would, it is necessary to provide them with linear plugs.

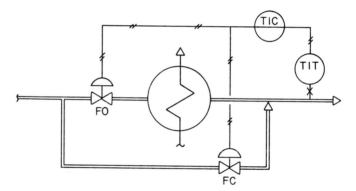

Fig. 10.11h   Exchanger by-pass control using two-way valves

It is logical for the reader to ask; why not always use this two two-way valve system, why consider three-way valves at all? The answer is basically; cost. The price of a three-way valve is about 65 percent of two two-ways, the installation cost is similarly lower and when positioners are required (the majority of cases), only one needs to be purchased, instead of two. There are some other considerations, which might tend to modify this impression of substantial cost savings. One such consideration is that the capacity of a three-way valve is the same as the capacity of a single-ported two-way valve, which is only 70 percent of the capacity of a double ported one. This could mean that instead of a 10-in. three-way unit, two 8-in. double ported two-way valves can be considered and therefore the cost advantage of a three-way installation is not as substantial as implied earlier. This is true if the comparison is based on double ported two-way valves, but the inherent leakage (about $\frac{1}{2}$ percent of full capacity) of these units makes them unsuitable for some installations. Therefore, where tight shutoff is required, only the three-way or the single-ported two-way valves can be considered, and their capacity is about the same.

As touched upon earlier, three-way valves are not recommended for high temperature or high-pressure differential service. In addition to the reasons noted earlier, the hollow plug design of three-way valves also contributes to this limitation because it is more subject to heat expansion and is more difficult to harden than the solid plugs.

To summarize, by-pass control is applied to circumvent the dynamic characteristics of heat exchangers and thereby improve their controllability. By-pass control can be achieved by the use of either one three-way or two two-way valves, and Table 10.11i summarizes the merits and drawbacks of using one or the other.

Table 10.11i
MERITS OF TWO-WAY vs THREE-WAY VALVES IN EXCHANGER
BY-PASS INSTALLATIONS

	One Three-Way Valve	Two Single-Seated Two-Way Valves	Two Double-Ported Two-Way Valves
Most economical	Yes		
Provides tight shutoff	Yes	Yes	
Applicable to service above 500°F		Yes	Yes
Applicable to differential temperature service above 300°F		Yes	Yes
Applicable to operation at high pressure and pressure differentials		Yes	Yes
Highest capacity for same valve size			Yes

Steam Heaters

The general discussion on loop components, accessories, sensor location and time lag considerations, presented in connection with liquid to liquid heat exchangers is also applicable here. The desirability of equal percentage valve trims is even more pronounced than it was on liquid to liquid exchangers because of the high rangeability required on most installations. The need for high rangeability is partially due to the variations in condensing pressure (valve back-pressure) with changes in process load. This can be best visualized by an example.

In Figure 10.11j both the high and the low load conditions are shown. When the steam flow demand is the greatest, the back-pressure is also the highest, leaving the lowest driving force (pressure drop) for the control valve. High flow and low pressure drop results in a large valve which might be throttled beyond its capability under low load conditions. In our example,

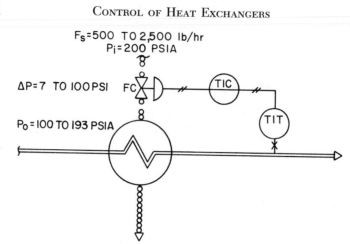

Fig. 10.11j   Feedback control of steam-heated exchanger

the back-pressure at low loads is only 100 PSIA, allowing some sixteen times greater pressure drop through the valve than at high loads. The ratio between the required valve coefficients for the high and low load conditions, represents the rangeability which the valve has to furnish.

$$\text{Rangeability} = S\frac{F_{s,\text{max}}}{F_{s,\text{min}}}\sqrt{\frac{[(P_1 - P_2)(P_1 + P_2)]_{\text{min}}}{[(P_1 - P_2)(P_1 + P_2)]_{\text{max}}}}$$
$$= 1.5 \times 5 \sqrt{\frac{100 \times 300}{7 \times 393}} = 25.5 \qquad 10.11(15)$$

The letter "S" with the numerical value of 1.5 in the above equation represents the safety factor which is applied in selecting the control valve. A rangeability requirement of this magnitude can create some control problems which can best be solved by the installation of a smaller control valve in parallel with the large one. If this is not done, the control quality will suffer for two reasons:

1. At low loads the valve will operate near to its clearance flow point where the flow versus lift curve changes abruptly, contributing to unstable or possibly on-off cycling valve operation.

2. For good control the system gain should not vary with changes in load, which an equal percentage control valve can guarantee only if control valve ΔP is not a function of load. This being the case the only way to guarantee constant system gain, is to install two valves for the two load conditions, both sized to maintain the same gain.

For a more detailed discussion on control valve rangeability, refer to Section 1.21.

Minimum Condensing Pressure

As it has been shown in connection with Figure 10.11j, the condensing pressure is a function of load, when the temperature is controlled by throttling the steam inlet. This at low loads and low operating temperatures can result in below atmospheric condensing pressures. If this occurs, then the condensing pressure will not be sufficient to discharge the condensate through the steam trap, which therefore accumulates inside the heater. As condensate accumulation progresses, more and more of the heat transfer area will be covered up, resulting in a corresponding increase in condensing pressure. When this pressure rises sufficiently to discharge the trap, the condensate is suddenly blown out and the effective heat transfer surface of the exchanger increases several fold, instantaneously. Such upsets of course make the control of temperature impossible and methods of improving this situation have to be considered. In the following paragraphs, the merits of some of these techniques will be discussed.

Control Valve in the Condensate Line

Mounting the control valve in the condensate line, as shown in Figure 10.11k, is sometimes proposed as a solution to minimum condensate pressure problems. An additional motivation to consider this technique is the cost advantage of purchasing a small condensate valve, instead of a larger one for steam service.

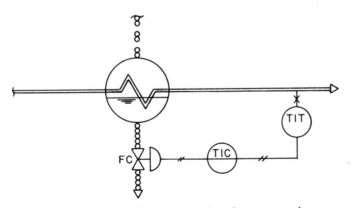

Fig. 10.11k   Controlling the rate of condensate removal

On the surface this appears to be a very convenient solution, since the throttling of the valve causes variations only in the condensate level inside the partially flooded heater and has no effect on the steam pressure, which stays constant. Therefore, there are no problems in condensate removal.

What is important to keep in mind is, that the criterion for a successful heater installation is not whether the accumulated condensate is removable, but whether it allows for precise temperature regulation. The answer is negative, the valve in the condensate line does not allow for accurate control of temperature. The reason for this is in the dynamics of the system.

When the load is decreasing, the valve is likely to close completely, before the condensate built up to a high enough level to match the new lower load with a reduced heat transfer area. In this direction, the process is slow, because steam has to condense before level can be affected. When the load increases, the process is fast because just a small change in control valve opening is sufficient to drain off enough condensate to expose an increased heat transfer surface. Having these "non-symmetrical" dynamics, control is bound to be poor. If the controller is tuned for the fast response speeds of the increasing load direction, substantial overshoot can result when load is decreasing, while if it is tuned for the slow part of the cycle, the overshoot and possible cycling occurs in the opposite direction.

For the above reasons, it is recommended that control valves be not mounted in the condensate lines on services where temperature control is important.

Pumping Traps

It is possible to prevent condensate accumulation in heaters operating at low condensing pressures, by the use of lifting traps. This device is illustrated in Figure 10.11l and it depends on an external pressure source for its energy source.

The unit is shown in its filling position, where the liquid head in the heater has opened the inlet check valve of the trap. Filling progresses until the condensate overflows into the bucket, which then sinks, closing the equalizer and opening the pressure source valve. As pressure builds in the trap, the

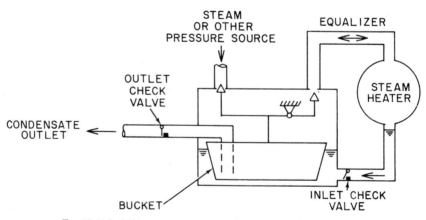

Fig. 10.11 l Lifting or pumping trap shown in its "filling" position

inlet check valve is closed and the outlet one opens when the pressure exceeds that of the condensate header. The discharge cycle follows during which the bucket is emptied. When near empty, the buoyant force raises the bucket which then closes off the steam valve and opens the equalizer. Once the pressure in the trap is lowered the condensate outlet check valve is closed by the header back pressure, and the inlet check is opened by the liquid head in the heater, which then is the beginning of another fill cycle.

The above described pumping trap guarantees condensate removal regardless of the minimum condensing pressure in the heater. If such a trap is placed on the exchanger illustrated in Figure 10.11j, it will make temperature control possible, even when the heater is under vacuum. This of course does not relieve the rangeability problems discussed earlier and the use of two valves in parallel might still be necessary.

Level Controllers

Because the low condensing pressure situation is a result of the combination of low load and high heat transfer surface area, it is possible to prevent the vacuum from developing by reducing the heat transfer area. One method of achieving this is shown in Figure 10.11m, where the steam trap has been replaced by a level control loop. With this instrumentation provided, it is possible to "adjust the size" of the heater by changing the level set-point, to match the process load. This technique gives good temperature control if the level setting is correctly made and if there are no sudden load variations in the system. The response to load variations is non-symmetrical just as it was outlined in connection with mounting the temperature control valve in the condensate line.

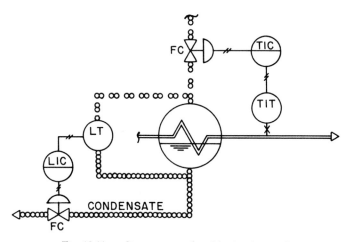

Fig. 10.11m   Steam trap replaced by level control

One disadvantage of this approach is the relatively high cost.

A continuous drainer trap, such as the one shown in Figure 10.11n serves the same purpose as the level control loop above. Its cost is substantially lower, but it is limited in the range within which the level setting can be varied and its control point is offset by load variations. For this reason, it is unlikely to be considered for installation on vertical heaters or reboilers, where the range of level adjustment can be substantial.

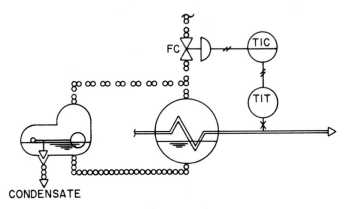

CONDENSATE

Fig. 10.11n   Continuous drainer trap

By-pass Control

Table 10.11i in the preceding text summarized some of the features of three-way valves, when installed to circumvent the transient characteristics of the cooler, by bypassing it. Figure 10.11o shows the same concept applied to a steam heater. The advantages and limitations of this system are the same

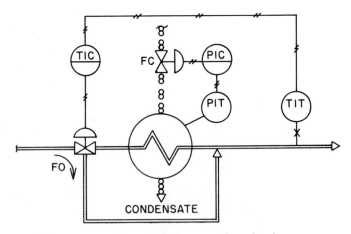

CONDENSATE

Fig. 10.11o   By-pass control on steam-heated exchanger

as discussed in connection with liquid-liquid exchangers, but there is one additional advantage. This has to do with the fact that the bypass created an additional degree of freedom, and therefore steam can now be throttled as a function of some other property. The logical decision is to adjust the steam feed so that it maintains the condensing pressure constant. This then eliminates problems associated with condensate removal. One should also realize that in case of full by-pass operation, the stagnant exchanger contents will be exposed to steam heat, and therefore, unless protected, it is possible to boil this liquid.

Table 10.11p summarizes some of the features of the various techniques discussed, which can be applied to combat problems created by low condensing pressures.

Table 10.11p
FEATURES OF HEATER CONTROL SYSTEMS

		Condition for Giving Precision Control		
	System Cost	Small or Slow Load Variations	Fast or Large Load Variations	Low (Vacuum) Condensing Pressures
Valve in condensate line throttled by temperature	Low	Questionable	No	No
Valve in steam line throttled by temperature	Medium	Yes	No	No
Two valves in steam line throttled by temperature	Medium	Yes	Yes	No
Valve in steam line throttled by temperature and condensate removed by drainer trap	Medium	Questionable	No	Yes
Valve in steam line throttled by temperature and condensate removed by pumping trap	Medium	Questionable	No	Yes
Valve in steam line throttled by temperature and condensate removed by level controller	High	Yes	No	Yes
Three-way valve controls temperature by throttling by-pass, and steam inlet is controlled to maintain condensing pressure	High	Yes	Yes	Yes

Condensers

The control of condensers and reboilers as part of larger systems is covered in Sections 10.3 and 10.4, for which reason only a brief discussion is presented here.

Depending on whether the control of condensate temperature or of condensing pressure is of interest, the systems shown in Figures 10.11q and r can be considered. Both of these throttle the cooling water flow through the condenser, causing a potential for high temperature rise which is acceptable only when the water is chemically treated against fouling. For good, sensitive control the water velocity through the condenser should be such that its residence time does not exceed one minute.

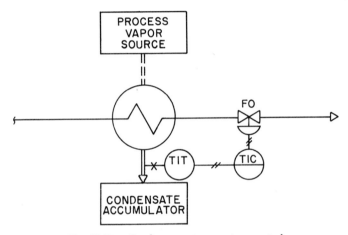

Fig. 10.11q   Condenser on temperature control

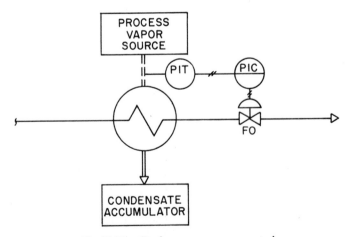

Fig. 10.11r   Condenser on pressure control

When it is not desirable to throttle the cooling water, the system illustrated in Figure 10.11s can be considered. Here the exposed condenser surface is varied to control the rate of condensation. Where non-condensables are present, a constant purge may serve to remove the inerts. One drawback of this system is the same "non-symmetricity" which has been discussed in connection with Figure 10.11k.

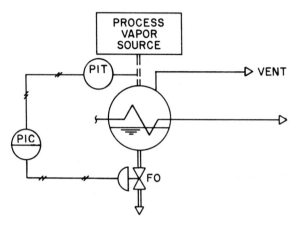

Fig. 10.11s   Condenser control by changing the wetted surface area

To reduce the problems associated with non-symmetric process dynamics, the "hot gas by-pass" systems shown in Figures 10.11t and 10.11u can be considered. In case of Figure 10.11t, the opening of the bypass valve, results

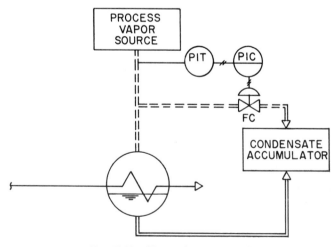

Fig. 10.11t   Hot gas by-pass control

in pressure equalization between condenser and accumulator causing partial flooding of the condenser, due to the relative elevations. When the condensing pressure is to be reduced, the valve closes, resulting in an increase of exposed condenser surface area. In order to expose more area, the condensate is transferred into the accumulator which can occur only if the accumulator vapor pressure has been lowered due to condensation. Therefore, the system speed in this direction is a function of condensate super-cooling. Increased super-cooling increases system response speed. Based on these considerations the unit might or might not be symmetrical in its dynamics. If high speed is desired, the controls shown in Figure 10.11u can be considered.

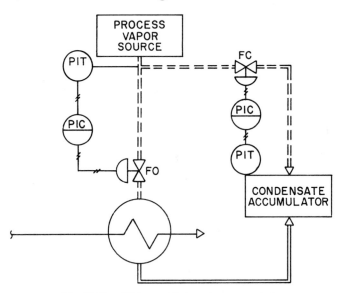

Fig. 10.11u   High-speed hot gas by-pass control

When the condensing temperature of the process fluid is low, water is no longer an acceptable cooling media. One standard technique of controlling a refrigerated condenser is illustrated in Figure 10.11v. Here the heat transfer area is set by the level control loop and the operating temperature is maintained by the pressure controller. When process load changes, it affects the rate of refrigerant vaporization which is compensated for by level controlled makeup. Usually the pressure and level settings are manually made, although there is no reason why these set points could not be automatically adjusted as a function of load, if required. (See Section 10.5 for refrigeration units.)

Reboilers and Vaporizers

As noted at the beginning of this section, in case of a steam heated reboiler, there is only one degree of freedom available and therefore only

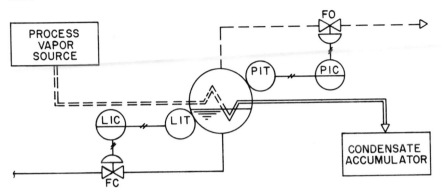

Fig. 10.11v   Condenser controls with refrigerant coolant

one controller can be installed without overdefining the system. This one controller is usually applied to adjust the rate of steam addition. As to the minimum condensing pressure considerations, the same applies here as has been discussed earlier in connection with liquid heaters. The Figures 10.11w and x show the two basic alternates for controlling the reboiler either to generate vapors at a controlled superheat temperature or to generate saturated vapors at a constant rate, set by the rate of heat input.

A number of more sophisticated alternates are discussed in Sections 10.3 and 10.4, where the possibilities of control by temperature difference, composition or by various cascade and computer methods are covered.

Cascade Control

The reader is referred for a general discussion of cascade control, to Section 7.8. Probably the most frequent use of cascade loops is in connection with heat transfer units. Cascade systems by definition consist of two control-

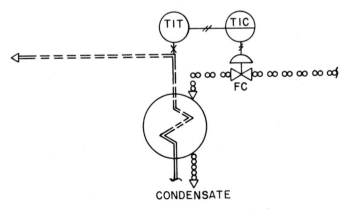

Fig. 10.11w   Temperature control of reboilers

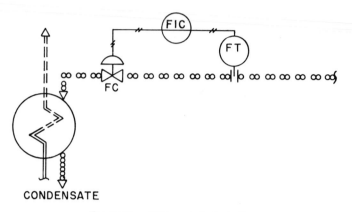

Fig. 10.11x   BTU control of reboilers

lers in series. The master in case of heat exchangers detects the process temperature and the slave is installed on a variable that may cause fluctuations in the process temperature. The master adjusts the slave set-point and the slave throttles the valve to maintain that set-point. It cannot be over-emphasized that a cascade loop controls *a single* temperature and the slave controller is there only to assist in achieving this. In other words, the cascade loop does not have two independent set points.

     Cascade loops are invariably installed to prevent outside disturbances from entering the process. An example of such would be the header pressure variations of a steam heater. The conventional single controller system (Figure 10.11j) cannot respond to a change in steam pressure until its effect is felt by the process temperature sensor. In other words, an error in the detected temperature has to develop before corrective action can be taken. The cascade loop in contrast, responds immediately, correcting for the effect of pressure change, before it could influence the process temperature (Figure 10.11y).

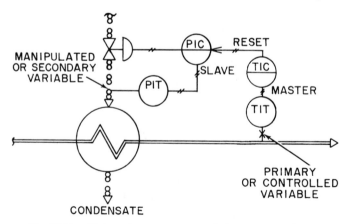

Fig. 10.11y   Temperature-pressure cascade loop on steam heater

The improvement in control quality due to cascading is a function of relative speeds and time lags. A slow primary (master) variable and a secondary (slave) variable which responds quickly to disturbances represent a desirable combination for this type of control. If the slave can quickly respond to fast disturbances then these will not be allowed to enter the process and thereby will not upset the control of the primary (master) variable. Some typical cascade applications for heat exchangers are illustrated on the following sketches.

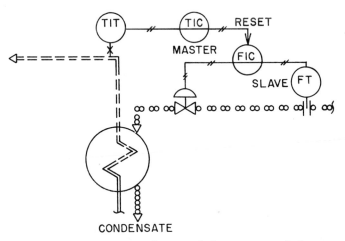

Fig. 10.11z   Temperature-flow cascade loop on steam reboiler

In Figures 10.11y and 10.11z, the controlled variable is temperature, while the manipulated variable is the pressure or flow of steam. The primary variable (temperature) is slow and the secondary (manipulated) variable is capable of quickly responding to disturbances. Therefore, if disturbances occur (sudden change in plant steam demand) upsetting the manipulated variable (steam pressure), it will be immediately sensed and corrective action will be taken by the secondary controller so that the primary variable (process temperature) will not be affected. As to the nature of the possible disturbances, they usually have to do with the properties of the heating or cooling media supply. These have been covered in connection with the discussion on system degrees of freedom at the beginning of this section, and it can be said that the use of cascade control on heat transfer equipment contributes to fast recovery from load changes or from other disturbances.

Feedforward Control

As discussed in Section 7.6, feedback control involves the detection of the controlled variable (temperature) and the counteracting of changes in its value relative to a set point, by the adjustment of a manipulated variable (coolant flow). This mode of control necessitates that the disturbance variable must first affect the controlled variable itself before correction can take place. Hence, the term "feedback" can imply a correction "back" in terms of time, a correction that should have taken place earlier, when the disturbance occurred.

In this manner of terminology, feedforward is a mode of control which responds to a disturbance such, that it instantaneously compensates for that error which the disturbance would have caused otherwise in the controlled variable, later in time.

Figure 10.11aa illustrates a steam heater under feedforward control. All variables that can affect the heat balance relationship, are measured and the manipulated variable (steam flow) is adjusted when upsets occur. The conservation of energy equation (equation 10.11(2)) describes the relationship between the four variables. A few computing modules allow this equation to be solved for W_s, such that any variations in W and T_1 (the load) are compensated for and therefore T_2 is maintained constant. Figure 7.6i illustrates an improved version of this system incorporating feedback trimming and dynamic lag-balancing features.

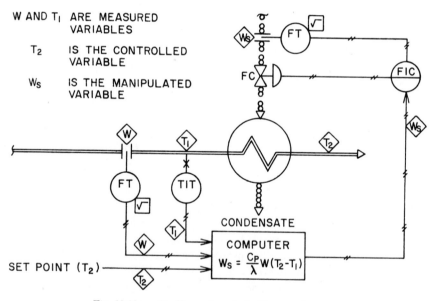

Fig. 10.11aa   Feedforward control of heat exchangers

The advantages of feed-forward control are similar to those of cascade because the load upset or supply disturbance is corrected for before its effect is felt by the controlled variable. As it can be expected, feedforward control contributes to stable, dampened response to load changes and to fast recovery from upsets.

In view of the remarkable improvements in the response to load and set-point changes, that can be attributed to the feedforward method of control, the reader is justified in questioning, why this technique should not be standardized on? This writer would answer that in two parts:

First, the cost of such installation is not always justified. Secondly, the implications of high precision control must be consistently accepted. This means that in order to keep T_2 within say $\pm\frac{1}{2}\%$ of full scale, the total accumulated error in flow measurements, temperature sensing, transmission, conversion, and computation should not exceed this amount. Therefore, in a computerized chemical plant, where the sensing elements have already been selected for precision consistent with the computer and where the inclusion of equation 10.11(2) in the computer program represents only a small added calculation, feedforward exchanger control could be standardized on. In conventional plants, the use of this technique is likely to be reserved for only the most critical heat transfer units, where the expense of additional detector and computing hardware can be justified by the resulting stable and accurate control.

Multipurpose Systems

In the earlier paragraphs the control of isolated heat transfer units has been discussed. In the majority of critical installations, the purpose of such systems is not limited to the addition or removal of heat, but to make use of both heating and cooling in order to maintain the process temperature constant. Such task necessitates the application of multipurpose systems, incorporating many of the features that have been individually discussed earlier.

Figure 10.11bb, for example, depicts a design which uses hot oil as its heat source and water as the means of cooling, arranged in a recirculating system. The points made earlier in connection with three-way valves, cascade systems, etc. also apply here, but there are a few additional considerations worth noting.

Probably the most important single feature of this design is that it operates on a "split range signal." This means that when the process temperature is above the desired set point, the valves will receive a dropping signal. While the value of this signal is between 9 and 15 PSIG the three-way valve is fully open to the exchanger by-pass and the two-way hot oil supply valve is partially open. If the reduction in the two-way valve opening is not sufficient to bring the process temperature down to set point, the signal further decreases, fully closing the two-way valve at 9 PSIG and beginning to open the cooler flow path through the three-way valve. At a 3 PSIG signal, the total cooling capacity of the system is applied to the recirculating stream, which in that case flows through the cooler without bypass.

The implications of such split-range operations should be fully realized:

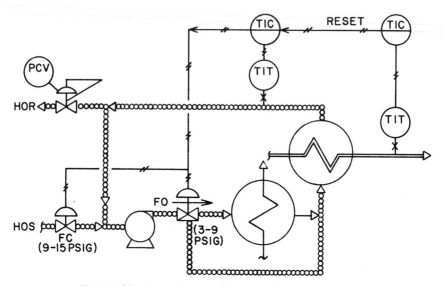

Fig. 10.11bb   Recirculating multi-purpose heat transfer system

1. At a signal level near 9 PSIG, the system can be unstable and cycling, because this is the point at which the three-way valve is just beginning to open to the cooler and the system might receive alternating slugs of cooling and heating due to the limited rangeability of the valves. (Zero flow is not enough, minimum flow is too much for the particular load condition.)

2. While the signal is in the 9–15 PSIG range, the cooler shell side becomes a reservoir of cold oil. This upsets the controls twice. Once when the three-way valve just opened to the cooler and once when the cold oil had been completely displaced and the oil outlet temperature from the cooler suddenly changed from that of the cooling water to some much higher value.

3. Most of these systems are "non-symmetrical" in that the process dynamics (lags and responses) are different for the cooling and heating phases.

To remedy the above problems, several steps can be and should be considered:

1. As shown in the illustration, a cascade loop should be used, so that upsets and disturbances in the circulating oil loop are prevented from entering the process and thereby from upsetting its temperature.

2. A slight overlapping of the two valve positioners is desirable, which offsets the beginning of cooling and the termination of heating phases, so that they will not both occur at 9 PSIG. The resulting sacrifice of heat energy is well justified by the improved control obtained.

3. In order to protect against the development of an extremely cold oil reservoir in the cooler, a set minimum continuous flow through this unit can be maintained.

In connection with the recirculating design shown in Figure 10.11bb, it is also important to realize that this is a flooded system and therefore when hot oil enters it, a corresponding volume of oil must be allowed to be removed. The pressure control valve shown serves this function. The same purpose can be achieved by elevational head on the return header, the important consideration being that whatever means are used, the path of least resistance for the oil, must be back to the pump suction to keep it always flooded and thereby prevent cavitation.

Most multipurpose systems represent a compromise of various degrees which is perfectly acceptable if the application engineer realizes the features that are lost and gained as the result of the compromise. Figure 10.11cc for

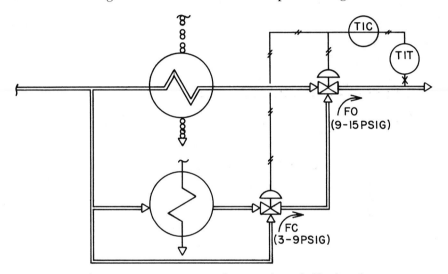

Fig. 10.11cc　Multi-purpose temperature control system utilizing the blending of process streams at differing temperatures

example illustrates a design where low cost and high response speed to load changes were the main considerations. This was achieved by the use of minimum hardware and by circumventing the transient characteristics of the exchangers. The price paid in this compromise involves the full use of utilities at all times, the development of hot and cold reservoirs and the necessity for supply disturbances to affect the process temperature before corrective action can be initiated. The reader will find it to be a valuable and educational exercise to arbitrarily select control system features that are desired and others that are of lesser consequence and then design a system which will satisfy those requirements.

Conclusions

The greatest danger in connection with heat exchanger control system design is over-simplification. Once it is realized that there are many factors entering the design problem, the good quality of control is half achieved. It is hoped that no reader of this section will put the book down and say, that "you control a cooler by throttling the water and control a heater by manipulating the steam to it." Having recognized a complex design challenge for what it is, a step by step analysis of desirable features will yield the desired results.

Some of the considerations that always pay to look into are—

The effect of supply disturbances on system performance.
The response speed of the system.
Rangeability considerations.
The quality of cooling water available.
Potential problems due to non-symmetrical dynamics and to low minimum condensing pressures.

Few generalizations are true in all cases, but there are a few which apply to the great majority of heat exchanger controls. In connection with the control valve used for example, it can be said that the use of a positioner and of equal percentage ports usually improves the performance. Similarly, it is true almost without exception, that cascade loops will give better control than conventional ones, and feedforward control will represent yet a further improvement.

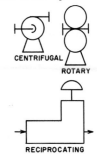

10.12 CONTROL OF PUMPS

Capacity control of pumps involves the incompressibility of liquids, because changes in the volume rate of flow throughout the system occur simultaneously and equally, and density is constant at constant temperature, regardless of pressure. Pump capacity may be affected by (1) a control valve in the discharge of a pump, (2) on-off switching, (3) varying the speed of the pump, or (4) stroke adjustment. Flow control by on-off switching provides only zero flow or full flow, whereas the other control methods provide continuously adjustable flows in the system. The applicability of these four methods of capacity control is related to the pump type, such as centrifugal, rotary or reciprocating. The possible types of capacity controls for the various pumps are summarized in Table 10.12a.

Table 10.12a
PUMP CONTROL METHODS

	Possible Types of Control	
Method of Control	*On-Off*	*Throttling*
On-Off Switch	Centrifugal, rotary or reciprocating	
Throttling Control Valve	Centrifugal or rotary	
Speed Control	Centrifugal, rotary or reciprocating	
Stroke Adjustment	Reciprocating	

Centrifugal Pumps

The centrifugal pump is the most common type of process pump, but its application is limited to liquids with viscosities up to 3,000 centistokes. The capacity-head curve is the operating line for the pump at constant speed and impeller diameter. Three types of curves are shown in Figure 10.12b, illustrating various relationships of capacity to discharge pressure. The capacity varies widely with changes in discharge pressure for all curves, but the shape of the curve determines the type of control which may be applied.

1513

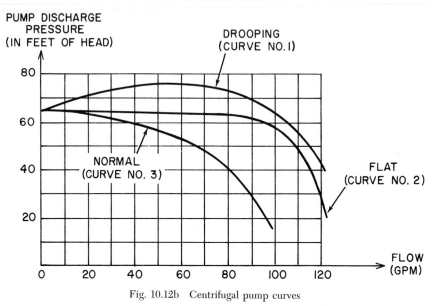

Fig. 10.12b   Centrifugal pump curves

For on-off switching control, curves #1 and #2 are satisfactory as long as the flow is above 100 GPM. Below this flow rate, curve #1 allows for two flows to correspond to the same head, and curve #2 may drop to zero flow to obtain a small head increase. Both are therefore unstable in this region. Curve #3 is stable for all flows and is best suited for throttling service where a wide range of flows is desired.

The following process loops illustrate typical control applications of centrifugal pumps, for both on-off and throttling control.

On-Off Level Control

Figure 10.12c illustrates the use of the level switch for on-off pump control. In order to prevent over-heating the motor, the number of pump starts should not exceed 15 per hour.

In connection with Figure 10.12c this can be checked by calculating the filling and emptying times. At a feed rate of 75 GPM, it takes 13 minutes to fill the 1,000-gallon volume between LSL and LSH, when the discharge pump is off. The pump capacity being 100 GPM, it takes 40 minutes to discharge this same volume. Therefore the pump will start approximately once an hour and run 75 percent of the time.

Therefore the 1,000-gallon tank is satisfactory. The switch prevents tank overflow and dry pumping, which would damage the seals. Since the pump starts once an hour, it will not overload the motor by excessive starts and stops. In this type of application, the capacity of the tank is the independent

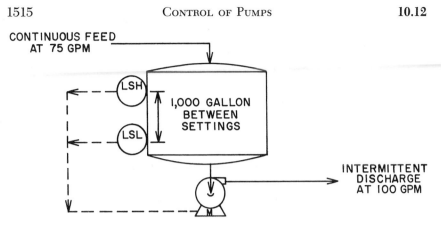

Fig. 10.12c   On-off level control

variable, since the pump capacity is usually selected at about 25 to 30 percent greater than the flow into the tank.

This design is modified when dual pumps are used. In this case, rather than to keep the spare pump idle, an alternator is interposed between the level switches and the motors. The alternator places the pumps in service in an alternating sequence. Thus, if two pumps are used, each will start half as many times per hour, permitting the tank size to be reduced. This application is shown in Figure 10.12d, where a cooling water return sump is illustrated. The minimum sump size is to be found allowing 10 starts per pump per hour. In this system, each pump is designed to handle the total feed flow into the sump by itself. The pumps do not operate together except when abnormally high flow rates are charged into the sump.

The design is for 10 starts per hour for each pump. There being two pumps, the sump capacity can allow 20 starts per hour or 3 minutes per empty

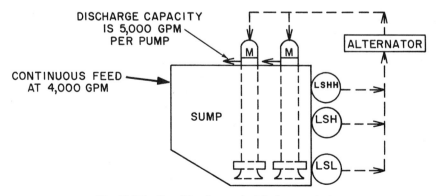

Fig. 10.12d   On-off level control of dual pump station

and refill cycle. Let x = sump capacity. The minimum sump capacity is calculated by allowing the sum of feed and discharge times to equal 3 minutes:

$$\frac{x}{4,000} + \frac{x}{1,000} = 3 \text{ minutes}$$

$$x = 2,400 \text{ gallons}$$

In addition to the reduction of sump volume, a further advantage is that a spare pump is available for emergency service. The purpose of a LSHH level switch is to put both pumps on line at once, should the continuous flow exceed 5,000 GPM, resulting in the level rising to this point. The combined pump flow of 10,000 GPM will then prevent over-flow of the sump. When the level drops to LSL, both pumps will be off, and the normal alternating cycle is resumed.

On-Off Flow Control

Illustrated in Figure 10.12e is a tandem pump arrangement which responds to varying flow demands on pump outlet. Pump I is normally operating

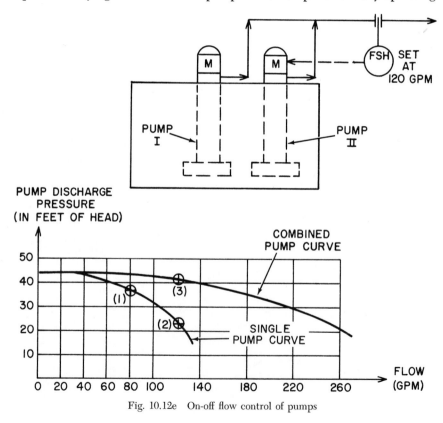

Fig. 10.12e   On-off flow control of pumps

at point (1) (at 80 GPM and 36 ft). When flow demand increases to 120 GPM, the head drops to 22 ft at point (2), and FSH starts pump II. The combined characteristic gives 120 GPM at 40 ft at point (3). In this control scheme, a wide range of flows is possible at high pump efficiency without serious loss in pressure head.

On-Off Pressure Control

A pressure switch may be used to start a spare pump in order to maintain pressure in a critical service when the operating pump fails. In this case, a low-pressure switch would be used to actuate the spare pump, piped in parallel with the first pump. A second function, as illustrated in Figure 10.12f, is to boost pressure. Pump I is normally operating at point (1). When the discharge pressure rises to 50 ft at point (2), the flow is reduced from 53 to 20 GPM. At this point the pressure switch will start pump II and close the by-pass

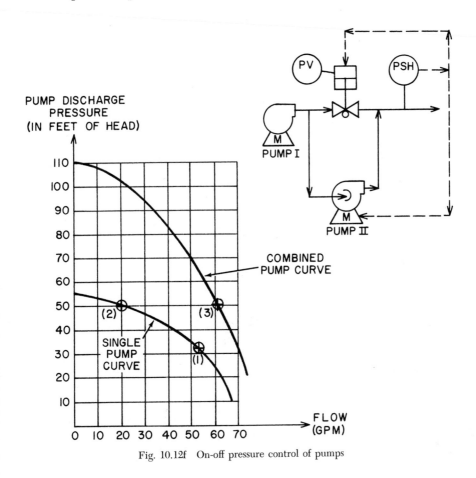

Fig. 10.12f   On-off pressure control of pumps

valve. The system will now operate at point (3) on the combined characteristic, delivering 60 GPM at 50 ft of head.

Throttling Control

Throttling control may be achieved by using control valves. In Figure 10.12g, design point (1) is near the maximum efficiency of the pump, and therefore, when throttling to point (2), the efficiency will drop, but the pump will still be stable.

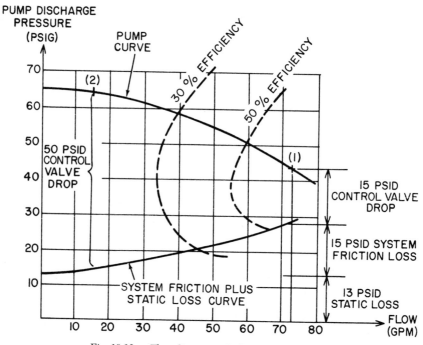

Fig. 10.12g   Throttling control of centrifugal pump

For good controllability, the control valve is usually sized to pass the design flow with a pressure drop equal to the system dynamic friction losses excluding the control valve but not less than 10 PSID minimum. (For more details on assigning sizing pressure drops to control valves, refer to Section 1.2) Pump flow is controlled by varying the pressure drop across the valve. This relationship is shown at point (1). When the flow is throttled to 15 GPM at point (2), the control valve must burn up a differential of 50 PSID. However, the pump must not be run at zero flow or over-heating will occur, and the fluid will vaporize, causing the pump to cavitate. To avoid this, a by-pass line can be provided with a back-pressure regulator. It is set at a pressure that will guarantee minimum flow as the pump is throttled toward zero flow.

A typical process loop on flow control is shown in Figure 10.12h. The rangeability of the control valve is assumed to be 25:1. (See Section 1.21 for details.) Thus, if the maximum flow required is 70 GPM through the flow control valve (Figure 10.12h), then about 3 GPM would be the minimum controllable flow. If lower flows are anticipated, then a second flow control valve should be installed in parallel with the first.

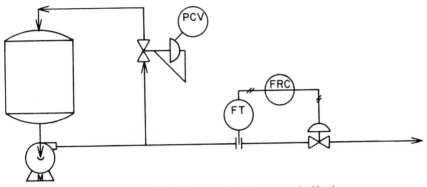

Fig. 10.12h   Throttling control with pressure kickback

The minimum flow needed through the pump to prevent vaporization can be calculated from a heat balance on the pump by assuming that the motor horsepower is converted to heat. If this flow were calculated to be 20 GPM, then the PCV in the by-pass line would be sized to pass 20 GPM with a corresponding set pressure of 63 PSIG. These same principles apply to process loops, where pump capacity is varied to maintain the level, pressure or temperature of other pieces of equipment.

Speed Variation

Flow control via speed control is less common than throttling with valves, because AC electric motors are constant-speed devices. If a turbine is considered, speed control is more convenient. An instrument air signal to the governor can control speed to within $\pm\frac{1}{2}$ percent of the set point. Throttle control on a gasoline engine may also alter speed but is used less frequently. In order to vary pump speeds with electric motors, it is usually necessary to use a variable-speed device in the power transmission train. This might consist of variable pulleys, gears, magnetic clutch, hydraulic coupling, etc., as covered in more detail in Section 2.2. In any case, variation of the pump speed generates a family of head-capacity curves, as shown in Figure 10.12i, where the volume flow is proportional to speed if the impeller diameter is constant. The head obtained is proportional to the square of the speed. The intersection of the system curve with the head curve determines the flow rate at points (1), (2) or (3).

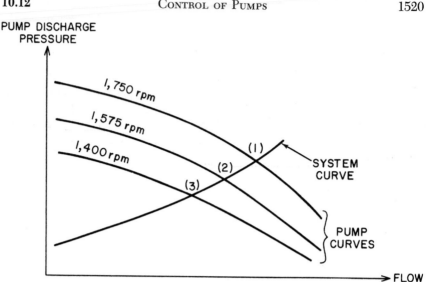

Fig. 10.12i   Centrifugal pump with speed variation

Rotary Pumps

The typical pump characteristics for a rotary pump, such as the gear, lobe, screw, or vane types, show a fairly constant capacity at constant speed with large changes in discharge pressure. This is shown in Figure 10.12j. These

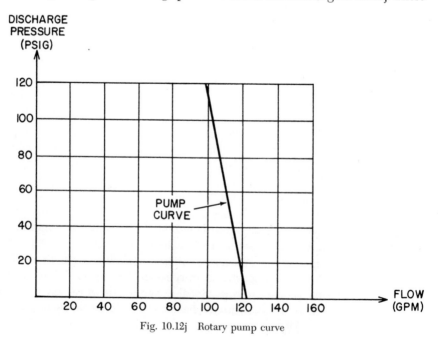

Fig. 10.12j   Rotary pump curve

pumps cover the viscosity range from less than 1 centipoise up to 500,000 centipoises. The usual application of this type of pump is for the highly viscous liquids and slurries which are beyond the capabilities of centrifugal pumps.

On-Off Control

The operation of rotary pumps with on-off control is similar to that with centrifugal pumps. The criterion for the maximum number of starts per hour must be checked carefully for motors on rotary pumps, since the starting torque may be very large, due to fluid viscosity, and so is the inertia load of the column of fluid in the piping, which accelerates under positive displacement each time the pump starts.

In slurry service, on-off control creates problems due to settling of solids, and it is therefore not recommended. Instead, a circulating loop is used with a pressure-controlled by-pass back to the feed tank. For example, intermittent flow to feed a centrifuge is obtained by opening an on-off valve via a signal from the cycle timer. Such a loop is shown in Figure 10.10e. The pressure controlled by-pass allows the normal pump flow to be maintained in the loop when the centrifuge feed valve is closed.

The on-off control in this case is applied to the fluid rather than to the pump motor.

Manual on-off control is often applied to rotary pumps in bulk storage batch transfer services with local level indication.

Safety and Throttling Controls

A safety relief valve is always provided on a rotary pump to protect the system and pump casing from excessive pressure should the discharge line be blocked while the pump is running. The relief valve may discharge to pump suction or to the feed tank. In cases when slurries and viscous materials may not be able to pass through the relief valve, a rupture disc is placed on the discharge line.

The output of a rotary pump may be continuously varied to suit process demands by use of a pressure-controlled by-pass, in combination with a flow control valve. The by-pass is necessary to accommodate changes in flow to the process, since the total flow through the by-pass plus to the process is constant at constant speed.

The capacity of a rotary pump is proportional to its speed, neglecting the small losses due to slippage. The flow can therefore be controlled by speed-modulating devices, which are fully described in Section 2.2.

In this case, no by-pass is needed. The rangeability is limited by the speed-control device, which for pulleys and magnetic drives is about 4 to 1. The response is slightly slower than with a control valve, due to the inertia of the system; however, this type of control would be favored when it is desired

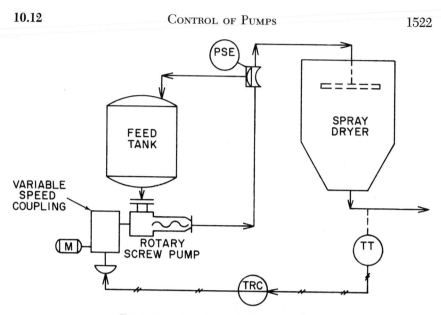

Fig. 10.12k   Speed control of rotary pumps

to avoid the use of control valves on slurry or gummy services. Such installation is shown in Figure 10.12k, where a screw pump feeds latex slurry to a spray dryer.

Reciprocating Pumps

Reciprocating pumps, such as the piston and diaphragm types, deliver a fixed volume of fluid per stroke. The control of these pumps is based on either changing the stroke length, changing the stroke speed, or varying the interval between strokes. In all cases, the discharge from these pumps is a pulsed flow, and for this reason it is not suited to control by throttling valves. In practice, the volume delivered per stroke is less than the full stroke displacement of the piston or diaphragm.

This hysteresis is due to high discharge pressures or high viscosity of the fluid pumped. Under these conditions, the check valves do not seat instantaneously. A calibration chart must therefore be drawn for the pump under actual operating conditions. A weigh tank or level-calibrated tank is usually the reference standard. Since the discharge is a pulsed flow, it must be totalized and divided by the time interval to get average flow rate for a particular speed and stroke setting. Metering accuracy is approximately ±1 percent of the actual flow with manual adjustment and ±1.5 percent with automatic positioning. Methods of stroke and speed adjustment are covered in detail in Sections 2.2 and 2.3, while other features of metering pumps are discussed in Section 5.17 of Volume I.

On-Off Control

Once the pump has been calibrated, it can be programmed via a timer or counter to deliver a known volume of fluid to a process. The pump may deliver one full stroke and then stop until the next electrical signal is received from the timer, or continuously charge a specified number of pump strokes and then be shut down by a counter. The flow may be smoothed by a pulsation dampener and the system and pump protected from over-pressure by a relief valve.

Throttling Control

Continuously variable flow control at constant speed may be accomplished by automatic adjustment of the stroke length. The range of flow control by stroke adjustment is zero to 100 percent. However, in order to maintain

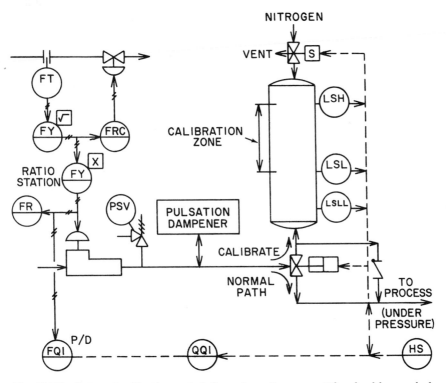

Fig. 10.12 1   Ratio and calibration controls for reciprocating pump. When level has reached LSH, the three-way valve returns to the "normal" path and nitrogen enters the tank to initiate discharge. When level drops to LSLL, discharge is terminated by venting off the nitrogen. Counter QQI is running while rising level is between LSL and LSH. Total count, when compared with known calibration volume, gives total error. Hand switch HS initiates calibration cycle by diverting the three-way valve to the "calibrate" path.

accuracy, the practical range is 10 to 100 percent of design flow. The flow
is related to stroke length through system calibration.

In some applications the reciprocating pump combines the measuring
and control functions, receiving no independent feedback to represent flow.
In other cases it is utilized as a final control element except that flow detection
is performed by an independent sensor. Figure 10.12l illustrates an installation
where the pump is both the measuring and the control device for ratio control,
and it is provided with automatic calibration capability.

Variable speed control is usually applied in multiple-head pumps where
all the pumps are coupled mechanically. A control signal may adjust all flows
in the same proportion simultaneously. The rangeability and accuracy of flow
control depends on the method of speed variation chosen.

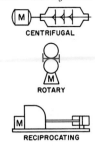

10.13 CONTROL OF COMPRESSORS

Compressors are gas-handling machines which perform the function of increasing gas pressure by confinement or by kinetic energy conversion. Methods of capacity control for the principal types of compressors are listed in Table 10.13a. The method of control to be employed is determined by the process requirements, type of driver, and cost.

Table 10.13a
CAPACITY CONTROL METHODS
OF COMPRESSORS

Compressor Type	Capacity Control Method
Centrifugal	Suction throttling Discharge throttling Variable inlet guide vanes Speed control
Rotary	By-passing Speed control
Reciprocating	On-off control Constant-speed unloading Speed control Speed control and unloading

The primary emphasis should be on the process requirements. After these have been met by use of the optimum control system, then the type of driver may be considered. Where the control involves variable speed, this is easily accomplished by using either a steam turbine, gas turbine, gasoline or diesel engine. The speed of these drivers can be regulated by throttling the fuel supply or the steam pressure. For constant-speed control, electric motors are well suited, since they are essentially constant-speed devices, while for variable speed they require expensive and inefficient variable-speed transmissions. For certain installations, the driver may be selected on the basis of overall plant material and energy balances. In these cases, the process may supply the heat or fuel which is used to run the steam or gas turbines.

The cost of the control must be considered, but this is usually subordinate to reliability requirements, especially in view of the key role of the compressor in most large, single-train processes. The critical nature of compressor controls extends to their lubrication and seal systems also, which must provide fail-safe, uninterrupted operation for long time periods.

Centrifugal Compressors

The centrifugal compressor is a machine that converts the momentum of gas into a pressure head. The head H, measured in feet, is equal to the torque rate $\tau\omega$, measured in foot-pounds force per hour, divided by the flow rate W, in pounds mass per hour, or,

$$H(ft) = \frac{\tau\omega(\text{ft-lbf/hr})}{W(\text{lbm/hr})} \qquad 10.13(1)$$

In addition, the head for a gas is related to the pressure ratio developed, as expressed by equation 10.13(2):

$$H = \frac{ZRT_I}{(n - 1/n)}[(P_D/P_I)^{(n-1/n)} - 1] \qquad 10.13(2)$$

Combining equations 10.13(1) and (2) gives

$$\frac{\tau\omega}{W} = \frac{ZRT_I}{(n - 1/n)}[(P_D/P_I)^{(n-1/n)} - 1] \qquad 10.13(3)$$

This equation is the basis for plotting the compressor curves and for understanding the operation of capacity controls. The nomenclature for the symbols in the equations is furnished at the end of this section.

In equation 10.13(3), the pressure ratio (P_D/P_I) varies inversely with mass flow (W). For a compressor running at constant speed (ω), constant inlet temperature (T_I), constant molecular weight (implicit in R), constant n, τ and Z, the discharge pressure may be plotted against weight flow as in Figure 10.13b (curve I), with the design point (1) being located in the maximum efficiency range at design flow and pressure.

Suction Throttling

The capacity of a centrifugal compressor can be controlled by placing a control valve in the suction line, thereby altering the inlet pressure (P_I). From equation 10.13(3) it can be seen that the discharge pressure will be altered for a given flow when P_I is changed, and a new compressor curve will be generated. This is illustrated in Figure 10.13b (curves II and III). Consider first that the compressor is operating at its normal inlet pressure (following curve I) and is intersecting the "constant pressure system" curve at point (1) with a design flow of 9,600 lbm/hr, at a discharge pressure of

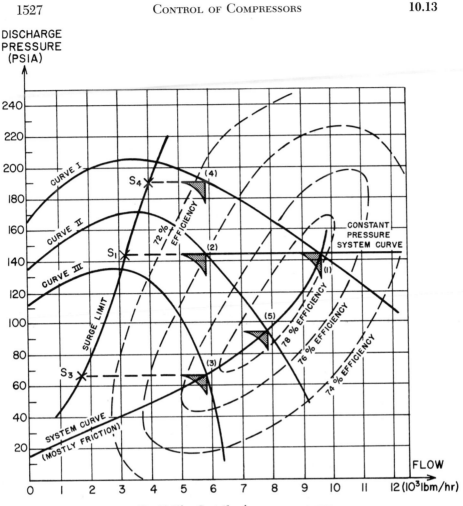

Fig. 10.13b   Centrifugal compressor curves

144 PSIA and 78 percent efficiency. If it is desired to change the flow to 5,900 lbm/hr, while maintaining the same discharge pressure, it would be necessary to shift the compressor from curve I to curve II. The new intersection with the "constant pressure system" curve is at the new operating point (2), at 74 percent efficiency. In order to shift from curve I to curve II, one must change the discharge pressure of 190 PSIA at the 5,900 lbm/hr flow on curve I, to 144 PSIA on curve II. If the pressure ratio is ten $(P_D/P_I = 10)$, then it would be necessary to throttle the suction by only $\Delta P_I = 46/10 = 4.6$ PSI, to achieve this shift.

It is also important to consider how close the operating point (2) is to the surge line. The surge line represents the low-flow limit for the compressor, below which its operation is unstable due to momentary flow reversals. Methods of surge control will be discussed later. At point (2) the flow is 5,900 lbm/hr and at the surge limit (S_1) it is 3,200 lbm/hr. Thus, the compressor is operating at 5,900/3,200 = 184 percent of surge flow. This may be compared with curve I at point (1), where prior to suction throttling the machine is operating at 9,600/3,200 = 300 percent of surge flow.

This same method of suction throttling may be applied in a "mostly friction system" also shown in Figure 10.13b. In order to reduce the flow from 9,600 lbm/hr to 5,900, it is necessary to alter the compressor curve to curve III, so that the intersection with the "mostly friction system curve" is at the new operating point (3), at 77 percent efficiency. In order to do this, one must change the discharge pressure from 190 PSIA (curve I) to 68 PSIA (on curve III). Thus, $\Delta P_D = 190 - 68 = 122$ PSI, and the amount of inlet pressure throttling for a machine, with a compression ratio of ten is $\Delta P_I = 122/10 = 12.2$ PSI. The corresponding surge flow is at 1,700 lbm/hr, which means that the compressor is operating at 5,900/1,700 = 347 percent of surge flow. Therefore, surge is less likely in a "mostly friction system" than in a "constant-pressure system" under suction throttling control.

Discharge Throttling

A control valve on the discharge of the centrifugal compressor may also be used to control its capacity. Referring to Figure 10.13b, if the flow is to be reduced from 9,600 lbm/hr at point (1) to 5,900 lbm/hr, the compressor must follow curve I and therefore operate at point (4), at 190 PSIA discharge pressure and 72 percent efficiency. However, the "mostly friction system" curve at this capacity requires only 68 PSIA discharge pressure. Therefore the 122 PSI excess pressure must be burned up in the discharge control valve. The surge flow (at S_4) is 4,000 lbm/hr, and the compressor is therefore operating at 5,900/4,000 = 148 percent of surge. Thus, surge is more likely to occur in a mostly friction system when using discharge throttling than when using suction throttling.

The various parameters involved in suction and discharge throttling of this sample compressor are compared in Table 10.13c.

Inlet Guide Vane

This method of control employs a set of adjustable guide vanes on the inlet to one or more of the compressor stages. By pre-rotation or counter-rotation of the gas stream relative to the impeller rotation, the stage is unloaded or loaded, thus lowering or raising the discharge head. The effect is similar to suction throttling as illustrated in Figure 10.13b (curves II and III), but less power is wasted because pressure is not throttled directly. Also, the control is two directional, since it may be used to raise as well as to lower the head. It is more complex and expensive than throttling valves, but may

Table 10.13c

COMPRESSOR PARAMETERS AS A FUNCTION
OF THROTTLING METHOD

	Control Valve ΔP(PSI)	Compressor Efficiency	Operation Above Surge By
Suction Throttling "Constant Pressure System"	4.6	74%	184%
Suction Throttling "Mostly Friction System"	12.2	77%	347%
Discharge Throttling "Mostly Friction System"	122	72%	148%

save 10 to 15 percent on power and is well suited for use on constant-speed machines in applications involving wide flow variations.

The guide vane effect on flow is more pronounced in constant discharge pressure systems. This can be seen in Figure 10.13b (curve II), where the intersection with the "constant pressure system" at point (2) represents a flow change from the normal design point (1) of $9{,}600 - 5{,}900 = 3{,}700$ lbm/hr, whereas the intersection with the "mostly friction system" at point (5) represents a flow change of only $9{,}600 - 7{,}800 = 1{,}800$ lbm/hr.

Variable Speed

The pressure ratio developed by a centrifugal compressor is related to tip speed by the following equation:

$$\psi u^2/2g = \frac{ZRT_I}{(n - 1/n)}[(P_D/P_I)^{(n-1/n)} - 1] \qquad 10.13(4)$$

From this relation the variation of discharge pressure with speed may be plotted for various percentages of design speed as shown in Figure 10.13d. The obvious advantage of speed control from a process viewpoint is that both suction and discharge pressures can be specified independently of the flow. The normal flow is shown at point (1) for 9,700 lbm/hr at 142 PSIA. If the same flow is desired at a discharge pressure of 25 PSIA, the speed is reduced to 70 percent of design, shown at point (2). In order to accomplish the same result through suction throttling with a pressure ratio of 10 to 1, the pressure drop across the valve would have to be $(142 - 25)/10 = 11.7$ PSI, with the attendant waste of power, due to throttling. This is in contrast to a power saving accomplished with speed control, since power input is reduced as the square of the speed.

One disadvantage of speed control is apparent in constant pressure systems where the change in capacity may be overly sensitive to relatively small speed changes. This is shown at point (3) where a 20 percent speed

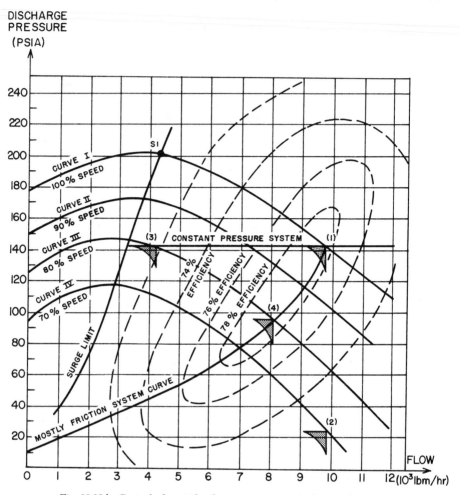

Fig. 10.13d   Control of centrifugal compressor capacity by speed variation

change gives a flow change of $(9,600 - 4,300)/9,600 = 55\%$. The effect is less pronounced in a "mostly friction system," where the flow change at point (4) is $(9,600 - 8,100)/9,600 = 16\%$.

Surge Control, Fixed Set Point

The design of compressor control systems is not complete without consideration of surge control, because this affects the stability of the machine. Surging begins at the positively sloped section of the compressor curve. In Figure 10.13d this occurs at S_1 on the 100 percent speed curve at 4,400 lbm/hr. This flow will insure safe operation for all speeds, but some power will be wasted at speeds below 100 percent because the surge limit decreases at reduced speeds. However, unless the compressor will be operating for long periods of time at reduced load below 4,400 lbm/hr, the power loss is usually neglected. Therefore, the flow through the compressor is maintained slightly above 4,400 lbm/hr by use of a flow-controlled by-pass back to the suction of the compressor. This is shown in Figure 10.13e for a compressor with suction throttling capacity control. Note that the surge control valve is in parallel with the capacity control valve. In this way it will afford positive protection against surge, regardless of the amount of throttling across the suction control valve. When the molecular weight of the gas varies widely, the surge controller may detect the motor current drawn, which is also related to surge.

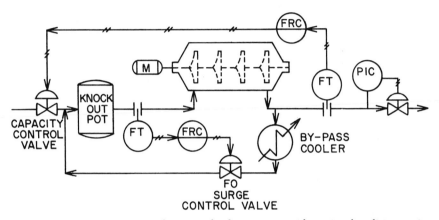

Fig. 10.13e   By-pass surge control on centrifugal compressor with suction-throttling capacity control

Anti-Surge Controller

Another method of surge control uses the ratio of the compressor pressure rise to the inlet flow rate to set the flow in the by-pass loop. This method saves considerable power at low flows, because the surge limit drops off markedly, as can be seen in Figures 10.13b and d. The surge limit curve is parabolic. Weight flow may be converted to volume flow by density correction and the surge curve may then be approximated by equation 10.13(5):

$$(P_D - P_I) = K_1 Q^2 (P_I/T_I) \qquad \text{10.13(5)}$$

Also, the head loss across an orifice plate in the suction line may be expressed by

$$h = K_2 Q^2 (P_I/T_I) \qquad \text{10.13(6)}$$

Combining 10.13(5) and (6) gives

$$(P_D - P_I) = K_3 h \qquad \text{10.13(7)}$$

Therefore, if the value of h is multiplied by K_3, this value may be compared with $(P_D - P_I)$ in the anti-surge controller. As long as hK_3 is less than $(P_D - P_I)$, the compressor will be outside the surge region. The control loop for this system is shown in Figure 10.13f.

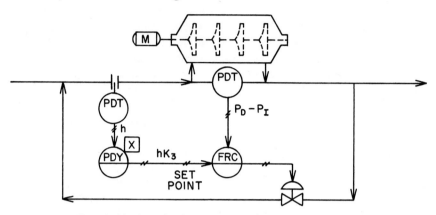

Fig. 10.13f   Centrifugal compressor with anti-surge controller

Because control is discontinuous, "reset windup" can be a problem. (See Section 4.3 for details.) It is therefore necessary to use a special controller suitable for such service. One technique used involves the detection of the controller output pressure by a switch, and when this signal reaches 15 PSIG, it automatically vents the reset bellows. This prevents controller saturation due to reset windup, and the functioning of the reset mode is reversed because now it repositions the controller proportional band to below the set point. Overshoot is eliminated by this technique, since measurement is allowed to enter the proportional band well before it reached the set point.

Rotary Compressors

Variable Speed

The rotary compressor is essentially a constant displacement variable discharge pressure machine. Common types are the lobe, sliding vane, and liquid ring. The characteristic curves for a lobe type unit are shown in Figure 10.13g. Since the unit is of the positive displacement type, the inlet flow will vary linearly with the speed. Curves I and II in Figure 10.13g show this. The small decrease in capacity at constant speed with increase in pressure is due to slip of the gas at impeller clearances. It is necessary to compensate for this by small speed adjustments as the discharge pressure varies. For example, when the compressor is operating at point (1) it delivers the design volume of 60 ACFM and $3\frac{1}{2}$ PSIG. In order to maintain the same flow when the discharge pressure is 7 PSIG, the speed must be increased from 1,420

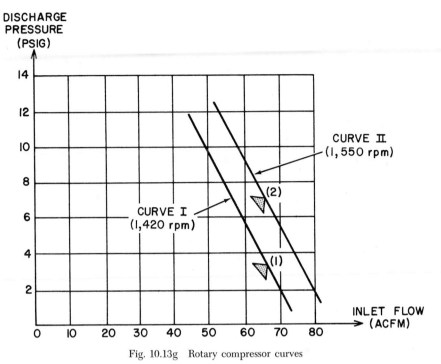

Fig. 10.13g   Rotary compressor curves

rpm to 1,550 rpm at point (2). This can be accomplished by a pneumatic flow controller on the discharge line manipulating the motor or drive speed.

By-pass and Suction Control

The discharge flow may be varied to suit process demands by a by-pass on pressure control. This is shown in Figure 10.13h. Excess gas is vented to the atmosphere as the temperature control valve closes. The temperature of

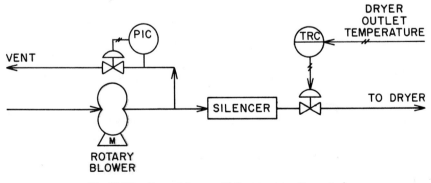

Fig. 10.13h   Rotary blower with by-pass capacity control

the outlet gas is controlled to prevent product degradation and provide the proper product dryness. In systems where the gas is not vented, it may be returned to the suction of the blower on pressure control.

Flow may also be varied by throttling the suction to the compressor, but this method is limited by the horsepower of the driver and by the temperature rise of the gas.

Suction Pressure Control

An important application of the liquid ring rotary compressor is in vacuum service. The suction pressure is often the independent variable and is controlled by bleeding gas into the suction on pressure control. This is shown in Figure 10.13i where suction pressure control is used on a rotary filter to maintain the proper drainage of liquor from the cake on the drum.

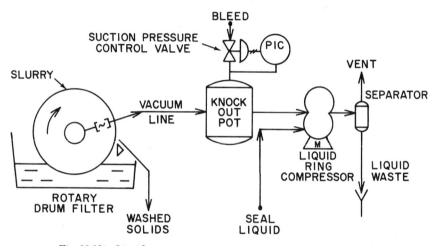

Fig. 10.13i Liquid ring rotary compressor on suction pressure control

Reciprocating Compressors

The reciprocating compressor is a constant volume, variable discharge pressure machine. A typical compressor curve is shown in Figure 10.13j for constant speed operation. The curve shows no variation in volumetric efficiency in the design pressure range, which may vary by 8 PSIG from unloaded to fully loaded.

The volumetric inefficiency is due to the clearance between piston end and cylinder end on the discharge stroke. The gas which is not discharged re-expands on the suction stroke, thus reducing the intake volume.

The relationship of speed to capacity is a direct ratio, since the compressor is a displacement type machine. The typical normal turndown with gasoline or diesel engine drivers is 50 percent of maximum speed, in order to maintain the torque within acceptable limits.

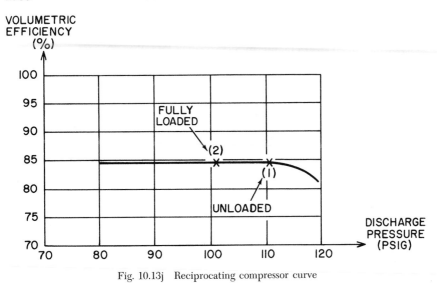

Fig. 10.13j   Reciprocating compressor curve

On-Off Control

For intermittent demand, where the compressor would waste power if run continuously, the capacity can be controlled by starting and stopping the motor. This can be done manually, or by use of pressure switches. Typical switch settings are: on at 140 PSIG, off at 175 PSIG. This type of control would suffice for processes where the continuous usage is less than 50 percent of capacity, as shown in Figure 10.13k, where an air mix blender uses a rapid series of high-pressure air blasts when the mixer becomes full. The high-pressure air for this purpose is stored in the receiver.

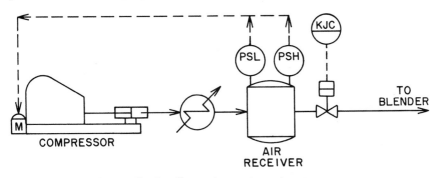

Fig. 10.13k   On-off control on reciprocating compressor

Constant Speed Unloading

In this type of control, the driver operates continuously, at constant speed, and the capacity is varied in discrete steps by holding suction valves open on the discharge stroke, or opening clearance pockets in the cylinder. The most common schemes are 3- and 5-step unloading techniques. The larger number of steps saves horsepower because it more closely matches the compressor output to the demand. In 3-step unloading, the capacity is either 100 percent, 50 percent or 0 percent of maximum flow. This is accomplished by the use of valve unloading in the double-acting piston. At 100 percent, both suction valves are closed during the discharge stroke. At 50 percent, one suction valve is open on the discharge stroke, wasting half the capacity of the machine. At 0 percent, both suction valves are held open on the discharge stroke, wasting the total machine capacity.

For 5-step unloading, a clearance pocket is employed in addition to suction valve control. The percentage capacity can be 100, 75, 50, 25 or 0 of maximum flow. This is shown in Figure 10.13l. At 100 percent, both suction valves and the clearance pocket are closed; at 75 percent, only the clearance pocket is open; at 50 percent only one suction valve is open on the discharge stroke; at 25 percent one suction valve and the clearance pocket are open; at 0 percent both suction valves are opened during the discharge stroke.

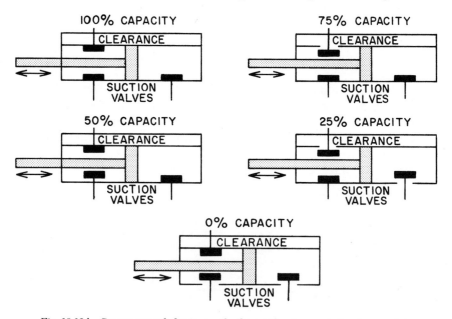

Fig. 10.13 l   Constant-speed, five-step unloading with valves and clearance pockets

The use of step unloading is most common where the driver is inherently a constant speed machine, such as an electric motor.

The flow sheet representation for this control method is shown in Figure 10.13m. The pressure controller signal from the air receiver operates a solenoid valve in the unloader mechanism. The action of the solenoid valve directs the power air to lift the suction valves and/or to open the clearance port.

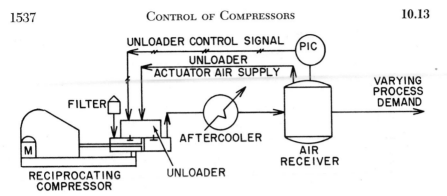

Fig. 10.13m   Constant-speed capacity control of reciprocating compressor

For 3-step unloading, two pressure switches are needed. The first switch loads the compressor to 50 percent if the pressure fell slightly below its design level, and the second switch loads the compressor to 100 percent if the pressure fell below the setting of the first switch.

For 5-step unloading, a pressure controller is usually substituted for the four pressure switches otherwise required, and the range between unloading steps is reduced to not more than 2 PSI deviation from the design level, keeping the minimum pressure within 8 PSI of design. Where exact pressure conditions must be met, a throttling valve is installed, which by-passes the gas from the discharge to the suction of the compressor. This device smoothes out pressure fluctuations and in some cases eliminates the need for a gas receiver in this service. This can prove economical in high-pressure services above 500 PSIG, where vessel costs become significant.

Variable Speed

This method of control is usually employed on gas turbines, steam turbines, gasoline and diesel engines which are easily adaptable to speed control by fuel throttling or steam regulation. By contrast, electric motors are usually constant-speed devices. When large flow changes are anticipated in the process, consideration must be given to loss of torque which occurs at reduced speed. The gas turbine is less susceptible to losses at reduced speed.

A patented control device known as the "Varsudi Controller" combines unloading control with speed control. This allows practically straight-line capacity control, while maintaining the speed within a range where the motor torque is close to the maximum. By sensing the suction and discharge pressure and relating these to the horsepower curve, the cylinder is unloaded when 3 percent overload occurs.

Lube and Seal Systems

A typical lube oil system is shown in Figure 10.13n. Dual pumps are provided to insure the uninterrupted flow of oil to the compressor bearings and seals. Panel alarms on low oil level, low oil pressure (or flow) and high

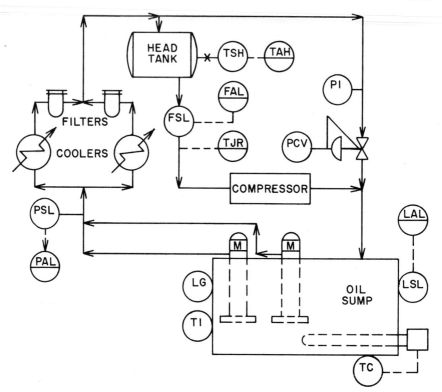

Fig. 10.13n   Compressor lube oil system for bearings and seals

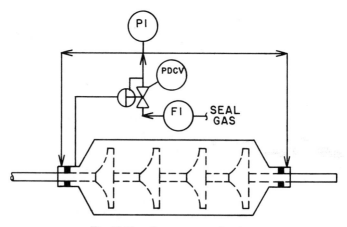

Fig. 10.13o   Compressor seal system

oil temperature are provided. A head tank provides oil for coasting down in case of a power failure. The design of these systems is critical, because failure of the oil supply could mean shutting down the entire process.

In cases where oil cannot be tolerated in contact with the process gas, an inert gas seal system may be used. This is shown in Figure 10.13o for a centrifugal compressor with balanced seals.

NOMENCLATURE

h = differential head (ft)
H = polytropic compressor head (ft)
$K_{1,2,3}$ = flow constant
n = polytropic coefficient
P_D = discharge pressure (PSIA)
P_I = inlet pressure (PSIA)
Q = volume flow rate (ACFH)
R = gas constant (1,544/molecular weight)
T_I = inlet temperature
u = rotor tip speed (ft/sec)
W = weight flow (lbm/hr)
Z = gas compressibility factor
τ = motor torque (ft-lbf)
ψ = head coefficient
ω = angular velocity (radians/hr)

10.14 EFFLUENT AND WATER TREATMENT CONTROLS

Not many processes used in the treatment of water and waste water are well suited to automatic process control. This is due to a notable absence of continuous analyzers reliable enough to measure all the necessary parameters of water quality. Chief among these are the lack of means to control coagulation and flocculation of water or the biological treatment of waste water. Both of these processes are, for the most part, an art, not a science, still requiring some human judgment to determine chemical application rates and process control parameters. There are only four control measurements commonly in use:

> pH
> ORP (oxidation-reduction potential)
> Residual chlorine
> Flow rate

Other properties which are measured to assist the operator controlling the system are

> Conductivity
> Alkalinity
> Temperature
> Suspended solids
> Dissolved oxygen
> Color

Water and waste water treatment consists of unit operations which may be classed as mechanical, chemical, biological and any combinations of these. The mechanical operations most often employed are screening, filtration and separation by gravity.

The majority of treatment processes employ chemicals in continuous rather than batch systems. Emphasis will be placed on this aspect under these general headings: (1) chemical oxidation, (2) chemical reduction, (3) neutralization, (4) precipitation and (5) biological control.

Critical design factors for all chemical treatment involve time, pH, concentration of contaminant(s) and chemical dosage rate.

Temperature is an effect in all chemical reactions, although most treatment processes presently employed are not substantially influenced by temperature variations. There are specific exceptions but, generally, temperature control is not practiced.

Chemical Oxidation

Treatment of water and waste water by chemical oxidation is employed in specific instances when the contaminant can either be destroyed, its chemical properties altered, or its physical form changed. Examples of chemicals which can be destroyed are cyanides and phenol. Sulfides can be oxidized to sulfates, thus changing their characteristics completely. Iron and manganese can be oxidized from the soluble ferrous or manganous state to the insoluble ferric or manganic state, respectively, permitting their removal by sedimentation. Strong oxidants are employed such as chlorine, chlorine dioxide, ozone, and potassium permanganate. Chlorine is preferred when it can be used because it is the least expensive and is readily available.

All chemical reactions of this type are pH dependent in relation to the time required for the reaction to proceed to completion, and for the desired end products. Residual oxidant or ORP measurement is used to control the process.

Oxidation-reduction implies a reversible reaction. Since these reactions are carried to completion and are not reversible, the term is misleading. In practice, control is by what may be called "electrode potential readings." An illustration is the oxidation of cyanide into cyanate with chlorine, according to the following reaction:

$$
\begin{array}{ll}
2Cl_2 & \text{Chlorine} \\
+ & \\
4NaOH & \text{Sodium hydroxide} \\
+ & \\
2NaCN & \text{Sodium cyanide} \\
\downarrow & \\
2NaCNO & \text{Sodium cyanate} \\
+ & \\
4NaCl & \text{Sodium chloride} \\
+ & \\
2H_2O & \text{Water}
\end{array}
\qquad 10.14(1)
$$

The electrode potential of the cyanide waste solution will be in the order of -200 to -400 millivolts. After sufficient chlorine has been applied to complete the reaction according to the equation above, the electrode potential will be in the order of $+200$ to $+400$ mv. The potential value will not increase until *all* cyanide has been oxidized. Control of pH is essential, with the minimum being 8.5. The reaction rate is faster at higher values.

Complete oxidation (destruction of cyanide) is a two-step reaction, the first being oxidation to the cyanate level described in equation 10.14(1). The end point of the reaction is again detected by electrode potential readings and will be in the order of $+700$ to $+800$ mv. The overall reaction is:

$$
\begin{array}{ll}
5Cl_2 & \text{Chlorine} \\
+ & \\
10NaOH & \text{Sodium hydroxide} \\
+ & \\
2NaCN & \text{Sodium cyanide} \\
\downarrow & \\
2NaHCO_3 & \text{Sodium bicarbonate} \\
+ & \\
N_2 & \text{Nitrogen} \\
+ & \\
4H_2O & \text{Water}
\end{array}
\qquad 10.14(2)
$$

While the process and process control are generally similar in all chemical oxidation reactions, the oxidation of cyanide is used as an example because it is unique in that it involves a two-step process, whereas the other applications cited involve only one. Because of this added complexity, the toxicity of cyanide, and the rigid requirements on waste discharge, a batch type treatment is recommended. This affords the assurance of complete treatment before discharge. In Figure 10.14a, chlorine is charged at a constant rate, with automatic pH control of caustic addition maintaining the batch pH at a value of 9.5.

When the ORP set point of $+750$ mv is reached on the ORP controller, a delay timer is actuated and the chemical feed systems are shut down. After a 30-minute delay, the tank contents are discharged if the potential reading is at or above the set point. If not, indicating that further reaction has been taking place in the delay period resulting in a subsequent drop in potential value, the system is reactivated and the cycle repeated. Additional, usually duplicate, tanks are required to receive the incoming waste while others are being used for treatment.

Continuous flow-through systems offer the advantage of less space requirements, but this is often offset in capital costs by the additional process equipment required. In the system, shown in Figure 10.14b, the two reaction steps previously mentioned are separated. In the first step, the ORP controller set point is approximately $+300$ mv. It controls the addition of chlorine to oxidize the cyanide into cyanate. pH is maintained at approximately 10. Reaction time is in the order of five minutes. Since the second step (that of oxidation of the cyanates), requires an additional amount of chlorine to be charged at nearly the same rate as in the first step, the chlorine flow rate signal is obtained by measuring the chlorine feed rate to the first step and

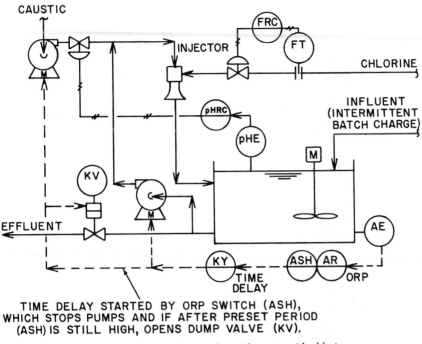

CAUSTIC

INJECTOR

CHLORINE

INFLUENT
(INTERMITTENT
BATCH CHARGE)

EFFLUENT

TIME
DELAY

ORP

TIME DELAY STARTED BY ORP SWITCH (ASH),
WHICH STOPS PUMPS AND IF AFTER PRESET PERIOD
(ASH) IS STILL HIGH, OPENS DUMP VALVE (KV).

Fig. 10.14a   Batch oxidation of cyanide waste with chlorine

multiplying it by a constant to control the chlorine feed rate in the second step. Caustic requirement is dependent solely on the chlorine rate (pH control is not necessary), and therefore the same signal can be used to adjust caustic feed. The ORP instrument which is sampling the final effluent can signal process failure if the potential level drops below approximately +750 mv.

It would appear that a feedback loop from the second ORP analyzer to the second chlorinator is desirable. Practice has not shown this to be necessary, because the ratioing accuracy available between first-stage chlorine rate and secondary addition (approximately 1:1) is sufficiently high. At worst, the system as shown will apply a little more chlorine than is actually required. Table 10.14c lists the set points and variables applicable to this oxidation process. Fixed flow rate systems are preferred to provide constant reaction times.

The use of residual chlorine analyzers is not applicable to this process, since the metal ions usually present in the waste and the intermediate products interfere with accurate determinations. They are used in processes where the presence of excess residual chlorine indicates a completed reaction. The set point is usually 1 milligram per liter or less.

Figure 10.14d is a schematic typical of systems with variable quality and variable flow rate. The chlorinator has two operators, one controlled by feed

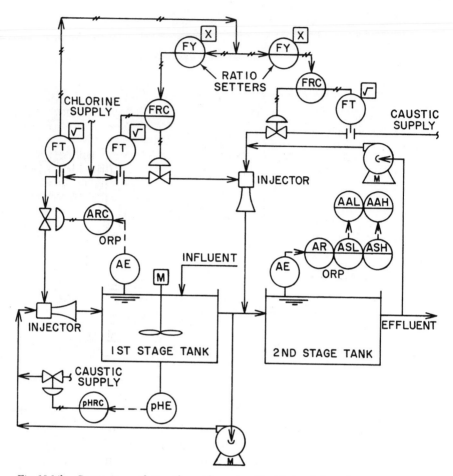

Fig. 10.14b Continuous oxidation of cyanide waste with chlorine. Influent here has continuous constant flow rate and variable quality.

Table 10.14c

SET POINTS AND PARAMETERS

Parameters	Process Steps	
	Cyanide to Cyanate	Destruction of Cyanate
pH	10–12	8.5–9.5
Reaction Time (minutes)	5	45
ORP (mv) Set Point	+300	+750
Maximum Concentration of Cyanide (Cyanate) That Can Be Treated	1,000 milligrams per liter	1,000 milligrams per liter

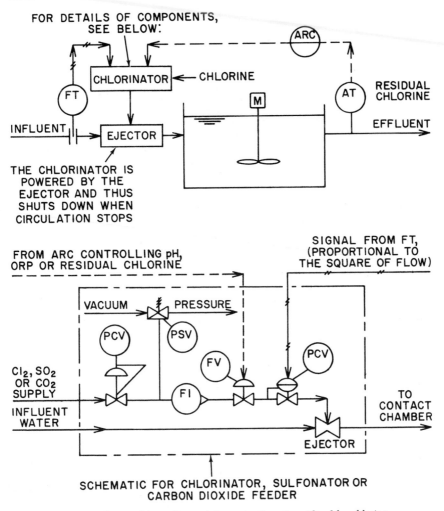

FOR DETAILS OF COMPONENTS,
SEE BELOW:

THE CHLORINATOR IS
POWERED BY THE
EJECTOR AND THUS
SHUTS DOWN WHEN
CIRCULATION STOPS

SCHEMATIC FOR CHLORINATOR, SULFONATOR OR
CARBON DIOXIDE FEEDER

Fig. 10.14d   Variable quality and flow rate of waste oxidized by chlorine

forward, the other by a feedback loop. Most reactions are completed within
five minutes, and except for cyanide treatment, most all other chemical
oxidation operations are carried out simultaneously with other unit operations,
such as coagulation and precipitation, which govern the pH value. Thus while
the pH value affects the rate of reaction, it is seldom controlled solely for
the oxidation process.

Chemical Reduction

Chemical reduction is quite similar to chemical oxidation, except that
reducing reactions are involved. Commonly used reductants are sulfur dioxide
and its sodium salts, such as sulfite, bisulfite, and metabisulfite. Ferrous iron

salts are infrequently used. Typical examples are reduction of hexavalent chromium, dechlorination, and deoxygenation.

Figure 10.14e is a schematic of a typical system for the reduction of highly toxic hexavalent chromium to the innocuous trivalent form according to the following reaction:

$$3SO_2 \quad \text{Sulfur dioxide}$$
$$+$$
$$2H_2CrO_4 \quad \text{Chromic acid}$$
$$\downarrow \qquad\qquad\qquad\qquad 10.14(3)$$
$$Cr_2(SO_4)_3 \quad \text{Chromic sulfate}$$
$$+$$
$$2H_2O \quad \text{Water}$$

Most hexavalent chrome wastes are acid, but the rate of reaction is much faster at very low pH values. For this reason pH control is essential. Sulfuric acid is preferred because it is cheaper than other mineral acids. The set point of the pH controller is approximately 2. As in the treatment of cyanide, the chemical reaction is not reversible, and the control of sulfur dioxide addition is by electrode potential level, using ORP instrumentation. The potential level of hexavalent chromium is $+700$ to $+1,000$ mv, while that of the reduced

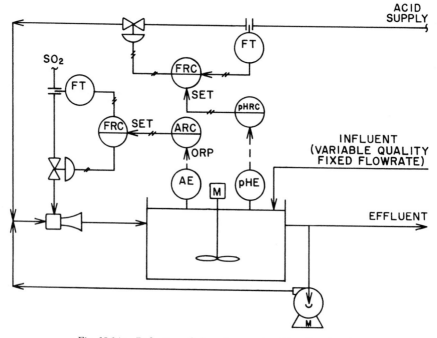

Fig. 10.14e   Reduction of chromium waste with sulfur dioxide

trivalent chrome is $+200$ to $+400$ mv. The set point on the ORP controller is approximately $+300$ mv.

The control system consists of feedback loops for both pH and ORP. The common and preferred design is for fixed flow systems.

The trivalent chromic sulfate is removed from solution by subsequently raising the pH to 8.5, at which point it will precipitate as chromic hydroxide. The control system for this step is identical with the one used in Figure 10.14k. Critical process control factors are summarized as follows:

Set Points and Parameters

Variable	Value
pH	2.0
ORP (mv) set point	$+300$
Reaction time (minutes)	10 at pH 2.0
	5 at pH 1.5

In the other examples cited, dechlorination and deoxygenation, control consists of adding the reducing agent in proportion to the oxidant concentration but maintaining a slight excess. In most cases, a slight excess of reducing agent is not detrimental. The pH value is not critical and is determined by other factors. Corrosion control is the most common of these factors.

Dechlorination to a fixed residual value is controlled, as illustrated in Figure 10.14d, except that sulfur dioxide is used instead of chlorine.

Neutralization

Strong alkalis react quickly and efficiently to neutralize strong acids. The simplicity of this fact is misleading in its consequences. Acid neutralization is a common requirement in waste water treatment, but few operations can be as complex. Essential information required for proper design includes: (1) flow rate and range of flow variations, (2) titratable acid content and variations in acid concentration, (3) rate of reaction and (4) discharge requirements for suspended solids, dissolved solids and pH range.

(The phenomena of pH and its detection is covered in Section 8.8 of Volume I.)

When the purpose of a control system is automatically to neutralize plant wastes, an understanding of the neutralization phenomenon is necessary. Figure 10.14f shows the pH values corresponding to the changing mixtures of a strong acid and a strong base. The slope of this pH curve is so great near neutrality (pH $=7$) that there is no likelihood of controlling such a system. Fortunately, plant effluents usually contain weak acids or bases which are neutralized by strong reagents. (The dotted pH curve shows that these are much easier processes to control.) The slope of the pH curve is affected by the ionization constants of the acid and base involved and by buffering.

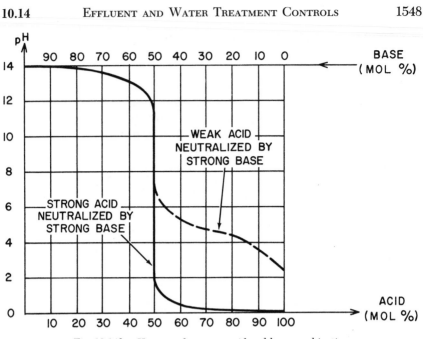

Fig. 10.14f   pH curve of a strong acid and base combination

Buffering compounds are those which contain no hydrogen or hydroxyl ions but are capable of suppressing the release of these ions from other solutes and thereby affect the solution acidity or alkalinity.

Neutralization control of plant wastes is difficult because of likely variation in—

> Acid or base contents by several decades.
> The type of acid (or base), thereby changing the applicable pH curve.
> Amount of buffering.
> Effluent from acidic to basic, which would require two reagents.
> Flow rates of effluent.

As a consequence, a reagent addition rangeability of several hundred to one may be required. This is accomplished by the use of two or more control valves in parallel. As shown in Figure 10.14g, the smaller valve has equal percentage characteristics and is throttled by a proportional controller. This is desirable to match the pH characteristics near neutrality with that of the valve. In order to maintain the relationship between pH and valve opening, the reset mode had to be eliminated. If pH measurement moves outside a preset "dead zone," this causes the second controller to make an adjustment in the opening of the large linear valve, thereby compensating for load changes. This second controller is provided with two control modes, with the integral action serving to bring the system back to set point after a load change.

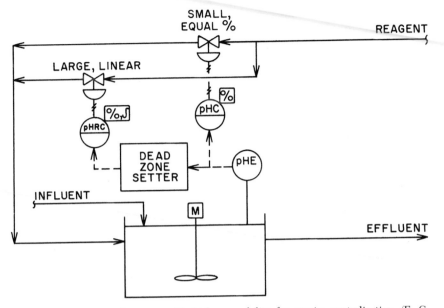

Fig. 10.14g Addition of reagent with high rangeability for precise neutralization. (F. G. Shinskey, *Process Control Systems*, McGraw-Hill Book Co., New York, 1967.)

It should be emphasized that pH is a measure of hydrogen ion activity and not acid concentration. A weak sulfuric acid solution will have a low pH value because of the high degree of hydrogen ion activity (disassociation), whereas some strong organic acids may show a pH value as high as 3 or 4.

An equalizing basin should be installed whenever possible ahead of the neutralizing system. This will tend to level out fluctuations in influent flow and concentration. This point cannot be overemphasized, because the lack of such a basin has been the cause of many failures. Pumping from an equalizing basin at a constant rate eliminates the need for high rangeability flow rate instrumentation. This, in combination with reduced variations in base or acid content, reduces the reagent feed range requirements. The obvious disadvantage is in the capital cost for large basins. Most systems are designed with as large an equalization tank as possible consistent with available space. Any equalization that can be installed will pay dividends.

Using acid wastes as an example, the maximum acid concentration and maximum flow rate will determine the capacity of the alkali feed system. The ratio between this and the requirement at minimum flow and minimum acid concentration will determine the range requirements of the reagent feed system. By using an equalization tank, a more dependable operation with lower investments in high rangeability equipment, can be expected.

It is essential that reaction rates be determined so that suitable reaction tank sizes (residence time) can be calculated. These rates are plotted by determining the total amount of alkali, in case of an acid waste, that is required to neutralize a sample of the waste, making sure the reaction has gone to completion. This amount is then added to a second sample as a single dose, and the pH rise vs time is plotted. Typical curves that can be expected are shown in Figure 10.14h. Sizing to provide at least 50 percent more holding time than that shown by the curve is recommended.

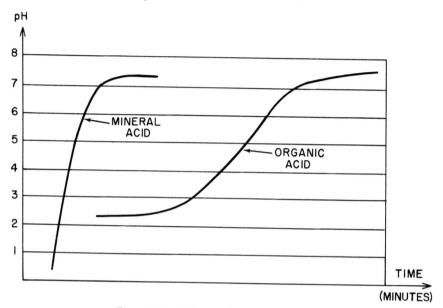

Fig. 10.14h   Acid neutralization reaction rates

The ideal aim is sufficient equalization to provide a homogeneous waste at a constant flow rate. To neutralize such a system, a lime feeder operating at a preset constant rate would suffice. Unfortunately, this seldom occurs, and provision must be made for reagent throttling. Figure 10.14i illustrates a system that can handle both flow and acid concentration changes.

The three treatment tanks and the final control elements for lime slurry feeding have identical capacities. Assuming a 10:1 range for each lime slurry control element, the range of the system is 30:1. Thus it can handle any combination of flow and acid concentration within that range.

The set point on the three pH controllers is at the final pH value desired. At periods of low flow, or when the acid content is low, treatment tank #1 can handle the entire requirement. Under other conditions, all three tanks may be required, with the first one or two satisfying a major portion of the reagent requirement, and the third serving a "polishing" or final trim function.

Two factors must be kept in mind in designing an acid neutralization

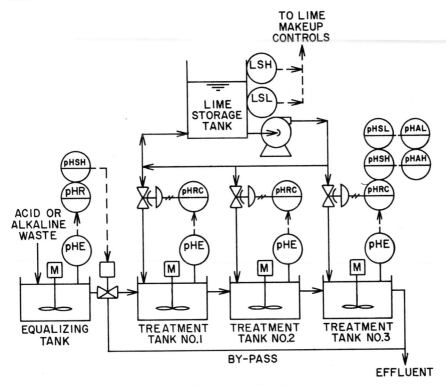

Fig. 10.14i   Acid waste neutralization system

control system: (1) pH is a logarithmic function, and (2) pH expresses hydrogen ion activity, not acid concentration.

Each tank in this system should be sized for a minimum of 50 percent of the total retention time determined at maximum flow rate. Where mixtures of acids are involved, the maximum time (not average) must be used.

Figure 10.14i has provision for the occasional case where incoming streams may be self-neutralizing. If this occurs, the treatment system is by-passed.

There are obviously many variations of this schematic that may be practical after a careful study of the waste characteristics. Tank sizes and number may be varied, the final control elements for lime feed can be of different sizes, and provision for recycle can be made if necessary.

Before design is started, regulations regarding discharge must be known. A common allowable range on pH values is between 6 and 8. There may also be limits on suspended or dissolved solids. The latter may be expressed as conductivity limits. A solution to these restrictions may be found in the choice of neutralizing chemicals. For instance, the resulting product of neutralizing sulfuric acid with lime (the cheapest alkali available) is calcium sulfate, a relatively insoluble product. This can be removed by sedimentation to reduce the quantity of suspended solids. Caustic soda reacts with sulfuric acid to form soluble sodium sulfate, requiring no subsequent sedimentation, but the effluent will contain substantial quantities of dissolved solids.

High maintenance costs of pH electrodes have been reported when lime is used as the reagent, due to the formation of calcium sulfate coatings on the electrodes. Daily maintenance may be needed. (See Section 8.8 in Volume I for details on devices for cleaning electrodes.)

In those systems where equalizing basins or other averaging techniques cannot be applied and accurate pH control is required, the concept of feedback control is insufficient, and feedforward schemes should be considered. Assuming that the reagent control valve is equal percentage in its characteristics, this can be used to advantage in a feedforward system to generate the forward loop pH function. Because the relationship between influent pH and reagent required to neutralize it is variable, feedback trimming of any feedforward loop is essential. A generalized feedforward-feedback model for the equal percentage valve position (not flow through it, but the signal it receives), represented by X, can be written as

$$X = (K_c)(\text{set-point} - \text{measurement}) + \log{(Fa)} \qquad 10.14(4)$$

where K_c is the proportional gain of the feedforward controller, F is the influent flow rate and a is the output signal from the feedback controller.

The nature of a neutralization system is such that the process gain (slope of the pH curve) is likely to be the highest at set point (around pH = 7) and decrease as deviation from set point increases. To compensate for this inverse relationship between process gain and deviation, the feedback controller is to be a non-linear one, with its gain directly varying (increasing) with deviation. With this controller, as shown in Figure 10.14j, the greater the deviation from set point, the larger the controller correction, but this does not create an unstable condition, because at high deviations the process gain is low.

If the reagent flow rangeability exceeds the capability of a single valve, the technique illustrated in Figure 10.14g can be applied.

Precipitation

Precipitation is the creation of insoluble materials by chemical reactions that provides treatment through subsequent liquid-solids separation. Typical of these operations is the removal of sulfates, removal of trivalent chromium, and softening of water with lime. Iron and manganese are removed by a variation of this process following the treatment discussed earlier, in connection with the chemical oxidation process. Lime softening is a common process and will be used as an example.

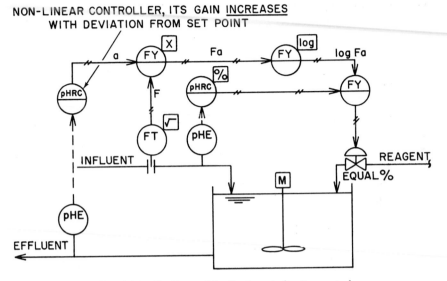

NON-LINEAR CONTROLLER, ITS GAIN <u>INCREASES</u>
WITH DEVIATION FROM SET POINT

Fig. 10.14j   Feedforward-feedback neutralization control

The reaction involved is

$$Ca(HCO_3)_2 \quad \text{Calcium bicarbonate}$$
$$+$$
$$Ca(OH)_2 \quad \text{Calcium hydroxide}$$
$$\downarrow$$
$$2CaCO_3 \quad \text{Calcium carbonate} \qquad 10.14(5)$$
$$+$$
$$2H_2O \quad \text{Water}$$

The calcium carbonate formed by this reaction is relatively insoluble and can be removed by gravity separation (settling). Typical settling time is 30 minutes or less, but most systems are designed for continuous operation with typical detention times of one hour.

Water treatment using this process is called *excess lime softening* derived from the application of lime in excess of that required for the reaction described in equation 10.14(5). Control consists of adding sufficient calcium hydroxide to maintain an excess hydroxide alkalinity of 10 to 50 milligrams per liter, as shown by equation 10.14(6):

$$2P = MO + 10 \text{ to } 50 \qquad 10.14(6)$$

where P is the phenolphthalein alkalinity and MO is the methyl orange alkalinity. This results in a pH value of 10 to 11, but pH control is not satisfactory for economical operations. This is an example where suitable analytical instrumentation is not available for continuous system control. If the quality of the untreated water is variable, operator control of lime dosage is essential. Pacing of manual dosage by feedforward control from flow rate is most frequently practiced.

A factor in the precipitation of calcium carbonate is a chemical phenomenon known as *crystal seeding*. This involves the acceleration of carbonate crystal formation by the presence of previously precipitated crystals. This is accomplished in practice by passing the water being treated through a "sludge blanket" in an upflow treatment unit shown schematically in Figure 10.14k. The resulting crystals of calcium carbonate are hard, dense and discrete, and they separate readily. When colloidal suspended material is also to be removed, which would be the case where surface waters are softened, a coagulant of aluminum or iron salts is also added to precipitate the colloids. Dosage is variable, depending on the quantity of suspended material. Application of both coagulant and calcium hydroxide is controlled by flow ratio modulation.

The resulting sludge, consisting of calcium carbonate, aluminum, or iron hydroxides, and the precipitated colloidal material are discharged to waste continuously. As previously noted, the presence of some precipitated carbonate is beneficial in order to remove all the sludge. An automatic sludge level control system is employed in Figure 10.14k to control the sludge level at an optimum.

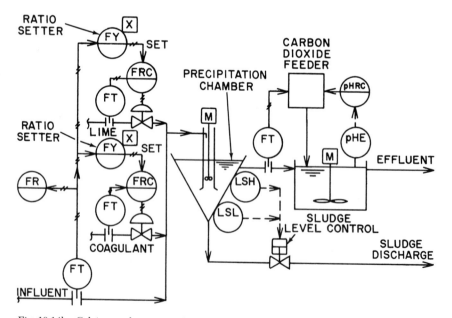

Fig. 10.14k   Calcium carbonate precipitation control system. Details of carbon dioxide feeder are shown in Figure 10.14d.

Water softened by the excess lime treatment is saturated with calcium carbonate and therefore is unstable. Stability is achieved by adding carbon dioxide to convert a portion of the carbonates into bicarbonate, according to the following equation:

$$CO_2 \quad \text{Carbon dioxide}$$
$$+$$
$$CaCO_3 \quad \text{Calcium carbonate} \qquad \text{10.14(7)}$$
$$+$$
$$H_2O \quad \text{Water}$$
$$\downarrow$$
$$Ca(HCO_3)_2 \quad \text{Calcium bicarbonate}$$

In contrast to the softening reaction, this process is suited to automatic pH control. Figure 10.14k shows this control system. The carbon dioxide feeder has two operators, one controlled by feedforward on influent flow, the other by feedback on effluent pH. The set point is in the order of 9.5 pH.

Fouling of the electrodes is likely to occur due to precipitation of crystallized calcium carbonate. Daily maintenance may be expected, unless automated cleaners are employed. The farther downstream (from the point of carbon dioxide application) the electrodes can be placed, consistent with acceptable loop time delays, the less will be the maintenance requirement.

Biological Control

Nearly every process for water or waste water treatment utilizes chlorination. It may be used either to reduce the possibility of pathogenic bacterial contamination of the receiving water or to prevent interference of biological growth with other processes. Disinfection of domestic waste and of potable water supplies is almost universally practiced. Control of biological slimes that interfere with heat exchange in cooling water systems, is readily accomplished with chlorine. The pulp and paper industry and the food industry typically use chlorination to prevent product deterioration.

Chlorine is a strong oxidant with many other uses, as was noted earlier in this section. It is a very effective bactericidal agent readily available at reasonable cost. Its effectiveness has been proved by over 50 years of use.

Several factors add to the complexity of chlorination systems for biological control. Materials in the water that can be chemically reduced cause an immediate reduction of the available active chlorine into an ineffective chloride form. Sufficient chlorine must be added to the water to account for this reaction and, in addition, provide a residual amount providing sufficient chlorine for subsequent reactions. The amount of chlorine involved in the initial reduction is called *chlorine demand*. A dosage greater than this amount provides excess chlorine which is called *residual chlorine.*

When nitrogenous material, particularly ammonia, is present in the water, the residual chlorine will be altered. Two kinds of residual chlorine are recognized: (1) Free residual is that remaining after the destruction with chlorine of ammonia or of certain organic nitrogen compounds. (2) Combined residual is produced by the reaction of chlorine with natural or added ammonia or with certain organic nitrogen compounds. This subject is dealt with adequately in "Water Treatment Plant Design," American Water Works Association Inc., 1969.

Since these two compounds (free residual chlorine and combined residual chlorine) are completely different in their ability to control bacterial organisms, it is important to differentiate which of the two forms of chlorine are involved in a process. Laboratory methods are available for both the measurement and differentiation of these two. Continuous analyzers suitable for control are also available to measure either the level of free residual chlorine or the level of total residual chlorine, meaning the combination of the two when both are present. The presence of residual chlorine in water does not ensure either disinfection or biological control. A bacteriological analysis requires several hours or even days. Through years of experience, it has been determined that measurement of residual chlorine can be a suitable inferential indicator of the effectiveness of biological control. For this reason, such processes are suitably controlled by analysis of residual chlorine.

An example of a typical control system employed in the disinfection of waste water is shown schematically in Figure 10.14l. The chemical feed system for applying chlorine must be of sufficient capacity to apply it both to satisfy the demand of the waste and to provide sufficient residual after the contact time required for the disinfection action. Typical of this for waste water treatment would be a chlorine demand of 5 milligrams per liter.

Another important factor is that the amount of residual chlorine will decline with time. There is no assurance that the initial residual concentration

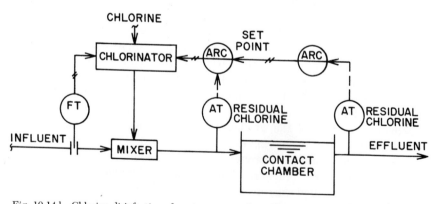

Fig. 10.14 l   Chlorine disinfection of waste water with variable flow rate and variable quality. Details of chlorinator are shown in Figure 10.14d.

will persist for the length of time required for disinfection to be accomplished. It is important that there be residual chlorine present during this entire time period. Typical design is for a contact chamber sized to permit retention for 30 minutes at maximum flow. Local regulations usually require that there be a minimum of 1 milligram per liter of available chlorine at the end of this contact time. It is not uncommon for flows to vary over a 6:1 range. To account for these variables and for the continuing chemical reactions, the control system automatically controls the residual after a short (approximately 5 minutes) fixed contact time. This is accomplished by building the 5-minute retention time into the sampling system. Feedback control, in conjunction with flow-proportioning feedforward control, establishes a constant residual value at the inlet of the contact chamber. Due to the variable rate at which residual decay occurs, assurance that the residual at the exit of the tank is maintained at a minimum value requires a second analyzer to change the set point on the controller as required.

The amount of residual chlorine decay varies widely due to variations

Table 10.14m

CONTROL INSTRUMENTATION APPLICABLE TO
VARIOUS EFFLUENT AND
WATER TREATMENT SYSTEMS

Treatment Process	Instrumentation			
	Flow	pH	ORP	Residual Chlorine
Chemical Oxidation of				
Cyanide		√	√	
Iron	√			√
Manganese	√			√
Hydrogen Sulfide	√			√
Chemical Reduction of				
Chromium		√	√	
Residual Chlorine	√			√
Precipitation of				
Chromium	√	√		
Iron	√	√		
Manganese	√	√		
Hardness	√			
Neutralization of				
Acid and Alkali	√	√		
Alkalinity (Recarbonation)	√	√		
Biological Control	√			√

in temperature, quality of waste water, and detention times. It may vary from a minimum of 0.5 to 5 milligrams per liter. To maintain an effluent residual of 1 milligram per liter, the set point on the controller for the chlorinator may be set anywhere from 1.5 to 6 milligrams per liter.

Chlorination of waste water for disinfection is unique in that it is usually the final process unit prior to discharge. For this reason detention time is provided as a part of the process. For most other biological control applications, other subsequent unit operations provide sufficient contact time, and the residual decay can be reasonably well predicted. A typical system is shown in Figure 10.14d.

Conclusions

Most water and waste water treatment systems are designed for continuous operation employing several process units. Many of the processes consist of chemical treatment. Relatively few of these are suited to automatic control because of the lack of reliable sensing devices. The development of *selective ion electrodes* suitable for continual analysis will, hopefully, increase the range of automatic control.

Some of the features of automated effluent and water treatment systems are summarized in Table 10.14m.

The natural laws governing chemical reactions dictate the design considerations. The most critical of these involve (1) pH, (2) reaction rates, (3) ratios (chemical dosage), (4) concentration and (5) temperature.

A discussion of coagulation and flocculation of water for suspended solids removal is conspicuously absent. It is because these have not yet been reduced from an art to a science. The same is true for biological waste treatment, although oxygen measurement (Volume I, Section 8.7) is coming into wider use for guidance to operator control.

10.15 ANALOG AND DIGITAL BLENDING SYSTEMS

Type of System:	(a) Analog Mechanical, (b) Analog Pneumatic, (c) Analog Electronic, (d) Digital Blending Note: In the feature summary below, the letters (a) to (d) refer to the listed designs.
Features Available:	All designs are available with indicating, recording or remote blend ratio adjustment features. In case of digital systems, signal conversion is usually required.

Ratio Adjustment Ranges:

	Linear	Square Root
(a)	0.1 to 3.0	0.5 to 1.7
(b)	0 to 3.0	0 to 1.7
(c)	0.3 to 3.0	0.6 to 1.7
(d)	0.001 to 1.999	

Precision:

	Accuracy	Repeatability
(a)	$\pm 2\%$	
(b)	$\pm 1\%$	0.25%
(c)	$\pm 0.5\%$	0.25%
(d)	$\pm 0.25\%$	0.1%

Controller Cost Ranges: (a) $400–$750, (b) $450–$850, (c) $600–$1,250, (d) $2,500–$5,000. In case of (d), signal-conditioning equipment is included and the price is given on a per flow stream basis.

Partial List of Suppliers: Fischer and Porter Co. (a and d); Foxboro Co. (b, c and d); Jordan Controls, Inc. (c); Moore Products Co. (b); Neptune Meter Co. (d); Robertshaw Controls Corp. (c); Taylor Instrument Cos. (a and c); Waugh Controls Corp. (d).

Blending systems are applied to a variety of materials in a number of industries: solvents, paints, reactor feeds, foams, fertilizers, soaps and liquid cleaners in the chemical industry; gasoline, asphalt, lube and fuel oils and distillates in the petroleum industry; wine, beer, candy, soups, ice cream mix and cake mixes in the food industry; cement, wire insulation, and asbestos products in the building industry.

These applications provide the processor with economic advantages by controlling the consumption of materials (costly components and additives can be blended more precisely) and by reducing investment in floor space and batching tanks (costly blend tanks are eliminated). Through the use of a continuous system, time lags of batch methods are eliminated, productivity is increased, manpower needs are reduced, and inventory can be in the form of component base stocks rather than as partially blended or finished products.

Technical advantages are provided by accurately controlling the quality of the product and by providing the flexibility to blend a variety of finished products with a minimum time required to change from one product to another.

These applications have a common denominator, which is the continuous control of the flow of each component with fixed ratios between components, so that when the streams are continuously combined to form the finished blend at a fixed through-put rate, the composition of the finished product is within specifications.

In this section, blending systems are described from the standpoint of control techniques. A number of typical blending systems are described to show the operating principles involved.

Various aspects of analog ratio control are also discussed in Sections 4.2, 7.6, 7.9 and 9.4.

Blending Methods

Automatic, continuous, in-line blending systems provide control of gases, liquids and solids in predetermined proportions at a desired total blend flow rate. The blending systems consist of flow transmitters (to detect controlled variables), ratio relays (to set proportions) and controllers (to complete the closed-loop control).

A two-stream blending system is illustrated in Figure 10.15a. The component "A" flow controller is set by the total blend flow controller, and the component "B" flow is ratioed to "A." It is also possible to have both blending components ratioed to the total blend flow, as shown in Figure 10.15b. In either case, the blending system maintains the blending ratio as well as the total flow rate. Incorporation of a preset totalizer with automatic system shutdown facilities provides batching capability as well.

Many of the commercially available blending systems provide system options such as flow alarm indications, system shutdown features, temperature

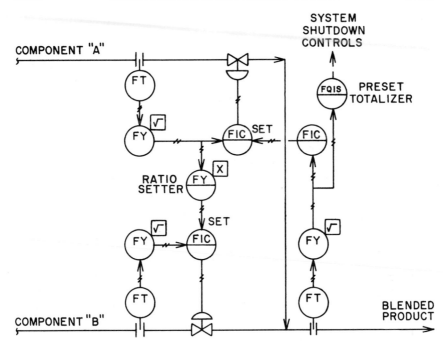

Fig. 10.15a   Analog rate blending of two components

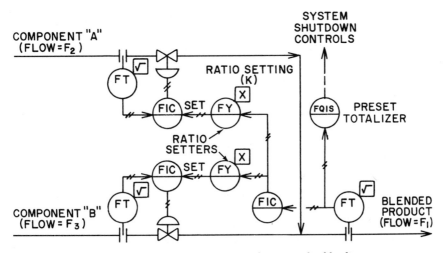

Fig. 10.15b   Both components directly ratioed in an analog blending system

compensation circuitries, scalers for conversion of transmitted signals to easily understood engineering units, pacing controls to slow down or shut down automatically and manual or automatic adjustment of blend rate and ratio. All of these systems share one type or another of a ratioing mechanism. The methods of blending will be examined here. The working principles of components are detailed in the latter part of this section.

Rate Blending

The blending system shown in Figure 10.15a is commonly found in the chemical industry for blending gas or liquid flows. A typical example is the manufacture of hydrochloric acid, which is done at fixed concentration and regulated flow rate by the absorption of anhydrous hydrogen chloride gas in water. The flow rate of hydrochloric acid from the absorption tower is measured to set the water flow thereby maintaining the desired through-put, and the anhydrous hydrogen chloride gas is ratioed to water flow to give constant concentration.

A system in which all the blend components are ratioed to the total blend flow is illustrated in Figure 10.15b. A well-known example of this application is in the continuous or semi-continuous charging of a batch reactor, where the recipe is given to set only the ratios of each ingredient and the total reactant charging rate. In the semi-continuous batch operation, a preset totalizer is utilized to terminate the charging operation when all ingredients have been charged. Numerous streams can be blended by incorporating additional ratio devices and related controls.

Totalizing Blending

When totalized flows are ratioed, the integrated quantity of each component (over a period of time) is controlled in a direct ratio to the total quantity of the blended product. A schematic diagram of a totalizing blending system is shown in Figure 10.15c. By this system, more precise control over the amount of each component is obtained than is possible with a rate-blending system. In the rate-blending system corrections are made to the ratio controller flow only after deviations have occurred (feedback control), and without correction for errors that have already occurred. In other words, the control system has no memory. By totalizing the flows and comparing the totals, the precise percentage of each component is ensured in the total blend.

Digital techniques have also been applied to digital blending systems, as shown in Figure 10.15d. Turbine meters or other pulse-generating devices can be employed to generate digital flow signals. A bi-directional counter is used to integrate the flow measurement and demand signals and to compare them in order to generate a corrective control signal whenever the counts in the bi-directional counter memory are not zero.

The totalizing-blending system (analog or digital) finds use in continuous

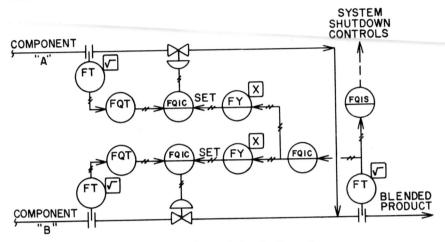

Fig. 10.15c   Analog totalizing blending system

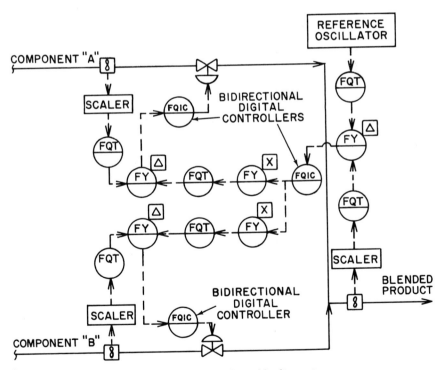

Fig. 10.15d   Digital totalizing blending system

in-line blending of petroleum products, such as gasoline and asphalts, where long runs and batching operations require precise control of ingredients to assure in-spec blending and uniform end-products. The accuracy of this system allows for on-line blending of many petroleum, chemical, food and cement products, with the product being sent directly to final shipping and storage containers. Multi-component blending systems can be obtained by adding more flow ratio controllers.

Optimizing Blending

Almost all of the commercially available ratio relays and controllers are able to accept remote set points. Therefore, the blend ratios and total blend flow rate can be automatically adjusted by a process variable. Thus an optimizing blending system has the added capability of automatically manipulating ratio settings and/or the total rate, based on certain criteria. A schematic of an optimizing blending system is shown in Figure 7.9d, where the blend analyzer is used to measure composition and to adjust the ratio setting while the level in some downstream tank is used to set the total flow rate.

In the hydrochloric manufacturing process, a densitometer can be used to detect the solution concentration and to correct the ratio settings, if deviations from set point occur. A chromatographic analysis or Reid vapor pressure measurement of gasoline provides automatic adjustment of blending ratio of components, such as butane, to give proper octane number.

Another application is reactor charge rate control of jacketed exthotheric reactors, based on the heat transfer coefficient and plant cooling capacity. For the selection of flow sensors, refer to Volume I.

Analog Blending

The heart of the analog blending system is the mechanism for ratio control. This is often a separate component, although it may be housed with the controller. As shown in Figure 10.15a, a blending system can be constructed by ratio controlling the blend components with the total blend flow. Thus the total blend rate is controlled together with the individual blend ratios. With incorporation of a preset totalizer, system shutdown can be initiated when batch blending is completed.

The ratio control relationship is derived with reference to the system shown in Figure 10.15b. It is assumed that C_1 and C_2 are the flow constants for the flow-measuring orifices. Then,

$$F_1 = C_1 \sqrt{P_1} \quad \text{or} \quad P_1 = (F_1/C_1)^2 \qquad 10.15(1)$$

and

$$F_2 = C_2 \sqrt{P_2} \quad \text{or} \quad P_2 = (F_2/C_2)^2 \qquad 10.15(2)$$

where F_1 = total flow rate,
    F_2 = component "A" flow rate,
    P_1 = output signal of total flow transmitter and
    P_2 = output signal of component "A" flow transmitter.

If K is the desired ratio setting,

$$P_2 = KP_1 \qquad\qquad 10.15(3)$$

By substituting equation 10.15(3) into (2),

$$F_2 = C_2 \sqrt{KP_1} \qquad\qquad 10.15(4)$$

and by substituting equation 10.15(1) into (4) and simplifying,

$$F_2 = (C_1/C_2)(\sqrt{K})(F_1) \qquad\qquad 10.15(5)$$

Using the same mathematical derivation, a ratioing system with linear input signals can be expressed as

$$F_2 = (C_1/C_2)(K)(F_1) \qquad\qquad 10.15(6)$$

A graphical representation of equation 10.15(6) is shown in Figure 10.15e, top, and the inverse-ratio controller characteristic is illustrated on the bottom. From equation 10.15(5) it can be seen that the square root input signals from flowmeters can be used as in Figure 10.15e, with the ratio flow control circuit based on a linear relationship, the only modification being the square root calibration of the ratio setting dial and of the indicating dials. The ratio setting dial is graduated and calibrated as the square root of the linear ratio. Most ratio control mechanisms provide for bias adjustments to change the basic characteristics shown in Figure 10.15e, to meet various process requirements. Figure 10.15f shows some of these biased relationships where the secondary flow has a preset minimum value or where it is kept zero until the primary flow reaches some value. Also refer to Figure 9.4d and text for a discussion of the limitations of biased ratio relays.

Mechanical Ratio Control

The mechanical ratio control system consists of a proportioning mechanism, pneumatic or electronic flow signal receivers and a case-mounted controller. The adjusting system resets the control point at a preset ratio by means of an adjustable mechanical linkage. The receivers are linked to a pen assembly, and output motion from the pen assembly operates the proportioning mechanism. Subsequent output motion from the proportioning mechanism positions the input lever of the pneumatic controller.

Figure 10.15g illustrates that each pen assembly spindle connects to one of the proportioning mechanism input levers through an adjustable link. Each input lever positions an internal crank arm assembly and raises or lowers one

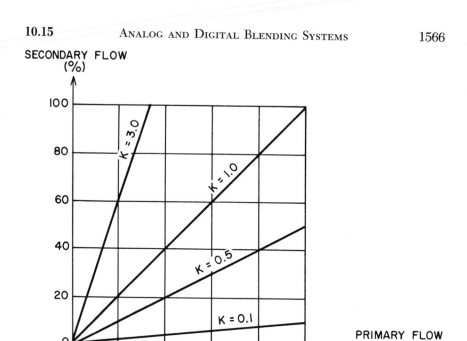

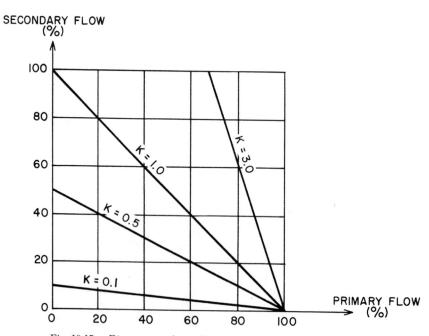

Fig. 10.15e   Direct ratio relationship (top) and inverse ratio relationship

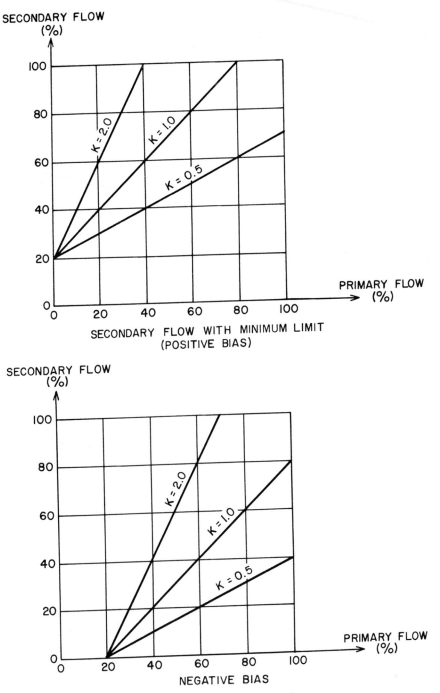

SECONDARY FLOW WITH MINIMUM LIMIT
(POSITIVE BIAS)

NEGATIVE BIAS

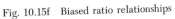

Fig. 10.15f   Biased ratio relationships

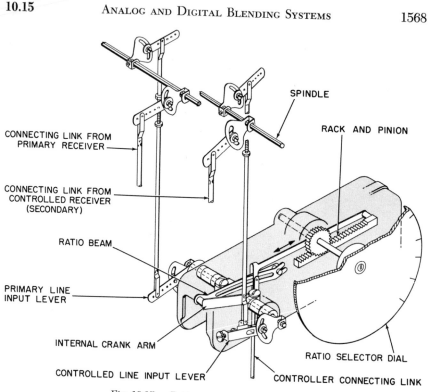

Fig. 10.15g   Ratio proportioning mechanism

end of the ratio beam. In turn, the link from the ratio beam connects to the input lever of the controller. The overall result is that controller output pressure changes whenever a receiver moves an input lever.

For remote adjustment of ratio set point, the manual set ratio mechanism is replaced by a pneumatic receiver. An external 3–15 PSIG set-point signal is used to position the receiver in proportion with the desired ratio. The ratio proportioning mechanism is precalibrated at the factory for the specific application, and all that is normally required is to check before use that the match marks are aligned and that the recorder pens and the transmitting meters are synchronized. Lubrication is seldom required, but the mechanism should be periodically inspected, cleaned and checked so that the proportioning mechanism operates frictionlessly.

The output accuracy of ±2 percent full scale can be obtained. Rangeability of this type of proportioning mechanism is about 40 to 1.

Pneumatic Ratio Control

The pneumatic ratio controllers contain no friction-producing mechanical links. The ratio relay modifies the input signal by means of the pneumatic circuit illustrated in Figure 10.15h. The primary variable signal is tubed through a fixed restriction (FO) into an adjustable area restriction. If the variable restriction valve is closed, the signal is not modified. This condition represents 100 percent ratio, because the controlled variable signal (the set point of the secondary flow controller) must equal the primary variable signal for the control circuit to be satisfied. If the adjustable area restriction is opened, the pressure between the two restrictions will drop until the flow through the fixed area restriction equals the flow through the adjustable area restriction. Thus the pressure is modified as a function of the opening of the adjustable area restriction. By calibrating the adjustable restriction in terms of percent ratio, the relay can be set for any desired ratio within its limits.

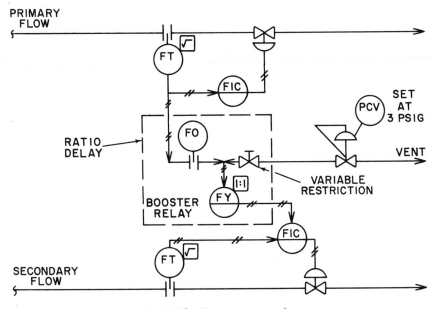

Fig. 10.15h   Pneumatic ratio relay

A booster relay should be used to ensure rapid transmission of the modified signal from the ratio relay to the controller. For applications in which the secondary variable set point will always be less than 100 percent of the primary, a 1:1 booster relay is recommended. In other cases a 2:1 or higher booster may be used.

The pneumatic set ratio circuit is identical to the manual set ratio relay, except that the adjustable restriction opening is set by a pneumatic diaphragm motor. This allows for continuous automatic adjustment of the ratio in accordance with a pneumatic signal received from a quality controller or other optimizing device. Incorporation of reset in the quality controller is recommended to eliminate the necessity for vernier adjustments to obtain the exact ratio and to compensate for the linearity limitations of the ratio unit.

The ratio relay should be calibrated at a specific ratio setting under actual operating conditions, even though the ratio setting is to be changed with operating conditions, to obtain maximum system accuracy. The accuracy (secondary flow set point) of ± 1 percent of full scale can be expected. Signal rangeabilities of 50 to 1 can be obtained, but system rangeability is determined by the flowmeters used. Most pneumatic ratio control systems are not designed to operate below 20 percent of full scale flow with square root signals or below 10 percent of full scale flow with linear signals.

Electronic Ratio Control

Electronic ratio control systems operate on the Wheatstone bridge principle, shown in Figure 10.15i. The bridge is said to be in a null, or balanced, condition when the ratios of resistance are such that $R_c/R_1 = R_f/R_2$, and no potential difference exists between points "A" and "B." If the ratio R_c/R_1 changes, then R_f/R_2 must also change by a like amount and in the same direction in order to maintain a null or balanced condition.

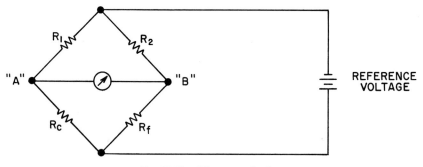

Fig. 10.15i   Wheatstone bridge

Figure 10.15j illustrates the operating principle of a Wheatstone bridge control system. Here the fixed resistors or bridge arms are replaced by potentiometers so that the ratios previously mentioned are easily varied. Assuming an initial balance, an increase or decrease in the setting of the command potentiometer (primary) causes an error signal to appear at the input of a servo amplifier which supplies power to the driven load. The direction of movement is dependent upon the polarity (or phase) of the error signal. The brush arm of the feedback potentiometer is mechanically geared to the driven load (secondary) and thus rotates until $R_f/R_2 = R_c/R_1$. At this point a null again exists (error signal equals zero) and positioning ceases. The actual position of each pot wiper, expressed as a fraction of its total possible travel may be written

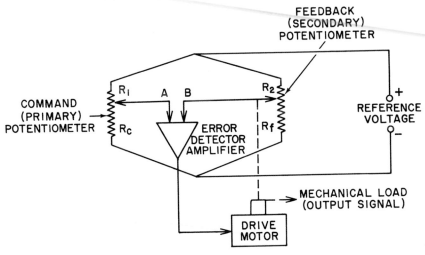

Fig. 10.15j   Wheatstone bridge operation

$$\frac{R_c}{R_c + R_1} \quad \text{and} \quad \frac{R_f}{R_f + R_2} \qquad 10.15(7)$$

At null, $R_c/(R_c + R_1) = R_f/(R_f + R_2)$. This is shown graphically in Figure 10.15k for a 0–100 percent movement of the command.

In this example the feedback signal is a measurement of actual movement or displacement, but when utilized in a flow ratio application, the feedback signal will be related to the secondary set point, and the command becomes the primary (wild) flow variable signal. Introduction of a fixed resistance in series with the command pot causes a change in the slope of the characteristic curve in Figure 10.15k. By making this additional resistance a potentiometer, as shown in Figure 10.15l, the full range travel of the feedback pot and of the driven load (secondary) can be limited to any desired degree for a 0–100 percent movement of the command. In Figure 10.15l the feedback pot position at null will be

$$\frac{R_f}{R_f + R_2} = \frac{R_c}{R_c + R_1 + R_3} = K_1\left(\frac{R_c}{R_c + R_1}\right) \qquad 10.15(8)$$

where $K_1 = (R_c + R_1)/(R_c + R_1 + R_3)$. The slope of the characteristic curve in Figure 10.15m is K_1, and its value is dependent upon the setting of R_3, which is usually calibrated to represent K_1, the ratio setting.

Electronic ratio control provides fast, accurate and adjustable ratios between input and output signals. Accuracy of ±0.5 percent of span is attainable.

Care should be exercised to provide a constant supply voltage and

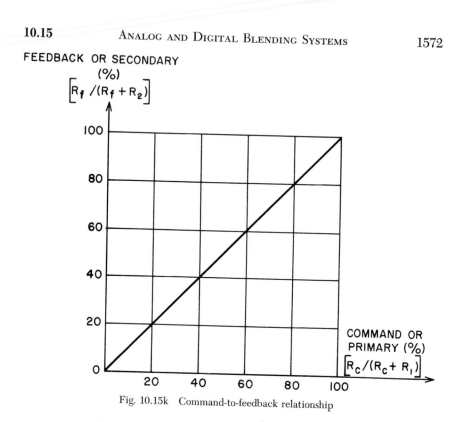

Fig. 10.15k   Command-to-feedback relationship

frequency. A change of 10 percent from the nominal voltage will cause a zero shift of as much as 0.5 percent of input value, and a change of 10 Hz over the range of 47 to 63 Hz will cause a zero shift of 0.25 percent of input value.

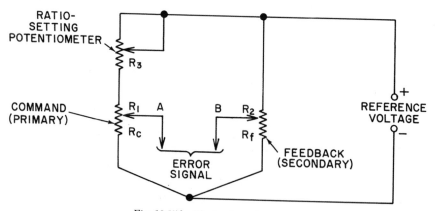

Fig. 10.15 l   Electronic ratio circuit

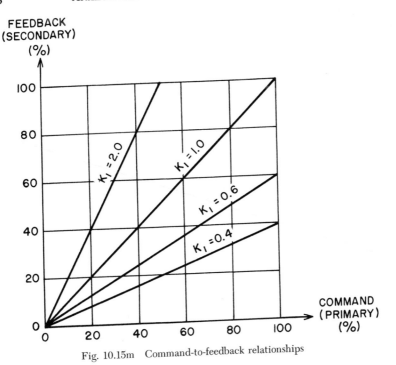

Fig. 10.15m   Command-to-feedback relationships

Ratio Dial Setting

The setting of the ratio relay is a function of the ranges of the transmitters. If the transmitters are measuring over the same range and in identical units, the graduations on the ratio dial represent the exact ratio between the primary and secondary flows. However, where maximum capacities and primary meter measurement units differ, the ratio selector dial setting must be calculated for each ratio desired. (For a review of general scaling procedures, refer to Section 9.1.)

Commercially available ratio control units are graduated to handle signals of the same characteristics (either linear or square root) and of the same units. The following equation is used to calculate ratio dial settings:

$$\text{Ratio dial setting} = \frac{(F_{pm})}{(F_{sm})}(R) \qquad\qquad 10.15(9)$$

where F_p = flow rate through primary,
$\quad F_s$ = flow rate through secondary,
$\quad F_{pm}$ = maximum capacity of primary flow transmitter,
$\quad F_{sm}$ = maximum capacity of secondary flow transmitter and
$\quad R$ = desired ratio of F_s/F_p.

If F_{pm} is 50 GPM and F_{sm} is 25 GPM, and it is desired to maintain the secondary at exactly 25 percent of the primary flow, then the ratio dial setting is $(50/25)(0.25) = 50\%$. In selecting the ratio dial settings, one should keep in mind the rangeability limitations of the flow transmitters. (See Chapter V of Volume I for details.)

Ratio Controller Tuning

A block diagram of a simple ratio control system is shown in Figure 10.15n. In the ratio control system, the set point of the secondary controller is directly related to the output of the primary flow transmitter. As the primary flow changes, the secondary controller assumes a new set point to maintain the desired ratio. (For a discussion of transfer functions and their use in feedback loops, refer to Sections 7.4 and 7.5.)

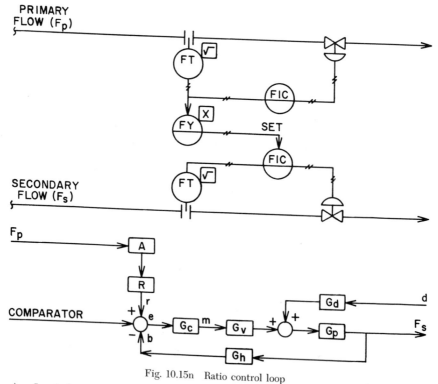

Fig. 10.15n   Ratio control loop

A = Input element
b = Feedback variable
d = Disturbance or load variable
e = Error (deviation) signal
F_p = Primary flow
F_s = Secondary flow
G_c = Controller transfer function

G_d = Disturbance or load transfer function
G_h = Feedback sensor transfer function
G_p = Process transfer function
G_v = Control valve transfer function
m = Manipulated variable
r = Reference (set-point) input
R = Desired ratio

If simple characteristics are assumed for the transfer functions in the block diagram, the overall system transfer function for set-point disturbances can be expressed as:

$$F_s = RF_p \left(\frac{T_s(s + 1)}{T_p(s + 1)} \right) \left(\frac{T_i(s + 1)}{(T_i T_s / K_c)s^2 + T_i(1/K_c + 1)(s + 1)} \right) \qquad 10.15(10)$$

where T_p = lag of primary flow measuring element,
$\quad\;\; T_s$ = lag of secondary flow measuring element,
$\quad\;\; T_i$ = integral time of controller and
$\quad\;\; K_c$ = controller gain.

As can be seen from equation 10.15(10), the best control of the controlled secondary variable (with primary variable changes) can be obtained when the lags T_p, T_s and T_i are minimum and the controller gain K_c is maximum, without creating instability. This statement is true not only for ratio loops, but for most feedback loops of all types.

Digital Blending System

The application of digital techniques to ratioing and blending may result in the total elimination of control system errors. This system continuously compares the total accumulated flows from each additive line with the total accumulated signal from a master oscillator. If there is a difference between these two values the corresponding control valve is repositioned to correct the deviation.

An overall digital blending system is illustrated in Figure 10.15o. The flow of each component is digitized by a turbine or displacement type flowmeter or by an analog-to-pulse generator, producing a pulse train whose frequency is proportional to flow rate. A standardizer is utilized to scale the transmitter output frequency to a common reference basis, such as 1,000 pulses per gallon, etc. This frequency is compared with a reference frequency produced by a numerically controlled frequency generator, which is commonly referred to as a *binary multiplier*. The inputs to the multiplier consist of a numeric quantity and a pulse frequency, and the output is a new pulse train whose frequency is the product of the two inputs. The multiplier produces two reference frequencies proportional to the manually set numeric ratio settings of K and $(1 - K)$.

Each digitized flow rate is compared with its corresponding demand signal generated by the ratio set module (binary multiplier). This comparison is performed by a *bi-directional binary counter*. The bi-directional counter counts in the positive direction on pulses from one input and in the negative direction on pulses from the other input. The set-point pulses produce "add" pulses and the measurement pulses produce "subtract" pulses. Hence, if the flow-generated pulses equal the demand pulses, the algebraic sum is zero and

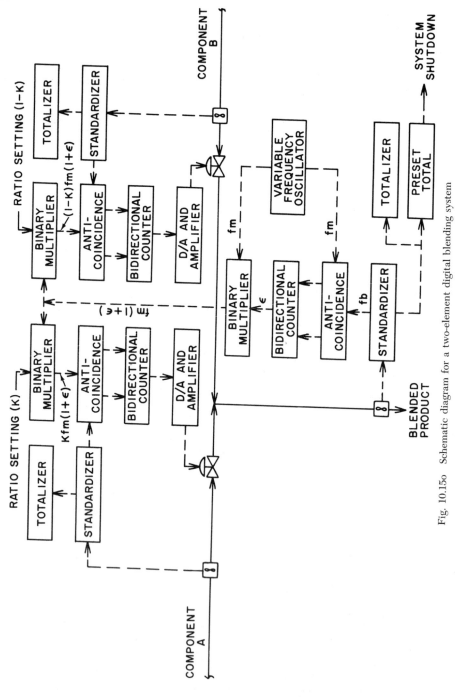

Fig. 10.15o Schematic diagram for a two-element digital blending system

1576

no change will occur in the binary memory, and no corrective action is taken. Should the rate from one input exceed that from the other, an error count will accumulate in the memory, causing the valve-control logic to generate a proportional correction. This correction signal, after conversion and amplification, positions the control valve. Thus, the quantitatively controlled flow rates of the blend components are maintained at the prescribed ratio.

For applications requiring precise control of total flow rate as well as of the blend ratios, a further digital control loop is provided, as shown in Figure 10.15o. Here a variable frequency oscillator is manually set so that its frequency is proportional to the desired total blend flow rate. This reference signal (f_m) together with the signal generated by the actual total blended flow rate (f_b) is synchronized by anticoincidence logic and accumulated in a bi-directional binary counter. The instantaneous counts (accumulation) of this counter are a measure of the difference between the total number of pulses generated by the reference oscillator and by the flowmeter, respectively. Thus

$$\epsilon = \Sigma f_m - \Sigma f_b \qquad\qquad 10.15(11)$$

where ϵ = instantaneous error accumulation in bi-directional counter,
  f_m = reference oscillator frequency and
  f_b = total blend flowmeter-generated frequency.

This instantaneous error (ϵ) serves as the numeric input into a binary multiplier whose input frequency is f_m. Therefore, the multiplier output frequency is ϵf_m. This output is then mixed with f_m in such a manner as to avoid time coincidence, and it thereby yields a pulse train having the average frequency of $f_m (1 + \epsilon)$. Thus if the blend operation produces a flow rate which is less than the sum of the constituent flow rates, or if the blend output flow rate must be controlled while keeping the blend ratios constant, then the error term (ϵ) provides the necessary augmentation to the total flow rate reference frequency. The frequency input to the ratio setting binary multipliers is $f_m(1 + \epsilon)$, and the resulting ratio demand outputs are

$$Kf_m(1 + \epsilon) \qquad \text{and} \qquad (1 - K)f_m(1 + \epsilon) \qquad 10.15(12)$$

for the two-component blending system.

This principle can also be used automatically to slow down the total blend flow rate, by substituting the master demand frequency (f_m) with a component flow frequency as an input to the master bi-directional counter. This feature is useful when one component may fall behind at startup, when a strainer is plugged or when a pump cannot meet the flow requirements. When this occurs, the component controller takes over the pacing from the master demand unit if a predetermined error has been accumulated, and adjusts the total flow rate to a value which the component can maintain. An alarm and automatic shutdown logic circuitry can be also incorporated to signal alarm conditions or automatically to shut down the system if any of the components fall below their preset minimum rates.

The total blended product requirement may be preset on a totalizer to initiate batch shutdown. Analyzers or optimizers can be added to adjust automatically the blend ratios or total blend flow rate as required.

The accuracy of the overall control system can exceed ±0.25 percent, with repeatability of better than 0.1 percent. The blend ratio setting can cover a range from 0.001 to 1.999, utilizing four-digit thumbwheels. In a digital blending system, the dynamic response is limited only by the control valve stroke speeds, since the control system itself has practically no deadtime.

Conclusions

Whether the blending system is designed by a user or is a package purchased from a manufacturer, the measuring and transmitting devices should be matched against the selected blending system in accuracy, rangeability and flexibility. It is inconsistent to install an accurate digital blending system with low-accuracy sensors.

In modern plants, the blending operation need not be costly, time-consuming and tedious. The instrument engineer can select from a wide range of flow measuring devices for liquids, solids and gases, and from analog or digital flow ratio controllers to design a continuous in-line blending system. The degree of his success will depend upon the proper selection of the suitable blending technique and of component hardware.

APPENDICES

H. D. Baumann, B. G. Lipták
and G. Platt

CONTENTS OF APPENDIX

A.1 INSTRUMENTATION FLOW SHEET SYMBOLS   1581
    Symbolizing Philosophy   1581
    Application Examples   1587

A.2 INTERLOCK LOGIC SYMBOLS   1592
    Logic Symbols   1592
    Logic Application   1597

A.3 ESTIMATING VALVE NOISE   1600
    Definitions   1600
    Estimating Method   1601
    Examples of Estimation   1603

A.1 INSTRUMENTATION FLOW SHEET SYMBOLS

This section describes the major elements of a method for symbolizing and identifying instruments on flow sheets and other documents. The symbols and identifications are based on the instrument functions. This method of representation identifies the means of measurement and the type of process control, but it leaves most details of instrumentation features to be determined from specifications or other documents.

Symbolizing Philosophy

Each instrument identification (or tag number) consists of a *functional* identification and a *loop* identification. A typical tag number is PRC-8, for a pressure recording controller (the functional identification is PRC), and the loop identification is the number 8. This tag number may include coded information such as plant area designation, flow sheet number, etc., resulting in multiple-digit identification numbers.

Table A.1a lists the meanings of the functional identification letters. The functional identification begins with a first letter denoting a measured or initiating variable. Readout or passive functional letters follow, describing the type of readout or display. The last letter represents the output function. They need follow no predetermined sequence, except that output letter C (control) precedes output letter V (valve); e.g., HCV, a hand-actuated valve. Modifying letters, if used, are interposed so that they are placed immediately following the letters they modify.

The functional identification is made according to the function and not according to the construction, e.g., a differential-pressure recorder used for flow measurement is an FR. The first letter of the functional identification follows the measured or initiating variable and not the manipulated variable, e.g., a control valve varying flow as commanded by a pressure controller is a PV, not an FV.

Each instrument loop should have a unique identification number. This number is common to all instruments of a loop. Because each instrument should have a unique identification, suffix letters A, B, C, etc. are used to distinguish among two or more instruments of similar function in a loop. With multipoint recorders, however, one may more conveniently use suffix numbers, such as TR-5-1, TR-5-2 and TR-5-3.

1581

Table A.1a

MEANINGS OF FUNCTIONAL INSTRUMENT IDENTIFICATION LETTERS

	First Letter		Succeeding Letters (3)		
	Measured or Initiating Variable (4)	Modifier	Readout or Passive Function	Output Function	Modifier
A	Analysis (5)		Alarm		
B	Burner Flame		User's Choice (1)	User's Choice (1)	User's Choice (1)
C	Conductivity (Electrical)			Control (13)	
D	Density (Mass) or Specific Gravity	Differential (4)			
E	Voltage (EMF)		Primary Element		
F	Flow Rate	Ratio (Fraction) (4)			
G	Gauging (Dimensional)		Glass (9)		
H	Hand (Manually Initiated)				High (7, 15, 16)
I	Current (Electrical)		Indicate (10)		
J	Power	Scan (7)			
K	Time or Time-Schedule			Control Station	
L	Level		Light (Pilot) (11)		Low (7, 15, 16)
M	Moisture or Humidity				Middle or Intermediate (7, 15)
N (1)	User's Choice		User's Choice	User's Choice	User's Choice
O	User's Choice (1)		Orifice (Restriction)		
P	Pressure or Vacuum		Point (Test Connection)		
Q	Quantity or Event	Integrate or Totalize (4)			
R	Radioactivity		Record or Print		
S	Speed or Frequency	Safety (8)		Switch (13)	
T	Temperature			Transmit	
U	Multivariable (6)		Multifunction (12)	Multifunction (12)	Multifunction (12)
V	Viscosity			Valve, Damper, or Louver (13)	
W	Weight or Force		Well		
X (2)	Unclassified		Unclassified	Unclassified	Unclassified
Y	User's Choice (1)			Relay or Compute (13, 14)	
Z	Position			Drive, Actuate or Unclassified Final Control Element	

Note: Numbers in parentheses refer to specific explanatory notes that follow.

Notes for Table A.1a

1. A *user's choice* letter is intended to cover unlisted meanings that will be used repetitively in a particular project. If used, the letter may have one meaning as a first letter and another meaning as a succeeding letter. The meanings need be defined only once in a legend, or otherwise, for that project. For example, the letter N may be defined as *modulus of elasticity* as a first letter and *oscilloscope* as a succeeding letter.

2. The *unclassified* letter X is intended to cover unlisted meanings that will be used only once or to a limited extent. If used, the letter may have any number of meanings as a first letter and any number of meanings as a succeeding letter. Except for its use with distinctive symbols, it is expected that the meanings will be defined outside a tagging balloon on a flow diagram. For example, XR-2 may be a *stress recorder*, XR-3 may be a *vibration recorder*, and XX-4 may be a *stress oscilloscope*.

3. The grammatical form of the succeeding-letter meanings may be modified as required. For example, *indicate* may be applied as *indicator* or *indicating, transmit* as *transmitter* or *transmitting*, etc.

4. Any first letter, if used in combination with modifying letters D (differential), F (ratio), or Q (integrate or totalize), or any combination of them, is construed to represent a new and separate measured variable, and the combination should be treated as a first-letter entity. Thus, instruments TDI and TI measure two different variables, namely, differential temperature and temperature. These modifying letters are used when applicable.

5. First letter A for *analysis* covers all analyses not listed in Table A.1a and not covered by a *user's choice* letter. It is expected that the type of analysis in each instance will be defined outside a tagging balloon on a flow diagram.

   Readily recognized self-defining symbols such as pH, O_2, and CO have been used optionally in the past in place of first-letter A.

6. Use of first-letter U for *multivariable* in lieu of a combination of first letters is optional.

7. The use of modifying terms *high, low, middle* or *intermediate*, and *scan* is preferred, but optional.

8. The term *safety* applies only to emergency protective primary elements and emergency protective final control elements. Thus, a self-actuated valve that prevents operation of a fluid system at a higher-than-desired pressure by bleeding fluid from the system is a back-pressure type PCV, even if the valve were not intended to be used normally. However, this valve would be a PSV if it were intended to protect against emergency conditions—i.e., conditions that are hazardous to personnel or equipment or both and that are not expected to arise normally.

   The designation PSV applies to all valves intended to protect against emergency pressure conditions, regardless of whether the valve construction and mode of operation place them in the category of the safety valve, relief valve, or safety relief valve. (For definitions of these terms, refer to Section 10.2 of Volume I.)

9. Passive function *glass* applies to instruments that provide an uncalibrated direct view of the process.

10. The term *indicate* applies only to the readout of an actual measurement. It does not apply to a scale for manual adjustment of a variable if there is no measurement input to the scale.

11. A *pilot light* that is part of an instrument loop is designated by a first letter followed by succeeding letter L. For example, a *pilot light* that indicates an expired time period may be tagged KL. However, if it is desired to tag a *pilot light* that is not part of a formal instrument loop, the *pilot light* may be designated in the same way or, alternatively, by a single letter L. For example, a running light for an electric motor may be tagged either EL, assuming that voltage is the appropriate measured variable, or XL, assuming that the light is actuated by auxiliary electric contacts of the motor starter, or simply L.

    The action of a *pilot light* may be accompanied by an audible signal.

12. Use of succeeding-letter U for *multifunction* instead of a combination of other functional letters is optional.

13. A device that connects, disconnects, or transfers one or more circuits may be either a *switch*, a *relay*, an on-off *controller*, or a *control valve*, depending on the application.

If the device manipulates a fluid process stream and is not a hand-actuated on-off block valve, it is designated as a *control valve*. For all applications other than fluid process streams, the device is designated as follows:

A *switch*, if it is actuated by hand.

A *switch* or an on-off *controller* if it is automatic and is the first such device in a loop. The term *switch* is generally used if the device is used for alarm, pilot light, selection, interlock or safety. The term *controller* is generally used if the device is used for normal operating control.

A *relay*, if it is automatic and is not the first such device in a loop, i.e., if it is actuated by a *switch* or an on-off *controller*.

14. It is expected that the functions associated with the use of succeeding-letter Y will be defined outside a balloon on a flow diagram

when it is convenient to do so. This need not be done when the function is self-evident, as for a solenoid valve in a fluid signal line.

15. Use of modifying terms *high, low,* and *middle* or *intermediate* correspond to values of the measured variable, not of the signal, unless otherwise noted. For example, a high-level alarm derived from a reverse-acting level transmitter signal is an LAH, even though the alarm is actuated when the signal falls to a low value. The terms may be used in combinations as appropriate.

16. The terms *high* and *low*, when applied to positions of valves and other open-close devices, are defined as follows: *high* denotes that the valve is in or approaching the fully open position, and *low* denotes in or approaching the fully closed position.

An instrument that performs two or more functions may be designated by all its functions. For example, a flow recorder FR with a pressure pen PR may be designated FR/PR or, alternatively, as UR a multivariable recorder. A common annunciator window for high- and low-temperature alarm may be shown as TAH/L.

Instrument relays perform various functions, such as computing, logic and signal conversion. The function of a relay represented on a diagram is usually clarified by placing one of the clarifying designations of Table A.1b outside the relay balloon.

Distinctive symbols (some of which are listed in Table A.1c) are used to represent instrumentation on flow diagrams and other documents. A circular balloon represents the instrument. The balloon may also be used to tag distinctive symbols, but this need not be done if the relationship of the distinctive symbol to the remainder of the loop is easily apparent. For example, an orifice plate or a control valve that is part of a system is not usually tagged on a diagram.

All identification letters are capitals for compatibility with automatic printing machines.

In general, one signal line suffices to represent the interconnections between two instruments on flow diagrams, even though they may be connected physically by more than one line. When needed to clarify the direction of flow of intelligence, directional arrowheads can be added to the signal lines.

If it is desired to simplify the representation of the instrumentation systems on a diagram, this can be done by omitting all instrumentation other than those representing the end functions needed for operation of the process. Thus, intermediate instruments—such as transmitters and signal converters—

RELAY FUNCTION SYMBOLS

Symbol	Function
1-0 or ON-OFF	Automatically connect, disconnect, or transfer one or more circuits, provided that this is not the first such device in a loop. (See note 13 on page 1583.)
Σ or ADD	Add or totalize (add and subtract), with two or more inputs.
Δ or DIFF.	Subtract (with two or more inputs)
$\pm$ $+$ $\boxed{-}$	Bias (single input)
AVG.	Average
% or 1:3 or 2:1 (typical)	Gain or attenuate (input:output), with single input
$\boxed{\times}$	Multiply (two or more inputs)
$\div$	Divide (two or more inputs)
$\boxed{\sqrt{}}$ or SQ. RT.	Extract square root
x^n or $x^{1/n}$	Raise to power
$f(x)$	Characterize
1:1	Boost
$\boxed{>}$ or HIGHEST (Measured Variable)	High-select. Select highest (higher) measured variable (not signal, unless so noted).
$\boxed{<}$ or LOWEST (Measured Variable)	Low-select. Select lowest (lower) measured variable (not signal, unless so noted).
REV.	Reverse
	Convert
a. E/P or P/I (typical)	For input/output sequences of the following:

Designation | Signal
E | Voltage
H | Hydraulic
I | Current (electrical)
O | Electromagnetic or sonic
P | Pneumatic
R | Resistance (electrical)

b. A/D or D/A	For input/output sequences of the following:

A | Analog
D | Digital

$\int$	Integrate (time integral)
D or d/dt	Derivative or rate
1/D	Inverse derivative

Note: The use of a box enclosing a symbol is optional. The box is intended to avoid confusion by setting off the symbol from other markings on a diagram.

Table A.1c
MISCELLANEOUS SYMBOLS

Instrument Line Symbols	*Power Supply Abbreviations*

All lines should be fine in relation to process piping lines.

Connection to process, or mechanical link, or instrument supply   ────────

Pneumatic signal   ─#─#─#─

The pneumatic signal symbol applies to a signal using any gas as the signal medium. If a gas other than air is used, the gas is identified by a note on the signal symbol or otherwise.

Electric signal   ─ ─ ─ ─ ─

Capillary tubing (filled system)   ─×─×─×─

Hydraulic signal   ─L─L─L─

Electromagnetic or sonic signal (without wiring or tubing)   ─∿─∿─∿─

Electromagnetic phenomena include heat, radio waves, nuclear radiation, and light.

Undefined signal   ─/─/─/─

The following abbreviations are suggested to denote the types of power supply. These designations may also be applied for purge fluid supplies.

AS	Air supply
ES	Electric supply
GS	Gas supply
HS	Hydraulic supply
NS	Nitrogen supply
SS	Steam supply
WS	Water supply

The power supply level may be added to the instrument supply line, e.g., AS 100 for a 100-PSIG air supply; ES 24DC for a 24-volt direct current supply, etc.

Instrument Symbol Balloons

APPROXIMATELY 7/16" DIAMETER

LOCALLY MOUNTED

MOUNTED ON MAIN BOARD

MOUNTED BEHIND THE BOARD

Instrument for two measured variables or single variable instrument with more than one function. Additional tangent balloons may be added as required.

can be eliminated from the diagram. Such simplification, if used, should be done consistently for a given type of drawing throughout a project.

The sequence in which the instruments of a loop are connected on a flow diagram should reflect the functional logic. This arrangement may differ from the actual connection sequence. Thus, a flow diagram may show instruments using electrical analog signals connected in parallel, regardless of whether the signal type is voltage or current.

Application Examples

The following examples of the symbol system are applied to typical industrial processes. No attempt has been made to show complete instrumentation. Primary flow elements, such as orifice plates, and control valves are not usually tagged on the flow diagram in actual practice, but they are tagged in the examples for illustrative purposes.

The symbol system can be either in its full form, which symbolizes each instrument in a loop, or in a simplified form. The two methods are compared in Figures A.1d and e. Figure A.1d shows that a gas is heated and temperature-controlled by a board-mounted controller, and the heating fluid is modulated by some type of control valve. The type of control signal is not specified, but records of gas flow, pressure, and outlet temperature, and a low-temperature alarm are required on the instrument board.

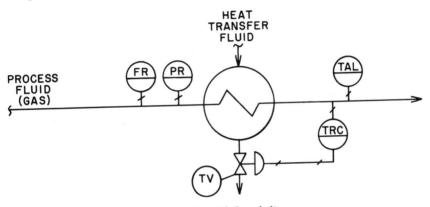

Fig. A.1d   Simplified symbolism

In Figure A.1e, all the instruments used are symbolized. The flow record is obtained by use of orifice plate, flow transmitter, square-root extractor mounted behind the board and a two-pen recorder on the board. The input to the pressure recorder is provided by a pressure transmitter, which measures the pressure on the downstream side of the orifice plate. The signals are pneumatic.

The gas outlet temperature is measured by a resistance-type element, mounted in a thermowell and connected to a board-mounted temperature-recording controller, with an electric output that throttles a ball-type control valve with a cylinder-type actuator and (by implication) with internal conversion from the electric signal to a fluid signal. The TRC has an integral low-temperature switch that actuates an alarm on the board.

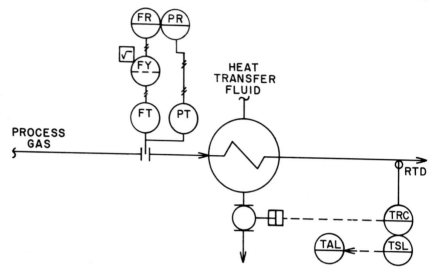

Figure A.1f provides a full symbolic description of a distillation process where the feed flow rate is not controlled, only measured and recorded. The heat input rate into the reboiler is proportioned to the feed rate by a multiplying relay (FY-1), which provides the set point for the control system of the hot oil flow.

The tower overhead is condensed, with instrumentation provided to maintain the column at constant pressure. The overhead product drawoff rate is flow controlled. The flow controller set point is adjusted by dividing relay (UY-2), whose inputs are the feed rate, as modified by the time-function relay (FY-3) and the output of the overhead product analyzer controller. This controller receives product analysis information from its transmitter, which also transmits to a dual (high/low) analysis switch which, in turn, actuates corresponding alarms.

Accumulator level is maintained by throttling the tower reflux, and an independent level switch actuates a common high/low accumulator level alarm.

Bottoms level in the tower is controlled by modulating product withdrawal, and local level indication is provided by separate gauge glass.

Temperature measurements at various points in the process are performed by a multipoint scanning recorder (TJR) and multipoint indicator (TI). Some of the points of TJR-4 have high- or low-temperature switches to actuate alarms. For example, overhead temperature is alarmed by TJSH-4-2 and TAH-4-2.

Figure A.1g illustrates an air cooling and humidifying system, where the

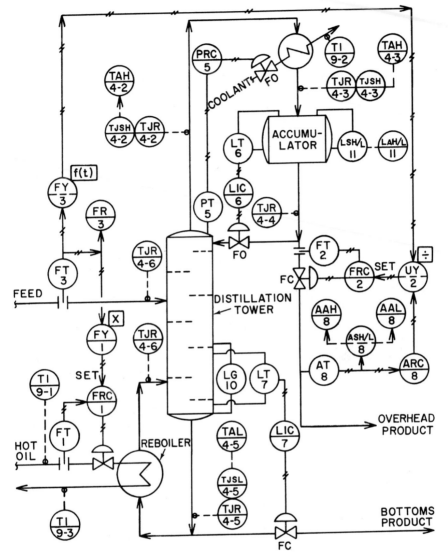

Fig. A.1f   Full symbolic description of distillation controls

ratio of makeup outdoor air to return air is manually adjusted by hand control valves. The combined streams enter a spray type washer. The cooled and humidified air is maintained at the proper humidity and temperature by the humidity controller resetting the temperature controller which, in turn, modulates the three-way valve, which varies the proportions of the cold water

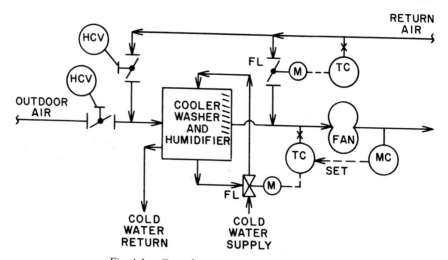

Fig. A.1g   Control system for air cooler-humidifier

and the recirculated warm water that are pumped to the spray nozzles. Return air is controlled to by-pass the washer in inverse proportion to its temperature.

Figure A.1h illustrates the control system on a chemical reactor where the feed of reactant A is flow-controlled.

The flows of A and B are ratio controlled, with the multiplying relay (FY-1) generating the set point for the reactant B controller. Reactor level is kept constant by throttling the discharge valve. If level is high, it automatically closes the reactant feed valves through solenoid valves UY-2 and UY-3 and actuates a high-level alarm. An alarm is also actuated on low reactor level.

The reaction is exothermic, and temperature is controlled by modulating the pressure of the coolant in the reactor jacket. This is done by the reactor temperature controller adjusting the set point of the jacket pressure controller, which controls the backpressure of steam generated by absorption of heat by the cooling water. High reactor temperature actuates an alarm, and, if the temperature gets very high, it closes the feed valves for reactants A and B and the steam backpressure valve, and it opens the water supply and return valves through solenoids UY-2 to UY-6. These valves can also be actuated manually through manual switch HS-7.

Coolant level is maintained constant in the jacket by throttling the water supply, and low jacket level actuates an alarm. Reactor pressure is controlled by throttling the inerts formed in the reaction, and the reactor is protected against hazardous overpressures by a rupture disc.

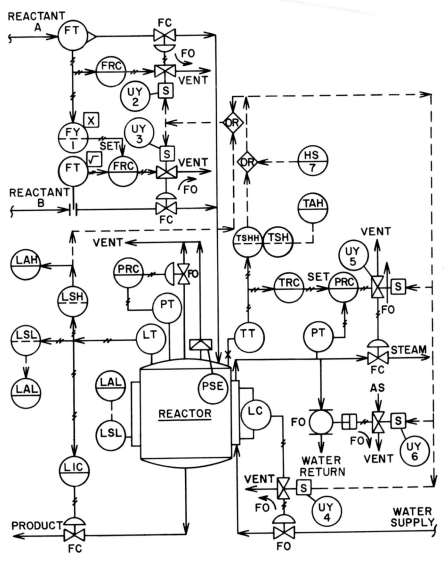

Fig. A.1h Chemical reactor control system

A.2 INTERLOCK LOGIC SYMBOLS

This section lists the symbols used to denote binary, i.e., on-off, process operations, and it illustrates a typical application of the symbols to an actual process.

This method of diagramming is applicable to any process which makes use of switching devices to initiate normal or emergency operations. The method is primarily process-based and describes operations in terms of the essential process functions. These functions can be carried out by any class of hardware, whether electric, pneumatic, fluidic, or other. The method is oriented toward the engineer who may have only a rudimentary knowledge of hardware circuit design but who knows what the process-sensing instruments are and how the process is supposed to operate. The hardware and circuit subfunctions needed to perform the process functions can then be detailed by the circuit designer as necessary to satisfy the instrument engineer's intentions.

The method is applicable whenever the operating requirements of the process need to be described to operating personnel, maintenance workers, designers, or others, and it is particularly useful for group discussions. It does not require these people to know how to read relatively complex and specialized circuit diagrams. However, where it is necessary to trace the actions of a circuit in detail, there is usually no substitute for the detailed circuit diagram.

A further discussion of logic circuits, their representation, optimization etc. is given in Sections 4.6 and 4.7.

Logic Symbols

A logic signal may correspond to either the existence or the non-existence of an instrument signal, depending on the particular type of hardware system and the circuit design philosophy that are selected.

Symbols shown in Tables A.2a and A.2b with three inputs, A, B, and C, are typical for the logic functions having any number of two or more inputs.

In the truth tables, 0 denotes the non-existence of the logic input or output signal or state given at the head of the column, and 1 denotes the existence of the logic input signal or state. D denotes the existence of the logic output signal or state as a result of appropriate logic inputs. The output states for a truth table are within the heavily outlined box.

1592

Table A.2a
BASIC LOGIC SYMBOLS FOR BINARY FUNCTIONS

Logic Function	Recommended Symbol	Acceptable Alternative Symbol	Definition and Truth Table	Example
AND	A → B →☐→ D C → If desired the inputs and outputs of all statements can be separated as noted below: A ⊢→ B ⊢→☐→⊦ D C ⊢→	A →⌐ B →⌐ D	Output D exists only if and while inputs A, B and C exist.	Operate pump if feed tank level is high and provided that discharge valve is open:

Truth table (AND):

		C	
A	B	0	1
0	0	0	0
0	1	0	0
1	0	0	0
1	1	0	D

HIGH TANK LEVEL →☐→ OPERATE PUMP
VALVE OPEN →

Logic Function	Recommended Symbol	Acceptable Alternative Symbol	Definition and Truth Table	Example
OR (Inclusive)	A → B →O→ D C →	A →⌐ B →⌐ D	Output D exists only if and while one or more of inputs A, B and C exist.	Stop compressor if cooling water pressure is low or bearing temperature is high:

Truth table (OR):

		C	
A	B	0	1
0	0	0	D
0	1	D	D
1	0	D	D
1	1	D	D

LOW WATER PRESSURE →
HIGH BEARING TEMPERATURE → O→ STOP COMPRESSOR

Table A.2a

Logic Function	Recommended Symbol	Acceptable Alternative Symbol	Definition and Truth Table	Example
			Output D exists only if and while one, and only one, of inputs A, B and C exists.	Alarm if one and only one of the tanks is available to use:
OR (Exclusive)				
			Output D exists only if and while two, and only two, of inputs A, B and C exist.	Operate feeder if two and only two mills are in service:
NOT			Output D exists only if and while input A does not exist (special case of NAND).	Close valve if pressure is not high:

Truth table (Exclusive OR, one):

A	B	C=0	C=1
0	0	0	D
0	1	D	0
1	0	D	0
1	1	0	0

Truth table (Exclusive OR, two):

A	B	C=0	C=1
0	0	0	0
0	1	0	D
1	0	0	D
1	1	D	0

Truth table (NOT):

A	D
0	1
1	0

Table A.2a

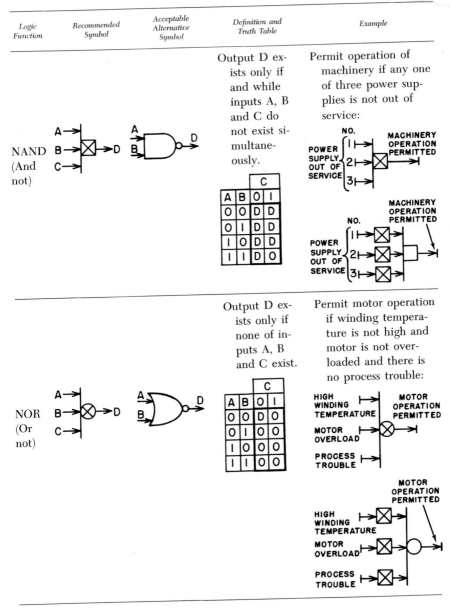

Logic Function	Recommended Symbol	Acceptable Alternative Symbol	Definition and Truth Table	Example
NAND (And not)			Output D exists only if and while inputs A, B and C do not exist simultaneously.	Permit operation of machinery if any one of three power supplies is not out of service:
NOR (Or not)			Output D exists only if none of inputs A, B and C exist.	Permit motor operation if winding temperature is not high and motor is not overloaded and there is no process trouble:

Table A.2b
MISCELLANEOUS BINARY LOGIC SYMBOLS

Function	Symbol	Description
Time Delay	A → [Td ( ) sec] → B	Output B will start to exist at a specified time after input A begins to exist, provided that A is maintained continuously for a specified time. B terminates when A does.
First Input Memory	A → [M First ( ) sec] → B	Output B exists as soon as there is an input, but continues to exist only for a specified time after first initiation of input, regardless of whether input continues to exist. Restoration of interrupted input during this time has no effect.
Last Input Memory	A → [M Last ( ) sec] → B	Output B exists as soon as there is an input, but continues to exist only for a specified time after last termination of input. Restoration of input during or after this time will reactivate the memory.
Maintained Memory	A → [M] → C B → [R] The R box can be above the M box. The following symbol should not be used. A → ○ → B	Memory is established and output C exists as soon as input A exists. C continues to exist, regardless of the subsequent state of A, until the memory is reset (meaning: terminated) by input B existing while A does not exist. (A overrides B). The action of the memory on loss of power supply is designated by letters as follows: M—Undefined. KM—Kept (retained) despite loss of power. LM—Lost if power is lost even momentarily. XM—Action on power loss is immaterial.
Break Point	90° ⟨	Denotes a break in the logic sequence and should be followed by a reference to the resumption point of the sequence. For reading convenience, the last statement preceding the break and the first statement following the break should use identical wording.

Table A.2b

Function	Symbol	Description
Resumption Point	90°	Denotes a resumption of the logic sequence that was broken elsewhere and should be preceded by a reference to the break point of the sequence.
Operating Status Summary	(OPTIONAL)	Sets off a summary of status of the operating system, wherever a summary is useful in the sequence. The use of this symbol is optional and its size is as required.
Special		This symbol is inscribed as necessary to describe any special function. The symbol size is as required.

Logic Application

The process illustrated in Figure A.2c must have high vacuum to perform satisfactorily. Vacuum is normally maintained by an air ejector, but in case of failure or overload of the air ejector the system pressure rises. This is sensed by pressure switch (PSH), which automatically starts a vacuum pump, provided that a hand-actuated control switch (HS) for the pump motor is in the AUTOMATIC position. This switch can also be used to start and stop the pump manually. However, the pump is not permitted to start or run if the discharge temperature, as sensed by temperature switch (TSH), is high or if the motor

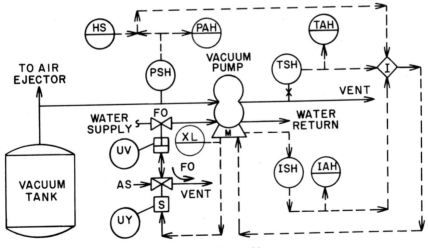

Fig. A.2c   Control system for stand-by vacuum pump

is overloaded and its circuit breaker is not manually reset. If the high pressure is maintained for ten minutes, high-pressure alarm (PAH) is actuated. High temperature is signaled by the alarm TAH. Pump motor overload is signaled by alarm IAH.

Whenever the pump is running, cooling water is automatically turned on. The water flow is controlled by the control valve UV, which is actuated by a solenoid valve (UY) which, in turn, is actuated by auxiliary contacts of the circuit breaker. The water is automatically turned off when the pump is stopped. The following instruments are on the instrument board:

HS   Manual control switch for pump operation. The switch has three momentary contact pushbuttons for START, AUTOMATIC and STOP.

PAH   Alarm actuated upon rise of pressure to abnormal value. However, this alarm is blocked for ten minutes after the occurrence of a high-pressure condition.

TAH   Alarm actuated if pump discharge temperature rises to abnormal value.

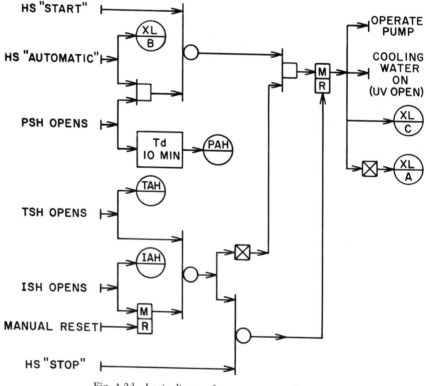

Fig. A.2d   Logic diagram for vacuum control system

XL-A　　Green pilot light denoting that the pump motor circuit breaker is not closed, i.e., that pump is not operating.

XL-B　　Amber pilot light denoting that the pump is ready for an automatic start.

XL-C　　Red pilot light denoting that the pump motor circuit breaker is closed, i.e., that pump is operating.

IAH　　Alarm actuated upon overload of pump motor.

The process functions required are described in the logic diagram in Figure A.2d. The hardware functions separately in Table A.2e.

Table A.2e
HARDWARE FUNCTIONING DESCRIPTION FOR VACUUM PUMP AND WATER VALVE COMPONENTS

Vacuum Pump	Motor Circuit Breaker Auxiliary Contacts	Solenoid Valve (UY) Coil	Control Valve (UV)		Cooling Water
			Actuator	Port	
Off	Closed	Energized	Pressurized	Closed	Off
On	Open	Deenergized	Vented	Open	On

A.3 ESTIMATING VALVE NOISE

Section 1.21 discusses the phenomena of noise, vibration and flowing velocity in control valves and also techniques for reducing noise and vibration. Methods of sizing control valves to satisfy these criteria are elaborated on in Section 1.2. The purpose of this section is therefore limited to outlining an approach to the approximation of control valve noise levels.

Definitions

Sound Intensity (I) The flow of a sound wave is a passage of sound energy. The power transmitted per unit of area in the direction of travel is defined as the sound intensity and is equal to

$$I = \frac{P^2}{\rho c} g \qquad\qquad A.3(1)$$

in a free homogeneous medium where ρ is the density, c is the velocity of sound, and P is the sound pressure.

Sound Power (W) If a sound source emits an amount of acoustic power W, then for a free field the sound intensity at any point a distance r from the source is

$$W = 4\, I\pi r^2 \qquad\qquad A.3(2)$$

Sound Power Level The sound power level of a sound source, in decibels, is 10 times the logarithm to the base 10 of the ratio of the sound power radiated by the source to a reference power. The reference power is usually taken as 1 micromicrowatt, or 10^{-12} watt.

Sound Pressure (P) The fluctuating part of the air pressure is called the sound pressure. Only frequencies above a few cycles per second are considered. The sound pressure is expressed in terms of its rms value unless otherwise stated.

Sound Pressure Level (SPL) The sound pressure level, in decibels, of a sound is 20 times the logarithm to the base 10 of the ratio of the pressure of this sound to the reference pressure. The reference pressure is usually taken as 0.0002 microbar. Refer also to equation 1.21(1). For the purposes of this discussion, the distance between the valve and the point of noise measurement is assumed to be 3 ft.

Estimating Method

Throttling noise produced by control valves has become a focal point of attention triggered in part by the Federal Government's enforcement of the Walsh-Healey Act which limits the noise level in an industrial location to 90 db for an 8-hour exposure, or 115 db for up to a 15-minute exposure.

As discussed in detail in Section 1.21, some of the major noise sources associated with control valves are:

> Due to mechanical vibration, i.e., resonance of the valve trim. This can be eliminated by a change in stem diameter, change in the mass of the plug or sometimes reversal in flow direction.
>
> Cavitation noise caused by the collapse of vapor bubbles in a liquid fluid stream. It can be avoided by use of a suitable trim or valve type (high C_f). For more details on cavitation, refer to Sections 1.2 and 1.21.
>
> Noise produced by fluid turbulence (aerodynamic noise) which is almost negligible with liquids, but highly important with gases due to the high velocity (above sonic) when passing the valve orifice at critical pressure drop. For velocity limiting at the valve outlet, refer to equations 1.21(2), (3) and (4).

The following is a mathematical way to estimate the aerodynamic noise level in decibels of any style control valve. This method is based on the basic research conducted by M. J. Lighthill[°] and yields accurate results considering the complexity of flow patterns in different styles of control valves.

This equation covers noise in conventional single stage orifice type valves. For major noise reduction employ the special valve designs described in Section 1.21. The sound pressure level for gas at or greater than critical pressure drop and at a distance of 3 ft from the valve is expressed by equation A.3(3):

$$SPL = 10 \log(9.2 \times 10^7 \eta P_1 P_2 C_v C_f) + G + T - A \qquad \text{A.3(3)}$$

where SPL = sound pressure level (db),
     η = efficiency factor (Figure A.3a),
     A = attenuation factor (db) (Table A.3b),
     G = gas property factor (db) (Table A.3c),
     P_1 = upstream pressure (PSIA),
     P_2 = downstream pressure (PSIA) and
     T = temperature correction factor (db) (Table A.3d).

The calculated results using this formula have been found to be accurate to ±5 db.

[°] "On Sound Generated Aerodynamically II, Turbulence as a Source of Sound," M. J. Lighthill, Proceedings of The Royal Society Of London, Vol. 222, Series A (1954).

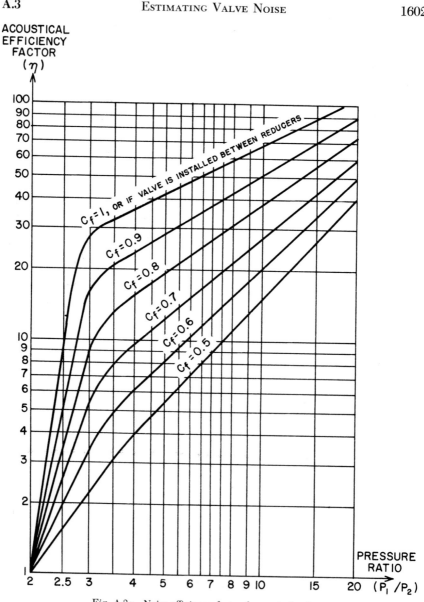

Fig. A.3a   Noise efficiency factor for control valves

In the Table A.3b, C_v is assumed to be $10d^2$ and $C_f = 0.8$. For more accurate results calculate A as defined by equation A.3(4):

$$A = 17 \log\left[m\left(\frac{38,000}{\sqrt{C_f C_v}} \right) \right] - 36 \qquad\qquad A.3(4)$$

Table A.3b
PIPE ATTENUATION FACTOR A (db)

Pipe Size (in.)	Pipe Schedule			Pipe Size (in.)	Pipe Schedule		
	40	80	160		40	80	160
1	46.8	48.9	51.4	6	39.2	42.2	45.9
1½	44.4	46.8	49.3	8	37.9	41.2	45.6
2	42.6	44.8	48.6	10	37.2	40.9	45.5
3	42.2	44.6	47.5	12	36.6	40.4	45.3
4	40.7	43.3	46.7				

Table A.3c
GAS PROPERTY FACTOR G

Gas	G(db)	Gas	G(db)
Saturated Steam	−0.3	Carbon Dioxide	−0.5
Superheated Steam	−0.3	Carbon Monoxide	0.2
Natural Gas	0.5	Helium	1.7
Hydrogen	0.2	Methane	−0.6
Oxygen	0	Nitrogen	0.1
Ammonia	−0.5	Propane	−1.5
Air	0	Ethylene	0.7
Acetylene	−0.5	Ethane	−1.2

Table A.3d
TEMPERATURE CORRECTION FACTOR-T

Flowing Gas Temperature(°F)	T(db)	Flowing Gas Temperature (°F)	T(db)
70	0	500	1.6
100	.4	700	2.0
200	.8	1,000	2.5
300	1.1	1,200	2.8

where A = pipe attenuation factor (db) (A ≥ 0) and
m = weight of pipe wall (lbm/ft^2).

The peak frequency is based on a Strouhal number of 0.2:

$$\text{Peak frequency} = \frac{17,000}{\sqrt{C_f \times C_v}} (\text{Hz}) \qquad \text{A.3(5)}$$

Examples of Estimation

Example 1: Determine the expected sound pressure level for a valve installation with the following conditions.

Valve type	Streamlined, reduced port, flow to close, angle.
Valve body size	2″
C_v	2.5
C_f	0.5 (Table 1.2)
Pipe	2″, schedule 80
Gas	Natural gas
P_1	3,800 PSIA
P_2	1,050 PSIA
P_1/P_2	3.6
η	3.3 (Figure A.3a)
A	44.8 (Table A.3b)
G	0.5 (Table A.3c)
T	0 at 70°F (Table A.3d)

From equation A.3(3),

$$
\begin{aligned}
\text{SPL} &= 10\log(9.2 \times 10^7 \times 3.3 \times 3,800 \times 1,050 \times 2.5 \times 0.5) \\
&\quad + 0.5 + 0 - 44.8 \\
&= 10\log(1.53 \times 10^{15}) - 44.3 = 151.85 - 44.3 \\
&= 107.55 \text{ db}
\end{aligned}
$$

Example 2: Determine the expected sound pressure level for a valve installation with the following conditions.

Valve type	Cage type, flow to open, full ported
Valve body size	4″
C_v	220 ($C_v = C_d \times d^2$, C_d, Table 1.2a)
C_f	0.85 (Table 1.2a)
Pipe	4″, schedule 40
Gas	Air
P_1	150 PSIA
P_2	50 PSIA
P_1/P_2	3
η	13 (Figure A.3a)
A	40.7 (Table A.3b)
G	0 (Table A.3c)
T	0 at 70°F (Table A.3d)

Therefore from equation A.3(3),

$$
\begin{aligned}
\text{SPL} &= 10\log(9.2 \times 10^7 \times 13 \times 150 \times 50 \times 220 \times 0.85) \\
&\quad + 0 + 0 - 40.7 \\
&= 10\log(1.68 \times 10^7) - 40.7 = 152.3 - 40.7 \\
&= 112.6 \text{ db}
\end{aligned}
$$

Example 3: What will be the noise level at a distance of 100 ft from the valve in Example 2? This is estimated by the use of equation A.3(6):

$$\text{SPL (at any distance)} = \text{SPL} - 20\log(\text{distance}/3) \qquad \text{A.3(6)}$$

Therefore, from equation A.3(6),

$$\text{SPL}_{100} = 112.6 - 20\log(100/3)$$
$$= 82.1 \text{ db}$$

Example 4: What will be sound pressure level at a 3 ft distance from the valve in Example 2, if it is throttled to 20 percent of its capacity? The decibel values to be subtracted due to throttling are obtained from Table A.3e:

$$\text{SPL} = 112.6 - 16$$
$$= 96.6 \text{ db}$$

Table A.3e
CORRECTION FOR
REDUCED CAPACITY

Percent Capacity (C_v)	Subtract (db)
10	22
20	16
30	13
40	10
50	8
60	7
70	5
80	3
90	2
100	0

The techniques utilized in examples 1 to 4 assume that the valve *outlet velocities are below sonic.* If this is not the case (equations 1.21(2), (3) and (4) are not satisfied) estimating equation A.3(7) should be used:

$$\text{SPL} = 3 + 10\log(P_2 \times 10^{15} \times \eta \times d^2) + G + T - A \qquad \text{A.3(7)}$$

where d = valve size in inches and
η = can be estimated as 10.

Example 5: Determine the expected sound pressure level for a valve installation with the following conditions.

Valve   2″ size
Pipe   2″, schedule 80
Gas   air at 70°F
P_2 20 PSIA
A   44.8 (Table A.3b)
G   0 (Table A.3c)
T   0 (Table A.3d)

Therefore, from equation A.3(7),

$$SPL = 3 + 10 \log(20 \times 10^{15} \times 10 \times 4) - 44.8$$
$$= 137.2 \text{ db}$$

SOURCES OF INFORMATION FOR THE APPENDIX

(Numbers refer to sections of Appendix)

1. Abstracted from Instrument Society of America Standard S5.1-1968, "Instrumentation Symbols and Identification."
2. Modified from "The Logic Diagram: A Communication Tool for Interlock Circuits," by George Platt, Instrument Society of America Paper 6.1-2-65, 1965.
3. Reproduced by permission of Masoneilan International.

INDEX

Abbreviations, 18
Abrasion-resistant materials, 274
Absolute value, analog computer, 1202
AC motor, speed changes, 315
Acceleration transducer, 646–7
Access, 720, 882
Accumulator, 919
  lag, 1242
Accuracy, 882
  analog, 4
  blending systems, 1564, 1574
  calculations, 1148
  computer modules, 1148
  digital, 9
    blending, 1578
  electronic ratio, 1571
  linear, 4
  non-linear, 6
  pneumatic ratio control, 1570
  system, 9
Action, 368
Actuator, 132, 221
  digital, 218
  drive sleeve, electric, 247
  electric, 222, 233
    gear operated, 247
    Quarter-Turn, 252
  electro-hydraulic, 236
    pneumatic, 232
  helical spline type, 225
  non-linearities, 136
  pneumatic, 223
  pneumo-hydraulic, 229
  quarter turn, 223
  rack and pinion, 223
    electric, 245
  rotary air motor, 231
  rotary electrical, 243
  sizing, 136
  solenoids, 233
  vane type, 228
A/D conversion channel, hybrid computer,
  1174
A/D converter characteristics, 970, 1075
Adaption, feedback parameters, 813
Adaption, self-, 816

Adaptive control, 720, 813
  equipment limitations, 817
  feedforward, 818
  flow, 815
  performance, 817
  programmed, 814
  self-, 816
Add, instruction, 919
Adder, 1157
  circuit, 1162
  electronic, 470
Adding amplifier input gains, 1162
Adding function, 1141
Adding relay, pneumatic, 467
Adding relays, 1141
Address, 882
Adiabatic process, 1370
Adjustment of control modes, 842
  drive speeds, 309
Advantages, computer, 1317
  control valves, 348
  regulators, 348
  simulation, 1268
Aerodynamic noise, 271
Aerospace valves, 213
Air cooler, 1590
Air failure, 142
Air filter regulators, 121
  fuel ratio, 1403
    limits, 1409
  humidifier, 1590
  locks, 144
  supply, panel, 552
Airsets, 121–2, 353
Alarm, zero voltage, 1083
Algebraic equations, analog computer, 1194
  loops, analog computations, 1196
Algol, 923–5
Algorithm, 931, 1070, 1128, 1215
  digital control, 931, 1215
  flow batching, 1293
  metering, 1293
  self-tuning, 1080
Alkalinity, measurement of, in water, 1540,
  1553
Alphanumeric codes, 920

1607

Alternator cost, 588
Aluminum tubing, 550
AM, amplitude modulation, 665
Ammonia in chlorination, 1556
  plant startup, 1423–4
  synthesis, 1433
Amplifier
  chopper stabilized, 1154
  guard, 952
  high gain, 1154
  magnetic, 306
  power, 307
  wide bandwidth, 1154
Amplitude modulation, 665
  analog computer, 1218
Amplitude stabilization, analog computer, 1218
Analog
  backup, 720, 883
  blending system, 1564
  computer, 457, 720, 886, 1140, 1151, 1179
    accuracy, 1148
    algebraic equations, 1194
    applications, 1179
    basic units, 1158
    calibration, 1149
    circuits, 1153
    computations, 1189
    control system, 1170
    design, 1171
    differential equations, 1189
    electronic, 1151
    furnaces, 1444
    integrator applications, 1160–1
    internal reflux, 1238
    iterative computations, 1197
    modules, 1153
    operation, 1170
    probability computation, 1199
    programming, 1180
    repetitive computations, 1196
    scaling, 1145, 1182, 1236
    simulation, 1179, 1199
    symbols, 1153, 1180
    time scale selection, 1185
  indicators, 567
  inputs, 967
  outputs, 964
  simulation circuit, reactor, 1263
  switch, 1154
  terminology, 719–35
Analog-to-digital converter, 967
Analog vs digital, 1120
Analyzer, furnace applications, 1436, 1441
  infrared, 1436
AND logic, 505, 517, 518, 720, 1593
Anti-reset wind-up, 406, 427, 453
Anti-surge control, compressor, 1532

Aperture time, 970
Apparent loudness, 265
Application, process computer, 1109
Application programs, 910
Arbitrary function generation, 1143
Area classifications, 553–4
Argument, 883
Arithmetic operations, Fortran, 925
Arithmetic unit, 883
Armature resistance speed control, 313
Armature voltage control, 313
Arrhenius equation, 1261
ASCII code, punched tape, 920
Aspirating effect, 360
Assembler, 883
Assembly language, 922, 1307, 1311
  computer, 916, 917
  pneumonic code, 922
  source program, 922
  translator, 917
Asynchronous computer, 883
Atmospheric dryers, 1453
Attenuation, 721, 883
Auctioneering device, 721
Automatic control terminology, 715
Automatic gain control, computer, 1221
Auxiliary fuel, boilers, 1393
Auxiliary memory, 997
Availability, 883
  computer, 1121
Averaging relay, 1141

Back-of-panel layout, 550
Backlash, 721
  function, computer simulation, 1208
Backup computer, 1100, 1121
  power system classification, 590
Baffle-nozzle error detector, 621
Balanced hybrid computers, 1227
Balanced plug, 359
Balanced resistance-capacitance filter, 950, 954
Balanced trim, 84
Ball and cage valve, 197
Ball check feeders, 333
Ball valve
  conventional, 161
  dead tight, 165
  fire safe, 163–4
  seals, 163–4
  sizing, 166
Band programming timers, 495
  reject filter, analog computer, 1219
Bandpass filter, analog simulation, 1220
Bang-bang control, 1292
  controller, 721, 756, 757
Baud, 884
Barrier repeater, 659

Basic Fortran, 925
Basic logic functions, 517
Batch
  controller circuit, 427, 453
  distillation control, 1256
  dryers, 1451
  reactor, 1287
    computer control, 928, 930, 1287
    control system, 928
Bath tub effect, 361, 1490
Battery, 603
  backup, 407, 556, 587–8
  charger, 603
    costs, 588
  cost, 587
  lead acid, 603
  nickel cadmium, 603
  sizing, 558
Baudot code, 585
BCD, 690
Belleville springs, 249
Bellows, 374
  seals, 100, 277, 285
Belt feeder, 335
Belt type volumetric feeders, 339
Bench range, 124
Bernoullis' theorem, 45, 115
Bias, 3
  force, 1141
Bidirectional binary counter, 1577
Binary
  coded decimal, 884
  counter, bidirectional, 1577
  multiplier, 1575
  number system, 917
  numbers, 917, 920
Bingham closure, 274
Biological control in water treatment, 1555
Biological treatment of waste water, 1540, 1555
Bipolar transistor, 996
Bit, 884, 918
Blender dryer, 1455
Blending, digital, 1575
  systems, 1559
Block diagram, 721, 778, 884
  algebra, 779
  closed loop, 781, 790
  conventions, 778
Boards, 535
Boiler
  air flow measurement, 1396
  combustion control, 1387
  components, 1386
  control, 1385
    furnace pressure, 1400
    interaction, 1386, 1408, 1419

Boiler (Continued)
  control, multiple fuel, 1390
  controller settings, 1388
  desuperheater spray control, 1416
  feedwater control, 1411
  flue gas analysis control, 1406
  fuel-air ratio control, 1403
    coal, 1394
    control, 1389
    positioner, 1390
    splitting control, 1390
    totalizing, 1390
    unmeasured, 1394
  interlocks, 1387
  limits, air-fuel, 1409
  load vs excess air, 1407
  master control, 1388
  once-through, 814
  one fuel uncontrolled, 1393
  performance, 1386
  safety, 1387
  shutdown, 1387
  spreader stoker control, 1394
  startup, 1387
  steam temperature control, 1416
  superheat control, 1416
  "swell" effect, 1411
Bolting materials, 96
Bonding cements for strain gauges, 677
Bonnets, 98
Boolean algebra, 527
  equivalents, 527
Booster, 391
  circuits, 423
  relay, 130, 391, 823
    in ratio control, 1569
Break frequency, 127
Break point, 721, 884
  computer, 1315
Breakaway torque, 248
Breakdown orifices, 263
Breakfront panels, 537, 538
Bridge-controlled oscillator, 690
Brine, 1373
Bubble point, 1319
Buffer, 884
Buffering, 1548
Bulb, 375, 379
  fittings, 379
  materials, 380
Bulk memory, 884
Bumpless transfer, 411
Buoyancy transmitter pneumatic, 633
Burner sequence control, 1387
  turndown, 1431
Bus transfer switch, 608–9
Butt welded valves, 95

Butterfly valve, 145
  dual range, 149, 150
  fish-tail, 151
  linings, 149, 150
  torque characteristics, 151–3
Butterfly vanes, 146
Byte, 919

Cage trims, 85, 110, 113
Cage valves, 84, 110
Calibrate, 885
Calibration, 13
  automatic, 1523
  certificate, 3
  computing system, 1149
  telemetry systems, 692
Cam characterized positioners, 125
  operated timers, 495
  programmer, 459, 460
Capacitance, 721, 744
  probe circuit, 651
  transmitter, 645–6
Capacitive-transfer multiplexer, 954
Capillary refrigerator, 1373
Capillary tubing, 380
Card, punched, 918
Card reader, 1078
Cascade, 402, 435
  control, 819, 1504
    ratio, 831
    station, single case, 435
Cascade
  instability, 822
    positioners, 823
    square root extractor, 823
  loop instability, 822
  primary, 819
  saturation problems, 824
  secondary, control modes, 822
    flow control, 821
    loops, 819, 821
    positioner, 821
    selection, 821
    temperature control, 822
    types, 821
    valve position, 821
  time lag distribution, 821
Cascaded, ddc, 1097
Catalytic reactions, 1267
Cathode ray tubes, 1078
Cavitation, 46, 49, 51, 113, 262
  methods to eliminate, 262
  pumps, 1518
  resistant materials, 263
Central control rooms, 536
Central processor, 885
Centrifugal compressors, 1526
  pump control, 1513–4
  pumps, 333

Centrifuge, automatic batch, 1480, 1482
  clarifier type, 1480
  continuous, 1481, 1484
  control, 1477
  filtration type, 1482
  sedimentation type, 1479
  selection, 1477
  semi-continuous, 1479
  sigma factor, 1479
C_f, 49, 57
Chamflex valve, 195
Characteristics of integrated circuits, 513, 515
  switching transistors, 465
Characterized ball valves, 167
  cams, 125
Charger, battery, 558
Charpy impact, 96
Checkout, computer, 1314
Checkout, panels, 562
Chemical oxidation in water treatment, 1541
Chemical plant telemetry, 681
Chemical reaction kinetics, 1259
Chemical reactor
  control, 1277
  dynamics, 790
  simulation, 1227
Chemical reduction in water treatment, 1541
Chimney effect, 1400
Chlorinator, 1544, 1557
Chlorine, 1541
  demand, definition, 1556
Choked flow condition, 47, 66
Circuit breakers, 555, 610
  diagrams, analog and hybrid, 1180
Circular chart recorder, 576
Clear, 885
Clearance flow, 258
Clock rate, 886
Closed-circuit television, 585
Closed loop, 721
  response, 789, 843
  tuning, 862
    fit 1 and 2, 849
Clutches, 248
Coagulation of water, 1540, 1544, 1558
Coal-air temperature control, 1394
Coal pulverizer compensation control, 1395
Coal pulverizer control, 1395
Coanda effect, 208
Cobol, 923–5
Code, 886
Coefficient, analog computer, 1168
  potentiometer, 1154, 1167
  value, analog computation, 1187
Cohen-coon, 850
Coincident-current, 988
Coking of tubes, 1438
Cold box, 281
Colloids, 1553

Color code for wiring, 560
  coded tubing, 550
  measurement, 1540
Colorimetric analyzers, furnaces, 1442
Column control, 1319
  analog computer, 1232
Combined residual chlorine, 1556
Combustible analyzers, furnaces, 1442
Combustion air flow control, 1399
  control, 1387
Command channels, hybrid computer, 1175
Common mode interference, 886, 938
  noise, 976
  rejection, 670, 957
    measurement, 957
Communication jack, 407
Comparator, 722, 886, 967, 1154
Comparator, analog computer, 1156
Compiler, computer, 886, 916
  language, 1307, 1311
Complementary value, analog computer
    simulation of, 1202
Complex roots, 776
  polynomials, 1194
Composition control in distillation, 1248
Compound DC motor, 311
Compression fittings, 550
Compression ratio, 1380
Compressor
  centrifugal, 1526
  control, 1525
    anti-surge system, 1531
    discharge throttling, 1528
    inlet guide vane, 1529
    rotary, 1532
    speed, 1529
    suction throttling, 1526
  lube oil systems, 1537
  overload, 1377
  reciprocating, 1534
  seal, oil free, 1537
  selective control, 835
  speed control, 1529
  surge, 1526, 1531
  throttling characteristics, 1529
  unloading, 1535
Computation, analog computer, 1189
  on-line, 1140
Computer
  advantages, 1317
  analog, 1140, 1151
    electronic, 1151
    inputs, 1073
  application, 1109
    chromatographs, 904
  assembly language, 916, 917, 922
  availability, 1121
  backup, 1100, 1121
  card reader, 1078

Computer (Continued)
  cascaded, 1097
  checkout, 1314
  codes, alphanumeric, 920
  compatibility, 392
  compiler, 916
  control, batch reactor, 928
    flow chart, 929
    furnace, 1445
    language, 916, 917
    levels, 1116
    summary, reactor, 1316
  core costs, 993
    memory, 986
      3-D, 988
  costs, 1079
  data entries, 1008
  ddc, 1069
  digital inputs, 1076
  direct digital control, 1069
  documentation, 913
  evaluation, 911
  furnace control programming, 1027
  hardware, 915
  heat transfer, 1147
  hybrid, 1151, 1172
  input-output interface, 1073
  installation, 900
    chemical, 900
    glass, 904
    manufacturing, 906
    petrochemical, 901
    petroleum, 902
    rubber, 905
  integrating amplifier, 982
  interface hardware, 961
  internal reflux, 1238
  I/O, 961, 1073
  languages, 915
  linear programming, 1041
  logic problem programming, 1024
  machine language, 915, 917, 919, 922
  management functions, 1132
  manual-automatic station, 1088
  manual stations, 1086
  mass flow problem programming, 1023-4
  measurement sampling rate, 1081
  memory, bulk storage, 1000
    cylindrical film, 995
    devices, 985
    disk, 999
    drum, 997
    film, 993
    integrated circuit, 996
    magnetic tape, 1001
    planar film, 993
    read only, 1004
    special, 1005
  modes, analog, 1159

Computer (*Continued*)
  multiple, 1097
  multiplexer, 916
  noise, 937
  operator's communication, 1086
    console, 916, 1091
    training, 912
  optimization sample problem, 1058
  optimizing programs, 1041
  pneumatic analog, 1140
  preamplifiers, 1075
  preset mode, 1087
  program form, action sheet, 1012
    equation sheet, 1012, 1014
    information sheet, 1016
    variable sheet, 1009
  project, critical path, 908
    planning, 907
  reader for tape, 1078
  real time clock, 1075
  reasonability, 1095
  reliability, 1079
  set control station, 435
  set-point stations, 978
  simulation, 1179, 1199, 1314
  software, 915
  stepping motor, 982
  storage devices, 985
  stored program, 915
  summary, 1109
  supervisory control, 982
  synchronous motor, 982
  terminology, 882
  tuning, conclusions,
    of controllers, 859
      example, 867
  types, 880
  wiring, 937
Computerized reactor control, 1287
Computing functions, 1234
Computing relays, 457, 464
Conclusions on controllers, 419
  on feeders, 340
Condenser control, 1503
Conduction angle, 301
Conductivity, measurement, 1540
Cone pulleys, 319
Console, 538
  operator's, 1091
Constraint equations, 909
  soft, 1051
  target, 1050
Contact closure input, 962
Contact materials, 488
Contact sense module, 887
Contacts, 484, 485
Continuous dryers, 1452
Continuous slope optimizing control, 839

Contoured plug, 108
Control
  accuracy, 722
  adaptive, 813
  algorithm, 925, 931
    feedforward, 933
    optimization, 935
    position, 931
    self-tuning, 935, 936
  algorithm, series, 934
    velocity, 932
  boards, 535
  boiler, 1385
  cascade, 819
  centers, 536
  compressors, 1525
  crystallizers, 1466
  devices for DC motors, 315
  dryers, 1449
  effluent, 1540
  feedback, 716
  feedforward, 717–8
  flow adapted, 815
  furnace, 1420
  heat exchanger, 1486
  language, computer, 915, 917, 926, 930
  levels, computer, 1116
  logic programming, 1307, 1311
  loop block diagram, 781
  mode, 402, 754
    adjustment, 842
    cascade secondary, 822
    inverse derivative, 757
    response to inputs, 756
    response to upsets, 796
    selection, 799
    setting ranges, 405
    tabulation, 755
    transfer functions, 792
  optimizing, 837
  output module, 887, 888
  panel, 535
  program, batch reactor, 930
    PID, 924
    temperature, 923
  pumps, 1513
  ratio, 722
  rooms, 536
  selective, 832
  software, 926
  states, computer, 1294
  station, self-synchronizing type, 437–8, 444
  steam boilers, 1385
  system, analog computer, 1170
    ratio, 826
    "tie-backs," 1419
  terminology, 715

Control (*Continued*)
valve accessories, 118
actuator, 132, 221
digital, 218
dynamics, 137
angle, 87
application, 256
ball, 161
ball and cage, 197
bellows seals, 100
bodies, 81
bonnets, 98
butterfly, 145
cage trim, 84
cavitation, 262
characteristics, 259
characterized ball, 167
clearance flow, 258
connections, 92
construction materials, 96
cryogenic, 281
diaphragm-operated cylinder type, 215
digital, 216
double-seated, 82, 84
elevation, 305
expandable element, 211
expansible diaphragm type, 206
tube type, 202
flow characteristics, 104
fluid interaction type, 208
for high pressure drops, 113
vacuum, 283
viscosity, 288
globe, 81
high pressure, 274
temperature, 276
jacketed, 280
leakage, 256
low temperature, 281
lubricators, 98
material selection, 96
noise, 113, 265, 1600
noise source, 267, 269
orientation table, 26–9
packings, 98, 99
pinch type, 171
plug type, 185
plugs, 108
positioned plug type, 212
positioners, 124
rangeability, 258, 1495
safe failure, 142
Saunders, 154
selection, 256
sizing, 40
sliding gate, 179
slurry, 288

Control
valve accessories
(*Continued*)
small flow, 90, 285
solid state diverting, 208
split body, 86
stroking times, 128
three-way, 89
trims, 104
velocity limits, 269
vibration, 265
wear, 113
y-style, 89
water treatment, 1540
Controlled variable, 736
limits, 834
Controller
alarm displays, 410
balancing, 411
components, 400
direct-connected field mounting, 449
displays, 408
electronic, 398
force balance, 429
functions, 398
high-density, 441
high-density scanning, 441
history, 398
inputs, 401
large case, 447
miniature, 427, 432
moment balance, 423, 428
mounting, 412
output load resistances, 405
meter, 410
pneumatic, 420
principle of operation, 423
response comparison, 796
scanning, 441
servicing, 414
setting, 842
boiler, 1388
closed-loop, 846
shelves, 413
switching, 401, 411
transfer functions, 792
tuning, 842, 865
by computer, 859
closed-loop, 843
criteria, 843
criteria, IAE, 843, 868
criteria, ISE, 843, 868
criteria, ITAE, 843, 868
criteria, tables, 853, 868
damped oscillation, 843
digital, 854
equations, 850, 851
example, 856

Controller
  tuning (*Continued*)
    integral criteria, 852
    limitations, 855
    non-symmetrical systems, 855
    open-loop, 850
    summary, 854
    ultimate method, 843
Conversion, degree of, 1284
  hybrid computer, 1172
Converter, 129, 392, 636, 655
  current-to-current, 658
  current-to-pneumatic, 129, 636
  D/A, 965
  digital to pneumatic, 218
  E/I, 657-8
  I/I, 658
  I/P, 636
  Mv/I, 656
  P/I, 656
  resistance-to-current, 659-60
  R/I, 659-60
Cooler controls, 1486
Cooling crystallizers, 1468-70
  of SCR's, 303-4
  regulator, 368
  water economizer, 1377
Copper tubing, 550
Core memory, 917, 986
  resident, 888
  switching times, 992
Corner frequency, 723
Correction time, 723
Corrosion resistance of materials, 264
Cost, alternators, engine-driven, 588
  batteries, 587
  battery charger, 588
  ddc, 1079
  inverter, static, 589
  pneumatic, vs electronics, 395
  programming, 1079
  telemetering, 662, 669
Counter, bidirectional, binary, 1577
Cracking furnaces, 1264
  controls, 1437
Critical damping, 723
Critical flow factor, 45
Critical flow terminology, 48
Critical path, computer project, 908
Critical point, 723
Critical pressure, 51
  ratio, 50, 51
  valve sizing method, 71
Critical speed, natural frequency, 1477
Critically damped, 752
Cross-ambient, 375
Crossbar switch, 972

Cryogenic valves, 89, 281
Crystal seeding, 1553
Crystallization process, 1466
Crystallizer
  classifying, 1468-70
  controlled growth magma, 1468-70
  controls, 1466
  cooling type, 1468-70
  degrees of freedom, 1468
  direct contact, 1471
  evaporator, 1471
  indirect heating, 1471
  reaction, 1475
  submerged combustion, 1471
  vacuum, 1473
Current limit, 600, 603
Current repeater, 659
  transmission signals, 649-50
Current-to-current converter, 658
Current-to-air converters, 129, 636
Cushion-loaded piston actuator, 133, 142
Cut-off frequency, 723
C_v, 41, 67
Cyanide, oxidation, 1541, 1544
Cybernetics, 723
Cycle time, 888
Cylinder actuators, 223
Cylindrical film memories, 995

D/A conversion channel, hybrid, 1175
  converter, 965, 1076
Dampeners, 272, 332, 661
Damper, 341
  characteristics, 1399
  motor, 245
Damping coefficient, 723
  ratio, 752
Dashpot, 723
Data break, 888
Data computer, 917
Data logging, program, 1006
Data processing, analog computers, 1222
DC motor, variable speed, 310
ddc, 1069
  advantages, 1079
  analog computer simulation, 1215
  analog inputs, 1073
  cascaded, 1097
  costs, 1079
  hierarchical, 1097
  implementation, 1081
  maximum number of loops, 1072
  reliability, 1079, 1082
  sampling frequency, 1081
Dead band, 723, 888
  analog computer simulation, 1205
Dead tight ball valves, 165

Dead time, 745, 861
Dead zone, 1051–2
  analog computer simulation, 1205
Debugging, computers, 1314
Dechlorination, 1547
Decibels, 265, 723, 950
Decoding error, 965
Decomposing a process, 1112
Decoupling of distillation control interaction,
    1252
Definitions
  actuator, 31, 37, 38
  bellows seal, 36
  bonnet, 34, 36
  control valves, 31, 32
  diaphragm actuator, 37, 38
  direct actuator, 39
  extension bonnet, 35, 36
  good control, 842
  isolating valve, 36
  noise, 1600
  packing box, 36
  plug guiding, 37
  radiation fin, 35, 36
  regulator, 31
  reverse actuator, 39
  seat, 37
  summary, 13
  three-way valves, 33
  transfer function, 776
  valve body, 31
    plugs, 34, 37
  yoke, 37
Degrees of freedom, 737, 1233, 1486
  boiler, 743
  distillation, 1320
  heat exchanger, 1486
  over-control, 742
  water heater, 739–40
Delatch solenoid, 234
DeMorgan's Theorem, 519
Derivative control, 796
Derivative mode definition, 754
Derivative non-interacting, 430–1
Derivative relay, 430–1
Derivative response, 427, 454, 723
Derivative time adjustment, 856
Descartes' rule of signs, 784
Detector nozzle, baffle-nozzle characteristics,
    621
Deviation display, 408, 441
Deviation scanning controllers, 441
Dew point, 1319
Diagnostic routine, 889
Dials, 567
Diaphragm
  actuators, 133

Diaphragm (Continued)
  operated cylinder type valve, 215
  pumps, 326
  (Saunders) valves, 154
  selection, 158
Differential equation, 746
  analog computer, 1189
  catalyst tank, 746
  chemical reactor, 748
  mass-spring dashpot, 749
  water heater, 747
Differential gap, 723–4, 757
Differential pressure transmitters pneumatic,
    627
Differential transformers, 643
Differentiating relay, pneumatic, 1145
Differentiator, electronic, 474
  pneumatic, 472
Digital
  actuator, 218
  attenuator, 1167
  backup, 723–4
  blending system, 1575
  control algorithm, 931
  control valve, 216
  electropneumatic positioners, 128, 217
  indicators, 573
  input, 961
  output, 963
Digital terminology, 719–20
Digital transmission signal, 652
Digital valve actuator, 218
Digital vs analog, 1120
Digital-to-analog conversion, 965
Digital-to-pneumatic transducer, 218
DIN flanges, 94
Diode, 499
  characteristics, 501
  switching circuits, 505
  symbols, 501
Direct-acting actuator, 135
Direct-actuated regulator, 365
Direct-calibrated charts, 578
Direct-connected, field mounting controllers,
    449
Direct digital control, 889, 1069
Direct digital controller, simulation, 1215
Direct instrumentation, analog computation,
    1189
Disc, 889
  file, computer, 918
  packs, 1000
Discharge coefficient, 8, 41
Disconnect plugs, 556
Discriminator, 665, 691
Displays, 401, 408
Distance-velocity lag, 724

Distillate control, fast response, 1242
Distillation column constraints, 1350–2
  dynamics, 1348
  feedforward control, 933
  loading, 1350–2
Distillation control
  analog computer, 1232
  analyzers, 1338
  batch, 1256
  boiling point sensors, 1338
  by feed composition, 1254
  by valve stem position, 1368
  chromatographs, 1338
  composition by temperature, 1335–6
    of two products, 1248
  computers, 1340
  condenser controls, 1332
  constant pressure, 1322
    separation, 1245
  degrees of freedom, 1320
  differential temperature, 1336
  elevated receiver, 1326
  feedback optimization, 1367
  feed conditions, 1329
  feedforward, 1245, 1342
  feed temperature, 1329
  furnace, 1331
  interaction, 1251
  liquid distillate and inerts, 1322
  loading constraints, 1362
  lowered condenser, 1326
  maximum recovery, 1246
  no inerts, 1324
  optimized, 1349
  optimum feed flow rate, 1363
  pressure, 1322
  reboiler, 1331
  reflux control, 1241, 1332
  side draw, 1254
  superdistillation, 1335–6
  temperature sensing, 1335–6
  thermosyphon reboiler, 1332
  vacuum operation, 1326
  valve position, 1367
  vapor distillate and inerts, 1323
  variable feed, 1333
  variables, 1320
Distillation equipment, 1319
Distillation optimization, 838, 1343
  bottom-to-feed ratio, 1347
  column constraint, 1353
  condenser constraint, 1350–2
  constraints, 1350–2
  equations, 1344, 1351, 1358, 1364
  expensive bottom product, 1357
    top product, 1355–6
  feed flow rate, 1363

Distillation optimization (Continued)
  internal reflux, 1347–8
  known prices, 1350
  limited feedstock, 1354
    market, 1354
  minimum utilities, 1349
  product prices unknown, 1349
  reboiler constraint, 1352, 1365
  unlimited feed, 1360–2
  unlimited market, 1360–2
Distortion coefficient, 76, 260
Disturbance response, 797
Disturbance variable, 352
Diverting valve, solid state, 208
Divider, 1164
  electronic, 465
Dividing functions, 1142
Division 0 (electric area), 395
DO loop, Fortran, 925–6
Documentation, analog and hybrid computer, 1187
Doping, 499
Double limiter, analog computer, 1205
Double packing, 101, 285
Double-acting piston actuator, 133, 142
Double-seated regulators, 369
  valves, 82, 84
Downcomer flooding, 1359
Draft elements, 627
Drift, 724
Drift vs temperature, 673
Drive sleeve type actuator, 247
Drive speed adjustment, 309
  controls, 247, 249
Driver, 889
Droop, 352, 359
  compensation, 360
Drooping resistor, 1073
Drum, 889
  programmer timers, 463, 496
Dryer
  atmospheric, 1453
  batch, 1451
  blender, 1456
  continuous, 1452, 1457
  control, 1449
    atmospheric dryers, 1453
    feedforward, 1463
    fluid bed, 1460
    heated cylinder, 1457–8
    rotary, 1457–8
    spray, 1462
    tumbling, 1457–8
    turbo, 1459
    vacuum, 1455
  fluid bed, 1460
    batch, 1456

Dryer (*Continued*)
  heated cylinder, 1457–8
  holdup, 1452
  rotary, 1457–8
  special batch, 1456
  spray, 1462
  tumbling, 1457–8
  turbo, 1459
  vacuum, 1455
Drying rate, 1450
DTL logic circuits, 513–5
Dual mode control, 1281, 1292
Dual range butterfly valves, 149–50
Dual range Saunders valve, 159–60
Dummy, 889
Dynamic breaking of DC motors, 315
  compensation by lead-lag, 809
  effect, 454
    pneumatic transmission, 638
  error, temperature, 1489
  functions, 1144
  gain, 724
  imbalance, 807
  model, 807
  response of positioners, 127
Dynamics
  distillation column, 1347–8
  example, 756
  on-off controller, 756
  PD controller, 764
  PI controller, 766
  PID controller, 768
  pneumatic components, 141
    controller, 756
  processes, 719, 736, 743
  proportional controller, 762
  valve, 137
    actuators, 137

EBCDIC, character code, 886
Economizer, cooling water, 1352
  expansion valve, 1359
  heat exchanger, 1374, 1380
Eddy current coupling, 315
  cooling, 317
Edge follower, electric programmer, 462
Effluent controls, 1540
E/I converter, 657–8
Electric actuators, 233
  drive sleeve, 247
  proportional control, 249
  rotary, 243
  rotating armature type, 252
Electric area classifications, 553–4
  DC motors, variable speed, 310
  energy throttling, 299

Electric area classifications (*Continued*)
  feeders, 555
  heater controls, 300
    resistance, 304
  interlock design, 516
    elements, 481, 498
Electric noise interference, 941
  proportional actuators, 251
  quarter-turn actuators, 252
  valve actuators, 222
Electrode potential, oxidation reduction, 1541
Electro-hydraulic actuators, 236
Electromagnetic relays, 484
Electromechanical relays, 482
  switching, 970, 1164, 1172
Electronic adder, 470
  computer, hybrid, 1172
  controllers, 388
    special, 415
  converters, 655
  differentiator, 474
  high-voltage limiter, 479
  integrator, 476
  inverter, 470
  low-voltage limiter, 479
  multiplier, 1164
  multiplier-divider, 465
  positioners, 127–8
  proportioner, 470
  ratio control system, 1570
  scaler, 470
  square root extractor, 477–8
  subtracter, 470
  transmission signals, 649
Electronic transmitters, 640
  vs pneumatic, 388
Electro-pneumatic actuator, 232
Electro-pneumatic converters, 636
  positioners, 127, 128
  transducers, 129
Electrostatic recording, 582
Electro-thermal relays, 482
Elementary reaction, 1260
Elevation, 626
Elutriation Leg, 1485
End connections on valves, 92
Equal loudness contours, 267
  percentage trims, 105, 259
Equalization basin sizing, 1485
Error
  message, computer, 1095
  random, 3
  rms, 1071
  signal, 727
  squared, 889–90
  systematic, 3
Event recorders, 581

Excess air, 1422, 1441
  vs boiler load, 1407
Excess lime softening, 1553
Exclusive OR, 725
Executive program, 889–90
Expandable element in-line valve, 211
Expansible diaphragm type valve, 206
  tube type valve, 202
Expansion of real roots, 775
  valves, refrigeration, 1376
  valves, special, 1377
Exponential functions, analog computer, 1165
Exponential stage, 725
Extension bonnets, 103, 282

Facsimile recording, 585
Fail-closed valves, 142
Fail-safe regulator, 381
Fail-safe valves, 121, 142
Fan suction dampers, 343
Feasibility study, 890
Feedback control, 716, 725, 802
  limitations, 803
  optimizing control, 839
Feed composition control of distillation, 1254
Feeder, 324
  ball check, 333
  belt, 335
  gravimetric, 335
  roll type, 337–8
  screw, 337–8
  volumetric, 337
Feedforward, adaptive control, 818
  adding feedback, 811
Feedforward computation, 1190
  control, 717–8, 805
    algorithm, 933
    dryers, 1463
    heat exchanger, 1507
    optimizing, 837
  dynamic model, 807
  input, 406
  load balancing, 805
  model, 806
  performance, 812
  signature curve, 811
Feed hoppers, 339
Feedwater control, boiler, 1411
  single element, 1412
  three element, 1414
  two element, 1412
Feedwater pump speed control, 1416
  system head curve, 1414
Feeler bulb fillings, 1377
Ferrite core, 986
Field-effect transistor, 996
  flux adjustment, 312
  mounted pneumatic controllers, 449

Fill-in-the-forms programs, 1006
Filter, analog simulations, 1220
  electronic, 725
  RC, 950, 1073
  regulators, 121–2
Filtering noise, 950
Filtration in water treatment, 1541
Final control elements, 19, 293
  value theorem, 771
Fire-safe ball valve, 163–4
Fired heater controls, 1427
  reboilers, 1425
Firing angle, 301
  controls, furnace, 1429
First-order filter time constant, 943
  lag, 725, 751, 861, 1210
    Laplace transform, 771, 1210
  model, 861
  process, 752
Fish-tail discs, 151
Fixed heads, 890
  point arithmetic, 890
  scale indicators, 568
Flame spray coatings, 677
Flangeless valve bodies, 94
Flanges, 92, 94
Flapper, 621
Flashing, 70, 72
Flat panels, 537–8
Flip-flop, 725, 890, 917
Floating control, 725, 755, 757
Float type refrigerator, 1374
Floating point arithmetic, 890
Flocculation of water, 1540, 1558
Flow, accuracy, 4
  adapted controller equation, 815
  calibration, 1523
  characteristics, 104, 259
  chart, computer control, 929
    process logic, 1296
    symbols, 1297
  loop, scan, 917
  ratio control, 826, 1435
  sheet symbols, 1581
Flowing velocity in valves, 265
Flue gas, percentage O_2 control, 1406
Fluid bed dryer, 1460
  batch, 1456
  clutches, magnetic, 318
  interaction valve, 208
Fluidic valve, 208
Flute type plugs, 286
Flying capacitor, 954, 1075
FM/FM telemetry, 669
FM, Frequency Modulation, 665
Force balance positioner, 125
  transmitter, 622, 641
  transmitters pneumatic, 634

Forced response, 725, 726
Form, computer control, 926, 927
Formica panel, 561
Fortran, 890, 915, 917, 923–5
  II, 923, 925
  IV, 923, 925
  constant, 925
  expressions, 925
  program, PID algorithm, 926
  subroutine, 925
  variable, 925
Fouling in water coolers, 1323
Four-pipe system, 391
Fractionation column, analog computer, 1232
Free residual chlorine, 1556
Freeze dryer, 1455
Frequency division multiplexing, 665
  domain, 787, 1201
  modulation, 665
    analog computer, 1218
  ranges of noise, 937
  response, 725, 726
  signals for transmission, 651
Frigorie, 1370
Front of panel layouts, 539–40
Fuel-air limiting, 1408
  ratio, 1403
Fuel cutback control, 1409
Full bore Saunders valve, 157–9
Full graphic panel, 546
Function generator, 457, 1143, 1154
  analog computer, 1163
  time, 459–60
Furnace air requirements, 1422
  controls, 1420
    advanced, 1442
    analog computer, 1444
    analyzers, 1441
    conclusions, 1448
    cracking, 1437
    digital computer, 1445
    excess air, 1441
    feedforward, 1442
    fired reboilers, 1425
    firing controls, 1429
    multiple fuel, 1431
    oxygen analyzers, 1441
    process heater, 1427
    pyrolysis, 1437
    reformer, 1433
    startup heaters, 1423
Furnace, cracking, 1264, 1437
  draft control, 1400
  efficiency, 1421
  fuel choices, 1422
Fuse, 301, 599, 610
  clearance time, 600
Fused disconnect switch, 555–6

Fusible resistor, 1073
Fusion type thermal system, 377

Gain, 725–6, 751, 861, 1141
  input to summing amplifiers, 1162
  loop, 727–8
  margin, 725–6
Galvanometer, 583
Gas distribution valve, 204
  flow pressure compensation, 1147
  industry regulators, 356–8
Gas regulator noise, 270
Gasket materials, 97
Gate circuit, 890
Gauge, 567
Gauge factors, 671, 673
Gear-operated electric actuators, 247
Generalized gas-sizing formula, 67
Globe body styles, 81
Graphic logic symbols, 521
  panel, 543
    back engraved plastic, 549
    painted, 546
    slide projector, 549
  symbols, 544
    analog and hybrid, 1181
  translation of logic, 521
Gravimetric feeder, 335
Gravity separation in water treatment, 1541
Gray code, 965
Greek alphabet, 18
Ground-loop in noise pickup, 947
  systems, 945
Grounding, 947
Guard bit, 890
Guide vane control, compressor, 1529
Guillotine valve, 179–82

Half duplex, 890
Handwheels, 118
Hard facing of trims, 111
Hardness of materials, 264
Hardware, computer, 915
Harmonic, 726
  oscillator, analog simulation, 1218
Harriott, 843, 846
Hazardous areas, 393
Head, 890
Heat exchanger, cascade control, 1504
  condensate level controller, 1499
    throttling, 1497–8
  control, 1486
    condenser, 1503
    two 2-way valves, 1494
  cooling water conservation, 1492
  degrees of freedom, 1486
  feedforward control, 1507
  furnace, 1420

Heat exchanger (*Continued*)
  heat and cool, 1509
  liquid-to-liquid, 1487
  minimum condensing pressure, 1497
  multipurpose systems, 1509
  pumping traps, 1497-8
  steam, 1495
  three-way valve, 1490
  three-way valve balancing, 1493
  valve rangeability, 1495
Heat pumps, 1370
  transfer computer, 1147
Heater controls, 1486
Heating regulator, 368
Heavy pattern butterfly valve, 148
Helical spline actuator, 225
Hermetically sealed relays, 491
Heterogeneous reactions, 1267
Hexadecimal code, 917, 918, 920
Hierarchical ddc, 1097
High-density instrumentation, 542
  mounting control stations, 441
High-pressure connections, 95
  drop valves, 113
  limiting relay, 477, 478
  packings, 275
  seals, 274
  selector relay, 477, 478
  valves, 274
High-temperature packing, 277
  valve, 276
    materials, 276
  viscosity valves, 288
  voltage limiter, 479
  selector, 479
Hill-climbing techniques, polynomials, 1193
HLL logic circuits, 513-5
Holding amplifier, 1077
Hollerith card code, 920, 921
Holograms, 585
Hot chamber thermal system, 376
  gas bypass, 1326, 1349, 1381, 1503
  vapor bypass, 1326, 1349, 1381
Hunting, 726
Hybrid computer, 1151
  applications, 1179, 1224
  concept, 1172
  hardware, 1172
Hybrid computer, interface, 1172
  software, 1176
Hybrid digital computers, 1225
Hydraulic hammer, 272
  snubber, 272
  variable-speed drive, 321
Hydrogen valves, 213
Hysteresis, 986, 993
  computer simulation, 1210

IAE, 843, 853, 868
Idler pulley, 320
Ignitron tube, 304
I/I converter, 658
Illegitimate error, 3
Illumination of control panels, 537
Implicit method, analog computation, 1191
Impulse, 726
Incipient cavitation point, 47
Incremental recorders, 579
  tape recorders, 584
Index of self-regulation, 850
  register, 891
Indicator, 566
  digital, 573
  movable-reference, 570
  movable scale, 571
  parallax, 570
  parametric indication, 572
  pointer, 570
Indicator terminology, 566
Indirect method, analog computation, 1190
Inductance-capacitance filter, 952
Inert buildup in condensers, 1326
Inertia, 744
Infeasibilities, 1047
Infrared analyzer, 1436
Initial conditions, analog computation, 1186
  error, 726
  valve theorem, 771
Initialize, 891
Inlet guide vane control, compressor, 1529
Input-output interface, computer, 961, 1073
  isolation, 306
Inspection of panels, 561
Instability in cascade loops, 822
Installation costs, 397
Installations of computers, 900
Installed valve characteristics, 260
Instructions, computer, 917
  machine language, 919
Instrument symbols, 1581
Instrumentation, electronic, 388
  pneumatic, 388
Integral action, 727
  control, 795, 799-800
  criteria in tuning, 852
Integral mode definition, 754
  time, 727, 755
Integrand, analog computer, 1190
Integrated circuit, 510
Integrated circuit characteristics, 513-5
Integrated circuit memories, 996
Integrated switching circuits, 512
Integrating amplifier, 982
Integrating relay, 1144
Integration speed, analog computation, 1186

Integrator, 1152
  analog computer, 1158
  electronic, 476
  mode switching, 1159
  pneumatic, 474
Interacting column control, 1251
Interaction, 427
Intercoolers, 1380
Interface, 891
  channels, hybrid computer, 1188
  hardware, 961
  hybrid computer, 1172
Interlock, boiler, 1387
  components, 481, 498
  design, 516
  optimization, 516, 523
  symbols, 1592
Internal reflex computer, 1238
Interrupts, 891, 963
Interval timers, 494
Intrinsic safety, 393, 414
Inverse derivative control mode, 430–1, 757, 800
  ratio controller characteristics, 1565
  transformation, 769
Inverter, 556
  analog computer, 1162
  circuit, 1162
  cost, 589
  electronic, 470
  mechanical, 607
  SCR, 607
Inverting relay, pneumatic, 467
I/O, batch reactors, 1290
  hardware, 1077
  interface computer, 961, 1073
I/P transducers, 129, 636
Iris diaphragm, 343
Iron, oxidation of, 1541, 1553
Irreversible reaction, 1259
ISE, 843, 853, 868
Isobaric process, 1370
Isolating valves, 552
Isothermal process, 1370
ITAE, 843, 853, 868
Iterative, computations, analog, 1197

Jacket, vacuum, 283
Jacketed pinch valve, 175
  valves, 89, 280

Karnough map, 524
Key punch, 891
Keyboard, operator's console, 1093
Kinescope recording, 585
Kinetic model, 1260
Kinetics, reaction, 1259

Knife gate valves, 179–82
K-values, 65

Lag, 727
  analog computer simulation, 1210
  function, 1144
  pneumatic transmission, 391, 454, 638
  time, 454
Laminated panels, 561
Language, computer, 915
  data logging, 1006
  fill-in-the-forms, 1006
  optimizing, 1041
  supervisory, 982, 1006
Laplace operator, 127
  theorems, 770
  transfer functions, 719, 727, 734, 1202
  transform, 769
    actuators, 139–40
    first-order lag, 771
    pairs, 771
Large case controllers, 447
  instruments, 539–40
Laser-based memories, 1005
Last input memory, 1596
Latching outputs, 964
  relays, 487
Layout of control rooms, 536
Lead acid batteries, 603
Lead-lag
  analog computer simulation, 1210
  function, 1144, 1348
  model, 810
  setting the unit, 809
Leakage in double-ported valves, 257
  seating forces required, 258
  temperature effects, 257
  through control valves, 256
Level-seeking trunk path, 1303–4
  transmitter pneumatic, 633
Life expectancy of batteries, 603
Lifting traps, 1499
Light pattern butterfly valve, 148
Lighthill's equation, 114
Lighting requirements of control rooms, 537
Limit stops, 121
  switches, 119
Limiter, analog computer simulation, 1202
Limiting relays, high and low pressure, 477–8
Line follower electric programmer, 462
Linear electric actuators, 233
  forcing, 727
  programming, 727, 891, 1041
  select, 991
  trims, 105
  variable differential transformer, 643
Linearity theorem, 770

Linearization of non-linear equations, 761, 780
Linings, 156
Liquid ring rotary compressor, 1533
Liquid-filled thermal systems, 375
Literature, analog computers, 1177
  hybrid computers, 1177
Live zero, 621
Load, instruction, 919
  variable, 727–8
Local panel cabinets, 539–40
Lock-up valve, 121
Log functions, analog computer, 1166
Logging typers, 1078
Logic diagram, batch reactor control, 929
  functions, 505, 517
  interlock design, 516
    elements, 481, 498
    synthesis, 516
Logic maps, 524
  operations, 505
  operation symbols, 505, 516, 518, 521
  redundancy, 523
  simplification, 524
  symbols, 521, 1592
  tables, 516
  translation, graphic, 521
Logical operation, 891
Log-modulus plot, 727–8
  program, 916
Loop gain, 727–8
Low-noise regulator design, 206
Low-pressure limiting relays, 477–8
  selector relay, 477–8
Low-recovery valves, 263
Low selectors, 835
Low-temperature valves, 281
Low-voltage limiter, 479
  selector, 479
Lube oil systems, 1537
Lubricators, 98
LVDT, 643

Machine language codes, 922
  computer, 915, 916, 919
Macro assembly language, 1307, 1311
Magnetic amplifiers, 306
Magnetic core, 891–2
  coupling, 315
  disc memory, 997, 999
  drum memory, 997
  field, 986
  fluid clutches, 318
  particle coupling, 317
  tape memories, 1001
Magnitude ratio, 728
Maintained memory, 1596
Maintenance,
  costs, 397

Maintenance (*Continued*)
  electronics, 395
  pneumatics, 395
  regulator, 382
Manganese, oxidation of, 1541, 1553
Manipulated variable, 736
  limits, 832
  selection, distillation, 1234
Manipulations of block diagrams, 779
Manual reset, 426
Masking, 891–2
Mass flow by orifice, 1147
Materials of construction for panels, 560
  valves, 96
Mathematical model, 861, 1053
Mathematical model, steady state, 806
Maximum recovery distillation, 1246
  value, computer detection, 1208
Meanings of identification letters, 1582
Measurement accuracy, 2
Mechanical pinch valves, 173
  ratio control system, 1565
  variable-speed drives, 319
Median selector, 835–6
Melamine panels, 561
Memory devices, computer, 985
  first input, 1596
  maintained, 1596
  protect, 892
Mercury arc rectifier, 304
  relays, 489
Merge, 892
Message typers, 1078
Meter factor, 1236
  relays, 417, 418, 483
Metering algorithm, 1293
  pump, 324
    diaphragm, 326
    installation, 330
    piston, 325
Meters, 566
Methane analyzer, 1436
Microprogramming, 892
Milliampere signals, 649, 650
Millivolt detectors, 649
  signals, 649
Millivolt-to-current converters, 656
Miniature instruments, 539–40
  recorders, 581
Minimum condensing pressures, 1497
  value, computer detection, 1208
Mixed fuel firing, 1431
  reactor model, 1261
Mixture density technique, 71
Mnemonic, 893
  symbols, 916
Mode control, analog computer, 1241
  switches, 1241

Model, dynamic, 807
  mathematical, 1053
  mixed reactor, 1261
  plug flow reactor, 1264
  process, 1233
  process reactor, 1259
  steady state, 806
  water heater, 806
Modulation, 729, 893
  of high-frequency carriers, 665
Molten metal valves, 89
Mosfet, 974
Motion balance positioner, 126
  transmitter, 626, 641–3
Motion transmitter, 440, 634
Motor gear boxes, 252
  speed controls, 249
Movable-reference indicators, 568
  scale indicators, 571
Mud feeders, 333
Multi-disc butterfly valve, 149, 150
Multiple address computer, 919
Multiple disc trim valve, 116
  fuel controls, 832
    furnace, 1431
  orifice cage valve, 114
    restrictions, 271
    in series, 115
  scales, 571
Multiplex, 893
Multiplexer, 916, 954, 1074
  crossbar, 972
  feature summary, 957
  pneumatic, 972
  rotary, 972
  three-wire, 976
  two-wire, 976
Multiplexing, 665, 669, 954, 970
  low-lever signals, 953
  systems, 956
Multiplication, software, 915
Multiplier, 1154, 1235
  analog computer, 1154
  electronic, 465
Multiplying and dividing relays, pneumatic, 464
  D/A converters, 1174
  functions, 1142
Multipoint recorder, 577, 579
Multi-port ball valves, 163, 164
  purpose regulators, 381
  stage pressure reduction, 361
Multi-velocity action, 729
Mv/I converter, 656

NAND logic, 505, 510, 512, 517, 520, 1594
Natural frequency, 729
  of control valves, 139–40
  vibration, 271

NEMA housing designations, 553–4
Neutral valve position, 249
Neutralization in water treatment, 1541, 1547
Newell principle, 1002
Newton-Raphson method, 1070
Newton's law, 744, 746, 749
Nichols plot, 727–8
No-break stand-by power supply, 591
Noise, 265, 360, 937, 976, 1600
  circuit arrangement, 941
  common impedance coupling, 939
  computer, 937
  conditions in various industries, 940
  conduction, 940
  contact resistance, 939
  control valves, 1600
  damage risk, 268
  definitions, 1600
  electrostatic coupling, 939
  estimating, valves, 1600
  filter, 649
    RC, 950, 1073
  inductive circuit, 938
  line transients, 940
  measuring equipment, 960
  reduction in valves, 113, 114, 265
  reducing, by impedance level, 943
  rejection, A/D converter, 956
    charge amplifier, 943
    electronic, 661
    grounding systems, 945
    unloading amplifier, 944
    voltage amplifiers, 943
  temperature compensation, 939
  treatment, 265
Nomenclature-control valves, 31
Non-destructive readout, 995
Non-ideal reactor, 1267
Non-interacting controller, 729
Non-interacting controller simulation, 1213
Non-linear controller, 1552
Non-Newtonian fluids, 64
Non-symmetrical process, 1497, 1498
Non-symmetrical systems, 855
NOR logic, 505, 510, 512, 517, 520, 729, 1594
Normal mode interference, 938
Normal mode noise, 976
Normalize, 893
Normalizing equations, 1146
NOT logic, 505, 507, 508, 729, 1594
NPN transistor, 501, 502
NPSH, 331
Nuclear reactor control, 835
Numerical approximation method, analog computer, 1196
Nyquist criterion, 787
Nyquist method, 723

Object program, 893
Octal code, 917, 918, 920
Off line, 893
Offset, 359, 426, 729, 795
Oil separator, 1376
Oil-free compressor seal, 1537
Once-through boiler, 814
One-to-one repeaters, 622
On line, 893
On-line application, analog computers, 1218
On-line computation, 1140
On-off controller, 416, 757, 756
On-off controller dynamics, 756
Open loop, 729
Open-loop gain, 127, 352, 361–2, 1157–8
Open-loop tuning, 847, 850
Operand, 893
Operation codes, 922
Operational amplifier, 1151, 1154
Operating profit, 909
Operator tuning, summary, 871
Operator's communication, computer, 1086
    console, computer, 893, 916, 979, 1091, 1314
        data entry, 1094
        displays, 1094
        keyboard, 1093
        keys, 1091
        operating sequence, 1094
Opposed centrifugal pumps, 333
Optical memories, 1005
    meter relay, 418
Optimization, control algorithm, 935
    distillation columns, 1343
    logic circuits, 516
    minimum operating cost, 1349
    primary convergence technique, 1045
    reactor, 1284
    sample, 1058
    secondary convergence, 1046
Optimizers, 837
Optimizing blending system, 1564
    control, 837
        continuous slope, 839
        feedback, 839
        feedforward, 837
        peak seeking, 841
        sampled data slope, 839
        steady state gain, 839
    distillation, 837
    programs, 1040
Orientation program, 1313
    table for control valves, 26–29
Orifice coefficients, 8
    mass flow, 1147
OR logic, 505, 507, 517, 518, 729–30, 1593
ORP in, 1541
    control of chromium treatment, 1544
    measurement, 1540

Oscillation of valve plugs, 272
Oscillator circuit, 604, 651
Oscillographs, 583
Oscilloscope, 567, 583
Output tracking, 407
Over-control, 742
Over-damped, 729, 730, 752
Overload controls, 249
Overload electric, 599
Overload protection, thermal, 248
Override control circuit, 451–2
Overshoot, 427, 453, 729–30
Over-temperature protection, 374
Over-voltage, 491
Oxygen, dissolved, 1540
Oxygen analyzers, furnaces, 1441
Ozone, 1541

Packing, 98
    and bonnet selection table, 277
    double, 285
    high-temperature, 277
Padé functions, 1213
Panel, 535
    air supply, 552
Panel check-out, 562
    formica, 561
    front layouts, 539–40
    inspection, 561
    laminated, 561
    materials of construction, 560
    melamine, 561
    piping and tubing, 550
    shipment, 564
    specification, 564
    steel, 560
    types, 537
    wiring, 553–4
Parabolic ball valves, 170
Parallax, 570
Parallel blade dampers, 341
    cascade action, 730
        processor, 1104
    fan interlocks, 1400
Parametric indication, 572
Parity check, 894
Partial differential equation computers, 1230
Partial differential equations, 1265
Partial fraction expansion, 775
PD control, 796, 800
PD controller dynamics, 764
PD controller settings, 846
Peak-seeking optimizing control, 841
Percent conversion, 1284
Perfectly mixed reactor, 1261
Perturbation, 730

Phase, 730
  angle, 730
    detection, 652
  margin, 730
  modulation, 665
    analog computer, 1218
  shift, 730
    detection, 652
pH control, 1541, 1548, 1553
  in chromium treatment, 1544
pH curve, 1548
pH detectors, 648
pH measurement, 1540
Phenol, destruction of, 1541
Phon, 265, 267
Photocell, 418
Photocoder, 644–5
Photoconductive crystal, 672
Photoelectric transducer, 644–5
Photographic recording, 582
PI control, 796, 799–800
PI controller dynamics, 766
PI controller settings, 846, 850, 851
P/I converters, 655
PID algorithm, 926
  control, 800
  controller dynamics, 768
    settings, 846, 850, 851
  control program, 924
  response, 798
Piezo-electric transducer, 646–7
Pilot controller, 449
Pilot-actuated regulator, 365
Pilot-operated regulator, 354
Pilot-operated solenoid, 234
Pinch valve, 171
  accurate closure, 177
  improved characteristics, 176
  jacketed, 175
  pneumatic, 175
  prepinched, 174
  reduced sleeve, 173
  shutter closure, 176
Pipe reducers, 52, 53
  strains, 258
Piping of panels, 550
Pirani gauge, 689
Piston actuator, 133, 141–2, 223
  pumps, 325
  ring seal, 149–50
Planar film, 993
Plant simulations, analog computer, 1199
Plastic tubing, 550
Plated-wire memories, 995
Plug, 104
  flow reactor, 1264
  for disconnect, 556
  stems, 112

Plug (Continued)
  valve, 185
    adjustable cylinder type, 188
    chamflex, 195
    eccentric rotating spherical segment, 195
      shaft design, 191
    expanding seat plate, 191
    overtravel seating design, 193
    retractable seat type, 193
    semispherical plug, 191
    V-ported, 185–6
Plunger pumps, 325
PL/I programming language-one, 923–5
PM, phase modulation, 665
P-N junctions, 500, 996
Pneumatic actuators, 223
  adding relay, 467
  analog computer, 1140
  computer, calibration, 1149
    internal reflux, 1238
  controller dynamics, 756
  differentiator, 472, 1145
Pneumatic heat transfer computer, 1147
  integrator, 474
  inverting relay, 467
  multiplexing, 972
  multiplying and dividing relays, 464
  pinch valve, 175
  positioners, 124
  proportioning relay, 470
  ratio control system, 1568
  scaling relay, 470
  square root extractor, 476
  subtracting relay, 467
  to electronic converters, 655
  transmission lags, 397, 454, 638
  transmitters, 619, 622, 625
  vs electronic, 388
Pneumo-hydraulic actuators, 229
PNP transistor, 501, 502
Poiseville-Hagen equation, 59
Polarized relays, 487
Pole, 730
Polyethelene tubing, 550
Polynomials, analog computer solution,
      1193
Portable recorders, 582
Ported plug, 109
Position control algorithm, 924, 931
  indicators for valve stems, 249
Positioned plug type valve, 212
Positioner, 124
  bypass, 126
  controller, 416
  digital-to-pneumatic, 217
  dynamic response, 127
Positive coefficient resistor, 493
Potassium permanganate, 1541

Potentiometer, 649
    analog computer, 1167
Potentiometric transmitter, 646, 647
Power amplifiers, 307
    distribution, 555
    factor constants, 306
    failure classification, 591
    pack, 237, 240
    relays, 483-4
    supply, 401
    transformer guard, 953
Preact control, 755
Preamplifiers, 1075
Precipitation in water treatment, 1541, 1544,
    1553
Precision, 3
Prepinched valves, 177
Preset mode, computer station, 1087
Pressure compensation, 1147
    control, reactor, 1282
    dampeners, 661
    gradients, 47, 57
Pressure recovery factors, 44, 49
    for reducers, 58
Pressure regulators, 351
    transducers, 671
    transmitter pneumatic, 622, 625
Priority interrupt, 894
Printers, 584
Procedureless switching, 437-8, 444
Process characteristics, 736
    computer, 879
        applications, 879, 1109
        cost, 880
        installations, 900
        limitations, 881
        summary, 879
    control, chemical reactors, 1277
        computerized reactor, 1287
        reactors, 1277
        reactor optimization, 1284
            pressure, 1282
        stability, 784
        terminology, 719, 720
    decomposition, 1112
    dynamics, 719, 736, 743
        capacitance, 744
        dead time, 745
Process dynamics, differential equations, 746
    first order, 751
    inertia, 744
    resistance, 744
    second order, 752
    transportation time, 745
    gain, 849
    heater controls, 1427
    logic flow charts, 1296

Process dynamics (Continued)
    model, 862, 1233
    reaction curve, 847
        kinetics, 1259
    reactor model, 1259
        simulation, 1259
    response, first-order, 751
        second-order, 752
    simulation, analog and hybrid, 1179, 1199,
        1210
    vaporizer controls, 1427
    variables, 736
Profile tracer pneumatic programmer, 460
Program, 895
    analog computer, 1180
    data logging, 1006
    fill-in-the-forms, 1006
    furnace control example, 1027
    hybrid computer, 1188
    logic problem example, 1024
    mass flow example, 1023-4
Program, optimizing computer, 1041
    orientation, 1313
    process executive, 1311
    supervisory, 1006
    system executive, 1311
    trace, 1315
    variable, analog computation, 1187
Programmed adaption, 814
Programmer, 457
    cam type, 459-60
    drum type, 463, 495
    electric line follower, 462
    pneumatic profile tracer, 460
    ramp and hold, 459-60
    step type, 463
    timers, 495
Programming, 910
    control logic, 1307-11
    linear, 1041
Properties of cryogenic fluids, 282
Proportional and integral control, 796
    band, 405
    band adjustment, 731, 754, 856
    control, 793, 799
    control, algorithm, 931
    controller dynamics, 762
    control of electric actuators, 249
Proportional derivative control algorithm, 932
    gain, 754
    integral control algorithm, 932
    mode, 381, 754
    response, 424
    sensitivity, 754
Proportioner, electronic, 470
Proportioning relay, pneumatic, 470
Pulley speed control, 319

Pulsation dampener, 332
Pulse coding, 665
  counting module, 895
  inputs, 962
  outputs, 964
  train, computer simulation, 1218
  upsets, 756
Pulverizer, 1394
Pump calibration, automatic, 1523
  cavitation, 1518
  control, 1513
    centrifugal, 1513–4
    flow, 1516
    level, 1513–4
    on-off, 1513–4, 1521
    pressure, 1517–8
    reciprocating, 1523
    rotary, 1519–20
    safety, 1521
    selective, 834
    speed variation, 1519–20, 1522
Pump control, throttling, 1517–8, 1523
  curve, centrifugal, 1513–4
Pumping traps, 1497–8
Punched card programmers, 496
Purge interlock, 1387
Pyrolysis furnace controls, 1437

Quarter share principle, 1154
Quarter-square multiplier, 1160
Quarter-turn
  actuator, 223, 252
  ball valves, 162
  valves, 145, 161, 185
Quick-change trims, 86
Quick-opening trims, 107

Rack and pinion actuators, 223
Radial vane dampers, 343
Radiation bonnets, 101
Radio telemetry, 662
Radio transmitters, 663
Radix, 895
Ramp and hold
  programmers, 459–60
  generator, 969
  response, 731
  upsets, 756
Random access, 895
Random error, 3
Random scanning, 974
Rangeability, 13, 105, 258
Rangeability
  definition, 260
  of speed controls, 313
  pump, 1522
  regulator, 361

Range elevation, 626
Range suppression, 626
Range controller, 805
  characteristics, 1565
  tuning, 1574
Range dial setting, 1573
Range relay, 131, 828
  pneumatic, 1568
Range setting determination, 828
Range station, 828
  remote set, 830
  scales, 828
Rate blending, 1562
Rate control, 755
Rate of reaction, 1259
Rate time, 731
Ratio control, 402, 826, 1235
  cascade, 831
  remote set, 830
  stability, 826
RC filter, noise, 950, 1073
Reaction curve, 847
Reaction kinetics, 1259
Reaction nozzle, 636
Reaction rate, 849, 1259
Reaction spring, 249
Reactor
  batch, 1287
  catalytic, 1267
  computer control summary, 1316
  control, 1277, 1590
    cascade, 822
    cascade temperature, 1279
    computerized, 1287
    derivative, 796
    flow charts, 1296, 1590
    integral, 795
    lags, 1279
    nuclear, 835
    PD, 796
    PI, 796
    proportional, 793
    temperature, 1278
  conversion control, 1285
  model, 1259
  non-ideal, 1267
  optimization, 1282
  plug flow, 1264
  pressure control, 1282
  process states, 1295
  reaction rate control, 1285
  recirculating temperature control, 1278
  response, 790
  sequence logic, 1294
  simulation, 1259
  simulation circuit, analog, 1263
  temperature control, 1292

Read-only memories, 1004
Real and distinct roots, 775
Real and repeated roots, 775
Real differentiation theorem, 770
Real integration theorem, 770
Real roots, polynomials, 1194
Real time clock, 1075
Real time operation, 895
Reboiler controls, 1331, 1504
   selective control, 835
Recipes, reactor control, 1294
Reciprocating compressor, 1534
   pumps, 1523
Recorder,
   cartesian coordinates, 576
   chart paper, 580
   circular chart, 576
   compact, 581
   curvilinear coordinates, 576, 578
   digital, 584
   drum type, 579
   event, 581
   incremental, 579
   magnetic, 584
   miniature, 581
   multipoint, 577, 579
   multi-range, 580
   multi-variable, 580
   operations, 581
   pen, 580
   polar coordinate, 576
   rectilinear coordinates, 576, 578
   scribing means, 580
   services, 582
   single-point, 578
   strip chart, 578
   stylus, 580
   telemetry, 691, 692
   time arcs, 577
   X-Y, 579
Recovery coefficients, 44, 49
Rectifier, 603
   tubes, 315
Reducers, 319, 320
Reed relays, 487
Reference voltage, 967
Reflux control, distillation, 1236
Reflux equations, 1347, 1348
Reflux internal, 1238
Reformer furnace controls, 1433
Refrigerant characteristics, 1372
Refrigerants, 1372
Refrigerated fluids, brines, 1373
Refrigeration controls, 1369
Refrigeration
   cycle, 1371
   expansion valves, 1376

Refrigeration (Continued)
   superheat control, 1376
   turndown ratio, 1381
   units, 1369
      conclusions, 1384
      economizer receiver, 1382
      high turn-down, 1381
      large industrial, 1381
      multi-stage, 1380
      of, 1370
      on-off industrial, 1378
Refrigerator,
   capillary, 1373
   float type, 1374
   household, 1373
   small industrial, 1375
Regulator, 348
   cooling, 368
   definitions, 352
   fail-safe, 381
   heating, 368
   installation, 362
   leakage, 371
   maintenance, 382
   multi-purpose, 381
   noise, 270, 360
   pilot-operated, 354
   pneumatically loaded, 356
   pressure, 351
   proportional mode, 381
   rangeability, 361
   safety, 362
   seating, 356-7
   selection, 358
   sizing, 361
   stability, 361-2
   temperature, 364
   thermal systems, 371
   vacuum, 354
   valve body, 368
   weight-adjusted, 372
   weight-loaded, 353
Rejection of common mode voltage, 670
Relative address, 895
Relay, 308, 481
   characteristics, 482, 484, 490, 491
   clapper type, 484
   contact configurations, 484, 485
   costs, 490, 491, 493
   functions, 482
   general purpose, 483, 484
   latching, 487
   mercury, 489
   miniature, 483-4
   (on-off) function, analog computer, 1205
   open contact, 490
   polarized, 487

Relay (*Continued*)
  power, 483, 484
  reed, 487
  reliability, 493
  sealed-contact, 491
  selection, 489, 490
  spark suppressors, 491
  special, 494
  structures, 484, 486
  symbols, 1585
  telephone type, 484
  time delay, 488
  types, 483
Reliability computer, 1079
  ddc, 1079
Remote-set control station, 494
Remote-set ratio station, 830
Remote-set signal conditioning, 944
Repeatability, 3, 13
Repeaters, one-to-one, 622
Repetitive computations, analog, 1196
Reset action, 755
Reset rate, 755
Reset response, 425
Reset time adjustment, 856
Reset wind-up, 406, 427, 453, 732, 800, 1281,
    1532
  prevention, 427, 453, 1281
Residence time, 1263
Residual chlorine
  analyzer, 1544, 1557
  decay, 1557
  definition, 1556
Resistance, 744
  detection, 649, 650, 651
Resistance-to-current converter, 659, 660
Resolution, 895, 896, 965
Resonance peak, 732
Response, 751
  closed-loop, 789
  first-order, 751
  PID controller, 798
  second-order, 752
  time, 732
    air tubing, 639
  to disturbance, 797
  to set-point change, 798
  various control modes, 756, 797, 798
Retransmission, 407
Retransmitting slidewire, 984
Reverse-acting
  actuator, 135
  controller, 732
  relay, 756
Reverse rotation of DC motors, 315
Reverse voltage, 501
Reversible reaction, 1259

Reversing relays, 131
Reynolds effect on discharge coefficient, 8
Reynolds number, 45
R/I converter, 659–60
Ring joint gaskets, 94
Rokide process, 677
Roll type feeder, 337–8
Root
  complex, 776
  mean-square, 1149
  real and distinct, 775
  real and repeated, 775
Rotameter transmitter pneumatic, 630
Rotary air motor
  actuator, 231
  compressor, 1532
  dryer, 1455, 1457–8
  electric actuators, 243
  multiplexers, 972
  pumps, 1519–20
  slide valve, 182
Routh criterion, 785–7
Routine, 895–6
RTL logic circuits, 513–5

Safety factors in valve sizing, 77
Safety failure, 142
Safety regulator, 362
Sample-and-hold, 970
Sampled data slope control, 839
Sampling, analog, 896
Sampling system, residual chlorine, 1557
Saturable core reactors, 305
Saturation, 732
  in cascade loops, 824
  selective control, 835–6
  voltage, 503
Saunders diaphragm selection, 158
Saunders valve, 154
  dual-range, 159–60
  full bore, 157–9
  linings, 156
  straight-through design, 157–8
  vacuum service, 154
Scale factor, 830, 896
  analog and hybrid, 1183
Scaler, electronic, 470
Scales, 567
Scaling, 1140, 1238, 1573
  analog computer variable, 1183
  example, 1146
  procedures, 1145
Scaling relay, pneumatic, 470
Scan cycles, computer, 1083
Scan rate, 962
Scanning, 974
  analog computer, polynomials, 1193

Scanning (*Continued*)
  controllers, 441
  stations, high-density, 441
Scooped-out plug valve, 288
Scotch yoke actuator, 223
SCR, 300, 416
  firing angle, 301
  fuses, 301
  rate of current rise, 301
  rate of voltage change, 303
Screening in water treatment, 1541
Screw feeder, 337–8
S domain, 770
Seals, oil-free, 1537
Seal rings for high pressure, 276
Seat materials, 258
Seating forces, 258
Secondary control modes, 822
Secondary convergence method for optimization, 1046
Secondary loop types, 821
Secondary valve orifice, 270
Second-order lags, 752
Second-order response, 752
Sectional linear programming, 1042
Selection of cascade secondaries, 821
Selective control, 832
  compressor, 835
  controlled variable, 834
  extremes, 835
  manipulated variable, 832
  median, 835–6
  multiple sensors, 835–6
  pump, 834
  reboiler, 835
  redundancy, 835–6
  saturation, 835–6
Selective ion probes, 648, 1558
Selector, high or low voltage, 479
Selector control circuit, 451–2
Selector relay, high and low pressure, 477–8
Self-adaption, 816
Self-contained regulator, 366
  temperature regulators, 378
Self-draining valve, 292
Self-energizing seals, 274
Self-powered feeders, 337
Self-regulation index, 850
Self-synchronizing control stations, 437–8, 444
Self-synchronizing regulators, 441
Self-tuning, control algorithm, 935, 1080
Semi-balanced valves, 369
Semiconductor doping, 499
Semiconductors, 499
Semi-graphic panels, 546
Separation efficiency, 1347
Sequence monitor, 896

Sequencing interlock components, 481, 498
Sequencing interlock design, 516
Sequential control, 897
Sequential scanning, 974
Series DC motor, 311
Service recorders, 582
Service routine, 897
Servo setpoint, 402
Servo system, 236
Servo-mechanism, 733
Set point, 401
  change response, 798
  ramping, 1293
  station, 978
    fail-safe, 979
    features, 980
    resolution, 982
Settling time, 733
Shielded-twisted wires, 948
Shielding of leads, 649
Shipment of panels, 564
Shipping by truck, 564
Shock waves, 270
Short circuits, 599
Shunt DC motor, 311
Sidedraw distillation control, 1252
Side-mounted handwheel, 118
Side-stream cooler and reboiler, 1320
Signal
  conditioners, 661
  conditioning, 975
    analog computer, 1220
  converters, 655
  delays, computer simulation, 1213
  electronic transmission, 649
  generation, computers, 1218
  ground, 946
  relay, 1154
  standardization, 390
  wiring, 661
Signature curve, feedforward, 811
Signum function, analog computer, 1205
Skewness, conversion channel, 1173
Silencers, 265
Silicone-controlled rectifier, 300, 415
Simulation, 733, 897
  advantages, 1268
  analog and hybrid computers, 1179, 1199
  analog computer, 1179, 1199
  circuit, analog, 1263
  computer, 1314
  control valve, 1216
  plug flow reactor, 1264
  process, 1199, 1210
  reactors, 1259
Simultaneous method of conversion, 967
Single-element feedwater control, 1412

Single-seated valves, 83, 368
Single-speed floating control, 757
Sinusoidal upsets, 756
Sizing data for control valves, 77
Sizing
    for critical flow and reducers, 57
    for high viscosity, 59, 63
    of control valves, 40
    of valve actuators, 136
    regulator, 361
Slab valve, 179–82
Sleeve materials, 173
Slide projectors, 549
Sliding gate
    plate and disc type, 182
    positioned disc type, 179–82
    valves, 179
Slimes biological, 1555
Slip-on flanges, 94
Sludge level, 1555
Slurry valves, 288
Small-flow valves, 90, 285
Snubbers, 661
Sockets, 379
Socket welded valves, 95
Soft constraint, 1051
Soft seats, 112, 257
Software, 897, 915
    checkout, 911
    documentation, 913
    fill-in-the-blanks, 927
    hybrid computer, 1176
    multiplication, 915
Solenoid, 119, 233
    lever type, 234
    pilot-operated, 234
    three-way, 236
Solid state
    diverting valve, 208
    logic interlocks, 498
    multiplexing system, 956, 974
    switching, 970
Sound
    intensity, 1600
    level, 266
    power, 1600
    pressure, 1600
Source language, 897
Source noise interference, 938
Spark suppression, 491
Speed changes on AC motors, 315
Speed control, 309
Speed transmitter pneumatic, 635
Split range, 124
    control, 1280
Split-body valves, 86
Spray dryer control, 1462

Spring and diaphragm actuators, 133
Spreader stoker, 1394
Square root extractor, 828, 1142, 1164
    electronic, 477, 478
    pneumatic, 476
Square wave, 608, 609
Squarer, 1154, 1164
Stability, 733, 784
    criteria, 784
    limitations of feedback control, 803
    regulator, 361–2
Stacked multiple disc valve, 116
Standard output pneumatic transmission
        systems, 621
Stand-by power supply, 555, 556, 587–8
    AC type, 592
    backup inverter, 598
    battery, 603
    battery chargers, 603
    bus transfer switch, 608–9
    delayed transfer, 596
    engines, 601, 602
    equipment failure, 595
    load failure, 599
    multicycle transfer, 610
    multiple sources, 595
    no-break transfer, 611–2
    no-break type, 591
    output redundancy, 598
    redundancy, 590
    redundant rectifiers, 592
    rotating equipment, 601
    specifications, 613
    static inverters, 603
    static rectifier, 601
    subcycle transfer, 610
    system, 610
    system redundancy, 611–2
    total redundancy, 598
    transfer time, 591
Stand-by power system classification, 590
Starting circuits for DC motors, 313
Startup heaters, 1423
Startup torque, low, 1374, 1377
Startup without overshoot, 427, 453
Static inverters, 603
Static logic switching, 498
Static rectifier, 601
Static switch, 608–9
Static test mode, analog computer, 1186
Static torque, 151–2
Statistical computations, analog, 1197
Status channels, hybrid computer, 1174
Status indication, computer, 1174
Status responding channels, hybrid, 1175
Status retaining channels, hybrid, 1175
Steady state model, 806

Steam boiler controls, 1385
Steam heater controls, 1495
Steam jackets, 89
Steel panel, 560
Stellite 6B, 263
Stem position indicators, 121, 249
Step response, 733–4
  first-order lag, 752
  lead-lag unit, 809
  second-order lag, 752
  test, modified, 863
Step type programmer, 463
Step upsets, 756
Stepped speed control, 319
Stepping motor, 982
Stepping switches, 487
Stirred tank model, 1261
Stoichiometric balance, 1468
Storage, 897
Storage devices, computer, 985
Store, instruction, 919
Straight-through Saunders valve, 157–8
Strain gauge, 671
  bonding cements, 677
  carrier materials, 677
  drift vs temperature, 674
Strain gauge
  dummy gauges, 680
  lead wires, 677
  materials, 673
  resistance vs temperature, 678
  temperature-compensated, 677
  transducer, 665
Strain slope change with temperature, 675
Strain transmission, 689
Strain-sensing alloys, 673
Strip chart recorder, 578
Stroking times, 128
Suboptimization, distillation, 1343
Subranging, 970
Subroutine, 897
Subtracter, electronic, 470
Subtracting relay, 1142
  pneumatic, 467
Successive approximation, 733–4, 897, 969
Suction throttling compressor control, 1526
Suction vane, 1381
Sulfides, oxidation of, 1541
Sulfur dioxide for chromium reduction, 1546
Sum of Cv's technique, 70
Superheat control valve, 1376
Supervisory control, 898
Supervisory programs, 910, 1006
Support programs, 911
Suppression, 626
Surge control, compressor, 1531
Surge line, compressor, 1526

Suspended solids, measurement, 1540
Swing-through butterfly, 146
Switch, electronic analog computer, 1160, 1161
Switching circuits, diodes, 505
Switching circuits, transistor, 507–8
Switching elements, 481, 498
Switching procedureless (of controllers),
    437–8, 444
Switching transistor characteristics, 504
Symbols
  block diagrams, 778
  flow chart, 1297
  flow sheet, 1581
  for logic operations, 505, 516, 518, 521
  instrumentation, 1581
  interlock, 1592
  logic, 1592
  relays, 1585
Synchronization circuit, 595
Synchronous computer, 898
Synchronous motor, 982
Syncro regulator, 440, 446
Synthesis of logic circuits, 516
Syphoning, 330
System ground, 946
System response, 751
System stability, 784
Systematic error, 3
Systematic variation, polynomials, 1193

Tabulation of control modes, 755
Tachometer, 652
Tag index, assembly language, 922
Tank sizing based on pumping rate, 1516
Tape
  magnetic, 918, 920
  printers, 585
  punched, 918, 920
  reader, 1078
  recorders, 584
Tare regulation, 634
Target constraints, 1050
Taylor's series expansion, 780, 1056
Teflon V-ring packings, 100
Telemetering, 662
Telemetry
  accuracy, 668
  antennas, 691–2
  applications, 680
  battery, 680, 691
  calibration, 689, 692
  chemical plant, 681
  control, 668
  conveyor chain, 681
  cost, 662, 669
  discriminator, 691
  dynamic testing, 670

Telemetry (*Continued*)
earth-moving equipment, 687
FM/FM, 669
gear train efficiency, 682
limitations of, 687
measurement, 668
methods, 664
multiple measurements, 669
noise, 669
power plants, 685
power sources, 680
receiver, 691
recorder, 691–2
redundant measurement, 689
sensing elements, 690
strain transmission, 689
system, 662
components, 690
temperature transmission, 689
textile mills, 681
total system operation, 691–2
transducers, 671, 688
transmitter, 668, 690
underground cable, 686
Teleprinters, 585
Teletypewriters, 585
Television, closed-circuit, 585
Temperature
control, computer, 924
control, heating and cooling, 1511
effect on strain gauge factor, 673
overshoot prevention, 1281
regulator, 364
transmission, 689
transmitter pneumatic, 631
Terminal, 554
identification, 559
strips, 559
Termination rack, 898
Terminology
computer related, 882
control valves, 31
for control theory, 715
Theorem
final value, 771
initial value, 771
linearity, 770
real differentiation, 770
real integration, 770
Thermal bulb, 375, 379
Thermal element lag, 1489
Thermal system, 371
fusion type, 377
hot chamber, 376
liquid-filled, 375
wax-filled, 377
Thermocouple signal conditioning, 941

Thermocouple wiring, 942
Thermostatic expansion, 1376
Thermowells, 379
Thin film, 993
Three-element feedwater control, 1414
Three-position controller, 734
Three-term controller, simulation, 1213
Three-way ball valves, 163, 164
Three-way globe valves, 89
Three-way solenoid pilot, 236
Three-way valves, 369
heat exchanger application, 1490
Three-wire multiplexing system, 955
Throttling electrical energy, 299
Thyristor, 595
Tight shut-off, 257
Time
constant, 751, 861
delay, 1596
relay, 488
thermal bulb, 1489
division multiplexing, 665, 669
domain, 769
function generators, 459, 460
lag, thermal element, 1489
proportional control, 416, 734
scale, analog computer, 1185
schedule controller, 734
Timers, 481
band programming, 495
cam-operated, 495
definitions, 494
drum programmers, 496
interval, 494
programming, 495
punched card, 496
Top-bottom guided valve, 83
Top-guided valve, 82
Top-mounted handwheel, 118
Top-of-scale value, analog variable, 1183
Toroid, magnetic, 917, 986
Torque characteristics, butterfly valves, 151, 152
Totalizing blending system, 1562
Track, 898
Transducer, 734
digital-to-pneumatic, 218
photoelectric, 644, 645
pressure, 671
strain gauge, 665
telemetry, 671, 688
Transfer function, 734, 776
analog computer, 1201
control modes, 792
filter, 1219
first-order lag, 777–8
general, 784

Transfer function (*Continued*)
  lead-lag, 809
  PID controller, 777–8, 1213
  representation, 1201
  second-order lag, 777, 778
  water tank, 782
Transfer time, 591
Transformer, 555
  differential, 643
Transient error, 807
  overvoltages, 306, 595
Transients, 489–90, 491, 492
Transistors, 308, 501
  switching circuits, 507–8
Transmission
  by resistance variation, 649–50
  distances and lag times, 392, 454, 639
  lag, 390, 454, 638
  signal, 390, 621, 649
    current, 649–50
    digital, 652
    electronic, 649
    frequency, 651
    phase angle, 652
    strain gauge, 651
  strain, 689
  temperature, 689
  wireless, 662
Transmitter
  buoyancy, 633
  capacitance, 645–6
  electronic, 640
  features, 653
  force, 634
    balance type, 622, 641
  level, 633
  motion, 634
    balance type, 626, 641–3
  piezoelectric, 646–7
  pneumatic, 619
  potentiometric, 646–7
  pressure, force balance, 625
  radio, 663
  rotameter type, 630
  speed, 635
  standard output range, 621
  telemetry, 690
  temperature, 631
  with voltage signals, 649
Transport lag, computer simulation, 1213
Transportation time, 745
Traps, pumping, 1497–8
Tray dryers, 1453
Trend recorders, 407, 543
Triangular wave, computer simulation, 1218
Trim, 104
  materials, 110, 111
  stability, 83

T-ring seal, 149
Triode vacuum tubes, 307
Truth tables, 516, 734
TTL logic circuits, 513–5
Tube fittings, 550
Tubes, vacuum, 307
Tubing, 380
  of panels, 550
Tumbling dryers, 1457–8
Tuning, 734, 898
  by computer, example, 867
  console, 1091
  of controllers, 842
Turbo dryer, 1459
Turndown, burner, 1431
Turndown, furnace, 1431
Turndown ratio, refrigerators, 1381
Twisted wires, 948
Two-element feedwater control, 1412
Two-pipe system, 432, 435
Two-position controller, 756, 757
Two-wire systems, 559
Type one servo, 735
Typewriter, 584
  logging, 1078

U-ball valves, 170
Ultimate gain, 843
Ultimate sensitivity tuning, 862
Ultimate time period, 843
Unbalance torque, 151, 152
Uncontrolled availability fuel, 1393
Underdamped, 752
Unit operations, 1273
Unit sensitivity, 260
Unloading, constant speed, 1535
Unloading, fire step, 1535
Unloading, two-stage, 1380
Update, 899
USASI, computer language, 925
USASI, Fortran, 925

Vacuum
  dryer, 1455
  jackets, 89, 283
  jets, 1329
  regulator, 326
  tubes, 307
  valves, 283
Valve
  accessories, 118
  actuator, 132, 221
    digital, 218
    dynamics, 137
  ball, 161
    and cage, 197
  bodies, 81
  body assembly, 81

Valve (*Continued*)
  bonnets, 98
  butterfly, 145
  cavitation, 262
  characteristics, 259
    installed, 260
    unit sensitivity, 260
  characterized ball, 256
  clearance flow, 258
  coefficients, 41, 44
  cryogenic, 281
  diaphragm-operated cylinder type, 215
  digital, 216
  discharge coefficient, 41
  expandable element, 211
  expansible diaphragm type, 206
    tube type, 202
  flow characteristics, 75
  fluid interaction type, 208
  for high pressure drops, 113
    high temperature, 276
  for vacuum, 283
  for viscous service, 288
  high-pressure, 274
  jacketed, 280
  leakage, 84, 256
  low-temperature, 281
  noise, 265, 1600
  opening indicators, 121
  pinch type, 171
  plug type, 185
  plug vibration, 272
  positioned plug type, 212
  positioners, 124
  pressure drop selection, 75
  rangeability, 258
  safe failure position, 142
  Saunders, 154
  seat materials, 258
  sizing
    equations, 42, 43
    for cavitating flow, 46, 51
    for critical flow, 46, 66
    for critical gas flow, 66
    for flashing, 70, 72
    for gases, 64
    for gas-liquid mixtures, 74
    for high viscosity, 59, 63
    for liquids, 45
    for non-Newtonian fluids, 64
    for pipe reducers, 52, 53, 56
    for toxic fluids, 213
    for superheated vapors, 65
    for vapors, 64
  sliding gate, 179
  slurry, 288
  small flows, 285
  soft seat, 257

Valve (*Continued*)
  solid state diverting, 208
  trim, 104
    material selection, 111
    velocity limits, 269
    vibration, 265
Vane type actuator, 228
Vapor-filled thermal system, 371
Vaporizer control, 1504
Variable orifice damper valve, 343
Variable pairing, 1252, 1256
Variable pitch pulley systems, 320
Variable pitch sheave, 320
Variable process, 736
Variable speed drive, 309, 321
  hydraulic, 321
Variable speed electric DC motors, 310
Varsud: controller, 1537
V-ball valves, 167
VCO, 690
Velocity
  constant, 735
  control algorithm, 932
  limit, 899
  limiting, simulation, 1216
  limits in valves, 269
Vena contracta effect, 47
Verify, 899
Vertical valve plug oscillation, 272
Vibrating feed hoppers, 340
Vibrating screw feeders, 339
Vibration, 265
Vibration, motors, 317
Vibration, natural frequency, 271
Vibration, valve plug, 272
Vibrationless valve design, 206
Video tape recording, 585
Viscosity correction factor, 61
Viscosity effect on sizing, 59, 63
Viscous service valves, 288
Voltage controlled oscillator, 690
Voltage to current converter, 657–8
Volume boosters, 130
Volume relays, 131
Volumetric feeder, 337
  belt type, 339
V-ring packing, 100

Waste treatment, 1540
  equalization basin, 1550
  neutralization, 1547
  oxidation, 1541
  reduction, 1546
Watchdog timer, 899
Water treatment controls, 1540
Wax-filled thermal system, 377
Weight-adjusted temperature regulator, 372
Weight-loaded regulator, 353

Weight transmitters, pneumatic, 634
Welding ends on valves, 95
Wet bulb temperature, 1450
Wheatstone bridge, 649–50, 1570
Wire color code, 560
Wire identification, 559
Wire lugs, 560
Wireless transmission, 662
Wiring
  computer, 937
  electric signal, 661
  noise pickup, 948
  panels, 553–4
  practices, 661
  rules, 950

Wobble plate, 427
Woven-wire memories, 995
Write, 899

X-Y recorders, 579

Y-style valves, 89

Zener diodes, 493, 605, 967
Zero voltage alarm, 1083
Ziegler and Nichols, 843, 850

$\Delta P_{allowed}$, 49, 66, 71
ΔP selection, 75
$3C$ equations, 851